PART FOUR
THE PRACTICE OF NAVIGATION

PART FOUR

THE PRACTICE OF NAVIGATION

Page

CHAPTER XXIII. The Practice of Marine Navigation 671
CHAPTER XXIV. Ship Weather Routing 684
CHAPTER XXV. Polar Navigation 700
CHAPTER XXVI. Lifeboat Navigation 724

CHAPTER XXIII

THE PRACTICE OF MARINE NAVIGATION

2301. Introduction.—In the preceding 22 chapters, dead reckoning, piloting, and celestial navigation are discussed separately. In this chapter the interrelationship of the various elements of navigation, including radionavigation, are discussed. However, the most important element of successful navigation cannot be acquired from this book—nor from any book or instructor. The *science* of navigation can be taught, but the *art* of navigation must be acquired. Modern navigation is a blending of the two—a *scientific art.* The truly successful navigator is one who supplements his knowledge with judgment, utilizing every opportunity to improve his judgment through experience. Even with knowledge and judgment, the navigator cannot expect to be fully reliable unless he is alert, constantly evaluating the situation as it develops, avoiding dangerous situations before they arise, or recognizing them if they do occur, and always keeping "ahead of the vessel." The elements of successful navigation, then, are *knowledge, judgment,* and *alertness.* To the person possessing these, navigation can be a pleasure. A person who tries to navigate without them is at best a doubtful asset. He may be a menace to his vessel and shipmates.

It is not wise to attempt to reduce navigation to a series of steps that can be followed mechanically. The methods and techniques to be used are those which are applicable to the type of vessel, the equipment available, the training and experience of the navigator and any assistants, the local situation, etc. The navigation of a small craft proceeding up the Choptank River, for instance, might be quite different from that of an ocean liner entering New York harbor. Both might differ from the navigation of a naval vessel approaching an assigned anchorage. It is important that a navigator make an "estimate of the situation" and use the methods and techniques that are best adapted to the conditions at hand.

The discussion that follows is generally applicable to any vessel under average conditions, but is written primarily for an average ship which might be planning and executing an ocean voyage.

2302. Advance preparation.—The initial planning for an ocean passage includes careful study of the *Sailing Directions (Planning Guide)* and inspection of charts of appropriate scale with the objective of determining the route, the conditions expected to be encountered, and the speed of advance required if the passage is to be completed by a predetermined time. This planning may utilize the services of ship weather routing (ch. XXIV).

As the planning progresses, the navigator obtains an overview of the passage. When the planning is completed, the navigator has developed the intended track based upon prudent consideration of: (1) the capabilities and limitations of navigational methods and systems to be employed; (2) navigational hazards; and (3) the possibility of equipment failure and human error. The detailed planning for the approach and entry at the destination is usually deferred until some time during the transit but before making landfall.

Before getting underway, the navigator should familiarize himself with his equipment. Any defective or questionable instruments should be repaired or replaced. The necessary charts and publications should be on hand. If the voyage is to extend beyond the time range of any publication, such as an almanac or tide tables, the volume for the next period should be included, or provision should be made to acquire it before the expiration date of the current volume. Charts and light lists should be checked to see that they have been corrected through the latest *Notice to Mariners.*

When all equipment is on hand and in suitable condition, the navigator should study his charts and publications. He should determine which soundings are in feet, which in fathoms, and whether other units are used. It is good practice to underline or circle with a colored pencil the statement of units as given on each chart. The various notes on the chart should be read, and applicable ones marked. The latitude and longitude scales should be observed and the units noted. The channels, currents, shoals, aids to navigation, and natural landmarks should be studied so that the general arrangement is familiar. Useful natural ranges should be located and marked. Where needed, turning bearings, danger angles, and danger bearings should be determined.

The tides and currents to be encountered should be determined from the tables and charts. The advice and warnings given in coast pilots or sailing directions should be read and pertinent parts marked or copied out. The light list should be studied, and the arcs of visual range for the usual height of eye drawn in. Characteristics, including shape and color of the structure, should be written on the chart, if not printed there, or in a notebook, to assist in identification. Useful radar targets, radiobeacons, Loran rates, etc., should be noted if equipment to utilize them is available. If a danger sounding is useful, it should be drawn in. The bottom configuration should be studied for distinctive features that will prove helpful in locating the position of the vessel, or keeping it in safe water. If foreign charts are to be used, the symbols should be understood.

The extent of the preliminary study depends somewhat upon the navigator's previous knowledge of the area. But however familiar he may be with local conditions, the navigator should not overlook the need for checking his equipment to be sure it is complete and up-to-date, nor to refresh his memory regarding critical items of information. The prudent navigator leaves nothing to chance and *assumes nothing that can be verified.*

In pilot waters with limited maneuvering space, the desired track might well be plotted in advance, and the predicted time between buoys, turns, etc., determined. Where repeated runs are made over the same routes, the entire track may be plotted in ink. Courses, distances between lights, visual range arcs, and other useful information might be prominently indicated. When this practice is followed, a positive routine should be set up to apply corrections and to bring these to the attention of all concerned.

2303. Getting underway.—Shortly before the ship gets underway the necessary charts, publications, and plotting equipment should be placed on the chart table. A check should be made to be sure that all marks (except those permanently plotted in ink or colored pencil) relating to a previous voyage have been erased from the charts. The navigator's binoculars should be checked to see that they are properly secured in their accustomed place on the bridge. The gyrocompasses should be started sufficiently in advance to insure proper operation, and should then be compared with the repeaters and the magnetic compass on the bridge. Gyro error should be determined before getting underway. The deviations of the magnetic compass may be unusually high because of cranes, cables, etc. A check should be made to see that the latest deviation tables are available, and that magnetic gear has not been left near the compass. Azimuth circles and peloruses should be in place and checked. The standard and emergency steering gear should be checked, as well as communication and signaling equipment. If practical, the mechanical log and electronic equipment such as radar, Loran, radio direction finder, and echo sounder should be started and checked. Radar and echo sounder errors should be determined prior to getting underway. The hand lead should be placed at a convenient location ready for immediate use. The anchor windlass should be tested. The sextant, chronometers, almanac, and tables should be checked to see that they are in their proper places. It is good practice for the navigator to prepare a check-off list to insure that nothing is overlooked. The checks should be made carefully by a responsible person who reports the results to the navigator.

Before getting underway the navigator should see that all navigational personnel are at their assigned stations and that each understands his duties. Aboard naval vessels a piloting team (normally composed of the navigator, plotter, recorder, bearing observers, echo sounder operator, and others) is usually employed when transiting restricted waters. Aboard merchant vessels in the same circumstances, the navigation may be performed by no more than the mate on watch and the helmsman. In any event, it is good practice to acquaint each person with the general plan of operation, for an informed person is less likely to make mistakes, and more likely to detect mistakes made by others.

2304. Leaving port.—In a harbor, the largest scale chart should be used for greatest accuracy and detail. The dead reckoning should be started as soon as the vessel steadies on its first course. If the desired track has not been plotted in advance, the dead reckoning is run ahead a short distance. In either event, the predicted time of arrival at the next turning bearing, or of passing the next aid to navigation is recorded on the chart. Predicted times of arrival at various points are of great importance in interpreting the information received and in avoiding dangerous situations. It is good practice to use *all* available information, and not rely solely upon a single aid. A good position should be maintained at all times. Fog may set in rapidly and without warning, obscuring landmarks before a round of bearings can be observed. Lights should be timed and identified by their characteristics. At a distance, the color and shape of buoys may not be apparent. Sometimes a sailboat can be mistaken for a buoy. Buoys may be out of position. Bearings and ranges on fixed objects are better than on floating aids which do not remain at fixed points. Soundings should be taken continuously in the vicinity of shoal water. It is good practice to check the compass and radar at convenient opportunities, as when on a range or passing between two headlands. Ranges are of great value for checking position or keeping on the desired track, and should be used whenever available.

By skillful navigation, one may be able to save many miles of steaming. However, it is possible to allow insufficient margin of safety. The navigator should always keep in mind the possibility of failure of some item of equipment, unexpected fog, or the need for maneuvering room if another vessel approaches too close. He should remember, too, that in pilot waters currents may be strong and variable.

A detailed record should be kept in a notebook. Entries should be made showing bearings and ranges, important soundings, all changes of course and speed, the times of passing important aids to navigation, and other pertinent information. The record should leave nothing in doubt, indicating whether bearings are true or by magnetic compass, whether soundings are in feet, fathoms, or meters. This record is useful in preparing the ship's log, providing guidance for future runs over the same area, establishing position if fog sets in, and in providing an acceptable record if the vessel experiences a mishap resulting in a later investigation.

The chart, also, should present a neat and intelligible record of the passage. Course lines and lines of position should be drawn boldly and neatly, and should be no longer than needed. Standard labels should be used wherever they contribute to an understanding of the plot. They should be so placed and worded that no doubt is left as to their applicability and meaning. If possible, lines and labels should not be drawn through chart symbols.

Outside the harbor, if the course is parallel to the coast, there may be advantages in remaining close enough to utilize major aids to navigation and other landmarks. However, a set toward the beach, particularly off the entrance to an estuary, can endanger the safety of a vessel. Many ships have grounded because a course was set too close to off-lying dangers.

2305. Taking departure.—When a vessel reaches the open sea and is about to leave the land astern, a last accurate position is obtained by means of landmarks available. This process is called **taking departure.** It marks the end of piloting and the beginning of the next phase of the navigation. The work of the navigator becomes less hurried, and fixes are obtained less frequently. Soundings become of less interest. The hand lead is secured. The position may be transferred from the chart to a plotting sheet. Courses and speeds will be maintained over relatively long periods. The sea routine begins. Even if the vessel is to follow the coast, it generally does so at such a distance that danger is some distance away, and fixing is an intermittent process rather than a continuous one.

2306. Navigation at sea, like piloting, varies somewhat from vessel to vessel depending upon the equipment available and the individual preferences of the navigator. A daily routine, called the **day's work,** is established by the navigator and carried out with such variations as dictated by circumstances. While details vary with the navigator, a typical minimum day's work is:

1. Plot of dead reckoning throughout the day.
2. Observation and reduction of celestial observations for a fix during morning twilight.
3. Winding of chronometers and determination of chronometer error.
4. Observation of the sun for a morning sunline (on or near the prime vertical if made at about the same time as 5).
5. Azimuth of the sun for a compass check, commonly made at about the same time as a morning sunline observation. This may be an amplitude observation at sunrise.
6. Observation of the sun at or near noon. This is crossed with a morning sunline, advanced, or with an observation of the moon or Venus to obtain a noon (ZT 1200) position.
7. Computation of the day's run (noon to noon, or midnight to midnight).
8. Observation of the sun during the afternoon (on or near the prime vertical if made at about the same time as 9). This is primarily for use with the advanced noon sunline, or with a moon or Venus line, if the skies are overcast during evening twilight.
9. Azimuth of the sun for a compass check. This is commonly made at about the same time as an afternoon sun observation, but may be an amplitude observation at sunset.

10. Computation of the time of sunset, sunrise, and twilight, and preparation of a list of stars and any planets in favorable positions for observation during each twilight period, with the approximate altitude and azimuth of each body.

11. Observation and reduction of celestial observations for a fix during evening twilight.

12. Computation of the time of moonrise and moonset (if required).

13. Use of Loran and any other available radionavigation aid on a regular schedule, as every hour.

The list of celestial bodies available for observation is customarily prepared with the aid of a star finder such as No. 2102–D (art. 2210) or volume I of Pub. No. 249. This list is particularly helpful during evening twilight, when one desires to know where to look for the brightest stars or planets before the general pattern of stars becomes visible. Some navigators list or make a simple plot of the *relative* azimuths of the bodies, to assist in locating them. The brightest bodies may be visible at about the time of sunset, or even a little before.

In general, it is good practice to observe the brightest bodies as they appear in the evening, while the horizon is clear and sharp, and the dimmest first in the morning, before they fade from view. Cloud cover permitting, the normal practice of navigators is to observe stars in the eastern sky first. During morning twilight the eastern horizon becomes clearly defined first due to the rising sun; the dim stars above this horizon fade from view first because of the brighter eastern sky. During evening twilight the eastern horizon becomes indistinct first due to the setting sun, dictating observation of stars above this horizon first. Also, during evening twilight the bright stars in the darker eastern sky will be visible first.

Several observations should be made of each body, each sight being taken quickly to avoid eye fatigue. In general, it is better to use one good observation than to average several of questionable accuracy. At least five or six bodies should be observed. If the four most favorably situated ones provide a good fix, additional sights need not be reduced, but if doubt remains, information for obtaining additional lines is available. It is better to observe bodies all around the horizon than in the same semicircle. Thus, three bodies separated by 120° are better than three separated by 60°, for in the former case any constant error in altitude will be neutralized.

If a comparing watch is used, it should be compared with the chronometer or a time tick every time celestial observations are made. The index correction should be determined each time the sextant is used. If the horizon is used for this purpose, the measurement should be made *before* evening twilight observations and *after* morning twilight observations, while the horizon is sharp. If the horizon is not equally sharp in all directions, the best part should be used.

When skies clear after a prolonged period of overcast, or when clouds threaten to obscure the heavens, additional observations should be made, if available. During the day a series of sunlines might be obtained and advanced to a common time, or the moon or Venus might be available at a favorable azimuth. Sometimes observations can be made during the night, either by use of moonlight to illuminate the horizon, or by dark-adapting the eyes. At this time the moon, and bodies having an azimuth nearly the same as the moon, should be avoided because of the probability of false horizons on the illuminated water.

Sights may be reduced by any reliable method. The one most widely used by mariners is Pub. No. 229, used in conjunction with the *Nautical Almanac.* If a check is needed, a good practice is to use a different method and a different almanac, so that mistakes will not be repeated.

Before the development of modern sight reduction methods, celestial navigation was largely a matter of determining latitude by observation of bodies on or near the celestial meridian (including Polaris) and longitude by observation of bodies on or near the prime vertical. Longitude was computed by time sight. Frequently, this method of navigation was inconvenient. Often it produced misleading results, as when a north-south "longitude" line was used instead of the true line of position which might differ in direction by as much as 30° or more. Errors were introduced when an incorrect longitude was used for solving a reduction to the meridian, or an incorrect latitude for solving a time sight of a body some distance from the prime vertical. The use of azimuth with a time sight was an improvement, but was not well adapted to observations of celestial bodies near the celestial meridian. The modern navigator is freed from these restrictions. He is able to obtain a *line of position* extending in the correct direction almost any time a celestial body can be observed. He places no special significance upon latitude and longitude lines, and solves all sights by a common method of sight reduction.

It is good practice to use a workbook for the various solutions made at sea. This provides a valuable record which may be of inestimable value in the future. Entries should be neat, orderly, and intelligible to another navigator. All original data and computations should be included. The use of standard work forms is recommended. They are considered adequate but, for sight reduction of celestial observations, there is merit in using a form which uses a single column, so that several sights can be reduced in parallel columns. The *best* form for anyone to use is one he thoroughly understands and finds logical and least confusing. If an alteration in a work form reduces the number of errors made, it is a desirable change. Because of the difference of opinion among marine navigators, and the tendency to follow mechanically an established form without fully understanding the principles involved, a work form standardized for all navigators is probably undesirable, although such is widely used by air navigators, who use celestial navigation somewhat intermittently. When one has established the work forms he desires to use, he can have a rubber stamp made, or have the forms reproduced by printing. The former is probably preferable because it permits use of a bound workbook. However, printed forms can be punched for retention in a looseleaf binder.

At sea it is good practice to run the dead reckoning from fix to fix, determining set and drift of the current at each fix. The use of single lines of position and current to establish estimated positions is a matter of judgment. The ability to predict the difference between dead reckoning positions and fixes, which ability may be developed when the need is not apparent, can serve as a valuable asset when fixes are not available. In the U. S. Navy, the best position available is recorded in the log at 0800, 1200, and 2000. A typical plot of part of a day's run at sea (omitting possible radionavigation fixes) is shown in figure 2306.

It is good practice to compare the gyro repeaters with the steering magnetic compass each half hour and after each change of course at sea, to detect any discrepancy which may arise through malfunction. In making the comparison, one should not overlook changes in variation and deviation. The master gyrocompass should be compared with its repeaters from time to time.

One of the duties of the navigator is to inform the captain of the expected time of crossing time zone boundaries. The change of time is usually made at a convenient whole hour near the time of crossing a boundary, or during the night. Aboard

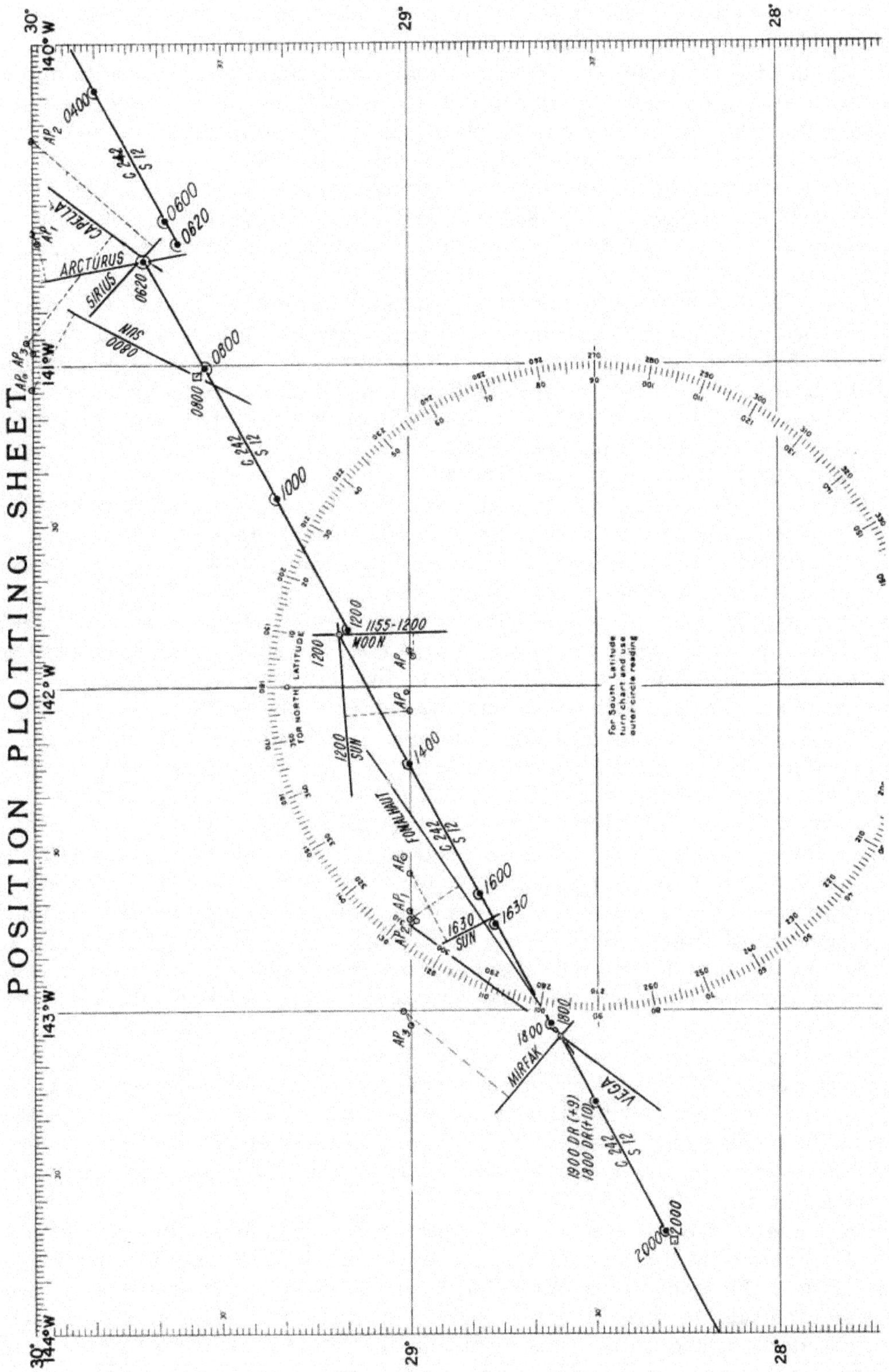

FIGURE 2306.—Typical minimum plot at sea (omitting possible radionavigation fixes).

some merchant ships the change is distributed equally through several watches, as 20 minutes during three consecutive watches.

It is common practice for the captain to maintain a **night order book.** Standing orders such as the conditions under which the captain is to be called, and the admonition to keep a sharp lookout, are usually given on the inside front cover. The orders for each night, if any, are recorded in order, over the captain's signature. They include items such as courses to be steered, speeds to be used, times and bearings of lights expected to be sighted, and any other pertinent navigational information. The navigator provides the captain with such information as he may require.

2307. Landfall.—After a voyage at sea, the first contact with land is of considerable importance. The accuracy with which one predicts the time and place of sighting land depends upon the accuracy of navigation. If consistent radionavigation fixes have been obtained at frequent intervals, and these positions are confirmed by a recent fix from celestial observations or other information, the prediction should be highly accurate. But if no fix has been available for several days, considerable doubt may surround the landfall.

Often the approximate distance offshore, if not the position, can be determined by means of soundings. Along most of their coasts the continents have a **continental shelf** (art. 3022) of relatively shoal water extending outward for a varying distance. A similar **island shelf** extends outward from many island groups. At the outer edge, called the **continental slope** (or **island slope**), a sharp increase in depth occurs. This is often at about the 100-fathom curve. Therefore, the crossing of this curve is often quite abrupt, and gives information on the distance offshore. The position of this and other depth curves may be indicated on the chart.

The place of making landfall has a definite relationship to the safety of the vessel, particularly in an area where shoaling is not uniform along the beach. For some time before making a landfall in such an area, it may be advisable to maintain both a dead reckoning and estimated position plot. The best obtainable position should be determined. Methods which are acceptable a thousand miles from land may not provide sufficiently exact data when a landfall is expected.

Only judgment, based upon existing circumstances, can determine the existence of a dangerous situation. If the water has shoaled to a dangerous degree, for instance, and the position of the vessel is seriously in doubt, one may have no recourse but to stand off or anchor and await daylight, improved visibility, or better information.

When contact is made with land, the first step should be to identify the point of contact. The *anticipated* point of making contact should be of assistance, but one should be alert to the possibility of similarly appearing land at other points within a reasonable distance on each side. The position of the vessel relative to land might be established even before land is sighted. Soundings, radio bearings, and radar may be used for this purpose.

2308. Entering port.—Before entering port, the navigator should have reliable information regarding the draft of his vessel. He should also have a reliable position relative to the land. Preparations for entering are similar to those for getting underway. The tide and tidal current tables, light list, coast pilot or sailing directions, and charts should all be broken out and studied so that one is familiar with conditions to be encountered. The *time* of entering might be selected to take advantage of favorable currents, and to arrive at the assigned berth at slack water. One should have a mental picture of what to expect when approaching from seaward under the anticipated conditions of lighting and visibility. The characteristics of all aids to navigation by day or night, as appropriate, and fog signals should be known or immediately available. In entering a strange port the navigator should carefully select the most suitable aids to use, with substitutes if these prove inadequate, or if there is any doubt as to their identity. Useful ranges, natural or artificial, should be noted. Danger bearings and danger circles should be drawn in and labeled, if this has not already been done. A danger sounding should be selected and drawn on the chart, if needed. Any shoal areas, wrecks, areas of unusually swift current, etc., should be noted.

The courses to be steered and the distance on each should be determined and recorded, or drawn and labeled on the chart. The identification of each turning point should be indicated. Definite courses should be steered, and changes made only when established positions indicate a departure from the planned track, or when necessitated by traffic. Course changes should occur at preselected points having definite identification. The position should not be permitted to be in doubt at any time, even in ports which are familiar to the navigator and considered easy to enter. Most avoidable groundings are caused by erroneous assumptions which should have been verified. The position should be checked frequently, using the most reliable information available. This may seem to be an unnecessary refinement, but in an emergency a position might be needed at a time when it cannot be obtained. When changes of course are ordered, it is good practice to indicate the amount and direction of change, or the new course, to avoid the possibility of having one's attention diverted at the moment the order should be given to check the swing or steady on the new course. In general, course changes are best made when a given aid to navigation or other landmark is abeam, or when the ship is on a range.

If it becomes necessary to pass between visible dangers without suitable marks for obtaining fixes, a track midway between dangers can be followed by eye more accurately than one closer to either side. If a vessel is to pass near reefs or shoals, it is sometimes possible to observe these from a position aloft, particularly if the sun is astern.

The actual navigation while entering port is similar to that when leaving port. A typical plot in pilot waters is shown in figure 2308. The entering of pilot waters should be accompanied by a mental reorientation and an increased alertness. The use of a local pilot, unless this is a mandatory requirement, is a matter which should be decided in each case. Whether or not a pilot is used, local harbor regulations should be followed, for the presence of a pilot does not relieve the master of his responsibility. One should not forget to note the time of entering the area where local or inland rules of the road apply.

Speed in the vicinity of wharves, construction work, dredges, small boats, etc., should be carefully controlled to avoid damage to them.

If the vessel anchors, the anchorage should be selected carefully, considering local regulations as well as suitability and safety, including the holding qualities of the bottom. If there is any doubt as to the depth of water, a boat might be sent in ahead to take soundings. If space is limited, the appraoch to the anchorage should be planned and executed carefully. As soon as the anchor is let go, the position should be determined accurately. Bearings of a number of prominent landmarks and lights should be measured and recorded, as a guide in determining whether or not the vessel drags anchor. A **drag circle** is helpful. Using the position of the anchor as its center, the drag circle is constructed with a radius equal to the scope

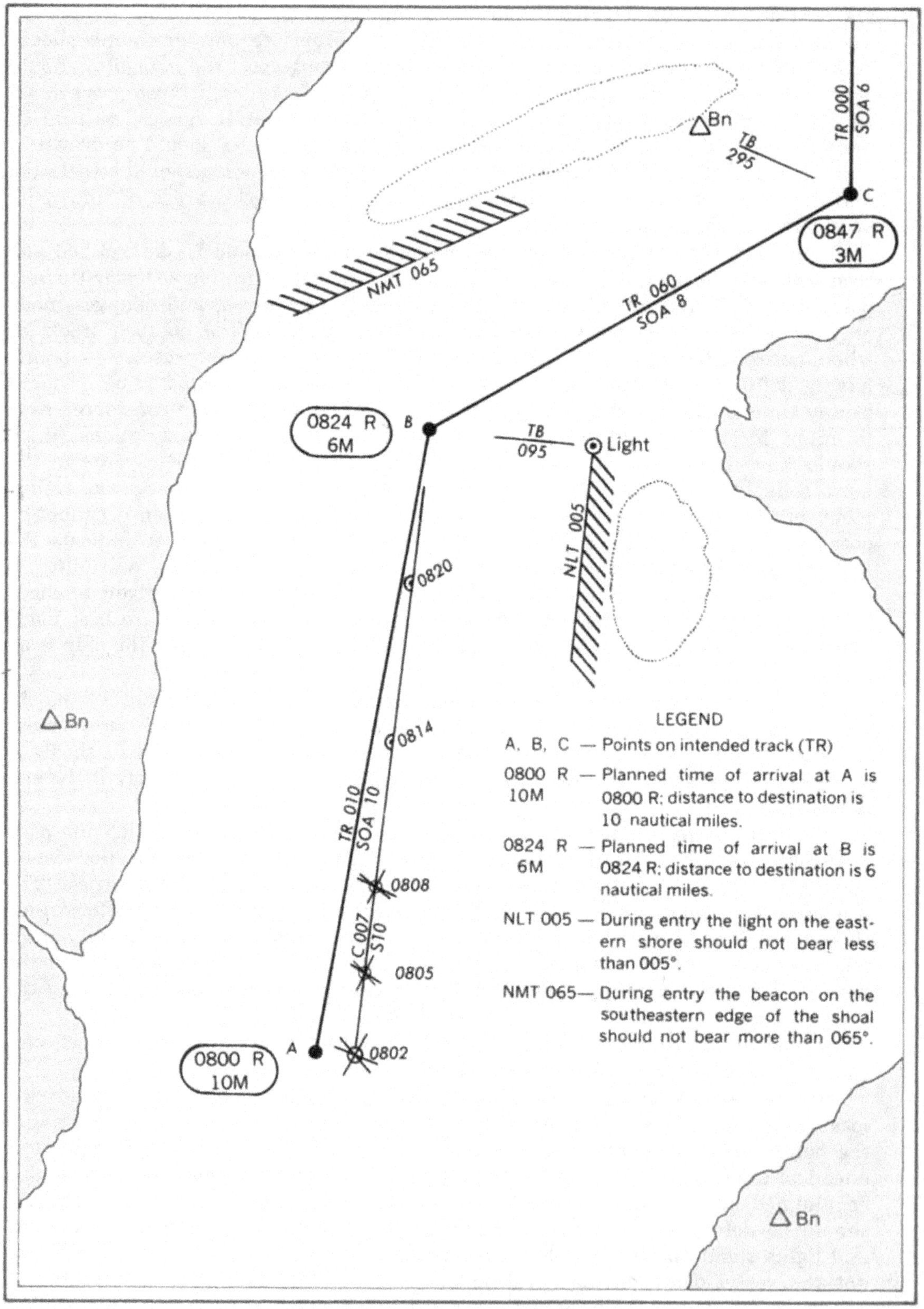

FIGURE 2308.—Typical plot in pilot waters.

plus the distance from the hawsepipe to the point of observation. A **swing circle**, with radius equal to the scope plus vessel's length, may also be helpful.

2309. Fog.—During periods of reduced visibility, the navigator's work is more difficult. At sea he is prevented from making celestial observations. Even when the fog is so shallow that celestial bodies are visible, the horizon is not available as a reference. An artificial-horizon sextant may prove of some value at such a time, but unless the sea is almost a flat calm, the results are likely to be less reliable than the dead reckoning. Radio aids to navigation are affected little by fog. Unless the vessel is approaching land, there is generally no cause for concern regarding the navigation, the principal danger being one of collision with other vessels. Usually the navigator merely waits for the fog to lift.

When a coast is approached, however, a wait may be impractical. The safety of the vessel requires reliable positional information. Along a coast where the shoaling is gradual, the echo sounder can be of great assistance in indicating the distance off. But along a coast having abrupt shoaling, the first indication of shallow water may be obtained so close to the beach that action to avoid grounding is not possible. If radio aids such as Loran, radio direction finder, and radar are available, they can provide useful information. If the vessel is near enough to a shore with steep cliffs, the echo of the vessel's whistle may provide indication of the distance off.

The decision of whether to enter a fogbound harbor should be made carefully. Once committed to the channel, the vessel may have no alternative but to continue on to the anchorage or wharf, for in some areas there is not room to turn back, and anchoring is unsafe. It is sometimes wiser to stand off or anchor for a few hours than to risk danger of grounding or collision.

If the decision is made to enter, one should be prepared for any reasonable eventuality. The proximity of danger and the presence of currents make necessary the maintenance of a good position at all times. Fog limits the number of objects that can be used for fixing position, and destroys the overall view of the area. The radio direction finder and radar, both shipborne and shore-based, have done much to reduce the hazard due to fog, but they have not eliminated it. The need for special precautions and increased vigilance is still present.

During periods of reduced visibility the practice of steering exact courses, with precise changes at definite points, is of great assistance in pilot waters. If the vessel is following a channel, each buoy should be located successively. If the fog is dense, this requires careful steering and attention to all details, such as indications of current, changes of wind, etc. If a single buoy is missed, consideration should be given to anchoring and waiting for improved visibility.

With the possible exception of radar, the most important navigational aid during fog in pilot waters is the echo sounder or hand lead. Continuous soundings, compared with the chart, can provide valuable information on the position and safety of the vessel. The decision as to whether to plot a line of soundings on transparent material, or along the edge of a piece of paper, and compare this with the chart is a matter of judgment. In general, the procedure is valuable when approaching a harbor or proceeding in an open part of a large bay, but in a channel or other restricted waters the method is not needed and might prove distracting.

During fog one should keep a sharp lookout for any objects that might appear momentarily through thin places in the fog. It is well to have a lookout stationed aloft, and another in the bow, for the visibility may vary with height.

The lookouts and all persons on the bridge should listen intently for fog signals. As soon as such a signal is heard, an effort should be made to identify its sources and determine its bearing. However, experience in the use of sound signals indicates that they are not wholly reliable. In particular, relative intensity of a sound is not a reliable indication of its distance, or whether the distance is increasing or decreasing. A signal may be totally inaudible in certain areas close to its source. Neither is its apparent direction always a correct indication of its actual direction. A fog signal may not be in operation when fog is present a short distance from a station but is unobserved from it. Transmission of sound through water is subject to uncertainties due principally to differences in density in different parts of the sea, causing the sound to be deflected.

It is well to remember that at reduced speed the relative effect of current is correspondingly greater, since the effect of current is proportional to time, not to the speed of the vessel.

2310. Navigation of small craft.—In principle, the navigation of small craft is the same as that of a large ship, but because of the shallower draft, greater maneuverability, and possible limitations of equipment of small craft, there are important differences. Small craft spend most of their time within sight of land, where navigation is largely a matter of piloting. They generally skirt the beach close enough to be able to reach safety in case of storm or fog, and since most of them are used primarily for pleasure, there is a natural tendency for the navigation to be a less continuous process than in larger craft.

The equipment carried and the type of navigation employed depend primarily upon the use of the craft and the preference of the user. If a rowboat, canoe, or small sailboat is to be used only close to the shore in good weather, "seaman's eye" might be sufficient for all navigational purposes. But if there is any possibility of the craft being out in a fog, or proceeding to greater distances from shore, fog-signalling apparatus, a compass, and some means of taking soundings should be carried.

A wide variety of equipment is available for yachts, and from this, suitable items can be selected. A minimum list should include a compass, pelorus, charts, plotting equipment (many types are available), means for determining speed or distance, log book, tide and tidal current tables, light list, coast pilot or sailing directions, hand lead, binoculars, flashlight, and fog-signal apparatus. A barometer and thermometer are useful.

Several items of electronic equipment, some of which are relatively inexpensive, are available for use in small craft, to aid in navigation and increase safety. The principal item of radio equipment, from the standpoint of safety, is a marine radiotelephone, which in addition to providing normal communication to other boats and the shore, permits the boat carrying it to call for help in distress, and assists in the location of the distressed vessel. The radio direction finder is a simple device requiring little power, an important factor on small craft. A multiband direction finder may be used as a second receiver in the broadcast and radiotelephone bands. Portable broadcast receivers permit reception of weather information on even the smallest boats. For larger craft, where ample power is available, radar and Loran may be good investments. In addition, every small craft should carry a corner reflector (art. 4207), so as more readily to reflect radar signals. In an emergency a metal bucket might be of some value as a reflector.

If the craft is to proceed out of sight of land for more than short intervals, celestial navigation equipment should be carried. This includes a sextant, an accurate timepiece, an almanac, sight reduction tables, and perhaps a star finder. If there is doubt as to advisability of including some item of equipment, the safer decision is to include it. It is better to have unused equipment than to risk danger of becoming lost because of lack of needed equipment.

The practice of navigation in small craft varies even more widely than the equipment carried. The variation extends from complete navigation similar to that of a large ocean vessel to no navigation other than by eye. The completeness of the navigation should fit the circumstances. There is an understandable tendency among small craft navigators of limited experience to underestimate the need for thorough and complete navigation. In general, it is good practice for the navigator of a small craft to establish the routine of always following definite courses from buoy to buoy or from landmark to landmark, so that the sudden onset of low visibility will not find him unable to proceed to safety without delay. He should change course at established points, maintain knowledge of his position at all times, and have reliable information on the deviation of his compass. There is a place in small craft navigation for a complete, accurate, neat plot. Where this is impractical because of heavy weather or limited plotting space, a careful log and dead reckoning by table 3 should be substituted.

The accounts given in yachting magazines, and the large number of calls for assistance received by the Coast Guard, indicate an inadequacy of the navigation of many small craft. Part of this is due to a lack of appreciation of the need for careful navigation. Much of it is due to lack of knowledge on the part of the small craft owner. The decision to omit some part of navigation should stem from knowledge, not ignorance. To the adequately informed, navigation can be part of the pleasure of yachting.

CHAPTER XXIV

SHIP WEATHER ROUTING

2401. Introduction.—*Ship weather routing* is a procedure whereby an optimum route is developed based on the forecasts of weather and seas and the ship's characteristics for a particular transit. Within specified limits of weather and sea conditions, the term optimum is used to mean maximum safety and crew comfort, minimum fuel consumption, minimum time underway, or any desired combination of these factors.

The ship routing agency, acting as an advisory service, attempts to avoid or reduce the effects of adverse weather and sea conditions on a ship by issuing initial route recommendations prior to sailing, recommendations for track changes while underway (diversions), and weather advisories to alert the commanding officer or master with respect to approaching unfavorable weather and sea conditions which cannot be effectively avoided by a diversion. Adverse weather and sea conditions are defined as those conditions which will cause a significant speed reduction or time loss, for example, speed reduced by one third or a lesser speed reduction causing a loss of at least 6 hours transit time.

The initial route recommendation is based on a survey of weather and sea forecasts between the point of departure and the destination and takes into account the hull type, speed capability, cargo, and loading conditions. The ship's progress is continually monitored and, if adverse weather and sea conditions are forecast along the ship's current track, a recommendation for a diversion is normally transmitted to the ship. By this process of initial route selection and continued monitoring of the ship's progress for possible change in relation to the forecast weather and sea conditions along a route, it is possible to maximize ship's speed and safety.

In providing optimum sailing conditions, the advisory service also attempts to reduce transit time by avoiding the adverse conditions which may be encountered on a shorter route; or if the forecasts permit, diverting to a shorter track to take advantage of favorable weather and sea conditions. The greatest potential advantage for this ship weather routing exists when: (1) the passage is relatively long, generally about 1,500 miles or more; (2) the waters are navigationally unrestricted so that there is a choice of routes (alternatively, navigational restrictions are limiting but at the same time offer possible protection from adverse weather); and (3) weather is a factor in determining the route to be followed.

The use of this advisory service in no way should relieve the commanding officer or master of responsibility for prudent seamanship and safe navigation. There is no intent by the routing agency to inhibit the exercise of professional judgement, capabilities, and the prerogatives of commanding officers and masters.

The purpose of this chapter is to acquaint the mariner with the basic philosophy and procedures of ship weather routing as an aid to his understanding of the ship routing agency's recommendations. This information should enhance his ability to determine the correct course of action when encountering a hostile environment.

2402. Development.—Today, ship weather routing uses modern weather forecasting techniques and computer procedures to provide optimum routes. It has only been with the recent advent of extended range forecasting and the development of selective climatology that a ship routing system has been possible. The ability to effectively advise ships to take advantage of favorable wind, seas, and ocean currents had been hampered previously by forecast limitations and the lack of an effective communications system with which to advise the commanding officer or master.

Prior to World War II all development work was in the area of data accumulation and climatology. Benjamin Franklin, as deputy postmaster general of the British Colonies in North America produced a chart of the Gulf Stream from information supplied by masters of New England whaling ships. This first mapping of the Gulf Stream helped improve the mail packet service between the British Colonies and England. In some passages the sailing time was reduced by as much as 14 days over routes previously sailed. In the mid-19th century, Lieutenant Matthew Fontaine Maury, USN, compiled large amounts of atmospheric and oceanographic data from ships' log books (art. 117). For the first time a climatology of ocean weather and currents of the world was available to the mariner. This information was used by Maury to develop seasonally recommended routes for sailing ships and early steam powered vessels in the latter half of the century. In many cases, Maury's charts were proved correct by the savings in transit time. On one trade route alone, the average transit time from New York to California around Cape Horn was reduced from 183 days to 139 days with the use of his recommended seasonal routes.

In the 1950's the concept of ship weather routing was put into operation by several private meteorological groups and by the U. S. Navy. By applying the available surface and upper air forecasts to transoceanic shipping, it was possible to effectively avoid much of the heavy weather while generally sailing shorter routes than previously.

Optimum Track Ship Routing (OTSR), the ship routing service of the U. S. Navy, utilizes short range and extended range forecasting techniques in the route selection and surveillance procedures. The short range dynamic forecasts of 3 to 5 days are derived from the meteorological primitive equations. These forecasts are computed twice daily from a data base of northern hemisphere surface and upper air observations, and include surface pressure, upper air constant pressure heights, and the spectral wave values. A significant increase in data input, particularly from derived satellite information over ocean areas, can be expected to extend the time period for which these forecasts are useful.

For extended range forecasting, generally 3 to 14 days, a computer searches a library of historical northern hemisphere surface pressure and 500 millibar analyses for an analogous weather pattern. This is an attempt at selective climatology by matching the current weather pattern with past weather patterns and providing a logical sequence-of-events forecast for the 10- to 14-day period following the dynamic forecast. It is performed for both the Atlantic and Pacific Oceans using climatological data for the entire period of data stored on tape (30 years in use in 1977). For longer ocean transits, monthly climatological values of wind, seas, fog, and ocean currents are used to further extend the time range.

Aviation was first in applying the principle of minimum time tracks (MTT) to a changing wind field. But the problem of finding an MTT for a specific flight is much simpler than for a transoceanic ship passage. This is because aircraft flight time is only hours, while the ship transit time is usually 7 to 14 days or more. Thus, marine minimum time tracks require significantly longer range forecasts to effectively develop a best weather-least cost route.

Automation has enabled ship routing agencies to develop realistic minimum time tracks. Computation of minimum time tracks for proposed transits or ships already underway makes use of:

1. a navigation system to compute route distance, time enroute, estimated times of arrival (ETA's), and to provide 6 hourly DR synoptic positions for the range of the dynamic forecasts for the ship's current track;

2. a surveillance system which surveys the environmental factors of wind, seas, fog, and ocean currents obtained from the dynamic and climatological fields;

3. an environmental constraint system imposed as part of the route selection and surveillance process. Constraints are the upper limits of wind and seas desired for the transit and are determined by the ship's loading, speed capability, and vulnerability. The constraint system is an important part of the route selection process and acts as a warning system when the weather and sea forecast along the present track is such that the ship is expected to transit an area which exceeds those predetermined limits;

4. ship performance curves, or speed curves, used to approximate ship's speed of advance (SOA) while transiting the forecast sea states.

Ship weather routing services are being offered as an aid to shipping by the governments of many nations. These include Japan, United Kingdom, Union of Soviet Socialist Republics, Netherlands, Federal Republic of Germany, and the United States. Also, several private firms provide the service to shipping industry clients.

2403. Ship and cargo considerations.—Ship and cargo characteristics have a significant influence on the application of ship weather routing. Ship size, speed capability, and type of cargo are important considerations in the route selection process prior to sailing and the surveillance procedure while underway. These ship characteristic factors help to identify the degree of vulnerability to potential adverse conditions and the ship's ability to effectively avoid the adverse weather and seas by diversion.

Generally, ships with higher speed capability and less cargo encumbrances will have shorter routes and be better able to maintain near normal SOA's than ships with lower speed capability or cargos which may generate unfavorable ship motions resulting from wind and sea conditions. Some routes are unique because of the type of ship or cargo. Avoiding one element of weather such as heavy head or beam seas to reduce pounding or rolling may be of prime importance. For example, a 20-knot ship with a deck cargo may be severely hampered in its ability to maintain a 20-knot SOA in any seas exceeding moderate head or beam seas because of the greater ship motions or other effects resulting from the deck load's characteristics. A similar ship without the deck load is not as vulnerable to these same or higher sea conditions and is able to nearly maintain the 20-knot SOA. In towing operations, a tug is more vulnerable to adverse weather and sea conditions, not only in consideration for the tow, but also because of its already limited speed capability and the difficulty in effectively avoiding adverse weather and sea conditions by a diversion.

Ship performance curves (speed curves) are used to estimate the ship's SOA while transiting the forecast sea states. The curves indicate the effect of head, beam, and following seas of various significant wave heights on the ship's speed. Figure 2403 is a performance curve prepared for an 18-knot vessel.

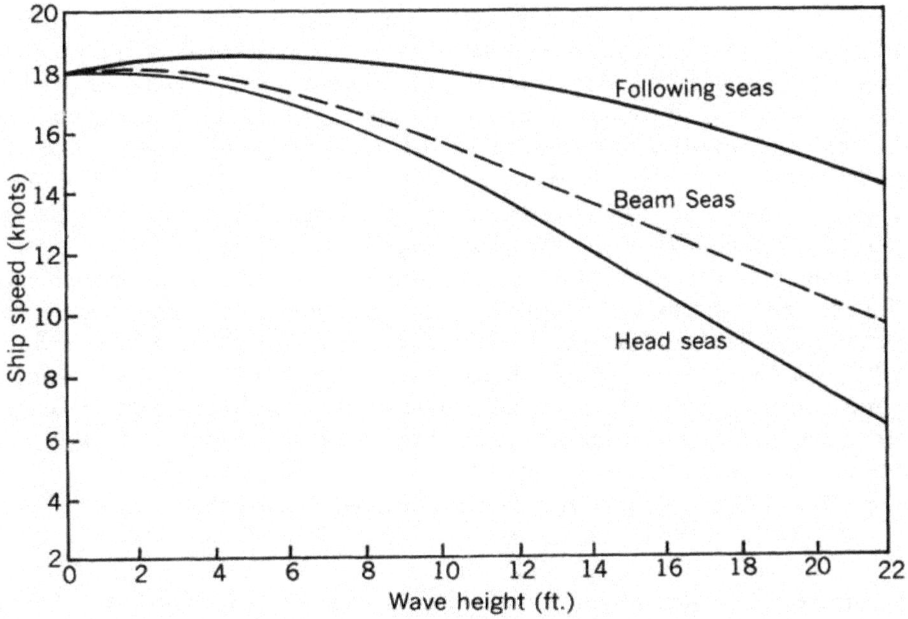

FIGURE 2403.—Performance curves for head, beam, and following seas (18-knot vessel).

With the aid of the speed curves it is possible to determine just how costly a diversion will be in terms of the required distance and time. A diversion may not be necessary where the duration of the adverse conditions is limited. In this case, it may be better to ride out the weather and seas knowing that a diversion, even if able to maintain the normal SOA, will not overcome the increased distance and time required by the diversion.

At other times, the diversion track is less costly because it avoids an area of adverse weather and sea conditions while being able to maintain normal SOA even though the distance to destination is increased. Based on input data for environmental conditions and ship's behavior, route selection and surveillance techniques seek to achieve the optimum balance between time and distance and acceptable environmental and seakeeping conditions. Although speed performance curves are an aid to the ship routing agency, the response by mariners to deteriorating weather and sea conditions is not uniform. Some reduce speed voluntarily or change heading sooner than others when unfavorable conditions are encountered. Certain waves with characteristics such that the ship's bow and stern are in successive crests and troughs present special problems for the mariner. Being nearly equal to the ship's length, such wavelengths may induce very dangerous stresses. The degree of hogging and sagging may be more apparent to the mariner than to the ship routing agency. Therefore, adjustment in course and speed for a more favorable ride may be initiated by the commanding officer or master when this situation is encountered.

2404. Environmental factors of importance to ship weather routing are those elements of the atmosphere and ocean that may produce a change in the status of a ship transit. In ship routing, consideration is given to wind, seas, fog, ice, and ocean currents. While all of the environmental factors are important for route selection and surveillance, optimum routing is normally considered attained if the effects of wind and seas can be optimized.

The effect of wind speed on ship performance is in some respects difficult to determine. In light winds (less than 20-knots), ships lose speed in headwinds and gain speed slightly in following winds. For higher wind speeds, ship speed is reduced in both head and following winds. This is due to the increased wave action and indicates the importance of sea conditions in determining ship performance. In dealing with wind, it is also necessary to know the ship's sail area. For example, high winds will have a greater adverse effect on a large, fully loaded container ship than a large, fully loaded tanker of similar length.

Wave height is the major factor affecting ship performance. Wave action is responsible for ship motions that reduce propeller thrust. The relationship of ship speed to wave direction and height is similar to that of wind. Head seas reduce ship speed, while following seas increase ship speed only slightly. In heavy seas, exact performance may be difficult to predict because of the adjustments to course and speed for shiphandling and comfort. Although the effect of sea and swell is much greater than wind, it is difficult to separate the two in ship routing.

In an effort to provide a more detailed description of the actual and forecast sea state, Fleet Numerical Oceanography Center, Monterey, California, uses the Spectral Ocean Wave Model (SOWM) for the U. S. Navy's Optimum Track Ship Routing service. The spectral wave model provides energy values from 12 different directions (30° sectors) and 15 frequency bands for wave periods from 6 to 26 seconds with the total wave energy being propagated throughout the grid system as a function of direction and frequency. It is based on the analyzed and forecast planetary boundary layer model wind fields and is produced for the Northern Hemisphere. For OTSR purposes, primary and secondary waves are derived from the spectral wave program, where the primary wave train has the principal energy (direction and frequency), and the secondary has to be 20 percent of the primary.

Fog, while not directly affecting ship performance, should be avoided as much as feasible in order to maintain normal speed in safe conditions. Extensive areas of fog during summertime can be avoided by selecting a lower latitude route than one based solely upon wind and seas. Although the route may be longer, transit time may be less due to not having to reduce speed in areas of very low visibility for safety considerations.

Ocean currents do not present a significant routing problem, but they can be a determining factor in route selection and diversion. This is especially true when the points of departure and destination are at relatively low latitude. The important considerations to be evaluated are the difference in distance between a great-circle route and a route selected for optimum current, with the expected increase in SOA from the following current, and the decreased probability of a diversion for weather and seas at the lower latitude. For example, it has proven beneficial to remain equatorward of approximately 22°N for westbound passages between the Canal Zone and southwest Pacific ports. For eastbound passages, if the maximum latitude on a great-circle track from the southwest Pacific to the Canal Zone is below 24°N, a route passing near the axis of the Equatorial Countercurrent is practical because the increased distance is offset by favorable current. Direction and speed of ocean currents are more predictable than wind and seas, but some variableness is to be expected. Major ocean currents can be disrupted for several days by very intense weather systems such as hurricanes and typhoons.

The problem of ice is twofold: floating ice (icebergs) and deck ice (art. 3825). If possible, areas of icebergs or pack ice should be avoided because of the difficulty of detection and the potential for disaster. Deck ice may be more difficult to contend with from a ship routing point of view because it is caused by freezing weather associated with a large weather system. While mostly a nuisance factor on large ships, it may cause significant problems with the stability of small ships.

Generally, the higher the latitude of a route, even in the summer, the greater are the problems with the environment. Certain operations should benefit from seasonal planning as well as optimum routing. For example, towing operations poleward of about 40° latitude should be avoided in non-summer months as much as possible.

2405. Synoptic weather considerations.—A ship routing agency should direct its forecasting skills to avoiding or limiting the effect of weather and seas associated with extratropical low pressure systems in the mid or higher latitudes and the tropical systems in low latitudes by route selection and diversion. Seasonal or monsoon weather is also a definite factor in route selection and diversion in certain areas of the world's oceans.

Despite the amount of attention and publicity given to tropical cyclones, midlatitude low pressure systems generally present more difficult problems to a ship routing agency. This is primarily due to the fact that major ship traffic is sailing in the latitudes of the migrating low pressure systems and the amount of potential exposure to intense weather systems, especially in winter, is much greater.

Low pressure systems weaker than gale intensity (winds less than 34 knots) are not a severe problem for most ships. However, a relatively weak system may generate prolonged periods of rough seas which are uncomfortable and may hamper normal work aboard ship. Ship weather routing can frequently limit rough conditions to short periods of time and provide more favorable conditions for most of the transit. Relatively small ships, tugs with tows, low powered ships, and ships with sensitive cargoes can be significantly affected by weather systems weaker than gale intensity. Use of a routing agency, or at least giving careful attention to the latest weather information, is considered very prudent and can be beneficial.

Gales (winds 34 to 47 knots) and storms (winds greater than 48 knots) are of such intensity that they generate very rough or high seas which force a reduction in speed in order to gain a more comfortable and safe ride. Because of the extensive geographic area covered by a well developed, intense, low pressure system, once ship's speed is reduced, the ability to improve the ship's situation is severely hampered. Thus, exposure to potential damage and danger is greatly increased. A recommendation for a diversion by a routing agency, well in advance of the intense weather and associated seas, will limit the duration of exposure of the ship to potential problems. If effective, ship speed will not be reduced and satisfactory progress will be maintained, even though the remaining distance to destination is increased. Overall transit time is usually shorter than if no track change had been made and the ship had remained in heavy weather. In some cases diversions are made to avoid adverse weather conditions and shorten the track at the same time. Significant savings can be the result.

In very intense low pressure systems with high winds and long duration over a long fetch, seas will be generated and propagated as swell over considerable distances. Even on a diversion, it is difficult to effectively avoid all unfavorable conditions. Generally, original routes for transoceanic passages issued by the U. S. Navy's ship routing service are equatorward of the 10% frequency isoline for gale force winds for the month of transit as interpreted from the *U. S. Navy Marine Climatic Atlas of the World.* These are shown in figures 2405a and 2405b for the

Pacific. To avoid the area of significant gale activity in the Atlantic from October to April, the latitude of transit is generally in the lower thirties.

The areas and seasons and, to some degree, the probability of development of tropical cyclones are fairly well defined in climatological publications. In long range planning considerable benefit can be gained by limiting the exposure to the potential hazards of tropical systems.

For routing in the North Pacific it has proven very helpful to provide routes which avoid the areas with the greatest probability of tropical cyclone formation. Avoiding existing tropical cyclones with a history of 24 hours or more of 6-hourly warnings is in most cases relatively straightforward. The concern is largely the price to pay in added time and distance to maintain satisfactory conditions. However, when transiting the tropical cyclone generating area the ship under routing may provide the first report of environmental conditions indicating that a new disturbnace is developing. In the eastern North Pacific the generating area for a high percentage of tropical cyclones is relatively compact (fig. 2405c). It has proven beneficial on certain ocean transits (Canal Zone to or from mid or western North Pacific ports) to remain equatorward of a line from lat. 9°N, long. 90°W to lat. 14°N, long. 115°W. In the western North Pacific, again depending on the point of departure of destination, it is advisable to hold north of 22°N when no tropical systems are known to exist (fig. 2405d).

In the Atlantic, it is considered prudent to sail near the axis of the Bermuda high or northward to avoid the area of formation of tropical cyclones. By remaining clear of these generating areas initially, it alleviates the problem of attempting to provide a diversion to a ship that may be in close proximity to a developing tropical system. Also, the distance and time lost will be much greater on a diversion than if the route had originally avoided the generating area. Avoiding any existing tropical cyclone takes precedence over avoiding the general area of potential development.

It has proven equally beneficial to employ similar considerations for routing in the monsoon areas of the Indian Ocean and the South China Sea. This is accomplished by providing routes and diversions that generally avoid the areas of high frequency of gale force winds and associated heavy seas as much as feasible. Ships are then able to remain in satisfactory conditions with only limited increases in route distance.

Depending upon the points of departure and destination, there are many combinations of routes that can be recommended when transiting the northern Indian Ocean (Arabian Sea, Bay of Bengal) and the South China Sea. For example, in the Arabian Sea during the summer monsoon, routes to and from the Red Sea, the western Pacific, and the eastern Indian Ocean should hold equatorward. Ships proceeding to the Persian Gulf during this period are held farther south and west, attempting to put the heaviest seas on the quarter or stern when transiting the Arabian Sea. Eastbound ships departing the Persian Gulf may proceed generally east southeast toward the Indian sub-continent then south, to pass north and east of the highest southwesterly seas in the Arabian Sea. Westbound ships out of the Persian Gulf for the Cape of Good Hope appear to have little choice in routes unless considerable distance is added to the transit by passing east of the highest seas. In the winter monsoon, routes to or from the Red Sea for the western Pacific and the Indian Ocean are held farther north in the Arabian Sea to avoid the highest seas. Ships proceeding to the Persian Gulf from the western Pacific and eastern Indian Ocean may hold more eastward when proceeding north in the Arabian Sea. Ships departing the Persian Gulf area will have considerably less difficulty than during

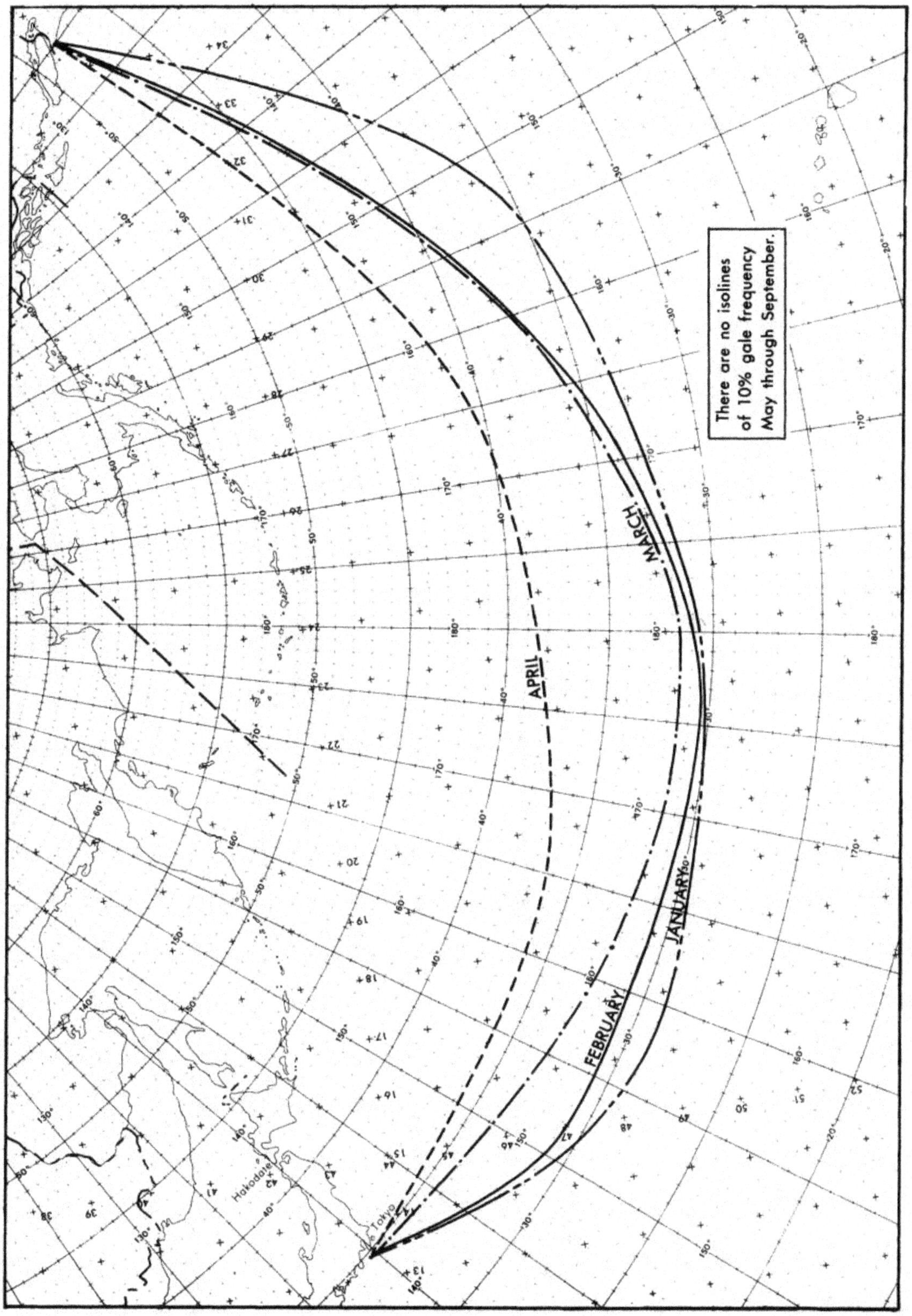

FIGURE 2405a.—Generalized 10% frequency isolines of gale force winds for January through April.

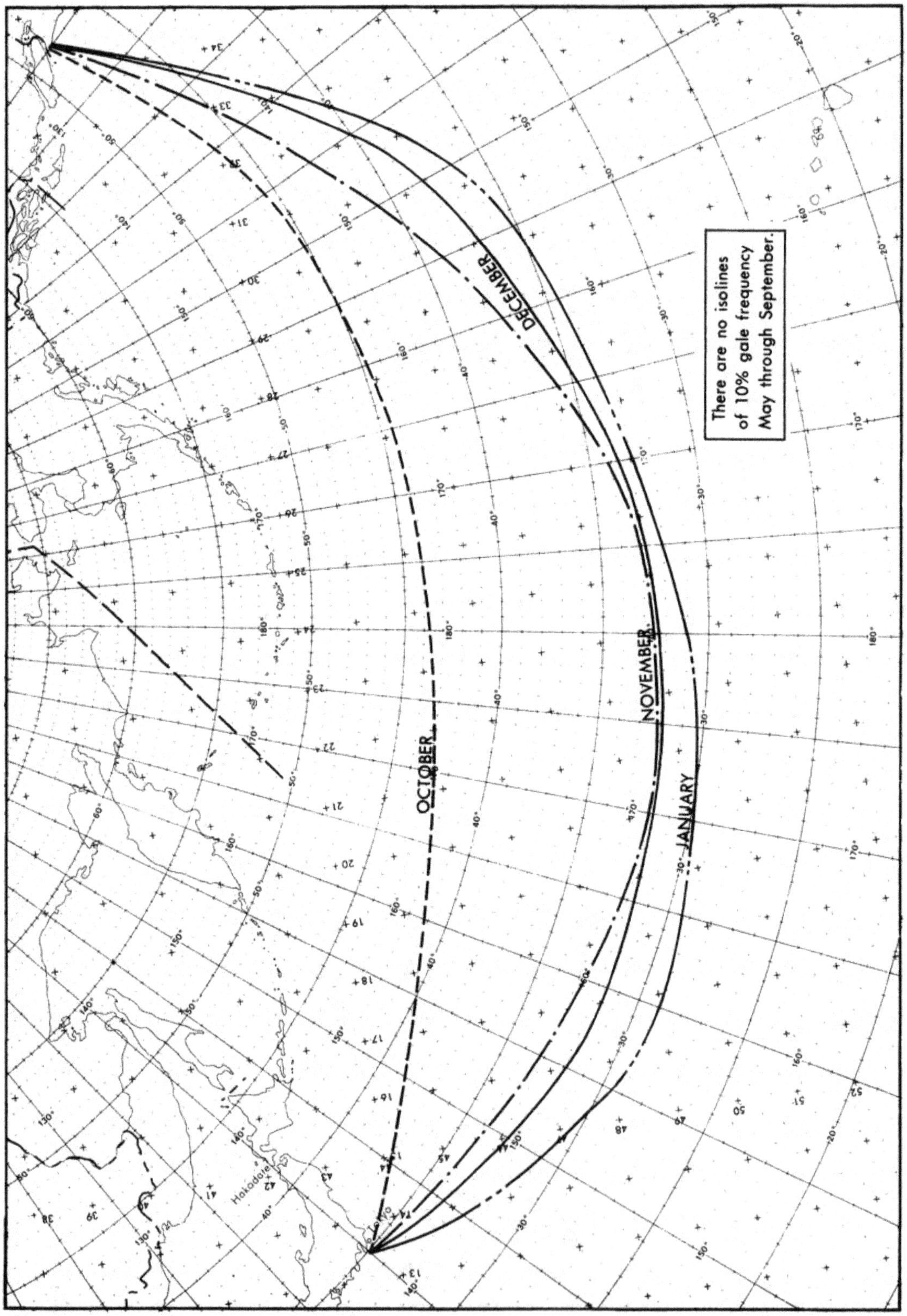

FIGURE 2405b.—Generalized 10% frequency isolines of gale force winds for October through April.

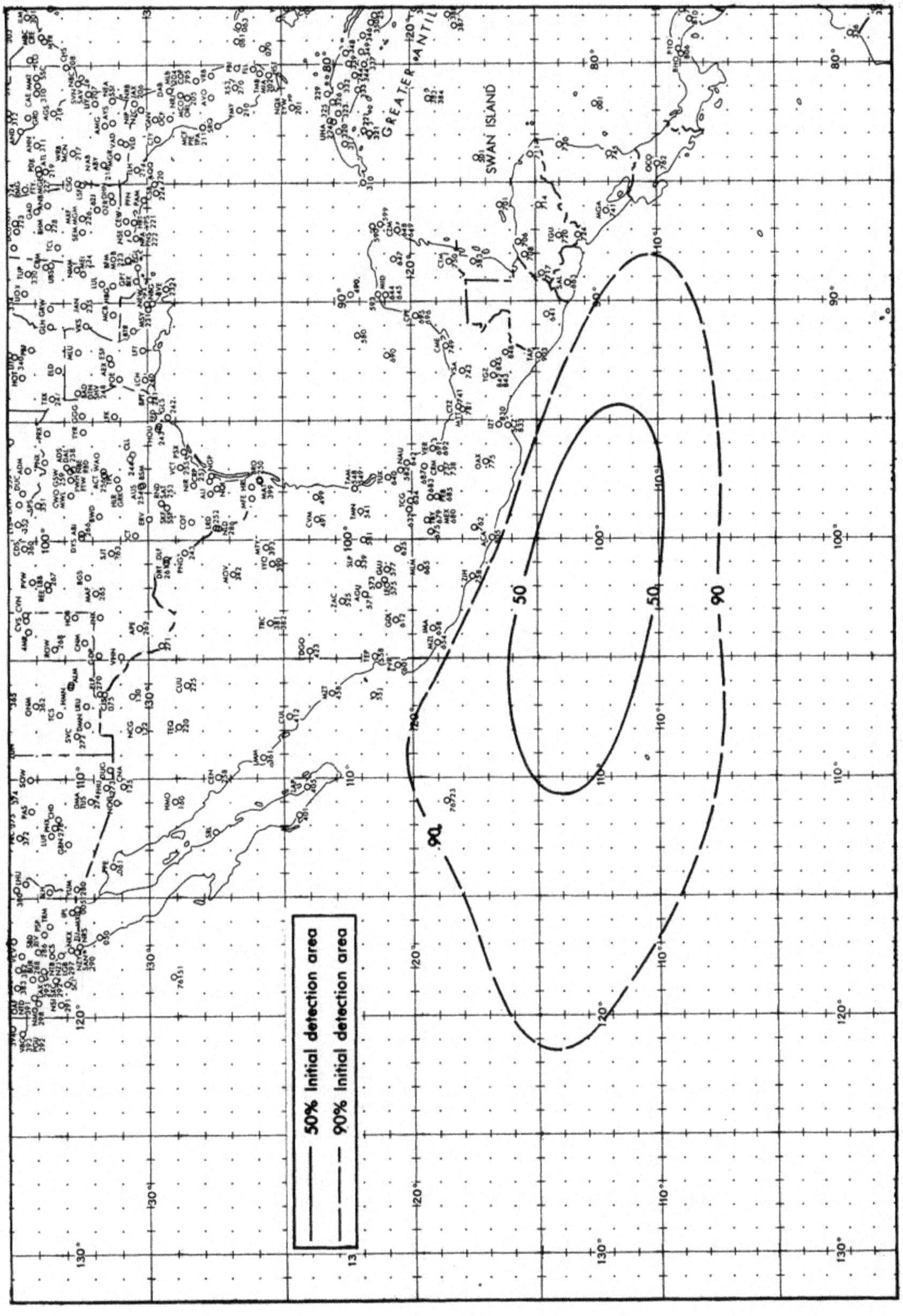

FIGURE 2405c.—Area of initial detection of high percentage of tropical cyclones which later developed to tropical storm or hurricane intensity, 1957–1974 (after Bowman and Renard).

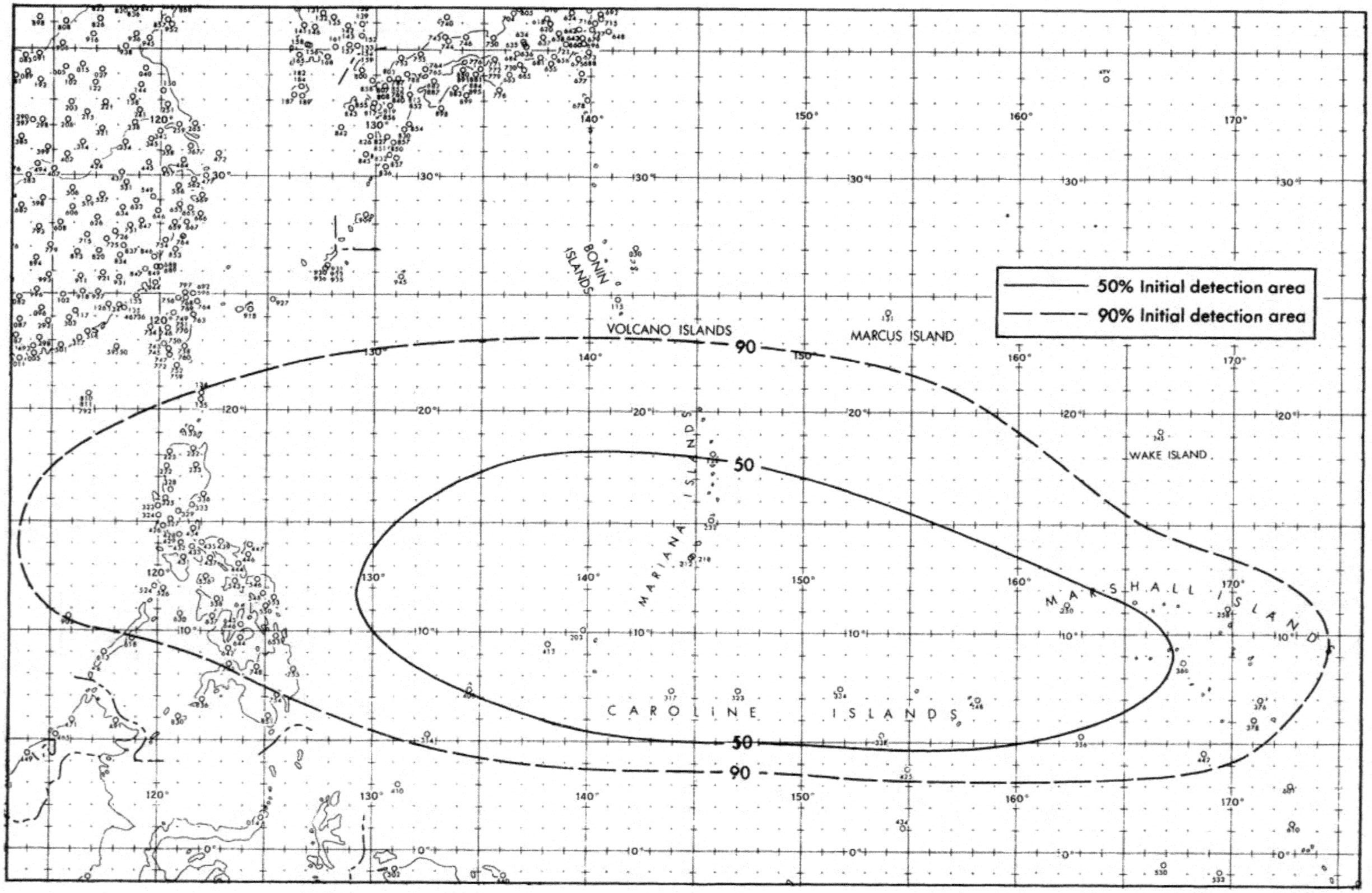

FIGURE 2405d.—Area of initial detection of high percentage of tropical cyclones which later developed to tropical storm or typhoon intensity, 1946–1973 (after Gray).

the summer monsoon. Similar considerations can be given when routing ships proceeding to and from the Bay of Bengal. Refer to sailing directions for recommended routes in the Indian Ocean.

In the South China Sea, transits via the Palawan Passage are given consideration when strong, opposing wind and seas are forecast. This is especially true during the winter monsoon.

During periods when the major monsoon flow is slack, ships can be diverted to the shortest track as conditions permit.

2406. Special weather and environmental considerations.—In addition to the synoptic weather considerations in ship weather routing, there are special environmental problems that can be avoided by following recommendations and advisories of ship routing agencies. These problems generally cover a smaller geographic area and are seasonal in nature but are still quite important to ship routing.

In the North Atlantic, because of heavy shipping traffic, frequent poor visibility in rain or fog, and restricted navigation, particularly east of Dover Strait, some mariners prefer to transit to or from the North Sea via Pentland Firth passing north of the British Isles rather than via the English Channel.

Weather routed ships generally avoid the area of dense fog with low visibility in the vicinity of the Grand Banks off Newfoundland and the area east of Japan north of 35°N. Fishing vessels in these two areas provide an added hazard to safe navigation. This condition exists primarily from June through August and sometimes into September. Arctic supply ships en route from the U. S. east coast to the Davis Strait-Baffin Bay area in the summer frequently transit via Cabot Strait and the Strait of Belle Isle where navigation aids are available and icebergs are generally grounded.

Icebergs are a definite hazard in the North Atlantic from late February through June and occasionally later. Although there is considerable variation between years in time and geographic area, the hazard of floating ice is frequently combined with restricted visibility in fog. International Ice Patrol reports and warnings are incorporated into the planning of routes to safely avoid dangerous iceberg areas. It is usually necessary to hold south of at least 45°N until well southeast of Newfoundland. The U. S. Navy Atlantic ship routing office at Naval Eastern Oceanography Center Norfolk maintains a safety margin of at least 100 miles from icebergs reported by the International Ice Patrol. In a severe winter, the Denmark Strait may be closed by fixed ice.

In the northern hemisphere winter, a strong high pressure system moving southeast out of the Rocky Mountains brings cold air down across Central America and the western Gulf of Mexico producing gale force winds in the Gulf of Tehuantepec. This *fall wind* is similar to the pampero, mistral, and bora of other areas of the world. A diversion or adjustment to ship's track can successfully avoid the highest seas associated with the Tehuantepecer. For transits between the Canal Zone and northwest Pacific ports, little additional distance is required to avoid this area (in winter) by remaining south of at least 12°N when crossing 97°W. While avoiding the highest seas some unfavorable swell conditions may be encountered south of this line. Transits between the Panama Canal and North American west coast ports can pass north near the coast of the Gulf of Tehuantepec to avoid the heavy seas during periods of gale conditions, but will still encounter high relative winds.

In the summer, the semi-permanent high pressure systems over the world's oceans produce strong equatorward flow along the west coasts of continents. This feature is most pronounced off the coast of California (U. S. west coast) and

Portugal in the Northern Hemisphere; Chile, western Australia, and southwest Africa in the Southern Hemisphere. Very rough seas are generated and are considered a definite factor in route selection or diversion when transiting these areas.

During the nonsummer months, a strong, cold outbreak off the land to relatively warmer water may occur which produces low pressure areas in the western Atlantic and Pacific. Occasionally, these are small but intense systems which develop rapidly and move much faster than normal, creating adverse weather and sea conditions for ships departing and arriving in these areas. Attention to coastal or area forecasts and the ship's own weather observations may be the first indication of this type of situation.

2407. Recommendations and advisories.—The recommendations and advisories issued by a ship routing agency are intended to provide the ship with timely route recommendations, diversions, and weather forecasts. It is by this method of direct involvement that the ship routing agency contributes to efficient shipping operations by avoiding or limiting the effect of adverse weather and sea conditions.

An *initial route recommendation* is issued to a ship or routing authority normally 48 to 72 hours prior to sailing and begins the process of surveillance. The surveillance procedure is a continual process and is maintained until the ship arrives at its destination. Initial route recommendations are a composite representation of experience, climatology, weather and seas forecasts, operational concerns, and the ship's characteristics. A *planning route* is a route recommendation that is intended to provide a best estimate of a realistic route for a specific transit period. Such routes are provided when estimated dates of departure (EDD's) are given to the routing agency well in advance of departure, usually a week to several months. Long range planning routes are based more on seasonal and climatological expectations than the current weather situation. While planning routes are an attempt at making extended range (more than a week) or long range (more than a month) forecasts, these recommendations are likely to have some revision near the time of departure to reflect the current synoptic weather pattern. An initial route recommendation is more closely related to the current synoptic weather patterns by using the latest dynamic forecasts than are the planning route recommendations. These, too, are subject to revision prior to sailing if weather and sea conditions warrant.

Adjustment of departure time is a recommendation for delay in departure, or early departure if feasible, and is intended to avoid or significantly reduce the adverse weather and seas forecast on the first portion of the route if sailing on the original EDD. The initial route is not revised, only the timing of the ship's transit through an area with currently unfavorable weather conditions. Adjusting the departure time is an effective method of avoiding a potentially hazardous situation where there is no optimum route for sailing at the originally scheduled time.

A *diversion* is an underway adjustment in track and is intended to avoid or limit the effect of adverse weather conditions forecast to be encountered along the ship's current track. Ship's speed is expected to be reduced by the encounter with the heavy weather. In most cases the distance to destination is increased in attempting to avoid the adverse weather, but this is partially overcome by being able to maintain near normal SOA. Diversions should also be recommended where satisfactory weather and sea conditions are forecast on a shorter track.

Adjustment of SOA is a recommendation for slowing or increasing the ship's speed as much as practicable in an attempt to avoid an adverse weather situation by adjusting the timing of the encounter. This is also an effective means of maintaining maximum ship operating efficiency and not diverting from the present ship's track. By adjusting the SOA, a major weather system can sometimes be avoided with no increase in distance. The development of fast ships (SOA greater than 30 knots) gives the ship routing agency the potential to "make the ship's weather" by adjusting the ship's speed and track for encounter with favorable weather conditions or ship response factors.

Evasion is a recommendation to the commanding officer or master to take independent action to avoid, as much as possible, a potentially dangerous weather system. The ship routing meteorologist may recommend a general direction for safe evasion but does not specify an exact track. The recommendation for evasion is an indication that the weather and sea conditions have deteriorated to a point where shiphandling and safety are the primary considerations and progress toward destination has been temporarily suspended or is at least of secondary consideration.

A *weather advisory* is a transmission sent to the ship advising the commanding officer or master of expected adverse conditions, their duration, and geographic extent. It is initiated by the ship routing agency as a service and an aid to the ship. The best example of a situation for which a forecast will be helpful is when the ship is currently in good weather but adverse weather is expected within 24 hours for which a diversion is not being recommended or, a diversion where adverse weather conditions are still expected. This type of advisory may include a synoptic weather discussion and a wind, seas, or fog forecast.

The ability of the routing agency to achieve optimum conditions for the ship is aided by the commanding officer or master adjusting course and speed where necessary for an efficient and safe ride. At times, the local sea conditions may dictate that the commanding officer or master take independent action.

2408. Southern Hemisphere.—Available data on which to base analyses and forecasts is generally very limited in the Southern Hemisphere. Weather information obtained from satellites offers the possibility of improvement in southern hemisphere forecast products.

Passages south of the Cape of Good Hope and Cape Horn should be timed to avoid heavy weather as much as possible since intense and frequent low pressure systems are common in these areas. In particular, near the southeast coasts of Africa and South America, intense low pressure systems form in the lee of relatively high terrain near the coasts of both continents. Winter transits south of Cape Horn are difficult since the time required at relatively higher latitudes is longer than the typical time interval between storms. Remaining equatorward of about 35°S as much as practicable will limit exposure to adverse conditions. If the frequency of lows passing these areas is one every three or four days, the probability of encountering heavy weather is high.

Tropical cyclones in the Southern Hemisphere present a significant problem because of the sparse surface and upper air observations from which forecasts can be made. Satellites provide the most reliable means by which to obtain accurate positions of tropical systems and also give the first indication of tropical cyclone formation.

In the Southern Hemisphere, OTSR and other ship weather routing services are available but are limited in their full application because of sparse data reports from which reliable short range and extended range forecasts can be produced. Strong climatological consideration is usually given to any proposed southern hemisphere transit. OTSR procedures (art. 2402) for the Northern Hemisphere can be instituted in the Southern Hemisphere whenever justified by basic data input and available forecast models.

2409. Communications.—A vital part of a ship routing service is communications between the ship underway and the routing agency. Reports from the ship give the progress and ability to proceed in existing conditions. Weather reports enrich the basic data collection on which analyses are based and forecasts derived in turn.

Despite all efforts to achieve the best forecasts possible, the quality of forecasts does not always warrant maintaining the route selected prior to departure until reaching destination. In the U. S. Navy's ship routing program, experience shows that one-third of the ships routed receive some operational or weather dependent change while underway.

The routing agency needs reports of the ship's position and the ability to transmit recommendations for track change or weather advisories to the ship. The ship needs send and receive capability for the required information. Information on seakeeping changes initiated by the ship is desirable in a coordinated effort to provide optimum transit conditions.

2410. Benefits.—The benefits of ship weather routing services are basically cost reduction and safety. The savings in operating costs are derived from reductions in transit time, heavy weather encounters, fuel consumption, cargo and hull damage, and more efficient scheduling of dockside activities. The savings are further increased by fewer emergency repairs, more efficient use of personnel, improved topside working conditions, lower insurance rates as preferred risks under weather routing, and ultimately extended ship operating life.

An effective routing service maximizes safety by greatly reducing the probability of severe or catastrophic damage to the ship and injury to crew members.

2411. Conclusion.—The success of ship weather routing is dependent upon the validity of the forecasts and the routing agency's ability to make appropriate route recommendations and diversions. Anticipated improvements in a routing agency's recommendations will come from advancements in meteorology, technology, and the application of ocean wave forecast models.

Advancements in mathematical meteorology coupled with the continued application of computers will extend the time range and skill of the dynamic and statistical forecasts.

Technological advancements in the areas of satellite and automated communications, and onboard ship response systems will increase the amount and type of information to and from the ship with fewer delays. An onboard ship response system will enable the commanding officer or master to effectively select the course and speed required to conserve fuel and provide the best ride while making satisfactory progress on the track recommended by the routing agency. Ship response and performance data included with the ship's weather report will provide the routing agency with real time information with which to ascertain the actual state of the ship. Being able to predict a ship's response in most weather and sea conditions will result in improved routing procedures.

Shipboard and anchored wave measuring devices contribute to the development of ocean wave analysis and forecast models. Shipboard seakeeping instrumentation with input of measured wave conditions and predetermined ship response data for the particular hull enables a master or commanding officer to adjust course and speed for actual conditions.

Modern ship designs, exotic (nonstandard) cargoes, and sophisticated transport methods require individual attention to each ship's areas of vulnerability. Any improvement in the description of sea conditions by ocean wave models will improve the output from ship routing and seakeeping systems.

Advanced planning of a proposed transit combined with the study of expected weather conditions, both before and during the voyage, as is done by ship routing agencies, and careful on board attention to seakeeping (with instrumentation if available) provide the greatest opportunity to achieve the goal of optimum environmental conditions for ocean transit.

References

Bijlsma, S. J. "On Minimal-Time Routing." Royal Netherlands Meteorological Institute, no. 94 (1975).

Bowman, W. N. and R. J. Renard. "Digest of Tropical Cyclones in the Eastern North Pacific." Environmental Prediction Research Facility, Monterey, California, Technical Paper (December 1975).

Bowman, W. N. and R. J. Renard. "The Climatology and Forecasting of Eastern North Pacific Tropical Cyclones." Environmental Prediction Research Facility, Monterey, California, Technical Paper 7–76 (July 1976).

Brand, S. and J. W. Bellock. "Typhoon Havens Handbook for the Western Pacific and Indian Oceans." Environmental Prediction Research Facility, Monterey, California, Technical Paper 5–76 (June 1976).

Cima, N. E. and W. F. Dupin. "Predicting Individual Ship Performance from Real Time Environmental Observations." Society of Naval Architects and Marine Engineers, (September 1974).

Clune, W. M. "Optimum Track Ship Routing at FNWC Monterey." *Mariner's Weather Log,* vol. 19, no. 1 (January 1975). National Oceanic and Atmospheric Administration.

Gray, W. M. "Tropical Cyclone Genesis in the Western North Pacific." Environmental Prediction Research Facility, Monterey, California, Technical Paper 16–75 (May 1975).

Harding, E. T. and W. J. Kotsch. *Heavy Weather Guide.* Annapolis, Maryland, United States Naval Institute, 1965.

Hoffman, D. "Fuel Conservation and Heavy Weather Avoidance." Lecture presented at Seminar on Maritime Fuel Conservation, May 8–9, 1975, at Webb Institute, Long Island, New York.

James, R. W. "Application of Wave Forecasts to Marine Navigation." U. S. Naval Oceanographic Office SP-1 (July 1957).

James, R. W. "Present Status of Ship Routing." Lecture presented at Interocean 70, (November 1970) Dusseldorf, Germany.

James, R. W. "Dangerous Waves Along North Wall of the Gulf Stream." *Mariner's Weather Log,* vol. 18, no. 3 (November 1974) National Oceanic and Atmospheric Administration.

Lazanoff, S. M. and N. M. Stevenson. "An Evaluation of a Hemispheric Operational Spectra Model." Fleet Numerical Weather Central, Monterey, California, Technical Note 75-3 (June 1975).

CHAPTER XXV

POLAR NAVIGATION

Polar Regions

2501. Introduction.—No single definition of the limits of the polar regions satisfies the needs of all who are interested in these areas. Astronomically, the parallels of latitude at which the sun becomes circumpolar (the Arctic and Antarctic Circles at about latitude 67°.5) are considered the lower limits. Meteorologically, the limits are irregular lines which, in the Arctic, coincides approximately with the tree line. For general purposes, the navigator may consider polar regions as extending from the geographical poles of the earth to latitude 70° (in the Arctic coinciding approximately with the northern coast of Alaska), with transitional **subpolar regions** extending for an additional 10° (in the Northern Hemisphere extending to the southern tip of Greenland).

This chapter deals primarily with marine navigation in high latitudes.

2502. Polar geography.—The north polar region, the **Arctic**, consists of an elongated central water area a little smaller than the United States, almost completely surrounded by land (fig. 2502a). Some of this land is high and rugged with permanent **ice caps**, but part of it is low and marshy when thawed. Underlying **permafrost**, permanently frozen ground, prevents adequate drainage, resulting in large numbers of lakes and ponds and extensive areas of **muskeg**, soft spongy ground with characteristic growths of certain types of moss and tufts of grass or sedge. There are also large areas of **tundra**, low treeless plains with vegetation consisting of mosses, lichens, shrubs, willows, etc., and usually having an underlying layer of permafrost. The northernmost point of land is Kap Morris Jessup, Greenland, about 380 nautical miles from the pole.

The central part of the **Arctic Ocean**, as the body of water is called, is a basin of about 12,000 feet average depth. However, the bottom is not level, having a number of seamounts and deeps. The greatest depth is probably a little more than 16,000 feet. At the North Pole the depth is 14,150 feet. Surrounding the polar basin is an extensive continental shelf, broken only in the area between Greenland and Svalbard (Spitsbergen). The many islands of the Canadian archipelago are on this shelf. The Greenland Sea, east of Greenland; Baffin Bay, west of Greenland; and the Bering Sea, north of the Aleutians, each has its independent basin. In a sense, the Arctic Ocean is an arm of the Atlantic, as shown in figure 2502a.

The south polar region, the **Antarctic**, is in marked contrast to the Arctic in physiographical features. Here a high, mountainous land mass about twice the area of the United States is surrounded by water (fig. 2502b). An extensive polar plateau covered with snow and ice is about 10,000 feet high. There are several mountain ranges with peaks rising to heights of more than 13,000 feet. The average height of Antarctica is about 6,000 feet, which is higher than any other continent. The height at the South Pole is about 9,500 feet. The barrier presented by land and tremendous **ice shelves** 500 to 1,000 feet thick prevent ships from reaching very high latitudes. Much of the coast of Antarctica is high and rugged, with few good harbors or anchorages.

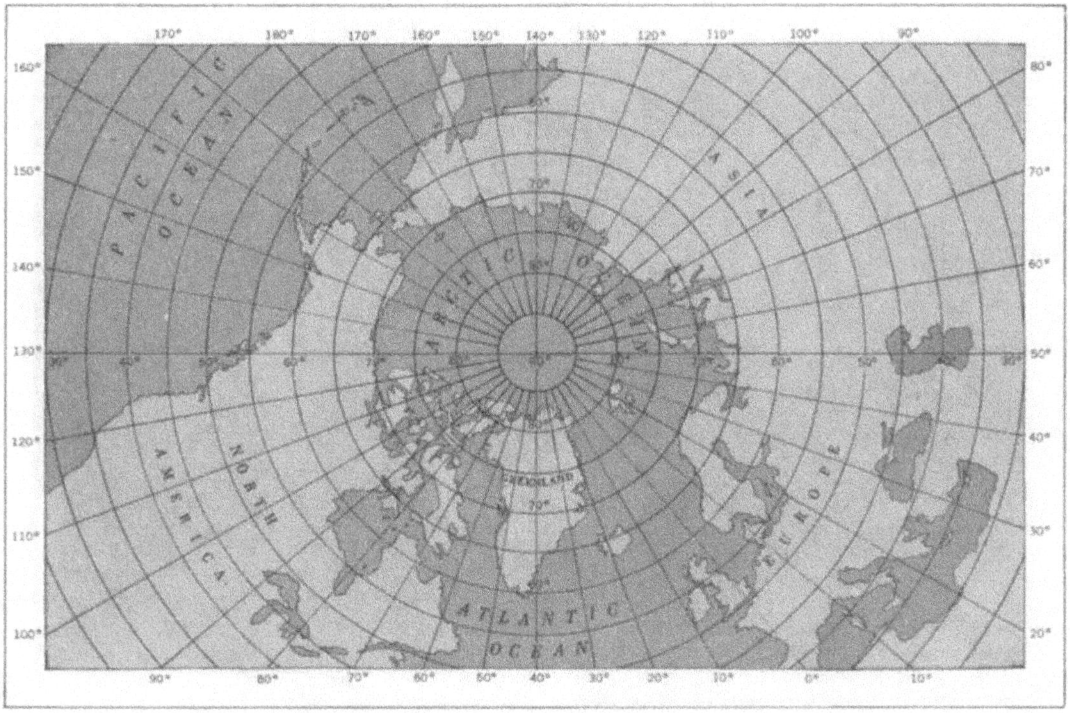

FIGURE 2502a.—The north polar region, or *Arctic.*

2503. Navigation in polar regions.—Special techniques have been developed to adapt navigation to the unique conditions of polar regions. These conditions are largely the result of (1) high latitude, and (2) meteorological factors.

2504. High-latitude effects.—Much of the thinking of the marine navigator is in terms of the "rectangular" world of the Mercator projection, on which the meridians are equally spaced, vertical lines perpendicular to the horizontal parallels of latitude. Directions are measured relative to the meridians, and are maintained by means of a magnetic or gyrocompass. A straight line on the chart is a rhumb line, the line used for ordinary purposes of navigation. Celestial bodies rise above the eastern horizon, climb to a maximum altitude often high in the sky as they cross the celestial meridian, and set below the western horizon. By this motion the sun divides the day naturally into two roughly equal periods of daylight and darkness, separated by relatively short transitional periods of twilight. The hour of the day is associated with this daily motion of the sun.

In polar regions conditions are different. Meridians all converge at the poles, which are centers of series of concentric circles constituting the parallels of latitude. The rapid convergence of the meridians renders the usual convention of direction inadequate for some purposes. A rhumb line is a curve which differs noticeably from a great circle, even for short distances. Even visual bearings cannot adequately be represented as rhumb lines. At the pole all directions are south or north, depending upon the pole. Direction in the usual sense is replaced by longitude.

At the pole the zenith and celestial pole coincide. Hence, the celestial horizon and celestial equator also coincide, and declination and computed altitude are the same. Therefore, celestial bodies change computed altitude only by changing declination. Stars circle the sky without noticeable change in altitude. Planets rise and

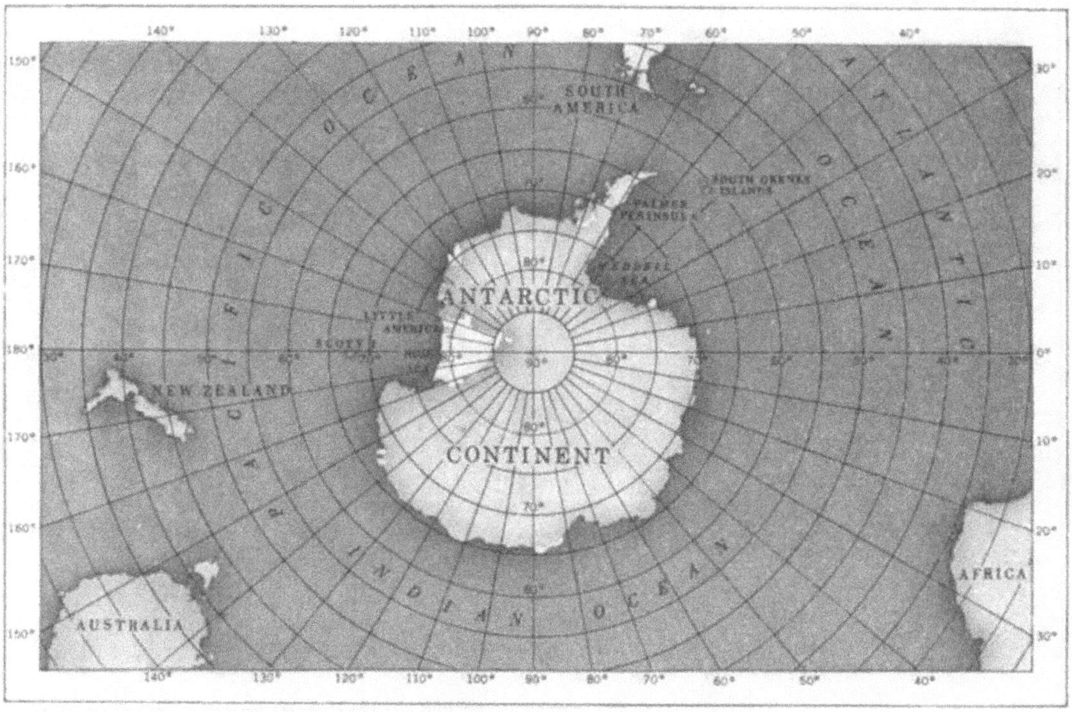

FIGURE 2502b.—The south polar region, or *Antarctic.*

set once each sidereal period (12 years for Jupiter, 30 years for Saturn). At the North Pole the sun rises about March 21, slowly spirals to a maximum altitude of about 23°27′ near June 21, slowly spirals downward to the horizon about September 23, and then disappears for another six months. At the South Pole a similar cycle takes place but during the opposite time of year. It requires about 32 hours for the sun to cross the horizon, during which time it circles the sky 1⅓ times. The twilight periods following sunset and preceding sunrise last for several *weeks.* The moon rises and sets about once each month. Only celestial bodies of north declination are visible at the North Pole; only bodies of south declination are visible at the South Pole.

The long polar night is not wholly dark. The full moon at this time rises relatively high in the sky. Light from the **aurora borealis** in the Arctic and the **aurora australis** in the Antarctic is often quite bright, occasionally exceeding that of the full moon. Even the planets and stars contribute an appreciable amount of light in this area where a snow cover provides an excellent reflecting surface.

All time zones, like all meridians, meet at the poles. *Local* time does not have its usual significance, since the hour of the day bears no relation to periods of light and darkness or to altitude of celestial bodies.

2505. Meteorological effects.—Polar regions are cold, but the temperature at sea is not as extreme as inland. The average winter temperature over the Arctic Ocean is (−)30°F to (−)40°F, with an extreme low value near (−)60°F. Colder temperatures have been recorded in Yellowstone National Park. During the summer the temperature remains above freezing over the ocean. Inland, extreme values are sometimes reached. At least one point on the Arctic Circle has experienced a temperature of 100°F. Few points on the Antarctic Continent have recorded temperatures above freezing, and the interior is probably the coldest part of the world.

Fog and clouds are common in polar regions, yet there is less precipitation than in some desert regions, since the cold air has small capacity for holding moisture. Very cold air over open water sometimes produces steaming of the surface, occasionally to a height of several hundred feet. This is called **frost smoke** or **sea smoke** (fig. 2505). When there is no fog or frost smoke, the visibility is often excellent. Sounds can sometimes be heard at great distances.

FIGURE 2505.—Frost smoke.

Sharp discontinuities or inversions in the temperature lapse rate sometimes produce a variety of mirages and extreme values of refraction. The sun has been known to rise several *days* before it was expected in the spring. False horizons are not uncommon.

Strong winds are common in the subarctic and in both the Antarctic and subantarctic. The belt of water surrounding Antarctica has been characterized as the stormiest in the world, being an area of high winds and high seas. Strong winds are not encountered over the Arctic Ocean.

In the polar and subpolar regions the principal hazard to ships is ice, both that formed at sea and land ice which has flowed into the sea in the form of glaciers.

Many low land areas are ice-free in summer. Ice is considered in more detail in chapter XXXVI.

When snow obliterates surface features, and the sky is covered with a uniform layer of cirrostratus or altostratus clouds, so that there are no shadows, the horizon disappears and earth and sky blend together, forming an unbroken expanse of white, without features. Landmarks cannot be distinguished, and with complete lack of contrast, distance is virtually impossible to estimate. This is called **arctic** (or **antarctic**) **whiteout.** It is particularly prevalent in northern Alaska during late winter and early spring.

2506. Miscellaneous.—The cold surface water of the Arctic Ocean flows outward between Greenland and Svalbard and is replaced by warmer subsurface water from the Atlantic. The surface currents depend largely upon the winds, and are generally quite weak in the Arctic Ocean. However, there are a number of well-established currents flowing with considerable consistency throughout the year. The general circulation in the Arctic is clockwise on the American side and around islands, and counterclockwise on the Asian side. Tidal ranges in this area are generally small. In the restricted waters of the upper Canadian-Greenland area both tides and currents vary considerably from place to place. In the Baffin Bay-Davis Strait area the currents are strong and the tides are high, with a great difference between springs and neaps. In the Antarctic, currents are strong, and the general circulation offshore is eastward or *clockwise* around the continent. Close to the shore, a weaker westerly or *counterclockwise* current may be encountered, but there are many local variations.

Since both magnetic poles are situated within the polar regions, the horizontal intensity of the earth's magnetic field is so low that the magnetic compass is of reduced value, and even useless in some areas. The magnetic storms centered in the auroral zones (art 4320) disrupt radio communications and alter magnetic compass errors. The frozen ground in polar regions is a poor conductor of electricity, another factor adversely affecting radio wave propagation.

2507. Summary of conditions in polar regions.—The more prominent characteristic features associated with large portions of the polar regions may be summarized as follows:

1. High latitude.
2. Rapid convergence of meridians.
3. Nearly horizontal diurnal motion of celestial bodies.
4. Long periods of daylight, twilight, and semidarkness.
5. Low mean temperatures.
6. Short, cool summers and long, cold winters.
7. High wind-chill factor.
8. Low evaporation rate.
9. Scant precipitation.
10. Dry air (low absolute humidity).
11. Excellent sound-transmitting conditions.
12. Periods of excellent visibility.
13. Extensive fog and clouds.
14. Large number and variety of mirages.
15. Extreme refraction and false horizons.
16. Winter freezing of rivers, lakes, and part of the sea.
17. Areas of permanent land and sea ice.
18. Areas of permanently frozen ground.

19. Large areas of tundra (Arctic).
20. Large areas of poor drainage, with many lakes and ponds (Arctic).
21. Large areas of muskeg (a grassy marsh when thawed) (Arctic).
22. Extensive auroral activity.
23. Large areas of low horizontal intensity of earth's magnetic field.
24. Intense magnetic storms.
25. Uncertain radio wave propagation.
26. Strong winds (Antarctic).
27. Frequent blizzards (Antarctic).
28. Large quantities of blowing snow.

Charts

2508. Projections.—In polar regions, as elsewhere, the chart is an important item of navigational equipment. The projections used for polar charts are considered in articles 321 and 322.

For ordinary navigation the Mercator projection has long been the overwhelming favorite of marine navigators, primarily because a rhumb line appears as a straight line on this projection. Even in high latitudes the mariner has exhibited an understandable partiality for Mercator charts, and these have been used virtually everywhere that ships have gone.

However, as the latitude increases, the superiority of the Mercator projection decreases, primarily because the value of the rhumb line becomes progressively less. At latitudes greater than 60° the decrease in utility begins to be noticeable, and beyond latitude 70° it becomes troublesome. In the clear polar atmosphere, visual bearings are observed at great distances, sometimes 50 miles or more. The use of a rhumb line to represent a bearing line introduces an error at any latitude, but at high latitudes this error becomes excessive.

Another objection to Mercator charts at high latitudes is the increasing rate of change of scale over a single chart. This results in distortion in the shape of land masses and errors in measuring distances.

At some latitudes the disadvantages of the Mercator projection outweigh its advantages. The latitude at which this occurs depends upon the physical features of the area, the configuration and orientation of land and water areas, the nature of the operation, and, mostly, upon the previous experience and personal preference of the mariner. Because of differences of opinion in this matter, a transitional zone exists in which several projections may be encountered. The wise high-latitude navigator is prepared to use any of them, since coverage of his operating area may not be adequate on his favorite projection.

2509. Adequacy.—Charts of most polar areas are generally inferior to those of other regions in at least three respects:

1. *Lack of detail.* Polar regions have not been surveyed with the thoroughness needed to provide charts of the accustomed detail. Relatively few soundings are available and many of the coastal features are shown by their general outlines only. Large areas are perennially covered by ice, which presents a changing appearance as the amount, position, and the character of the ice change. Heavy covers of ice and snow prevent accurate determination of surface features of the earth beneath. Added to this is the similarity between adjacent land features where the hundreds of points and fiords in a rugged area or the extensive areas of treeless, flat coastal land in another look strikingly alike. The thousands of shallow lakes and ponds along a flat coastal plain lack distinctive features.

2. *Inaccuracy.* Polar charts are based upon incomplete surveys and reports of those who have been in the areas. These reports are less reliable than in other areas because icebergs are sometimes mistaken for islands, ice-covered islands are mistaken for grounded icebergs, shorelines are not easy to detect when snow covers both land and attached sea ice, inlets and sounds may be completely obscured by ice and snow, and meteorological conditions may introduce inaccuracy in determination of position. Consequently, many features are inaccurately shown in location, shape, and size, and there are numerous omissions. Isogonic lines, too, are based upon incomplete information, resulting in less than desired accuracy.

3. *Coverage.* Relatively few nautical charts of polar regions are available, and the limits of some of these are not convenient for some operations. As in other areas, charts have been made as the need has arisen. Hence, large-scale charts of some areas are completely lacking. Aeronautical charts are sometimes quite helpful, as they often show more detail of land areas than do the nautical charts. However, aeronautical charts do not show soundings.

2510. Polar grid.—Because of the rapid convergence of the meridians in polar regions, the true direction of an oblique line near the pole may vary considerably over a relatively few miles. The meridians are radial lines meeting at the poles, instead of being parallel, as they appear on the familiar Mercator chart.

Near the pole the convenience of parallel meridians is attained by means of a **polar grid.** On the chart a number of lines are printed parallel to a selected reference meridian, usually that of Greenwich. On transverse Mercator charts the fictitious meridians may serve this purpose. Any straight line on the chart makes the same angle with all grid lines. On the transverse Mercator projection it is therefore a **fictitious rhumb line.** On any polar projection it is a close approximation to a great circle. If north along the reference meridian is selected as the reference direction, all parallel grid lines can be considered extending in the same direction. The constant direction relative to the grid lines is called **grid direction.** North along the Greenwich meridian is usually taken as grid north in both the Northern and Southern Hemispheres.

The value of grid directions is indicated in figure 2510. In this figure *A* and *B* are 400 miles apart. The true bearing of *B* from *A* is 023°, yet at *B* this bearing line, if continued, extends in true direction 163°, a change of 140° in 400 miles. The grid direction at any point along the bearing line is 103°.

When north along the Greenwich meridian is used as grid north, interconversion between grid and true directions is quite simple. Let G represent a grid direction and T the corresponding true direction. Then for the Arctic,

$$G=T+\lambda W.$$

That is, in the Western Hemisphere, in the Arctic, grid direction is found by *adding* the longtitude to the true direction. From this it follows that

$$T=G-\lambda W,$$

and in the Eastern Hemisphere

$$G=T-\lambda E,$$
$$T=G+\lambda E.$$

In the Southern Hemisphere the signs (+ or −) of the longitude are reversed in all formulas.

If a magnetic compass is used to follow a grid direction, variation and convergency can be combined into a single correction called **grid variation** or **grivation.** It

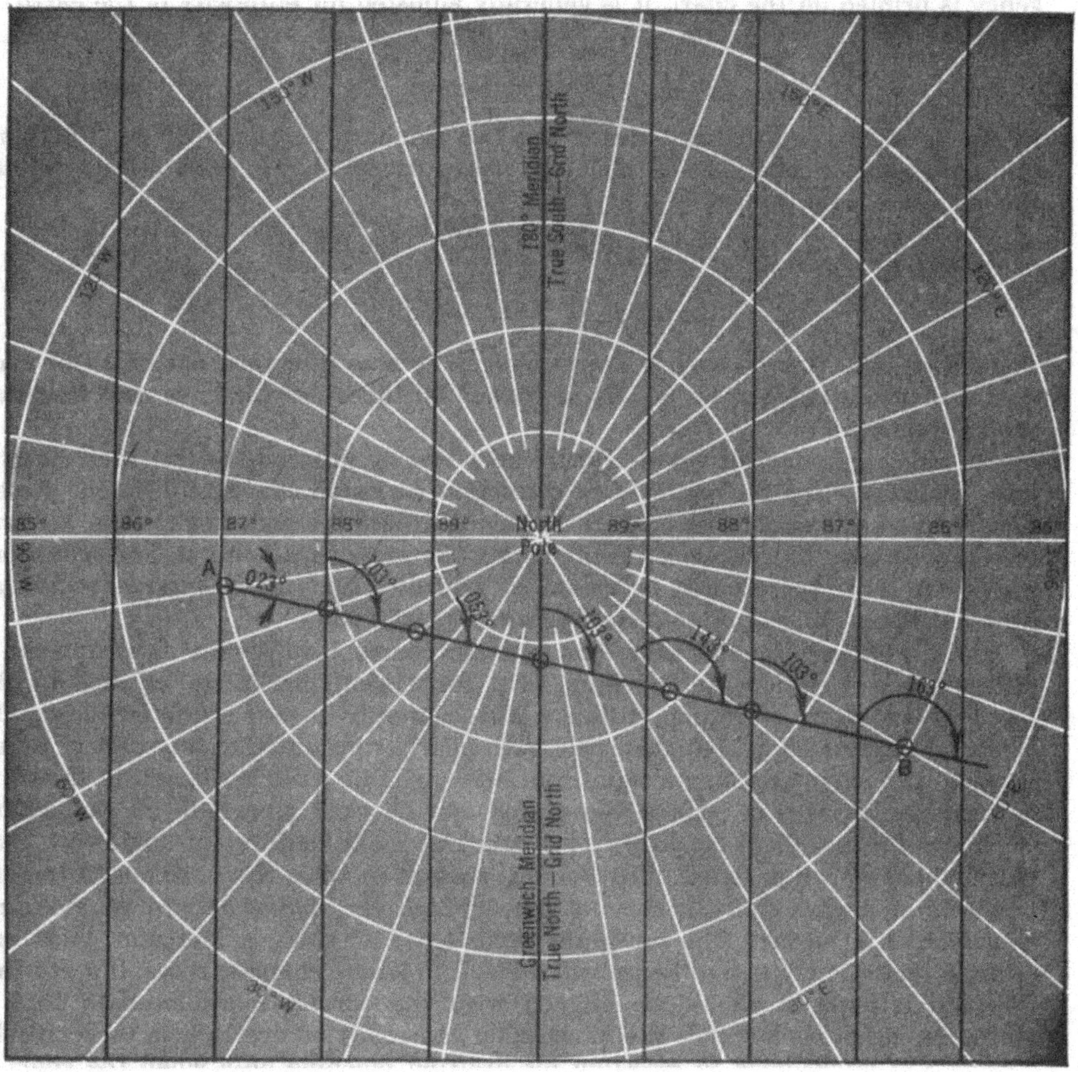

FIGURE 2510.—Polar grid navigation.

is customary to show lines of equal grivation on polar charts rather than lines of equal variation. Defense Mapping Agency Hydrographic Topographic Center chart 43 shows the **isogrivs** (lines of equal grivation) for the north and south polar areas.

With one modification the grid system of direction can be used in any latitude. Meridians 1° apart make an angle of 1° with each other where they meet at the pole. The **convergency** is one, and the 360° of longitude cover all 360° around the pole. At the equator the meridians are parallel and the convergency is zero. Between these two limits the convergency has some value between zero and one. On a sphere it is equal to the sine of the latitude. For practical navigation this relationship can be used on the spheroidal earth. On a simple conic or Lambert conformal chart a constant convergency is used over the entire chart, and is known as the **constant of the cone.** On a simple conic projection it is equal to the sine of the standard parallel. On a Lambert conformal projection it is equal (approximately) to the sine of the latitude midway between the two standard parallels. When convergency is printed on the chart, it is generally adjusted for ellipticity of the earth. If K is the constant of the cone,

$$K = \sin \tfrac{1}{2} (L_1 + L_2),$$

where L_1 and L_2 are the latitudes of the two standard parallels. On such a chart, grid navigation is conducted as explained above, except that in each of the formulas the longitude is multiplied by K:

$$G = T + K\lambda W,$$
$$T = G - K\lambda W,$$
$$G = T - K\lambda E,$$
$$T = G + K\lambda E.$$

Thus, a straight line on such a chart changes its true direction, not by 1° for each degree of longitude, but by $K°$. As in higher latitudes, convergency and variation can be combined.

In using grid navigation one should keep clearly in mind the fact that the grid lines are parallel *on the chart.* Since distortion varies on charts of different projections, and on charts of conic projections having different standard parallels, *the grid direction between any two given points is not the same on all charts.* For operations which are to be coordinated by means of grid directions, *it is important that all charts showing the grid be on a single graticule* (art. 303).

2511. Plotting on polar charts, as on other charts, involves the measurement of distance and direction. On a chart with converging meridians, as one on the Lambert conformal projection, distance is measured by means of the latitude scale, as on a Mercator chart, but this scale is so nearly constant that any part of it can be used without introducing a significant error. A mile scale is sometimes shown in or near the margin of such a chart, and can be used anywhere on that chart.

Since the meridians converge, a straight line makes a different angle with each meridian, as shown in figure 2510. For this reason, compass roses are not customarily shown on such a chart. If they do appear, *each one applies only to the meridian on which it is located.* The navigator accustomed to using a Mercator chart can easily forget this point, and hence will do well to ignore compass roses. If a drafting machine is used, it should be aligned with the correct meridian each time a measurement is made. Since this precaution can easily be overlooked, especially by a navigator accustomed to resetting his drafting machine only when the chart is moved, and since the resulting error may be too small to be apparent but too large to ignore, it is good practice to discard this instrument when the Mercator chart is replaced by one with converging meridians, unless positive steps are taken to prevent error.

The most nearly fool-proof and generally the most satisfactory method of measuring directions on a chart with converging meridians is to use a protractor, or some kind of plotter combining the features of a protractor and straightedge (fig. 2511a).

If a course is to be measured, the mid meridian of each leg should be used, as shown in figure 2511a. If a bearing is to be measured, the meridian nearest the point at which the bearing was determined should be used, as shown in figure 2511b. Thus, in the usual case of determining the bearing of a landmark from a ship, the meridian nearest the ship should be used. In using either of the plotters shown in figures 2511a or 2511b, note that the center hole is placed over the meridian used, the straightedge part is placed along the line to be drawn or measured, and the angle is read on the protractor at the same meridian which

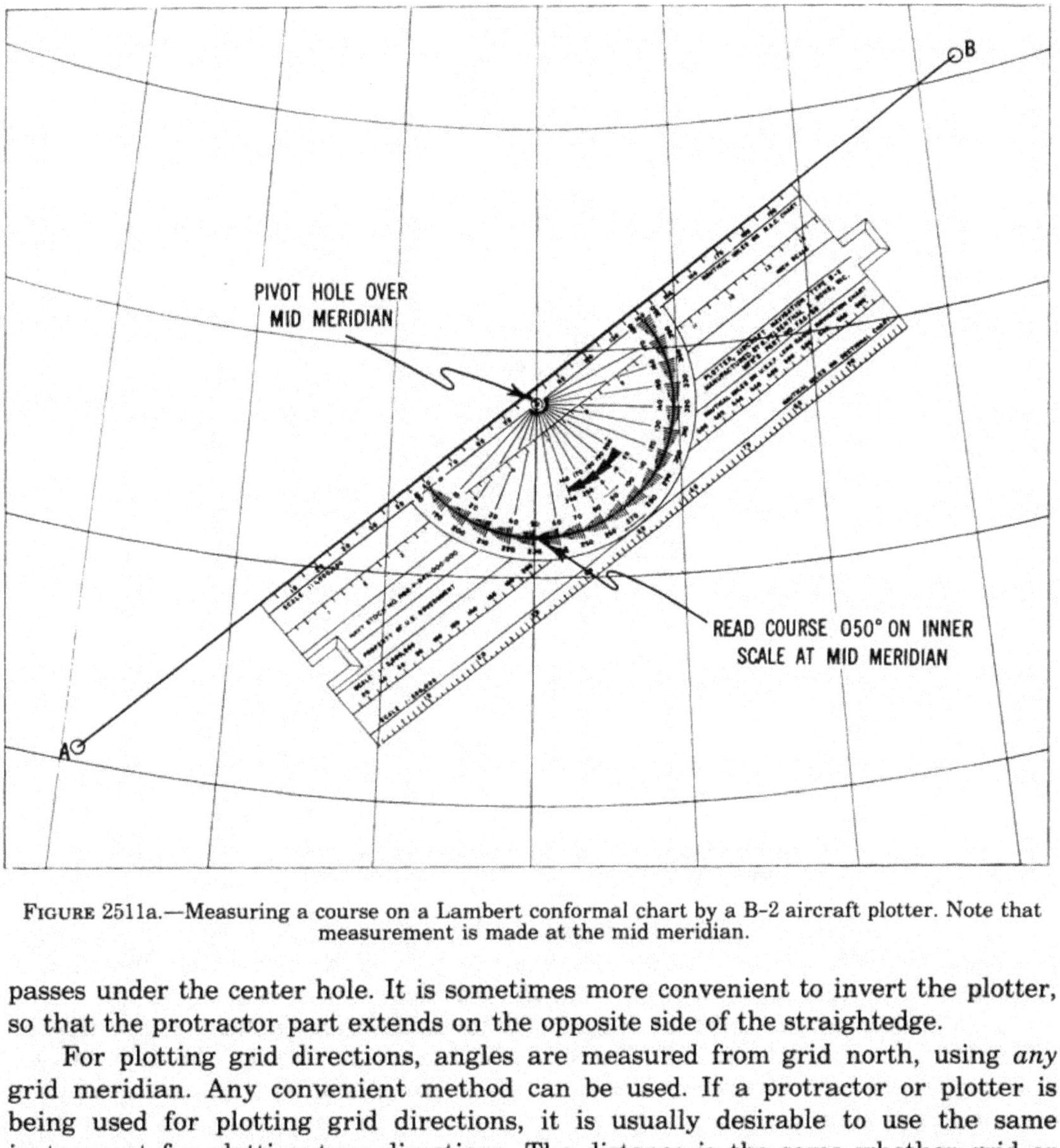

FIGURE 2511a.—Measuring a course on a Lambert conformal chart by a B-2 aircraft plotter. Note that measurement is made at the mid meridian.

passes under the center hole. It is sometimes more convenient to invert the plotter, so that the protractor part extends on the opposite side of the straightedge.

For plotting grid directions, angles are measured from grid north, using *any* grid meridian. Any convenient method can be used. If a protractor or plotter is being used for plotting grid directions, it is usually desirable to use the same instrument for plotting true directions. The distance is the same whether grid or true directions are used.

Dead Reckoning

2512. Polar dead reckoning.—In polar regions, as elsewhere, dead reckoning involves measurement of direction and distance traveled, and the use of this information for determination of position.

Direction is normally determined by a compass, but in polar regions both magnetic and gyrocompasses are subject to certain limitations not encountered elsewhere. However, the navigator who thoroughly understands the use of these instruments in high latitudes can get much useful information from them. The polar navigator should not overlook the value of radar tracking or visual tracking for determining direction of motion. This is discussed in article 2515.

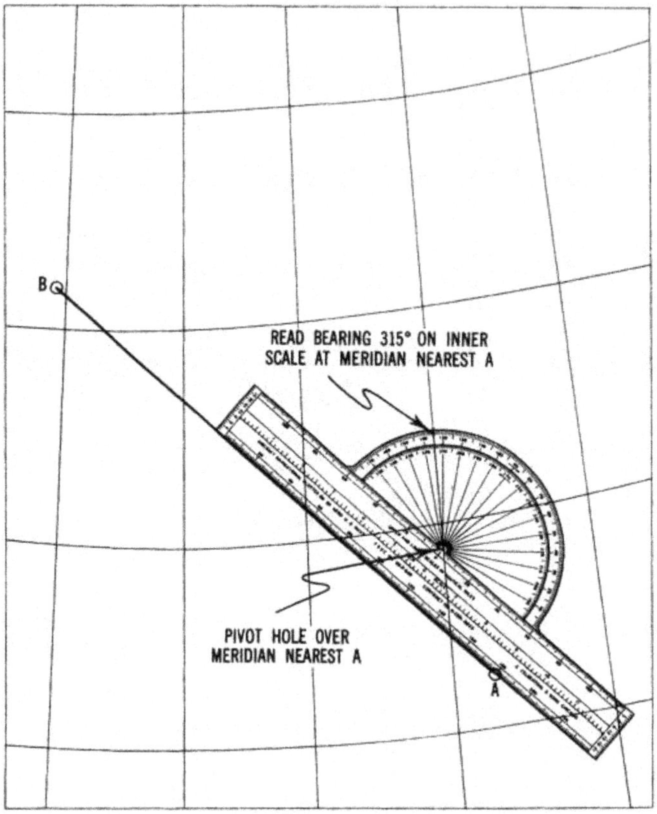

FIGURE 2511b.—Measuring a bearing on a Lambert conformal chart by AN plotter. Note that measurement is made at the meridian nearest the ship.

Speed or distance is normally measured by log or engine revolution counter, but these methods are not entirely suitable when the ship is operating in ice. The problem of determining speed or distance in ice is discussed in article 2515.

2513. The magnetic compass depends for its directive force upon the horizontal intensity of the magnetic field of the earth. As the magnetic poles are approached, this force becomes progressively weaker until at some point the magnetic compass becomes useless as a direction-measuring device. In a marginal area it is good practice to keep the magnetic compass under almost constant scrutiny, as it is somewhat erratic in dependability and its errors may change rapidly. Frequent compass checks by celestial observation or any other method available are wise precautions. A log of compass comparisons and observations is useful in predicting future reliability.

The magnetic poles themselves are somewhat elusive, since they participate in the normal diurnal, annual, and secular changes in the earth's field, as well as the more erratic changes caused by magnetic storms. Measurements indicate that the north magnetic pole moves within an elongated area of perhaps 100 miles in a generally north-south direction and somewhat less in an east-west direction. Normally, it is at the southern end of its area of movement at local noon and at the northern extremity twelve hours later, but during a severe magnetic storm this motion is upset and becomes highly erratic. Because of the motions of the poles, they are sometimes regarded as *areas* rather than points. There is some evidence to support the belief that several secondary poles exist, although such alleged poles may be anomalies (local attractions), possibly of intermittent or temporary existence. Various severe anomalies have been located in polar areas and others may exist.

The continual motion of the poles may account, at least in part, for the large diurnal changes in variation encountered in high latitudes. Changes as large as 10° have been reported.

Measurements of the earth's magnetic field in polar regions are neither numerous nor frequent. The isogonic lines in these areas are close together, resulting in rapid change in short distances in some directions, and their locations are imperfectly known. As a result, charted variation in polar regions is not of the same order of accuracy as elsewhere.

The decrease in horizontal intensity encountered near the magnetic poles, as well as magnetic storms, affects the deviation. Any deviating magnetic influence remaining after adjustment, which is seldom perfect, exerts a greater influence as the directive force decreases. It is not uncommon for residual deviation determined in moderate latitudes to increase 10- or 20-fold in marginal areas. Interactions between correctors and compass magnets exert a deviating influence that may increase to a troublesome degree in high latitudes. The heeling magnet, correcting for both permanent and induced magnetism, is accurately located only for one magnetic latitude. Near the magnetic pole its position might be changed, but this may induce sufficient magnetism in the Flinders bar to more than offset the change in deviation due to the change in the position of the heeling magnet. The relatively strong vertical intensity may render the Flinders bar a stronger influence than the horizontal field of the earth. When this occurs, the compass reading remains nearly the same on any heading.

Another effect of the decrease in the directive force of the compass is a greater influence of frictional errors. This, combined with an increase in the period of the compass, results in greatly increased sluggishness in its return to the correct reading after being disturbed. For this reason the compass performs better in a smooth sea free from ice than in an ice-infested area where its equilibrium is frequently upset by impact of the vessel against ice.

Magnetic storms affect the magnetism of a ship as well as that of the earth. Changes in deviation of as much as 45° have been reported during severe magnetic storms, although it is possible that such large changes may be a combination of deviation and variation changes.

The area in which the magnetic compass is of reduced value cannot be stated in specific terms. A magnetic compass in an exposed position performs better than one in a steel pilot house. The performance of the compass varies considerably with the type of compass, sensitiveness and period, thoroughness of adjustment, location on the vessel, and magnetic properties of the vessel. It also varies with local conditions.

In a very general sense the magnetic compass can be considered of reduced reliability when the horizontal intensity is less than 0.09 oersted, erratic when the field is less than 0.06 oersted, and useless when it is less than 0.03 oersted. The extent of these areas in the Northern Hemisphere is indicated in figure 2513. Similar areas extend around the south magnetic pole, which is located at latitude 68°S, longitude 139°E, not far from the eastern shore of the Ross Sea. DMAHTC chart 33 shows lines of equal horizontal intensity in the north and south polar regions, respectively. However, the effectiveness of the magnetic compass is influenced also by local conditions. A compass on a vessel making a voyage through the islands of the Canadian archipelago has been reported to give fair indication of direction in certain small areas where the horizontal intensity is less than 0.02 oersted, yet to be useless at some places where the horizontal intensity is greater than 0.04 oersted.

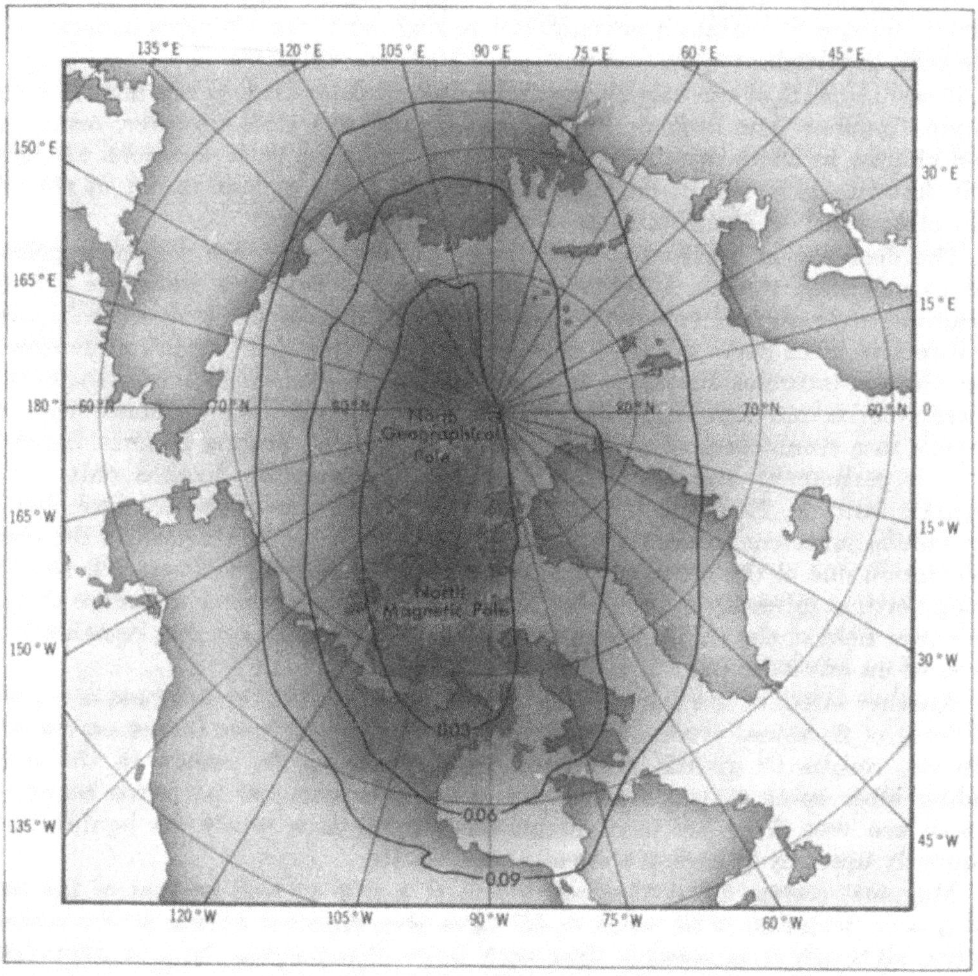

FIGURE 2513.—Arctic areas in which the magnetic compass is of reduced value. Inside the curves representing the 0.09, 0.06, and 0.03 oersted values of horizontal intensity the compass can be considered of reduced reliability, erratic, and useless, respectively.

Despite its various limitations, the magnetic compass is a valuable instrument in much of the polar regions, where the gyrocompass is also of reduced reliability. With careful adjustment, frequent checks, and a record of previous behavior, the polar navigator can get much useful service from his instrument.

When a compass is subjected to extremely low temperatures, there is danger of the liquid freezing. Sufficient heat to prevent this can normally be obtained from the compass light, which should not be turned off during severe weather.

2514. The gyrocompass depends for its operation upon the rotation of the earth about its axis. Its maximum directive force is at the equator, where the axis of the compass is parallel to the axis of the earth. As the latitude increases, the angle between these two axes increases. At the *geographical* poles the gyrocompass has no directive force.

The gyrocompass is generally reliable to latitude 70°. At higher latitudes the disturbing effect of imperfections in compass or adjustment is magnified. Latitude adjustment becomes critical. Speed error increases as the speed of the vessel approaches the rotational speed of the earth. Ballistic deflection error becomes large and the compass is slow to respond to correcting forces. Frequent changes of course and speed, often necessary when proceeding through ice, introduce errors which are slow to settle out. The impact of the vessel against ice deflects the gyrocompass, which does not return quickly to the correct reading.

The error increases and becomes more erratic as the vessel proceeds to higher latitudes. Extreme errors as large as 27° have been reported at latitudes greater than 82°. The gyrocompass probably becomes useless at about latitude 85°. At latitude 70° the gyro error should be determined frequently, perhaps every four hours, by means of celestial bodies when these are available. As the error increases and becomes more erratic, with higher latitude, it should be determined more frequently. In heavy ice at extreme latitudes an almost constant check is desirable. The gyro and magnetic compasses should be compared frequently and a log kept of the results of these comparisons and the gyro error determinations.

Most gyrocompasses are not provided with a latitude correction setting above 70°. Beyond this, correction can be made by either of two methods: (1) set the latitude and speed correctors to zero and apply a correction from a table or diagram obtainable from the manufacturer of the compass, or constructed as explained in article 639; or (2) use an equivalent latitude and speed setting. Both of these methods have proved generally satisfactory, although the second is considered superior to the first because it at least partly corrects for errors introduced by a change in course. In certain types of gyrocompasses, facilities for their operation in a high latitude mode up to about 86° and as directional gyros even to the poles is provided.

2515. Distance and direction in ice.—In ice-free waters, distance or speed is determined by some form of log or by engine revolution counter. In the presence of ice, however, most logs are inoperative or inaccurate due to clogging by the ice. Engine revolution counters are not accurate speed indicating devices when a ship is forcing its way through ice. With experience, one can estimate the speed in relation to ice, or a correction can be applied to speed by engine revolution counter. At best, however, these methods are seldom of the desired accuracy.

If ranges and bearings of a land feature can be determined either visually or by radar, course and speed of the vessel or distance traveled over the ground can be determined by tracking the landmark and plotting the results. The feature used need not be identified. Ice can be used if it is grounded or attached to the shore. Course and speed or distance through the water can be determined by tracking a floating iceberg or other prominent floating ice feature. However, an error may be introduced by this method if the effect of wind and current upon the floating feature is different than upon the ship.

2516. Tide, current, and wind.—Relatively little is known of tides and currents in the polar regions. The tables do not extend to these areas, but some information is given in the sailing directions. In general, tidal ranges are small, and the water in most anchorages is relatively deep.

Currents in many coastal areas are strong and somewhat variable. When a vessel is operating in ice, the current is often difficult to determine because of frequent changes in course and speed of the vessel and inaccuracies in the measurement of direction and distance traveled.

In the vicinity of land, and in the whole antarctic area, winds are variable in direction, gusty, and often strong. Offshore, in the Arctic Ocean, the winds are not strong and are steadier, but ships rarely operate in this area. The wind in polar regions, as elsewhere, has two primary navigational effects upon vessels. First, its direct effect is to produce leeway. When a vessel is operating in ice, the leeway may be different from that in open water. It is well to determine this effect for one's own vessel. The second effect is to produce wind currents in the sea.

2517. Keeping the dead reckoning.—Because of the lack of facilities for fixing the position of a vessel in polar regions, accurate dead reckoning is even more important than elsewhere. The problem is complicated by the fact that the *elements* of dead reckoning, direction and distance, are usually known with less certainty than in lower latitudes. This only heightens the need for *keeping* the dead reckoning with all the accuracy obtainable. This may usually be accomplished by careful hand plotting on the available charts or plotting sheets.

Piloting

2518. Piloting in high latitudes is basically no different from that elsewhere. However, in polar regions piloting is the primary method of marine navigation. As previously indicated, dead reckoning is difficult and generally less accurate than in lower latitudes.

Piloting is associated with proximity to land and shoal water. A ship in polar regions is seldom far from land, and the areas are not so accurately surveyed that the navigator can be sure that uncharted shoals are not nearby.

Piloting is characterized by an alertness not required when a vessel is far from danger of grounding. Nowhere is this alertness more necessary than in polar regions. Added to the usual reasons for constant vigilance are the uncertainties of charted information and the lack of detail, as discussed in article 2509.

2519. Natural landmarks are plentiful in some areas, but their usefulness is restricted by the difficulty in identifying them, or locating them on the chart. Along many of the coasts the various points and inlets bear a marked resemblance to each other. The appearance of a coast is often very different when many of its features are obliterated by a heavy covering of snow or ice than when it is ice-free.

2520. Bearings are useful, but have limitations. When bearings on more than two objects are taken, they may fail to intersect at a point because the objects may not be charted in their correct relation to each other. Even a point fix may be considerably in error geographically if all of the objects used are shown in correct relation to each other, but in the wrong position on the earth. However, in restricted waters it is usually more important to know the position of the vessel relative to nearby land and shoals than its latitude and longitude. The bearing and distance of even an unidentified or uncharted point are valuable.

When a position is established relative to nearby landmarks, it is good practice to use this to help establish the identity and location of some prominent feature a considerable distance ahead, so that this feature, in turn, can be used to establish future positions.

In high latitudes it is not unusual to make use of bearings on objects a considerable distance from the vessel. Because of the rapid convergence of the meridians in these areas, such bearings are not correctly represented by straight lines on a Mercator chart. If this projection is used, the bearings should be corrected in the same manner that radio bearings are corrected (using table 1), since both can be considered great circles. Neither visual nor radio bearings are corrected when plotted on a Lambert conformal or polar steorographic chart.

2521. Soundings are so important in polar regions that echo sounders are customarily operated continuously while the vessel is underway. It is good practice to have at least two such instruments, preferably those of the recording type and having a wide flexibility in the range of the recorder. In few parts of the polar regions have enough soundings been obtained and made available to charting agencies to permit adequate portrayal of the bottom configuration. However, since depth of water is a primary consideration in avoiding an unwanted grounding, a constant watch should be maintained to avoid unobserved shoaling.

Polar regions have relatively few shoals, but in some areas, notably along the Labrador coast, a number of pinnacles and ledges rise abruptly from the bottom. These constitute a real danger to vessels, since they are generally not surrounded by any apparent shoaling. In such an area, or when entering an unknown harbor or any area of questionable safety, it is good practice to send one or more small craft ahead with portable sounding gear.

In very deep water, of the order of 1,000 fathoms or more, the echo returned from the bottom is sometimes masked by the sound of ice coming in contact with the hull, but this is generally not a problem when the bottom is close enough to be menacing.

The hand lead is of little value to a ship underway in ice, because the ice generally prevents its effective use unless the vessel is stopped.

If a ship becomes **beset** by ice, so that steerage way is lost and the vessel drifts with the ice, it may be in danger of grounding as the ice moves over a shoal. Hence, it is important that soundings be continued even when beset. If necessary, a hole should be made in the ice and a hand lead used. A vessel with limited means for freeing itself may prudently save such means for use only when there is danger of grounding.

Useful information on the depth of water in the vicinity of a ship can sometimes be obtained by watching the ice. A stream of ice moving faster than surrounding ice, or a stretch of open water in loose pack ice often marks the main channel through shoal water. A patch of stationary ice in the midst of moving ice often marks a shoal.

Knowledge of earth formations may also prove helpful. The slope of land is often an indication of the underwater gradient. Shoal water is often found off low islands, spits, etc., but seldom near a steep shore. Where glaciation has occurred, the moraine deposits are likely to have formed a bar some distance offshore. Submerged rocks and pinnacles are more likely to be encountered off a rugged shore than near a low, sandy beach.

2522. Anchorages.—Because good anchorages are not plentiful in high latitudes, there is an understandable temptation to be less demanding in their selection. This is dangerous practice, for in polar regions some of the requirements are accentuated. The factors to be considered are:

1. *Holding quality of the bottom.* In polar regions a rocky bottom or one with only fair to poor holding qualities is not uncommon. Sometimes the bottom is steep or irregular. Since the nature of the bottom is seldom adequately shown on charts, a wise precaution is to sample the bottom, and sound in the vicinity before anchoring.

2. *Adequacy of room for swing.* Because high winds are frequent along polar shores, sometimes with little or no warning, long scopes of anchor chain are

customarily used. Some harbors are otherwise suitable, but allow inadequate room for swing of the vessel at anchor, or even for its yaw in a high wind. If a vessel is to anchor in an unsurveyed area, the area should first be adequately covered by small boats with portable sounding gear to detect any obstructions.

3. *Protection from wind and sea.* In polar regions protection from wind is probably the most difficult requirement to meet. Generally, high land is accompanied by strong wind blowing directly down the side of the mountains. Polar winds are extremely variable, both in direction and speed. Shifts of 180° accompanied by an increase in speed of more than 50 knots in a few minutes have been reported. It is important that ground tackle be in good condition and that maximum-weight anchors be used. All available weather reports should be obtained and a continuous watch kept on the local weather. Whenever a heavy blow might reasonably be anticipated, the main engines should be kept in an operating condition and on a standby status. Heavy seas are seldom a problem.

4. *Availability of suitable exit in event of extreme weather.* In ice areas it is important that a continuous watch be kept to prevent blocking of the entrance by ice, or actual damage to the vessel by floating ice. However, in an unsurveyed area it may be dangerous to shift anchorage without first sounding the area. It is a wise precaution to do this in advance. Unless the vessel is immediately endangered by ice, it is generally safer to remain at anchor with optimum ground tackle and use of engines to assist in preventing dragging, than to proceed to sea in a high wind, especially in the presence of icebergs and growlers, and particularly during darkness.

5. *Availability of objects for position determination.* The familiar polar problem of establishing a position by inaccurately charted or inadequately surveyed landmarks is accentuated when an accurate position is desired to establish the position of an anchor. Sometimes a trial and error method is needed, and it may be necessary to add landmarks located by radar or visual observation. Because of chart inadequacy, the suitability of an anchorage, from the standpoint of availability of suitable landmarks, cannot always be adequately predicted before arrival.

An unsurveyed harbor should be entered with caution at slow speed, with both the pilot house and engine room force alerted to possible radical changes in speed or course with little or no warning. The anchor should be kept ready for letting go on short notice and should be adequately attended. An engine combination providing full backing power should be maintained.

2523. Sailing directions for high latitudes contain a wealth of valuable information acquired by those who have previously visited the areas. However, since high latitudes have not been visited with the frequency of other areas, and since they are inadequately surveyed, the sailing directions for polar areas are neither as complete nor as accurate as for other areas, and information on unvisited areas is completely lacking. Until traffic in high latitudes increases and the sailing directions for these areas incorporate the additional information obtained, unusual caution should accompany their use. Each vessel that enters polar regions can help correct this condition by recording accurate information and sending it to the Defense Mapping Agency Hydrographic/Topographic Center or its counterpart in other countries.

Electronics and Navigation

2524. Propagation.—In general, radio wave propagation in high latitudes follows the same principles that apply elsewhere, as described in chapter XL. However, certain anomalous conditions occur, and although these are but imperfectly understood, and experience to date has not always seemed consistent, there is much that has been established. An understanding of these conditions is important if maximum effective use is to be made of electronics in high latitudes. Such anomalous conditions are discussed in chapters XL and XLIII.

2525. Radar.—In polar regions, where fog and long periods of continuous daylight or darkness reduce the effectiveness of both celestial navigation and visual piloting, and where other electronic aids are generally not available, radar is particularly valuable. Its value is further enhanced by the fact that polar seas are generally smooth, resulting in relatively little oscillation of the shipborne antenna. When ice is not present, relatively little sea return is encountered from the calm sea.

However, certain limitations attend the use of radar in polar regions. Similarity of detail along the polar shore is even more apparent by radar than by visual observation. Lack of accurate detail on charts adds to the difficulty of identification. Identification is even more of a problem when the shoreline is beyond the radar horizon and accurate contours are not shown on the chart. When an extensive ice pack extends out from shore, accurate location of the shoreline is extremely difficult.

Good training and extensive experience are needed to interpret accurately the returns in polar regions where ice may cover both land and sea. A number of icebergs close to a shore may be too close together to be resolved, giving an altered appearance to a shoreline, or they may be mistaken for off-lying islands. The shadow of an iceberg or pressure ridge and the lack of return from an open lead in the ice may easily be confused. Smooth ice may look like open water. In making rendezvous, one might inadvertently close on an iceberg instead of a ship.

As with visual bearings, radar bearings need correction for convergency unless the objects observed are quite close to the ship.

2526. Long-range and worldwide radionavigation aids, such as Loran-C, Omega, and the Navy Navigation Satellite System, are particularly useful in the Arctic because of the scarcity of aids of shorter range. Such short range aids as may be in existence are subject to damage or failure by ice or storms, or other causes. Ice and storm damage may be widespread and require considerable time to repair. Isolated damage may exist for a long time without being discovered and reported.

2527. Other electronic aids are virtually nonexistent in polar regions.

The *radio direction finder* is useful when the few transmitting stations are within range. One of the principal uses of RDF in polar regions is to assist in locating other vessels, for rendezvous or other purposes. This is particularly true in an area of many icebergs, where radar may not distinguish between ships and icebergs.

Consol is available in the Norwegian Sea between Norway and Greenland.

The *echo sounder* is highly useful, as indicated in article 2521, and is operated continuously in high latitudes.

Sonar is useful primarily for detecting ice, particularly growlers. Since about ½ to ⅞ of the ice is under water, its presence can sometimes be detected by sonar when it is overlooked by radar or visual observation.

Celestial Navigation

2528. Celestial navigation in high latitudes.—Of the various types of navigation, celestial is perhaps least changed in polar regions. However, certain special considerations are applicable.

Because of the limitations of other forms of navigation, as discussed earlier in this chapter, celestial navigation provides the principal means of determining geographical position. However, as indicated in article 2520, position relative to nearby dangers is usually of more interest to the polar navigator than geographical position. Since ships in high latitudes are seldom far from land, and since celestial navigation is attended by several limitations, discussed in article 2529, its use in marine navigation is generally confined to the following applications:

1. navigation while proceeding to and from polar regions;
2. checking the accuracy of dead reckoning;
3. checking the accuracy of charted positions of landmarks, shoals, etc; and
4. providing a directional reference, either by means of a celestial compass or by providing a means of checking the magnetic or gyrocompass.

Although its applications are limited, celestial navigation is important in high latitudes. Application 3 above, and application 4, even more so, can be of great value to the polar navigator.

2529. Celestial observations.—The best celestial fixes are usually obtained by star observations during twilight. As the latitude increases, these periods become longer, providing additional time for observation. But with this increase comes longer periods when the sun is just below the horizon and the stars have not yet appeared. During this period, which in the extreme condition at the pole lasts for several *days,* no celestial observations may be available. The moon is sometimes above the horizon during this period and bright planets, notably Venus and Jupiter, may be visible. With practice, the brighter stars can be observed when the sun is 2° to 3° below the horizon.

Beyond the polar circles the sun remains above the horizon without setting during part of the summer. The length of this period increases with latitude. At Thule, Greenland, about 10° inside the Arctic Circle, the sun remains above the horizon for four months. During this period of continuous daylight the sun circles the sky, changing azimuth about 15° each hour. A careful observation, or the average of several observations, each two hours provides a series of running fixes. An even better check on position is provided by making hourly observations and establishing the most probable position at each observation. Sometimes the moon is above the horizon, but within several days of the new or full phase it provides lines of position nearly parallel to the sunlines and hence of limited value in establishing fixes.

During the long polar night the sun is not available and the horizon is often indistinct. However, the long twilight, a bright aurora, and other sources of polar light (art. 2504) shorten this period. By adapting their eyes to darkness, some navigators can make reasonably accurate observations throughout the polar night. The full moon in winter remains above the horizon more than half the time and attains higher altitudes than at other seasons.

In addition to the long periods of darkness in high latitudes, other conditions are sometimes present to complicate the problem of locating the horizon. During daylight the horizon is frequently obscured by low fog, frost smoke, or blowing snow, yet the sun may be clearly visible. Hummocked sea ice is sometimes a problem, particularly at low heights of eye. Nearby land or an extensive ice foot can also be troublesome. Extreme conditions of abnormal refraction are not uncommon in high latitudes, sometimes producing false horizons and always affecting the refraction and dip corrections.

Because of these conditions, it is advisable to be provided with an artificial-horizon sextant (art. 1513). This instrument is generally not used aboard ship because of the excessive acceleration error encountered as the ship rolls and pitches. However, in polar regions there is generally little such motion and in the ice there may be virtually none. Some practice is needed to obtain good results with an artificial-horizon sextant, but these results are sometimes superior to those obtainable with a marine sextant, and when some of the conditions mentioned above prevail, the artificial-horizon sextant may provide the only means of making an observation. Better results with this instrument can generally be obtained if the instrument is hung from some support, as it generally is when used in aircraft.

An artificial horizon (art. 1512) can sometimes be used effectively, even an improvised one, as by placing heavy lubricating oil in a bucket.

It is sometimes possible to make better observations by artificial-horizon sextant or artificial horizon from a nearby cake of ice than from the ship.

Clouds and high fog are frequent in high latitudes, but it is not uncommon, particularly in the Antarctic, for the fog to lift for brief periods, permitting an alert navigator to obtain observations.

As the latitude increases, an error of time has less effect upon altitude. At the equator an error of 4 seconds in time may result in an error in the location of the position line of as much as 1 mile. At latitude 60° a position error of this magnitude cannot occur unless the timing error is 8 seconds. At 70° nearly 12 seconds are needed, and at 80° about 23 seconds are needed for such a position error.

Polaris is of diminished value in high northern latitudes because of its high altitude. At high latitudes the second correction to observed altitude (a_1) becomes greater. The almanac makes no provision for applying this beyond latitude 68°. Bodies at high altitudes are not desirable for azimuth determination, but if Polaris is used, the use of the actual azimuth given at the bottom of the Polaris tables of the *Nautical Almanac* is of increased importance because of its larger variation from 000° in high latitudes. No azimuth is provided beyond latitude 65°.

In applying a sextant altitude correction for dip of the horizon, one should use height of eye *above the ice at the horizon,* instead of height above water. The difference between ice and water levels at the horizon can often be estimated by observing ice near the vessel.

2530. Low-altitude observations.—Because of large and variable refraction at low altitudes, navigators customarily avoid observations below some minimum, usually 5° to 15°, if higher bodies can be observed. In polar regions low-altitude observations are often the only ones available. The sun, moon, and planets remain low in the sky for relatively long periods, their diurnal motion being nearly horizontal. The only lower limit is that imposed by the horizon itself. In fact, good observations can sometimes be made without a sextant by noting the time at which either the upper or lower limb is tangent to the horizon. To such an observation sextant altitude corrections are applied as for a marine sextant without an index correction.

Correction of low-altitude observations made by marine sextant is discussed in article 1625. If a bubble or other artificial-horizon sextant is used, corrections are made as for higher altitudes, being careful to use the refraction value corrected for temperature, or to make a separate correction for air temperature. In addition, a correction for atmospheric pressure (tab. 24) is applied if of sufficient size to be of importance.

Solution of low-altitude observations is discussed in article 2023.

2531. Abnormal refraction and dip.—Tables of refraction correction are based upon a standard atmosphere. Variations in this atmosphere result in changes in the refraction, and since the atmosphere is seldom exactly standard, the mean refraction is seldom the same as shown in the tables. Variations from standard conditions are usually not great enough to be troublesome.

In polar regions, however, it is normal for the atmosphere to differ considerably from the standard, particularly near the surface. This affects both refraction and dip, as indicated in article 1605. Outside polar regions, variations in refraction seldom exceed 2′ or 3′, although extreme values of more than 30′ have been encountered. In polar regions refraction variations of several minutes are not uncommon and an extreme value of about 5° has been reported. This would produce an error of 300 miles in a line of position. The sun has been known to rise as much as ten *days* before it was expected.

Most celestial observations in polar regions produce satisfactory results, but the high-latitude navigator should be on the alert for abnormal conditions, since they occur more often than elsewhere, and have greater extreme values. A wise precaution is to apply corrections for air temperature (tab. 23) and atmospheric pressure (tab. 24), particularly for altitudes of less than 5°.

Abnormal dip affects the accuracy of celestial observations equally at any altitude, if the visible horizon is used. Such errors may be avoided in any *one* of four ways:

1. The artificial-horizon sextant may be used, as indicated in article 2529.

2. When stars are available, three stars may be observed at azimuth intervals of approximately 120°, (or four at 90° intervals, five at 72°, etc.). Any error in dip *or refraction* will alter the size of the enclosed figure, but will not change the location of its center unless the dip or refraction error varies in different directions. The stars should preferably be at the same altitude.

3. The altitude of a single body may be observed twice, facing in opposite directions. The sum of the two readings differs from 180° by twice the sum of the index and dip corrections (also personal and instrument corrections, if present). This method assumes that dip is the same in both directions, an assumption that is usually approximately correct. Also, the method requires that the arc of the sextant be sufficiently long and the altitude of the body sufficiently great to permit observation of the back sight in the opposite direction. In making such observations, it is necessary that allowance be made for the change of altitude between readings. This may be done by taking a direct sight, a back sight, and then another direct sight at equal intervals of time, and using the average of the two direct sights.

4. A correction for the difference between air and sea temperatures (art. 814, vol. II) may be applied to the sextant altitude. This will often provide reasonably good results. However, there is considerable disagreement in the manner in which temperature is to be measured, and in the factor to use for any given difference. Therefore, the validity of this correction is not fully established.

There is still much to be learned regarding refraction and even with all known precautions, results may occasionally be unsatisfactory.

2532. Sight reduction in polar regions is virtually the same as elsewhere. Computation can be made by nearly any method. In Pub. No. 214, tabulations are not extended below an altitude of 5°, but this method can be used for lower altitudes, which are not uncommon in polar regions, by selecting an assumed position some distance away, in the general direction of the body. Thus, if the altitude is 2°, an assumed position 3° (180 miles) nearer the body (4° is a better choice to allow for possible error in the dead reckoning and for adjustment for a convenient assumed position) should result in a computed altitude of 5° or more. This method will result in an unusually long altitude intercept, but the error introduced will be negligible *if* the assumed position is in the direction of the body, and the chart used is one on which a straight line is a close approximation to a great circle. A Lambert conformal chart is satisfactory for this purpose. An example of such a solution is given in article 2023.

One special method of considerable interest is conveniently applicable only within about 5° of the pole, a higher latitude than is usually attainable by ships. This is the method of using the pole as the assumed position. At this point the zenith and pole coincide and hence the celestial equator and celestial horizon also coincide, and the systems of coordinates based upon these two great circles of the celestial sphere become identical. The declination is computed altitude, and GHA replaces azimuth. A "toward" altitude intercept is plotted along the upper branch of the meridian over which the body is located, and an "away" intercept is plotted in the opposite direction, along the lower branch. Such a line or its AP is advanced or retired in the usual manner. This method is a special application of the meridian altitude sometimes used in lower latitudes. Beyond the limits of this method the meridian altitude can be used in the usual manner (art. 2025) without complications and with time of transit being less critical. However, table 29, for reduction to the meridian, extends only to latitude 60°.

2533. Plotting lines of position from celestial observations.—Lines of position from celestial observations in polar regions are plotted as elsewhere, using an assumed position, altitude intercept, and azimuth. If a Mercator chart is used, the error introduced by using rhumb lines for the azimuth line (a great circle) and line of position (a small circle) is accentuated. This can be overcome by using a chart on a more favorable projection.

If a chart with nonparallel meridians, such as the Lambert conformal, is used, the true azimuth should be plotted by protractor or plotter and measured at the meridian of the assumed position. On a chart having a grid overprint the true azimuth can be converted to grid azimuth, using the longitude of the assumed position, and the direction measured from any grid line. This method involves an additional step, with no real advantage.

Lines of position from high-altitude observations, to be plotted as circles with the geographical position as the center (art. 2024), should not be plotted on a Mercator chart because of the rapid change of scale, resulting in distortion of the circle as plotted on the chart.

Lines of position are advanced or retired as in any latitude. However, the movement of the line is no more accurate than the estimate of the direction and distance traveled, and in polar regions this estimate may be of less than usual accuracy. In addition to his problem of estimated direction of travel, the polar navigator may encounter difficulty in accurately plotting the direction determined. If an accurate gyrocompass is used, the ship follows a rhumb line, which is accurately shown only on a Mercator chart. If a magnetic compass is used, the rapid change in variation may be a disturbing factor. If the ship is in ice, the course line may be far from straight.

Because of the various possible sources of error involved, it is good practice to avoid advancing or retiring lines for a period longer than about two hours. When the sun is the only body available, best results can sometimes be obtained by making an observation every hour, retiring the most recent line one hour and advancing for one hour the line obtained two hours previously. The present position is then obtained by dead reckoning from the running fix of an hour before. Another technique is to advance the one or two previous lines to the present time for a running fix. A third method is to drop a perpendicular from the dead reckoning or estimated position to the line of position to obtain a new estimated position, from which a new dead reckoning plot is carried forward to the time of the next observation. A variation of this method is to evaluate the relative accuracy of the new line of position and the dead reckoning or estimated position run up from the previous position and take some point *between* them, halfway if no information is available on which to evaluate the relative accuracies. None of these techniques is suitable for determining set and drift of the current.

2534. Rising, setting, and twilight data are tabulated in the almanacs to latitude 72°N and 60°S. Within these limits the times of these phenomena are determined as explained in chapter XIX.

Beyond the northern limits of these tables the values can be obtained from a series of graphs given near the back of the *Air Almanac.* These graphs are shown in appendix G. For high latitudes, graphs are used instead of tables because graphs give a clearer picture of conditions, which may change radically with relatively little change in position or date. Under these conditions interpolation to practical precision is simpler by graph than by table. In those parts of the graph which are difficult to read, the times of the phenomena's occurrence are themselves uncertain, being altered considerably by a relatively small change in refraction or height of eye. The use of the graphs is explained in chapter XIX.

General

2535. Ice.—Several references have been made to ice. The almost constant presence of large quantities of ice is one of the distinctive features of polar regions, and is one of the primary considerations in any operations in these areas. The subject of ice in the sea is covered in chapter XXXVI.

2536. Knowledge of polar regions.—Operations in polar regions are attended by hazards and problems not encountered elsewhere. Lack of knowledge, sometimes accompanied by fear of the unknown, has prevented navigation in these areas from being conducted with the same confidence with which it is pursued in more familiar areas. As experience in high latitudes has increased, much of the mystery surrounding these areas has been dispelled, and operations there have become more predictable.

Before entering polar regions, the navigator will do well to acquaint himself with the experience of those who have preceded him into the areas and under the conditions he anticipates. This information can be found in a growing literature composed of the accounts of explorers, reports of previous operations in high latitudes, articles in professional journals, and several books on operations in polar regions. Some of it is given in various volumes of sailing directions.

The search for knowledge should not be confined to navigation. The wise polar navigator will seek information on living conditions, survival, geography, ice, climate and weather, and operational experience of others who have been to the same area. As elsewhere, knowledge and experience are valuable.

2537. Planning, important in any operation, is vital to the success of polar navigation. The first step to adequate planning is the acquisition of full knowledge, as discussed in article 2536. No item, however trivial, should escape attention. The ship should be provided with all the needed charts, publications, and special navigational material. All available data and information from previous operations in the area should be studied. Key personnel should be adequately instructed in polar navigation prior to departure or while en route to the polar regions. Forecasts on anticipated ice and weather conditions should be obtained before departure and after getting under way. All equipment should be put in top operating condition. All material should be carefully inspected for completeness and condition. The navigator should make certain that all items of equipment are familiar to those who will use them. This is particularly true of items not generally used at sea, such as charts on an unfamiliar projection, or a bubble sextant. Do not *assume* anything that can be *known.* On the adequacy and thoroughness of the advanced planning and preparation, perhaps more than anything else, will depend the success of polar navigation.

CHAPTER XXVI

LIFEBOAT NAVIGATION

Before Emergency Arises

2601. Introduction.—The methods and techniques used in lifeboat navigation are those available at the time. With full equipment, lifeboat navigation differs little from that aboard ship. More often, however, it is a matter of improvising equipment from available materials, and developing procedures from a knowledge of basic principles. Ingenuity is often essential. The officer who navigates by blindly "following the steps" may be of little more value in a lifeboat devoid of familiar navigational equipment than the man who has never set foot on the bridge of a ship. The wise officer becomes thoroughly familiar with the theory of navigation: the celestial triangle, the circle of equal altitude, and the other basic principles involved. He should be able to identify the most useful stars, and know how to solve his sights by any widely used method, because his favorite method may not be available. He should be able to construct a plotting sheet with a protractor, and use distress signaling equipment. Familiarity with the coordinates (latitude and longitude) of land points in the area of operations, ability to interpret wind and weather signs, knowledge of the ocean currents, and skill in handling a small boat are parts of the practical navigator's basic education which assume their greatest importance in an emergency. For the navigator prepared with such knowledge, and a determination to succeed, the situation is never hopeless. *Some* method of navigation is always available.

2602. Emergency navigation kit.—In time of national emergency, the prudent navigator will provide each lifeboat with a kit containing the equipment which it is practical to carry for emergency navigational purposes (art. 2603). Even in peacetime it is good practice to have one such kit permanently located in the chart house or the wheel house so that it can be quickly transferred to a lifeboat when needed.

The least preparation made should be a check-off list of items to be assembled if time permits, so that nothing will be overlooked. Such a list can be helpful even if one or more emergency kits have been provided. The list should be kept in a prominent place on the bridge or near the lifeboats, perhaps framed under glass. All officers should be familiar with its location and should be acquainted with the location and identity of each item listed.

Junior officers or reliable crew members should be assigned the duty of bringing to their stations, during abandon ship drill, emergency navigational equipment not permanently stowed in the boats. A senior officer should then check each item against the equipment check-off list to ascertain that nothing has been overlooked.

2603. Equipment.—If practicable, full navigational equipment should be provided. As many as possible of the items in the following list should be included. All of these except a timepiece, and possibly a sextant and radio, can be kept in the emergency navigation kit recommended in article 2602.

1. **Notebook** suitable for use as a deck log and for performing computations. Several items of information should be written in this notebook in advance, so as to be available when and if needed. Such items include the latitude and longitude of

various places in the area of operation; any desired information on currents and weather; declination and SHA of several widely scattered stars, with any needed information on identifying them; desired notes and tables from this chapter and elsewhere; any desired general information, such as a list of poisonous fish and those items which may prove useful for survival. This section of the notebook should be brief and the items limited to those most essential in time of emergency.

2. **Charts and other plotting materials.** A pilot chart is most suitable for lifeboat use, both for plotting and as a source of information on variation of the compass, shipping lanes, currents, winds, and weather. Charts for both summer and winter seasons should be included. During World War II pilot charts were printed on waterproof material suitable for use in a lifeboat. **Plotting sheets** (art. 323) are useful but not essential if charts are available. The plotting sheets should cover the latitudes in which the ship operates. **Universal plotting sheets** (art. 324) may be preferred, particularly if the latitude coverage is large. Several **maneuvering boards,** and several sheets of **cross-section paper** (preferably with ten squares per inch) should be included, as these have many uses.

3. **Plotting equipment.** Pencils, erasers, straightedge, protractor, dividers and compasses (not essential, but useful), and a knife or pencil sharpener should be included. Preferably, the straightedge and protractor should be combined in a single device constituting some kind of **plotter** (art. 605). A ruler graduated in inches and fractions may be useful.

4. **Timepiece.** A good watch is needed if longitude is to be determined astronomically. This watch should be waterproof or kept in a waterproof container which permits reading and winding of the watch without exposing it to the elements. The watch should be wound regularly and a record kept of its error and rate of change. Even if one or more such watches are available, the possibility of taking along the chronometers should not be overlooked. The optimum timepiece is a quartz crystal watch.

5. **Sextant.** A marine sextant should be taken along if possible. However, since this may be impractical, a lifeboat sextant, or materials for constructing one, should be provided. The relatively inexpensive plastic sextants should be more than adequate for lifeboat use. They should be protected from the direct rays of the sun when not in use. Several commercially manufactured lifeboat sextants have been made available, particularly during wartime. A lifeboat sextant can be made of wood or other rigid material, two small mirrors, and a pivot. The graduations of the arc should be double those of a compass rose (an angle of 5° should be labeled 10°, etc.). It is not necessary to provide a vernier, or means of adjusting the sextant, since accuracy of 0°.1 is satisfactory for lifeboat use.

6. **Almanac.** A *Nautical Almanac* for the current year is desirable. In an emergency an almanac for another year can be used for stars and the sun without serious error by lifeboats standards, if suitable adjustment is made (art. 2617). Some form of long-term almanac, as that given in appendix H, might well be copied or pasted in the notebook suggested as item 1, above.

7. **Tables.** Some form of table will be needed for reducing celestial observations. The most suitable is one that does not require much space. If a table of trigonometric functions (either logarithmic or natural) is provided, formulas should be included with them. It is not wise to trust the memory for such vital information. A set of tables similar to Pub. No. 214 can be made at 5° intervals of the arguments. Only one page is needed for each latitude entry (5°) if declination is limited to about 30° (sufficient for bodies of the solar system and many stars), entries are given to the

nearest 0°.1 for altitudes and 1° for azimuth, and the delta (Δ) values are omitted. **Traverse tables** and others given in this chapter are useful. Volume II provides tables of trigonometric functions (logarithmic and natural), formulas, a long-term almanac, traverse tables, and a sight reduction table (table 35). Due to its contents, volume II would be very useful in a lifeboat.

8. **Compass.** Each lifeboat is required to carry a magnetic compass. A deviation table for each compass should be made while in port, with magnetic material in its normal place. It would be well to check the accuracy of each table periodically.

9. **Flashlight.** A flashlight is required to be carried in each lifeboat. The batteries should be replaced from time to time, as necessary. Extra batteries and bulbs might well be carried.

10. **Portable radio.** If a portable radio is available, be sure it is included. Whether this is one of the transmitting-receiving sets approved by the Federal Communications Commission for lifeboat use, or merely a small receiver of limited range owned by a crew member, do not overlook it, as it may be used as a radio direction finder.

2604. Position of ship.—A knowledge of the position of the vessel at the time it is abandoned is of great importance. The officer on watch on the bridge should never permit himself to become careless in the matter of keeping a mental note of the approximate position of the vessel. During wartime, or whenever the possibility of abandoning ship might reasonably be anticipated, the radio operator should be provided with a list of advance dead reckoning positions.

Abandoning Ship

2605. Before lowering boats.—The period between the decision to abandon ship and the actual leaving of the vessel is a highly important one. It is also a period of mental strain and possible confusion. The degree to which the crew can be prepared for the ordeal ahead depends upon the amount of time available and the thoroughness of the preparation that has been made. If there has been advance warning of the possibility of the decision, certain preparations can be made before the decision is reached. If time permits, after the decision to abandon ship has been made, the radio operator should send a final distress message, giving the ship's position and any other pertinent information. It will be important later to know whether an acknowledgement of receipt of the message was received. Any available time can be wisely used to check the navigational equipment in each boat and assemble missing items. There may be time to make a last minute check of position of the ship, position of any nearby land, set and drift of current, present and forecast weather, watch error, and date. These items should be written down. Perhaps the chart can be taken along. Equipment should be properly secured before lowering the boats. In a rough sea it may be desirable to lower the sextant, chronometer, and radio into the boat after it is afloat.

2606. Establishing command.—The identity of the person in command of each boat, and the over-all commander, should be firmly established. Almost invariably this will be the senior officer present. In a lifeboat, perhaps more than in any other circumstances, strong leadership is required if the confidence of the crew is to be maintained. The officer whom the crew respects as a *man,* admires as a *seaman,* and recognizes as a *gentleman* will have little or no trouble with discipline and cooperation of all on board.

Morale is a prime consideration, and it grows in importance with the passage of time. The person in command should be recognized as the final authority in all matters, but it is important that he give to each person an opportunity to be heard, and that he keep all hands fully informed of the bad as well as of the good. Decisions will be more acceptable if the crew has been informed of each consideration as it arises, and so has been somewhat prepared. Complete fairness and impartiality are essential.

2607. Estimate of the situation.—Perhaps the first item which should engage the attention of the person in command, after the lifeboat has cleared the stricken vessel, is the questioning of each person aboard to collect all the useful information available. It is well to determine what is known regarding the position of the ship, ocean currents, weather, astronomy, navigation, seamanship, sailing, etc. Find out who owns watches and what each owner knows about the error and rate of his watch. Establish a routine for winding and comparing them. No useful skill or knowledge should be overlooked; all should be fully considered in making the important decision of whether to remain in the vicinity of the disaster in the hope of rescue, or to attempt to reach land or a more heavily traveled shipping lane.

This decision of whether to stay or leave may be the most important one of the entire experience. Until comparatively recent times there was no problem. Because there was virtually no hope of assistance, the lifeboat crew had to rely upon itself. Since the development of modern communication and rescue facilities, however, it is often wiser to remain than to complicate the rescue problem by increasing the area to be searched.

The decision should not be made until careful consideration has been given to all factors, nor should it be delayed longer than necessary. Considerations vary with the circumstances, but certainly the following should be included:

Was a distress message sent before the ship was abandoned? Did it include the position of the ship? How accurate was the position? Is there any reasonable doubt that the message was received? If no message was sent, how soon will the ship be missed? What rescue facilities are available? How far away are they and how long will it be before help arrives? How conspicuous is the lifeboat? What facilities are available for attracting attention, either visually or by radar? How proficient is the crew in using such equipment? Is a radio transmitter available? What is the probable running time to the nearest land in several directions, considering the prevailing winds and currents, the motive power available (art. 2614), and the ability of the crew to use it? How long will the fresh water and rations last, and will they be sufficient to sustain the crew in the physical exertion required?

If the decision is to stay, how will the crew occupy its time, remembering the increased morale problem with an idle crew? How will position be maintained, or regained if the boat drifts? Would it be practical to wait two or three days, perhaps, in the hope of rescue, and then to set out for land if help does not come?

If the decision is to leave, where should the boat head? How soon can a well-traveled shipping lane be reached? In time of war, where is the enemy and where are friends? How large and conspicuous is the land in each direction, considering the low height of eye in a lifeboat? It may be better to head for conspicuous land 500 miles away than for a small, low island 200 miles away, particularly if the latter is in a direction of unfavorable winds or currents, or takes the boat farther away from shipping lanes.

Avoid, if possible, a hasty decision that will later be recognized as unwise. Discuss the matter thoroughly with the crew, and when the decision is made, inform them of the reason for it. Do this in a manner that will invite their confidence and support. Inform them of the best estimate of the situation.

2608. Selecting the route.—It is not always desirable to head directly for the objective. A longer route with favorable winds and currents may be quicker. A longer route by way of shipping lanes may enhance the possibility of rescue.

With clear skies, latitude can be found with relatively crude equipment. But unless accurate Greenwich time is available, longitude cannot be found astronomically, even with the best equipment; nor is a nonastronomical method likely to be available. In the absence of reliable longitude information, it is better to head for a point at the latitude of the destination but so far east or west of it that no reasonable doubt will exist as to the direction of land when that latitude is reached. The distance of the point from the destination depends upon the degree of uncertainty of the longitude, remembering that this uncertainty is likely to increase with time. This method of "parallel sailing" was used for centuries before a method of determining or "discovering" longitude at sea was developed.

If the objective has a considerable extent in a north-south direction, the need for a final east-west leg is less critical, and in attempting to reach a continent or very large island, one need not consider it at all. In the absence of better information, an east or west course should be selected from the outset, since most large land masses of the earth are oriented in a general north-south direction.

2609. Keeping boats together.—If more than one boat is launched, every effort should be made to keep them together. While the person in charge of each boat is responsible for decisions regarding his boat, considerable advantage is to be gained by keeping the boats together and recognizing one person, logically the senior officer present, as the over-all commander. Since navigational equipment and skill probably will differ widely from boat to boat, the benefits of any accurate navigation can be shared by all if the boats are close together. Other knowledge can be exchanged, equipment shared, and rations distributed equitably. It may be wise to shift some personnel among the boats, perhaps on a periodic basis, either to effect a better balance of skill and knowledge, or for morale purposes.

2610. Lookout.—Always there is the possibility of sighting another vessel. Hence, a lookout should be posted at all times. This becomes of even greater importance when approaching land, or if the location of all land along the route is not known. If it is possible to rig a metal object high in the boat, this should be done to enhance the possibility of detection by radar.

Dead Reckoning

2611. Importance of dead reckoning.—Of the various kinds of navigation, dead reckoning alone is always available in some form. It should never be neglected, but in a lifeboat it is of more than average importance. A close check should be kept on the direction and distance made good, and all disturbing elements such as wind and current should be carefully evaluated. Long voyages have been successfully completed by this method alone, and landfalls have been made with surprising accuracy. This is not meant to minimize the importance of other methods of determining position, but with the methods generally available in a lifeboat, one may well find that, during the first few days, his dead reckoning positions are more accurate than those determined by other methods. If the means of determining direction and distance—the elements of dead reckoning—are accurate, it might be well to make an adjustment to the dead reckoning only after consistent indication of the magnitude and direction of its error. The dropping of the dead reckoning at each uncertain "fix" is at best a questionable procedure. The conflicting information likely to be available calls for careful analysis and good judgment on the part of the navigator.

2612. Deck Log.—From the beginning a careful log should be kept. The date and time of abandoning ship should be the first entry, followed by navigational information available, and the various important decisions and the reasons for them. Since the conservation of paper may be important, record only the essentials of the important items, but do not overlook the recording in considerable detail of the selection of a commanding officer, changes in command, deaths, missing persons, and navigational information.

The best determination of the position of abandoning ship should be recorded, followed by a full account of courses, distances, positions, winds, currents, and leeway. No important navigational information should be left to memory if it can be recorded.

2613. Direction.—As one of the elements of dead reckoning, direction is an important item. As indicated in article 2603, a deviation table for each lifeboat compass should be determined in port, and checked periodically. At the first convenient opportunity after abandoning ship the accuracy should be checked on the course to be followed.

If an almanac, accurate Greenwich time, and the necessary tables are available, the azimuth of any celestial body can be computed and this value compared with the azimuth as measured by the compass. If it is difficult to observe the compass azimuth, select a body dead ahead and note the compass heading. The difference between computed and observed azimuths is compass error. This is of more immediate value than deviation, but if the latter is desired, it can be determined by applying to the compass error the variation, from the pilot chart.

Several unique astronomical situations occur, permitting determination of azimuth without computation:

Polaris is always within 2° of true north for observers between the equator and latitude 60°N. When this star is directly above or below the celestial pole, its azimuth is exactly north at any latitude. This occurs approximately when the *trailing* star of either *Cassiopeia* (ε *Cassiopeiae*) or the Big Dipper (Alkaid) is directly above or directly below Polaris (fig. 2621). When a line through the trailing stars and Polaris is horizontal, the maximum correction should be applied. Below latitude 50° this can be considered 1°; and between 50° and 65°, 2°. If *Cassiopeia* is to the *right* of Polaris, the azimuth is 001° (or 002°), and if to the left, 359° (or 358°). The *south* celestial pole is located approximately at the intersection of a line through the longer axis of the Southern Cross with a line from the northernmost star of *Triangulum Australe* perpendicular to the line joining the other two stars of the triangle. No conspicuous star marks this spot (figs. 2205–2208).

Meridian transit. Any celestial body bears due north or south at meridian transit, either upper or lower. This is the moment of maximum (or minimum) altitude of the body. However, since the altitude at this time is nearly constant during a considerable change of azimuth, the instant of meridian transit may be difficult to determine. If time and an almanac are available, and the longitude is known, the time of transit can be computed.

Body on prime vertical. If any method is available for determining when a body is on the prime vertical (due east or west), the compass azimuth at this time can be observed. Table 25 provides this information. Any body on the celestial equator (declination 0°) is on the prime vertical at the time of rising or setting. For the sun this occurs at the time of the equinoxes (art. 1419). The star Mintaka (δ *Orionis*), the *leading* star of *Orion's* belt, has a declination of approximately 0°.3S and can be considered on the celestial equator. For an observer near the equator, such a body is always nearly east or west. Because of refraction and dip, the azimuth should be noted when the center of the sun or a star is a little more than one sun diameter (half a degree) *above* the horizon. The moon should be observed when its *upper* limb is *on* the horizon.

Body at rising or setting. Except for the moon, the azimuth angle (art. 1430) of a body is almost the same at rising as at setting, except that the former is toward the *east* and the latter toward the *west.* If the azimuth is measured *both* at rising and setting, true south (or north) is midway between the two observed values, and the difference between this value and 180° (or 000°) is the compass error. Thus, if the compass azimuth of a body is 073° at rising, and 277° at setting, true south (180°) is at

$$\frac{073^\circ+277^\circ}{2}=175^\circ$$

by compass, and the compass error is 5°E. This method may be in error if the boat is moving rapidly in a north or south direction. If the declination and latitude are known, the true azimuth of any body at rising or setting can be determined by means of a diagram on the plane of the celestial meridian (art. 1432) or by computation (art. 2125). For this purpose the body (except the moon) should be considered as rising or setting when its center is a little more than one sun diameter (half a degree) above the horizon, because of refraction and dip.

The direction of the sun in relation to the hands of a watch is sometimes advocated, but the limitations of this method are too great to permit general application.

A simple nonastronomical method can be used for determining the *deviation.* An object that will float but not drift rapidly before the wind is thrown overboard. The boat is then steered as steadily as possible in the *opposite* direction to that desired. At a distance of perhaps half a mile, or more if the floating object is still clearly in view, the boat is turned around in the smallest practicable radius, and headed back toward the floating object. The *magnetic* course is midway between the course toward the object and the *reciprocal* of the course away from the object. Thus, if the boat is on compass course 151° while heading away from the object, and 337° while returning, the magnetic course is midway between 337° and 151°+180°=331°, or

$$\frac{337^\circ+331^\circ}{2}=334^\circ.$$

Since 334° magnetic is the same as 337° by compass, the deviation on this heading is 3°W.

If a compass is not available, any celestial body can be used to steer by, if its diurnal apparent motion is considered. A reasonably straight course can be steered by noting the direction of the wind, the movement of the clouds, the direction of the waves, or by watching the wake of the boat. A line can be secured to the side of the boat at a point amidships or forward. The line should tend parallel to the centerline of the boat if on a straight course. The angle between the centerline and the wake is an indication of the amount of leeway. The accuracy of the towed-object or wake method is affected adversely by a cross sea.

A body having a declination the same as the latitude of the destination is over the destination once each day, at the time when its hour angle is the same as the longitude, measured westward through 360°. At this time it should be dead ahead if the boat is following the great circle leading directly through the destination.

2614. Motive power.—A lifeboat is equipped with one or more of the following means of locomotion: oars, hand-operated propeller, motor, sail. Inflatable rafts usually have no means of locomotion other than paddles. Of these, only sail offers a practical means of travel over an extended period of time. Men living in an open boat, perhaps on reduced rations, should not attempt to expend their strength on hand locomotion, except for short periods. Likewise, the comparatively small fuel supply in a motorboat should be hoarded jealously. It may be desperately needed later, as for landing through a surf, preventing the boat from drifting onto a rocky coast, or making the land when a strong current is carrying the boat past an island.

A sail should be rigged, for in it lies the best hope of reaching distant land. If the standard lifeboat sail is not available, a substitute can usually be devised, using the boat cover, or even clothing, and oars.

2615. Distance can be determined directly between accurate fixes, but generally it is found by means of speed and elapsed time. A loaded lifeboat will not travel fast, under normal conditions. With fair wind and weather it may make good a speed of about two knots through the water. Hence the importance of wind and current. The navigator used to observing the sea from a high bridge usually overestimates his speed in a lifeboat, where he is only a few feet from the water. With practice, his ability should improve.

Speed may be determined by using a form of **chip log.** Attach a long line to a heavy, floating object. Put one knot in the line twelve or fifteen fathoms from the object, and another just ten fathoms (or any convenient distance) from the first. Stream the device over the side and let the line run out freely, noting the elapsed time between passage of the two knots through the hand. A variation of this is the **Dutchman's log.** A floating object is thrown overboard at the bow, and the elapsed time required for a known length along the centerline to pass it is noted. If a line is attached to the object, it may be used many times. With either variation, it is well to tie the bitter end of the line to the boat, to minimize danger of losing the whole device overboard.

With either the chip or Dutchman's log, the speed is determined by the formula:

$$S=\frac{60 \text{ seconds per minute} \times 60 \text{ minutes per hour} \times \text{feet between marks}}{6{,}000 \text{ feet per mile} \times \text{seconds of elapsed time}}.$$

This is equal to:

$$S=\frac{3{,}600 \times \text{feet between marks}}{6{,}000 \times \text{seconds of elapsed time}}=\frac{0.6 \times \text{feet between marks}}{\text{seconds of elapsed time}}.$$

Since the feet between marks is constant, a convenient number can be selected.

Thus, if the length is 16⅔ feet, the formula becomes

$$S = \frac{10}{\text{seconds of elapsed time}}.$$

If the elasped time is ten seconds, the boat is traveling at one knot; if five seconds, at two knots; if eight seconds, at 1¼ knots, etc.

If a watch is not available, a simple pendulum may be devised to time the interval. A piece of string with a weight attached, of a length of 9.8 inches (to the center of gravity of the weight), will, when suspended, make a complete swing (back *and* forth) once every second. For a pendulum 39.1 inches long the period is two seconds. With practice, time can be estimated with fair accuracy.

It is not always possible to head directly along the course to the destination, because of adverse winds. It is better to make good progress in the general direction desired than none at all, and much better on morale. However, at times conditions, may be so adverse that it will be best to drop sail until the wind shifts or abates. At such a time a sea anchor should be streamed to minimize loss of precious mileage, and, in severe conditions, to keep the boat headed into the sea.

2616. Position by dead reckoning.—Plotting can be done directly on a pilot chart or plotting sheet. If this proves too difficult, or if an independent check is desired, some form of mathematical reckoning may be useful. Table 2616, a simplified traverse table, can be used for this purpose. This is a critical-type table, various factors being given for limiting values of certain angles. To find the difference or change of latitude,

ANGLE	0°	18°	31°	41°	49°	56°	63°	69°	75°	81°	87°	90°
FACTOR	1.0	0.9	0.8	0.7	0.6	0.5	0.4	0.3	0.2	0.1	0.0	

TABLE 2616.—Simplified traverse table.

in minutes, enter the table with course angle, reckoned from north or south toward the east or west. Multiply the distance run, in miles, by the factor. To find the departure, in miles, enter the table with the *complement* of the course angle. Multiply the distance run, in miles, by the factor. To convert departure to difference of longitude, in minutes, enter the table with mid-latitude. Divide the departure by the factor.

Example.—A lifeboat travels 26 miles on course 205°, from L41°44′N, λ56°21′W.

Required.—Latitude and longitude of the point of arrival.

Solution.—The course angle is 205°−180°=S25°W, and the complement is 90°−25°=65°. The factors corresponding to these angles are 0.9 and 0.4, respectively. The difference of latitude is 26×0.9=23′ (to the nearest minute) and the departure is 26×0.4=10 mi. Since the course is in the southwestern quadrant, in the Northern Hemisphere, the latitude of the point of arrival is 41°44′N−23′=41°21′N. The factor corresponding to the mid-latitude 41°32′N is 0.7. The difference of longitude is 10÷0.7=14′. The longitude of the point of arrival is 56°21′W+14′=56°35′W.

Answer.—L41°21′N, λ56°35′W.

Celestial Navigation

2617. Celestial coordinates.—Almanac information, particularly declination and Greenwich hour angle of bodies, is important to celestial navigation. If the current *Nautical Almanac* is available, there is no problem. If the only copy available is for a previous year, it can be used for the sun, Aries, and stars without serious error, by lifeboat standards. However, for greater accuracy, proceed as follows: For declination of the sun, enter the almanac with a time that is *earlier* than the correct time by 5^h49^m times the number of years between the date of the almanac and the correct date, adding 24^h for each February 29 that occurs between the dates. If the date is February 29, use March 1 and reduce by one the number of 24^h periods added. For GHA of the sun or Aries determine the value for the correct time, adjusting the minutes and tenths of arc to agree with that at the time for which the declination is determined. Since the adjustment never exceeds half a degree, care should be used when the value is near a whole degree, to prevent the value from being in error by 1°. Appendix H is a long-term almanac giving values of GHA ♈, and GHA and declination of the sun. Instructions for its use are included in the appendix. A reproduction of this almanac might profitably be included in the navigational kit mentioned in article 2602.

If no almanac is available, a rough approximation of the declination of the sun can be obtained as follows: Count the days from the given date to the *nearer* solstice (June 21 or December 22). Divide this by the number of days from that solstice to the equinox (March 21 or September 23), using the equinox that will result in the given date being between it and the solstice. Multiply the result by 90°. Enter table 2616 with the angle so found, and extract the factor. Multiply this by 23°.45 to find the declination.

Example 1.—The date is August 24.

Required.—The approximate declination of the sun.

Solution.—The number of days from the given date to the nearer solstice (June 21) is 64. There are 94 days between June 21 and September 23. Dividing and multiplying by 90°,

$$\frac{64}{94}\times 90^\circ = 61^\circ.3.$$

The factor from table 2616 is 0.5. The declination is 23°.45×0.5=11°.7. It is known to be north because of the date.

Answer.—Dec. 11°.7N.

The accuracy of this solution can be improved by considering the factor of table 2616 as the value for the mid-angle between the two limiting ones (except that 1.00 is correct for 0° and 0.00 is correct for 90°), and interpolating to one additional decimal. In this instance the interpolation would be between 0.50 at 59°.5 and 0.40 at 66°. The interpolated value is 0.47, giving a declination of 11°.0N. Still greater accuracy can be obtained by using a table of natural cosines instead of table 2616. By natural cosine the value is 11°.3N.

If the latitude is known, the declination of any body can be determined by observing a meridian altitude. In a lifeboat it is usually best to make a number of observations shortly before and after transit, plot the values on cross-section paper, letting the ordinate (vertical scale) represent altitude, and the abscissa (horizontal scale) the time. The altitude is found by fairing a curve or drawing an arc of a circle through the points, and taking the highest value. A meridian altitude problem is then solved in reverse.

Example 2.—The latitude of a lifeboat is 40°16′S. The sun is observed on the meridian, bearing north. The *observed* altitude is 36°29′.

Required.—Declination of the sun.

Solution.—The zenith distance is 90° − 36°29′ = 53°31′. The sun is 53°31′ north of the observer, or 13°15′ north of the equator. Hence, the declination is 13°15′N.

Answer.—Dec. 13°15′N.

The GHA ♈ can be determined approximately by considering it equal to GMT (in angular units) on September 23. To find GHA ♈ on any other date, add 1° for each day following September 23. The value is approximately 90° on December 22, 180° on March 21, and 270° on June 21. The values so found can be in error by as much as several *degrees,* and so should not be used if better information is available. An approximate check is provided by the great circle through Polaris, Caph (the leading star of *Cassiopeia*), and the eastern side of the square of *Pegasus.* When this great circle coincides with the meridian, LHA ♈ is approximately 0°. The hour angle of a body is equal to its SHA plus the hour angle of Aries.

If an error of as much as 4°, or a little more, is acceptable, the GHA of the sun can be considered equal to GMT± 180° (12^h). For more accurate results, one can make a table of the equation of time from the *Nautical Almanac* perhaps at five- or ten-day intervals, and include this in the emergency navigation kit mentioned in article 2602. The equation of time is applied according to its sign to GMT±180° to find GHA.

2618. Altitude measurement.—If a sextant is available, either one from the pilot house or an emergency-type instrument, altitudes are measured in the usual manner. The sextant should be shielded as much as possible from wind and spray. If the sea is rough, the observer should brace himself against the mast and make his observation when on the crest of a wave, when the horizon is least likely to be obscured by nearby waves. It is usually good practice to make a number of observations and average both the altitudes and times, or plot on cross-section paper the altitudes versus time, using any convenient time and the corresponding altitude for solving the observation.

The improvisations which may be made in the absence of a sextant are so varied that in virtually any circumstances the application of a little ingenuity and some effort will produce a device for measuring altitude. The results obtained with any improvised method will be approximate at best, but if a number of observations are averaged, the accuracy should be improved. Almost always a measurement, however approximate, is better than an estimate. Two general classes of improvisation are available:

1. *By circle.* Any circular scale, such as a maneuvering board, compass rose, protractor, or plotter can be used to measure altitude or zenith distance directly. This is the principle of the ancient astrolabe (art. 124). A maneuvering board or compass rose is usually handled best by mounting it on a flat board. A protractor or plotter may be so mounted or used directly. There are a number of variations of the technique of using such a device. Some of them are:

A peg or nail is placed at the center of the circle and perpendicular to it. A weight is hung from the 90° graduation, and a string for holding the device is attached at the 270° graduation. When it is held with the weight acting as a plumb bob, the 0°−180° line is horizontal (fig. 2618a). In this position the board is turned in azimuth until it is in line with the sun. The intersection of the shadow of the center peg with the arc of the circle indicates the altitude of the center of the sun.

The weight and loop can be omitted and pegs placed at the 0° and 180° points of the circle. While one observer sights along the line of pegs to the horizon, an assistant notes the altitude.

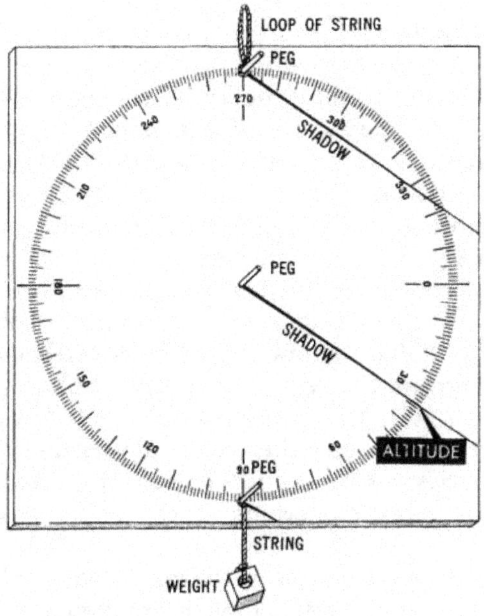

FIGURE 2618a.—Improvised astrolabe; shadow method. Pegs and board shown tilted for clarity.

The weight can be attached to the center pin, and the three pins (0°, center, 180°) aligned with the celestial body. The reading is made at the point where the string holding the weight crosses the scale. The reading thus obtained is the zenith distance unless the graduations are labeled to indicate altitude. This method, illustrated in figure 2618b, is used for bodies other than the sun.

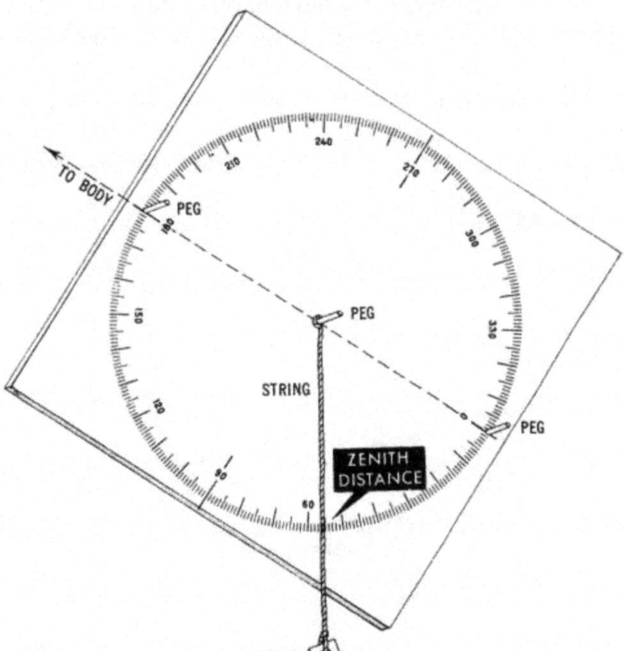

FIGURE 2618b.—Improvised astrolabe; direct sighting method. Pegs and board shown tilted for clarity.

Whatever the technique it is good practice to reverse the device for half the readings of a series, to minimize errors of construction. Generally, the circle method produces more accurate results than the right triangle method, described below.

2. *By right triangle.* The principle of the ancient cross-staff can be used to establish one or more right triangles, which can be solved by measurement of the angle representing the altitude, either directly or by reconstructing the triangle. Another way of determining the altitude is to measure two of the sides of the triangle and divide one by the other to determine one of the trigonometric functions. This procedure, of course, requires a source of information on the values of trigonometric functions corresponding to various angles. If the cosine is found, table 2616 can be used. The tabulated factors can be considered correct to one additional decimal for the value midway between the limited values (except that 1.00 is the correct value for 0° and 0.00 is the correct value for 90°) without serious error by lifeboat standards. Interpolation can then be made between such values. By either protractor or table, most devices can be graduated in advance so that angles can be read directly. There are many variations of the right triangle method. Some of these are:

Two straight pieces of wood can be attached to each other in such a way that the shorter one can be moved along the longer, the two always being perpendicular to each other. The shorter piece is attached at its center. One end of the longer arm is held to the eye. The shorter arm is moved until its top edge is in line with the celestial body, and its bottom edge is in line with the horizon. Thus, two right triangles are used (the third sides being the slant distances between the ends of the arms) each representing half the altitude (fig. 2618c). For low altitudes, only one of the triangles is used, the long arm being held in line with the horizon. The length of half the short arm, divided by the length of that part of the long arm between the eye and the intersection with the short arm, is the tangent of half the altitude (the whole altitude if only one right triangle is used). The cosine can be found by dividing that part of the long arm between the eye and the intersection with the short arm by the slant distance from the eye to one end of the short arm. Graduations consist of a series of marks along the long arm indicating settings for various angles. The device should be inverted for alternate readings of a series.

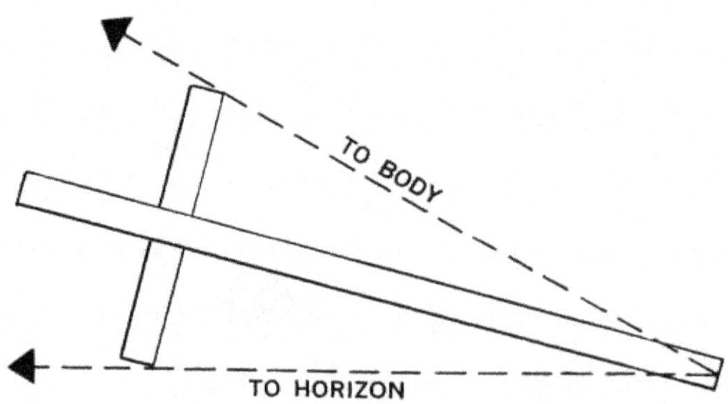

FIGURE 2618c.—Improvised cross-staff.

A rule or any stick can be held at arm's length. The top of the rule is placed in line with the celestial body being observed, and the top of the thumb is placed in line with the horizon. The rule is held vertical. The length of rule above the thumb, divided by the distance from the eye to the top of the thumb is the tangent of the angle observed. The cosine can be found by dividing the distance from the eye to the top of the thumb by the distance from the eye to the top of the rule. If the rule is tilted toward the eye until the minimum of rule is used, the distance from the eye to the middle of the rule is substituted for the distance from the eye to the top of the thumb, *half* the length of the rule above the thumb is used, and the angle found is multiplied by two. Graduations consist of marks on the rule or stick indicating various altitudes. For the average observer each inch of rule will subtend an angle of about 2°.3, assuming an eye-to-ruler distance of 25 inches. This relationship is good to a maximum altitude of about 20°. The accuracy of this relationship for a specific observer can be checked by comparing the measurement against known angles in the sky. Angular distances between stars can be computed by sight reduction methods, including Pub. No. 229, by using the declination of one star as the latitude of the assumed position, and the difference between the hour angles (or SHA's) of the two bodies as the local hour angle. The angular distance is the complement of the computed altitude. The angular distances between some well-known star pairs are: end stars of *Orion's* belt, 2°.7; pointers of the Big Dipper, 5°.4, Rigel to *Orion's* belt, 9°.0; eastern side of the great square of *Pegasus,* 14°.0; Dubhe (the pointer nearer Polaris) and Mizar (the second star in the Big Dipper, counting from the end of the handle), 19°.3.

The angle between the lines of sight from each eye is, at arm's length, about 6°. By holding a pencil or finger horizontal, and placing the head on its side, one can estimate an angle of about 6° by closing first one eye and then the other, and noting how much the pencil or finger appears to move in the sky.

The length of the shadow of a peg or nail mounted perpendicular to a horizontal board can be used as one side of an altitude triangle. The other sides are the height of the peg and the slant distance from the top of the peg to the end of the shadow. The height of the peg, divided by the length of the shadow, is the tangent of the altitude of the center of the sun. The length of the shadow divided by the slant distance is the cosine. Graduations consist of a series of concentric circles indicating various altitudes, the peg being at the common center. The device is kept horizontal by floating it in a bucket of water. Half the readings of a series are taken with the board turned 180° in azimuth.

Two pegs or nails can be mounted perpendicular to a board, with a weight hung from the one farther from the eye. The board is held perpendicular and the two pegs aligned with the body being observed. The finger is then placed over the string holding the weight, to keep it in position as the board is turned on its side. A perpendicular is dropped from the peg nearer the eye, to the string. The altitude is the acute angle nearer the eye. For alternate readings of a series, the board should be inverted. Graduations consist of a series of marks indicating the position of the string at various altitudes.

As the altitude decreases, the triangle becomes smaller. At the celestial horizon it beocomes a straight line. No instrument is needed to measure the altitude when either the upper or lower limb is tangent to the horizon, as the "sextant" altitude is then 0°.

2619. Sextant altitude corrections.—If altitudes are measured by a marine sextant, the usual sextant altitude corrections apply (ch. XVI). If the center of the sun or moon is observed, either by sighting at the center or by shadow, the lower-limb corrections should be applied, as usual, and an additional correction of (−)16′ applied. If the upper limb is observed, use (−)32′. If a weight is used as a plumb bob, or if the *length* of a shadow is measured, omit the dip (height of eye) correction.

If the almanac is not available for making corrections, each source of error can be corrected separately, as follows:

Index correction. If a sextant is used, the index correction should be determined and applied to all observations, or the sextant adjusted to eliminate index error.

Refraction is given to the nearest minute of arc in table 2619. The value for a horizon observation is 34′. If the nearest 0°.1 is sufficiently accurate, as with an improvised method of observing altitude, a correction of 0°.1 should be applied for altitudes between 5° and 18°, and no correction applied for greater altitudes. Refraction applies to all observations, and is always a minus (−) correction.

ALT.	5°		6°		7°		8°		10°		12°		15°		21°		33°		63°		90°
REFR.		9′		8′		7′		6′		5′		4′		3′		2′		1′		0	

TABLE 2619.—Refraction.

Dip, in minutes of arc, is approximately equal to the square root of the height of eye, in feet. The correction applies to all observations in which the horizon is used as the horizontal reference. It is always a minus (−) correction. If 0°.1 accuracy is used, no dip correction is needed for lifeboat heights of eye.

Semidiameter. The semidiameter of either the sun or moon does not differ greatly from 16′. The correction does not apply to other bodies or to observations of the center of the sun and moon, by whatever method, including shadow. The correction is plus (+) if the lower limb is observed, and minus (−) if the upper limb is observed.

Parallax. For lifeboat accuracy, parallax is applied to observations of the moon only. An approximate value, in minutes of arc, can be found by multiplying 57′ by the factor from table 2616, entering that table with altitude. For more accurate results the factors can be considered correct to one additional decimal for the altitude midway between the limiting values (except that 1.00 is correct for 0° and 0.00 is correct for 90°), and the values for other altitudes can be found by interpolation. This correction is always plus (+).

For observations of celestial bodies on the horizon, the total correction for zero height of eye is:

Sun. Lower limb: (−)18′, upper limb: (−)50′.
Moon. Lower limb: (+)39′, upper limb: (+)7′.
Planet or star. (−)34′.

Dip should be added algebraically to these values.

Since the "sextant" altitude is zero, the "observed" altitude is equal to the total correction.

2620. Sight reduction.—If any tables designed for sight reduction, such as Pub. No. 229, are available, they should be safeguarded to prevent loss or damage. If trigonometric tables and the necessary formulas are available, they will serve the purpose. Speed in solution is seldom a factor in a lifeboat. A slow method might actually be an asset, from a morale standpoint, as it will provide occupation for a limited time for at least one crew member. The tables and formulas given in volume II would be useful. If tables but no formulas are available, carefully determine the mathematical knowledge possessed by the crew. Someone may be able to provide the missing information. If the formulas are available, but no tables, approximate natural values of the various trigonometric functions can be obtained graphically by the method explained in article 139 of volume II. Graphical solution of the navigational triangle can be made by the orthographic method explained in article 1432. A maneuvering board might prove helpful in the graphical solution for either trigonometric functions or altitude and azimuth. Very careful work will be needed for useful results by either method.

Unless full navigational equipment is available, better results might be obtained by making separate determinations of latitude and longitude.

2621. Latitude determination.—Several methods are available for determining latitude, and in none of them is accurate time needed.

Meridian altitude. Latitude can be determined by means of a meridian altitude of any body, if its declination is known. The method is explained in article 2103. If accurate time, knowledge of the longitude, and an almanac are available, the observation can be made at the correct moment, as determined in advance. However, if any of these is lacking, or if an *accurate* altitude-measuring instrument is unavailable, better procedure is to make a number of altitude observations before and after meridian transit. A plot is then made of altitude versus time, if cross-section paper is available, and the highest (or lowest, for lower transit) altitude is scaled from a curve faired through the plotted points. At lifeboat speeds this procedure is not likely to introduce a significant error. The time used for plotting the observations need not be accurate, as *elapsed* time between observations is all that is needed, and this is not of critical accuracy. Thus, even a watch that has run down and then been rewound can be used without resetting. Any altitudes that are not consistent with others of the series should be discarded.

Polaris. Latitude by Polaris is explained in article 2027. In a lifeboat, only the first correction is of practical significance. If suitable tables are not available, this correction can be estimated. The trailing star of *Cassiopeia (ε Cassiopeiae)* and Polaris have almost exactly the same SHA. The trailing star of the Big Dipper (Alkaid) is nearly opposite Polaris and *ε Cassiopeiae.* These three stars, *ε Cassiopeiae,* Polaris, and Alkaid, form a line through the pole (approximately). When this line is horizontal, there is no correction. When it is vertical, the maximum correction of 56′ applies. It should be added to the observed altitude if Alkaid is at the top, and subtracted if *ε Cassiopeiae* is at the top. For any other position, estimate the angle this line makes with the vertical (fig. 2621), and multiply the maximum correction (56′) by the factor from table 2616, adding if Alkaid is higher than *ε Cassiopeiae,* and subtracting if it is lower. For more accurate results, the factor from table 2616 can be considered accurate to one additional decimal for the mid-value between those tabulated (except that 1.00 is correct for 0° and 0.00 for 90°). Other values can be found by interpolation.

Length of day. The length of the day varies with latitude. Hence, latitude can be determined if the elapsed time between sunrise and sunset can be observed. Correct the observed length of day by *adding* 1^m for each 15′ of longitude traveled toward the east and *subtracting* 1^m for each 15′ of longitude traveled toward the west. The latitude determined by length of day is the value for the time of meridian transit. Since meridian transit occurs approximately midway between sunrise and sunset, half the interval may be observed and doubled. If a sunrise and sunset table is not available, the length of daylight can be determined graphically by means of a diagram on the plane of the celestial meridian (art. 1432). A maneuvering board is useful for this purpose. This method cannot be used near the time of the equinoxes, and is of little value near the equator. The moon can be used if moonrise and

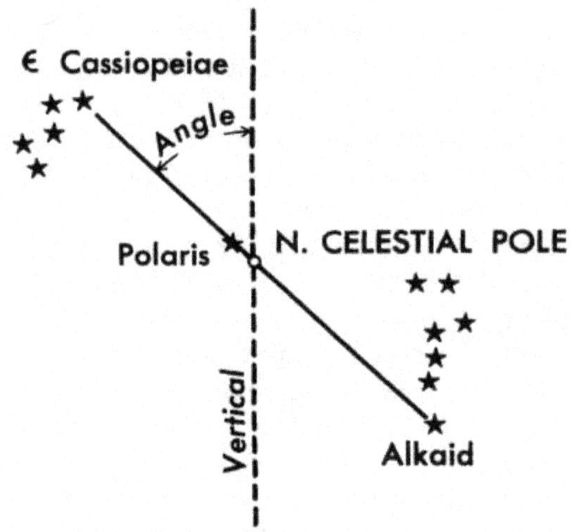

FIGURE 2621.—Relative positions of ε *Cassiopeiae*, Polaris, and Alkaid with respect to the north celestial pole.

moonset tables are available, but with the moon the half-interval method is of insufficient accuracy, and allowance should be made for the longitude correction.

Body in zenith. The declination of a body in the zenith is equal to the latitude of the observer. If no means are available for measuring the altitude, the position of the zenith may possibly be estimated in a calm sea by lying in the lifeboat and looking skyward. The accuracy of the results depends upon the ability to estimate the position of the zenith. Use of a plumb bob may help.

Variation of the compass can occasionally be used for determining latitude, as explained in article 2622.

2622. Longitude determination.—Unlike latitude, longitude requires accurate Greenwich time for its determination by astronomical means. All such methods consist of noting the Greenwich time at which a phenomenon occurs locally. In addition, a table indicating the time of occurrence of the same phenomenon at Greenwich, or equivalent information, is needed.

Time of transit. When a body is on the local celestial meridian, its GHA is the same as the longitude of the observer if in west longitude, or $360°-\lambda$ in east longitude. Thus, if the GMT of local transit is determined and a table of Greenwich hour angles (or time of transit of the Greenwich meridian) is available, longitude can be computed. If only the equation of time is available, the method can be used with the sun. This is the reverse of the problem of finding the time of transit of a body (art. 2104). The time of transit is not always apparent. If a curve is made of altitude versus time, as suggested in article 2621, the time corresponding to the highest altitude is used in the determination of longitude. Under some conditions it may be preferable to observe an altitude before meridian transit and then again after meridian transit, when the body has returned to the same altitude as at the first observation. Meridian transit occurs midway between these two times. A body in the zenith is on the celestial meridian. If accurate azimuth measurement is available, note the time when the azimuth is 000° or 180°.

Sunrise and sunset. The difference between the observed GMT of sunrise or sunset and the LMT tabulated in the almanac is the longitude in time units, which can then be converted to angular measure. If the *Nautical Almanac* is used, this information is tabulated for each third day only. Greater accuracy can be obtained if interpolation is used for determining intermediate values. Moonrise or moonset can be used if the tabulated LMT is corrected for longitude (art. 1812). Planets and stars can be used if the means are available for determining the time of rising or setting. This can be determined by computation (art 1909) or, approximately, by means of a diagram on the plane of the celestial meridian (art. 1432).

Either of these methods can be used in reverse to set a watch that has run down, or to check the accuracy of a watch, if the longitude is known. In the case of a meridian transit the time need not be determined at the instant of transit. The watch is started and the altitude is then measured several times before and after transit, or at equal altitudes. The times of these observations are noted and from them the time of meridian transit is determined. The difference between this time and the correct time of transit can then be used as a correction to reset the watch. If a watch runs down and cannot be reset from other timepieces, the correct time should be determined at the first opportunity, if the longitude accuracy is likely to deteriorate.

Variation of the compass. If the deviation of the compass is known accurately and an accurate azimuth can be observed, it is possible to determine the variation. If this is compared with the variation shown on the pilot chart, an approximate line of position can be determined. Since in many areas these lines run in a generally north-south direction, this may be an indication of the longitude. However, if the line has a large east-west component, it should be considered as any other such line of position, rather than as a longitude line. In some areas it is more nearly a latitude line. The accuracy of the method depends upon the accuracy with which the variation can be determined, and the spacing between adjacent isogonic lines.

Time sight. If altitude of a celestial body is available, including zero "sextant" altitude at rising or setting (art. 2619), longitude can be found by time sight (art. 2106).

Approaching Land

2623. Signs of land.—There are a number of signs which may indicate that the lifeboat is approaching land.

The sky will sometimes indicate a break in the open sea. A small fixed cloud, when surrounding ones are in motion or absent, will usually be over or close to land. At high latitudes, a light-colored reflection in the sky might be over an ice area; a light green reflection in the tropical sky might indicate a shallow lagoon. Such indications may be even more apparent on the under side of a uniform cloud layer.

Birds most often fly away from land at dawn and toward it at dusk. A large number of birds may indicate the nearness of land.

Swell, properly interpreted, may be used as a guide to land. Consecutive swells travel parallel until they reach an island and then "bend" around it. Eddies are formed where the distorted swell meets beyond the island. This eddy line may be used as a bearing to land, sometimes at a considerable distance.

The color of the sea may act as a guide in finding land as the open sea generally appears dark blue or dark green, and a lighter shade indicates shallow water, which may be near land.

The sound of the surf is often heard while still a considerable distance from land. Other sounds may also be heard at great distances.

Odors, as from burning wood, sometimes carry a long way out to sea.

Sounds and odors may be particularly helpful in periods of reduced visibility.

2624. Distance off.—At sea in a lifeboat the navigator is handicapped by his limited range of visibility. Distance to the horizon, in nautical miles, is given approximately by the formula $1.15\sqrt{h}$, h being the height of eye in feet. Thus, distance in miles is approximately 1⅐ times the square root of the height in feet. At an eye height of nine feet, the horizon is about 3½ miles away. A loaded Victory ship, whose greatest mast height is about 81 feet above the waterline, could be seen $1.15\sqrt{81}$ or 10.35 miles by an observer at zero height of eye. At a height of eye of nine feet the top of the mast should break the horizon when the ship is about 13.8 miles off.

If the height of an object above the horizon, or the distance between points on it is known, a simple proportion can be solved to determine the distance off by use of the cross-staff (art. 2618) or a similar device. To do this, align the two ends of the crosspiece with top and bottom, or two ends, of the object. The ratio of the length of the crosspiece to the length from this piece to the eye is the same as the ratio of the height (or length) of the object to its distance from the observer (fig. 2624). Thus, if the crosspiece is 18 inches and the intercepted length of the long piece is 31 inches, the distance to an island 1½ miles wide in the line of sight is found from the proportion

$$\frac{18}{31}=\frac{1.5}{D}, \text{ or } \frac{D}{1.5}=\frac{31}{18}$$

$$D=\frac{1.5\times 31}{18}=2.6 \text{ miles.}$$

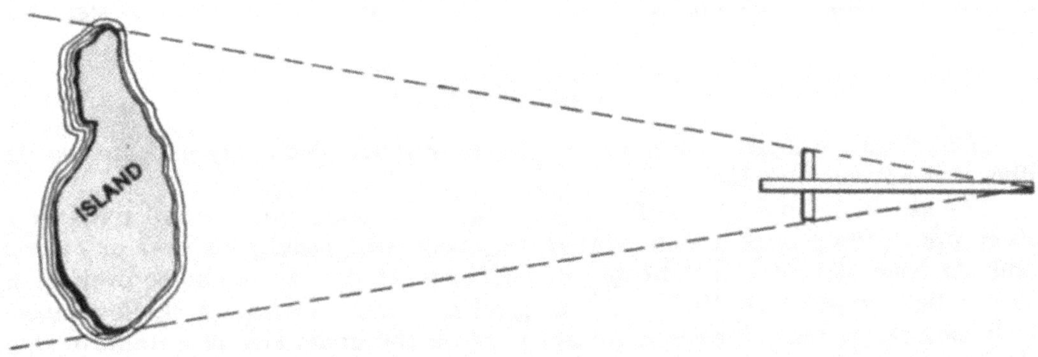

FIGURE 2624.—Using the cross-staff to measure distance.

In this proportion the two parts of either fraction must be expressed in the same units if results are to be obtained without a conversion factor. Thus, both 18 and 31 are expressed in inches, and both 1.5 and 2.6 are in miles. For small or distant objects the crosspiece may be too long. In this case replace it with a shorter one, use half or less of it, or substitute some other device such as a rule held at arm's length. In the case of a height, only the visible part of the object is used if the horizon is between the observer and the object.

A variation of this method can produce approximate results rather quickly. Hold a pencil, stick, or finger vertical at arm's length. Close one eye and align the vertical member with one end of an object such as an island. Open the closed eye and close the other one. Estimate the distance the vertical member appears to move against the background. The distance of the background object is ten times the amount of apparent movement, in the same units. The actual ratio varies somewhat among individuals and can be determined by comparing the length of the outstretched arm with the distance between eyes—or by practice on objects of known size at known distances. For vertical objects hold the extended member horizontal and bend the head until it, also, is horizontal.

2625. Beaching the boat.—The beaching of a lifeboat may be one of the most dangerous parts of the entire experience. The approach to an island should be made on the lee side, if possible, and every effort should be made to attract the attention of any inhabitants so that advice on the best place to land, and perhaps assistance, might be obtained. If no help is available, sail parallel to the coast to study the terrain and determine the safest place to beach the boat. A lagoon or other sheltered area may be available. It may be necessary to delay the landing overnight to make a complete study of the terrain and to beach the boat by daylight. Surf appears less rough from the sea than from land. High spray indicates a rough surf.

If a steering oar is available, the rudder should be unshipped before the boat is brought in, as the steering oar will provide better control in the surf zone. The sea anchor should be used to lessen the possibility of broaching and capsizing. Storm oil should be used, if available, to reduce the roughness of the surf. It is possible that the course can be altered somewhat while heading in to the beach, to take advantage of a better opening, but care should be taken to avoid broaching. Additional information on handling a boat in a surf can be found in nearly any book on seamanship.

2626. Ashore.—Once the boat has been safely beached, the problem remains to lead the survivors to civilization. Perhaps the land will be heavily populated and the boat met by local people, or the way to safety may be indicated by a road or trail. But the boat may be beached at a deserted place where there are no signs of life.

Many of the methods used to determine position at sea may also be used ashore, and usually with greater accuracy due to the absence of motion.

Problems

2613a. The compass azimuth of the sun is 126° at rising and 252° at setting.

Required.—Compass error.

Answer.—CE 9°W.

2613b. A life preserver is thrown overboard from a lifeboat and the boat headed away on course 355°. At a distance of half a mile the boat turns and heads back for the life preserver. The return course is 169°. The variation is 5°W.

Required.—(1) True course back to the life preserver.

(2) Deviation on this heading.

Answers.—(1) TC 167°, (2) D 3°E.

2615. The two knots in the log line of an improvised chip log of a lifeboat are 16⅔ feet apart. The elapsed time between passage of the knots through the hands of the observer is 4 seconds.

Required.—Speed of the lifeboat.

Answer.—S 2.5 kn.

2616. A lifeboat travels 18 miles on course 110°, from lat. 35°15′S, long. 82°31′W.

Required.—Latitude and longitude of the point of arrival.

Answer.—L 35°20′S, λ 82°11′W.

2617a. The date is November 15.

Required.—The approximate declination of the sun, without reference to an almanac.

Answer.—Dec. 18.°8′S.

2617b. The latitude of a lifeboat is 22°47′N. A star is observed on the meridian, bearing north. The observed altitude is 66°50′.

Required.—Declination of the star.

Answer.—Dec. 45°57′N.

2617c. The GMT is 1000, October 15.

Required.—Approximate GHA ♈, without reference to an almanac.

Answer.—GHA ♈ 172°.

2624. Approaching land, the navigator wishes to determine his distance from a lighthouse situated on the coast. He holds a rule at arm's length and finds that ⅝ inch of the rule appears the same height as the top of the lighthouse above water. He estimates the distance from his eye to the rule as 24 inches, and the height of the top of the lighthouse as 150 feet above water.

Required.—Distance to the lighthouse.

Answer.—D 0.9 mi.

PART FIVE
NAVIGATIONAL SAFETY

PART FIVE

NAVIGATIONAL SAFETY

Page

CHAPTER XXVII. Navigational Safety 747

CHAPTER XXVIII. Oceanic Soundings and Hydrographic Reports 756

CHAPTER XXIX. Position Reporting Systems 765

CHAPTER XXVII

NAVIGATIONAL SAFETY

2701. Introduction.—The discussion of navigational safety in this chapter *supplements* the discussions of safety matters in other chapters. For example, the *fallibility of floating aids to navigation* is discussed in chapter IV.

Where applicable, the mariner should always refer to the more detailed information given in the publications corrected by *Notice to Mariners,* e.g. Pubs. Nos. 117A and 117B, *Radio Navigational Aids.*

Published Warnings

2702. ***Notice to Mariners*** is published weekly by the Defense Mapping Agency Hydrographic/Topographic Center and is prepared jointly with the National Ocean Service and the U. S. Coast Guard. It is published to advise mariners of important matters affecting navigational safety, including new hydrographic discoveries, changes in channels and aids to navigation, etc. (U. S. Code Title 10, Secs. 7391 and 7392 and Title 44, Sec. 1336 refers). Besides keeping mariners informed generally, the information published in *Notice to Mariners* is particularly designed to simplify the correction aboard oceangoing ships of charts, sailing directions, light lists, and other publications produced by the Defense Mapping Agency Hydrographic/Topographic Center, the National Ocean Service, and the U. S. Coast Guard.

All corrections listed affect the safety of navigation, and it is the responsibility of users to decide which of their charts and publications require correction. In the interest of navigational safety, suitable records of *Notice to Mariners* corrections should be maintained to facilitate the updating of charts and publications prior to their use. Because of the sometimes transitory nature of aids to navigation, depths, and port information, local area sources should be consulted whenever possible.

Information for *Notice to Mariners* is contributed by the following agencies: Defense Mapping Agency Hydrographic/Topographic Center (Department of Defense) for waters outside the territorial limits of the United States; National Ocean Service (National Oceanic and Atmospheric Administration, Department of Commerce), which is charged with the surveys and charting of the coasts and harbors of the United States and its territories; the U. S. Coast Guard (Department of Transportation) which is responsible for the safety of life at sea and the establishment and operation of aids to navigation; and the Corps of Engineers, U. S. Army (Department of Defense) which is charged with the improvement of rivers and harbors of the United States. In addition, important contributions are made by foreign hydrographic offices and cooperating observers of all nationalities.

Notice to Mariners, relating to the *Great Lakes* and tributary waters west of Montreal, Canada, is published weekly by the U. S. Coast Guard. These Notices contain selected items from *Local Notices to Mariners* and other reported marine information and are intended primarily for use in correcting Great Lakes charts and related publications.

Mariners are requested to cooperate in the corrective maintenance of navigational charts and publications by reporting all discrepancies between published information and conditions actually observed or encountered, and by recommending appropriate additions, deletions, or improvements. A convenient reporting form is provided in the back of most issues of *Notice to Mariners.*

Information affecting National Ocean Service charts and publications, or concerning coasts and waters of the United States and its territories should be sent to the Director, National Ocean Service. Deficiencies in aids to navigation, either visual or electronic, on the coasts or waters of the United States or its territories should be reported to the nearest Coast Guard District Office. Suggestions regarding these aids should also be sent to the Coast Guard District Office.

Notice to Mariners No. 1 contains *important information* on a variety of subjects, amplifying information not usually found on charts or navigational publications. This information is promulgated as *Special Notice to Mariners Paragraphs* once each year in the interest of safe navigation. Additional items considered of interest to the mariner, including a selected listing of firing, bombing, and exercise areas, are also included in this Notice.

The *Summary of Corrections* published by the Defense Mapping Agency Hydrographic/Topographic Center contains corrections to charts, *Sailing Directions,* and *United States Coast Pilots* previously published in *Notice to Mariners.* The weekly *Notice to Mariners* should be referred to for corrective material after the last *Notice to Mariners* published in the *Summary.*

2703. The *Local Notice to Mariners* is issued by each U. S. Coast Guard District to disseminate important information affecting navigational safety within the District. This Notice reports changes to and deficiencies in aids to navigation maintained by and under the authority of the U. S. Coast Guard and such other marine information as new charts, channel depths, naval operations, regattas, etc. Since temporary information, known or expected to be of short duration, is not included in the weekly *Notice to Mariners* published by the Defense Mapping Agency Hydrographic/Topographic Center, the appropriate *Local Notice to Mariners* may be the only source of such information. Small craft using the Intracoastal Waterway and other waterways and small harbors that are not normally used by oceangoing vessels need it to keep chart and related publications up-to-date.

If still significant at the time of publication, the U. S. Coast Guard's *Broadcast Notice to Mariners* (art. 2705) is included in the *Local Notice to Mariners.*

The *Local Notice to Mariners* is published as often as required; usually weekly. It may be obtained, free of charge, by making application to the appropriate Coast Guard District Commander. Vessels operating in ports and waterways in several districts will have to obtain the *Local Notice to Mariners* from each district in order to be fully informed.

2704. The *Daily Memorandum* provides navigators of ships in port with printed copy of HYDROLANT/NAVAREA IV warnings or HYDROPAC/NAVAREA XII warnings (art. 2707 and 2708) broadcast in the past 24 hours or since the previous working day.

The *Daily Memorandum* is published by the Defense Mapping Agency Hydrographic/Topographic Center each working day in two editions: the Atlantic Edition and the Pacific Edition, both prepared at Washington, D.C.

The *Daily Memorandum* is sent to fleet operating bases, naval stations, custom houses, shipping company offices, etc., where it may be picked up by navigational personnel of vessels in port.

Radio Navigational Warning Systems

2705. Short range or local warnings.—One of the two distinct types of radio navigational warning systems is the short range or local radio navigational warnings for mariners plying nearby waters. Short range or local warnings are intended primarily to help local and coastal traffic. Usually, local or short range warnings are broadcast from a single coast station, frequently by voice as well as radiotelegraph. Examples of short range navigational warning systems are the radio navigational warning broadcasts of the various U. S. Coast Guard Districts. Most maritime nations broadcast short range warnings to notify local mariners of dangers in coastal and inland waters. Schedules and working particulars of many such stations throughout the world can be found in Pubs. Nos. 117A and 117B, *Radio Navigational Aids.*

Broadcast Notice to Mariners are originated by the U. S. Coast Guard and broadcast from U. S. Coast Guard, U. S. Navy, and some commercial radio stations to report deficiencies and changes in aids to navigation.

2706. Long range radio navigational warnings are primarily intended to assist mariners on the high seas by giving them navigational safety information in ports, harbors, coastlines, or areas at sea in major ocean areas. Such warnings are usually broadcast by means of radiotelegraph and radioteleprinter from several widely dispersed radio stations with sufficient power to ensure their availability to all ships in the oceanic area. The HYDROLANT and HYDROPAC system (art. 2707) of the Defense Mapping Agency Hydrographic/Topographic Center, for the general Atlantic and Pacific areas, is typical of a long range radio navigational warning system. Long range warning systems are discussed in more detail in Pubs. Nos. 117A and 117B, *Radio Navigational Aids,* and volume V of *Admiralty List of Radio Signals.*

2707. The U.S. long range radio navigational warning system provides coverage outside NAVAREAS IV and XII (art. 2708) by HYDROLANT and HYDROPAC messages originated by the Defense Mapping Agency Hydrographic/Topographic Center. The particulars of this system are discussed in detail in Pubs. Nos. 117A and 117B, *Radio Navigational Aids.*

HYDROLANTS and HYDROPACS are restricted to the more important marine incidents or navigational changes for which a delay in disseminating the information to mariners would adversely affect navigational safety. Many of these warnings are temporary in nature. Others might remain in force for long periods of time and ultimately be superseded by a numbered paragraph in *Notice to Mariners.*

Printed copies of HYDROLANTS and HYDROPACS are published each working day in the appropriate edition of the *Daily Memorandum.* The text of effective HYDROLANTS and HYDROPACS issued during a week is printed in the weekly *Notice to Mariners.*

It is important that the mariner retains the radio warning until the printed *Notice to Mariners* is received.

2708. NAVAREA Warnings, containing information which may affect the safety of navigation on the high seas, are broadcast in accordance with international obligations. The Defense Mapping Agency Hydrographic/Topographic Center is responsible for disseminating navigational information for ocean areas designated as NAVAREAS IV and XII of the Worldwide Navigational Warning Service (fig. 2708).

As is the case with HYDROLANTS and HYDROPACS, warnings for NAVAREAS IV and XII may be superseded by a numbered paragraph in *Notice to Mari-*

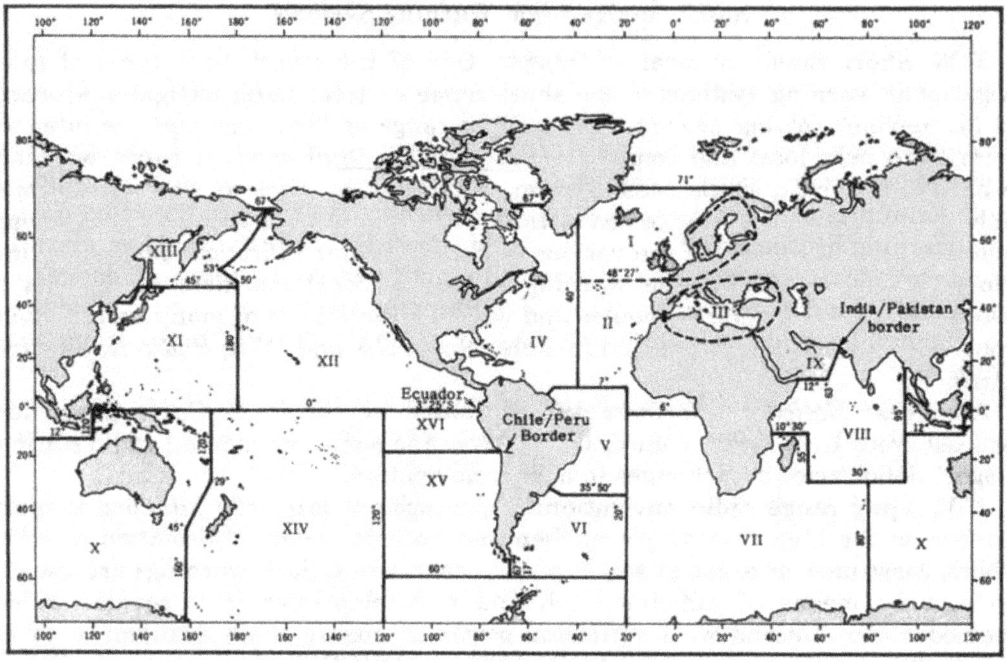

FIGURE 2708.—NAVAREAS of the Worldwide Navigational Warning Service.

ners. Printed copies are published each working day in the appropriate edition of the *Daily Memorandum.* The text of effective warnings for NAVAREAS IV and XII is printed in the weekly *Notice to Mariners.*

2709. Special Warnings are broadcast by U. S. Navy and U. S. Coast Guard radio stations primarily for the dissemination of official U. S. Government proclamations affecting shipping. Special Warnings are published in all editions of the *Daily Memorandum* and in the weekly *Notice to Mariners.* Upon issuance the text of all effective Special Warnings are published in *Notice to Mariners No. 1,* annually.

Safety Information

2710. Traffic separation schemes.—To increase the safety of navigation, particularly in areas of high shipping density, routes incorporating traffic separation have, with the approval of the International Maritime Organization (IMO), been established in certain areas of the world. In the interest of safe navigation, it is recommended that through traffic use these routes, as far as circumstances permit, by day and by night and in all weather conditions. The routes, which are intended for use by all vessels are governed by Rule 10 of the *International Regulations for Preventing Collisions at Sea, 1972* (72 COLREGS). Rule 10 states the following:

(a) This Rule applies to traffic separation schemes adopted by the Organization.

(b) A vessel using a traffic separation scheme shall:

(i) proceed in the appropriate traffic lane in the general direction of traffic flow for that lane;

(ii) so far as practicable keep clear of a traffic separation line or separation zone;

(iii) normally join or leave a traffic lane at the termination of the lane, but when joining or leaving from either side shall do so at as small an angle to the general direction of traffic flow as practicable.

(c) A vessel shall so far as practicable avoid crossing traffic lanes, but if obliged to do so shall cross as nearly as practicable at right angles to the general direction of traffic flow.

(d) Inshore traffic zones shall not normally be used by through traffic which can safely use the appropriate traffic lane within the adjacent traffic separation scheme. However, vessels of less than 20 meters in length and sailing vessels may under all circumstances use inshore traffic zones.

(e) A vessel other than a crossing vessel or a vessel joining or leaving a lane shall not normally enter a separation zone or cross a separation line except:

(i) in cases of emergency to avoid immediate danger;

(ii) to engage in fishing within a separation zone.

(f) A vessel navigating in areas near the terminations of traffic separation schemes shall do so with particular caution.

(g) A vessel shall so far as practicable avoid anchoring in a traffic separation scheme or in areas near its terminations.

(h) A vessel not using a traffic separation scheme shall avoid it by as wide a margin as is practicable.

(i) A vessel engaged in fishing shall not impede the passage of any vessel following a traffic lane.

(j) A vessel of less than 20 meters in length or a sailing vessel shall not impede the safe passage of a power-driven vessel following a traffic lane.

(k) A vessel restricted in her ability to maneuver when engaged in an operation for the maintenance of safety of navigation in a traffic separation scheme is exempted from complying with this Rule to the extent necessary to carry out the operation.

(l) A vessel restricted in her ability to maneuver when engaged in an operation for the laying, servicing or picking up of a submarine cable, within a traffic separation scheme, is exempted from complying with this Rule to the extent necessary to carry out the operation.

2711. Offshore oil-well structures.—Caution should be exercised when navigating in waters contiguous to the U. S. and its territories, particularly in the Gulf of Mexico; Santa Barbara Channel, California; and Cook Inlet, Alaska, in order to avoid collision with oil structures and their associated mooring piles, anchor and mooring buoys, etc.

The identification of the oil-well structures is discussed in *Notice to Mariners No. 1.* Due to the number of oil wells in the Gulf of Mexico, Shipping Safety Fairways have been established, and while adherence to these fairways is not mandatory, mariners should take advantage of the safer passage ways made available.

The most accurate information regarding the position of oil-well structures is contained in the latest editions of the pertinent National Ocean Service charts. (Additionally, an annually up-dated listing of Gulf of Mexico oil-well structures is published by the U. S. Coast Guard. Mariners desiring to receive issues of the publication should write to the District Commander (OAN), Eighth Coast Guard District, Hale Boggs Federal Building, 500 Camp Street, New Orleans, Louisiana 70130, requesting to be placed on their mailing list.)

Corrective information concerning the establishment, change, or discontinuance of the oil-well structures (with the exception of mobile drilling rigs) is published in the weekly *Notice to Mariners.* Movements of mobile drilling rigs and seismic survey operations are generally covered in HYDROLANT/NAVAREA IV or HYDROPAC/NAVAREA XII radio navigational warning broadcasts (art. 2707 and 2708).

2712. Danger from submarine cables and pipelines.—Submarine cables or pipelines pass beneath various navigable waterways throughout the world. Installation of new submarine cables and pipelines may be reported in *Notice to Mariners;* their locations may or may not be charted. Where feasible, warning signs are often erected to warn the mariners of their existence.

In view of the serious consequences resulting from damage to submarine cables and pipelines, mariners should take special care when anchoring, fishing or engaging in underwater operations near areas where these cables or pipelines may exist or have been reported to exist.

Certain cables carry high voltages; many pipelines carry natural gas under high pressure, or petroleum products. Electrocution, fire, or explosion with injury or loss of life or a serious pollution incident could occur if they are penetrated.

Vessels fouling a submarine cable or pipeline should attempt to clear without undue strain. Anchors or gear that cannot be cleared should be slipped; no attempt should be made to cut a cable or pipeline.

2713. Artificial obstructions to navigation.—**Disposal areas** are designated by the Corps of Engineers for depositing dredged material where existing depths indicate that the deposits will not cause sufficient shoaling to create a danger to surface navigation. The areas are shown on National Ocean Service charts without blue tint, and soundings and depth curves are retained.

Dumping grounds are areas established by federal regulation in which dumping of dredged material and other nonbuoyant objects is prohibited or in which such dumping is allowed with the permission of and under the supervision of the Corps of Engineers.

Spoil areas are for the purpose of depositing dredged material, usually near and parallel to dredged channels; *they are usually a hazard to navigation.* Spoil areas are usually charted from survey drawings from Corps of Engineers after-dredging surveys, though they may originate from private or other government agency surveys. Spoil areas are tinted blue on the chart and labeled, and all soundings and depth contours are omitted. *Navigators of even the smallest craft should avoid crossing spoil areas.*

Fish havens are established by private interests, usually sport fishermen, to simulate natural reefs and wrecks that attract fish. The reefs are constructed by dumping assorted junk, ranging from old trolley cars and barges to scrap building material in areas which may be of very small extent or may stretch a considerable distance along a depth contour; old automobile bodies are a commonly used material. The Corps of Engineers must issue a permit, specifying the location and depth over the reef, before such a reef may be built. However, the reef builders adherence to permit specifications can be checked only with a wire drag. Fish havens are outlined and labeled on the charts, but soundings and depth contours are usually retained and blue tinting is seldom used. *Navigators should be cautious about passing over fish havens or anchoring in their vicinity.*

Fishtrap areas are areas established by the Corps of Engineers in which traps may be built and maintained according to established regulations. The fish stakes which may exist in these areas are obstructions to navigation and may be dangerous. The limits of fishtrap areas and a cautionary note are usually charted. *Navigators should avoid these areas.*

2714. Net bottom clearances.—It is becoming increasingly evident that economic pressures are causing mariners to navigate through waters of barely adequate depth, with net bottom clearances (also known as **under-keel clearances**) being finely assessed from the charted depths, predicted tide levels, and depths recorded by echo sounders.

It cannot be too strongly emphasized that even charts based on modern surveys may not show all sea-bed obstructions or the least depths, and actual tide levels may be appreciably lower than those predicted.

In many ships an appreciable correction must be applied to shoal soundings recorded by echo sounders due to the horizontal distance between the transducers. This separation correction, which is the amount by which recorded depths exceed true depths, increases with decreasing depths to a maximum equal to half the distance apart of the transducers; at this maximum the transducers are aground. Mariners should have available a table of true and recorded depths.

Other appreciable corrections include the effects of water density, trim, list, squat (bodily sinkage and change of trim), heaving, rolling, and pitching.

The effects of water density, trim, and list upon bottom clearance are easily calculated assuming at-rest conditions. Squat, heaving, rolling, and pitching are dynamic phenomena and are more complex to predict singly or in combination.

Squat, which includes bodily sinkage and change of trim, is a result of the pressure distribution on the hull caused by the relative motion of water and hull. The effect begins to increase significantly at depth-to-draft ratios less than 2.5. It increases rapidly with speed and is augmented in narrow channels.

The effect of heaving, pitching, and rolling on net bottom clearance can be quite significant in critical combinations of wave height and period, ship dimensions, and angle of incidence. Passages over bars in exposed areas require adequate allowance for these factors.

2715. Controlling depths.—The **controlling depth** of a channel is the least depth within the limits of the channel; it restricts the safe use of the channel to drafts of less than that depth. The **centerline controlling depth** of a channel applies only to the channel centerline; lesser depths may exist in the remainder of the channel. The **mid-channel controlling depth** of a channel is the controlling depth of only the middle half of the channel. **Federal project depth** is the design dredging depth of a channel constructed by the Corps of Engineers, U. S. Army; the project depth *may or may not be the goal of maintenance dredging* after completion of the channel, and, for this reason, project depth must not be confused with controlling depth.

Depths alongside wharves usually have been reported by owners or operators of the waterfront facilities, and may not have been verified by government surveys. Since these depths may be subject to change, local authorities should be consulted for the latest controlling depths.

In general, the *Coast Pilot* gives the project depths for deep-draft ship channels maintained by the Corps of Engineers. The latest controlling depths are usually shown on the charts and published in *Notice to Mariners.* For other channels, the latest controlling depths available at the time of publication are given.

The most authoritative source for channel depths in U. S. ports is the field office of the Corps of Engineers in the vicinity. Addresses and telephone numbers are given in *Notice to Mariners No. 1.*

2716. Use of foreign charts.—Caution should be exercised in the use of foreign charts not maintained by *Notice to Mariners* published by the Defense Mapping Agency Hydrographic/Topographic Center. Foreign charts are sometimes referred to in *Sailing Directions* published by this Center. The mariner is advised that when such foreign charts are used for navigation, it is his responsibility to maintain the charts by the *Notice to Mariners* of the country producing the charts.

The mariner is warned that the buoyage system, shapes, and colors used by other countries often have a different significance than the U. S. system.

2717. Nautical chart symbols and abbreviations.—The standard symbols and abbreviations approved for use on all regular nautical charts published by the Defense Mapping Agency Hydrographic/Topographic Center and the National Ocean Service are contained in Chart No. 1, *United States of America Nautical Chart Symbols and Abbreviations.* The mariner should insure that a copy of the latest edition of Chart No. 1 is readily available.

The symbols and abbreviations used by other countries often vary from those used by the United States. Charts produced by the Defense Mapping Agency Hydrographic/Topographic Center will show the colors, lights, and other characteristics in use for the area of the individual chart. Certain reproductions of foreign charts published and distributed by this Center may also show the shapes and other distinctive features that will vary from those illustrated in Chart No. 1. Mariners who acquire and use foreign charts and reproductions of such charts are advised to procure the symbol sheet of the foreign chart agency.

2718. Unverified information.—Information received by the Defense Mapping Agency Hydrographic/Topographic Center and the National Ocean Service from various sources concerning depths, dangers, currents, facilities, and other subjects, which has not been verified by government surveys or inspections, is often included in *Sailing Directions* and *Coast Pilots;* such unverified information is qualified as "reported," and should be regarded with caution.

2719. International Ice Patrol.—Beginning in February or March and ending in August or September of each year depending upon ice conditions, the International Ice Patrol conducts its annual service of guarding the southeastern, southern, and southwestern limits of the regions of icebergs in the vicinity of the Grand Banks of Newfoundland for the purpose of informing passing ships of the extent of ice in this dangerous region. Ice Patrol Bulletins are sent via CW, voice and radiofacsimile broadcasts. Details are given in Pub. No. 117A, *Radio Navigational Aids.*

2720. Firing danger areas.—Firing and bombing practice takes place either occasionally or regularly in numerous areas established for those purposes along the coasts of practically all maritime countries.

In view of difficulty in keeping these areas up-to-date on the charts, and since the responsibility to avoid accidents rests with the authorities using the areas for firing or bombing practice, these areas will not as a rule be shown on Defense Mapping Agency Hydrographic/Topographic Center charts. The National Ocean Service charts show firing and bombing practice areas as defined by the *Code of Federal Regulations* in United States waters.

Any aid to navigation that may be established to mark a danger area, any target, fixed or floating, that may constitute a danger to navigation, will be shown on the appropriate charts.

Danger areas that will be in force for some length of time will be published in *Notice to Mariners.*

Warning signals, usually consisting of red flags or red lights, are customarily displayed before and during practice, but the absence of such warnings cannot be accepted as evidence that a practice area does not exist.

Vessels should be on the lookout for local warnings and signals, and should whenever possible, avoid passing through an area in which practice is in progress, but if compelled to do so should endeavor to clear it at the earliest possible moment.

CHAPTER XXVIII

OCEANIC SOUNDINGS AND HYDROGRAPHIC REPORTS

General

2801. Introduction.—Maritime shipping continues to increase with the growth in volume of domestic and international commerce, and seagoing vessels continue to increase in number, size, and speed. Not only does this growth result in constant demands for expansion and improvement of harbor, port, and navigation facilities, but it also necessitates improvements in the quality, quantity, and display of all forms of navigation information.

Although both government agencies and private institutions operate as many oceanographic and hydrographic ships as their resources permit, the magnitude of requirements for information far exceeds the collection capabilities of these comparatively few vessels. It is not possible for any hydrographic institute to conduct a continuous worldwide survey. Consequently, the Defense Mapping Agency Hydrographic/Topographic Center and the National Ocean Service depend to a very great extent upon reports from voluntary seagoing observers for information pertaining to navigation and oceanography. Information from these reports and other sources is evaluated and used in the improvement, production, correction, and maintenance of charts and publications. After careful analysis of a report and comparison with all other data concerning the same area or subject, appropriate action is taken. If the report is of sufficient importance to affect the immediate safety of navigation, the information will be broadcast as a navigational warning. Each bit of information, no matter how trivial it may seem, is coordinated with other reports and used in some way in the compilation, construction, and correction of charts and publications. It is only through the constant flow of new information that charts and publications can be kept accurate and up-to-date.

Discrepancies are sometimes found in source material, mainly as a result of comparison with the reports submitted by mariners who have recently visited the area. Often, errors in basic source material are of the type that would affect the safety of navigation. Several confirming ship reports will usually reveal these errors so that corrective action can be taken. The greater the volume of confirming ship reports, the greater is the accuracy of the finished product.

2802. Marine reports.—Frequently the most valuable information is recent information reported by a mariner, who records the information in the greatest detail possible and reports it promptly.

Depending on the type of report, certain items of information are absolutely essential for a correct evaluation. An example is the state of tide. State of tide is of paramount importance in a report of near shore shoals, but of no interest with respect to shoals reported beyond the continental shelf.

Several reasons have been found for hesitancy on the part of the mariner to report his observations. He is frequently in doubt as to what information, and in what detail, to report. He often believes that the data that he reports will be inconsequential to the Defense Mapping Agency Hydrographic/Topographic Center since "they already receive the most recent information available, and they are already aware of this data," as one mariner put it.

Excellence in grammar is unnecessary. Reports, or supplemental information, such as port plans, obtained locally, may even be in a foreign language. Observations should be reported in the language of mariners since they will be evaluated by ex-mariners, who will make full and appropriate use of the information.

Reports by letter are just as acceptable as those prepared on regular forms. In some instances, a letter report will permit greater flexibility in reporting details, conclusions, or recommendations concerning the observation. When using the regular report forms, one should not hesitate to use additional sheets to complete the details of his observation. One should never be reluctant to report in detail.

The following general suggestions are offered as an aid to making reports that will be of maximum value.

1. The geographical position included in the report may be used in the correction of charts. Accordingly, it should be fixed by the most accurate method available. If practicable, the position should be verified by additional means.

2. The report should state the method by which the position was fixed, so that the degree of accuracy can be established.

3. When reporting the position of an object or condition that is not shown on the chart but is within sight of charted objects, the simplest and most accurate method is to express the position in terms of bearings and distances from charted objects.

4. Should geographical coordinates be used to report position, they should be made as precise as circumstances permit. Either tenths of a minute or seconds, depending upon the scale of the chart, should be included. Unfortunately, coordinates of all charts for a given area are not always in exact agreement. Accordingly, one should refer to the chart by number and include the edition number and the date of the printing being used. Both are shown in the bottom margin.

5. When describing the sectors in which a light is either visible or obscured, the limiting bearings from the ship toward the light should be given. Although this is just the reverse of the form used for locating objects, it is the standard method used by the hydrographic institutes of practically all countries.

6. All bearings used in reports should be *true* bearings, expressed in degrees. Should magnetic bearings be used, for any reason, such use should be stated in the report.

7. A report prepared by one person should, if practicable, be checked by another.

In most cases marine information can be adequately reported on one of various forms printed by the Defense Mapping Agency Hydrographic/Topographic Center and shown in *Guide to Marine Observing and Reporting* (art. 2804) or by the reporting sheet in the weekly *Notice to Mariners*. However, in some cases it is both more convenient and more valuable to annotate information directly on the affected chart and mail the chart to DMAHTC.

As an example, new construction, such as port facilities, may be drawn on the chart in cases where a written report would be inadequate. Another example would be a chart showing the trackline and pertinent soundings through a critical passage or strait when the ship's draft is close to the controlling depth of the water in the passage or strait. Information, such as times, state of tide, draft, and method by which fixes were obtained should also be included.

Whenever it is necessary to send a chart to amplify or explain a report, the Defense Mapping Agency Hydrographic/Topographic Center, upon request, will replace the chart free of charge on a one-for-one basis.

2803. Urgent reports by radio.—The *International Convention for the Safety of Life at Sea* (1960), which is applicable to all U. S. flag ships, requires: "The master of every ship which meets with dangerous ice, or dangerous derelict, or any other direct danger to navigation, or a tropical storm, or encounters sub-freezing air temperatures associated with gale force winds causing severe ice accretion on superstructures, or winds of force 10 or above on the Beaufort scale for which no storm warning has been received, is bound to communicate the information by all the means at his disposal to ships in the vicinity, and also to the competent authorities at the first point on the coast with which he can communicate."

The master must first warn ships in the vicinity and must then report the danger to competent authorities at the first point on the coast with which radio contact can be made, using relay procedure if necessary. The report should be broadcast first on 500 kHz prefixed by the Safety Signal "TTT TTT TTT." This should be followed by transmission of the message to the proper authorities ashore.

Details on reporting via radio are contained in chapters 4 and 5 of Pubs. Nos. 117A and 117B, *Radio Navigational Aids.*

2804. *Guide to Marine Observing and Reporting,* Pub. 606, prepared jointly by the Naval Oceanographic Office, the Defense Mapping Agency Hydrographic/Topographic Center, U. S. Coast Guard, and the National Oceanic and Atmospheric Administration, provides detailed guidance for submitting hydrographic and oceanographic reports. Where appropriate, this guidance includes check lists of key questions as a means of insuring that no essential fact will be missing from a report.

Oceanic Soundings

2805. Soundings.—Relatively little is known of the surface features of the nearly 71 percent of the earth covered by water. However, enough has been learned to indicate that the unseen topography beneath the oceans has all the features common to that above water. It is known that there are submerged mountains extending to greater heights above their surroundings than do the Rockies, and depressions deeper than the Grand Canyon.

While many of the general features are known, details are lacking. In any given area, a very large number of accurately located soundings are needed to provide sufficient information for mapping the ocean floor. If sufficient information is available, the relief can be depicted on bathymetric charts by means of contours. A simplified chart of this type is shown in the upper part of figure 2805. The lower part of the figure is a block diagram of the area shown on the chart. Only a relatively small part of the oceans has been sounded sufficiently to provide the detailed information needed for such a chart, mainly narrow strips along coasts, i.e., the continental shelves. In these areas, the soundings have the necessary accuracy and density to portray underwater relief.

As long as oceanic soundings could be made only by a vessel stopping and lowering a weight, a process which might require several hours for a single sounding in very deep water, it was impractical for most vessels to obtain very much depth information at sea. With the development of the echo sounder, however, this situation has changed. With a recording echo sounder, a ship can obtain a profile along its track from continent to continent without slowing, using about a yard of

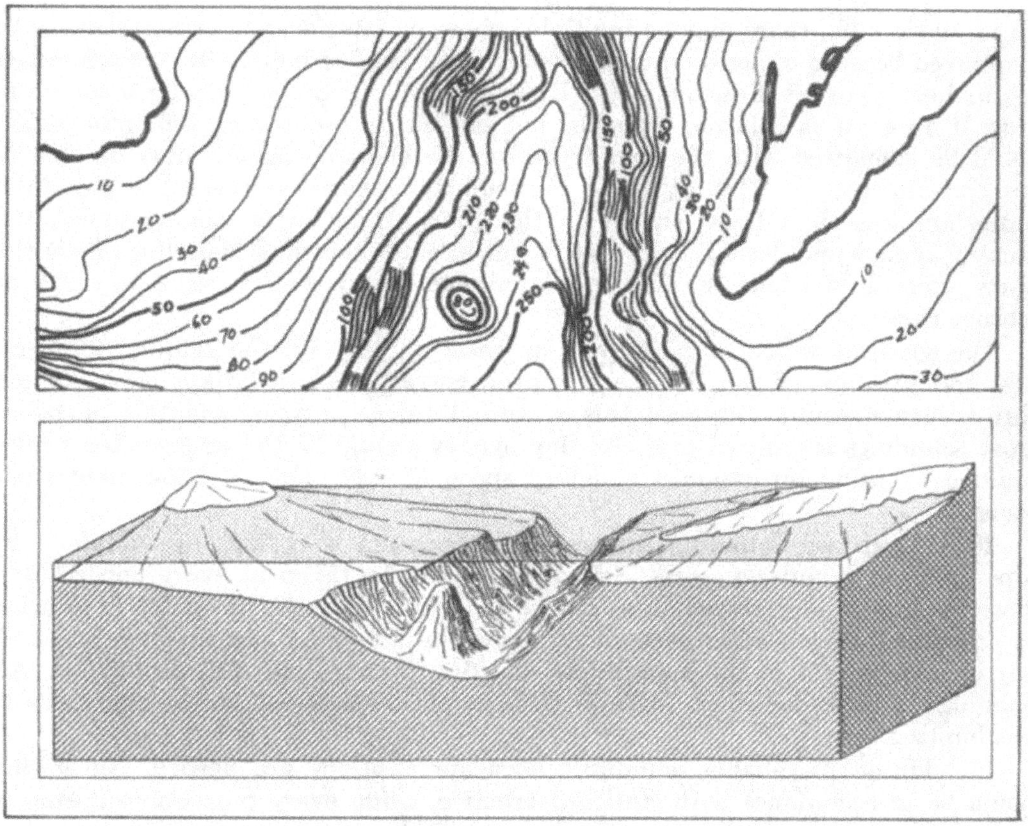

FIGURE 2805.—Contour lines and hachures (top) may be used to show underwater relief (bottom).

recording paper per day. Such information, if reliable, is of great assistance to charting agencies in preparing more adequate charts of the ocean areas.

2806. Sounding equipment.—While lead lines and sounding machines have been used at sea, almost all deep-sea soundings are now taken by echo sounder (art. 619). If a depth recording device is available, it should be used, as the profile thus produced is a better indication of the bottom than even the most closely spaced visual readings.

All echo sounding equipment is subject to certain errors unless the operator has a clear understanding of the operating characteristics and limitations of the instrument. The routine checks recommended by the manufacturer should be made at every change of the watch, or oftener. In addition, the operator should be alert for certain possible errors peculiar to his instrument. A close watch should be kept on the proper functioning of the stylus, recorder speed, the zero adjustment, and the frequency of the electric current. The percentage error in the recorded depth is the same as that of the electric current frequency. Thus, at 3,000 fathoms, the error of a 60-cycle echo sounder is 100 fathoms if the actual frequency is in error by two cycles.

2807. Evaluating results.—Inaccurate results may be worse than no information at all. Therefore, every effort should be made to obtain reliable data. Particularly, soundings which conflict with known or charted depths should be carefully analyzed. Even when the equipment is operating correctly, false returns might be received due to sources external to the vessel. A shoal "phantom bottom" may be due to marine life, there may be multiple echoes or interference, or no return may be received because of aeration of the water or suspended matter in it. Such errors are further discussed in article 3504. Unusual local conditions may be a source of error. If an error is believed probable, but no source is detected, full information should be submitted with the soundings, for the charting agency may be able to interpret the results. This action is particularly important where the measured depths are less than those shown on the chart. If no error can be found, the charting agency may have no alternative but to enter the shoal soundings upon the charts affected, and take the first opportunity to send a survey vessel to verify or disprove them.

The speed at which sound travels in water varies with the salinity, temperature, and pressure. When these are known, corrections can be applied to obtain more accurate results. However, this is normally done only for scientific purposes. Those soundings submitted to a charting agency should be the uncorrected values obtained by using an assumed standard speed of 4,800 feet or 1,500 meters per second.

2808. Deep sea sounding lines.—Many deep sea soundings are obtained by ships proceeding between ports. Soundings should be taken at every opportunity. Those taken in well-surveyed areas can be of assistance to the navigator in locating his position. If they conflict with values shown on the chart, and no error is found, they should be sent to the appropriate charting agency, with full particulars. All soundings in areas for which little depth information is shown on the chart should be submitted.

In addition to reliable soundings, accurate positions are needed. Navigation should be in accordance with standard practice, using every practicable means to reduce error and provide frequent checks on position.

When two or more ships are operating together, they should steam on parallel courses about five miles apart. Each ship should collect and record its own navigational data for subsequent submission to the appropriate charting agency.

2809. Investigating small areas.—If a feature of particular interest, such as an isolated shoal or a seamount, is found or reported in the vicinity of the vessel, a service can be rendered by conducting a further investigation in the vicinity of the feature. Two methods are in common use for this purpose:

Radial. A system of radial lines 20° apart are laid out from a central control point, preferably at the center of the feature to be investigated. These are extended outward for a distance of about 30 miles, and the ends of alternate ones are connected, as shown in figure 2809a. These form a series of course lines as shown.

Parallel. A north-south, east-west square is laid out with perhaps 60-mile sides, the center of the feature of interest being at the center of the square. A series of course lines are drawn parallel to one side of the square, at intervals of about 5 miles. The ends of alternate parallel course lines are connected, as shown in figure 2809b.

During such an investigation, by either method, the best control of position can usually be obtained by anchoring a buoy, if practicable, at the center of the area. In some instances, several buoys might be used. Any rig having buoyancy adequate to support the necessary length of anchor cable is satisfactory. The type generally used consists of a steel drum or mooring buoy with a weight attached to a cable, in the case of a large buoy, or piano wire if the buoy is small and of insufficient buoyancy to support a cable. A chain is not generally used. Buoys of this type have been successfully anchored in depths to 2,500 fathoms. The position of the buoy is

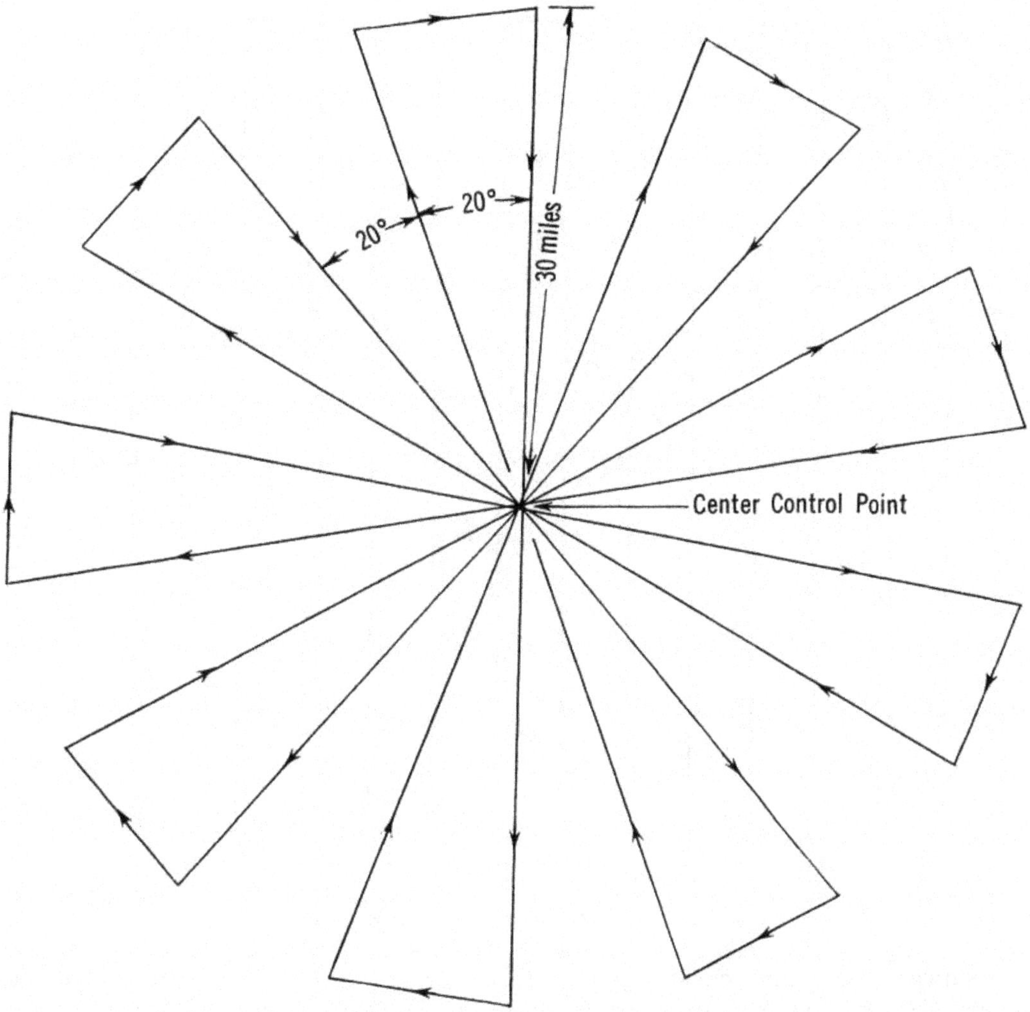

FIGURE 2809a.—Radial course line pattern.

determined as accurately as practicable, using celestial navigation, Loran, or whatever means are available. Position of the vessel is determined relative to the buoy or buoys, using visual or radar bearings and ranges at intervals of half an hour or less. Beyond this range, the best available means are used. A balloon with a suspended radar reflector might be attached to the buoy to extend its range of usefulness. The securing line of the balloon should be at least 400 feet long, if practicable.

Sonar ranging, if available, should be used to assist in the location of shoal areas.

Sounding Reports

2810. Records.—Today much of the data processing necessary for compilation of nautical or bathymetric charts is accomplished by computers. Consequently, many of the laborious and time-consuming aspects of sounding report preparation have been eliminated. Computers, if given the proper information, can quickly and efficiently perform all the routine data manipulation tasks formerly done manually

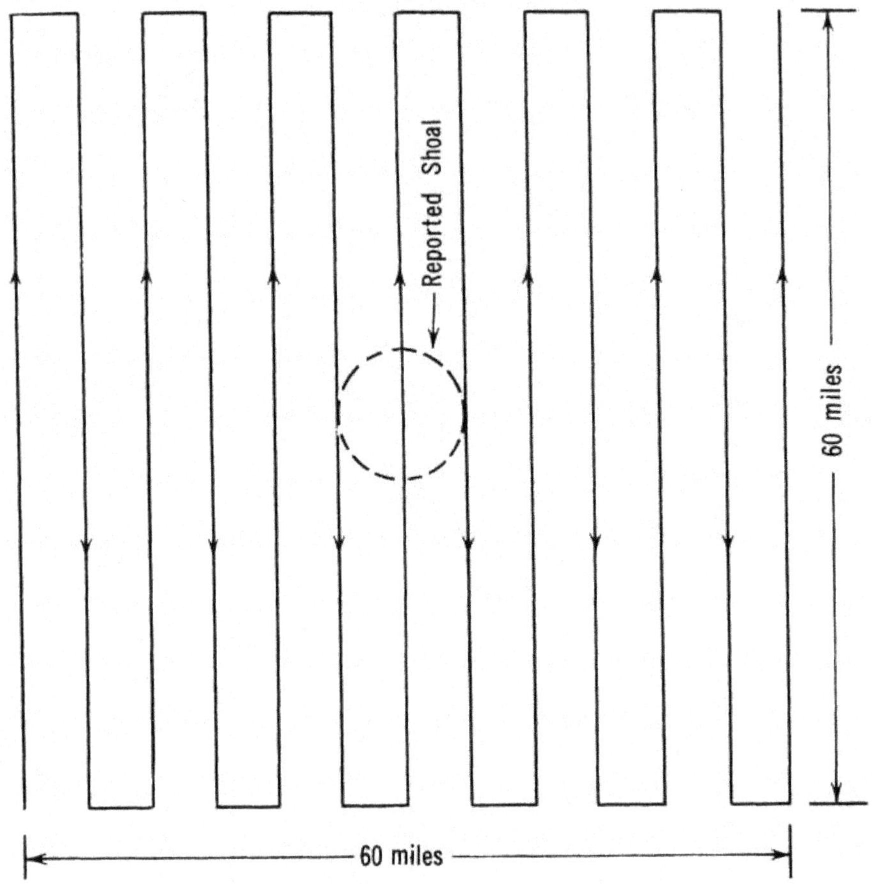

FIGURE 2809b.—Parallel course line pattern.

by oceanographers and cartographers. However, it is still imperative that the mariner prepare his report accurately and completely, insuring that the basic elements of depth and position, both correlated with time, are included. Described below are the report records that best present these basic elements.

1. **Echograms.** Depth is best depicted by the echogram itself, a continuous analog record that serves not only as a report but also provides verification of the shipboard interpretation. There are many formats for echogram paper, but the essential information needed on an echogram is the following:

(i) *Ship Name.* Record at the beginning and end of each roll of echogram or portion thereof.

(ii) *Date.* Annotate at least once each day at 1200 and when starting and stopping echo sounder.

(iii) *Time.* The event marker should be activated and annotated with the correct time at the beginning of the echogram, at least once each watch thereafter, and at the end of the echogram.

(iv) *Time Zones.* Greenwich mean time (GMT) should be used if practicable. In the event local time zones are used, annotate echogram whenever clocks are reset and identify time zone in use. Ambiguity of time zone is the most common cause of difficulty in relating a sounding trace to a ship's position.

(v) *Phase or Scale Changes.* Clearly label all depth phase (or depth scale) changes and the exact time they occur. Annotate the upper and lower limits of echograms if necessary.

(vi) *Transducer Depth.* The depth of the transducer beneath the surface and whether it is allowed for in the trace is necessary to compute true depth, especially in shoal areas. A specimen echo sounding record is shown in figure 2810a.

2. **Navigation Log.** In the past a smooth plotted track supplemented by the unadjusted plot was an essential part of a sounding report. The computer, with its tremendous capacity for data processing, has relieved the navigator of this monotonous and time-consuming task so that today only the navigation log is necessary. However, it is still important that the navigation log be accurate and contain all of the following information:

(i) Date
(ii) Time (GMT)
(iii) Latitude and longitude
(iv) Type of navigational fix
(v) Course
(vi) Speed
(vii) Remarks

Figure 2810b illustrates a typical navigation log.

When the above information has been collected and properly annotated, it should be sent to the appropriate charting agency, usually the Defense Mapping Agency Hydrographic/Topographic Center. The commanding officer's or master's forwarding letter should indicate the type of sounding system, any difficulties encountered, and pertinent remarks regarding estimated reliability of the data. Areas where sounding data are most needed are outlined on chart 5103, *Bathymetric Data Requirements,* which should be inspected prior to a voyage.

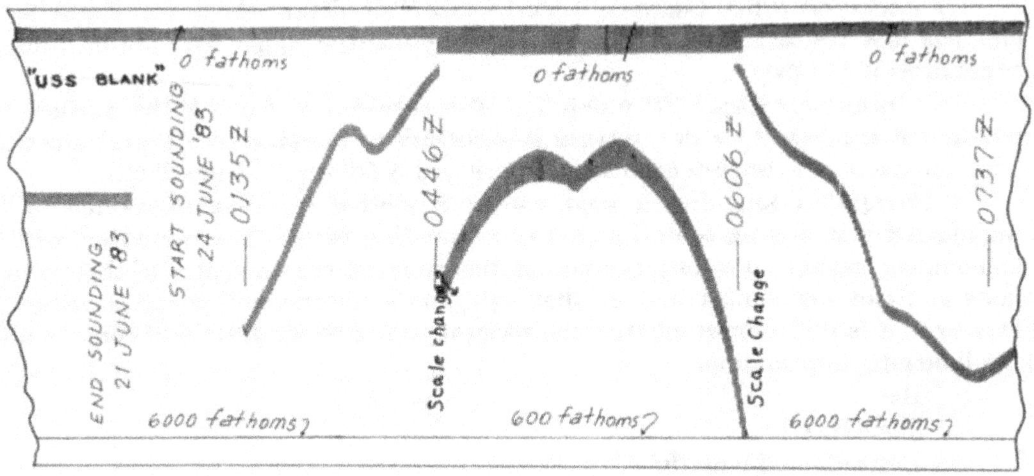

FIGURE 2810a.—Specimen echo sounding record.

NAVIGATION LOG							REMARKS
DATE	TIME (GMT)	LAT.	LONG.	NAV. FIX	COURSE	SPEED	
11/2/83	0221	29°41'N	124°10'E	LORAN	093°	12.3	
	0340				097°	12.3	CHANGE COURSE
	0400	29°40'N	124°35'E	NOON FIX	097°	12.3	
	0728	29°35'N	125°22'E	LORAN	097°	12.3	
	0810				VARIOUS	8.2	REDUCE SPEED – MANUVERING TO AVOID FISHING BOATS
	0826	29°34'N	125°35.5'E	LORAN	097°	12.3	RESUME COURSE AND SPEED
	1011	29°32'N	125°56'E	EVENING STARS	097°	12.3	
	1620	29°23'N	127°22'E	LORAN	102°	12.4	CHANGE COURSE
	2230	29°06.2'N	128°48.5'E	RADAR / STAR	102°	12.5	
	2305				102°	10.1	REDUCE SPEED

FIGURE 2810b.—Typical navigation log.

CHAPTER XXIX

POSITION REPORTING SYSTEMS

2901. Introduction.—Several thousand merchant vessels are at sea at one time. These vessels have the proven potential for early arrival at a scene of distress. The purpose of a position reporting system is to make possible maximum efficiency in coordinating assistance by merchant vessels at a scene of distress in order to save life and property at sea.

It is important that information be readily available to Search and Rescue (SAR) coordinators immediately upon occurrence of an emergency so that potential assistance can be obtained effectively and with the least delay to those offering and needing aid. Establishing communications is sometimes difficult even when automatic alarms are used and determination of SAR capabilities and intentions of vessels is time-consuming, unless the essential information has been made readily available beforehand by their participation in a position reporting system.

Regulation 10, chapter V of the *Convention on Safety of Life at Sea* (SOLAS 1960) obligates the master of any vessel at sea who becomes aware of a distress incident to attempt to render assistance. He must proceed and assist until aware that other aid is at hand or until released by the distressed unit. Other international treaties and conventions impose the same requirement. Position reporting systems permit determination of the most appropriate early assistance, provide the means for a timely resolution of distress cases, and enable vessels responding to distress calls to continue their passage with a minimum amount of delay.

Recommendation 47 of IMCO's SOLAS 1960 Conference reads as follows:

"The Conference recommends that Contracting Governments should encourage all ships to report their positions when traveling in areas where arrangements are made to collect these positions for Search and Rescue (SAR) use. Each Government should arrange that such messages shall be free of cost to the ship concerned."

There are currently many position reporting systems in operation throughout the world. The particulars of each system are given in publications of the International Maritime Organization (IMO), formerly known as the Inter-Governmental Maritime Consultative Organization (IMCO).

Masters of vessels making offshore passages are requested by the U. S. Coast Guard to *always* participate in the AMVER System, and to participate in the *other* systems whenever sailing within the areas covered by them.

2902. The Automated Mutual-Assistance Vessel Rescue System (AMVER), operated by the United States Coast Guard, is an international maritime mutual assistance program which provides important aid to the development and coordination of search and rescue (SAR) efforts in many offshore areas of the world. Merchant ships of all nations making offshore passages are encouraged to voluntarily send movement (sailing) reports and periodic position reports to the AMVER Center at Coast Guard, New York via selected radio stations. Information from these reports is entered into an electronic computer which generates and maintains dead reckoning positions for the vessels. Characteristics of vessels which are valuable for determining SAR capability are also entered into the computer from available sources of information.

Information concerning the predicted location and SAR characterstrics of each vessel known to be within the area of interest is made available upon request to recognized SAR agencies of any nation or vessels needing assistance. Predicted locations are only disclosed for reasons related to marine safety.

Messages sent within the AMVER System are at no cost to the ship or owner. Benefits to shipping include: (1) improved chances of aid in emergencies, (2) reduced number of calls for assistance to vessels not favorably located, and (3) reduced time lost for vessels responding to calls for assistance. An AMVER participant is under no greater obligation to render assistance during an emergency than a vessel who is not participating.

All AMVER messages should be addressed to Coast Guard, New York regardless of the station to which the message is delivered, except those sent to Canadian stations which should be addressed to AMVER Halifax or AMVER Vancouver to avoid incurring charges to the vessel for these messages.

In addition to the information calculated from sail plans and position reports, the AMVER Center stores data on the characteristics of merchant vessels. This data, reflecting SAR capability, includes the following; vessel name; international call sign; nation of registry; owner or operator; type of rig; type of propulsion; gross tonnage; length; normal cruising speed; radio schedule; medium, high, and very high frequency radio facilities; radio telephone installed; surface search radar installed; doctor normally carried. Vessels can assist the AMVER Center in keeping this data accurate by sending a complete report by message, letter, or by completing a SAR Information Questionnaire (fig. 2905) available from AMVER, and then sending corrections as the characteristics change. The corrections may easily be included in regular AMVER reports as remarks.

For AMVER participants bound for U.S. ports there is an additional benefit. AMVER participation via messages which include the necessary information is considered to meet the requirements of 33 CFR 161 (Notice of arrival).

2903. AMVER System communications network.—An extensive radio station communications network supports AMVER system and provides two routes for assistance messages as well as for AMVER messages: coast radio stations and Ocean Station Vessel radio facilities. Propagation conditions, location of vessel, and message density will normally determine which station may be best contacted to establish communications. To insure that no charge is applied, all AMVER messages should be passed through specified radio stations. Those which currently accept AMVER messages and apply to coastal station, ship station, or landline charge are listed in each issue of the *AMVER Bulletin* (art. 2904) together with respective call sign, location, frequency bands, and hours of guard. Although AMVER messages may be sent through other stations, the Coast Guard cannot reimburse the sender for any charges applied.

2904. The *AMVER Bulletin*, published quarterly by Commander, Atlantic Area, United States Coast Guard, Governors Island, New York, New York, 10004, provides information on the operation of the AMVER System of general interest to the mariner. It also provides up-to-date information on the AMVER communications network (art. 2903) and Radio Wave Propagation Charts which indicate recommended frequencies for contacting U.S. coast radio stations participating in the AMVER System, according to the time of day and the season of the year.

AMVER-Automated Mutual-Assistance Vessel Rescue System

AUTOMATED MUTUAL-ASSISTANCE VESSEL RESCUE SYSTEM
SEARCH AND RESCUE INFORMATION QUESTIONNAIRE

(Please Print)

NAME OF VESSEL ______________________ CALL SIGN ______________

PREVIOUS NAME OF VESSEL ______________________ PREVIOUS CALL SIGN ______________

MANAGER'S COMPLETE NAME AND ADDRESS: NATION OF REGISTRY______________

______________________ LENGTH (FEET) ______________

______________________ GROSS TONNAGE ______________

______________________ SERVICE SPEED (KNOTS) ______________

______________________ COASTAL RADIO STATION NORMALLY USED ______________

TYPE OF RIG (Circle One)

A Academic or Training
B Cable
C Cargo, Dry
D Dredge
E Pipe-Layer
F Fishing or Whaling
H Hospital
I Icebreaker
K Car Carrier, RO-RO
L Log, Lumber
N Naval
O Ore, Dry Bulk
P Passenger
R Refrigerated Cargo
S Salvage, Tug or Tender
T Tanker
U LASH
V Van, Container
W Weather Station
X Miscellaneous: Research, Survey, Ferry, etc.

TYPE OF PROPULSION (Circle One)

DE Diesel-Electric
GT Gas Turbine
MV Oil or Gas
NR Nuclear Reactor
SR Steam Reciprocating
SS Steam Turbine
TE Turbo-Electric

COMMUNICATIONS/NAVIGATION EQUIPMENT (Circle All That Apply)

R Surface Radar
T Radiotelephone (2182 kHz)
M INMARSAT
V VHF-FM (156.8 MHz)
X Medium Frequency (405-535 kHz)
Z High Frequency (4000-25110 kHz)
S Single Sideband

MEDICAL CAPABILITIES (Circle All That Apply)

Doctor Normally Carried	YES	NO
Paramedic or Physician's Assistant Normally Carried	YES	NO

RADIO WATCH SCHEDULE (Circle one)

H24 24-Hour Continuous Service
H16 16-Hour Service (ITU Schedule)
H8 8-Hour Service (ITU Schedule)
HX 8-Hour Service (Unspecified Schedule)
N No CW Operation (Radiotelephone Only)

DATE ______________ YOUR NAME AND TITLE ______________________

COMMENTS: ______________________

If you are not on the mailing list of the AMVER quarterly Magazine, the AMVER BULLETIN, and wish to receive it at no cost, please fill in your vessel s mailing address.

FIGURE 2905.—SAR Information Questionnaire.

2905. AMVER participation.—Instructions guiding participation in the AMVER System are available in the following languages: Chinese, Danish, Dutch,

English, French, German, Greek, Italian, Japanese, Korean, Norwegian, Polish, Portuguese, Russian, Spanish and Swedish. The *AMVER Users Manual* is available from: Commander, Atlantic Area, U.S. Coast Guard, Governors Island, N.Y. 10004; Commander Pacific Area, U.S. Coast Guard, Government Island, Alameda, CA 94501; and at U.S. Coast Guard District Offices, Marine Safety Offices, Marine Inspection Offices and Captain of the Port Offices in major U.S. Ports. Requests for instructions should state the language desired if other than English.

Search and Rescue Operation procedures are contained in the *Merchant Ship Search and Rescue Manual (MERSAR)* published by the International Maritime Organization (IMO). U.S. flag vessels may obtain a copy of *MERSAR* from local Coast Guard Marine Safety Offices and Marine Inspection Offices or by writing to U.S. Coast Guard (G-OSR), Washington, DC 20593. Other flag vessels may purchase *MERSAR* directly from IMO.

The Coast Guard conducts and/or coordinates search and rescue operations for surface vessels and aircraft that are in distress or overdue.

In connection with a vessel's first AMVER-plotted voyage, the master is requested to complete a questionnaire (fig. 2905) providing the radio watch schedule, available medical and communications facilities, and other useful characteristics. Stored in the AMVER computer, this information can be electronically processed with great speed in an emergency at the same time that a position is calculated.

Any vessel of any nation departing on an offshore passage of 24 hours duration or greater is encouraged to become a participant in the AMVER System by sending appropriate AMVER messages in types of formats. The messages may be transmitted at any convenient time as long as the information is accurate and the data corresponds to the time specified. For example, the information may be estimated for a short time in the future, for the present, or for a short time past.

There are five types of AMVER Reports—Sailing Plan, Departure, Arrival, Position, and Deviation Reports.

AMVER participants need to be familiar with the five types of reports. AMVER permits sailing plan and departure information to be combined into a single report. AMVER accepts sailing plan information separately—for example, several days prior to departure.

Only the above five types of AMVER messages require specific formats. (See DMAHTC PUBS. 117A and 117B). Other messages relating to a vessel's AMVER participation or data, such as facts on her SAR capabilities, may also be sent via the AMVER communications network.

Additional information concerning the AMVER System may be obtained by writing to Commandant, U.S. Coast Guard, Washington, DC 20590, or by writing or visiting Commander, Atlantic Area, U.S. Coast Guard, Governors Island, New York, NY 10004.

The AMVER System is coordinated in the Pacific regions by Commander, Pacific Area, U.S. Coast Guard, Government Island, Alameda, CA 94501.

Other countries such as Canada are a formal part of the AMVER System and provide radio stations for relay of AMVER reports as well as coordinating rescue efforts in certain regions. Applicable instuctions have been promulgated by official publications of the participating countries.

2906. AMVER Reporting Required.—U.S. Maritime Administration regulations effective August 1, 1983, state that certain U.S. flag vessels and foreign flag "War Risk" vessels must report and regularly update their voyages to the AMVER Center. This reporting is required of the following: (a) U.S. flag vessels of 1,000 tons or greater, operating in foreign commerce; (b) foreign flag vessels of 1,000 gross tons or greater, for which an Interim War Risk Insurance Binder has been issued under the provisions of Title XII, Merchant Marine Act, 1936.

2907. AMVER plot information.—The information stored in the computer can be used to provide several types of display according to the needs of controllers at Rescue Coordination Centers. The surface picture (SURPIC) can be displayed as a **Radius SURPIC** (fig. 2907a). When requesting a Radius SURPIC, the controller specifies the date and time, a latitude and longitude to mark the center (P), the radius (in nautical miles) that the SURPIC should cover (R), whether the names of all ships are desired (or only those with doctors or just those heading either east or west).

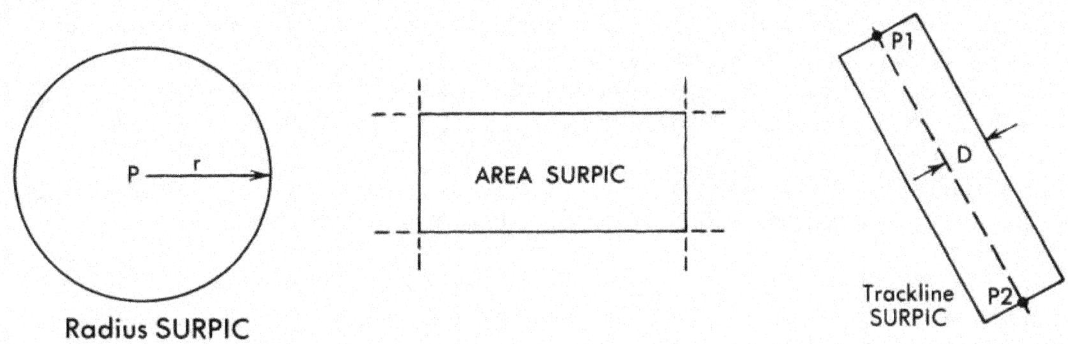

FIGURE 2907a.—Radius SURPIC, AREA SURPIC, and Trackline SURPIC.

A Radius SURPIC may be requested for any radius from 1 to 999 miles. A sample request is as follows:

"REQUEST 062100Z RADIUS SURPIC OF DOCTOR-SHIPS WITHIN 800 MILES OF 43.6N 030.2W FOR MEDICAL EVALUATION M/V SEVEN SEAS."

The **AREA SURPIC** (fig. 2907a) is obtained by the controller specifying the date, time, and two latitudes and two longitudes. The controller can limit the ships to be listed as with the Radius SURPIC. The computer responds with a listing of vessels within the boundaries specified.

There is no maximum or minimum size limitation on an AREA SURPIC. A sample AREA SURPIC request is as follows:

"REQUEST 151300Z AREA SURPIC OF WESTBOUND SHIPS FROM 43N TO 31N LATITUDE AND FROM 130W TO 150W LONGITUDE FOR SHIP DISTRESS M/V EVENING SUN LOCATION 37N, 140W."

The **Trackline SURPIC** (fig. 2907a) is obtained by the controller specifying the date and time, two points (P–1 and P–2), whether the trackline should be rhumb line or great circle, what the half-width (D) coverage should be (in miles), and whether all ships are desired (or only doctor ships, or just those east or westbound). The half-width (D) specified should not exceed 100 miles. When received, the SURPIC will list ships in order from P–1 to P–2.

There is no maximum or minimum distance between P–1 and P–2.

A sample Trackline SURPIC request is as follows:

"REQUEST 310100Z GREAT CIRCLE TRACKLINE SURPIC OF ALL SHIPS WITHIN 50 MILES OF A LINE FROM 20.1N 150.2W TO 21.5N 158.0W FOR AIRCRAFT PRECAUTION."

A **Location Vessel** is not a SURPIC, as such. It is used to determine the location of a specific ship. It permits a controller to determine the position of an AMVER participant wherever located.

A sample Location Vessel request is as follows:

"REQUEST PRESENT POSITION, COURSE, AND SPEED OF M/V SOLID STATE/HIND."

A Radius SURPIC as it would be received by a rescue center, listing all ships within a 200-mile radius of 26.2N, 179.9W, is shown in figure 2907b.

Name	*Call sign*	*Position*	*Course*	*Speed*	*SAR data*	*Destination and ETA*	
CHILE MARU	JAYU	26.2N 179.9E	C294	12.5K	H 1 6 R T X Z	KOBE	11
CPA 258 DEG. 012 MI. 032000Z							
WILYAMA	LKBD	24.8N 179.1W	C106	14.0K	H X R T V X Z	BALBOA	21
CPA 152 DEG. 092 MI. 032000Z							
PRES CLEVELAND	WITM	25.5N 177.0W	C284	19.3K	H 2 4 R D T X Z S	YKHAMA	08
CPA WILL PASS WITHIN 10 MI 040430Z							
AENEAS	GMRT	25.9N 176.9E	C285	16.0K	H 8 R N V X Z	YKHAMA	10
CPA 265 DEG. 175 MI. 03200Z							

FIGURE 2907b.—Radius SURPIC as received by a rescue center.

2908. Uses of AMVER plot information.—An example of the use of a Radius SURPIC is depicted in figure 2908. In this situation rescue authorities believe that a ship in distress, or her survivors, will be found in the rectangular area. The Rescue Coordination Center requests a listing of all eastbound ships within 100 miles of a carefully chosen position. Once this list is received by the Rescue Coordination Center a few moments later, the names and call letters of those ships chosen to assist in the search can be passed to a powerful commercial radio station nearby for inclusion in their next regularly scheduled TRAFFIC LISTS (normally broadcast every 2 hours). These ships will be notified that rescue authorities are waiting to contact them on a given working frequency.

Each ship contacted may be asked to sail a rhumb line between two specified points, one at the beginning of the search area and one at the end. By carefully assigning ships to areas of needed coverage, very little time need be lost from the sailing schedule of each cooperating ship. Those ships joining the search would report their positions every few hours to the Rescue Coordination Center, together with weather data and any significant sightings. In order to achieve saturation coverage, a westbound SURPIC at the eastern extremity of the search area would be used.

The Trackline SURPIC is most commonly used as a precautionary measure for aircraft. Rarely, if ever, is a major airliner forced to ditch at sea anymore. But occasions frequently arise where a plane loses the services of one or more of its

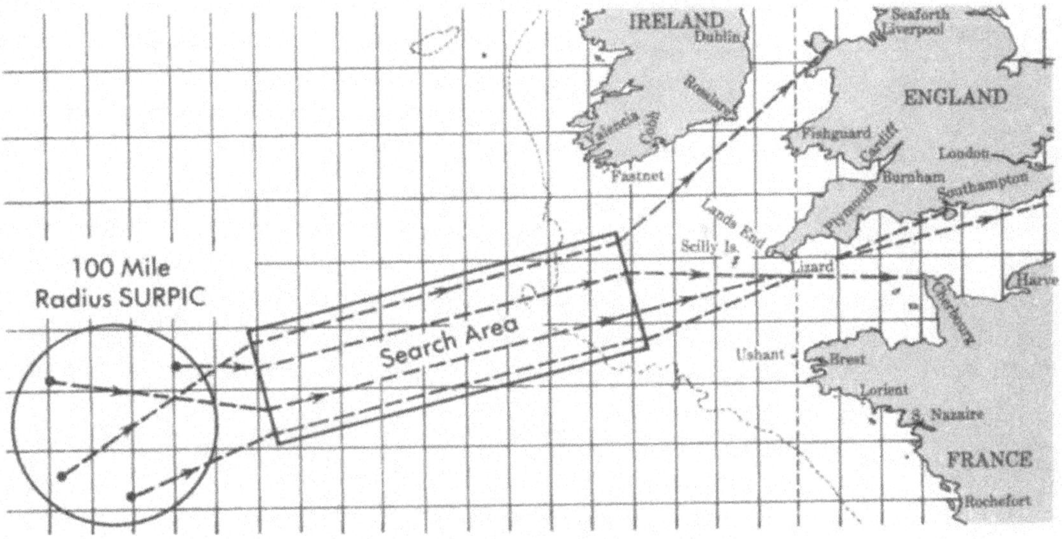

FIGURE 2908.—Use of Radius SURPIC.

engines. A Trackline SURPIC provided from the point of difficulty to the destination provides the pilot with the added assurance of knowing the positions of vessels beneath him. SURPIC's have been used successfully to save the lives of pilots of small aircraft.

PART SIX
OCEANOGRAPHY

PART SIX

OCEANOGRAPHY

	Page
CHAPTER XXX. The Oceans	775
CHAPTER XXXI. Tides and Tidal Currents	789
CHAPTER XXXII. Ocean Currents	815
CHAPTER XXXIII. Ocean Waves	824
CHAPTER XXXIV. Breakers and Surf	834
CHAPTER XXXV. Sound in the Sea	838
CHAPTER XXXVI. Ice in the Sea	843

CHAPTER XXX

THE OCEANS

3001. Introduction.—Oceanography is the application of the sciences to the phenomena of the oceans. It includes a study of their forms; physical, chemical, geological, and biological features; and phenomena. Thus, it embraces the widely separated fields of geography, geology, chemistry, physics, and biology. Many subdivisions of these sciences, such as sedimentation, ecology (biological relationship between organisms and their environment), bacteriology, biochemistry, hydrodynamics, acoustics, and optics, have been extensively studied in the oceans.

The oceans cover 70.8 percent of the surface of the earth. The Atlantic covers 16.2 percent, the Pacific 32.4 percent (3.2 percent more than the land area of the entire earth), the Indian Ocean 14.4 percent, and marginal and adjacent areas (of which the largest is the Arctic Ocean) 7.8 percent. Their extent alone makes them an important subject for study. However, greater incentive lies in their use for transportation, their influence upon weather and climate, and their potentiality as a source of power, food, freshwater, and mineral and organic substances.

3002. History of oceanography.—The earliest studies of the oceans were concerned principally with problems of navigation. Information concerning tides, currents, soundings, ice, and distances between ports was needed as ocean commerce increased. According to Posidonius, a depth of 1,000 fathoms had been measured in the Sea of Sardinia as early as the second century BC. About the middle of the 19th century, the Darwinian theories of evolution gave a great impetus to the collection of marine organisms, since it is believed by some that all terrestrial forms have evolved from oceanic ancestors. Later, the serious depletion of many fisheries called for investigation of the relation of the economically valuable organisms to the physical characteristics of their environment, especially in northwestern Europe and off Japan. Still later, the growing use of the oceans in warfare, particularly after the development of the submarine, required that much effort be expended in problems of detection and attack, resulting in the study of many previously neglected scientific aspects of the sea.

Oceanographic exploration. Exploration of the seas was primarily geographical until the 19th century, although the accumulated observations of seafarers, as recorded in the early charts and sailing directions, often included data on tides, currents, and other oceanographic phenomena. The great voyages of discovery, particularly those beginning in 1768 with Captain Cook, and continued by such commanders as La Pérouse, Bellingshausen, and Wilkes, included scientists in their complements. However, scientific work on the oceans at this period was severely limited by lack of suitable instruments for probing conditions below the surface. Meanwhile, Lieutenant Matthew Fontaine Maury, USN, working in the forerunner of the U. S. Navy Hydrographic Office in Washington, developed to a high degree of perfection the analysis of log-book observations. His first results, published in 1848, were of great importance to ship operations in the recommendation of favorable sailing routes, and they stimulated international cooperation in the fields of oceanography and marine meteorology.

In the rapid advances in technology after 1850, oceanographic instrumentation problems were not neglected, with the result that the British Navy in 1872–76 was able to send HMS *Challenger* around the world on the first purely deep-sea oceanographic expedition ever attempted. Her bottom samples, as analyzed by Sir John Murray, laid the foundation of geological oceanography, and 77 of her seawater samples, analyzed by C. R. Dittmar, proved for the first time that various constituents of the salts in seawater are everywhere in virtually the same proportions.

Since that time, the coastal waters and fishing banks of many nations have been extensively studied, and numerous vessels of various nationalities have conducted work on the high seas. Notable among these have been the American *Albatross* from 1882 to 1920; the Austrian *Pola* in the Mediterranean and Red Seas between 1890 and 1896; the Danish *Dana,* which during its voyages of 1920–22 discovered the breeding place of the European eels in the Sargasso Sea; the American *Carnegie* in 1927–29; the German *Meteor* in the Atlantic from 1928 to 1938; and the British *Discovery II* in the Antarctic between 1930 and 1939. Notable also were the drifts of the Norwegian vessels *Fram* and *Maud* in the arctic ice pack from 1893 to 1896 and 1918 to 1925, respectively; the attempt by Sir George Hubert Wilkins to operate under the ice in the British submarine *Nautilus* in 1931; and the Russian station set up at the North Pole in 1937, which made observations from the drifting pack ice.

At the same time, investigations pursued ashore provided the theoretical basis for the explanation of ocean currents, under the leadership of Helland-Hansen in Norway and Ekman and the Bjerknes in Sweden, while Martin Knudsen in Denmark worked out the precise details of the relationship between chlorinity, salinity, and density, enabling the theories to be verified by field observations.

During World War II, basic investigations were interrupted while work on purely military applications of oceanography was carried out. Deep-sea expeditions were renewed by the Swedish *Albatross* after the war, followed by the Danish *Galathea,* the second British *Challenger* (built in 1931), and *Discovery II* in the Antarctic, and vessels of the American Scripps Institution in the Pacific. Oceanographic work was carried out by Americans and Russians in the Arctic.

3003. Origin of the oceans.—Although many leading geologists still disagree with the conclusion that the structure of the continents is fundamentally different from that of the oceans, there is a growing body of evidence in support of the theory that the rocks underlying the ocean floors are more dense than those underlying the continents. According to this theory, all the earth's crust floats on a central liquid core, and the portions that make up the continents, being lighter, float with a higher freeboard. Thus, the thinner areas, composed of heavier rock, form natural basins where water has collected.

The shape of the oceans is constantly changing due to **continental drift.** The surface of the earth may be conceived as consisting of several "plates." These plates are joined along **fracture** or **fault lines.** There is constant and measurable movement of these plates.

The origin of the water in the oceans is also controversial. Although some geologists have postulated that all the water existed as vapor in the atmosphere of the primeval earth, and that it fell in great torrents of rain as soon as the earth cooled sufficiently, another school holds that the atmosphere of the original hot earth was lost, and that the water gradually accumulated as it was given off in steam by volcanoes or worked to the surface in hot springs.

Most of the water on the earth's crust is now in the oceans—about 328,000,000 cubic statute miles, or about 85 percent of the total. The mean depth of the ocean is 2,075 fathoms, and the total area is 139,000,000 square statute miles.

3004. Oceanographic chemistry may be divided into three main parts: the chemistry of (1) seawater, (2) marine sediments, and (3) organisms living in the sea. The first is of particular interest to the navigator.

Chemical properties of seawater are determined by analyzing samples of water obtained at various places and depths. Samples from below the surface are obtained by means of metal bottles designed for this purpose. The open bottles are attached at suitable intervals to a wire lowered into the sea. When they reach the desired depths, a metal ring or **messenger** is dropped down the wire. When the messenger arrives at the first bottle, it causes the bottle to close, trapping a sample of the water at that depth, and releasing a second messenger which travels on down the wire. The process is repeated at each bottle until all are closed, when they are hauled up and each bottle detached as it comes within reach. Of the various types devised, the **Nansen bottle** is the most widely known. It is equipped with a removable frame for attaching a thermometer.

3005. Physical properties of seawater are dependent primarily upon salinity, temperature, and pressure. However, factors like motion of the water and the amount of suspended matter affect such properties as color and transparency, conduction of heat, absorption of radiation, etc.

3006. Salinity is the amount of dissolved solid material in the water when carbonate has been converted to oxide, bromide and iodide to chloride, and organic material oxidized. It is usually expressed as parts per thousand (by weight), under certain standard conditions. This is not the same as **chlorinity**, which is equal approximately to the amount of chlorine, with bromides and oxides converted to chloride. (Actually the chlorine content is about 1.00045 times the chlorinity as determined by standard procedures.) The two have been found to be related empirically by the formula:

$$\text{salinity} = 0.03 + 1.805 \times \text{chlorinity}.$$

Historically the determination of salinity was a slow and difficult process, while chlorinity could be determined easily and accurately by titration with silver nitrate. It was customary to determine chlorinity and compute salinity by the formula given above. By this process, salinity could be determined with an error not exceeding 0.02 parts per thousand. Salinity can now be measured directly using a **salinometer** which measures changes in conductivity. Salinity generally varies between about 33 and 37 parts per thousand, the average being about 35 parts per thousand. However, when the water has been diluted, as near the mouth of a river or after a heavy rainfall, the salinity is somewhat less; and in areas of excessive evaporation, the salinity may be as high as 40 parts per thousand. In certain confined bodies of water, notably the Great Salt Lake in Utah, and the Dead Sea in Asia Minor, the salinity is several times this maximum. Chlorinity accounts for about 55 percent of salinity, the average being about 19 parts per thousand.

3007. Temperature in the ocean varies widely, both horizontally and with depth. Maximum values of about 90°F are encountered at the surface in the Persian Gulf in summer, and the lowest possible values of about 28°F (the usual minimum freezing point of seawater) occur in polar regions and near the ocean bottom everywhere, including the Tropics. Pub. No. 225, *World Atlas of Sea Surface Temperatures,* shows in detail the average sea surface temperatures for each month.

The following tabulation gives the percentage distribution of temperatures for the world for the months of February and August, as derived from this source:

Surface temperature	*Percentage of area of ocean*	
°F	*February*	*August*
<35	12.0	13.1
35–40	6.5	3.3
40–45	4.0	3.0
45–50	4.5	5.0
50–55	4.0	6.5
55–60	5.0	6.0
60–65	5.5	6.3
65–70	8.0	7.0
70–75	10.0	10.4
75–80	17.5	16.5
80–85	23.0	22.7
85–90	0.0	0.2

The vertical distribution of temperature in the sea nearly everywhere shows a decrease of temperature with depth. Since colder water is denser (assuming the same salinity), it sinks below warmer water. This results in a temperature distribution just opposite to that of the earth's crust, where temperature increases with depth below the surface of the ground.

In general, in the sea there is usually a mixed layer of isothermal water below the surface, where the temperature is the same as that of the surface. This layer is caused by two physical processes: wind mixing, and convective overturning as surface water cools and becomes more dense. The layer is best developed in the Arctic and Antarctic regions and seas like the Baltic and Sea of Japan during the winter, where it may extend to the bottom of the ocean. In the Tropics, the wind-mixed layer may exist to a depth of 125 meters. The layer may exist throughout the year. Below this layer is a zone of rapid temperature decrease, called the **thermocline,** to the temperature of the deep oceans. At a depth greater than 200 fathoms, the temperature everywhere is below 60°F, and in the deeper layers, fed by cooled waters that have sunk from the surface in the Arctic and Antarctic, temperatures as low as 28°5F exist.

In the colder regions the cooling creates the convective overturning and isothermal water in the winter; but in the summer a seasonal thermocline is created as the upper water becomes warmer.

A typical curve of temperature at various depths is shown in figure 3503a. Temperature at any desired depth can be determined by means of a **reversing thermometer** attached to a Nansen bottle (art. 3004). When the bottle closes, the thermometer measures the temperature to within 0°04F, thus providing a reading for a particular time and point. Instruments with **thermistors** (devices that utilize the change in conductivity of a semiconductor with change in temperature) are commonly used to measure temperature. The STD (salinity-temperature-depth) is an instrument that provides continuous signals as it is lowered from the vessel; temperature is determined by means of a thermistor, salinity by conductivity, and depth by pressure. Continuous records of temperature were first obtained by an instrument called a **bathythermograph,** invented by Spilhaus in 1938. This device functioned to a depth of 75 meters.

The mechanical bathythermograph has been replaced almost entirely by the **expendable bathythermograph (XBT),** which uses a thermistor. The XBT is connected to the vessel by a fine wire. The wire is coiled inside the probe and as the probe freefalls in the ocean, the wire pays out. Depth is determined by elapsed time and a known sink rate. Depth range is determined by the amount of wire stored in the probe; the most common model has a depth range of 500 meters. At the end of the drop, the wire breaks and the probe falls to the ocean bottom. One instrument of this type is dropped from an aircraft, the data being relayed to the aircraft from a buoy to which the wire of the XBT is attached.

3008. Pressure.—In oceanographic work, pressure is generally expressed in units of the centimeter-gram-second system. The basic unit of this system is 1 dyne per square centimeter. This is a very small unit, one million constituting a practical unit called a bar, which is nearly equal to 1 atmosphere. Atmospheric pressure is often expressed in terms of **millibars,** 1,000 of these being equal to 1 bar. In oceanographic work, water pressure is commonly expressed in terms of **decibars,** 10 of these being equal to 1 bar. One decibar is equal to nearly 1½ pounds per square inch. This unit is convenient because it is very nearly the pressure exerted by 1 meter of water. Thus, the pressure in decibars is approximately the same as the depth in meters, the unit of depth customarily used in oceanographic research. In terms more familiar to the mariner, the pressure at various depths is as follows:

Depth in fathoms	*Pressure in pounds per square inch*
1,000	2,680
2,000	5,390
3,000	8,100
4,000	10,810
5,000	13,520

The increase in pressure with depth is nearly constant because water is only slightly compressible.

Although virtually all of the physical properties of seawater are affected to a measurable extent by pressure, the effect is not as great as those of salinity and temperature. Pressure is of particular importance to submarines, directly because of the stress it induces on the materials of the craft, and indirectly because of its effect upon buoyancy.

3009. Density is mass per unit volume. Oceanographers use the centimeter-gram-second system, in which density is expressed as grams per cubic centimeter. The ratio of the density of a substance to that of a standard substance under stated conditions is called **specific gravity.** By definition, the density of distilled water at 4°C (39°.2 F) is 1 gram per milliliter (approximately 1 gram per cubic centimeter). Therefore, if this is used as the standard, as it is in oceanographic work, density and specific gravity are virtually identical numerically.

The density of seawater depends upon salinity, temperature, and pressure. At constant temperature and pressure, density varies with salinity or, because of the relationship between this and chlorinity, with the chlorinity. A temperature of 32°F and atmospheric pressure are considered standard for density determination. The effects of thermal expansion and compressibility are used to determine the density at other temperatures and pressures. The density at a particular pressure affects the buoyancy of submarines. It is also important in its relation to ocean currents.

The greatest changes in density of seawater occur at the surface, where the water is subject to influences not present at depths. Here density is decreased by precipitation, run-off from land, melting of ice, or heating. When the surface water becomes less dense, it tends to float on top of the more dense water below. There is little tendency for the water to mix, and so the condition is one of stability. The density of surface water is increased by evaporation, formation of sea ice, and by cooling. If the surface water becomes more dense than that below, it causes convective mixing. The more dense surface water sinks and mixes with less dense water below. The resultant layer of water is of intermediate density. This process continues until the density of the mixed layer becomes less than that of the water below. The convective circulation established as part of this process can create very deep uniform mixed layers. If the surface water becomes sufficiently dense, it sinks all the way to the bottom. If this occurs in an area where horizontal flow is unobstructed, the water which has descended spreads to other regions, creating a dense bottom layer. Since the greatest increase in density occurs in polar regions, where the air is cold and great quantities of ice form, the cold, dense polar water sinks to the bottom and then spreads to lower latitudes. In the Arctic Ocean region, the cold, dense water is confined by the Bering Strait and the underwater ridge from Greenland to Iceland to Europe. In the Antarctic, however, there are no similar geographic restrictions and large quantities of very cold, dense water formed there flow to the north along the ocean bottom. This process has continued for a sufficiently long period of time that the entire ocean floor is covered with this dense water, thus explaining the layer of cold water at great depths in all the oceans.

In some respects, oceanographic processes are similar to those occurring in the atmosphere (ch. XXXVIII). The convective circulation in the ocean is somewhat similar to that in the atmosphere. Water masses having nearly uniform characteristics are analogous to airmasses.

3010. Compressibility.—Seawater is nearly incompressible, its coefficient of compressibility being only 0.000046 per bar under standard conditions. This value changes slightly with changes of temperature or salinity. The effect of compression is to force the molecules of the substance closer together, causing it to become more dense. Even though the compressibility is low, its total effect is considerable because of the amount of water involved. If the compressibility of seawater were zero, sea level would be about 90 feet higher than it now is.

3011. Viscosity is resistance to flow. Seawater is slightly more viscous than freshwater. Its viscosity increases with greater salinity, but the effect is not nearly as marked as that occurring with decreasing temperature. The rate is not uniform, becoming greater as the temperature decreases. Because of the effect of temperature upon viscosity, an incompressible object might sink at a faster rate in warm surface water than in colder water below. However, for most objects, this effect may be more than offset by the compressibility of the object.

The actual relationships existing in the ocean are considerably more complicated than indicated by the simple explanation given above, because of turbulent motion within the sea. The disturbing effect is called **eddy viscosity.**

3012. Specific heat is the amount of heat required to raise the temperature of a unit mass of a substance a stated amount. In oceanographic work, specific heat is stated, in centimeter-gram-second units, as the number of calories needed to raise 1 gram of the substance 1°C. Specific heat at constant pressure is usually the quantity desired when liquids are involved, but occasionally the specific heat at constant volume is required. The ratio of these two quantities has a direct relationship to the speed of sound in seawater.

The specific heat of seawater decreases slightly as salinity increases. However, it is much greater than that of land. The ocean is a giant sink and source for heat. It can absorb large quantities of heat with very little change in temperature. This is partly due to the high specific heat of water and partly due to mixing in the ocean that distributes the heat throughout a layer. Land has a lower specific heat and, in addition, all heat is lost or gained from a thin layer at the surface. This accounts for the greater temperature range of land and the atmosphere above it, resulting in monsoons (art. 3809) and the familiar land and sea breezes of tropical and temperate regions (art. 3813).

3013. Thermal expansion.—One of the more interesting differences between salt and freshwater relates to thermal expansion. Saltwater continues to become more dense as it cools to the freezing point; freshwater reaches maximum density at 4°C and then expands (becomes less dense) as the water cools to 0°C and freezes. This means that the convective mixing of freshwater stops at 4°C; freezing proceeds very rapidly beyond that point. The rate of expansion with increased temperature is greater in seawater than in freshwater. Thus, at temperature 15°C (59°F), and atmospheric pressure, the coefficient of thermal expansion is 0.000151 per degree Celsius for freshwater and 0.000214 per degree Celsius for water of 35 parts per thousand salinity. The coefficient of thermal expansion increases not only with greater salinity, but also with increased temperature and pressure. At 35 parts per thousand, the coefficient of surface water increases from 0.000051 per degree Celsius at 0°C (32°F) to 0.000334 per degree Celsius at 30°C (86°F). At a constant temperature of 0°C (32°F) and a salinity of 34.85 parts per thousand, the coefficient increases to 0.000276 per degree Celsius at a pressure of 10,000 decibars (at a depth of approximately 10,000 meters).

3014. Thermal conductivity.—In water, as in other substances, one method of heat transfer is by conduction. Freshwater is a poor conductor of heat, having a coefficient of thermal conductivity of 0.00139 calories per second per centimeter per degree Celsius. For seawater it is slightly less but increases with greater temperature or pressure.

However, if turbulence is present, which it nearly always is to some extent in the ocean, the processes of heat transfer are altered. The effect of turbulence is to increase greatly the rate of heat transfer. The "eddy" coefficient used in place of the still-water coefficient is so many times larger, and so dependent upon the degree of turbulence that the effects of temperature and pressure are not important.

3015. Electrical conductivity.—Water without impurities is a very poor conductor of electricity. However, when salt is in solution in water, the salt molecules are ionized (art. 4007) and therefore are carriers of electricity. (What is commonly called freshwater has many impurities and is a good conductor of electricity; only *pure* distilled water is a poor conductor of electricity.) Hence, the electrical conductivity of seawater is directly proportional to the number of salt molecules in the water. For any given salinity, the conductivity increases with an increase in temperature.

3016. Radioactivity.—Although the amount of radioactive material in seawater is very small, this material is present in marine sediments to a greater extent than in the rocks of the earth's crust. This is probably due to precipitation of radium or other radioactive material from the water. The radioactivity of the top layers of sediment is less than that of deeper layers. This may be due to absorption of radioactive material in the soft tissues of marine organisms.

3017. Refractive index (art. 1606) of seawater increases as salinity becomes greater, or as temperature decreases. Since it varies with frequency of the radiant energy, the "D line" of sodium is usually used as the standard for comparison.

3018. Surface tension of water in dynes per square centimeter is approximately equal to $75.64-0.144T+0.0399Cl$, where T is temperature in degrees Celsius (centigrade) and Cl is the chlorinity of the water in parts per thousand. As indicated by the last term, the surface tension increases with chlorinity, and is therefore a little more for seawater than for freshwater. However, the presence of impurities causes it to be somewhat less than indicated by the formula.

3019. Transparency of seawater varies with the number, size, and nature of particles suspended in the water, as well as with the nature and intensity of illumination. The rate of decrease of light energy with depth is called the "extinction coefficient." The earliest method of measuring transparency was by means of a **Secchi disk,** a white disk 30 centimeters (a little less than 1 foot) in diameter. This was lowered into the sea, and the depth at which it disappeared was recorded. In coastal waters the depth varies from about 5 to 25 meters (16 to 82 feet). Offshore, the depth is usually about 45 to 60 meters (148 to 197 feet). The greatest recorded depth at which the disk has disappeared is 66 meters (217 feet), in the Sargasso Sea.

Although the Secchi disk still affords a simple method of measuring transparency, more exact methods have been devised.

3020. Color.—The color of seawater varies considerably. Water of the Gulf Stream is a deep indigo blue, while a similar current off Japan was named Kuroshio (Black Stream) because of the dark color of its water. Along many coasts the water is green. In certain localities a brown or brownish-red water has been observed. Colors other than blue are caused by biological sources, such as plankton, or by suspended sediments from river runoff.

Offshore, some shade of blue is common, particularly in tropical or subtropical regions. It is due to scattering of sunlight by minute particles suspended in the water, or by molecules of the water itself. Because of its short wavelength, blue light is more effectively scattered than light of longer waves. Thus, the ocean appears blue for the same reason that the sky does (art. 3816). The green color often seen near the coast is a mixture of the blue due to scattering of light and a stable soluble yellow pigment associated with phytoplankton. Brown or brownish-red water receives its color from large quantities of certain types of **algae**, microscopic plants in the sea or from river runoff.

3021. Marine geology is a branch of oceanography dealing with bottom relief, particularly the characteristics of ocean basins and the geological processes that brought them into being and tend to alter them, as well as with marine sediments.

3022. Bottom relief.—Compared to land, relatively little is known of relief below the surface of the sea. Until recent years, the sea has proved an effective barrier to acquisition of knowledge of features below its surface. Although soundings of 1,000 fathoms were probably made as early as the second century BC (art. 3002), the number of deep sea soundings by means of a weight lowered to the bottom had been relatively few. The process was a time-consuming one requiring special equipment. Several hours were needed for a single sounding. Since the development of an effective echo sounder (art. 619) in 1922, the number of deep sea soundings has greatly increased. Later, a recording echo sounder was developed to permit the continuous tracing of a **bottom profile.** This has assisted materially in the acquisition of knowledge of bottom relief. By this means, many mountain ranges, and other features have been discovered. Although the main features are becoming known, a great many details are yet to be learned.

Along most of the coasts of the continents, the bottom slopes gradually downward to a depth of about 100 fathoms or somewhat less, where it falls away more rapidly to greater depths. This **continental shelf** (fig. 3022a) averages about 30 miles in width, but varies from nothing to about 800 miles, the widest part being off the Siberian arctic coast. A similar shelf extending outward from an island or group of islands is called an **island shelf.** At the outer edge of the shelf, the steeper slope of 2° to 4° is called the **continental slope,** or the **island slope,** according to whether it surrounds a continent or group of islands. The shelf itself is not uniform, but has numerous hills, ridges, terraces, and canyons, the largest being comparable in size to the Grand Canyon.

The relief of the ocean floor is comparable to that of land. Both have steep, rugged mountains, deep canyons, rolling hills, plains, etc. Most of the ocean floor is considered to be made up of a number of more-or-less circular or oval depressions called **basins,** surrounded by walls (sills) of lesser depth.

Undersea features (figs. 3022a and 3022b) are defined as follows:

Archipelagic apron or *apron.*—A gentle slope with a generally smooth surface on the sea floor, particularly as found around groups of islands or seamounts.

Bank.—An elevation of the sea floor located on a shelf and over which the depth of water is relatively shallow but sufficient for safe surface navigation.

Basin.—A depression of variable extent and more-or-less circular or oval in form.

Borderland or *continental borderland.*—A region adjacent to a continent, normally occupied by or bordering a shelf, that is highly irregular with depths well in excess of those typical of a shelf.

Canyon.—A relatively narrow, deep depression with steep slopes, the bottom of which generally grades downward.

Cone.—See FAN.

Continental borderland.—See BORDERLAND.

Continental margin.—The zone separating the emergent continent from the deep sea bottom, generally consisting of the rise, slope, and shelf.

Continental rise.—A gentle slope rising toward the foot of the continental slope. See RISE.

Continental shelf.—See SHELF.

Cordillera.—An entire mountain system including all the subordinate ranges, interior plateaus, and basins.

Escarpment or *scarp.*—An elongated and comparatively steep slope of the sea floor, separating flat or gently sloping areas.

Fan or *cone.*—A gently sloping, fan-shaped feature normally located near the lower termination of a canyon.

Fracture zone.—An extensive linear zone of unusually irregular topography of the sea floor characterized by large seamounts, steep-sided or asymmetrical ridges, troughs, or escarpments.

Gap.—A depression cutting transversely across a ridge or rise.

Hill.—A small elevation rising generally less than 200 meters from the sea floor.

Hole.—A small depression of the sea floor.

Knoll.—An elevation rising less than 1,000 meters from the sea floor and of limited extent across the summit.

Levee.—An embankment bordering either one or both sides of a seachannel or the low-gradient seaward part of a canyon or valley.

Moat.—An annual depression that may not be continuous, located at the base of many seamounts or islands.

Mountains.—A well delineated subdivision of a large and complex positive feature, generally part of a cordillera.

Peak.—An individual pointed top on a ridge or a complex seamount.

Plain.—A flat, gently sloping or nearly level region of the sea floor.

Plateau.—A comparatively flat-topped elevation of the sea floor of considerable extent across the summit and usually rising more than 200 meters on at least one side.

Province.—A region composed of a group of similar bathymetric features whose characteristics are markedly in contrast with surrounding areas.

Range.—A series of ridges or seamounts, generally parallel.

Reef.—An offshore consolidated rock hazard to navigation with a least depth of 20 meters (or 10 fathoms) or less.

Ridge.—A long, narrow elevation of the sea floor with steep sides.

Rise.—A long, broad elevation that rises gently and generally smoothly from the sea floor.

Saddle.—A low part on a ridge or between seamounts.

Seachannel.—A long, narrow, U-shaped, or V-shaped, shallow depression of the sea floor, usually occurring on a gently sloping plain or fan.

Seamount.—An elevation rising 1,000 meters or more from the sea floor, and of limited extent across the summit.

Shelf or *continental shelf.*—A zone adjacent to a continent or around an island, and extending from the low waterline to the depth at which there is usually a marked increase of slope to greater depth.

Shoal.—An offshore hazard to navigation with a least depth of 20 meters (or 10 fathoms) or less, composed of unconsolidated material.

Sill.—The low part of the ridge or rise separating ocean basins from one another or from the adjacent sea floor.

Slope or *continental slope.*—The declivity seaward from a shelf into greater depth.

Spur.—A subordinate elevation, ridge, or rise projecting from a larger feature.

Tablemount or *Guyot.*—A seamount having a comparatively smooth, flat top.

Terrace or *bench.*—A bench-like feature bordering an undersea feature.

Trench.—A long, narrow and deep depression of the sea floor, with relatively steep sides.

Trough.—A long depression of the sea floor, normally wider and shallower than a trench.

Valley.—A relatively shallow, wide depression with gentle slopes, the bottom of which generally grades continuously downward. This term is used for features that do not have canyon-like characteristics in any significant part of their extent.

The term **deep** may be used for a very deep part of the ocean, generally that part deeper than 3,000 fathoms.

The average depth of water in the oceans is 2,075 fathoms (12,450 feet), as compared to an average height of land above the sea of about 2,750 feet. The greatest known depth is 35,800 feet, in the Marianas Trench in the Pacific. The highest known land is Mount Everest, 29,002 feet. About 23 percent of the ocean is shallower than 10,000 feet, about 76 percent is between 10,000 and 20,000 feet, and a little more than 1 percent is deeper than 20,000 feet.

3023. Marine sediments.—The ocean floor is composed of material deposited there through the years. This material consists principally of (1) earth and rocks washed into the sea by streams and waves, (2) volcanic ashes and lava, and (3) the

FIGURE 3022a.—Ocean basin features.

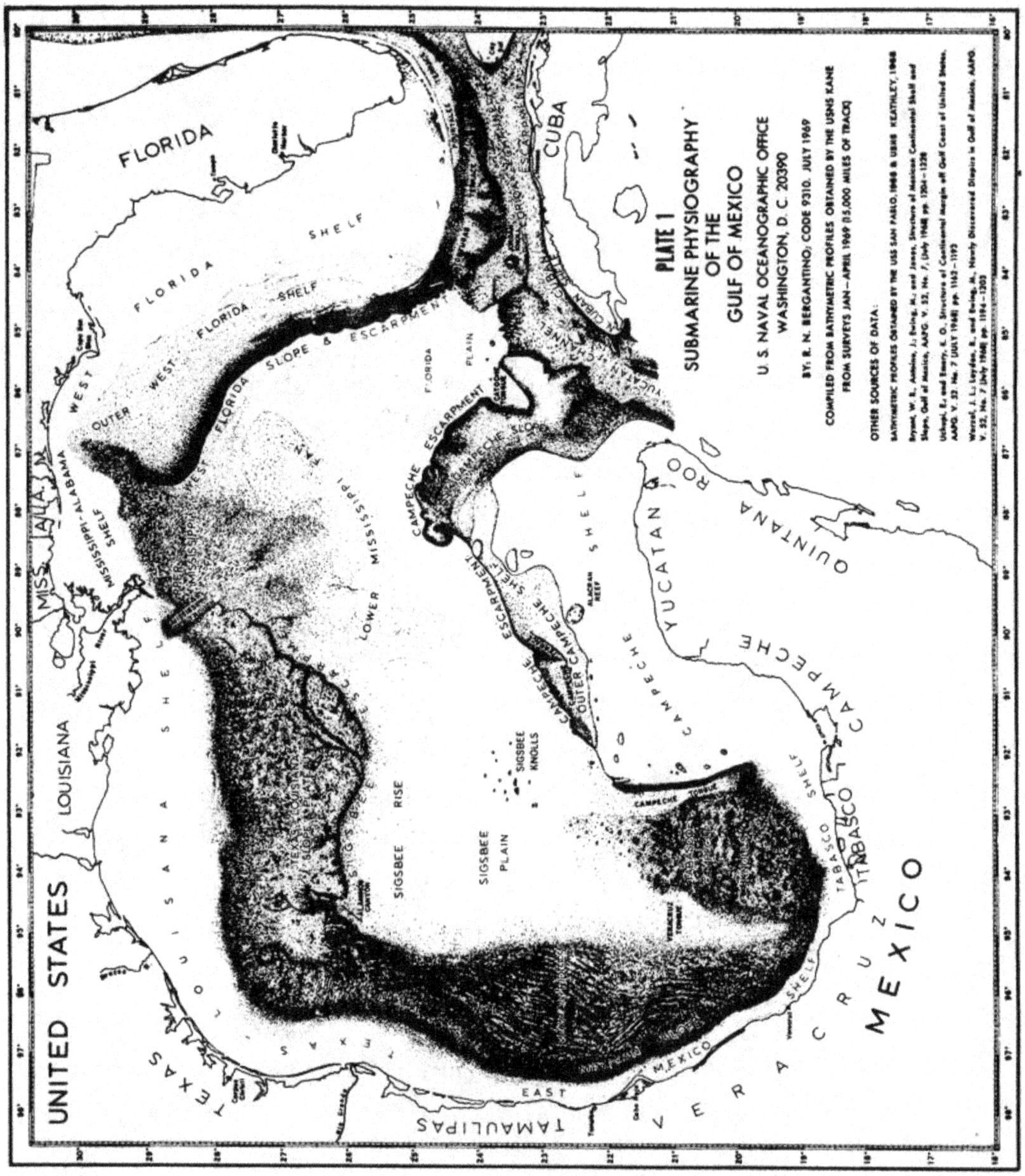

FIGURE 3022b.—Ocean basin features.

remains of marine organisms. Lesser amounts of land material are carried into the sea by glaciers, blown out to sea by wind, or deposited by chemical means. This latter process is responsible for the manganese nodules that cover some parts of the ocean floor. In the ocean, the material is transported by ocean currents, waves, and ice. Near shore the material is deposited at the rate of about 3 inches in 1,000 years, while in the deep water offshore the rate is only about half an inch in 1,000 years. Marine deposits in water deep enough to be relatively free from wave action are subject to little erosion. Recent studies have shown that some bottom currents are strong enough to move sediments. There are **turbidity currents**, similar to land slides, that move large masses of sediments. Turbidity currents have been known to rip apart large transoceanic cables on the ocean bottom. Because of this and the slow rate of deposit, marine sediments provide a better geological record than does the land.

Marine sediments are composed of individual particles of all sizes from the finest clay to large boulders. In general, the inorganic deposits near shore are relatively coarse (sand, gravel, shingle, etc.), while those in deep water are much finer (clay). In some areas the siliceous remains of marine organisms or the calcareous deposits (of either organic or inorganic origin) are sufficient to predominate on the ocean floor.

A wide range of colors is found in marine sediments. The lighter colors (white or a pale tint) are usually associated with coarse-grained quartz or limestone deposits. Darker colors (red, blue, green, etc.) are usually found in mud having a predominance of some mineral substance, such as an oxide of iron or manganese. Black mud is often found in an area that is little disturbed, such as at the bottom of an inlet or in a depression without free access to other areas.

Marine sediments are studied primarily by means of bottom samples. Samples of surface deposits are obtained by means of a **snapper** (for mud, sand, etc.) or "dredge" (usually for rocky material). If a sample of material below the bottom surface is desired, a "coring" device is used. This device consists essentially of a tube driven into the bottom by weights or explosives. A sample obtained in this way preserves the natural order of the various layers. Samples of more than 100 feet in depth have been obtained by means of coring devices. The bottom sample obtained by the mariner, by arming his lead with tallow or soap (art. 618), is an incomplete indication of bottom surface conditions.

3024. Synoptic oceanography.—Bathythermograph and sea surface temperature observations are reported directly to the Fleet Numerical Oceanography Center, Monterey, California. These synoptic reports are then analyzed by computer to determine ocean thermal conditions at any point in the Northern Hemisphere.

References

Crease, J. "The Origin of Ocean Currents." *Journal of the Institute of Navigation* (British), vol. 5, no. 3 (July 1952).

Day, A., Rear Admiral. "Navigation and Hydrography." *Journal of the Institute of Navigation* (British), vol. 6, no. 1 (January 1953).

Deacon, G. E. R. "Oceanographical Research and Navigation." *Journal of the Institute of Navigation* (British), vol. 4, no. 3 (July 1951).

Defant, A. *Physical Oceanography.* (2 vols.) New York, Pergamon, 1961.

Marmer, H. A. *The Scope of Oceanography.* James Johnstone Memorial Volume. Liverpool, University Press of Liverpool, 1934.

National Research Council. *Physics of the Earth—Oceanography.* Bulletin no. 85, Chapter V. Washington, The National Academy of Sciences, 1932.

Satow, P. G. "Some Problems of Underwater Navigation." *Journal of the Institute of Navigation* (British), vol. 4, no. 3 (July 1951).

Shepard, F. P. *Submarine Geology.* New York, Harper, 1948.

Sverdrup, H. U., M. W. Johnson, and R. H. Fleming. *The Oceans, Their Physics, Chemistry and General Biology.* New York, Prentice-Hall, 1942.

CHAPTER XXXI

TIDES AND TIDAL CURRENTS

General

3101. The tidal phenomenon is the periodic motion of the waters of the sea due to differences in the attractive forces of various celestial bodies, principally the moon and sun, upon different parts of the rotating earth. It can be either a help or hindrance to the mariner—the water's rise and fall may at certain times provide enough depth to clear a bar and at others may prevent him from entering or leaving a harbor. The flow of the current may help his progress or hinder it, may set him toward dangers or away from them. By understanding this phenomenon and by making intelligent use of predictions published in tide and tidal current tables and of descriptions in sailing directions, the mariner can set his course and schedule his passage to make the tide serve him, or at least to avoid its dangers.

3102. Tide and current.—In its rise and fall, the tide is accompanied by a periodic horizontal movement of the water called **tidal current.** The two movements, tide and tidal current, are intimately related, forming parts of the same phenomenon brought about by the tide-producing forces of the sun and moon, principally.

It is necessary, however, to distinguish clearly between tide and tidal current, for the relation between them is not a simple one nor is it everywhere the same. For the sake of clearness and to avoid misunderstanding, it is desirable that the mariner adopt the technical usage: **tide** for the vertical rise and fall of the water, and **tidal current** for the horizontal flow. The tide rises and falls, the tidal current floods and ebbs. In British usage, tidal current is called **tidal stream.**

3103. Cause.—It is often said of science that the ability to predict a natural event is indicative of understanding. Since tides are the most accurately predictable oceanographic phenomena, one could easily assume that physical oceanographers truly understand them. Unfortunately, this is not true; significant gaps remain. An examination of the details of this apparent contradiction gives insight into one of the most exciting areas of oceanography—ocean tides.

To facilitate this examination, it will be desirable, first of all, to discuss the fundamental tide-generating forces and the theoretical equilibrium tide they try to produce. The principal tide-generating forces on the surface of the earth result from the differential gravitational forces of the moon and sun. The moon is the main tide-generating body. Due to its greater distance, the effect of the sun is only 46 percent of the effect due to the moon. After the theoretical equilibrium tide produced by the sun and moon is discussed in this article, the actual tide as observed in nature is described in article 3104. Observed tides will differ considerably from the tides predicted by equilibrium theory since size, depth, and configuration of the basin or waterway, friction, landmasses, inertia of watermasses, Coriolis acceleration, and other factors are neglected in this theory. Nevertheless, equilibrium theory will be sufficient to describe the magnitude and distribution of the main tide-generating forces across the surface of the earth.

Tide-Generating Forces

Newton's universal law of gravitation governs both the orbits of celestial bodies and the tide-generating forces which occur on these bodies. The force of gravitational attraction between any two masses, m_1 and m_2, is given by

$$F=\frac{Gm_1m_2}{d^2},$$

where d is the distance between the two masses and G is a constant which depends upon the units employed. This law assumes that m_1 and m_2 are point masses. Newton was able to show that homogeneous spheres could be treated as point masses when determining their orbits. However, when computing differential gravitational forces, the actual dimensions of the masses must be taken into account.

Using the law of gravitation, it is found that the orbits of two point masses are conic sections about the barycenter of the two masses (art. 1407). If either one or both of the masses are homogeneous spheres instead of point masses, the orbits are the same as the orbits which would result if all of the mass of the sphere were concentrated at a point at the center of the sphere. In the case of the earth-moon system, both the earth and the moon describe elliptical orbits about their barycenter if the simplifying assumption is made that both bodies are homogeneous spheres and the gravitational forces of the sun and other planets are neglected. The earth-moon barycenter is located at a distance of 0.74 R_E from the earth's center, where R_E is the radius of the earth. This is approximately three-fourths of the distance from the center of the earth to the surface of the earth along the line connecting the centers of the earth and moon (fig. 3103a).

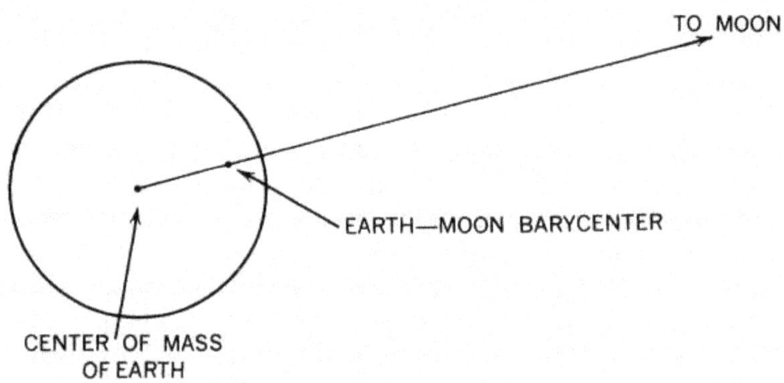

FIGURE 3103a.—Earth-moon barycenter.

Thus, the center of mass of the earth describes a very small ellipse about the earth-moon barycenter whereas the center of mass of the moon describes a much larger ellipse about the same barycenter. If the gravitational forces of the other bodies of the solar system are neglected, Newton's law of gravitation also predicts that the earth-moon barycenter will describe an orbit which is approximately elliptical about the barycenter of the sun-earth-moon system. This barycentric point lies inside the sun (fig. 3103b).

The *differences* in gravitational attraction of various celestial bodies, principally the moon and sun, upon different parts of the rotating and revolving earth are the fundamental tide-generating forces. These *differential gravitational* forces are described here by considering first the effects of the moon only. The results will be

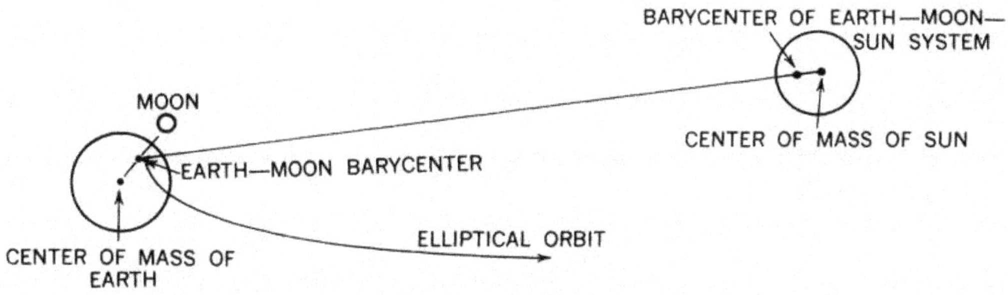

FIGURE 3103b.—Orbit of earth-moon barycenter (not to scale).

general and can be applied directly to describe the differential gravitational forces of the sun.

Earth-Moon System

When determining the orbit of the earth's center of mass about the earth-moon barycenter, the gravitational force exerted by the moon on the earth is given by

$$F = \frac{GM_E M_M}{d^2_M},$$

where M_E and M_M are the masses of the earth and moon, and d_M is the distance between their centers of mass. Acceleration and force are related by Newton's second law of motion, $F=ma$. The acceleration, a, of a mass, m, is then $a=F/m$. The terms "acceleration" and "force per unit mass" may be used interchangeably. Combining Newton's law of gravitation and his second law of motion, the acceleration of the earth's center of mass about the earth-moon barycenter is

$$a_c = \frac{GM_M}{d^2_M}.$$

In determining the direction and magnitude of tide-generating forces of the moon, the simplest case to treat is that of the sublunar point and its antipode on the earth. These two points are labeled P_1 and P_2 in figure 3103c.

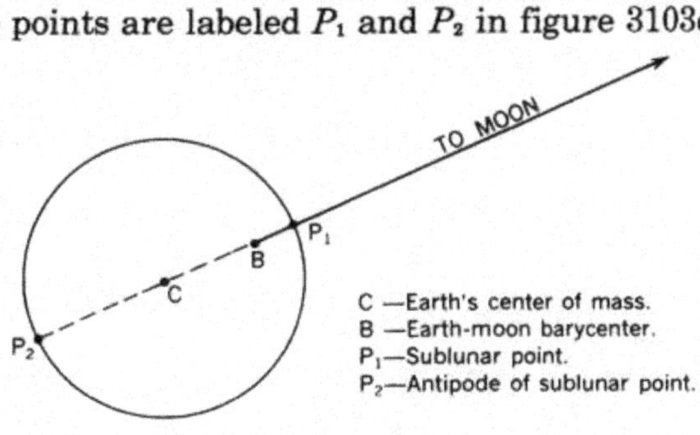

FIGURE 3103c.—Sublunar point and antipode.

The acceleration at the point P_1 due to the gravitational attraction of the moon is

$$a_1 = \frac{GM_M}{(d_M - R_E)^2},$$

where R_E is the radius of the earth. The acceleration at the point P_2 due to the gravitational attraction of the moon is

$$a_2 = \frac{GM_M}{(d_M + R_E)^2}.$$

The differential acceleration at P_1, that is, the acceleration at P_1 relative to the acceleration of the center of the earth, is

$$a_1 - a_c = \frac{GM_M}{(d_M - R_E)^2} - \frac{GM_M}{d^2_M}.$$

With some simplification, this expression becomes

$$a_1 - a_c = \frac{2GM_MR_E}{d^3_M}.$$

This is the differential acceleration or differential force per unit mass at the sublunar point P_1. In a similar manner, the differential acceleration at the antipode, point P_2, is found to be

$$a_2 - a_c = \frac{-2GM_MR_E}{d^3_M}.$$

Both the accelerations and the resulting differential accelerations or differential forces per unit mass are shown in figure 3103d.

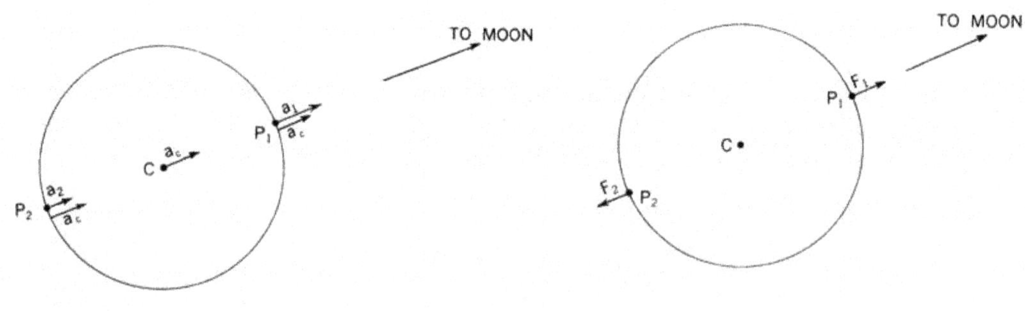

Differential gravitational force per unit mass at sublunar point P_1 is $a_1 - a_c$.

Differential gravitational force per unit mass at antipode, point P_2, is $a_2 - a_c$.

F_1 and F_2 represent the differential forces per unit mass at the sublunar point and the antipode, points P_1 and P_2, where
$F_1 = a_1 - a_c$
$F_2 = a_2 - a_c$

FIGURE 3103d.—Differential gravitational forces per unit mass at sublunar point and antipode.

Note that the differential gravitational force per unit mass at the antipode, point P_2, is negative. Thus, at both the sublunar point and the antipode, the moon's differential gravitational forces are vertical and directed away from the center of the earth.

A more complicated situation occurs when the moon's gravitational forces act on points of the earth's surface other than the sublunar point and the antipode. To find the differential accelerations at the sublunar point and the antipode, it was only necessary to take the algebraic difference between the accelerations at the surface and the acceleration of the earth's center. At other points on the surface, however, both the magnitude and direction of the moon's gravitational forces differ from that at the earth's center (figure 3103e). To obtain the differential accelerations at the other points, it is necessary to subtract the accelerations vectorially: $\bar{F}_D = \bar{a}_M - \bar{a}_c$, where $\bar{F}_D$ is the differential acceleration or differential force per unit mass at any point on the surface of the earth, $\bar{a}_c$ is the acceleration of the earth's center of mass, and $\bar{a}_M$ is the acceleration or force per unit mass at the point under consideration due to the moon's gravitational force acting at that point (figure 3103f).

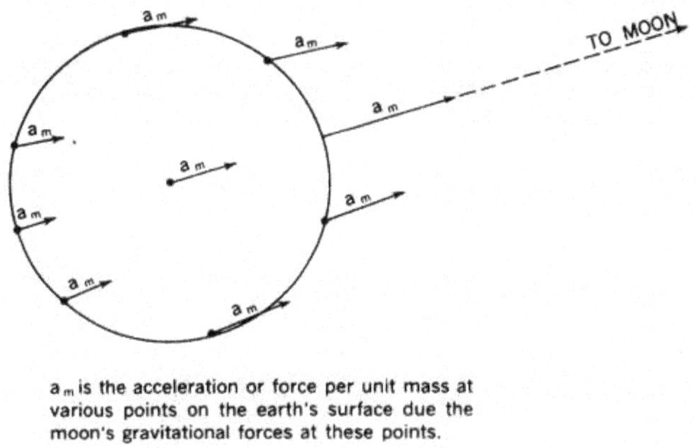

FIGURE 3103e.—Forces per unit mass on earth's surface due to moon's gravitational forces. Only the acceleration at the earth's center and the accelerations along one great circle through the sublunar point and the antipode are shown.

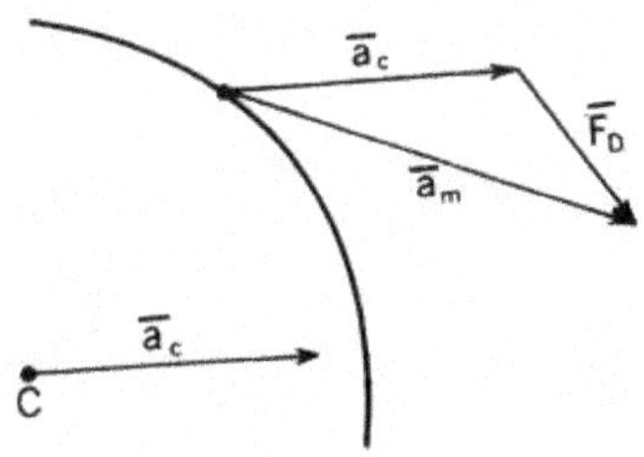

FIGURE 3103f.—Differential force per unit mass, $\bar{F}_D$, is the vector difference $\bar{a}_M - \bar{a}_C$.

The relative effects of $\bar{a}_M$ and $\bar{a}_C$ at the center of the earth and at various points along one great circle through the sublunar point and the antipode are shown in figure 3103g. The resultant differential forces per unit mass are shown in figure 3103h.

If it is assumed that the entire surface of the earth is covered with a uniform layer of water, the differential forces may be resolved into components perpendicular and parallel to the surface of the earth (fig. 3103i) to determine their effect.

The components of these differential forces which are perpendicular to the earth's surface have the effect of changing the weight of the mass on which they are acting. These vertical components do not contribute to the tidal effect. The horizontal components which are parallel to the earth's surface, although small, have the effect of moving the water in a horizontal direction toward the sublunar and antipodal points until an equilibrium position is found. The horizontal components of the differential forces are the principal tide-generating forces. These are also called **tractive forces.** Figure 3103j shows the tractive forces across the surface of the earth.

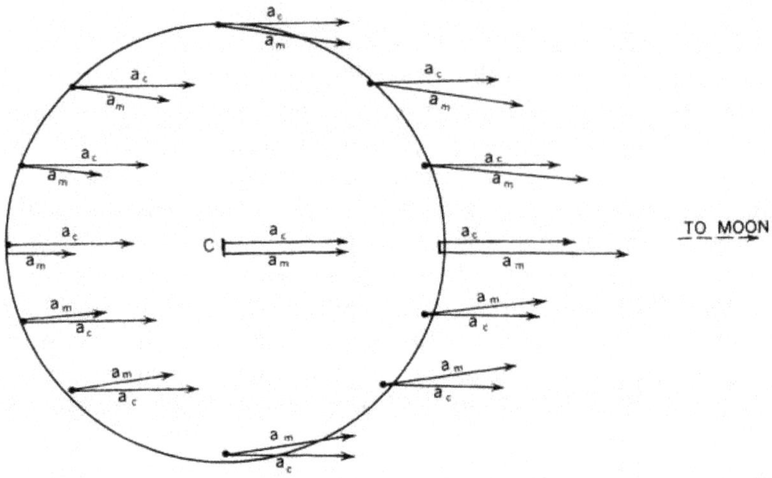

FIGURE 3103g.—Accelerations due to the moon's gravitational forces, $\bar{a}_M$, compared to the acceleration of the center of the earth, $\bar{a}_C$. Comparisons are made at the earth's center and at various points along a great circle connecting the sublunar point and the antipode. The effects will be the same along all great circles connecting these two points.

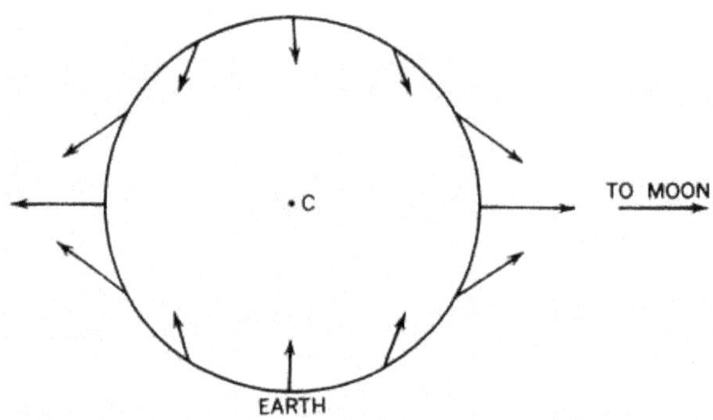

FIGURE 3103h.—Differential forces along a great circle connecting the sublunar point and antipode.

The magnitudes of the horizontal and vertical components are

$$F_H = \frac{3}{2}\,\frac{GM_M R_E}{d^3_M}\sin 2A$$

$$F_V = \frac{GM_M R_E}{d^3_M}(3\cos^2 A - 1),$$

where A is the angle at the center of the earth between the line connecting the sublunar point and the antipode and the line from the center of the earth to the point under consideration (fig. 3103i). Thus, it can be seen that the horizontal component, which is the tide-generating force, is zero when the angle A is zero (sublunar point and antipode). It is also zero when the angle A is 90°. This corresponds approximately to the great circle connecting observers for which the moon is setting. The maximum value of F_H occurs when the angle A is 45°.

Equilibrium will be reached when a bulge of water has formed at the sublunar and antipodal points such that the tractive forces due to the moon's differential gravitational forces on the mass of water covering the surface of the earth are just balanced by the earth's gravitational attraction (fig. 3103k).

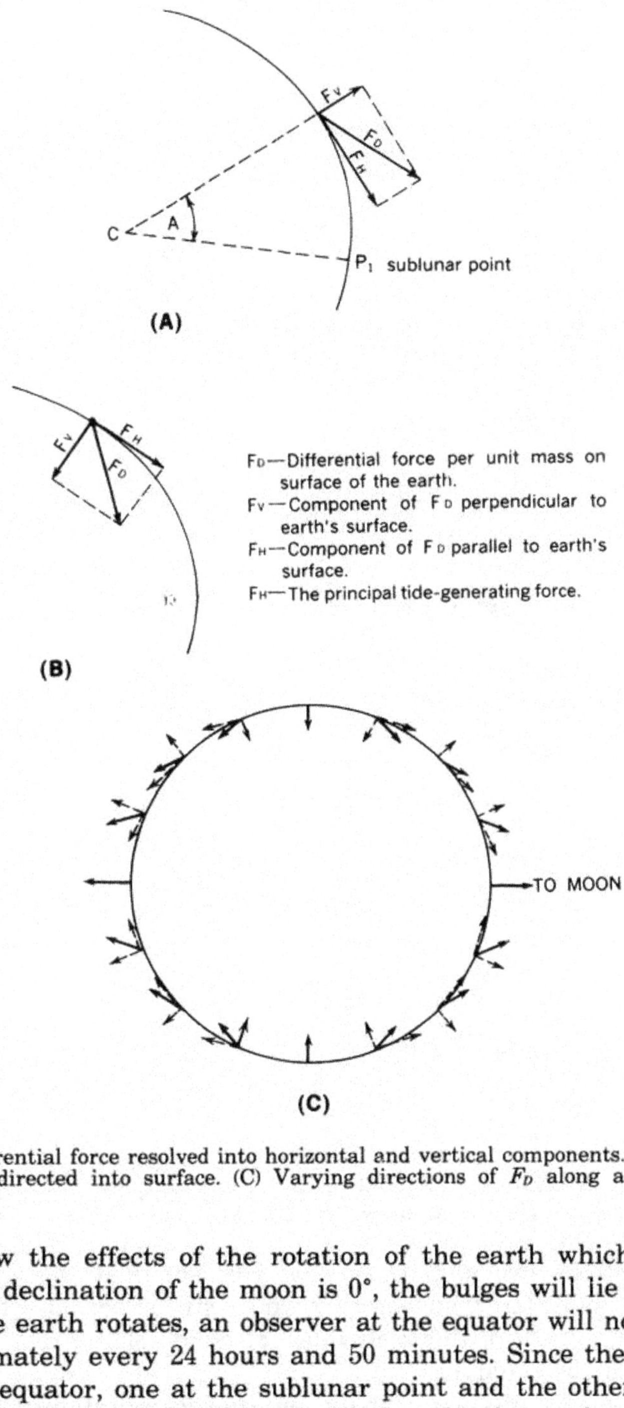

FIGURE 3103i.—Differential force resolved into horizontal and vertical components. (A) F_D directed out of surface. (B) F_D directed into surface. (C) Varying directions of F_D along a great circle through sublunar point.

Consider now the effects of the rotation of the earth which were previously neglected. If the declination of the moon is 0°, the bulges will lie on the equator of the earth. As the earth rotates, an observer at the equator will note that the moon transits approximately every 24 hours and 50 minutes. Since there are two bulges of water on the equator, one at the sublunar point and the other at the antipode, the observer will also see two high tides during this interval with one high tide occurring when the moon is overhead and another high tide 12 hours 25 minutes later when the observer is located at the antipode. He will also experience two low tides, one between each high tide. The range of these equilibrium tides at the equator will be less than 1 meter.

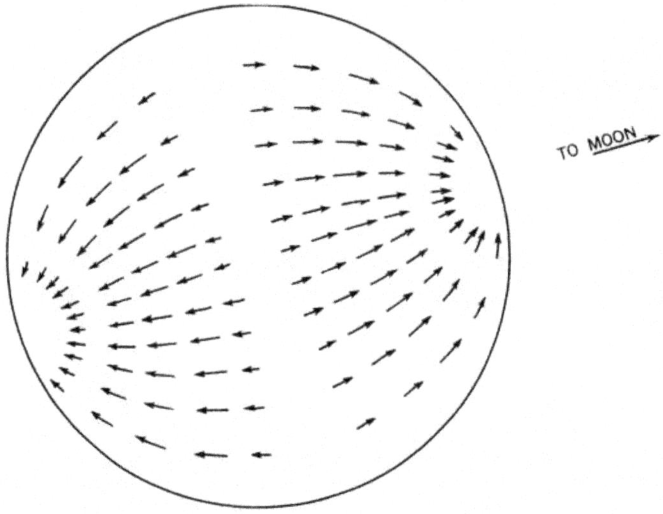

FIGURE 3103j.—Tractive forces across the surface of the earth. Tractive forces are zero at the sublunar and antipodal points and along the great circle halfway between these two points. Tractive forces are maximum along the small circles located 45° from the sublunar point and the antipode.

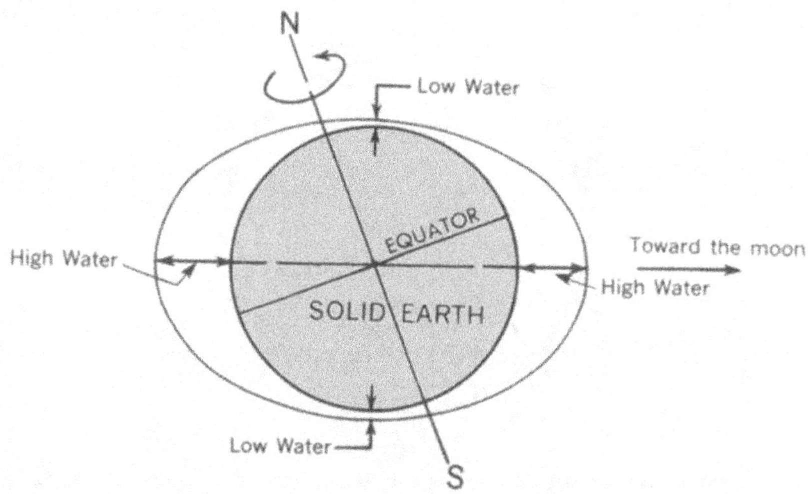

FIGURE 3103k.—Theoretical equilibrium configuration due to moon's differential gravitational forces. One bulge of the water envelope is located at the sublunar point, the other bulge at the antipode.

The heights of the two high tides should be equal at the equator. At points north or south of the equator, an observer would still experience two high and two low tides, but the heights of the high tides, although still equal, would not be as great as they are at the equator.

The effects of the declination of the moon are shown in figure 3103l.

The preceding paragraphs addressed the tide-generating forces due to the differential gravitational forces of the moon on the earth. For the sublunar point, the

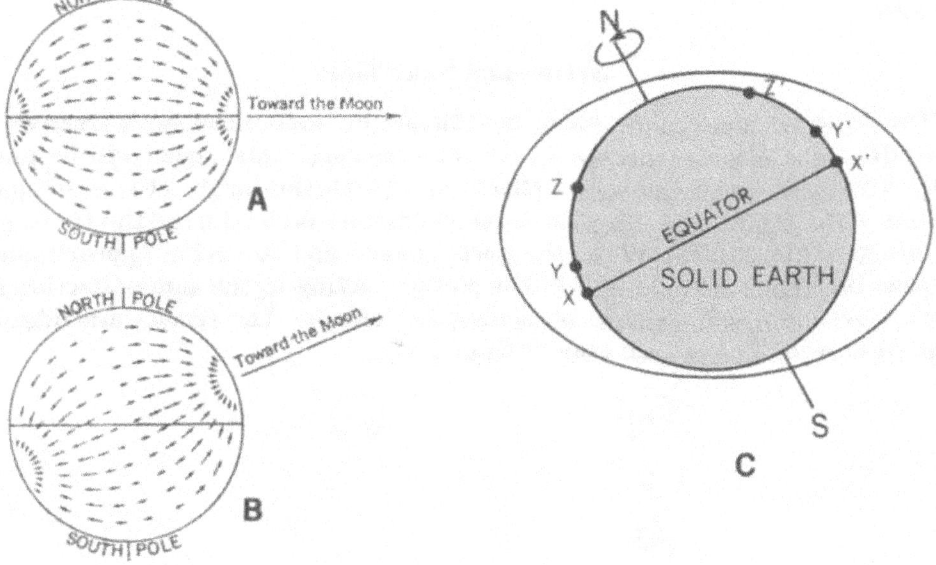

FIGURE 31031.—Effects of the declination of the moon. (A) When the moon is in the plane of the equator, the forces are equal in magnitude at the two points on the same parallel of latitude and 180° apart in longitude. (B) When the moon is at north (or south) declination, the forces are unequal at such points and tend to cause an inequality in the two high waters and the two low waters of a day. (C) Observers at points X, Y, and Z experience one high tide when moon is on their meridian, then another high tide 12 hours 25 minutes later when at X', Y', and Z'. The second high tide is the same at X' as at X. High tides at Y' and Z' are lower than high tides at Y and Z.

force per unit mass was found to be

$$\frac{2GM_M R_E}{d^3_M}.$$

In a similar manner, the differential gravitational force per unit mass due to the sun at the subsolar point is found to be

$$\frac{2GM_S R_E}{d^3_S},$$

where M_S is the mass of the sun and d_S is the distance between the centers of mass of the sun and earth. To find the relative effects of the sun and moon, the ratio of the expressions can be used. This ratio is

$$\left(\frac{M_S}{M_M}\right)\left(\frac{d_M}{d_S}\right)^3.$$

The numerical value of this ratio is 0.46. Thus, the effect of the moon is approximately two and one-quarter times greater than the effect of the sun even though the moon's mass is but a fraction of the sun's. This is due to the fact that the differential forces vary inversely as the *cube* of the distance. Thus the moon's smaller mass is offset by its much shorter distance to the earth.

The preceding discussion pertaining to the effects of the moon is equally valid when discussing the effects of the sun, taking into account that the magnitude of the solar effects are smaller than the lunar effects. Hence, the tides will also vary according to the sun's declination and its varying distance from the earth. A second envelope of water representing the equilibrium tides due to the sun would resemble the envelope shown in figure 3103k except that the heights of the high tides would be smaller.

Spring and Neap Tides

The combined lunar-solar effect is obtained by adding the sun's tractive forces vectorially to the moon's tractive forces. The resultant tidal bulge will be predominantly lunar with modifying solar effects upon both the height of the tide and the direction of the tidal bulge. Special cases of interest occur during the times of new and full moon (fig. 3103m). With the earth, moon, and sun lying approximately on the same line, the tractive forces of the sun are acting in the same direction as the moon's tractive forces (modified by declination effects). The results are tides called *spring tides* whose ranges are greater than average.

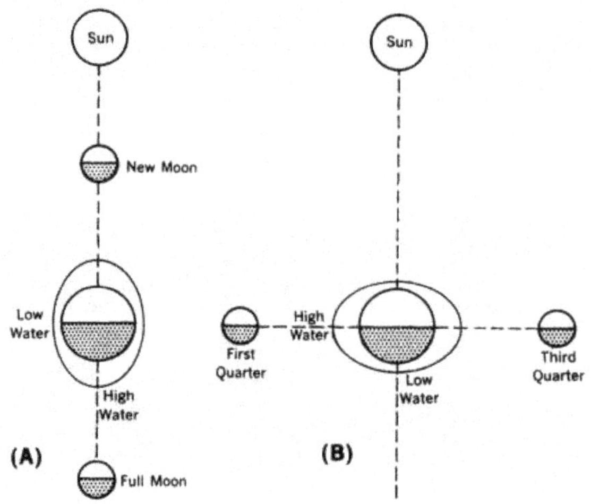

FIGURE 3103m.—(A) Spring tides occur at times of new and full moon. Range of tide is greater than average since solar and lunar tractive forces act in same direction. (B) Neap tides occur at times of first and third quarters. Range of tide is less than average since solar and lunar tractive forces act at right angles.

Another case of interest occurs when the moon is at first and third quarters. At those times, the tractive forces of the sun are acting at approximately right angles to the moon's tractive forces (fig. 3103m). The results are tides called *neap tides* whose ranges are less than average.

With the moon in positions between quadrature and new and full moon, the effect of the sun is to cause the tidal bulge to either lag or precede the moon (fig. 3103n). These effects are called *priming* and *lagging* the tides.

Tide

3104. General features.—Tide is the periodic rise and fall of the water accompanying the tidal phenomenon. At most places it occurs twice daily. The tide rises until it reaches a maximum height, called **high tide** or **high water,** and then falls to a minimum level called **low tide** or **low water.**

The rate of rise and fall is not uniform. From low water, the tide begins to rise slowly at first but at an increasing rate until it is about halfway to high water. The rate of rise then decreases until high water is reached and the rise ceases. The falling tide behaves in a similar manner. The period at high or low water during

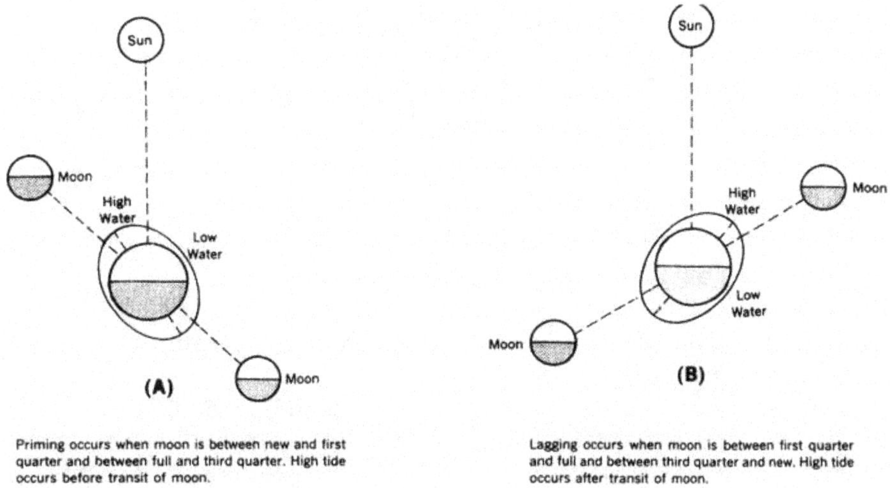

FIGURE 3103n.—Priming and lagging the tides.

which there is no sensible change of level is called **stand.** The difference in height between consecutive high and low waters is the **range.**

Figure 3104 is a graphical representation of the rise and fall of the tide at New York during a 24-hour period. The tide curve has the general form of a sine curve.

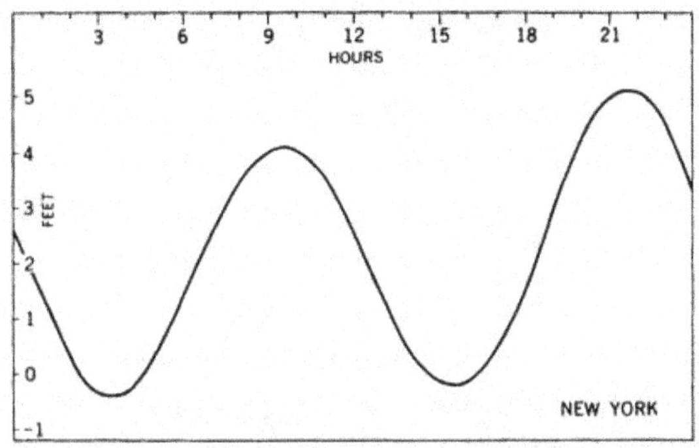

FIGURE 3104.—The rise and fall of the tide at New York, shown graphically.

3105. Types of tide.—A body of water has a natural period of oscillation that is dependent upon its dimensions. None of the oceans appears to be a single oscillating body, but rather each one is made up of a number of oscillating basins. As such basins are acted upon by the tide-producing forces, some respond more readily to daily or diurnal forces, others to semidiurnal forces, and others almost equally to both. Hence, tides at a place are classified as one of three types—**semidiurnal, diurnal,** or **mixed**—according to the characteristics of the tidal pattern occurring at the place.

In the **semidiurnal** type of tide, there are two high and two low waters each tidal day, with relatively small inequality in the high and low water heights. Tides on the Atlantic coast of the United States are representative of the semidiurnal type, which is illustrated in figure 3105a by the tide curve for Boston Harbor.

In the **diurnal** type of tide, only a single high and single low water occur each tidal day. Tides of the diurnal type occur along the northern shore of the Gulf of Mexico, in the Java Sea, the Gulf of Tonkin (off the Vietnam-China coast), and in a

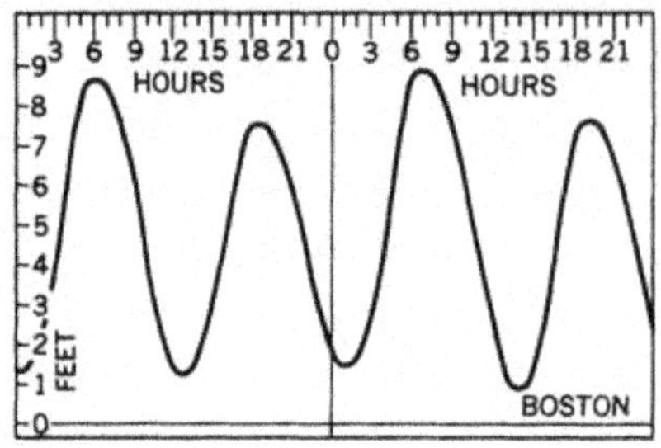

FIGURE 3105a.—Semidiurnal type of tide.

few other localities. The tide curve for Pei-Hai, China, illustrated in figure 3105b, is an example of the diurnal type.

In the **mixed** type of tide, the diurnal and semidiurnal oscillations are both important factors and the tide is characterized by a large inequality in the high water heights, low water heights, or in both. There are usually two high and two low waters each day, but occasionally the tide may become diurnal. Such tides are prevalent along the Pacific coast of the United States and in many other parts of the world. Examples of mixed types of tide are shown in figure 3105c. At Los Angeles, it is typical that the inequalities in the high and low waters are about the same. At Seattle the greater inequalities are typically in the low waters, while at Honolulu it is the high waters that have the greater inequalities.

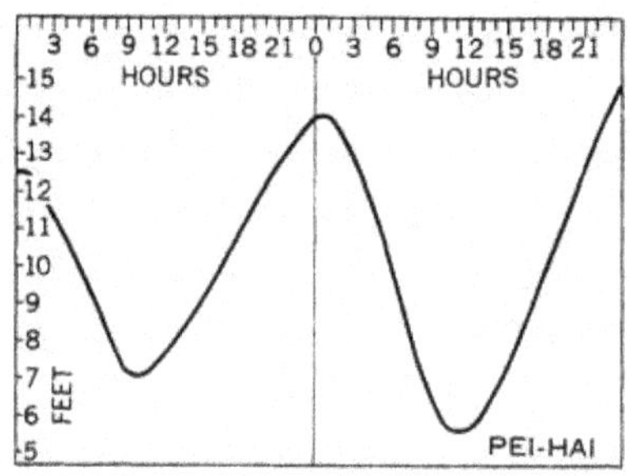

FIGURE 3105b.—Diurnal type of tide.

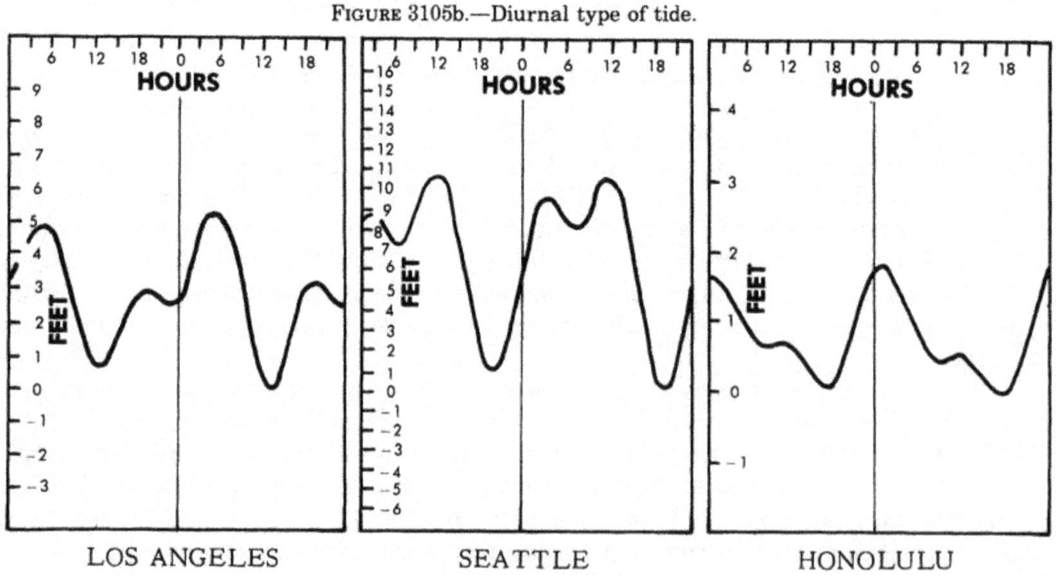

FIGURE 3105c.—Mixed type of tide.

3106. Solar tide.—The natural period of oscillation of a body of water may accentuate either the solar or the lunar tidal oscillations. Though it is a general rule that the tides follow the moon, the relative importance of the solar effect varies in different areas. There are a few places, primarily in the South Pacific and the Indonesian areas, where the solar oscillation is the more important, and at those places the high and low waters occur at about the same time each day. At Port Adelaide, Australia (fig. 3106), the solar and lunar semidiurnal oscillations are equal and nullify one another at neaps (art. 3108).

3107. Special effects.—As a progressive wave enters shallow water its speed is decreased. Since the trough is shallower than the crest, its retardation is greater, resulting in a steepening of the wave front. Therefore, in many rivers, the duration of rise is considerably less than the duration of fall. In a few estuaries, the advance of the low water trough is so much retarded that the crest of the rising tide overtakes the low, and advances upstream as a churning, foaming wall of water called a **bore.** Bores that are large and dangerous at times of large tidal ranges may be mere ripples at those times of the month when the range is small. Examples occur in the Petitcodiac River in the Bay of Fundy, and at Haining, China, in the Tsientang Kaing. The tide tables indicate where bores occur.

Other special features are the **double low water** (as at Hoek Van Holland) and the **double high water** (as at Southampton, England). At such places there is often a slight fall or rise in the middle of the high or low water period. The practical effect is to create a longer period of stand at high or low tide. The tide tables direct attention to these and other peculiarities where they occur.

3108. Variations in range.—Though the tide at a particular place can be classified as to type, it exhibits many variations during the month (fig. 3106). The range of the tide varies in accordance with the intensity of the tide-producing force, though there may be a lag of a day or two (**age of tide**) between a particular astronomic cause and the tidal effect.

Thus, when the moon is at the point in its orbit nearest the earth (at *perigee*), the lunar semidiurnal range is increased and **perigean** tides occur; when the moon is farthest from the earth (at *apogee*), the smaller **apogean** tides occur. When the moon and sun are in line and pulling together, as at new and full moon, **spring** tides occur (the term *spring* has nothing to do with the season of year); when the moon and sun oppose each other, as at the quadratures, the smaller **neap** tides occur.

When certain of these phenomena coincide, the great **perigean spring** tides, the small **apogean neap** tides, etc., occur.

These are variations in the semidiurnal portion of the tide. Variations in the diurnal portion occur as the moon and sun change declination. When the moon is at its maximum semi-monthly declination (either north or south), **tropic** tides occur in which the diurnal effect is at a maximum; when it crosses the equator, the diurnal effect is a minimum and **equatorial** tides occur.

It should be noted that when the range of tide is increased, as at spring tides, there is more water available only at *high* tide; at *low* tide there is less, for the high waters rise higher and the low waters fall lower at these times. There is more water at neap low water than at spring low water. With tropic tides, there is usually more depth at one low water during the day than at the other. While it is desirable to know the meanings of these terms, the best way of determining the height of the tide at any place and time is to examine the tide predictions for the place as given in the tide tables. Figure 3108 illustrates variations in the ranges

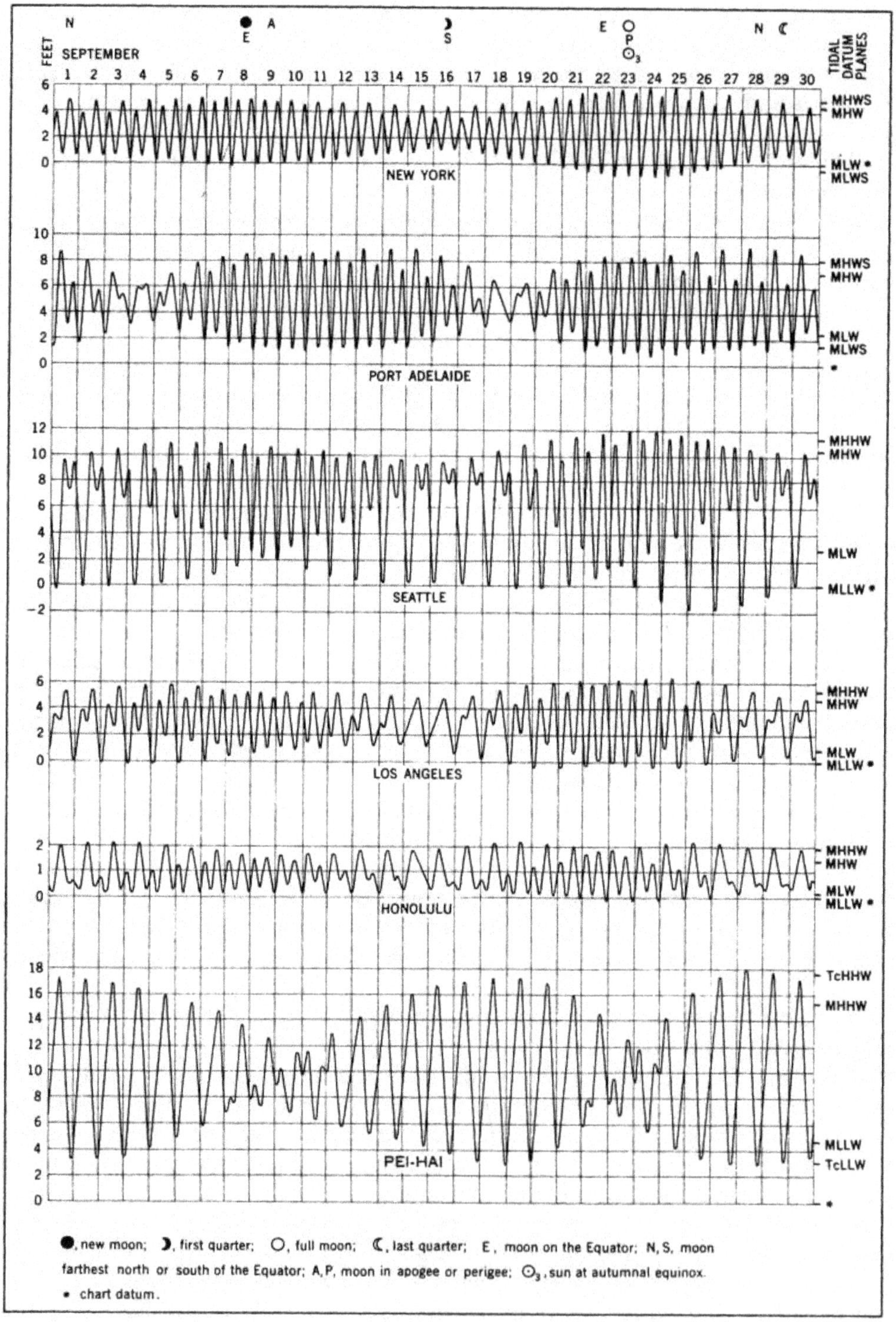

FIGURE 3106.—Tidal variations at various places during a month.

and heights of tides in a locality such as the Indian Ocean where predicted and observed water levels are referenced to a chart sounding datum that will always cause them to be additive relative to the charted depth.

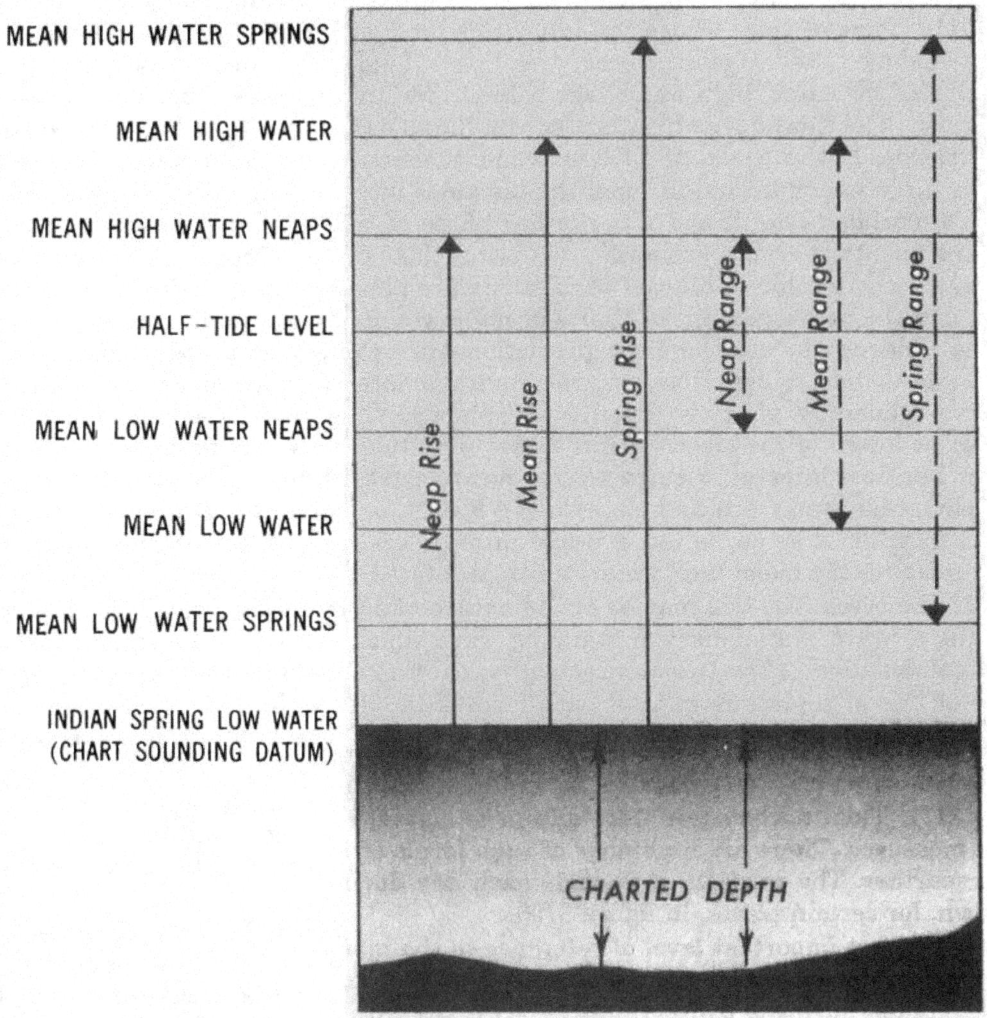

FIGURE 3108.—Variations in the ranges and heights of tide in a locality where the chart sounding datum is Indian Spring Low Water.

3109. Tidal cycles.—Tidal oscillations go through a number of cycles. The shortest cycle, completed in about 12 hours and 25 minutes for a semidiurnal tide, extends from any phase of the tide to the next recurrence of the same phase. During a **lunar day** (averaging 24 hours and 50 minutes) there are two highs and two lows (two of the shorter cycles) for a semidiurnal tide. The moon revolves around the earth with respect to the sun in a **synodical month** of about 29½ days, commonly called the **lunar month.** The effect of the phase variation is completed in one-half a synodical month or about 2 weeks as the moon varies from new to full or full to new. The effect of the moon's declination is also repeated in one-half of a **tropical month** of 27⅓ days or about each 2 weeks. The cycle involving the moon's distance requires an **anomalistic month** of about 27½ days. The sun's declination and distance cycles are respectively a half year and a year in length. An important lunar cycle, called the **nodal period,** is 18.6 years (usually expressed in round figures as 19 years). For a tidal value, particularly a range, to be considered a true mean, it must be either based upon observations extended over this period of time or adjusted to take account of variations known to occur during the cycle.

3110. Time of tide.—Since the lunar tide-producing force has the greater effect in producing tides at most places, the tides "follow the moon." Because of the rotation of the earth, high water lags behind meridian passage (upper and lower) of the moon. The **tidal day,** which is also the **lunar day,** is the time between consecutive transits of the moon, or 24 hours and 50 minutes on the average. Where the tide is largely semidiurnal in type, the **lunitidal interval**—the interval between the moon's meridian transit and a particular phase of tide—is fairly constant throughout the month, varying somewhat with the tidal cycles. There are many places, however, where solar or diurnal oscillations are effective in upsetting this relationship, and the newer editions of charts of many countries now omit intervals because of the tendency to use them for prediction even though accurate predictions are available in tide tables. However, the lunitidal interval may be encountered. The interval generally given is the average elapsed time from the meridian transit (upper or lower) of the moon until the next high tide. This may be called **mean high water lunitidal interval** or **corrected** (or **mean**) **establishment.** The **establishment of the port, high water full and change (HWF &C),** or **vulgar** (or **common**) **establishment,** sometimes given, is the average interval on days of full or new moon, and approximates the mean high water lunitidal interval.

In the ocean, the tide may be of the nature of a progressive wave with the crest moving forward, a stationary or standing wave which oscillates in a seesaw fashion, or a combination of the two. Consequently, caution should be used in inferring the time of tide at a place from tidal data for nearby places. In a river or estuary, the tide enters from the sea and is usually sent upstream as a progressive wave, so that the tide occurs progressively later at various places upstream.

3111. Tidal datums.—A **tidal datum** is a level from which heights and depths are measured. There are a number of such levels of reference that are important to the mariner. The relation of the tide each day during a month to these datums is shown, for certain places, in figure 3106.

The most important level of reference to the mariner is the datum of soundings on charts (art. 511). Since the tide rises and falls continually while soundings are being taken during a hydrographic survey, the tide should be observed during the survey so that soundings taken at all stages of the tide can be reduced to a common **chart sounding datum.** Soundings on charts show depths below a selected low water datum (occasionally mean sea level), and tide predictions in tide tables show heights above the same level. The depth of water available at any time is obtained by adding the height of the tide at the time in question to the charted depth, or by subtracting the predicted height if it is negative.

By international agreement, the level used as chart datum should be just low enough so that low waters do not go far below it. At most places, however, the level used is one determined from a mean of a number of low waters (usually over a 19-year period); therefore, some low waters can be expected to fall below it. The following are some of the datums in general use.

The highest low water datum in considerable use is **mean low water (MLW),** which is the average height of all low waters at a place. About half of the low waters fall below it. **Mean low water springs (MLWS),** usually shortened to **low water springs,** is the average level of the low waters that occur at the times of spring tides. **Mean lower low water (MLLW)** is the average height of the lower low waters of each tidal day. **Tropic lower low water (TcLLW)** is the average height of the lower low waters (or of the single daily low waters if the tide becomes diurnal) that occur when the moon is near maximum declination and the diurnal effect is most pronounced. This datum is not in common use as a tidal reference. **Indian spring low water (ISLW)** sometimes called **Indian tide plane** or **harmonic tide plane,** is a low water datum that includes the spring effect of the semi-diurnal portion of the tide and the tropic effect of the diurnal portion. It is about the level of lower low water of mixed tides at the time that the moon's maximum declination coincides with the time of new or full moon. **Mean lower low water springs** is the average level of the lower of the two low waters on the days of spring tides. Some still lower datums used on charts are determined from tide observations and some are determined arbitrarily and later referred to the tide. Most of them fall close to one or the other of the following two datums. **Lowest normal low water** is a datum that approximates the average height of monthly lowest low waters, discarding any tides disturbed by storms. **Lowest low water** is an extremely low datum. It conforms generally to the lowest tide observed, or even somewhat lower. Once a tidal datum is established, it is sometimes retained for an indefinite period, even though it might differ slightly from a better determination from later observations. When this occurs, the established datum may be called **low water datum, lower low water datum,** etc. These datums are used in a limited area and primarily for river and harbor engineering purposes. Examples are *Boston Harbor Low Water Datum* and *Columbia River Lower Low Water Datum.*

In some areas where there is little or no tide, such as the Baltic Sea, **mean sea level (MSL)** is used as chart datum. This is the average height of the surface of the sea for all stages of the tide over a 19-year period. This may differ slightly from **half-tide level,** which is the level midway between mean high water and mean low water.

Inconsistencies of terminology are found among charts of different countries and between charts issued at different times. For example, the spring effect as defined here is a feature of only the semidiurnal tide, yet it is sometimes used synonymously with tropic effect to refer to times of increased range of a diurnal tide. Such inconsistencies are being reduced through increased international cooperation.

Large-scale charts usually specify the datum of soundings and may contain a tide note giving mean heights of the tide at one or more places on the chart. These heights are intended merely as a rough guide to the change in depth to be expected under the specified conditions. They should not be used for the prediction of heights on any particular day. Such predictions should be obtained from *tide tables* (arts. 1203–1206).

3112. High water datums.—Heights of land features are usually referred on nautical charts to a high water datum. The one used on charts of the United States, its territories, and possessions, and widely used elsewhere, is **mean high water (MHW),** which is the average height of all high waters over a 19-year period. Any other high water datum in use on charts is likely to be higher than this. Other high water datums are **mean high water springs (MHWS),** which is the average level of the high waters that occur at the time of spring tides; **mean higher high water (MHHW),** which is the average height of the higher high waters of each tidal day; and **tropic higher high water (TcHHW),** which is the average height of the higher high waters (or the single daily high waters if the tide becomes diurnal) that occur when the moon is near maximum declination and the diurnal effect is most pronounced. A reference merely to "high water" leaves some doubt as to the specific level referred to, for the height of high water varies from day to day. Where the range is large, the variation during a 2-week period may be considerable.

As there are periodic and apparent secular trends in sea level, a specific 19-year cycle (the **National Tidal Datum Epoch**) is issued for all United States datums. The National Tidal Datum Epoch officially adopted by the National Ocean Service is presently 1941 through 1959. The Epoch will be reviewed for consideration for revision at 25-year intervals.

3113. Observations and predictions.—Since the tide at different places responds differently to the tide-producing forces, the nature of the tide at any place can be determined most accurately by actual observation. The predictions in tide tables and the tidal data on nautical charts are based upon observations.

Tides are usually observed by means of a continuously recording gage. A year of observations is the minimum length desirable for determining the **harmonic constants** used in prediction. For establishing mean sea level and the long-time changes in the relative elevations of land and sea, as well as for other special uses, observations have been made over periods of 20, 30, and even 120 years at important locations. Observations for a month or less will establish the *type* of tide and suffice for comparison with a longer series of a similar type to determine tidal differences and constants.

Mathematically, the variations in the lunar and solar tide-producing forces, such as those due to changing phase, distance, and declination, are considered as separate constituent forces, and the **harmonic analysis** of observations reveals the response of each constituent of the tide to its corresponding force. At any one place this response remains constant and is shown for each constituent by **harmonic constants** which are in the form of a phase angle for the time relation and an amplitude for the height. Harmonic constants are used in making technical studies of the tide and predictions on computers and mechanical **tide predicting machines.** Most published tide predictions are made by computer.

3114. Tide tables are published annually by most of the maritime nations of the world. They consist primarily of two parts. One contains predictions of the time and height of each high and low water for every day of the year for many important ports called **reference stations.** The other part contains tidal differences and ratios for thousands of other places, called **subordinate stations,** and specifies the reference station to which the differences are to be applied in order to obtain time and height of tide for any day at the subordinate station. The type of tide at a subordinate station is the same as at its reference station. The use of tide tables is explained in articles 1203-1206.

3115. Meteorological effects.—The foregoing discussion of tide behavior assumes normal weather conditions. The level of the sea is affected by wind and atmospheric pressure. In general, onshore winds raise the level and offshore winds lower it, but the amount of change varies at different places. During periods of low atmospheric pressure, the water level tends to be higher than normal. For a stationary low, the increase in elevation can be found by the formula

$$R_0 = 0.0325(1010 - P),$$

in which R_0 is the increase in elevation in feet, and P is the atmospheric pressure in millibars. This is equal approximately to 1 centimeter per millibar depression, or

1 foot (13.6 inches) per inch depression. For a moving low, the increase in elevation is given by the formula

$$R = \frac{R_0}{1 - \frac{C^2}{gh}},$$

in which R is the increase in elevation in feet, R_0 is the increase in feet for a stationary low, C is the rate of motion of the low in feet per second, g is the acceleration due to gravity (32.2 feet per second per second), and h is the depth of water in feet.

Where the range of tide is very small, the meteorological effect may sometimes be greater than the normal tide.

Tidal Current

3116. Tidal and nontidal currents.—Horizontal movement of the water is **current**. It may be classified as "tidal" and "nontidal." **Tidal current** is the periodic horizontal flow of water accompanying the rise and fall of the tide, and results from the same cause. **Nontidal current** is any current not due to the tidal movement. Nontidal currents include the permanent currents in the general circulatory system of the oceans as well as temporary currents arising from meteorological conditions. The current experienced at any time is usually a combination of tidal and nontidal currents.

In navigation, the effect of the tidal current is often of more importance than the changing depth due to the tide, and many mariners speak of "the tide," when they have in mind the flow of the tidal current.

3117. General features.—Offshore, where the direction of flow is not restricted by any barriers, the tidal current is **rotary;** that is, it flows continuously, with the direction changing through all points of the compass during the tidal period. The tendency for the rotation in direction has its origin in the deflecting force of the earth's rotation, and unless modified by local conditions, the change is clockwise in the Northern Hemisphere and counterclockwise in the Southern Hemisphere. The speed usually varies throughout the tidal cycle, passing through two maximums in approximately opposite directions, and two minimums about halfway between the maximums in time and direction. Rotary currents can be depicted as in figure 3117a, by a series of arrows representing the direction and speed of the current at each hour. This is sometimes called a **current rose.** Because of the elliptical pattern formed by the ends of the arrows, it is also referred to as a **current ellipse.**

In rivers or straits, or where the direction of flow is more or less restricted to certain channels, the tidal current is **reversing;** that is, it flows alternately in approximately opposite directions with an instant or short period of little or no current, called **slack water,** at each reversal of the current. During the flow in each direction, the speed varies from zero at the time of slack water to a maximum, called **strength of flood** or **ebb,** about midway between the slacks. Reversing currents can be indicated graphically, as in figure 3117b, by arrows that represent the speed of the current at each hour. The flood is usually depicted above the slack waterline and the ebb below it. The tidal current curve formed by the ends of the arrows has the same characteristic sine form as the tide curve. (In illustrations for certain purposes, as in figures 3118b and 3120b, it is convenient to omit the arrows and show only the curve.)

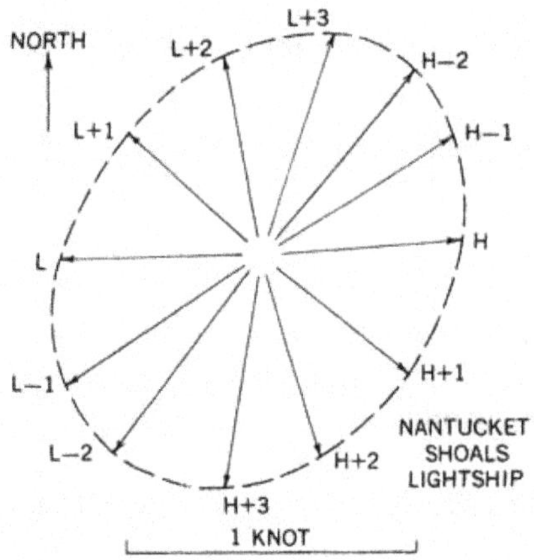

FIGURE 3117a.—Rotary tidal current. Times are hours before and after high and low tide at Nantucket Shoals. The bearing and length of each arrow represents the hourly direction and speed of the current. See figure 3120a.

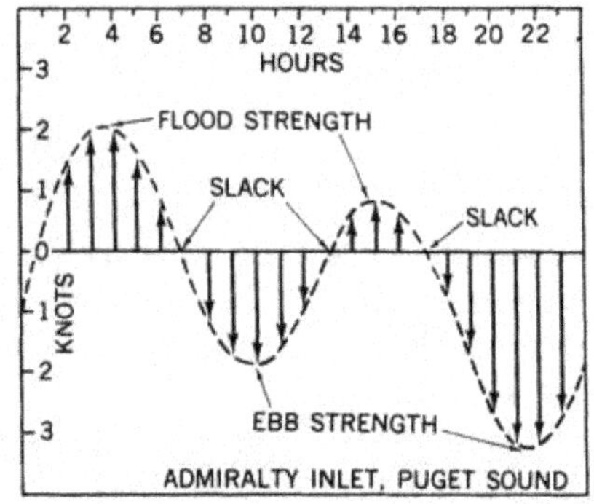

FIGURE 3117b.—Reversing tidal current. (Such graphs may show only the curved pattern without the arrows, as in figures 3118b and 3120b.) See figure 3120b.

A slight departure from the sine form is exhibited by the reversing current in a strait, such as East River, New York, that connects two tidal bodies of water. The tides at the two ends of a strait are seldom in phase or equal in range, and the current, called **hydraulic current**, is generated largely by the continuously changing difference in height of water at the two ends. The speed of a hydraulic current varies nearly as the square root of the difference in height. The speed reaches a maximum more quickly and remains at strength for a longer period than shown in figure 3117b, and the period of weak current near the time of slack is considerably shortened.

The current *direction* or **set** is the direction *toward* which the current flows. The *speed* is sometimes called the **drift**. The term "velocity" is often used as the equivalent of "speed" when referring to current, although strictly "velocity" implies direction as well as speed. The term "strength" is also used to refer to speed, but more often to greatest speed between consecutive slack waters. The movement toward shore or upstream is the **flood**, the movement away from shore or downstream is the **ebb**. In a purely semidiurnal type of current unaffected by nontidal flow, the flood and ebb each last about 6 hours and 13 minutes. But if there is either diurnal inequality or nontidal flow, the durations of flood and ebb may be quite unequal.

3118. Types of tidal current.—Tidal currents may be of the **semidiurnal, diurnal,** or **mixed** type; corresponding to a considerable degree to the type of tide at the place, but often with a stronger semidiurnal tendency.

The tidal currents in tidal estuaries along the Atlantic coast of the United States are examples of the semidiurnal type of reversing current. At Mobile Bay entrance they are almost purely diurnal. At most places, however, the type is mixed to a greater or lesser degree. At Tampa and Galveston entrances there is only one flood and one ebb each day when the moon is near its maximum declination, and two floods and two ebbs each day when the moon is near the equator. Along the Pacific coast of the United States there are generally two floods and two ebbs every day, but one of the floods or ebbs has a greater speed and longer duration than the other, the inequality varying with the declination of the moon. The inequalities in the current often differ considerably from place to place even within limited areas, such as adjacent passages in Puget Sound and various passages between the Aleutian Islands. Figure 3118a shows several types of reversing current. Figure 3118b shows how the flood disappears as the diurnal inequality increases at one station.

Offshore rotary currents that are purely semidiurnal repeat the elliptical pattern (fig. 3117a) each tidal cycle of 12 hours and 25 minutes. If there is considerable diurnal inequality, the plotted hourly current arrows describe a set of two ellipses of different sizes during a period of 24 hours and 50 minutes, as shown in figure 3118c, and the greater the diurnal inequality, the greater the difference between the sizes of the two ellipses. In a completely diurnal rotary current, the smaller ellipse disappears and only one ellipse is produced in 24 hours and 50 minutes.

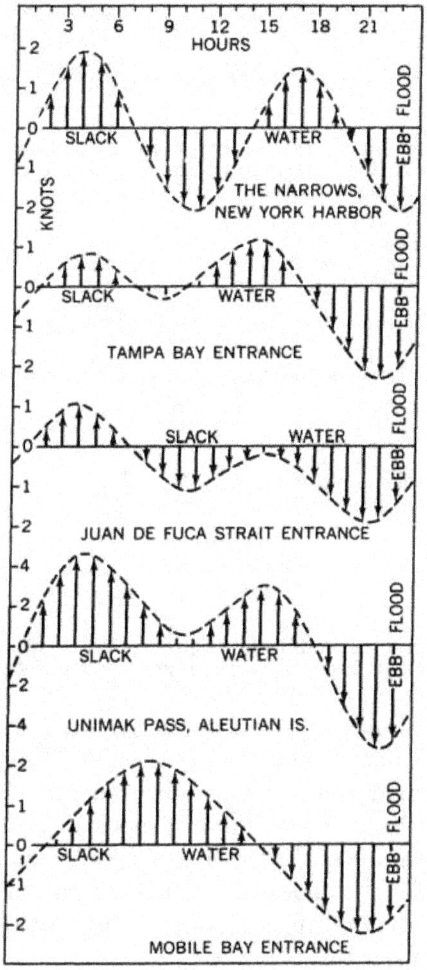

FIGURE 3118a.—Several types of reversing current. The pattern changes gradually from day to day, particularly for mixed types, passing through cycles somewhat similar to that shown for tides in figure 3106.

3119. Variations and cycles.—Tidal currents have periods and cycles similar to those of the tides (art. 3109) and are subject to similar variations, but flood and ebb of the current do not necessarily occur at the same times as the rise and fall of the tide. The relationship is explained further in article 3121.

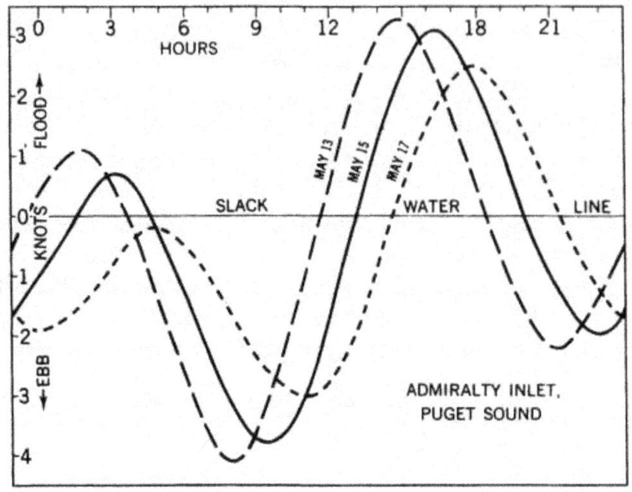

FIGURE 3118b.—Changes in a current of the mixed type. Note that each day as the inequality increases, the morning slacks draw together in time until on the 17th the morning flood disappears. On that day the current ebbs throughout the morning.

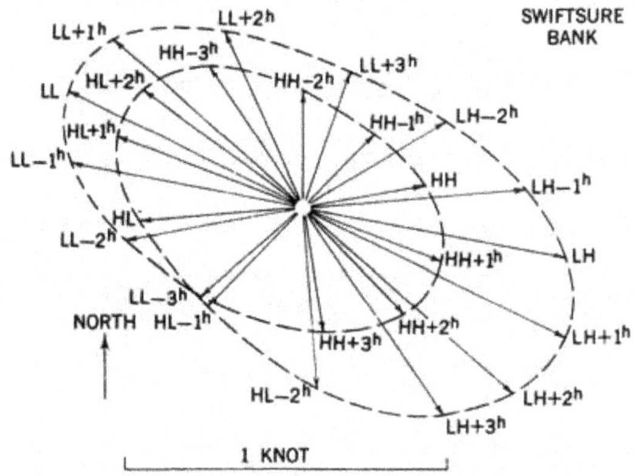

FIGURE 3118c.—Rotary tidal current with diurnal inequality. Times are in hours referred to tides (higher high, lower low, lower high, and higher low) at Swiftsure Bank.

The speed at strength increases and decreases during the 2-week period, month, and year with the variations in the range of tide. Thus, the stronger **spring** and **perigean currents** occur near the times of new and full moon and near the times of the moon's perigee, or at times of spring and perigean tides (art. 3108); the weaker **neap** and **apogean currents** occur at the times of neap and apogean tides; **tropic currents** with increased diurnal speeds or with larger diurnal inequalities in speed occur at times of tropic tides; and **equatorial currents** with a minimum diurnal effect occur at times of equatorial tides; etc.

As with the tide, a *mean value* represents an average obtained from a 19-year series. Since a series of current observations is usually limited to a few days, and seldom covers more than a month or two, it is necessary to adjust the observed values, usually by comparison with tides at a nearby place, to obtain such a mean.

3120. Effect of nontidal flow.—The current existing at any time is seldom purely tidal, but usually includes also a nontidal current that is due to drainage, oceanic circulation, wind, or other cause. The method in which tidal and nontidal currents combine is best explained graphically, as in figures 3120a and 3120b. The pattern of the tidal current remains unchanged, but the curve is shifted from the point or line from which the currents are measured in the direction of the nontidal current and by an amount equal to it. It is sometimes more convenient graphically merely to move the line or point of origin in the opposite direction.

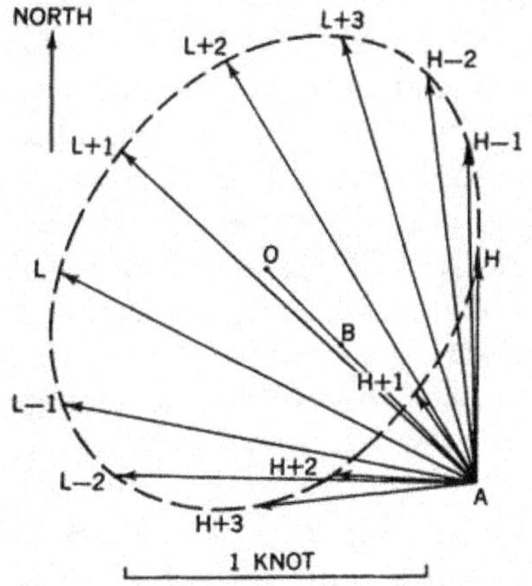

FIGURE 3120a.—Effect of nontidal current on the rotary tidal current of figure 3117a. If the nontidal current is northwest at 0.3 knot, it may be represented by *BO*, and all hourly directions and speeds will then be measured from *B*. If it is 1.0 knot, it will be represented by *AO* and the actual resultant hourly directions and speeds will be measured from *A*, as shown by the arrows.

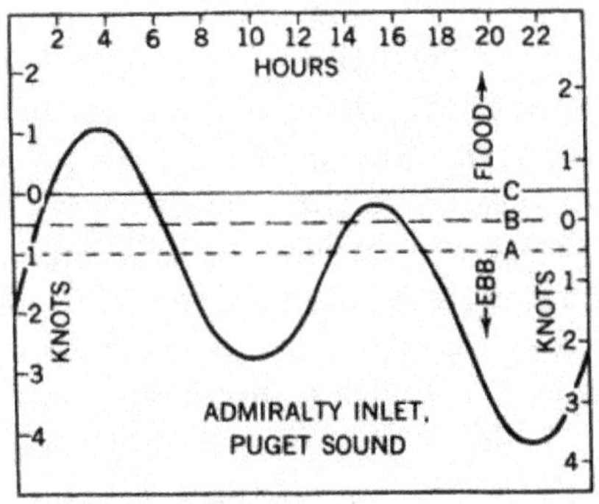

FIGURE 3120b.—Effect of nontidal current on the reversing tidal current of figure 3117b. If the nontidal current is 0.5 knot in the ebb direction, the ebb is increased by moving the slack water line from position *A* up 0.5 knot to position *B*. Speeds will then be measured from this broken line as shown by the scale on the right, and times of slack are changed. If the nontidal current is 1.0 knot in the ebb direction, as shown by line *C*, the speeds are as shown on the left, and the current will not reverse to a flood in the afternoon; it will merely slacken at about 1500.

Thus, the speed of the current flowing in the direction of the nontidal current is increased by an amount equal to the magnitude of the nontidal current, and the speed of the current flowing in the opposite direction is decreased by an equal amount. In figure 3120a, a nontidal current is represented both in direction and speed by the vector *AO*. Since this is greater than the speed of the tidal current in the opposite direction, the point *A* is outside the ellipse. The direction and speed of the combined tidal and nontidal currents at any time is represented by a vector from *A* to that point on the curve representing the given time, and can be scaled from the graph. The strongest and weakest currents may no longer be in the directions of the maximum and minimum of the tidal current. In a reversing current (fig. 3120b), the effect is to advance the time of one slack and to retard the following one. If the speed of the nontidal current exceeds that of the reversing tidal current, the resultant current flows continuously in one direction without coming to a slack. In this case, the speed varies from a maximum to a minimum and back to a maximum in each tidal cycle. In figure 3120b, the horizontal line *A* represents slack water if only tidal currents are present. Line *B* represents the effect of a 0.5-knot nontidal ebb, and line *C* the effect of a 1.0-knot nontidal ebb. With the condition shown at *C* there is only one flood each tidal day. If the nontidal ebb were to increase to approximately 2 knots, there would be no flood, two maximum ebbs and two minimum ebbs occurring during a tidal day.

3121. Relation between time of tidal current and time of tide.—At many places where current and tide are both semidiurnal, there is a definite relation between times of current and times of high and low water in the locality. Current atlases and notes on nautical charts often make use of this relationship by presenting for particular locations the direction and speed of the current at each succeeding hour after high and low water at a place for which tide predictions are available.

In localities where there is considerable diurnal inequality in tide or current, or where the type of current differs from the type of tide, the relationship is not constant, and it may be hazardous to try to predict the times of current from times of tide. Note the current curve for Unimak Pass in the Aleutians in figure 3118a. It shows the current as predicted in the tidal current tables. Predictions of high and low waters in the tide tables might have led one to expect the current to change from flood to ebb in the late morning, whereas actually the current continued to run flood with some strength at that time.

Since the relationship between times of tidal current and tide is not everywhere the same, and may be variable at the same place, one should exercise extreme caution in using general rules. The belief that slacks occur at local high and low tides and that the maximum flood and ebb occur when the tide is rising or falling most rapidly may be approximately true at the seaward entrance to, and in the upper reaches of, an inland tidal waterway. But generally this is not true in other parts of inland waterways. When an inland waterway is extensive or its entrance constricted, the slacks in some parts of the waterway often occur midway between the times of high and low tide. Usually in such waterways the relationship changes from place to place as one progresses upstream, slack water getting progressively closer in time to the local tide maximum until at the head of tidewater (the inland limit of water affected by a tide) the slacks occur at about the times of high and low tide.

3122. Relation between speed of current and range of tide.—The variation in the speed of the tidal current from place to place is not necessarily consistent with the range of tide. It may be the reverse. For example, currents are weak in the Gulf of Maine where the tides are large, and strong near Nantucket Island and in Nantucket Sound where the tides are small.

At any one place, however, the speed of the current at strength of flood and ebb varies during the month in about the same proportion as the range of tide, and one can use this relationship to determine the relative strength of currents on any day.

3123. Variation across an estuary.—In inland tidal waterways the *time* of tidal current varies across the channel from shore to shore. On the average, the current turns earlier near shore than in midstream, where the speed is greater. Differences of half an hour to an hour are not uncommon, but the difference varies and the relationship may be nullified by the effect of nontidal flow.

The *speed* of the current also varies across the channel, usually being greater in midstream or midchannel than near shore, but in a winding river or channel the strongest currents occur near the concave shore. Near the opposite (convex) shore the currents are weak or may eddy.

3124. Variation with depth.—In tidal rivers the subsurface current acting on the lower portion of the hull may differ considerably from the surface current. An appreciable subsurface current may be present when the surface movement appears to be practically slack, and the subsurface current may even be flowing with appreciable speed in the opposite direction to the surface current.

In a tidal estuary, particularly in the lower reaches where there is considerable difference in density from top to bottom, flood usually begins earlier near the bottom than at the surface. The differences may be an hour or two or as little as a few minutes, depending upon the estuary, the location in the estuary, and freshet conditions. Even when the freshwater runoff becomes so great as to prevent the surface current from flooding, it may still flood below the surface. The difference in time of ebb from surface to bottom is normally small but subject to variation with time and location.

The ebb speed at strength usually decreases gradually from top to bottom, but the speed of flood at strength often is stronger at subsurface depths than at the surface.

3125. Observations.—Observations of the current are made by means of a current meter or current pole and log line. In the past, most successful meters required a vessel and observers in continual attendance, as is necessary with the pole and line. Because of the difficulty and expense of such observations, they usually covered only a period of a day or two at a place. Observations of a month are the exception, and longer series were obtained only where ship and observers were available because of other duties, such as at lightships, where observations have been continued over a number of years.

Newer meters have been and are being developed that are suspended from a buoy and that record either in the buoy or send speed and direction impulses by radio to a base station on ship or land. With them, the period of observation has been increased so that in some recent surveys of United States harbors, the minimum period of observation was 1 week, with observations at several stations being continued over a period of 1 to 6 months.

3126. Tidal current tables and other sources of information.—The navigator should not attempt to predict currents without specific information for the locality in which he is interested. Such information is contained in various forms in many navigational publications.

Tidal current tables, issued annually, list daily predictions of the times and strengths of flood and ebb currents, and of the times of intervening slacks or minima. Due to lack of observational data, coverage is considerably more limited than for the tides. The tidal current tables do include supplemental data by which tidal current predictions can be determined for many places in addition to those for which daily predictions are given. The predictions are made by computers, using current harmonic constants that are obtained by analyzing current observations in the same manner as for tides (art. 3113). The use of tidal current tables is explained in articles 1207–1210.

Sailing directions and **coast pilots** issued by maritime nations include general descriptions of current behavior in various localities throughout the world.

Tidal current charts. A number of important harbors and waterways are covered by sets of tidal current charts showing graphically the hourly current movement.

Tidal Current Diagrams are a series of monthly diagrams used with the tidal current charts. The diagrams directly indicate the chart to use and the speed correction factor to apply to each chart.

The use of tables and charts for tide and current predictions is discussed in chapter XII.

References

Doodson, A. T., and H. D. Warburg. *Admiralty Manual of Tides.* London, H. M. Stationery Office, 1941.

Marmer, H. A. *The Tide.* New York, Appleton, 1926.

Schureman, Paul. *Manual of the Harmonic Analysis and Prediction of Tides.* Rev. ed. U.S. Coast and Geodetic Survey Special Publication No. 98, Washington, U.S. Govt. Print. Off., 1940.

Schureman, Paul. *Tide and Current Glossary.* National Ocean Survey. Washington, U.S. Govt. Print. Off., 1975.

U.S. Coast and Geodetic Survey. *Manual of Current Observations.* Special Publication No. 215. Rev. ed. Washington, U.S. Govt. Print. Off., 1950.

U.S. Coast and Geodetic Survey. *Manual of Tide Observations.* Publication 30–1. Washington, U.S. Govt. Print. Off., 1965.

National Ocean Survey. *Tidal Current Charts* and *Tidal Current Diagrams.* Washington, several published periodically.

National Ocean Survey. *Tide Tables* and *Tidal Current Tables.* Washington, U.S. Govt. Print. Off., several volumes of each published annually.

CHAPTER XXXII

OCEAN CURRENTS

3201. Introduction.—The movement of water comprising the oceans is one of the principal sources of discrepancy between dead reckoning and actual positions of vessels. Water in essentially horizontal motion is called a **current**, the direction *toward* which it moves being the **set**, and its speed the **drift**. A well-defined current extending over a considerable region of the ocean is called an **ocean current**.

A **periodic current** is one the speed or direction of which changes cyclically at somewhat regular intervals, as a tidal current. A **seasonal current** is one which has large changes in speed or direction due to seasonal winds. A **permanent current** is one which experiences relatively little periodic or seasonal change.

A **coastal current** flows roughly parallel to a coast, outside the surf zone, while a **longshore current** is one parallel to a shore, inside the surf zone, and generated by waves striking the beach at an angle. Any current some distance from the shore may be called an **offshore current**, and one close to the shore an **inshore current**.

A **surface current** is one present at the surface, particularly one that does not extend more than a relatively few feet below the surface. A **subsurface current** is one which is present below the surface only.

There is evidence to indicate that the strongest ocean currents consist of relatively narrow, high-speed streams that follow winding, shifting courses. Often associated with these currents are secondary **countercurrents** flowing adjacent to them but in the opposite direction, and somewhat local, roughly circular, **eddy currents**. A relatively narrow, deep, fast-moving current is sometimes called a **stream current**, and a broad, shallow, slow-moving one a **drift current**.

3202. Causes of ocean currents.—Although man's knowledge of the processes which produce and maintain ocean currents is far from complete, he does have a general understanding of the principal factors involved. The primary generating forces are wind and the density differences in the water. In addition, such factors as depth of water, underwater topography, shape of the basin in which the current is running, extent and location of land, and deflection by the rotation of the earth all affect the oceanic circulation.

3203. Wind currents.—The stress of wind blowing across the sea causes the surface layer of water to move. This motion is transmitted to each succeeding layer below the surface, but due to internal friction within the water, the rate of motion decreases with depth. The current is called **Ekman wind current** or simply **wind current**. Although there are many variables, it is generally true that a steady wind for about 12 hours is needed to establish such a current.

A wind-driven current does not flow in the direction of the wind, being deflected by Coriolis force (art. 3803), due to rotation of the earth. This deflection is toward the *right* in the Northern Hemisphere, and toward the *left* in the Southern Hemisphere. The Coriolis force is greater in higher latitudes, and is more effective in deep water. In general, the difference between wind direction and surface wind-current direction varies from about 15° along shallow coastal areas to a maximum of 45° in the deep oceans. As the motion is transmitted to successive deeper layers, the Coriolis force continues to deflect the current. At several hundred fathoms the current may flow in the opposite direction to the surface current. This shift of current directions with depth combined with the decrease in velocity with depth is called the **Ekman spiral.**

The speed of the current depends upon the speed of the wind, its constancy, the length of time it has blown, and other factors. In general, however, about 2 percent of the wind speed, or a little less, is a good average for deep water where the wind has been blowing steadily for at least 12 hours.

3204. Currents related to density differences.—As indicated in article 3009, the density of water varies with salinity, temperature, and pressure. At any given depth, the differences in density are due to differences in temperature and salinity. When suitable information is available, a map showing geographical density distribution at a certain depth could be drawn, with lines connecting points of equal density. These **isopycnic lines,** or lines connecting points at which a given density occurs at the same depth, would be similar to isobars on a weather map (art. 3826), and would serve an analogous purpose, showing areas of high density and those of low density. In an area of high density, the water surface is lower than in an area of low density, the maximum difference in height being of the order of 1 to 2 feet in 40 miles. Because of this difference, water tends to flow from an area of higher water (low density) to one of lower water (high density), but due to rotation of the earth, it is deflected toward the right in the Northern Hemisphere, and toward the left in the Southern Hemisphere. Thus, a circulation is set up similar to the cyclonic and anticyclonic circulation in the atmosphere. The greater the density gradient (rate of change with distance), the faster the related current.

3205. Oceanic circulation.—A number of ocean currents flow with great persistence, setting up a circulation that continues with relatively little change throughout the year. Because of the influence of wind in creating current (art. 3203), there is a relationship between this oceanic circulation and the general circulation of the atmosphere (art. 3803). The oceanic circulation is shown in figure 3205, with the names of the major ocean currents. Some differences in opinion exist regarding the names and limits of some of the currents, but those shown are representative. The spacing of the lines is a general indication of speed, but conditions vary somewhat with the season. This is particularly noticeable in the Indian Ocean and along the South China coast, where currents are influenced to a marked degree by the monsoons (art. 3809).

3206. Atlantic Ocean currents.—The trade winds (art. 3805), which blow with great persistence, set up a system of **equatorial currents** which at times extends over as much as 50° of latitude, or even more. There are two westerly flowing currents conforming generally with the areas of trade winds, separated by a weaker, easterly flowing countercurrent.

The **North Equatorial Current** originates to the northward of the Cape Verde Islands and flows almost due west at an average speed of about 0.7 knot.

The **South Equatorial Current** is more extensive. It starts off the west coast of Africa, south of the Gulf of Guinea, and flows in a generally westerly direction at an average speed of about 0.6 knot. However, the speed gradually increases until it may reach a value of 2.5 knots or more off the east coast of South America. As the current approaches Cabo de São Roque, the eastern extremity of South America, it divides, the southern part curving toward the south along the coast of Brazil, and the northern part being deflected by the continent of South America toward the north.

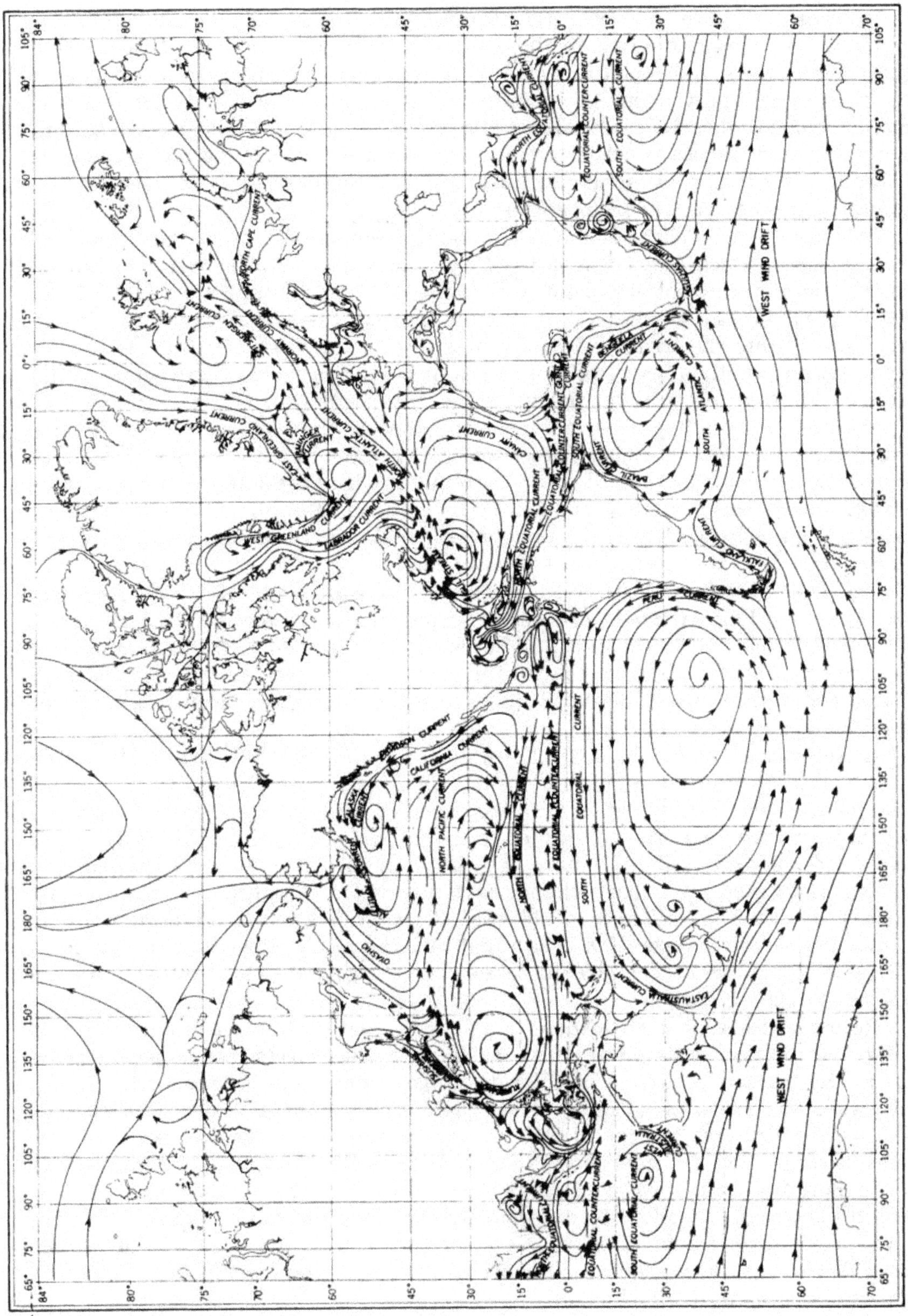

FIGURE 3205.—Major surface currents of the world (northern hemisphere winter).

Between the North and South Equatorial Currents a weaker **Equatorial Countercurrent** sets toward the east in the general vicinity of the doldrums (art. 3804). This is fed by water from the two westerly flowing equatorial currents, particularly the South Equatorial Current. The extent and strength of the Equatorial Countercurrent changes with the seasonal variations of the wind. It reaches a maximum during July and August, when it extends from about 50° west longitude to the Gulf of Guinea. During its minimum, in December and January, it is of very limited extent, the western portion disappearing altogether.

That part of the South Equatorial Current flowing along the northern coast of South America which does not feed the Equatorial Countercurrent unites with the North Equatorial Current at a point west of the Equatorial Countercurrent. A large part of the combined current flows through various passages between the Windward Islands, into the Caribbean Sea. It sets toward the west, and then somewhat north of west, finally arriving off the Yucatan peninsula. From there, some of the water curves toward the right, flowing some distance off the shore of the Gulf of Mexico, and part of it curves more sharply toward the east and flows directly toward the north coast of Cuba. These two parts reunite in the Straits of Florida to form the most remarkable of all ocean currents, the **Gulf Stream.** Off the southeast coast of Florida this current is augmented by a current flowing along the northern coasts of Puerto Rico, Hispaniola, and Cuba. Another current flowing eastward of the Bahamas joins the stream north of these islands.

The Gulf Stream follows generally along the east coast of North America, flowing around Florida, northward and then northeastward toward Cape Hatteras, and then curving toward the east and becoming broader and slower. After passing the Grand Banks, it turns more toward the north and becomes a broad drift current flowing across the North Atlantic. That part in the Straits of Florida is sometimes called the **Florida Current.**

A tremendous volume of water flows northward in the Gulf Stream. It can be distinguished by its deep indigo-blue color, which contrasts sharply with the dull green of the surrounding water. It is accompanied by frequent squalls. When the Gulf Stream encounters the cold water of the Labrador Current, principally in the vicinity of the Grand Banks, there is little mixing of the waters. Instead, the junction is marked by a sharp change in temperature. The line or surface along which this occurs is called the **cold wall.** When the warm Gulf Stream water encounters cold air, evaporation is so rapid that the rising vapor may be visible as frost smoke (art. 3814). The stream carries large quantities of gulfweed from the Tropics to higher latitudes.

Recent investigations have shown that the current itself is much narrower and faster than previously supposed, and considerably more variable in its position and speed. The maximum current off Florida ranges from about 2 to 4 knots. To the northward the speed is generally less, and decreases further after the current passes Cape Hatteras. As the stream meanders and shifts position, eddies sometimes break off and continue as separate, circular flows until they dissipate. Boats in the Bermuda Race have been known to be within sight of each other and be carried in opposite directions by different parts of the same current. As the current shifts position, its extent does not always coincide with the area of warm, blue water. When the sea is relatively smooth, the edges of the current are marked by ripples.

Information is not yet available to permit prediction of the position and speed of the current at any future time, but it has been found that tidal forces apparently influence the current, which reaches its daily maximum speed about 3 hours after transit of the moon. The current generally is faster at the time of neap tides than at spring tides. When the moon is over the equator, the stream is narrower and faster than at maximum northerly or southerly declination. Variations in the trade winds (art. 3805) also affect the current.

As the Gulf Stream continues eastward and northeastward beyond the Grand Banks, it gradually widens and decreases speed until it becomes a vast, slow-moving current known as the **North Atlantic Current,** in the general vicinity of the prevailing westerlies (art. 3807). In the eastern part of the Atlantic it divides into the **Northeast Drift Current** and the **Southeast Drift Current.**

The Northeast Drift Current continues in a generally northeasterly direction toward the Norwegian Sea. As it does so, it continues to widen and decrease speed. South of Iceland it branches to form the **Irminger Current** and the **Norway Current.** The Irminger Current curves toward the north and northwest to join the East Greenland Current southwest of Iceland. The Norway Current continues in a northeasterly direction along the coast of Norway. Part of it, the **North Cape Current,** rounds North Cape into the Barents Sea. The other part curves toward the north and becomes known as the **Spitsbergen Current.** Before reaching Svalbard (Spitsbergen), it curves toward the west and joins the cold **East Greenland Current** flowing southward in the Greenland Sea. As this current flows past Iceland, it is further augmented by the Irminger Current.

Off Kap Farvel, at the southern tip of Greenland, the East Greenland Current curves sharply to the northwest following the coastline. As it does so, it becomes known as the **West Greenland Current.** This current continues along the west coast of Greenland, through Davis Strait, and into Baffin Bay. Both East and West Greenland Currents are sometimes known by the single name **Greenland Current.**

In Baffin Bay the Greenland Current generally follows the coast, curving westward off Kap York to form the southerly flowing **Labrador Current.** This cold current flows southward off the coast of Baffin Island, through Davis Strait, along the coast of Labrador and Newfoundland, to the Grand Banks, carrying with it large quantities of ice (ch. XXXVI). Here it encounters the warm water of the Gulf Stream, creating the cold wall. Some of the cold water flows southward along the east coast of North America, inshore of the Gulf Stream, as far as Cape Hatteras. The remainder curves toward the east and flows along the northern edge of the North Atlantic and Northeast Drift Currents, gradually merging with them.

The Southeast Drift Current curves toward the east, southeast, and then south as it is deflected by the coast of Europe. It flows past the Bay of Biscay, toward southeastern Europe and the Canary Islands, where it continues as the **Canary Current.** In the vicinity of the Cape Verde Islands, this current divides, part of it curving toward the west to help form the North Equatorial Current, and part of it curving toward the east to follow the coast of Africa into the Gulf of Guinea, where it is known as the **Guinea Current.** This current is augmented by the Equatorial Countercurrent and, in summer, it is strengthened by monsoon winds. It flows in close proximity to the South Equatorial Current, but in the opposite direction. As it curves toward the south, still following the African coast, it merges with the South Equatorial Current.

The clockwise circulation of the North Atlantic leaves a large central area having no well-defined currents. This area is known as the **Sargasso Sea,** from the large quantities of sargasso or gulfweed encountered there.

That branch of the South Equatorial Current which curves toward the south off the east coast of South America follows the coast as the warm, highly-saline **Brazil Current,** which in some respects resembles the Gulf Stream. Off Uruguay, it encounters the colder, less-salty Falkland Current and the two curve toward the east to form the broad, slow-moving, **South Atlantic Current,** in the general vicinity of the prevailing westerlies (art. 3807). This current flows eastward to a point west of the Cape of Good Hope, where it curves northward to follow the west coast of Africa as the strong **Benguela Current,** augmented somewhat by part of the Agulhas Current flowing around the southern part of Africa from the Indian Ocean. As it continues northward, the current gradually widens and slows. At a point east of St. Helena Island it curves westward to continue as part of the South Equatorial Current, thus completing the counterclockwise circulation of the South Atlantic. The Benguela Current is augmented somewhat by the **West Wind Drift,** a current which flows easterly around Antarctica. As the West Wind Drift flows past Cape Horn, that part in the immediate vicinity of the cape is called the **Cape Horn Current.** This current rounds the cape and flows in a northerly and northeasterly direction along the coast of South America as the **Falkland Current.**

3207. Pacific Ocean currents follow the general pattern of those in the Atlantic. The **North Equatorial Current** flows westward in the general area of the northeast trades, and the **South Equatorial Current** follows a similar path in the region of the southeast trades. Between these two, the weaker **Equatorial Countercurrent** sets toward the east, just north of the equator.

After passing the Mariana Islands, the major part of the North Equatorial Current curves somewhat toward the northwest, past the Philippines and Taiwan. Here it is deflected further toward the north, where it becomes known as the **Kuroshio,** and then toward the northeast past the Nansei Shoto and Japan, and on in a more easterly direction. Part of the Kuroshio, called the **Tsushima Current,** flows through Tsushima Strait, between Japan and Korea, and the Sea of Japan, following generally the northwest coast of Japan. North of Japan it curves eastward and then southeastward to rejoin the main part of the Kuroshio. The limits and volume of the Kuroshio are influenced by the monsoons (art. 3809), being augmented during the season of southwesterly winds, and diminished when the northeasterly winds are prevalent.

The Kuroshio (Japanese for "Black Stream") is so named because of the dark color of its water. It is sometimes called the **Japan Current.** In many respects it is similar to the Gulf Stream of the Atlantic. Like that current, it carries large quantities of warm tropical water to higher latitudes, and then curves toward the east as a major part of the general clockwise circulation in the Northern Hemisphere. As it does so, it widens and slows. A small part of it curves to the right to form a weak clockwise circulation west of the Hawaiian Islands. The major portion continues on between the Aleutians and the Hawaiian Islands, where it becomes known as the **North Pacific Current.**

As this current approaches the North American continent, most of it is deflected toward the right to form a clockwise circulation between the west coast of North America and the Hawaiian Islands. This part of the current has become so broad that the circulation is generally weak. A small part near the coast, however, joins the southern branch of the Aleutian Current, and flows southeastward as the **California Current.** The average speed of this current is about 0.8 knot. It is strongest near land. Near the southern end of Baja (Lower) California, this current curves sharply to the west and broadens to form the major portion of the North Equatorial Current.

During the winter, a weak countercurrent flows northwestward along the west coast of North America from southern California to Vancouver Island, inshore of the southeasterly flowing California Current. This is called the **Davidson Current.**

Off the west coast of Mexico, south of Baja California, the current flows southeastward, as a continuation of part of the California Current, during the winter. During the summer, the current in this area is northwestward, as a continuation of the Equatorial Countercurrent, before it turns westward to help form the North Equatorial Current.

As in the Atlantic, there is in the Pacific a counterclockwise circulation to the north of the clockwise circulation. Cold water flowing southward through the western part of Bering Strait between Alaska and Siberia is joined by water circulating counterclockwise in the Bering Sea to form the **Oyashio.** As the current leaves the strait, it curves toward the right and flows southwesterly along the coast of Siberia and the Kuril Islands. This current brings quantities of sea ice, but no icebergs. When it encounters the Kuroshio, the Oyashio curves southward and then eastward, the greater portion joining the Kuroshio and North Pacific Current. The northern portion continues eastward to join the curving Aleutian Current.

As this current approaches the west coast of North America, west of Vancouver Island, part of it curves toward the right and is joined by water from the North Pacific Current, to form the California Current. The northern branch of the Aleutian Current curves in a counterclockwise direction to form the **Alaska Current,** which generally follows the coast of Canada and Alaska. When it arrives off the Aleutian Islands, it becomes known as the **Aleutian Current.** Part of it flows along the southern side of these islands to about the 180th meridian, where it curves in a counterclockwise direction and becomes an easterly flowing current, being augmented by the northern part of the Oyashio. The other part of the Aleutian Current flows through various openings between the Aleutian Islands, into the Bering Sea. Here it flows in a general counterclockwise direction, most of it finally joining the southerly flowing Oyashio, and a small part of it flowing northward through the eastern side of the Bering Strait, into the Arctic Ocean.

The South Equatorial Current, extending in width between about 4°N latitude and 10°S, flows westward from South America to the western Pacific. After this current crosses the 180th meridian, the major part curves in a counterclockwise direction, entering the Coral Sea, and then curving more sharply toward the south along the east coast of Australia, where it is known as the **East Australia Current.** In the Tasman Sea, northeast of Tasmania, it is augmented by water from the West Wind Drift, flowing eastward south of Australia. It curves toward the southeast and then the east, gradually merging with the easterly flowing West Wind Drift, a broad, slow-moving current that circles Antarctica.

Near the southern extremity of South America, most of this current flows eastward into the Atlantic, but part of it curves toward the left and flows generally northward along the west coast of South America as the **Peru Current** or **Humboldt Current.** Occasionally a set directly toward land is encountered. At about Cabo Blanco, where the coast falls away to the right, the current curves toward the left, past the Galapagos Islands, where it takes a westerly set and constitutes the major portion of the South Equatorial Current, thus completing the counterclockwise circulation of the South Pacific.

During the northern hemisphere summer, a weak northern branch of the South Equatorial Current, known as the **Rossel Current**, continues on toward the west and northwest along both the southern and northeastern coasts of New Guinea. The southern part flows through Torres Strait, between New Guinea and Australia, into the Arafura Sea. Here, it gradually loses its identity, part of it flowing on toward the west as part of the South Equatorial Current of the Indian Ocean, and part of it following the coast of Australia and finally joining the easterly flowing West Wind Drift. The northern part of the Rossel Current curves in a clockwise direction to help form the Pacific Equatorial Countercurrent. During the northern hemisphere winter, the Rossel Current is replaced by an easterly flowing current from the Indian Ocean.

3208. Indian Ocean currents follow generally the pattern of the Atlantic and Pacific but with differences caused principally by the monsoons (art. 3809) and the more limited extent of water in the Northern Hemisphere. During the northern hemisphere winter, the **North Equatorial Current** and **South Equatorial Current** flow toward the west, with the weaker, easterly flowing **Equatorial Countercurrent** flowing between them, as in the Atlantic and Pacific (but somewhat south of the equator). But during the northern hemisphere summer, both the North Equatorial Current and the Equatorial Countercurrent are replaced by the **Monsoon Current,** which flows eastward and southeastward across the Arabian Sea and the Bay of Bengal. Near Sumatra, this current curves in a clockwise direction and flows westward, augmenting the South Equatorial Current and setting up a clockwise circulation in the northern part of the Indian Ocean.

As the South Equatorial Current approaches the coast of Africa, it curves toward the southwest, part of it flowing through the Mozambique Channel between Madagascar and the mainland, and part flowing along the east coast of Madagascar. At the southern end of this island the two join to form the strong **Agulhas Current,** which is analogous to the Gulf Stream.

A small part of the Agulhas Current rounds the southern end of Africa and helps form the Benguela Current. The major portion, however, curves sharply southward and then eastward to join the West Wind Drift. This junction is often marked by a broken and confused sea. During the northern hemisphere winter the northern part of this current curves in a counterclockwise direction to form the **West Australia Current**, which flows northward along the west coast of Australia. As it passes Northwest Cape, it curves northwestward to help form the South Equatorial Current. During the northern hemisphere summer, the West Australia Current is replaced by a weak current flowing around the western part of Australia as an extension of the southern branch of the Rossel Current.

3209. Polar currents.—The waters of the North Atlantic enter the Arctic Ocean between Norway and Svalbard. The currents flow easterly north of Siberia to the region of the Novosibirskiye Ostrova, where they turn northerly across the North Pole and continue down the Greenland coast to form the **East Greenland Current.** On the American side of the arctic basin, there is a weak, continuous clockwise flow centered in the vicinity of 80°N, 150°W. A current north through Bering Strait along the American coast is balanced by an outward southerly flow along the Siberian coast, which eventually becomes part of the Oyashio. Each of the main islands or island groups in the Arctic, as far as is known, seems to have a clockwise nearshore circulation around it. The Barents Sea, Kara Sea, and Laptev Sea each have a weak counterclockwise circulation. A similar but weaker counterclockwise current system appears to exist in the East Siberian Sea.

In the Antarctic, the circulation is generally from west to east in a broad, slow-moving current extending completely around Antarctica. This is called the **West Wind Drift**, although it is formed partly by the strong westerly wind in this area and partly by density differences. This current is augmented by the Brazil and Falkland Currents in the Atlantic, the East Australia Current in the Pacific, and the Agulhas Current in the Indian Ocean. In return, part of it curves northward to form the Cape Horn, Falkland, and most of the Benguela Currents in the Atlantic, the Peru Current in the Pacific, and West Australia Current in the Indian Ocean.

3210. Ocean currents and climate.—Many of the ocean currents exert a marked influence upon the climate of the coastal regions along which they flow. Thus, warm water from the Gulf Stream, continuing as the North Atlantic, Northeast Drift, and Irminger Currents, arrives off the southwest coast of Iceland, warming it to the extent that Reykjavik has a higher average winter temperature than New York City, far to the south. Great Britain and Labrador are about the same latitude, but the climate of Great Britain is much milder because of the difference of temperature of currents. The west coast of the United States is cooled in the summer by the California Current, and warmed in the winter by the Davidson Current. As a result of this condition, partly, the range of monthly average temperature is comparatively small.

Currents exercise other influences besides those on temperature. The pressure pattern is affected materially, as air over a cold current contracts as it is cooled, and that over a warm current expands. As air cools above a cold ocean current, fog is likely to form. Frost smoke (art. 3814) is most prevalent over a warm current which flows into a colder region. Evaporation is greater from warm water than from cold water.

In these and other ways, the climate of the earth is closely associated with the ocean currents, although other factors, such as topography and prevailing winds, are also important.

References

Stream Drift Chart of the World—January. Defense Mapping Agency Hydrographic/Topographic Center Pilot Charts (various editions).

Stream Drift Chart of the World—July. Defense Mapping Agency Hydrographic/Topographic Center Pilot Charts (various editions).

Sverdrup, H. U., M. W. Johnson, and R. H. Fleming. *The Oceans, Their Physics, Chemistry and General Biology.* New York, Prentice-Hall, 1942.

CHAPTER XXXIII

OCEAN WAVES

3301. Introduction.—Undulations of the surface of the water, called **waves**, are perhaps the most widely observed phenomenon at sea, and possibly the least understood by the average seaman. The mariner equipped with a knowledge of the basic facts concerning waves is able to use them to his advantage, and either avoid hazardous conditions or operate with a minimum of danger if such conditions cannot be avoided.

3302. Causes of waves.—Waves on the surface of the sea are caused principally by wind, but other factors, such as submarine earthquakes, volcanic eruptions, and the tide, also cause waves. If a breeze of less than 2 knots starts to blow across smooth water, small wavelets called **ripples** form almost instantaneously. When the breeze dies, the ripples disappear as suddenly as they formed, the level surface being restored by surface tension of the water. If the wind speed exceeds 2 knots, more stable **gravity waves** gradually form, and progress with the wind.

While the generating wind blows, the resulting waves may be referred to as **sea.** When the wind stops or changes direction, the waves that continue on without relation to local winds are called **swell.**

Unlike wind and current, waves are not deflected appreciably by the rotation of the earth, but move in the direction in which the generating wind blows. When this wind ceases, friction and spreading cause the waves to be reduced in height, or **attenuated,** as they move across the surface. However, the reduction takes place so slowly that swell continues until it reaches some obstruction, such as a shore.

The Fleet Numerical Oceanography Center, Monterey, California, produces synoptic analyses and predictions of ocean wave heights using a spectral numerical model. The wave information consists of heights and directions for different periods and wavelengths. The model generates and propagates wave energy. Verification has been very good. Information from the model is provided to the U.S. Navy on a routine basis and is a vital input to the Optimum Track Ship Routing program (ch. XXIV).

3303. Wave characteristics.—Ocean waves are very nearly in the shape of an inverted **cycloid,** the figure formed by a point inside the rim of a wheel rolling along a level surface. This shape is shown in figure 3303a. The highest parts of waves are called **crests,** and the intervening lowest parts, **troughs.** Since the crests are steeper and narrower than the troughs, the mean or still water level is a little lower than halfway between the crests and troughs. The vertical distance between trough and crest is called **wave height,** labeled H in figure 3303a. The horizontal distance between successive crests, measured in the direction of travel, is called **wavelength,** labeled L. The time interval between passage of successive crests at a stationary point is called **wave period (P).** Wave height, length, and period depend upon a number of factors, such as the wind speed, the length of time it has blown, and its **fetch** (the straight distance it has traveled over the surface). Table 3303 indicates the relationship between wind speed, fetch, length of time the wind blows, wave height, and wave period in deep water.

Fetch	BEAUFORT NUMBER																											Fetch
	3			4			5			6			7			8			9			10			11			
	T	H	P	T	H	P	T	H	P	T	H	P	T	H	P	T	H	P	T	H	P	T	H	P	T	H	P	
10	4.4	1.8	2.1	3.7	2.6	2.4	3.2	3.5	2.8	2.7	5.0	3.1	2.5	6.0	3.4	2.3	7.3	3.9	2.0	8.0	4.1	1.9	10.0	4.2	1.8	10.0	5.0	10
20	7.1	2.0	2.5	6.2	3.2	2.9	5.4	4.9	3.3	4.7	7.0	3.8	4.2	8.6	4.3	3.9	10.0	4.4	3.5	12.0	5.0	3.2	14.0	5.2	3.0	16.0	5.9	20
30	9.8	2.0	2.8	8.3	3.8	3.3	7.2	5.8	3.7	6.2	8.0	4.2	5.8	10.0	4.6	5.2	12.1	5.0	4.7	15.8	5.5	4.4	18.0	6.0	4.1	19.8	6.3	30
40	12.0	2.0	3.0	10.3	3.9	3.6	8.9	6.2	4.1	7.8	9.0	4.6	7.1	11.2	4.9	6.5	14.0	5.4	5.8	17.7	5.9	5.4	21.0	6.3	5.1	22.5	6.7	40
50	14.0	2.0	3.2	12.4	4.0	3.8	11.0	6.5	4.4	9.1	9.8	4.8	8.4	12.2	5.2	7.7	15.7	5.6	6.9	19.8	6.3	6.4	23.0	6.7	6.1	25.0	7.1	50
60	16.0	2.0	3.5	14.0	4.0	4.0	12.0	6.8	4.6	10.2	10.3	5.1	9.6	13.2	5.5	8.7	17.0	6.0	8.0	21.0	6.5	7.4	25.0	7.0	7.0	27.5	7.5	60
70	18.0	2.0	3.7	15.8	4.0	4.1	13.5	7.0	4.8	11.9	10.8	5.4	10.5	13.9	5.7	9.9	18.0	6.4	9.0	22.5	6.8	8.3	26.5	7.3	7.8	29.5	7.7	70
80	20.0	2.0	3.8	17.0	4.0	4.2	15.0	7.2	4.9	13.0	11.0	5.6	12.0	14.5	6.0	11.0	18.9	6.6	10.0	24.0	7.1	9.3	28.0	7.7	8.6	31.5	7.9	80
90	23.6	2.0	3.9	18.8	4.0	4.3	16.5	7.3	5.1	14.1	11.2	5.8	13.0	15.0	6.3	12.0	20.0	6.7	11.0	25.0	7.2	10.2	30.0	7.9	9.5	34.0	8.2	90
100	27.1	2.0	4.0	20.0	4.0	4.4	17.5	7.3	5.3	15.1	11.4	6.0	14.0	15.5	6.5	12.8	20.5	6.9	11.9	26.5	7.6	11.0	32.0	8.1	10.3	35.0	8.5	100
120	31.1	2.0	4.2	22.4	4.1	4.7	20.0	7.8	5.4	17.0	11.7	6.2	15.9	16.0	6.7	14.5	21.5	7.3	13.1	27.5	7.9	12.3	33.5	8.4	11.5	37.5	8.8	120
140	36.6	2.0	4.5	25.8	4.2	4.9	22.5	7.9	5.8	19.1	11.9	6.4	17.6	16.2	7.0	16.0	22.0	7.6	14.8	29.0	8.3	13.9	35.5	8.8	13.0	40.0	9.2	140
160	43.2	2.0	4.9	28.4	4.2	5.2	24.3	7.9	6.0	21.1	12.0	6.6	19.5	16.5	7.3	18.0	23.0	8.0	16.4	30.5	8.7	15.1	37.0	9.1	14.5	42.5	9.6	160
180	50.0	2.0	4.9	30.9	4.3	5.4	27.0	8.0	6.2	23.1	12.1	6.8	21.3	17.0	7.5	19.9	23.5	8.3	18.0	31.5	9.0	16.5	38.5	9.5	16.0	44.5	10.0	180
200				33.5	4.3	5.6	29.0	8.0	6.4	25.4	12.2	7.1	23.1	17.5	7.7	21.5	23.5	8.5	19.3	32.5	9.2	18.1	40.0	9.8	17.1	46.0	10.3	200
220				36.5	4.4	5.8	31.1	8.0	6.6	27.2	12.3	7.2	25.0	17.9	8.0	22.9	24.0	8.8	20.9	34.0	9.6	19.1	41.5	10.1	18.2	47.5	10.6	220
240				39.2	4.4	5.9	33.1	8.0	6.8	29.0	12.4	7.3	26.8	17.9	8.2	24.4	24.5	9.0	22.0	34.5	9.8	20.5	43.0	10.3	19.5	49.0	10.8	240
260				41.9	4.4	6.0	34.9	8.0	6.9	30.5	12.6	7.5	28.0	18.0	8.4	26.0	25.0	9.2	23.5	34.5	10.0	21.8	44.0	10.6	20.9	50.5	11.1	260
280				44.5	4.4	6.2	36.8	8.0	7.0	32.4	12.9	7.8	29.5	18.0	8.5	27.7	25.0	9.4	25.0	35.0	10.2	23.0	45.0	10.9	22.0	51.5	11.3	280
300				47.0	4.4	6.3	38.5	8.0	7.1	34.1	13.1	8.0	31.5	18.0	8.7	29.0	25.0	9.5	26.3	35.0	10.4	24.3	45.0	11.1	23.2	53.0	11.6	300
320							40.5	8.0	7.2	36.0	13.3	8.2	33.0	18.0	8.9	30.2	25.0	9.6	27.6	35.5	10.6	25.5	45.5	11.2	24.5	54.0	11.8	320
340							42.4	8.0	7.3	37.6	13.4	8.3	34.2	18.0	9.0	31.6	25.0	9.8	29.0	36.0	10.8	26.7	46.0	11.4	25.5	55.0	12.0	340
360							44.2	8.0	7.4	38.8	13.4	8.4	35.7	18.1	9.1	33.0	25.0	9.9	30.0	36.5	10.9	27.7	46.5	11.6	26.6	55.0	12.2	360
380							46.1	8.0	7.5	40.2	13.5	8.5	37.1	18.2	9.3	34.2	25.5	10.0	31.3	37.0	11.1	29.1	47.0	11.8	27.7	55.5	12.4	380
400							48.0	8.0	7.7	42.2	13.5	8.6	38.8	18.4	9.5	35.6	26.0	10.2	32.5	37.0	11.2	30.2	47.5	12.0	28.9	56.0	12.6	400
420							50.0	8.0	7.8	43.5	13.6	8.7	40.0	18.7	9.6	36.9	26.5	10.3	33.7	37.5	11.4	31.5	47.5	12.2	29.6	56.5	12.7	420
440							52.0	8.0	7.9	44.7	13.7	8.8	41.3	18.8	9.7	38.1	27.0	10.4	34.8	37.5	11.5	32.5	48.0	12.3	30.9	57.0	12.9	440
460							54.0	8.0	8.0	46.2	13.7	8.9	42.8	19.0	9.8	39.5	27.5	10.6	36.0	37.5	11.7	33.5	48.5	12.5	31.8	57.5	13.1	460
480							56.0	8.0	8.1	47.8	13.7	9.0	44.0	19.0	9.9	41.0	27.5	10.8	37.0	37.5	11.8	34.5	49.0	12.6	32.7	57.5	13.2	480
500							58.0	8.0	8.2	49.2	13.8	9.1	45.5	19.1	10.1	42.1	27.5	10.9	38.3	38.0	11.9	35.5	49.0	12.7	33.9	58.0	13.4	500
550										53.0	13.8	9.3	48.5	19.5	10.3	44.9	27.5	11.1	41.0	38.5	12.2	38.2	50.0	13.0	36.5	59.0	13.7	550
600										56.3	13.8	9.5	51.8	19.7	10.5	47.7	27.5	11.3	43.6	39.0	12.5	40.3	50.0	13.3	38.7	60.0	14.0	600
650													55.0	19.8	10.7	50.3	27.5	11.6	46.4	39.5	12.8	43.0	50.0	13.7	41.0	60.0	14.2	650
700													58.5	19.8	11.0	53.2	27.5	11.8	49.0	40.0	13.1	45.4	50.5	14.0	43.5	60.5	14.5	700
750																56.2	27.5	12.1	51.0	40.0	13.3	48.0	51.0	14.2	45.8	61.0	14.8	750
800																59.2	27.5	12.3	53.8	40.0	13.5	50.6	51.5	14.5	47.8	61.5	15.0	800
850																			56.2	40.0	13.8	52.5	52.0	14.6	50.0	62.0	15.2	850
900																			58.2	40.0	14.0	54.6	52.0	14.9	52.0	62.5	15.5	900
950																						57.2	52.0	15.1	54.0	63.0	15.7	950
1000																						59.3	52.0	15.3	56.3	63.0	16.0	1000

TABLE 3303.—Minimum Time (T) in hours that wind must blow to form waves of H significant height (in feet) and P period (in seconds). Fetch in nautical miles.

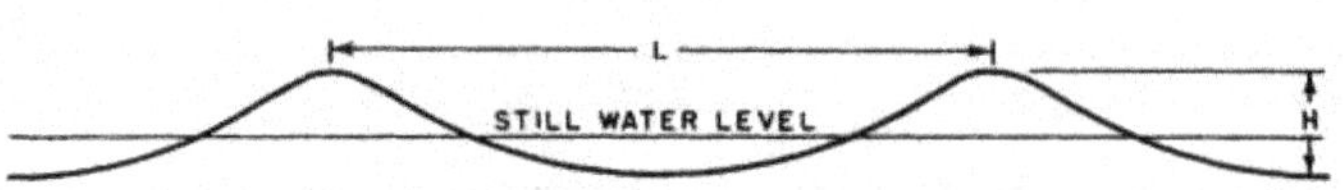

FIGURE 3303a.—A typical sea wave.

If the water is deeper than one-half the wavelength (L), this length in feet is theoretically related to period (P) in seconds by the formula

$$L=5.12P^2.$$

The actual value has been found to be a little less than this for swell, and about two-thirds the length determined by this formula for sea. When the waves leave the generating area and continue as free waves, the wavelength and period continue to increase, while the height decreases. The rate of change gradually decreases.

The speed (S) of a free wave in deep water is nearly independent of its height or steepness. For swell, its relationship in knots to the period (P) in seconds is given by the formula

$$S=3.03P.$$

The relationship for sea is not known.

The theoretical relationship between speed, wavelength, and period is shown in figure 3303b. As waves continue on beyond the generating area, the period, wavelength, and speed remains the same. Because the waves of each period have different speeds they tend to sort themselves by periods as they move away from the generating area. The longer period waves move at a greater speed and move ahead. At great enough distances from a storm area the waves will have sorted themselves into packets based on period.

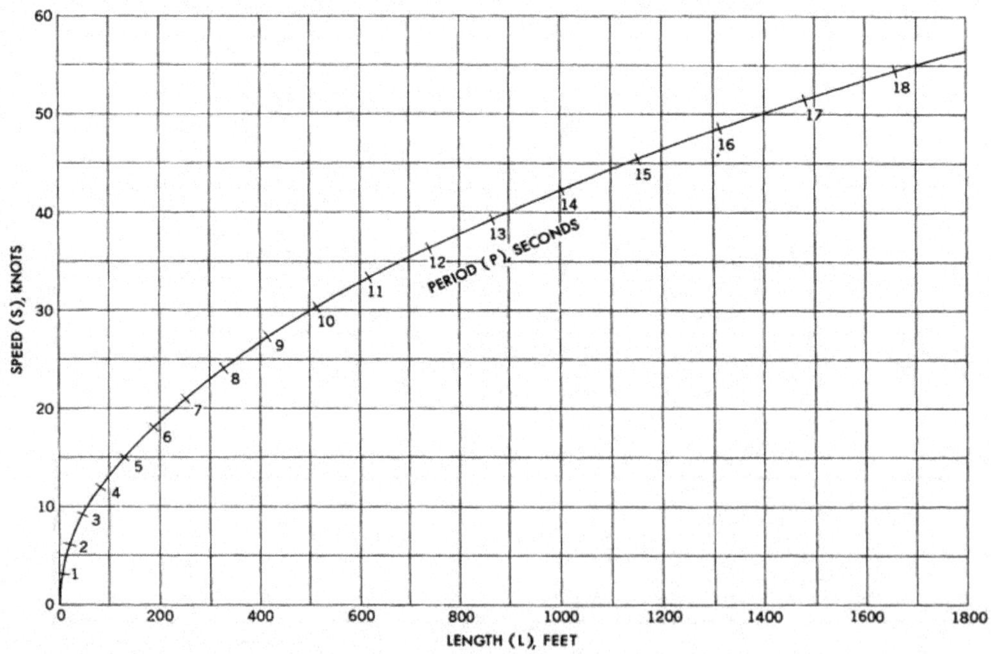

FIGURE 3303b.—Relationship between speed, length, and period of waves in deep water, based upon the theoretical relationship between period and length.

All the waves are attenuated as they propagate but the short period waves attenuate faster so that at a long distance from a storm only the longer waves remain.

The time needed for a wave system to travel some distance is *double* that which would be indicated by the speed of individual waves. This is because the front wave gradually disappears and transfers its energy to succeeding waves. The process is followed by each front wave in succession, at such a rate that the wave *system* advances at a speed which is just *half* that of *individual* waves. This process can be seen in the bow wave of a vessel. The speed at which the wave system advances is called **group velocity.**

Because of the existence of many independent wave systems at the same time, the sea surface acquires a complex and irregular pattern. Also, since the longer waves outrun the shorter ones, the resulting interference adds to the complexity of the pattern. The process of interference, illustrated in figure 3303c, is duplicated many times in the sea, being the principal reason that successive waves are not of the same height. The irregularity of the surface may be further accentuated by the presence of wave systems crossing at an angle to each other, producing peak-like rises.

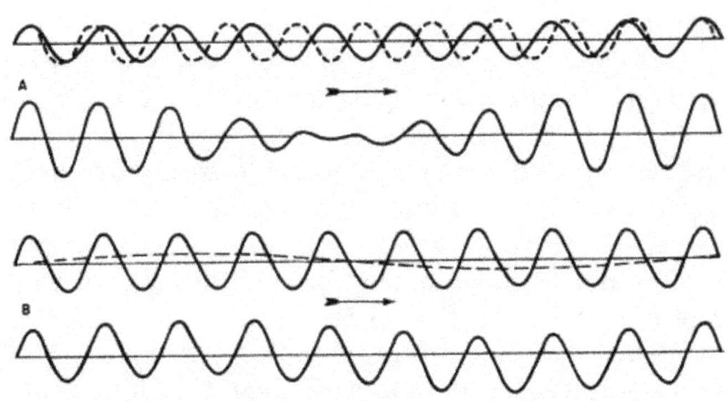

FIGURE 3303c.—Interference. The upper part of *A* shows two waves of equal height and nearly equal length traveling in the same direction. The lower part of *A* shows the resulting wave pattern. In *B* similar information is shown for short waves and long swell.

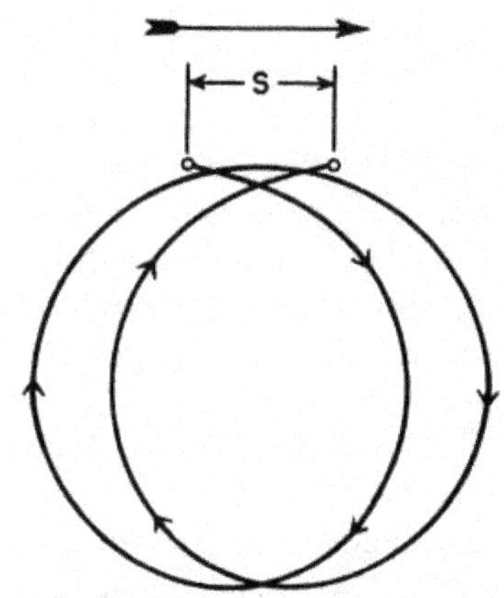

FIGURE 3304.—Orbital motion and displacement, *s*, of a particle on the surface of deep water during two wave periods.

In reporting average wave heights, the mariner has a tendency to neglect the lower ones. It has been found that the reported value is about the average for the highest one-third. This is sometimes called the "significant" wave height. The approximate relationship between this height and others, is as follows:

Wave	*Relative height*
Average	0.64
Significant	1.00
Highest 10 percent	1.29
Highest	1.87

3304. Path of water particles in a wave.—As shown in figure 3304, a particle of water on the surface of the ocean follows a somewhat circular orbit as a wave passes, but moves very little in the direction of motion of the wave. The common wave producing this action is called an **oscillatory wave.** As the crest passes, the particle moves forward, giving the water the appearance of moving with the wave. As the trough passes, the motion is in the opposite direction. The radius of the circular orbit decreases with depth, approaching zero at a depth equal to about half the wavelength. In shallower water the orbits become more elliptical, and in very shallow water, as at a beach, the vertical motion disappears almost completely.

Since the speed is greater at the top of the orbit than at the bottom, the particle is not at exactly its original point following passage of a wave, but has moved slightly in the direction of motion of the wave. However, since this advance is small in relation to the vertical displacement, a floating object is raised and lowered by passage of a wave, but moved little from its original position. If this were not so, a slow moving vessel might experience considerable difficulty in making way against a wave train. In figure 3304 the forward displacement is greatly exaggerated.

3305. Effects of currents on waves.—A following current increases wavelengths and decreases wave heights. An opposing current has the opposite effect, decreasing the length and increasing the height. A strong opposing current may cause the waves to break. The extent of wave alteration is dependent upon the ratio of the still-water wave speed to the speed of the current.

Moderate ocean currents running at oblique angles to wave directions appear to have little effect, but strong tidal currents perpendicular to a system of waves have been observed to completely destroy them in a short period of time.

3306. The effect of ice on waves.—When ice crystals form in seawater, internal friction is greatly increased. This results in smoothing of the sea surface. The effect of pack ice is even more pronounced. A vessel following a lead through such ice may be in smooth water even when a gale is blowing and heavy seas are beating against the outer edge of the pack. Hail is also effective in flattening the sea, even in a high wind.

3307. Waves and shallow water.—When a wave encounters shallow water, the movement of the individual particles of water is restricted by the bottom, resulting in reduced wave speed. In deep water wave speed is a function of period. In shallow water, the wave speed becomes a function of depth. The shallower the water the slower is the wave speed. As the wave speed slows, the period remains the same so the wavelength becomes shorter. Since the energy in the waves remains the same, the shortening of wavelengths results in increased heights. This process is called **shoaling.** If the wave approaches the shoal at an angle, each part is slowed successively as the depth decreases. This causes a change in direction of motion or **refraction,** the wave tending to become parallel to the depth curves. The effect is similar to the refraction of light and other forms of radiant energy (art. 1606).

As each wave slows, the next wave behind it, in deeper water, tends to catch up. As the wavelength decreases, the height generally becomes greater. The lower part of a wave, being nearest the bottom, is slowed more than the top. This may cause the wave to become unstable, the faster-moving top falling or **breaking.** Such a wave is called a **breaker,** and a series of breakers, **surf.** This subject is covered in greater detail in chapter XXXIV.

Swell passing over a shoal but not breaking undergoes a decrease in wavelength and speed, and an increase in height. Such **ground swell** may cause heavy rolling if it is on the beam and its period is the same as the period of roll of a vessel, even though the sea may appear relatively calm. Figure 3307 illustrates the approximate alteration of the characteristics of waves as they cross a shoal.

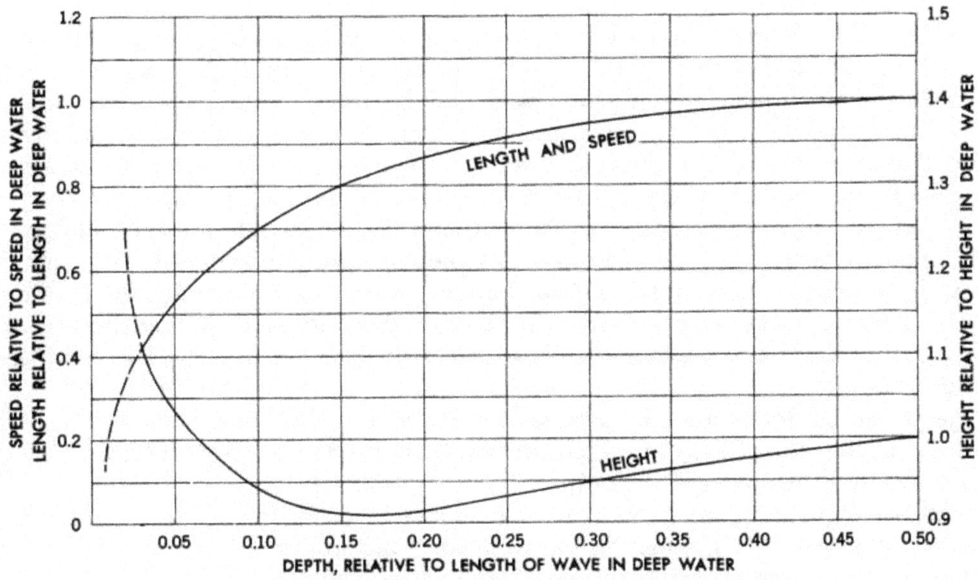

FIGURE 3307.—Alteration of the characteristics of waves as they cross a shoal.

3308. Energy of waves.—The potential energy of a wave is related to the vertical distance of each particle from its still-water position, and therefore moves with the wave. In contrast, the kinetic energy of a wave is related to the speed of the particles, being distributed evenly along the entire wave.

The amount of kinetic energy in even a moderate wave is tremendous. A 4-foot, 10-second wave striking a coast expends more than 35,000 horsepower per mile of beach. For each 56 miles of coast, the energy expended equals the power generated at Hoover Dam. An increase in temperature of the water in the relatively narrow **surf zone** in which this energy is expended would seem to be indicated, but no pronounced increase has been measured. Apparently, any heat that may be generated is dissipated to the deeper water beyond the surf zone.

3309. Wave measurement aboard ship.—With suitable equipment and adequate training, one can make reasonably reliable measurements of the height, length, period, and speed of waves. However, the mariner's estimates of height and length usually contain relatively large errors. There is a tendency to underestimate the heights of low waves, and overestimate the heights of high ones. There are numerous accounts of waves 75 to 80 feet high, or even higher, although waves more than 55 feet high are very rare. Wavelength is usually underestimated. The motions of the vessel from which measurements are made perhaps contribute to such errors.

Height. Measurement of wave height is particularly difficult. A microbarograph (art. 3705) can be used if the wave is long enough to permit the vessel to ride up and down with it. If the waves are approaching from dead ahead or dead astern, this requires a wavelength at least twice the length of the vessel. For most accurate results the instrument should be placed at the center of roll and pitch, to minimize the effects of these motions. Wave height can often be estimated with reasonable accuracy by comparing it with freeboard of the vessel. This is less accurate as wave height and vessel motion increase. If a point of observation can be found at which the top of a wave is in line with the horizon when the observer is in the trough, the wave height is equal to height of eye. However, if the vessel is rolling or pitching, this height at the moment of observation may be difficult to determine. The highest wave ever reliably reported was 112 feet observed from the USS *Ramapo* in 1933.

Length. The dimensions of the vessel can be used to determine wavelength. Errors are introduced by perspective and disturbance of the wave pattern by the vessel. These errors are minimized if observations are made from maximum height. Best results are obtained if the sea is from dead ahead or dead astern.

Period. If allowance is made for the motion of the vessel, wave period can be determined by measuring the interval between passages of wave crests past the observer. The correction for the motion of the vessel can be eliminated by timing the passage of successive wave crests past a patch of foam or a floating object at some distance from the vessel. Accuracy of results can be improved by averaging several observations.

Speed can be determined by timing the passage of the wave between measured points along the side of the ship, if corrections are applied for the direction of travel of the wave and the speed of the ship.

The length, period, and speed of waves are interrelated by the relationships indicated in article 3303. There is no definite mathematical relationship between wave height and length, period, or speed.

3310. Tsunamis are ocean waves produced by sudden, large-scale motion of a portion of the ocean floor or the shore, as by volcanic eruption, earthquake (sometimes called **seaquake** if it occurs at sea), or landslide. If they are caused by a submarine earthquake, they are usually called **seismic sea waves.** The point directly above the disturbance, at which the waves originate, is called the **epicenter.** Either a tsunami or a storm tide (art. 3311) that overflows the land is popularly called a **tidal wave,** although it bears no relation to the tide.

If a volcanic eruption occurs below the surface of the sea, the escaping gases cause a quantity of water to be pushed upward in the shape of a dome or mound. The same effect is caused by the sudden rising of a portion of the bottom. As this water settles back, it creates a wave which travels at high speed across the surface of the ocean.

Tsunamis are a series of waves. Near the epicenter, the first wave may be the highest. At greater distances, the highest wave usually occurs later in the series, commonly between the third and the eighth wave. Following the maximum, they again become smaller, but the tsunami may be detectable for several days.

In deep water the wave height of a tsunami is probably never greater than 2 or 3 feet. Since the wavelength is usually considerably more than 100 miles, the wave is not conspicuous at sea. In the Pacific, where most tsunamis occur, the wave period varies between about 15 and 60 *minutes,* and the speed in deep water is more than 400 knots. The approximate speed can be computed by the formula

$$S=0.6\sqrt{gd}=3.4\sqrt{d},$$

where S is the speed in knots, g is the acceleration due to gravity (32.2 feet per second per second), and d is the depth of water in feet. This formula is applicable to any wave in water having a depth of less than half the wavelength. For most ocean waves it applies only in shallow water, because of the relatively short wavelength.

When a tsunami enters shoal water, it undergoes the same changes as other waves. The formula indicates that speed is proportional to depth of water. Because of the great speed of a tsunami when it is in relatively deep water, the slowing is relatively much greater than that of an ordinary wave crested by wind. Therefore, the increase in height is also much greater. The size of the wave depends upon the nature and intensity of the disturbance. The height and destructiveness of the wave arriving at any place depend upon its distance from the epicenter, topography of the ocean floor, and the coastline. The angle at which the wave arrives, the shape of the coastline, and the topography along the coast and offshore all have their effect. The position of the shore is also a factor, as it may be sheltered by intervening land, or be in a position where waves have a tendency to converge, either because of refraction or reflection, or both.

Tsunamis 50 feet in height or higher have reached the shore, inflicting widespread damage. On April 1, 1946, seismic sea waves originating at an epicenter near the Aleutians spread over the entire Pacific. Scotch Cap Light on Unimak Island, 57 feet above sea level, was completely destroyed. Traveling at an average speed of 490 miles per hour, the waves reached the Hawaiian Islands in 4 hours and 34 minutes, where they arrived as waves 50 feet above the high water level, and flooded a strip of coast more than 1,000 feet wide at some places. They left a death toll of 173, and property damage of $25 million. Less destructive waves reached the shores of North and South America, and Australia, 6,700 miles from the epicenter.

After this disaster, a tsunami warning system was set up in the Pacific, even though destructive waves are relatively rare (averaging about one in 20 years in the Hawaiian Islands).

In addition to seismic sea waves, earthquakes below the surface of the sea may produce a longitudinal wave that travels upward toward the surface, at the speed of sound. When a ship encounters such a wave, it is felt as a sudden shock which may be of such severity that the crew thinks the vessel has struck bottom.

3311. Storm tides.—In relatively tideless seas like the Baltic and Mediterranean, winds cause the chief fluctuations in sea level. Elsewhere, the astronomical tide usually masks these variations. However, under exceptional conditions, either severe extra-tropical storms or tropical cyclones can produce changes in sea level that exceed the normal range of tide. Low sea level is of little concern except to shipping, but a rise above ordinary high-water mark, particularly when it is accompanied by high waves, can result in a catastrophe.

Although, like tsunamis, these **storm tides** or **storm surges** are popularly called **tidal waves**, they are not associated with the tide. They consist of a single wave crest and hence have no period or wavelength.

Three effects in a storm induce a rise in sea level. The first is wind stress on the sea surface, which results in a piling-up of water (sometimes called "wind set-up"). The second effect is the convergence of wind-driven currents, which elevates the sea surface along the convergence line. In shallow water, bottom friction and the effects of local topography cause this elevation to persist and may even intensify it. The low atmospheric pressure that accompanies severe storms causes the third effect, which is sometimes referred to as the "inverted barometer." An inch of mercury is equivalent to about 13.6 inches of water (art. 3115) and the adjustment of the sea surface to the reduced pressure can amount to several feet at equilibrium (art. 3912).

All three of these causes act independently, and if they happen to occur simultaneously, their effects are additive. In addition, the wave can be intensified or amplified by the effects of local topography. Storm tides may reach heights of 20 feet or more, and it is estimated that they cause three-fourths of the deaths attributed to hurricanes.

3312. Standing waves and seiches.—Previous articles in this chapter have dealt with **progressive waves** which appear to move regularly with time. When two systems of progressive waves having the same period travel in opposite directions across the same area, a series of **standing waves** may form. These appear to remain stationary.

Another type of standing wave, called a **seiche** (sāsh), sometimes occurs in a confined body of water. It is a long wave, usually having its crest at one end of the confined space, and its trough at the other. Its period may be anything from a few minutes to an hour or more, but somewhat less than the tidal period. Seiches are usually attributed to strong winds or differences in atmospheric pressure.

3313. Tide waves.—As indicated in chapter XXXI, there are, in general, two regions of high tide separated by two regions of low tide, and these regions move progressively westward around the earth as the moon revolves in its orbit. The high tides are the crests of these **tide waves,** and the low tides are the troughs. The wave is not noticeable at sea, but becomes apparent along the coasts, particularly in funnel-shaped estuaries. In certain river mouths or estuaries of particular configuration, the incoming wave of high water overtakes the preceding low tide, resulting in a high-crested, roaring wave which progresses upstream in one mighty surge called a **bore.**

3314. Internal waves.—Thus far, the discussion has been confined to waves on the surface of the sea, the boundary between air and water. **Internal waves,** or **boundary waves,** are created below the surface, at the boundaries between water strata of different densities. The density differences between adjacent water strata in the sea are considerably less than that between sea and air. Consequently, internal waves are much more easily formed than surface waves, and they are often much larger. The maximum height of wind waves on the surface is about 60 feet, but internal wave heights as great as 300 feet have been encountered.

Internal waves are detected by a number of observations of the vertical temperature distribution, using recording devices such as the bathythermograph (art. 3007). They have periods as short as a few minutes, and as long as 12 or 24 hours, these greater periods being associated with the tides.

A slow-moving ship operating in a freshwater layer having a depth approximating the draft of the vessel may produce short-period internal waves. This may occur off rivers emptying into the sea or in polar regions in the vicinity of melting ice. Under suitable conditions, the normal propulsion energy of the ship is expended in generating and maintaining these internal waves and the ship appears to "stick" in the water, becoming sluggish and making little headway. The phenomenon, known as **dead water,** disappears when speed is increased by a few knots.

The full significance of internal waves has not been determined, but it is known that they may cause submarines to rise and fall like a ship at the surface, and they may also affect sound transmission in the sea.

3315. Waves and ships.—The effects of waves on a ship vary considerably with the type of ship, its course and speed, and the condition of the sea. A short vessel has a tendency to ride up one side of a wave and down the other side, while a larger vessel may tend to ride *through* the waves on an even keel. If the waves are of such length that the bow and stern of a vessel are alternately in successive crests and successive troughs, the vessel is subject to heavy sagging and hogging stresses, and under extreme conditions may break in two. A change of heading may reduce the danger. Because of the danger from sagging and hogging, a small vessel is sometimes better able to ride out a storm than a large one.

If successive waves strike the side of a vessel at the same phase of successive rolls, relatively small waves can cause heavy rolling. The effect is similar to that of swinging a child, where the strength of the push is not as important as its timing. The same effect, if applied to the bow or stern in time with the pitch, can cause heavy pitching. A change of either heading or speed can reduce the effect.

A wave having a length twice that of a ship places that ship in danger of falling off into the trough of the sea, particularly if it is a slow-moving vessel. The effect is especially pronounced if the sea is broad on the bow or broad on the quarter. An increase of speed reduces the hazard.

3316. Use of oil for modifying the effects of breaking waves.—Oil has proved effective in modifying the effects of breaking waves, and has proved useful to vessels at sea, whether making way or stopped, particularly when lowering or hoisting boats. Its effect is greatest in deep water, where a small quantity suffices if the oil can be made to spread to windward. In shallow water where the water is in motion over the bottom, oil is less effective but of some value.

The heaviest oils, notably animal and vegetable oils, are the most effective. Crude petroleum is useful, but its effectiveness can be improved by mixing it with animal and vegetable oils. Gasoline or kerosene are of little value. Oil spreads slowly. In cold weather it may need some thinning with petroleum to hasten the process and produce the desired spread before the vessel is too far away for the effect to be useful.

At sea, best results can be expected if the vessel drifts or runs slowly before the wind, with the oil being discharged on both sides from waste pipes or by other convenient methods. If a sea anchor is used, oil can be distributed from a container inserted within it for this purpose. If such a container is not available, an oil bag can be fastened to an endless line rove through a block on the sea anchor. This permits distribution of oil to windward, and provides a means for hauling the bag aboard for refilling. If another vessel is being towed, the oil should be distributed from the towing vessel, forward and on both sides, so that both vessels will be benefited. If a drifting vessel is to be approached, the oil might be distributed from both sides of the drifting vessel or by the approaching vessel, which should distribute it to leeward of the drifting vessel so that the vessel will drift into it. If the vessel being approached is aground, the procedure best suiting the circumstances should be used.

If oil is needed in crossing a bar to enter a harbor, it can be floated in ahead of the vessel if a flood current is running. A considerable amount may be needed. During slack water a hose might be trailed over the bow and oil poured freely through it if no more convenient method is available. With an ebb current oil is of little use, unless it can be distributed from another vessel or in some other manner from the opposite side of the bar.

CHAPTER XXXIV

BREAKERS AND SURF

3401. Introduction.—The purpose of this chapter is to acquaint the navigator with the oceanographic factors affecting the safe navigation through the surf zone to the beach.

3402. Refraction.—As explained in article 3307, wave speed is slowed in shallow water, causing **refraction** if the waves approach the beach at an angle. Along a perfectly straight beach, with uniform shoaling, the wave fronts tend to become parallel to the shore. Any irregularities in the coastline or bottom contours, however, affect the refraction, causing irregularity. In the case of a ridge perpendicular to the beach, for instance, the shoaling is more rapid, causing greater refraction towards the ridge. The waves tend to align themselves with the bottom contours. Waves on both sides of the ridge have a component of motion toward the ridge. This **convergence** of wave energy toward the ridge causes an increase in wave or breaker height. A submarine canyon or valley perpendicular to the beach, on the other hand, produces **divergence**, with a decrease in wave or breaker height. These effects are illustrated in figure 3402. Bends in the coastline have a similar effect, convergence occurring at a *point,* and divergence if the coast is *concave* to the sea. Points act as focal areas for wave energy and experience large breakers. Concave bays have small breakers because the energy is spread out as the waves approach the beach.

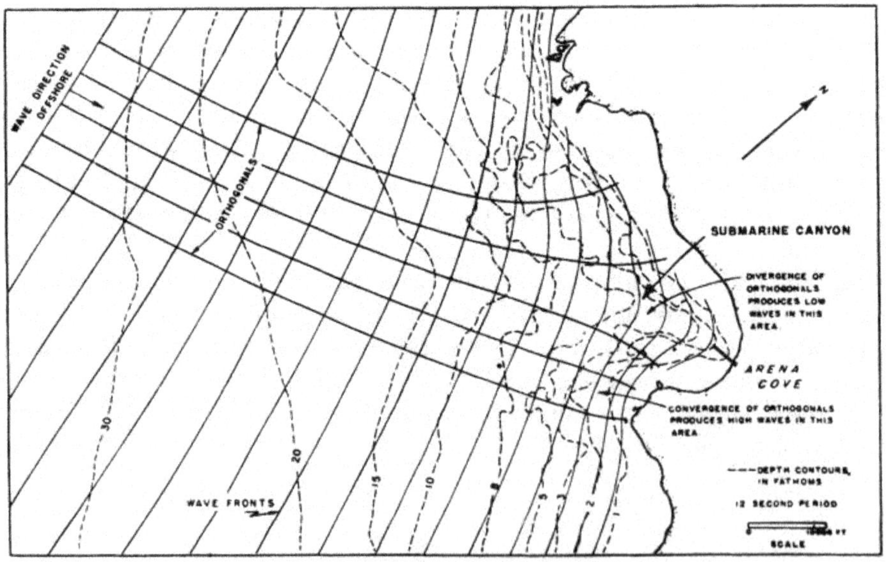

Courtesy of Robert L. Wiegel, Council on Wave Research, University of California.

FIGURE 3402.—The effect of bottom topography in causing wave convergence and wave divergence.

Under suitable conditions, currents also cause refraction. This is of particular importance at entrances of tidal estuaries. When waves encounter a current running in the opposite direction, they become higher and shorter. This results in a choppy sea, often with breakers. When waves move in the same direction as current, they decrease in height, and become longer. Refraction occurs when waves encounter a current at an angle.

Refraction diagrams, useful in planning amphibious operations, can be prepared with the aid of nautical charts or aerial photographs. When computer facilities are available, complex computer programs are used to determine refraction diagrams, quickly and accurately.

3403. Breakers and surf.—In deep water, swell generally moves across the surface as somewhat regular, smooth undulations (ch. XXXIII). When shoal water is reached, the wave period remains the same, but the speed decreases. The amount of decrease is negligible until the depth of water becomes about one-half the wavelength, when the waves begin to "feel" bottom. There is a slight decrease in wave height, followed by a rapid increase, if the waves are traveling perpendicular to a straight coast with a uniformly sloping bottom. As the waves become higher and shorter, they also become steeper, and the crest becomes narrower. When the speed of individual particles at the crest becomes greater than that of the wave, the front face of the wave becomes steeper than the rear face. This process continues at an accelerating rate as the depth of water decreases. At some point the wave may become unstable, toppling forward to form a **breaker.**

There are three general classes of breakers. A **spilling breaker** breaks gradually over a considerable distance. A **plunging breaker** tends to curl over and break with a single crash. A **surging breaker** peaks up, but surges up the beach without spilling or plunging. It is classed as a breaker even though it does not actually break. The type of breaker is determined by the steepness of the beach and the steepness of the wave before it reaches shallow water, as illustrated in figure 3403.

Longer waves break in deeper water, and have a greater breaker height. The effect of a steeper beach is also to increase breaker height. The height of breakers is less if the waves approach the beach at an acute angle. With a steeper beach slope there is greater tendency of the breakers to plunge or surge. Following the **uprush** of water onto a beach after the breaking of a wave, the seaward **backrush** occurs. The returning water is called **backwash.** It tends to further slow the bottom of a wave, thus increasing its tendency to break. This effect is greater as either the speed or depth of the backwash increases. The still water depth at the point of breaking is approximately 1.3 times the average breaker height.

Surf varies with both position along the beach and time. A change in position often means a change in bottom contour, with the refraction effects discussed in article 3402. At the same point, the height and period of waves vary considerably from wave to wave. A group of high waves is usually followed by several lower ones. Therefore, passage through surf can usually be made most easily immediately following a series of higher waves.

Since surf conditions are directly related to height of the waves approaching a beach, and the configuration of the bottom, the state of the surf at any time can be predicted if one has the necessary information and knowledge of the principles involved. Height of the sea and swell can be predicted from wind data, and information on bottom configuration can generally be obtained from the nautical chart. In addition, the area of lightest surf along a beach can be predicted if details of the bottom configuration are available.

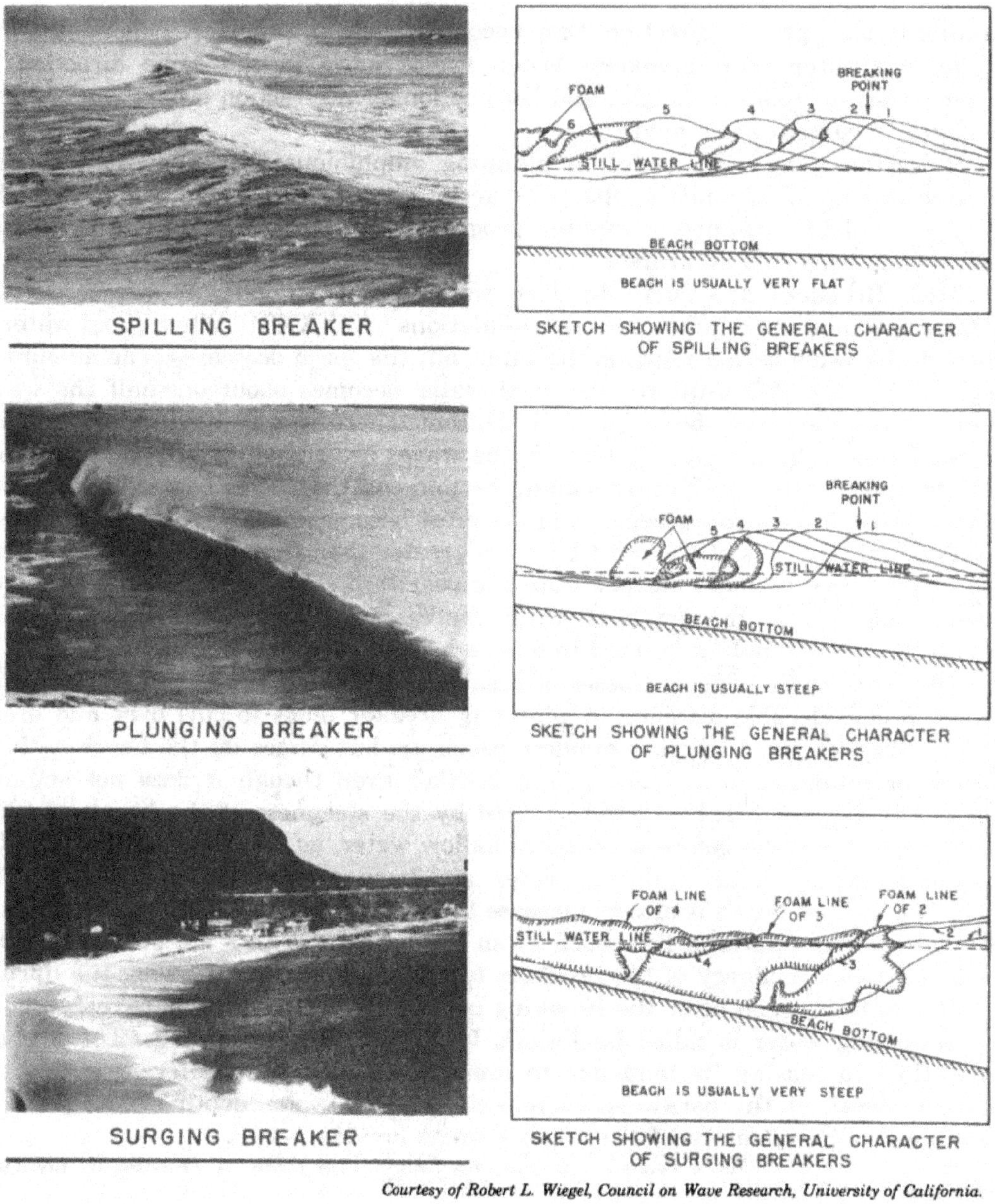

Courtesy of Robert L. Wiegel, Council on Wave Research, University of California.

FIGURE 3403.—The three types of breakers.

3404. Currents in the surf zone.—In and adjacent to the surf zone, currents are generated by waves approaching the bottom contours at an angle, and by irregularities in the bottom.

Waves approaching at an angle produce a **longshore current** parallel to the beach, within the surf zone. Longshore currents are most common along straight beaches. Their speeds increase with increasing breaker height, decreasing wave period, increasing angle of breaker line with the beach, and increasing beach slope. Speed seldom exceeds 1 knot, but sustained speeds as high as 3 knots have been recorded. Longshore currents are usually constant in direction. They increase the danger of landing craft broaching to.

As explained in article 3402, wave fronts advancing over nonparallel bottom contours are refracted to cause convergence or divergence of the energy of the waves. Energy concentrations, in areas of convergence, form barriers to the returning backwash, which is deflected *along* the beach to areas of less resistance. Backwash accumulates at weak points, and returns seaward in concentrations, forming **rip currents** through the surf. At these points the large volume of returning water has a retarding effect upon the incoming waves, thus adding to the condition causing the rip current. The waves on one or both sides of the rip, having greater energy and not being retarded by the concentration of backwash, advance faster and farther up the beach. From here, they move *along* the beach as **feeder currents.** At some point of low resistance, the water flows seaward through the surf, forming the **neck** of the rip current. Outside the breaker line the current widens and slackens, forming the **head.** The various parts of a rip current are shown in figure 3404.

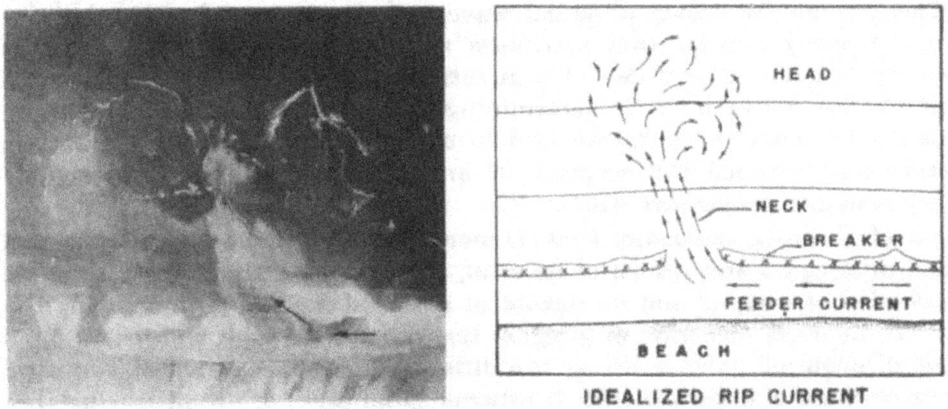

Courtesy of Robert L. Wiegel, Council on Wave Research, University of California.

FIGURE 3404.—A rip current (left) and a diagram of its parts (right).

Rip currents may also be caused by irregularities in the beach face. If a beach indentation causes an uprush to advance farther than the average, the backrush is delayed and this in turn retards the next incoming **foam line** (the front of a wave as it advances shoreward after breaking) at that point. The foam line on each side of the retarded point continues in its advance, however, and tends to fill in the retarded area, producing a rip current.

3405. Beach sediments.—In the surf zone, large amounts of sediment are suspended in the water. When the water motion decreases, the sediments are deposited as sand. The water motion can be either waves or currents. Promontories or points are rocky because the large breakers scour the points and small sediments are suspended in the water and carried away. Bays have sandy beaches because of the small wave conditions.

In the winter when storms create larger breakers and surf, the waves erode the beaches and carry the particles offshore where offshore sand bars form; sandy beaches tend to be narrower. In the summer the waves gradually move the sand back to the beaches and the offshore sand bars decrease; sandy beaches tend to be wider.

Longshore currents move large amounts of sand along the coast. These currents deposit sand on the upcurrent side of a jetty and erode the beach on the downcurrent side.

CHAPTER XXXV

SOUND IN THE SEA

3501. Underwater sound and the navigator.—The clarity with which the noises associated with weighing anchor, propelling a ship, and other underwater motions are heard below the waterline and near the skin of a vessel is an indication of the high sound-transmitting qualities of seawater. Water is a better conductor of sound than is air because it absorbs less energy from the sound. There are several ways in which underwater sound can be used in navigation.

The *direction* of travel of sound waves can be measured either by means of **binaural hearing** (hearing with two "ears"), or by equipment which has directional characteristics similar to those of a directional antenna used in radio (art. 4103). Either method can be used for determining the direction from which general noise is coming, but only the latter is used in active sonar equipment for determining direction and distance by reception of an echo from a directional signal, in a manner similar to radar (art. 4201).

Distance can be determined by (1) measuring the elapsed time between transmission of a signal and return of its echo, (2) measuring the elapsed time between transmission of a signal and its receipt at a second station, (3) measuring the time *difference* between reception of a signal transmitted through water and one transmitted through air, (4) measuring the difference in phase between two signals or change of phase of a signal when it returns as an echo, or (5) measuring the angle at which an echo is received from a signal produced at another place. The first method is used in active sonar and echo sounding equipment (art. 619). The fourth and fifth methods were used in early forms of echo sounders.

3502. Sources of sound in the ocean.—Underwater sounds intended for navigational use are produced in one of three basic ways: (1) by percussion, as the striking of a bell, gong, or the bottom of the vessel; (2) by oscillator, as the vibration of a diaphragm; (3) by explosion, as by small bomb or depth charge. Certain man-made noises ordinarily produced in water, such as those due to operation of the main engines of a vessel, can be detected by an appropriate listening device.

In addition, many noises are made by animals living in the ocean. Certain shrimp, great numbers of which inhabit some areas, make a snapping noise with their claws. Some fish make a noise by stridulating (scraping). When shellfish are being eaten, a sound is emitted as the shells are broken by the teeth of the fish which are feeding. Grunting noises are made by many kinds of fish, usually by means of their swim bladders. Porpoises produce sounds of a high pitch. Sounds of various frequency and amplitude are produced by other forms of marine life. Where sound-producing marine life is very abundant, it interferes with detection of man-made sounds, requiring a high signal-to-noise ratio. The effect is similar to that of a high atmospheric noise level in radio.

3503. Speed of sound in seawater.—Three variables govern the speed (S) of sound in a fluid. They are density (ρ), compressibility (β), and the ratio between the specific heats of the fluid at constant pressure and at constant volume (γ). The following formula is sufficiently accurate for most navigational purposes:

$$S=\sqrt{\frac{\gamma}{\rho\beta}}.$$

Density and specific heat are discussed in articles 3009 and 3012, respectively. Compressibility refers to the relative change in volume for a given change in pressure. The compressibility of water is low, and consequently the speed of sound in water is high. The specific heat ratio enters the formula because the energy of a sound impulse is briefly transformed into heat, and then reconverted (with slight loss) into kinetic energy. The ratio rarely exceeds 1.02 in seawater and is commonly taken as unity.

For atmospheric pressure 29.92 inches of mercury, temperature 60°F, and salinity 34.85 parts per thousand, the density of seawater is 64 pounds per cubic foot and the compressibility approximately 0.0000435 per atmosphere (1 atmosphere equals 14.696 pounds per square inch). Using these values and 32.174 feet per second per second (the acceleration of gravity at latitude 45°) and 144 square inches per square foot, and taking γ equal to unity, one obtains:

$$S=\sqrt{\frac{1.0\times 32.174\times 14.696\times 144}{64\times 0.0000435}}=4945 \text{ ft./sec.}$$

The same formula can be used to determine the speed of sound in air. For atmospheric pressure 29.92 and temperature 60°F, the density of air is 0.0764 pound per cubic foot and, since air is a gas, the compressibility is the reciprocal of the pressure. Taking γ equal to 1.4, one obtains:

$$S=\sqrt{\frac{1.4\times 32.174\times 14.696\times 144}{0.0764\times 1}}=1117 \text{ ft./sec.}$$

The speed of sound in water is approximately 4.5 times its speed in air.

An increase in temperature decreases both density and compressibility, resulting in an increase in the speed of sound. In seawater, an increase in pressure or salinity produces a slight increase in density and a larger decrease in compressibility, resulting in a net increase in the speed of sound. Thus, in seawater, an increase in temperature, pressure, or salinity results in greater speed of sound. Of the three, temperature has the greatest influence on the speed of sound in seawater in the upper layers. At depth, pressure, and in coastal areas, changes in salinity, may have the greatest effect.

Normally, the change of these three elements is much more rapid in a vertical direction than in a horizontal direction. The change with depth varies with location. With respect to temperature, much of the ocean is considered to consist of three layers, a **mixed layer** influenced greatly by the temperature of the air above it, a **thermocline** of rapidly decreasing temperature, and a nearly uniform **deep-water layer.** Typical curves showing change of temperature and salinity with depth are shown in figure 3503a. The increase of pressure with depth is almost uniform, the pressure at 10,000 feet being approximately twice that at 5,000 feet, and 10 times that at 1,000 feet. A typical curve of speed of sound with depth is shown in figure 3503b. In this case there is little or no mixed layer and the temperature decreases rapidly from the surface; therefore, the sound velocity also decreases rapidly. Below the range of temperature decrease, the pressure effect becomes the primary factor and sound velocity starts to increase. Note that the minimum sound velocity is at 2,400 feet. This would be the depth of the deep sound channel.

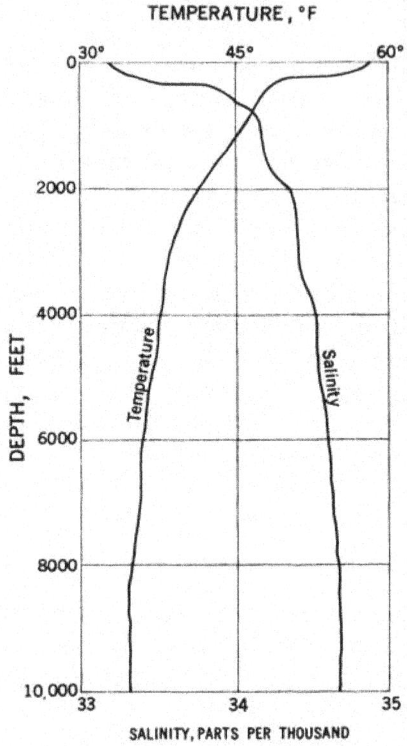

FIGURE 3503a.—Variation of temperature and salinity with depth at one locality.

Study of transmission of sound from underwater explosions indicates that near the explosion the speed of sound may be somewhat higher than expected, probably due to increased pressure caused by the disturbance. This effect extends over such a short distance that it is insignificant in ordinary underwater sound transmission.

3504. Reflection of underwater sound waves.—In water, as in air, sound is reflected by obstructions in the form of solid objects or sharp discontinuities. Thus, sound is reflected from the bottom, the shore, hulls of ships, the surface of the water, etc. It is this reflecting energy that is used in echo sounders (art. 619) to determine depth, and in sonar equipment used for echo ranging.

Reflecting properties of various substances differ markedly. Rock reflects almost all of the sound that strikes its surface, while soft mud absorbs or is penetrated by sound. Thus, in echo sounding, a layer of soft mud over rock may result in two echoes, indicating two depths.

Fish and even tiny sea animals also reflect sound. As a result, echo sounders are widely used among fishermen to locate schools of fish. In deep water it is not unusual for an echo sounder to receive an echo from a depth of about 200 fathoms, although the depth is shallower somewhat at night. This **phantom bottom** or **deep scattering layer**, which is undoubtedly the source of many erroneous shoal sounding reports, is due to large numbers of tiny marine animals, or other marine life.

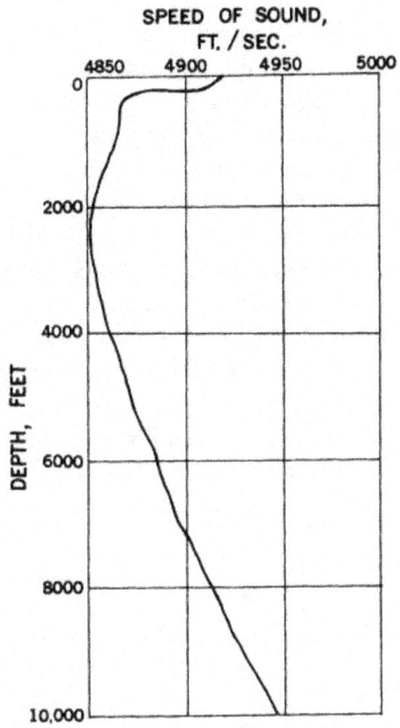

FIGURE 3503b.—Typical variation of speed of sound with depth in the ocean.

A sharp discontinuity within the water causes reflection of sound. Thus, an echo sounder may detect the boundary between a layer of freshwater overlying saltwater, a condition which might occur near the mouth of a river.

Sharp, distinct echoes denoting precise depths are difficult to obtain over roughsurfaced bottoms. Therefore, considerable discretion should be exercised in evaluating soundings taken over bottoms possessing a high degree of relief.

3505. Refraction of underwater sound waves.—The laws of refraction as applied to light (art. 1609) and radio waves (art. 4006) apply also to sound. Because of differences of velocity of sound in seawater, an advancing sound wave is refracted toward the area of slower sound velocity. If sound is traveling vertically downward, as in echo sounding, the effect of refraction is relatively slight because the layers of water in which velocity differs are approximately horizontal, and when the direction of travel of the sound is normal to the refracting surface or layer, there is no refraction.

If a sound beam is transmitted outward from a source, it will start at a particular sound velocity but the sound velocity will either increase or decrease as the beam moves into water of different temperatures, salinity, or pressure. The beam will refract toward the region where the sound velocity is slower. In a mixed surface layer the temperature is isothermal so the sound velocity increases with depth due to the pressure effect. A sound beam in that layer would be refracted upward to the surface where it would reflect off the surface. Sound beams can be trapped in the mixed layer and create a "surface sound duct." If a beam penetrates below the mixed layer into the thermocline it is in a region where sound velocity decreases rapidly with depth due to the temperature decrease. In the thermocline sound beams are refracted sharply downward. Sonar ranges can be very short in the thermocline layer. This is the region of the "shadow zone."

With typical distribution of sound velocity with depth, as shown in figure 3503b, sound velocity decreases with depth until a minimum is reached at some level below the surface, and below this it increases. In figure 3503b minimum velocity occurs at about 2,400 feet. In the Tropics this level of minimum velocity may be as deep as 6,000 feet, and in polar regions it may be at the surface. This level is referred to as the **deep sound channel.** Sound produced at any level tends to be refracted to the level of minimum speed, and to remain there, for as it attempts to leave this level, it is refracted back toward it, as shown in figure 3505. This, of course, does not refer to sound traveling vertically. If a sound is produced at this level, as by the explosion of a bomb or depth charge, the sound waves start to move outward as expanding spheres, but most of the rays are refracted back toward the minimum speed level. Because of this effect, such a sound may travel great distances with relatively little decrease in intensity. Listening gear placed at this level has detected sounds produced thousands of miles away.

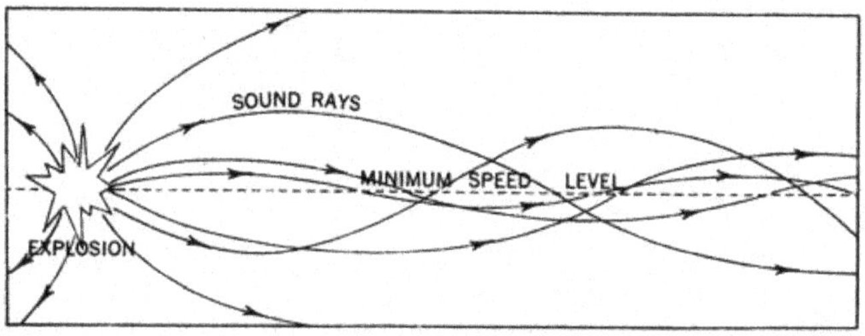

FIGURE 3505.—Transmission of sound rays along the minimum sound level.

Sound beams that penetrate the deep sound channel without being trapped continue on to regions of increased sound velocity. If the water is deep enough these beams will be refracted upwards towards the surface. If this occurs, the energy converges near the surface at ranges of about 30 nautical miles. This is the **convergence zone.** Convergence zone detection is significant in modern sonar applications.

3506. Attenuation of sound.—As sound is transmitted from a source, the energy is lost or *attenuated* due to reflection, spreading, and absorption. A sound beam reflected from the bottom or surface loses energy; although the sound energy is concentrated near the source, as the range increases the same energy is spread over a sphere whose radius is the range. The rate of absorption is a function of frequency; high frequency sound is absorbed more than sound of lower frequency.

CHAPTER XXXVI

ICE IN THE SEA

3601. Ice and the navigator.—*Sea ice* has posed a problem to the polar navigator since antiquity. Pytheas of Massalia (art. 104) sighted a strange substance which he described as "neither land nor air nor water" floating upon and covering the northern sea over which the summer sun barely set. Pytheas named this lonely region Thule, hence Ultima Thule (farthest north or land's end). Thus began over 20 centuries of polar exploration.

Ice is of direct concern to the navigator because it restricts and sometimes controls his movements, it affects his dead reckoning by forcing frequent and sometimes inaccurately determined changes of course and speed, it affects his piloting by altering the appearance or obliterating the features of landmarks and by rendering difficult the establishment and maintenance of aids to navigation, it affects his use of electronics by its effect upon propagation of radio waves and the changes it produces both in surface features and radar returns from such features, it affects his celestial navigation by altering the refraction and obscuring his horizon and celestial bodies either directly or by the weather it influences, and it affects his charts by introducing various difficulties to the hydrographic surveyor.

Because of his direct concern with ice, the prospective polar navigator will do well to acquaint himself with its nature and extent in the area he expects to navigate. To this end he should consult the sailing directions for the area, and whatever other literature may be available to him, including reports of previous operations in the same area.

3602. Formation of ice.—As it cools, water contracts until the temperature of maximum density is reached. Further cooling results in expansion. The maximum density of freshwater occurs at a temperature of 4°.0C (39°.2F), and freezing takes place at 0°C (32°F). The addition of salt lowers both the temperature of maximum density and, to a lesser extent, that of freezing. The relationships are shown in figure 3602. The two lines meet at a salinity of 24.7 parts per thousand, at which maximum density occurs at the freezing temperature of −1°.3C (29°.61F). At this and greater salinities, the density increases right down to the freezing point. At a salinity of 35 parts per thousand, the approximate average for the oceans, the freezing point is −1°.88C (28°.6F).

As the density of surface seawater increases with decreasing temperature, density currents are induced bringing warmer, less dense water to the surface. If the polar seas consisted of water with constant salinity, the entire water column would have to be cooled to the freezing point in this manner before ice would begin to form. This is not the case, however, in the polar regions where the vertical salinity distribution is such that the surface waters are underlaid at shallow depth by waters of higher salinity. In this instance density currents form a shallow mixed layer which subsequently cannot mix with the deep layer of warmer but saltier water. Ice will then begin forming at the water surface when density currents cease and the surface water reaches its freezing point. In shoal water, however, the mixing process can be sufficient to extend the freezing temperature from the

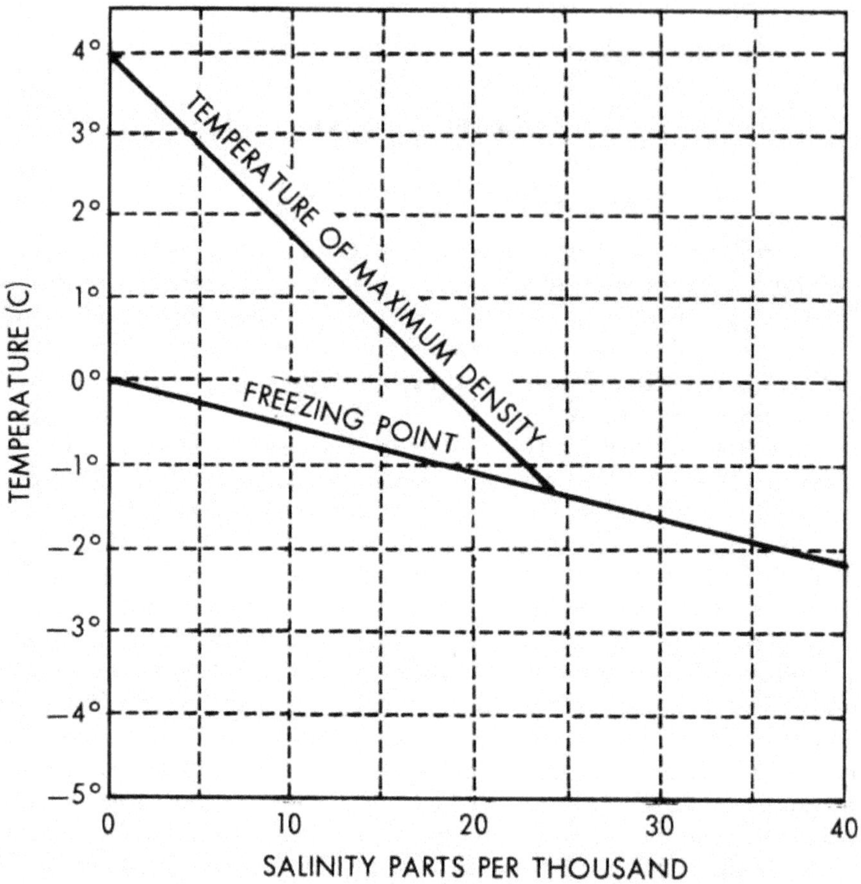

FIGURE 3602.—Relationship between temperature of maximum density and freezing point for water of varying salinity.

surface to the bottom. Ice crystals can, therefore, form at any depth in this case. Because of their decreased density, they tend to rise to the surface unless they form at the bottom and attach themselves there. This ice, called **anchor ice,** may continue to grow as additional ice freezes to that already formed.

3603. Ice of land origin is formed on land by the freezing of freshwater or the compacting of snow as layer upon layer adds to the pressure on that beneath.

Under great pressure ice becomes slightly plastic and is forced outward and downward along an inclined surface. If a large area is relatively flat, as on the antarctic plateau, or if the outward flow is obstructed, as on Greenland, an **ice cap** forms and remains winter and summer. The thickness of these ice caps range from nearly 1 kilometer on Greenland to as much as 4.5 kilometers on the Antarctic Continent. Where ravines or mountain passes permit flow of the ice, a **glacier** is formed. This is a mass of snow and ice which continuously flows to lower levels, exhibiting many of the characteristics of rivers of water. The flow may be more than 30 meters per day, but is generally much less. When a glacier reaches a comparatively level area, it spreads out. When a glacier flows into the sea, the buoyant force of the water breaks off pieces from time to time, and these float away as **icebergs.** Icebergs may be described as dome shaped, sloping or pinnacled (fig. 3603a), tabular (fig. 3603b), glacier, or weathered.

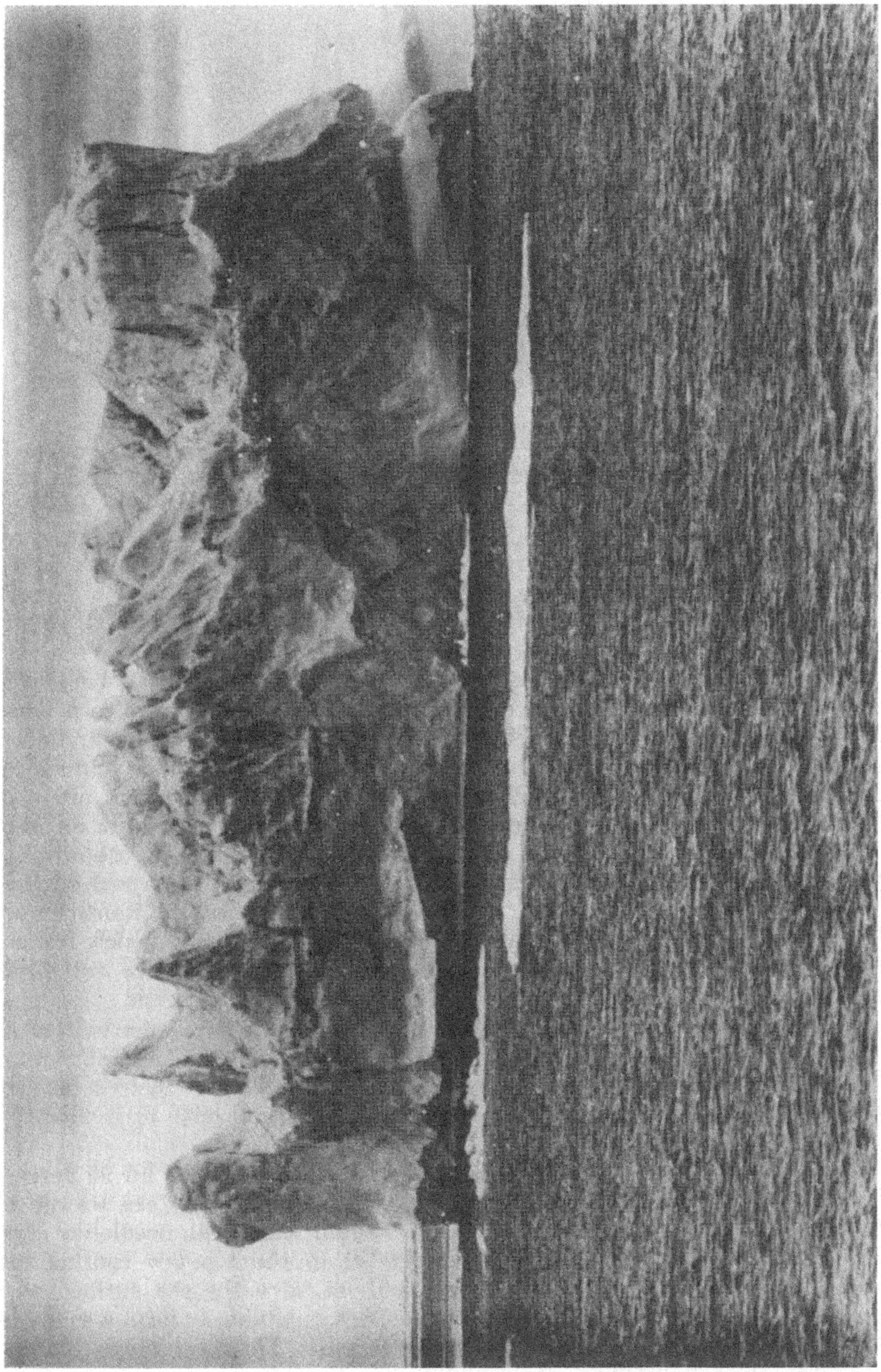

FIGURE 3603a.—Pinnacled iceberg.

FIGURE 3603b.—A tabular iceberg.

An iceberg seldom melts uniformly because of lack of uniformity in the ice itself, differences in the temperature above and below the waterline, exposure of one side to the sun, strains, cracks, mechanical erosion, etc. The inclusion of rocks, silt, and other foreign matter further accentuates the differences. As a result, changes in equilibrium take place, which may cause the berg to tilt or capsize. Parts of it may break off or **calve,** forming separate smaller bergs. A relatively large piece of floating ice generally extending 1 to 5 meters above the sea surface and normally about 100 to 300 square meters in area is called a **bergy bit.** A smaller piece of ice but one large enough to inflict serious damage to a vessel is called a **growler** because of the noise it sometimes makes as it bobs up and down in the sea. Growlers extend less than 1 meter above the sea surface and normally occupy an area of about 20 square meters. Bergy bits and growlers are usually pieces calved from icebergs, but they may be formed by the melting of an iceberg. The principal danger from icebergs is their tendency to break or shift position, and possible underwater extensions, called **rams,** which are usually formed due to the more intensive melting or erosion of the unsubmerged portions. Rams may also extend from a vertical **ice cliff,** also known as an **ice front,** which forms the seaward face of a massive ice sheet or floating glacier; or from an **ice wall** which is the ice cliff forming the seaward margin of a glacier which is aground.

As strange as it may seem, icebergs may be helpful to the mariner in some ways. The melt water found on the surface of icebergs is a source of freshwater, and in the past some daring seamen have made their vessels fast to icebergs which, because they are affected more by currents than the wind, have proceeded to tow them out of the ice pack.

3604. Sea ice forms by the freezing of seawater and accounts for 95 percent of all ice encountered. The first indication of the formation of **new sea ice** (up to 10 centimeters in thickness) is the development of small individual, needlelike crystals of ice, called **spicules,** which become suspended in the top few centimeters of seawater. These spicules, also known as **frazil ice,** give the sea surface an oily appearance. **Grease ice** is formed when the spicules coagulate to form a soupy layer on the surface giving the sea a matte appearance. The next stage in sea ice formation occurs when **shuga,** an accumulation of spongy white ice lumps a few centimeters across, develops from grease ice. Upon further freezing, and depending upon wind exposure, seas, and salinity, shuga and grease ice develop into **nilas,** an elastic crust of high salinity up to 10 centimeters in thickness with a matte surface or into **ice rind,** a brittle, shiny crust of low salinity with a thickness up to approximately 5 centimeters. A layer of 5 centimeters of freshwater ice is brittle but strong enough to support the weight of a heavy man. In contrast, the same thickness of newly formed sea ice will support not more than about 10 percent of this weight, although its strength varies with the temperatures at which it is formed; very cold ice supports a greater weight than warmer ice. As it ages, sea ice becomes harder and more brittle.

New ice may also develop from **slush** which is formed when snow falls into seawater which is near its freezing point, but colder than the melting point of snow. The snow does not melt but floats on the surface, drifting with the wind into beds. If the temperature then drops below the freezing point of the seawater, the slush freezes quickly into a soft ice similar to shuga.

Sea ice is exposed to several forces, including currents, wave motion, tides, wind, and temperature differences. In its early stages, its plasticity permits it to conform readily to virtually any shape required by the forces acting upon it. As it becomes older, thicker, more brittle, and exposed to the influence of wind and wave action, new ice usually separates into circular pieces from 30 centimeters to 3 meters in diameter and up to approximately 10 centimeters in thickness with raised edges due to individual pieces striking against each other. These circular pieces of ice are called **pancake ice** (fig. 3604a) and may break into smaller pieces with strong wave motion. Any single piece of relatively flat sea ice less than 20 meters across is called an **ice cake.** With continued low temperatures individual ice cakes and pancake ice will, depending on wind or wave motion, either freeze together to form a continuous sheet or unite into pieces of ice 20 meters or more across. These larger pieces are then called **ice floes** which may further freeze together to form an ice covered area greater than 10 kilometers across known as an **ice field.** In wind sheltered areas thickening ice usually forms a continuous sheet before it can develop into the characteristic ice cake form. When sea ice reaches a thickness of between 10 to 30 centimeters it is referred to as **grey** and **grey-white ice,** or collectively as **young ice,** and is the transition stage between nilas and **first-year ice.** First-year ice usually attains a thickness of between 30 centimeters and 2 meters in its first winter's growth.

Sea ice may grow to a thickness of 10 to 13 centimeters within 48 hours, after which it acts as an insulator between the ocean and the atmosphere progressively slowing its further growth. However, sea ice may grow to a thickness of between 2 to 3 meters in its first winter. Ice which has survived at least one summer's melt is classified as **old ice.** If it has survived only one summer's melt it is referred to as **second-year ice.** Because it is thicker and less dense than first-year ice, it stands higher out of the water. Old ice which has attained a thickness of 3 meters or more and has survived at least two summers' melt is known as **multiyear ice** and is almost salt free. Old ice may often be recognized by a bluish tone to its surface color in contrast to the greenish tint of first-year ice.

Greater thicknesses in both first and multiyear ice are attained through the deformation of the ice resulting from the movement and interaction of individual floes. Deformation processes occur after the development of new and young ice and are the direct consequence of the effects of winds, tides, and currents. These processes transform a relatively flat sheet of ice into **pressure ice** which has a readily observed roughness in its surface. **Bending,** which is the first stage in the formation of pressure ice, is the upward or downward motion of thin and very

FIGURE 3604a.—Pancake ice, with an iceberg in the background.

plastic ice. **Tenting** occurs when bending produces an upward displacement of ice forming a flat sided arch with a cavity beneath. More frequently, however, **rafting** takes place as one piece of new and young ice overrides another. When pieces of first-year ice are piled haphazardly over one another forming a wall or line of broken ice, referred to as a **ridge**, the process is known as **ridging.** Pressure ice with topography consisting of numerous mounds or hillocks is called **hummocked ice,** each mound being called a **hummock.**

The motion of adjacent floes is seldom equal. The rougher the surface, the greater is the effect of wind, since each piece extending above the surface acts as a sail. Some ice floes are in rotary motion as they tend to trim themselves into the wind. Since ridges extend below as well as above the surface, the deeper ones are influenced more by deep water currents. When a strong wind blows in the same direction for a considerable period, each floe exerts pressure on the next one, and as the distance increases, the pressure becomes tremendous. Ridges on sea ice are generally about 1 meter high and 5 meters deep, but under considerable pressure may attain heights of 30 meters and depths of 150 meters in extreme cases.

The alternate melting and growth of sea ice, combined with the continual motion of various floes that results in separation as well as consolidation, causes widely varying conditions within the ice cover itself. The mean areal density, or **concentration,** of pack ice in any given area is expressed in *oktas* (eights). Concentrations range from: **open water** (total concentration of all ice does not exceed ⅛), **very open pack** (⅛ to less than ⅜ concentration), **open pack** (⅜ to less than ⅝ concentration), **close pack** (⅝ to less than ⅞ concentration), **very close pack** (⅞ to less than 8/8 concentration), to **compact** or **consolidated pack** (8/8 or complete coverage). The extent to which an ice cover of varying concentrations can be penetrated by a vessel varies from place to place and with changing weather conditions. With a concentration of 1 to 2 oktas in a given area, an unreinforced vessel can generally navigate safely, but the danger of receiving heavy damage is always present. When the concentration increases to between 2 and 4 oktas, the area becomes only occasionally accessible to an unreinforced vessel depending upon the vagaries of wind and current. With concentrations of 4 to 6 oktas, the area becomes accessible only to ice strengthened vessels which on occasion will require icebreaker assistance. Navigation in areas with concentrations of 6 oktas or more should only be attempted by modern icebreakers.

Within the ice cover, openings may develop resulting from a number of deformation processes. Long, jagged cracks may appear first in the ice cover or through a single floe. When these cracks part and reach lengths of a few meters to many kilometers, they are referred to as **fractures.** If they widen further to permit passage of a ship, they are called **leads.** In winter, a thin coating of new ice may cover the water within a lead, but in summer the water usually remains ice-free until a shift in the movement forces the two sides together again. Before this occurs, lateral motion generally occurs between the floes, so that they no longer fit and unless the pressure is extreme, numerous large patches of open water remain. These nonlinear shaped openings enclosed in ice are called **polynyas.** Polynyas may contain small fragments of floating ice and may be covered with miles of new and young ice.

Sea ice which is formed in situ from seawater or by the freezing of pack ice of any age to the shore and which remains attached to the coast, to an ice wall, to an ice front, or between shoals is called **fast ice.** The width of this fast ice varies considerably and may extend for a few meters or several hundred kilometers. In bays and other sheltered areas, fast ice, often augmented by annual snow accumulations and the seaward extension of land ice, may attain a thickness of over 2 meters above the sea surface. When a floating sheet of ice grows to this or a greater thickness and extends over a great horizontal distance, it is called an **ice shelf.** Massive ice shelfs where the ice thickness reaches several hundred meters are found in both the Arctic and Antarctic.

The majority of the icebergs found in the Antarctic do not originate from glaciers as those found in the Arctic, but are calved from the outer edges of broad expanses of shelf ice. Icebergs formed in this manner are called **tabular icebergs,** having a boxlike shape with horizontal dimensions measured in kilometers, and heights above the sea surface approaching 60 meters. The largest Antarctic ice shelves are found in the Ross and Weddell Seas. The expression "tabular iceberg" is not applied to bergs which break off from arctic ice shelves; similar formations there are called **ice islands.** These originate when shelf ice, such as that found on the northern coast of Greenland and in the bays of Ellesmere Island, breaks up. As a rule, arctic ice islands are not as large as the tabular icebergs found in the Antarctic. They attain a thickness of up to 55 meters and on the average extend 5 to 7 meters above the sea surface. Both tabular icebergs and ice islands possess a nearly level, but gently rolling surface. Because of their deep draft, they are influenced much more by current than wind. Both the United States and the U.S.S.R. have used arctic ice islands as floating scientific platforms from which polar research has been conducted.

3605. Thickness of sea ice.—Sea ice has been observed to grow to a thickness of almost 3 meters during its first year. However, the thickness of first-year ice that has not undergone deformation does not generally exceed 2 meters. In coastal areas where the melting rate is less than the freezing rate, the thickness may increase during succeeding winters, being augmented by compacted and frozen snow, until a maximum thickness of about 3.5 to 4.5 meters may eventually be reached. Old sea ice may also attain a thickness of over 4 meters in this manner, or when summer melt water from its surface or from snow cover runs off into the sea and refreezes under the ice where the seawater temperature is below the freezing point of the fresher melt water.

The growth of sea ice is dependent upon a number of meteorological and oceanographic parameters. Such parameters include air temperature, initial ice thickness, snow depth and density, wind speed, seawater salinity and density, and the specific heats of sea ice and seawater. Investigations, however, have shown that the most influential parameters affecting sea ice growth are air temperature, wind speed, snow depth and density, and initial ice thickness. Many complex equations have been formulated to predict ice growth using these five parameters. However, except for the first two, these parameters are not routinely observed for remote polar locations.

In the early 1940's a Russian geographical scientist, N. N. Zubov, formulated an ice growth equation as a function of air temperature alone and based on his empirical observations of ice formation along portions of the northern Russian arctic coast. Air temperatures are translated into accumulated frost degree days from which theoretical ice thicknesses are calculated using the equation:

$$T_j = \frac{-50 + \sqrt{2500 + 32 \sum_{i=1}^{j} DDi}}{2},$$

where T_j is the ice thickness in centimeters on day j and DDi is the frost degree day accumulation in degrees Celsius on day i.

A **frost degree day** is defined as a day with a mean temperature of 1° below an arbitrary base. The base most commonly used is the freezing point of freshwater (0°C). If, for example, the mean temperature on a given day is 5° below freezing, then five frost degree days are collected for that day. These frost degree days are then added to those collected the very next day to obtain an accumulated value, which is then added to the number of degree days collected the following day. This process is repeated daily throughout the ice growing season. Temperatures usually fluctuate above and below freezing for several days before remaining below freezing. Therefore, frost degree day accumulations are initiated on the first day of the period when temperatures remain below freezing. The relationship between frost degree day accumulations and theoretical ice growth curves at Point Barrow, Alaska is shown in figure 3605a. Similar curves for other arctic stations are contained in publications available from the U. S. Naval Oceanographic Office. Figure 3605b graphically depicts the relationship between accumulated frost degree days (°C) and ice thickness in centimeters.

During the winter the ice usually becomes covered with snow which insulates the ice beneath and tends to slow down its rate of growth. This thickness of snow cover varies considerably from region to region as a result of differing climatic conditions. Its depth may also vary widely within very short distances in response to variable winds and ice topography. While this snow cover persists, almost 90 percent of the incoming radiation is reflected back to space. Eventually, however, the snow begins to melt as the air temperature rises above 0°C in early summer and the resulting freshwater forms puddles on the surface. These puddles absorb about 90 percent of the incoming radiation and rapidly enlarge as they melt the surrounding snow or ice. Eventually the puddles penetrate to the bottom surface of the floes and are known as **thawholes.** This slow process is characteristic of ice in

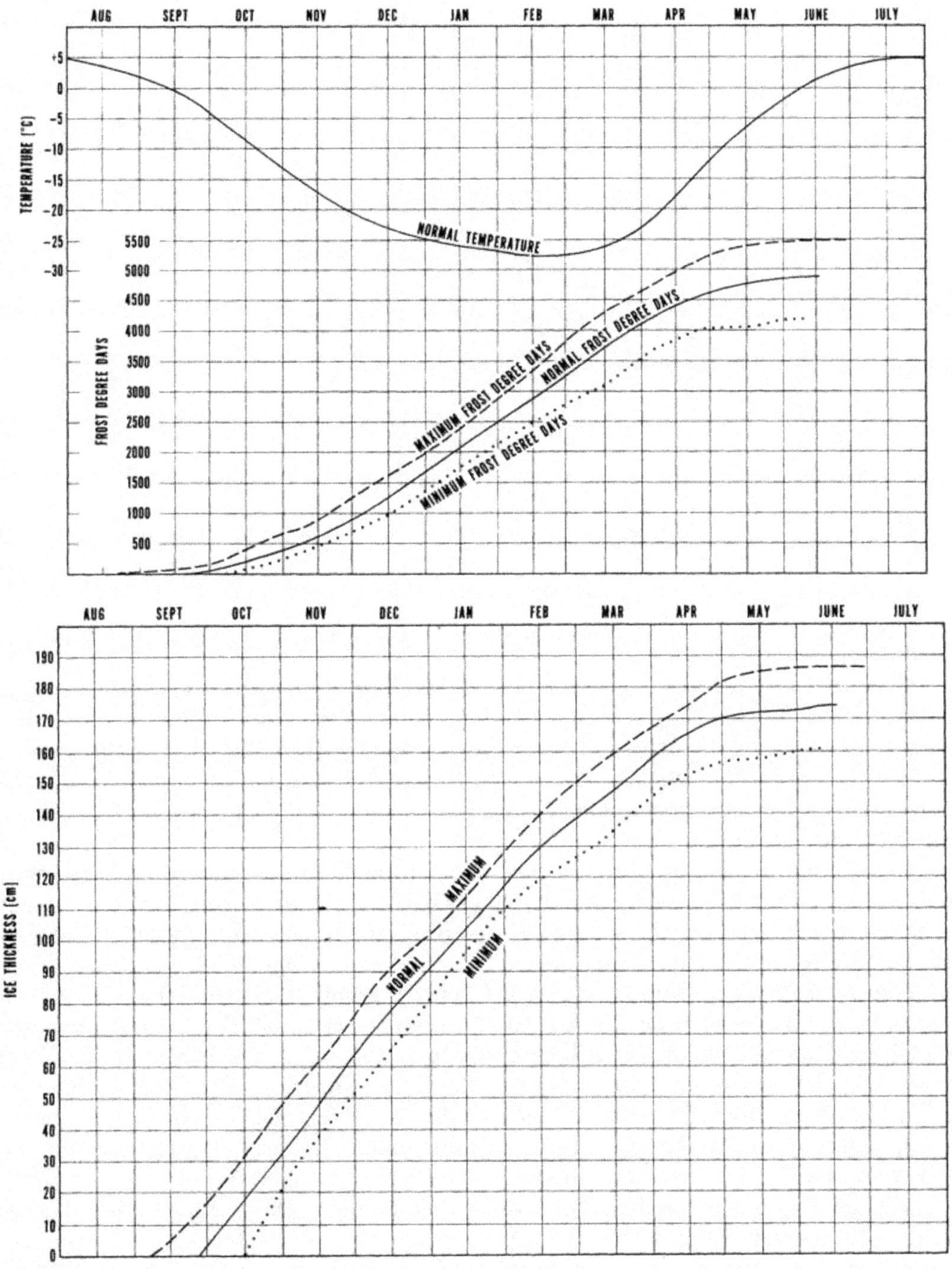

FIGURE 3605a.—Relationship between accumulated frost degree days and theoretical ice thickness at Point Barrow, Alaska.

the Arctic Ocean and seas where movement is restricted by the coastline or islands. Where ice is free to drift into warmer waters (e.g., the Antarctic, East Greenland, and the Labrador Sea) decay is accelerated in response to wave erosion as well as warmer air and sea temperatures.

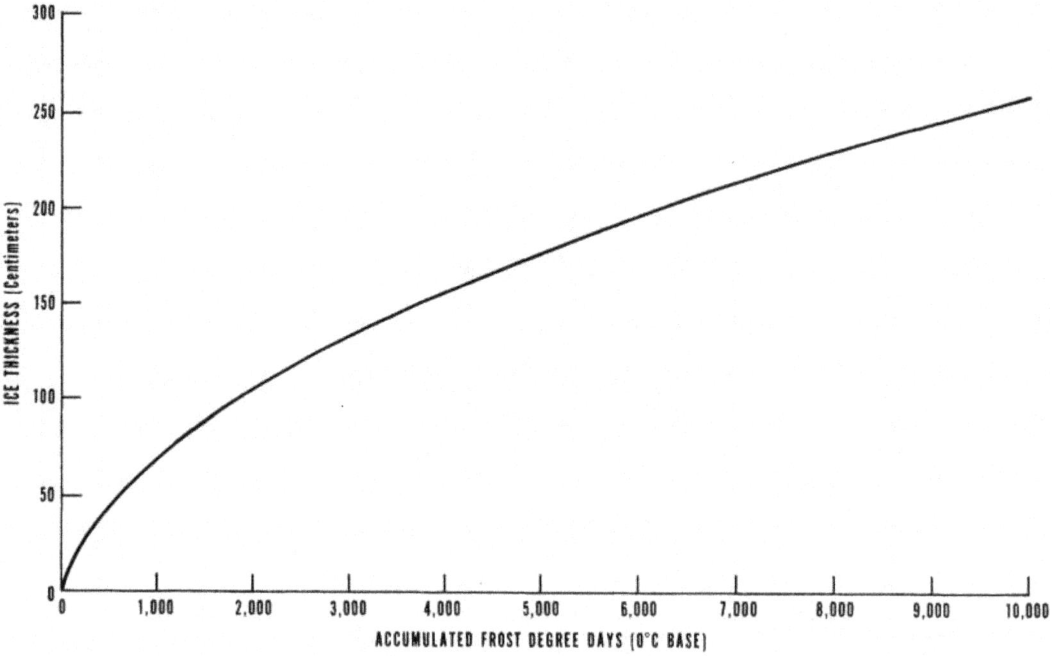

FIGURE 3605b.—Relationship between accumulated frost degree days (°C) and ice thickness (cm).

3606. Salinity of sea ice.—Sea ice forms first as salt-free crystals near the surface of the sea. As the process continues, these crystals are joined together and, as they do so, small quantities of brine are trapped within the ice. On the average, new ice 15 centimeters thick contains 5 to 10 parts of salt per thousand. With lower temperatures, freezing takes place faster. With faster freezing, a greater amount of salt is trapped in the ice.

Depending upon the temperature, the trapped brine may either freeze or remain liquid, but because its density is greater than that of the pure ice, it tends to settle down through the pure ice. As it does so, the ice gradually freshens, becoming clearer, stronger, and more brittle. At an age of 1 year, sea ice is sufficiently fresh that its melt water, if found in **puddles** of sufficient size, and not contaminated by spray from the sea, can be used to replenish the freshwater supply of a ship. However, ponds of sufficient size to water ships are seldom found except in ice of great age, and then much of the melt water is from snow which has accumulated on the surface of the ice. When sea ice reaches an age of about 2 years, virtually all of the salt has been eliminated. Icebergs contain no salt, and uncontaminated melt water obtained from them is fresh.

The settling out of the brine gives sea ice a honeycomb structure which greatly hastens its disintegration when the temperature rises above freezing. In this state, when it is called **rotten ice,** much more surface is exposed to warm air and water, and the rate of melting is increased. In a day's time, a floe of apparently solid ice several inches thick may disappear completely.

3607. Density of ice.—The density of freshwater ice at its freezing point is 0.917. Newly formed sea ice, due to its salt content, is more dense, 0.925 being a representative value. The density decreases as the ice freshens (art. 3606). By the time it has shed most of its salt, sea ice is less dense than freshwater ice, because ice formed in the sea contains more air bubbles. Ice having no salt but containing air to the extent of 8 percent by volume (an approximately maximum value for sea ice) has a density of 0.845.

The density of land ice varies over even wider limits. That formed by freezing of freshwater has a density of 0.917, as stated above. Much of the land ice, however, is formed by compacting of snow. This results in the entrapping of relatively large quantities of air. **Névé**, a snow which has become coarse grained and compact through temperature change, forming the transition stage to glacier ice, may have an air content of as much as 50 percent by volume. By the time the ice of a glacier reaches the sea, its density approaches that of freshwater ice. A sample taken from an iceberg on the Grand Banks had a density of 0.899.

When ice floats, part of it is above water and part is below the surface. The percentage of the mass below the surface can be found by dividing the average density of the ice by the density of the water in which it floats. Thus, if an iceberg of density 0.920 floats in water of density 1.028 (corresponding to a salinity of 35 parts per thousand and a temperature of −1°C, or 30°F), 89.5 percent of its mass will be below the surface. That is, about nine-tenths of the mass will be below the surface, and only about one-tenth will be above the surface.

The height to draft ratio for a blocky or tabular iceberg probably varies fairly closely about 1:5. This average ratio was computed for icebergs south of Newfoundland by considering density values and a few actual measurements, and by seismic means at a number of locations along the edge of the Ross Ice Shelf near Little America Station. It was also substantiated by density measurements taken in a nearby hole drilled through the 256-meter thick ice shelf. The height to draft ratios of icebergs become significant when determining their drift (art. 3609).

3608. Drift of ice.—Although surface currents have some effect upon the drift of pack ice, the principal factor is wind. Due to Coriolis force (art. 3803), ice does not drift in the direction of the wind, but varies from approximately 18° to as much as 90° from this direction, depending upon the force of the surface wind and the ice thickness. In the Northern Hemisphere, this drift is to the *right* of the direction toward which the wind blows, and in the Southern Hemisphere it is toward the *left.* Although early investigators computed average angles of approximately 28° or 29° for the drift of close multiyear pack ice, large drift angles were usually observed with low rather than high wind speeds. The relationship between surface wind speed, ice thickness, and drift angle, shown in figure 3608, was derived theoretically for the drift of consolidated pack under equilibrium (a balance of forces acting on the ice) conditions, and shows that the drift angle increases with increasing ice thickness and decreasing surface wind speed. A slight increase also occurs with higher latitude.

Since the cross-isobar deflection of the surface wind over the oceans is approximately 20°, the deflection of the ice varies from approximately along the isobars to as much as 70° to the right of the isobars, with low pressure on the left and high pressure on the right in the Northern Hemisphere. The positions of the low and high pressure areas are, of course, reversed in the Southern Hemisphere. The drift angles that are given in figure 3608 may be used for all ice concentrations and polar latitudes.

The rate of drift, compiled from observations of ice drift along the northern Russian coast bordering the Chukchi Sea, is presented in table 3608. Rates are given as a percentage of the surface wind speed and depend upon the roughness of the surface and the concentration of the ice. Percentages vary from approximately one quarter of 1 percent to almost 8 percent of the surface wind speed measured

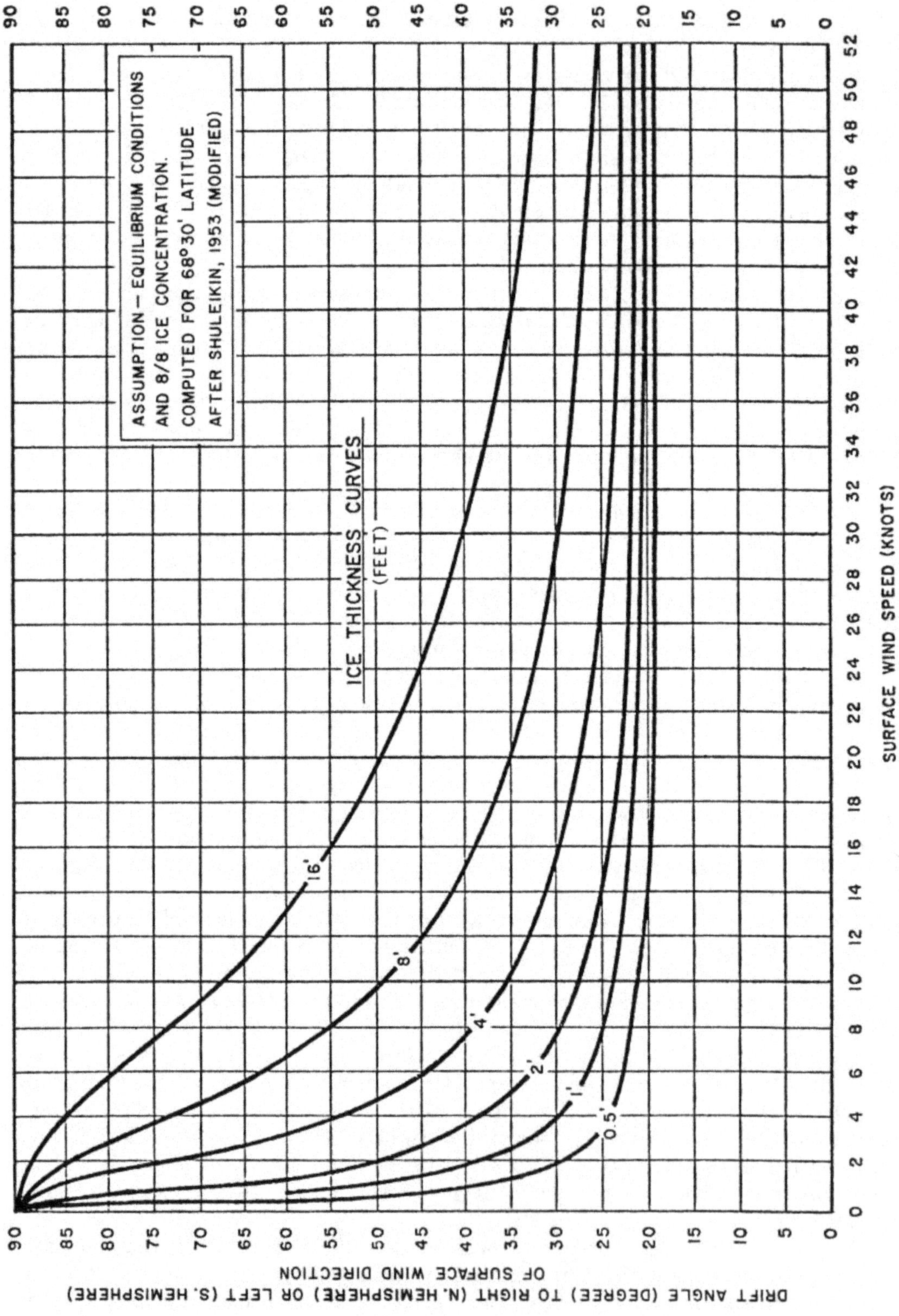

FIGURE 3608.—Ice drift direction for varying wind speed and ice thickness.

approximately 6 meters above the ice surface. Low concentrations of heavily ridged or hummocked floes drift faster than high concentrations of lightly ridged or hummocked floes with the same wind speed. From table 3608 it can be seen that sea ice of 6 to 7 okta concentrations and six tenths hummocking or close multiyear ice will drift at approximately 2 percent of the surface wind speed. Additionally, the response factors of 1 and 4 okta ice concentrations respectively, are approximately three times and twice the magnitude of the response factor for 7 okta ice concentrations with the same extent of surface roughness. Although a maximum ice drift to surface wind speed ratio of approximately 8 percent is indicated by table 3608, isolated ice floes have been observed to drift as fast as 10 percent to 12 percent of strong surface winds.

EXTENT OF RIDGING & HUMMOCKING (TENTHS)	ICE CONCENTRATION (OKTAS) 1	2	3	4	5	6	7
1	0.88	0.75	0.63	0.50	0.39	0.33	0.26
2	1.75	1.50	1.25	1.05	0.85	0.65	0.53
3	2.64	2.33	2.05	1.75	1.44	1.13	0.85
4	3.53	3.13	2.69	2.30	1.86	1.45	1.08
5	4.40	3.90	3.40	2.90	2.40	1.90	1.40
6	5.28	4.65	4.06	3.50	2.94	2.35	1.73
7	6.66	5.48	4.79	4.10	3.41	2.73	2.04
8	7.03	6.25	5.55	4.75	3.95	3.18	2.36
9	7.93	7.05	6.10	5.20	4.35	3.50	2.61

TABLE 3608.—Rate of the wind drift of sea ice (given as a percent of the surface wind speed) for varying ice concentration and surface roughness.

The rates with which sea ice drifts have been quantified through empirical observation. The drift angle, however, has been determined theoretically for 8 okta ice concentrations. This relationship presently is extended to the drift of all ice concentrations due to the lack of basic knowledge of the dynamic forces that act upon and result in redistributions of sea ice in the polar regions.

3609. Iceberg drift.—Icebergs extend a considerable distance below the surface and have relatively small "sail areas" compared to their subsurface mass. Therefore, the near-surface current is thought to be primarily responsible for drift; however, observations have shown that wind can be the dominant force that governs iceberg drift at a particular location or time. Also, the current and wind may contribute nearly equally to the resultant drift.

Two other major forces which act on a drifting iceberg are the Coriolis force and, to a lesser extent, the pressure gradient force which is caused by gravity owing to a tilt of the sea surface and is important only for iceberg drift in a major current. Near-surface currents are generated by a variety of factors such as horizontal pressure gradients owing to density variations in the water, rotation of the earth, gravitational attraction of the moon, and slope of the sea surface. Wind not only acts directly on an iceberg, but also indirectly by generating waves and a surface current in about the same direction as the wind. Because of inertia, an iceberg may continue to move from the influence of wind for some time after the wind stops or changes direction.

The relative influence of currents and winds on the drift of an iceberg varies according to the direction and magnitude of the forces acting on its sail area and subsurface cross-sectional area. The resultant force therefore involves the proportions of the iceberg above and below the sea surface in relation to the velocity and depth of the current, and the velocity and duration of the wind. Studies tend to show that, generally, where strong currents prevail, the current is dominant. In regions of weak currents, however, winds that blow for a number of hours in a steady direction materially affect the drift of icebergs. Generally, it can be stated that currents tend to have a greater effect on deep-drafted icebergs, while winds tend to have a greater effect on shallow-drafted icebergs.

As icebergs waste through melting, erosion, and calving, observations indicate the height to draft ratio may approach 1:1 during their last stage of decay when they are referred to as valley, winged, horned, or spired icebergs. The height to draft ratios found for icebergs in their various stages are presented in table 3609a. Since wind tends to have a greater effect on shallow than deep-drafted icebergs, the wind can be expected to exert increasing influence on iceberg drift as wastage increases.

Iceberg type	*Height to draft ratio*
Blocky or tabular	1:5
Rounded or domed	1:4
Picturesque or Greenland (sloping)	1:3
Pinnacled or ridged	1:2
Horned, winged, valley, or spired (weathered)	1:1

TABLE 3609a.—Height to draft ratios for various types of icebergs.

Simple equations which precisely define iceberg drift cannot be formulated at present because of the uncertainty in the water and air drag coefficients associated with iceberg motion. Values for these parameters not only vary from iceberg to iceberg, but they probably change for the same iceberg over its period of wastage.

Present investigations utilize an analytical approach facilitated by computer calculations in which the air and water drag coefficients are varied within reasonable limits. Combinations of these drag values are then used in several increasingly complex water models that try to duplicate observed iceberg trajectories. The results indicate that with a wind generated current, Coriolis force, and a uniform wind, but without a gradient current, small and medium icebergs will drift with the percentages of the wind as given in table 3609b. The drift will be to the right in the Northern Hemisphere and to the left in the Southern Hemisphere.

Wind Speed (knots)	*Ice Speed/Wind Speed (percent)*		*Drift Angle (degrees)*	
	Small Berg	*Med. Berg*	*Small Berg*	*Med. Berg*
10	3.6	2.2	12	69
20	3.8	3.1	14	55
30	4.1	3.4	17	36
40	4.4	3.5	19	33
50	4.5	3.6	23	32
60	4.9	3.7	24	31

TABLE 3609b.—Drift of iceberg as percentage of wind speed.

When gradient currents are introduced, trajectories vary considerably depending on the magnitude of the wind and current and whether they are in the same or opposite direction. When a 1-knot current and wind are in the same direction, drift is to the right of both wind and current with drift angles increasing linearly from approximately 5° at 10 knots to 22° at 60 knots. When the wind and a 1-knot current are in opposite directions, drift is to the left of the current with the angle increasing from approximately 3° at 10 knots, to 20° at 30 knots, and to 73° at 60 knots. As a limiting case for increasing wind speeds, drift may be approximately normal (to the right) to the wind direction. This indicates that the wind generated current is clearly dominating the drift. In general, the various models used demonstrated that a combination of the wind and current was responsible for the drift of icebergs.

3610. Extent of ice in the sea.—When an area of sea ice, no matter what form it takes or how it is disposed, is described, it is referred to as **pack ice.** In both polar regions the pack ice is a very dynamic feature with wide deviations in its areal extent dependent upon changing oceanographic and meteorological phenomena. In winter the arctic pack extends over the entire Arctic Ocean and for a varying distance outward from it; the limits receding considerably during the warmer summer months. Each year a large portion of the ice from the Arctic Ocean moves outward between Greenland and Spitsbergen, into the North Atlantic, and is replaced by new ice. Relatively little of the arctic pack ice is more than 10 years old. An example of the variance possible in the outer limit of the arctic ice pack is shown in figure 3610a where the average positions of the maximum and minimum extents of sea ice are plotted.

Ice covers a large portion of the antarctic waters and is probably the greatest single factor contributing to the isolation of the Antarctic Continent. During the austral winter (June through September), ice completely surrounds the continent, forming an almost impassable barrier that extends northward on the average to about 54°S in the Atlantic and to about 62°S in the Pacific. Disintegration of the pack ice during the austral summer months of December through March allows the limits of the ice edge to recede considerably opening some coastal areas of the Antarctic to navigation. The mean maximum and mean minimum positions of the antarctic ice limit are shown in figure 3610b.

Historical information on sea conditions for specific localities and time periods can be found in publications of the U. S. Naval Oceanographic Office and the Defense Mapping Agency Hydrographic/Topographic Center. Such publications include sailing directions, forecasting guides, and ice atlases.

3611. Ice in the North Atlantic. Sea level glaciers exist on a number of landmasses bordering the northern seas, including Alaska, Greenland, Svalbard (Spitsbergen), Zemlya Frantsa-Iosifa (Franz Josef Land), Novaya Zemlya, and Severnaya Zemlya (Nicholas II Land). Except in Greenland, the rate of calving is relatively slow, and the few icebergs produced melt near their points of formation. Many of those produced along the coasts of Greenland, however, are eventually carried into the shipping lanes of the North Atlantic, where they constitute a major menace to ships.

Generally the majority of icebergs produced along the east coast of Greenland remain near their source of origin. However, a small number of bergy bits, growlers, and small icebergs are transported from this region by the East Greenland Current around Kap Farvel at the southern tip of Greenland and then northward by the West Greenland Current into Davis Strait to the vicinity of 67°N. Relatively few of these icebergs menace shipping but some are carried to the south and southeast of Kap Farvel by a counterclockwise current gyre centered near 57°N and 43°W.

The main source of the icebergs encountered in the North Atlantic is the west coast of Greenland between 67°N and 76°N where approximately 7,500 icebergs are

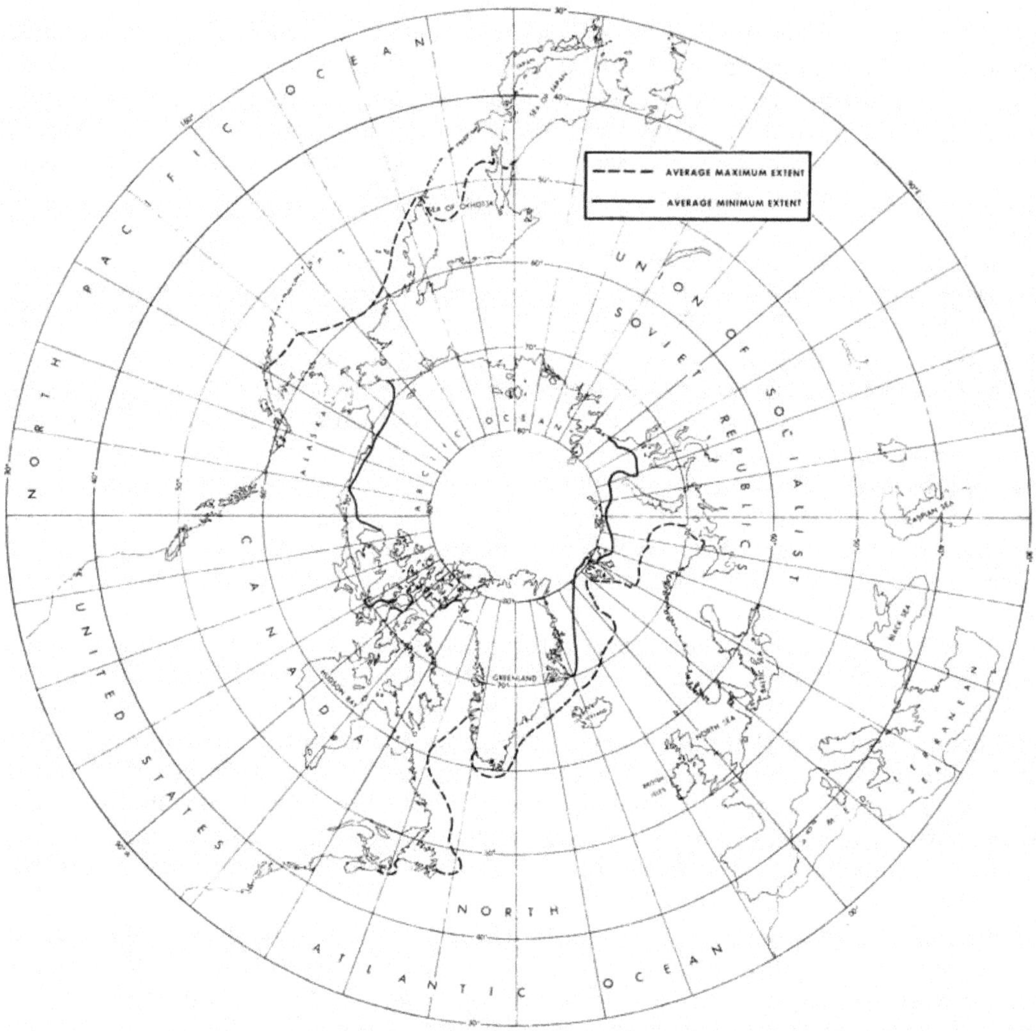

FIGURE 3610a.—Average maximum and minimum extent of arctic sea ice.

formed each year. In this area there are about 100 low lying coastal glaciers, 20 of them being the principal producers of icebergs. Of these 20 major glaciers, 2 located in Disko Bugt between 69°N and 70°N are estimated to contribute 28 percent of all icebergs appearing in Baffin Bay and the Labrador Sea. The West Greenland Current carries icebergs from this area northward and then westward until they encounter the south flowing Labrador Current. West Greenland icebergs generally spend their first winter locked in the Baffin Bay pack ice; however, a large number can also be found within the sea ice extending along the entire Labrador coast by late winter. During the next spring and summer, when they are freed by the break up of the pack ice, they are transported further southward by the Labrador Current. The general drift patterns of icebergs that are prevalent in the eastern portion of the North American Arctic are shown in figure 3611a. Observations over a 69-year period show that an average of 365 icebergs per year reach latitudes south of 48°N, with approximately 10 percent of this total carried south of the Grand Banks (43°N) before they melt. Icebergs may be encountered during any part of the year, but in the Grand Banks area they are most numerous during spring. The maximum

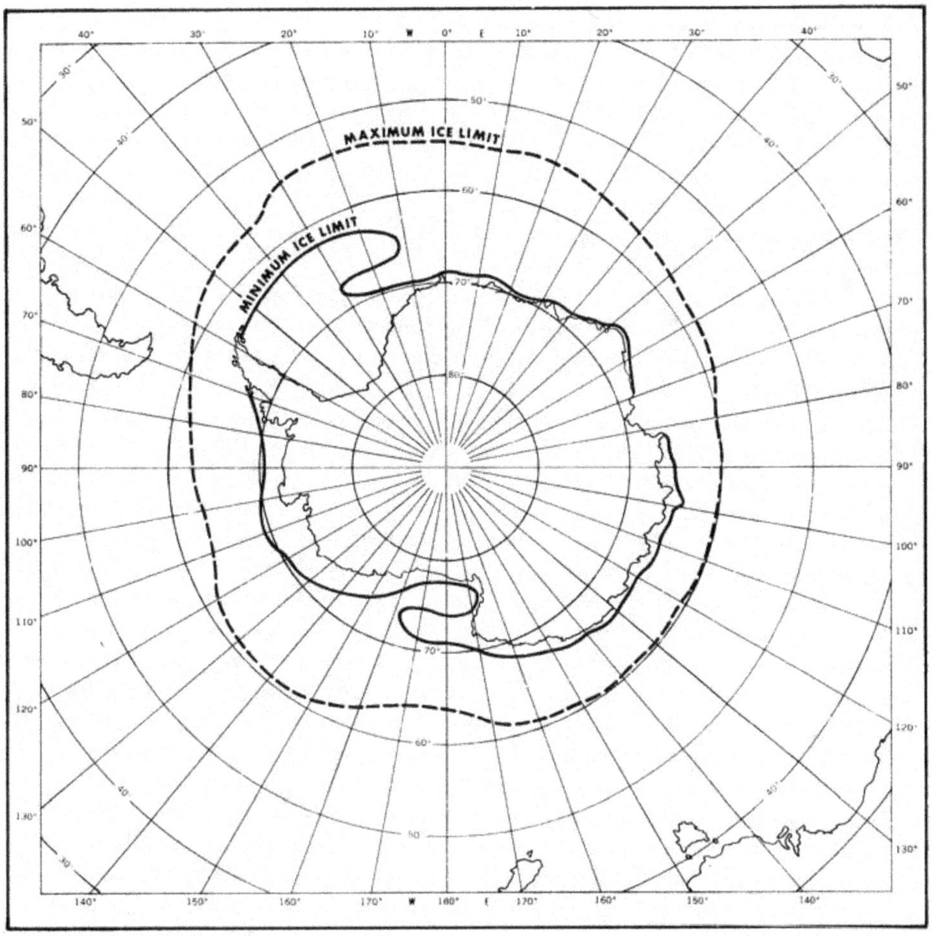

FIGURE 3610b.—Average maximum and minimum extent of antarctic sea ice.

monthly average of iceberg sightings occur during April, May, and June, with May having the highest average of 124.

The variation from average conditions is considerable. More than 1,587 icebergs have been sighted south of latitude 48°N in a single year (1972), while in 1966 not a single iceberg was encountered in this area. In the years of 1940 and 1958 only one iceberg was observed south of 48°N. Although this variation has not been fully explained, it is apparently related to wind conditions, the distribution of pack ice in Davis Strait, and to the amount of pack ice off Labrador. It has been suggested that the distribution of the Davis Strait-Labrador Sea pack ice influences the effectiveness of this ice in holding back the icebergs. According to this theory, when pack ice is heavy along the Labrador coast, the icebergs are forced well offshore, where warmer water causes them to melt before they reach the North Atlantic shipping lanes; but when the pack ice is not sufficient for this, the icebergs drift closer to shore, where there is colder water which prolongs their existence.

Average iceberg and pack ice limits in this area during April, May, and June are shown in figures 3611b, 3611c, and 3611d. Icebergs have been observed in the vicinity of Bermuda, the Azores, and within 400 to 500 kilometers of Great Britain.

FIGURE 3611a.—General drift pattern of icebergs.

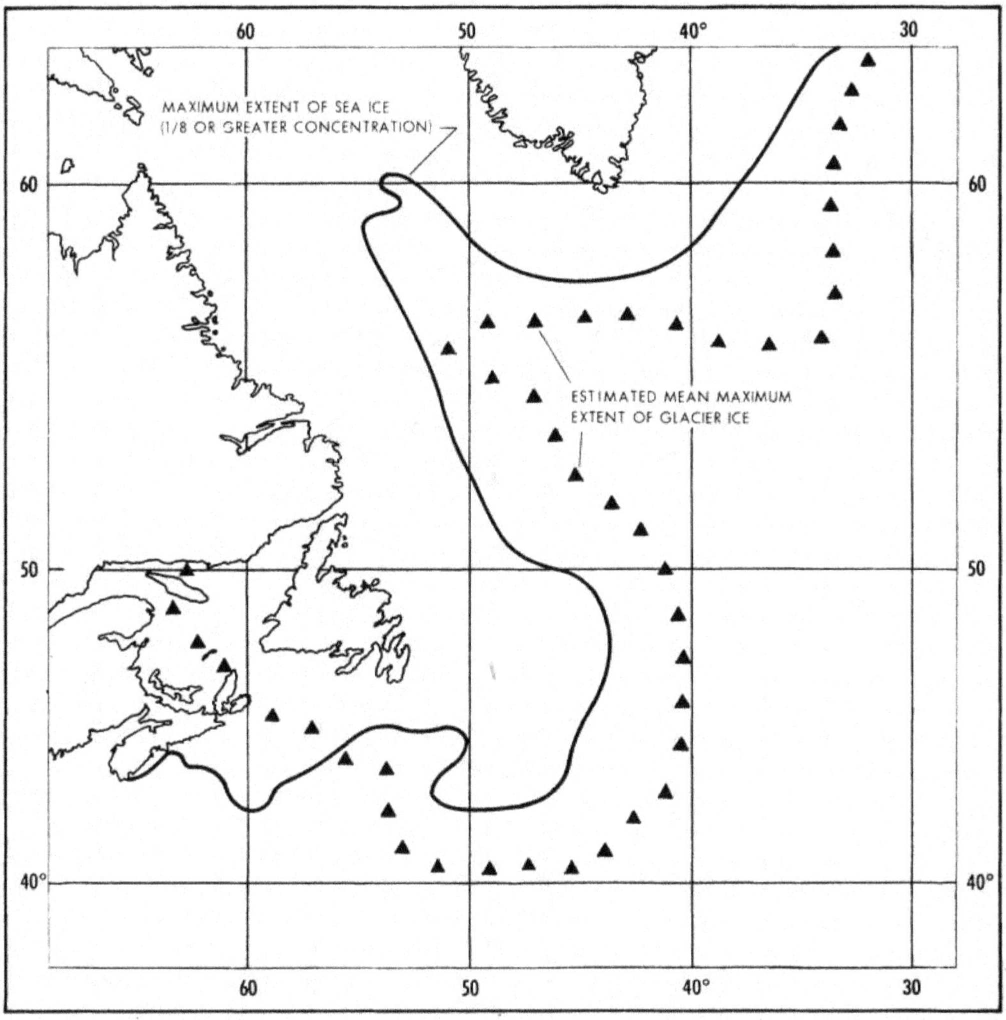

FIGURE 3611b.—Average iceberg and pack ice limits during the month of April.

Pack ice may also be found in the North Atlantic, some having been brought south by the Labrador Current and some coming through Cabot Strait after having formed in the Gulf of St. Lawrence.

3612. The International Ice Patrol was established in 1914 by the *International Convention for the Safety of Life at Sea,* held in 1913 as a result of the sinking of the SS *Titanic* in 1912. On its maiden voyage this vessel struck an iceberg and sank with the loss of 1,513 lives. In accordance with the agreement reached at the convention, this patrol is conducted by the U. S. Coast Guard, which is responsible for the observations and dissemination of information concerning ice conditions in the North Atlantic. Information on ice conditions for the Gulf of St. Lawrence and the coastal waters of Newfoundland and Labrador, including the Strait of Belle Isle to west of Belle Isle itself, is provided by the Canadian Ministry of Transport between the months of December through late June. Ice data for these areas are obtained from the Ice Operations Officer located at Dartmouth, Nova Scotia via Sidney or Halifax marine radio.

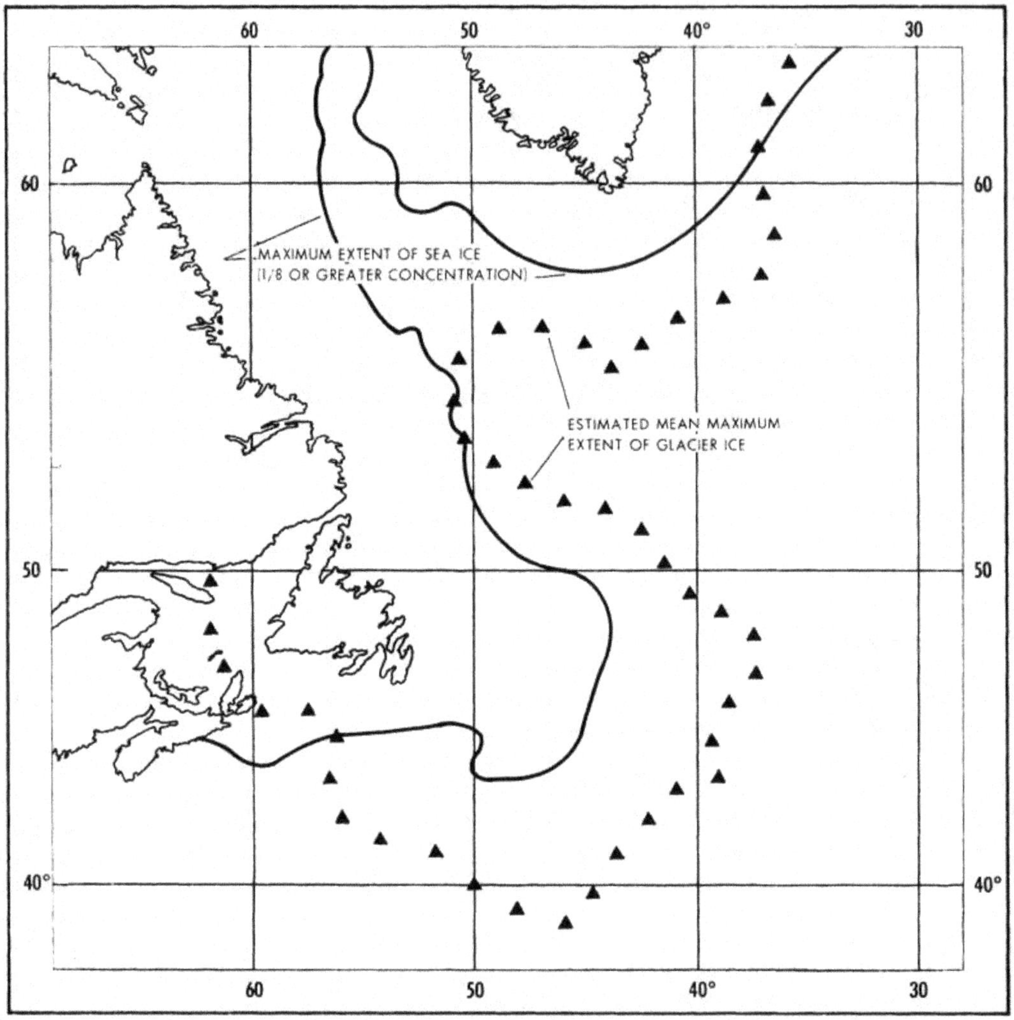

FIGURE 3611c.—Average iceberg and pack ice limits during the month of May.

During each ice season, aerial reconnaissance surveys are made in the vicinity of the Grand Banks of Newfoundland to determine the southeastern, southern, and southwestern limit of the seaward extent of icebergs. During the war years of 1916–18 and 1941–45 the patrol was suspended. Aircraft were added to the patrol force following World War II, and today perform the majority of the work. Reports of ice sightings are also requested and collected from ships transiting the Grand Banks area. When reporting ice, vessels are requested to detail the type of ice (icebergs or sea ice) sighted, its position, concentration and thickness (for sea ice), and size and shape (for icebergs).

In addition to ice reports, masters who do not issue routine weather reports, are urged to make sea surface temperature and weather reports to the Ice Patrol every 6 hours within latitudes 40° to 52°N and longitudes 38° to 58°W.

Operations of the Ice Patrol are directed from the U. S. Coast Guard Base, Avery Point, Groton, Connecticut. Regularly scheduled bulletins are issued by the Ice Patrol twice daily during the ice season by radio and landline communications from Boston, Massachusetts. When icebergs are sighted outside the known limits of ice, special broadcasts are issued from St. Johns, Newfoundland, between those regularly scheduled. Iceberg positions in the ice bulletins are updated for drift at 12-hour intervals. A radio-facsimile chart is also broadcast once a day throughout the ice season. The Ice Patrol, in addition to patrolling possible iceberg areas, conducts oceanographic surveys, maintains up-to-date records of the currents in its area of operation to aid in predicting the drift of icebergs, and studies sea ice conditions in general.

3613. Ice detection.—Safe navigation in the polar seas depends on a number of factors, not the least of which is accurate knowledge of the location and amount of sea ice that lies between the mariner and his destination. Sophisticated electronic equipment such as radar, sonar, and the visible, infrared, and microwave radiation remote sensors on board earth orbiting satellites have joined forces with the polar traveler's own eyesight and, in some cases, hearing to aid him in detecting ice in the sea.

As a ship proceeds into higher latitudes, the first ice it encounters is likely to be in the form of icebergs, because such large pieces require a longer time to disintegrate. Icebergs can easily be avoided if detected soon enough. The distance at which an iceberg can be seen visually depends upon meteorological visibility, height of the iceberg, source and condition of lighting, and the observer. On a clear day with excellent visibility, a large iceberg, due to its brilliant luster, might be sighted at a distance of almost 35 kilometers. With a low-lying haze around the horizon, this distance may be reduced by one-half. In light fog or drizzle this distance is further reduced from 1,850 meters to 5.5 kilometers.

In a dense fog an iceberg may not be perceptible at a distance of over 100 meters or until it is close aboard where it will appear in the form of a luminous, white object if the sun is shining; or as a dark, somber mass with a narrow streak of blackness at the waterline if the sun is not shining. If the layer of fog is not too thick, an iceberg may be sighted from aloft sooner than from a point lower in the vessel, but this fact should not be considered justification for omitting a bow lookout. The diffusion of light in a fog will produce a blink, or area of whiteness, above and at the sides of an iceberg which will appear to increase the apparent size of its mass.

On dark, clear nights icebergs may be seen at a distance of from 1,850 meters to 4 kilometers, appearing either as white or black objects with an occasional light spot where a wave breaks against it. Under such conditions of visibility growlers are a greater menace to vessels, and the vessel's speed should be reduced and a sharp lookout maintained.

The moon may either help or hinder, depending upon its phase and position relative to ship and iceberg. A full moon in the direction of the iceberg interferes with its detection, while light from one in the opposite direction may produce a blink which renders the iceberg visible for a greater distance, possibly as much as 5.5 kilometers. A clouded sky at night, through which the moonlight is intermittent, also renders ice detection difficult. A night sky with heavy passing clouds may also dim or obscure any object which has been sighted, and fleecy cumulus and cumulonimbus clouds often may give the appearance of blink from icebergs.

If an iceberg is in the process of disintegration, its presence may be detected by the cracking sound as a piece breaks off, or by the thunderous roar as a large piece falls into the water. The appearance of smaller pieces of ice in the water often indicates the presence of an iceberg nearby. In calm weather such pieces may form

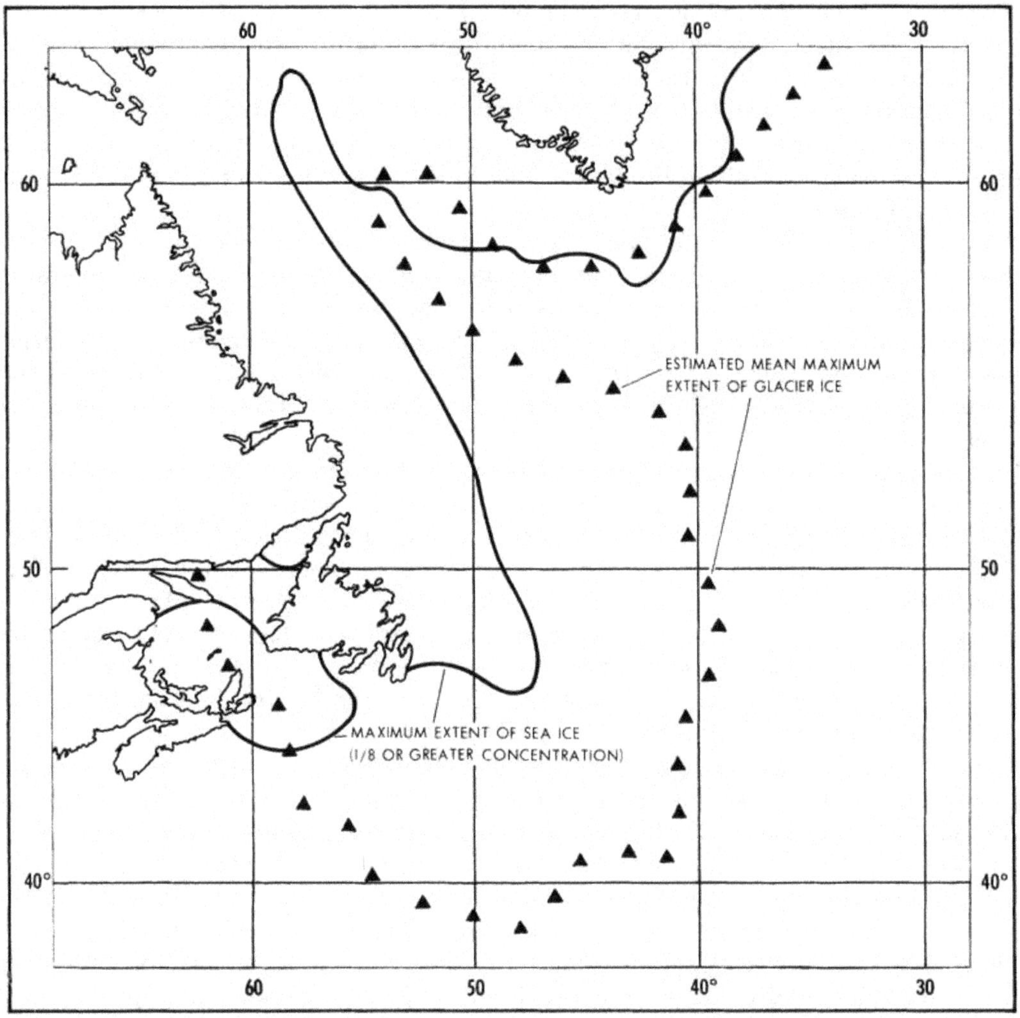

FIGURE 3611d.—Average iceberg and pack ice limits during the month of June.

a curved line with the parent iceberg on the concave side. Some of the pieces broken from an iceberg are themselves large enough to be a menace to ships.

As the ship moves closer towards areas known to contain sea ice, one of the most reliable signs that pack ice is being approached is the absence of swell or wave motion in a fresh breeze or a sudden flattening of the sea, especially from leeward. The observation of icebergs in itself is not a good indication that pack ice will be encountered soon, since icebergs may be found at great distances from pack ice. If the sea ice is approached from windward, it is usually compacted and the edge will be sharply defined. However, if it is approached from leeward, the ice is likely to be loose and somewhat scattered, often in long narrow arms.

Another reliable sign of the approach of pack ice, not yet in sight, is the appearance of a pattern, or **sky map,** on the horizon or on the underside of distant, extensive cloud areas, created by the varying amounts of light reflected from different materials on the sea or earth's surface. A bright white glare, or **snow blink,** will be observed above a snow covered surface. When the reflection on the underside of clouds is caused by an accumulation of distant ice, the glare is a little less bright and is referred to as an **ice blink.** A relatively dark pattern is reflected on the underside of clouds when it is over land that is not snow covered. This is known as a **land sky.** The darkest pattern will occur when the clouds are above an open water area, and is called a **water sky.** A mariner experienced in recognizing the sky maps detailed above will find them useful in avoiding ice or searching out openings which may permit his vessel to make progress while proceeding through an ice field.

Another indication of the presence of sea ice is the formation of thick bands of fog over the ice edge as moisture condenses from warm air as it passes over the colder ice. An abrupt change in air or sea temperature or seawater salinity is not a reliable sign of the approach of icebergs or pack ice. However, a drop in the seawater temperature to 1°.1C may be an indication that a ship is within 90 kilometers of pack ice.

The presence of certain species of animals and birds can also indicate that pack ice is in close proximity. The sighting of walruses, seals, or polar bears in the Arctic should warn the mariner that pack ice is close at hand. In the Antarctic, the usual precursors of sea ice are penguins, terns, fulmars, petrels, and skuas. The mariner will do well to observe the habits of all species encountered, for the information gained will be useful on subsequent journeys.

When visibility becomes limited, radar can prove to be an invaluable tool for the polar mariner. Although many icebergs will be observed visually on clear days before there is a return on the radarscope, radar under bad weather conditions will detect the average iceberg at a range of about 15 to 18 kilometers. The intensity of the return is a function of the nature of the iceberg's exposed surface (slope, surface roughness); however, it is unusual to find an iceberg which will not produce a detectable echo.

Large, vertical-sided tabular icebergs of the antarctic and arctic ice islands are usually detected by radar at ranges of 28 to 55 kilometers, with ranges of 68.5 kilometers having been reported.

Whereas a large iceberg is almost always detected by radar in time to be avoided, a growler large enough to be a serious menace to a vessel may be lost in the sea return and escape detection. If an iceberg or growler is detected by radar, tracking is sometimes necessary to distinguish it from a rock, islet, or another ship.

Radar can be of great assistance to one experienced in interpreting the radarscope. Smooth sea ice, like smooth water, returns little or no echo, but small floes of rough, hummocky sea ice capable of inflicting damage to a ship can be detected in a smooth sea at a range of about 4 to 6 kilometers. The return may be similar to sea return, but the same echoes appear at each sweep. A lead in smooth ice is clearly visible on a radarscope, even though a thin coating of new ice may have formed in the opening. A light covering of snow obliterating many of the features to the eye has little effect upon a radar return. The ranges at which ice can be detected by radar are somewhat dependent upon refraction, which is sometimes quite abnormal in polar regions. Adequate training and experience are essential if full benefit is to be realized from radar.

Echoes from the ship's whistle or horn will sometimes indicate the presence of icebergs. Such echoes can give an indication of direction. If the time interval between the sound and its echo is measured, the distance in meters can be determined by multiplying the number of seconds by 168. However, echoes are not a reliable indication because only those pieces of ice with large vertical areas facing the ship return enough echo to be heard. Also, echoes might be received from land or a fog bank.

At relatively short ranges, sonar is sometimes helpful in locating ice. The initial detection of icebergs may be made at a distance of about 5 kilometers or more, but usually considerably less. Growlers may be detected at a distance of 900 meters to 2.5 kilometers, and even smaller pieces may be detected in time to avoid them. Since one-half to seven-eighths of the mass of an iceberg may lie below the surface, the underwater portion presents a better target than the portion above water.

Ice in the polar regions is best detected and observed from the air either from aircraft or by satellite remote sensing surveillance systems. Fixed-winged aircraft have been utilized extensively for obtaining detailed aerial ice reconnaissance information since the early 1930's and will no doubt continue to provide this invaluable service for many years to come. Some ships, particularly icebreakers, proceeding into high latitudes carry helicopters, which are invaluable in locating ice and determining the relative navigability of different portions of the ice pack. If these helicopters, their support vessels, or aircraft flying aerial reconnaissance can be contacted by radio, much useful information will be obtained from them. Ice reports from personnel at arctic and antarctic coastal shore stations can also prove valuable to the polar mariner.

The enormous ice reconnaissance capabilities of meteorological satellites were confirmed within hours of the launch by the National Aeronautics and Space Administration (NASA) of the first experimental meteorological satellite, TIROS I, on April 1, 1960. Although this satellite was placed in an equatorial orbit, it was able to detect ice covered waters in the Gulf of St. Lawrence-Newfoundland region. With the advent of the polar-orbiting meteorological satellites during the mid and late 1960's, the U. S. Navy initiated an operational satellite ice reconnaissance program which could, depending upon solar illumination, observe on a daily basis ice and its movement in any region of the globe. With the further addition of improved sensors such as high resolution infrared and visible scanning radiometers (SR); very high resolution radiometers (VHRR), also in the visible and infrared spectrum; and microwave systems; detailed global satellite ice data were made available under all weather and lighting conditions with resolutions in some cases below 100 meters. Examples of satellite imagery of ice covered waters are shown in figures 3613a and 3613b.

Utilizing portable Automatic Picture Transmission (APT) equipment, which can easily be installed aboard ships or aircraft, visible and infrared radiation data transmitted from operational satellites can be collected during a satellite's passage overhead. In this manner ice data from the satellite's scanning radiometers can be received by APT stations anywhere in the polar regions during the time they are in the line of sight of the satellite. Portable APT equipment is generally small and inexpensive, usually consisting of a receiver with a camera pack, a scanning radiometer adapter, a tape recorder for later data playback, and an omnidirectional antenna for ship and aircraft use. A printed display is available with the addition of a mini-computer that also enhances the image and a small printer to display the digitized data. General information relating to operational satellites, various APT systems, types and modes of satellite data available, and transmission frequencies and times can be obtained from the National Environmental Satellite, Data, and Information Service, National Oceanic and Atmospheric Administration, Washington, D.C.

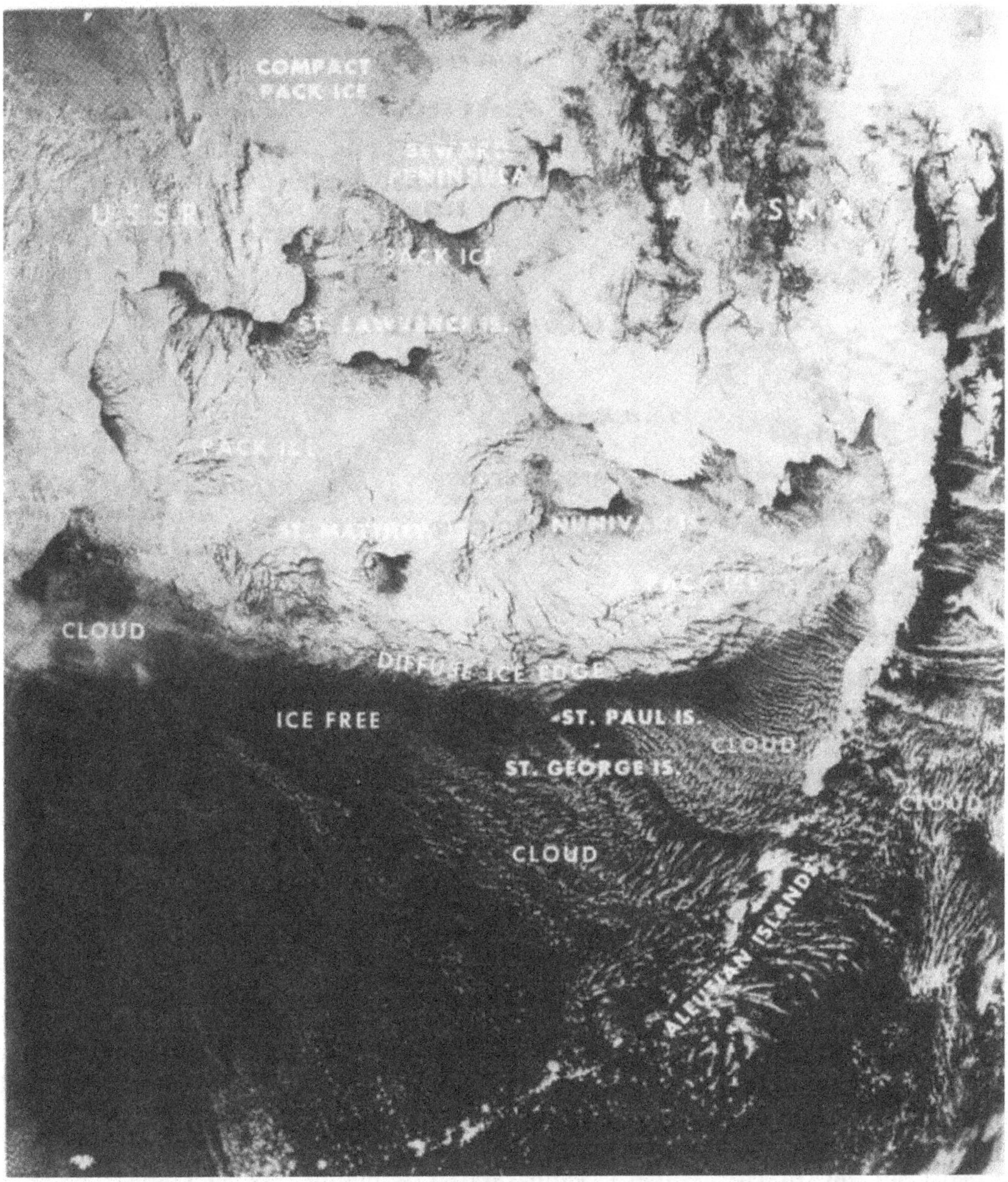

FIGURE 3613a.—Example of satellite imagery with a resolution of 0.9 kilometer.

3614. Operations in ice.—Operations in the polar regions necessarily require considerable advanced planning and many more precautionary measures than those taken prior to a typical open ocean voyage. The crew, large or small, of a polar-bound vessel should be thoroughly indoctrinated in the fundamentals of polar operations, utilizing the best information sources available. The subjects covered should include training in shiphandling in ice, polar navigation, effects of low temperatures on materials and equipment, damage control procedures, communications problems inherent in polar regions, polar meteorology, sea ice terminology, ice observing and reporting procedures (including classification and codes) and polar

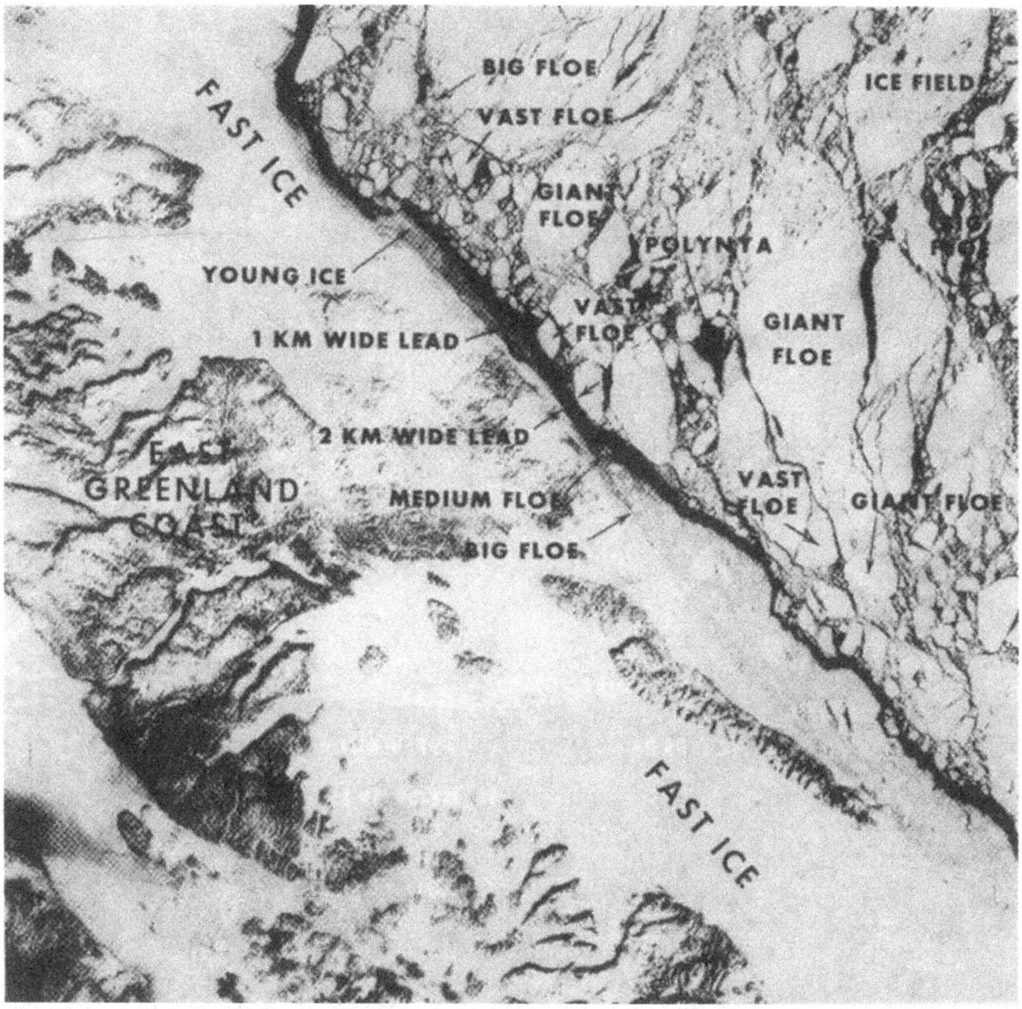

FIGURE 3613b.—Example of satellite imagery with a resolution of 80 meters.

survival. Training materials should consist of reports on previous arctic and antarctic voyages, sailing directions, ice atlases, training films on polar operations, and U. S. Navy service manuals detailing the recommended procedures to follow during high latitude missions. Information relating to sources of information can be obtained from the Director, Naval Oceanography Command, Bay St. Louis, Mississippi, and from the Office of Polar Programs, National Science Foundation, Washington, D.C.

The preparation of a vessel for polar operations is of extreme importance and the considerable experience gained from previous operations should be drawn upon to bring the ship to optimum operating condition. At the very least, operations conducted in ice infested waters require that the vessel's hull and propulsion system undergo certain modifications.

The bow and waterline of the forward part of the vessel should be heavily reinforced. Similar reinforcement should also be considered for the propulsion spaces of the vessel. Cast iron propellers and those made of a bronze alloy do not possess the strength necessary to operate safely in ice. Therefore, it is strongly recommended that propellers made of these materials be replaced by those fabricated from steel. Other desirable features are the absence of vertical sides, deep placement of the propellers, a blunt bow, metal guards to protect propellers from ice damage, and lifeboats for 150 percent of personnel aboard. The complete list of desirable features depends upon the area of operations, types of ice to be encountered, length of stay in the vicinity of ice, anticipated assistance by icebreakers, and possibly other factors. Strength requirements and the minimum thicknesses deemed necessary for the vessel's frames and additional plating to be used as reinforcement, as well as other procedures needed to outfit a vessel for ice operations, can be obtained from the American Bureau of Shipping. For a more definitive and complete guide to the ice strengthening of ships, the mariner may desire to consult the procedures outlined in *Rules for Ice Strengthening of Ships,* from the Board of Navigation, Helsinki, Finland.

Equipment necessary to meet the basic needs of the crew and to insure the successful and safe completion of the polar voyage should not be overlooked. A minimum list of essential items should consist of polar clothing and footware, food, vitamins, medical supplies, fuel, storage batteries, antifreeze, explosives, detonators, fuses, meteorological supplies, and survival kits containing sleeping bags, trail rations, firearms, ammunition, fishing gear, emergency medical supplies, and a repair kit.

Whatever the nature of the vessel, it will be subjected to various hazards which may cause damage. Its safety depends largely upon the thoroughness of advance preparations, the alertness and skill of its crew, and their ability to make repairs if damage is incurred. Spare propellers, rudder assemblies, and patch materials, together with the equipment necessary to effect emergency repairs of structural damage should be carried. Examples of repair materials needed include quick setting cement, oakum, canvas, timbers, planks, pieces of iron of varying shapes, welding equipment, clamps, and an assortment of nuts, bolts, washers, screws, and nails.

Ice and snow accumulation on portions of the vessel poses a definite safety hazard. Therefore, mallets, hammers, and scrapers to aid in the removal of heavy accumulations of ice, together with a supply of snow shovels and stiff brooms for snow removal should be provided.

Navigation in polar waters is, even under optimum conditions, difficult and, during poor conditions, almost impossible. Environmental conditions encountered in the high latitudes such as fog, storms, compass anomalies, atmospheric effects, and, of course, ice, hinder polar operations. Also, deficiencies in the reliability and detail of hydrographic and geographical information presented on polar navigation charts coupled with a distinct lack of reliable bathymetry, current, and tidal data add to the problems of polar navigation. Much work is being carried out in the polar regions to improve the geodetic control, triangulation, and quality of hydrographic and topographic information necessary for accurate polar charts; however, until this massive task is completed, the only resource open to the polar navigator, especially during periods of poor environmental conditions, is to rely upon the basic principles of navigation and adapt them to unconventional methods when abnormal situations arise. A guide to polar navigation is presented in chapter XXV.

Upon the approach to pack ice, a careful decision is needed to determine the best action. Often it is possible to go around the ice, rather than through it. Unless the pack is quite loose, this action usually gains rather than loses time. When skirting an ice field or an iceberg, do so to windward, if a choice is available, to avoid projecting tongues of ice or individual pieces that have been blown away from the main body of ice.

When it becomes necessary to enter pack ice, a thorough examination of the distribution and extent of the ice conditions should be made beforehand from the highest possible location. Aircraft (particularly helicopters) and direct satellite readouts are of great value in determining the nature of the ice to be encountered. The most important features to be noted include the location of open water such as leads and polynyas which may be manifested by water sky, icebergs, and the presence or absence of both ice under pressure and rotten ice. Some protection may be offered the propeller and rudder assemblies by trimming the vessel down by the stern slightly (at no time more than 60 to 90 centimeters) prior to entering the ice; however, this precaution usually impairs the maneuvering characteristics of most vessels not specifically built for icebreaking.

Selection of the point of entry into the pack should be undertaken with great care; and if the ice boundary consists of closely packed ice or ice under pressure, it is advisable to skirt the edge until a more desirable point of entry is located. Seek areas with low ice concentrations, areas of rotten ice or those containing navigable leads, and if possible enter from leeward on a course perpendicular to the ice edge. It is also advisable to take into consideration the direction and force of the wind, and the set and drift of the prevailing currents when determining the point of entry and the course followed thereafter. Due to wind induced wave action, ice floes close to the periphery of the ice pack will take on a bouncing motion which can be quite hazardous to the hull of thin-skinned vessels. In addition, keep in mind that pack ice will drift slightly to the right of the lee of the true wind in the Northern Hemisphere and to the left of the lee in the Southern Hemisphere (art. 3608), and that leads opened by the force of the wind will appear perpendicular to the wind direction. If a suitable entry point cannot be located due to less than favorable conditions, one should be patient. Unfavorable conditions generally improve over a short period of time by a change in the wind, tide, or sea state.

Having entered the pack, *always work with the ice, not against it, and keep moving,* but do not rush the work of negotiating the pack. Patience may pay big dividends. Respect the ice but do not fear it. Proceed at slow speed at first, staying in open water or in areas of weak ice if possible. The vessel's speed may be safely increased after it has been ascertained how well it handles under the varying ice conditions encountered. Remember that it is always better to make good progress in the *general* direction desired than to fight large thick floes in the *exact* direction to be made good. However, avoid the temptation to proceed far to one side of the intended track; it is almost always better to back out and seek a more penetrable area. During those situations when it becomes necessary to back, always do so with extreme caution.

Ice conditions may change rapidly while a vessel is working in pack ice, necessitating some quick maneuvering. It must never be forgotten that conventional vessels, even though ice strengthened, are not built for ice navigation. The vessel should be conned to first attempt to place it in leads or polynyas, giving due consideration to wind conditions. The age, thickness, and size of ice which can be broken depends upon the type, size, strength, and shaft horsepower of the vessel employed. If contact with an ice floe is unavoidable, never strike it a glancing blow. This maneuver may cause the ship to veer off in a direction which will swing the stern into the ice. If possible, seek weak spots in the floe and hit it head-on at slow speed. Unless the ice is rotten or very young, do not attempt to break through the floe, but rather make an attempt to swing it aside as speed is slowly increased. Keep clear of corners and projecting points of ice, but do so without making sharp turns which may throw the stern against the ice, resulting in a damaged propeller, propeller shaft, or rudder. The use of full rudder, in non-emergency situations, is not recommended because it may swing either the stern or mid-section of the vessel into the ice. Keep a sharp watch on the propellers and rudder, fending off with long ice poles pieces of ice which might damage these vital parts. Stop the propellers only if ice cannot be avoided.

Offshore winds may open relatively ice free navigable coastal leads, but such leads should not be entered without benefit of icebreaker escort. If it becomes necessary to enter coastal leads, narrow straits, or bays, an alert watch should be maintained since a shift in the wind may force drifting ice down upon the vessel. An increase in wind on the windward side of a prominent point, grounded iceberg, or land ice tongue extending into the sea will similarly endanger a vessel. It will always be wiser to seek out leads toward the windward side of the main body of the ice pack. In the event that the vessel is under imminent danger of being trapped close to shore by pack ice, immediately attempt to orient the vessel's bow seaward. This will help to take advantage of the little manuevering room available in the open water areas found between ice floes. Work carefully through these areas, easing the ice floes aside while maintaining a close watch on the general movement of the ice pack.

If the vessel is completely halted by pack ice, it is best to keep the rudder amidships and the propellers turning at slow speed. The wash of the propellers may help to clear ice away from the stern, making it possible to back down safely. When the vessel is stuck fast as is the case when the bow is forced up onto a massive ice floe, an attempt first should be made to free the vessel by going full speed astern. If this manuever proves ineffective, it may be possible to get the vessel's stern to move slightly, thereby causing the bow to shift, by shifting the rudder from one side to the other while going full speed ahead. Another attempt at going astern should then free the vessel. The vessel may also be freed by either transferring water from the ballast tanks causing the vessel to list, or by alternately flooding and emptying the fore and aft tanks. Men wielding crowbars may also be able to split the ice at the pressure points. If all these methods fail, the utilization of deadmen (2- to 4-meter lengths of timber buried in holes out in the ice and to which a vessel is moored) and ice anchors (a stockless, single-fluked hook embedded in the ice) may be helpful. With a deadman or ice anchors attached to the ice astern, the vessel may be warped off the ice by winching while the engines are going full astern. If all the foregoing methods fail, explosives placed in holes cut nearly to the bottom of the ice approximately 10 to 12 meters off the beam of the vessel and detonated while the engines are working full astern should succeed in freeing the vessel. A vessel may also be sawed out of the ice if the ambient air temperature is above the freezing point of seawater.

When a vessel becomes so closely surrounded by ice that all steering control is lost and it is unable to move, it is **beset.** It may then be carried by the drifting pack into shallow water or areas containing thicker ice or icebergs with their accompanying dangerous underwater projections. If ice forcibly presses itself against the hull, the vessel is said to be **nipped,** whether or not damage is sustained. When this occurs, the gradually increasing pressure may be capable of holing the vessel's bottom. When a vessel is beset or nipped, freedom may be achieved through the careful manuevering procedures, the physical efforts of the crew, or by the use of explosives similar to those previously detailed. Under severe conditions the mariner's best ally may be patience since there will be many times when nothing can be done to improve the vessel's plight until there is a change in meteorological conditions. It is a time to preserve fuel and perform any needed repairs to the vessel and its engines. Damage to the vessel while it is beset is usually attributable to collisions or pressure exerted between the vessel's hull, propellers, or rudder assembly and the sharp corners of ice floes. These collisions can be minimized greatly by attempting to align the vessel in such a manner as to insure that the pressure from the surrounding pack ice is distributed as evenly as possible over the hull. This is best accomplished when medium or large ice floes encircle the vessel.

In the vicinity of icebergs, either in or outside of the pack ice, a sharp lookout should be kept and all icebergs given a wide berth. The commanding officers and masters of all vessels, irrespective of their size, should treat all icebergs with due respect. The best locations for lookouts are generally in a crow's nest rigged in the foremast or housed in a shelter built specifically for a bow lookout in the eyes of a vessel. Telephone communications between these sites and the navigation bridge on larger vessels will prove invaluable. It is dangerous to approach close to an iceberg of any size because of the possibility of encountering underwater extensions, and because icebergs that are disintegrating may suddenly capsize or readjust their masses to new positions of equilibrium. In periods of low visibility the utmost caution is needed at all times. Vessel speed should be reduced and the watch prepared for quick maneuvering. Radar becomes an effective tool in this case, but does not negate the need for trained lookouts.

Since icebergs may have from eight to nine-tenths of their masses below the water surface, their drift (art. 3608) is generally influenced more by currents than winds, particularly under light wind conditions. The drift of pack ice, on the other hand, is usually dependent upon the wind. Under these conditions, icebergs within the pack may be found moving at a different rate and in a different direction from that of the pack ice. In regions of strong currents, icebergs should always be given a wide berth because they may often travel upwind at great speeds under the influence of contrary currents, wreaking heavy pack in their paths and endangering those vessels that are unable to work clear. In these situations, open water will generally be found leeward of the iceberg, with piled up pack ice to windward. Where currents are weak and a strong wind predominates, similar conditions will be observed as the wind driven ice pack overtakes an iceberg and piles up to windward with an open water area lying to leeward.

Under ice submarine operations require knowledge of prevailing and expected sea ice conditions to ensure maximum operational efficiency and safety. The most important ice features are the frequency and extent of downward projections (bummocks and ice keels) from the underside of the **ice canopy** (pack ice and enclosed water areas from the point of view of the submariner), the distribution of thin ice areas through which submarines can attempt to surface, and the probable location of the outer pack edge where submarines can remain surfaced during emergencies to rendezvous with surface ship or helicopter units.

Bummocks are the subsurface counterpart of hummocks, and **ice keels** are similarly related to ridges. When the physical nature of these ice features is considered, it is apparent that ice keels may have considerable horizontal extent whereas individual bummocks can be expected to have little horizontal extent. In shallow water lanes to the Arctic Basin such as the Bering Strait and the adjoining portions of the Bering Sea and Chukchi Sea, deep bummocks and ice keels may leave little vertical leeway for submarine passage. Widely separated bummocks may be circumnavigated but make for a hazardous passage. Extensive ice areas with numerous bummocks or ice keels which cross the lane, however, may effectively block passage to the Arctic Basin.

Bummocks and ice keels extend downward approximately five times their vertical extent above the ice surface. Therefore, observed ridges of approximately 10 meters may extend as much as 50 meters below sea level. Owing to the direct relation of the frequency and vertical extent between these surface features and their subsurface counterparts, aircraft reconnaissance of ice conditions over a planned submarine cruise track should be conducted before under ice operations are commenced.

Skylights are defined as thin places (usually less than 1 meter thick) in the ice canopy and appear from below as relatively light translucent patches in dark surroundings. The undersurface of a skylight is usually flat; not having been subjected to great pressure although the ice canopy may have a concentration of nearly 8 oktas. Skylights are called large if big enough for a submarine to attempt to surface through them; that is, have a linear extent of at least 120 meters. Skylights smaller than 120 meters are called small. An ice canopy along a submarine's track that contains a number of large skylights or other features such as leads and polynyas which permit a submarine to surface more frequently than 10 times per 56 kilometers is called **friendly ice.** An ice canopy containing no large skylights or other features which permit a submarine to surface is called **hostile ice.**

For a more comprehensive guide to operations in ice, it is recommended that the mariner refer to *Polar Operations,* by Captain Edwin A. MacDonald, USN (Ret.), published by the United States Naval Institute, Annapolis, Maryland.

3615. Great Lakes ice.—Large vessels have been navigating the Great Lakes since the early 1760's. This large expanse of navigable water has since become one of the world's busiest waterways. Due to the northern geographical location of the Great Lakes Basin and its susceptibility to arctic outbreaks of polar air during winter, the formation of ice plays a major role, albeit a disruptive one, in this region's economically vital marine industry. Because of the relatively large size of the five Great Lakes, the ice cover which forms on them is affected by the wind and currents to a greater degree than that on smaller lakes. The Great Lakes northern location results in a long ice growth season which in combination with the effect of wind and current imparts to their ice covers some of the characteristics and behavior of an arctic ice pack. For these reasons, this article is being included in this chapter on ice in the sea.

Since the five Great Lakes extend over a distance of approximately 800 kilometers in a north-south direction, each lake is influenced by varying degrees of meteorological parameters. These parameters, in combination with the fact that each lake also possesses differing hydrometeorological characteristics, materially affects the extent and distribution of their respective ice covers. The largest, deepest, and most northern of the five Great Lakes is Lake Superior. Ice not under pressure, especially along this lake's northern shores, can attain a thickness of between 70 to 100 centimeters which is equivalent to medium first-year ice. Winds and currents acting upon the ice have been known to cause ridging with heights approaching 10 meters. The great depth of Lake Superior, however, provides it with a large heat storage capacity which hinders the growth of ice somewhat, particularly during the period of initial ice formation. During a normal winter, it can be

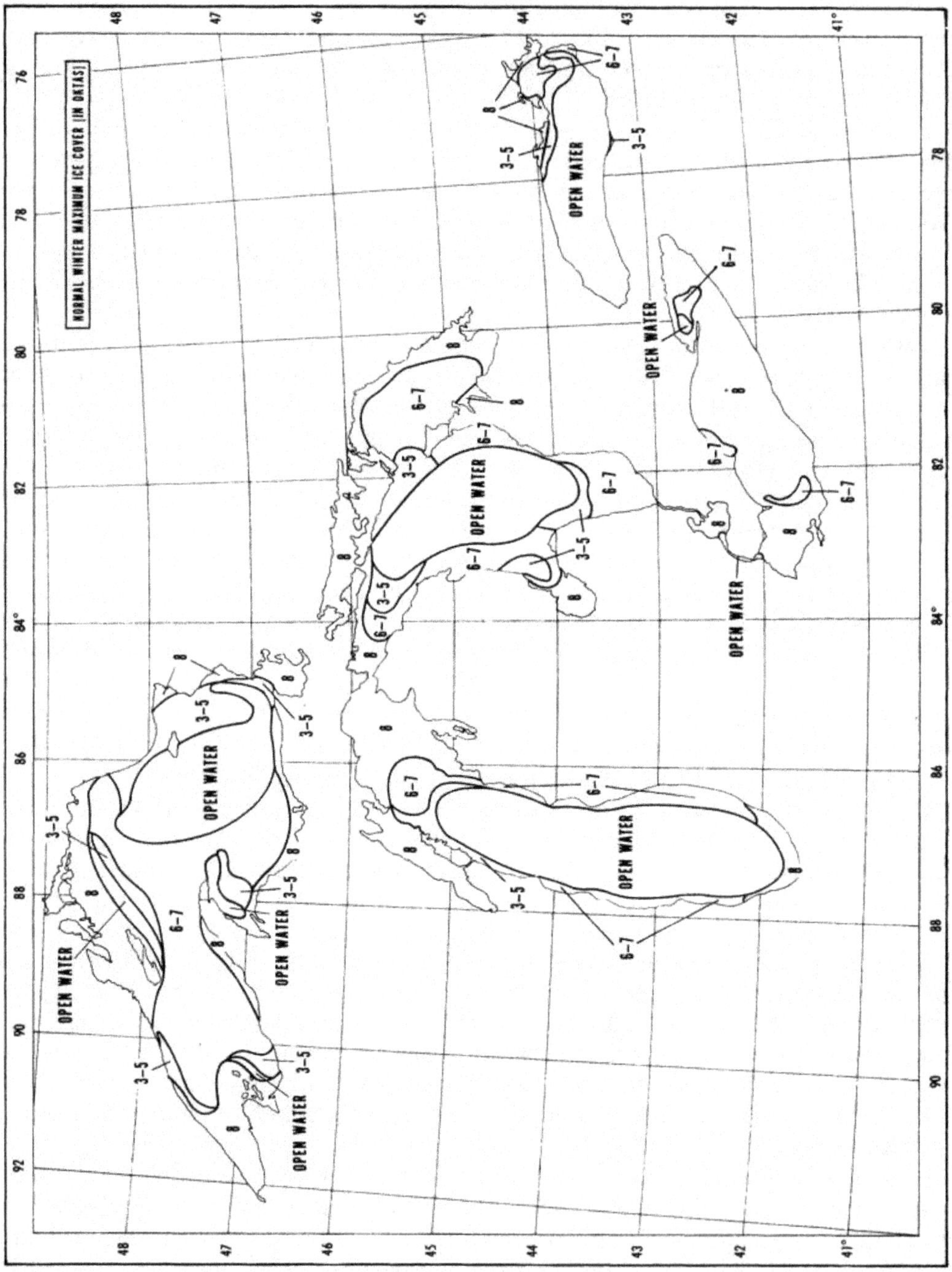

FIGURE 3615a.—Great Lakes maximum ice cover during a normal winter.

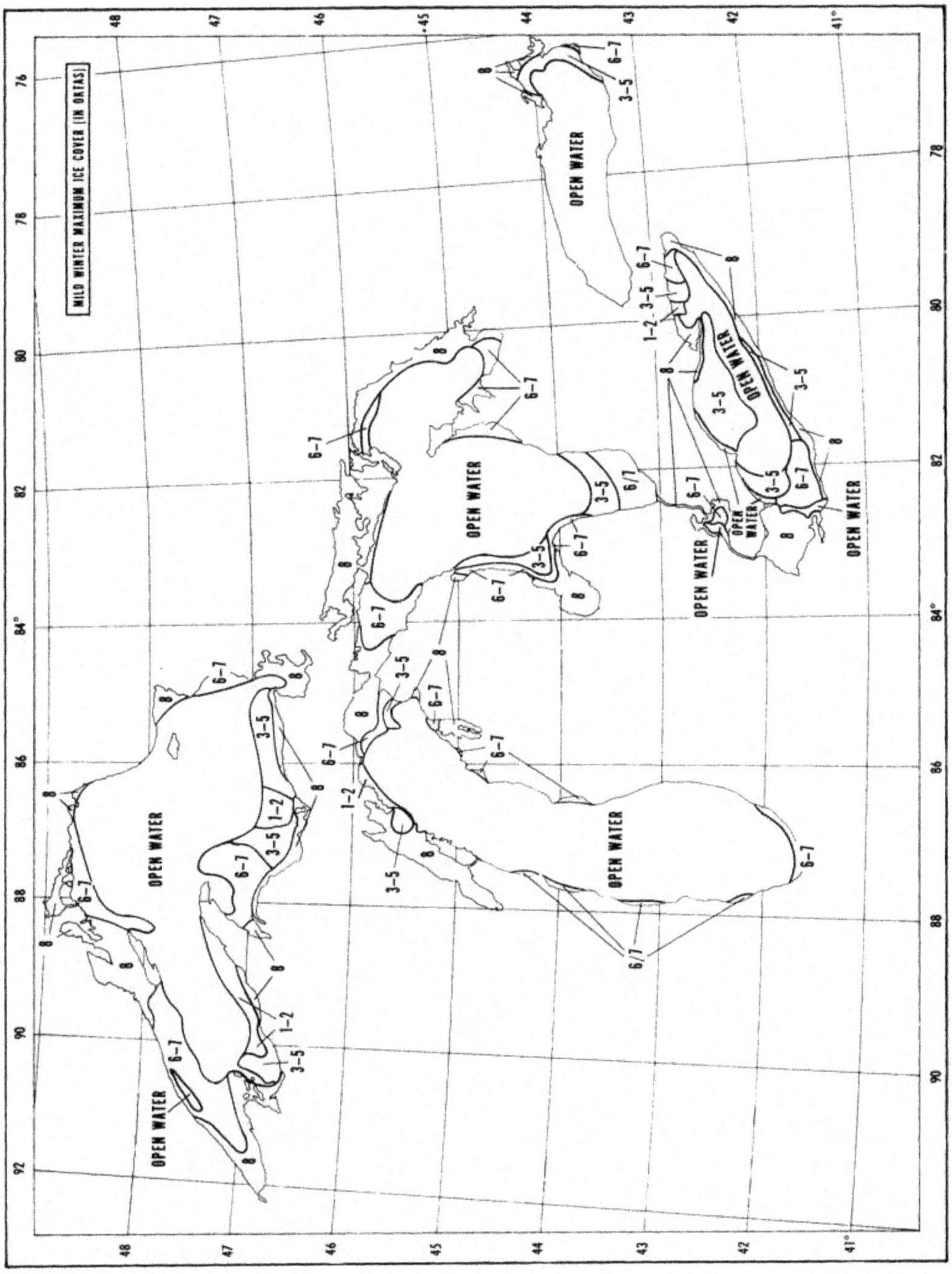

FIGURE 3615b.—Great Lakes maximum ice cover during a mild winter.

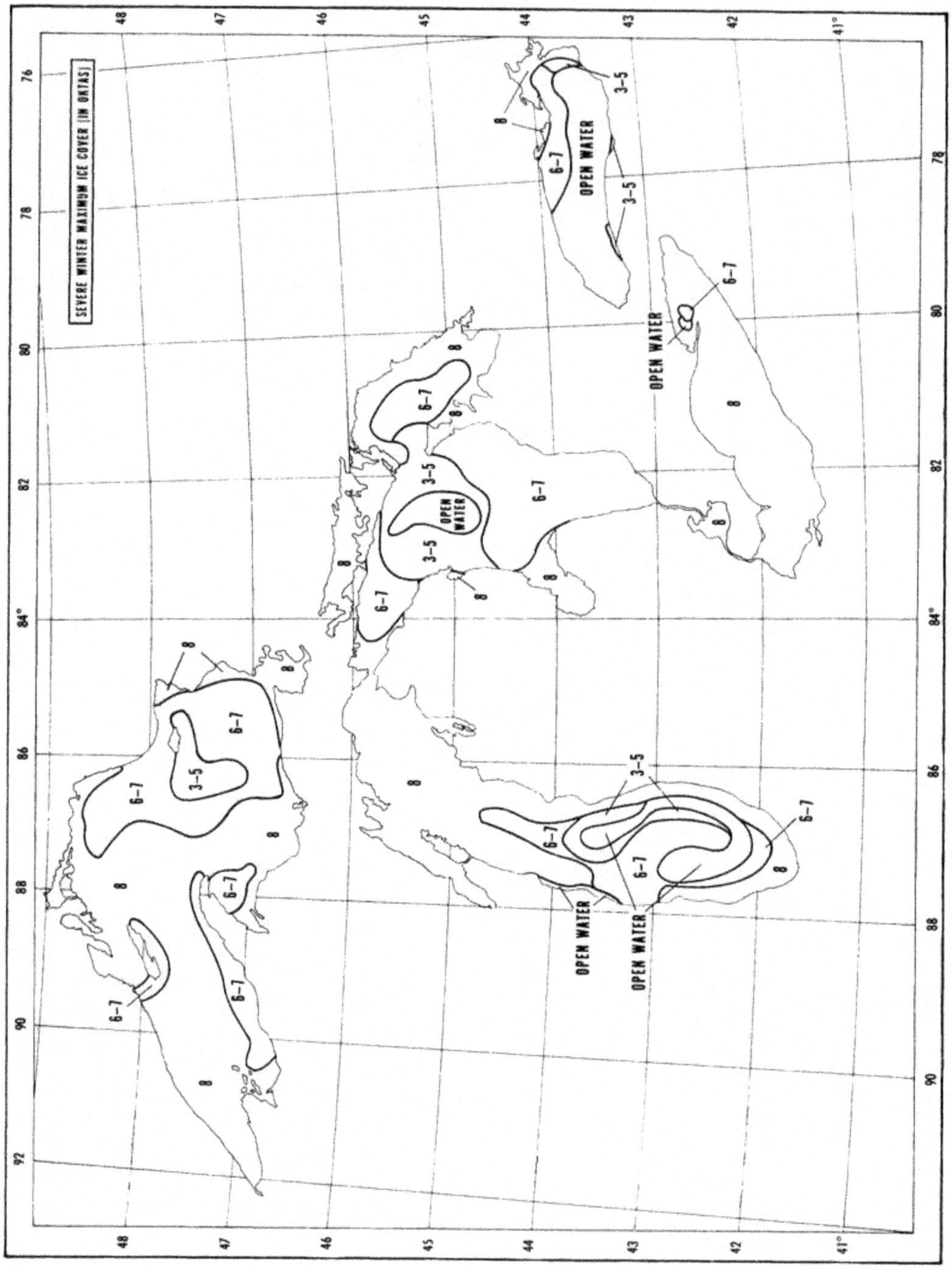

FIGURE 3615c.—Great Lakes maximum ice cover during a severe winter.

expected that 60 percent of the surface area of Lake Superior will become covered by ice. This value increases to 95 percent during a severe winter and decreases to 40 percent during a mild winter. Under average conditions, ice which presents an obstacle to navigation on Lake Superior appears during the last week of December along both the north and south shores with the maximum extent of ice cover occurring between March 30 and April 10.

Lake Michigan extends in a north-south direction for approximately 480 kilometers and possesses the third largest mean depth of the Great Lakes. Its north-south alignment causes ice to accumulate in the northern portions of the lake initially and then grow in a southerly direction as the winter progresses. Ice thickness ranges from an average of 10 to 20 centimeters (grey to grey-white ice) in the southern portion to 50 to 80 centimeters (thin to medium first-year ice) in the northern portion. The ice cover becomes a hazard to navigation in the northern sector of Lake Michigan during the last week of December with the average date of maximum extent of ice cover ranging from March 10 in the Chicago area to March 28 at the northern end of Green Bay. During a severe winter, 80 percent of Lake Michigan's surface area will be covered by ice. This value reduces to 40 percent for a normal winter and to only 10 percent during a mild winter.

Ice formation on Lake Huron generally commences along the northeastern and western shorelines during the last week in December of each year. The deep north central basin of the lake does not, except during severe winters, generally acquire even a partial ice cover. The ice on Lake Huron will on the average consist predominantly of thin first-year ice (30–70 centimeters), with some medium first-year ice (70–120 centimeters) forming during a severe winter. The time interval between dates of the average maximum ice cover extent for this lake ranges from March 11 in the extreme southern portion to March 28 in the northern sector. The percent of lake surface area that will become ice covered is 60 percent for a normal winter, 40 percent for a mild winter, and 80 percent for a severe winter.

Lake Erie is the shallowest of the Great Lakes with a mean depth of just under 20 meters. Because of its shallowness, this lake is greatly influenced by seasonal temperature changes and will accumulate a considerable ice cover over a short period of time. Ice will begin to form first in the very shallow western portion of the lake during mid-December. During its growth period, the lake ice is acted upon by the prevailing winds and currents which concentrate it at the northeastern end of the lake. Generally, Lake Erie's ice cover is made up of a combination of grey-white ice (15–30 centimeters) and thin first-year ice (30–70 centimeters). The average dates on which the maximum extent of ice cover is attained on Lake Erie varies from March 5 in the western sector to March 15 for the northeastern portion of the lake. Since it reacts rapidly to the change in the seasons, Lake Erie will attain an ice cover that blankets 95 to 100 percent of its surface area during a normal winter. During a mild winter, the ice will occupy an area covering 50 percent of the surface area.

Lake Ontario has the smallest surface area of the five Great Lakes, but the second greatest mean depth, second only to Lake Superior. Like Lake Superior, its large mean depth gives Lake Ontario a large heat storage capacity which, in combination with its small surface area, causes the lake to respond slowly to changing meteorological conditions. This, in turn, produces the smallest amount of ice cover found on the Great Lakes. Ice will begin forming during mid-December in the northeastern section of the lake, and wind and current conditions similar to those found on Lake Erie will confine the majority of the ice cover to that section of the lake. The majority of the ice formed will consist of thin first-year ice (30–70 centimeters) with a small concentration of grey-white ice (15–30 centimeters). The date at which the ice cover on Lake Ontario reaches its maximum extent varies on the average from March 10 to March 20. During a mild winter only 8 percent of the lake surface area will be covered by ice. This value increases to 15 percent for a normal winter and to 25 percent for a severe winter.

The maximum ice cover distribution attained by each of the Great Lakes for a normal, mild, and severe winter are shown in figures 3615a, 3615b, and 3615c. It should be noted that although the average maximum ice cover distributions for all five lakes appear on a single chart, they occur during the average time periods detailed above for each lake.

Information concerning analyses, forecasts, and climatology of Great Lakes ice can be obtained from the Great Lakes Environmental Research Laboratory or the National Weather Service Forecast Office, both located in Ann Arbor, Michigan.

3616. Ice observing, reporting, and forecasting.—Advance knowledge of ice conditions to be encountered and knowledge of how these conditions will change over specified time periods are invaluable in both the planning and operational phases of a voyage undertaken in polar regions. Typical ice support services offered to the polar navigator generally include analyses of current ice conditions; short-range (24 to 48 hour), weekly (5 to 7 day), and long range (15 to 30 day) ice forecasts; seasonal long-range (60 to 90 day) ice outlooks which are ordinarily updated by 15 to 30 day forecasts as the season progresses; and ship weather routing service (ch. XXIV) through ice infested waters. Generally an ice analysis or forecast will depict the current or expected configuration and location of the pack ice edge, ice concentrations within the pack itself, and the locations of features such as leads, polynyas, fast ice, and areas of open water.

The single most important input into an ice forecast of any duration is an accurate, current, ice analysis based on the latest ice observations available. As stated previously, ice in the sea can be observed from vessels, fixed winged aircraft, helicopters, and by earth orbiting satellites, as well as by personnel at arctic and antarctic coastal shore stations. By means of modern communications networks, the reports of these observations are relayed to the offices of both federal government agencies and private commercial companies.

Federal agencies providing both ice observing and forecasting services on an operational basis include the Department of Defense, the Department of Transportation, and the Department of Commerce. The Department of Defense, principally through its Department of the Navy, provides ice support services to all U. S. and allied military units. Additional support may be provided to commercial concerns in some instances with the prior approval of the Chief of Naval Operations. Specific information relating to ice observation; forecasting freeze-up, ice growth, movement, concentration, and break-up; ship weather routing services; as well as the methods used to disseminate ice information can be obtained from Director, Naval Oceanography Command, Bay St. Louis, Mississippi. The Department of Transportation, through the U. S. Coast Guard, provides ice breaker support for polar operations, and the International Ice Patrol provides the ice support services detailed in article 3612. Ice forecasts for Alaskan waters are provided to commercial interests by the National Weather Service Forecast Offices in Anchorage and Fairbanks, Alaska. Inquiries concerning assistance available from the Coast Guard while conducting programs in ice should be sent to Commandant, U. S. Coast Guard Headquarters, Washington, D.C.

A network of polar orbiting meteorological satellites and a worldwide communications system provide the National Oceanic and Atmospheric Administration (NOAA) with great quantities of detailed ice information from all areas of the world. Inquiries concerning the products and services available to the polar navigator should be directed to either the National Weather Service or the National Environmental Satellite, Data, and Information Service, NOAA, Department of Commerce, Washington, D.C. Listings of consulting meteorologists and oceanographers, and firms employing such personnel, that provide ice support services to individuals or commercial organizations can be usually located in the professional directories of the leading meteorology and oceanography journals.

Mariners operating in and around sea ice can do much to contribute to the overall effectiveness of many of the services provided them. To assist these programs it is essential that all vessels and aircraft operating in ice areas submit as detailed an ice report as possible to interested agencies. Several codes are now available for reporting ice conditions. The code normally used by those trained to make meteorological observations, but not specifically sea ice observations, consists of a five-character group appended to the World Meteorological Organization (WMO) weather reporting codes FM 21–V, FM 22–V, and FM 23–V. These codes are completely described in volume I of World Meteorological Organization, *Manual on Codes,* WMO No. 306, available from the Secretariat of the World Meteorological Organization, Geneva, Switzerland. A more complete and detailed reporting code (ICEOB) has been in use in the United States since 1972. This code and the procedures for its use are contained in the *Ice Observers Manual,* U. S. Naval Oceanographic Office Technical Note 3700–49–76 (unpublished manuscript).

These codes make use of special nomenclature which has been precisely defined by the World Meteorological Organization in several languages. *Sea-Ice Nomenclature,* (WMO No. 259.TP.145) contains the nomenclature along with photography of most ice features. This publication is very useful for those who plan to submit ice condition reports.

In addition, *Guide to Meteorological Instruments and Observing Practices* (WMO No. 8.TP.31) contains a chapter on ice development, dynamic processes, and observing procedures and should also prove extremely useful.

The mariner who regularly sends as complete an ice report as possible contributes substantially to an increase in the knowledge of synoptic ice conditions and therefore to the accuracy and timeliness of subsequent ice analyses and forecasts.

PART SEVEN
WEATHER

PART SEVEN

WEATHER

	Page
CHAPTER XXXVII. Weather Observations	883
CHAPTER XXXVIII. Weather Elements	903
CHAPTER XXXIX. Tropical Cyclones	931

CHAPTER XXXVII

WEATHER OBSERVATIONS

3701. Introduction.—Weather forecasts are generally based upon information acquired by observations made at a large number of stations. Ashore, these stations are located so as to provide adequate coverage of the area of interest. Most observations at sea are made by mariners, wherever they happen to be. Since the number of observations at sea is small compared to the number ashore, marine observations are of importance in areas where little or no information is available from other sources. Results of these observations are recorded in the deck log (art. 3719), or other appropriate form. Data recorded by designated vessels are sent by radio to centers ashore, where they are plotted, along with other observations, to provide data for drawing synoptic charts (art. 3826). These charts are used to make forecasts. Complete weather information gathered at sea is mailed to the appropriate meteorological services for use in the preparation of weather atlases and in marine climatological studies.

A special effort should be made to provide routine synoptic reports when transiting those areas where few ships are available to make and report weather observations. Such effort is particularly important when vessels are transiting the tropical regions. A vessel's synoptic weather report may be the first indication of a developing tropical cyclone.

In many instances, the analysis of the surface weather map and subsequent forecasts can be no better than the weather reports received.

A knowledge of *weather elements* and the instruments used to measure them is therefore of importance to the mariner who hopes to benefit from weather *broadcasts.*

3702. Atmospheric pressure measurement.—The sea of air surrounding the earth exerts a pressure of about 14.7 pounds per square inch on the surface of the earth. This **atmospheric pressure,** sometimes called **barometric pressure,** varies from place to place, and at the same place it varies with time.

Atmospheric pressure is one of the basic elements of a meteorological observation. When the pressure at each station is plotted on a synoptic chart, lines of equal atmospheric pressure, called **isobars,** are drawn to indicate the areas of high and low pressure and their centers. These are useful in making weather predictions, because certain types of weather are characteristic of each type area, and the wind patterns over large areas are deduced from the isobars.

Atmospheric pressure is measured by means of a **barometer.** A **mercurial barometer** does this by balancing the weight of a column of air against that of a column of mercury. The **aneroid barometer** has a partly evacuated, thin-metal cell which is compressed by atmospheric pressure, the amount of the compression being related to the pressure.

Early mercurial barometers were calibrated to indicate the height, usually in inches or millimeters, of the column of mercury needed to balance the column of air above the point of measurement. While the units **inches of mercury** and **millimeters of mercury** are still widely used, many modern barometers are calibrated to indicate the centimeter-gram-second unit of pressure, the **millibar**, which is equal to 1,000 dynes per square centimeter. A **dyne** is the force required to accelerate a mass of one gram at the rate of one centimeter per second per second. A reading in any of the three units of measurement can be converted to the equivalent reading in either of the other units by means of table 14, or the conversion factors given in appendix D. However, the pressure reading should always be reported in millibars.

3703. The mercurial barometer was invented by Evangelista Torricelli in 1643. In its simplest form it consists of a glass tube a little more than 30 inches in length and of uniform internal diameter; one end being closed, the tube is filled with mercury, and inverted into a cup of mercury. The mercury in the tube falls until the column is just supported by the pressure of the atmosphere on the open cup, leaving a vacuum at the upper end of the tube. The height of the column indicates atmospheric pressure, greater pressures supporting higher columns of mercury.

The mercurial barometer is subject to rapid variations in height, called **pumping,** due to pitch and roll of the vessel and temporary changes in atmospheric pressure in the vicinity of the barometer. Because of this, the care required in the reading of the instrument, its bulkiness, and its vulnerability to physical damage, the mercurial barometer has been replaced at sea by the aneroid barometer.

3704. The aneroid barometer (fig. 3704) measures atmospheric pressure by means of the force exerted by the pressure on a partly evacuated, thin-metal element called a **sylphon cell (aneroid capsule).** A small spring is used, either internally or externally, to partly counteract the tendency of the atmospheric pressure to crush the cell. Atmospheric pressure is indicated directly by a scale and a pointer connected to the cell by a combination of levers. The linkage provides considerable magnification of the slight motion of the cell, to permit readings to higher precision than could be obtained without it.

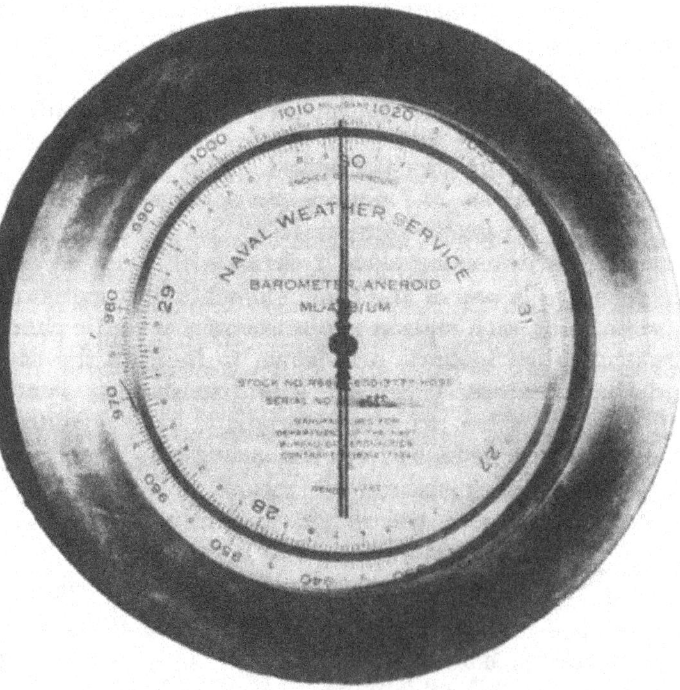

FIGURE 3704.—An aneroid barometer.

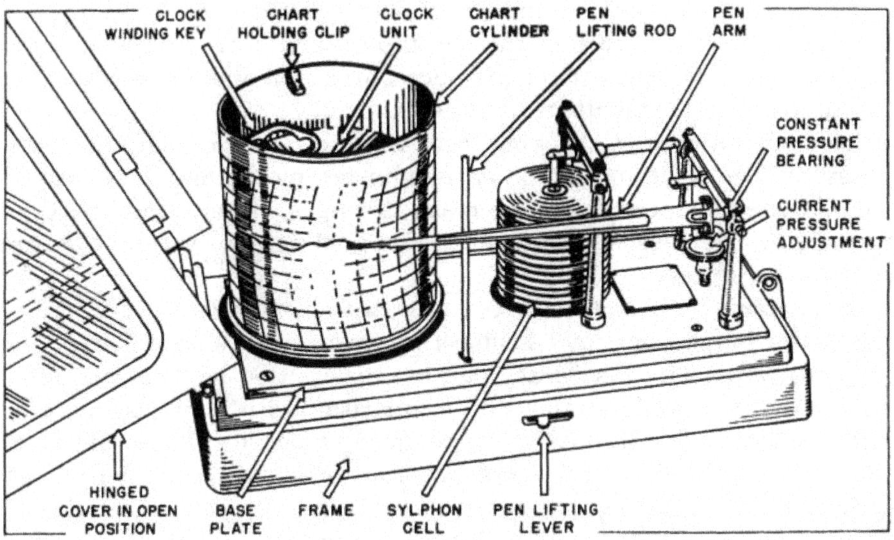

FIGURE 3705.—A barograph.

An aneroid barometer should be mounted permanently. Prior to installation, the barometer should be carefully set to station pressure (art. 3706). An adjustment screw is provided for this purpose. The error in the reading of the instrument is determined by comparison with a mercurial barometer or a standard precision aneroid barometer. If a qualified meteorologist is not available to make this adjustment, it is good practice to remove only one-half the apparent error. The case should then be tapped gently to assist the linkage to adjust itself, and the process repeated. If the remaining error is not more than half a millibar (0.015 inch), no attempt should be made to remove it by further adjustment. Instead, a correction should be applied to the readings. The accuracy of this correction should be checked from time to time.

3705. The barograph (fig. 3705) is a recording barometer. Basically, it is the same as a nonrecording aneroid barometer except that the pointer carries a pen at its outer end, and the scale is replaced by a slowly rotating cylinder around which a prepared chart is wrapped. A clock mechanism inside the cylinder rotates the cylinder so that a continuous line is traced on the chart to indicate the pressure at any time.

A **marine microbarograph** is a precision barograph with greater magnification of deformations due to pressure changes, and a correspondingly expanded chart. It is designed to maintain its precision through the varied and exacting conditions encountered in shipboard use. Two sylphon cells are used, one being mounted over the other in tandem. Minor fluctuations due to shocks or vibrations are eliminated by damping. Since oil-filled dashpots are used for this purpose, the instrument should not be inverted.

Ship motions are compensated by damping and spring loading which make it possible for the microbarograph to be tilted up to 22° without varying more than 0.3 millibars from true reading.

The barograph is usually mounted on a shelf or desk in a room open to the atmosphere, and in a location which minimizes the effect of the ship's vibration. Shock-absorbing material such as sponge rubber is placed under the instrument to minimize the transmission of shocks.

The pen should be checked and the inkwell filled each time the chart is changed, every week in the case of the barograph, and each 4 days in the case of the microbarograph. The dashpots of the microbarograph should be kept filled with dashpot oil to within three-eights inch of the top.

Both instruments require checking from time to time to insure correct indication of pressure. The position of the pen is adjusted by a small knob provided for this purpose. The adjustment should be made in stages, eliminating half the apparent error, tapping the case to insure linkage adjustment to the new setting, and then repeating the process.

3706. Adjustment of barometer readings.—Atmospheric pressure as indicated by a barometer or barograph may be subject to several errors, as follows:

Instrument error. Any inaccuracy due to imperfection or incorrect adjustment of the instrument can be determined by comparison with a precision instrument. The National Weather Service provides a comparison service. In certain U. S. ports a Port Meteorological Officer carries a portable precision aneroid barometer for barometer comparisons on board ships which participate in the cooperative observation program of the National Weather Service. The portable barometer is compared with station barometers before and after each ship visit. If a barometer is taken to a National Weather Service shore station, the comparison can be made there. The correct sea-level pressure can be obtained by telephone. The shipboard barometer should be corrected for height, as explained below, before comparison with this telephoned value. If there is reason to believe that the barometer is in error, it should be compared with a standard, and if an error is found, the barometer should be adjusted to the correct reading, or a correction applied to all readings.

Height error. The atmospheric pressure reading at the height of the barometer is called the **station pressure** and is subject to a height correction in order to make it a sea level pressure reading. Isobars adequately reflect wind conditions and geographic distribution of pressure only when they are drawn for pressure at constant height (or the varying height at which a constant pressure exists). On synoptic charts it is customary to show the equivalent pressure at sea level, called **sea level pressure.** This is found by applying a correction to station pressure. The correction, given in table 11, depends upon the height of the barometer and the average temperature of the air between this height and the surface. The outside air temperature taken aboard ship is sufficiently accurate for this purpose. *This is an important correction which should be applied to all readings of any type barometer.*

Gravity error. Mercurial barometers are calibrated for standard sea-level gravity at latitude 45°32′40″. If the gravity differs from this amount, an error is introduced. The correction to be applied to readings at various latitudes is given in table 12. *This correction does not apply to readings of an aneroid barometer or microbarograph.* Gravity also changes with height above sea level, but the effect is negligible for the first few hundred feet, and so is not needed for readings taken aboard ship.

Temperature error. Barometers are calibrated at a standard temperature of 32°F. The liquid of a mercurial barometer expands as the temperature of the mercury rises, and contracts as it decreases. The correction to adjust the reading of the instrument to the true value is given in table 13. *This correction is to be applied to readings of mercurial barometers only.* Modern aneroid barometers are compensated for temperature changes by the use of different metals having unequal coefficients of linear expansion.

3707. Wind measurement consists of determination of the direction *from* which the wind is blowing, and the speed of the wind. Wind direction is measured by a **wind vane,** and wind speed by an **anemometer.**

A wind vane consists of a device pivoted on a vertical shaft, with more surface area on one side of the pivot than on the other, so that the wind exerts more force on one side, causing the smaller end to point into the wind.

In its simplest form, an anemometer consists of a number of cups mounted on short horizontal arms attached to a longer vertical shaft which rotates as the wind blows against the cups. The speed at which the shaft rotates is directly proportional to the wind speed.

Several types of wind speed and direction recorders are available.

If no anemometer is available, wind speed can be estimated by its effect upon the sea and objects in its path, as explained in article 3709.

3708. True and apparent wind.—An observer aboard a vessel proceeding through still air experiences an **apparent wind** which is from dead ahead and has an apparent speed equal to the speed of the vessel. Thus, if the actual or **true wind** is zero and the speed of the vessel is 10 knots, the apparent wind is from dead ahead at 10 knots. If the true wind is from dead ahead at 15 knots, and the speed of the vessel is 10 knots, the apparent wind is $15+10=25$ knots from dead ahead. If the vessel makes a 180° turn, the apparent wind is $15-10=5$ knots from dead astern.

In any case, the apparent wind is the vector sum (art. 118, vol. II) of the true wind and the *reciprocal* of the vessel's course and speed vector. Since wind vanes and anemometers measure *apparent* wind, the usual problem aboard a vessel equipped with an anemometer is to convert this to true wind. There are several ways of doing this. Perhaps the simplest is by the graphical solution illustrated in the following example:

Example 1.—A ship is proceeding on course 150° at a speed of 17 knots. The apparent wind is from 40° off the starboard bow, speed 15 knots.

Required.—The relative direction, true direction, and speed of the true wind.

Solution (fig. 3708a).—Starting at the center of a manuevering board or other suitable form, draw a line in the relative direction *from* which the apparent wind is blowing. Locate point 1 on this line, at a distance from the center equal to the speed of the apparent wind (2:1 scale is used in figure 3708a). From point 1, draw a line vertically *downward.* Locate point 2 on this line at a distance from point 1 equal to the speed of the vessel in knots, to the same scale as the first line. The relative direction of the true wind is *from* point 2 (120°) toward the center, and the speed of the true wind is the distance of point 2 from the center, to the same scale used previously (11 kn.). The true direction of the wind is the relative direction plus the true heading, or $120°+150°=270°$.

Answers.—True wind from 120° relative, is 270° true, at 11 knots.

A quick solution can be made without an actual plot, in the following manner: On a maneuvering board, label the circles 5, 10, 15, 20, etc., from the center, and draw vertical lines tangent to these circles. Cut out the 5:1 scale and discard that part having graduations greater than the maximum speed of the vessel. Keep this equipment for all solutions. (For durability, the two parts can be mounted on cardboard or other suitable material.) To find true wind, spot in point 1 by eye. Place the zero of the 5:1 scale on this point and align the scale (inverted) by means of the vertical lines. Locate point 2 at the speed of the vessel as indicated on the 5:1 scale. It is always vertically *below* point 1. Read the relative direction and the speed

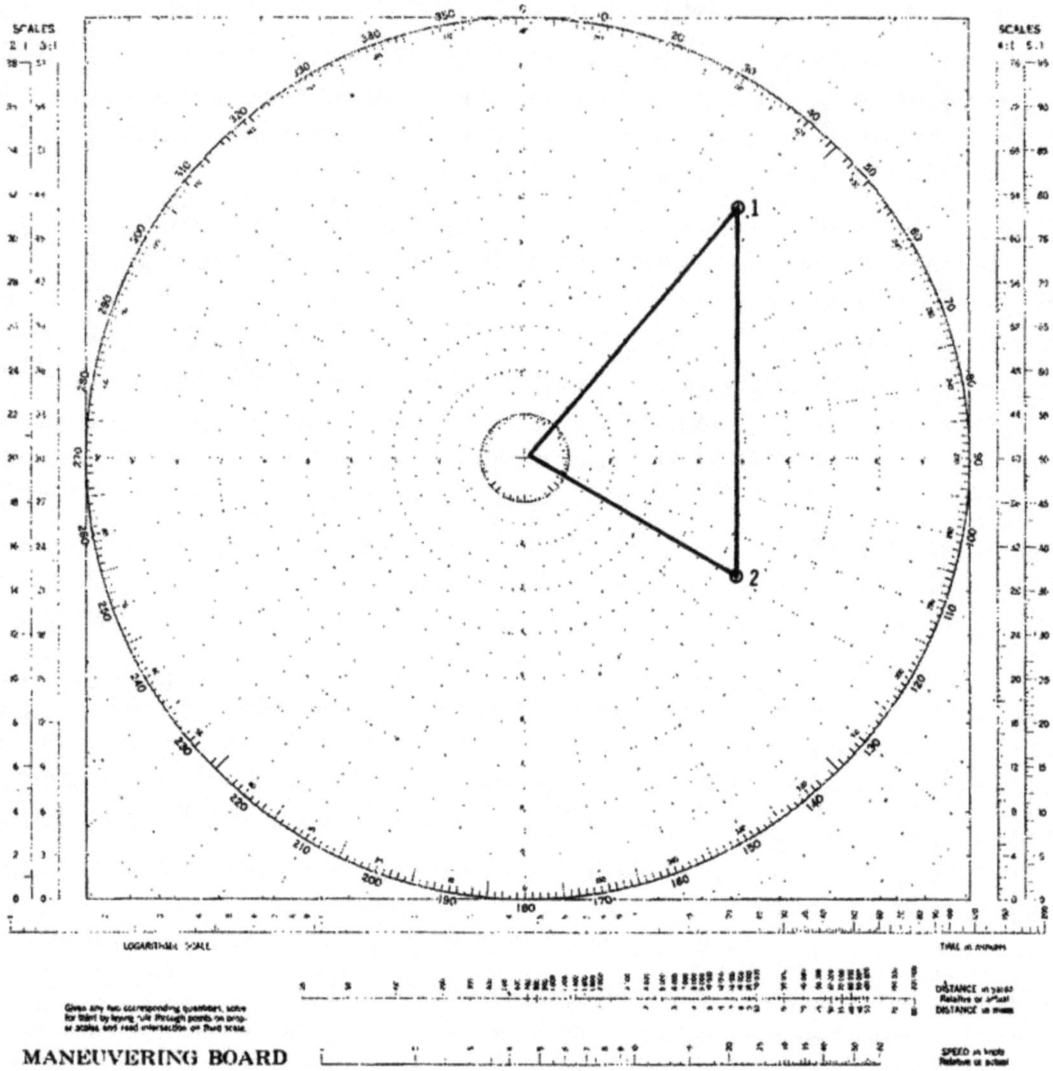

FIGURE 3708a.—Finding true wind by maneuvering board.

of the true wind using eye interpolation if needed. The National Weather Service distributes a wind vector computer called a *Shipboard Wind Plotter* (fig. 3708b). Solution by means of this plotter is illustrated in the following example:

Example 2.—A ship is proceeding on course 270° at a speed of 14.5 knots. The apparent wind is from 40° off the starboard bow, speed 20 knots.

Required.—The relative direction, true direction, and speed of the true wind by Shipboard Wind Plotter.

Solution (fig. 3708b).—The true direction of the apparent wind is determined by adding the apparent wind direction to the ship's heading if the wind is from off the starboard bow and subtracting the apparent wind direction if the wind is from off the port bow. In this example, the true direction of the apparent wind is 310°. In this solution the red arrowhead is considered the top of the plotter. Set ship's course, 270°, to the top of the plotter by rotating the protractor disk to set 270° at the red arrow. Using a convenient linear scale, measure vertically downward from

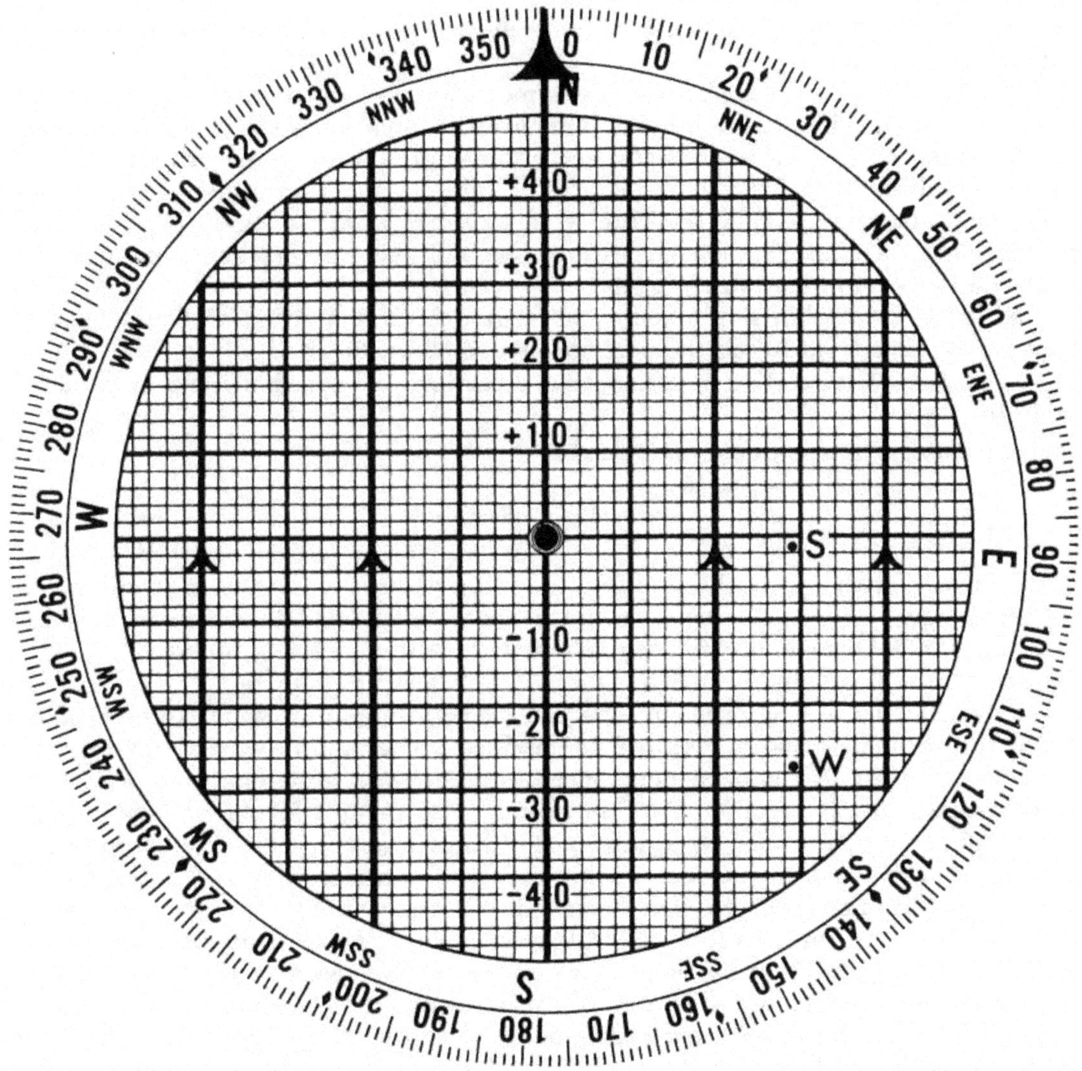

FIGURE 3708b.—Finding true wind by Shipboard Wind Plotter.

the center peg of the plotting board a distance equivalent to 14.5 knots. Mark this point "S" for ship. Rotate the protractor disk of the plotting board until 310° is at the red arrowhead at the top of the plotting board. Using the same linear scale as for ship's speed, plot vertically downward from the center peg of the plotting board a distance equivalent to 20 knots. Mark this point "W." Rotate the protractor disk until the "S" is vertically above the "W," using the vertical lines on the plotting board to line up the two points. Read the true wind direction at the top of the plotting board. The distance between points "S" and "W" is the true wind speed, using the same scale as in plotting points "S" and "W."

Answers.—True wind direction is 357°, true wind speed is 13 knots.

Such problems can be solved by the use of true directions and a regular vector solution, but the use of relative directions simplifies the plot because that component of the apparent wind due to the vessel's motion is always parallel (but reversed) to the vessel's motion, and the apparent wind is always *forward* of the true wind.

A tabular solution based upon the same principle can be made by means of table 10. The entering values for this table are the apparent wind speed *in units of ship's speed,* and the difference between the heading and the apparent wind direction. The values taken from the table are the relative direction (right or left) of the true wind, and the speed of the true wind *in units of ship's speed.* If a vessel is proceeding at 12 knots, 6 knots constitutes one-half (0.5) unit, 12 knots one unit, 18 knots 1.5 units, 24 knots two units, etc.

Example 3.—A ship is proceeding on course 270° at a speed of 10 knots. The apparent wind is from 10° off the port bow, speed 30 knots.

Required.—The relative direction, true direction, and speed of the true wind by table 10.

Solution.—The apparent wind speed is

$$\frac{30}{10} = 3.0 \text{ ship's speed units.}$$

Enter table 10 with 3.0 and 10° and find the relative direction of the true wind to be 15° off the port bow (345° relative), and the speed to be 2.02 times the ship's speed, or $2.02 \times 10 = 20$ knots, approximately. The true direction is $345° + 270° = 255°$.

Answers.—True wind from 345° relative, 255° true, at 20 knots.

By variations of this problem, one can find the apparent wind from the true wind, the course or speed required to produce an apparent wind from a given direction or speed, or the course and speed to produce an apparent wind of a given speed from a given direction. Such problems arise in aircraft carrier operations.

Wind speed determined by appearance of the sea (art. 3709) is the speed of the true wind. The sea also provides an indication of the direction of the true wind, because waves move in the same direction as the generating wind, not being deflected by earth rotation (art. 3302). If a wind vane is used, the direction of the apparent wind thus determined can be used with the speed of the true wind to determine the direction of the true wind by vector diagram. If a maneuvering board is used, draw a circle about the center equal to the speed of the true wind. From the center, plot the ship's vector (true course and speed). From the end of this vector draw a line in the direction in which the apparent wind is blowing (reciprocal of the direction from which it is blowing) until it intersects the speed circle. This line is the apparent wind vector, its length denotes the speed. A line from the center of the board to the end of the apparent wind vector is the true wind vector. The reciprocal of this vector is the direction from which the true wind is blowing. If the true wind speed is less than the speed of the vessel, two solutions are possible. If solution is by table 10, the true speed, in units of ship's speed, is found in the column for the direction of the apparent wind. The number to the left is the relative direction of the true wind. The number on the same line in the side columns is the speed of the apparent wind in units of ship's speed. Again, two solutions are possible if true wind speed is less than ship's speed.

3709. Wind and the sea.—The action of the wind in creating ocean currents and waves is discussed in chapters XXXII and XXXIII, respectively. There is a relationship between the speed of the wind and the state of the sea in the immediate vicinity of the wind. This is useful in predicting the sea conditions to be anticipated when future wind speed forecasts are available. It can also be used to estimate the speed of the wind, which may be desirable when an anemometer is not available.

Wind speeds are usually grouped in accordance with the **Beaufort scale** named after Admiral Sir Francis Beaufort, who devised it in 1806. As adopted in 1838, Beaufort numbers ranged from 0, calm, to 12, hurricane. The Beaufort scale, with certain other pertinent information, is given in appendix V. The appearance of the sea at different Beaufort scale numbers from 0 through 10 is shown in appendix W.

3710. Temperature is the intensity or degree of heat. It is measured in degrees. Several different temperature scales are in use.

On the **Fahrenheit (F)** scale pure water freezes at 32° and boils at 212°.

On the **Celsius (C)** scale commonly used with the metric system, the freezing point of pure water is 0° and the boiling point is 100°. This scale, has been known by various names in different countries. In the United States it was formerly called the **centigrade** scale. The Ninth General Conference of Weights and Measures, held in France in 1948, adopted the name Celsius to be consistent with the naming of other temperature scales after their inventors, and to avoid the use of different names in different countries. On the original Celsius scale, invented in 1742 by a Swedish astronomer named Anders Celsius, the numbering was the reverse of the modern scale, 0° representing the boiling point of water, and 100° its freezing point.

Absolute zero is considered to be the lowest possible temperature, at which there is no molecular motion and a body has no heat. For some purposes, it is convenient to express temperature by a scale at which 0° is absolute zero. This is called **absolute** temperature. If Fahrenheit degrees are used, it may be called **Rankine (R)** temperature; and if Celsius, **Kelvin (K)** temperature. The Kelvin scale is more widely used than the Rankine. Absolute zero is at (−) 459°.69F or (−) 273°.16C.

Temperature by one scale can be converted to that at another by means of the relationship that exists between the scales. Thus,

$$C=\frac{5}{9}(F-32),$$

$$F=\frac{9}{5}C+32$$

and

$$K=C+273^{\circ}.16C.$$

A temperature of (−) 40° is the same by either the Celsius or Fahrenheit scale. Similar formulas can be made for conversion of other temperature scale readings. Table 15 gives the equivalent values of Fahrenheit, Celsius, and Kelvin temperatures.

The intensity or degree of heat (temperature) should not be confused with the *amount* of heat. If the temperature of air or some other substance is to be increased (the substance made hotter) by a given number of degrees, the amount of heat that must be added is dependent upon the amount of the substance to be heated. Also, equal amounts of different substances require the addition of unequal amounts of heat to effect equal increase in temperature because of their difference of specific heat (art. 3012). Units used for measurement of amount of heat are the **British thermal unit (BTU)**, the amount of heat needed to raise the temperature of 1 pound of water 1° Fahrenheit; and the **calorie**, the amount of heat needed to raise the temperature of 1 gram of water 1° Celsius.

3711. Temperature measurement is made by means of a **thermometer.** Most thermometers are based upon the principle that materials expand with increase of temperature, and contract as temperature decreases. In its most usual form a thermometer consists of a bulb filled with mercury and connected to a tube of very small cross-sectional area. The mercury only partly fills the tube. In the remainder is a vacuum created during construction of the instrument. The air is driven out by boiling the mercury, and the top of the tube is then sealed by a flame. As the mercury expands or contracts with changing temperature, the length of the mercury column in the tube changes. Temperature is indicated by the position of the top of the column of mercury with respect to a scale etched on the glass tube or placed on the thermometer support.

Temperature measuring equipment should be placed in a shelter which protects it from mechanical damage and direct rays of the sun. The shelter should have louvered sides to permit free access of air. Aboard ship, the shelter should be placed in an exposed position as far as practicable from metal bulkheads. On vessels where shelters are not available, the temperature measurement should be made in shade at an exposed position on the windward side.

Sea surface temperature observations are used in the forecasting of fog and furnish important information about the development and movement of tropical cyclones. Commercial fishermen are interested in the sea surface temperature as an aid in locating certain species of fish. There are several methods of determining seawater temperature. These include engine room intake readings, condenser intake readings, thermister probes attached to the hull, and readings from buckets recovered from over the side. Although the condenser intake method is not a true measure of surface water temperature, the error is generally small. Measurement should be made near the entrance of the intake.

If the temperature of the water at the surface is desired, a sample should be obtained by bucket, preferably a canvas bucket, from a forward position well clear of any discharge lines. The sample should be taken immediately to a place where it is sheltered from wind and sun. The water should then be stirred with the thermometer, keeping the bulb submerged, until an essentially constant reading is obtained.

A considerable variation in sea surface temperature can be experienced in a relatively short distance of travel. This is especially true when crossing major ocean currents such as the Gulf Stream and the Kuroshio. Significant variations also occur where large quantities of freshwater are discharged from rivers.

3712. Humidity is the condition of the atmosphere with reference to its water vapor content. **Relative humidity** is the ratio (stated as a percentage) of the pressure of water vapor present in the atmosphere to the saturation vapor pressure at the same temperature.

As air temperature decreases, the relative humidity increases. At some point, saturation takes place, and any further cooling results in condensation of some of the moisture. The temperature at which this occurs is called the **dew point,** and the moisture deposited upon natural objects is called **dew** if it forms in the liquid state, or **frost** if it forms in the frozen state.

The same process causes moisture to form on the outside of a container of cold liquid, the liquid cooling the air in the immediate vicinity of the container until it reaches the dew point. When moisture is deposited on man-made objects, it is usually called **sweat.** It occurs whenever the temperature of a surface is lower than the dew point of air in contact with it. It is of particular concern to the mariner because of its effect upon his instruments, and possible damage to his ship or its cargo. Lenses of optical instruments may sweat, usually with such small droplets that the surface has a "frosted" appearance. When this occurs, the instrument is said to "fog" or "fog up," and is useless until the moisture is removed. Damage is often caused by corrosion or direct water damage when pipes sweat and drip, or when the inside of the shell plates of a vessel sweat. Cargo may sweat if it is cooler than the dew point of the air. One of the principal problems of preserving ships of the reserve fleet is the protection against moisture. An important step is the draining of all water, sealing of compartments, and drying of the air.

Clouds and fog form by "sweating" of minute particles of dust, salt, etc., in the air. Each particle forms a nucleus around which a droplet of water forms. If air is completely free from solid particles on which water vapor may condense, the extra moisture remains in the vapor state, and the air is said to be **supersaturated.**

Relative humidity and dew point are measured by means of a **hygrometer.** The most common type, called a **psychrometer,** consists of two thermometers mounted together on a single strip of material, as shown in figure 3712a. One of the thermometers is mounted a little lower than the other, and has its bulb covered with muslin. When the muslin covering is thoroughly moistened and the thermometer well ventilated, evaporation cools the bulb of the thermometer, causing it to indicate a lower reading than the other. A **sling psychrometer,** illustrated in figure 3712a, is ventilated by whirling the thermometers. **Dry-bulb temperature** is indicated by the uncovered **dry-bulb thermometer,** and **wet-bulb temperature** is indicated by the muslin-covered **wet-bulb thermometer.** The difference between these two temperatures, and the dry-bulb temperature, are used to enter **psychrometric tables** to find the relative humidity (tab. 16) and dew point (tab. 17). If the wet-bulb temperature is above freezing, reasonably accurate results can be obtained by a psychrometer consisting of wet- and dry-bulb thermometers mounted so that air can circulate freely around them without special ventilation. This type of installation is common aboard ship.

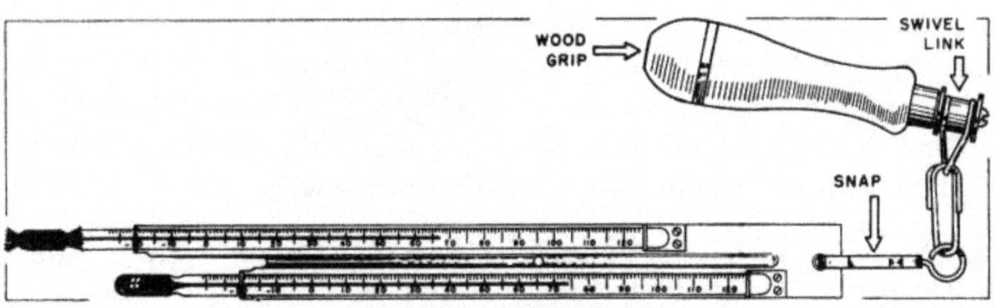

FIGURE 3712a.—A sling psychrometer.

Example.—The dry-bulb temperature is 65°F and the wet-bulb temperature is 61°F.

Required.—(1) Relative humidity, (2) dew point.

Solution.—The difference between readings is 4°. Entering table 16 with this value and a dry-bulb temperature of 65°, the relative humidity is found to be 80 percent. From table 17 the dew point is found to be 58°.

Answers.—(1) Relative humidity 80 percent, (2) dew point 58°.

Also in use aboard many ships is the **electric psychrometer** (fig. 3712b). This is a handheld, battery operated instrument with two mercury thermometers for obtaining dry- and wet-bulb temperature readings. It consists of a plastic housing that holds the thermometers, batteries, motor, and fan. Although the electric psychrometer is constructed primarily of non-corrodible materials, prolonged exposure to the marine environment will shorten the life of the instrument.

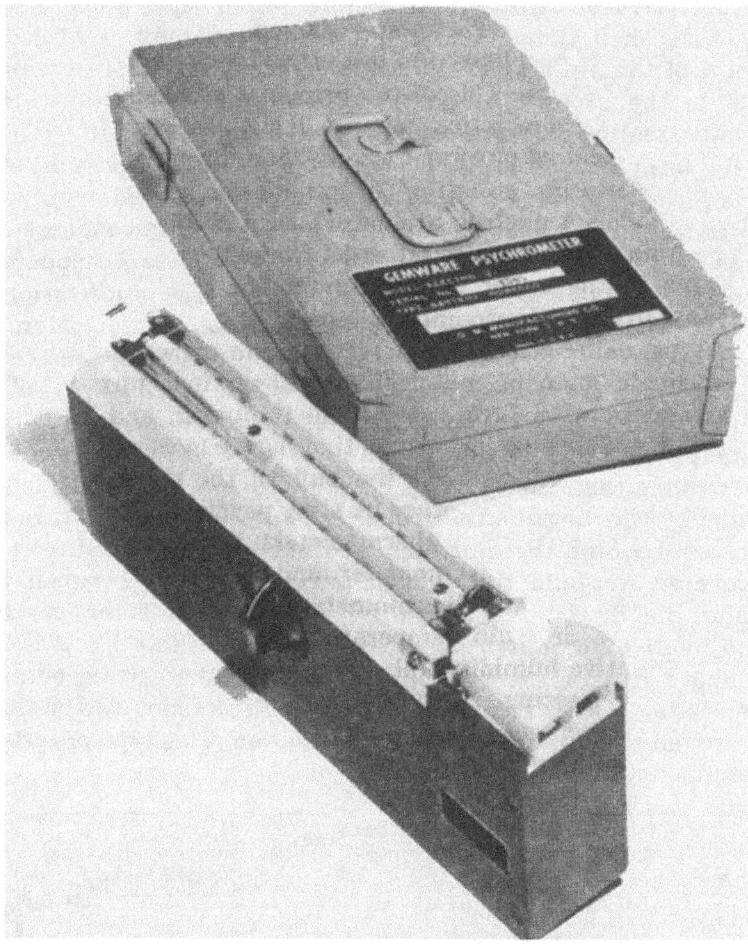

FIGURE 3712b.—Electric psychrometer.

3713. Clouds are visible assemblages of numerous tiny droplets of water, or ice crystals, formed by condensation of water vapor in the air, with the bases of the assemblages above the surface of the earth. **Fog** is a similar assemblage in contact with the surface of the earth.

The shape, size, height, thickness, and nature of a cloud depend upon the conditions under which it is formed. Therefore, clouds are indicators of various processes occurring in the atmosphere. The ability to recognize different types and a knowledge of the conditions associated with them are useful in predicting future weather.

Although the variety of clouds is virtually endless, they may be classified according to general type. Clouds are grouped generally into three "families" according to some common characteristic. **High clouds** are those having a mean lower level above 20,000 feet. They are composed principally of ice crystals. **Middle clouds** have a mean level between 6,500 and 20,000 feet. They are composed largely of water droplets, although the higher ones have a tendency toward ice particles. **Low clouds** have a mean lower level of less than 6,500 feet. These clouds are composed entirely of water droplets.

Within these 3 families are 10 principal cloud types. The names of these are composed of various combinations and forms of the following basic words, all from Latin:

Cirrus, meaning "curl, lock, or tuft of hair."

Cumulus, meaning "heap, a pile, an accumulation."

Stratus, meaning "spread out, flatten, cover with a layer."

Alto, meaning "high, upper air."

Nimbus, meaning "rainy cloud."

Individual cloud types recognize certain characteristics, variations, or combinations of these. The 10 principal cloud types are:

High clouds. **Cirrus (Ci)** are detached high clouds of delicate and fibrous appearance, without shading, generally white in color, and often of a silky appearance (figs. 3713a and 3713d). Their fibrous and feathery appearance is due to the fact that they are composed entirely of ice crystals. Cirrus appear in varied forms such as isolated tufts; long, thin lines across the sky; branching, feather-like plumes; curved wisps which may end in tufts, etc. These clouds may be arranged in parallel bands which cross the sky in great circles and appear to converge toward a point on the horizon. This may indicate, in a general way, the direction of a low pressure area. Cirrus may be brilliantly colored at sunrise and sunset. Because of their height, they become illuminated before other clouds in the morning, and remain lighted after others at sunset. Cirrus are generally associated with fair weather, but if they are followed by lower and thicker clouds, they are often the forerunner of rain or snow.

Cirrocumulus (Cc) are high clouds composed of small white flakes or scales, or of very small globular masses, usually without shadows and arranged in groups of lines, or more often in ripples resembling those of sand on the seashore (fig. 3713b). One form of cirrocumulus is popularly known as "mackerel sky" because the pattern resembles the scales on the back of a mackerel. Like cirrus, cirrocumulus are composed of ice crystals and are generally associated with fair weather, but may precede a storm if they thicken and lower. They may turn gray and appear hard before thickening.

Cirrostratus (Cs) are thin, whitish, high clouds (fig. 3713c) sometimes covering the sky completely and giving it a milky appearance and at other times presenting, more or less distinctly, a formation like a tangled web. The thin veil is not sufficiently dense to blur the outline of sun or moon. However, the ice crystals of which the cloud is composed refract the light passing through in such a way that halos (art. 3818) may form with the sun or moon at the center. Figure 3713d shows cirrus thickening and changing into cirrostratus. In this form it is popularly known as "mares' tails." If it continues to thicken and lower, the ice crystals melting to form water droplets, the cloud formation is known as altostratus. When this occurs, rain may normally be expected within 24 hours. The more brushlike the cirrus when the sky appears as in figure 3713d, the stronger the wind at the level of the cloud.

Middle clouds. **Altocumulus (Ac)** are middle clouds consisting of a layer of large, ball-like masses that tend to merge together. The balls or patches may vary in thickness and color from dazzling white to dark gray, but they are more or less regularly arranged. They may appear as distinct patches (fig. 3713e) similar to

FIGURE 3713a.—Cirrus.

FIGURE 3713b.—Cirrocumulus.

FIGURE 3713c.—Cirrostratus.

FIGURE 3713d.—Cirrus and cirrostratus.

cirrocumulus (fig. 3713b) but can be distinguished by the fact that individual patches are generally larger, and show distinct shadows in some places. They are often mistaken for stratocumulus (fig. 3713i). If this form thickens and lowers, it may produce thundery weather and showers, but it does not bring prolonged bad weather. Sometimes the patches merge to form a series of big rolls that resemble ocean waves, but with streaks of blue sky (fig. 3713f). Because of perspective, the rolls appear to run together near the horizon. These regular parallel bands differ from cirrocumulus in that they occur in larger masses with shadows. These clouds move in the direction of the short dimension of the rolls, as do ocean waves. Sometimes altocumulus appear briefly in the form shown in figure 3713g, usually before a thunderstorm. They are generally arranged in a line with a flat horizontal base, giving the impression of turrets on a castle. The turreted tops may look like miniature cumulus and possess considerable depth and great length. These clouds usually indicate a change to chaotic, thundery skies.

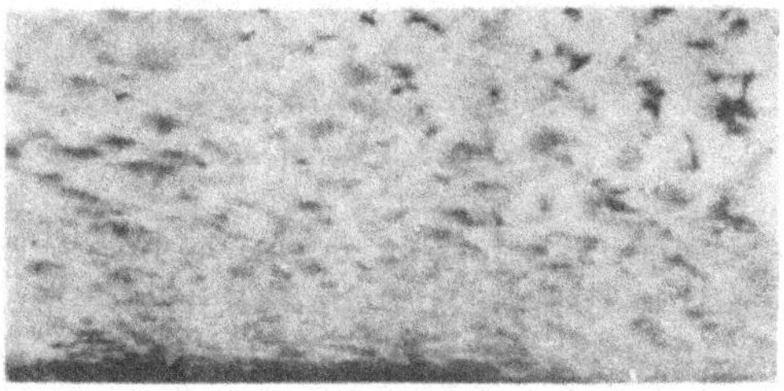

FIGURE 3713e.—Altocumulus in patches.

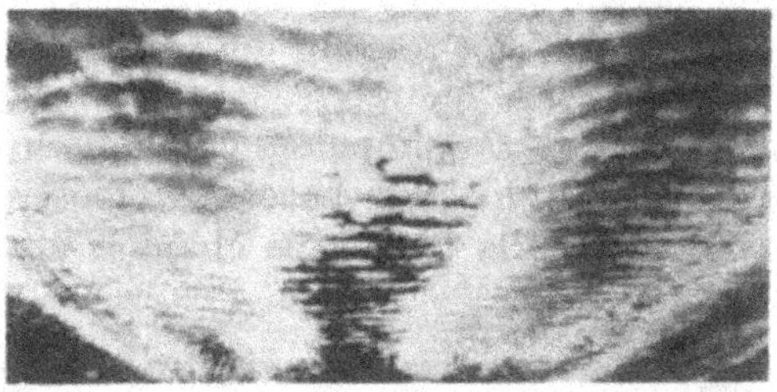

FIGURE 3713f.—Altocumulus in bands.

FIGURE 3713g.—Turreted altocumulus.

Altostratus (As) are middle clouds having the appearance of a grayish or bluish, fibrous veil or sheet (fig. 3713h). The sun or moon, when seen through these clouds, appears as if it were shining through ground glass, with a corona (art. 3819) around it. Halos are not formed. If these clouds thicken and lower, or if low, ragged "scud" or rain clouds (nimbostratus) form below them, continuous rain or snow may be expected within a few hours.

Low clouds. **Stratocumulus (Sc)** are low clouds composed of soft, gray, roll-shaped masses (fig. 3713i). They may be shaped in long, parallel rolls similar to altocumulus (fig. 3713f), moving forward with the wind. The motion is in the

FIGURE 3713h.—Altostratus.

direction of their short dimension, like ocean waves. These clouds, which vary greatly in altitude, are the final product of the characteristic daily change that takes place in cumulus clouds. They are usually followed by clear skies during the night.

Stratus (St) is a low cloud in a uniform layer (fig. 3713j) resembling fog. Often the base is not more than 1,000 feet high. A veil of thin stratus gives the sky a hazy appearance. Stratus is often quite thick, permitting so little sunlight to penetrate that it appears dark to an observer below it. From above, it looks white. Light mist may descend from stratus. Strong wind sometimes breaks stratus into shreds called "fractostratus."

Nimbostratus (Ns) is a low, dark, shapeless cloud layer, usually nearly uniform, but sometimes with ragged, wet-looking bases. Nimbostratus is the typical rain cloud. The precipitation which falls from this cloud is steady or intermittent, but not showery.

FIGURE 3713i.—Stratocumulus.

FIGURE 3713j.—Stratus.

Cumulus (Cu) are dense clouds with *vertical development* (clouds formed by rising air which is cooled as it reaches greater heights). They have a horizontal base

FIGURE 3713k.—Cumulus.

FIGURE 3713l.—Cumulonimbus.

and dome-shaped upper surface, with protuberances extending above the dome. Cumulus appear in small patches, and never cover the entire sky. When the vertical development is not great, the clouds appear in patches resembling tufts of cotton or wool, being popularly called "woolpack" clouds (fig. 3713k). The horizontal bases of such clouds may not be noticeable. These are called "fair weather" cumulus because they always accompany good weather. However, they may merge with altocumulus, or may grow to cumulonimbus before a thunderstorm. Since cumulus are formed by updrafts, they are accompanied by turbulence, causing "bumpiness" in the air. The extent of turbulence is proportional to the vertical extent of the clouds. Cumulus are marked by strong contrasts of light and dark.

Cumulonimbus (Cb) is a massive cloud with great vertical development, rising in mountainous towers to great heights (fig. 3713l). The upper part consists of ice crystals, and often spreads out in the shape of an anvil which may be seen at such distances that the base may be below the horizon. Cumulonimbus often produces showers of rain, snow, or hail, frequently accompanied by thunder. Because of this, the cloud is often popularly called a "thundercloud" or "thunderhead." The base is horizontal, but as showers occur it lowers and becomes ragged.

3714. Cloud height measurement.—At sea, cloud heights are often determined by estimate. This is a difficult task, particularly at night.

The height of the base of clouds formed by vertical development (any form of cumulus), if formed in air that has risen from the surface of the earth, can be determined by psychrometer, because the height to which the air must rise before condensation takes place is proportional to the difference between surface air temperature and the dew point. At sea, this difference multiplied by 236 gives the height in feet. That is, for every degree difference between surface air temperature and the dew point, the air must rise 236 feet before condensation will take place. Thus, if the dry-bulb temperature is 80°F, and the wet-bulb temperature is 77°F, the dew point (from tab. 17) is 76°F, or 4° lower than the surface air temperature. The height of the cloud base is $4 \times 236 = 944$ feet.

3715. Visibility measurement.—Visibility is the extreme horizontal distance at which prominent objects can be seen and identified by the unaided eye. It is usually measured directly by the human eye. Ashore, the distances of various buildings, trees, lights, and other objects are measured and used as a guide in estimating the visibility. At sea, however, such an estimate is difficult to make with accuracy. Other ships and the horizon may be of some assistance.

Ashore, visibility is sometimes measured by a **transmissometer**, a device which measures the transparency of the atmosphere by passing a beam of light over a known short distance, and comparing it with a reference light.

3716. Upper air observations.—Upper air information provides the third dimension to the weather map. Unfortunately, the equipment necessary to obtain such information is quite expensive, and the observations are time consuming. Consequently, the network of observing stations is quite sparse compared to that for surface observations, particularly over the oceans and in isolated land areas. Where facilities exist, upper air observations are made by means of unmanned balloons in conjunction with theodolites, radiosondes, radar, and radio direction finders.

3717. Storm detection radar.—During World War II, it was found that certain radar equipment gave an indication of weather fronts (art. 3811) and precipitation areas. It was of particular value near hurricanes and typhoons. Since the close of that war a great amount of work has been done in perfecting radar equipment for use in weather observation. It has proved of immense value in detecting, tracking, and interpreting weather activity out to a distance of as much as 400 miles from the observing station.

3718. Automated weather stations and buoy systems provide regular transmissions of meteorological and oceanographic information by radio. They are generally used at isolated and relatively inaccessible locations from which weather and ocean data are of great importance. Depending on the type of system used, the elements usually measured include wind direction and speed, atmospheric pressure, air and sea surface temperature, spectral wave data, and a temperature profile from the sea surface to a predetermined depth.

3719. Recording observations.—Instructions for recording weather observations aboard vessels of the U.S. Navy are given in NAVOCEANCOMINST 3144.1 (series), *Manual for Ship's Surface Weather Observations.* Instructions for recording observations aboard merchant vessels are given in Weather Service Observing Handbook No. 1, *Marine Surface Observations.*

Problems

3708a. A ship is proceeding on course 180° at a speed of 22 knots. The apparent wind is from 70° off the port bow, speed 20 knots.

Required.—The relative direction, true direction, and speed of the true wind by maneuvering board or the National Weather Service plotter.

Answers.—True wind from 231° relative, 051° true, at 24.3 knots.

3708b. A ship is proceeding on course 050° at a speed of 13.5 knots. The apparent wind is from broad on the starboard bow, speed 20 knots.

Required.—The relative direction, true direction, and speed of the true wind by table 10.

Answers.—True wind from 086° relative, 136° true, at 14.3 knots.

3708c. A ship is proceeding on course 020° at a speed of 16 knots. The true wind is estimated to be from 110° on the port bow, speed 10 knots.

Required.—The relative direction, true direction, and speed of the apparent wind by maneuvering board or the National Weather Service plotter.

Answers.—Apparent wind from 323° relative, 343° true, at 15.6 knots.

3708d. A ship is proceeding on course 190° at a speed of 14 knots. The true wind is estimated to be from broad on the starboard quarter, speed 20 knots.

Required.—The relative direction, true direction, and speed of the apparent wind by table 10.

Answers.—Apparent wind from 090° relative, 280° true, at 14.0 knots.

3708e. The true wind has been determined to be from 210°, speed 12 knots. The captain of an aircraft carrier desires an apparent wind of 30 knots from 10° on the port bow for launching aircraft.

Required.—The course and speed of the aircraft carrier.

Answers.—C 235°, S 18.6 kn. (The required apparent wind could also be produced by C 005°, S 40.5 kn.)

3708f. A ship is proceeding on course 255° at a speed of 15 knots. The wind vane indicates the apparent wind is broad on the starboard beam. From the appearance of the sea the navigator estimates the speed of the true wind as Beaufort 5 (19 knots).

Required.—(1) Relative and true directions of the true wind, (2) speed of the apparent wind. Use the maneuvering board.

Answers.—(1) True wind from 142° relative, 037° true; (2) apparent wind speed 11.6 knots.

3708g. A ship is proceeding on course 135° at a speed of 18 knots. The wind vane indicates the apparent wind is 40° on the starboard bow. From the appearance of the sea the navigator estimates the speed of the true wind as Beaufort 6 (24.5 knots).

Required.—(1) Relative and true directions of the true wind, (2) speed of the apparent wind. Use table 10.

Answers.—(1) True wind from 069° relative, 204° true; (2) apparent wind speed 36 knots.

3708h. A ship is proceeding on course 330° at a speed of 20 knots. The wind vane indicates the apparent wind is 30° on the port bow. From the appearance of the sea the navigator estimates the speed of the true wind as Beaufort 4 (13.5 knots).

Required.—(1) Relative and true directions of the true wind, (2) speed of the apparent wind. Solve first by maneuvering board and then by table 10.

Answers.—Graphical solution: (1) true wind from 199° relative, 169° true or from 282° relative, 252° true; (2) apparent wind speed 8.5 knots or 26.3 knots. Table 10 solution: (1) true wind from 197° relative, 167° true or from 283° relative, 253° true; (2) apparent wind speed 8.0 knots or 26.0 knots.

3712. The dry-bulb temperature is 41°F and the wet-bulb temperature is 35°F.

Required.—(1) Relative humidity, (2) dew point.

Answers.—(1) Relative humidity 53 percent, (2) dew point 26°.

3714a. The dry-bulb temperature is 72°F and the wet-bulb temperature is 58°F.

Required.—Height of the base of cumulonimbus clouds formed in air which has risen from the surface of the sea.

Answer.—Height 5,900 feet.

References

Kotsch, W. J. *Weather for the Mariner.* Annapolis, United States Naval Institute, 1970.

National Weather Service, *Marine Surface Observations,* Weather Service Observing Handbook No. 1. Washington, U. S. Govt. Print. Off., 1974.

Royal Meteorology Office. *Meteorology for Mariners.* London, Her Majesty's Stationery Office, 1967.

CHAPTER XXXVIII

WEATHER ELEMENTS

3801. Introduction.—Weather is the state of the earth's atmosphere with respect to temperature, humidity, precipitation, visibility, cloudiness, etc. In contrast, the term **climate** refers to the prevalent or characteristic meteorological conditions of a place or region.

All weather may be traced ultimately to the effect of the sun on the earth, including the lower portions of the atmosphere. Most changes in weather involve large-scale, approximately horizontal, motion of air. Air in such motion is called **wind.** This motion is produced by differences of atmospheric pressure, which are attributable both to differences of temperature and the nature of the motion itself.

The weather is of considerable interest to the mariner. The wind and state of the sea affect dead reckoning. Reduced horizontal visibility limits piloting. The state of the atmosphere affects electronic navigation and radio communication. If the skies are overcast, visual celestial observations are not available; and under certain conditions refraction and dip are disturbed. When wind was the primary motive power, knowledge of the areas of favorable winds was of great importance. This consideration led Matthew Fontaine Maury, more than a century ago, to seek information from ships' logs to establish speed and direction of prevailing winds over the various trade routes of the world. The information thus gathered was shown on pilot charts. By means of these charts, the mariner could select a suitable route for a favorable passage. Even power vessels are affected considerably by wind and sea. Less fuel consumption and a more comfortable passage are to be expected if wind and sea are moderate and favorable. Pilot charts are useful in selecting suitable routes.

Optimizing ship speed and safety by taking advantage of favorable wind and sea conditions has become practicable with the advent of extended range forecasting techniques. These techniques and the procedures used in ship weather routing are discussed in chapter XXIV.

3802. The atmosphere is a relatively thin shell of air, water vapor, dust, smoke, etc., surrounding the earth. The air is a mixture of transparent gases and, like any gas, is elastic and highly compressible. Although extremely light, it has a definite weight which can be measured. A cubic foot of air at standard sea-level temperature and pressure weighs 1.22 ounces, or about 1/817th part of the weight of an equal volume of water. Because of this weight, the atmosphere exerts a pressure upon the surface of the earth, amounting to about 15 pounds per square inch.

As altitude increases, less atmosphere extends upward, and pressure decreases. With less pressure, the density decreases. More than three-fourths of the air is concentrated within a layer averaging about 7 statute miles thick, called the **troposphere.** This is the region of most "weather," as the term is commonly understood.

The top of the troposphere is marked by a thin transition zone called the **tropopause,** immediately above which is the **stratosphere.** Beyond this lie several other layers having distinctive characteristics. The average height of the tropopause ranges from about 5 miles or less at high latitudes to about 10 miles at low latitudes.

The **standard atmosphere** is a conventional vertical structure of the atmosphere characterized by a standard sea-level pressure of 29.92 inches of mercury (1013.25 millibars) and a sea-level temperature of 59°F (15°C), the rate of temperature decreases with height (i.e., **standard lapse rate**) being a uniform 3°6F (2°C) per thousand feet to 11 kilometers (36,089 feet) and thereafter a constant temperature of (−)69°7F (−56°5C).

With the aid of *weather satellite observations,* meteorologists are continually learning more of the atmospheric processes in the troposphere and stratosphere as they affect weather at sea. In recent years research has indicated that the **jet stream** is an important entity in relation to the sequence of weather. The jet stream refers to relatively strong (≥60 knots) quasi-horizontal winds usually concentrated within a restricted layer of the atmosphere. Although jet stream winds can occur at any level and geographic location and from any direction, the term is most often associated with mid-latitude winds with maximum speeds from 270° (±45°). Such winds, called **polar jet stream winds,** average 90 knots maximum, but speeds up to 200 knots may occur in the winter season.

3803. General circulation of the atmosphere.—The heat required for warming the air is supplied originally by the sun. As radiant energy from the sun arrives at the earth, about 29 percent is reflected back into space by the earth and its atmosphere, 19 percent is absorbed by the atmosphere, and the remaining 52 percent is absorbed by the surface of the earth, much of which is reradiated back into space. This earth radiation is in comparatively long waves relative to the short-wave radiation from the sun, since it emanates from a cooler body. Long-wave radiation, being readily absorbed by the water vapor in the air, is primarily responsible for the warmth of the atmosphere near the earth's surface. Thus, the atmosphere acts much like the glass on the roof of a greenhouse. It allows part of the incoming solar radiation to reach the surface of the earth, but is heated by the terrestrial radiation passing outward. Over the entire earth and for long periods of time, the total outgoing energy must be equivalent to the incoming energy (minus any converted to another form and retained), or the temperature of the earth, including its atmosphere, would steadily increase or decrease. In local areas, or over relatively short periods of time, such a balance is not required, and in fact does not exist, resulting in changes such as those occurring from one year to another in different seasons and in different parts of the day.

As shown in figure 1419b, the more nearly perpendicular the rays of the sun strike the surface of the earth, the more heat energy per unit area is received at that place. Physical measurements show that in the Tropics more heat per unit area is received than is radiated away, and that in polar regions the opposite is true. Unless there were some process to transfer heat from the Tropics to polar regions, the Tropics would be much warmer than they are, and the polar regions would be much colder. Atmospheric motions bring about the required transfer of heat. The oceans also participate in the process, but to a lesser degree.

If the earth had a uniform surface and did not rotate on its axis, with the sun following its normal path across the sky (solar heating increasing with decreasing latitude), a simple circulation would result, as shown in figure 3803a. However, the surface of the earth is far from uniform, being covered with an irregular distribution of land of various heights, and water; the earth rotates about its axis once in approximately 24 hours, so that the portion heated by the sun continually changes;

and the axis of rotation is tilted so that as the earth moves along its orbit about the sun, seasonal changes occur in the exposure of specific areas to the sun's rays, resulting in variations in the heat balance of these areas. These factors, coupled with others, result in constantly changing large-scale movements of air. For example, the rotation of the earth exerts an apparent force, known as **Coriolis force,** which diverts the air from a direct path between high and low pressure areas. The diversion of the air is toward the right in the Northern Hemisphere and toward the left in the Southern Hemisphere. At some distance above the surface of the earth, the wind tends to blow along the isobars, being called the **geostrophic wind** if the isobars are straight (great circles), and **gradient wind** if they are curved. Near the surface of the earth, friction tends to divert the wind from the isobars toward the center of low pressure. At sea, where friction is less than on land, the wind follows the isobars more closely.

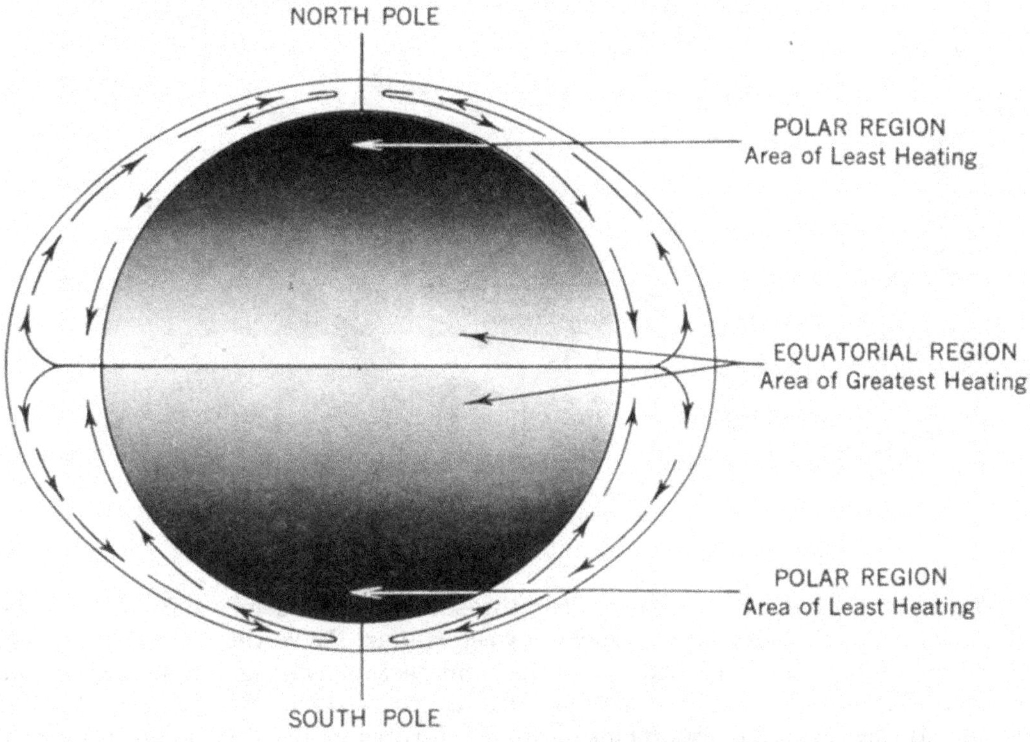

FIGURE 3803a.—Ideal atmospheric circulation for a uniform and nonrotating earth.

A simplified diagram of the general circulation pattern is shown in figure 3803b. Figures 3803c and 3803d give a generalized picture of the world's pressure distribution and wind systems as actually observed.

The change in pressure with horizontal distance is called **pressure gradient.** It is maximum along a normal (perpendicular) to the isobars. A force results which is called **pressure gradient force** and is always directed from high to low pressure. Speed of the wind as illustrated in figures 3803c and 3803d is approximately proportional to this pressure gradient.

3804. The Doldrums.—The belt of low pressure at the surface near the equator occupies a position approximately midway between high pressure belts at about latitude 30° to 35° on each side. Except for significant intradiurnal changes, the atmospheric pressure along the equatorial low is almost uniform. With minimal pressure gradient, wind speeds are light and directions are variable. Hot, sultry days are common. The sky is often overcast, and showers and thundershowers are relatively frequent; in these disturbed areas, brief periods of strong wind occur.

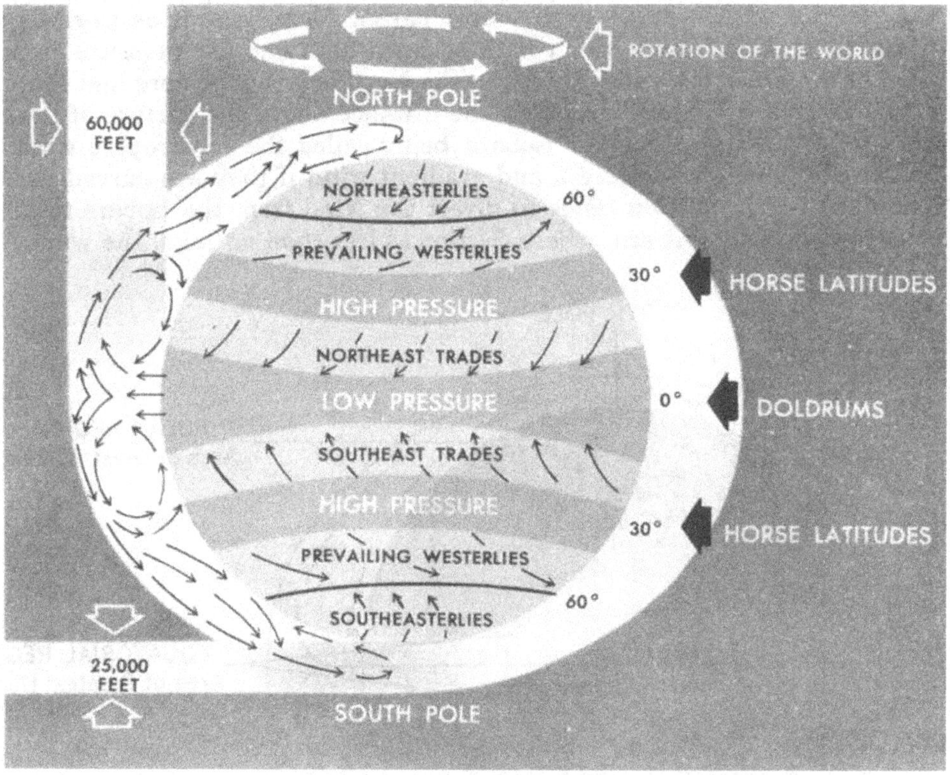

FIGURE 3803b.—Simplified diagram of the general circulation of the atmosphere.

The area involved is a thin belt near the equator, the eastern part in both the Atlantic and Pacific being wider than the western part. However, both the position and extent of the belt vary with longitude and season. During all seasons in the Northern Hemisphere, the belt is centered in the eastern Atlantic and Pacific; however, there are wide excursions of the doldrums regions at longitudes with considerable landmass. On the average, the position is at 5°N, frequently called the **meteorological equator.**

3805. The trade winds at the surface blow from the belts of high pressure toward the equatorial belts of low pressure. Because of the rotation of the earth, the moving air is deflected toward the west. Therefore, the trade winds in the Northern Hemisphere are from the northeast and are called the **northeast trades,** while those in the Southern Hemisphere are from the southeast and are called the **southeast trades.** The trade-wind directions are best defined over eastern ocean areas.

The trade winds are generally considered among the most constant of winds, blowing for days or even weeks with little change of direction or speed. However, at times they weaken or shift direction, and there are regions where the general pattern is disrupted. A notable example is found in the island groups of the South Pacific, where the trades are practically nonexistent during January and February. Their best development is attained in the South Atlantic and in the South Indian

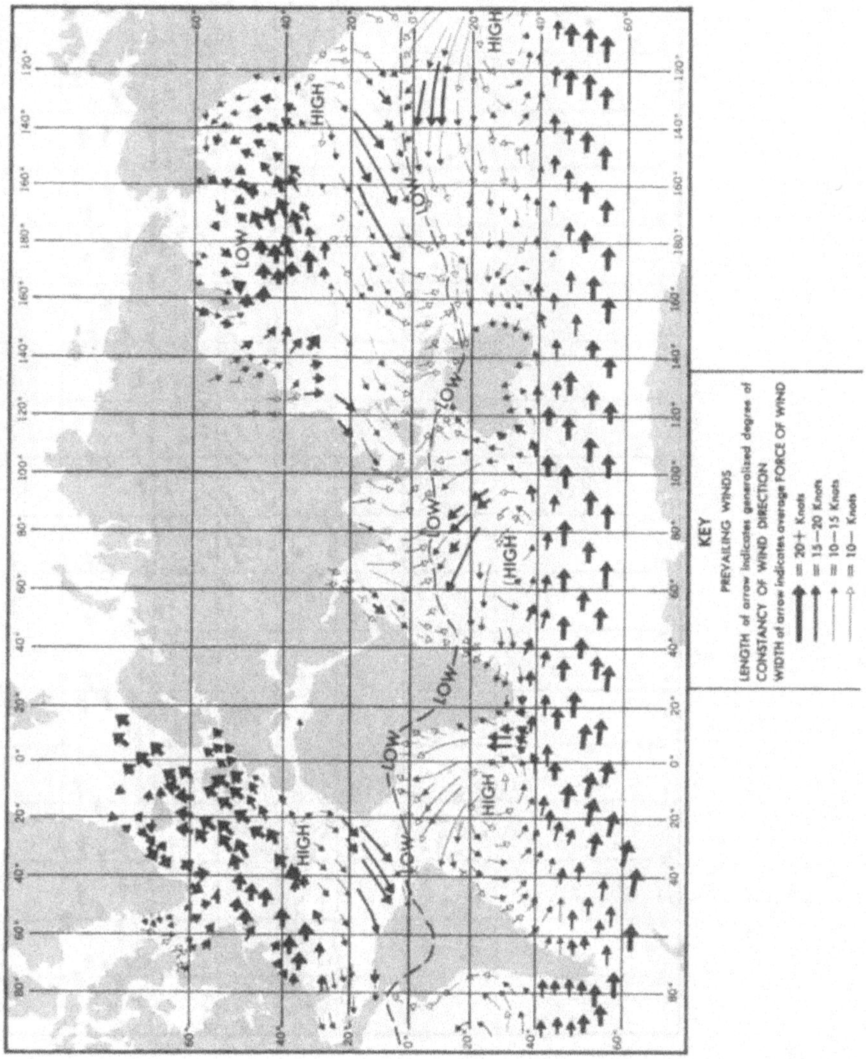

FIGURE 3803c.—Generalized pattern of actual surface winds in January and February.

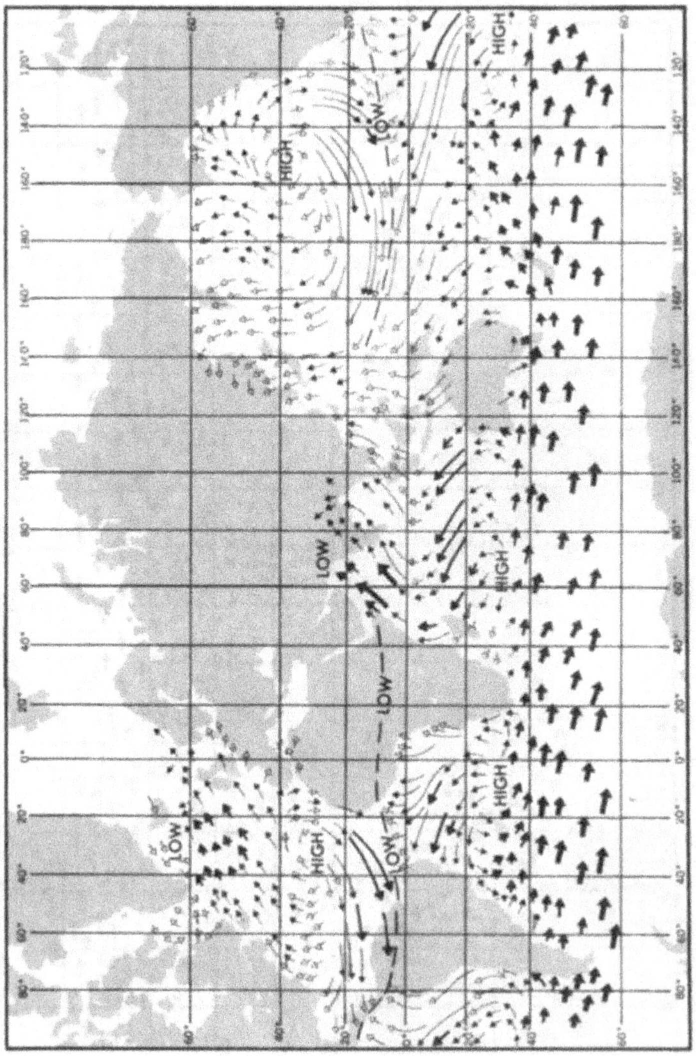

FIGURE 3803d.—Generalized pattern of actual surface winds in July and August. (See key with figure 3803c).

Ocean. In general, they are stronger during the winter than during the summer season.

In July and August, when the belt of equatorial low pressure moves to a position some distance north of the equator, the southeast trades blow across the equator, into the Northern Hemisphere, where the earth's rotation diverts them toward the right, causing them to be southerly and southwesterly winds. The "southwest monsoons" of the African and Central American coasts have their origin partly in such diverted southeast trades.

Cyclones (art. 3812) from the middle latitudes rarely enter the regions of the trade winds, although tropical cyclones (ch. XXXIX) originate within these areas.

3806. The horse latitudes.—Along the poleward side of each trade-wind belt, and corresponding approximately with the belt of high pressure in each hemisphere, is another region with weak pressure gradients and correspondingly light, variable winds. These are called the **horse latitudes.** The weather is generally good although low clouds are common. Compared to the doldrums, periods of stagnation in the horse latitudes are less persistent, being of a more intermittent nature. The difference is due primarily to the fact that rising currents of warm air in the equatorial low carry large amounts of moisture which condenses as the air cools at higher levels, while in the horse latitudes the air is apparently descending and becoming less humid as it is warmed at lower heights.

3807. The prevailing westerlies.—On the poleward side of the high pressure belt in each hemisphere the atmospheric pressure again diminishes. The currents of air set in motion along these gradients toward the poles are diverted by the earth's rotation toward the east, becoming southwesterly winds in the Northern Hemisphere and northwesterly in the Southern Hemisphere. These two wind systems are known as the **prevailing westerlies** of the temperate zones.

In the Northern Hemisphere this relatively simple pattern is distorted considerably by secondary wind circulations, due primarily to the presence of large landmasses. In the North Atlantic, between latitudes 40° and 50°, winds blow from some direction between south and northwest during 74 percent of the time, being somewhat more persistent in winter than in summer. They are stronger in winter, too, averaging about 25 knots (Beaufort 6) as compared with 14 knots (Beaufort 4) in the summer.

In the Southern Hemisphere the westerlies blow throughout the year with a steadiness approaching that of the trade winds (art. 3805). The speed, though variable, is generally between 17 and 27 knots (Beaufort 5 and 6). Latitudes 40°S to 50°S (or 55°S) where these boisterous winds occur, are called the **roaring forties.** These winds are strongest at about latitude 50°S.

The greater speed and persistence of the westerlies in the Southern Hemisphere are due to the difference in the atmospheric pressure pattern, and its variations, from that of the Northern Hemisphere. In the comparatively landless Southern Hemisphere, the average yearly atmospheric pressure diminishes much more rapidly on the poleward side of the high pressure belt, and has fewer irregularities due to continental interference, than in the Northern Hemisphere.

3808. Winds of polar regions.—Partly because of the low temperatures near the geographical poles of the earth, the surface pressure tends to remain higher than in surrounding regions. Consequently, the winds blow outward from the poles, and are deflected westward by the rotation of the earth, to become **northeasterlies** in the Arctic, and **southeasterlies** in the Antarctic. Where the polar easterlies meet the prevailing westerlies, near 50°N and 50°S on the average, a discontinuity in temperature and wind exists. This discontinuity is called the **polar front.** Here the warmer low-latitude air ascends over the colder polar air creating a zone of cloudiness and precipitation.

In the Arctic, the general circulation is greatly modified by surrounding landmasses. Winds over the Arctic Ocean are somewhat variable, and strong surface winds are rarely encountered.

In the Antarctic, on the other hand, a high central landmass is surrounded by water, a condition which augments, rather than diminishes, the general circulation. The high pressure, although weaker than in the horse latitudes, is stronger than in the Arctic, and of great persistence especially in eastern Antarctica. The cold air from the plateau areas moves outward and downward toward the sea and is deflected toward the west by the earth's rotation. The katabatic winds (art. 3813) remain strong throughout the year, frequently attaining hurricane force near the base of the mountains. These are some of the strongest surface winds encountered anywhere in the world, with the possible exception of those in well-developed tropical cyclones (ch. XXXIX).

3809. Modifications of the general circulation.—The general circulation of the atmosphere as described in articles 3803–3808 is greatly modified by various conditions.

The high pressure in the horse latitudes is not uniformly distributed around the belts, but tends to be accentuated at several points, as shown in figures 3803b and 3803c. These **semipermanent highs** remain at about the same places with great persistence.

Semipermanent lows also occur in various places, the most prominent ones being west of Iceland, and over the Aleutians (winter only) in the Northern Hemisphere, and in the Ross Sea and Weddell Sea in the antarctic areas. The regions occupied by these semipermanent lows are sometimes called the graveyards of the lows, since many lows move directly into these areas and lose their identity as they merge with and reinforce the semipermanent lows. The low pressure in these areas is maintained largely by the migratory lows which stall there, with topography also important, especially in Antarctica.

Another modifying influence is land, which undergoes greater temperature changes than does the sea. During the summer, a continent is warmer than its adjacent oceans. Therefore, low pressures tend to prevail over the land. If a climatological belt of high pressure encounters a continent, its pattern is distorted or interrupted, whereas a belt of low pressure is intensified over the same area. In winter, the opposite effect takes place, belts of high pressure being intensified over land and those of low pressure being weakened.

The most striking example of a wind system produced by the alternate heating and cooling of a landmass is the **monsoons** (seasonal wind) of the China Sea and Indian Ocean. A portion of this effect is shown in figures 3809a and 3809b. In the summer (fig. 3809a), low pressure prevails over the warm continent of Asia, and relatively higher pressure prevails over the adjacent sea. Between these two systems the wind blows in a nearly steady direction. The lower portion of the pattern is in the Southern Hemisphere, extending to about 10° south latitude. Here the rotation of the earth causes a deflection to the left, resulting in southeasterly winds. As they cross the equator, the deflection is in the opposite direction, causing them to curve toward the right, becoming southwesterly winds. In the winter (fig. 3809b), the positions of high and low pressure areas are interchanged, and the direction of flow is reversed.

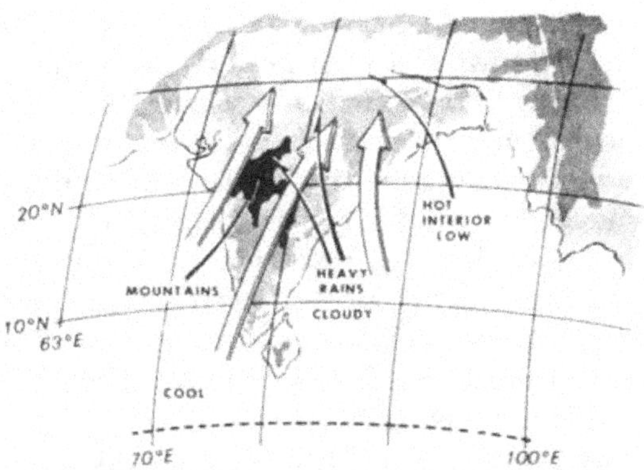

FIGURE 3809a.—The summer monsoon.

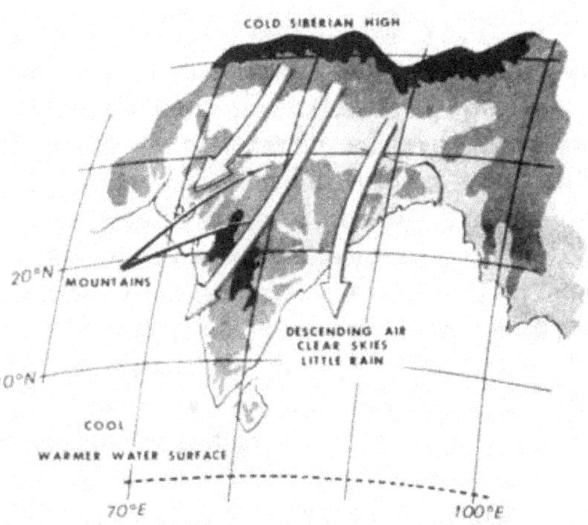

FIGURE 3809b.—The winter monsoon.

In the China Sea the summer monsoon blows from the southwest, usually from May to September. The strong winds are accompanied by heavy squalls and thunderstorms, the rainfall being much heavier than during the winter monsoon. As the season advances, squalls and rain become less frequent. In some places the wind becomes a light breeze which is unsteady in direction, or stops altogether, while in other places it continues almost undiminished, with changes in direction or calms being infrequent. The winter monsoon blows from the northeast, usually from October to April. It blows with a steadiness similar to that of the trade winds, often attaining the speed of a moderate gale (28–33 knots). Skies are generally clear during this season, and there is relatively little rain.

The general circulation is further modified by winds of cyclonic origin (art. 3812), and various local winds (art. 3813).

3810. Airmasses.—Because of large differences in physical characteristics of the earth's surface, particularly the oceanic and continental contrasts, the air overlying these surfaces acquires differing values of temperature and moisture. The processes of radiation and convection in the lower portions of the troposphere act in differing characteristic manners for a number of well-defined regions of the earth. The air overlying these regions acquires characteristics common to the particular area, but contrasting to those of other areas. Each distinctive part of the atmosphere, within which common characteristics prevail over a reasonably large area, is called an **airmass.**

Airmasses are named according to their source regions. Four such regions are generally recognized: (1) *equatorial* (*E*), the doldrum area between the north and south trades; (2) *tropical* (*T*), the trade wind and lower temperate regions; (3) *polar* (*P*), the higher temperate latitudes; and (4) *arctic or antarctic* (*A*), the north or south polar regions of ice and snow. This classification is a general indication of relative temperature, as well as latitude of origin.

Airmasses are further classified as maritime (*m*) or continental (*c*), depending upon whether they form over water or land. This classification is an indication of the relative moisture content of the airmass. Tropical air, then, might be designated maritime tropical (*mT*) or continental tropical (*cT*). Similarly, polar air may be either maritime polar (*mP*) or continental polar (*cP*). Arctic/antarctic air, due to the predominance of landmasses and ice fields in the high latitudes, is rarely maritime arctic (*mA*). Equatorial air is found exclusively over the ocean surface and is designated neither (*cE*) nor (*mE*), but simply (*E*).

A third classification sometimes applied to tropical and polar airmasses indicates whether the airmass is *warm* (*w*) or *cold* (*k*) relative to the underlying surface. Thus, the symbol *mTw* indicates maritime tropical air which is warmer than the underlying surface, and *cPk* indicates continental polar air which is colder than the underlying surface. The *w* and *k* classifications are primarily indications of stability (i.e., change of temperature with increasing height). If the air is cold relative to the surface, the lower portion of the airmass is being heated, resulting in instability (temperature markedly decreases with increasing height) as the warmer air tends to rise by convection. Conversely, if the air is warm relative to the surface, the lower portion of the airmass is cooled, tending to remain close to the surface. This is a stable condition (temperature increases with increasing height).

Two other types of airmasses are sometimes recognized. These are *monsoon* (*M*), a transitional form between *cP* and *E;* and *superior* (*S*), a special type formed in the free atmosphere by the sinking and consequent warming of air aloft.

3811. Fronts.—As airmasses move within the general circulation, they travel from their source regions and invade other areas dominated by air having different characteristics. Such a process leads to a zone of separation between the two airmasses. The gradients of thermal and moisture properties are maximized in the zone. Since the zone or discontinuity is so thin as to approach a sheet when viewed on a small scale map, it is called a **frontal surface.** The intersection of a frontal surface and a horizontal plane in a line is called a **front,** although the term "front" is commonly used as a short expression for "frontal surface" when this will not introduce an ambiguity.

Indicative of the differences in the motion of adjacent airmasses, the front takes a wave like character, hence the term "frontal wave."

Before the formation of frontal waves, the isobars (lines of equal atmospheric pressure) tend to run parallel to the fronts. As a wave is formed, the pattern is distorted somewhat, as shown in figure 3811a. In this illustration, colder air is north of warmer air. In figures 3811a–3811d isobars are drawn at 4-millibar intervals.

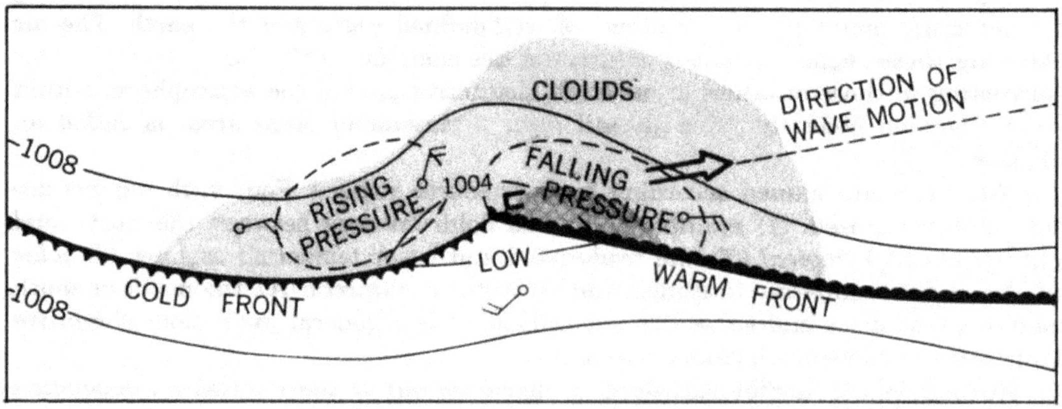

FIGURE 3811a.—First stage in the development of a frontal wave (top view).

The wave tends to travel in the direction of the general circulation, which in the temperate latitudes is usually in a general easterly and slightly poleward direction.

Along the leading edge of the wave, warmer air is replacing colder air. This is called the **warm front.** The trailing edge is the **cold front,** where colder air is underrunning and displacing warmer air.

The warm air, being less dense, tends to ride up over the colder air it is replacing. The slope is gentle, varying between 1:100 and 1:300. Partly because of the replacement of cold, dense air with warm, light air, the pressure decreases. Since the slope is gentle, the upper part of a warm frontal surface may be many hundreds of miles ahead of the surface portion. The decreasing pressure, indicated by a "falling barometer," is often an indication of the approach of such a wave. In a slow-moving, well-developed wave, the barometer may begin to fall several *days* before the wave arrives. Thus, the amount and nature of the change of atmospheric pressure between observations, called **pressure tendency,** is of assistance in predicting the approach of such a system.

The advancing cold air, being more dense, tends to cut under the warmer air at the cold front, lifting it to greater heights. The slope here is such that the upper-air portion of the cold front is behind the surface position relative to its motion. The slope generally ranges from 1:25 to 1:100, being steeper than the warm front. After a cold front has passed, the pressure increases—a "rising barometer."

In the first stages, these effects are not marked, but as the wave continues to grow, they become more pronounced, as shown in figure 3811b. As the amplitude of the wave increases, pressure near the center usually decreases, and the "low" is said to "deepen." As it deepens, its forward speed generally decreases.

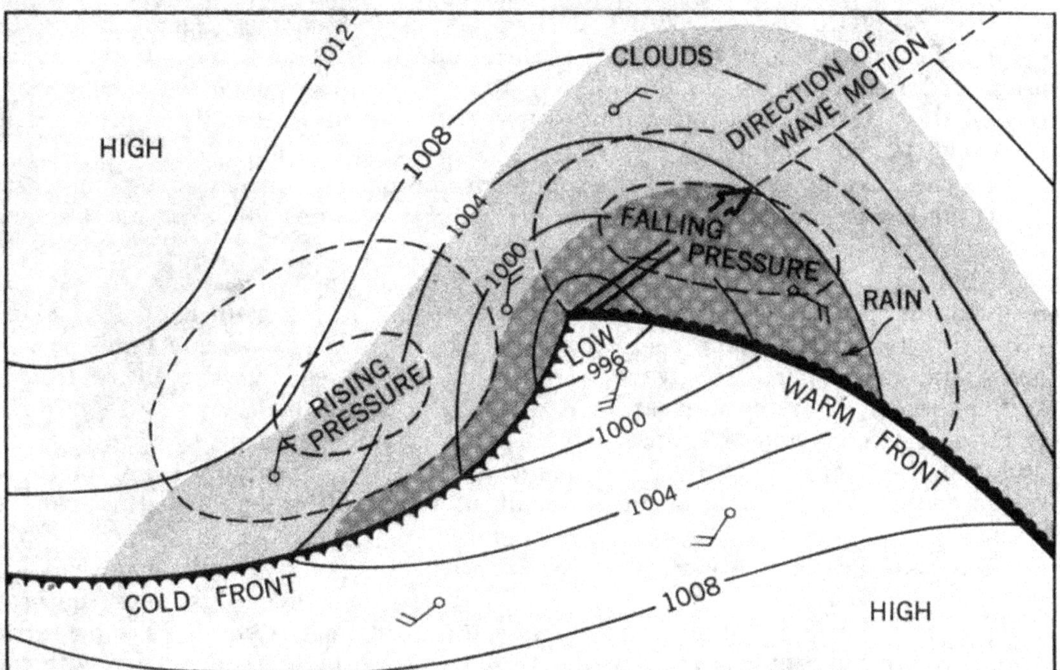

FIGURE 3811b.—A fully developed frontal wave (top view).

The approach of a well-developed warm front (i.e., when the warm air is *mT*) is usually heralded not only by falling pressure, but also by a more-or-less regular sequence of clouds. First, cirrus appear. These give way successively to cirrostratus, altostratus, altocumulus, and nimbostratus. Brief showers may precede the steady rain accompanying the nimbostratus.

As the warm front passes, the temperature rises, the wind shifts clockwise (in the Northern Hemisphere), and the steady rain stops. Drizzle may fall from low-lying stratus clouds, or there may be fog for some time after the wind shift. During passage of the **warm sector** between the warm front and the cold front, there is little change in temperature or pressure. However, if the wave is still growing and the low deepening, the pressure might slowly decrease. In the warm sector the skies are generally clear or partly cloudy, with cumulus or stratocumulus clouds most frequent. The warm air is usually moist, and haze or fog may often be present.

As the faster moving, steeper cold front passes, the wind shifts clockwise in the Northern Hemisphere (counterclockwise in the Southern Hemisphere), the temperature falls rapidly, and there are often brief and sometimes violent showers, frequently accompanied by thunder and lightning. Clouds are usually of the convective type. A cold front usually coincides with a well-defined **wind-shift line** (a line along which the wind shifts abruptly from southerly or southwesterly to northerly or northwesterly in the Northern Hemisphere and from northerly or northwesterly to southerly or southwesterly in the Southern Hemisphere). At sea a series of brief showers accompanied by strong, shifting winds may occur along or some distance (up to 200 miles) ahead of a cold front. These are called **squalls** (in common nautical use the term squall may be additionally applied to any severe local storm accompanied by gusty winds, precipitation, thunder, and lightning), and the line along which they occur is called a **squall line.**

Because of its greater speed and steeper slope, which may approach or even exceed the vertical near the earth's surface (due to friction), a cold front and its associated weather passes more quickly than a warm front. After a cold front passes, the pressure rises, often quite rapidly, the visibility usually improves, and the clouds tend to diminish.

As the wave progresses and the cold front approaches the slower moving warm front, the low becomes deeper and the warm sector becomes smaller. This is shown in figure 3811c.

Finally, the faster moving cold front overtakes the warm front (fig. 3811d), resulting in an **occluded front** at the surface, and an **upper front** aloft (fig. 3811e). When the two parts of the cold airmass meet, the warmer portion tends to rise above the colder part. The warm air continues to rise until the entire frontal system dissipates. As the warmer air is replaced by colder air, the pressure gradually rises, a process called "filling." This usually occurs within a few days after an occluded front forms. Finally, there results a **cold low,** or simply a low pressure system across which little or no gradient in temperature and moisture can be found.

The sequence of weather associated with a low depends greatly upon location with respect to the path of the center. That described above assumes that the observer is so located that he encounters each part of the system. If he is poleward of the path of the center of the low, the abrupt weather changes associated with the passage of fronts are not experienced. Instead, the change from the weather characteristically found ahead of a warm front to that behind a cold front takes place gradually, the exact sequence being dictated somewhat by distance from the center, as well as severity and age of the low.

Although each low follows generally the pattern given above, no two are ever exactly alike. Other centers of low pressure and high pressure and the airmasses

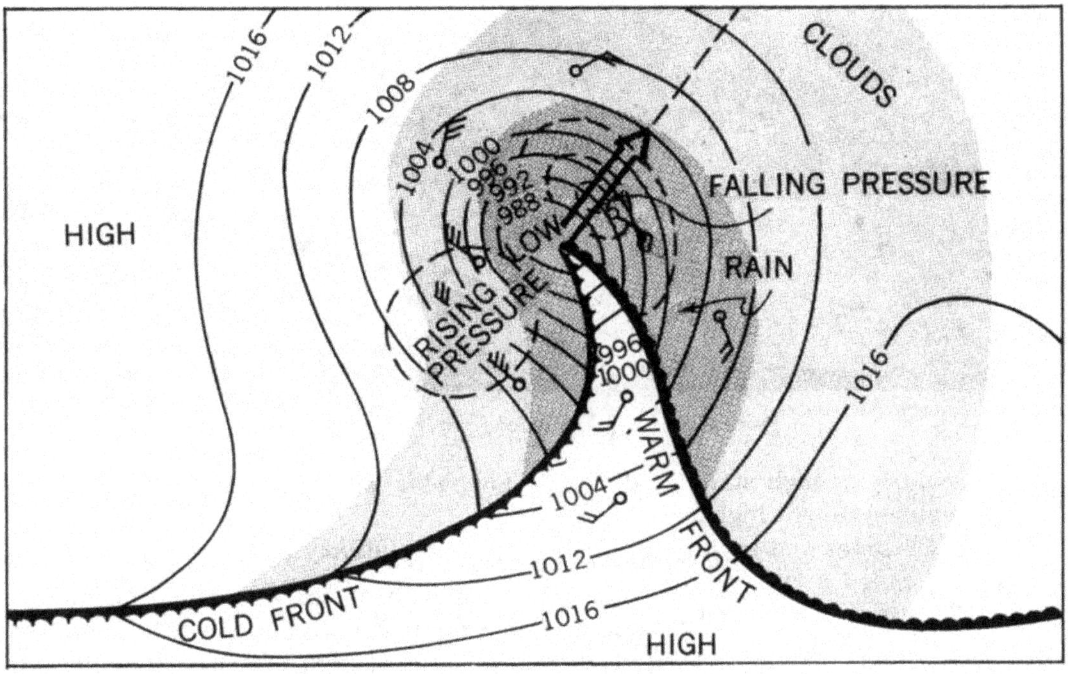

FIGURE 3811c.—A frontal wave nearing occlusion (top view).

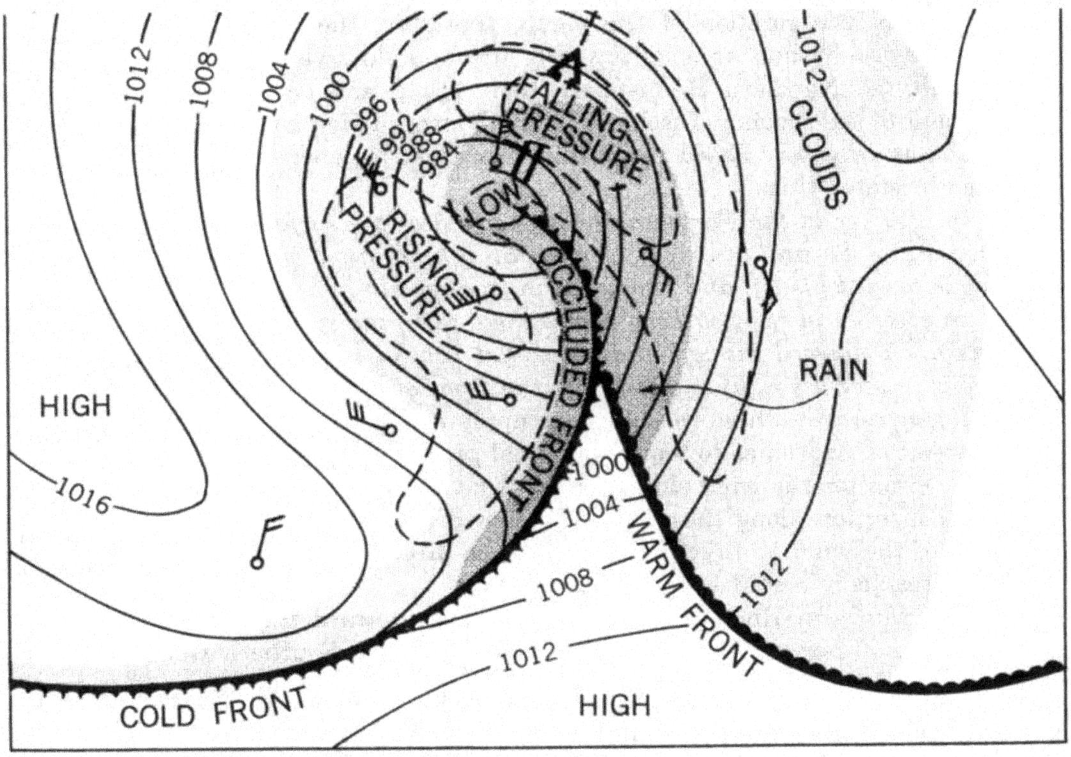

FIGURE 3811d.—An occluded front (top view).

associated with them, even though they may be 1,000 miles or more away, influence the formation and motion of individual low centers and their accompanying weath-

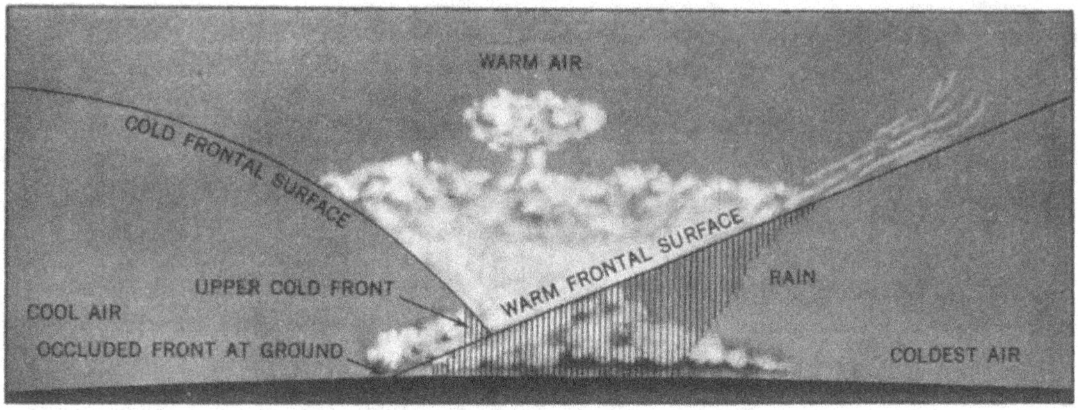

FIGURE 3811e.—An occluded front (cross section).

er. Particularly, a high stalls or diverts a low. This is true of temporary highs as well as semipermanent highs.

3812. Cyclones and anticyclones.—An area of relatively low pressure, generally circular, is called a **cyclone.** Its counterpart for high pressure is called an **anticyclone.** These terms are used particularly in connection with the winds associated with such centers. Wind tends to blow from an area of high pressure to one of low pressure, but due to rotation of the earth, they are deflected toward the right in the Northern Hemisphere and toward the left in the Southern Hemisphere (art. 3803).

Because of the rotation of the earth, therefore, the circulation tends to be counterclockwise around areas of low pressure and clockwise around areas of high pressure in the Northern Hemisphere (figs. 3811c and 3811d), the speed being proportional to the spacing of isobars. In the Southern Hemisphere, the direction of circulation is reversed. Based upon this condition, a general rule (**Buys Ballot's Law**) can be stated thus:

If an observer in the Northern Hemisphere faces the surface wind, the center of low pressure is toward his right, somewhat behind him; and the center of high pressure is toward his left and somewhat in front of him.

If an observer in the Southern Hemisphere faces the surface wind, the center of low pressure is toward his left and somewhat behind him; and the center of high pressure is toward his right and somewhat in front of him.

In a general way, these relationships apply in the case of the general distribution of pressure, as well as to temporary local pressure systems.

The reason for the wind shift along a front is that the isobars have an abrupt change of direction along these lines, as shown in figures 3811a–3811d. Since the direction of the wind is directly related to the direction of isobars, any change in the latter results in a shift in the wind direction.

In the Northern Hemisphere, the wind shifts toward the *right* (clockwise or veering) when either a warm or cold front passes. In the Southern Hemisphere, the shift is toward the *left* (counterclockwise or backing). When an observer is on the poleward side of the path of a frontal wave, wind shifts are reversed (i.e., backing in the Northern Hemisphere and veering in the Southern Hemisphere).

In an anticyclone, successive isobars are relatively far apart, resulting in light winds. In a cyclone, the isobars are more closely spaced. With a steeper pressure gradient, the winds are stronger.

Since an anticyclonic area is a region of outflowing winds, air is drawn into it from aloft. Descending air is warmed, and as air becomes warmer, its capacity for holding uncondensed moisture increases. Therefore, clouds tend to dissipate. Clear skies are characteristic of an anticyclone, although scattered clouds and showers are sometimes encountered.

In contrast, a cyclonic area is one of converging winds. The resulting upward movement of air results in cooling, a condition favorable to the formation of clouds and precipitation. More or less continuous rain and generally stormy weather are usually associated with a cyclone.

Between the two hemispheric belts of high pressure associated with the horse latitudes, called **subtropical anticyclones** (art. 3806), cyclones form only occasionally over certain areas at sea, generally in summer and fall only. These **tropical cyclones** are usually quite violent, being known under various names according to their location. They are discussed in chapter XXXIX.

In the areas of the prevailing westerlies (art. 3807) in temperate latitudes, migratory cyclones (**lows**) and anticyclones (**highs**) are a common occurrence. These are sometimes called **extratropical cyclones** and **extratropical anticyclones** to distinguish them from the more violent tropical cyclones. Formation occurs over sea and land. The lows intensify as they move poleward; the highs weaken as they move equatorward. In their early stages, cyclones are elongated, as shown in figure 3811a, but as their life cycle proceeds, they become more nearly circular (figs. 3811b–3811d).

3813. Local winds.—In addition to the winds of the general circulation (arts. 3803–3808) and those associated with migratory cyclones and anticyclones (art. 3812), there are numerous local winds which influence the weather in various places.

The most common of these are the **land** and **sea breezes**, caused by alternate heating and cooling of land adjacent to water. The effect is similar to that which causes the monsoons (art. 3809), but on a much smaller scale, and over shorter periods. By day the land is warmer than the water, and by night it is cooler. This effect occurs along many coasts during the summer. Between about 0900 and 1100 the temperature of the land becomes greater than that of the adjacent water. The lower levels of air over the land are warmed, and the air rises, drawing in cooler air from the sea. This is the **sea breeze.** Late in the afternoon, when the sun is low in the sky, the temperature of the two surfaces equalizes and the breeze stops. After sunset, as the land cools below the sea temperature, the air above it is also cooled. The contracting cool air becomes more dense, increasing the pressure near the surface. This results in an outflow of winds to the sea. This is the **land breeze,** which blows during the night and dies away near sunrise. Since the atmospheric pressure changes associated with this cycle are not great, the accompanying winds generally do not exceed gentle to moderate breezes. The circulation is usually of limited extent, reaching a distance of perhaps 20 miles inland, and not more than 5 or 6 miles offshore, and to a height of a few hundred feet. In the doldrums and subtropics, this process is repeated with great regularity throughout most of the year. As the latitude increases, it becomes less prominent, being masked by winds of migratory cyclones and anticyclones (art. 3812). However, the effect often may be present to reinforce, retard, or deflect stronger prevailing winds.

Varying conditions of topography produce a large variety of local winds throughout the world. Winds tend to follow valleys, and to be deflected from high banks and shores. In mountain areas wind flows in response to temperature distribution and gravity. An **anabatic wind** is one that blows up an incline, usually as a result of surface heating. A **katabatic wind** is one which blows down an incline. There are two types, *foehn* and *fall wind.*

A dry wind with a downward component, warm for the season, is called a **foehn.** The foehn occurs when horizontally moving air encounters a mountain barrier. As it blows upward to clear the barrier, it is cooled below the dew point, resulting in loss of moisture by cloud formation and perhaps rain. As the air continues to rise, its rate of cooling is reduced because the condensing water vapor gives off heat to the surrounding atmosphere. After crossing the mountain barrier, the air flows downward along the leeward slope, being warmed by compression as it descends to lower levels. Thus, since it loses less heat on the ascent than it gains during descent, and since it loses moisture during ascent, it arrives at the bottom of the mountains as very warm, dry air. This accounts for the warm, arid regions along the eastern side of the Rocky Mountains and in similar areas. In the Rocky Mountain region this wind is known by the name **chinook.** It may occur at any season of the year, at any hour of the day or night, and have any speed from a gentle breeze to a gale. It may last for several days, or for a very short period. Its effect is most marked in winter, when it may cause the temperature to rise as much as 20°F to 30°F within 15 minutes, and cause snow and ice to melt within a few hours. On the west coast of the United States, a foehn wind, given the name **Santa Ana,** blows through a pass and down a valley by that name in Southern California. This wind may blow with such force that it endangers small craft immediately off the coast.

A cold wind blowing down an incline is called a **fall wind.** Although it is warmed somewhat during descent, as is the foehn, it remains cold relative to the surrounding air. It occurs when cold air is dammed up in great quantity on the windward side of a mountain and then spills over suddenly, usually as an overwhelming surge down the other side. It is usually quite violent, sometimes reaching hurricane force. A different name for this type wind is given at each place where it is common. The **tehuantepecer** of the Mexican and Central American coast, the **pampero** of the Argentine coast, the **mistral** of the western Mediterranean, and the **bora** of the eastern Mediterranean are examples of this type wind.

Many other local winds common to certain areas have been given distinctive names.

A **blizzard** is a violent, intensely cold wind laden with snow mostly or entirely picked up from the ground, although the term is often used popularly to refer to any heavy snowfall accompanied by strong wind. A **dust whirl** is a rotating column of air about 100 to 300 feet in height, carrying dust, leaves, and other light material. This wind, which is similar to a waterspout at sea (art. 3824), is given various local names such as **dust devil** in southwestern United States and **desert devil** in South Africa. A **gust** is a sudden, brief increase in wind speed followed by a slackening, or the violent wind or squall that accompanies a thunderstorm. A puff of wind or a light breeze affecting a small area, such as would cause patches of ripples on the surface of water, is called a **cat's paw.**

3814. Fog is a cloud (art. 3713) whose base is low enough to restrict visibility. Fog is composed of droplets of water, or ice crystals (**ice fog**) formed by condensation or crystallization of water vapor in the air.

Radiation fog forms over low-lying land on clear, calm nights. As the land radiates heat and becomes cooler, it cools the air immediately above the surface. This causes a **temperature inversion** to form, the temperature for some distance upward *increasing* with height. If the air is cooled to its dew point (art. 3712), fog forms. Often, cooler and more dense air drains down surrounding slopes to heighten the effect. Radiation fog is often quite shallow, and is usually densest at the surface. After sunrise the fog may "lift," as shown in figure 3814, and gradually dissipate, usually being entirely gone by noon. At sea the temperature of the water undergoes little change between day and night, and so radiation fog is seldom encountered more than 10 miles from shore.

Advection fog forms when warm, moist air blows over a colder surface and is cooled below its dew point. This type, most commonly encountered at sea, may be quite dense and often persists over relatively long periods. Advection fog is common over cold ocean currents. If the wind is strong enough to thoroughly mix the air, condensation may take place at some distance above the surface of the earth, forming low stratus clouds (art. 3713) rather than fog.

Off the coast of California, seasonal winds create an offshore current which displaces the warm surface water, causing an upwelling of colder water. Moist Pacific air is transported along the coast in the same wind system and is cooled by the relatively cold water. Advection fog results. In the coastal valleys, fog is sometimes formed when moist air blown inland during the afternoon is cooled by radiation during the night.

When very cold air moves over warmer water, wisps of visible water vapor may rise from the surface as the water "steams," as shown in figure 2505. In extreme cases this **frost smoke**, or **arctic sea smoke**, may rise to a height of several hundred feet, the portion near the surface constituting a dense fog which obscures the horizon and surface objects, but usually leaves the sky relatively clear.

Haze consists of fine dust or salt particles in the air, too small to be individually apparent, but in sufficient number to reduce horizontal visibility and cast a bluish or yellowish veil over the landscape, subduing its colors and making objects appear indistinct. This is sometimes called **dry haze** to distinguish it from **damp haze**, which consists of small water droplets or moist particles in the air, smaller and more scattered than light fog. In international meteorological practice, the term "haze" is used to refer to a condition of atmospheric obscurity caused by dust and smoke.

Mist is synonymous with **drizzle** in the United States but is often considered as intermediate between haze and fog in its properties.

A mixture of smoke and fog is called **smog.**

3815. Mirage.—As explained in article 1606, light is refracted as it passes through the atmosphere. When refraction is normal, objects appear slightly elevated, and the visible horizon is farther from the observer than it otherwise would be. Since the effects are uniformly progressive, they are not apparent to the observer. When refraction is not normal, some form of **mirage** may occur. A mirage is an optical phenomenon in which objects appear distorted, displaced (raised or lowered), magnified, multiplied, or inverted due to varying atmospheric refraction which occurs when a layer of air near the earth's surface differs greatly in density from surrounding air. This may occur when there is a rapid and sometimes irregular change of temperature or humidity with height.

If there is a temperature inversion (increase of temperature with height), particularly if accompanied by a rapid decrease in humidity, the refraction is greater than normal. Objects appear elevated, and the visible horizon is farther away. Objects which are normally below the horizon become visible. This is called **looming.** If the upper portion of an object is raised much more than the bottom

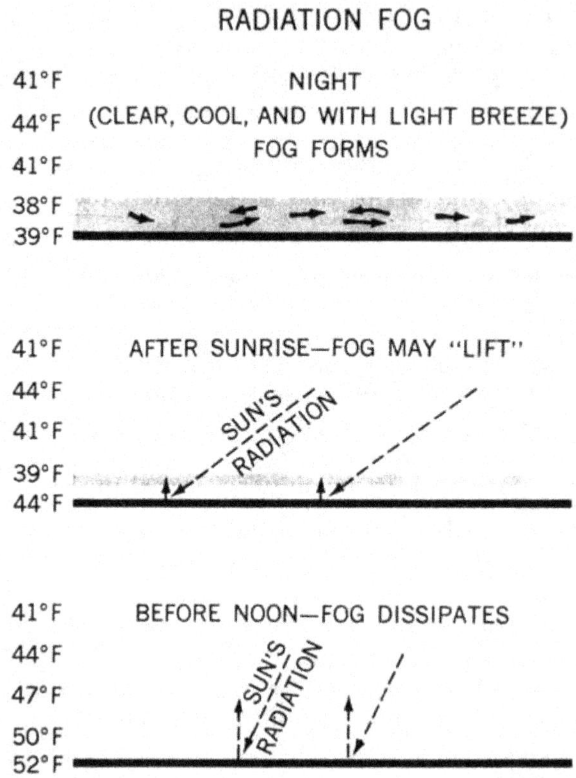

FIGURE 3814.—Formation and dissipation of radiation fog.

part, the object appears taller than usual, an effect called **towering.** If the lower part of an object is raised more than the upper part, the object appears shorter, an effect called **stooping.** When the refraction is greater than normal, a **superior mirage** may occur. An inverted image is seen above the object, and sometimes an erect image appears over the inverted one, with the bases of the two images touching. Greater than normal refraction usually occurs when the water is much colder than the air above it.

If the temperature decrease with height is much greater than normal, refraction is less than normal, or may even cause bending in the opposite direction. Objects appear lower than normal, and the visible horizon is closer to the observer. This is called **sinking.** Towering or stooping may occur if conditions are suitable. When the refraction is reversed, an **inferior mirage** may occur. A ship or an island appears to be floating in the air above a shimmering horizon, possibly with an inverted image beneath it. Conditions suitable to the formation of an inferior mirage occur when the surface is much warmer than the air above it. This usually requires a heated landmass, and therefore is more common near the coast than at sea.

When refraction is not uniformly progressive, objects may appear distorted, taking an almost endless variety of shapes. The sun when near the horizon is one of the objects most noticeably affected. A **fata morgana** is a complex mirage characterized by marked distortion, generally in the vertical. It may cause objects to appear towering, magnified, and at times even multiplied.

3816. Sky coloring.—White light is composed of light of all colors. Color is related to wavelength, the visible spectrum varying from about 0.000038 to 0.000076 centimeters (art. 4003). The characteristics of each color are related to its wavelength (or frequency). Thus, the shorter the wavelength, the greater the amount of bending when light is refracted. It is this principle that permits the separation of light from celestial bodies into a **spectrum** ranging from red, through orange, yellow, green, and blue, to violet, with long-wave infrared (black light) being slightly outside the visible range at one end and short-wave ultraviolet being slightly outside the visible range at the other end. Light of shorter wavelength is scattered and diffracted more than that of longer wavelength.

Light from the sun and moon is white, containing all colors. As it enters the earth's atmosphere, a certain amount of it is scattered. The blue and violet, being of shorter wavelength than other colors, are scattered most. Most of the violet light is absorbed in the atmosphere. Thus, the scattered blue light is most apparent, and the sky appears blue. At great heights, above most of the atmosphere, it appears black.

When the sun is near the horizon, its light passes through more of the atmosphere than when higher in the sky, resulting in greater scattering and absorption of blue and green light, so that a larger percentage of the red and orange light penetrates to the observer. For this reason the sun and moon appear redder at this time, and when this light falls upon clouds, they appear colored. This accounts for the colors at sunset and sunrise. As the setting sun approaches the horizon, the sunset colors first appear as faint tints of yellow and orange. As the sun continues to set, the colors deepen. Contrasts occur, due principally to difference in height of clouds. As the sun sets, the clouds become a deeper red, first the lower clouds and then the higher ones, and finally they fade to a gray.

When there is a large quantity of smoke, dust, or other material in the sky, unusual effects may be observed. If the material in the atmosphere is of suitable substance and quantity to absorb the longer wave red, orange, and yellow radiations, the sky may have a greenish tint, and even the sun or moon may appear green. If the green light, too, is absorbed, the sun or moon may appear blue. A **green moon** or **blue moon** is most likely to occur when the sun is slightly below the horizon and the longer wavelength light from the sun is absorbed, resulting in green or blue light being cast upon the atmosphere in front of the moon. The effect is most apparent if the moon is on the same side of the sky as the sun.

3817. Rainbows.—The familiar arc of concentric colored bands seen when the sun shines on rain, mist, spray, etc., is caused by refraction, internal reflection, and diffraction of sunlight by the drops of water. The center of the arc is a point 180° from the sun, in the direction of a line from the sun, through the observer. The radius of the brightest rainbow is 42°. The colors are visible because of the difference in the amount of refraction of the different colors making up white light, the light being spread out to form a spectrum (art. 3816). Red is on the outer side and blue and violet on the inner side, with orange, yellow, and green between, in that order from red.

Sometimes a secondary rainbow is seen outside the primary one, at a radius of about 50°. The order of colors of this rainbow is reversed. On rare occasions a faint rainbow is seen on the same side as the sun. The radius of this rainbow and the order of colors are the same as those of the primary rainbow.

A similar arc formed by light from the moon (a lunar rainbow) is called a **moonbow.** The colors are usually very faint. A faint, white arc of about 39° radius is occasionally seen in fog opposite the sun. This is called a **fogbow,** although its origin is controversial, some considering it a halo (art. 3818).

3818. Halos.—Refraction, or a combination of refraction and reflection, of light by ice crystals in the atmosphere (cirrostratus clouds, art. 3713) may cause a **halo** to appear. The most common form is a ring of light of radius 22° or 46° with the sun or moon at the center. Occasionally a faint, white circle with a radius of 90° appears around the sun. This is called a **Hevelian halo.** It is probably caused by refraction and internal reflection of the sun's light by bipyramidal ice crystals. A halo formed by refraction is usually faintly colored like a rainbow (art. 3817), with red nearest the celestial body, and blue farthest from it.

A brilliant rainbow-colored arc of about a quarter of a circle with its center at the zenith, and the bottom of the arc about 46° above the sun, is called a **circumzenithal arc.** Red is on the outside of the arc, nearest the sun. It is produced by the refraction and dispersion of the sun's light striking the top of prismatic ice crystals in the atmosphere. It usually lasts for only about 5 minutes, but may be so brilliant as to be mistaken for an unusually bright rainbow. A similar arc formed 46° *below* the sun, with red on the upper side, is called a **circumhorizontal arc.** Any arc tangent to a heliocentric halo (one surrounding the sun) is called a **tangent arc.** As the sun increases in elevation, such arcs tangent to the halo of 22° gradually bend their ends toward each other. If they meet, the elongated curve enclosing the circular halo is called a **circumscribed halo.** The inner edge is red.

A halo consisting of a faint, white circle through the sun and parallel to the horizon is called a **parhelic circle.** A similar one through the moon is called a **paraselenic circle.** They are produced by reflection of sunlight or moonlight from vertical faces of ice crystals.

A **parhelion** (plural *parhelia*) is a form of halo consisting of an image of the sun at the same altitude and some distance from it, usually 22°, but occasionally 46°. A similar phenomenon occurring at an angular distance of 120° (sometimes 90° or 140°) from the sun is called a **paranthelion.** One at an angular distance of 180°, a rare occurrence, is called an **anthelion,** although this term is also used to refer to a luminous, colored ring or **glory** sometimes seen around the shadow of one's head on a cloud or fog bank. A parhelion is popularly called a **mock sun** or **sun dog.** Similar phenomena in relation to the moon are called **paraselene** (popularly a **mock moon** or **moon dog**), **parantiselene,** and **antiselene.** The term *parhelion* should not be confused with *perihelion,* that orbital point nearest the sun when the sun is the center of attraction (art. 1407).

A **sun pillar** is a glittering shaft of white or reddish light occasionally seen extending above and below the sun, usually when the sun is near the horizon. A phenomenon similar to a sun pillar, but observed in connection with the moon, is called a **moon pillar.** A rare form of halo in which horizontal and vertical shafts of light intersect at the sun is called a **sun cross.** It is probably due to the simultaneous occurrence of a sun pillar and a parhelic circle.

3819. Corona.—When the sun or moon is seen through altostratus clouds (art. 3713), its outline is indistinct, and it appears surrounded by a glow of light called a **corona.** This is somewhat similar in appearance to the corona seen around the sun during a solar eclipse (art. 1424). When the effect is due to clouds, however, the glow may be accompanied by one or more rainbow-colored rings of small radii, with the celestial body at the center. These can be distinguished from a halo by their much smaller radii and also by the fact that the order of the colors is reversed, red being on the inside, nearest the body, in the case of the halo, and on the outside, away from the body, in the case of the corona.

A corona is caused by diffraction of light by tiny droplets of water. The radius of a corona is inversely proportional to the size of the water droplets. A large corona indicates small droplets. If a corona decreases in size, the water droplets are becoming larger and the air more humid. This may be an indication of an approaching rainstorm.

The glow portion of a corona is called an **aureole.**

3820. The green flash.—As light from the sun passes through the atmosphere, it is refracted. Since the amount of bending is slightly different for each color, separate images of the sun are formed in each color of the spectrum. The effect is similar to that of imperfect color printing in which the various colors are slightly out of register. However, the difference is so slight that the effect is not usually noticeable. At the horizon, where refraction is maximum, the greatest difference, which occurs between violet at one end of the spectrum and red at the other, is about 10 seconds of arc. At latitudes of the United States, about 0.7 second of time is needed for the sun to change altitude by this amount when it is near the horizon. The red image, being bent least by refraction, is first to set and last to rise. The shorter wave blue and violet colors are scattered most by the atmosphere, giving it its characteristic blue color (art. 3816). Thus, as the sun sets, the green image may be the last of the colored images to drop out of sight. If the red, orange, and yellow images are below the horizon, and the blue and violet light is scattered and absorbed, the upper rim of the green image is the only part seen, and the sun appears green. This is the **green flash.** The shade of green varies, and occasionally the blue image is seen, either separately or following the green flash (at sunset). On rare occasions the violet image is also seen. These colors may also be seen at sunrise, but in reverse order. They are occasionally seen when the sun disappears behind a cloud or other obstruction.

The phenomenon is not observed at each sunrise or sunset, but under suitable conditions is far more common than generally supposed. Conditions favorable to observation of the green flash are a sharp horizon, clear atmosphere, a temperature inversion (art. 3814), and an attentive observer. Since these conditions are more frequently met when the horizon is formed by the sea than by land, the phenomenon is more common at sea. With a sharp sea horizon and clear atmosphere, an attentive observer may see the green flash at as many as 50 percent of sunsets and sunrises, although a telescope may be needed for some of the observations.

Duration of the green flash (including the time of blue and violet flashes) of as long as 10 seconds has been reported, but such length is rare. Usually it lasts for a period of about ½ second to 2½ seconds with about 1¼ seconds being average. This variability is probably due primarily to changes in the index of refraction (art. 1606) of the air near the horizon.

Under favorable conditions, a momentary green flash has been observed at the setting of Venus and Jupiter. A telescope improves the chances of seeing such a flash from a planet, but is not a necessity.

3821. Crepuscular rays are beams of light from the sun passing through openings in the clouds, and made visible by illumination of dust in the atmosphere along their paths. Actually, the rays are virtually parallel, but because of perspective, appear to diverge. Those appearing to extend downward are popularly called **backstays of the sun,** or **sun drawing water.** Those extending upward and across the sky, appearing to converge toward a point 180° from the sun, are called **anticrepuscular rays.**

3822. The atmosphere and radio waves.—Radio waves traveling through the atmosphere exhibit many of the properties of light, being refracted, reflected, diffracted, and scattered. These and other effects are discussed in chapter XL.

3823. Atmospheric electricity.—Various conditions induce the formation of electrical charges in the atmosphere. When this occurs, there is often a difference of electron charge between various parts of the atmosphere, and between the atmosphere and earth or terrestrial objects. When this difference exceeds a certain minimum value, depending upon the conditions, the static electricity is discharged, resulting in phenomena such as lightning or St. Elmo's fire.

Lightning is the discharge of electricity from one part of a thundercloud (art. 3713) to another, from one such cloud to another, or between such a cloud and the earth or a terrestrial object.

Enormous electrical stresses build up within thunderclouds, and between such clouds and the earth. At some point the resistance of the intervening air is overcome. At first the process is a progressive one, probably starting as a brush discharge (St. Elmo's fire) and growing by ionization. The breakdown follows an irregular path along the line of least resistance. A hundred or more individual discharges may be necessary to complete the path between points of opposite polarity. When this "leader stroke" reaches its destination, a heavy "main stroke" immediately follows in the opposite direction. This main stroke is the visible lightning, which may be tinted any color, depending upon the nature of the gases through which it passes. The illumination is due to the high degree of ionization of the air, which causes many of the atoms to be in excited states and emit radiation.

Thunder, the noise that accompanies lightning, is caused by the heating and ionizing of the air by lightning, which results in rapid expansion of the air along its path and the sending out of a compression wave. Thunder may be heard at a distance of as much as 15 miles, but generally does not carry that far. The elapsed time between the flash of lightning and reception of the accompanying sound of thunder is an indication of the distance, because of the difference in travel time of light and sound. Since the former is comparatively instantaneous, and the speed of sound is about 1,117 feet per second, the approximate distance in nautical miles is equal to the elapsed time in seconds, divided by 5.5. If the thunder accompanying lightning cannot be heard due to its distance, the lightning is called **heat lightning,** a phenomenon not unusual during continental "hot spells."

St. Elmo's fire is a luminous discharge of electricity from pointed objects such as the masts and yardarms of ships, lightning rods, steeples, mountain tops, blades of grass, human hair, arms, etc., when there is a considerable difference in the electrical charge between the object and the air. It appears most frequently during a storm. An object from which St. Elmo's fire emanates is in danger of being struck by lightning, since this type discharge may be the initial phase of the leader stroke. Throughout history those who have not understood St. Elmo's fire have regarded it with superstitious awe, considering it a supernatural manifestation. This view is reflected in the name **corposant** (from "corpo santo," meaning "body of a saint") sometimes given this phenomenon.

The **aurora** is a luminous glow appearing in varied forms in the thin atmosphere high above the earth in middle and high latitudes due to radiation emissions from gases in the high atmosphere.

3824. Waterspouts.—A waterspout is a small, whirling storm over the ocean or inland waters. Its chief characteristic is a funnel-shaped cloud; when fully developed it extends from the surface of the water to the base of a cumulus type cloud (fig. 3824). The water in a spout is mostly confined to its lower portion, and may be either salt spray drawn up by the sea surface, or freshwater resulting from condensation due to the lowered pressure in the center of the vortex creating the spout. The air in waterspouts may rotate clockwise or counterclockwise, depending on the manner of formation. They are found most frequently in tropical regions, but are not uncommon in higher latitudes.

FIGURE 3824.—Waterspouts.

There are two types of waterspouts: those derived from violent convective storms over land moving seaward, called **tornadoes**, and those formed over the sea and which are associated with fair or foul weather. The latter type is most common, lasts a maximum of 1 hour, and has variable strength. Many waterspouts are no stronger than dust whirlwinds, which they resemble; at other times they are strong enough to destroy small craft or to cause damage to larger vessels, although modern ocean-going vessels have little to fear from this type.

Waterspouts vary in diameter from a few feet to several hundred feet, and in height from a few hundred feet to several thousand feet. Sometimes they assume fantastic shapes; in early stages of development an hour glass shape between cloud and sea is common. Since a waterspout is often inclined to the vertical, its actual length may be much greater than indicated by its height.

3825. Deck ice.—Ships traveling through regions where the air temperature is below freezing may acquire thick deposits of ice as a result of salt spray freezing on the rigging or deck areas (fig. 3825). This accumulation of ice is called **ice accretion.** Also, precipitation may freeze to the superstructure and exposed areas of the vessel, increasing the load of ice.

On small vessels in heavy seas and freezing weather, deck ice may accumulate very rapidly and increase the topside weight to such an extent as to reduce seriously the stability of the vessel.

FIGURE 3825.—Deck ice.

3826. Forecasting weather.—The prediction of weather at some future time is based upon an understanding of weather processes, and observations of present conditions. Thus, one learns that when there is a certain sequence of cloud types (art. 3713), rain usually can be expected to follow within a certain period. If the sky is cloudless, more heat will be received from the sun by day, and more heat will be radiated outward from the warm earth by night than if the sky is overcast. If the wind is in such a direction that warm, moist air will be transported over a colder surface, fog can be expected. A falling barometer indicates the approach of a "low," probably accompanied by stormy weather. Thus, before meteorology passed from an "art" to "science," many individuals learned to interpret certain atmospheric phenomena in terms of future weather, and to make reasonably accurate forecasts for short periods into the future.

With the establishment of weather observation stations, continuous and accurate weather information became available. As such observations expanded, and communication facilities improved, knowledge of simultaneous conditions over wider areas became available. This made possible the collection of these "synoptic" reports at civilian and military forecast centers.

The individual observations are made at government-operated stations on shore, and aboard vessels at sea. Observations aboard merchant ships at sea are made and transmitted on a voluntary and cooperative basis. The various national meteorological services supply shipmasters with blank forms, printed instructions, and other materials essential to the making, recording, and interpreting of observations. Any shipmaster can render a particularly valuable service by reporting all unusual or non-normal weather occurrences.

Symbols and numbers are used to indicate on a **synoptic chart,** popularly called a **weather map,** the conditions at each observation station. Isobars are drawn through lines of equal atmospheric pressure, fronts are located and symbolically marked (fig. 3826), areas of precipitation and fog are indicated, etc.

Ordinarily, weather maps for surface observations are prepared every 6 (sometimes 3) hours. In addition, synoptic charts for selected heights are prepared every 12 (sometimes 6) hours. Knowledge of conditions aloft is of value in establishing the three-dimensional structure and motion of the atmosphere as input to the forecast.

With the advent of the digital computer, highly sophisticated numerical models have been developed to analyze and prognosticate weather patterns. The civil and military weather centers prepare and disseminate vast numbers of weather charts (analyses and prognoses) daily to assist local forecasters in their efforts to provide users with accurate, predicted weather parameters. It must be remembered that in any area, the accuracy of forecasted parameters decreases with the length of the forecast period. Thus, a 12-hour forecast is likely to be more reliable than a 24-hour forecast. Long term forecasts for 2 weeks or a month in advance are limited to general statements. For example, a prediction is made as to which areas will have temperatures above or below normal, and how precipitation will compare with normal, but no attempt is made to state that rainfall will occur at a certain time and place.

Forecasts are issued for various areas. The national meteorological services of most maritime nations, including the United States, issue forecasts for ocean areas and warnings of the approach of storms. The efforts of the various nations are coordinated through the World Meteorological Organization.

3827. Dissemination of weather information is carried out in a number of ways. Forecasts are widely broadcast by commercial and government radio stations, and printed in newspapers. Shipping authorities on land are kept informed by telegraph and telephone. Visual storm warnings are displayed in various ports, and storm warnings are broadcast by radio.

Through the use of codes, a simplified version of synoptic weather charts is transmitted to various stations ashore and afloat. Rapid transmission of completed maps has been made possible by the development of facsimile transmitters and receivers. This system is based upon detailed scanning, by a photoelectric detector, of properly illuminated black and white copy. The varying degrees of light intensity are converted to electric energy which is transmitted to the receiver and converted back to a black and white presentation.

Complete information on dissemination of weather information by radio is provided in *Selected Worldwide Marine Weather Broadcast* published jointly by the National Weather Service and the Naval Oceanography Command. This publication lists broadcast schedules and weather codes. Information on day and night visual storm warnings is given in the various volumes of sailing directions.

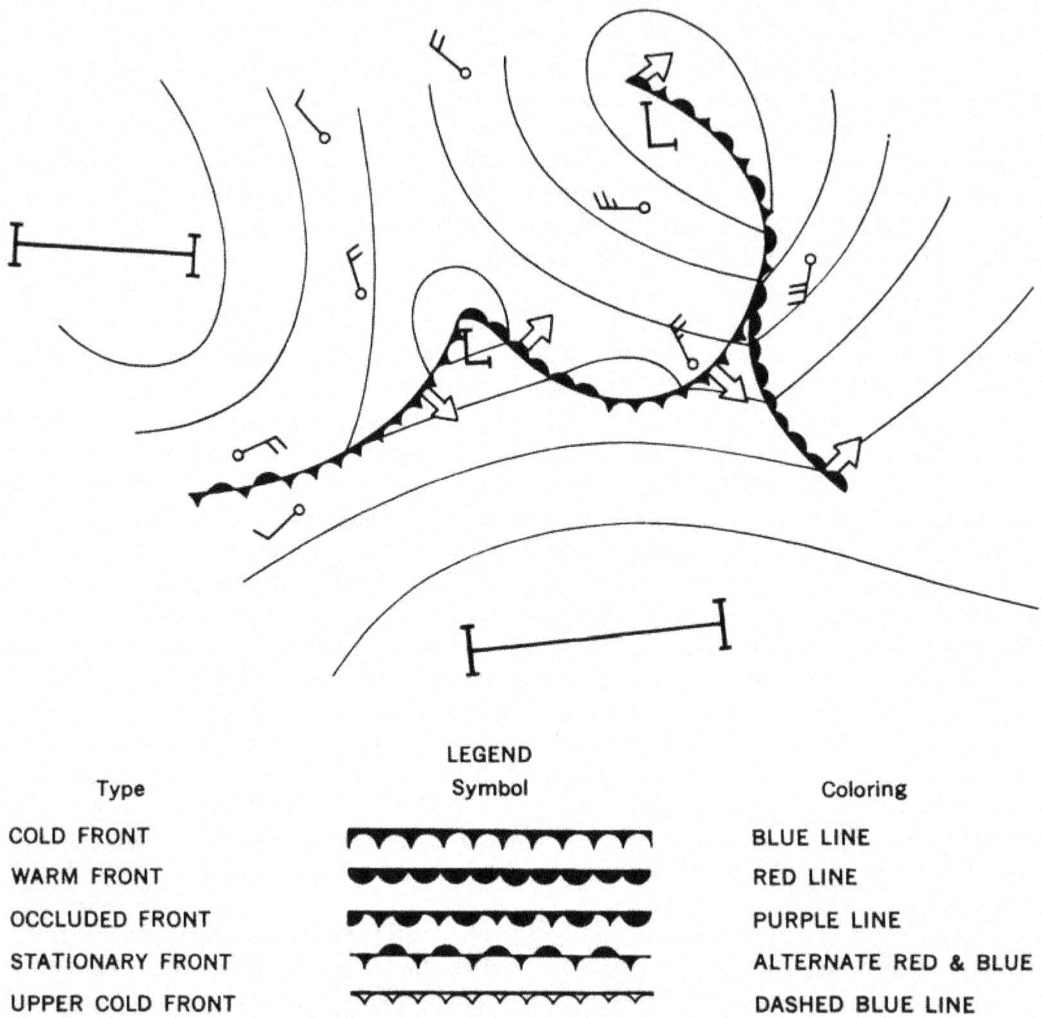

FIGURE 3826.—Designation of fronts on weather maps.

3828. Interpreting the weather.—The factors which determine weather are numerous and varied. Ever-increasing knowledge regarding them makes possible a continually improving weather service. However, the ability to forecast is acquired through study and long practice, and therefore the services of a trained meteorologist should be utilized whenever available.

The value of a forecast is increased if one has access to the information upon which it is based, and understands the principles and processes involved. It is sometimes as important to know the various types of weather that *might* be experienced as it is to know which of several possibilities is *most likely* to occur.

At sea, reporting stations are unevenly distributed, sometimes leaving relatively large areas with incomplete reports, or none at all. Under these conditions, the locations of highs, lows, fronts, etc., are imperfectly known, and their very existence may even be in doubt. At such times the mariner who can interpret the observations made from his own vessel may be able to predict weather for the next several hours more reliably than a trained meteorologist some distance away with incomplete information for the area of concern.

Knowledge of the various relationships given in chapters XXXVII, XXXVIII, and XXXIX is of value, but only the more elementary principles are presented. Further information can be obtained from meteorological publications such as those listed at the ends of the weather chapters. The information obtained from these references will provide a background for proper interpretation of individual weather experiences. If one uses every opportunity to observe and interpret weather sequences, knowledge and skill can be developed that will serve as a valuable supplement to information given in weather broadcasts, or to supply information for areas not covered by such broadcasts.

3829. Influencing the weather.—Meteorological activities are devoted primarily to understanding weather processes and predicting future weather. However, as knowledge regarding cause-and-effect relationships increases, the possibility of being able to induce certain results by artificially producing the necessary conditions becomes greater. The most promising results to date have been in inducing or increasing precipitation on a local scale by "seeding" supercooled clouds with powdered dry ice or silver iodide smoke. The effectiveness of this procedure on a larger scale is still controversial. Experiments in decreasing the intensity of severe tropical cyclones (i.e., hurricanes, typhoons), have been carried out but an operational method is still many years away.

References

American Meteorological Society. *Compendium of Meteorology.* Boston, American Meteorological Society, 1951.

American Meteorological Society. *Glossary of Meteorology.* Boston, American Meteorological Society, 1959 (with corrections, 1970).

Anderson, R. K. et al. *Applications of Meteorological Satellite Data in Analysis and Forecasting,* ESSA TR NESS51. Washington, D.C., National Oceanic and Atmospheric Administration, National Environmental Satellite Service, 1974.

Atkinson, G. D. *Forecasting Guide to Tropical Meteorology,* TR 240, Air Weather Service, U. S. Air Force, 1971.

Berry, F. A. Jr., E. Bollay, and N. R. Beers. *Handbook of Meteorology.* New York, McGraw-Hill, 1945.

Byers, H. R. *General Meteorology,* 4th ed. New York, McGraw-Hill, 1974.

Donn, W. L. *Meteorology,* 4th ed. New York, McGraw-Hill, 1975.

Haltiner, G. J., and F. L. Martin. *Dynamical and Physical Meteorology,* New York, McGraw-Hill, 1957.

Kotsch, W. J. *Weather for the Mariner,* Annapolis, Maryland, Naval Institute Press, 1970.

Neuberger, H., and J. Cahir. *Principles of Climatology, A Manual on Earth Science.* New York, Holt, Rinehart and Winston, Inc., 1969.

Petterssen, S. *Introduction to Meteorology,* 3rd ed. New York, McGraw-Hill, 1969.

Riehl, H. *Tropical Meteorology.* New York, McGraw-Hill, 1954.

Trewartha, G. T. *An Introduction to Climate,* 4th ed. New York, McGraw-Hill, 1968.

U. S. Department of Commerce. *Mariners Weather Log.* National Oceanic and Atmospheric Administration, Environmental Data Service. (periodical).

U. S. Department of the Navy. *Numerical Environmental Products Manual.* NAVAIR 50-1G-522. Naval Weather Service Command, 1975.

U. S. Department of the Navy. *U. S. Navy Marine Climatic Atlas of the World,* vol. VIII, The World. NAVAIR 50-1C-54, Naval Weather Service Command, 1969.

U. S. Department of the Navy. *Aerographer's Mate 1 and C,* NAVEDTRA 10362-B. Naval Education and Training Command, 1974.

U. S. Department of the Navy. *Aerographer's Mate 3 and 2,* NAVEDTRA 10363-E, 1976.

Williams, J., J. J. Higginson, and J. D. Rohrbough. *Sea and Air, The Marine Environment.* Annapolis, Maryland, Naval Institute Press, 1973.

World Meteorological Organization. *The Preparation and Use of Weather Maps by Mariners,* TN72. Secretariat, World Meteorological Organization, Geneva, Switzerland, 1966.

CHAPTER XXXIX

TROPICAL CYCLONES

3901. Introduction.—A **tropical cyclone** is a cyclone (art. 3812) originating in the Tropics or subtropics. Although it generally resembles the extratropical cyclone originating in higher latitudes, there are important differences, the principal one being the concentration of a large amount of energy into a relatively small area. Tropical cyclones are infrequent in comparison with middle- and high-latitude storms, but they have a record of destruction far exceeding that of any other type of storm. Because of their fury, and the fact that they are predominantly oceanic, they merit the special attention of all mariners, whether professional or amateur.

Rarely does the mariner who has experienced a fully developed tropical cyclone at sea wish to encounter a second one. He has learned the wisdom of avoiding them if possible. The uninitiated may be misled by the deceptively small size of a tropical cyclone as it appears on a weather map, and by the fine weather experienced only a few hundred miles from the reported center of such a storm. The rapidity with which the weather can deteriorate with approach of the storm, and the violence of the fully developed tropical cyclone, are difficult to visualize if they have not been experienced.

On his second voyage to the New World, Columbus encountered a tropical storm. Although his vessels suffered no damage, this experience proved valuable during his fourth voyage when his vessels were threatened by a fully developed hurricane. Columbus read the signs of an approaching storm from the appearance of a southeasterly swell, the direction of the high cirrus clouds, and the hazy appearance of the atmosphere. He directed his vessels to shelter. The commander of another group, who did not heed the signs, lost most of his ships; more than 500 men in their crews perished.

3902. Definitions.—Tropical cyclone is the general term for cyclones originating in the Tropics or subtropics. These cyclones are classified by form and intensity as follows:

Tropical disturbance is a discrete system of apparently organized convection—generally 100 to 300 miles in diameter—having a nonfrontal migratory character, and having maintained its identity for 24 hours or more. It may or may not be associated with a detectable perturbation of the wind field. It has no strong winds and no closed isobars i.e., isobars that completely enclose the low. (In successive stages of intensification, the tropical cyclone may be classified as a tropical disturbance, tropical depression, tropical storm, and hurricane or typhoon.)

Tropical depression has one or more closed isobars and some rotary circulation at the surface. The highest sustained (1-minute mean) surface wind speed is 33 knots.

Tropical storm has closed isobars and a distinct rotary circulation. The highest sustained (1-minute mean) surface wind speed is 34 to 63 knots.

Hurricane or **typhoon** has closed isobars, a strong and very pronounced rotary circulation, and a sustained (1-minute mean) surface wind speed of 64 knots or higher.

3903. Areas of occurrence.—Tropical cyclones occur almost entirely in six rather distinct areas, four in the Northern Hemisphere and two in the Southern Hemisphere as shown in figure 3903. The name by which the tropical cyclone is commonly known varies somewhat with the locality, as follows:

North Atlantic. A tropical cyclone with winds of 64 knots or greater is called a **hurricane.**

Eastern North Pacific. The name **hurricane** is used as in the North Atlantic.

Western North Pacific. A fully developed storm with winds of 64 knots or greater is called a **typhoon** or, locally in the Philippines, a **baguio.**

North Indian Ocean. A tropical cyclone with winds of 34 knots or greater is called a **cyclonic storm.**

South Indian Ocean. A tropical cyclone with winds of 34 knots or greater is called a **cyclone.**

Southwest Pacific and Australian Area. The name **cyclone** is used as in the South Indian Ocean. A severe tropical cyclone originating in the Timor Sea and moving southwest and then southeast across the interior of northwestern Australia is called a **willy-willy.**

Tropical cyclones have not been observed in the South Atlantic or in the South Pacific east of 140°W.

3904. Origin, season, and frequency of occurrence of the tropical cyclones in the six areas are as follows:

North Atlantic tropical cyclones can affect the entire North Atlantic Ocean in any month. However, they are mostly a threat south of about 35°N from June through November; August, September, and October are the months of highest incidence (tab. 3904). About 9 or 10 tropical cyclones (tropical storms and hurricanes) form each season; 5 or 6 reach hurricane intensity (winds of 64 knots and higher). A few hurricanes have generated winds estimated as high as 200 knots. Early- and late-season storms usually develop west of 50°W; during August and September, this spawning ground extends to the Cape Verde Islands. These storms usually move westward or westnorthwestward at speeds of less than 15 knots in the lower latitudes. After moving into the northern Caribbean or Greater Antilles regions, they will usually either move toward the Gulf of Mexico or recurve and accelerate in the North Atlantic. Some will recurve after reaching the Gulf of Mexico, while others will continue westward to landfall (fig. 3904).

Eastern North Pacific season is from June through October, although a storm can form in any month. An average of 15 tropical cyclones (tropical storms and hurricanes) form each year with about 6 reaching hurricane strength. The most intense storms are often the early- and late-season ones; these form close to the coast and far south. Midseason storms form anywhere in a wide band from the Mexican-Central American coast to the Hawaiian Islands. August and September are the months of highest incidence. These storms differ from their North Atlantic counterparts in that they are usually smaller in size. However, they can be just as intense.

Western North Pacific. More tropical cyclones form in the tropical western North Pacific than anywhere else in the world. More than 25 (tropical storms and typhoons) develop each year, and about 18 become typhoons. These typhoons are the largest and most intense tropical cyclones in the world. Each year an average of five generate maximum winds over 130 knots; circulations covering more than 600 miles in diameter are not uncommon. Most of these storms form east of the Philippines, and move across the Pacific toward the Philippines, Japan, and China;

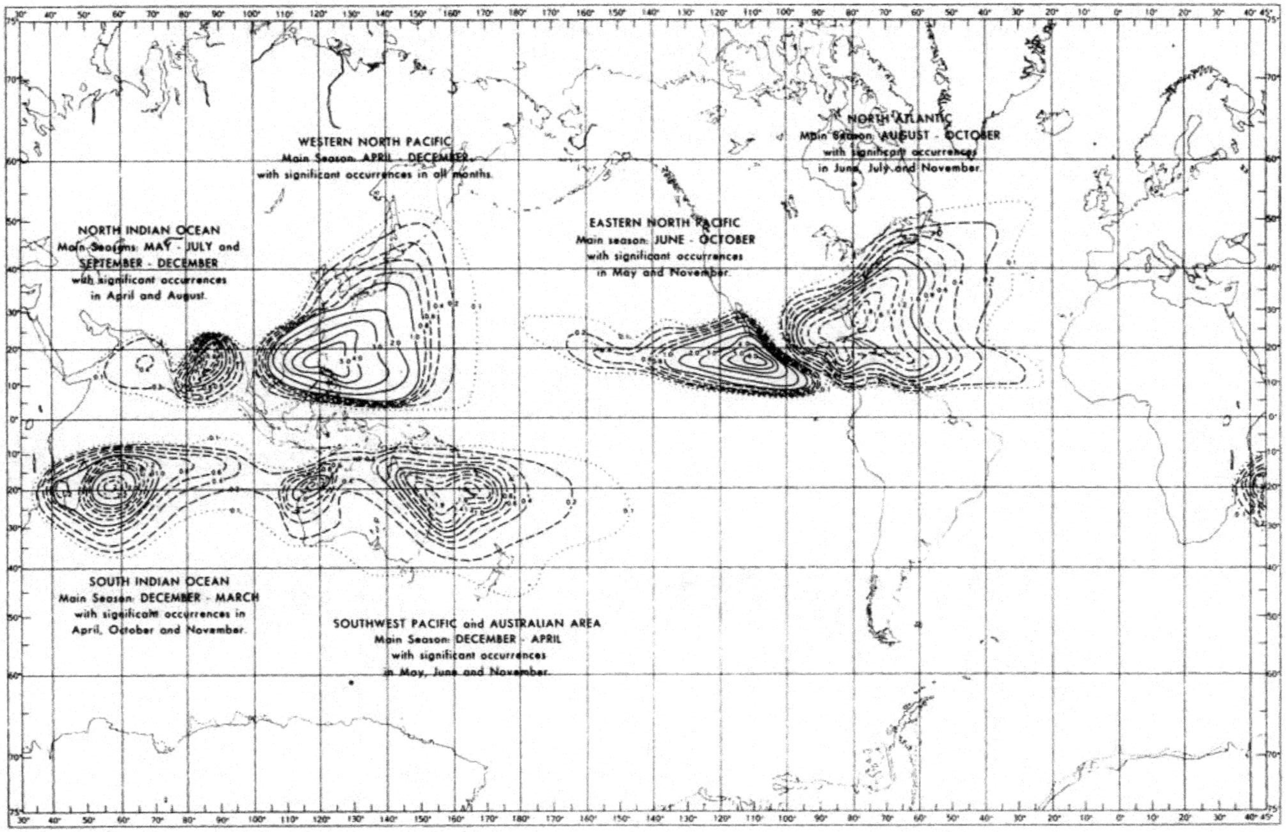

FIGURE 3903.—Areas in which tropical cyclones occur. The average number of tropical cyclones per 5° square has been analyzed for this figure. The main season for intense tropical storm activity is also shown for each major basin.

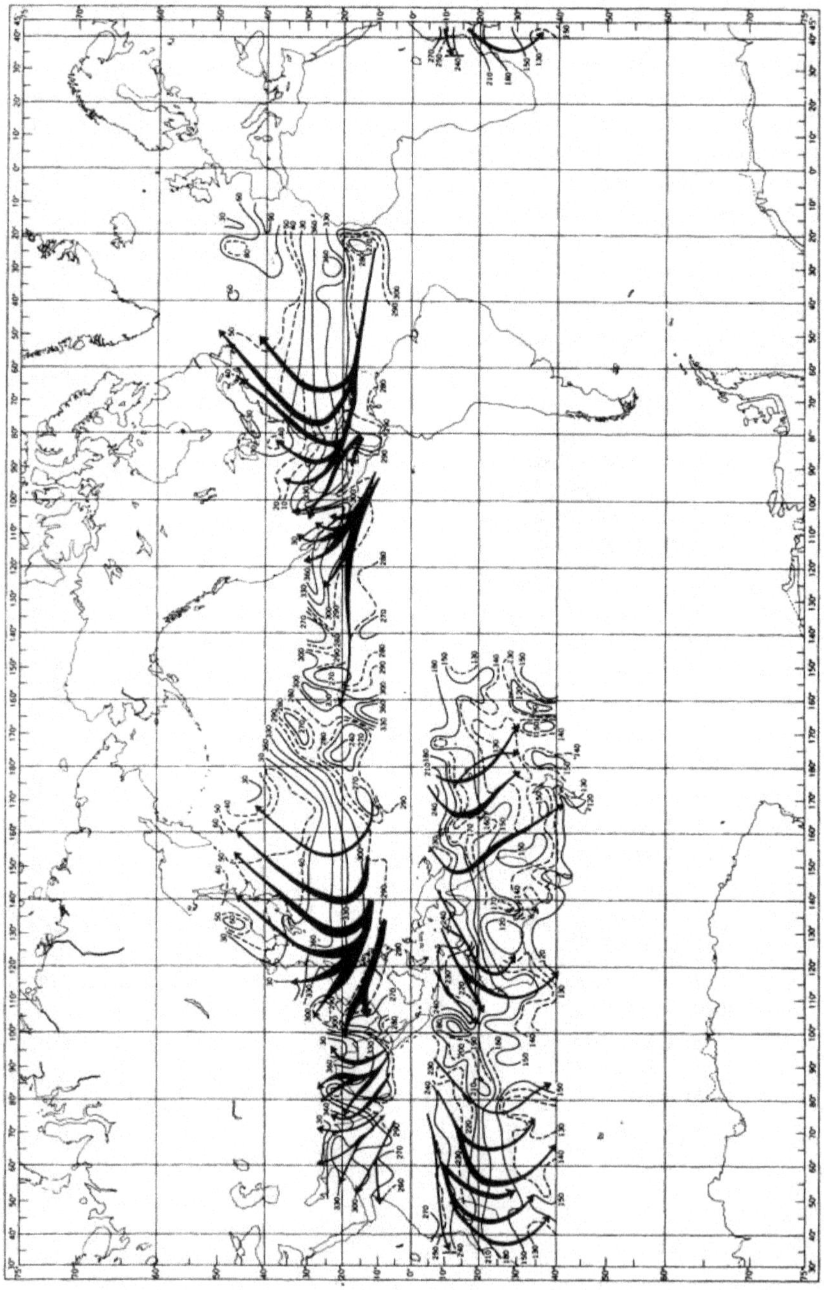

FIGURE 3904.—Storm tracks. The width of the arrow indicates the approximate frequency of storms; the wider the arrow the higher the frequency. Isolines on the base map show the resultant direction toward which storms moved. Data for the entire year has been summarized for this figure.

a few storms form in the South China Sea. The season extends from April through December. However, tropical cyclones are more common in the off-season months in this area than anywhere else. The peak of the season is July through October, when nearly 70 percent of all typhoons develop. There is a noticeable seasonal shift in storm tracks in this region. From July through September, storms move north of the Philippines and recurve, while early- and late-season typhoons move on a more westerly track through the Philippines before recurving (fig. 3904).

AREA AND STAGE	JAN	FEB	MAR	APR	MAY	JUN	JUL	AUG	SEP	OCT	NOV	DEC	ANNUAL
NORTH ATLANTIC													
TROPICAL STORMS	*	*	*	*	0.1	0.4	0.3	1.0	1.5	1.2	0.4	*	4.2
HURRICANES	*	*	*	*	*	0.3	0.4	1.5	2.7	1.3	0.3	*	5.2
TROPICAL STORMS AND HURRICANES	*	*	*	*	*0.2*	*0.7*	*0.8*	*2.5*	*4.3*	*2.5*	*0.7*	*0.1*	*9.4*
EASTERN NORTH PACIFIC	JAN	FEB	MAR	APR	MAY	JUN	JUL	AUG	SEP	OCT	NOV	DEC	ANNUAL
TROPICAL STORMS	*	*	*	*	*	1.5	2.8	2.3	2.3	1.2	0.3	*	9.3
HURRICANES	*	*	*	*	0.3	0.6	0.9	2.0	1.8	1.0	*	*	5.8
TROPICAL STORMS AND HURRICANES	*	*	*	*	*0.3*	*2.0*	*3.6*	*4.5*	*4.1*	*2.2*	*0.3*	*	*15.2*
WESTERN NORTH PACIFIC	JAN	FEB	MAR	APR	MAY	JUN	JUL	AUG	SEP	OCT	NOV	DEC	ANNUAL
TROPICAL STORMS	0.2	0.3	0.3	0.2	0.4	0.5	1.2	1.8	1.5	1.0	0.8	0.6	7.5
TYPHOONS	0.3	0.2	0.2	0.7	0.9	1.2	2.7	4.0	4.1	3.3	2.1	0.7	17.8
TROPICAL STORMS AND TYPHOONS	*0.4*	*0.4*	*0.5*	*0.9*	*1.3*	*1.8*	*3.9*	*5.8*	*5.6*	*4.3*	*2.9*	*1.3*	*25.3*
SOUTHWEST PACIFIC AND AUSTRALIAN AREA	JAN	FEB	MAR	APR	MAY	JUN	JUL	AUG	SEP	OCT	NOV	DEC	ANNUAL
TROPICAL STORMS	2.7	2.8	2.4	1.3	0.3	0.2	*	*	*	0.1	0.4	1.5	10.9
HURRICANES	0.7	1.1	1.3	0.3	*	*	0.1	0.1	*	*	0.3	0.5	3.8
TROPICAL STORMS AND HURRICANES	*3.4*	*4.1*	*3.7*	*1.7*	*0.3*	*0.2*	*0.1*	*0.1*	*	*0.1*	*0.7*	*2.0*	*14.8*
SOUTHWEST INDIAN OCEAN	JAN	FEB	MAR	APR	MAY	JUN	JUL	AUG	SEP	OCT	NOV	DEC	ANNUAL
TROPICAL STORMS	2.0	2.2	1.7	0.6	0.2	*	*	*	*	0.3	0.3	0.8	7.4
HURRICANES	1.3	1.1	0.8	0.4	*	*	*	*	*	*	*	0.5	3.8
TROPICAL STORMS AND HURRICANES	*3.2*	*3.3*	*2.5*	*1.1*	*0.2*	*	*	*	*	*0.3*	*0.4*	*1.4*	*11.2*
NORTH INDIAN OCEAN	JAN	FEB	MAR	APR	MAY	JUN	JUL	AUG	SEP	OCT	NOV	DEC	ANNUAL
TROPICAL STORMS	0.1	*	*	0.1	0.3	0.5	0.5	0.4	0.4	0.6	0.5	0.3	3.5
CYCLONES [1]	*	*	*	0.1	0.5	0.2	0.1	*	0.1	0.4	0.6	0.2	2.2
TROPICAL STORMS AND CYCLONES [1]	*0.1*	*	*0.1*	*0.3*	*0.7*	*0.7*	*0.6*	*0.4*	*0.5*	*1.0*	*1.1*	*0.5*	*5.7*

* Less than .05 [1] Winds ≥ 48 Kts.

Monthly values cannot be combined because single storms overlapping two months were counted once in each month and once in the annual.

TABLE 3904.—Monthly and annual average number of storms per year for each area.

North Indian Ocean tropical cyclones develop in the Bay of Bengal and Arabian Sea during the spring and fall. Tropical cyclones in this area form between latitudes 8°N and 15°N, except from June through September, when the little activity that does occur is confined north of about 15°N. These storms are usually short-lived and weak; however, winds of 130 knots have been encountered. They often develop as perturbations along the Intertropical Convergence Zone (ITCZ); this inhibits summertime development since the ITCZ is usually over land during this monsoon season. However, it is sometimes displaced southward, and when this occurs, storms will form over the monsoon-flooded plains of Bengal. On the average, six cyclonic storms form each year. These include two storms that generate winds of 48 knots or greater. Another 10 tropical cyclones never develop beyond tropical depressions. The Bay of Bengal is the area of highest incidence. However, it is not unusual for a storm to move across southern India and reintensify in the Arabian

Sea. This is particularly true during October—the month of highest incidence during the tropical cyclone season. It is also during this period that torrential rains from these storms dumped over already rain-soaked areas cause disastrous floods.

South Indian Ocean. Over the waters west of 100°E to the east African coast, an average of 11 tropical cyclones (tropical storms and hurricanes) form each season, and about 4 reach hurricane intensity. The season is from December through March, although it is possible for a storm to form in any month. Tropical cyclones in this region usually form south of 10°S. The latitude of recurvature usually migrates from about 20°S in January to around 15°S in April. After crossing 30°S, these storms sometimes become intense extratropical lows.

Southwest Pacific and Australian Area. These tropical waters spawn an annual average of 15 tropical cyclones (tropical storms and hurricanes), 4 of which reach hurricane intensity. The season extends from about December through April, although storms can form in any month. Activity is widespread in January and February, and it is in these months that tropical cyclones are most likely to affect Fiji, Samoa, and the other eastern islands.

Tropical cyclones usually form in the waters from 105°E to 160°W, between 5° and 20°S. Storms affecting northern and western Australia often develop in the Timor or Arafura Sea, while those that affect the east coast form in the Coral Sea. These storms are often small, but can develop winds in excess of 130 knots. New Zealand is sometimes reached by decaying Coral Sea storms; occasionally, it is reached by an intense hurricane. In general, tropical cyclones in this region move southwestward and then recurve southeastward (fig. 3904).

3905. Hurricane formation was once believed to result from an intensification of convective forces which produce the cumulonimbus towers of the doldrums. This view of hurricane generation held that surface heating caused warm moist air to ascend convectively to levels where condensation produced cumulonimbus clouds, which, after an inexplicable drop in atmospheric pressure, coalesced and were spun into a cyclonic motion by Coriolis force.

This hypothesis left much to be desired. Although some hurricanes develop from disturbances beginning in the doldrums (art. 3804), very few reach maturity in that region. Also, the high incidence of seemingly ideal convective situations does not match the low incidence of Atlantic hurricanes. Finally, the hypothesis did not explain the drop in atmospheric pressure, so essential to development of hurricane-force winds.

There is still no exact understanding of the triggering mechanism involved in hurricane generation, the balance of conditions needed to generate hurricane circulation, and the relationships between large- and small-scale atmospheric processes. But scientists today, treating the hurricane system as an atmospheric heat engine, present a more comprehensive and convincing view.

They begin with a starter mechanism in which either internal or external forces intensify the initial disturbance. The initial disturbance becomes a region into which low-level air from the surrounding area begins to flow, accelerating the convection already occurring inside the disturbance. The vertical circulation becomes increasingly well organized as water vapor in the ascending moist layer is condensed (releasing large amounts of heat energy to drive the wind system) and as the system is swept into a counterclockwise cyclonic spiral. But this incipient hurricane would soon fill up because of inflow at lower levels unless the chimney in which converging air surges upward is provided the exhaust mechanism of high-altitude winds.

These high-altitude winds (fig. 3905) pump ascending air out of the cyclonic system into a high-altitude anticyclone, which transports the air well away from the disturbance before sinking occurs. Thus, a large scale vertical circulation is set up in which low-level air is spiraled up the cyclonic twisting of the disturbance, and, after a trajectory over the sea, returned to lower altitudes some distance from the storm. This pumping action—and the heat released by the ascending air—may account for the sudden drop of atmospheric pressure at the surface, which produces the steep pressure gradient along which winds reach hurricane proportions.

It is believed that the interaction of low-level and high-altitude wind systems determines the intensity the hurricane will attain. If less air is pumped out than converges at low levels, the system will fill and die out. If more is pumped out than flows in, the circulation will be sustained and will intensify.

Research has shown that any process which increases the rate of low-level inflow is favorable for hurricane development, provided the inflowing air carries sufficient heat and moisture to fuel the hurricane's power system. It has also been shown that air above the developing disturbance at altitudes between 20,000 and 40,000 feet increases 1° to 3° in temperature about 24 hours before the disturbance develops into a hurricane. But it is not known whether low-level inflow and high-level warming *cause* hurricanes. They could very well be measurable symptoms of another effect which actually triggers the storm's increase to hurricane intensity.

The view of hurricanes as atmospheric engines is necessarily a general one. The exact role of each contributor is not completely understood. The engine seems to be both inefficient and unreliable; a myriad of delicate conditions must be satisfied for the atmosphere to produce a hurricane. Their relative infrequency indicates that many a potentially healthy hurricane ends early as a misfiring dud of a disturbance, somewhere over the sea.

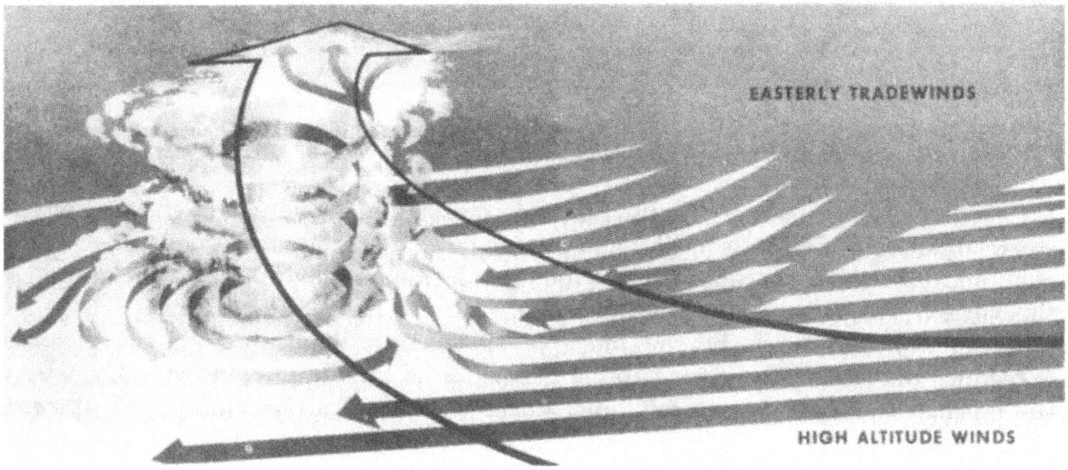

Figure 3905.—Pumping action of high altitude winds.

3906. Portrait of a hurricane.—In the early life of the hurricane, the spiral covers an area averaging 100 miles in diameter with winds of 64 knots and greater, and spreads gale-force winds over a 400-mile diameter. The cyclonic spiral (fig. 3906) is marked by heavy cloud bands from which torrential rains fall, separated by areas of light rain or no rain at all. These spiral bands ascend in decks of cumulus and cumulonimbus clouds to the convective limit of cloud formation, where condensing water vapor is swept off as ice-crystal wisps of cirrus clouds. Thunderstorm electrical activity is observed in these bands, both as lightning and as tiny electrostatic discharge.

In the lower few thousand feet, air flows in through the cyclone, and is drawn upward through ascending columns of air near the center. The size and intensity decrease with altitude, the cyclonic circulation being gradually replaced above 40,000 feet by an anticyclonic circulation centered hundreds of miles away—the enormous high-altitude pump which is the exhaust system of the hurricane heat engine.

FIGURE 3906.—Cutaway view of a hurricane greatly exaggerated in vertical dimension. Actual hurricanes are less than 50,000 feet high and may have a diameter of several hundred miles.

At lower levels, where the hurricane is more intense, winds on the rim of the storm follow a wide pattern, like the slower currents around the edge of a whirlpool; and, like those currents, these winds accelerate as they approach the center of the vortex. The outer band has light winds at the rim of the storm, perhaps no more than 25 knots; within 30 miles of the center, winds may have velocities exceeding 130 knots. The inner band is the region of maximum wind velocity, where the storm's worst winds are felt, and where ascending air is chimneyed upward, releasing heat to drive the storm. In most hurricanes, these winds reach 85 knots and more than 170 knots in the more memorable ones.

In the hurricane, winds flow toward the low pressure in the warm, comparatively calm core. There, converging air is whirled upward by convection, the mechanical thrusting of other converging air, and the pumping action of high-altitude circulations. This spiral is marked by the thick cloud walls curling inward toward the storm center, releasing heavy precipitation and enormous quantities of heat energy. At the center, surrounded by a band in which this strong vertical circulation is greatest, is the **eye** of the hurricane.

The eye, like the spiral rainbands, is unique to the hurricane; no other atmospheric phenomenon has this calm core. On the average, eye diameter is about 14 miles, although diameters of 25 miles are not unusual. From the heated tower of maximum winds and cumulonimbus clouds, winds diminish rapidly to something less than 15 miles per hour in the eye; at the opposite wall, winds increase again, but come from the opposite direction because of the cyclonic circulation of the storm. This transformation of storm into comparative calm, and calm into violence from another quarter is spectacular. The eye's abrupt existence in the midst of opaque rainsqualls and hurricane winds, the intermittent bursts of blue sky and sunlight through light clouds in the core of the cyclone, and the galleried cumulus and cumulonimbus clouds are unforgettable.

That is how an average hurricane is structured. But every hurricane is individual, and the more or less orderly circulation described here omits the extreme variability and instability within the storm system. Pressure and temperature gradients fluctuate wildly across the storm as the hurricane maintains its erratic life in the face of forces which will ultimately destroy it. If it is an August storm, its average life expectancy is 12 days; if a July or November storm, it lives an average of 8 days.

3907. Life of a tropical cyclone.—Reports from ships in the vicinity of an **easterly wave** (a westward-moving trough of low pressure embedded in deep easterlies) indicate that the atmospheric pressure in the region has fallen more than 5 millibars in the past 24 hours. This is cause for alarm because in the Tropics pressure varies little; the normal diurnal pressure change is only about 3 millibars. Satellite pictures indicate thickening middle and high clouds, squalls are reported ahead of the easterly wave, and wind reports indicate a cyclonic circulation is forming. The former easterly wave—now classified a *tropical disturbance*—is moving westward at 10 knots under the canopy of a large high-pressure system aloft. Sea surface temperatures in the vicinity are in the mid-80°F range.

Within 48 hours winds increase to 25 knots near the center of definite circulation, and central pressure has dropped below 1000 millibars. The disturbance is now classified as a *tropical depression.* Soon the circulation extends out to 100 miles and upward to 20,000 feet. Winds near the center increase to gale force, central pressure falls below 990 millibars, and towering cumulonimbus clouds shield a developing eye; a *tropical storm* has developed.

Satellite photographs now reveal a tightly organized tropical cyclone, and reconnaissance reports indicate maximum winds of 80 knots around a central pressure of 980 millibars; a *hurricane* has developed. A ship to the right (left in the Southern Hemisphere) of the hurricane's center (looking toward the direction of storm movement) reports a 30-foot sea. The hurricane is fast maturing; it continues westward.

A few days later the hurricane reaches its peak. The satellite photographs a textbook picture (fig. 3907), as 120-knot winds roar around a 940-millibar pressure center; hurricane-force winds extend 50 miles in all directions, and seas are reported up to 40 feet. There is no further deepening now, but the hurricane begins to expand. In 2 days, gales extend out to 200 miles, and hurricane winds out to 75 miles. Then the hurricane slows and begins to recurve; this turning marks the beginning of the end.

FIGURE 3907.—Satellite photograph of hurricane.

The hurricane accelerates, and, upon reaching the temperate latitudes, it begins to lose its tropical characteristics. The circulation continues to expand, but now cold air is intruding (cold air, cold water, dry air aloft, and land aid in the decay of a tropical cyclone). The warm core survives for a few more days before the transformation to a large extratropical low-pressure system is complete.

Not all tropical cyclones follow this ideal pattern. Most falter in the early stages, some dissipate over land, and others remain potent for several weeks.

The lowest-sea-level pressure ever recorded was 877 millibars in typhoon Ida, on September 24, 1958. The observation was taken by a reconnaissance aircraft dropsonde some 750 miles east of Luzon, Philippines. This observation was obtained again in typhoon Nora on October 6, 1973. The lowest barometric reading of record for the United States is 892.3 millibars obtained during a hurricane at Lower Matecumbe Key, Florida in September 1935. In hurricane Camille in 1969, a 905 millibar pressure was measured by reconnaissance aircraft. During a 1927 typhoon, the S. S. *Sapoeroea* recorded a pressure of 886.6 millibars, the lowest sea-level pressure reported from a ship. Pressure has been observed to drop more than 33 millibars per hour, with a pressure gradient amounting to a change of 3.7 millibars per mile.

3908. The marine weather broadcast is the most important tool the mariner has for avoiding the tropical cyclone. This broadcast, covering all tropical areas, provides information about the tropical cyclone's present location, maximum winds and seas, and future condition.

The U. S. Navy, the National Oceanic and Atmospheric Administration, and the U. S. Air Force have developed a highly effective surveillance system for the tropical cyclone areas of the world. Routine and special weather reports (from land stations, ships at sea, aircraft; daytime weather satellite reports; radar reports from land stations; special reports from ships at sea; and the specially instrumented weather reconnaissance aircraft of National Oceanic and Atmospheric Administration and the U. S. Air Force) enable accurate detection, location, and tracking of tropical cyclones. International cooperation is good. In addition to improved satellites permitting nighttime surveillance, data buoys provide another new source of information for the protection of the mariner.

The tropical warning services have three principal functions:

1. the collection and analysis of the necessary observational data;
2. the preparation of timely and accurate forecasts and warnings; and
3. the rapid and efficient distribution of advisories, warnings, and all other pertinent information.

To provide timely and accurate information and warnings regarding tropical cyclones, the oceans have been divided into overlapping geographical areas of responsibility.

For detailed information on the areas of responsibility of the countries participating in the international forecasting and warning program, and radio aids, refer to *Selected Worldwide Marine Weather Broadcasts,* published jointly by the Naval Oceanography Command and the National Weather Service.

Although the areas of forecasting responsibility are fairly well defined for the Department of Defense, the international and domestic civilian system provides many overlaps and is dependent upon qualitative factors. For example, when a tropical storm or hurricane is traveling westward and crosses 35°W longitude, the continued issuance of forecasts and warnings to the general public, shipping interests, etc., becomes the responsibility of the National Hurricane Center of the National Weather Service at Miami, Florida. When a tropical storm or hurricane crosses 35°W longitude traveling from west to east, the National Hurricane Center ceases to issue formal public advisories, but will issue marine bulletins on any dangerous tropical cyclone in the North Atlantic, if it is of importance or constitutes a threat to shipping and other interests. These advisories are included in National Weather Service Marine Bulletins broadcast to ships over radio station NAM Norfolk, Virginia. (*See Selected Worldwide Marine Weather Broadcasts*). Special advisories may be issued at any time.

In the eastern Pacific (east of longitude 140°W), responsibility for the issuance of tropical storm and hurricane advisories and warnings for the general public, merchant shipping, and other interests rests with the National Weather Service Eastern Pacific Hurricane Center, San Francisco, California. The Department of Defense responsibility rests with the Naval Western Oceanography Center, Pearl Harbor, Hawaii. Formal advisories and warnings are issued daily and are included in the marine bulletins broadcast by radio stations KFS, NMC, and NMQ.

In the central Pacific (between the 180th meridian and longitude 140°W), the civilian responsibility rests with the National Weather Service Central Pacific Hurricane Center, Honolulu, Hawaii. Department of Defense responsibility rests with the Naval Western Oceanography Center, Pearl Harbor. Formal tropical storm and hurricane advisories and warnings are issued daily and are included in the marine bulletins broadcast by radio station NMO and NRV.

Tropical cyclone information messages generally contain position of the storm, intensity, direction and speed of movement, and a description of the area of strong winds. Also included is a forecast of future movement and intensity. When the storm is likely to affect any land area, details on when and where it will be felt, and data on tides, rain, floods, and maximum winds are also included. Figure 3908 provides an example of a marine advisory issued by the National Hurricane Center.

The U. S. Navy's Oceanography Command Center/Joint Typhoon Warning Center (JTWC) in Guam, has a primary area of responsibility for all U. S. tropical storm and typhoon advisories and warnings from the 180th meridian westward to the mainland of Asia. A secondary area of responsibility extends westward to longitude 90°E. Whenever a tropical cyclone is observed in the western North Pacific area, serially numbered warnings, bearing an immediate precedence are broadcast from the Oceanography Command Center/JTWC at 0000, 0600, 1200, and 1800 GMT.

The responsibility for issuing gale and storm warnings for the Indian Ocean, Arabian Sea, Bay of Bengal, Western Pacific, and South Pacific rests with many countries. In general, warnings of approaching tropical cyclones which may be hazardous will include the following information: storm type, central pressure given in millibars, windspeed observed within the storm, storm location, speed and direction of movement, the extent of the affected area, visibility, and the state of the sea, as well as any other pertinent information received. All storm warning messages commence with the international call sign "TTT."

These warnings are broadcast on prespecified radio frequency bands immediately upon receipt of the information and at specific intervals thereafter. Generally, the broadcast interval is every 6 to 8 hours, depending upon receipt of new information.

Bulletins and forecasts are excellent guides to the present and future behavior of the tropical cyclone, and a plot should be kept of all positions.

3909. The passage of a tropical cyclone at sea is an experience not soon to be forgotten.

NOAA/NATIONAL HURRICANE CENTER MARINE ADVISORY NUMBER 13 HURRICANE LADY 0400Z SEPTEMBER 21 19--.

HURRICANE WARNINGS ARE DISPLAYED FROM KEY LARGO TO CAPE KENNEDY. GALE WARNINGS ARE DISPLAYED FROM KEY WEST TO JACKSONVILLE AND FROM FLORIDAY BAY TO CEDAR KEY.

HURRICANE CENTER LOCATED NEAR LATITUDE 25.5 NORTH LONGITUDE 78.5 WEST AT 21/0400Z. POSITION EXCELLENT ACCURATE WITHIN 10 MILES BASED ON AIR FORCE RECONNAISSANCE AND SYNOPTIC REPORTS.

PRESENT MOVEMENT TOWARD THE WEST NORTHWEST OR 285 DEGREES AT 10 KT. MAX SUSTAINED WINDS OF 100 KT NEAR CENTER WITH GUSTS TO 160 KT.
MAX WINDS OVER INLAND AREAS 35 KT.
RAD OF 65 KT WINDS 90 NE 60 SE 80 SW 90 NW QUAD.
RAD OF 50 KT WINDS 120 NE 70 SE 90 SW 120 NW QUAD.
RAD OF 30 KT WINDS 210 NE 210 SE 210 SW 210 NW QUAD.
REPEAT CENTER LOCATED 25.5N 78.3W at 21/0400Z.

12 HOUR FORECAST VALID 21/1600Z LATITUDE 26.0N LONGITUDE 80.5W.
MAX WINDS OF 100 KT NEAR CENTER WITH GUSTS TO 160 KT.
MAX WINDS OVER INLAND AREAS 65 KT.
RADIUS OF 50KT WINDS 120 NE 70 SE 90 SW 120 NW QUAD.
24 HOUR FORECAST VALID 22/0400Z LATITUDE 26.0N LONGITUDE 83.0W.
MAX WINDS OF 75 KT NEAR CENTER WITH GUSTS TO 120 KT.
MAX WINDS OVER INLAND AREAS 45 KT.
RADIUS OF 50 KT WINDS 120 NE 120 SE 120 SW 120 NW QUAD.

STORM TIDE OF 9 TO 12 FT SOUTHEAST FLA COAST GREATER MIAMI AREA TO THE PALM BEACHES.

NEXT ADVISORY AT 21/1000Z.

FIGURE 3908.—Example of marine advisory issued by National Hurricane Center.

An early indication of the approach of such a storm is the presence of a long swell. In the absence of a tropical cyclone, the crests of swell in the deep waters of the Atlantic pass at the rate of perhaps eight per minute. Swell generated by a hurricane is about twice as long, the crests passing at the rate of perhaps four per minute. Swell may be observed several days before arrival of the storm.

When the storm center is 500 to 1,000 miles away, the barometer usually rises a little, and the skies are relatively clear. Cumulus clouds, if present at all, are few in number and their vertical development appears suppressed. The barometer usually appears restless, **pumping** up and down a few hundredths of an inch.

As the tropical cyclone comes nearer, a cloud sequence begins which resembles that associated with the approach of a warm front in middle latitudes (art. 3811). Snow-white, fibrous "mare's tails" (cirrus) appear when the storm is about 300 to 600 miles away. Usually these seem to converge, more or less, in the direction from which the storm is approaching. This convergence is particularly apparent at about the time of sunrise and sunset.

Shortly after the cirrus appears, but sometimes before, the barometer starts a long, slow fall. At first the fall is so gradual that it only appears to alter somewhat the normal daily cycle (two maxima and two minima in the Tropics). As the rate of fall increases, the daily pattern is completely lost in the more or less steady fall.

The cirrus becomes more confused and tangled, and then gradually gives way to a continuous veil of cirrostratus. Below this veil, altostratus forms, and then stratocumulus (art. 3713). These clouds gradually become more dense, and as they do so, the weather becomes unsettled. A fine, mist-like rain begins to fall, interrupted from time to time by showers. The barometer has fallen perhaps a tenth of an inch.

As the fall becomes more rapid, the wind increases in gustiness, and its speed becomes greater, reaching a value of perhaps 22 to 40 knots (Beaufort 6–8). On the horizon appears a dark wall of heavy cumulonimbus (art. 3713), the **bar** of the storm. Portions of this heavy cloud become detached from time to time and drift across the sky, accompanied by rainsqualls and wind of increasing speed. Between squalls, the cirrostratus can be seen through breaks in the stratocumulus.

As the bar approaches, the barometer falls more rapidly and wind speed increases. The seas, which have been gradually mounting, become tempestuous. Squall lines, one after the other, sweep past in ever increasing number and intensity.

With the arrival of the bar, the day becomes very dark, squalls become virtually continuous, and the barometer falls precipitously, with a rapid increase in wind speed. The center may still be 100 to 200 miles away in a fully developed tropical cyclone. As the center of the storm comes closer, the ever-stronger wind shrieks through the rigging and about the superstructure of the vessel. As the center approaches, rain falls in torrents. The wind fury increases. The seas become mountainous. The tops of huge waves are blown off to mingle with the rain and fill the air with water. Objects at a short distance are not visible. Even the largest and most seaworthy vessels become virtually unmanageable, and may sustain heavy damage. Less sturdy vessels may not survive. Navigation virtually stops as safety of the vessel becomes the prime consideration. The awesome fury of this condition can only be experienced. Words are inadequate to describe it.

If the eye of the storm passes over the vessel, the winds suddenly drop to a breeze as the wall of the eye passes. The rain stops, and the skies clear sufficiently to permit the sun to shine through holes in the comparatively thin cloud cover. Visibility improves. Mountainous seas approach from all sides, apparently in complete confusion. The barometer reaches its lowest point, which may be 1½ or 2 inches below normal in fully developed tropical cyclones. As the wall on the opposite side of the eye arrives, the full fury of the wind strikes as suddenly as it ceased, but from the opposite direction. The sequence of conditions that occurred during approach of the storm is reversed, and passes more quickly, as the various parts of the storm are not as wide in the rear of a storm as on its forward side.

Typical cloud formations associated with a hurricane are shown in figure 3909.

3910. Locating the center of a tropical cyclone.—If intelligent action is to be taken to avoid the full fury of a tropical cyclone, early determination of its location and direction of travel relative to the vessel is essential. The bulletins and forecasts are an excellent general guide, but they are not infallible and may be sufficiently in error to induce a mariner in a critical position to alter course so as to unwittingly increase the danger to his vessel. Often it is possible, using only those observations made aboard ship, to obtain a sufficiently close approximation to enable the vessel to maneuver to the best advantage.

As stated in article 3909, the presence of an exceptionally long swell is usually the first visible indication of the existence of a tropical cyclone. In deep water it approaches from the general direction of origin (the position of the storm center *when the swell was generated*). However, in shoaling water this is a less reliable indication because the direction is changed by refraction, the crests being more nearly parallel to the bottom contours.

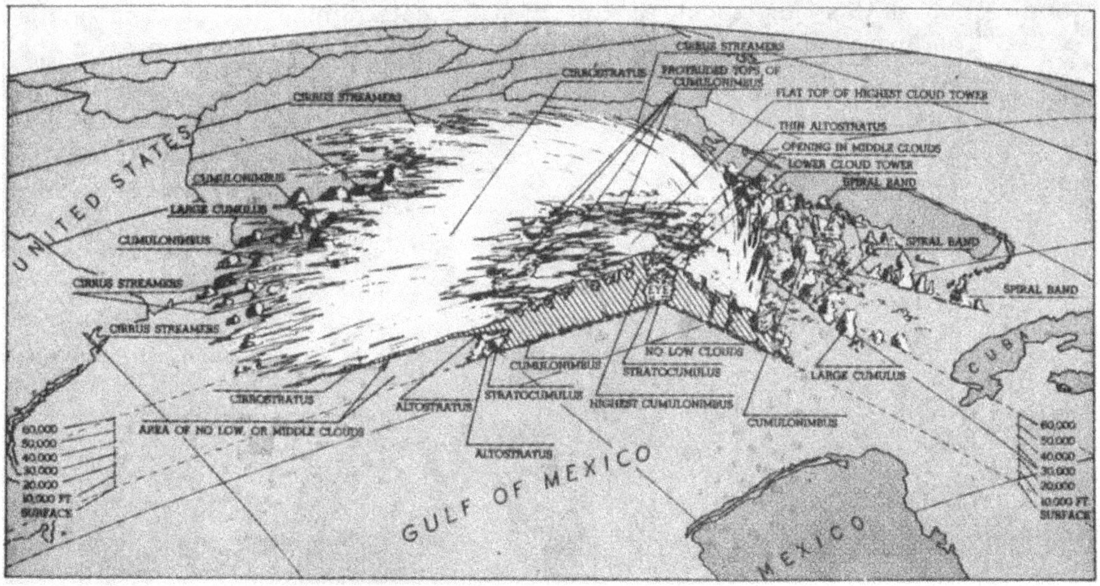

FIGURE 3909.—Typical hurricane cloud formations.

When the cirrus clouds appear, their point of convergence provides an indication of the direction of the storm center. If the storm is to pass well to one side of the observer, the point of convergence shifts slowly in the direction of storm movement. If the storm center will pass near the observer, this point remains steady. When the bar (art. 3909) becomes visible, it appears to rest upon the horizon for several hours. The darkest part of this cloud is in the direction of the storm center. If the storm is to pass to one side, the bar appears to drift slowly along the horizon. If the storm is heading directly toward the observer, the position of the bar remains fixed. Once within the area of the dense, low clouds, one should observe their direction of movement, which is almost exactly along the isobars, with the center of the storm being 90° from the direction of cloud movement (left of direction of movement in the Northern Hemisphere, and right in the Southern Hemisphere).

The winds are probably the best guide to the direction of the center of a tropical cyclone. The circulation is cyclonic (art. 3812), but because of the steep pressure gradient near the center, the winds there blow with greater violence and are more nearly circular than in extratropical cyclones.

According to Buys Ballot's law (art. 3812) an observer who faces into the wind has the center of the low pressure on his right in the Northern Hemisphere, and on his left in the Southern Hemisphere, and in each case somewhat behind him. If the wind followed circular isobars exactly, the center would be exactly 8 points, or 90°, from dead ahead when facing into the wind. However, the track of the wind is usually inclined somewhat toward the center, so that the angle from dead ahead varies between perhaps 8 and 12 points (90° to 135°). The inclination varies in different parts of the same storm. It is least in front of the storm, and greatest in the rear, since the actual wind is the vector sum of that due to the pressure gradient and the motion of the storm along the track. A good average is perhaps 10 points in front, and 11 or 12 points in the rear. These values apply when the storm center is still several hundred miles away. Closer to the center, the wind blows more nearly along the isobars, the inclination being reduced by one or two points at the wall of the eye. Since wind direction usually shifts temporarily during a squall, its direction at this time should not be used for determining the position of the center. The approximate relationship of wind to isobars and storm center in the Northern Hemisphere is shown in figure 3910a.

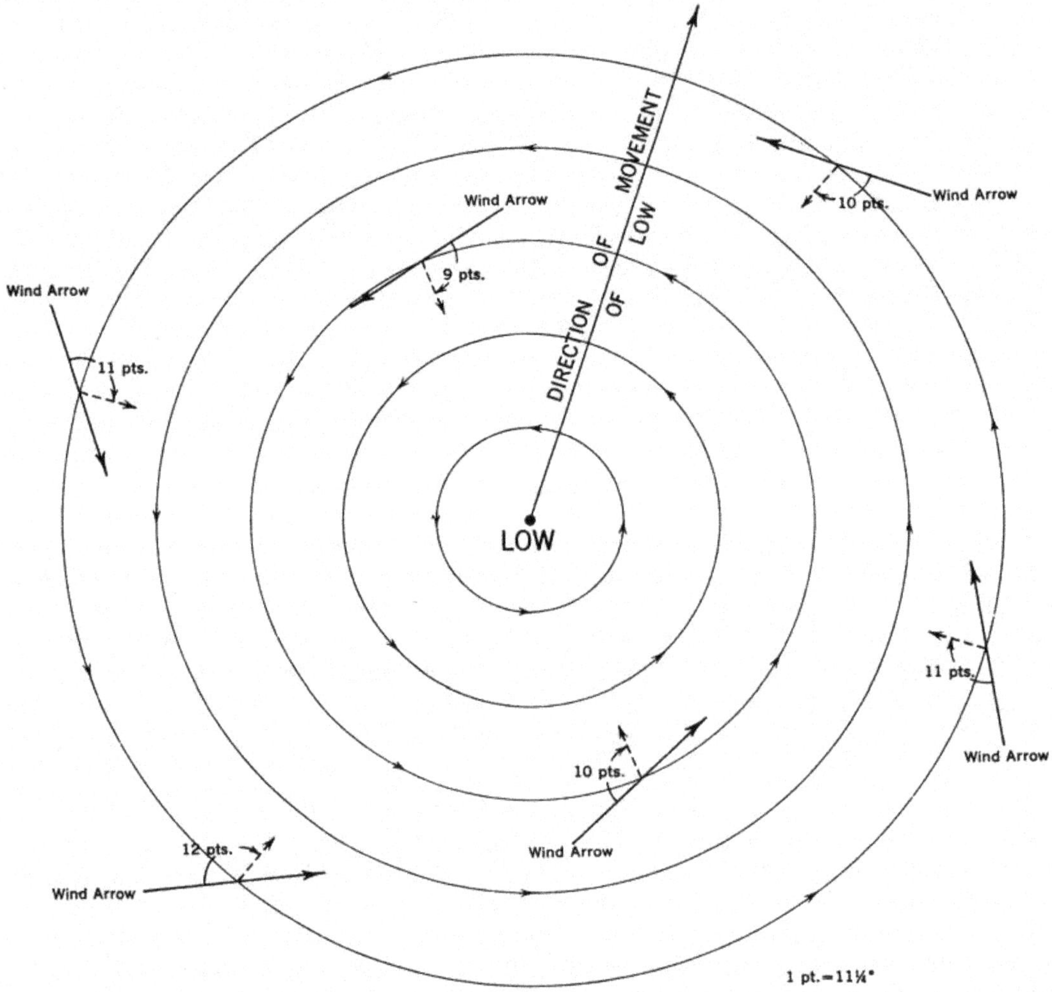

FIGURE 3910a.—Approximate relationship of wind to isobars and storm center in the Northern Hemisphere.

When the center is within radar range, it might be located by this equipment. However, since the radar return is predominantly from the rain, results can be deceptive, and other indications should not be neglected. Figure 3910b shows a radar PPI presentation of a tropical cyclone. If the eye is out of range, the spiral bands (fig. 3910b) may indicate its direction from the vessel. Tracking the eye or upwind portion of the spiral bands enables determining the direction and speed of movement; this should be done for at least 1 hour because the eye tends to oscillate. The tracking of individual cells, which tend to move tangentially around the eye, for 15 minutes or more, either at the end of the band or between bands, will provide an indication of the wind speed in that area of the storm.

Distance from the storm center is more difficult to determine than direction. Radar is perhaps the best guide. However, the rate of fall of the barometer is some indication.

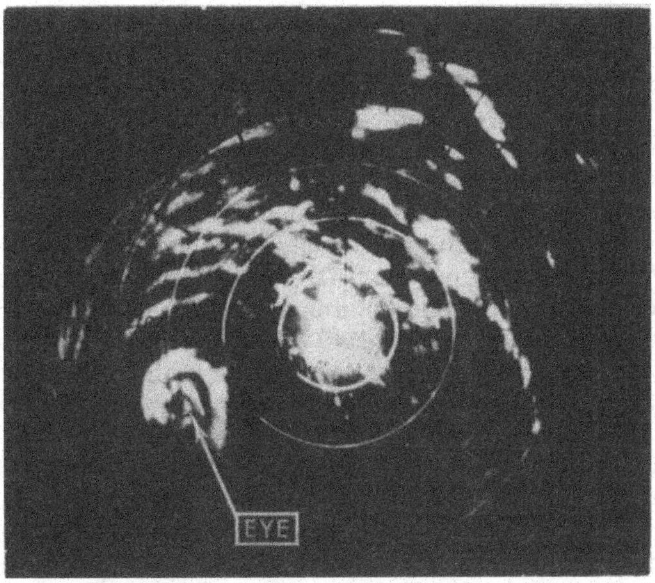

FIGURE 3910b.—Radar PPI presentation of a tropical cyclone.

3911. Maneuvering to avoid the storm center.—The safest procedure with respect to tropical cyclones is to avoid them. If action is taken sufficiently early, this is simply a matter of setting a course that will take the vessel well to one side of the probable track of the storm, and then continuing to plot the positions of the storm center, as given in the weather bulletins, revising the course as needed.

However, such action is not always possible. If one finds himself within the storm area, the proper action to take depends in part upon his position relative to the storm center and its direction of travel. It is customary to divide the circular area of the storm into two parts. In the Northern Hemisphere, that part to the *right* of the storm track (facing in the direction *toward* which the storm is moving) is called the **dangerous semicircle.** It is considered dangerous because (1) the actual wind *speed* is greater than that due to the pressure gradient alone, since it is augmented by the forward motion of the storm, and (2) the *direction* of the wind and sea is such as to carry a vessel into the path of the storm (in the forward part of the semicircle). The part to the left of the storm track is called the **navigable semicircle.** In this part, the wind is decreased by the forward motion of the storm, and the wind blows vessels away from the storm track (in the forward part). Because of the greater wind speed in the dangerous semicircle, the seas are higher than in the navigable semicircle. In the Southern Hemisphere, the dangerous semicircle is to the left of the storm track, and the navigable semicircle is to the right of the storm track.

A plot of successive positions of the storm center should indicate the semicircle in which a vessel is located. However, if this is based upon weather bulletins, it is not a reliable guide because of the lag between the observations upon which the bulletin is based and the time of reception of the bulletin, with the ever present possibility of a change in the direction of motion of the storm. The use of one's radar eliminates this lag, but the return is not always a true indication of the center. Perhaps the most reliable guide is the wind. Within the cyclonic circulation, a *veering* wind (one changing direction to the right in the Northern Hemisphere and to the left in the Southern Hemisphere) indicates the vessel is probably in the dangerous semicircle, and a *backing* wind (one changing in a direction opposite to a veering wind) indicates the vessel is probably in the navigable semicircle. However, if a vessel is underway, its motion should be considered. If it is outrunning the storm or pulling rapidly toward one side (which is not difficult during the early stages of a storm, when its speed is low), the opposite effect occurs. This should usually be accompanied by a rise in atmospheric pressure, but if motion of the vessel is nearly along an isobar, this may not be a reliable indication. If in doubt, the safest action is usually to stop long enough to determine definitely the semicircle. The loss in valuable time may be more than offset by the minimizing of the possibility of taking the wrong action and increasing the danger to the vessel. If the wind direction remains steady (for a vessel which is stopped), with increasing speed and falling barometer, the vessel is in or near the path of the storm. If it remains steady with decreasing speed and rising barometer, the vessel is on the storm track, behind the center.

The first action to take if one finds himself within the cyclonic circulation, is to determine the position of his vessel with respect to the storm center. While the vessel can still make considerable way through the water, a course should be selected to take it as far as possible from the center. If the vessel can move faster than the storm, it is a relatively simple matter to outrun the storm if sea room permits. But when the storm is faster, the solution is not as simple. In this case, the vessel, if ahead of the storm, will approach nearer to the center. The problem is to select a course that will produce the greatest possible minimum distance. This is best determined by means of a relative movement plot, as shown in the following example solved on a maneuvering board.

Example.—A tropical cyclone is estimated to be moving in direction 320° at 19 knots. Its center bears 170°, at an estimated distance of 200 miles from a vessel which has a maximum speed of 12 knots.

Required.—(1) The course to steer at 12 knots to produce the greatest possible minimum distance between the vessel and the storm center.

(2) The distance of the storm center at nearest approach.

(3) Elapsed time until nearest approach.

Solution (fig. 3911).—Consider the vessel remaining at the center of the plot throughout the solution, as on a radar PPI.

(1) Plot point *C* at a distance of 200 miles (scale 20:1) in direction 170° from the center of the diagram, to locate the position of the storm center relative to the vessel. From the center of the diagram, draw *RA,* the speed vector of the storm center, in direction 320°, speed 19 knots (scale 2:1). From *A* draw a line tangent to the 12-knot speed circle (labeled 6 at scale 2:1) *on the side opposite the storm center.* From the center of the diagram draw a perpendicular to this tangent line, locating point *B.* The line *RB* is the required speed vector for the vessel. Its direction, 011°, is the required course.

(2) The path of the storm center *relative to the vessel,* will be along a line from *C* in the direction *BA,* if both storm and vessel maintain course and speed. The point of nearest approach will be at *D,* the foot of a perpendicular from the center of the diagram. This distance, at scale 20:1, is 187 miles.

(3) The length of the vector *BA* (14.8 knots) is the speed of the storm with respect to the vessel. Mark this on the lowest scale of the nomogram at the bottom of the diagram. The relative distance *CD* is 72 miles, by measurement. Mark this (scale 10:1) on the middle scale at the bottom of the diagram. Draw a line between the two points and extend it to intersect the top scale at 29.2 (292 at 10:1 scale). The

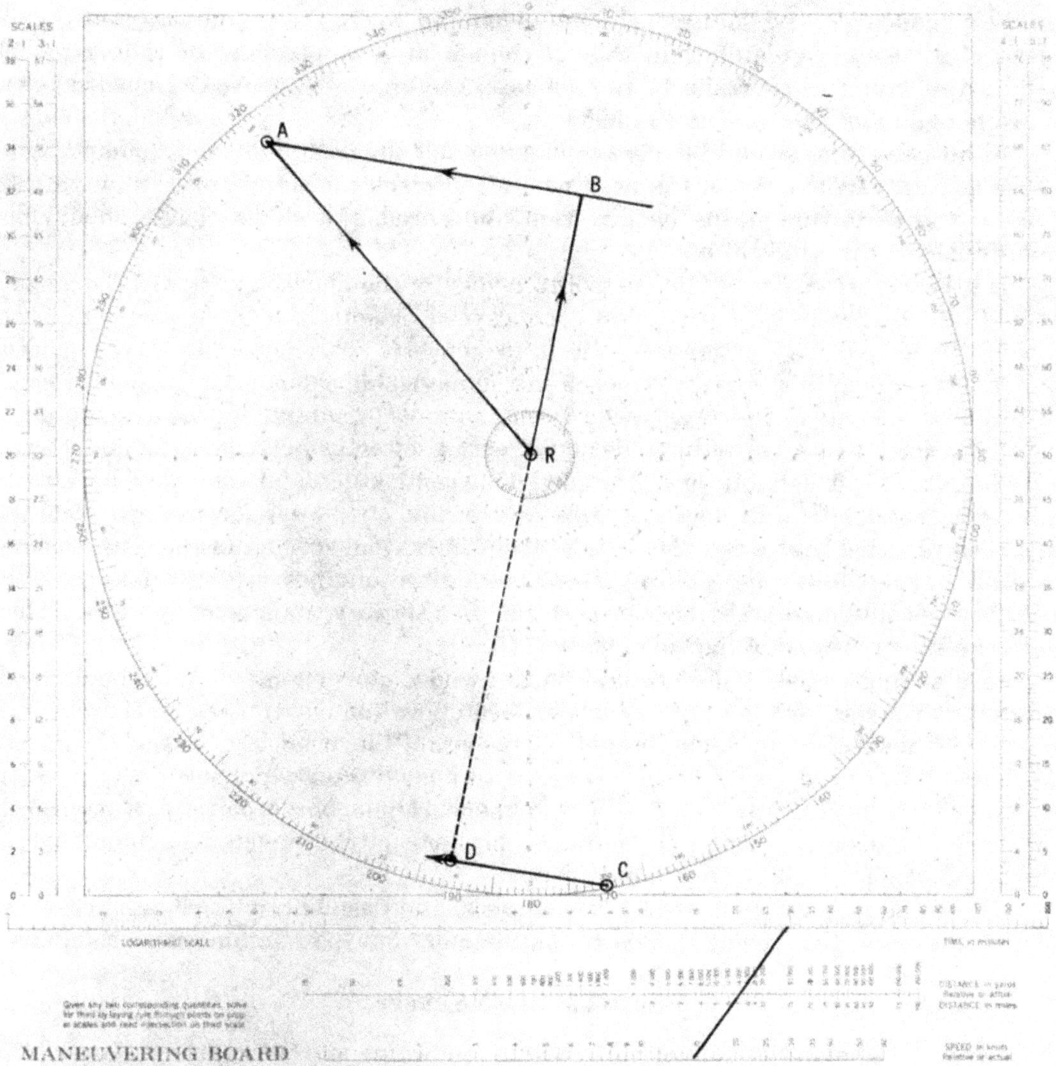

FIGURE 3911.—Solution to determine course for avoiding storm center.

elapsed time is therefore 292 minutes, or 4 hours 52 minutes, or 5 hours, approximately.

Answers.—(1) C 011°, (2) D 187 mi., (3) t 5^h (approximately).

The storm center will be dead astern at its nearest approach.

As a very general rule, for a vessel in the Northern Hemisphere, safety lies in placing the wind on the starboard bow in the dangerous semicircle and on the starboard quarter in the navigable semicircle. If on the storm track ahead of the storm, the wind should be put about 2 points on the starboard quarter until the vessel is well within the navigable semicircle, and the rule for that semicircle then followed. A study of figure 3910a should indicate why these headings are desirable. In the Southern Hemisphere the same rules hold, but with respect to the port side. With a faster than average vessel, the wind can be brought a little farther aft in each case. However, as the speed of the storm increases along its track, the wind should be brought farther forward. If land interferes with what would otherwise be the best maneuver, the solution should be altered to fit the circumstances. If the speed of a vessel is greater than that of the storm, it is possible for the vessel, if behind the storm, to overtake it. In this case, the only action usually needed is to slow enough to let the storm pull ahead.

In all cases, one should be alert to changes in the direction of movement of the storm center, particularly in the area where the track normally curves toward the pole. If the storm maintains its direction and speed, the ship's course should be maintained as the wind shifts.

If it becomes necessary for a vessel to heave to, the characteristics of the vessel should be considered. A power vessel is concerned primarily with damage by direct action of the sea. A good general rule is to heave to with head to the sea in the dangerous semicircle or stern to the sea in the navigable semicircle. This will result in greatest amount of headway away from the storm center, and least amount of leeway toward it. If a vessel handles better with the sea astern or on the quarter, it may be placed in this position in the navigable semicircle or in the rear half of the dangerous semicircle, but *never* in the forward half of the dangerous semicircle. It has been reported that when the wind reaches hurricane speed and the seas become confused, some ships ride out the storm best if the engines are stopped, and the vessel is permitted to seek its own position. In this way, it is said, the ship rides *with* the storm instead of fighting *against* it.

In a sailing vessel, while attempting to avoid a storm center, one should steer courses as near as possible to those prescribed above for power vessels. However, if it becomes necessary for such a vessel to heave to, the wind is of greater concern than the sea. A good general rule always is to heave to on whichever tack permits the shifting wind to draw aft. In the Northern Hemisphere this is the starboard tack in the dangerous semicircle and the port tack in the navigable semicircle. In the Southern Hemisphere these are reversed.

While each storm requires its own analysis, and frequent or continual resurvey of the situation, the general rules for a steamer may be summarized as follows:

NORTHERN HEMISPHERE

Right or dangerous semicircle.—Bring the wind on the starboard bow (045° relative), hold course and make as much way as possible. If obliged to heave to, do so with head to the sea.

Left or navigable semicircle.—Bring the wind on the starboard quarter (135° relative), hold course and make as much way as possible. If obliged to heave to, do so with stern to the sea.

On storm track, ahead of center.—Bring the wind 2 points on the starboard quarter (about 160° relative), hold course and make as much way as possible. When well within the navigable semicircle, maneuver as indicated above.

On storm track, behind center.—Avoid the center by the best practicable course, keeping in mind the tendency of tropical cyclones to curve northward and eastward.

SOUTHERN HEMISPHERE

Left or dangerous semicircle.—Bring the wind on the port bow (315° relative), hold course and make as much way as possible. If obliged to heave to, do so with head to the sea.

Right or navigable semicircle.—Bring the wind on the port quarter (225° relative), hold course and make as much way as possible. If obliged to heave to, do so with stern to the sea.

On storm track, ahead of center.—Bring the wind 2 points on the port quarter (about 200° relative), hold course and make as much way as possible. When well within the navigable semicircle, maneuver as indicated above.

On storm track, behind center.—Avoid the center by the best practicable course, keeping in mind the tendency of tropical cyclones to curve southward and eastward.

Whenever a tropical cyclone is encountered, the wise procedure is to begin preparing the vessel for heavy weather in sufficient time to permit thorough preparation, so that damage may be minimized. One should be particularly careful to keep free surfaces of liquids to a minimum.

It is possible, particularly in temperate latitudes after the storm has recurved, that the dangerous semicircle is the left one in the Northern Hemisphere (right one in the Southern Hemisphere). This can occur if a large high lies north of the storm and causes a tightening of the pressure gradient in the region.

Typhoon Havens Handbook for the Western Pacific and Indian Oceans is published by the Naval Environmental Prediction Research Facility, Monterey, California, as an aid to commanders and commanding officers of ships in evaluating a typhoon situation and to assist them in deciding whether to sortie, to evade, or to remain in port to take shelter within a specific harbor.

3912. Effects.—The high winds of a tropical cyclone inflict widespread damage when such a storm leaves the ocean and crosses land. Aids to navigation may be blown out of position or destroyed. Craft in harbors, unless they are properly secured, drag anchor or are blown against obstructions. Ashore, trees are blown over, houses are damaged, power lines are blown down, etc. The greatest damage usually occurs in the dangerous semicircle a short distance from the center, where the strongest winds occur. As the storm continues on across land, its fury subsides faster than it would if it had remained over water.

Wind instruments are usually incapable of measuring the 175- to 200-knot winds of the more intense hurricanes; if the instrument holds up, often the supporting structure gives way.

Wind gusts, which are usually 30 to 50 percent higher than sustained winds, add significantly to the destructiveness of the tropical cyclone. Many tropical cyclones that reach hurricane intensity develop winds of more than 90 knots sometime during their lives, but few develop winds of more than 130 knots.

Tropical cyclones have produced some of the world's heaviest rainfalls. While average amounts range from 6 to 10 inches, totals near 100 inches over a 4-day period have been observed. A 24-hour world's record of 73.62 inches fell at Reunion Island during a tropical cyclone in 1952. Forward movement of the storm and land topography have a considerable influence on rainfall totals. Torrential rains can occur when a storm moves against a mountain range; this is common in the Philippines and Japan, where even weak tropical depressions produce considerable rainfall. A 24-hour total of 46 inches was recorded in the Philippines during a typhoon in 1911. As hurricane Camille crossed southern Virginia's Blue Ridge Mountains in August of 1969, there was nearly 30 inches of rain in about 8 hours. This caused some of the most disastrous floods in the state's history.

Flooding is an extremely destructive by-product of the tropical cyclone's torrential rains. Whether an area will be flooded depends on the physical characteristics of the drainage basin, rate and accumulation of precipitation, and river stages at the time the rains begin. When heavy rains fall over flat terrain, the countryside may lie underwater for a month or so, and while buildings, furnishings, and underground powerlines may be damaged, there are usually few fatalities. In mountainous or hill country, disastrous floods develop rapidly and can cause a great loss of life.

There have been occasional reports in tropical cyclones of waves greater than 40 feet in height, and numerous reports in the 30- to 40-foot category. However, in tropical cyclones, strong winds rarely persist for a sufficiently long time or over a large enough area to permit enormous wave heights to develop. The direction and speed of the wind changes more rapidly in tropical cyclones than in extratropical storms. Thus, the maximum duration and fetch length for any wind condition is often less in tropical cyclones than in extratropical storms of similar intensity, and the waves accompanying any given local wind conditions are generally not so high as those expected, with similar local wind conditions, in the high-latitude-type storms. In hurricane Camille, significant waves of 43 feet were recorded; an extreme wave height reached 72 feet.

Exceptional conditions may arise when waves of certain dimensions travel within the storm at a speed equal to the storm's speed, thus, in effect, extending the duration and fetch of the wave and significantly increasing its height. This occurs most often to the right of the track in the Northern Hemisphere (left of the track in the Southern Hemisphere). Another condition that may give rise to exceptional wave heights is the intersection of waves from two or more distinct directions. This may lead to a zone of confused seas in which the heights of some waves will equal the sum in each individual wave train. This process can occur in any quadrant of the storm and so it should not be assumed that the highest waves will always be encountered to the right of the storm track in the Northern Hemisphere (left of the track in the Southern Hemisphere).

When these waves move beyond the influence of the generating winds, they become swell. They are recognized by their smooth, undulating form, in contrast to the steep, ragged crests of the winds' waves. This swell, particularly that generated by the right side of the storm, can travel a thousand miles or more and may produce tides 3 or 4 feet above normal along several hundred miles of coastline.

When a tropical cyclone moves close to a coast, wind often causes a rapid rise in water level, and along with the falling pressure may produce a storm **surge.** This surge is usually confined to the right of the track in the Northern Hemisphere (left of the track in the Southern Hemisphere) and to a relatively small section of the coastline. It most often occurs with the approach of the storm, but in some cases, where a surge moves into a long channel, the effect may be delayed. Occasionally, the greatest rise in water is observed on the opposite side of the track, when northerly winds funnel into a partially landlocked harbor. The surge could be 3 feet or less, or it could be 20 feet or more, depending on the combination of all the factors involved.

There have been reports of a "hurricane wave," described as a "wall of water," which moves rapidly toward the coastline. Authenticated cases are rare, but some of the world's greatest natural disasters have occurred as a result of this wave, which may be just a rapidly rising and abnormally high storm surge. In India, such a disaster occurred in 1876, between Calcutta and Chittagong, and drowned more than 100,000 persons.

Along the coast, particularly, greater damage may be inflicted by water than by the wind. There are at least four sources of water damage. First, the unusually high seas generated by the storm winds pound against shore installations and craft in their way. Second, the continued blowing of the wind toward land causes the water level to increase perhaps 3 to 10 feet above its normal level. This **storm tide**, which may begin when the storm center is 500 miles or even farther from the shore, gradually increases until the storm passes. The highest storm tides are caused by a slow-moving tropical cyclone of large diameter, because both of these effects result in greater duration of wind in the same direction. The effect is greatest in a partly enclosed body of water, such as the Gulf of Mexico, where the concave coastline does not readily permit the escape of water. It is least on small islands, which present little obstruction to the flow of water. Third, the furious winds which blow around the wall of the eye create a ridge of water called a **storm wave**, which strikes the coast and often inflicts heavy damage. The effect is similar to that of a **seismic sea wave**, caused by an earthquake in the ocean floor. Both of these waves are popularly called **tidal waves**. Storm waves of 20 feet or more have occurred. About 3 or 4 feet of this is due to the decrease of atmospheric pressure, and the rest to winds. Like the damage caused by wind, that due to high seas, the storm surge and tide, and the storm wave is greatest in the dangerous semicircle, near the center. The fourth source of water damage is the heavy rain that accompanies a tropical cyclone. This causes floods that add to the damage caused in other ways.

There have been many instances of tornadoes occurring within the circulation of tropical cyclones. Most of these have been associated with tropical cyclones of the North Atlantic Ocean and have occurred in the West Indies and along the gulf and Atlantic coasts of the United States. They are usually observed in the forward semicircle or along the advancing periphery of the storm. These tornadoes are usually short-lived and less intense than those that occur in the midwestern United States.

When proceeding along a shore recently visited by a tropical cyclone, a navigator should remember that time is required to restore aids to navigation which have been blown out of position or destroyed. In some instances the aid may remain but its light, sound apparatus, or radiobeacon may be inoperative. Landmarks may have been damaged or destroyed.

References

Australia Bureau of Meteorology. *Occurrence of Tropical Depressions and Cyclones in the Northeastern and Northwestern Australian Regions* (annual summary 1957–1962). Maribyrnong, Department of Supply, Central Drawing Office.

Australia Bureau of Meteorology. *Tropical Cyclones in the Northern Australian Regions* (annual summary 1962–1969). Maribyrnong, Department of Supply, Central Drawing Office.

Brand, S., and J. W. Blelloch. *Typhoon Havens Handbook for the Western Pacific and Indian Oceans.* Technical Paper 5-76, Naval Environmental Prediction Research Facility, Monterey, California, 1976.

Chin, P. C. *Tropical Cyclones in the Western Pacific and China Sea Area from 1884 to 1953.* Hong Kong, Royal Observatory, 1958.

Crutcher, H. L., and R. G. Quayle. *Mariners Worldwide Climatic Guide to Tropical Storms at Sea,* Naval Weather Service Command, U. S. Govt. Print. Off., 1974.

Cry, G. W. "Tropical Cyclones of the North Atlantic Ocean: Tracks and Frequencies of Hurricanes and Tropical Storms, 1871–1963," Technical Paper No. 55, Washington, U. S. Govt. Print. Off., 1965.

DeAngelis, R. M. "North Pacific Hurricanes: Timid or Treacherous?" *Mariners Weather Log,* vol. 11, No. 6 (November 1967), pp. 193–200.

Dunn, G. E., and B. I. Miller. *Atlantic Hurricanes.* Louisiana State University Press, 1960.

Environmental Science Services Administration. *Hurricane the Greatest Storm on Earth.* Washington, U. S. Govt. Print. Off., 1967.

Harding, E. T., and Wm. J. Kotsch. *Heavy Weather Guide.* Annapolis, U. S. Naval Institute, 1965.

Harris, D. L. "Wave Patterns in Tropical Cyclones." *Mariners Weather Log,* vol. 6, No. 5 (September 1962), pp. 156–160.

Harris, D. L. "Characteristics of the Hurricane Storm Surge," Technical Paper No. 48. Washington, U. S. Govt. Print. Off., 1963.

Hodge, W. T. "North Pacific Typhoons Where and When Are They Most Frequent?" *Mariners Weather Log,* vol. 9, No. 3 (May 1965), pp. 73–76.

India Meteorological Department. *Tracks of Storms and Depressions in the Bay of Bengal and the Arabian Sea, 1877–1960.* New Delhi, India, Meteorological Department, 1964.

Joint Typhoon Warning Center. *Annual Typhoon Reports 1959–1973.* Guam, U. S. Fleet Weather Central/Joint Typhoon Warning Center.

Mauritius Meteorological Department. *Annual Report of the Meteorological Department, 1950–1966.* Port Louis, Mauritius Government Printer.

New Zealand Meteorological Service. *Annual Meteorological Summary, 1957–1966.* Laucala Bay, Suva, Fiji Government Press.

Nimitz, Chester W., et al. "Typhoon Doctrine." U. S. Naval Institute Proceedings, vol. 82, No. 1 (January 1956), pp. 83–93.

U. S. Weather Bureau. *Climatological and Oceanographic Atlas for Mariners,* Vol. I North Atlantic Ocean. Washington, U. S. Govt. Print. Off., 1959.

U. S. Weather Bureau. *Climatological and Oceanographic Atlas for Mariners,* Vol. II North Pacific Ocean. Washington, U. S. Govt. Print. Off., 1961.

PART EIGHT
ELECTRONICS AND NAVIGATION

PART EIGHT

ELECTRONICS AND NAVIGATION

	Page
CHAPTER XL. Radio Waves	957
CHAPTER XLI. Radio Direction Finding	972
CHAPTER XLII. Radar Navigation	983
CHAPTER XLIII. Radionavigation Systems	1011
CHAPTER XLIV. Satellite Navigation	1066
CHAPTER XLV. Doppler Navigation	1118
CHAPTER XLVI. Introduction to Inertial Navigation	1129

CHAPTER XL

RADIO WAVES

4001. Source of radio waves.—All matter is made up of tiny particles called **atoms.** Each atom has a central **nucleus** composed principally of subatomic particles called **protons** and **neutrons.** One or more **electrons** revolve around the nucleus in orbits resembling those of planets around the sun (art. 1407). The number and arrangement of the particles constituting an atom of each element of matter determine the properties of that element. Electrons, each having a mass of only about 1/1,840 that of a proton or neutron, are kept in their orbits principally by means of an attractive electrical force, each electron carrying one negative "charge" and each proton one positive "charge." Like charges repel and unlike charges attract. This electrical attraction is additional to the gravitational attraction existing between all particles in the universe. The neutron is electrically neutral.

Under suitable conditions, some electrons become detached from their atoms. An excess or deficiency of electrons in a nonconductor is called **static electricity.** A substance which provides a path for electron movement with relatively little resistance is called a **conductor.** A *flow* of electrons along such a conductor constitutes an **electric current,** although the current direction is conventionally considered to be opposite to the direction of flow of the electrons. A **direct current** flows continuously in the same direction. If the strength of the current varies rhythmically but does not change direction, the current is said to be **pulsating.** If the direction of flow periodically reverses, an **alternating current** results.

In addition to its electrical and gravitational forces, a moving electron is accompanied by a **magnetic force.** As long as the flow is steady, the magnetic force is constant. If a conductor is in the region of influence or **field** of magnetism, there is no noticeable effect unless the strength of the field is changing, or relative motion exists between the conductor and the field, when an **induced current** flows in the conductor. The extent to which a substance has electrons free to move under suitable influence determines its value as a conductor. One which offers great resistance to such flow is called an **insulator.**

In a suitable electrical system, an electric charge creates a magnetic field which builds up to a maximum. If the electric current is then discontinued, the magnetic field collapses. This change in the strength of the magnetic field induces an electric current in the conductor, but in the opposite direction to the original current. This current creates a new magnetic field, and the cycle repeats. Thus, an alternating current is produced, the strength increasing to a maximum in one direction, decreasing to zero, increasing to a maximum in the opposite direction, and again decreasing to zero to start a new cycle. This cycle is repeated many times each second, the number depending upon the characteristics of the system. Such a system is called an **oscillating circuit.**

A relatively small amount of energy is dissipated as heat in overcoming the resistance of the circuit. The remainder continues to oscillate between electric and magnetic fields. The build-up and collapse of each field occurs at about the speed of light, which is approximately 186,000 statute miles (300,000,000 meters) per second. If a relatively long period of time is available for the cycle to occur, the energy is fully transferred before the next step occurs. However, if the cycle is speeded until the time needed for each field to build up or collapse is more than about one-half cycle, some of the energy becomes detached and is radiated into space, through which it travels at about the speed of light. This **electromagnetic radiation,** having both electrical and magnetic properties, is known as **radio waves,** if of a frequency suitable for radio communication.

4002. Radio wave terminology.—The build-up and collapse of the electric and magnetic fields are proportional to the *sine* of the portion of the cycle completed, as shown in figure 4002. This representation has led to the use of the term "wave" when referring to electromagnetic propagation. The highest point on the curve (in the direction considered positive) is the **crest,** and the lowest point the **trough.** Either point may be called the **peak,** considered positive or negative if a distinction is desired. The displacement of a peak from zero is called the **amplitude.** The forward side of any wave is called the **wave front.** For a nondirectional antenna each wave proceeds outward as an expanding sphere (or hemisphere).

One **cycle** is a complete sequence of values, as from crest to crest. The distance traveled by the energy during one cycle is the **wavelength,** usually expressed in metric units (meters, centimeters, etc.). The number of cycles repeated during unit time (usually one second) is the **frequency.** This is given in **hertz** (cycles per second). A **kilohertz (kHz)** is 1,000 cycles per second. A **megahertz (MHz)** is 1,000,000 cycles per second. Wavelength and frequency are inversely proportional. The approximate value of either may be found by dividing 300,000,000 by the other quantity, if wavelength is expressed in meters and frequency in hertz. Thus, if the frequency is 1,500 kilohertz (1,500,000 cycles per second), the wavelength is

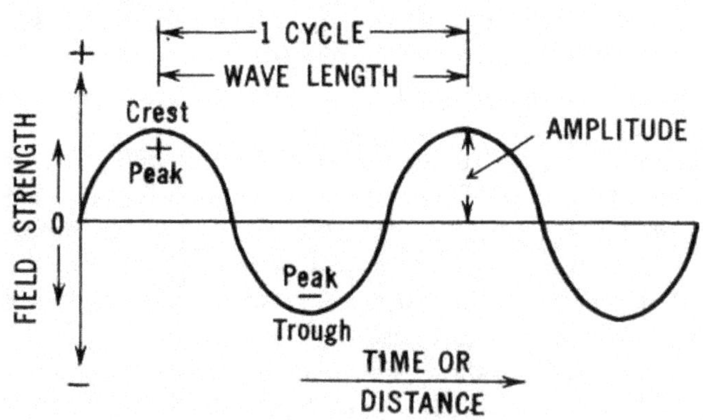

FIGURE 4002.—Radio wave terminology.

$$\frac{300,000,000}{1,500,000} = 200 \text{ meters.}$$

If the wavelength is 10 centimeters (0.1 meter), the frequency is about

$$\frac{300,000,000}{0.1} = 3,000,000,000$$

cycles per second or 3 **gigahertz.** A more precise value for the speed of propagation in air is 299,708,000 meters per second. This is equivalent to 186,230 statute miles, or 161,829 nautical miles, per second. The exact value varies slightly with density of the medium through which the wave travels, and frequency. The speed in a vacuum is a little more than that in air.

The **phase** of a wave is the amount by which the cycle has progressed from a specified origin. For most purposes it is stated in circular measure, a complete cycle being considered 360°. Generally, the origin is not important, principal interest being the phase relative to that of some other wave. Thus, two waves having crests one-fourth cycle apart are said to be 90° "out of phase." If the crest of one wave occurs at the trough of another, the two are 180° out of phase.

4003. Electromagnetic spectrum.—The entire range of electromagnetic radiation frequencies is called the **electromagnetic spectrum.** The range of frequencies suitable for radio transmission, called the **radio spectrum,** extends from 10 kilohertz to 300,000 megahertz, approximately. For convenience, it is divided into a number of **bands,** as shown in table 4003. Below the radio spectrum, but overlapping it, is the **audio frequency** band, extending from 20 to 20,000 hertz, approximately. This is the range of frequencies that can be heard by the human ear. Above the radio spectrum are heat and infrared, the visible spectrum (light in its various colors), ultraviolet, X-rays, gamma rays, and cosmic rays. These are included in table 4003. Waves shorter than 30 centimeters are usually called **microwaves.**

Band	Abbreviation	Range of frequency	Range of wavelength
Audio frequency	AF	20 to 20,000 Hz	15,000,000 to 15,000 m
Radio frequency	RF	10 kHz to 300,000 MHz	30,000 m to 0.1 cm
Very low frequency	VLF	10 to 30 kHz	30,000 to 10,000 m
Low frequency	LF	30 to 300 kHz	10,000 to 1,000 m
Medium frequency	MF	300 to 3,000 kHz	1,000 to 100 m
High frequency	HF	3 to 30 MHz	100 to 10 m
Very high frequency	VHF	30 to 300 MHz	10 to 1 m
Ultra high frequency	UHF	300 to 3,000 MHz	100 to 10 cm
Super high frequency	SHF	3,000 to 30,000 MHz	10 to 1 cm
Extremely high frequency	EHF	30,000 to 300,000 MHz	1 to 0.1 cm
Heat and infrared*		10^6 to 3.9×10^8 MHz	0.03 to 7.6×10^{-5} cm
Visible spectrum*		3.9×10^8 to 7.9×10^8 MHz	7.6×10^{-5} to 3.8×10^{-5} cm
Ultraviolet*		7.9×10^8 to 2.3×10^{10} MHz	3.8×10^{-5} to 1.3×10^{-6} cm
X-rays*		2.0×10^9 to 3.0×10^{13} MHz	1.5×10^{-5} to 1.0×10^{-9} cm
Gamma rays*		2.3×10^{12} to 3.0×10^{14} MHz	1.3×10^{-8} to 1.0×10^{-10} cm
Cosmic rays*		$> 4.8\times10^{15}$ MHz	$< 6.2\times10^{-12}$ cm

*Values approximate.

TABLE 4003.—Electromagnetic spectrum.

4004. Polarization.—As indicated in article 4001, radio waves have both electric and magnetic fields. The two fields are conceived as having direction associated with the orientation of the vibrations. The direction of the electric component of the field is called the **polarization** of the electromagnetic field. Thus, if the electric component is vertical, the wave is said to be "vertically polarized," and if horizontal, "horizontally polarized." A wave traveling through space may be polarized in any direction. One traveling along the surface of the earth is always vertically polarized because the earth, a conductor, short-circuits any horizontal component. The magnetic field and the electric field are always mutually perpendicular.

4005. Reflection.—When radio waves strike a surface, they are reflected in the same manner as light waves, if conditions are favorable. Radio waves of all frequencies are reflected by the surface of the earth. The strength of the **reflected wave**

depends upon **grazing angle** (the angle between the incident ray and the horizontal), type of polarization, frequency, reflecting properties of the surface, and divergence of the reflected ray. Lower frequency results in greater penetration. At very low frequencies usable radio signals can be received some distance below the surface of the sea.

A change of phase takes place when a wave is reflected from the surface of the earth. The amount of the change varies with the conductivity of the earth and the polarization of the wave, reaching a maximum of 180° for a horizontally polarized wave reflected from seawater (considered to have infinite conductivity). When **direct waves** (those traveling from transmitter to receiver in a relatively straight line, without reflection) and reflected waves arrive at a receiver, the total signal is the vector sum of the two. If the signals are in phase, they reinforce each other, producing a stronger signal. If there is a phase difference, the signals tend to cancel each other, the cancellation being complete if the phase difference is 180° and the two signals have the same amplitude. This interaction of waves is called **wave interference.** A phase difference may occur because of the change of phase of a reflected wave, or because of the longer path followed by it. The second effect decreases with greater distance between transmitter and receiver, for under these conditions the difference in path lengths is smaller. At lower frequencies there is no practical solution to interference caused in this way. For VHF and higher frequencies the condition can be improved by elevating the antenna, if the wave is vertically polarized. Also, interference at higher frequencies can be more nearly eliminated because of the greater ease of beaming the signal to avoid reflection.

Reflections may also occur from mountains, trees, and other obstacles. Such reflection is negligible for lower frequencies, but becomes more prevalent as frequency increases. In radio communication it can be reduced by using directional antennas, but this solution is not always available for navigational systems.

Various reflecting surfaces occur in the atmosphere. At high frequencies, reflections take place from rain. At still higher frequencies, reflections are possible from clouds, particularly rain clouds. Reflections may even occur at a sharply defined boundary surface between airmasses, as when warm, moist air flows over cold, dry air. When such a surface is roughly parallel to the surface of the earth, radio waves may travel for greater distances than normal. A somewhat similar condition is described in article 4006. The principal source of reflection in the atmosphere is the ionosphere (arts. 4007, 4008).

4006. Refraction of radio waves is similar to that of light waves (art. 1606). Thus, as a signal passes from air of one density to that of a different density, the direction of travel is altered. The principal cause of refraction in the atmosphere is the difference in temperature and pressure occurring at various heights and in different airmasses.

Refraction occurs at all frequencies, but at those below 30 MHz the effect is small as compared with ionospheric effects (art. 4008), diffraction (art. 4009), and absorption (art. 4010). At higher frequencies, refraction in the lower layer of the atmosphere extends the **radio horizon** to a distance about 15 percent greater than the visible horizon. The effect is the same as if the radius of the earth were about one-third greater than it is, and there were no refraction.

Sometimes the lower portion of the atmosphere becomes stratified with horizontal layers of air having certain characteristics, resulting in nonstandard temperature and moisture changes with height. If there is a marked temperature inversion (art. 3814) or a sharp decrease in water vapor content with increased height, a horizontal radio **duct** may be formed. High frequency radio waves traveling horizontally within the duct are refracted to such an extent that they remain within the duct, following the curvature of the earth for phenomenal distances. This is called **super-refraction.** Maximum results are obtained when both transmitting and receiving antennas are within the duct. There is a lower limit to the frequency affected by ducts. It varies from about 200 MHz to more than 1,000 MHz.

At night, surface ducts may occur over land due to cooling of the surface. At sea, surface ducts about 50 feet thick may occur at any time in the trade wind belt. Surface ducts 100 feet or more in thickness may extend from land out to sea when warm air from the land flows over the cooler ocean surface. Elevated ducts from a few feet to more than 1,000 feet in thickness may occur at elevations of 1,000 to 5,000 feet, due to the settling of a large airmass. This is a frequent occurrence in Southern California and certain areas of the Pacific Ocean.

Refraction effects associated with the ionosphere are discussed in article 4008.

A bending in the horizontal plane occurs when a groundwave (art. 4008) crosses a coast at an oblique angle. This is due to a marked difference in the conducting and reflecting properties of the land and water over which the wave travels. The effect is known as **coastal refraction** or **land effect.**

4007. The ionosphere.—Since an atom normally has an equal number of negatively charged electrons and positively charged protons, it is electrically neutral. An **ion** is an atom or group of atoms which has become electrically charged, either positively or negatively, by the loss or gain of one or more electrons.

Loss of electrons may occur in a variety of ways. In the atmosphere, ions are usually formed by collision of atoms with rapidly moving particles, or by the action of cosmic rays or ultraviolet light. In the lower portion of the atmosphere, recombination soon occurs, leaving a small percentage of ions. In thin atmosphere far above the surface of the earth, however, atoms are widely separated and a large number of ions may be present. The region of numerous positive and negative ions and unattached electrons is called the **ionosphere.** The extent of ionization depends upon the kinds of atoms present in the atmosphere, the density of the atmosphere, and the position relative to the sun (time of day and season). After sunset, ions and electrons recombine faster than they are separated, decreasing the ionization of the atmosphere.

An electron can be separated from its atom only by the application of greater energy than that holding the electron. Since the energy of the electron depends primarily upon the kind of an atom of which it is a part, and its position relative to the nucleus of that atom, different kinds of radiation may cause ionization of different substances.

In the outermost regions of the atmosphere the density is so low that oxygen exists largely as separate atoms, rather than combining as molecules as it does nearer the surface of the earth. At great heights the energy level is low and ionization from solar radiation is intense. This is known as the **F layer.** Above this level the ionization decreases because of the lack of atoms to be ionized. Below this level it decreases because the ionizing agent of appropriate energy has already been absorbed. During daylight, two levels of maximum F ionization can be detected, the F_2 layer at about 125 statute miles above the surface of the earth, and the F_1 layer at about 90 statute miles. At night, these combine to form a single F layer.

At a height of about 60 statute miles the solar radiation not absorbed by the F layer encounters, for the first time, large numbers of oxygen *molecules.* A new maximum ionization occurs, known as the **E layer.** The height of this layer is quite constant, in contrast with the fluctuating F layer. At night the E layer becomes weaker by two orders of magnitude.

Below the E layer a weak **D layer** forms at a height of about 45 statute miles, where the incoming radiation encounters ozone (O_3) for the first time. The D layer is the principal source of absorption of HF waves, and of reflection of LF and VLF waves during daylight.

4008. The ionosphere and radio waves.—When a radio wave encounters a particle having an electric charge, it causes that particle to vibrate. The vibrating particle absorbs electromagnetic energy from the radio wave and reradiates it. The net effect is a change of polarization and an alteration of the path of the wave. That portion of the wave in a more highly ionized region travels faster, causing the wave front to tilt and the wave to be directed toward a region of less intense ionization.

Refer to figure 4008a, in which a single layer of the ionosphere is considered. Ray 1 enters the ionosphere at such an angle that its path is altered, but it passes on through and proceeds outward into space. As the angle with the horizontal decreases, a critical value is reached where the ray (2) is bent or reflected back toward the earth. As the angle is still further decreased, as at 3, the return to earth occurs at a greater distance from the transmitter.

A wave reaching a receiver by way of the ionosphere is called a **skywave.** This expression is also appropriately applied to a wave reflected from an airmass boundary. In common usage, however, it is generally associated with the ionosphere. The wave which travels along the surface of the earth is called a **groundwave.** At angles greater than the critical angle, no skywave signal is received. Therefore, there is a minimum distance from the transmitter at which skywaves can be received. This is called the **skip distance,** shown in figure 4008a. If the groundwave extends out for less distance than the skip distance, a **skip zone** occurs, in which no signal is received.

The critical radiation angle depends upon the intensity of ionization, and the frequency of the radio wave. As the frequency increases, the angle becomes smaller. At frequencies greater than about 30 MHz virtually all of the energy penetrates through or is absorbed by the ionosphere. Therefore, at any given receiver there is a maximum usable frequency if skywaves are to be utilized. The strongest signals are received at or slightly below this frequency. There is also a lower practical frequency beyond which signals are too weak to be of value. Within this band the optimum frequency can be selected to give best results. It cannot be too near the maximum usable frequency because this frequency fluctuates with changes of intensity within the ionosphere. During magnetic storms the ionosphere density decreases. The maximum usable frequency decreases, and the lower usable frequency increases. The band of usable frequencies is thus narrowed. Under extreme conditions it may be completely eliminated, isolating the receiver and causing a radio **blackout.**

Skywave signals reaching a given receiver may arrive by any of several paths, as shown in figure 4008b. A signal which undergoes a single reflection is called a "one-hop" signal, one which undergoes two reflections with a ground reflection between is called a "two-hop" signal, etc. A "multihop" signal undergoes several reflections. The layer at which the reflection occurs is usually indicated, also, as "one hop E," "two hop F," etc.

Because of the different paths and phase changes occurring at each reflection, the various signals arriving at a receiver have different phase relationships. Since the density of the ionosphere is continually fluctuating, the strength and phase

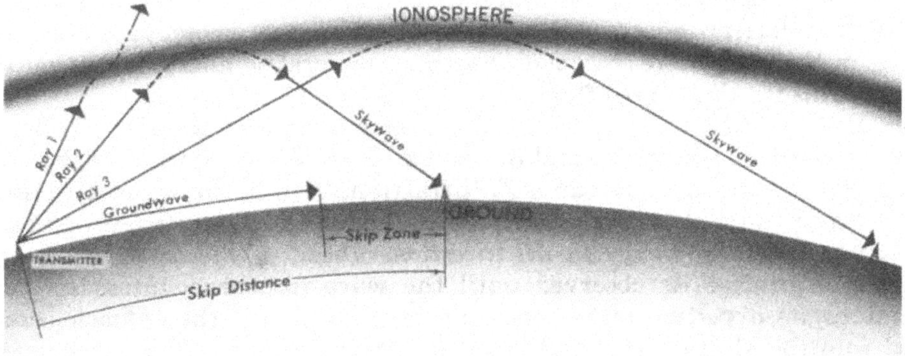

FIGURE 4008a.—The effect of the ionosphere on radio waves.

relationships of the various signals may undergo an almost continuous change. Thus, the various signals may reinforce each other at one moment and cancel each other at the next, resulting in fluctuations of the strength of the total signal received. This is called **fading.** This phenomenon may also be caused by interaction of components within a single reflected wave, or changes in its strength due to changes in the reflecting surface. Ionospheric changes are associated with fluctuations in the radiation received from the sun, since this is the principal cause of ionization. Signals from the F layer are particularly erratic because of the rapidly fluctuating conditions within the layer itself.

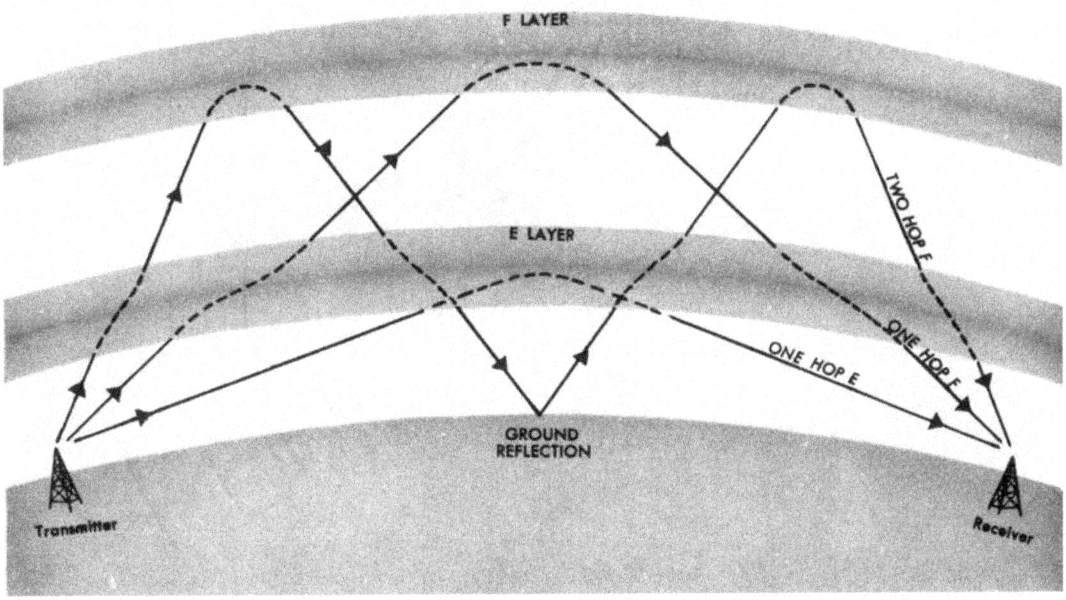

FIGURE 4008b.—Various paths by which a skywave signal might be received.

The maximum distance at which a one-hop-E signal can be received is about 1,400 miles. At this distance the signal leaves the transmitter in approximately a horizontal direction. A one-hop-F signal can be received out to about 2,500 miles. At low frequencies groundwaves extend out for great distances.

A skywave may undergo a change of polarization during reflection from the ionosphere, accompanied by an alteration in the direction of travel of the wave. This is called **polarization error.** Near sunrise and sunset, when rapid changes are occurring in the ionosphere, reception may become erratic and polarization error a maximum. This is called **night effect.**

4009. Diffraction.—When a radio wave encounters an obstacle, its energy is reflected or absorbed, causing a **shadow** beyond the obstacle. However, some energy does enter the shadow area because of **diffraction.** This is explained by **Huygens' principle,** which states that *every point on the surface of a wave front is a source of radiation, transmitting energy in all directions ahead of the wave.* No noticeable effect of this principle is observed until the wave front encounters an obstacle, which intercepts a portion of the wave. From the edge of the obstacle, energy is radiated into the shadow area, and also outside of the area. The latter interacts with energy from other parts of the wave front, producing alternate bands in which the secondary radiation reinforces or tends to cancel the energy of the primary radiation. Thus, the practical effect of an obstacle is a greatly reduced signal strength in the shadow area, and a disturbed pattern for a short distance outside the shadow area. This is illustrated in figure 4009.

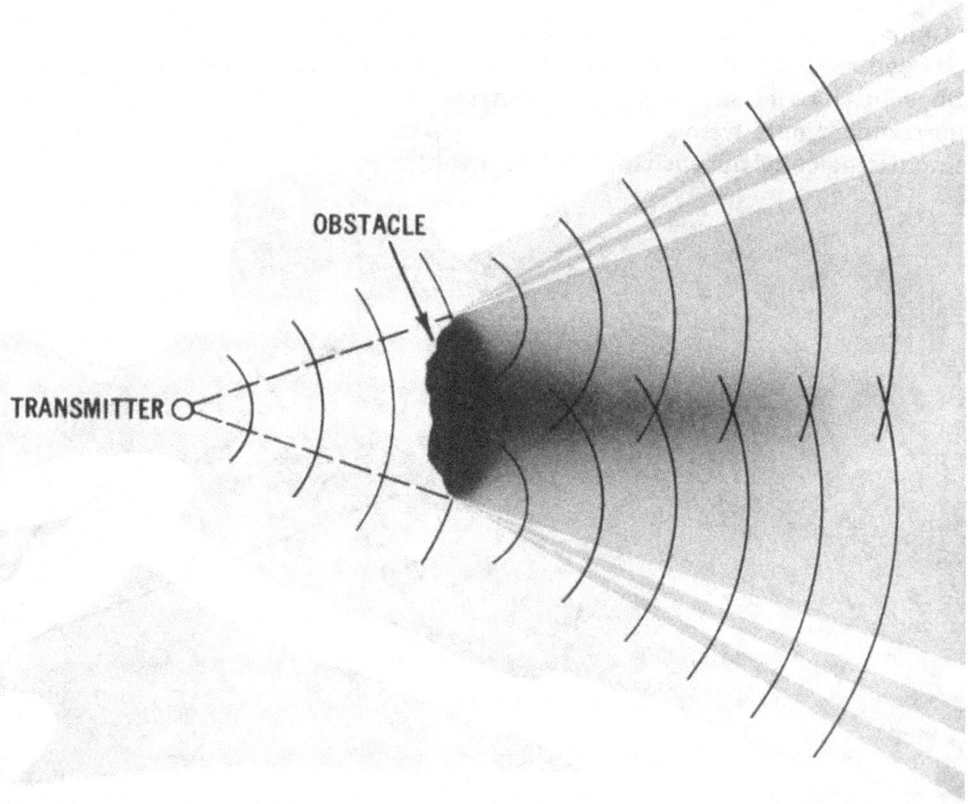

FIGURE 4009.—Diffraction.

The amount of diffraction is inversely proportional to the frequency, being greatest at very low frequencies.

4010. Absorption and scattering.—The amplitude of a radio wave expanding outward through space varies inversely with distance. That is, it gets weaker with increased distance. The decrease of strength with distance is called **attenuation.** Under certain conditions the attenuation is greater than in free space.

A wave traveling along the surface of the earth loses a certain amount of energy to the earth. The wave is diffracted downward and absorbed by the earth. As a result of this absorption, the remainder of the wave front tilts downward, resulting in further absorption by the earth. Attenuation is greater over a surface that is a poor conductor. Relatively little absorption occurs over seawater, which is an excellent conductor at low frequencies, and low frequency groundwaves travel great distances over water.

A skywave suffers an attenuation loss in its encounter with the ionosphere. The amount depends upon the height and composition of the ionosphere, as well as the frequency of the radio wave. Maximum ionospheric absorption occurs at about 1,400 kHz.

In general, atmospheric absorption increases with frequency, being a problem only at SHF and EHF. At these frequencies, attenuation is further increased by **scattering** due to reflection by oxygen, water vapor, water droplets, and rain in the atmosphere.

4011. Noise.—Unwanted signals in a receiver are called **interference.** The intentional production of such interference to obstruct communication is called **jamming.** Unintentional interference is called **noise.**

Noise may originate within the receiver. **Hum** is usually the result of induction from neighboring circuits carrying alternating current. **Microphonic noise** is the result of vibration of elements in an electron tube. Irregular crackling or sizzling sounds may be caused by poor contacts or faulty components within the receiver. Electron movement in normal components causes some noise. This source sets the ultimate limit of sensitivity (art. 4018) that can be achieved in a receiver. It is the same at any frequency.

Noise originating outside the receiver may be either man-made or natural. Man-made noises originate in electrical appliances, motor and generator brushes, ignition systems, and other sources of sparks which transmit electromagnetic signals that are picked up by the receiving antenna.

Natural noise is caused principally by discharge of static electricity in the atmosphere. This is called **atmospheric noise, atmospherics,** or **static.** An extreme example is a thunderstorm. An exposed surface may acquire a considerable charge of static electricity. This may be caused by friction of water or solid particles blown against or along such a surface. It may also be caused by splitting of a water droplet which strikes the surface, one part of the droplet requiring a positive charge and the other a negative charge. These charges may be transferred to the surface. The charge tends to gather at points and ridges of the conducting surface, and when it accumulates to a sufficient extent to overcome the insulating properties of the atmosphere, it discharges into the atmosphere. Under suitable conditions this becomes visible and is known as **St. Elmo's fire,** which is sometimes seen at mastheads, the ends of yardarms, etc.

Atmospheric noise occurs to some extent at all frequencies, but decreases with higher frequencies. Above about 30 MHz it is not generally a problem.

Since most of the noise occurs at low frequencies, it travels great distances and the accumulation may reach troublesome proportions at these frequencies, particularly during the summer in mountainous regions.

4012. Antenna characteristics.—Antenna design and orientation have a marked effect upon radio wave propagation. For a single-wire antenna, strongest signals are transmitted along the perpendicular to the wire, and virtually no signal in the direction of the wire. For a vertical antenna, the signal strength is the same in all horizontal directions. Unless the polarization undergoes a change during transit, the strongest signal *received* from a vertical transmitting antenna occurs when the receiving antenna is also vertical.

For lower frequencies the radiation of a radio signal takes place by interaction between the antenna and the ground. For a vertical antenna, efficiency increases with greater length of the antenna. For a horizontal antenna, efficiency increases with greater distance between antenna and ground. Near-maximum efficiency is attained when this distance is one-half wavelength. This is the reason for elevating low frequency antennas to great heights. However, at the lowest frequencies, the required height becomes prohibitively great. At 10 kHz it would be about 8 nautical miles for a half-wavelength antenna. Therefore, lower frequency antennas are inherently inefficient. This is partly offset by the greater range of a low frequency signal of the same transmitted power as one of higher frequency.

At higher frequencies, the ground is not used, both conducting portions being included in a **dipole** antenna. Not only can such an antenna be made efficient, but it can also be made sharply directive, thus greatly increasing the strength of the signal transmitted in a desired direction.

The power received is inversely proportional to the square of the distance from the transmitter, assuming there is no attenuation due to absorption or scattering.

4013. Range.—The range at which a usable signal is received depends upon the power transmitted, the sensitivity of the receiver, frequency, route of travel, noise level, and perhaps other factors. For the same *transmitted* power, both the groundwave and skywave ranges are greatest at the lowest frequencies, but this is somewhat offset by the lesser efficiency of antennas for these frequencies. At higher frequencies, only direct waves are useful, and the effective range is greatly reduced. Attenuation, skip distance, ground reflection, wave interference, condition of the ionosphere, atmospheric noise level, and antenna design all affect the distance at which useful signals can be received.

4014. Frequency and radio wave propagation.—Frequency is an important consideration in radio wave propagation, as indicated previously. The following summary indicates the principal effects associated with the various frequency bands, starting with the lowest and progressing to the highest usable radio frequency.

Very low frequency (VLF, 10 to 30 kHz). The VLF signals propagate between the bounds of the ionosphere and the earth and are thus guided around the curvature of the earth to great distances with low attenuation and excellent stability. Diffraction is maximum. Because of the long wavelength, large antennas are needed, and even these are inefficient, permitting radiation of relatively small amounts of power. Magnetic storms have little effect upon transmission because of the efficiency of the "earth-ionosphere waveguide." During such storms, VLF signals may constitute the only source of radio communication over great distances. However, interference from atmospheric noise may be troublesome. Signals may be received from below the surface of the sea. The characteristics of VLF propagation are discussed in more detail in the coverage of the Omega Navigation System in chapter XLIII.

Low frequency (LF, 30 to 300 kHz). As frequency is increased to the LF band, diffraction decreases, there is greater attenuation with distance, and range for a given power output falls off rapidly. However, this is partly offset by more efficient transmitting antennas, which can be of a size practical for use aboard ship. LF signals are most stable within groundwave distance of the transmitter. A wider bandwidth permits pulsed signals at 100 kHz. This allows separation of the stable groundwave pulse from the variable skywave pulse up to 1,500 km, and up to 2,000 km for overwater paths. The frequency for Loran-C (ch. XLIII) is in the LF band. This band is also useful for radio direction finding (ch. XLI), and time dissemination (ch. XVIII).

Medium frequency (MF, 300 to 3,000 kHz). Groundwaves provide dependable service, but the range for a given power is reduced greatly, varying from about 400 miles at the lower portion of the band to about 15 miles at the upper end for a transmitted signal of 1 kilowatt. These values are influenced, however, by the power of the transmitter, the directivity and efficiency of the antenna, and the nature of the terrain over which signals travel. Elevating the antenna to obtain direct waves may improve the transmission. At the lower frequencies of the band, skywaves are available both day and night. As the frequency is increased, ionospheric absorption increases to a maximum at about 1,400 kHz. At higher frequencies the absorption decreases, permitting increased use of skywaves. Since the ionosphere changes with the hour, season, and sunspot cycle, the reliability of skywave signals is variable. By careful selection of frequency, one can obtain ranges of as much as 8,000 miles with 1 kilowatt of transmitted power, using multihop signals. However, the frequency selection is critical. If it is too high, the signals penetrate the ionosphere and are lost in space. If it is too low, signals are too weak. In general, skywave reception is equally good by day or night, but lower frequencies are needed at night. The standard broadcast band for commercial stations (535 to 1,605 kHz) is in the MF band.

High frequency (HF, 3 to 30 MHz). As with higher medium frequencies, the groundwave range of HF signals is limited to a few miles, but the elevation of the antenna may increase the direct-wave distance of transmission. Also, the height of the antenna does have an important effect upon skywave transmission because the antenna has an "image" within the conducting earth. The distance between antenna and image is related to the height of the antenna, and this distance is as critical as the distance between elements of an antenna system. Maximum usable frequencies (art. 4008) fall generally within the HF band. By day this may be 10 to 30 MHz, but during the night it may drop to 8 to 10 MHz. The HF band is widely used for ship-to-ship and ship-to-shore communication.

Very high frequency (VHF, 30 to 300 MHz). Communication is limited primarily to the direct wave, or the direct wave plus a ground-reflected wave. Elevating the antenna to increase the distance at which direct waves can be used results in increased distance of reception, even though some wave interference between direct and ground-reflected waves is present. Diffraction is much less than with lower frequencies, but is most evident when signals cross sharp mountain peaks or ridges. Under suitable conditions, reflections from the ionosphere are sufficiently strong to be useful, but generally they are unavailable. There is relatively little interference from atmospheric noise in this band. Reasonably efficient directional antennas are possible with VHF. The VHF band is much used for communication.

Ultra high frequency (UHF, 300 to 3,000 MHz). Skywaves are not used in the UHF band because the ionosphere is not sufficiently dense to reflect the waves, which pass through it into space. Groundwaves and ground-reflected waves are used, although there is some wave interference. Diffraction is negligible, but the radio horizon extends about 15 percent beyond the visible horizon, due principally to refraction. Reception of UHF signals is virtually free from fading and interference by atmospheric noise. Sharply directive antennas can be produced for transmission in this band, which is widely used for ship-to-ship and ship-to-shore communication.

Super high frequency (SHF, 3,000 to 30,000 MHz). In the SHF band, also known as the *microwave* or as the *centimeter wave band,* there are no skywaves, transmission being entirely by direct and ground-reflected waves. Diffraction and interference by atmospheric noise are virtually nonexistent. Highly efficient, sharply directive antennas can be produced. Thus, transmission in this band is similar to that of UHF, but with the effects of shorter waves being greater. Reflection by clouds, water droplets, dust particles, etc., increases, causing greater scattering, increased wave interference, and fading. The SHF band is used for marine navigational radar.

Extremely high frequency (EHF, 30,000 to 300,000 MHz). The effects of shorter waves are more pronounced in the EHF band, transmission being free from wave interference, diffraction, fading, and interference by atmospheric noise. Only direct and ground-reflected waves are available. Scattering and absorption in the atmosphere are pronounced and may produce an upper limit to the frequency useful in radio communication.

4015. Regulation of frequency use.—While the characteristics of various frequencies are important to the selection of the most suitable one for any given purpose, these are not the only considerations. Confusion and extensive interference would result if every user had complete freedom of selection. Some form of regulation is needed. The allocation of various frequency bands to particular uses is a matter of international agreement. Within the United States, the Federal Communications Commission has responsibility for authorizing use of particular frequencies. In some cases a given frequency is allocated to several widely separated transmitters, but only under conditions which minimize interference, as during daylight hours. Interference between stations is further reduced by the use of **channels,** each of a narrow band of frequencies. That is, assigned frequencies are separated by an arbitrary band of frequencies that are not authorized for use. In the case of radio aids to navigation, ship communications, etc., bands of several channels are allocated, permitting selection of band and channel by the user.

4016. Kinds of radio transmission.—A series of waves transmitted at constant frequency and amplitude is called a **continuous wave (CW).** This cannot be heard except at the very lowest radio frequencies, when it may produce, in a receiver, an audible hum of high pitch.

Although a continuous wave may be used directly, as in radio direction finding (art. 4101) or Decca (art. 4344), it is more commonly modified in some manner. This is called **modulation.** When this occurs, the continuous wave serves as a **carrier wave** for information. Any of several types of modulation may be used.

In **amplitude modulation (AM)** the amplitude of the carrier wave is altered in accordance with the amplitude of a **modulating wave,** usually of audio frequency, as shown in figure 4016a. In the receiver the signal is **demodulated** by removing the modulating wave and converting it back to its original form. This form of modulation is widely used in voice radio, as in the standard broadcast band of commercial broadcasting.

If the *frequency* instead of the *amplitude* is altered in accordance with the amplitude of the impressed signal, as shown in figure 4016a, **frequency modulation (FM)** occurs. This is used for FM broadcasts and the sound portion of television broadcasts.

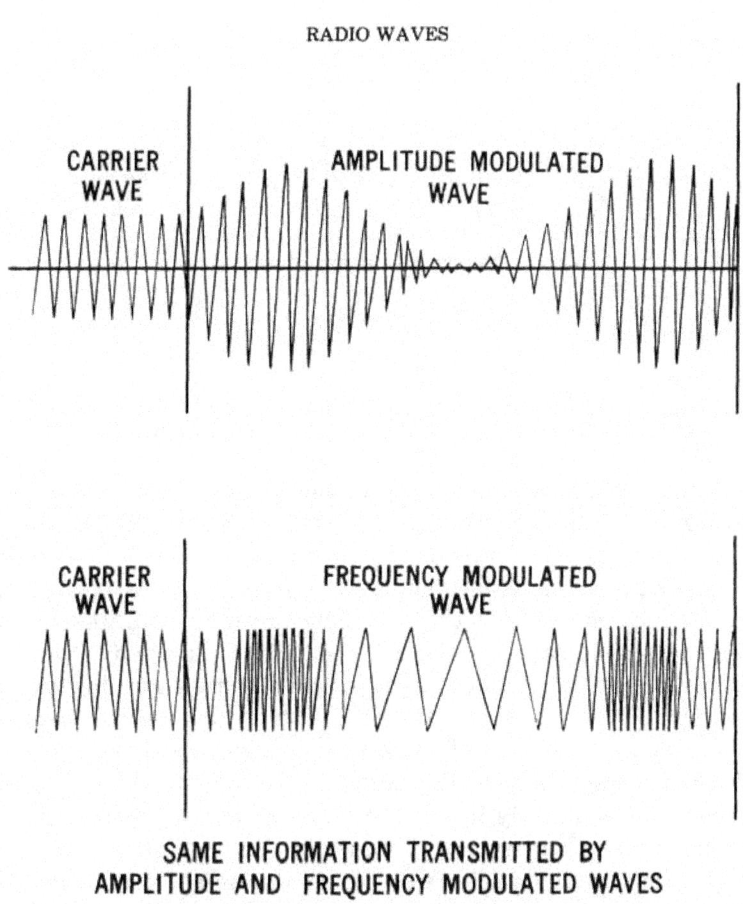

FIGURE 4016a.—Amplitude modulation (upper figure) and frequency modulation (lower figure) by the same modulating wave.

Pulse modulation (PM) is somewhat different, there being no impressed modulating wave. In this form of transmission, very short bursts of carrier wave are transmitted, separated by relatively long periods of "silence," during which there is no transmission. This type of transmission, illustrated in figure 4016b, is used in some common radio navigational aids, including radar (art. 4201) and Loran-C (art. 4329).

FIGURE 4016b.—Pulse modulation.

4017. Transmitters.—A radio **transmitter** consists essentially of (1) a **power supply** to furnish direct current, (2) an **oscillator** to convert direct current into radio-frequency oscillations (the carrier wave), (3) a device to control the generated signal, (4) an **amplifier** to increase the output of the oscillator. For some transmitters a **microphone** is needed with a **modulator** and final amplifier to modulate the carrier wave. In addition, an **antenna** and **ground** (for lower frequencies) are needed to produce electromagnetic radiation. These components are illustrated diagrammatically in figure 4017.

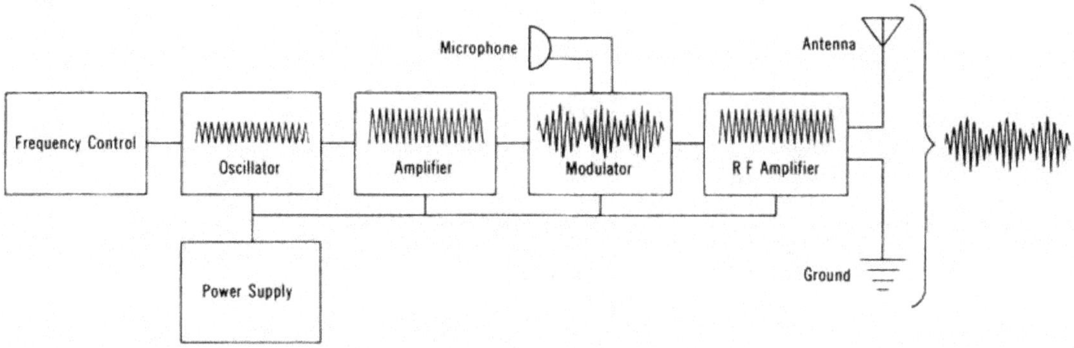

FIGURE 4017.—Components of a radio transmitter.

4018. Receivers.—When a radio wave passes a conductor, a current is induced in that conductor. A radio **receiver** is a device which accepts the power thus generated in an antenna, and transforms it into usable form. It should be able to select signals of a single frequency (actually a narrow band of frequencies) from among the many which may reach the receiving antenna. If necessary, the receiver should be able to demodulate the signal, and it should always provide adequate amplification. The output of a receiver may be presented audibly by earphones or loudspeaker; or visually on a dial, cathode-ray tube (art. 4019), counter, or other display. Thus, the useful reception of radio signals requires three components: (1) an antenna, (2) a receiver, and (3) a display unit.

Radio receivers differ mainly in (1) **frequency range,** the range of frequencies to which they can be tuned; (2) **selectivity,** the ability to confine reception to signals of the desired frequency and avoid others of nearly the same frequency; (3) **sensitivity,** the ability to amplify a weak signal to usable strength against a background of noise; (4) **stability,** the ability to resist drift from conditions or values to which set; and (5) **fidelity,** the completeness with which the essential characteristics of the original signal are reproduced. Receivers may have additional features such as an automatic frequency control, automatic noise limiter, etc.

Some of these characteristics are interrelated. For instance, if a receiver lacks selectivity, signals of a frequency differing slightly from those to which the receiver is tuned may be received. This condition is called **spillover,** and the resulting interference is called **crosstalk.** If the selectivity is increased sufficiently to prevent spillover, it may not permit receipt of a great enough band of frequencies to obtain the full range of those of the desired signal. Thus, the fidelity may be reduced.

A **transponder** is a transmitter-receiver capable of accepting the challenge of an interrogator and automatically transmitting an appropriate reply.

4019. The cathode-ray tube is a useful device for presenting certain types of information. This tube, with its associated controls, is often called an **oscilloscope,** or **scope** for short. In television receivers it is usually called the **picture tube.**

The essential components of a cathode-ray tube are shown in figure 4019. At the left is a **cathode** which serves as a source of electrons. In this usage it is called an **electron gun.** The electrons are collected and focused into a beam by a **focusing anode,** and then speeded up by an **accelerating anode.** If there were no other controls, the beam of electrons would travel the remainder of the length of the tube and strike the enlarged, curved surface of the tube **face** at its center, approximately. The inside of the face is coated with a material known as a **phosphor** (such as zinc sulphide or calcium sulphide) which become luminous (**phosphorescent**) where a beam of electrons impinges upon it. If the beam is sharply focused, a dot of light appears at the point of impact.

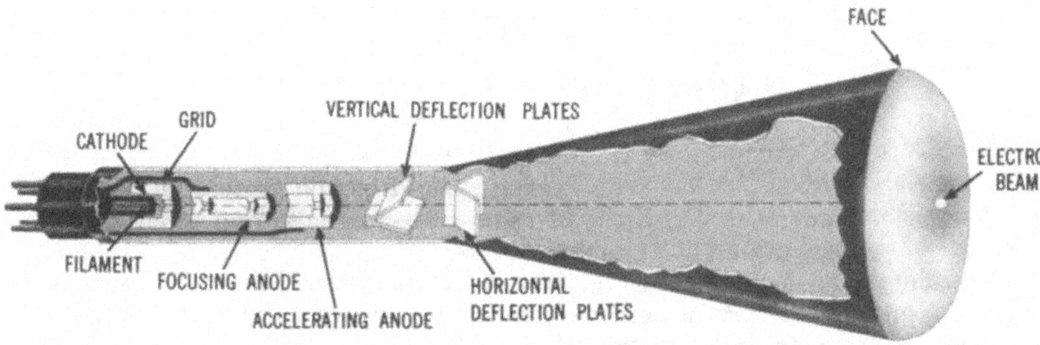

FIGURE 4019.—A cathode-ray tube.

By means of the **vertical deflection plates,** the beam is bent upward or downward. This is accomplished by impressing electric charges on these plates. The beam, being negatively charged, is repelled by the negative plate and attracted by the positive plate. If an alternating current is used, the strength and polarity of the electric charge on each plate changes continually, causing the beam to be deflected alternately up and down. This results in vertical motion of the spot of light on the face of the tube. If the motion is sufficiently rapid, a vertical *line* appears on the face of the tube. This is true not only because of the persistency of vision within the eye, but also because the tube face does not immediately fade when the stream of electrons is moved to another point. This visible line is called a **trace,** and the motion of the dot in producing it, a **sweep.** A horizontal trace can be made by means of the **horizontal deflection plates** which operate in a manner similar to that of the vertical deflection plates.

If both sets of plates are energized at the same time, the spot of light can be moved to various places on the face of the tube. If two alternating currents are properly synchronized, the spot can be made to trace repeatedly some pattern, such as a sine wave. It is generally desirable to have one trace repeated in accordance with a prearranged plan, having the deflection such that motion in one direction across the face of the tube is relatively slow, and that in the opposite direction is very fast, so that the return of the spot to a starting point is almost instantaneous. Such a return is called **flyback,** and the faint trace that may be visible is called a **retrace.** The position of the spot along the trace can be used as a measurement of elapsed time since the spot was at some reference point. This is usually accomplished by having a received signal impress a momentary charge on the *other* set of deflecting plates, causing a deflection of the trace as the spot is momentarily moved to one side of the line; or by causing the received signal to intensify the spot, causing it to glow brighter.

By suitable controls, the trace can be divided into two or more parts, made to rotate, or take any of a great variety of motions and shapes.

In a **dark trace tube** the spot appears dark on a lighter background.

The cathode-ray tube has many applications in electronic navigational equipment.

CHAPTER XLI

RADIO DIRECTION FINDING

4101. Introduction.—Medium frequency radio direction finders on board ships enable measurement of the bearings of radio transmissions from other ships, aircraft, shore stations, marine radiobeacons, aeronautical radiobeacons, and the coastal stations of the radio communications network.

Depending upon the design of the **radio direction finder (RDF)**, the bearings of the radio transmissions are measured as relative bearings or as both relative and true bearings. In one design, the true bearing dial is manually set with respect to the relative bearing dial in accordance with the ship's heading. In another design, the true bearing dial is rotated electrically in accordance with a course input from the gyrocompass. In some of the earlier designs, the RDF is mounted over the ship's compass so as to permit the bearings to be read directly from the compass card.

Whatever means is used to read the bearings, corrections for errors of radio bearings (art. 4104) must be applied. In some designs two distinct means are employed for automatic error compensation. An electrical compensating system provides for automatic compensation of errors which are symmetrically distributed, fore-and-aft or athwartship. A mechanical compensating system provides for automatic compensation of residual errors up to a limiting value. The mechanical compensator may consist of a stationary cam which is cut for each calibration (art. 4107), a movable roller, and associated linkage for causing the goniometer (art. 4103) pointer to lead or lag the actual goniometer setting.

When plotting radio bearings on a Mercator chart, conversion angle (art. 403, vol. II) must be applied to convert the radio bearing (great-circle direction) to the equivalent rhumb line bearing. Conversion angles are given in table 1.

4102. Radiobeacons established to be of primary usefulness to mariners are known as **marine radiobeacons;** beacons established to be of primary usefulness to airmen are known as **aeronautical radiobeacons;** other beacons established for both classes of user are sometimes known as **aeromarine radiobeacons.** The most common type of marine radiobeacon transmits radio waves of approximately uniform strength in all directions. These *omnidirectional* beacons are known as **circular radiobeacons.**

Directional radiobeacons transmit radio waves within a narrow sector. Compared with the circular radiobeacon, the transmissions from the directional radiobeacons have relatively short range. The rotating loop radiobeacon is discussed in article 4111.

Most United States and Canadian radiobeacons are grouped together on the same operating frequency and are assigned a specific sequence of transmission within this group. This reduces station interference and undesirable retuning. Normally the stations operate in groups of six, each station in a group of **sequenced radiobeacons** using the same frequency and transmitting for 1 minute in its proper sequence. A few radiobeacons transmit for 1 minute with 2 minutes of silence, and some radiobeacons transmit continuously without interruption.

Except for **calibration radiobeacons,** radiobeacons operate during all periods either sequenced or continuously, regardless of weather conditions.

Simple combinations of dots and dashes are used for station identification. These combinations and the duration of the dots, dashes, and spaces are chosen for ease of identification. Where applicable, the Morse equivalent character or characters are shown in conjunction with the station characteristic. All radiobeacons superimpose the characteristic on a carrier which is on continuously during the period of transmission. This extends the usefulness of marine radiobeacons to an airborne or marine user of an **automatic radio direction finder (ADF).** Users of the "aural null" type radio direction finder will notice no change in quality of service. A 10-second dash is incorporated in the characteristic signal to enable the user of the aural null type of radio direction finder to refine his bearing.

Aeronautical radiobeacons are sometimes used by marine navigators for determining lines of position when marine radiobeacons are not available. Since it is not possible to predict the extent to which land effect (art. 4104) may render mariners' observations of the bearings of these beacons unreliable, their inclusion in Pubs. Nos. 117A and 117B, *Radio Navigational Aids,* does not imply that the beacons have been found reliable for marine use. Those aeronautical radiobeacons included in Pubs. Nos. 117A and 117B may be useful to the marine navigator who recognizes their limitations. These aeronautical aids become less trustworthy, so far as marine applications are concerned, as they are situated farther inland or when high land intervenes between them and the coast.

Pubs. Nos. 117A and 117B, *Radio Navigational Aids,* include in the details of many radiobeacons located at or near light stations a statement of the distance and bearing of the radiobeacon transmitting antenna from the light tower. Use should be made of this information when calibrating (art. 4107) the radio direction finder.

4103. Direction measurement at the receiving site is accomplished by means of a directional antenna. Nearly all antennas have some directional properties, but in the usual antenna used for radio communication, these properties are not sufficiently critical for navigational use.

A widely used directional antenna is in the form of a loop. Suppose a transmitted radio signal encounters such a loop oriented in the direction of travel of the radio signal, as shown in figure 4103a. If the diameter of the loop is half the wavelength, the crest of one wave arrives at one side of the loop at the same time that the trough arrives at the opposite side, as shown. Thus, the currents induced in the two sides reinforce each other, causing maximum output from the antenna. A short time later, as the wave continues to move past the antenna, the crest reaches the other side of the loop, and a new trough reaches the approach side. A maximum current now flows in the opposite direction. Therefore, with the antenna in this orientation, an alternating current flows in the loop. If the loop diameter is less than half a wavelength, the current is less than maximum.

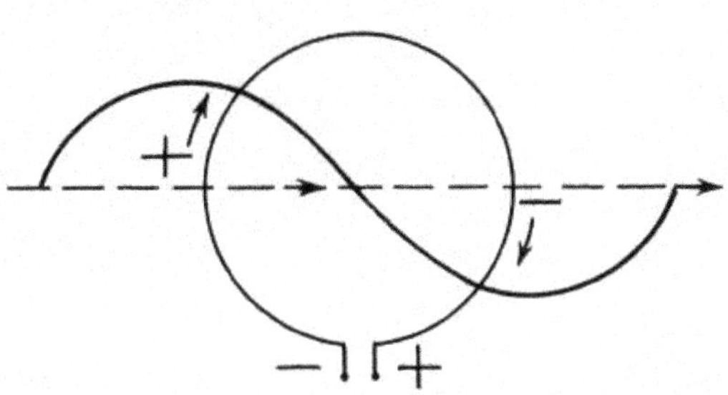

FIGURE 4103a.—Principle of the loop antenna.

If the antenna is rotated 90°, the alternate crests and troughs arrive at both sides at the same time, tending to cause currents to flow in opposite directions around the loop. Under these conditions the two parts cancel each other, resulting in zero antenna output. This condition is called a **null.**

As the antenna is rotated, its output varies with the angle relative to the direction of motion of the radio signal. This condition is illustrated in figure 4103b. The length of a line from the center to the outer edge of the shaded area represents the strength of the antenna output at that bearing, relative to the direction of motion of the radio wave. Thus, when it is in line, *with either side of the loop toward the approaching signal,* the output is maximum, and at 90° it is minimum. Since the change with bearing is most rapid near the region of minimum signal, this is the portion used for determination of direction.

Because of the characteristics of the simple loop antenna, a 180° **ambiguity** exists. That is, a signal approaching from either of two directions 180° apart would cause the same antenna output. This ambiguity can be resolved by using a vertical **sense antenna** in connection with a loop. The output from this wire, if the direction of motion of the signal is horizontal, is the same in all directions. Therefore, the polar diagram of its output is a circle, with the same polarity in all directions. If this output is exactly equal to the maximum of the loop, it will cancel the output from one side and double that from the other, since the polarity in the two sides is opposite. The resulting diagram of antenna output is shown in figure 4103c. With this arrangement, a single minimum exists, permitting the determination of which of the two reciprocal bearings is correct, thereby removing the ambiguity. The loop antenna is then used for making the reading. This is the type of equipment commonly used with a radio direction finder.

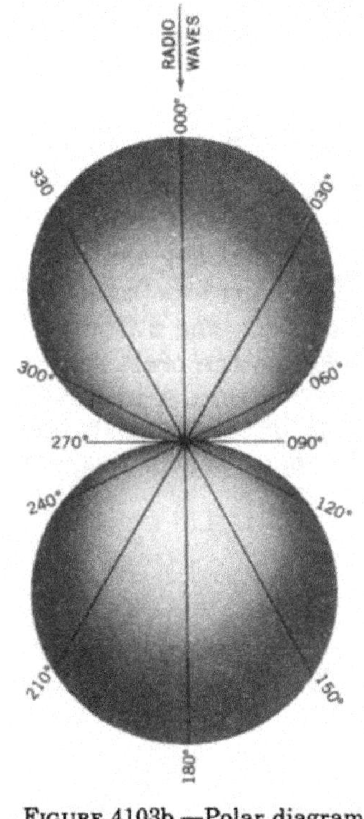

FIGURE 4103b.—Polar diagram of output of loop antenna.

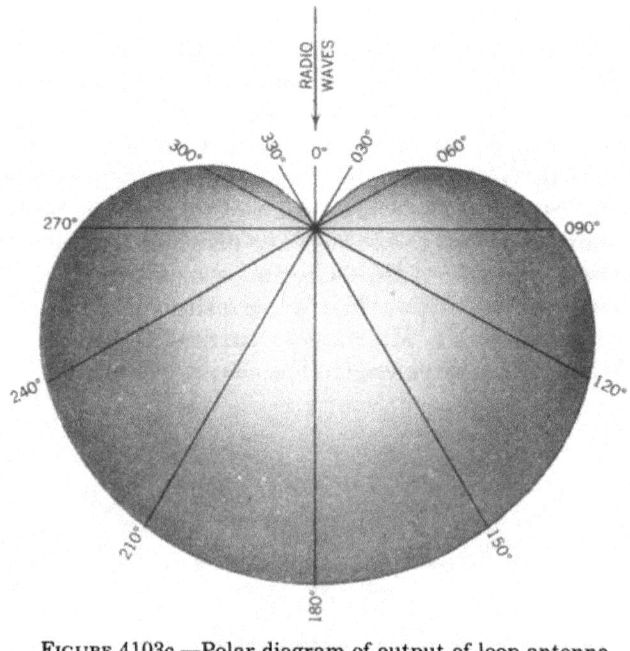

FIGURE 4103c.—Polar diagram of output of loop antenna with vertical sense antenna.

Two variations of the loop antenna are also used in radio direction finders. In one of these, the **crossed loop** type, two loops are rigidly mounted in such manner that one is rotated 90° with respect to the other. The *relative* output of the two antennas is related to the orientation of each with respect to the direction of travel of the radio wave, and is measured by a device called a **goniometer.** This is the type antenna used in an automatic direction finder. In the other variation, the **rotating loop** type, a single loop is kept in rapid rotation by means of a motor. The antenna output is shown on a cathode-ray tube, and the resulting display shows the direction of the signal.

4104. Errors of radio bearings.—Bearings obtained by radio direction finder are subject to certain errors, as follows:

Quadrantal error. When radio waves arrive at a receiver, they are influenced somewhat by the environment. An erroneous radio direction finder bearing results from currents induced in the direction finder antenna by re-radiation from the structural features of the vessel's superstructure and distortion of the radio wave front due to the physical dimensions and contour of the vessel's hull. This quadrantal error is a function of the *relative* bearing, normally being maximum for bearings broad on the bow and broad on the quarter. Its value for various bearings can be determined, and a **calibration table** made (art. 4107).

Coastal refraction. As indicated in article 4006, a radio wave crossing a coastline at an oblique angle undergoes a change of direction due to difference in conducting and reflecting properties of land and water. This is sometimes called **land effect.** It is avoided by not using, or regarding as of doubtful accuracy, bearings of waves which cross a shoreline at an oblique angle. Bearings within 15° to 20° of being parallel to a shoreline should not be trusted. If the transmitter is near the coast, negligible error is introduced because of the short distance the waves travel before undergoing refraction.

Polarization error. As indicated in article 4008, the direction of travel of radio waves may undergo an alteration during the confused period near sunrise or sunset, when great changes are taking place in the ionosphere. This error is sometimes called **night effect.** The error can be minimized by averaging several readings, but any radio bearings taken during this period should be considered of doubtful accuracy.

Reciprocal bearings. Unless a radio direction finder has a vertical sensing wire (art. 4103), there is a possible 180° ambiguity in the reading. If such an error is discovered, one should take the reciprocal of the *uncorrected* reading, and apply the correction for the new direction. If there is doubt as to which of the two possible directions is the correct one, one should wait long enough for the bearing to change appreciably and take another reading. The transmitter should draw *aft* between readings. If the reciprocal is used, the station will appear to have drawn *forward.* A reciprocal bearing furnished by a direction finder station should not be used because the quadrantal error is not known, either on the given bearing or its reciprocal.

4105. Accuracy of radio bearings.—In general, good radio bearings should not be in error by more than 2° for distances under 150 nautical miles. However, conditions vary considerably, and skill is an important factor. By observing the technical instructions for the equipment and practicing frequently when results can be checked by visual observation or by other means, one can develop skill and learn to what extent radio bearings can be relied upon under various conditions.

Other factors affecting accuracy include the errors discussed in article 4104, range, the condition of the equipment, and the accuracy of the calibration (art. 4107). Errors in bearing can result if the selectivity (art. 4106) of a radio direction finder is poor.

4106. Factors affecting maximum range.—The service range of a radiobeacon is determined by the strength of the radiated signal. Field strength requirements for a given service range vary with latitude, being higher in the southern latitudes. The actual useful range may vary considerably from the service range with different types of radio direction finders and during varying atmospheric conditions.

Sensitivity is a measure of the ability of a receiver to detect transmissions. All direction finder receivers do not have the same sensitivity. Some will detect a radiobeacon signal at its rated range whereas others will not detect the same signal until such time that the distance to the beacon has decreased. For example, radio direction finders having a sensitivity of 75 microvolts per meter on the radiobeacon band should be capable of detecting a signal whose intensity is 75 microvolts per meter or more. A radio direction finder having a sensitivity of 120 microvolts per meter will be unable to receive a radiobeacon signal rated at 75 microvolts per meter at 100 miles until 56 miles from the radiobeacon. At 56 miles the signal strength of the transmitted signal is 120 microvolts per meter, which is equal to the sensitivity of the receiver. It follows that the sensitivity of a radio direction finder determines the degree to which the full range capability of the radiobeacon system can be utilized.

Selectivity is a measure of the ability of a receiver to choose one frequency and reject all others. The selectivity varies with the type of receiver and its condition. The transmitted radiobeacon signal is comprised of a band of frequencies, 286.000 kHz to 287.020 kHz for example. A radio direction finder capable of accepting only this narrow band of frequencies would be ideal. If a radio direction finder accepts a wide band of frequencies (280 to 292 kHz) when tuned to 286 kHz, it will admit more noise and signals than desired. This additional interference may reduce the usefulness of the desired signal, and effectively decrease the maximum range of reception of the radiobeacon.

4107. Calibration.—The reliability of a radio direction finder is largely dependent upon the accuracy of the calibration. A good initial calibration not only increases the reliability of operation but also reduces the need for repeated recalibration, provided the superstructure and rigging of the vessel are not altered. Correct radio direction finder calibration compensates for errors caused by induced currents and vessel configuration, i.e., quadrantal error (art. 4104), also known as **direction finder error.**

Proper preliminary procedures for this calibration include accurately aligning the vessel's pelorus fore-and-aft, providing adequate communications between the pelorus and the radio direction finder, developing a plan for coordination between the calibrator and conning officer, determining that the equipment is in proper operating condition, and determining that metal booms, cranes, antennas, etc., are in their normal positions.

While the vessel is enroute to the calibration site, it may be advisable to take a number of bearings on the station to be used for calibration and on one or more other charted stations whose relative bearings can be ascertained. If it is feasible to take an RDF bearing on such a station when it bears broad on the bow or quarter, the magnitude of the quadrantal error can be estimated. Then the electrical error compensator, if installed, should be set to correct the error. It is also desirable to take RDF bearings while the vessel is enroute to the calibration site on a station at 135°, 225°, and 315° from the vessel's heading. These additional readings serve to check the setting of the electrical error compensator and to indicate whether additional adjustment of the compensator is required. Generally, if the magnitudes of the bearing errors on 135°, 225°, and 315° are less than the initial error used to adjust the compensator, the setting should be left unaltered.

The source of the radio signals used in calibration may be a radiobeacon operating on schedule or on request.

Sequenced radiobeacons cannot broadcast at any time other than on their assigned operating minute for the purpose of enabling vessels to calibrate their radio direction finders without causing interference. Special radio direction finder calibration transmitters of short range are operated at certain localities to provide calibration service during specified periods or on request. These stations with information as to position, frequency, characteristic, times of service, requests for use, etc., are listed in Pubs. Nos. 117A and 117B, *Radio Navigational Aids* and the *Light List.*

The position given for the antenna is the point from which the radiobeacon signal is emitted.

The calibration must be made on approximately the same frequency or frequencies as will be used to take RDF bearings because the direction finder error for several frequencies is not likely to be the same. It is believed that one calibration curve or table is satisfactory for the normal radiobeacon frequency (285 to 325 kHz); but the instructions issued by the manufacturer of the particular radio direction finder to be calibrated should be studied in this respect.

The usual method of calibration is to obtain a series of simultaneous radio and visual bearings on a transmitter. This can be done while a vessel swings at anchor or more quickly by steaming in a circle at the greatest distance compatible with clear visual observation of the transmitter, preferably over 1 mile. The simultaneous bearings should be observed and recorded at least every 10°, preferably every 5°.

The difference between a radio bearing and a simultaneously observed visual bearing, using the same reference, is the direction finder error. The error is negative if the visual bearing is greater than the radio bearing; the correction is positive.

The quadrantal error being maximum, generally, when the station is broad on the bow or quarter, setting of the electrical error compensator for a correction equal to the error measured when the visual bearing is 45° or 315° should provide marked reduction in errors corresponding to visual bearings of 45°, 135°, 225°, and 315°. If the vessel is swung again and another set of visual and RDF bearings are observed, the errors computed from this swing should comprise primarily the residual nonsymmetrical error components. This residual error can be corrected by cutting a cam for the mechanical error compensator.

The radio direction finder should be recalibrated after any changes have been made in the set or its surroundings, whenever there is reason to believe that the previous calibration has become inaccurate, and also at periodic intervals.

While RDF bearings are being taken, other radio antennas on board must be in the same condition as they were when the calibration was made; movable parts of the ship's superstructure such as booms, davits, wire rigging, etc., must be secured in the positions which they occupied when the radio direction finder was calibrated. Unusual cargoes such as large quantities of metals and extraordinary conditions of loading may cause errors.

4108. Using radio bearings.—A bearing obtained by radio, like one determined in any other manner, provides means for establishing a line of position. By heading in the direction from which the signal is coming, one can proceed toward, or **home** on, the transmitter. In thick weather one should avoid heading directly toward the source of radiation unless he has reliable information to indicate that he is some distance away. In 1934 the Nantucket Lightship was rammed and sunk by a ship homing on its radiobeacon.

Due to the many factors which enter into the transmission and reception of radio signals, a mariner cannot practically estimate his distance from a radiobeacon either by the strength of the signals received or by the time at which the signals were first heard. Mariners should give this fact careful consideration in approaching radiobeacons. When approaching a lightship, large navigational buoy, ocean station vessel, or a station on a submarine site, on radio bearings, the risk of collision will be avoided by insuring that the radio bearing does not remain constant.

It should be borne in mind that most lightships and large buoys are anchored to a very long scope of chain and, as a result, the radius of their swinging circle is considerable. The charted position is the location of the anchor. Furthermore, under certain conditions of wind and current, they are subject to sudden and unexpected sheers which are certain to hazard a vessel attempting to pass close aboard.

Radio waves, like light, travel along great circles. Except in high latitudes, visual bearings can usually be plotted as straight lines on a Mercator chart, without significant error. Radio bearings, however, are often observed at such positions with respect to the transmitter that the use of a rhumb line is not satisfactory. Under these conditions it is customary to apply the **conversion angle** (art. 403, vol. II) as a correction to the observed angle, to find the equivalent rhumb line. Such a correction is not needed when a bearing is plotted on a gnomonic chart or one on which a straight line is a good approximation of a great circle. In other situations, a correction may be necessary.

If the transmitter and receiver are on the same meridian, or are both on the equator, no correction is needed because rhumb lines and great circles coincide under these conditions. The size of the correction increases with degree of departure from these conditions, and with greater distance between transmitter and receiver.

Conversion angles are given in table 1. This table is used to convert great circle to rhumb line directions. If the difference of longitude is not more than 4°.5, and the mid-latitude between transmitter and receiver is not more than 85°, the first part of the table should be used. The simplifying assumptions used in the computation of this part of the table do not introduce a significant error within the limits of the table.

The sign of the correction can be determined by referring to the rules given at the bottom of each page of table 1. These follow from the fact that the great circle is nearer the pole than the rhumb line.

Before taking bearings on a commercial broadcasting station, the mariner should consider the following:

1. The operating frequency of the commercial station may differ widely from the frequency for which the radio direction finder is calibrated.
2. The broadcast antenna may be remote from the broadcast station.
3. The commercial stations are usually inland.

Accordingly, the use of commercial broadcasting stations to obtain a direction finder bearing is not recommended. If these stations are used, the mariner should recognize the limitations of the bearings obtained.

4109. Radio direction finder stations.—Radio direction finder stations are stations equipped with special apparatus for determining the direction of radio signals transmitted by ships and other stations. The bearings taken by radio direction finder stations, and reported to ships, are corrected for all determinable errors except conversion angle.

The bearings are normally accurate within 2° for distances under 150 nautical miles. The best bearings are obtained on ships whose signals are steady, clear, and strong. Therefore, the ship's transmitter should be finely tuned to the frequency of the transmitter. If the ship's transmitter is not finely tuned, it is difficult for the station to obtain bearings sufficiently accurate for navigational purposes.

Where bearing lines intersect an intervening coastline at an oblique angle or cross high intervening land, errors of from 4° to 5° may be expected due to refraction (art. 4104). However, the sectors in which such refraction may be expected is normally known by station personnel. Such sectors may not be included in the published sectors of calibration or are indicated as **sectors of uncertain calibration.**

The **sector of calibration** of a radio direction finder station is the sector about the receiving coil of the station in which the deviation of radio bearings is known. In Pubs. Nos. 117A and 117B, *Radio Navigational Aids,* the sectors are measured clockwise from 0° (true north) to 360° and are given looking from the station to seaward. Bearings which do not lie within the sector of calibration of a station should be considered unreliable.

4110. Distance finding stations.—At some locations a radio signal is synchronized with a sound signal which may be transmitted through either air or water or both. The travel time of the radio signal is negligible compared to that of the sound signal. Consequently, the difference in time between reception of the two signals is proportional to the distance from the station. The distance in nautical miles is equal to the number of seconds of time interval divided by 5½ if the sound travels through air, or by 1¼ if through water (or multiplied by 0.18 or 0.8, respectively). The distance so found is from the origin of the *sound* signal, which might differ somewhat from that of the radio signal. Table 4110 can be used for finding the distance in nautical miles from a sound signal source.

Interval in seconds	Distance in nautical miles from sound signal source	
	Air	Submarine
1	0.18	0.8
2	.36	1.6
3	.54	2.4
4	.72	3.2
5	.90	4.0
6	1.08	4.8
7	1.26	5.6
8	1.44	6.4
9	1.62	7.2
10	1.80	8.0
20	3.60	16.0
30	5.40	24.0
40	7.20	
50	9.00	
60	10.80	

TABLE 4110.—Table for finding distance from a sound source.

The speed of sound travel is influenced by a number of conditions making it impracticable to state a factor that will give exact results under all conditions. The results obtained by the methods described may be accepted as being accurate to within 10 percent of the distance.

Ordinarily, the sound signals do not operate during the transmission period of the radio signal in clear weather. The methods in use employ, as a rule, distinctive signals to indicate the point of synchronization. Methods of synchronizing the signals vary and are described or illustrated in official announcements regarding them. It is essential to note carefully the point of synchronization used so that no error will be made through taking time on the wrong signal or the wrong part of it.

An example of the synchronized signals is shown in figure 4110. In this example, the beginning of the 10-second radio dash and the beginning of the 5-second fog signal blast are synchronized. The observer may use as the time interval the interval from the time of hearing the beginning of the long radio dash to the instant of hearing the beginning of the long blast of the fog signal.

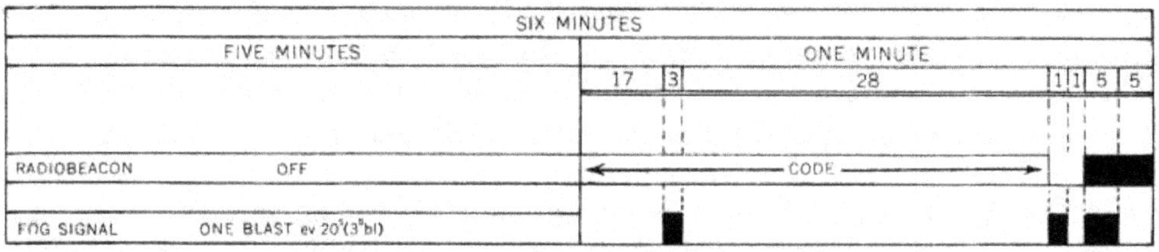

FIGURE 4110.—Synchronized radio and fog signals.

In observing air signals it is usually sufficient to use a watch with second hand, although a stopwatch is helpful. For submarine signals where the interval is shorter and a time error correspondingly more important, it is essential that a stopwatch or other timing device be used. Where the radiobeacon and submarine signals are not received at the same point on the vessel, means of instant communication between two observers should be available or synchronized stopwatches provided for each.

In the case of some sound signals a series of short radio dashes is transmitted at intervals following the synchronizing point, so that by counting the number of such short dashes heard after the distinctive radio signal and before hearing the corresponding distinctive sound signal, the observer obtains his distance, in miles equal to the number of dashes counted, from the sound signal apparatus unless stated otherwise.

Ships not equipped with an RDF receiver can take advantage of the distance finding feature of a radiobeacon station, if equipped with a radio receiver capable of receiving the transmission. In the case of obtaining distance from a radiobeacon station which is synchronized with a submarine sound signal, the ship must also be equipped with a device for picking up submarine sound signals.

4111. Rotating loop radiobeacon.—The rotating loop radiobeacon *used in Japanese waters* consists of a rotating loop transmitter having directional properties by which an observer in a ship can obtain his bearing from the beacon without the use of a radio direction finder. Any radio receiving set capable of being tuned to the radiobeacon frequency can be used. The only other equipment required is a reliable stopwatch with a sweep second hand. Stopwatches and clocks with dials graduated in degrees can be used, from which bearings can be read off directly without any mathematical calculation.

During each revolution of the beacon, the signals received by the observer will rise and fall in intensity, passing through a maximum and a minimum twice each minute. The positions of minimum intensity, which occur at intervals of 30 seconds from one another, are very sharp and can be accurately observed. These are, therefore, used for navigation purposes.

The beacon may be regarded as having a line or beam of minimum intensity which rotates at a uniform speed of 360° in 1 minute (i.e. 6° in 1 second) based on the true meridian as starting point. Therefore, if the observer can (1) identify the beacon and (2) measure the number of seconds which this minimum beam takes to reach his position starting from the true meridian, this number multiplied by six will give his true bearing from the beacon or its reciprocal.

Each transmission (fig. 4111) from the beacon lasts for 4 minutes, and automatically starts again at the end of the silent period. Each transmission consists of two parts: (1) the identification signal of the station set at a slow speed for the first minute, commencing when the minimum beam is true east and west and followed by a long dash of about 12 seconds duration and (2) the signal group commencing when the minimum beam is approaching the true meridian and consisting of (i) the north starting signal, which is Morse V followed by two dots (··· — ··); (ii) a long dash of about 12 seconds duration; (iii) the east starting signal, which is Morse B followed by two dots (— ··· ··); and (iv) a long dash for about 42 seconds.

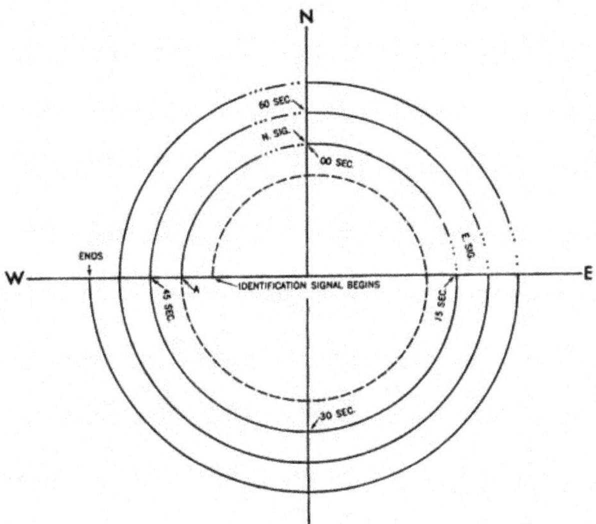

FIGURE 4111.—Transmissions from rotating loop radiobeacon.

The navigation signals are repeated during the remainder of the transmission and signals cease when the minimum beam is in the east-west position.

4112. Using the rotating loop radiobeacon.—Procedures for using the rotating loop radiobeacon are as follows:

1. Set stopwatch to zero.
2. Listen for identification signal.
3. When the first long dash begins (at *A* of figure 4111) standby for the north starting signal (Morse V followed by two dots); start stopwatch exactly at beginning of long dash (00^s of figure 4111), counting "one"-"two" with the two preceding dots and "three" for the start of the stopwatch.

4. Listen for minimum and record its exact time by stopwatch.

5. Multiply the number of seconds by 6° for bearing.

6. Determine whether bearing is reciprocal or direct.

7. If the north signal is faint, use the east signal but add 90° to the bearing.

The true bearing established is *from* the beacon. Therefore, if conversion angle (art 403, vol. II) is to be used it should be applied to the true bearing of the observer *from* the beacon, as in correcting bearings obtained by shore direction finder stations for conversion angle.

The following precautions should be observed:

1. Stopwatch should be started exactly at the beginning of the long dash.

2. The time of the minimum should be observed to the nearest fifth of a second, if possible.

3. Be sure to determine whether the bearing is the direct bearing to the beacon or its reciprocal.

4. If the east signal is used, be sure to add 90°.

5. The stopwatch should be checked before use. This can be done by checking the time by stopwatch of the complete revolution of the beacon transmission. A comparatively large bearing error can result from inaccurate timing.

CHAPTER XLII

RADAR NAVIGATION

4201. Introduction.—*Radar navigation* involves the use of radio waves, usually in the centimeter band, to determine the direction and distance of an object reflecting the waves to the sender.

A more detailed discussion of this part of radionavigation is given in Pub. No. 1310, *Radar Navigation Manual.*

4202. Radar, a term derived from **ra**dio **d**etection **a**nd **r**anging, is applied to electronic equipment designed to determine distance by measuring the time required for a radio signal to travel from a transmitter to a "target" and return, either as a reflected "echo" (**primary radar**) or as a retransmitted signal from a transponder (art. 4018) triggered by the original signal (**secondary radar**). Since primary radar uses a directional antenna, the direction of the target is also determined, but with somewhat less accuracy than the distance.

In the usual design for marine navigation applications, the radio signal is *pulse modulated* (art. 4016). Signals are generated in a transmitter by a timing circuit so that energy leaves the antenna in very short bursts or "pulses." During transmission of a pulse, the antenna is connected to the transmitter but not the receiver. As soon as the pulse leaves, an electronic switch disconnects the antenna from the transmitter and connects it to the receiver. Another pulse is not transmitted until after the preceding one has had time to travel to the most distant target within range, and return. Since the interval between pulses is long compared with the length of a pulse, strong signals can be provided with low *average* power. The duration or length of a single pulse is called **pulse length, pulse duration,** or **pulse width.**

From the receiver, the return signal goes to the indicator. This consists of a cathode-ray tube (art. 4019) and appropriate circuits. Many types of displays have been devised, a number of them to meet specialized requirements. For navigational use, the earliest type of display was the **A-scope.** The principle of this scope is illustrated in figure 4202a. At *A* a pulse leaves the antenna of a ship, and a vertical deflection appears at the start of the horizontal trace on the scope face. At *B* the pulse has traveled some distance outward from the antenna. A short horizontal line appears after the vertical deflection on the scope face. The length of this line is directly proportional to the distance traveled by the pulse. At *C* the pulse encounters a target with a reflecting surface. At *D* the original pulse has moved on beyond the target, but part of its energy has been reflected back toward the transmitter. At *E* the echo has arrived back at the transmitting craft, causing a vertical deflection of the horizontal trace. The height of this deflection is directly proportional to the strength of the returning signal. At *F* the echo has proceeded on past the transmitting ship, and the trace is completed.

This sequence is repeated a great many times, perhaps 1,000 per second, the rate being called the **pulse repetition rate (PRR)** or **pulse recurrence rate.** The start of each trace is synchronized with transmission of the signal so that each trace is a repetition of the previous one, if slight changes in relative positions of

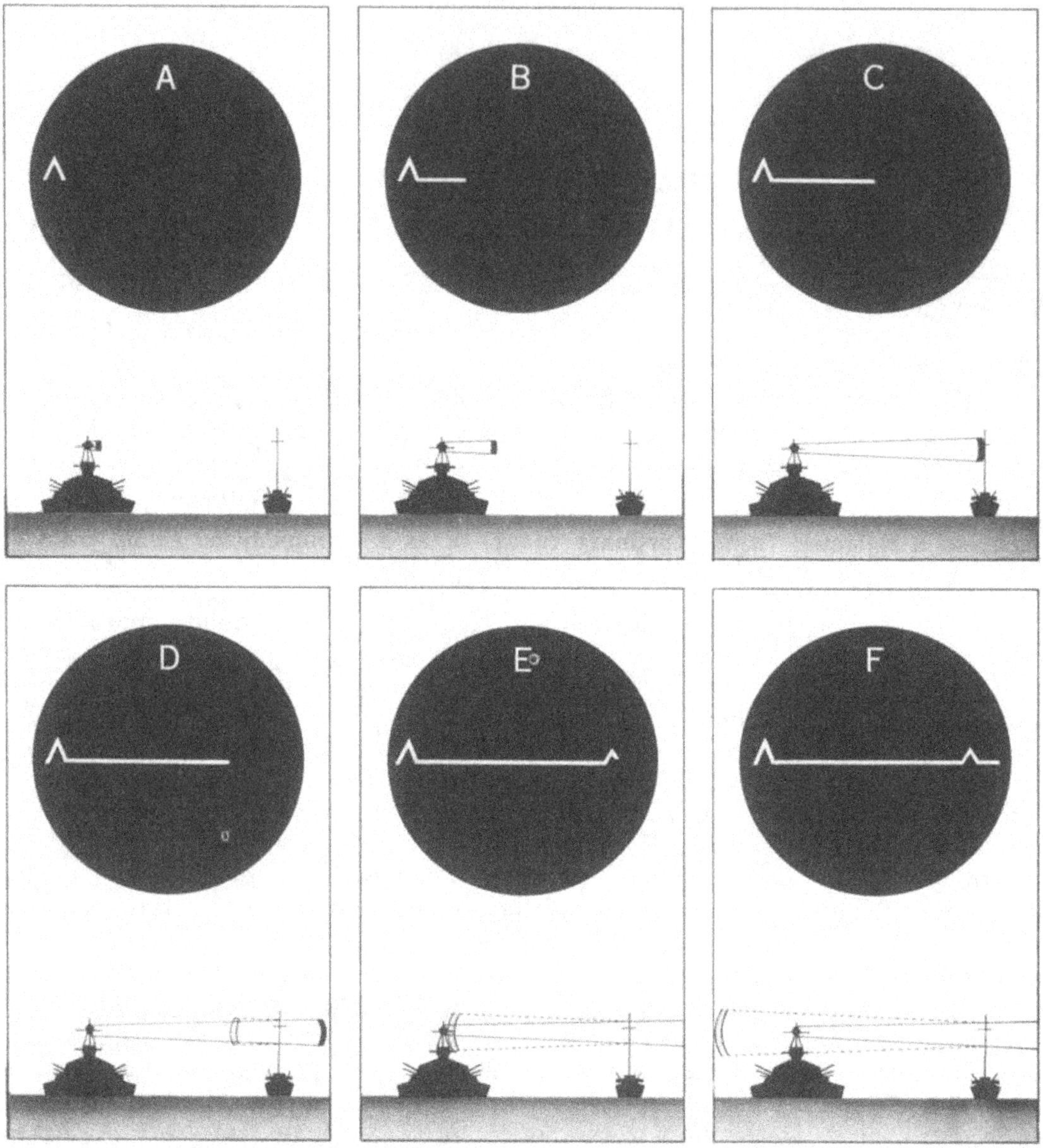

FIGURE 4202a.—A-scope.

transmitting ship, target, and antenna orientation are neglected. Therefore, the trace and all deflections appear as a continuous line. The distance between leading edges of the vertical deflections, or "pips," is directly proportional to range. A change of range alters the positions of the second pip. The orientation of the antenna is an indication of direction. A pip appears only when the antenna is pointed toward the target.

The type of presentation now most commonly used for navigational radar is called the **plan position indicator (PPI).** On this presentation the sweep starts at the center of the tube face and moves outward along a radial line which rotates in synchronization with the antenna. Instead of being deflected, the trace glows with greater intensity (brightness) at the appropriate places. Because of the persistence of the tube face coating, the glow continues after the trace rotates on past the target, resulting in a maplike presentation on the scope. This presentation is shown in figure 4202b.

On a PPI, the range of a target is proportional to the distance of its echo signal from the center of the scope. This may be measured by a series of visible concentric circles at established distances from the center, or by means of an adjustable ring (**variable range marker**) synchronized with a counter. Bearing is indicated by the direction of an echo signal from the center of the scope. To facilitate measurement of direction, a movable, radial, guide line or **cursor** is provided, and a compass rose is placed around the outside of the scope. In the "heading-upward" presentation, relative bearings are indicated, the top of the scope representing the direction of the ship's head. In this *unstabilized* presentation, the orientation changes with changes in ship's heading. In the *stabilized* "north-upward" presentation, gyro north is always at the top, regardless of the heading. True bearings are indicated if there is no gyro error. On this type presentation a radial line is customarily provided at the heading of the vessel.

Provision may be made for offsetting the center of the PPI presentation from the center of the tube face, to permit large-scale observation of distant targets in one direction. With "true motion" radar, the center of the PPI continues to represent the same geographical position until reset. The actual motion of all moving objects, including one's own vessel, appears on the PPI, instead of the relative movement usually shown.

Other modifications have been devised. In some installations a **repeater** duplicates the presentation, making the information available at a distance from the radar.

4203. The radar beam.—The pulses of energy as emitted from a feedhorn at the focal point of a reflector or as emitted and radiated directly from the slots of a slotted waveguide antenna would, for the most part, form a single lobe-shaped pattern of radiation if emitted in free space. Figure 4203a shows this free space radiation pattern, including the undesirable minor lobes or side lobes associated with practical antenna design. Because of the large differences in the various dimensions of the radiation pattern, figure 4203a is necessarily distorted.

Although the radiated energy is concentrated or focused into a relatively narrow main beam by the antenna, similar to a beam of light from a flashlight, there is no clearly defined envelope of the energy radiated. Although the energy is concentrated along the axis of the beam, the strength of the energy decreases rapidly in directions away from the beam axis. The power in watts at points in the beam is inversely proportional to the square of the distance. Therefore, the power at 3 miles is only one-ninth of the power at 1 mile in a given direction. The field intensity in volts at points in the beam is inversely proportional to the distance. Therefore, the voltage at 2 miles is only one-half the voltage at 1 mile in a given direction. With the rapid decrease in the amount of radiated energy in directions away from the axis and in conjunction with the rapid decreases of this energy with distance, it follows that practical limits of power of voltage may be used to define the dimensions of the radar beam or to establish its envelope of useful energy.

The three-dimensional radar beam is normally defined by its *horizontal* and *vertical beam widths.* Beam width is the angular width of a radar beam between points within which the field strength or power is greater than arbitrarily selected lower limits of field strength or power.

There are two limiting values, expressed either in terms of field intensity or power ratios, used conventionally to define beam width. One convention defines

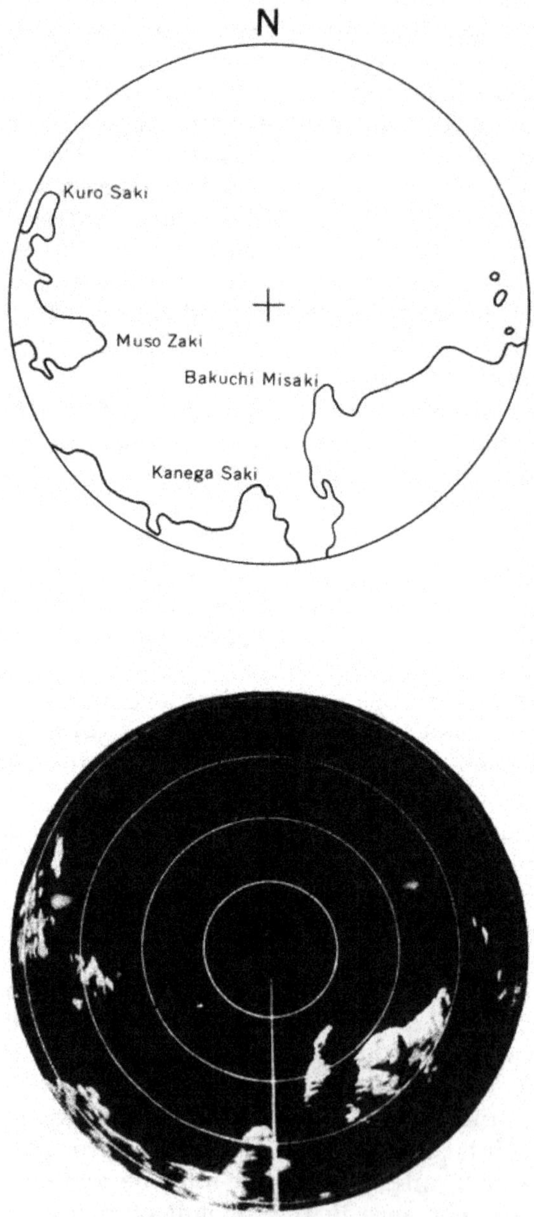

HONSHU, NW COAST—APPROACH TO MAIZURU KO

Radar position: 35° 34'.1 N., 135° 19'.6 E.
Range ring interval: 1 mile

(Source: Japanese Sailing Directions)

FIGURE 4202b.—Plan position indicator display.

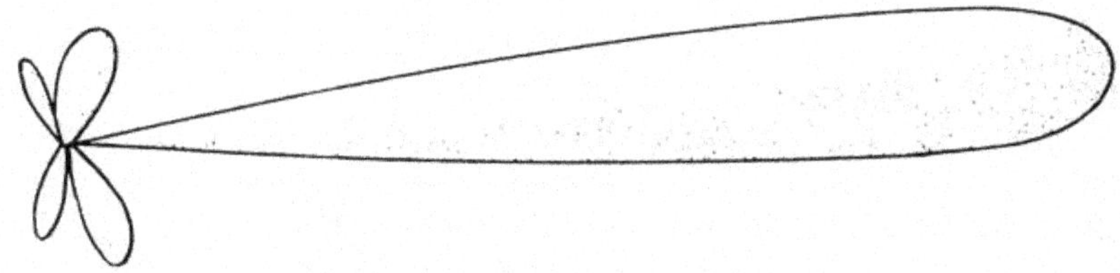

FIGURE 4203a.—Free space radiation pattern.

beam width as the angular width between points at which the field strength is 71 percent of its maximum value. Expressed in terms of power ratio, this convention defines beam width as the angular width between **half-power points.** The other convention defines beam width as the angular width between points at which the field strength is 50 percent of its maximum value. Expressed in terms of power ratio, the latter convention defines beam width as the angular width between **quarter-power points.**

The half-power ratio is the most frequently used convention. Which convention has been used in stating the beam width may be identified from the decibel (dB) figure normally included with the specifications of a radar set. Half-power and 71 percent field strength correspond to −3 dB; quarter-power and 50 percent field strength correspond to −6 dB.

The radiation diagram shown in figure 4203b depicts relative values of power in the same plane existing at the same distances from the antenna or the origin of the radar beam. Maximum power is in the direction of the axis of the beam. Power values diminish rapidly in directions away from the axis. The beam width in this case is taken as the angle between the half-power points.

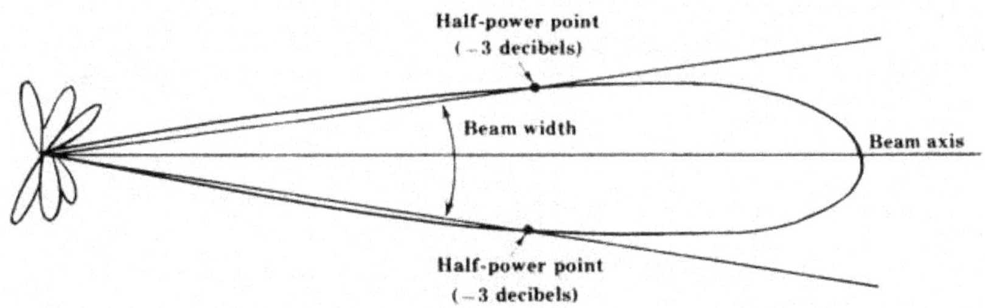

FIGURE 4203b.—Radiation diagram.

For a given amount of transmitted power, the main lobe of the radar beam extends to a greater distance at a given power level with greater concentration of power in narrower beam widths. To increase maximum detection range capabilities, the energy is concentrated into as narrow a beam as is feasible. Because of practical considerations related to target detection and discrimination, only the horizontal beam width is quite narrow, typical values being between about 0°65 to 2°0. The vertical beam width is relatively broad, typical values being between about 15° to 30°.

The beam width is dependent upon the frequency or wavelength of the transmitted energy, antenna design, and the dimensions of the antenna.

For a given antenna size (antenna aperture), narrower beam widths are obtained when using shorter wavelengths. For a given wavelength, narrower beam widths are obtained when using larger antennas.

The slotted waveguide antenna has largely eliminated the side-lobe problem.

With radar waves being propagated in the vicinity of the surface of the sea, the main lobe of the radar beam, as a whole, is composed of a number of separate lobes as opposed to the single lobe-shaped pattern of radiation as emitted in free space. This phenomenon is the result of interference between radar waves directly transmitted and those waves which are reflected from the surface of the sea. The vertical beam widths of navigational radars are such that during normal transmission, radar waves will strike the surface of the sea at points from near the antenna to the radar horizon, depending upon antenna height and vertical beam width. The indirect waves (fig. 4203c) reflected from the surface of the sea may, on rejoining the direct waves, either reinforce or cancel the direct waves depending upon whether they are in phase or out of phase with the direct waves, respectively. Where the direct and indirect waves are exactly in phase, i.e., the crests and troughs of the waves coincide, hyperbolic lines of maximum radiation known as *lines of maxima* are produced. Where the direct and indirect waves are exactly of opposite phase, i.e., the trough of one wave coincides with the crest of the other wave, hyperbolic lines of minimum radiation known as *lines of minima* are produced. Along directions away from the antenna, the direct and indirect waves will gradually come into and pass out of phase, producing lobes of useful radiation separated by regions within which, for practical purposes, there is no useful radiation. Except for the fact that the phase of the indirect wave is reversed on being reflected from the surface of the sea, points on the lines of minima would correspond to differences in the lengths of the paths of the direct and indirect waves of an odd number of half wavelengths. Because of the reversal of the phase of the indirect wave on being reflected from the surface of the sea, points on the lines of minima correspond to differences in the lengths of the paths of the direct and indirect waves of an even number of half wavelengths.

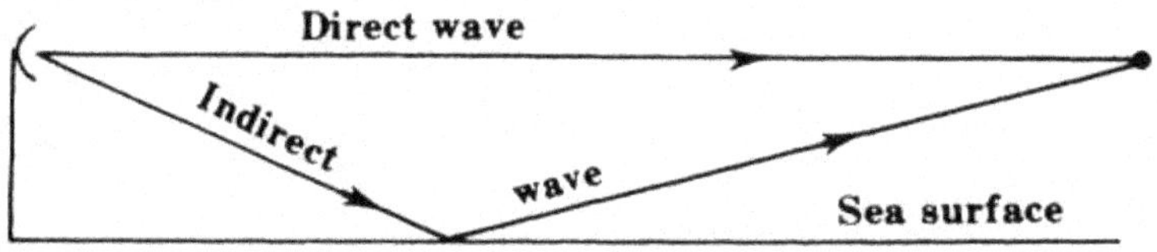

FIGURE 4203c.—Direct and indirect waves.

Figure 4203d shows the lower region of the interference pattern of a representative navigational radar. Since the first line of minima is at the surface of the sea, the first region of minimum radiation or energy is adjacent to the sea's surface.

From this figure it should be obvious that if energy is to be reflected from a target, the target must extend somewhat above the radar horizon, the amount of extension being dependent upon the reflecting properties of the target.

A **vertical-plane coverage diagram** (fig. 4203d) is used by radar designers and analysts to predict regions in which targets will and will not be detected.

Of course, on the small page of a book it would be impossible to illustrate the coverage of a radar beam to scale with antenna height being in feet and the lengths of the various lobes of the interference pattern being in miles. In providing greater clarity of the presentation of the lobes, non-linear graduations of the arc of the vertical beam width are used.

The lengths of the various lobes shown in figures 4203d and 4203e should be given no special significance with respect to the range capabilities of a particular radar set. As with other coverage diagrams, the lobes are drawn to connect points

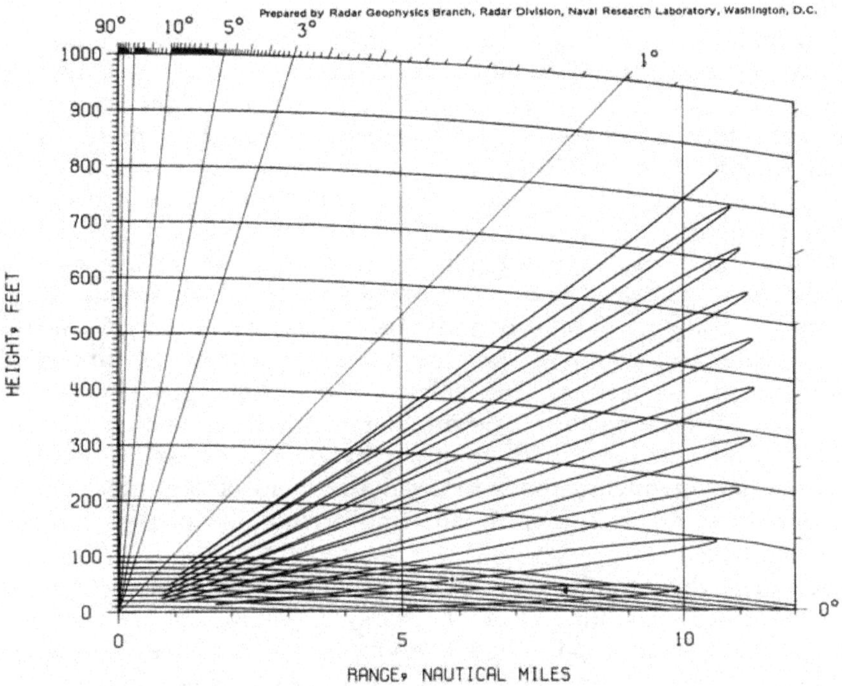

FIGURE 4203d.—Vertical-plane coverage diagram (3050 MHz, antenna height 125 feet, wave height 4 feet).

of equal field intensities. Longer and broader lobes may be drawn connecting points of equal, but lesser, field intensities.

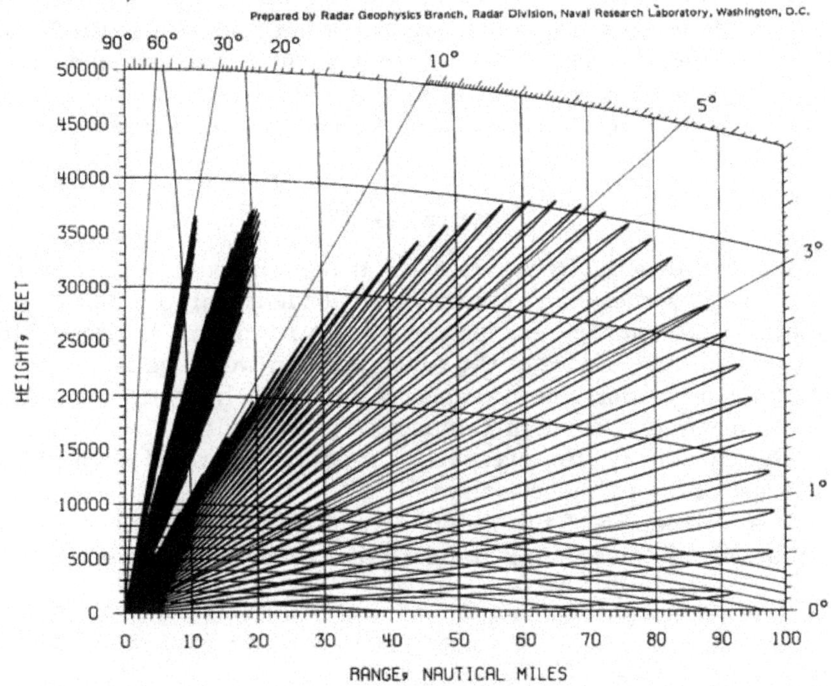

FIGURE 4203e.—Vertical-plane coverage diagram (1000 MHz, vertical beam width 10°, antenna height 80 feet, wave height 0 feet).

The vertical plane coverage diagram as shown in figure 4203e, while not representative of navigational radars, does indicate that at the lower frequencies the interference pattern is coarser than the patterns for higher frequencies. This particular diagram was constructed with the assumption that the free space useful range of the radar beam was 50 nautical miles. From this diagram it is seen that the ranges of the useful lobes are extended to considerably greater distance because of the reinforcement of the direct radar waves by the indirect radar waves. Also, the elevation of the lowest lobe is higher than it would be for a higher frequency. Figure 4203e also illustrates the vertical view of the undesirable side lobes associated with practical antenna design. In examining these radiation coverage diagrams, the reader should keep in mind that the radiation pattern is three-dimensional.

Antenna height as well as frequency or wavelength governs the number of lobes in the interference pattern. The number of the lobes and the fineness of the interference pattern increase with antenna height. Increased antenna height as well as increases in frequency tends to lower the lobes of the interference pattern.

The pitch and roll of the ship radiating do not affect the structure of the interference pattern.

Diffraction is the bending of a wave as it passes an obstruction. Because of diffraction there is some illumination of the region behind an obstruction or target by the radar beam. Diffraction effects are greater at the lower frequencies. Thus, the radar beam of a lower frequency radar tends to illuminate more of the shadow region behind an obstruction than the beam of a radar of higher frequency or shorter wavelength.

Attenuation is the scattering and absorption of the energy in the radar beam as it passes through the atmosphere. It causes a decrease in echo strength. Attenuation is greater at the higher frequencies or shorter wavelengths.

While reflected echoes are much weaker than the transmitted pulses, the characteristics of their return to the source are similar to the characteristics of propagation. The strengths of these echoes are dependent upon the amount of transmitted energy striking the targets and the size and reflecting properties of the targets.

Refraction

If the radar waves actually traveled in straight lines or rays, the distance to the horizon grazed by these rays would be dependent only on the height of the antenna, assuming adequate power for the rays to reach this horizon. Without the effects of refraction, the distance to the **radar horizon** would be the same as that of the geometrical horizon for the antenna height. Like light rays, radar rays are subject to bending or refraction in the atmosphere resulting from travel through regions of different density. However, radar rays are refracted slightly more than light rays because of the frequencies used.

Where h is the height of the antenna in feet, the distance, d, to the radar horizon in nautical miles, assuming standard atmospheric conditions, may be found as follows:

$$d = 1.22\sqrt{h}.$$

With the distances to the geometrical and optical horizons being $1.06\sqrt{h}$ and $1.15\sqrt{h}$, respectively, the radar horizon exceeds the geometrical horizon by 15 percent and the optical horizon by 6 percent. Thus, like light rays in the standard atmosphere, radar rays are bent or refracted slightly downwards approximating the curvature of the earth.

The distance to the radar horizon does not in itself limit the distance from which echoes may be received from targets. Assuming that adequate power is transmitted, echoes may be received from targets beyond the radar horizon if their reflecting surfaces extend above it. Note that the distance to the radar horizon is the distance at which the radar rays graze the surface of the earth.

In the preceding discussion, standard atmospheric conditions were assumed. The *standard atmosphere* is a hypothetical vertical distribution of atmospheric temperature, pressure, and density which is taken to be representative of the atmosphere for various purposes. While the atmospheric conditions at any one locality during a given season may differ considerably from standard atmospheric conditions, the slightly downward bending of the light and radar rays may be described as the typical case.

Although the formula for the distance to the radar horizon ($d=1.22\sqrt{h}$) is based upon a wavelength of 3 centimeters, this formula may be used in the computation of the distance to the radar horizon for other wavelengths used with navigational radar. The value so determined should be considered only as an approximate value because the mariner generally has no means of knowing what are the actual refraction conditions.

In calm weather with no turbulence when there is an upper layer of warm, dry air over a surface layer of cold, moist air, a condition known as **super-refraction** may occur. The effect of super-refraction is to increase the downward bending of the radar rays and thus increase the ranges at which targets may be detected. Super-refraction occurs often in the tropics when a warm land breeze blows over cooler ocean currents.

If a layer of cold, moist air overrides a shallow layer of warm, dry air, a condition known as **sub-refraction** may occur. The effect of sub-refraction is to bend the radar rays upward and thus decrease the maximum ranges at which targets may be detected.

Sub-refraction also affects minimum ranges and may result in failure to detect low lying targets at short range. This condition may occur in polar regions when cold airmasses move over warm ocean currents.

Most radar operators are aware that at certain times they are able to detect targets at extremely long ranges, but at other times they cannot detect targets within visual ranges, even though their radars may be in top operating condition in both instances. These phenomena occur during extreme cases of super-refraction. Energy radiated at angles of 1° or less may be trapped in a layer of the atmosphere called a **surface radio duct.** In the surface radio duct, the radar rays are refracted downward to the surface of the sea, reflected upward, refracted downward again within the duct, and so on continuously. The energy trapped by the duct suffers little loss; thus, targets may be detected at exceptionally long ranges. Surface targets have been detected at ranges in excess of 1,400 miles with relatively low-powered equipment. There is a great loss in the energy of the rays escaping the duct, thus reducing the chances for detection of targets above the duct.

Ducting sometimes reduces the effective radar range. If the antenna is below a duct, it is improbable that targets above the duct will be detected. In instances of extremely low-level ducts when the antenna is above the duct, surface targets lying below the duct may not be detected. The latter situation does not occur very often.

Although ducting conditions can happen any place in the world, the climate and weather in some areas make their occurrence more likely. In some parts of the world, particularly those having a monsoonal climate, variation in the degree of ducting is mainly seasonal, and great changes from day to day may not take place. In other parts of the world, especially those in which low barometric pressure areas recur often, the extent of nonstandard propagation conditions varies considerably from day to day.

4204. Minimum and maximum range.—Since the receiver is disconnected during transmission of a signal, no echo can be received during this period. As a result, there is a minimum range at which objects can be detected. The shortest pulses are about 0.05 microsecond in duration, or approximately 49 feet long. Since the time measurement is of the *round trip* as the signal travels to the target and the echo returns, the range is *half* the distance corresponding to the measured time interval. Therefore, a minimum range of about 25 feet is theoretically possible with a pulse of 0.05 microsecond. However, the *practical* minimum range is somewhat greater because of receiver recovery time, sea return, side-lobe echoes, and the vertical beam width.

1. *Recovery time* of the receiver extends the minimum range at which a target can be detected beyond the range determined by the pulse length.

2. *Sea return* or echoes received from waves may clutter the indicator within and beyond the minimum range established by the pulse length and recovery time.

3. When operating near land or large targets, *side-lobe echoes* (echoes from targets detected in the side-lobes of the antenna beam pattern) may clutter the indicator and prevent detection of close targets, without regard to the direction in which the antenna is trained.

4. Small surface targets may escape detection when close due to the limits of effective radiation established by the *vertical beam width.*

The *maximum range* is limited by the frequency, peak power, pulse length, pulse repetition rate, beam width, receiver sensitivity, antenna rotation rate, curvature of the earth, target characteristics, and weather.

1. Lower frequencies (longer wavelengths) generally provide longer detection ranges because the attenuation of the higher frequencies is greater, regardless of weather.

2. The range capability increases with increase in peak power. Doubling the peak power increases the range capability by about 25 percent.

3. As the pulse length is increased, the range capability also increases due to the greater amount of energy transmitted.

4. The pulse repetition rate (PRR) determines the maximum *measurable* range of the radar. Ample time must be allowed between pulses for an echo to return from any target located within the maximum workable range of the system. Otherwise, echoes returning from the more distant targets are blocked by succeeding transmitted pulses. This necessary time interval determines the highest PRR that can be used. The PRR must be high enough, however, that sufficient pulses hit the target, and enough echoes are returned to the radar.

5. The detection range increases as the radar beam becomes more concentrated.

6. The more sensitive receivers provide greater detection ranges.

7. The detection range increases as the antenna rotation rate decreases. For a radar set having a PRR of 1000 pulses per second, a horizontal beam width of 2°.0, and an antenna rotation rate of 6 RPM (1 revolution in 10 seconds or 36 scanning degrees per second), there is one pulse transmitted each 0°.036 of rotation. There are

56 pulses transmitted during the time required for the antenna to rotate through its beam width:

$$\frac{\text{Beam Width}}{\text{Degrees per Pulse}} = \frac{2^{\circ}\!.0}{0^{\circ}\!.036} = 56 \text{ pulses.}$$

With an antenna rotation rate of 15 RPM (1 revolution in 4 seconds or 90 scanning degrees per second), there is only one pulse transmitted each 0°.090 of rotation. There are only 22 pulses transmitted during the time required for the antenna to rotate through its beam width:

$$\frac{\text{Beam Width}}{\text{Degrees per Pulse}} = \frac{2^{\circ}\!.0}{0^{\circ}\!.090} = 22 \text{ pulses.}$$

From the foregoing it is apparent that at the higher antenna rotation rates, the maximum ranges at which targets, particularly small targets, may be detected are reduced.

8. Since radar operates in the higher frequencies that are essentially line-of-sight, the maximum range is limited by the curvature of the earth. The radar horizon, at which rays from the transmitting antenna graze the surface of the earth, is at a distance about 15 percent greater than that of the visible horizon (tab. 8). Under conditions of abnormal refraction (art. 4203), both visible and radar horizons may be extended to greater distances. Due to curvature of the earth, maximum range increases as either antenna or target height increases.

9. There are several target characteristics which will enable one target to be detected at a greater range than another, or for one target to produce a stronger echo than another target of similar size:

Height. Since radar wave propagation is almost line-of-sight, the height of the target is of prime importance. If the target does not rise above the radar horizon, the radar beam cannot be reflected from the target. Because of the interference pattern in the radar beam, the target must rise somewhat above the radar horizon.

Size. Up to certain limits, targets having larger reflecting areas return stronger echoes than targets having smaller reflecting areas. Should a target be wider than the horizontal beam width, the strength of the echoes is not increased on account of the greater width of the target because the area not exposed to the radar beam at any instant cannot, of course, reflect an echo. Since the vertical dimensions of most targets are small compared to the vertical beam width of marine navigational radars, the beam width limitation is not normally applicable to the vertical dimensions. However, there is a vertical dimension limitation in the case of sloping surfaces or stepped surfaces. In this case, only the projected vertical area lying within the distance equivalent of the pulse length can return echoes at any instant.

Aspect of a target is its orientation to the axis of the radar beam. With change in aspect, the effective reflecting area may change, depending upon the shape of the target. The nearer the angle between the reflecting area and the beam axis is 90°, the greater is the strength of the echo returned to the antenna.

Shape. Targets of identical shape may give echoes of varying strength, depending on aspect. Thus a flat surface at right angles to the radar beam, such as the side of a steel ship or a steep cliff along the shore, reflects very strong echoes. As the aspect changes, this flat surface tends to reflect more of the energy of the beam away from the antenna, and may given rather weak echoes. A concave surface tends to focus the radar beam back to the antenna while a convex surface tends to scatter the energy. A smooth conical surface does not reflect energy back to the antenna. However, echoes may be reflected to the antenna if the conical surface is rough.

Texture of the target may modify the effects of shape and aspect. A smooth texture tends to increase the reflection qualities, but unless the aspect and shape of the target are such that the reflection is focused directly back to the antenna, the smooth surface will give a poor radar echo because most of the energy is reflected in another direction. On the other hand, a rough surface tends to break up the reflection and improves the strength of echoes returned from those targets whose shape and aspect normally give weak echoes.

Composition. The ability of various substances to reflect radar pulses depends on the intrinsic electrical properties of those substances. Thus metal and water are good reflectors. Ice is a fair reflector, depending on aspect. Land areas vary in their reflection qualities depending on the amount and type of vegetation and the rock and mineral content. Wood and fiber glass boats are poor reflectors. It must be remembered that all of the characteristics interact with each other to determine the strength of the radar echo, and no factor can be singled out without considering the effects of the others.

10. The usual effects of weather are to reduce the ranges at which targets can be detected and to produce unwanted echoes on the radarscope which may obscure the returns from important targets or from targets which may be dangerous to one's ship.

The wind produces waves which reflect unwanted echoes (art. 4206) which appear on the PPI as *sea return.* Sea return is normally greater in the direction from which the wind and seas are coming. Rain, hail, and snow storms all may return echoes which appear on the PPI as a blurred or cluttered area. In addition to masking targets which are within the storm area, heavy precipitation may absorb some of the strength of the pulse and decrease maximum detection range.

4205. Radarscope interpretation.—Radar may serve the navigator as a very valuable tool if its characteristics and limitations are understood. Although determining position through observation of the range and bearing of a charted, isolated, and well defined object having good reflecting properties is relatively simple, this task still requires that the navigator have an understanding of the characteristics and limitations of his radar. The more general task of using radar in observing a shoreline where the radar targets are not so obvious or well defined requires considerable expertise which may be gained only through an adequate understanding of the characteristics and limitations of the radar being used.

Although the plan position indicator does provide a chartlike presentation when a landmass is being scanned, the image painted by the sweep is not a true representation of the shoreline. But with practice, one can acquire considerable skill in interpreting the signals appearing on the radarscope. Some of the factors to be kept in mind in interpretation are the following:

Resolution in range. In part *A* of figure 4205a, a transmitted pulse has arrived at the second of two targets of insufficient size or density to absorb or reflect all of the energy of the pulse. While the pulse has traveled from the first to the second target, the echo from the first has traveled an equal distance in the opposite direction. At *B* the transmitted pulse has continued on beyond the second target, and the two echoes are returning toward the transmitter. The distance between leading edges of the two echoes is *twice* the distance between targets. The correct distance will be shown on the scope, which is calibrated to show *half* the distance traveled out and back. At *C* the targets are closer together and the pulse length has been increased. The two echoes merge, and on the scope will appear as a single, large target. At *D* the pulse length has been decreased, and the two echoes appear separated. The ability of a radar to separate targets close together on the same bearing is called **resolution in range.** It is related primarily to pulse length, the minimum distance between targets that can be distinguished as separate ones being *half* the pulse length. This (half the pulse length) is the apparent depth or thickness of a target presenting a flat perpendicular surface to the radar beam. Thus, several ships close together may appear as an island. Echoes from a number of small boats, piles, breakers, or even large ships close to the shore may blend with echoes from the shore, resulting in an incorrect indication of the position and shape of the shoreline.

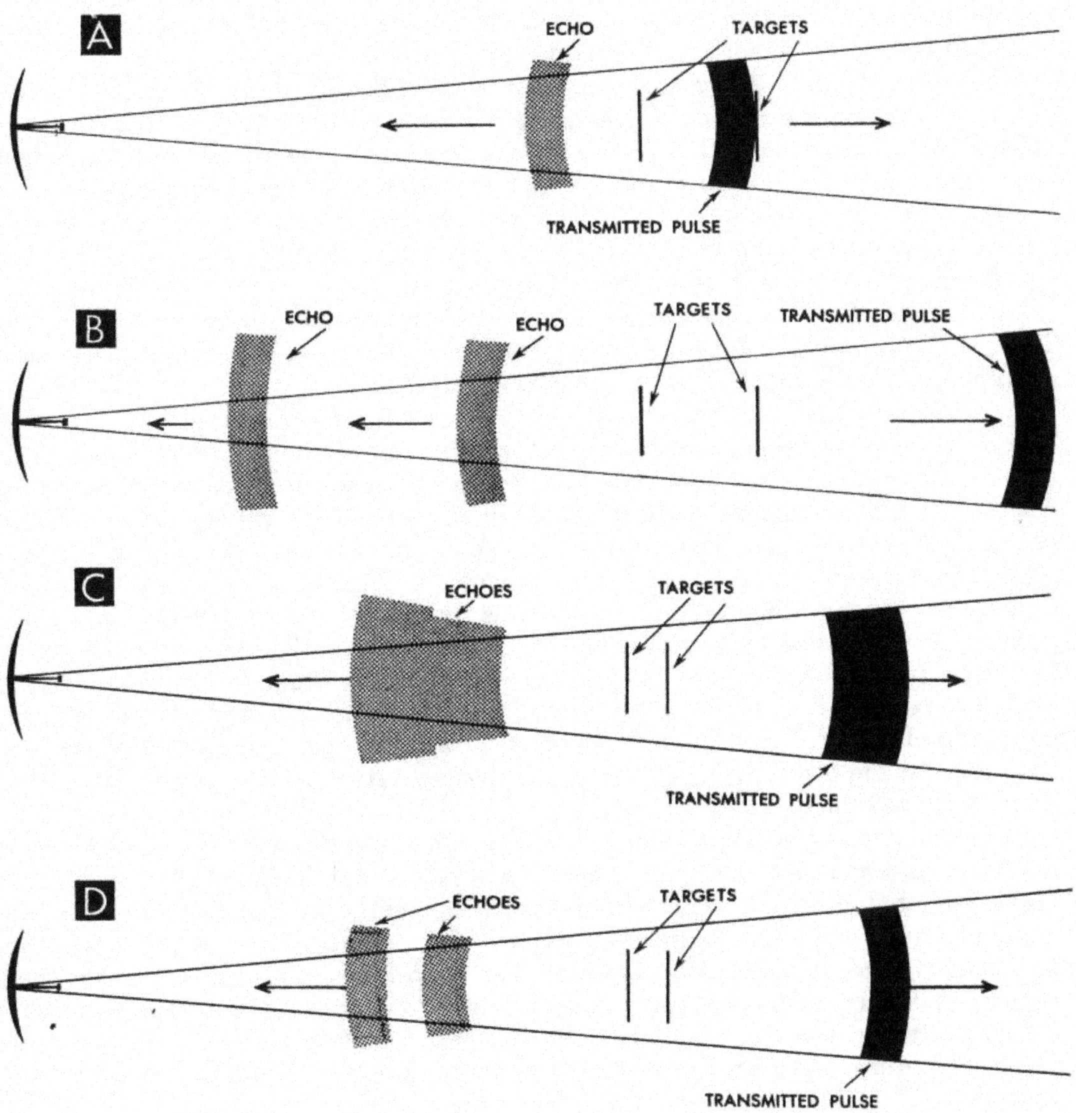

FIGURE 4205a.—Resolution in range.

Resolution in bearing is similar to that in range. A pulse proceeds outward along a narrow sector. As the beam rotates, energy is returned during the entire time that a target is "illuminated," the same as with a searchlight. A vertical target such as a mast is "seen" over the arc in which there is sufficient illumination to render it visible. On a radar PPI a target appears widened by an amount equal to the effective **beam width,** half the effective beam width being added to each side. Thus, the echoes from two or more targets close together at the same range may merge to form a single, wider echo. The ability to separate such targets is called **resolution in bearing.** In angular units it is dependent primarily upon beam width, a narrower beam having a higher resolution. In terms of distance between targets, range is also important, resolution increasing as range decreases.

Height of antenna and target. If the radar horizon (art. 4203) is between the transmitting vessel and the target, the lower part of the target will not be visible. A large vessel may appear as a small craft, a shoreline may appear at some distance inland. Areas within shadows may not be visible at all.

Reflecting quality of target. Echoes from several targets of the same size may be quite different in appearance. A metal surface is a better reflector of radio waves than a wooden surface. A surface perpendicular to the beam returns a stronger echo than a nonperpendicular one. For this reason, a gently sloping beach may not be visible. A vessel encountered broadside returns a stronger echo than one heading toward or away from the radar vessel. In some instances, the strength of an echo can be increased by means of a corner reflector (art. 4207).

Frequency. As the frequency is increased, reflections occur from smaller targets. Thus, a 10-centimeter radar generally penetrates fog, rain, snow, etc., while a 3-centimeter radar receives returns from such obstacles.

Radarscope interpretation is complicated somewhat by the presence of unwanted signals from atmospheric noise, sea return, precipitation, etc. Collectively, this is called **clutter.** Generally, it is strongest near the vessel and gradually decreases with increased range, because of reduced sea return. Strong echoes can sometimes be detected by reducing the volume or "gain" of the receiver (not the image intensity of the indicator), so that weaker signals will not appear. Even when the amplitude of the clutter is about the same as that of desired signals, the latter can sometimes be detected by watching the scope during several rotations of the antenna. At each rotation the signals from targets remain at about the same place, and of about the same magnitude, while those from waves, noise, etc., fluctuate, appearing different on each revolution. Floating ice or a small boat may not be detected at any range if the waves are high. A rough surface returns a stronger echo than a smooth surface.

Sometimes a signal appears on a radar screen when there is no visible object at the point indicated, and no apparent source of the signal. This is called a **ghost.** It may be due to faulty operation of the radar set or to an actual echo returned from a discontinuity in the atmosphere. Sometimes such discontinuities reflect light, also, producing images or apparent images similar to mirages and of seeming apparent reality. A similar condition occasionally occurs in the sea. This phenomenon is undoubtedly the basis of many reports of strange objects sighted visually or by radar. Sometimes such apparent objects exhibit incredible speed or maneuverability.

The major problem is that of determining which features in the vicinity of the shoreline are actually reflecting the echoes "painted" on the radarscope. Particularly in cases where a low lying shore is being scanned, there may be considerable uncertainty.

An associated problem is the fact that certain features on the shore will not return echoes, even if they have good reflecting properties, simply because they are blocked from the radar beam by other physical features or obstructions. This factor in turn causes the chartlike image painted on the scope to differ from the chart of the area.

If the navigator is to be able to interpret the chartlike presentation on his radarscope, he must have at least an elementary understanding of the characteristics of radar propagation, the characteristics of his radar set, the reflecting properties of different types of radar targets, and the ability to analyze his chart to make an estimate of just which charted features are most likely to reflect the transmitted pulses or to be blocked from the radar beam. Although contour lines on the chart topography aid the navigator materially in the latter task, experience gained during clear weather comparison of the visual cross-bearing plot and the radarscope presentation is invaluable.

Observing experience gained during good weather transits of confined waters can be invaluable during later inclement weather transits, particularly if the navigator gives close attention to the identification of those features the reliable identification of which would be of material aid in effecting a safe transit.

On relative and true motion displays, landmasses are readily recognizable because of the generally steady brilliance of the relatively large areas painted on the PPI. Also land should be at positions expected from knowledge of the ship's navigational position. On relative motion displays, landmasses move in directions and at rates opposite and equal to the actual motion of the observer's ship. Individual pips do not move relative to one another. On true motion displays, landmasses do not move on the PPI if there is accurate compensation for set and drift. Without such compensation, i.e., when the true motion display is sea-stabilized, only slight movements of landmasses may be detected on the PPI.

Although landmasses are readily recognizable, the primary problem is the identification of specific features so that such features can be used for fixing the position of the observer's ship. Identification of specific features can be quite difficult because of various factors, including distortion resulting from beam width and pulse length, and uncertainty as to just which charted features are reflecting the echoes. The following hints may be used as an aid in identification:

1. *Sandspits* and *smooth, clear beaches* normally do not appear on the PPI at ranges beyond 1 or 2 miles because these targets have almost no area that can reflect energy back to the radar. Ranges determined from these targets are not reliable. If waves are breaking over a sandbar, echoes may be returned from the surf. Waves may, however, break well out from the actual shoreline, so that ranging on the surf may be misleading when a radar position is being determined relative to the shoreline.

2. *Mud flats* and *marshes* normally reflect radar pulses only a little better than a sandspit. The weak echoes received at low tide disappear at high tide. *Mangroves* and other *thick growth* may produce a strong echo. Areas that are indicated as swamps on a chart, therefore, may return either strong or weak echoes, depending on the density and size of the vegetation growing in the area.

3. When *sand dunes* are covered with vegetation and are well back from a low, smooth beach, the apparent shoreline determined by radar appears as the line of the dunes rather than the true shoreline. Under some conditions, sand dunes may return strong echo signals because the combination of the vertical surface of the vegetation and the horizontal beach may form a sort of corner reflector.

4. *Lagoons* and *inland lakes* usually appear as blank areas on a PPI because the smooth water surface returns no energy to the radar antenna. In some instances, the sandbar or reef surrounding the lagoon may not appear on the PPI because it lies too low in the water.

5. *Coral atolls* and *long chains of islands* may produce long lines of echoes when the radar beam is directed perpendicular to the line of the islands. This indication is especially true when the islands are closely spaced. The reason is that the spreading resulting from the width of the radar beam causes the echoes to blend into continuous lines. When the chain of islands is viewed lengthwise, or obliquely, however, each island may produce a separate pip. *Surf breaking on a reef* around an atoll produces a ragged, variable line of echoes.

6. One or two *rocks* projecting above the surface of the water, or waves breaking over a reef, may appear on the PPI. When an object is submerged entirely and the sea is smooth over it, no indication is seen on the PPI.

7. *If the land rises in a gradual, regular manner from the shoreline,* no part of the terrain produces an echo that is stronger than the echo from any other part. As a result, a general haze of echoes appears on the PPI, and it is difficult to ascertain the range to any particular part of the land.

Land can be recognized by plotting the contact. Care must be exercised when plotting because, as a ship approaches or goes away from a shore behind which the land rises gradually, a plot of the ranges and bearings to the land may show an "apparent" course and speed.

8. Blotchy signals are returned from *hilly ground* because the crest of each hill returns a good echo although the valley beyond is in a shadow. If high receiver gain is used, the pattern may become solid except for the very deep shadows.

9. *Low islands* ordinarily produce small echoes. When thick *palm trees* or other foliage grow on the island, strong echoes often are produced because the horizontal surface of the water around the island forms a sort of corner reflector with the vertical surfaces of the trees. As a result, *wooded islands* give good echoes and can be detected at a much greater range than barren islands.

With the appearance of a small pip on the PPI, *its identification as a ship* can be aided by a process of elimination. A check of the navigational position can overrule the possibility of land. The size of the pip can be used to overrule the possibility of land or precipitation, both usually having a massive appearance on the PPI. The rate of movement of the pip on the PPI can overrule the possibility of aircraft.

Having eliminated the foregoing possibilities, the appearance of the pip on the PPI at a medium range as a bright, steady, and clearly defined image indicates a high probability that the target is a steel ship.

The pip of a ship target may brighten at times and then slowly decrease in brightness. Normally, the pip of a ship target fades from the PPI only when the range becomes too great.

Although PPI displays are approximately chartlike when landmasses are being scanned by the radar beam, there may be sizable areas missing from the display because of certain features being blocked from the radar beam by other features. A shoreline which is continuous on the PPI display when the ship is at one position may not be continuous when the ship is at another position and scanning the same shoreline. The radar beam may be blocked from a segment of this shoreline by an obstruction such as a promontory. An indentation in the shoreline, such as a cove or bay, appearing on the PPI when the ship is at one position may not appear when the ship is at another position nearby. Thus, **radar shadow** alone can cause considerable differences between the PPI display and the chart presentation. This effect in conjunction with beam width and pulse length distortion of the PPI display can cause even greater differences.

The pips of ships, rocks, and other targets close to shore may merge with the shoreline image on the PPI. This merging is due to the distortion effects of horizontal beam width and pulse length. Target images on the PPI always are distorted angularly by an amount equal to the effective horizontal beam width. Also, the target images always are distorted radially by an amount at least equal to one-half the pulse length (164 yards per microsecond of pulse length).

Figure 4205b illustrates the effects of ship's position, beam width, and pulse length on the radar shoreline. Because of beam width distortion, a straight, or nearly straight, shoreline often appears crescent-shaped on the PPI. This effect is greater with the wider beam widths. Note that this distortion increases as the angle between the beam axis and the shoreline decreases.

Figure 4205c illustrates the distortion effects of radar shadow, beam width, and pulse length. View A shows the actual shape of the shoreline and the land behind it. Note the steel tower on the low sand beach and the two ships at anchor close to shore. The heavy line in view B represents the shoreline on the PPI. The dotted lines represent the actual position and shape of all targets. Note in particular:

1. The low sand beach is not detected by the radar.
2. The tower on the low beach is detected, but it looks like a ship in a cove. At closer range the land would be detected and the cove-shaped area would begin to fill in; then the tower could not be seen without reducing the receiver gain.
3. The radar shadow behind both mountains. Distortion owing to radar shadows is responsible for more confusion than any other cause. The small island does not appear because it is in the radar shadow.
4. The spreading of the land in bearing caused by beam width distortion. Look at the upper shore of the peninsula. The shoreline distortion is greater to the west because the angle between the radar beam and the shore is smaller as the beam seeks out the more westerly shore.
5. Ship No. 1 appears as a small peninsula. Her pip has merged with the land because of the beam width distortion.
6. Ship No. 2 also merges with the shoreline and forms a bump. This bump is caused by pulse length and beam width distortion. Reducing receiver gain might cause the ship to separate from land, provided the ship is not too close to the shore. The Fast Time Constant (FTC) control could also be used to attempt to separate the ship from land.

4206. Recognition of unwanted echoes.—The navigator must be able to recognize various abnormal echoes and effects on the radarscope so as not to be confused by their presence.

Indirect or **false echoes** are caused by reflection of the main lobe of the radar beam off ship's structures such as stacks and kingposts. When such reflection does occur, the echo will return from a legitimate radar contact to the antenna by the same indirect path. Consequently, the echo will appear on the PPI at the bearing of the reflecting surface. As shown in figure 4206a, the indirect echo will appear on the PPI at the same range as the direct echo received, assuming that the additional distance by the indirect path is negligible.

Characteristics by which indirect echoes may be recognized are summarized as follows:

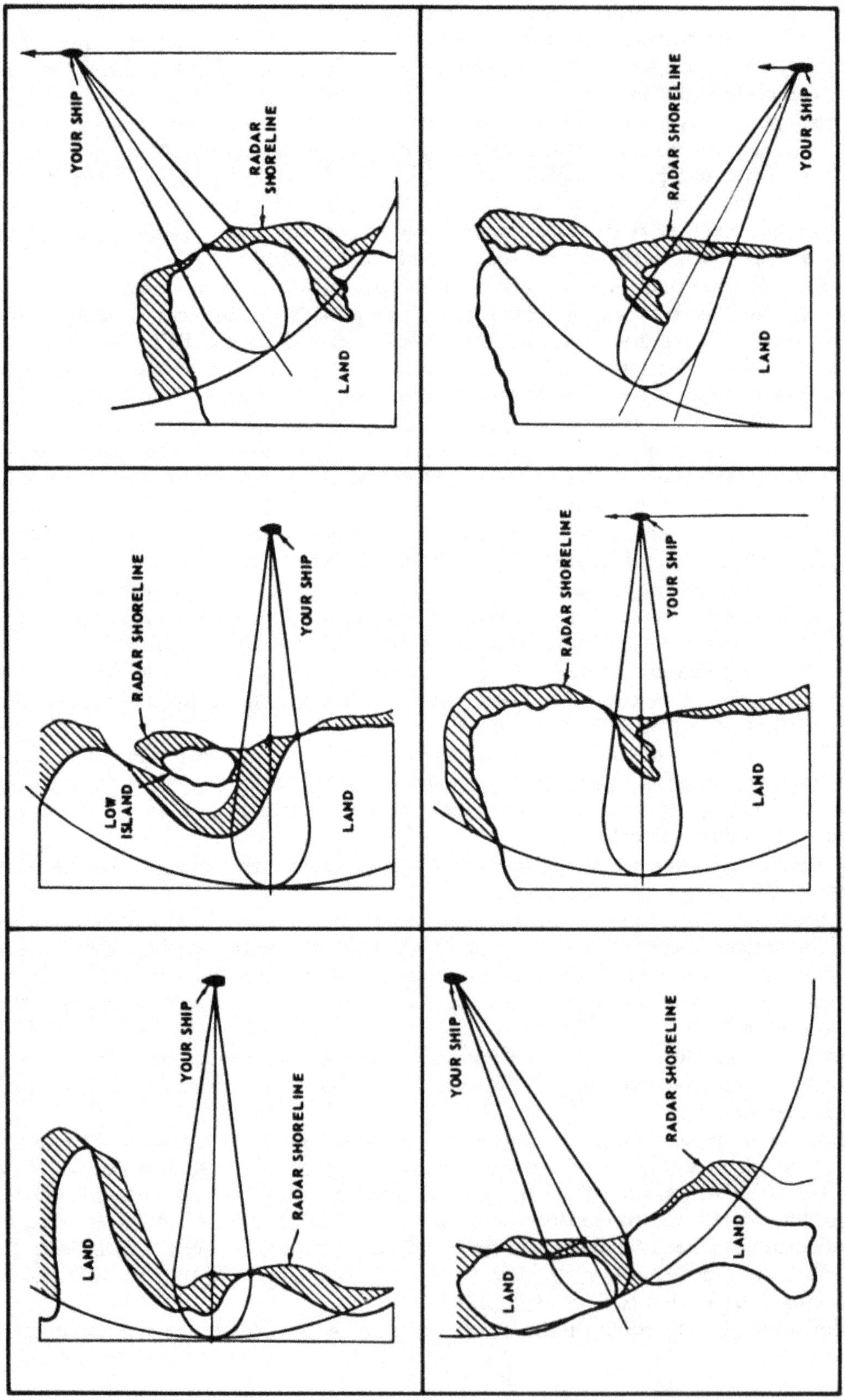

FIGURE 4205b.—Effects of ship's position, beam width, and pulse length on radar shoreline.

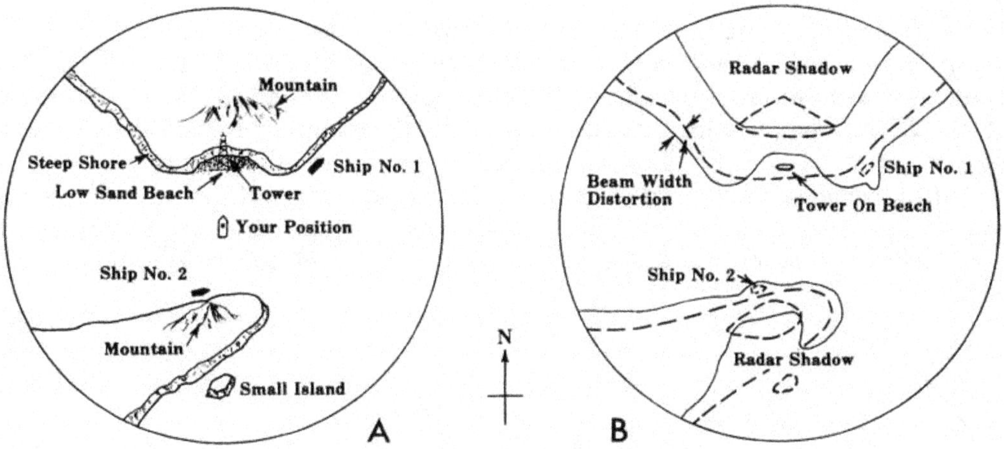

FIGURE 4205c.—Distortion effects of radar shadow, beam width, and pulse length.

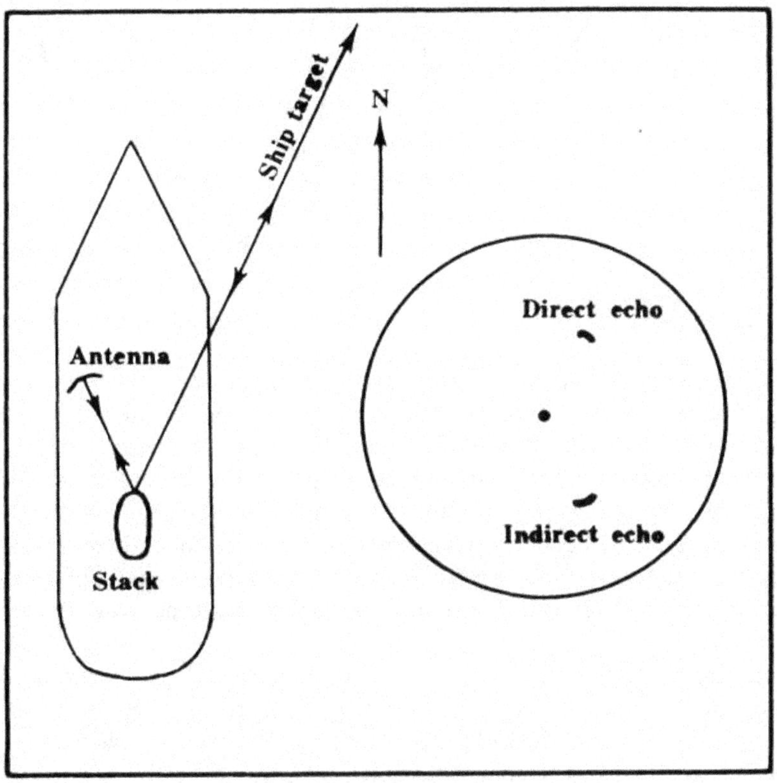

FIGURE 4206a.—Indirect echo.

1. The indirect echoes will usually occur in shadow sectors.
2. They are received on substantially constant bearings although the true bearing of the radar contact may change appreciably.
3. They appear at the same ranges as the corresponding direct echoes.
4. When plotted, their movements are abnormal, usually.
5. Their shapes may indicate that they are not direct echoes.

Side-lobe effects are readily recognized in that they produce a series of echoes (fig. 4206b) on each side of the main lobe echo at the same range as the latter.

Semicircles or even complete circles may be produced. Because of the low energy of the side-lobes, these effects will normally occur only at the shorter ranges. The effects may be minimized or eliminated through use of the gain and anti-clutter controls. Slotted wave guide antennas have largely eliminated the side-lobe problem.

Multiple echoes may occur when a strong echo is received from another ship at close range. A second or third or more echoes may be observed on the radarscope at double, triple, or other multiples of the actual range of the radar contact (fig. 4206c).

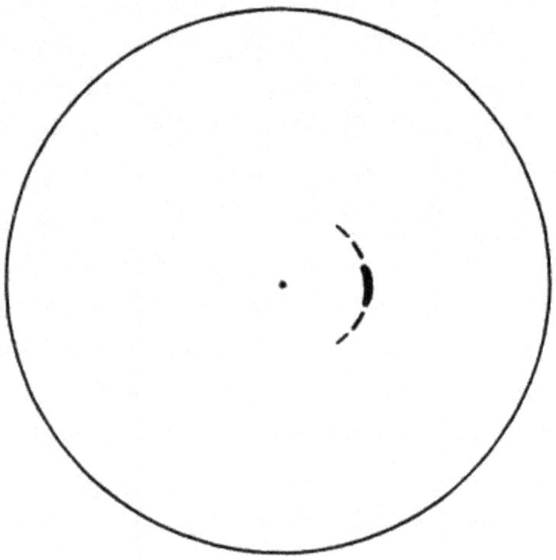

FIGURE 4206b.—Side-lobe effects.

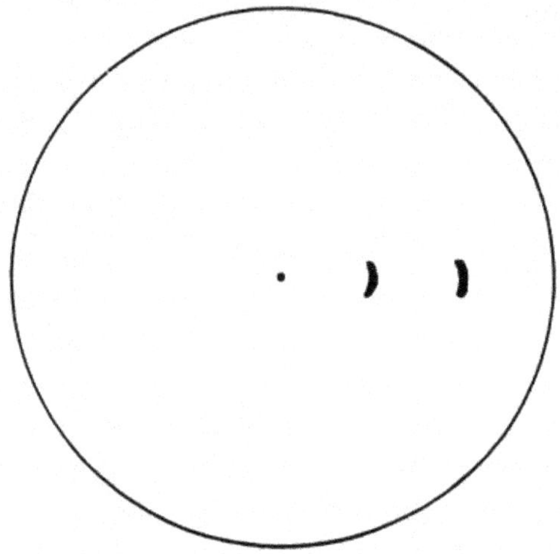

FIGURE 4206c.—Multiple echoes.

Second-trace echoes (multiple-trace echoes) are echoes received from a contact at an actual range greater than the radar range setting. If an echo from a distant target is received after the following pulse has been transmitted, the echo will appear on the radarscope at the correct bearing but not at the true range. Second-trace echoes are unusual except under abnormal atmospheric conditions, or conditions under which super-refraction is present. Second-trace echoes may be recognized through changes in their positions on the radarscope on changing the pulse repetition rate (PRR); their hazy, streaky, or distorted shape; and the erratic movements on plotting.

As illustrated in figure 4206d, a target pip is detected on a true bearing of 090° at a distance of 7.5 miles. On changing the PRR from 2000 to 1800 pulses per second, the same target is detected on a bearing of 090° at a distance of 3 miles (fig. 4206e). The change in the position of the pip indicates that the pip is a second-trace echo. The actual distance of the target is the distance as indicated on the PPI plus half the distance the radar wave travels between pulses.

Electronic interference effects, such as may occur when in the vicinity of another radar operating in the same frequency band as that of the observer's ship, is usually seen on the PPI as a large number of bright dots either scattered at random or in the form of dotted lines extending from the center to the edge of the PPI.

Interference effects are greater at the longer radar range scale settings. The interference effects can be distinguished easily from normal echoes because they do not appear in the same places on successive rotations of the antenna.

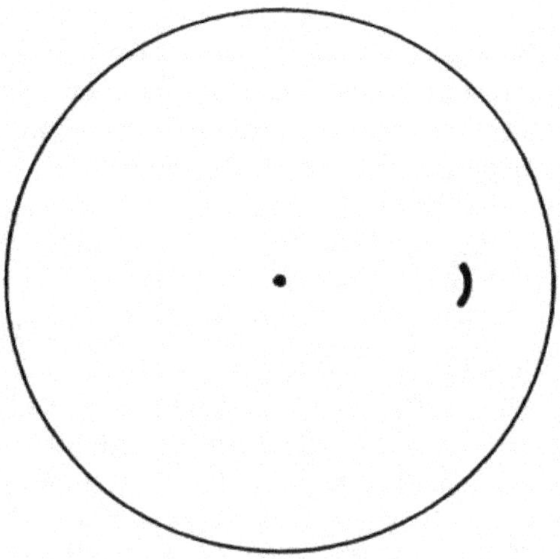

FIGURE 4206d.—Second-trace echo on 12-mile range scale.

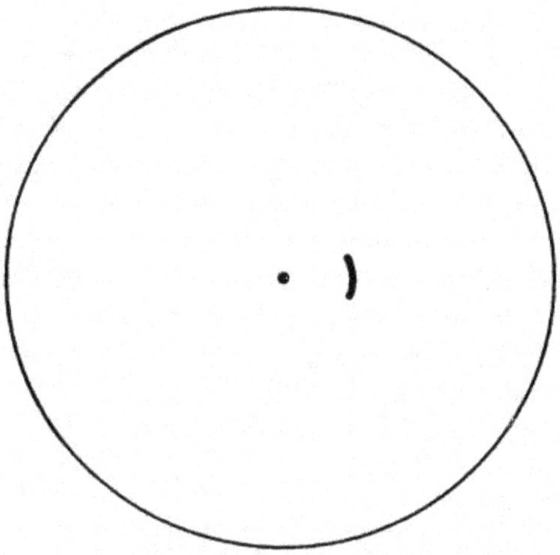

FIGURE 4206e.—Position of second-trace echo on 12-mile range scale after changing PRR.

Stacks, masts, samson posts, and other structures may cause a reduction in the intensity of the radar beam beyond these obstructions, especially if they are close to the radar antenna. If the angle at the antenna subtended by the obstruction is more than a few degrees, the reduction of the intensity of the radar beam beyond the obstruction may be such that a **blind sector** is produced. With lesser reduction in the intensity of the beam beyond the obstructions, **shadow sectors** can be produced. Within these shadow sectors, small targets at close range may not be detected while larger targets at much greater ranges may be detected.

Spoking appears on the PPI as a number of spokes or radial lines. Spoking is easily distinguished from interference effects because the lines are straight on all range-scale settings and are lines rather than a series of dots.

The spokes may appear all around the PPI, or they may be confined to a sector. Should the spoking be confined to a narrow sector, the effect can be distinguished from a Ramark signal (art. 4207) of similar appearance through observation of the steady relative bearing of the spoke in a situation where the bearing of the Ramark signal should change.

The appearance of spoking is indicative of need for equipment maintenance.

The PPI display may appear as alternately normal and dark sectors. This **sectoring** phenomenon is usually due to the automatic frequency control being out of adjustment.

The appearance of **serrated range rings** is indicative of need for equipment maintenance.

After the radar set has been turned on, the display may not spread immediately to the whole of the PPI because of static electricity inside the CRT. Usually, the **static electricity effect**, which produces a distorted PPI display, lasts no longer than a few minutes.

Hour-glass effect appears as either a constriction or expansion of the display near the center of the PPI. The expansion effect is similar in appearance to the expanded center display. This effect, which can be caused by a nonlinear time base or the sweep not starting on the indicator at the same instant as the transmission of the pulse, is most apparent when in narrow rivers or close to shore.

The echo from an **overhead power cable** appears on the PPI as a single echo always at right angles to the line of the cable. If this phenomenon is not recognized, the echo can be wrongly identified as the echo from a ship on a steady bearing. Avoiding action results in the echo remaining on a constant bearing and moving to the same side of the channel as the ship altering course. This phenomenon is particularly apparent for the power cable spanning the Straits of Messina.

4207. Aids to radar navigation.—Various aids to radar navigation have been developed to aid the navigator in identifying radar targets and for increasing the strength of the echoes received from objects which otherwise are poor radar targets.

Buoys and small boats, particularly those boats constructed of wood, are poor radar targets. Weak fluctuating echoes received from these targets are easily lost in the sea clutter on the radarscope. To aid in the detection of these targets, radar reflectors, of the corner reflector type, may be used. The corner reflectors may be mounted on the tops of buoys, or the body of the buoy may be shaped as a corner reflector.

Each **corner reflector,** shown in figure 4207a, consists of three mutually perpendicular flat metal surfaces. A radar wave on striking any of the metal surfaces or plates will be reflected back in the direction of its source, i.e., the radar antenna. Maximum energy will be reflected back to the antenna if the axis of the radar beam makes equal angles with all the metal surfaces. Frequently, corner reflectors are assembled in clusters to insure receiving strong echoes at the antenna.

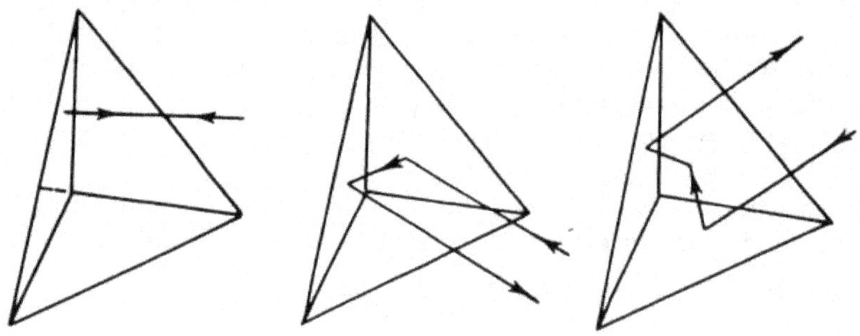

FIGURE 4207a.—Corner reflectors.

Although radar reflectors are used to obtain stronger echoes from radar targets, other means are required for more positive identification of radar targets. **Radar beacons** are transmitters operating in the marine radar frequency band which produce distinctive indications on the radarscopes of ships within range of these beacons. There are two general classes of these beacons: racon which provides both bearing and range information to the target, and ramark which provides bearing information only. However, if the ramark installation is detected as an echo on the radarscope, the range will be available also.

Racon is a radar transponder which emits a characteristic signal when triggered by a ship's radar. The signal may be emitted on the same frequency as that of the triggering radar, in which case it is superimposed on the ship's radar display automatically. The signal may be emitted on a separate frequency, in which case to receive the signal the ship's radar receiver must be capable of being tuned to the beacon frequency or a special receiver must be used. In either case, the PPI will be blank except for the beacon signal. However, the only racons in service are "in band" beacons which transmit in one of the marine radar bands, usually only the 3-centimeter band.

The racon signal appears on the PPI as a radial line originating at a point just beyond the position of the radar beacon or as a Morse code signal (fig. 4207b) displayed radially from just beyond the beacon.

Ramark is a radar beacon which transmits either continuously or at intervals. The latter method of transmission is used so that the PPI can be inspected without any clutter introduced by the ramark signal on the scope. The ramark signal as it appears on the PPI is a radial line from the center. The radial line may be a continuous narrow line, a broken line (fig. 4207c), a series of dots, or a series of dots and dashes.

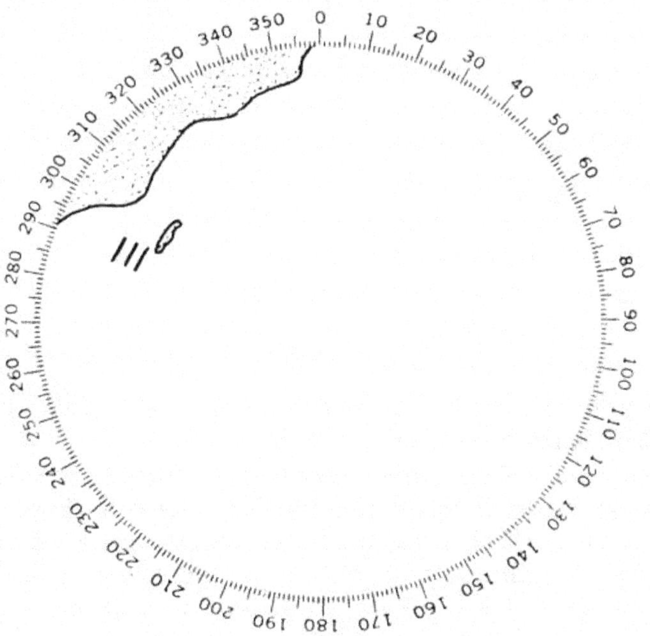

FIGURE 4207b.—Coded racon signal.

4208. Radar navigation.—Radar provides a means of establishing position, or keeping a vessel in safe water during periods of reduced visibility, or at considerable distance from shore, when other methods may not be available. Since both range and bearing can be obtained, a single identifiable object is needed. However, if a visual bearing is available, it should be more reliable than one obtained by radar. Since radar range is usually more accurate than radar bearing, a fix by two or more ranges is generally preferable to one obtained by two bearings or by range and bearing. However, accurate range requires reliable identification of the part of the target returning the echo. This is not always apparent when natural objects are used.

In addition to the usual methods of piloting, radar is adapted to several methods of somewhat limited application. If a single prominent target is available in an operating area, a series of concentric circles and radial lines can be drawn on the chart and suitably labeled. If bearing and distance are measured frequently, an almost continuous fix can be obtained by spotting in the positions by eye. If a polar plot is made on a piece of transparent material to the same scale as the chart, the ranges and bearings of a number of points can be plotted in quick succession, and

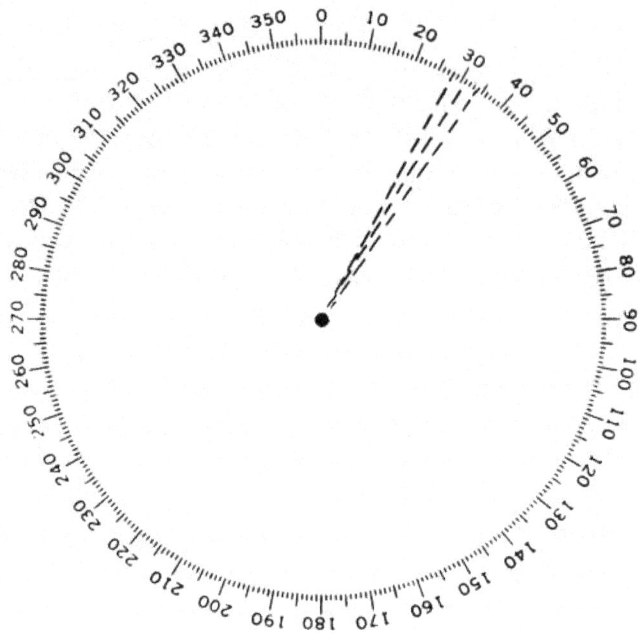

FIGURE 4207c.—Ramark signal appearing as a broken radial line.

the transparent material fitted to the chart by trial and error. The center of the plot is then the position of the radar.

In that the navigator of a radar-equipped vessel always must be prepared to use radar as his primary means of navigation in pilot waters, during the planning for a transit of these waters it behooves him to study the navigational situation with respect to any special techniques which can be employed to enhance the use of radar. The effectiveness of such techniques usually is dependent upon adequate preparation for their use, including special constructions on the chart or the preparation of transparent chart overlays.

The correlation of the chart and the PPI display during a transit of confined waters frequently can be aided through the use of a transparent chart overlay on which properly scaled concentric circles are inscribed as a means of simulating the fixed range rings on the PPI. By placing the center of the concentric circles at appropriate positions on the chart, the navigator is able to determine by rapid inspection, and with close approximation, just where the pips of certain charted features should appear with respect to the fixed range rings on the PPI when the vessel is at those positions. This technique compensates for the difficulty imposed by viewing the PPI at one scale and the chart at another scale. Through study of the positions of various charted features with respect to the simulated fixed range rings on the transparency as the center of the simulated rings is moved along the intended track, certain possibilities for unique observations may be revealed.

By placing the center of the properly scaled simulated range ring transparency over the observer's most probable position on the chart, the identification of echoes is aided. The positions of the range rings relative to the more conspicuous objects aid in establishing the most probable position. With better positioning of the center of the simulated rings, more reliable identification is obtained.

By placing the simulated range ring transparency over the chart so that the simulated rings have the same relationship to charted objects as the actual range rings have to the corresponding echoes, the observer's position is found at the center of the simulated range rings. Under some conditions, there may not be enough suitable objects and corresponding echoes to correlate with the range rings to obtain the desired accuracy.

This method of fixing should be particularly useful aboard small craft with limited navigational personnel, equipment, and plotting facilities. It should serve to overcome difficulties associated with unstabilized displays and lack of a variable range marker.

Preferably, radar fixes obtained through measuring the range and bearing to a single object should be limited to small, isolated fixed objects which can be identified with reasonable certainty. In many situations, this may be the only reliable method which can be employed. If possible, the fix should be based upon a radar range and visual gyro bearing because radar bearings are less accurate than visual gyro bearings. A primary advantage of the method is the rapidity with which a fix can be obtained. A disadvantage is that the fix is based upon only two intersecting lines of position, a bearing line and a range arc, obtained from observations of the same object. Identification mistakes can lead to disaster. *If the fix is based upon a floating aid, it should be treated with considerable caution.*

Generally, fixes obtained from radar bearings are less accurate than those obtained from intersecting range arcs. The accuracy of fixing by this method is greater when the center bearings of small, isolated, radar-conspicuous objects can be observed. The rapidity of the method affords a means for initially determining an approximate position for subsequent use in more reliable identification of objects for fixing by means of two or more ranges.

Fixing by tangent bearings is one of the least accurate methods. The use of tangent bearings with a range measurement can provide a fix of reasonably good accuracy. As illustrated in figure 4208a, the tangent bearing lines intersect at a range from the island observed less than the range as measured because of beam width distortion. Right tangent bearings should be decreased by an estimate of half the horizontal beam width. Left tangent bearings should be increased by the same amount. The fix is taken as that point on the range arc midway between the bearing lines.

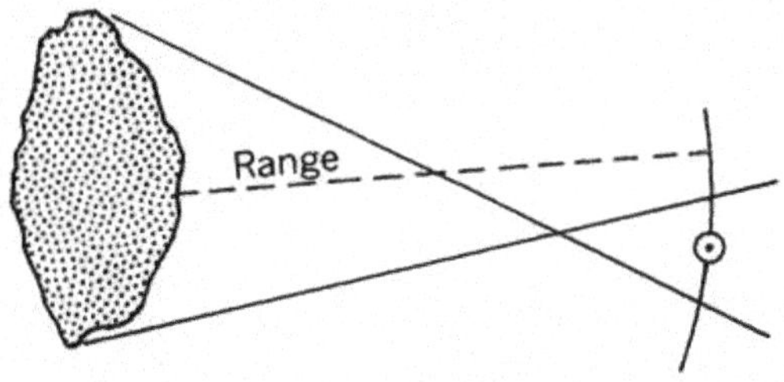

FIGURE 4208a.—Fixing by tangent bearings and radar range.

It is frequently quite difficult to correlate the left and right extremities of the island as charted with the island image on the PPI. Therefore, even with compensation for half of the beam width, the bearing lines usually will not intersect at the range arc.

In many situations, the more accurate radar fixes are determined from nearly simultaneous measurements of the ranges to two or more fixed objects. Preferably, at least three ranges should be used for the fix. The number of ranges which it is feasible to use in a particular situation is dependent upon the time required for identification and range measurements. In many situations, the use of more than three range arcs for the fix may introduce excessive time lag.

If the most rapidly changing range is measured first, the plot will indicate less progress along the intended track than if it were measured last. Thus, less lag in the radar plot from the vessel's actual position is obtained through measuring the most rapidly changing range last.

Similar to a visual cross-bearing fix, the accuracy of the radar fix is dependent upon the angles of cut of the intersecting lines of position (range arcs). For greater accuracy, the objects selected should provide range arcs with angles of cut as close to 90° as is possible. In cases where two identifiable objects lie in opposite or nearly opposite directions, their range arcs, even though they may intersect at a small angle of cut or may not actually intersect, in combination with another range arc intersecting them at an angle approaching 90°, may provide a fix of high accuracy (fig. 4208b). The near tangency of the two range arcs indicates accurate measurements and good reliability of the fix with respect to the distance off the land to port and starboard.

Small, isolated, radar-conspicuous fixed objects afford the most reliable and accurate means for radar fixing when they are so situated that their associated range arcs intersect at angles approaching 90°.

Figure 4208c shows a fix obtained by measuring the ranges to three radar-conspicuous objects well situated. The fix is based solely upon range measurements because radar ranges are more accurate than radar bearings even when small objects are observed. Note that in this rather ideal situation, a point fix was not obtained. Due to inherent radar errors, any point fix should be treated as an accident dependent upon plotting errors, the scale of the chart, etc.

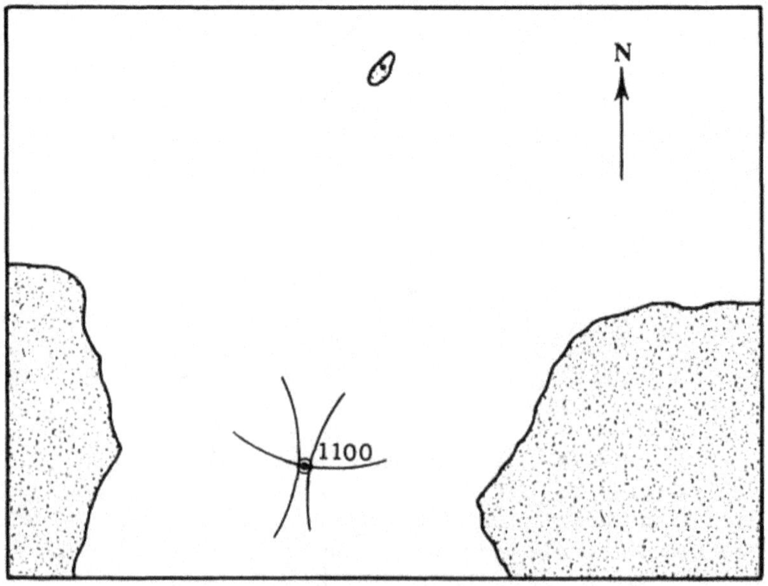

FIGURE 4208b.—Fixing by range arcs.

Although observed radar bearings were not used in establishing the fix as such, the bearings were useful in the identification of the radar-conspicuous objects.

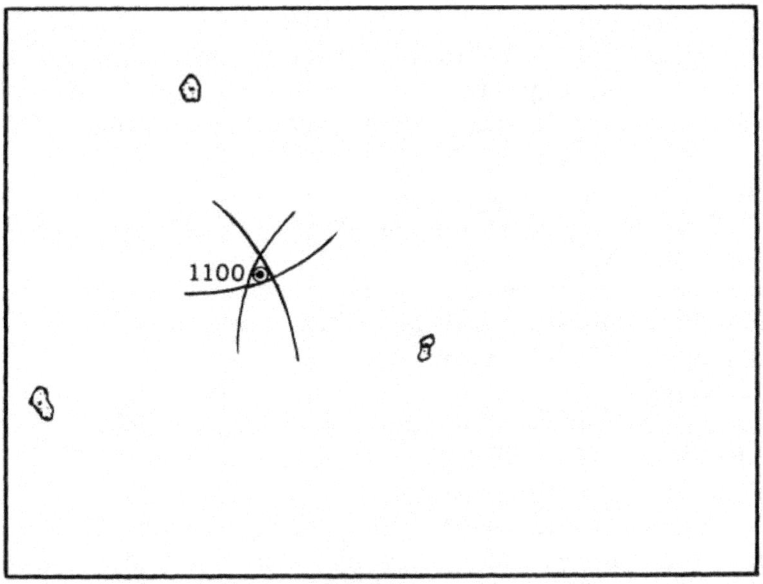

FIGURE 4208c.—Fix by small, isolated radar-conspicuous objects.

As the ship travels along its track, the three radar-conspicuous objects still afford good fixing capability until such time as the angles of cut of the range arcs have degraded appreciably. At such time, other radar-conspicuous objects should be selected to provide better angles of cut. Preferably, the first new object should be selected and observed before the angles of cut have degraded appreciably. Incorporating the range arc of the new object with range arcs of objects which have provided reliable fixes affords more positive identification of the new object.

While fixing by means of intersecting range arcs, the usual case is that two or more small, isolated, and conspicuous objects, which are well situated to provide good angles of cut, are not available. The navigator must exercise considerable skill in radarscope interpretation to estimate which charted features are actually displayed. If initially there are no well defined features displayed and there is considerable uncertainty as to the vessel's position, the navigator may observe the radar bearings of features tentatively identified as a step towards their more positive identification. If the cross-bearing fix does indicate that the features have been identified with some degree of accuracy, the estimate of the vessel's position obtained from the cross-bearing fix can be used as an aid in subsequent interpretation of the radar display. With better knowledge of the ship's position, the factors affecting the distortion of the radar display can be used more intelligently in the course of more accurate interpretation of the radar display.

Frequently there is at least one object available which, if correctly identified, can enable fixing by the range and bearing of a single object method. A fix so obtained can be used as an aid in radarscope interpretation for fixing by two or more intersecting range arcs.

The difficulties which may be encountered in radarscope interpretation during a transit may be so great that accurate fixing by means of range arcs is not obtainable. In such circumstances, range arcs having some degree of accuracy can be used to aid in the identification of objects used with the range and bearing method.

With correct identification of the object observed, the accuracy of the fix obtained by the range and bearing to single object method usually can be improved through the use of a visual gyro bearing instead of the radar bearing. Particularly during periods of low visibility, the navigator should be alert for visual bearings of opportunity.

Although the best method or combination of methods for a particular situation must be left to the good judgment of the experienced navigator, factors affecting method selection include:

1. The general need for redundancy—but not to such extent that too much is attempted with too little aid or means in too little time.
2. The characteristics of the radar set.
3. Individual skills.
4. The navigational situation, including the shipping situation.
5. The difficulties associated with radarscope interpretation.
6. Angles of cut of the lines of position.

Beam compasses (art. 602) shown in figure 4208d can be a very valuable radar navigation tool. If the navigator does not have such tool readily available, he may find himself in a situation where his only observations are radar ranges, but he cannot strike the range arcs to fix his position because of the limited spread of the points of his compasses.

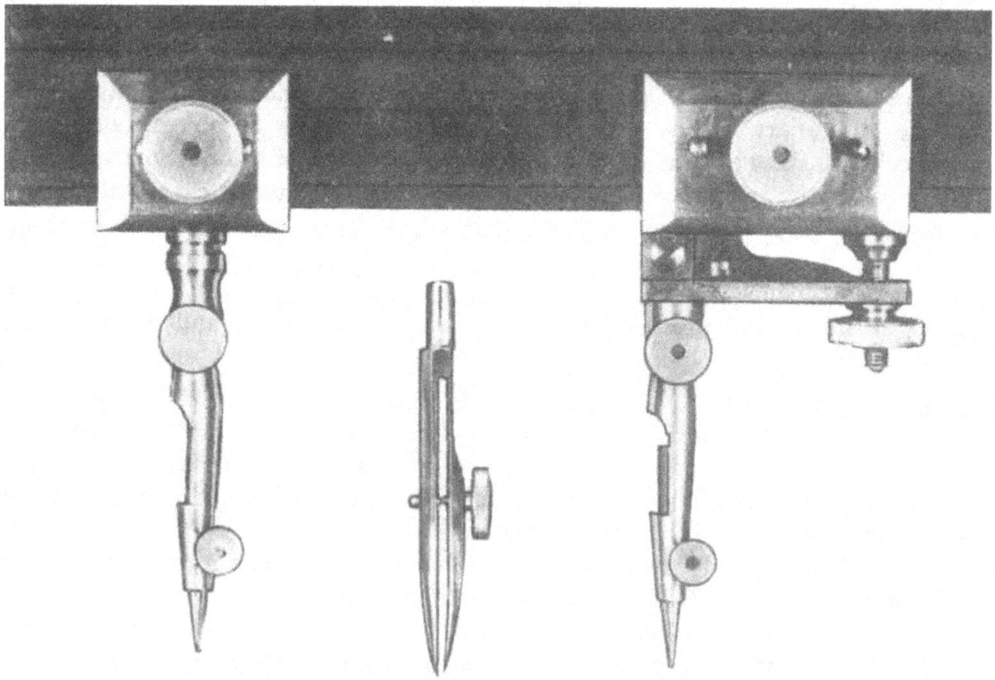

FIGURE 4208d.—Beam compasses.

CHAPTER XLIII

RADIONAVIGATION SYSTEMS

General

4301. Introduction.—Radionavigation systems are characterized with respect to their transmissions by the radiofrequency of the carrier wave, the power, and the signal type. These systems are also characterized by how the signals are analyzed upon reception, the nature of the lines of position they supply, their range or area of coverage, their availability to the user, their passivity, and their accuracy.

As discussed in article 4014, the frequency and power of the radio wave transmission are among the factors determining the effective range of a system. However, very low frequency (VLF) signals travel very long distances with relatively small amounts of power radiated. Although low frequency (LF) transmissions generally show greater attenuation with distance than VLF signals, the 100 kHz Loran-C (art. 4329) transmissions in particular, provide reliable groundwave reception at distances of about 800 to 1,200 nautical miles. The Decca Navigator System (art. 4344), also operating in the LF band at 70 to 130 kHz but at considerably less power than Loran-C, provides reliable groundwave reception at distances of more than 400 nautical miles during daylight under good reception conditions. At night this reception may be reduced significantly by skywave contamination of the groundwave. The very high frequency (VHF) transmission, as from navigational satellites (ch. XLIV), and the super high frequency (SHF) transmissions, as from navigational radar (ch. XLII), are limited to the line-of-sight.

The signal type affects how signals are analyzed upon reception. In general, the receivers of pulse modulated systems determine time intervals; the receivers of continuous wave (CW) systems compare phases. From the point of view of reception, these systems are designated as **time difference measuring systems** and **phase comparison systems.** This distinction is made despite the fact that both systems detect time intervals. The phase comparison consists in fact in determining the time shift of one wave with respect to another, taking as a unit of time the inverse of the frequency.

Using Loran-C as a specific example, however, the pulse type signal is processed in the receiver by both time difference and phase comparison methods.

The difference in the ranges of reliable reception of Loran-C and Decca signals is also due to the pulse modulation of the Loran-C signal. Multiple pulse formats raise the average power and thus increase the range. Other advantages to be derived from pulse modulation include:

1. freedom from lane ambiguities,
2. means to discriminate between groundwaves and skywaves,
3. suitable coding techniques can be used for groundwave and skywave discrimination even if the skywave of an earlier pulse coincides with the groundwave of a later pulse, and
4. the envelope of the pulse permits identification of a specific carrier wave cycle of a coherent radiofrequency signal for phase measurement purposes.

The use of continuous wave transmissions for wide area coverage from a single complex of stations results in two serious problems: (1) ambiguities in position determination, and (2) difficulties in discriminating against skywave contamination. Also, where high accuracy coverage of long coastlines and vast ocean areas is required, and where relatively few suitable sites for transmitting stations are available, continuous wave systems are generally unsatisfactory because of the limitations of the maximum permissible baseline lengths of the high accuracy continuous wave systems.

According to the nature of the lines of position provided, radionavigation systems may be described as **hyperbolic, ranging, azimuthal,** or **composite.**

Radionavigation systems may be described as **short range** if their positioning capability is limited to coastal regions, or the maximum useful range is limited to making landfall. Examples are radar and the radio direction finder. The systems may be described as **medium range** if their positioning is generally limited to ranges permitting reliable positioning for about 1 day prior to making landfall; Decca is an example.

Those radionavigation systems providing positioning capability on the high seas in areas well beyond the positioning capability of the medium range systems are described as **long range systems.** Examples of these systems are Loran–C and Omega. When the latter system is fully implemented, it may be described as a **worldwide** or **global system.** The Navy Navigation Satellite System is an example of the worldwide or global system. The worldwide or global system provides positioning capability wherever the observer may be located on navigable waters.

The maximum range of a radionavigation system depends upon the frequency, transmitter power, signal-to-noise ratio in the service area, receiver sensitivity, and losses over the signal path. The signal-to-noise ratio, often expressed in decibels (app. D), is a more realistic indicator of range capability than the field intensity of the groundwave.

Radionavigation systems may also be classified with respect to availability. Loran–C may be described as a **continuous system** because the user within the effective range of the system can use it at any time to determine his position. Omega signals as transmitted in the 10-second commutation pattern (tab. 4312) may be considered as continuously available. Therefore, for practical purposes positioning can be effected continuously. When using the Navy Navigation Satellite System (ch. XLIV) near the equator, the time between fixes may be well in excess of 1 hour because one of the satellites in near polar orbit is not available for observations. Therefore, the Navy Navigation Satellite is an example of a system that is *not* continuous.

Those systems whose operation does not require the user to transmit signals and can be used simultaneously by an unlimited number of users are described as *nonsaturable.* **Saturable** systems are limited to a single user or a limited number of users on a time-shared basis.

Systems whose operation does not require the user to transmit a signal may be described as **passive.**

Due to varying accuracy requirements among users, accuracy (app. Q) classifications, such as high, medium, and low, have limited value. In this chapter Loran–C is described as a system of high accuracy when the groundwave is used; except for differential Omega (art. 4321), Omega is described as a system of medium accuracy; Consol is described as a system of low accuracy.

4302. Distance-difference measurement.—If synchronized signals from two stations are transmitted, the *difference* in distance from the stations can be measured, either by means of the elapsed time interval between the arrival of the two signals, or by measurement of the phase difference between the signals.

Refer to figure 4302. Let *M* and *S* be two stations. Synchronization is achieved by letting the signals of *M*, the **master**, control those of *S*, the **slave** or **secondary.** Circles *M*1, *M*2, *M*3, etc., are units of distance from *M*; and circles *S*1, *S*2, *S*3, etc., are units of distance from *S*. If both signals are transmitted at the same instant, they will arrive together at any point along a line equidistant from the two stations. This **centerline** is the perpendicular bisector of the **baseline** joining the two stations. On a sphere, both centerline and baseline are great circles.

The centerline is the zero time difference line. If the *M* signal arrives at some point before the *S* signal, the time difference can be found by subtracting the *M* signal travel time (or distance) from the *S* signal travel time (or distance). If a line is drawn through all intersections at which units of distance from *S* are greater by *one* than those from *M* (*S*8, *M*7; *S*7, *M*6; *S*6, *M*5; etc.), a curve is formed, as shown at "+1" in figure 4302. A similar curve labeled "−1" is formed if all points at which units of distance from *S* are *less* by *one* than those from *M* (*M*8, *S*7; *M*7, *S*6; etc.) are connected. The minus sign indicates that the *M* signal arrives (−)1 time unit *before* the *S* signal, or $S-M=(-)1$. On a plane surface, such curves are **hyperbolas** (art 134, vol. II) because they connect points of equal *difference* of distance between two fixed points. On a sphere, such curves are called **spherical hyperbolas.** On the spheroidal earth the curves are called **spheroidal hyperbolas** and differ somewhat from spherical hyperbolas.

Other, more sharply curving hyperbolas are formed by connecting lines of greater time (distance) difference, as at (+)2, (−)2, (+)3, (−)3, etc. The maximum difference occurs along the **baseline extensions** beyond the transmitters. This difference depends upon the distance between stations. A pattern of all positive readings can be obtained by delaying the start of the *S* signal until the *M* signal is received at *S*, or longer. Suppose the *S* signal is transmitted 10 units *after* the *M* signal. The *M* signal for a baseline six units long will already have traveled four units beyond *S* when the *S* signal leaves the transmitter. Therefore, the reading along the baseline extension from *S* is (+)4, or 10 *more* than shown in figure 4302. By the time the *S* signal arrives at the master transmitter, the *M* signal will be at 10 (the delay) plus 6 (the number of units between *M* and *S*) units, or 16 units away. Therefore, the reading along the baseline extension beyond *M* is $10+6=(+)16$. Similarly it can be shown that all other readings are also increased by (+)10.

Each hyperbola is a line of position for the associated distance-difference measurement. Intersecting families of hyperbolas, as derived from three or more stations, is known as a hyperbolic **lattice.**

Hyperbolic Radionavigation Systems

4303. Hyperbolic radionavigation systems are so called because of the hyperbolic lattices (art. 4302) formed in their signal fields and the associated hyperbolic lines of position supplied. With appropriate modifications hyperbolic systems can be operated in the circular or ranging mode with circular lines of position.

Hyperbolic systems provide the greatest potential capability to date for coverage of large areas with the smallest number of transmitting sites, and the best compromise between accuracy and coverage. Also, the hyperbolic mode of operation enables simpler receiver design because time differences can be measured with

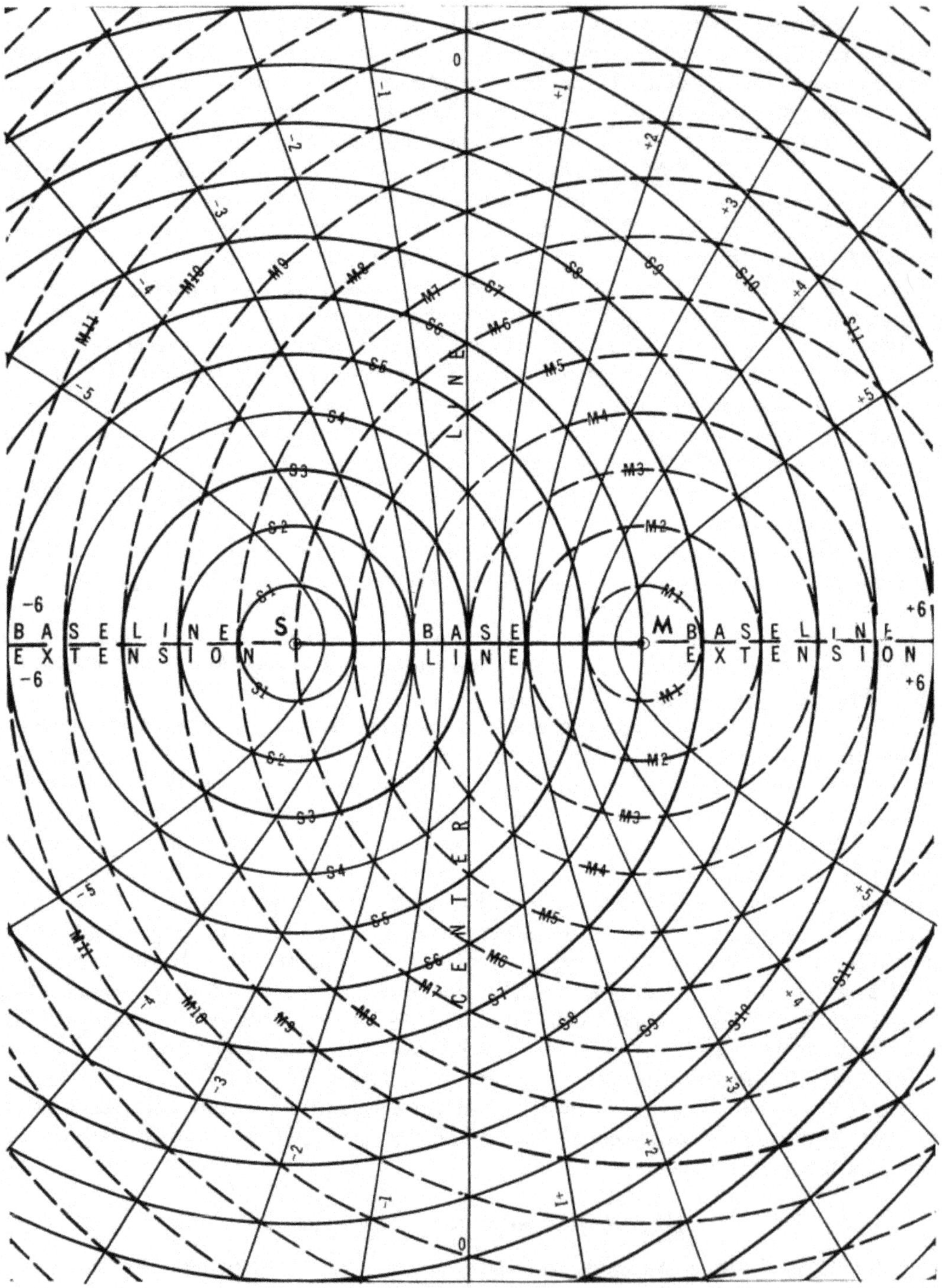

FIGURE 4302.—Hyperbolic lattice.

sufficient precision without the need for a precision oscillator. Since the precision control equipment is located at the transmitting sites, user equipment can be designed to meet different requirements at reasonable cost. The systems are continuous, non-saturable, and passive (art. 4301).

The disadvantage of a hyperbolic system in needing two stations to provide a single family of hyperbolas (figure 4303) can be partly overcome by using a group or **chain** of stations that operates to form a lattice of intersecting families of hyperbolas.

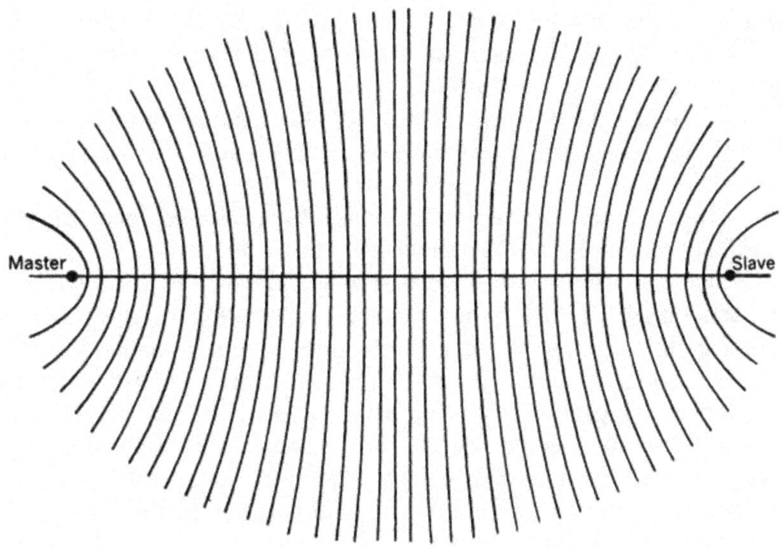

FIGURE 4303.—A family of hyperbolas. Each hyperbola is a line of position for the associated time-difference measurement.

The accuracy (app. Q) of the fix provided by a hyperbolic radionavigation system is dependent upon the accuracy of each line of position used to obtain the fix and the angle of cut of the lines of position. The accuracy of each line of position depends upon the following factors:

1. the precision with which the difference between the times of arrival of two signals can be measured;
2. the synchronization of the transmitting stations;
3. the accuracy of propagation predictions;
4. operational or receiver accuracy;
5. user's position relative to the transmitting stations; and
6. lattice table and charting accuracies, including the accuracies of the positions of the transmitting stations.

Since the velocity of propagation of radio energy is approximately 1 foot per nanosecond, for accuracies on the order of tens or hundreds of feet, measurements must be made to tens or hundreds of nanoseconds. If the time differences are to be converted to lines of position accurately, propagation conditions must be reliably predictable to tens or hundreds of nanoseconds. Receiver accuracy is dependent upon signal-to-noise ratio, operator skill, and instrumentation. The user's position relative to the transmitting stations governs the **gradient** or spacing between consecutive lines of position per unit time difference, as 1 microsecond. If the gradient is high a relatively small time-difference error will result in a relatively high position error. Lines are most closely spaced, giving highest accuracy, along the baseline between stations. As the distance between consecutive lines increases, the accuracy decreases, being so low along the baseline extensions that the use of this part of the lattice is normally avoided.

The undesirable divergence of the hyperbolic lines of position varies with the length of the baseline. However, each hyperbola becomes more nearly a straight line (great circle) as distance from the baseline increases. At a distance from the center of the baseline of five times the length of the baseline, the departure of the hyperbola from a great circle becomes very small. Thus, if the baseline is very short, as in Consol (art. 4354), the system can be considered directional (azimuthal) beyond a distance of a few miles from the station.

The use of time-difference measurements to establish lines of position serves to minimize the effects on position accuracy of errors caused by propagation anomalies.

4304. The ranging mode, often called the **Range-Range** or **Rho-Rho** mode, is that method of operation in which the times for the signals to travel from each transmitting station to the receiver are measured rather than their differences. Each time measurement (range measurement) provides a circular line of position.

The feasibility of ranging mode is dependent upon a stable frequency source, propagation predictability, and a stable time reference within the receiver. The range to a station is calculated from the difference in the known time of transmission and the measured time of arrival of the signal at the receiver. The user's time reference must be initially synchronized to the time standard at the transmitting station. Since perfect synchronization is, in general, not feasible, calibration of the user's time standard is required.

While the gradient (art. 4303) in the hyperbolic mode degrades with increasing distance from the transmitting stations, the gradient in the ranging mode is a constant equal to the propagation velocity. Therefore, the ranging mode overcomes the *geometric dilution* associated with the hyperbolic mode at extended ranges. Thus, the ranging mode can be used to increase the coverage area.

For a given service area and the associated transmitting station locations, the ranging mode may provide better system geometry, i.e. better angles of cut of the intersecting circles forming the lattice. Greater freedom in selecting stations for better system geometry is afforded by not having to use a master station, two intersecting lines of position being obtainable from two slave (secondary) stations. The ranging mode of operation can be used to extend the coverage area since the user must be within range of only two transmitting stations. Also the coverage area is extended by overcoming the geometric dilution of the hyperbolic mode.

When three stations can be received, the ranging mode is enhanced by redundant information since only two stations are needed. This redundancy serves to increase fix accuracy. With these circular lines of position, the fix determined by the intersection of one pair will, in general, be inconsistent with a fix determined by the intersection of a second pair. Since the inconsistency will be due to errors in the system, including drift in the time standard, the third (redundant) measurement can be used to estimate the errors. If the estimate is correct the three lines of position will intersect at a common point.

Omega Navigation System

4305. The Omega Navigation System is a worldwide continuous radionavigation system of medium accuracy which provides hyperbolic lines of position through phase comparisons of VLF (10–14 kHz) continuous wave signals transmitted on a common frequency on a time shared basis. The fully implemented system is comprised of only eight transmitting stations.

4306. Omega Navigation System principles.—The navigator using what may be described as a standard Omega receiver determines his position within a hyperbolic lattice formed by contours of constant phase differences in the signal field of the transmitting stations. If the design of the receiver includes a very stable oscillator, the navigator may employ a circular or ranging lattice to determine his position. However, this mode of operation requires that the local oscillator be synchronized, both in frequency and time, with the oscillator at the transmitting station.

With the Omega receiver being used in the hyperbolic mode, the measured phase difference of phase synchronized and time shared signals received at a particular position from two stations transmitting on the same frequency depends solely upon how much farther the position is from one transmitting station than the other. The term **phase synchronized** indicates that the phases of all transmissions are maintained synchronized at their sources. This is accomplished through the use of cesium frequency standards. Phase synchronization between transmitter and receiver oscillator is not required in the hyperbolic mode of operation because the phase error is cancelled on taking the difference. The term "time shared" indicates that the signals from the stations are not transmitted simultaneously but sequentially in a 10-second commutation pattern (tab. 4312). Time sharing or sequential transmission is necessary because of the use of the single frequency.

Although the signals from the transmitting stations are actually transmitted sequentially in accordance with the commutation pattern, simultaneous transmission can be used to explain the basic principles. Unless ionospheric characteristics change suddenly within the period of the sequence, the same phase-difference measurement can be obtained by a stationary observer, theoretically, whether the transmissions are simultaneous or sequential. When the observer is in motion, the use of sequential transmission for practical considerations, including station identification and simplicity of receiver design, does introduce an error because the difference in phase between the local oscillator and one station will be measured at a position different from that at which the measurement is made between the local oscillator and the second station. However, at the relatively low speeds of marine applications this error is inconsequential.

As shown in figure 4306, phase synchronized signals are considered as being transmitted from stations *X* and *Y* simultaneously. At the instant the phase of the signal transmitted from station *Y* is 180° at a point one-half wavelength ($\frac{1}{2}\lambda$) from *Y*, the phase of the signal transmitted from station *X* is 180° at a point one-half wavelength from *X*. When the phase of the signal transmitted from station *Y* has increased to 360° (or returned to 0°) at a point one wavelength from station *Y*, the signal transmitted from station *X* will have increased to 360° (or returned to 0°) at a point one wavelength from station *X* at the same instant, etc.

Theoretically, the phase of the signal received from each station can be compared with a locally generated reference signal, having the same wavelength and synchronized in phase with the phase of the signal being transmitted from each station. Each such comparison will indicate the fractional part of a wavelength in excess of an integral number of wavelengths that the receiver is distant from a station.

In actual practice (sequential transmission), the receiver first measures the difference in phase between the locally generated signal and the phase of the signal received from that station of the pair the letter designation of which appears first

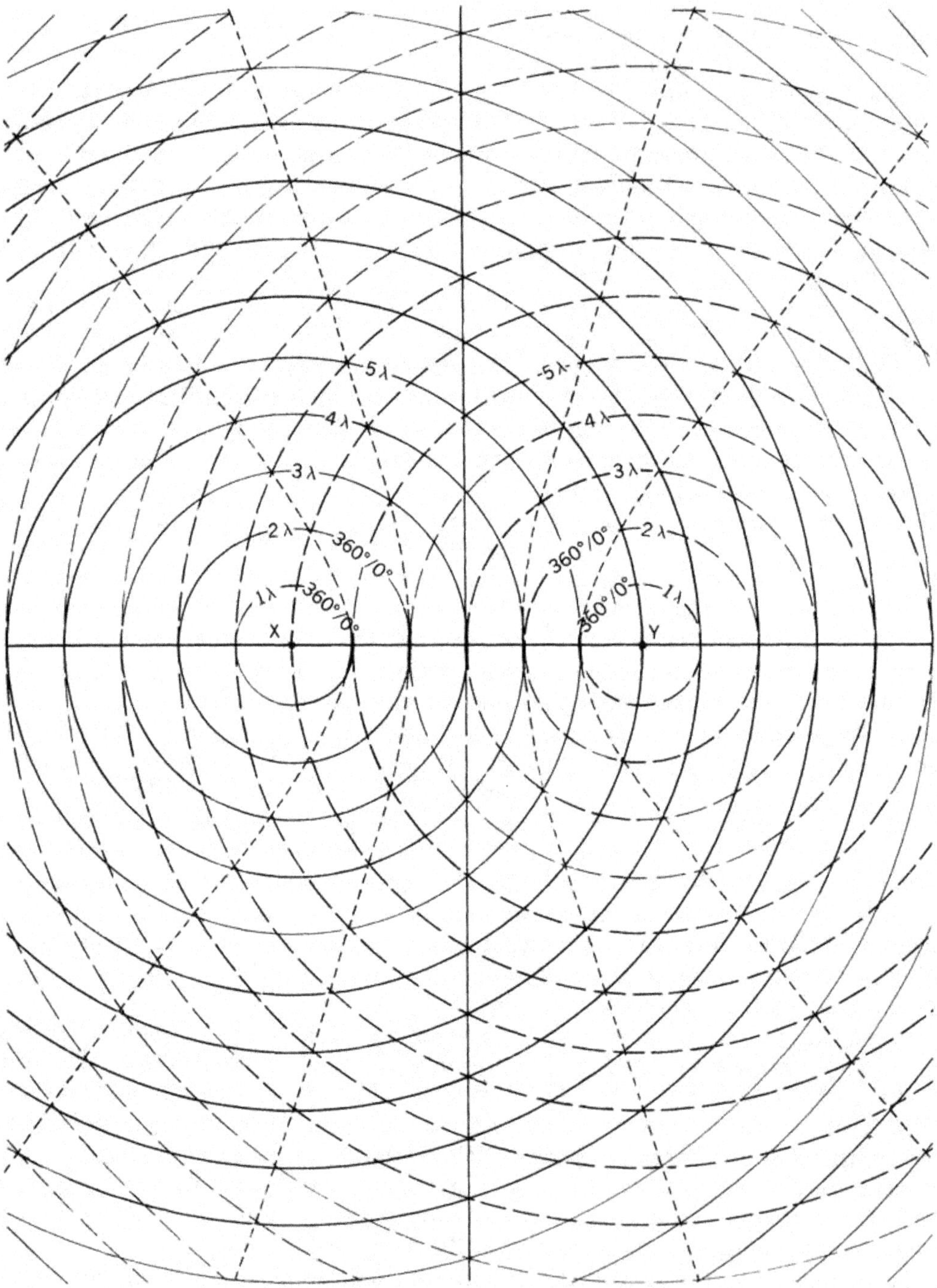

FIGURE 4306.—Phase synchronized signals.

in the alphabet. The difference, the phase angle, is stored for future reference. Then the receiver measures the difference in phase between the locally generated signal and the phase of the signal received from the other station of the pair. This phase angle is subtracted from the phase angle previously obtained using the station the letter designation of which precedes that of the second station in the alphabet. This phase-difference measurement represents a difference in distance of the receiver from the two stations. Insofar as the observer is stationary, the storing of the first phase-angle measurement for future (within 10 seconds) comparison with the second measurement is comparable to a simultaneous transmission format. The same phase-difference measurement will be observed at all points which have the same difference in distance from the two transmitting stations. The locus of such points is a contour of constant phase (isophase contour) which is fixed on the surface of the earth with respect to the positions of the two transmitting stations.

By repeating measurements for another pair of stations, a second isophase contour or line of position is obtained. The intersection of the lines of position establishes the position of the observer.

4307. Omega lanes.—With the stations transmitting at 10.2 kHz, the wavelength is approximately 16 nautical miles. Thus, identical phase angles are repeated at 16-mile intervals. However, on the baseline between two transmitting stations a specific difference in the phase angles of the radio waves received from the two transmitting stations repeats itself every one-half wavelength or about every 8 miles.

As is obvious from inspection of figure 4307, isophase contours pass through all points which are an integral number of wavelengths distant from the two transmitting stations, *X* and *Y*. At points any integral number of wavelengths plus a half wavelength from the stations, the phase-difference is also zero. Thus, isophase contours pass through points on the baseline a half wavelength apart.

Since a representative baseline in the Omega system is about 5,000 nautical miles in length, or about 300 wavelengths at 10.2 kHz, specific phase-angle differences in the 10.2 kHz signal field are repeated about 600 times on the baseline. Hence, at 10.2 kHz there are about 600 points on the baseline at which the phase-angle differences are zero.

Contours connecting points in the signal field at which the phase-angle differences are zero are the constant phase or isophase contours plotted on Omega charts and tabulated in the Omega lattice tables. At 10.2 kHz each such contour is spaced at intervals of about 8 nautical miles on the baseline. Away from the baseline there is a small divergence of the hyperbolic contours. The exact spacing of the hyperbolic contours on the baseline as printed on a chart is governed by the value used for the phase velocity, v. For 10.2 kHz charted contours, $c/v = 0.9974$, where c is the group velocity (299,792.5 km/sec). The group velocity, c, refers to the velocity at which the energy is being transferred. This cannot exceed relativistic limits. The phase velocity, v, refers to the transfer of a point of reference, a phase difference, rather than a physical quantity and, therefore, is not subject to relativistic velocity limitations.

The area lying between two zero phase-difference contours is known as an **Omega lane.** Thus, on the baseline between two stations transmitting at 10.2 kHz the lane width is about 8 nautical miles. Each such pair of stations transmitting at 10.2 kHz produces a pattern of about 600 lanes. Away from the baseline the lane width for 10.2 kHz transmissions increases gradually. When the two transmitting stations and the receiver form an equilateral triangle (the receiver being 4,300 nautical miles from the baseline), the lane width will have increased to about 12 nautical miles.

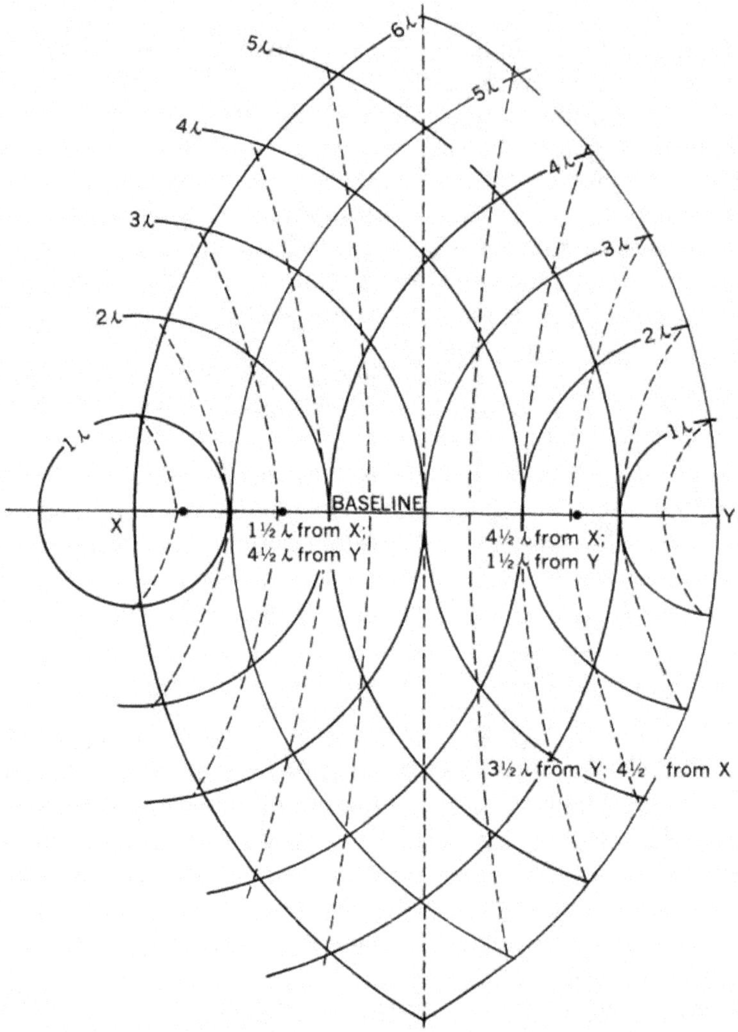

FIGURE 4307.—Isophase contours.

In the ranging mode of operation, the lane is the area lying between two zero phase-difference contours within which there is a 360° change in phase of the CW wave received from a single transmitting station as the receiver is moved from one isophase contour to the other. Thus, in the ranging mode, lane width at 10.2 kHz is constant at about 16 nautical miles.

4308. Omega line of position.—In the normal hyperbolic mode, the phase-difference readout from the receiver (with the propagation correction applied) only indicates the isophase contour (LOP) corresponding to the observer's position within a lane. When the stations are transmitting at 10.2 kHz, the observer must know the lane in which he is located or his position within an accuracy of 4 miles, if on the baseline.

The difference in phase angle measurements for a pair of stations (with the propagation correction applied) establishes the percentage value of the lane defining the LOP. For example, the phase angle for station *A* of figure 4308 is 25 percent of a cycle (25 centicycles or 25 cecs) and the phase angle for station *B* is 75 percent of a cycle (75 centicycles or 75 cecs). The difference, 50 percent of a cycle or 50 cecs, is numerically equal to the percentage value of the lane defining the LOP. That is, the LOP is defined by 50 percent of the lane or 50 centilanes (50 cels).

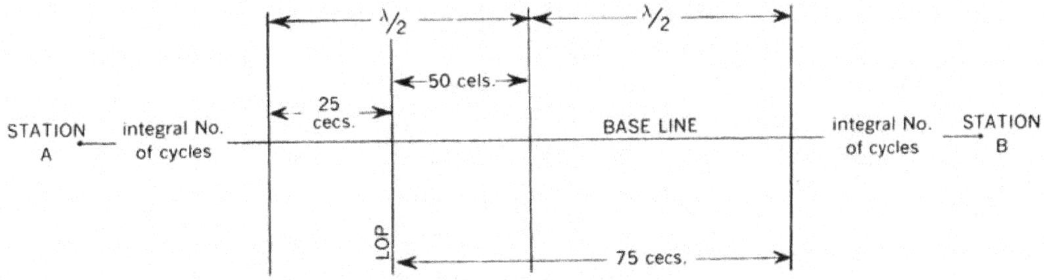

FIGURE 4308.—Omega line of position.

4309. Lane identification.—Normally, the lane counter of the receiver, which is set on departure from a known position, provides the lane. If the lane counting capability of the equipment is lost for any reason, such as temporary equipment failure, the navigator usually can determine the lane he is in by dead reckoning or other navigational means.

Should there be any ambiguity with respect to the lane, the multiple Omega frequencies can be used to determine the lane by obtaining fixes on two or more pairs of stations if the user has a multi-channel receiver. Any two stations can be used as a pair rather than a particular master and slave as in the Loran concept.

If one of the Omega frequencies were 3.4 kHz, the wavelength would be 48 nautical miles; the lane width on the baseline would be 24 nautical miles or three times the lane width at 10.2 kHz. The difference between the 13.6 kHz and 10.2 kHz Omega frequencies is 3.4 kHz. By observing the difference in the phase-difference readings at 10.2 kHz and 13.6 kHz, it is possible to form a *coarse lane* corresponding to a frequency of 3.4 kHz, neglecting the slight difference in propagational velocities at 10.2 kHz and 13.6 kHz. With the coarse lane so formed, the navigator now need only to know his position with an accuracy of 12 miles to identify the coarse lane without ambiguity. Since the 3.4 kHz coarse lane is formed by three 10.2 kHz *fine lanes* as shown in figure 4309, the correct fine lane can be identified by means of the equivalent of the phase-difference reading for 3.4 kHz. The equivalent phase reading is obtained by subtracting the 10.2 kHz phase reading from the 13.6 kHz phase reading and then applying a 3.4 kHz correction. *The use of this technique is dependent upon the availability of a multi-channel receiver and propagation corrections for 10.2 kHz and 3.4 kHz.* With the correct 10.2 kHz lane so identified without ambiguity, it is then possible to determine a more exact line of position within the fine lane by means of a phase reading at 10.2 kHz.

4310. Identification of coarse lane.—Since the equivalent of the 3.4 kHz coarse lane is formed by three 10.2 kHz fine lanes, the first step is to inspect the 10.2 kHz isophase contours on the chart to determine which lane numbers are multiples of three, or which numbers on the contours can be divided exactly by three. Having identified the coarse lane or lanes in which the observer might be located, the next step is to apply the equivalent of the phase-difference reading for 3.4 kHz. If the correct coarse lane has been selected, the isophase contour (LOP) as plotted within the coarse lane will lie within 12 miles of the DR position, assuming that such position is not in error by more than 12 miles. The LOP will establish in which of the three fine lanes the observer is located. If the LOP as plotted within the coarse

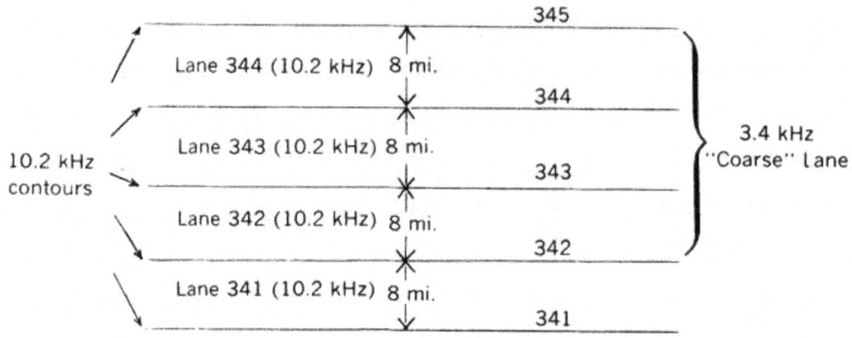

FIGURE 4309.—The coarse lane.

lane does not lie within 12 miles of the DR position, an adjacent coarse lane should be selected for use in identification of the fine 10.2 kHz lane.

The lane resolution process can be extended further by using even lower difference frequencies. By taking the difference of 11.333 kHz and 10.2 kHz Omega frequencies to obtain a difference frequency of 1.133 kHz, a coarse lane having a width of 72 nautical miles on the baseline can be used.

4311. Station locations and coverage.—The Omega stations, their letter designators, and approximate locations are given in figure 4311. The Omega coverage diagram, not shown here but generally similar to the Loran-C coverage diagram (fig. 4340), provides the repeatable accuracy for different regions expressed in terms of d_{rms} (arts. Q7 and Q8). The accuracy is determined using a standard deviation of 4 centilanes for each of the two lines of position making up the fix.

4312. Characteristics of the Omega transmission.—Omega is a VLF, continuous wave, time shared radionavigation system with all stations transmitting at the same frequencies. The stations always transmit in the same order, with the length of each transmission varying between 0.9, 1.0, 1.1, and 1.2 seconds from station to station. The order and lengths of the station transmissions at 10.2 kHz, in accordance with the Omega station identification code, are contained in table 4312.

Station	*A*	*B*	*C*	*D*	*E*	*F*	*G*	*H*
Length of transmission in seconds	0.9	1.0	1.1	1.2	1.1	0.9	1.2	1.0

TABLE 4312.—Omega commutation pattern.

4313. Signal format.—As shown in figure 4313, the order of transmission is such that when station *A* transmits at 10.2 kHz, stations *G* and *H* transmit at 11.33 kHz and 13.6 kHz, respectively, for the same time duration that station *A* transmits at 10.2 kHz. When station *B* transmits at 10.2 kHz, stations *A* and *H* transmit at 13.6 kHz and 11.33 kHz, respectively, for the same time duration that station *B* transmits at 10.2 kHz, and so on until the whole sequence of transmissions has been completed. Since there is a time interval of 0.2 second between transmissions, the entire cycle of the commutation pattern is repeated every 10 seconds. The start of one of the segments of the 10-second cycle is sufficiently close to Coordinated Universal Time (art. 1805) to permit station identification while listening to UTC.

The 0.2-second interval between transmissions rules out the possibility of overlapping of the signals received from different stations. The 0.2-second interval between transmissions also eases the requirement for perfect alignment of the

Letter Designation	*Station Name*	*Approximate Location* *General*	*Latitude*	*Longitude*
A	OMEGA Norway	Norwegian Coast at Arctic Circle	66°N	13°E
B	OMEGA Liberia	Monrovia, Liberia	6°N	11°W
C	OMEGA Hawaii	Oahu, Hawaii	21°N	158°W
D	OMEGA North Dakota	La Moure, North Dakota	46°N	98°W
E	OMEGA Reunion	Reunion Island, France	21°S	55°E
F	OMEGA Argentina	Golfo Nuevo, Argentina	43°S	65°W
G	OMEGA Australia	Victoria, Australia	38°S	147°E
H	OMEGA Japan	Tsushima, Japan	35°N	129°E

FIGURE 4311.—Omega station locations.

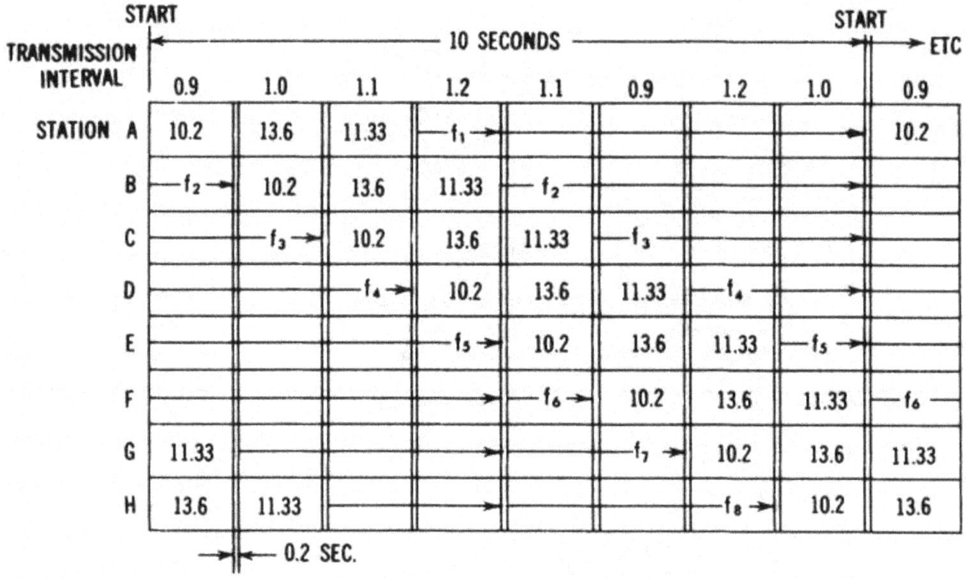

FIGURE 4313.—Signal format.

receiver commutator. A 0.1-second error in setting the commutator cycle has negligible effect.

4314. Identification of the Omega signal.—Depending upon the observer's distance from the various transmitters, the Omega signals are of differing amplitude or strength. The relative strengths of the signals received and the time sequence of transmission can be used to identify the Omega signal.

By the signal strength method of identification, the various stations can be recognized either by the relative sound levels or by the heights of the signals on an oscilloscope. In identifying the stations aurally, the signals from the nearer stations will sound relatively loud, whereas the signals from the more distant stations will be relatively weak, or they may not be heard. On an oscilloscope display, the amplitudes of the signals from the nearer stations will be relatively large whereas the amplitudes of the signals from the more distant stations will be relatively small, or the signals may not be seen. It should be kept in mind, however, these methods depend upon observing or listening to the entire signal format. The transmission from any single transmitter can be distorted by propagation effects, and unless all the signals are taken into consideration, an erroneous lane count can be established. Also, "long path" signals, or signals which travel the longer of the two possible great-circle paths from the transmitter, degrade this technique.

Each of the eight transmissions during the 10-second period of transmitting the complete sequence of signals is called a *time segment.* The first transmission in the sequence is segment A, the second is segment B, etc. (the time segment designation should not be confused with the station designations). A particular station can be identified by observing the relative time of its transmissions in the segment sequence. The 13.6 kHz transmission from station B occurs during time segment C, the 13.6 kHz transmission from station C occurs during time segment D (the commutation cycle in the receiver must be synchronized so that this condition occurs). By reference to standard time transmissions (WWV, WWVH, etc.), the start of the sequence of transmissions can be determined. Through identification of the ·rious time segments, the various stations can be identified.

Some Omega receivers are designed for automatically adjusting the receiver's internal timing circuits to agree with the Omega transmission pattern.

4315. Characteristics of Omega propagation.—The propagation of very low frequency (VLF) electromagnetic waves in the region between the lower portion of the ionosphere and the surface of the earth may be described in much the same manner as the propagation of higher frequency waves in conventional waveguides. The basic quantities in the description are certain electromagnetic waves that satisfy (1) the conditions on the electric and magnetic fields at the material boundaries, i.e., the earth's surface and the lower ionosphere; and (2) the equations for propagating waves everywhere within the region between these boundaries. These waves can be described by "the natural modes of propagation," or simply "modes." The behavior of the VLF wave may be discussed in terms of the modes.

There are three parameters that indicate how a mode will propagate in the earth-ionosphere waveguide: its attenuation rate, excitation factor, and phase velocity. The first and second parameters are measures of the energy lost by the mode per unit length and how strongly the source generates the mode in comparison to other modes, respectively; the third parameter is the mode's velocity of propagation (phase velocity). The modes are usually ordered by increasing attenuation rates, so that normally mode 1 has the lowest rate. For frequencies in the 10–14 kHz band, the attenuation rates for the second and higher modes are so large that only the first mode is of any practical importance at very long distances. However, since mode 2 is more strongly excited than mode 1 by the type of transmitters used in the Omega system, both modes must be considered at intermediate distances. In the near field region of a transmitter, other higher-order modes may have to be taken into account.

Another consideration of some importance is the fact that the modes have different phase velocities. In particular, mode 2 has a higher phase velocity than mode 1. Thus, as these modes propagate outward from the transmitter, they move in and out of phase with one another, so that the strength of the vertical electric field of the signal displays "dips" or "nulls" at several points on a radial from the transmitter. These nulls gradually disappear, however, as mode 2 attenuates, so that the strength behaves in a smooth and regular manner at long distances (where mode 1 dominates). However, within about 450 nautical miles of a transmitter the modal interference is such that the use of Omega is not generally reliable.

Since the degree of modal interference is also dependent upon factors other than proximity to the transmitter, e.g. path geomagnetic latitude and bearing, the minimum distance for reliable use is variable. For applications sensitive to spatial irregularities, such as lane resolution, separations greater than 450 miles may be required. Lesser separations may be adequate for daylight path propagation at 10.2 kHz. As a warning, the Omega LOPs depicted on charts are dashed within 450 nautical miles of a station.

Since the characteristics of the Omega signal are largely determined by the electromagnetic properties of the lower ionosphere and the surface of the earth, any change in these properties along a propagation path will generally affect the behavior of these signals. Of course, the changes will not all produce the same effect. Some will lead to small effects due to a relatively insensitive relationship between the signal characteristics and the corresponding properties. For Omega signals, one of the most important properties in this category is the effective height of the ionosphere. This height is about 90 kilometers (km) at night, but decreases quite rapidly to about 70 km soon after sunrise due to the ionization produced by solar radiation. As expected, it increases again after sunset.

The phase velocity of mode 1 is critically dependent on this height, increasing as the height decreases or vice versa. Since the phase of the Omega signal at long distances is inversely proportional to this velocity, it will readily increase or decrease in step with the effective height. In other words, the phase will be advanced or retarded as the effective height changes. One therefore expects regular diurnal changes in the phase from day to day, and studies have shown that this is indeed the case, although the exact shape and magnitude of this diurnal variation depends on several factors, including the geographic position of the receiver and transmitter and the orientation of the path relative to the boundary between the day and night hemispheres.

In general, however, if an initially all-night path is partially illuminated by the sun, starting at the transmitter, the phase will decrease, i.e. be advanced, from the night value to the nominal day value; while the opposite change will take place on an all-day path that moves into the night hemisphere starting at the transmitter. Moreover, the transition from night to day values of phase, or vice versa, will be relatively long for east-west paths, and will become shorter as the propagation path approaches a north-south orientation. Clearly, these changes will have an important effect on uncompensated LOPs. Since one wavelength is the distance covered by a wave (traveling with a given phase velocity) in one period, the LOPs constructed from the measured phase-difference will be squeezed together in going from all day to all night conditions over the propagation path, and stretched out for night to day transitions, relative to the *charted* LOPs.

Finally, while the diurnal variation in phase represents the major time-variation in the characteristics of the Omega signal at long distances, the presence of a boundary between the day and night hemispheres may produce an additional variation that arises in the following manner. In the night hemisphere, both mode 1 and mode 2 are usually present. In the day hemisphere, however, only mode 1 is usually present. Hence, as the signal passes from the night to the day hemisphere, mode 2 will be converted into the daytime mode 1 at the day-night boundary. This resultant mode 1 may then interfere with the nighttime mode 1 that passes unchanged into the day hemisphere. Thus, some additional variation in the characteristics may be present due to such interference.

4316. Geophysical parameters.—Effects less pronounced than those associated with diurnal phase shift are produced by various geophysical parameters including:

ground conductivity. Very high attenuation is incurred with propagation over freshwater ice caps.

earth's magnetic field. Propagation towards magnetic west is attenuated more than propagation towards magnetic east.

solar activity. See Sudden Ionospheric Disturbances (art. 4317) and Polar Cap Absorption (art. 4318).

latitude. The height of the ionosphere varies with latitude, being slightly higher over the higher latitudes than over the lower latitudes during the summer.

solar zenith angle. The height of the ionosphere varies with the solar zenith angle. For example, the ionosphere is higher in the summer than in the winter of the Northern Hemisphere.

4317. Sudden Ionospheric Disturbances (**SIDs**) occur when there is a very sudden and large increase in X-ray flux emitted from the sun, usually during a solar flare. SIDs also occur during flares called "X-ray flares" that produce large X-ray flux, but which have no components in the visible light spectrum. The effect, which is restricted to sunlit propagation paths, causes a phase advance and is known as a **sudden phase anomaly** (**SPA**). The SID effects are related to the solar zenith angle, and consequently, occur mostly in lower latitude regions. Usually there is a fast advance over a period of 5 to 10 minutes followed by a recovery over a period of about 30 to 60 minutes. Significant SIDs could cause position errors of about 2 to 3 miles.

4318. The polar cap disturbance (**PCD**) is an ionospheric disturbance which is in no way dependent on the ice cap in the polar region. It is the result of the focusing effect the earth's magnetic field has on particles released from the sun during a solar proton event. The effect concentrates high-energy particles in the region of the magnetic pole with the result that normal VLF transmission is disrupted.

The effect on radio waves is known as **polar cap absorption** (**PCA**). Historically, PCDs produced large or total absorption of high frequency radio waves crossing the polar region, hence the term polar cap absorption. The amount of transmission anomaly depends on how much of the total transmission path actually crosses the region of the magnetic pole and its associated auroral zone. A transmission path which is entirely outside the Arctic region will be unaffected by the PCD. Although the occurrence of a PCD is random and unpredictable, the probability of PCD incidence increases during periods of high solar activity. The *Omega Propagation Correction Tables* make no allowance for this phenomenon which is random in occurrence, although the frequency of occurrence increases during those years of peak solar activity.

The PCDs, often called PCA events (PCAs), may persist for a week or more although a duration of only a few days is more common. HYDROLANT/HYDROPAC messages are originated by the Defense Mapping Agency Hydrographic/Topographic Center if significant PCDs are detected.

The position error that will be experienced for a given LOP will depend upon the transmitting stations used and the effect of the PCD on each signal. In some cases the effect will tend to be cancelled out if the navigator is using the hyperbolic mode and has chosen station pairs whose transmission paths through the auroral zone are similar. However, no cancellation will be realized if the ranging mode is used. Using circular LOPs in this case, the position error will depend upon the severity of the PCD and the propagation path geometry involved. Phase shifts of as much as 50 cecs for a given signal have been observed. At 10.2 kHz in the ranging mode, this would correspond to a position error of approximately 8 nautical miles.

4319. Arctic paths.—The predicted propagation corrections include allowance for propagation over regions of very poor conductivity such as Greenland and parts of Iceland. Little data are available for these areas, hence even the best estimates are uncertain. In particular, rather rapid attenuation of the signal with position occurs as one passes into the "shadow" of the Greenland ice cap.

4320. Auroral zones.—Paths which intercept the auroral zones surrounding the north and south geomagnetic poles are known to affect the phase of Omega signals. Auroral effects are believed to arise from electron precipitation in the higher regions of the ionosphere which serves as a source of ionization for the D-region of the ionosphere. Although the visual auroral zone is generally oval in shape, it is now thought that the shape of the effective region producing ionization of the D-region is circular about the geomagnetic poles. Thus, auroral effects occur in a circular band between 60° and 80° north and south geomagnetic latitude. In terms of phase velocity, the nighttime auroral variation is about four times that of the daytime variation. The actual effect on Omega signals for paths intersecting this region is a decrease in phase, i.e. a phase advance with respect to nominal values.

4321. Spatial and dispersive correlation.—Over very long transmission paths from a transmitting station to various points within a limited area, approximately the same errors will be accumulated along segments of the different paths. It follows that within this limited area, two Omega receivers will exhibit similar errors. This spatial correlation results in Omega having excellent rendezvous capability. **Differential Omega,** in which a propagation correction as accurately determined at a fixed site is broadcast to Omega users within the immediate area (radius up to about 300 nautical miles), makes use of this spatial correlation.

Over the same propagation path, fluctuations on 10.2 kHz are usually closely related to those on 13.6 kHz. This dispersive correlation generally reduces errors associated with the use of the two frequencies in lane resolution.

4322. The stability of propagation has been demonstrated by long term monitoring of Omega signals at fixed sites by the Naval Electronics Laboratory Center (NELC) and others. The monitoring verifies that the standard deviation of an LOP derived from a station pair is on the order of 4 cels. This monitoring also indicates that the phase-difference measurement is dependent upon the time of day. During the day the required propagation corrections are relatively small but undergo more variations than the required nighttime corrections; during the nighttime, the required corrections are relatively large but more nearly constant. During the daytime the actual and charted Omega lanes differ than during the night.

4323. Propagation prediction.—The practicality of the Omega Navigation System is dependent upon the fact that the radio signals in the VLF band have very good phase stability over extremely long distances. The accuracy of the system is dependent upon this inherent stability and the predictability of the phase variations along the propagation path. Long term studies indicate that the Omega signal phase, which varies diurnally in response to diurnal changes in the ionosphere, can be predicted with enough accuracy to enable positioning to within 1 to 2 nautical miles d_{rms} (app. Q).

The model for propagation predictions, based on theoretical and empirical physical principles, is revised periodically to account for changes in solar activity and other propagation anomalies. It is necessary to update the propagation correction tables about every 2 years. Through force fitting, the tabular corrections are adjusted in accordance with local prediction errors determined by monitoring phase data at the transmitting stations and other sites.

4324. The *Omega Propagation Correction Tables* contain the necessary data for correcting Omega receiver readouts, affected by the prevailing propagation conditions, to the standard conditions on which all Omega hyperbolic charts and tables are based. Propagation corrections are *always* necessary.

The corrections are presented in the single station (range) mode so that the navigator only need acquire the tables for the stations and areas desired. Each table contains propagation corrections for the station and area listed on the cover.

The publication number, pertinent suffix followed by the letter C, and the designator of the single stations for which they are computed fully identify each propagation correction table. Using Pub. No. 224 (109–C) D shown in figure 4324a as an example, the 224 designates an Omega publication; the first digit of the suffix (109) identifies the frequency as 10.2 kHz (2 denotes 3.4 kHz); the last two digits of the suffix identify the area of coverage of the table as 09 (fig. 4324b); the letter C indicates that the table is a PPC table as opposed to a lattice table; and the station designator (D) completes the full identification of the table.

Pub. 224 (109-C) D

OMEGA PROPAGATION
CORRECTION TABLES

FOR 10.2 kHz

AREA 09

NORTHWEST PACIFIC

STATION D (NORTH DAKOTA)

Prepared and published by the
DEFENSE MAPPING AGENCY HYDROGRAPHIC CENTER
Washington, D. C. 20390
Reprinted 1976

For sale by authorized Sales Agents of the Defense Mapping Agency Hydrographic Center
DMA STOCK NO. OMPUB224109CD

FIGURE 4324a.—*Omega Propagation Correction Tables.*

Propagation corrections are tabulated for each grid point of a grid constructed for 4° intervals for latitudes between 0° and 45°. The longitude interval is increased to 6° for latitudes between 45° and 60°, and to 8° for latitudes between 60° and 80°. Additionally, propagation corrections for the north and south polar areas between 70° and 90° are tabulated at intervals of 4° of latitude and 20° of longitude.

The corrections for a given coverage area are arranged in order of increasing west longitude and increasing north latitude as shown in figure 4324c. The numbers in the centers of the small quadrangles of this figure are the page numbers of the table in which the corrections are tabulated. Also, the centers of these quadrangles are the grid points at which sets of corrections are computed. As shown in figure 4324d, each set contains a matrix of propagation correction values arranged horizontally by Greenwich mean time (GMT) and vertically in semimonthly periods.

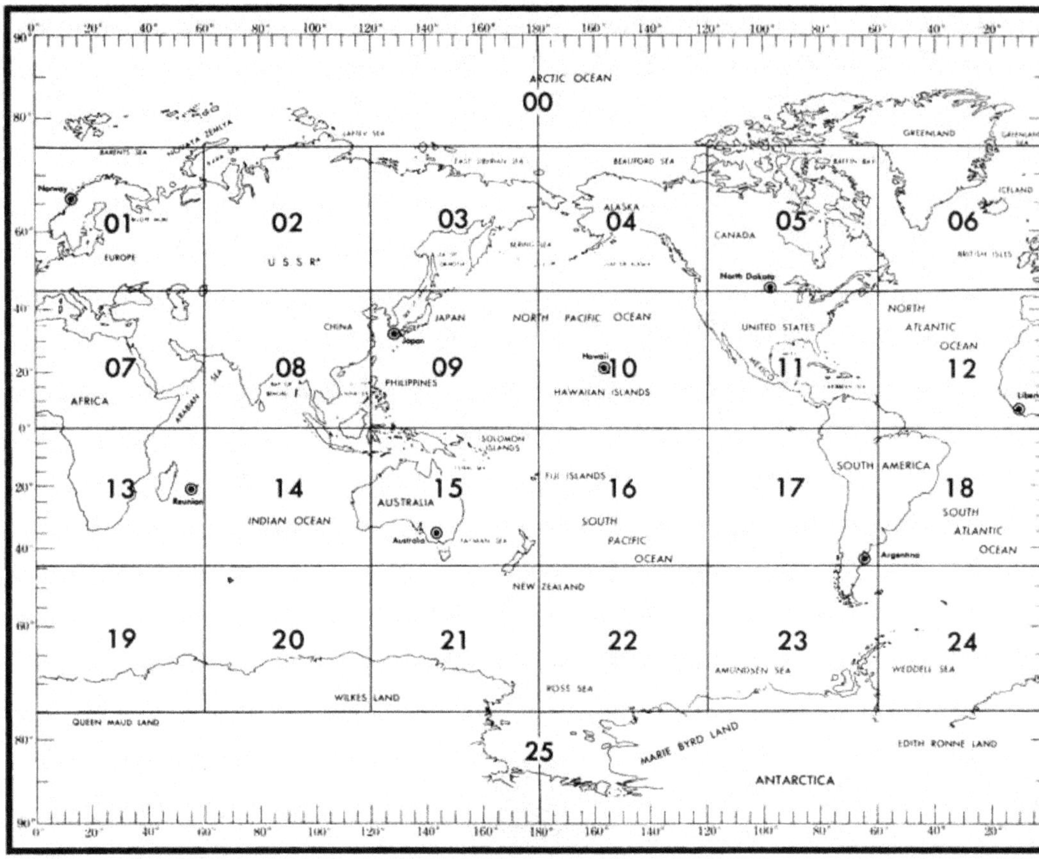

FIGURE 4324b.—Omega table areas.

4325. The propagation corrections effect a conversion from the prevailing propagation conditions to the standard conditions on which the Omega charts and lattice tables are based.

The values as extracted from the tables are in units of centicycles (cecs). The centicycle (cec) is equal to 1 percent of a cycle. Note, however, that the difference in cecs of the phase readings of a station pair is numerically equal to the percentage value of the lane defining the LOP, i.e., the difference in cecs is numerically equal to the number of cels defining the LOP within the lane.

To obtain the propagation correction for a station pair, the correction for each station is determined first. Following such determinations, the correction for the station the designator of which follows the designator of the other station in the alphabet is subtracted algebraically from the propagation correction of the other station. The net value is added algebraically to the phase-difference readout on the receiver for that station pair.

Although extracting the correction data for each station of a pair from the pertinent tables and forming the propagation correction for the station pair are simple processes, the navigator may prefer to construct graphs in advance for a particular area. Since the corrections are for semimonthly periods, the graphs can be used during a 2-week period. These graphs may be constructed for single stations, pairs of stations, or both. As an example, the following corrections are extracted from the pertinent tables:

Propagation Corrections

GMT	*Station A*	*Station B*	*Hyperbolic A-B*
07	−71	−(−36)	−35
08	−71	−(−36)	−35
09	−14	−(−36)	22
10	−05	−(−25)	30
11	−02	−(−08)	06
12	−01	−(−02)	01
13	00	00	00
14	01	−(03)	−02
15	00	−(04)	−04

Note that since B follows A in the alphabet, the correction for station B is subtracted algebraically from the correction for station A.

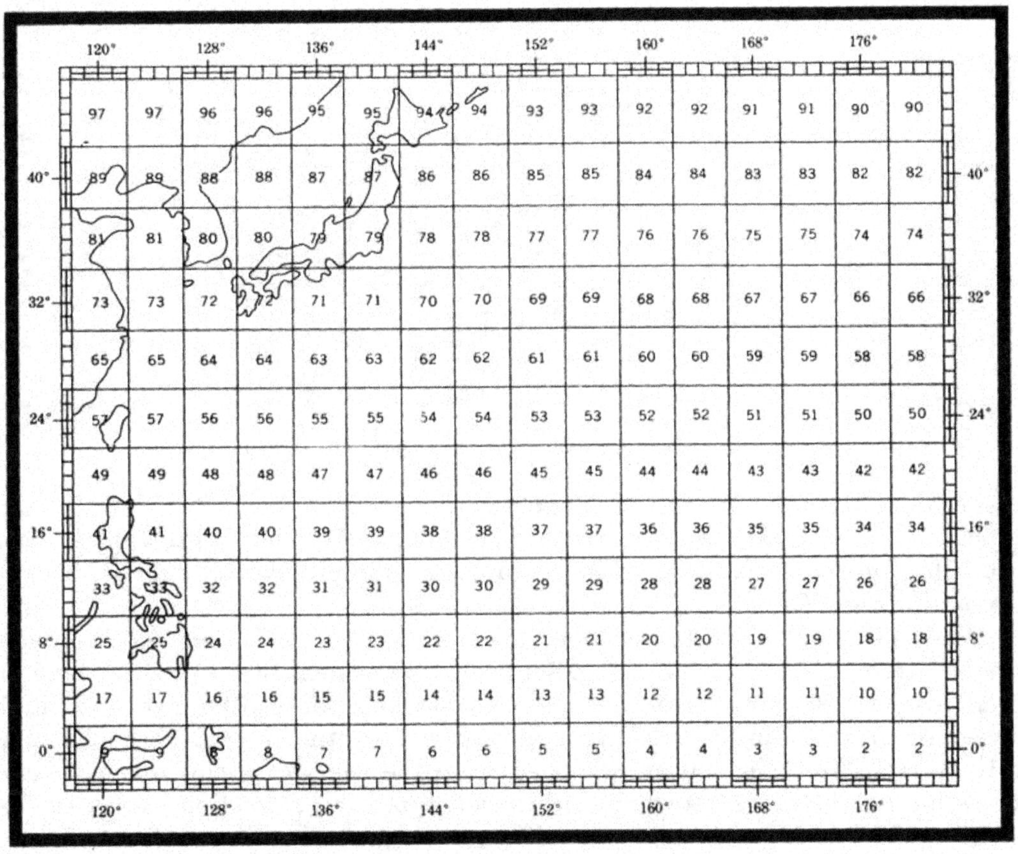

FIGURE 4324c.—Index to corrections for a given coverage area.

As illustrated in figure 4325a, a simple graph can be constructed which simplifies any interpolation that might be required and depicts any rapid variations in the corrections indicating uncertainty in the validity of the correction. In this graph propagation changes occur quite rapidly about 0830Z. Station A (Norway) corrections only are graphed in figure 4325b. These corrections for 10 January are for a grid point (lat. 16°N, long. 40°W) at which the sun rises at about the same time as

LOCATION 20.0 N 160.0 E
STATION D NORTH DAKOTA

DATE	GMT 00	01	02	03	04	05	06	07	08	09	10	11	12	13	14	15	16	17	18	19	20	21	22	23	24
1-15 JAN	-19	-26	-35	-46	-58	-71	-87	-98	-101	-103	-103	-103	-103	-104	-103	-90	-80	-68	-54	-28	-7	-9	-8	-13	-19
16-31 JAN	-15	-23	-33	-42	-54	-68	-84	-97	-101	-102	-103	-103	-103	-104	-102	-88	-78	-65	-50	-25	-5	-7	-5	-7	-15
1-14 FEB	-8	-19	-28	-38	-49	-63	-80	-95	-100	-102	-103	-103	-103	-104	-98	-85	-73	-61	-45	-20	-3	-4	-2	-3	-8
15-29 FEB	-3	-12	-22	-33	-45	-59	-76	-92	-99	-102	-103	-103	-103	-103	-92	-78	-68	-56	-38	-13	-2	-2	1	1	-3
1-15 MAR	1	-7	-18	-28	-41	-55	-72	-90	-99	-102	-103	-103	-103	-100	-85	-73	-63	-50	-32	-7	-1	0	3	4	1
16-31 MAR	4	-2	-12	-24	-36	-50	-68	-87	-98	-101	-103	-103	-103	-93	-79	-69	-58	-44	-24	-2	0	2	5	6	4
1-15 APR	7	2	-8	-19	-31	-45	-63	-84	-97	-101	-102	-103	-99	-84	-73	-63	-51	-37	-16	0	2	5	7	8	7
16-30 APR	8	4	-4	-15	-27	-41	-58	-80	-96	-100	-102	-103	-92	-79	-67	-58	-46	-30	-10	1	3	6	8	9	8
1-15 MAY	9	7	0	-11	-23	-37	-54	-77	-94	-99	-102	-99	-85	-73	-63	-53	-40	-26	-6	2	4	8	9	10	9
16-31 MAY	10	8	2	-8	-20	-33	-49	-72	-92	-99	-101	-93	-81	-69	-60	-50	-37	-23	-3	2	5	8	10	11	10
1-15 JUN	11	9	3	-5	-17	-30	-46	-68	-89	-98	-100	-90	-78	-67	-59	-49	-37	-21	-3	2	6	9	11	11	11
16-30 JUN	11	9	4	-4	-16	-29	-43	-66	-88	-97	-100	-90	-78	-67	-60	-49	-37	-21	-3	2	5	9	11	11	11
1-15 JUL	11	9	4	-5	-16	-29	-44	-66	-88	-97	-100	-91	-79	-69	-60	-51	-37	-23	-5	2	5	8	10	11	11
16-31 JUL	10	9	3	-6	-18	-31	-46	-68	-89	-98	-101	-96	-83	-72	-62	-51	-42	-25	-7	2	5	8	10	11	10
1-15 AUG	10	7	1	-10	-22	-35	-51	-74	-92	-99	-102	-101	-87	-76	-66	-55	-45	-29	-10	1	4	7	9	10	10
16-31 AUG	8	5	-3	-15	-27	-40	-57	-79	-95	-100	-102	-103	-93	-80	-69	-57	-48	-33	-13	1	3	6	8	9	8
1-15 SEP	6	1	-9	-20	-33	-47	-65	-86	-97	-101	-102	-103	-99	-84	-74	-63	-51	-36	-15	0	2	4	7	7	6
16-30 SEP	3	-4	-15	-27	-39	-54	-73	-91	-99	-102	-103	-103	-103	-90	-77	-65	-55	-39	-19	0	0	3	5	5	3
1-15 OCT	-2	-11	-22	-33	-46	-61	-80	-95	-100	-102	-103	-103	-103	-96	-82	-71	-59	-44	-24	-1	-1	1	3	3	-2
16-31 OCT	-7	-18	-28	-39	-52	-67	-85	-97	-101	-102	-103	-103	-103	-100	-85	-76	-63	-49	-28	-3	-3	-1	1	-1	-7
1-15 NOV	-14	-24	-34	-44	-57	-72	-89	-99	-101	-103	-103	-103	-103	-103	-92	-79	-69	-55	-36	-7	-5	-4	-3	-6	-14
16-30 NOV	-19	-27	-36	-48	-60	-75	-90	-99	-102	-103	-103	-103	-103	-104	-96	-85	-73	-61	-43	-11	-7	-7	-6	-11	-19
1-15 DEC	-21	-29	-38	-49	-61	-75	-91	-99	-102	-103	-103	-103	-103	-104	-100	-88	-77	-66	-48	-18	-8	-9	-8	-16	-21
16-31 DEC	-20	-28	-38	-48	-60	-74	-89	-99	-102	-103	-103	-103	-103	-104	-102	-91	-79	-68	-52	-26	-8	-10	-9	-15	-20

LOCATION 20.0 N 160.0 E
STATION H JAPAN

DATE	GMT 00	01	02	03	04	05	06	07	08	09	10	11	12	13	14	15	16	17	18	19	20	21	22	23	24
1-15 JAN	-9	-8	-7	-7	-8	-9	-11	-21	-36	-43	-46	-47	-48	-48	-48	-48	-48	-48	-48	-48	-39	-21	-9	-10	-9
16-31 JAN	-9	-8	-7	-7	-7	-8	-10	-18	-33	-43	-46	-47	-48	-48	-48	-48	-48	-48	-48	-48	-39	-21	-8	-10	-9
1-14 FEB	-9	-7	-6	-6	-7	-8	-9	-15	-30	-41	-45	-47	-48	-48	-48	-48	-48	-48	-48	-48	-37	-19	-8	-10	-9
15-29 FEB	-8	-6	-6	-5	-6	-7	-9	-14	-27	-39	-44	-47	-47	-48	-48	-48	-48	-48	-48	-48	-33	-14	-8	-10	-8
1-15 MAR	-7	-6	-5	-5	-5	-7	-8	-12	-25	-38	-44	-46	-47	-48	-48	-48	-48	-48	-48	-48	-28	-9	-9	-9	-7
16-31 MAR	-7	-5	-4	-4	-5	-6	-8	-11	-23	-36	-43	-46	-47	-48	-48	-48	-48	-48	-48	-46	-22	-6	-9	-9	-7
1-15 APR	-6	-5	-4	-4	-5	-6	-8	-11	-21	-34	-43	-46	-47	-48	-48	-48	-48	-48	-48	-41	-15	-7	-10	-8	-6
16-30 APR	-5	-4	-4	-4	-4	-6	-7	-10	-19	-33	-42	-45	-47	-48	-48	-48	-48	-48	-48	-35	-8	-9	-9	-7	-5
1-15 MAY	-5	-4	-3	-3	-4	-5	-7	-10	-18	-31	-41	-45	-47	-48	-48	-48	-48	-48	-48	-28	-6	-9	-9	-7	-5
16-31 MAY	-5	-4	-3	-3	-4	-5	-7	-9	-17	-29	-39	-45	-47	-47	-48	-48	-48	-48	-48	-24	-6	-10	-9	-7	-5
1-15 JUN	-5	-4	-3	-3	-4	-5	-7	-9	-15	-27	-38	-44	-46	-47	-48	-48	-48	-48	-48	-22	-7	-10	-9	-7	-5
16-30 JUN	-5	-4	-3	-3	-4	-5	-7	-9	-15	-26	-37	-44	-46	-47	-48	-48	-48	-48	-48	-23	-7	-10	-9	-7	-5
1-15 JUL	-5	-4	-3	-3	-4	-5	-7	-9	-14	-26	-38	-44	-46	-47	-48	-48	-48	-48	-48	-24	-6	-10	-9	-7	-5
16-31 JUL	-5	-4	-3	-3	-4	-5	-7	-9	-15	-27	-38	-44	-46	-47	-48	-48	-48	-48	-48	-29	-6	-9	-9	-7	-5
1-15 AUG	-5	-4	-3	-3	-4	-5	-7	-9	-16	-29	-40	-45	-47	-47	-48	-48	-48	-48	-48	-34	-8	-9	-10	-7	-5
16-31 AUG	-6	-4	-4	-4	-4	-5	-7	-10	-19	-32	-41	-45	-47	-48	-48	-48	-48	-48	-48	-38	-11	-8	-10	-7	-6
1-15 SEP	-6	-5	-4	-4	-5	-6	-8	-11	-22	-36	-43	-46	-47	-48	-48	-48	-48	-48	-48	-40	-15	-7	-10	-8	-6
16-30 SEP	-6	-5	-4	-4	-5	-7	-8	-13	-26	-39	-44	-46	-47	-48	-48	-48	-48	-48	-48	-42	-18	-7	-9	-8	-6
1-15 OCT	-7	-5	-5	-5	-6	-7	-9	-16	-31	-41	-45	-47	-48	-48	-48	-48	-48	-48	-48	-43	-20	-7	-9	-9	-7
16-31 OCT	-7	-6	-6	-6	-7	-8	-10	-20	-34	-43	-46	-47	-48	-48	-48	-48	-48	-48	-48	-45	-24	-9	-9	-9	-7
1-15 NOV	-8	-7	-6	-6	-7	-9	-12	-23	-37	-44	-46	-47	-48	-48	-48	-48	-48	-48	-48	-47	-28	-12	-9	-10	-8
16-30 NOV	-9	-7	-7	-7	-8	-9	-13	-25	-38	-44	-46	-47	-48	-48	-48	-48	-48	-48	-48	-47	-30	-13	-9	-10	-9
1-15 DEC	-9	-8	-7	-7	-8	-9	-13	-25	-38	-44	-46	-47	-48	-48	-48	-48	-48	-48	-48	-48	-35	-18	-9	-10	-9
16-31 DEC	-9	-8	-7	-7	-8	-9	-12	-24	-38	-44	-46	-47	-48	-48	-48	-48	-48	-48	-48	-48	-37	-19	-9	-10	-9

FIGURE 4324d.—Extracts from *Omega Propagation Correction Tables.*

at station A. Hence, propagation changes over most of the 2,300 nautical mile propagation path from northern Norway to 16°N, 40°W occur quite rapidly about 0830Z. The prudent procedure would be to use a different station pair at this time for obtaining an LOP.

When forming the 3.4 kHz coarse lane from the 10.2 kHz and 13.6 kHz transmissions from a station pair and determining in which 10.2 kHz fine lane the observer is located by means of the equivalent 3.4 kHz phase-difference reading, the phase-difference reading at 10.2 kHz is subtracted from the phase difference reading at 13.6 kHz. The correction as extracted from the *Omega Propagation Correction Tables* for 10.2 kHz is subtracted algebraically from the correction extracted from the tables for 13.6 kHz. The algebraic difference of the corrections so found for each station of the pair is then applied to the phase difference reading.

4326. Omega plotting charts.—Several series of charts have either been produced or are under development for use with the Omega system. The 7600 and 7700 series, designed to give global coverage on the Mercator projection with a scale of 1:2,187,400 (1 inch = 30 nautical miles), are plotting sheets overprinted with Omega zero phase-difference contours for standard propagation conditions.

On these plotting charts the LOP for a corrected receiver reading of a station pair is constructed by graphical linear interpolation between the charted LOPs.

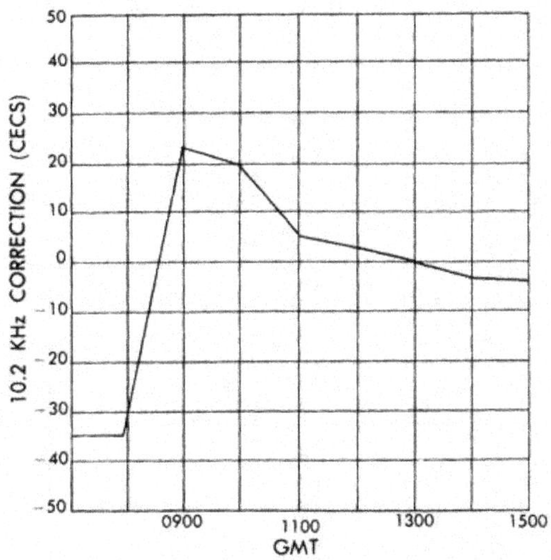

FIGURE 4325a.—Diurnal effects.

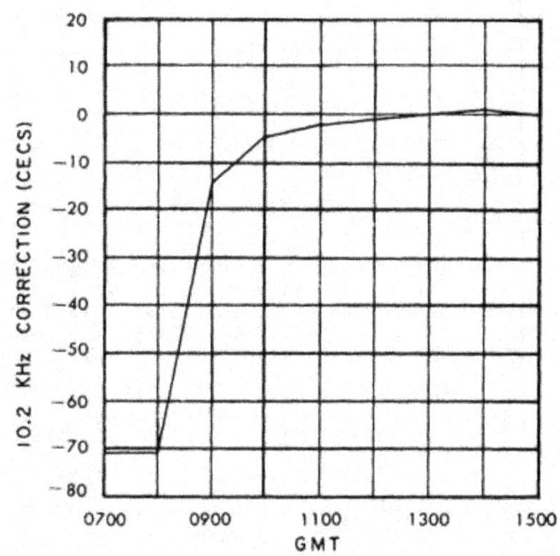

FIGURE 4325b.—Diurnal effects.

Notice to Mariners are not issued for changes to hydrographic information on these plotting charts. The mariner should consult the nautical charts covering the same area for hydrographic information.

4327. Lattice tables.—It is not feasible to provide the great number of large-scale charts required for every application of the Omega System. The counterpart to the Omega chart is the lattice table (Omega charting coordinate table) contained in the Pub. 224 Series. Through use of this table, Omega lines of position can be plotted on a suitable plotting sheet or chart having a scale as large as 1:800,000.

Because all Omega frequencies are integral multiples of a common base frequency (1.133 kHz), there is no real need to publish lattice tables for frequencies other than 10.2 kHz. For instance, every third 10.2 kHz hyperbola is also a 3.4 kHz hyperbola; every fourth 13.6 kHz hyperbola is also a 3.4 kHz hyperbola. The tables are computed for hyperbolic lanes at a frequency of 10.2 kHz. Every 10.2 kHz hyperbola, exactly divisible by three (897,900,903, etc.) is also a 3.4 kHz hyperbola.

The numbering of the lanes established by a station pair increases in the direction away from that station of the pair the designator of which appears first in the alphabet; the 10.2 kHz lane at the perpendicular bisector of the baseline of the pair is assigned a value of 900. A fictitious minimum lane count is inserted in the lattice computations to provide this 10.2 kHz lane count of 900 lanes on the perpendicular bisector of the baseline.

For each coverage area as shown in figure 4324b there is a separate Omega Table (lattice table) for each station pair that can be used in that area.

The publication number, pertinent suffix, and station pair fully identify each lattice table. Using Pub. No. 224 (109), Pair D–H, as an example, the 224 designates an Omega publication; the first digit of the suffix (109) identifies the frequency as 10.2 kHz; the last two digits of the suffix identify the area of coverage of the table as area 09; the station pair (D–H) completes the full identification of the table.

As shown in figure 4327, the columns are headed by lane values at intervals of one lane in most cases. Lane values corresponding to the baseline extensions are so marked.

Because Omega isophase contours fan out in all directions, it is sometimes necessary to tabulate the latitude at which the contours intersect meridians; at

T		D-H 1090		D-H 1091		D-H 1092		D-H 1093		D-H 1094		T
Lat ° ′		° ′	Δ	° ′	Δ	° ′	Δ	° ′	Δ	° ′	Δ	Long ° ′
0		162 16.2E	-121	162 04.0E	-121	161 51.9E	-122	161 39.6E	-122	161 27.4E	-122	
1	N	162 18.7E	-120	162 06.5E	-121	161 54.4E	-121	161 42.3E	-121	161 30.1E	-121	
2	N	162 20.7E	-120	162 08.7E	-120	161 56.6E	-120	161 44.5E	-120	161 32.5E	-120	
3	N	162 22.4E	-119	162 10.4E	-119	161 58.5E	-119	161 46.4E	-120	161 34.4E	-120	
4	N	162 23.7E	-119	162 11.8E	-119	161 59.9E	-119	161 47.9E	-119	161 36.0E	-119	
5	N	162 24.6E	-118	162 12.8E	-118	162 00.9E	-118	161 49.0E	-118	161 37.1E	-118	
6	N	162 25.1E	-117	162 13.4E	-117	162 01.6E	-118	161 49.7E	-118	161 37.9E	-118	
7	N	162 25.2E	-116	162 13.6E	-117	162 01.8E	-117	161 50.0E	-117	161 38.3E	-117	
8	N	162 24.9E	-116	162 13.3E	-116	162 01.6E	-116	161 49.9E	-117	161 38.2E	-117	
9	N	162 24.2E	-115	162 12.6E	-115	162 01.0E	-116	161 49.3E	-116	161 37.7E	-116	
10	N	162 23.0E	-115	162 11.5E	-115	161 59.9E	-115	161 48.3E	-115	161 36.7E	-115	
11	N	162 21.4E	-114	162 09.9E	-115	161 58.3E	-115	161 46.8E	-115	161 35.3E	-115	
12	N	162 19.2E	-114	162 07.8E	-114	161 56.3E	-114	161 44.8E	-114	161 33.3E	-114	
13	N	162 16.6E	-113	162 05.3E	-114	161 53.8E	-114	161 42.3E	-114	161 30.9E	-114	
14	N	162 13.5E	-113	162 02.2E	-113	161 50.8E	-114	161 39.3E	-114	161 27.9E	-114	
15	N	162 09.8E	-113	161 58.5E	-113	161 47.2E	-113	161 35.8E	-113	161 24.5E	-113	
16	N	162 05.6E	-112	161 54.3E	-112	161 43.0E	-112	161 31.7E	-112	161 20.4E	-112	
17	N	162 00.8E	-112	161 49.5E	-112	161 38.2E	-112	161 27.0E	-112	161 15.7E	-112	
18	N	161 55.3E	-112	161 44.1E	-112	161 32.9E	-111	161 21.7E	-112	161 10.4E	-112	
19	N	161 49.2E	-111	161 38.1E	-111	161 26.9E	-112	161 15.6E	-112	161 04.4E	-111	
20	N	161 42.4E	-111	161 31.3E	-111	161 20.1E	-111	161 09.0E	-111	160 57.8E	-111	
21	N	161 34.9E	-110	161 23.8E	-111	161 12.7E	-111	161 01.6E	-111	160 50.4E	-111	
22	N	161 26.7E	-110	161 15.6E	-110	161 04.5E	-111	160 53.4E	-111	160 42.3E	-111	
23	N	161 17.6E	-110	161 06.6E	-110	160 55.5E	-111	160 44.3E	-111	160 33.3E	-110	
24	N	161 07.7E	-110	160 56.6E	-110	160 45.6E	-110	160 34.5E	-111	160 23.3E	-111	
25	N	160 56.9E	-110	160 45.9E	-110	160 34.8E	-111	160 23.6E	-110	160 12.6E	-110	
26	N	160 45.1E	-110	160 34.1E	-110	160 23.0E	-111	160 11.9E	-110	160 00.8E	-110	
27	N	160 32.3E	-110	160 21.3E	-110	160 10.1E	-111	159 59.0E	-110	159 48.0E	-110	
28	N	160 18.4E	-110	160 07.3E	-110	159 56.2E	-111	159 45.1E	-111	159 34.0E	-110	
29	N	160 03.3E	-110	159 52.2E	-111	159 41.0E	-111	159 29.9E	-111	159 18.8E	-110	
30	N	159 46.9E	-110	159 35.8E	-111	155 24.6E	-111	159 13.4E	-111	159 02.3E	-111	
31	N	159 29.2E	-111	159 18.0E	-111	159 06.8E	-112	158 55.6E	-111	158 44.4E	-111	
32	N	159 10.0E	-112	158 58.8E	-112	158 47.5E	-112	158 36.3E	-112	158 25.1E	-112	
33	N	158 49.3E	-112	158 37.9E	-113	158 26.6E	-112	158 15.3E	-112	158 04.1E	-113	
34	N	158 26.8E	-113	158 15.4E	-114	158 04.0E	-113	157 52.6E	-113	157 41.3E	-113	
35	N	158 02.4E	-114	157 50.9E	-114	157 39.5E	-114	157 28.0E	-114	157 16.5E	-115	
36	N	157 36.1E	-116	157 24.4E	-115	157 12.9E	-115	157 01.4E	-115	156 49.7E	-116	
37	N	157 07.5E	-117	156 55.8E	-116	156 44.1E	-116	156 32.4E	-117	156 20.7E	-117	
38	N	156 36.5E	-118	156 24.7E	-118	156 12.9E	-118	156 01.0E	-119	155 49.1E	-118	
39	N	156 03.0E	-119	155 51.0E	-120	155 38.9E	-120	155 26.8E	-120	155 14.8E	-120	
40	N	155 26.5E	-121	155 14.3E	-122	155 02.0E	-122	154 49.7E	-122	154 37.4E	-122	
41	N	154 46.8E	-124	154 34.3E	-125	154 21.7E	-125	154 09.2E	-124	153 56.8E	-125	
42	N	154 03.5E	-127	153 50.7E	-128	153 37.8E	-128	153 25.1E	-127	153 12.2E	-128	
43	N	153 16.1E	-131	153 03.0E	-131	152 49.9E	-131	152 36.7E	-132	152 23.5E	-132	
44	N	152 24.4E	-134	152 10.9E	-135	151 57.3E	-135	151 43.7E	-135	151 30.1E	-136	
45	N	151 27.5E	-139	151 13.5E	-140	150 59.4E	-140	150 45.3E	-140	150 31.2E	-141	
T		D-H 1090		D-H 1091		D-H 1092		D-H 1093		D-H 1094		T

FIGURE 4327.—Extract from Omega Table.

other times it is necessary to tabulate the longitudes at which the contours intersect parallels of latitude. The tables are arranged so that the latitude always appears to the left of the longitude.

For the entire area of coverage, points on the isophase contours are tabulated at intervals of 1° of latitude or longitude from the equator to 60° north and south. From 60° latitude to 80° latitude, the points are tabulated at intervals of 1° of latitude or 2° of longitude. From 80° latitude to the pole, the points are tabulated at intervals of 1° of latitude or 5° of longitude.

Close to the transmitting stations where the hyperbolic contours curve sharply, additional points are tabulated at intervals of 15 minutes of arc. The spacing of the points is such that the navigator can safely use a straight line between any two adjacent tabulated points as an approximation of the hyperbola. Within approximately *20 miles* of a transmitting station, the curvature of the hyperbolic contours is excessive. The navigator is *cautioned* that *straight line segments* will introduce *appreciable errors.* A plot of three consecutive points will indicate the amount of curvature and the actual contour. When operating within *450 nautical miles* from a transmitting station, *caution* should be exercised because transmissions from the station may not be reliable. The use of the signals when within 450 nautical miles of a transmitting station should be avoided if possible.

4328. Interpolation.—The usual situation is that the LOP corresponding to a corrected receiver reading lies *within* a lane, i.e., the LOP usually lies between adjacent zero phase-difference contours for standard propagation conditions. Since the lattice table tabulations are points on these contours, interpolation is usually required.

To facilitate linear interpolation in the lattice tables, the rate of change of latitude or longitude per lane is tabulated immediately to the right of each tabulated point on the zero phase-difference contour or lane boundary. This Δ value is only an average rate of change for a two-lane band. It is assumed to be correct at the tabulated point of the middle hyperbola. For this reason it is always necessary to interpolate from the nearest tabulated lane value, T. The difference between the corrected receiver reading and tabular lane is multiplied by Δ. This product is added algebraically to the value of latitude or longitude tabulated under the column headed by T. Either Δ or the difference between the corrected receiver reading and the tabulated lane value (T) can be negative.

The Δ value is in units of minutes of arc, the last digit being tenths of minutes of arc. However, the decimal is not shown in the tabulation.

As the baseline extensions are approached, accurate interpolation is impossible. In these areas, values for Δ have been omitted.

Example.—The 0400Z DR position of a ship on 5 February is latitude 21°36′N, longitude 161°15′E. The 0400Z Omega reading for station pair D–H is as follows:

0400Z Pair D–H 1090.15.

Required.—The 0400Z Omega LOP.

Solution.—The appropriate predicted propagation correction (PPC) must be applied to the observed receiver reading to obtain the corrected receiver reading. Therefore, enter the PPC tables for the pertinent stations with the DR position. Select the PPC for the closest geographic intersection to the ship's DR position for the proper time and date. Combine the PPC's for the individual stations comprising the pair by algebraically subtracting the correction for the second station alphabetically of the pair from the correction for the first station. Apply the correction to the uncorrected receiver reading. Enter the lattice table in the T column nearest to the value of the corrected receiver reading and with the latitudes or longitudes closest to and on each side of the DR position. Extract the corresponding longitudes or latitudes for these two points. Interpolate if necessary. Plot the two points thus obtained. Connect these two points with a straight line to determine the Omega LOP.

PPC table excerpts for Stations D and H are as follows (see figure 4324d):

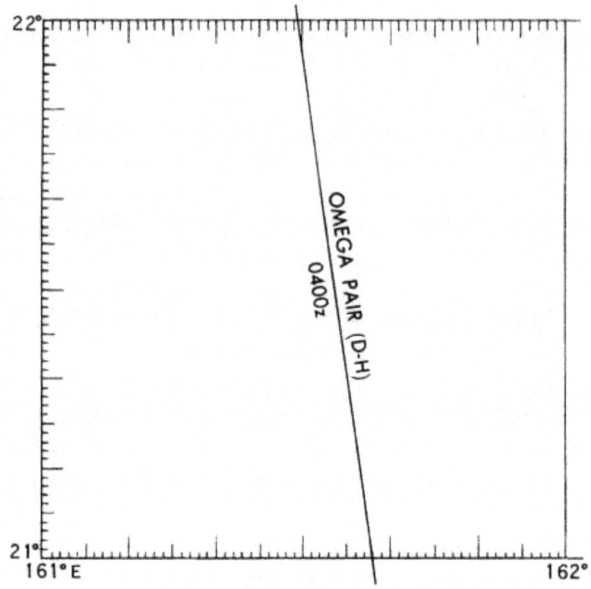

FIGURE 4328.—Plot of Omega line of position.

PPC for Station D	−.49
PPC for Station H	−.07
PPC for Pair D–H=(−.49)−(−.07)	−.42
Pair D–H lattice table excerpt (see figure 4327):	
Receiver reading (uncorrected)	1090.15
PPC for Pair D–H	−.42
Corrected receiver reading	1089.73
T	1090.00
Diff	−.27

Latitude	*Tabulated longitude*	Δ
21°N	161°34′.9E	−11.0
22°N	161°26′.7E	−11.0

Longitude change (diff.×Δ)	*Interpolated longitude*
(−.27)×(−11.0)= +3′.0	161°37′.9E
(−.27)×(−11.0)= +3′.0	161°29′.7E

Answer.—The LOP is plotted through the following positions: latitude 21°N, longitude 161°37′.9E, and latitude 22°N, longitude 161°29′.7E (see figure 4328).

Loran-C

4329. Loran-C is a long range hyperbolic radionavigation system of high accuracy which processes a pulsed LF (100 kHz) signal by both the time difference and phase comparison methods. Ranges of 800 to 1,200 nautical miles are obtainable when using the groundwave, depending upon transmitter power, signal-to-noise ratio in the service area, receiver sensitivity, and losses over the signal path. Wide coverage areas are made possible by the low propagation losses of LF groundwaves, by the use of pulse groups to increase the average power without any increase in the peak power, and by the use of long baselines which, as compared to shorter baselines, serve to increase the total area where the angles of cut of the lines of position are 15° or greater.

The normal propagation mode is the groundwave. The skywave propagation mode can be used with reduced accuracy beyond the area of groundwave reception. One-hop skywaves may be received at ranges up to about 2,300 nautical miles.

Although it is designed and operated for use in the hyperbolic mode, Loran-C in the Range-Range mode (art. 4304) can be used to extend the coverage area within which accurate fixes are obtainable.

The low frequency (90–110 kHz) signals used in the Loran-C system do not suffer from the high propagation losses over land associated with medium frequency (MF) and high frequency (HF) signals. The MF and HF signals also suffer loss of propagation predictability due to natural and man-made features whose sizes are significant fractions of a wavelength. Higher frequency signals are range limited to line-of-sight. Thus, the center frequency of 100 kHz was selected for Loran-C to benefit from the stable propagation characteristics and long range of the LF band.

4330. Loran-C chains are comprised of a master transmitting station, two or more secondary (slave) transmitting stations and, if necessary, system area monitor stations. The transmitting stations are located so that the signals from the master and at least two secondary stations can be received throughout the desired coverage or service area. For convenience, the master station is designated by the letter *M* and the secondary stations are designated *W, X, Y,* or Z. Thus, a particular master-secondary pair and the time difference (TD) which it produces can be referred to by the letter designations of both stations or just that of the secondary, e.g. MX time difference or TDX (fig. 4330).

4331. The Loran-C signal format.—The transmitting stations of a Loran-C chain transmit groups of pulses at a specified **group repetition interval (GRI).** Each pulse has a 100 kHz carrier and is of the shape shown in figure 4331a. The shape is such that 99 percent of the radiated energy is contained between the frequencies of 90 and 110 kHz. For each chain a minimum GRI is selected of sufficient length so that it contains time for transmission of the pulse group from each station (10,000 microseconds for the master and 8,000 microseconds for each secondary) plus time between each pulse group so that signals from two or more stations cannot overlap in time anywhere in the coverage area (fig. 4331b). Thus, with respect to the time of arrival of the master, a secondary station will delay its own transmissions for a specified time, called the **secondary coding delay.** The minimum GRI is therefore a direct function of the number of stations and the distance between them. A GRI for the chain is then selected so that adjacent chains do not cause mutual (cross-rate) interference. Possible values for GRI are listed in table 4331a. The GRI is defined to begin coincident with the start of the first pulse of the master group.

Each station transmits one pulse group per GRI. The master pulse group consists of eight pulses spaced 1,000 microseconds apart, and a ninth pulse 2,000 microseconds after the eighth. Secondary pulse groups contain eight pulses spaced 1,000 microseconds apart. Multiple pulses are used so that more signal energy is available at the receiver, improving significantly the signal-to-noise ratio without having to increase the peak transmitted power capability of the transmitters. The master's ninth pulse is used for visual identification of the master, and for blink. **Blink,** used to warn users that there is an error in the transmissions of a particular station, is accomplished by turning the ninth pulse on and off in a specified code as shown in table 4331b. The secondary station of the unusable pair also blinks by turning its first two pulses on and off. Most modern receivers automatically detect secondary station blink only, as this is sufficient to trigger alarm indicators.

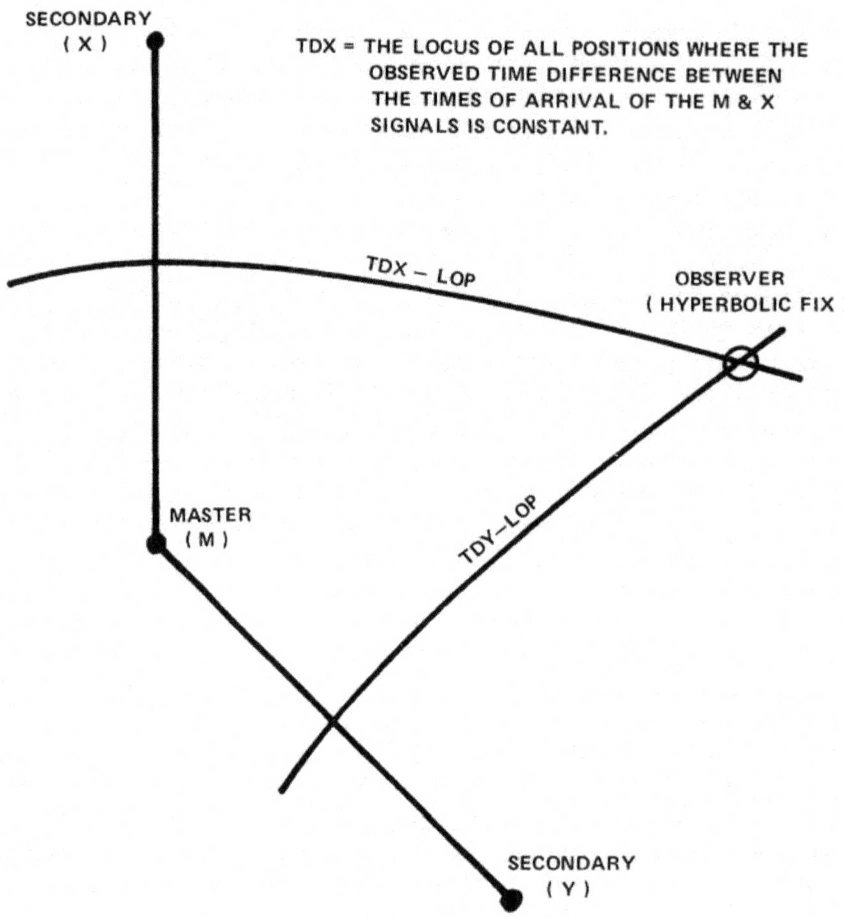

FIGURE 4330.—Loran-C chain.

The rate structure for Loran-C is limited in theory to GRIs of 00010 to 99990 microseconds in 10 microsecond steps. In actual practice the GRIs are between 40000 and 99990 microseconds with limits placed on rates actually selected. The designation of a Loran-C rate is by the first four digits of the specific GRI. This is a newly expanded rate structure. The old rate structure consisted of those GRIs with old rate designations shown in parentheses (thus rate SS7 is now referred to as 9930).

4332. Avoiding skywave contamination.—A skywave may arrive at a receiver as little as 35 microseconds (μs) or as much as 1,000 μs after the groundwave. In the first case, the skywave will overlap its own groundwave while in the second case the skywave will overlap the groundwave of the succeeding pulse. Either case will cause distortion of the received signal in the form of fading and pulse shape changes. Large positional errors would result if these conditions were not accounted for in the selection of the Loran-C signal format, and the design of the receivers.

The early arriving skywave is overcome by making time of arrival measurements on the first part of the pulse. This ability is enhanced by the fast-rising pulse (fig. 4331a), achieving high power prior to the arrival of the skywaves. The shape of the pulse also allows the receiver to identify one particular cycle of the 100-kHz carrier. This is essential to prevent whole cycle ambiguities in the time-difference

9999		8999	7999		6999	5999		4999	
9998		8998	7998		6998	5998		4998	
9997		8997	7997		6997	5997		4997	
•		•	•		•	•		•	
•		•	•		•	•		•	
•		•	•		•	•		•	
9991		8991	7991		6991	5991		4991	
9990	(SS1)	8990	7990	(SL1)	6990	5990	(SH1)	4990	(S1)
9989		8989	7989		6989	5989		4989	
•		•	•		•	•		•	
•		•	•		•	•		•	
•		•	•		•	•		•	
9971		8971	7971		6971	5971		4971	
9970	(SS3)	8970	7970	(SL3)	6970	5970	(SH3)	4970	(S3)
9969		8969	7969		6969	5969		4969	
•		•	•		•	•		•	
•		•	•		•	•		•	
•		•	•		•	•		•	
9931		8931	7931		6931	5931		4931	
9930	(SS7)	8930	7930	(SL7)	6930	5930	(SH7)	4930	(S7)
9929		8929	7929		6929	5929		4929	
•		•	•		•	•		•	
•		•	•		•	•		•	
•		•	•		•	•		•	
9000		8000	7000		6000	5000		4000	

TABLE 4331a.—Group repetition intervals (GRI in tens of microseconds).

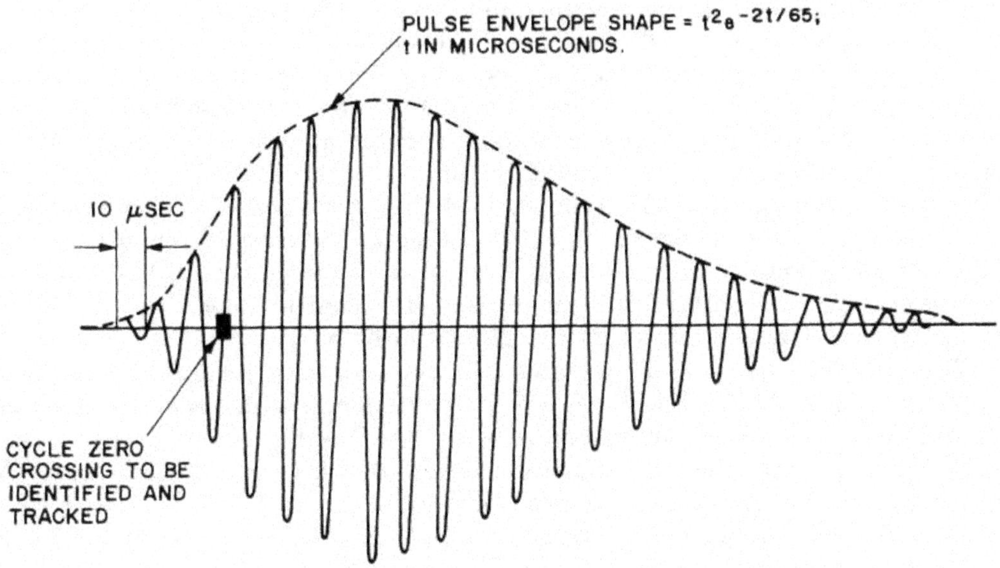

FIGURE 4331a.—Loran-C pulse.

measurement and allows the high accuracy of the phase measurement system to be achieved.

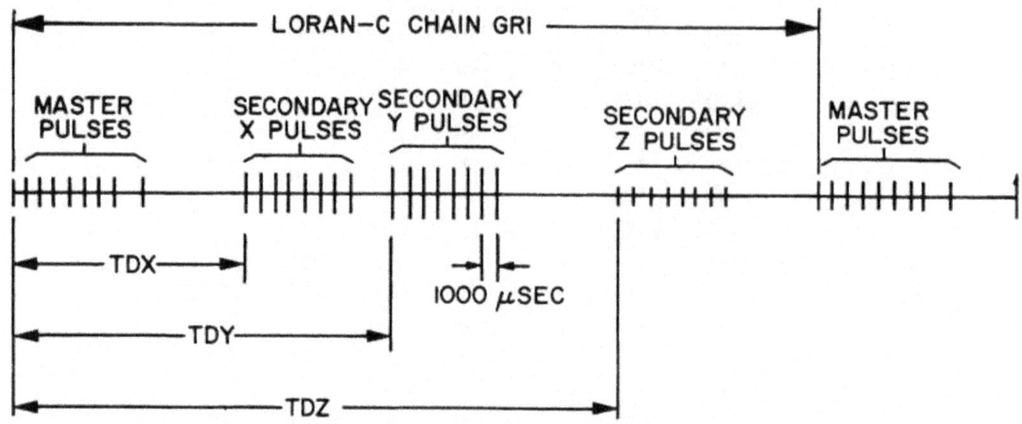

FIGURE 4331b.—Loran-C signal format.

To prevent the long-delay skywaves from affecting the time-difference measurement, the phase of the 100-kHz carrier is changed in each pulse of a group in accordance with a predetermined pattern. The phase codes for Loran-C are shown in table 4331c. The different phase codes for the master and secondary signals also allow automatic receivers to use the code for master and secondary station identification.

4333. Synchronization control.—All transmitting stations are equipped with cesium frequency standards. The high stability and accuracy of these standards permit each station to derive its own time of transmission without reference to another station.

The objective for control of a Loran-C chain is to maintain constant the observed time difference (TD) of each master-secondary pair throughout the coverage area. Frequency offsets in the cesium standards and changes in propagation conditions can cause the observed TD to vary. Therefore, one or more **system area monitor (SAM)** stations with precision receiving equipment are established in the coverage area to monitor continuously the TDs of the master-secondary pairs. In some cases a transmitting station is suitably located and performs the SAM function. A control TD is established through calibration (art. 4334). When the observed TD varies from the control TD by one-half of the prescribed control tolerance, the SAM directs a change in the timing of the secondary station to remove the error. The control tolerance is plus or minus 200 nanoseconds or better. If the observed TD differs from the control TD by more than the control tolerance, "blink" (art. 4331) is ordered to advise users that the TD is unusable.

The Loran-C system as it operates has maintained a record of 99.7 percent availability, not including scheduled off-air maintenance, which reduces that figure to about 99 percent. New equipment is being developed which will permit on-air maintenance, and also improve the system availability, with a goal of better than 99.7 percent, including all interruptions of service.

4334. System calibration.—When a Loran-C chain is established (or when a secondary is added) and periodically thereafter, chain calibration is conducted. The purpose of the calibration is to record the Loran-C time differences at a number of known geographical points in the coverage area. This information is then used to:

1. verify the initial chain synchronization to ensure that the chain performs as advertised;

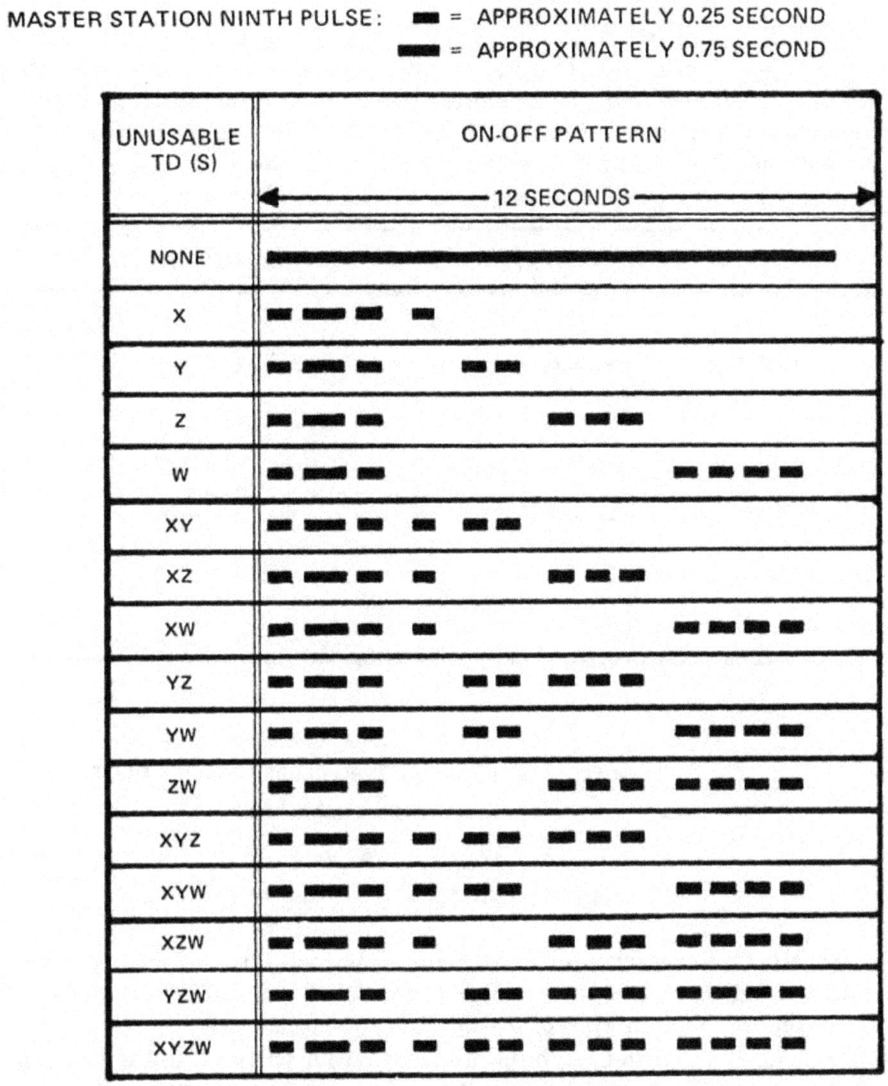

SECONDARY STATION FIRST TWO PULSES:

TURNED ON (BLINKED) FOR APPROXIMATELY 0.25 SECONDS EVERY 4.0 SECONDS. ALL SECONDARIES USE SAME CODE, AUTOMATICALLY RECOGNIZED BY MOST MODERN LORAN–C RECEIVERS.

TABLE 4331b.—Loran-C blink code.

2. establish the control time differences, which are then used as the reference for measuring synchronization control;

3. ensure the accuracy of existing control time differences;

4. provide survey data for accurate charting and for use in determining surface conductivities.

In performing a calibration, an effort is made to distribute the monitor sites uniformly over the coverage area. Each site is visited for a minimum period of 4 hours, during which time the Loran-C signals are monitored precisely and the average value for the period is determined. The transmitting stations and the station for which the control time differences are to be established are also required to conduct precise monitoring. The geographic location of the monitor site (if not already known) is determined at the same time by use of satellite positioning. The calibration data is then reduced to establish or check the control time differences, and to estimate surface conductivities throughout the coverage area. This improves the ability to predict the time differences at particular locations.

	MASTER	EACH SECONDARY
GRI A	++−−+−+− +	+++++−−+
GRI B	+−−+++++ −	+−+−++−−

NOTE: (+) INDICATES ZERO DEGREE CARRIER PHASE
(−) INDICATES 180° CARRIER PHASE
LORAN–C INTERVALS A&B ALTERNATE IN TIME

TABLE 4331c.—Loran-C phase codes.

4335. Receiver characteristics.—A "true" low-cost Loran-C receiver which will be useful to the limits of the U. S. Coast Guard's advertised coverage area for the Coastal Confluence Zone has the following characteristics:

1. It acquires the Loran-C signals automatically, without the use of an oscilloscope.
2. It accomplishes cycle matching on all pulses to take advantage of the maximum accuracy of the system.
3. It automatically tracks the signals once they have been acquired.
4. It displays two time-difference readings.

There are many Loran-C receivers on the market, exhibiting varying degrees of performance. A few of these receivers are known to meet all of the above requirements. There is no guarantee that all Loran-C receivers will accurately measure the received Loran-C signals out to the limits of the advertised coverage area, since most of them use pulse envelope matching techniques and little or no cycle matching. Some receivers perform all of the desired functions except that they depend on the operator using an oscilloscope for signal acquisitions. In a "noisy" environment (e.g. where the atmospheric or man-made noise level is high compared to that of the desired Loran-C signals), this will be extremely difficult, again limiting the area in which that receiver can be used.

4336. Interference filters.—Like all radionavigation systems, Loran-C can be adversely affected by interference. The effect of interference is to make it difficult to acquire the Loran-C signals or to make the readings fluctuate more than usual or both. Most manufacturers provide tuneable filters which can be used to minimize such interference. If a user is to remain always in the same general area (about 100-mile radius from a center point) it will probably never be necessary to readjust these filters once they are properly set by the manufacturer or his local representative.

If the user is to travel great distances, it will be necessary for him to learn how to readjust the filters. This is not a difficult task following some initial training.

4337. Signal acquisition.—To acquire the Loran-C signals it is important to enter into the receiver the GRI of the local Loran-C chain. Usually, the GRI will remain unchanged as long as the vessel stays on the same coast of the United States.

The speed at which the receiver will find the Loran-C signals depends upon the signal strength and how much noise is present. In some receivers, the user can expedite the process by preselecting the approximate Loran-C readings he expects to read. Most modern receivers will be automatically tracking within 5 minutes of initial turn-on, and will continue to track until the receiver is turned off. If a user is at a known location (at a pier and ready for departure, for example) it will be obvious to him when the receiver is providing the correct information. In any event, most receivers show some type of an alarm which remains lighted until the receiver is tracking properly.

Initially acquiring Loran-C signals when arriving from far out (several hundred miles or more) at sea is a more difficult problem than the one of a vessel at a pier where the Loran-C readings are known. Thus, the receiver may take more time to acquire the signals. When first entering a Loran-C coverage area, the receiver should be checked frequently to ensure that all the alarms are extinguished. Sometimes, due to weak signals and high noise, the receiver alarms will go out even though the receiver is not tracking precisely. However, as the vessel continues to enter the stronger signal area, the receiver will automatically recognize that it has made an error and will give the user an alarm light. This should occur well before the vessel enters the CCZ.

4338. Lattice tables provide the coordinates necessary for the construction of straight line representations of segments of the hyperbolic lines of position. Except when the user of the Loran-C system is within about 20 nautical miles of a transmitting station, the straight line segment joining any two adjacent tabulated points can be used without appreciable error. Should there be doubt about the accuracy of the line of position with regard to curvature, the amount of error can be resolved by plotting an adjoining straight line segment of the line of position.

Pub. 221, *Loran-C Table,* is a series of lattice tables published by the Defense Mapping Agency Hydrographic/Topographic Center. In this series there is a separate lattice table for each station pair of a chain. Each table is identified by the publication number (221), pertinent suffix, and station pair. For example, Pub. 221 (2013) Pair 7960–X is the lattice table for the 7960–X pair in the Gulf of Alaska chain.

Points on hyperbolas separated by 10 microseconds of time difference are tabulated in the lattice tables (fig. 4338a) at intervals of whole degrees of latitude or longitude except in areas close to transmitting stations. In such areas, points are tabulated at intervals of whole degrees or quarter degrees, depending upon the degree of curvature of the line. A separate column is given for each tabulated reading, at 10-microsecond intervals. An auxiliary tabulation labeled Δ (delta) gives the change in longitude or latitude (to 0ʹ.01) for a 1-microsecond change in the time-difference reading. Also, points on the baseline extension are tabulated in a column headed by the time-difference reading on the baseline extension.

Tabulated readings are for groundwaves. Skywave readings are corrected to the equivalent groundwave readings before entering the tables. A groundwave reading is designated T_G, and a skywave reading T_S. If a groundwave is matched with a skywave, the reading is labeled T_{GS} if the groundwave is from the master station, and T_{SG} if from the secondary (slave) station.

The Loran-C lattice table is entered with the groundwave reading in microseconds, and the latitude *or* longitude. For a line running in a generally north-south direction, the table is entered with the latitude, and the corresponding longitude is taken from the table. For an east-west line, the table is entered with longitude, and latitude is taken from the table. Two such points are thus determined and plotted, one on each side of the dead reckoning position. The straight line connecting them is an approximation of a small segment of the line of position.

Interpolation is usually required. For each latitude *or* longitude argument, the lattice table is entered with the tabulated time-difference (T) *nearest* to the groundwave reading. The longitudes *or* latitudes extracted are then corrected by the amounts of the products of (T_G-T) and Δ, the change in longitude *or* latitude for 1-microsecond change in T. Careful attention must be given to the signs of Δ and (T_G-T). The sign of Δ is found by inspection. The sign of (T_G-T) is found by *always* algebraically subtracting T from T_G.

Example 1.—The 1530 DR position of a ship is lat. 48°35ʹN, long. 30°17ʹW. Loran-C readings are obtained, as follows:

1530	7930–X	T_G 29523.8
1530	7930–Z	T_G 48635.7

Required.—The 1530 Loran-C fix.

Solution.—Enter the lattice tables (fig. 4338a) in the T (time difference) column nearest to the value of T_G and with the latitudes or longitudes closest to and on each side of the DR position. Extract the corresponding longitudes or latitudes for these two points. Interpolate if necessary. Plot the two points thus obtained. Connect these two points with a straight line to determine a segment of the hyperbolic line of position. The intersection of two lines for different pairs determines the Loran-C fix (fig. 4338b).

Pair	*7930–X*	*Latitude*	*Tabulated longitude*	Δ	*Longitude change* $(T_G-T)\times(\Delta)$	*Interpolated longitude*
T	29520					
T_G	29523.8	48°N	30°24ʹ.3W	+22	(+3.8)×(+.22)= +0ʹ.8	30°25ʹ.1W
T_G-T	+3.8	49°N	30°21ʹ.5W	+21	(+3.8)×(+.21)= +0ʹ.8	30°22ʹ.3W

Pair	*7930–Z*	*Longitude*	*Tabulated latitude*	Δ	*Latitude change* $(T_G-T)\times(\Delta)$	*Interpolated latitude*
T	48640					
T_G	48635.7	30°W	48°24ʹ.3N	+20	(−4.3)×(+.20)= −0ʹ.9	48°23ʹ.4N
T_G-T	−4.3	31°W	48°47ʹ.5N	+19	(−4.3)×(+.19)= −0ʹ.8	48°46ʹ.7N

Loran-C lines of position are plotted through the following positions: lat. 48°N, long. 30°25ʹ.1W; lat. 49°N, long. 30°22ʹ.3W; and lat. 48°23ʹ.4N, long. 30°W; lat. 48°46ʹ.7N, long. 31°W.

240 GF 7930-X

T Lat ° ′	29500 ° ′	Δ	29520 ° ′	Δ	29540 ° ′	Δ	29560 ° ′	Δ	29580 ° ′	Δ	T Long ° ′
25 N	31 39.6W	36	31 46.8W	35	31 53.9W	35	32 01.1W	36	32 08.3W	36	
26 N	31 36.1W	35	31 43.3W	35	31 50.4W	35	31 57.5W	35	32 04.5W	35	
27 N	31 32.8W	34	31 39.7W	34	31 46.8W	35	31 53.8W	35	32 00.8W	34	
28 N	31 29.4W	34	31 36.3W	34	31 43.2W	34	31 50.0W	34	31 57.0W	34	
29 N	31 25.9W	34	31 32.8W	33	31 39.5W	33	31 46.3W	34	31 53.1W	34	
30 N	31 22.4W	33	31 29.1W	33	31 35.8W	33	31 42.6W	33	31 49.3W	33	
31 N	31 19.0W	32	31 25.5W	32	31 32.1W	32	31 38.7W	33	31 45.4W	33	
32 N	31 15.4W	32	31 21.9W	32	31 28.4W	32	31 34.9W	32	31 41.4W	32	
33 N	31 11.9W	31	31 18.2W	31	31 24.6W	31	31 31.0W	32	31 37.5W	31	
34 N	31 08.3W	31	31 14.6W	31	31 20.8W	31	31 27.1W	31	31 33.4W	31	
35 N	31 04.7W	30	31 10.9W	30	31 17.0W	30	31 23.2W	31	31 29.4W	30	
36 N	31 01.1W	30	31 07.1W	30	31 13.1W	30	31 19.2W	30	31 25.3W	30	
37 N	30 57.5W	29	31 03.4W	29	31 09.3W	29	31 15.3W	29	31 21.3W	29	
38 N	30 53.8W	29	30 59.6W	28	31 05.4W	29	31 11.3W	29	31 17.2W	29	
39 N	30 50.2W	28	30 55.9W	28	31 01.6W	28	31 07.2W	28	31 13.0W	28	
40 N	30 46.6W	27	30 52.1W	27	30 57.6W	27	31 03.3W	28	31 08.9W	27	
41 N	30 42.9W	27	30 48.4W	27	30 53.8W	27	30 59.3W	27	31 04.8W	27	
42 N	30 39.3W	26	30 44.7W	26	30 50.0W	26	30 55.3W	26	31 00.7W	26	
43 N	30 35.7W	26	30 41.0W	26	30 46.2W	26	30 51.4W	25	30 56.6W	26	
44 N	30 32.3W	25	30 37.4W	25	30 42.5W	25	30 47.5W	25	30 52.6W	25	
45 N	30 28.9W	24	30 33.9W	24	30 38.8W	24	30 43.7W	24	30 48.7W	25	
46 N	30 25.7W	24	30 30.5W	23	30 35.3W	24	30 40.1W	24	30 45.0W	24	
47 N	30 22.6W	23	30 27.3W	23	30 31.9W	23	30 36.6W	23	30 41.4W	23	
48 N	30 19.7W	22	30 24.3W	22	30 28.9W	22	30 33.4W	22	30 37.9W	22	
49 N	30 17.2W	22	30 21.5W	21	30 25.9W	22	30 30.4W	22	30 34.8W	22	

7930-Z GR 197

T Lat ° ′	48600 ° ′	Δ	48620 ° ′	Δ	48640 ° ′	Δ	48660 ° ′	Δ	48680 ° ′	Δ	T Long ° ′
	47 51.8N	21	47 56.0N	21	48 00.3N	21	48 04.5N	21	48 08.7N	21	29 W
	48 16.2N	20	48 20.3N	20	48 24.3N	20	48 28.3N	20	48 32.3N	20	30 W
	48 39.9N	19	48 43.7N	19	48 47.5N	19	48 51.3N	19	48 55.1N	19	31 W
	49 02.7N	18	49 06.3N	18	49 09.9N	18	49 13.6N	18	49 17.2N	17	32 W
	49 24.7N	17	49 28.1N	17	49 31.5N	17	49 35.0N	17	49 38.4N	17	33 W
	49 45.9N	16	49 49.2N	16	49 52.4N	16	49 55.7N	16	49 58.9N	16	34 W
	50 06.4N	15	50 09.5N	15	50 12.5N	15	50 15.6N	15	50 18.7N	15	35 W
	50 26.1N	14	50 29.1N	14	50 32.0N	14	50 34.9N	14	50 37.8N	14	36 W
	50 45.2N	13	50 48.0N	13	50 50.8N	13	50 53.5N	13	50 56.2N	13	37 W
	51 03.6N	13	51 06.2N	13	51 08.9N	13	51 11.5N	12	51 14.0N	13	38 W
	51 21.3N	12	51 23.9N	12	51 26.3N	12	51 28.8N	12	51 31.3N	12	39 W
	51 38.5N	11	51 40.8N	11	51 43.2N	11	51 45.6N	11	51 47.9N	11	40 W
	51 55.0N	11	51 57.3N	11	51 59.6N	11	52 01.8N	10	52 04.0N	11	41 W
	52 11.1N	10	52 13.2N	10	52 15.3N	10	52 17.5N	10	52 19.6N	10	42 W
	52 26.6N	10	52 23.6N	10	52 30.6N	10	52 32.7N	10	52 34.7N	9	43 W

FIGURE 4338a.—Extracts from lattice tables.

Answer.—lat. 48°32′.7N, long. 30°23′.6W.

When it is possible to match two groundwaves, never match a groundwave with a skywave. Under the best conditions for matching groundwaves to skywaves, the value obtained may be uncertain by an amount equivalent to several miles in position. As is the case with matching two skywaves, the error may be very large when the user is within 200 microseconds of the baseline extensions.

When receiving a skywave signal from one Loran-C station and strong groundwave and skywave signals from a second Loran-C station, the practical procedure is to match the skywave from the first station with the groundwave from the second station. This situation would be encountered when the receiver is located far from the first station and near to the second station. The use of an all skywave match in this situation may cause large errors because of large uncertainties in the skywave correction for the near station.

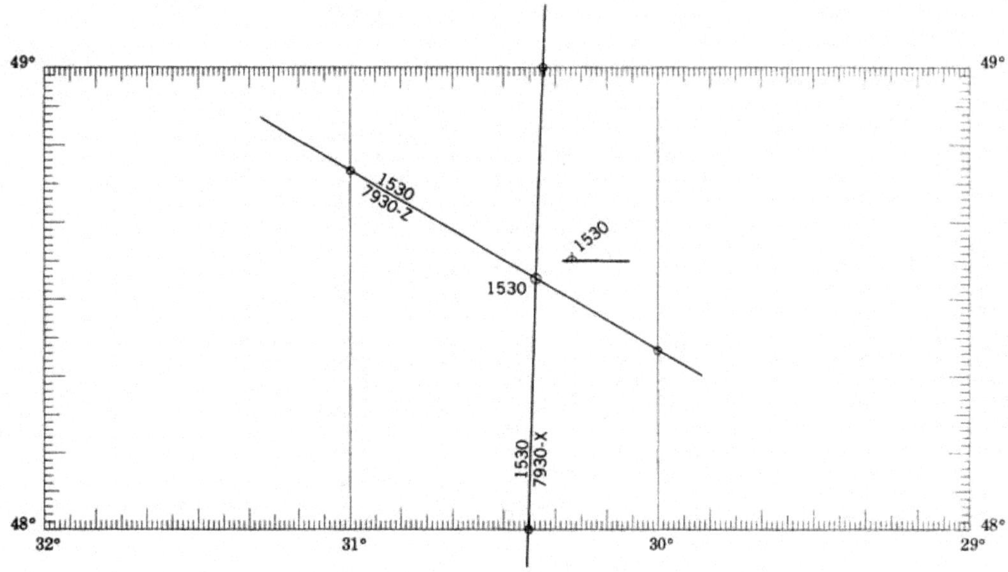

FIGURE 4338b.—Loran-C fix.

Example 2.—The 2130 DR position of a ship is lat. 48°35′N, long. 46°45′W. Loran-C readings are obtained as follows:

2130	7930–X	T_S 33114.3
2130	7930–Z	T_{SG} 54632.2

Required.—The 2130 Loran-C fix.

Solution.—The observed skywave reading (T_S) must be corrected to an equivalent groundwave reading (T_G). Preceding each time-difference table are conventional "Skywave Correction" tables for daytime (ionosphere height=73 kilometers) and for nighttime (h=91 km.). Enter the pertinent table (fig. 4338c) (in this example, nighttime—since the Loran-C readings are taken at 2130) with the DR position and obtain the skywave correction. Interpolate if necessary. Apply the correction to the observed T_S to determine T_G.

If a skywave and a groundwave are matched, the observed reading (T_{SG} or T_{GS}) must also be corrected to an equivalent groundwave reading (T_G). Special "Groundwave to Skywave Correction" tables (fig. 4338d) precede the time-difference tables. Enter the appropriate special correction table (in this example, nighttime) and obtain the special correction. To determine T_G, apply this correction to the observed T_{SG} or T_{GS} reading.

After T_G is determined, the remainder of the solution follows that of example 1. Extracts from the lattice table are given in figure 4338e.

Pair 7930–X		*Pair 7930–X*	
T_S	33114.3	T	33120
Conventional Corr	+8.1	T_G	33122.4
T_G	33122.4	T_G-T	+2.4

Latitude	*Tabulated longitude*	Δ	*Longitude change* $(T_G-T)\times(\Delta)$	*Interpolated longitude*
48°N	46°59′.5W	+40	(+2.4)×(+.40)=+1′.0	47°00′.5W
49°N	46°20′.5W	+38	(+2.4)×(+.38)=+0′.9	46°21′.4W

SKYWAVE CORRECTION

7930-X GF

NIGHTTIME (h = 91 km.)

Longitude—59°W to 45°W

LATITUDE NORTH	59	58	57	56	55	54	53	52	51	50	49	48	47	46	45	LATITUDE NORTH
89																89
88																88
87																87
86																86
85																85
84																84
83									0.0	0.0	0.0					83
82	0.0	0.0	0.0	0.0	0.0	0.0	0.0	0.0	0.1	0.1	0.1	0.0	0.0	0.0	0.0	82
81	0.2	0.2	0.2	0.2	0.3	0.3	0.3	0.3	0.2	0.2	0.2	0.2	0.1	0.1	0.1	81
80	0.4	0.5	0.5	0.5	0.5	0.5	0.4	0.4	0.4	0.3	0.3	0.3	0.2	0.2	0.2	80
			ADD					ADD					ADD			
79	0.7	0.7	0.7	0.7	0.7	0.6	0.6	0.6	0.5	0.5	0.5	0.4	0.4	0.4	0.3	79
78	1.0	1.0	1.0	0.9	0.9	0.9	0.9	0.8	0.8	0.8	0.7	0.7	0.6	0.6	0.5	78
77	1.3	1.3	1.3	1.3	1.3	1.2	1.2	1.2	1.1	1.1	1.1	1.0	1.0	0.9	0.8	77
76	1.7	1.7	1.7	1.7	1.7	1.7	1.7	1.6	1.6	1.6	1.5	1.5	1.4	1.3	1.3	76
75	2.3	2.3	2.3	2.3	2.3	2.3	2.3	2.3	2.2	2.2	2.2	2.1	2.0	2.0	1.9	75
74	2.9	3.0	3.0	3.1	3.1	3.1	3.1	3.1	3.1	3.1	3.0	3.0	2.9	2.8	2.7	74
73	3.8	3.9	4.0	4.0	4.1	4.2	4.2	4.2	4.2	4.2	4.2	4.1	4.1	4.0	3.9	73
72	4.9	5.0	5.2	5.3	5.4	5.5	5.6	5.7	5.7	5.7	5.7	5.7	5.6	5.5	5.4	72
71	6.2	6.4	6.7	6.9	7.1	7.3	7.4	7.6	7.7	7.7	7.8	7.8	7.7	7.6	7.5	71
70	7.8	8.1	8.5	8.8	9.2	9.5	9.8	10	10	10	10	10	10	10	10	70
			ADD					ADD					ADD			
69	9.7	10	11	11	12	12	13	13	13	14	14	14	14	14	14	69
68	12	13	14	14	15	16	17	17	18	18	19	19	19	19	19	68
67	15	16	17	18	19	20	21	22	23	24	25	25	26	26	26	67
66	17	19	20	22	24	26	27	29	31	32	33	34	35	35	35	66
65	20	22	25	27	30	32	35	38	40	43	45	47	48	49	49	65
64	23	26	29	32	36	40	44	48	53	57	61	65				64
63	26	29	33	38	42	48	54	61								63
62	28	32	36	42	48	56	64				DO NOT USE					62
61	29	33	38	45	52	61										61
60	29	33	38	45	52	62					SKYWAVES					60
			ADD													
59	27	31	36	42	50	58					IN THIS AREA					59
58	25	29	33	38	44	52	60									58
57	22	25	29	33	38	44	50	57	65							57
56	19	22	25	28	32	36	40	45	50	56	61	65				56
55	16	18	21	23	26	29	32	35	39	42	45	48	50	51	51	55
54	13	15	17	19	21	23	25	27	30	32	34	35	37	37	37	54
53	11	12	14	15	17	18	20	21	23	24	26	27	27	28	28	53
52	9.0	10	11	12	13	14	15	17	18	19	19	20	21	21	21	52
51	7.3	8.0	8.8	9.6	10	11	12	13	14	14	15	15	16	16	16	51
50	5.8	6.4	7.0	7.6	8.2	8.8	9.4	10	11	11	11	12	12	12	12	50
			ADD					ADD					ADD			
49	4.6	5.0	5.5	5.9	6.4	6.8	7.3	7.7	8.1	8.4	8.7	9.0	9.1	9.2	9.3	49
48	3.6	3.9	4.3	4.6	5.0	5.3	5.6	5.9	6.2	6.5	6.7	6.9	7.0	7.0	7.1	48
47	2.8	3.1	3.3	3.6	3.8	4.1	4.3	4.6	4.8	4.9	5.1	5.2	5.3	5.4	5.4	47
46	2.2	2.4	2.6	2.7	2.9	3.1	3.3	3.5	3.6	3.8	3.9	4.0	4.0	4.1	4.1	46
45	1.6	1.8	1.9	2.1	2.2	2.4	2.5	2.6	2.7	2.8	2.9	3.0	3.0	3.1	3.1	45
44	1.2	1.3	1.5	1.6	1.7	1.8	1.9	2.0	2.1	2.1	2.2	2.3	2.3	2.3	2.3	44
43	0.9	1.0	1.1	1.2	1.2	1.3	1.4	1.5	1.5	1.6	1.6	1.7	1.7	1.7	1.7	43
42	0.6	0.7	0.8	0.8	0.9	0.9	1.0	1.1	1.1	1.2	1.2	1.2	1.2	1.2	1.3	42
41	0.4	0.4	0.5	0.5	0.6	0.6	0.7	0.7	0.8	0.8	0.8	0.9	0.9	0.9	0.9	41
40	0.1	0.2	0.2	0.3	0.3	0.4	0.4	0.5	0.5	0.5	0.5	0.6	0.6	0.6	0.6	40
			ADD					ADD					ADD			
39	0.0	0.0	0.0	0.1	0.1	0.1	0.2	0.2	0.2	0.3	0.3	0.3	0.3	0.3	0.3	39
38				0.0	0.0	0.0	0.0	0.0	0.0	0.0	0.1	0.1	0.1	0.1	0.1	38
37											0.0	0.0	0.0	0.0	0.0	37
36																36
35																35
34																34
33																33
32																32
31																31
30																30
	59	58	57	56	55	54	53	52	51	50	49	48	47	46	45	

Longitude—59°W to 45°W

NIGHTTIME

LXV

FIGURE 4338c.—Conventional skywave correction table.

7930-Z **GROUNDWAVE TO SKYWAVE CORRECTION** GR

NIGHTTIME (h = 91 km.)

Groundwave from Slave (R) to First-hop Skywave from Master (G)

Longitude—59°W to 45°W

LATITUDE NORTH	59	58	57	56	55	54	53	52	51	50	49	48	47	46	45	LATITUDE NORTH
69																69
68																68
67																67
66																66
65																65
64																64
63																63
62																62
61																61
60																60
59																59
58																58
57																57
56																56
55																55
		ADD														
54	68	69	71	73												54
53	65	66	68	69	71	72										53
52	63	64	65	66	67	68	70	71	72	73						52
51	61	62	63	64	65	65	66	67	68	68	69	69	70	70	70	51
50	60	60	61	62	62	63	63	64	65	65	66	66	66	66	66	50
					ADD						ADD					
49	59	59	60	60	60	61	61	62	62	63	63	63	63	63	63	49
48	58	58	58	59	59	59	60	60	60	61	61	61	61	61	61	48
47	57	57	57	58	58	58	58	59	59	59	59	59	59	59	59	47
46	56	56	57	57	57	57	57	58	58	58	58	58	58	58	58	46
45	56	56	56	56	56	56	57	57	57	57	57	57	57	57	57	45
		ADD						ADD						ADD		
44	55	55	56	56	56	56	56	56	56	56	56	56	56	56	56	44
43	55	55	55	55	55	55	55	56	56	56	[illegible]	56	56	56	56	43
42	55	55	55	55	55	55	55	55	55	55	55	55	55	55	55	42
41	54	55	55	55	55	55	55	55	55	55	55	55	55	55	55	41
40	54	54	54	54	54	54	55	55	55	55	55	55	55			40
					ADD											
39			54	54	54	54	54	54	54	54						39
38																38
37																37
36																36
35																35
34																34
33																33
32																32
31																31
30																30
	59	58	57	56	55	54	53	52	51	50	49	48	47	46	45	

Longitude—59°W to 45°W

Groundwave from Slave (R) to First-hop Skywave from Master (G)

NIGHTTIME

XLVII

FIGURE 4338d.—Special groundwave to skywave correction table.

328 GF 7930-X

T Lat ° ′	33100 ° ′	Δ	33120 ° ′	Δ	33140 ° ′	Δ	33160 ° ′	Δ	33180 ° ′	Δ	T Long ° ′
34 N	54 29.0W	56	54 40.5W	57	54 51.9W	57	55 03.5W	58	55 15.1W	58	
35 N	54 01.1W	55	54 12.3W	56	54 23.6W	56	54 35.0W	57	54 46.5W	57	
36 N	53 32.6W	54	53 43.5W	55	53 54.7W	55	54 05.8W	56	54 17.2W	56	
37 N	53 03.4W	53	53 14.1W	54	53 25.1W	54	53 36.0W	55	53 47.2W	55	
38 N	52 33.4W	52	52 44.1W	53	52 54.8W	53	53 05.6W	54	53 16.5W	54	
39 N	52 02.9W	51	52 13.3W	52	52 23.8W	52	52 34.3W	53	52 45.0W	53	
40 N	51 31.5W	50	51 41.7W	51	51 52.0W	51	52 02.4W	51	52 12.8W	52	
41 N	50 59.5W	49	51 09.4W	49	51 19.5W	50	51 29.5W	50	51 39.8W	51	
42 N	50 26.5W	48	50 36.3W	48	50 46.0W	49	50 56.0W	49	51 05.9W	50	
43 N	49 52.8W	47	50 02.3W	47	50 11.8W	47	50 21.5W	48	50 31.2W	48	
44 N	49 18.3W	45	49 27.5W	46	49 36.7W	46	49 46.1W	47	49 55.6W	47	
45 N	48 42.9W	44	48 51.8W	44	49 00.8W	45	49 09.9W	45	49 19.1W	46	
46 N	48 06.6W	43	48 15.2W	43	48 23.9W	43	48 32.8W	44	48 41.6W	44	
47 N	47 29.5W	41	47 37.8W	41	47 46.2W	42	47 54.7W	42	48 03.3W	42	
48 N	46 51.5W	39	46 59.5W	40	47 07.7W	40	47 15.8W	40	47 24.1W	41	
49 N	46 12.9W	38	46 20.5W	38	46 28.3W	38	46 36.1W	39	46 44.0W	39	
50 N	45 33.6W	36	45 40.9W	36	45 48.3W	37	45 55.7W	37	46 03.3W	37	
51 N	44 53.8W	34	45 00.8W	34	45 07.7W	35	45 14.8W	35	45 21.9W	35	
52 N	44 13.9W	32	44 20.5W	32	44 27.0W	33	44 33.7W	33	44 40.4W	33	
53 N	43 34.2W	30	43 40.4W	30	43 46.5W	30	43 52.8W	31	43 59.0W	31	

7930-Z GR 97

T Lat ° ′	45700 ° ′	Δ	45720 ° ′	Δ	45740 ° ′	Δ	45760 ° ′	Δ	45780 ° ′	Δ	T Long ° ′
	45 57.7N	16	47 01.0N	16	47 04.3N	15	47 07.6N	15	47 10.8N	16	44 W
	47 36.2N	14	47 39.2N	14	47 42.1N	14	47 45.0N	14	47 47.8N	14	45 W
	48 10.9N	13	48 13.6N	13	48 16.2N	12	48 18.7N	12	48 21.3N	12	46 W
	48 41.8N	11	48 44.2N	11	48 46.5N	11	48 48.8N	11	48 51.1N	11	47 W
	49 08.9N	10	49 11.0N	10	49 13.1N	10	49 15.2N	10	49 17.2N	10	48 W
	49 32.1N	9	49 34.0N	9	49 36.0N	9	49 37.9N	9	49 39.8N	9	49 W
	49 37.3N	9	49 39.2N	9	49 41.1N	9	49 43.0N	9	49 44.9N	9	49 15W
	49 42.3N	9	49 44.2N	9	49 46.0N	9	49 47.9N	9	49 49.8N	9	49 30W
	49 47.0N	9	49 48.9N	9	49 50.8N	9	49 52.6N	8	49 54.4N	9	49 45W
	49 51.6N	8	49 53.4N	9	49 55.2N	9	49 57.1N	9	49 58.8N	8	50 W
	49 56.0N	8	49 57.7N	8	49 59.4N	8	50 01.3N	9	50 03.1N	8	50 15W
	50 00.0N	8	50 01.7N	8	50 03.5N	8	50 05.3N	8	50 07.0N	8	50 30W
	50 03.9N	8	50 05.6N	8	50 07.4N	8	50 09.1N	8	50 10.9N	8	50 45W
	50 07.6N	8	50 09.3N	8	50 11.0N	8	50 12.7N	8	50 14.5N	8	51 W
	50 11.0N	8	50 12.7N	8	50 14.4N	8	50 16.1N	8	50 17.9N	8	51 15W
	50 14.3N	8	50 15.9N	8	50 17.7N	8	50 19.4N	8	50 21.1N	8	51 30W
	50 17.4N	8	50 19.0N	8	50 20.7N	8	50 22.4N	8	50 24.1N	8	51 45W
	50 20.2N	8	50 21.9N	8	50 23.6N	8	50 25.2N	8	50 26.9N	8	52 W
	50 22.8N	8	50 24.5N	8	50 26.2N	8	50 27.9N	8	50 29.5N	8	52 15W
	50 25.3N	8	50 27.0N	8	50 28.7N	8	50 30.4N	8	50 32.0N	8	52 30W

FIGURE 4338e.—Extracts from lattice tables.

Pair 7930–Z		*Pair 7930–Z*	
T_{SG}	45632.2	T	45700
Special Corr	+62	T_G	45694.2
T_G	45694.2	$T_G - T$	−5.8

Longitude	*Tabulated latitude*	Δ	*Latitude change* $(T_G - T) \times (\Delta)$	*Interpolated latitude*
46°W	48°10′.9N	+13	(−5.8)×(+.13)=−0′.8	48°10′.1N
47°W	48°41′.8N	+11	(−5.8)×(+.11)=−0′.6	48°41′.2N

Answer.—Following the procedure for plotting lines of position as shown in example 1, the 2130 Loran-C fix is lat. 48°31′.3N, long. 46°40′.2W.

Loran-C lattices on charts.—The Defense Mapping Agency Hydrographic/Topographic Center provides Loran-C lattices on charts which are within the Loran-C system's coverage area.

Loran-C lattices are for use with groundwaves. Skywave corrections are shown on the chart at projection intersections (fig. 4338f) where skywave contamination is expected. The navigator must determine if the receiver is tracking:

(1) groundwave signals from both master and secondary stations (T_G), or

(2) skywave signals from both master and secondary stations (T_S), or

(3) groundwave signal for master and skywave signal for secondary (T_{GS}), or

(4) skywave signal for master and groundwave signal for secondary (T_{SG}).

LORAN-C RATES ON THIS CHART

Additional loran rates in this area may be provided by other charts and publications listed in Pub. P2V10.

Skywave corrections for this chart are:

9970 X − 02 D SG + 38 D
9970 X − 03 N SG + 54 N
9970 Y − 02 D SG + 38 D
9970 Y *

* Do not use skywaves in this area.

FIGURE 4338f.—Loran-C skywave corrections as shown on charts.

When the navigator determines that the signals being used are both skywave then the proper correction at the nearest projection intersection is chosen based on time of fix, (N for night and D for day), and the station pairs being used. This correction will reduce the skywave signals to a groundwave reading for plotting. If the navigator determines that the received signals are of mixed propagation modes (skywave for one and groundwave for the other) a special correction must be used to reduce the received signals to a groundwave reading. These special corrections are also located at the projection intersections and prefixed by either GS or SG to indicated the order of mix. A station pair designator followed by an asterisk indicates that skywaves should not be used for this pair in that area. Additional skywave correction data is located in the DMAHTC *Loran-C Rate Tables* for each station pair.

4339. Propagation prediction.—The accuracy of the lattice tables or of the lattices overprinted on the nautical chart is dependent upon knowledge of the LF signal transit time between a transmitting station and receiver. The signal propagation velocity in free-space is well known; however, there exists an additional time (or phase) delay compared to free-space propagation time when the signal propagates over the earth's surface. Compensation for this additional delay is necessary in order to obtain maximum positioning accuracy. The compensation for these delays, in the form of **Secondary Phase Factor (SF)** corrections for signal transmission over all seawater paths and **Additional Secondary Phase Factor** (**ASF**) corrections for transmission over land paths, is generally derived from both analytical and empirical models for predicting phase delay.

The Loran-C lattices as tabulated in the tables or overprinted on the nautical chart normally include compensation for Secondary Phase Factor. The lattices overprinted on nautical charts covering areas within the Coastal Confluence Zone may include compensation for Additional Secondary Phase Factor and will be stated on the chart when included. **Loran-C Correction Tables** tabulate the ASF in the Coastal Confluence Zone and when not included on the chart should be applied algebraically to the receiver time difference reading to increase the accuracy of the lines of position.

4340. Accuracy.—The *repeatable* accuracy (art. Q8) to be expected of a Loran-C fix for 1:3 and 1:10 signal-to-noise ratios is expressed in terms of $2d_{rms}$ (art. Q7) in figures 4343a and 4343b. The error is stated as 1,500 feet. The probability is given as 95 percent. The standard deviation of each of the two intersecting lines of position establishing the fix is 0.1 μs.

The repeatable accuracy is affected by systematic and random errors (arts. Q3 and Q4). The systematic errors are largely due to errors in propagation prediction (art. 4339).

4341. Ranging mode.—With the user's receiver appropriately modified to enable time measurements with respect to a local time reference, the Loran-C system can be operated in the ranging or Range-Range mode, which is discussed in more detail in article 4304.

4342. Loran-D, designed for military tactical use, is a lower power, shorter range version of Loran-C. Since the stations are readily transportable, the system provides the potential for filling any gaps in the higher power Loran-C coverage.

4343. The U. S. Coastal Confluence Zone (**CCZ**) as defined by the *National Plan for Navigation,* promulgated by the Secretary of Transportation, is that area of water extending outward from the shore for 50 nautical miles or to the 100-fathom curve, whichever is farther from the shore. The inner boundary is the harbor entrance. It is the area where transoceanic traffic converges and interport traffic exists.

After extensive study, the U. S. Coast Guard recommended and the Secretary of Transportation approved in 1974 the selection of Loran-C as the government sponsored navigation system for the CCZ.

Within the limits of the CCZ, users of "true" Loran-C receivers (art. 4335) should obtain a repeatable accuracy (art. Q8) of 0.25 nautical mile with a probability of 95 percent. That is, only 1 fix in 20 should deviate from the mean of a large number of positions that could be established by the Loran-C system at a given place and time by more than 0.25 nautical mile.

Loran-C coverage in the CCZ is shown in figures 4343a and 4343b.

LORAN-C COVERAGE DIAGRAM

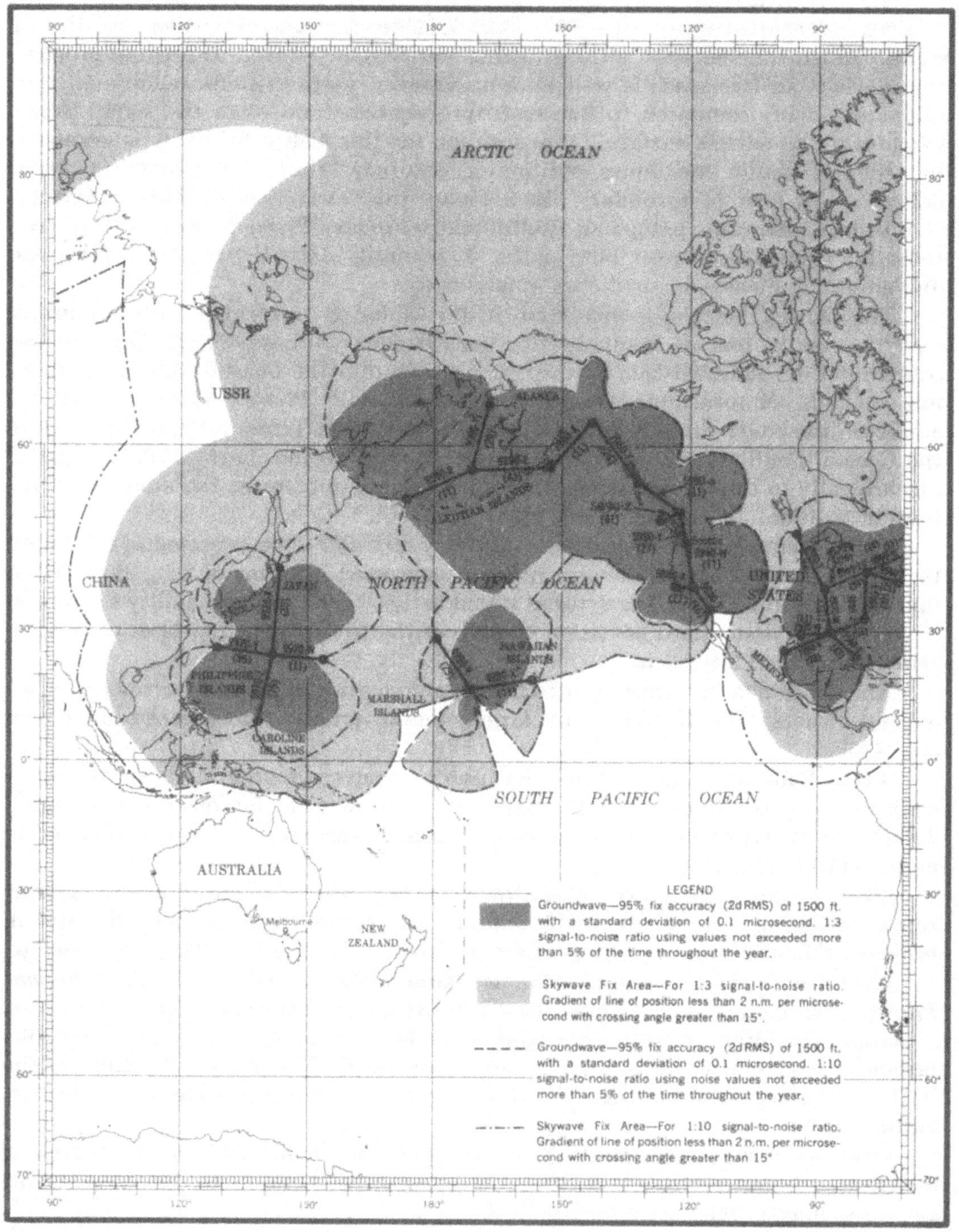

FIGURE 4343a.—Loran-C coverage diagram.

LORAN-C COVERAGE DIAGRAM

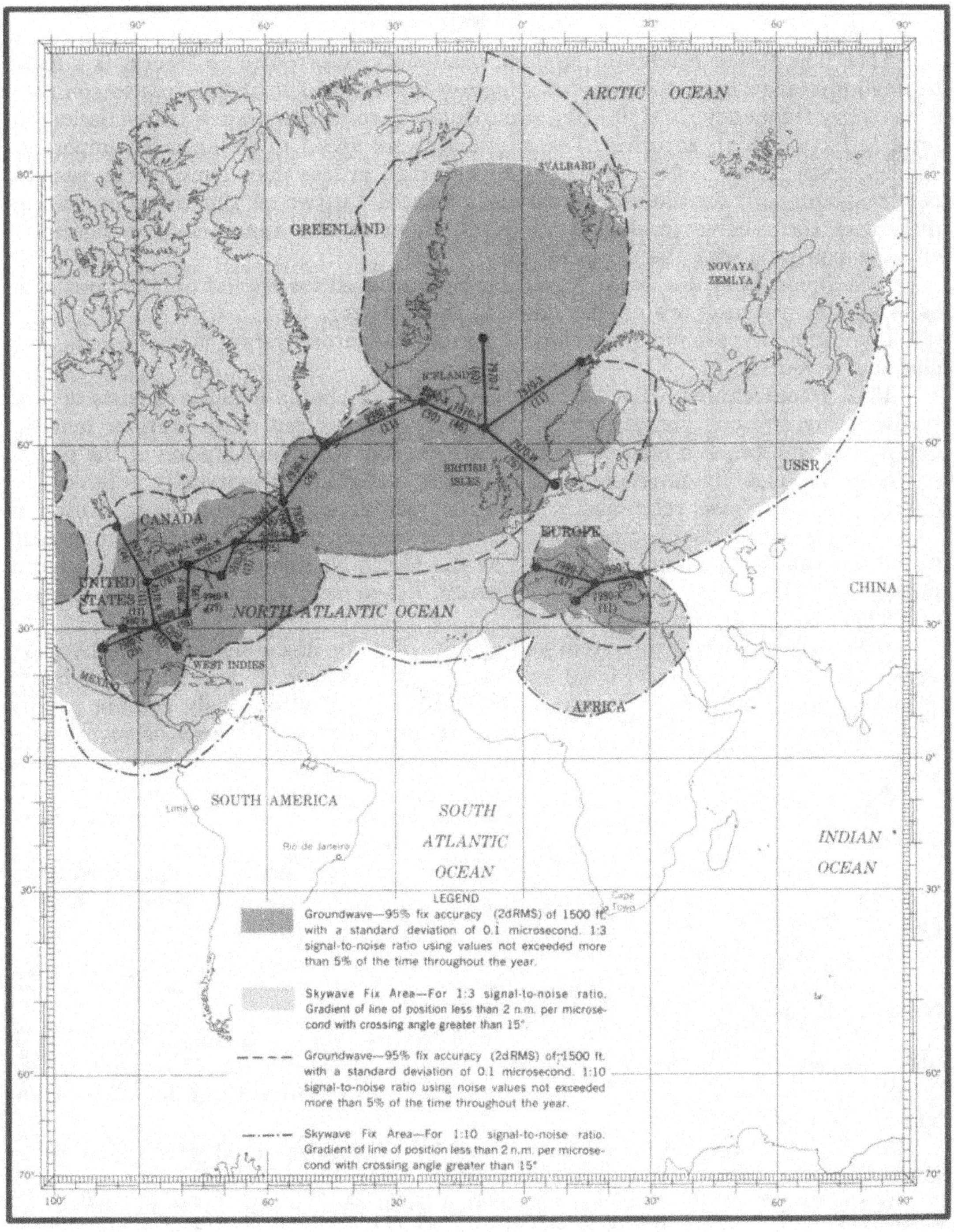

Figure 4343b.—Loran-C coverage diagram.

Decca

4344. The Decca Navigator System, commonly referred to as **Decca,** is a short to medium range hyperbolic radionavigation system which utilizes phase comparisons of low frequency (70–130 kHz), unmodulated, continuous-wave transmissions to provide fixes of high accuracy. The simplicity and speed in fixing is an important characteristic of the system. A fix can be obtained in less than 1 minute by reading two relevant position coordinate values indicated by two of the three coordinate meters of the receiver and then referring them to the appropriate navigational chart overprinted with the Decca lattice.

Developments of the Decca Navigator System used for special applications such as in narrow channels are not discussed in this chapter.

Like other hyperbolic systems, the Decca Navigator System is *continuous, passive,* and *non-saturable.*

4345. Decca chains.—Each of the several Decca chains usually consists of four transmitting stations operating in three pairs: master/red slave station, master/green slave station, and master/purple slave station. The slave station of the pair is so called because its function is to transmit a signal harmonically related and having a fixed phase relationship with the master signal. Each slave station is located about 60 to 120 nautical miles from the master station. Information on each of the several Decca chains established by the Decca Navigator Company Limited is presented in *The Decca Navigator Operating Instructions and Marine Data Sheets,* published by that company.

4346. Decca Navigator System principles.—For the discussion of Decca here, let us assume that two transmitting stations, master station *A* and slave station *B*, separated by a distance *S*, known as the baseline, simultaneously radiate a pure continuous wave of frequency *f*, and that at any point in the area of coverage of these stations, both transmissions can be received separately and their phases compared.

Let us also assume that the phase of the wave transmitted from *B* is so adjusted that at station *A* the two waves are in phase.

Figure 4346a shows the relative positions of the two waves of equal frequency at two successive instants. The continuous line shows the wave transmitted from *A*, and the broken line the wave transmitted from *B*.

It can readily be seen that:

1. At master station *A* the two waves are always in phase and remain so at all points on the baseline extension to the left of *A*.
2. At slave station *B* the wave emitted by that station is and remains out of phase in relation to the wave emitted by master station *A*. The same phase difference prevails as between these two waves at all points of the baseline extension to the right of *B*.
3. Along the baseline between stations *A* and *B*, the points at which both waves are in phase are separated by a distance equal to half the wavelength.

If the length of the baseline (fig. 4346a) were equal to an integral number of halfwavelengths corresponding to the frequency *f*, at station *B* and at all points on the baseline extension to the right of that station, the two waves would always be in phase.

Let us assume that such is the case for the pair of stations shown in figure 4346b, and that the earth is a plane within the area of coverage.

At any point *Q*, two fields exist simultaneously radiated respectively by *A* and *B*. If we consider a point Q_1 in the bisector of the baseline, distance Q_1A is equal to

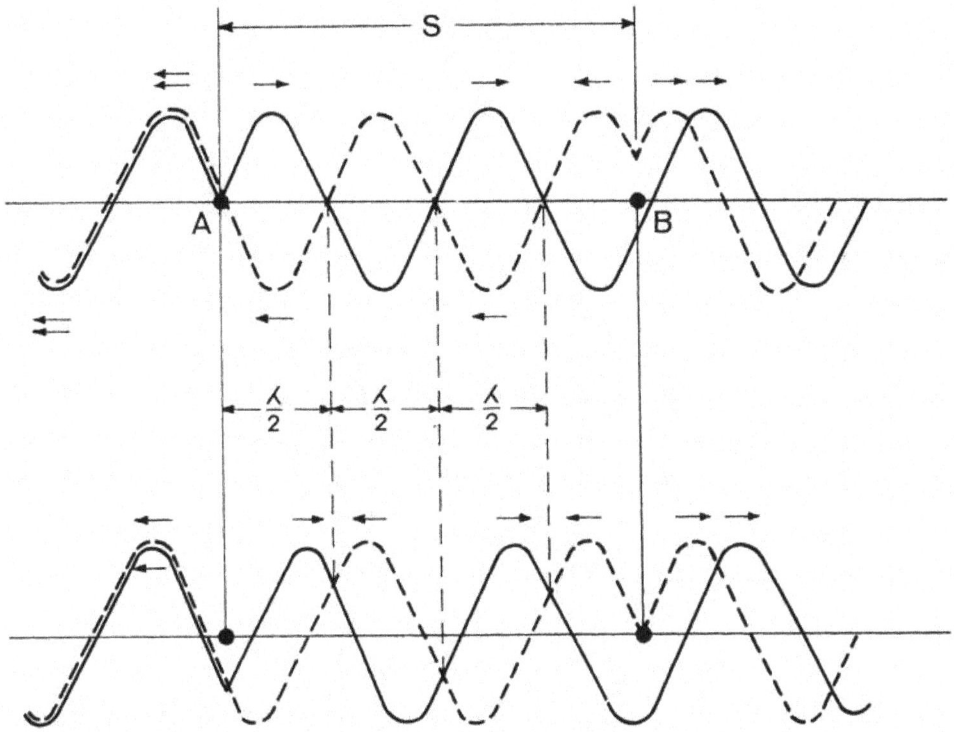

FIGURE 4346a.—Relative positions of two waves of equal frequency at two successive instants.

distance Q_1B, and if the currents in the transmitting antennas A and B are in phase, it is clear that the two fields produced at Q_1 will also be in phase. At a point slightly to the right of Q_1, the distance to antenna A increases and that to antenna B decreases. The two fields are therefore no longer in phase. The more the point is moved to the right of Q_1, the more the difference in phase increases. The phase difference may thus reach 360°, which is equivalent to a phase difference of 0°, and the two fields are once more in phase. This occurs in the case of a point Q_2 which is farther removed from A than from B by one wavelength.

If we now plot the hyperbola having A and B as its foci and passing through point Q_2, all points on the hyperbola (according to the actual definition of a hyperbola) will be one wavelength farther away from A than from B. Consequently, for all points on the hyperbola, both fields are in phase.

If point Q_3 is two wavelengths farther from A than from B, the phase difference between the two fields is 720° or twice 360°, i.e. 0° again. This is true for all points of the hyperbola passing through Q_3.

In practice the length of the baseline is not an integral number of wavelengths. The fixed relationship that the phase of the wave emitted by the slave station bears to that of the wave emitted by the master station is such that the waves are in-phase at the master station as shown in figure 4346a. Neither detail invalidates the basic principle.

The fixed phase relationship is maintained by **phase locking**, i.e. keeping the frequency emitted by the slave station at a constant phase angle relative to the stable frequency received from the master station.

Since it is impossible in practice to separate two waves of identical frequency at a receiver in order to measure their phase difference, as they would then combine

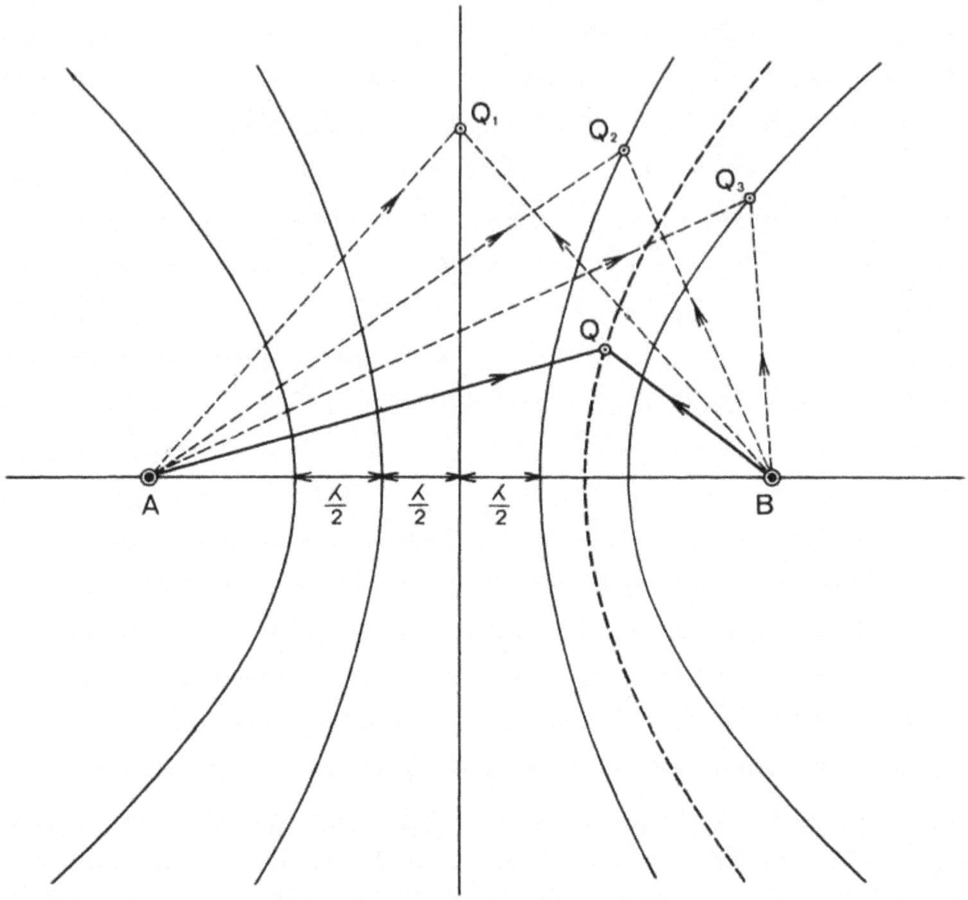

FIGURE 4346b.—Hyperbolas defined by phase-difference measurements.

into a single wave, recourse is made to a technical device which achieves the effect of transmitting signals of equal frequency from the master and slave stations. The effect of having waves sent from both stations to arrive at the phasemeter on an identical frequency, free from interference or wave distortion, is achieved by assigning harmonically-related values to the two frequencies actually transmitted so that multiplying circuits in the receiver can derive from each a common harmonic, the comparison frequency. As shown in table 4346, the master station transmits a signal of $6f$, the red slave $8f$, the green slave $9f$, and the purple slave $5f$, where the fundamental frequency, f, which is not transmitted, lies between 14.00 and 14.33 kHz. Thus the master and red slave signals are multiplied to a common **comparison frequency** of $24f$. Geometrically the system functions as if the common frequency $24f$ were radiated from the two stations. The master/green and master/purple pairs function similarly by using the frequency values and multiplication factors shown in table 4346.

	Master	*Red*	*Green*	*Purple*
Transmission frequency (kHz)	85.000	113.333	127.500	70.833
Harmonic	6*f*	8*f*	9*f*	5*f*
Phase comparison frequency		340.000	255.000	425.000
Harmonic		24*f*	18*f*	30*f*
Lane width on baseline (meters)		440.735	587.647	352.588
Comparison frequency for lane identification (1*f*)		14.166	14.166	14.166
Zone width on baseline (km)		10.56	10.56	10.56

TABLE 4346.—Transmission and comparison frequencies for a chain using a fundamental frequency of 14.166 kHz. The fundamental frequency *f* varies from chain to chain. The lane width on the baseline is for a propagation velocity of 299,700 kilometers per second.

4347. The Decca lattice.—Each hyperbola of the Decca lattice overprinted on the navigational chart in the color of its respective slave station is an in-phase or zero phase-difference hyperbola. The area between adjacent hyperbolas of zero phase difference is called a **lane.**

Each hyperbola *as charted* is based upon a velocity of propagation of the radio waves, assuming an all seawater path. As discussed in article 4339, an appropriate correction may be required when land intervenes.

4348. Decometer readings.—Since the phase-measuring equipment is sensitive to phase differences of the order of 3° to 3°.5, the number of hyperbolas that can be derived per lane is limited in practice to 100. Accordingly, the Decometer gives the fractional value of the lane in hundredths. Although the whole lane numbers are also given, the lane number must be initially set to the proper lane as established by other means. This is due to the lane ambiguity associated with the phase comparison method. As the vessel moves from one lane boundary to another, the fractional pointer of the Decometer makes one revolution.

The complete reading of the red, green, or purple Decometer consists of the Decometer color, the zone letter, the lane number, and the fractional value of the lane. The **zone** is a group of adjacent lanes used for lane identification purposes. Each zone contains 24 red, 18 green, or 30 purple lanes and is about 10 kilometers wide measured along the baseline. As is the case with the lane number, the zone letter must be initially set by manual means.

Since the zones are about 6 miles wide on the baseline, about 20 miles wide at 100 miles from the stations, and about 50 miles wide at the edge of the service area, there is seldom any problem in setting the zone indication on the Decometer.

The lanes in each family of hyperbolas are numbered differently: red 0–23, green 30–47, purple 50–79. These numbers recur in zones denoted by letters *A* to *J*. In cases where the master and slave are more than 10 zones apart, the zone lettering after *J* starts again at *A*.

4349. Lane identification within a known zone is obtained by a process essentially the same as that of determining the fractional value of a lane. As the vessel moves from one zone boundary to another, the lane identification pointer of the Decometer makes one revolution against a scale marked in lane numbers.

The lane identification transmissions emitted by Decca chains are classified as V1, V2, or Multipulse (MP). Most Decca chains emit either MP only or MP and V2 combined; a few chains emit V1 or V2 only. Decca Mark V receivers can receive V1 or V2 transmissions only; Mark 12 receivers can receive both MP and V-type transmissions; and Mark 21 receivers can receive only MP transmissions. Thus, when operating within the coverage area of chains emitting V-type transmissions, the users of Mark 21 receivers must ascertain the correct lane number by other means.

At distances in excess of 250 nautical miles from the stations of chains employing MP lane identification and 150 nautical miles in the case of stations employing V-type lane identification, skywave contamination may be of sufficient intensity to cause the lane identification meter to malfunction, resulting in *lane slip.* The effect is most pronounced at night or during the darker part of twilight. Lane slip or incorrect lane identification can also result from interruption or disturbance of the Decca transmission, incorrect initial referencing of the receiver, or from snow static or electrical storms.

Any interruption or disturbance of the normal transmissions of Decca stations is broadcast as a Decca Warning by the coast radio station in the vicinity.

The Decca Navigator Operating Instructions and Marine Data Sheets provide comprehensive information on lane identification according to receiver type.

In addition to providing initial lane identification within a known zone, lane identification provides an independent check on the lane counting process at any time.

4350. Chain numbers.—Each Decca chain is identified by a number from 0 to 10 used to denote a group of basic frequencies and a letter of *A* to *F* used to denote one of the six master frequencies in each group, except for group 10 which contains the *A, B,* and *C* frequencies only. Thus, there are 63 frequencies available for use with the several chains of the system. Mark V receivers can receive only the *A, B,* or *C* frequencies in groups 1 to 9, whereas Mark 12 and Mark 21 receivers can receive all 63 frequencies. Details are given in Pubs. Nos. 117A and 117B, *Radio Navigational Aids,* and volume 5 of *Admiralty List of Radio Signals.*

4351. Inter-chain fixing is positioning by means of intersecting hyperbolas obtained from different multipulse type chains in situations where a better angle of cut is obtained thereby. This method of operation is particularly advantageous at the longer ranges when the user is unable to resolve zone ambiguity. In such operation the reading from the second chain is obtained from the lane identification readout. Although the readout precision is only 0.1 lane, it is adequate for the purpose.

The reliability of the multipulse lane identification is the main factor permitting interchain fixing, since the method would not be feasible if, on switching to the second chain, the correct lane number could not be ascertained immediately. Special charts overprinted with the appropriate pair of families of hyperbolas are available for certain areas where there is overlapping coverage from multipulse type chains. The transmitters for the two chains are not phase locked, and therefore it is not possible to use combinations of transmitters from the two chains to provide new families of hyperbolas.

4352. Accuracy and coverage.—The *predictable* accuracy (art. Q8) to be expected of a Decca fix in a particular coverage area is given in diagrams in *The Decca Navigator Operating Instructions and Marine Data Sheets* and NP 275(5a), an appendix to volume 5 of the *Admiralty List of Radio Signals.* The accuracy is expressed in terms of d_{rms} (art. Q7). The assumed standard deviation of each line of position is 0.02 mean lane. The probability is given as 68 percent. The diagrams in the Decca publication present the predictable accuracy according to time and season.

The predictable accuracy is affected by both random and fixed (systematic) errors. The random errors (art. Q4) are largely due to skywave interference with the groundwave, resulting in incorrect Decometer readings. The fixed errors (art. Q3) are largely due to errors in propagation prediction (art. 4339). The Decca lattice as overprinted on the navigational chart does not reflect compensation for differences in the conductivities along the actual paths of the signals and the conductivity along an all seawater path. Although the fixed errors are generally small, there are certain regions (e.g. certain coastal waters) where consideration should be given to them when fixing the vessel's position. *The Decca Navigator Operating Instructions and Marine Data Sheets* include diagrams showing the known corrections which should be applied to the Decca readings. However, it should not be assumed that a fixed error does not exist when a correction is not given. The corrections as given on the diagrams may be too sparse to permit a valid assessment of the correction to apply to the Decometer reading. In such case not too much reliance should be placed on the absolute or predictable accuracy (art. Q8).

The repeatable accuracy of the system ranges from a few tens of meters by daytime in areas where the geometry is favorable to a few nautical miles in the presence of skywave interference by night at the limit of the range.

The coverage of a Decca chain is determined primarily by the effects of skywave contamination of the groundwave rather than factors of signal-to-noise ratio or system geometry, and varies with the low frequency propagation conditions characteristic of different regions of the world. Ranges of 175 nautical miles by night and 350 nautical miles by day may be taken as representative when the transmission paths lie over seawater.

4353. Decca Track Plotter.—It is possible to use an automatic track plotter with the Decca receivers, displaying and recording the track made good as a pen trace. This enables the vessel to follow any desired or intended track.

The hyperbolic lattice is represented on a special chart as a rectilinear inverse lattice so that, for example, the pen moves in response to the red pattern and the chart to the green. The pen therefore indicates the position of the vessel within the lattice and traces a continuous record of the **track made good.** Since the hyperbolas are represented as straight parallel lines intersecting at right angles, a certain amount of distortion occurs, but provision is made for reducing this distortion to an acceptable level. However, if a geographically straight track appears on the chart as a curve, steering the vessel so as to keep the pen on the curve will result in the intended track being made good.

Consol

4354. Consol is a long range azimuthal radionavigation system of low accuracy operated primarily for air navigation. Although not sufficiently accurate for coastal navigation or making landfall, the system can be useful to the marine navigator as an aid to ocean navigation.

A Consol station consists basically of a medium frequency (MF) radio transmitter with three antennas in line, equally spaced apart at a distance of the order of three times the wavelength of the transmitted frequency. The three antennas are fed with signals in such a manner that radial patterns of alternate dot and dash sectors are formed, separated by the **equisignal,** the whole pattern rotating through one sector width within a transmission cycle. An observer will therefore hear the equisignal between the dot and dash signals once per transmission cycle. A count of the dots and dashes will give the angular position within a sector. This sector has to be identified either by direction finding methods or some other form of navigation.

Since the radial sectors are not formed close to the transmitting site, there is a minimum range limitation within which the system cannot be used. This is usually taken to be 25 to 30 nautical miles from the site. There is also a sector 30° on either side of the baseline extension which is not usable because of pattern distortion.

The system is described as azimuthal even though it is basically a hyperbolic system. As discussed in article 4303, a system can be considered directional (azimuthal) beyond a distance of a few miles from the station if the baseline is very short.

Sometimes Consol is classified as a radiobeacon because of the frequency of operation and being azimuthal.

A modified form of Consol called **Consolan** was developed in the United States. In this system only two antennas are used.

In the U.S.S.R. a further modification of Consol is in use. This system, called BPM5, uses five antennas in the form of a cross to obtain narrower dot and dash sectors.

The main advantage of Consol is that the signal can be received on a standard communications receiver.

4355. The signal from the Consol transmitter consists of two parts, the transmission of the call sign, sometimes with a continuous wave signal, and the transmission of the rotating pattern (navigational period). The duration of the total transmitting cycle is not the same for all stations, but the navigational period has been standardized at 30 seconds. If the call sign only is sent in addition to the navigational cycle, the total signal duration is 40 seconds. When the continuous wave signal is incorporated, the total signal duration is 60 seconds. The call sign and the continuous wave signal are transmitted only from the center antenna in the Consol system. They can be used for coarse tuning or direction finding purposes.

The signals can be received by any medium frequency receiver operating in the maritime radionavigation band and suitable for receiving A1 telegraphy. The range and accuracy of the signal is affected by receiver selectivity, increasing as the selectivity increases, although above 100 Hz there is little increase in either range or accuracy. Some experience in the use of the system is found to be necessary to obtain the best results, especially if the signal is weak or there is excessive background noise.

Some special adaptors and receivers have been produced for use with the Consol system. These do aid in the counting of the dots and dashes. One such adaptor uses a meter indication in which dots give a deflection to one side of the center position and dashes give a deflection to the other side. An automatic Consol receiver displays a digital readout of the dot and dash count. This receiver can compensate, within certain limits, for missing characters when the equisignal condition is masked by noise.

4356. Method of use.—Procedures for using Consol and Consolan and the tables for converting dot and dash counts to true bearings *from* the station are given in Pubs. Nos. 117A and 117B, *Radio Navigational Aids,* and volume 5 of *Admiralty List of Radio Signals.* These bearings must be corrected for conversion angle (tab. 1) before plotting on the Mercator chart.

Figure 4356 shows what the observer will obtain in the receiver, depending on his position within a dash or dot sector. At the beginning of the keying cycle, the equisignal lines will be as shown, with the dash and dot sectors on either side.

During the navigational period, these lines will rotate as shown, pushing the various sectors in front of them. There are 60 dots or dashes in each sector.

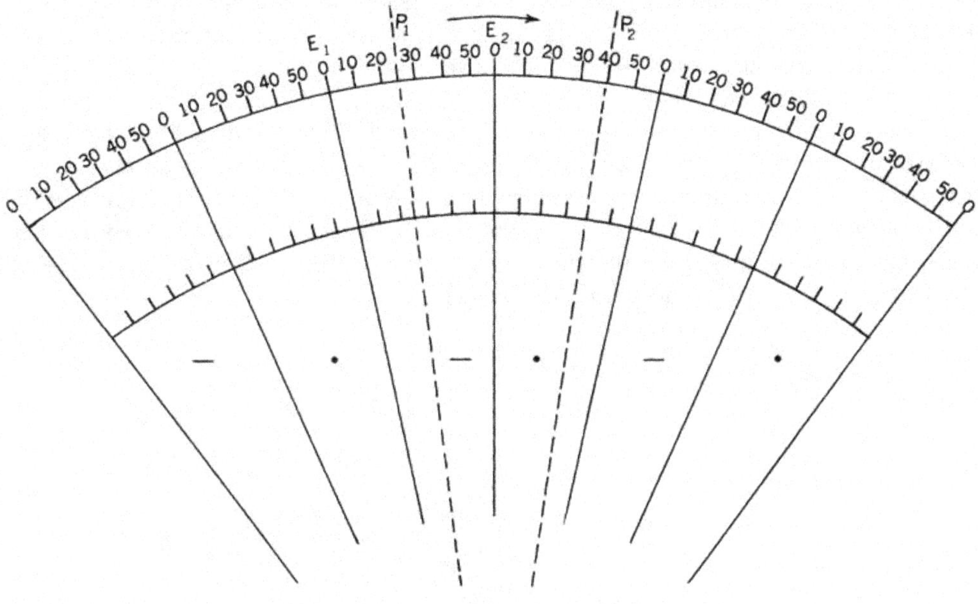

FIGURE 4356.—Signal reception.

Consider first an observer at P_1 within a dash sector. As the pattern rotates a number of dashes will be heard before the equisignal condition. As the pattern continues to rotate, the equisignal will resolve into dots. By counting the number of dashes, the angular displacement from the equisignal line E_1 can be obtained.

Then consider an observer at P_2 within a dot sector. As the pattern rotates, a number of dots will be heard before the equisignal, followed by a number of dashes. In this case, the count of the number of dots will give the angular displacement with respect to the equisignal line E_2.

Due to the fact that the dots and dashes merge into the equisignal, it is difficult, in the equisignal, to distinguish between the dot and dash signals. In theory the total number of dashes and dots heard by the observers at P_1 and P_2 should in each case be 60. In practice this is not the case, there being a period of confusion around the equisignal time. In order to obtain an accurate assessment of the true count, the total number of dots and dashes heard during one navigational period is subtracted from the theoretical total count, that is 60. Half the resultant figures is then added to the original count of dots or dashes.

As an example, take the case where the theoretical dash count should be 25, but the actual count is 23 followed by a dot count of 33. This gives a total dot and dash count of 56. This is subtracted from the total theoretical count of 60 giving a figure of 4. Half this, 2, is added to the dash count of 23 to obtain the assessment of the true count of 25 from the equisignal line. This count is then referred to a Consol chart of tables to give the bearing from the station within the dash sector. The sector itself is identified either by radio direction finding or other navigational means.

4357. Accuracy.—The number of dot and dash sectors, and therefore the accuracy of the Consol line of position is dependent on the baseline length, $n\lambda$, between the inner and each outer antenna. Since the wavelength is fixed for each station, the number of sectors will therefore depend upon n. For any given value of n, there will be $8n$ equisignal lines in the full 360° coverage. For practical purposes the value of n is usually taken as three. In this case the equisignal lines will be at angles of 0°, ±9°.6, ±19°.5, ±30°, ±41°.8, and ±90° from the normal to the baseline through the center antenna. In the last sector on each side, about 34° wide, the rotation speed of the equisignal line is not constant, especially at the beginning of the navigational cycle when it will be a maximum. The readings in these sectors are therefore normally taken as unreliable. In the remaining sectors the widths vary from about 10° to about 14°, with an average of 12°. Hence, if there are 60 dot and dash elements in each sector, the resolution will be 12°/120=0°.1. By increasing n, the number of wavelengths in the baseline, the width of the unuseable sector is reduced and the resolution accuracy increased.

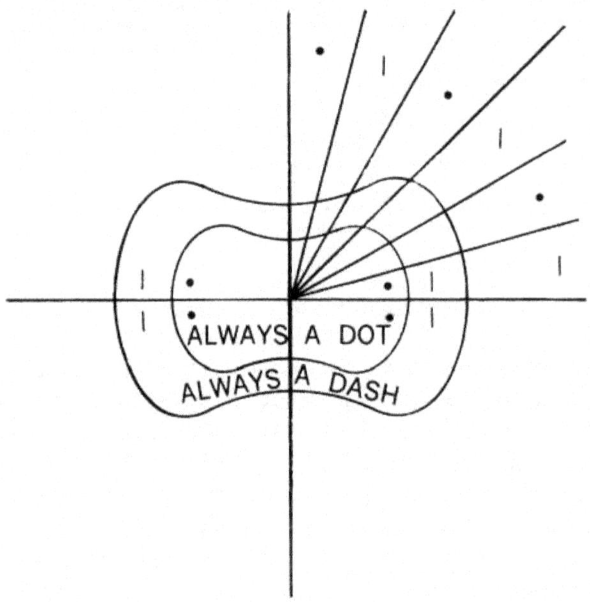

FIGURE 4357.—Fixed equisignal curves at close range.

The minimum distance below which the system cannot be used is due to two factors, the presence of fixed equisignal curves (fig. 4357) and the fact that at close range the equisignal lines are hyperbolic rather than radial or azimuthal. For the first factor it can be shown that where n (in wavelengths) is distance between the center and outer antennas, and λ is the wavelength, the minimum distance is given by

$$D=\left(\frac{16n^2-1}{8}\right)\lambda.$$

Hence, if n is 3 and λ is 1 kilometer, D is 18 kilometers, so that at 20 kilometers and above, the fixed equisignal lines will not adversely affect bearings.

For the second factor or the case where the hyperbolas have not yet become asymptotic with the radial lines, the error in bearing is given by

$$\epsilon = \frac{n^2\lambda^2}{4D^2} \sin 2\theta,$$

where θ is the bearing. The bearing error ϵ always has the same sign in each quadrant as the error is always on the same side of the radial line. The error is maximum when bearing θ is 45° from the perpendicular bisectors. Table 4357a gives some values for the errors for different ranges and different bearings.

Table 4357a shows that the error is reduced with increase in range. At 50 kilometers the mean error is 2 minutes of arc; however, at the higher angles, this will be increased to 3 minutes of arc. When this error is combined with the minimum resolution error of 0°1, the minimum error becomes 8 minutes of arc. This minimum error will vary with the observer's position within the sector and his range and bearing from the station. This error is due entirely to the system itself, and will be the minimum system error. There are other factors which will affect the accuracy of the measured bearing.

Distance (D) (km)	*Error in minutes of arc for angle of*					*Mean error in useable sector (minutes of arc)*
	0°	15°	30°	45°	60°	
20	0	10	17	19	17	12
30	0	4	7.5	8.5	7.5	4.5
40	0	2	4	5	4	2.5
50	0	1.5	3	3	3	2
60	0	1	2	2	2	1
70	0	1	1	1.5	1	1
80	0	1	1	1	1	1

TABLE 4357a.—Values of error for different ranges and different bearings.

The overall accuracy of the system is dependent on a number of factors in addition to the system errors described above. These include asymmetrical phasing errors in the signals from the inner and outer antennas, phasing and amplitude errors in the signals from the outer antennas, propagation errors, and incorrect functioning of the receiver.

Phasing errors affect the width of the equisignal zone or its position, but these errors are normally corrected by a monitoring station.

Propagation conditions can introduce errors which cannot be corrected at the transmitting site. These are the errors which have the most effect on bearing accuracies. These errors can include:

1. those caused by a curved propagation path, giving a fixed bearing error;
2. those caused by a difference in propagation conditions between the outer antennas and the angular error which increases with bearing; and
3. those due to phase differences in groundwave and skywave path lengths.

At any receiving point, it is the stronger of the groundwave and skywave signals that will be detected. During daylight groundwaves predominate up to a range of about 150 nautical miles, while skywaves predominate at ranges greater than about 500 nautical miles. Between these two ranges is an area where skywave errors could cause problems.

In addition to the signal errors, it is also possible for the receiver on the vessel to introduce errors. If a loop antenna is used and it is set to a minimum signal condition as may be used for obtaining a direction finder bearing using the first part of the transmission, and readings are then taken during the navigational period of the transmission, large errors could result since the groundwave signal would be at a minimum while the skywave signal would be at a maximum, thus "swamping" the groundwaves. Therefore, it is essential that either the loop antenna be aligned for maximum signal during the navigational period or that a vertical antenna be used.

An automatic gain control in a receiver can cause difficulty in determining the equisignal. This determination is based upon the relative magnitudes of the two signals, while the function of the automatic gain control is to equalize the signal amplitudes if possible. Therefore, the observer should insure that the automatic gain control is not in operation during the period of counting the dots and dashes.

In general it is found that the bearing error increases with range, both day and night. The systematic errors are usually small except over rough ground. Skywave systematic errors, although large at some ranges, are usually smaller than the random errors. Table 4357b shows typical random errors of count under different conditions. The probability is given as 95 percent.

Day range (nautical miles)		Night range (nautical miles)				
Over land	Over sea	0–150	250	350	450	Over 550
4	2	2	8	14	10	4

TABLE 4357b.—Typical random errors of count under different conditions for a probability of 95 percent.

Table 4357c shows the 2 σ (art. Q6) errors of Consol lines of position.

Angle from normal	*Day range (nautical miles)*					*Night range (nautical miles)*		
	Over sea			*Over land*				
	250	*500*	*1000*	*250*	*500*	*100*	*300–1000*	*1500*
On normal	1.5	3	6	3	6	0.5	10	18
60°	3	6	12	6	12	1	20	36
75°	6	12	24	12	24	2	40	72

TABLE 4357c.—Errors of Consol lines of position in terms of 2 σ.

If a number of counts are taken sequentially and the mean value is used to obtain the bearing, the error should be less than shown in table 4357c.

4358. The operational range of the system is a function of the power transmitted from the three antennas. The total power from the transmitter is divided among the three antennas during the navigational period, the power from the center antenna being greater than that from the outer antennas by a factor of k. This factor then has an influence on the range. However, as k is increased the available power from the outer antennas is decreased, making it more difficult to count the dots and dashes in the presence of noise. For optimum operation of the system, the factor k is normally four.

The distance at which the signals can be used properly depends on propagation conditions as well as the ratio of signal to outside interference or to receiver noise.

Figure 4358 shows the field strength at different ranges in the case of a 1 kW, 275 kHz transmitter. Curves *a, b,* and *c* refer to groundwave propagation during daylight hours when skywaves at this frequency are heavily absorbed in the upper atmosphere. At night, however, the skywaves are reflected and can be received at great distances. It is much more difficult to predict field strengths under these conditions as the reflected waves are subject to variable absorption. However, curve *d* shows a typical mean level. It can be seen that, except at short ranges, the skywave signals received at night are in general stronger than the groundwaves received during the day.

The operational range of a Consol station depends not only on its output power and the time of day, but also on the receiver noise figure and outside interference levels. As an average, the maximum range is considered to be between 500 and 1,200 nautical miles by day and between 900 and 1,500 nautical miles by night.

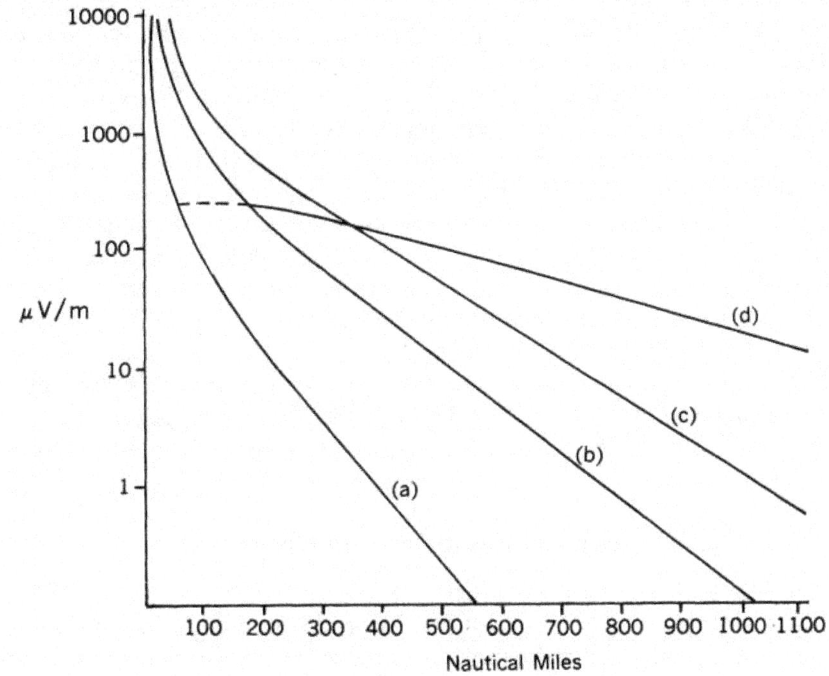

FIGURE 4358.—Typical field strength curves.

CHAPTER XLIV

SATELLITE NAVIGATION

4401. Introduction.—The Navy Navigation Satellite System (NAVSAT) is the operational satellite navigation system of the United States. It is all-weather, worldwide, and passive. The system's accuracy is better than 0.1 nautical mile anywhere in the world, on land and sea. It is available to all civilian users of the world. The system is used primarily for the navigation of surface ships and submarines; it has some application in air navigation. It is also used in hydrographic surveying and geodetic position determination. The system is also known as TRANSIT.

The NAVSAT system utilizes the doppler shift of radio signals transmitted from the satellite to measure the relative velocity between the satellite and the navigator. Knowing the satellite orbit precisely, the navigator's absolute position can be accurately determined from this time rate of change of range to the satellite. The satellites also transmit timing signals which provide time automatically. Frequency accuracy is better than 1 part in 10 billion for precise determination of the doppler shift; the time is given in Coordinated Universal Time (UTC) to within 200 microseconds (usually 20-50 microseconds).

NAVSAT was conceived and developed by the Applied Physics Laboratory of the Johns Hopkins University for the U. S. Navy. The operation of the system is under the control of the U. S. Navy Astronautics Group with headquarters at Point Mugu, California.

System Configuration and Operation

4402. The Navy Navigation Satellite System (fig. 4402a) consists of a constellation of orbiting satellites, a network of tracking stations that continuously monitor the satellites and update the information they transmit, and any number of user equipments composed of receivers and computers.

Each navigation satellite is in a nominally circular polar orbit at an approximate altitude of 600 nautical miles. The orbital planes of the satellites intersect at the earth's axis of rotation and are spaced apart in longitude. Thus, the orbital paths cross at the North and South Poles (fig. 4402b). Although the orbital planes remain nearly fixed in space, the satellites appear to traverse the longitudinal meridians as the earth rotates beneath them. There are usually five satellites operating in the system, and these provide navigation fixes anywhere on the earth on nearly an hourly basis. Five satellites in orbit provide redundancy; the minimum constellation for system operation is four. This redundancy allows for an unexpected failure of a satellite and the relatively long period of time desired to schedule, prepare for launch, and orbit a replacement satellite on an economical basis (not an emergency basis). This redundancy also provides for turning off a satellite when (on rare occasions) its orbital plane precesses near another satellite's plane, or when the timing (phasing) of several satellites in their orbits are temporarily such that many satellites pass nearly simultaneously near one of the poles.

Each satellite orbits the earth in approximately 106 minutes. Throughout its useful life, each satellite continuously transmits the following phase-modulated data

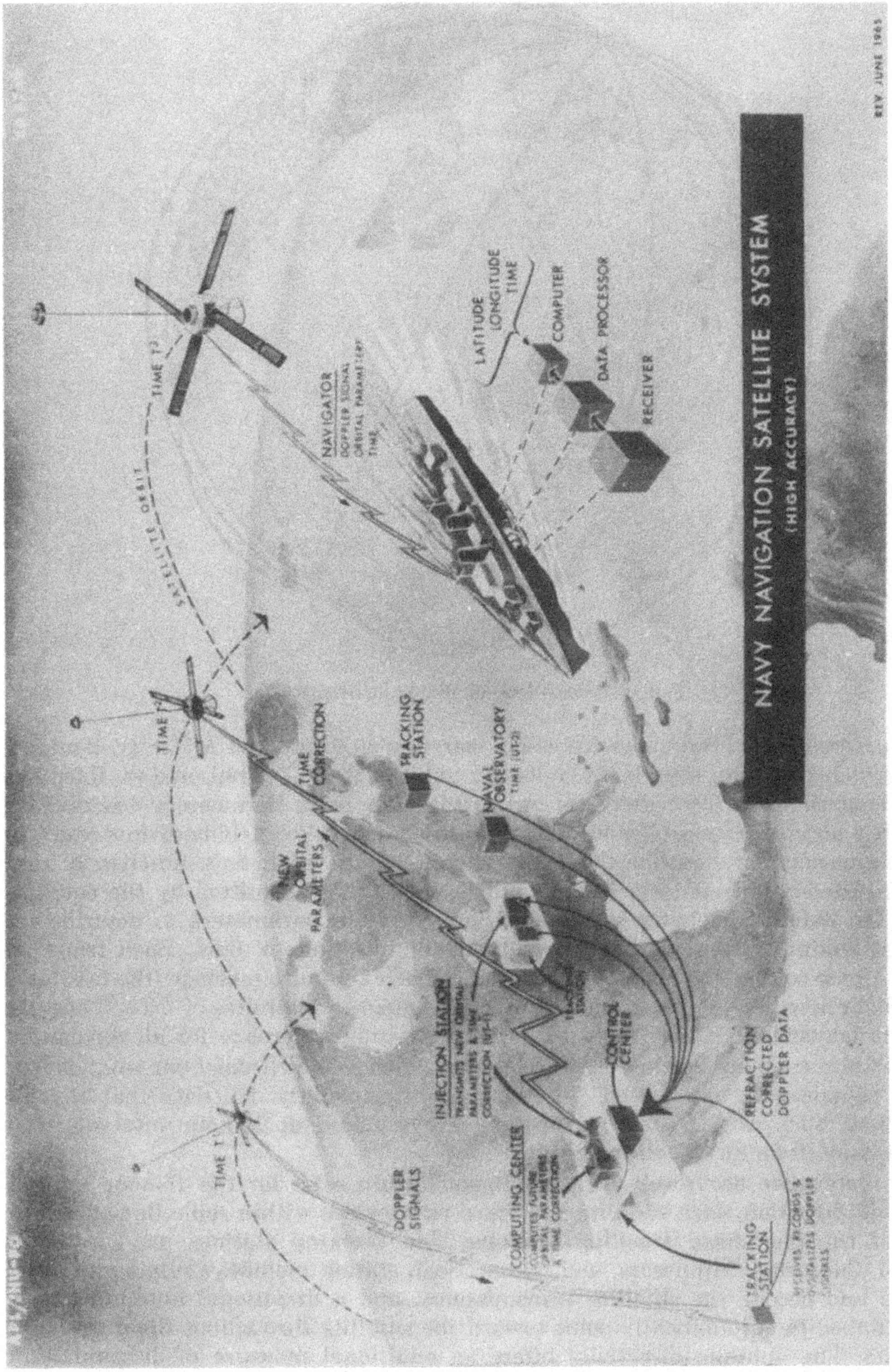

Courtesy of Applied Physics Laboratory, The Johns Hopkins University.

FIGURE 4402a.—Navy Navigation Satellite System.

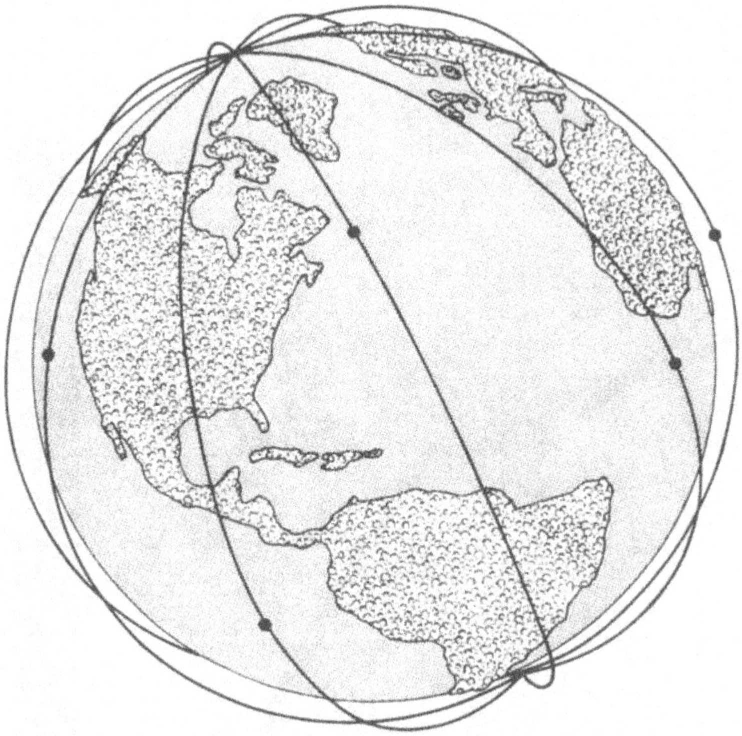

FIGURE 4402b.—Satellite orbital distribution.

as two messages on two radio frequency carriers (150 and 400 MHz): (1) 2-minute mark synchronization signals, (2) a 400 Hz audible "beep" signal, and (3) fixed and variable parameters describing its own orbit. The fixed parameters describe the satellite's approximate orbit and typically are used only for a 16-hour interval. The variable parameters describe the fine structure of the orbit as a function of time and are correct only for the time at which they are transmitted by the satellite. Thus, the satellite memory stores sufficient variable parameters to describe the orbit at 2-minute intervals between subsequent injections of data. Each transmission is timed so that the end of the 78th bit of each 2-minute message (the last bit of the synchronization signal), coincides with the integral 2 minutes of UTC. Thus, the satellite transmissions also serve as an accurate time reference for all navigators.

All data transmitted that does not change, such as synchronization and identification signals, etc., are "wired" into the satellite memory. All data that changes with time, such as the orbit parameters, are replaced at 16-hour intervals by a transmission from an **injection station.**

To determine accurately its present and future orbit for the 16-hour interval after data injection, each satellite is tracked as it passes within radio line-of-sight of each of the four fixed **tracking stations.** The tracking stations are located in Hawaii, California, Minnesota, and Maine. Each station includes equipments which receive and decode the satellite transmissions, and a directional antenna that is programmed to automatically point toward the satellite throughout the duration of the pass. The antenna directivity offers an additional measure of discrimination against spurious signals from local transmitters and ensures tracking of the selected satellite during those instances when two satellites converge within radio line-of-sight.

Programming data for pointing the station antennas either originate at the **central computing center** and are routed through the **control center** to the tracking station, or are locally derived at the tracking station. Just before the satellite time-of-rise, the antenna at the tracking station is pointed to acquire the satellite signals. As the satellite rises above the horizon, the antenna continues to follow the pass, enabling the frequency tracking loop of the radio receiver at the station to "lock onto" the signals. The receiver and data processing equipment decode and record the satellite message. The doppler signal is digitized and sent with satellite time measurements, via the control center, to the central computing center.

The central computing center continually accepts satellite data inputs from the four tracking stations. Periodically, to obtain the orbital parameters for a satellite, the central computing center computes an orbit for each satellite that best fits the doppler curves obtained from all tracking stations. Then, using the computed orbital shape, the central computing center extrapolates the position of the satellite at each integral 2 minutes in Coordinated Universal Time for the next 16 hours subsequent to data injection. The data, together with commands and time correction data for the satellite, and antenna-pointing orders for the injection station antennas, are supplied to the four injection stations located in Hawaii, California, Minnesota, and Maine via the control center.

The injection stations, after receiving and verifying the incoming message from the central computing center, store the message until it is needed for transmission to the satellite. Just before a satellite's time-of-rise, the injection station antenna is pointed to acquire, "lock on", and track the satellite throughout the pass. As soon as the receiving equipment at the injection station receives and locks onto the satellite signals, the injection station transmits the new injection data and commands to the satellite. Transmission to the satellite is on a frequency different from those used by the satellite, and the bit rate is much higher; thus, injection is completed in a matter of seconds and does not disturb use of the satellite for navigation.

The next integral 2-minute transmission by the satellite during the pass contains part of the newly injected data. In the injection station, this read-back is compared with the data that the satellite should be transmitting as a check for injection errors. Because most of the newly injected data (the variable parameters) will not be transmitted until the appropriate time during the satellite orbit, the initial read-back from the satellite includes parity check data. These data provide for error detection of the variable parameters so that the injection station can verify that the parameters were received correctly. If no errors are detected, injection is complete. If one or more errors are noted, injection is repeated at 2-minute intervals (updating the variable parameters as necessary) until the satellite transmission is verified as being correct or until the satellite is no longer available for data injection.

Once data injection is complete, the satellite continues to transmit its normal 2-minute messages. Any time corrections for the satellite clock and any commands for the satellite (such as changeover to the standby oscillator, cease transmission, etc.) also are performed during the period of data injection. These precautions ensure that the navigation equipment, which depends on accurate satellite data for determining its position, is provided the best possible data from each satellite. Any time that the satellite is within radio line-of-sight of the navigation equipment and has a maximum elevation at time of closest approach between 15° and 75°, the satellite transmission can be used to compute the exact position on earth of the navigator, although good data can be received frequently when the satellite is not within these elevation requirements.

4403. Navy Navigation Satellites are launched into nearly exact polar orbits from Vandenberg Air Force Base, California by four-stage, solid-fuel Scout rockets (fig. 4403). These polar orbits are nearly circular at nominal altitudes of 600 nautical miles. The orbits are circular in order to attenuate acceleration and deceleration characteristics of elliptical orbits, and polar to reduce the precession of orbital planes which results in eventual overlap.

Although successive models may differ, each satellite contains: (1) receiver equipment to accept injection data and operational commands from the ground, (2) a decoder for digitizing the data, (3) switching logic and memory banks for sorting and storing the digital data, (4) control circuits to cause the data to be read out at specific times in the proper format, (5) an encoder to translate the digital data to phase modulation, (6) ultrastable 5 MHz oscillators, and (7) 1.5-watt transmitters to broadcast the 150- and 400-MHz oscillator-regulated frequencies that carry the data to earth.

4404. Configuration.—The satellites (sometimes called operational, OSCAR, or NOVA satellites) weigh 116 pounds and have their antenna and solar cell panels configured for a nominal circular polar orbit at an altitude of about 600 nautical miles. Each satellite is an octagonal cylinder 18 inches wide and 12 inches high, and has four solar cell panels that extend from four of eight faces (fig. 4404a). These four panels fold down around the fourth stage of the Scout rocket during launch, and are held by wires which also serve as a "yo-yo" despin mechanism after the fourth stage has fired. When released, the blades are erected into the position shown by swinging "door hinges." Subsequently, the external configuration is modified by extending a 100-foot boom from the top of the satellite. The boom has a 3-pound weight at the end in order to achieve gravity-gradient stabilization, so that the bottom of the satellite containing the antennas always points toward the ground.

A follow-on version of the operational satellites, designated NOVA (figure 4404b) was launched in 1981. NOVA contains a number of improved features; however, it is not possible to distinguish between OSCAR and NOVA except for a minimum of 3 to 4 dB signal enhancement from the improved antenna and transmitters.

4405. Stabilization.—The operational satellite and the last stage of the Scout rocket are despun soon after achieving orbit by the yo-yo technique (art. 4404). The yo-yo immediately reduces the spin to a few percent of the spin rate of the fourth stage (spin-stabilized) Scout rocket. This small residual spin is removed by incorporating long, thin rods of a magnetic material which exhibit substantial magnetic hysteresis in the solar panel blades. These hysteresis rods are inductively magnetized by the earth's magnetic field in alternate directions as the satellite spins. The hysteresis loss involved in this process takes energy from the satellite rotation and shows up as heat energy in the rods. This process slows the rotation rate to zero within less than a day. When the spin rate reaches zero, an electromagnet is activated along the vertical axis to align itself with the earth's magnetic field line. The hysteresis rods in the solar cell blades provide damping for the resulting magnetic attitudes stabilization. Within another day, the librations (oscillations) of the satellite about the local magnetic field have damped out sufficiently so that the satellite axis is always within 10° of the local magnetic field direction.

Courtesy of Applied Physics Laboratory, The Johns Hopkins University.

FIGURE 4403.—Scout launch of Navy Navigation Satellite.

As the satellite passes over the north magnetic pole, it is vertical, with the bottom side of the satellite pointing toward the earth. At this point a command from the ground turns the electromagnet off, and the 100-foot boom is extended. Thus, during its time in orbit, the satellite always points its antennas toward the earth. This attitude is maintained because of the difference in the level of gravity at the satellite and at the end of the boom. In this "gravity-gradient" stabilization mode, the hysteresis rods provide enough damping to bring the librations below 10° within a day or two. This orientation makes it possible for the satellite to send most of its transmitted power toward the earth, and for circular polarization to be used so that no loss results from Faraday rotation in the ionosphere. In addition, the transmitter antenna shapes the beam so that more power is transmitted at large

FIGURE 4404a.—Navy Navigation Satellite.

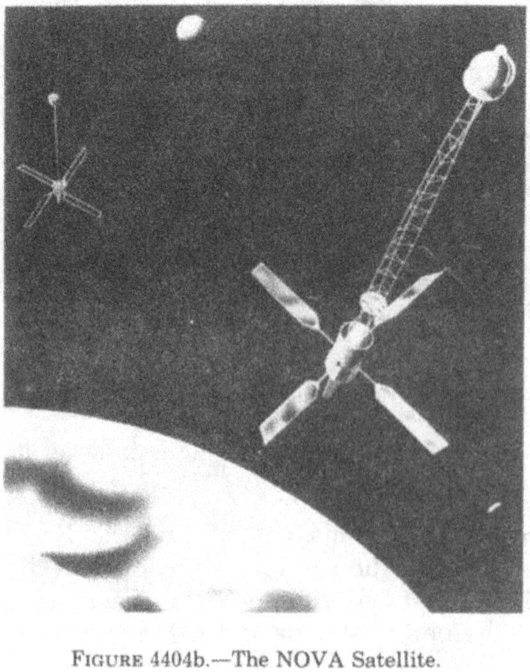

FIGURE 4404b.—The NOVA Satellite.

angles, so that at any moment the received power anywhere above the satellite's horizon is about the same.

4406. Power and thermal design.—The power system is based on solar cells that charge a set of sealed nickel-cadmium batteries and provide power during the dark portion of each orbit. The solar cells initially provide 30 watts of power when the satellite is launched. After 5 years of exposure to the radiation present in the space environment, 25 watts are still available, which is sufficient to power the satellite. A major problem with near-earth satellite design is caused by the variation in sunlight as the orbit plane precesses with respect to the terminator. A polar satellite near the earth has two extended periods each year during which the satellite is continuously in sunlight. This is often referred to as the **dawn orbit** since it corresponds to the time when the satellite passes overhead at approximately local sunrise (and, of course, sunset). At the other extreme, there are two periods each year during which the sun is nearly in the orbital plane; during this period, the satellite is in darkness for 33 percent of each orbit. This is known as the **noon orbit** because the satellite passes overhead near the time of local noon (and, of course, midnight). The satellite design assures that both the thermal balance and the power balance are acceptable under these two extreme situations.

For a gravity-gradient stabilized satellite, it is true that when the satellite is in constant sunlight (dawn orbit) the sun always "looks" at the side of the satellite whereas for the minimum sunlight case (noon orbit) the sun "looks" at the top and part of the edge of the satellite. Accordingly, by carefully choosing the ratio of the top area to the projected area of the sides, the total thermal input during a full orbit is approximately the same in both the dawn and noon orbits. Similarly, by canting the antenna blades on which the solar cells are mounted to the proper angle, the total power input from the solar cells is made approximately the same in both the dawn and noon orbit cases.

In the operational satellite the use of this technique has made it possible to provide automatic temperature control that is accurate to a few degrees by means of thermostats and heaters and, further, to have acceptable temperature limits even if the automatic temperature control system fails.

4407. Packaging techniques.—The operational navigation satellites are relatively simple from a mechanical standpoint. There are no tape recorders or television cameras and, in fact, no moving parts other than command relays, which are operated very infrequently. However, they do contain some 35,000 magnetic (memory) cores and 6,200 other electronic components. In order to reliably package this equipment within 140 lbs (a Scout rocket compatible payload), it is necessary to use modern packaging techniques. Welded "cordwood" construction with a package density of 50,000 parts per cubic foot is used wherever applicable. Of the 46,000 permanent electrical joints in the electronics, 40,000 are welds and only 6,000 are solder connections. Generally, except in telemetry functions which can be lost without destroying the operational usefulness, redundant wiring and redundant solder connections are used. The use of mechanical connectors (plugs) in electrical circuits is kept to an absolute minimum, and, where plugs must be used in critical circuitry, complete redundancy is provided. In most places where plugs would normally be used, wire-wrapped connections are used.

4408. Memory organization.—The information on satellite position required for navigation is stored in a magnetic core memory and transmitted as phase modulation on the two basic stable frequencies used to generate the doppler shift. The modulation pattern is quite symmetrical so as not to introduce an error in the measurement of doppler. The specific modulation patterns that are to be interpret-

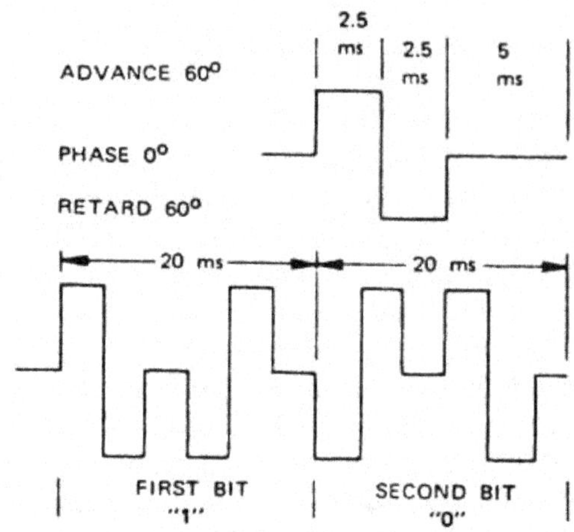

THE PHASE OF THE DOPPLER SIGNAL IS ADVANCED AND THEN RETARDED TO REPRESENT ONE POLARITY: RETARDED AND THEN ADVANCED FOR THE REVERSE POLARITY. EACH HALF-BIT IS TRANSMITTED TWICE. THE SECOND TIME IN REVERSE POLARITY.

"1" = ADVANCE RETARD SPACE
RETARD ADVANCE SPACE

"0" RETARD ADVANCE SPACE
ADVANCE RETARD SPACE

BIT RATE ≈ 50/SECOND

FIGURE 4408.—Phase modulation waveforms.

ed as "zero" or "one" are shown in figure 4408.

The memory, which is read out every 2 minutes, contains 156 words of 39 bits each, plus an additional 19 bits. The great majority of these words are not required for navigation but disseminate other information. The first two words are simply a fixed pattern of "zeros" and "ones" used to recognize the start of a message and to establish synchronization of the ground equipment with the satellite transmissions. Thereafter, the words with specific significance for navigation are a total of 19 words divided into 2 sets, an initial set of 8 so-called "ephemeral words" (consisting of word numbers 8, 14, 20, 26, 32, 38, 44, and 50) and a set of 11 fixed parameters that are changed only by injection.

Because of the departure of the earth's gravity field from that of an ideal spheroid, atmospheric drag, solar photon pressure, attraction by the sun and moon, etc., the satellite orbit cannot be given accurately by algebraic equations. However, there is an algebraic description using 11 orbit parameters which affords a good approximation and is simple to compute in navigation equipment. The deviations of the actual orbit from this 11-parameter algebraically described orbit are small and require relatively little satellite memory to store a complete set, each 2 minutes for 16 hours. The appropriate deviations of the actual orbit are transmitted each integral 2 minutes of UTC and are the eight ephemeral words mentioned above.

After injection, the main memory is simply read out serially every 2 minutes until a new injection takes place—all main memory words being unchanged during the interval between injections (usually about 12 hours). However, the ephemeral words must be advanced on every 2-minute memory readout, and are not stored in the main memory. Specifically, consider the memory readout that lasts from t

minutes to $t+2$ minutes. The eight ephemeral words will contain the orbit deviations appropriate to the times $t-6$, $t-4$, $t-2$, t, $t+2$, $t+4$, $t+6$, and $t+8$, respectively. Thus, on each successive readout, the ephemeral words are moved upward (advanced) so that word number 8 in the previous readout is discarded and replaced by the previous word 14, the old word 14 is replaced by the old word 20, etc., and finally a new word 50 is transmitted. This new word is transferred from a separate memory in the satellite known as the ephemeral memory, which is filled at the time of injection. On each memory readout, a single 39-bit word is "transferred across" to fill word 50 in the main memory. The ephemeral memory contains 480 words, and since a fresh word is used every 2 minutes, the ephemeral memory is used up in $480\times2=960$ minutes, or 16 hours. Thus, a new injection must be made within 16 hours to prevent the ephemeral memory from running out. It should be noted that each word in the ephemeral memory is read out 8 successive times, first as word 50, then as word 44, etc., finally as word 8 and then is discarded. Notice also that a single 2-minute memory readout gives orbit information for a full 14-minute interval, spanning the time of readout symmetrically.

4409. Timing.—The orbit readout rate is controlled by counting down from the basic stable oscillator. However, in spite of the excellent stability of the satellite oscillator, there are long-term drifts that slowly change the basic oscillator frequency during the life of the satellite. Thus, to keep the memory readout period accurate at integral 2 minutes of UTC, it is necessary to modify the countdown of the oscillator. This is done by using the special bit of each 39-bit word in the main memory as a signal to determine whether or not to suppress a single count in the countdown process. This is done at a point where a single count has a value of $10\mu s$. Thus, a total variation of $156\times10\mu s=1.56$ ms in each 2 minutes is available. With only this level of adjustment, time correction may not be sufficient for a 12 hour period. Thus, a fine adjustment is available by inserting an appropriate count suppression signal in the ephemeral memory. Each ephemeral memory correction bit is used only once and then discarded.

Use of the System

4410. Equipment developed for Navy Navigation Satellite System use includes the AN/BRN-3, AN/SRN-9, AN/WRN-5, and AN/SRN-19 radionavigation sets. The AN/PRR-14 Geoceiver was developed for using NAVSAT for geodetic position determination. Several commercial radionavigation sets are also available for use with NAVSAT.

AN/BRN-3 Radionavigation Set

4411. AN/BRN-3 Radionavigation Set was developed and deployed in the 1960's as a result of several programs and U. S. Navy contracts. Active development of major units was initiated by the Applied Physics Laboratory of The Johns Hopkins University in late 1960, with requests for technical proposals from industrial suppliers. In about 1 year the first prototype had been developed, tested, and delivered to the Applied Physics Laboratory to be used in the development of the total Navy Navigation Satellite System, and to be used as a standard for procurement control of follow-on prototype and production units. By 1964, the total system had matured and was operational with regular service being provided to the fleet ballistic missile (FBM) submarine forces while on patrol.

The AN/BRN-3 is a complex, multi-unit radionavigation aid whose purpose is to compute accurate ship's position using satellite orbital data and measurements of doppler shift to the received satellite reference signals. The position data are provided to the Navigation Control Console (NCC) via the Navigation Data Assimilation Computer (NAVDAC) for the calibration and adjustment of the Ship's Inertial Navigation System (SINS) so that continuous navigation is possible with accuracies sufficient for POLARIS or POSEIDON fleet ballistic missile (FBM) targeting. All units are designed for the special military environments prevailing in the FBM submarine application.

During the production years the Sperry Rand Corporation was designated the prime contractor for AN/BRN-3 procurement, deployment, and maintenance in the fleet. Applied Physics Laboratory of The Johns Hopkins University was retained and continues to provide technical services. The principal components of the system are: (1) radio receiver and RF amplifier/power supply, (2) RF antenna, (3) data processor/computer, (4) tape punch and reader or magnetic tape unit, and (5) typewriters. Figure 4411a shows a typical FBM submarine installation.

The AN/BRN-3 Radionavigation Set, in conjunction with the SINS and other navigational aids, has a primary function of providing continuous submarine navigation with global, all-weather coverage and high accuracies sufficient to provide position references for POLARIS and POSEIDON missile targeting. The Navy Navigation Satellite System and the AN/BRN-3 provide highly accurate position fixes (at intermittent times dependent upon satellite availability in the local area and at the discretion of the submarine commander) from which a calibration of the SINS, relative to a standard earth coordinate system, can be made to indicate the compensating adjustments necessary to remove the effects of inertial system anomalies and drifts.

In addition to FBM submarine installations, the AN/BRN-3 has been used for special missions aboard surface ships. Several units are installed in training centers, instrumentation ships, and research centers.

In addition, each operational station of the navigation satellite ground support subsystem uses the AN/BRN-3 receiver for satellite tracking and orbit determination rather than for navigation. The AN/BRN-3 has been used as a standard for performance comparison with other alternative navigation sets and for performance evaluation of newly launched satellites prior to their acceptance into official Navy operational service.

The AN/BRN-3 receives navigation satellite signals on the 400 and 150 MHz channels from which it measures doppler shift versus time and recovers "message" data from the modulation. It receives ship's velocity information and position estimates from SINS, timing from the ship's clock, and operational commands from the ship's navigation center. Data other than the satellite signal-derived information are received via the ship's NAVDAC computer. The AN/BRN-3 computes from these inputs the ship's position at sea to a high accuracy and provides the information for printout at the Navigation Control Center (NCC).

In addition, the AN/BRN-3 computes from its data memory (which is updated by new message data recovered during each satellite pass reception interval actually taken), predictions, and information for the selection of usable future satellite passes from those that will be available within a time interval surrounding the next desired SINS calibration time. The prediction is based on a constellation of up to eight navigation satellites and the ship's estimated position. The information pro-

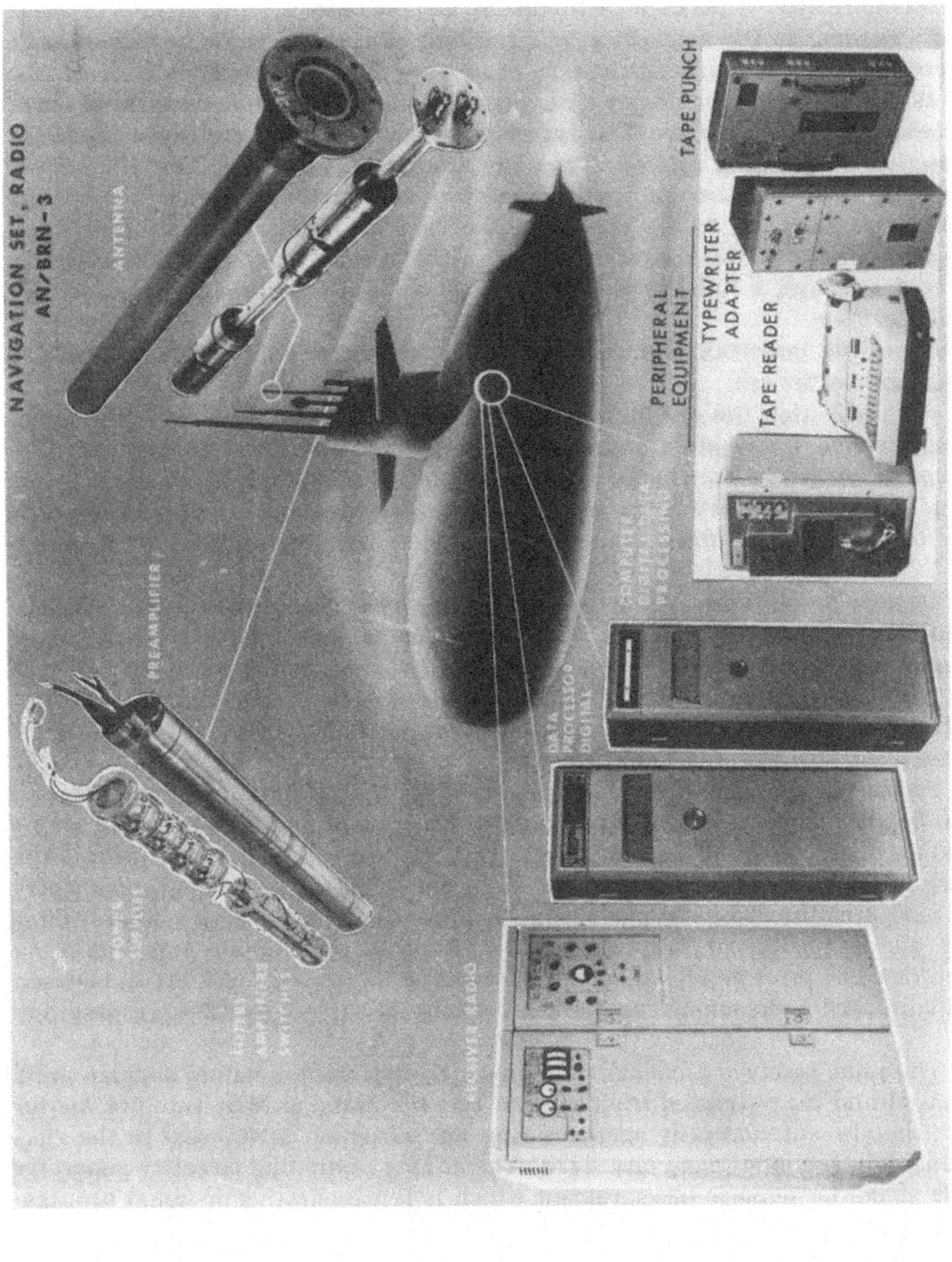

Courtesy of Applied Physics Laboratory, The Johns Hopkins University.

FIGURE 4411a.—AN/BRN-3 Radionavigation Set.

vided includes predictions of time of satellite rise (when the radio signal can be received) and the value of the doppler shift.

From a model of the total system, the AN/BRN-3 computes forcing functions for the enhancement of its own performance. It generates driving signals to assist the radio receiver in the acquisition and tracking of satellite radio signals so as to minimize the loss of information due to temporary failure to coherently track the signal as a result of its temporary disappearance due to spray and wavewash over the antenna at the sea surface. The system is decision directed and self-adaptive in the selection of operating mode and receiver tracking bandwidths. The AN/BRN-3 has self-diagnostic features for routine maintenance, pre-pass readiness checks to avoid abortive surfacing by the submarine for satellite reception, and for performance enhancement of the operational mode. The normal operation is completely automated although a full complement of controls and monitors are provided for manual operation.

Some of the important features and design characteristics of the AN/BRN-3 Radionavigation Set are:

1. For navigation the system uses iterative least squares curve fitting routines to match doppler information measured from received signals to doppler information that is predicted for the best estimate of the prevailing satellite-navigator dynamic geometry. A very large number of measurements are made on received signals to generate, in effect, a smoothed curve of observed doppler shift versus time.

2. Doppler measurements are made using the set's local clock to determine the time interval in microseconds that is required for a preset digital doppler shift counter to overflow when the doppler shift is offset about a reference frequency of 55 kHz. Each doppler information sample therefore constitutes a nominal short (1 second) count, the actual interval varying between about 0.8 to 1.4 seconds for satellites at an altitude of 600 nautical miles.

3. Provisions are made in the data processor for the validation and qualification of each doppler measurement sample and for selective grouping of qualified points to obtain a distribution versus time during the satellite pass which will provide the greatest efficiency in total time spanned and accuracy of the fix computation. Entry to the navigation fix computation routines requires 165 doppler data samples. After editing and grouping, 70 fully qualified data points are required for final fix computation—35 prior and 35 following the time of closest approach (TCA) between the satellite and submarine. This data processing is subject to software programming.

4. The radio receiver combines in analog circuitry the prevailing doppler shifts as received and reconstructed from each of the 400 and 150 MHz satellite carrier signals, thereby automatically compensating for refraction errors due to the electron density in the ionosphere (art. 4429). The analog combining circuitry generates a single signal for doppler measurement which is representative of signal propagation through nonrefracted vacuum. Due to the scaling factors employed, the magnitude of doppler shift presented for measurement is equivalent to that which would occur if the satellite transmitted a single carrier frequency at 687.5 MHz. The circuitry also provides an analog derivation of a refraction signal. This signal is a measure of the departure from exact mathematical coherency between the 400 and 150 MHz carrier signals as received. The refraction signal is used as a quality indicator—if the frequency of the refraction signal exceeds a threshold of 10 Hz for

a sustained interval, the accompanying quantitative doppler data is deleted and not accepted to the navigation computing routines.

5. Data processing and computing routines are provided for the calibration of the local clock of the navigation set against epochs recovered from the satellite modulation and message formats.

6. One phase modulation decoder is provided in the radio receiver for the recovery of satellite timing, satellite orbital data, cross satellite alert, and other system data that are encoded in the 6,103 bits per 120 second repeatable satellite format which is transmitted by the satellite on each of the 400 and 150 MHz carrier signals. The phase modulation decoder is preceded by a proportional predetection combiner of the 500 kHz IF signals of the respective 400 and 150 MHz receiver channels.

7. After phase modulation detection of the combined satellite signals, information decoding is performed by the computing processor using majority voting and piecewise fitting routines to obtain at least one complete message from the several that occur within the interval of satellite signal reception.

8. There is an automatic search and satellite signal acquisition mode for the carrier tracking circuits of the radio receiver with the controls provided by the computer and data processor based upon predictions from a local memory of orbital parameters for up to eight satellites and the ship's inertial position data. Optional manual controls are also available at the receiver.

9. The system has built-in test features for the special analysis of the radio spectrum received by the antenna which can be used to identify and measure certain interfering and jamming signals at the sea surface.

10. The system has self-test features for equipment malfunction diagnosis and routine preventive maintenance covering the entire AN/BRN-3 system beginning at the antenna input terminals to the radio receiver. Software programs establish and control operational readiness tests prior to submarine surfacing while in a pre-pass period. The diagnostic analysis can extend to the smallest modular unit level.

The navigation accuracies that are obtained at sea depend in part upon the accuracies of inputs describing the ship's motion in the interval of the satellite pass when doppler data was measured.

The statistical accuracies achieved at a non-navigating fixed site, such as Station 110 located at Applied Physics Laboratory, using a typical operational software program, produce a circular probable error (CEP) of 0.017 nautical mile as shown in figure 4411b.

Integrated Doppler Tracking Equipment

4412. Developmental AN/SRN-9 equipment.—In the early stages of the development by the Applied Physics Laboratory of receiving equipment for use in the integrated doppler count method of navigation, the technical approach was centered around a single-frequency system. It was recognized that the use of a single-frequency system operating at the higher frequencies, i.e., 400 MHz, would result in a navigation error as large as 1 nautical mile because of the refraction effect of the ionosphere (art. 4429). The elimination of the requirements for a 150 MHz phase-locked receiver, for a more complex antenna with dual preamplifiers, and for refraction correction equipment appeared desirable in terms of the resultant equipment simplification and lower cost. The single-frequency system was built in breadboard form at the Laboratory, and the feasibility of the system demonstrated in mid-1961.

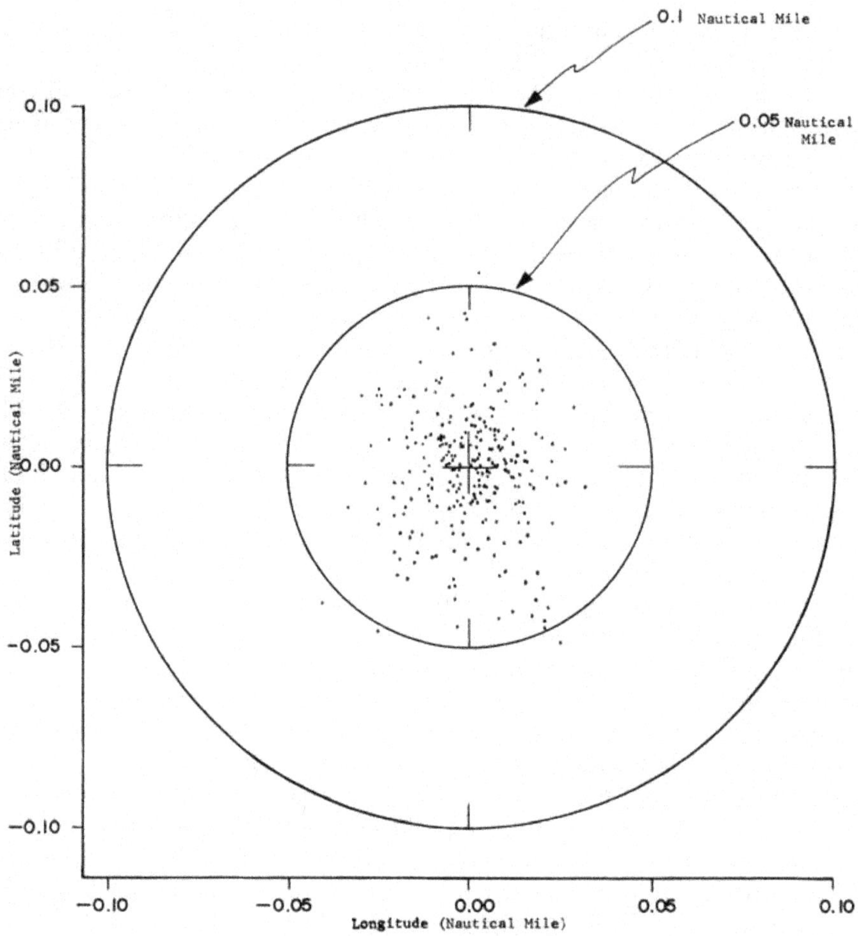

Courtesy of Applied Physics Laboratory, The Johns Hopkins University.

FIGURE 4411b.—Accuracy results at Station 110 for October 1974.

The design of a two-frequency system was begun by the Laboratory about the same time the single-frequency system reached its breadboard stage. This design effort disclosed that since the two received frequencies are always in constant ratio within a few parts in 10^8 (the order of the refraction effect) the second receiver need not be a phase-locked receiver, but could be merely slaved to the 400 MHz phase-locked receiver. The two-frequency system design was developed and tested as an engineering model and subsequently developed into a prototype form designated XN-5 (fig. 4412). In the period between 1964 and 1967, a total of 23 prototype AN/SRN-9 (XN-5) sets were produced by the Laboratory and placed in service, primarily aboard attack aircraft carriers and oceanographic ships.

Basic to the design of both systems is the stable oscillator. Any bias in measuring frequency that is maintained over a pass (as opposed to point-to-point noise within a pass) produces a proportional error in position. The assumption is made, therefore, that the frequency of the local oscillator is an unknown (art. 4427). This assumption requires that the measurements and computations needed for a navigation fix be arranged to eliminate the value of the frequency of the oscillator. When this elimination is done properly, the only stability required is five parts in 10^{11}

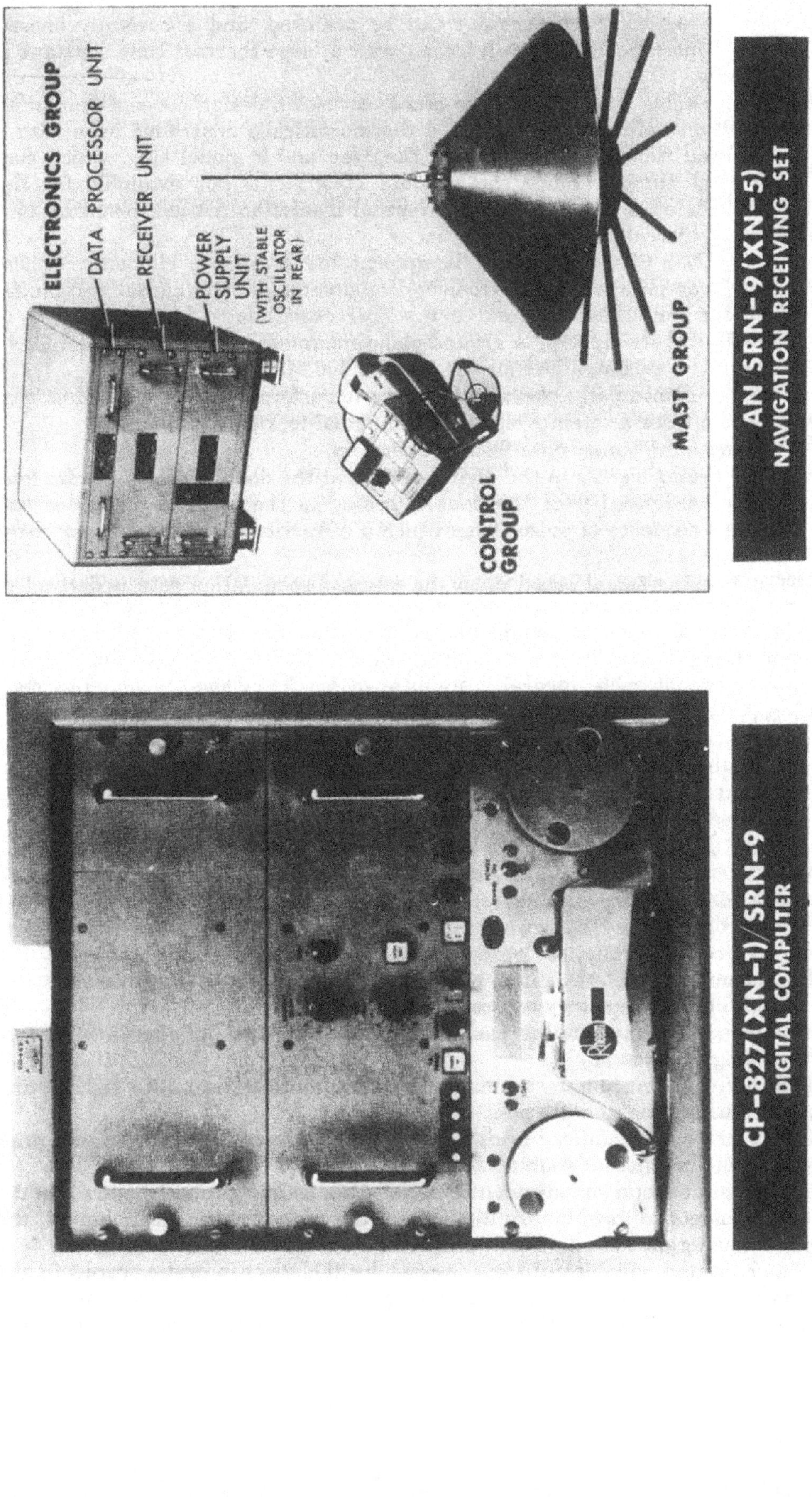

Courtesy of Applied Physics Laboratory, The Johns Hopkins University.

FIGURE 4412.—Developmental equipment.

over a 2-minute period. Such stability can be achieved, and a carefully chosen crystal in a thermostatically controlled oven with a large thermal time constant is entirely adequate.

The (XN-5) stable 5 MHz reference oscillator uses a design concept similar to those used in the satellite oscillator, i.e., a thermostatically controlled oven with a very long thermal time constant between the oven and a monel slug, which contains the critical circuits. Since the vacuum of space is not available for the earthbound oscillator, a great amount of thermal insulation is used, resulting in a relatively large physical size.

The AN/SRN-9 (XN-5) receiving equipment has five basic elements: (1) the antenna and preamplifiers, (2) the receiver-demodulator, (3) the digital section, (4) the control group (output section), and (5) the 5 MHz oscillator.

The antenna is a whip over a ground-plane mounted on the superstructure of the ship, along with preamplifiers for the 150 and 400 MHz signals.

The receiver-demodulator contains circuitry to perform the following functions:

1. selectively track a satellite signal after manual lock-on;

2. demodulate the binary data from the carriers;

3. provide timing signals to the digital section at the doublet (half bit) rate (one every 9.83 ms) as derived from the doublet coding in the satellite messages; and

4. produce a sequence of pulses from which a refraction corrected doppler count is obtained.

A precise timing signal based upon the message modulation rate is derived in an internal clock in the receiving equipment. This synchronized internal clock controls the decoding, printing, and doppler count gating operations with an accuracy of better than 0.2 ms. Because the operational satellites transmit the end of message work, two at each integral 2 minutes of UTC ($\pm 200 \mu s$), adequate time information is obtained from the satellite for navigation and doppler gating.

The digital section contains shift registers for accumulating the doppler count and for storing the serial binary data decoded from the satellite messages.

The digital section also contains an output register and the necessary counting and control logic to organize the satellite messages into words and digits (output format control). It also programs the data and other timing signals to the output terminals. The message data are extracted in four-bit groups (i.e., excess-three binary coded decimal format). Control signals are available to take all data (every word) or select only every sixth word (all that is necessary) for normal navigation.

From the control group, the navigator can monitor the operation of the equipment. In operation, the navigator remotely tunes the 400 MHz receiver from whence he obtains all necessary control functions.

In summary, for any satellite pass the following sequence of events will occur in the receiving equipment:

1. The receiver-demodulator is manually locked onto the satellite signals, and phase tracks during the satellite pass.

2. The receiver-demodulator begins decoding the binary data based on an arbitrary association of adjacent doublets.

3. The digital section monitors the decoded data, and properly pairs the demodulated doublets to form binary bits. When the proper pairing is achieved, the digital section energizes the bit synchronization line.

4. The counting and control logic is reset by the synchronization word in the satellite data format. The first time the synchronization word is received after bit synchronization, the digital section outputs a synchronization pulse. A 2 minute

UTC pulse is also generated each time the synchronization sequence (011111111111111111111111111110) is received.

5. The counting and format control logic in the digital section governs the handling of the binary data from the satellite, and the accumulation and output of the doppler count.

6. The 2-minute doppler count and satellite message data are printed out in decimal form on the control group printer.

7. Whenever an interrupt in the satellite signal occurs, bit synchronization must be reestablished.

4413. Functional description.—The navigation program currently used in conjunction with the CP-827 computer has the following functional capabilities:

1. Processes data from the SRN-9 receiver in an "on line" mode (the receiver and computer are electrically interfaced, and satellite data collected by the receiver will be transferred and stored by the computer).

2. Formats and prints the receiver data on the Central Group.

3. Majority-votes satellite message data from several messages and modifies this data in the case of injection of new message data into the satellite.

4. Edits doppler data to 7°5 provided at least four doppler counts are received. Additionally checks for monotonically increasing doppler counts and zero doppler counts.

5. Provides navigator's motion description in the form of constant course and speed or maneuvering with various options: distance north (DN) and distance east (DE) per 2 minutes, latitude, longitude, or range and bearing at the 2-minute time-marks.

6. Provides automatic satellite alert computation for the satellite just tracked. Alert computation for a single satellite may be performed at any time using message data from a previous satellite pass.

7. Renavigates satellite pass with different dopplers or different motion inputs.

8. A test satellite message can be navigated for test purposes.

4414. Accuracy.—The AN/SRN-9 (XN-5)/CP-827 produces an error at a fixed site of less than 50 meters (CEP). A bull's-eye plot of fixed site error is shown in figure 4414. The fixed site error is also given in table 4414. The difference in computed mean position and reference position is less than 17 meters. The shipboard accuracy at sea depends upon precise inputs of ship motion during a satellite pass. When SINS motion inputs are used, errors less than 0.1 nautical mile are obtained.

COMPUTED MEAN POSITION	
Latitude:	39°09′.8124
Longitude:	−76°53′.8408

FIX REFERENCE POSITION
Latitude: 39°09′.8214
Longitude: −76°53′.8410
Antenna Height: 106.00 meters

DIFFERENCE OF MEAN REFERENCE POSITION	
D. Lat. −0′.0090	−0.0090 mi.
DLo 0′.0002	0.0002 mi.
Mean Radial Error about the Mean Position:	0.0284 mi.
DEVIATION ABOUT THE COMPUTED MEAN POSITION	
Latitude:	0.0215 mi.
Longitude:	0.0226 mi.
Root Sum Square:	0.0312 mi.
Circular Probable Error (CEP):	0.0260 mi.

TOTAL NUMBER OF PASSES USED: 32.

TABLE 4414.—Fixed site error of AN/SRN-9(XN-5)/CP-827.

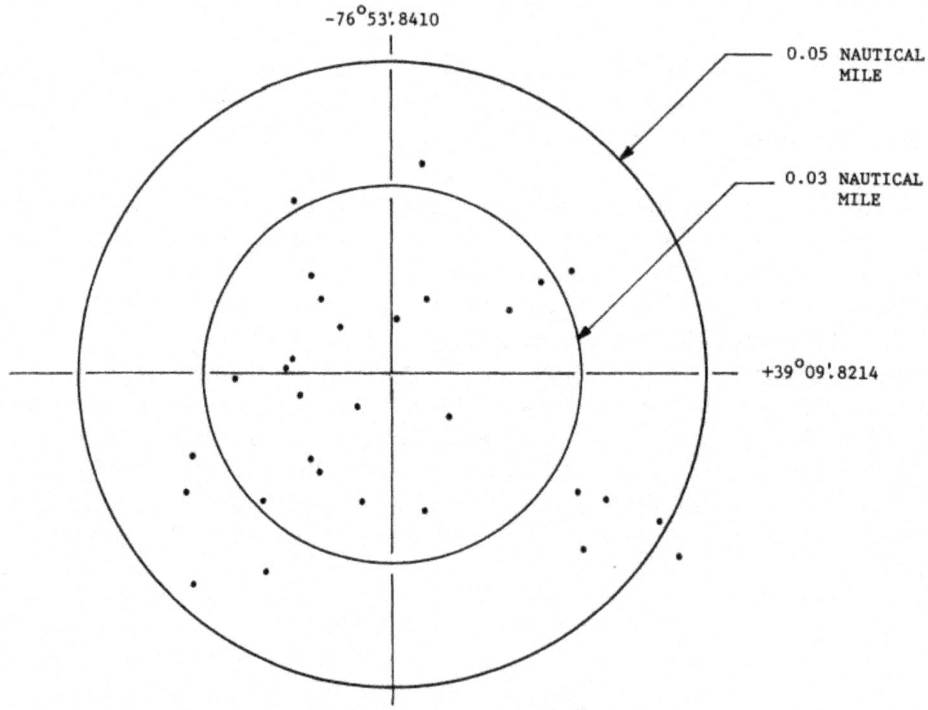

FIGURE 4414.—Bull's-eye plot of AN/SRN-9(XN-5)/CP827 navigation fixes at the Applied Physics Laboratory of The Johns Hopkins University.

Current Navy Navigation Satellite Equipment

4415. The AN/WRN-5 Radionavigation Set, a self contained navigation set consisting of preamplifier and receiver/computer assembly, is capable of automatic operation and utilizes short-count doppler data to provide navigation in a submarine environment. A synchro-to-digital interface allows interconnection with the SINS for the automatic inputting of ship's motion data. The set is fully militarized and uses the standard UHF submarine antenna for receiving 150 and 400 MHz signals from the Navy Navigation Satellites to obtain navigation fixes. The set is shown in figure 4415.

4416. Functional description.—The AN/WRN-5 Radionavigation Set consists of (1) antenna (uses the submarine's UHF antenna), (2) preamplifier, and (3) receiver assembly. The receiver assembly consists of the following subassemblies:

1. power supply,
2. radio receiver,
3. expanded data processor,
4. RF test signal generator,
5. memory loader (tape cassette),
6. 5 MHz frequency standard,
7. remote video display,
8. software program.

The AN/WRN-5 set weighs 150 pounds and uses 350 watts of power.

4417. Design characteristics.—The AN/WRN-5 Radionavigation Set employs 150 and 400 MHz phase-locked tracking loops, and has the following features:

1. automatic acquisition,
2. short count (23 second nominal interval) 400 MHz and 150 MHz doppler data,
3. integral test set,
4. integral 5 MHz oscillator,
5. integral computer with 16K-word, 16-bit memory,
6. teletype interface,
7. SINS interface,
8. navigation fix display,
9. minimum exposure time,
10. latest navigation program.

The AN/WRN-5 set interfaces with a remote video display, printer, and teletype.

4418. Accuracy.—Fixed-site accuracy of the AN/WRN-5 is shown in tables 4418a, 4418b, and figure 4418. The evaluation was conducted during four modes of operation: (1) the normal mode-automatic acquisition, dual channel; (2) single channel, 400 MHz only; (3) single channel, 150 MHz only; and (4) dual channel, short exposure. In the short exposure mode, approximately 4 minutes of data were taken including at least 2 minutes before the time of closest approach. In a few cases where lock was momentarily lost on one channel, data were taken for longer periods of time to allow a refraction corrected fix to be made.

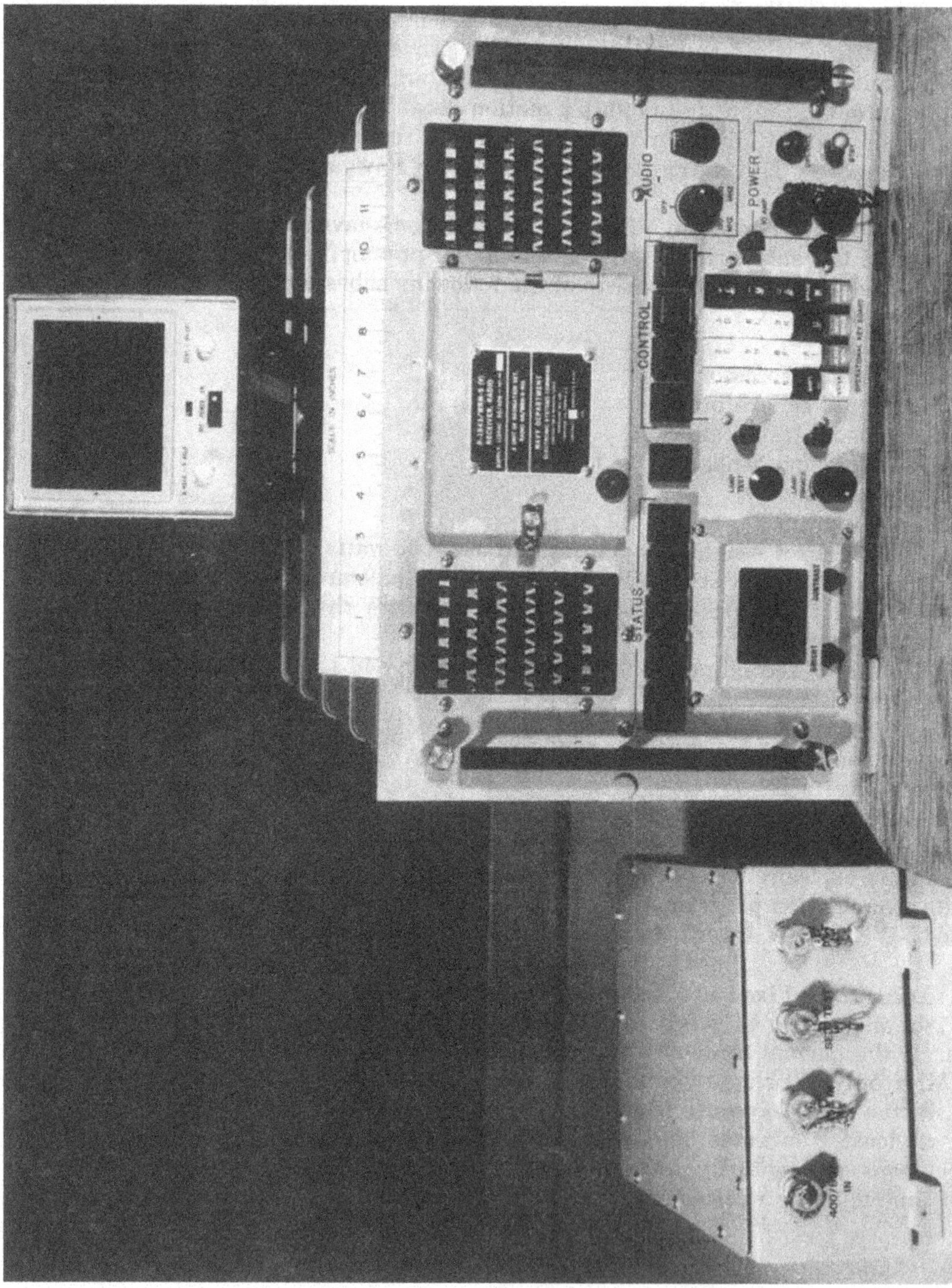

Courtesy of Applied Physics Laboratory, The Johns Hopkins University.

FIGURE 4415.—AN/WRN-5 Radionavigation Set.

DUAL CHANNEL, AUTOMATIC ACQUISITION	
Number of Passes:	39 (ECA** 10° – 70°)
Computed Mean Position*—	
Latitude:	39°09′.8158
Longitude:	–76°53′.8410
Difference of Mean Reference Position—	
Latitude:	–0.0055 mi.
Longitude:	0.0007 mi.
Mean Radial Error about Mean Position:	0.0204 mi.
Deviation about Computed Mean—	
Latitude:	0.0167 mi.
Longitude:	0.0161 mi.
Root Sum Square:	0.0232 mi.
Circular Probable Error (CEP):	0.0193 mi.

SINGLE CHANNEL, 400 MHz	
Number of Passes:	10 (ECA** 10° – 70°)
Computed Mean Position*—	
Latitude:	39°09′.8144
Longitude:	–76°53′.8390
Difference of Mean Reference Position—	
Latitude:	0.0070 mi.
Longitude:	0.0016 mi.
Mean Radial Error about Mean Position:	0.0491 mi.
Deviation about Computed Mean—	
Latitude:	0.0195 mi.
Longitude:	0.0477 mi.
Root Sum Square:	0.0515 mi.
Circular Probable Error (CEP)	0.0396 mi.

*Reference Position—Latitude: 39°09′.8214; Longitude: –76°53′.8410.
**ECA—Elevation at Closest Approach.

TABLE 4418a.—AN/WRN-5 fix results.

4419. AN/SRN-19 (XN-1) Radionavigation Set.—The radionavigation sets available for use with the NAVSAT, such as the AN/WRN-5, are rather sophisticated. They receive data on two satellite frequencies; they are more accurate; and they are relatively expensive. To meet the need for a low-cost simplified satellite navigation set for use aboard the smaller naval vessels, particularly destroyers, the Applied Physics Laboratory designed the AN/SRN-19 Radionavigation Set as a single channel integral doppler set that includes all of the units necessary to compute and display a navigation fix in one chassis and an external antenna/preamplifier unit. Developed in response to Advanced Development Objective (ADO) 3411, the AN/SRN-19 (XN-1)

SINGLE CHANNEL, 150 MHz	
Number of Passes:	14 (ECA** 10° – 70°)
Computed Mean Position*—	
Latitude:	39°09'.8391
Longitude:	–76°54'.0482
Difference of Mean Reference Position—	
Latitude:	0.0177 mi.
Longitude:	–0.1612 mi.
Mean Radial Error about Mean Position:	0.2676 mi.
Deviation about Computed Mean—	
Latitude:	0.2662 mi.
Longitude:	0.2175 mi.
Root Sum Square:	0.3438 mi.
Circular Probable Error (CEP):	0.2848 mi.

DUAL CHANNEL, SHORT EXPOSURE	
Number of Passes:	20 (ECA** 10° – 70°)
Computed Mean Position*—	
Latitude:	39°09'.7683
Longitude:	–76°53'.8468
Difference of Mean Reference Position—	
Latitude:	–0.0530 mi.
Longitude:	–0.0045 mi.
Mean Radial Error about Mean Position:	0.1042 mi.
Deviation about Computed Mean—	
Latitude:	0.1393 mi.
Longitude:	0.0847 mi.
Root Sum Square:	0.1630 mi.
Circular Probable Error (CEP):	0.1319 mi.

*Fix Reference Position—Latitude: 39°09'.8214; Longitude: –76°53'.8410.
**ECA—Elevation at Closest Approach.

TABLE 4418b.—AN/WRN-5 fix results.

will enhance a vessel's navigation capabilities by providing regular position fixes and dead reckoning (between satellite fixes) anywhere in the world, day and night, in any weather. Design features of the set are such that it is expected to cost less than $10,000 in production quantities. It is intended for use on surface ships, especially those with an Anti Submarine Warfare mission, where the considerably higher cost of the AN/WRN-5 Radionavigation Set cannot be accommodated and where the extreme accuracy of the higher priced set is not required.

4420. Functional description.—Figure 4420 shows the AN/SRN-19 Radionavigation Set with its antenna/preamplifier and printer, which is optional. The set consists of a 400 MHz receiver; a 5 MHz oscillator; an 8,192 word, 16-bit data

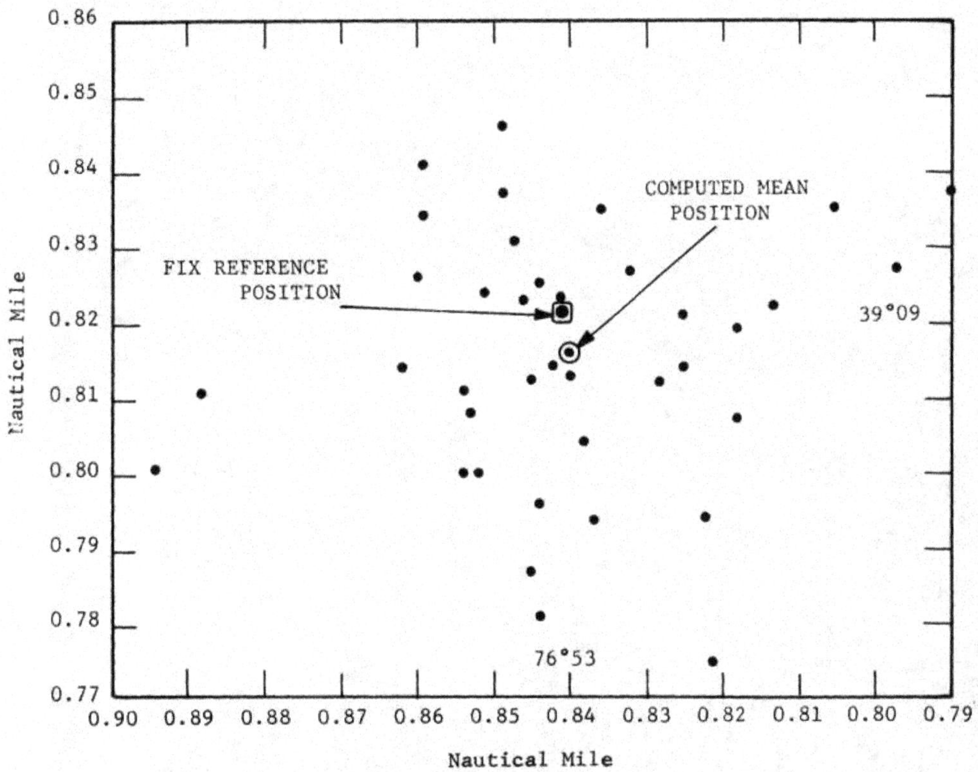

FIGURE 4418.—AN/WRN-5 receiver accuracy, dual channel operation.

processor; power supplies; and 2 synchro-to-digital units mounted in its chassis. A display, keyboard, and a cassette recorder are mounted on the front panel of the chassis. Controls and monitoring devices are also on the front panel.

The receiver automatically searches for and locks onto a satellite's 400 MHz signal, and extracts from the signal two types of information. One is the reconstructed doppler shift of the satellite signal which results from the relative motion between the navigator and the satellite transmitting the signal. The other type of information is obtained by demodulation of the satellite carrier which is phase modulated with a message describing the satellite position in inertial space. These data along with the information on ship's motion during the satellite pass are used by the navigation program to compute the ship's position.

The navigation program is permanently programmed on Read-Only-Memory (ROM) chips in the data processor. This program provides a continuous dead reckoning position using ship's course and speed through the two synchro-to-digital converters and updates the DR position periodically with a satellite navigation fix. Ship's position information in the form of time, latitude, and longitude is displayed on the AN/SRN-19 front panel and updated every 5 seconds. AN/SRN-19 time is kept by processing once per second interrupts to the data processor. The interrupts are provided by a countdown of the 5 MHz oscillator. This "clock" is initialized by the operator and corrected thereafter by satellite data.

The operator uses the keyboard to initialize the navigation program and to select special purpose routines that compute and display the desired information, e.g., ship's speed and heading, and satellite fix results. The cassette tape recorder records ship's position every 15 minutes for playback later at a shore facility.

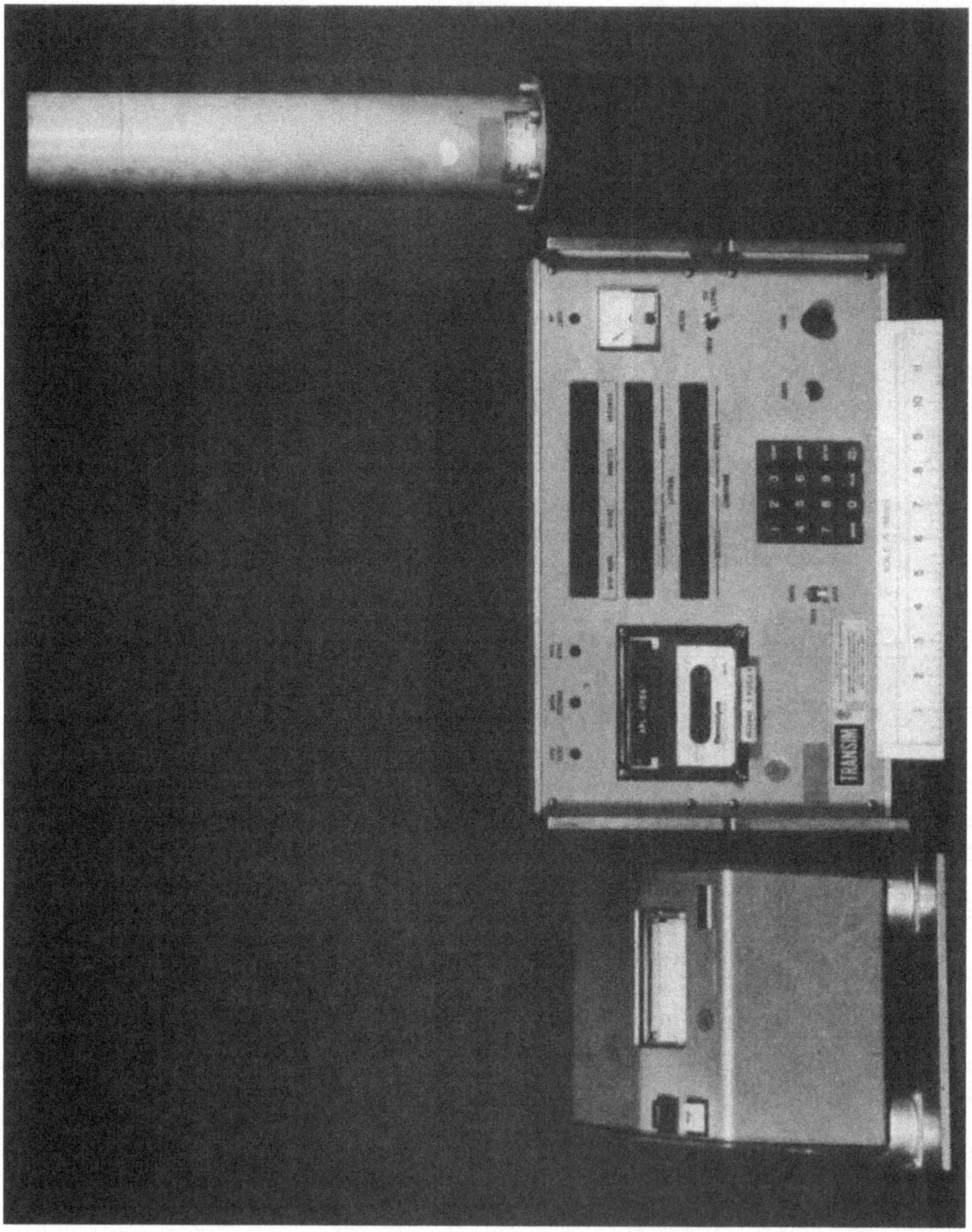

Courtesy of Applied Physics Laboratory, The Johns Hopkins University.

FIGURE 4420.—AN/SRN-19 (XN-1) Radionavigation Set.

The navigation program has the following capabilities:

1. Updates present latitude and longitude once per second by dead reckoning on ship's true speed and heading. The ship's position on the front panel display is refreshed every 5 seconds. True speed and heading are obtained by summing vectorially the speed and heading of the ship with respect to the water with the set and drift of the current in the area. The heading and speed of the ship with respect to the water are obtained automatically from the ship's Mark 19 Gyrocompass and electromagnetic underwater log through the AN/SRN-19 synchro-to-digital converters. The operator has the option of manually entering the set and drift of the ocean currents if they are known. If the ship does not have a Mark 19 Gyrocompass and electromagnetic log, the ship's heading and speed with respect to the water may be entered manually. The dead reckoning program continues to run once per second regardless of any other programs that may be executing (i.e., great circle program, satellite fix program, or alert program).

2. When the receiver locks onto a satellite signal, the program recognizes the beginning of a 2-minute data transmission interval, collects and majority votes the satellite message data, and computes a satellite fix provided there is sufficient satellite data. The fix is used to update automatically the dead reckoning position provided the following conditions are met:

i. The satellite elevation at closest approach is between 10° and 80°.

ii. The fix computation converges in five or less iterations and the total correction is less than 30 nautical miles.

If the above conditions are not met, the operator may manually enter the fix in the data processor to update the DR position.

The special purpose routines of the navigation program include the following:

1. A great circle program which will compute and display the great-circle distance and bearing from the ship's present position to any destination that has been entered by the operator.

2. A satellite alert program which computes and displays the next expected satellite rise time, and the elevation at closest approach of the satellite just tracked. A new alert is computed each time when the display button is depressed.

3. A keyboard/display program which is used to enter and display various parameters of interest.

4421. Accuracy.—The AN/SRN-5 (XN-1) has a fixed-site CEP of less than 120 meters. Figure 4421 shows plots of 24 satellite fixes taken at dockside aboard the USS *Forest Sherman* (DD-931) at Norfolk, Virginia.

During the period 24 April to 26 April 1974, an AN/SRN-19 was installed for a technical evaluation. The dockside RMS error for 24 navigation fixes was 0.076 nautical mile. The design goal was 0.1 nautical mile RMS at dockside. During at-sea tests, automatic inputs of ship's course and speed were used. Based on eight satellite fixes obtained from the two at-sea tests, the RMS radial fix error of the AN/SRN-19 was 0.13 nautical mile. The design goal was 0.3 nautical mile RMS.

The DR velocity error is dependent upon the particular ship's motion measuring devices that are used. Fifty-three satellite navigation fixes were computed by the AN/SRN-19 located at the Applied Physics Laboratory with a 1-knot error intentionally inserted into the DR system. The RMS radial error of these 53 passes was 0.31 nautical mile.

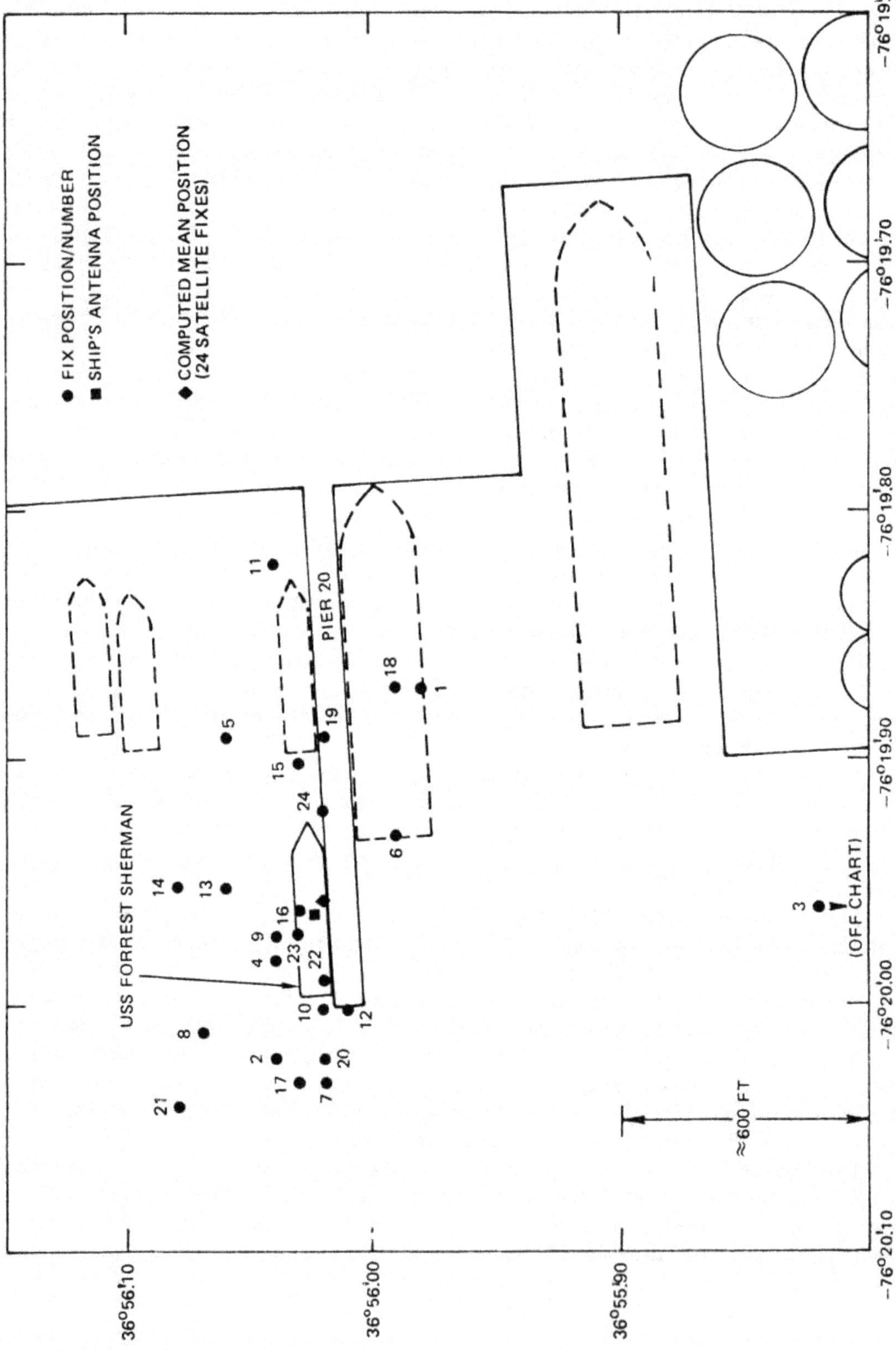

FIGURE 4421.—AN/SRN-19 (XN-1) Dockside fixes and area plan view.

4422. AN/PRR-14 Geoceiver.—The basic purpose of the Geoceiver conceived at the Applied Physics Laboratory is to receive and record data for geodetic position determination. Secondary uses are for satellite orbit determination and shipboard navigation. The Geoceiver makes integrated doppler measurements from satellites radiating on 324 and 162 MHz frequency channels, as well as from the Navy navigation satellites radiating on the 400 and 150 MHz frequency.

4423. Functional description.—The received frequency pair may be selected manually or automatically. When in the automatic mode, the receiver alternately searches for the frequency pairs until satellite signals are acquired. Data processed by the Geoceiver—time, integrated doppler, and integrated refraction correction—are stored on punched paper tape which are subsequently transmitted to a computing center. Provision is also included for processing the Navy navigation satellite message with the doppler and refraction correction data to a computer for real-time navigation applications.

The main receiving unit of the Geoceiver is shown in figure 4423.

4424. Design characteristics.—The Geoceiver design specifications include the following:

1. double conversion 400/324 MHz channel,
2. double conversion slaved 150/162 MHz channel,
3. integrated doppler frequency measurements timed by a receiver clock (30-second or 60-second integration intervals),
4. receiver clock calibrated against received Navy navigation satellite time,
5. separate integrated refraction data from 150/162 MHz channel,
6. time, doppler, and refraction correction data recorded on punched paper tape with additional data entered from the receiver control panel,
7. Navy navigation satellite message, time, doppler, and refraction correction data available on a computer output to provide real-time navigation,
8. automatic search and acquisition of satellite signals with manual over-ride,
9. rapid resynchronization with satellite message and time if carrier tracking is lost for up to a minute,
10. built-in self-test and troubleshooting aids,
11. portable; small in size, weight and power consumption.

A suitable external computer is required for real-time navigation applications.

4425. Accuracy.—Table 4425 indicates single pass survey capabilities of the Geoceiver by using satellite orbits derived from tracked data as opposed to navigation results obtained from projected orbits.

Equipment	Satellite	No. Passes	Mean Error (Meters)		Standard Deviation (Meters)	
			Long.	Lat.	Long.	Lat.
Breadboard	30120	44	6.2	−5.1	8.7	6.1
Geoceiver	30130	53	8.0	−3.3	13.7	5.7
(Station 391)	30140	50	7.6	−3.1	8.6	4.9
	30180	44	8.0	−5.7	9.3	6.4
Prototype	30120	32	5.7	−4.3	12.5	7.5
Geoceiver	30130	37	8.9	−3.8	6.6	3.5
(Station 137)	30140	42	6.3	−4.2	7.6	3.9
	30180	42	6.7	−5.0	10.8	6.9

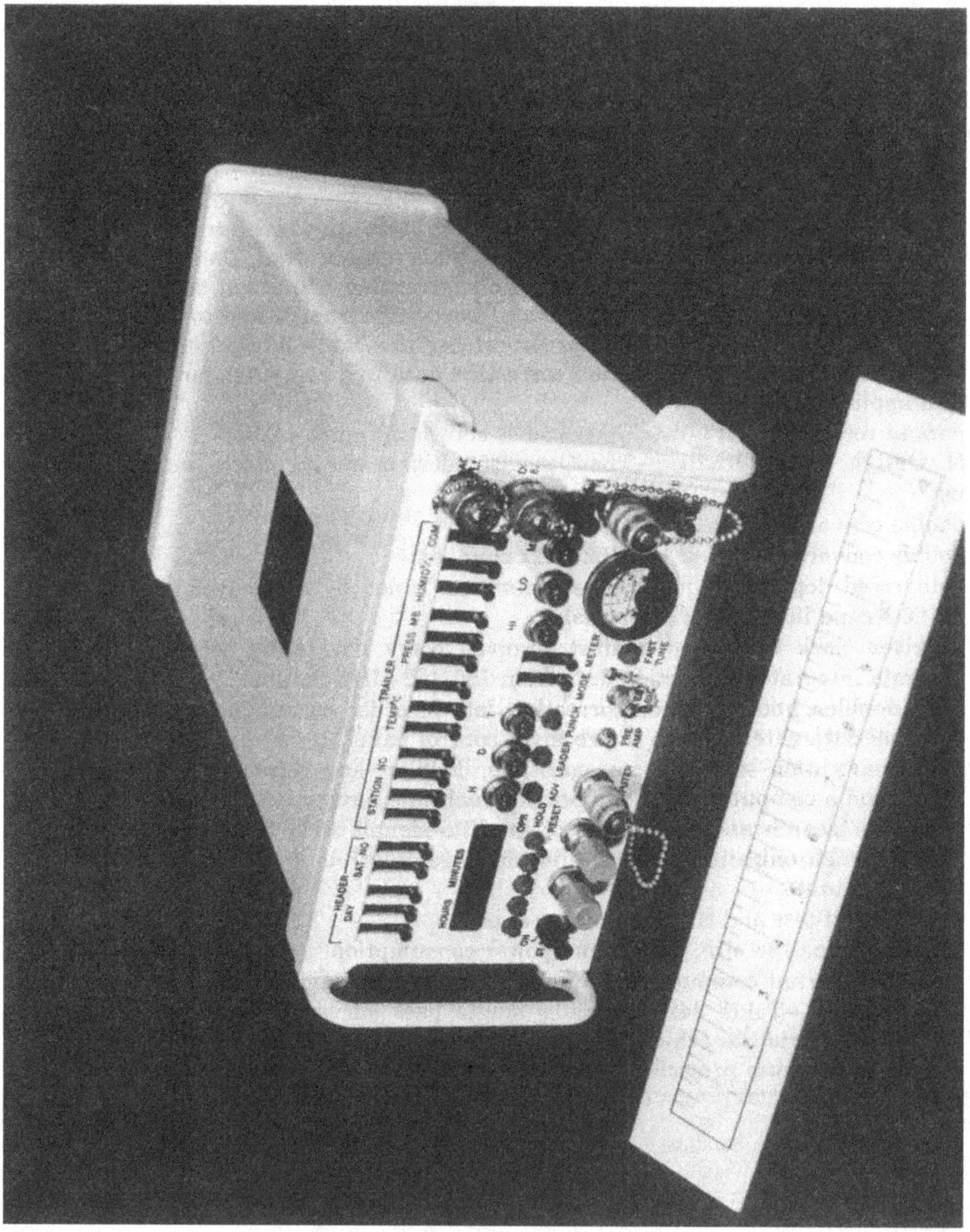

Courtesy of Applied Physics Laboratory, The Johns Hopkins University.

FIGURE 4423.—AN/PRR Receiver Unit.

Equipment	Satellite	No. Passes	Mean Error (Meters)		Standard Deviation (Meters)	
			Long.	Lat.	Long.	Lat.
Breadboard	30120	7	13.8	−1.9	13.2	9.6
Geoceiver	30130	16	8.0	−2.9	9.0	3.9
(Station 136)	30140	14	7.2	−2.8	7.3	6.7
	30180	18	9.7	−3.3	11.4	5.7

NOTES: 1. Obtained from NWL Technical Report TR-2338 dated September 1969.
2. Data obtained with antenna common to both the breadboard and prototype Geoceivers.

TABLE 4425.—Geoceiver survey data statistics for single pass solutions.

Basic Doppler Principles

4426. Design considerations.—A satellite navigation system is truly all weather when radio frequencies are utilized which are not sensitive to weather or atmospheric noise, and when frequencies are combined such as to obtain independence from ionospheric variations. The use of orbiting satellites makes the system inherently worldwide. Other attributes of a satellite navigation system depend upon other system design considerations such as the question of whether to utilize radio sextant techniques analogous to celestial navigation, whether to utilize simultaneous transmissions from multiple orbiting satellites in analogy with Loran or Omega, or whether to use an altogether new method which is more natural to artificial satellites than established techniques, e.g., taking advantage of the high speed of the satellite by using the doppler shift.

NAVSAT through its use of range rate by measurement of the doppler shift of satellite radio transmissions provides the following advantages:

1. There is no need for a large receiving antenna array as required by radio sextants.

2. Only one satellite need be utilized to obtain a position at a given time so that the complexity of maintaining cooperative satellites in orbit for simultaneous signal reception is avoided.

3. Doppler measurements are inherently open loop, that is, there is no need for transponding signals back and forth between the satellite and the observer as would be necessary in measuring range directly. Therefore, NAVSAT is passive in that there are no radio transmissions from the observer's station.

4427. Basic doppler navigation principles.—Measurement of the satellite doppler curve in the vicinity of zero doppler shift allows a determination of the time and the line-of-sight range (slant range) to the closest point of approach of the satellite to the observer. This measurement enables determining the observer's position relative to the satellite in a coordinate system which is local to the satellite at the time of closest approach and whose axes are parallel and perpendicular to the direction the satellite travels over the earth at that time. Knowing the satellite's orbit, its subtrack over the earth's surface can be determined. Knowing the minimum slant range to the satellite and the altitude of the satellite, the distance from this subtrack can be found. This can be thought of as the cross-track coordinate of the observer relative to the satellite. Knowing the exact time when this minimum slant range occurred, the position of the satellite along its earth subtrack is known which can be considered the along-track coordinate of the observer relative to the satellite. Knowledge of the satellite orbit in detail tells how to transform the observer's relative along-track and cross-track coordinates to latitude and longitude.

While this method is adequate to determine the observer's position, other information in the doppler shift can be used to advantage. If the whole of the doppler shift is used, sufficient redundant information is obtained to overwhelm any error due to random noise such as receiver and atmospheric noise. Furthermore, one can tell which side of the satellite subtrack the navigator is on by utilizing the fact that the earth's rotation produces a slight difference between the doppler shift as received east of the satellite orbit and west of the orbit.

Effective utilization of the whole of the doppler shift as received by the observer also allows the navigated position to be essentially free of any momentary receiver "drop out," short term interference, etc., and some important potential system errors; for example, the error caused by an error in the frequency standard of the equipment. (The satellite frequency is maintained to an accuracy of better than one part in 10 billion but the navigator's local oscillator frequency can be in substantial error in some of the less expensive navigator equipments.) Figure 4427 shows the time dependence of the change in the received doppler shift for a set on the earth's equator which is changed by 1 kilometer in its along-track and cross-track components, respectively. It can be seen that the change in the along-track component produces a symmetric frequency change. Also, it can be seen that the change in cross-track component produces an anti-symmetric frequency change so that with doppler data only near the time of closest approach (zero relative time in figure 4427) along-track and cross-track components can be identified unambiguously. However, if a frequency standard error (local oscillator error) exists also, data only near closest approach will not allow unambiguous identification of both frequency and along-track changes. On the other hand, if the whole of the doppler shift is considered, the two errors can be easily differentiated since a change in frequency produces a constant frequency error, while a change in the along-track component produces a time dependent doppler error, the error being a maximum near the time of closest approach.

4428. Effect of random measurement errors.—Profiles of over-all system navigation error can be constructed assuming that only random doppler (frequency) errors are present. Figure 4428 shows a typical navigation error profile as the ground distance to the satellite subtrack changes. As is to be expected, the along-track error becomes relatively large when the satellite is on the horizon since it is not above the horizon sufficiently long to accurately establish separation of along-track error from the set's local oscillator error. The cross-track error becomes large when the satellite is nearly overhead since there is relatively small change in slant range to the satellite for a change in cross-track component when the satellite is overhead.

The error profiles like those in figure 4428 are pessimistic since they consider the navigation errors when many independent measurements of frequency are made, e.g., one each second. Modern techniques dictate that coherent measurements be taken so that any error in frequency which has an average error of zero is

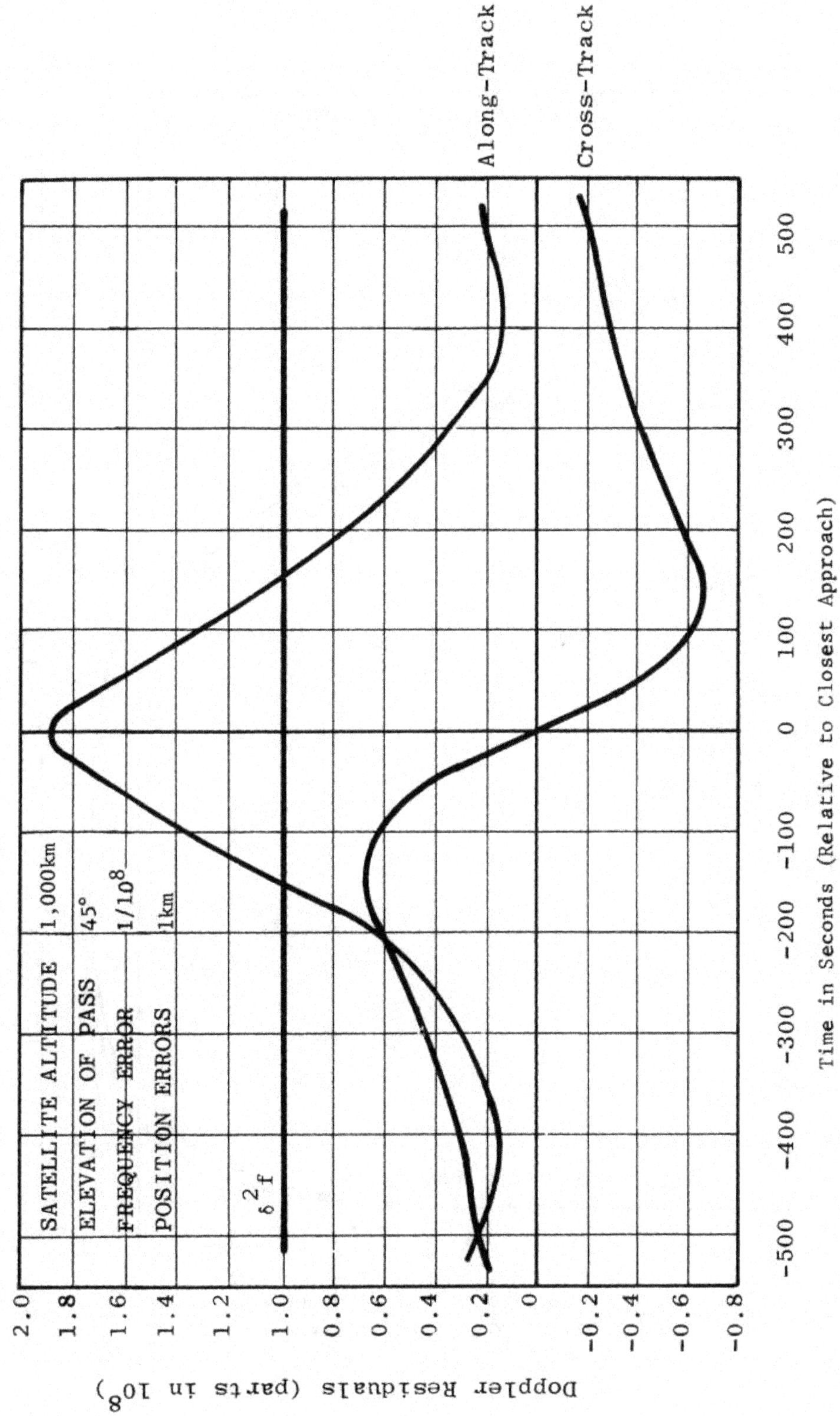

FIGURE 4427.—Two position parameters.

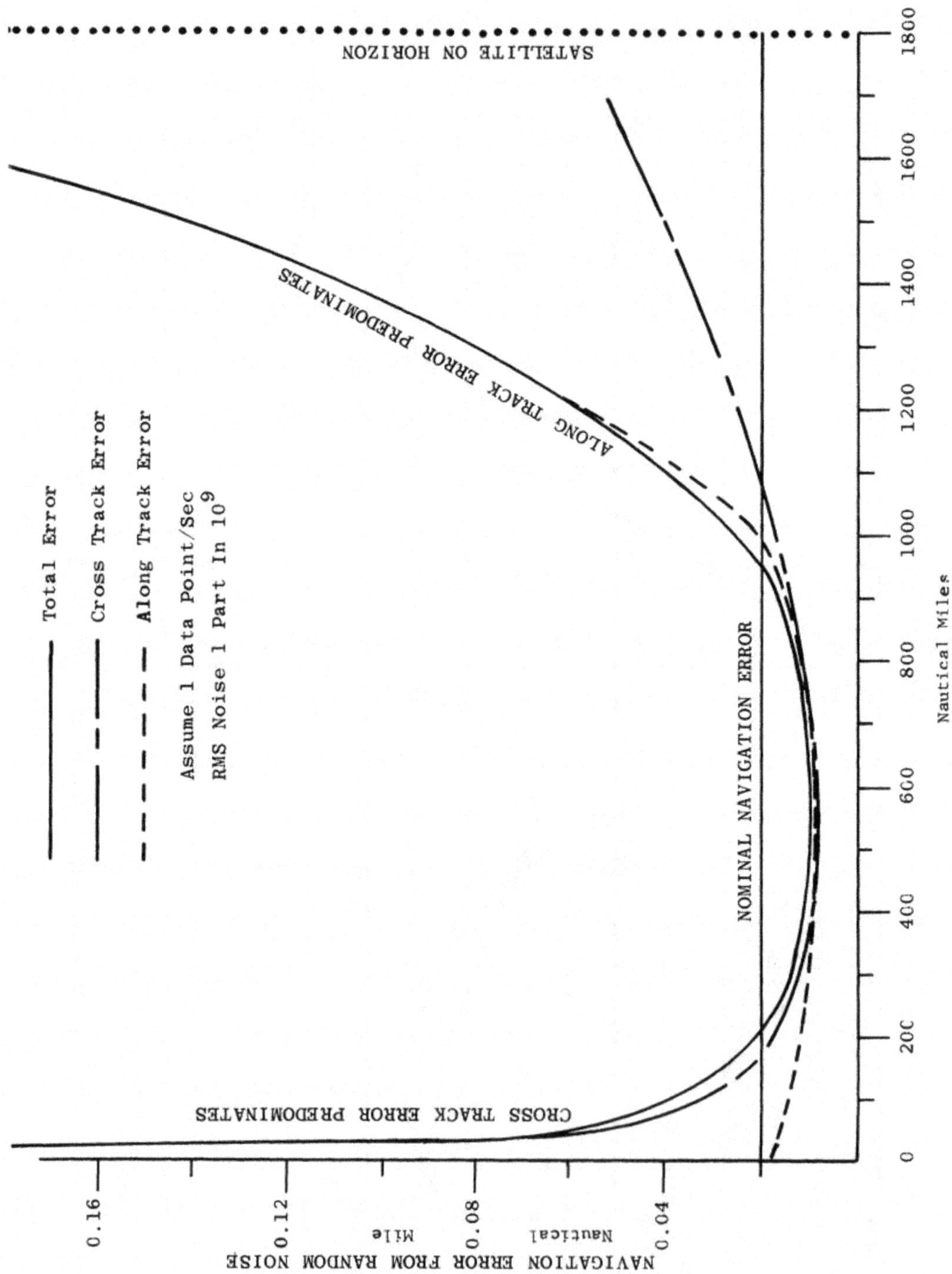

FIGURE 4428.—Ground range to satellite subtrack.

suppressed. With such techniques, the errors due to random noise are suppressed more than the square root of the number of measurements, but probably still less than the total length of time of reception itself. In other words, noise can be suppressed somewhere between a factor of 15 min. (average pass duration) $\times 60$ sec. $=900$ and

$$\sqrt{15\times 60}=30,$$

dependent upon other considerations.

One manner in which coherent measurements can be very conveniently utilized is by not measuring frequency per se but to maintain a running count of the number of zero crossings of the received signal to obtain the coherent integral of the doppler shift. Utilizing integral doppler data in this way yields more potential accuracy and can be easier to implement in the navigator's equipment. In doing this, the equipment does become more sensitive to intermittent signal dropouts, etc., so that the more reliable navigator's equipments utilize more sophisticated methods than this. However, several of the less expensive equipments, such as the AN/SRN-9 (art. 4412), utilize explicitly this technique, where the integral doppler value is read each time the satellite 2-minute time marker is received.

4429. Non-random error sources.—For all practical purposes, random-frequency errors in the doppler data will not cause significant navigation errors. This is not true of systematic errors. The major sources of error which have been found are related to ionospheric refraction, errors in knowledge of the gravity field acting on the navigation satellites, and errors in knowledge of ship's course and speed through the water.

A major potential source of error in all satellite radio navigation systems is ionospheric refraction. Navigation errors can result from the effect of refraction on the measurement of the doppler shift and from the errors in the satellite's orbit if refraction is not accurately accounted for in the satellite tracking. Since the doppler shift in the presence of the ionosphere is basically the time rate of change of the electromagnetic path length, the doppler shift is altered from what it would be in a vacuum. However, it can be shown that the ionosphere is **dispersive** (the amount of refraction is dependent upon the frequency) and for frequencies significantly above any ionospheric resonance frequency (usually less than 30 MHz) the dependence of the doppler shift with frequency is accurately known. For frequencies above 100 MHz, the dependence is inversely proportional to the frequency to very high accuracy. The refraction contribution can be eliminated by the proper mixing of the received doppler shift from two harmonically related frequencies (150 and 400 MHz for NAVSAT) to yield an accurate estimate of the vacuum doppler shift.

In addition, the two received doppler frequencies can be mixed to yield an experimentally measured effect of the ionosphere, and the probable error caused by the ionospheric contributions can be studied. Such studies show that the principal navigation error due to refraction is in the cross-track component; it typically can be several tenths of a nautical mile, and can be as large as 1 nautical mile. Unlike the mixing of local oscillator frequency error with the navigated along-track error, where use of the whole of the doppler curve allows both the along-track component and local oscillator frequency to be error determined, the ionospheric error has a time dependence during the pass of the satellite above the navigator's horizon which mimics the time dependence of the doppler signal for a cross-track error. In this case, the cross-track component and a parametric fit to ionospheric parameters does not yield negligible navigation errors. Consequently, except for the simplest navigator equipment, two frequencies are received to yield a navigation position via the vacuum doppler data. Figure 4429a shows the ionospheric contribution at 150 MHz which can be compared with the cross-track error of figure 4427.

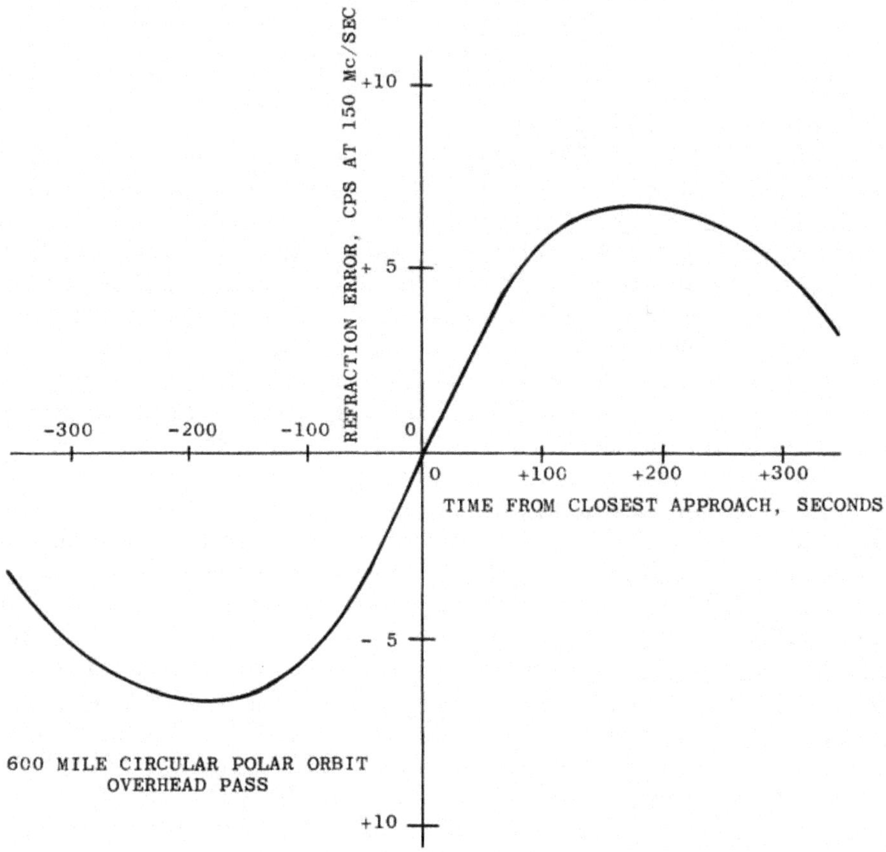

FIGURE 4429a.—First order theoretical refraction error.

Any error in the navigator's knowledge of the satellite position during the time doppler data is used feeds directly (and in nearly a one-to-one manner) into the navigation error. Consequently, every effort is made to reduce the navigation satellite orbit error to a minimum. The major contributor to this error is inadequate knowledge of the gravity field of the earth, i.e., the gravity forces acting on the satellite. In 1960 it was possible to determine the orbit of a satellite and then predict the orbit ahead in time for 1 day (for use by navigators) to an accuracy of 0.5 to 1.0 nautical mile due to errors in the gravity forces. Since that time extensive geodetic research utilizing research satellites, similar to the navigation satellites with added tracking capability by optical means, has resulted in a significant improvement in knowledge of the earth's gravity field and consequent improvement in navigation accuracy.

The technique for determining the gravity field is to accumulate a large amount of tracking data (primarily doppler and optical) as observed on a large number of satellites in orbits of varying inclinations. A gravity model is explicitly "parameterized" in terms of unknown coefficients in an expansion of the earth's gravity field. An over-all least squares fit of the tracking data is then made considering all of the satellite orbit parameters, all of the Legendre expansion coefficients of the gravity field, and the location of the tracking stations as unknowns. The gravity field which gradually emerged from such worldwide studies exhibited a complexity far beyond that ever suspected.

It is difficult to summarize the complex gravity field as it is known today. One fairly common method is to present contours of geoid heights. Such heights represent the height that water would rise or sink to—in seeking its own level—over and above that water level which would occur if the gravity field were a spherical mass (suitably modified for the equatorial bulge due to the earth's spinning). Consequently, such geoidal contour maps can indicate the departure of the earth's gravity field from what it was thought to be prior to the advent of satellite geodesy. Such a contour plot is shown in figure 4429b, which portrays the gravity field as known in 1968. With modern gravity models for the earth, satellite tracking error has been reduced from about ¼ mile to about 20 meters with commensurate reduction in the navigation error (orbit prediction error) caused by errors in the gravity forces acting on the satellite. As knowledge of the gravity field improves, errors due to this source will be further reduced.

The remaining large potential source of error is the error in the navigator's motion during the receipt of doppler data. Since the doppler shift is determined by the relative motion between satellite and navigator, any error in the navigator's motion will appear as an apparent error in the doppler data used to find the navigator's position. On the average this error causes about 400 yards navigation error for every knot of error in velocity. Navigator velocity error can become appreciable in aircraft navigation applications, and special techniques are used to bound this error in aircraft equipments. This error can become significant for high speed surface vessels and submarines in the most demanding applications. For such cases, inertial systems are usually used to determine vessel speed and aid in dead reckoning between satellite passes.

Other potential sources of error are usually negligible except for the most demanding applications. Sources of errors such as abnormal tropospheric refraction (e.g., navigation near a weather front); neglect of the variable height of the sea surface (e.g., ignore the geoidal height, figure 4429b, as an altitude correction of the navigator's antenna height); and even wander of the geographic position of the spin axis of the earth (the so called Chandler wobble) are errors which can and should be accounted for when highest accuracy is desired. With all such corrections, navigation accuracy rivals world surveying accuracy (art. 4422). For this reason, satellite navigation has brought attention to some large mapping errors. A good discussion of current navigation accuracy capability can be obtained in the bull's-eye plots shown for the navigation equipment discussed (figs. 4411b, 4414, and tab. 4425).

Integral Doppler Navigation

4430. AN/SRN-9 integral doppler navigation.—It is beyond the scope of this chapter to present details of the more sophisticated navigation techniques, e.g., the AN/BRN-3 coupled to SINS. However, the essentials and many of the details are similar for utilizing integral doppler data in the AN/SRN-9. Therefore, the navigation procedure is given for this equipment in more detail to provide one detailed example of satellite navigation.

The navigation fix obtained from the AN/SRN-9 is based upon the shift in frequency (doppler frequency shift) that occurs whenever the relative distance between a radio transmitter and receiver is changing. Such a change can be

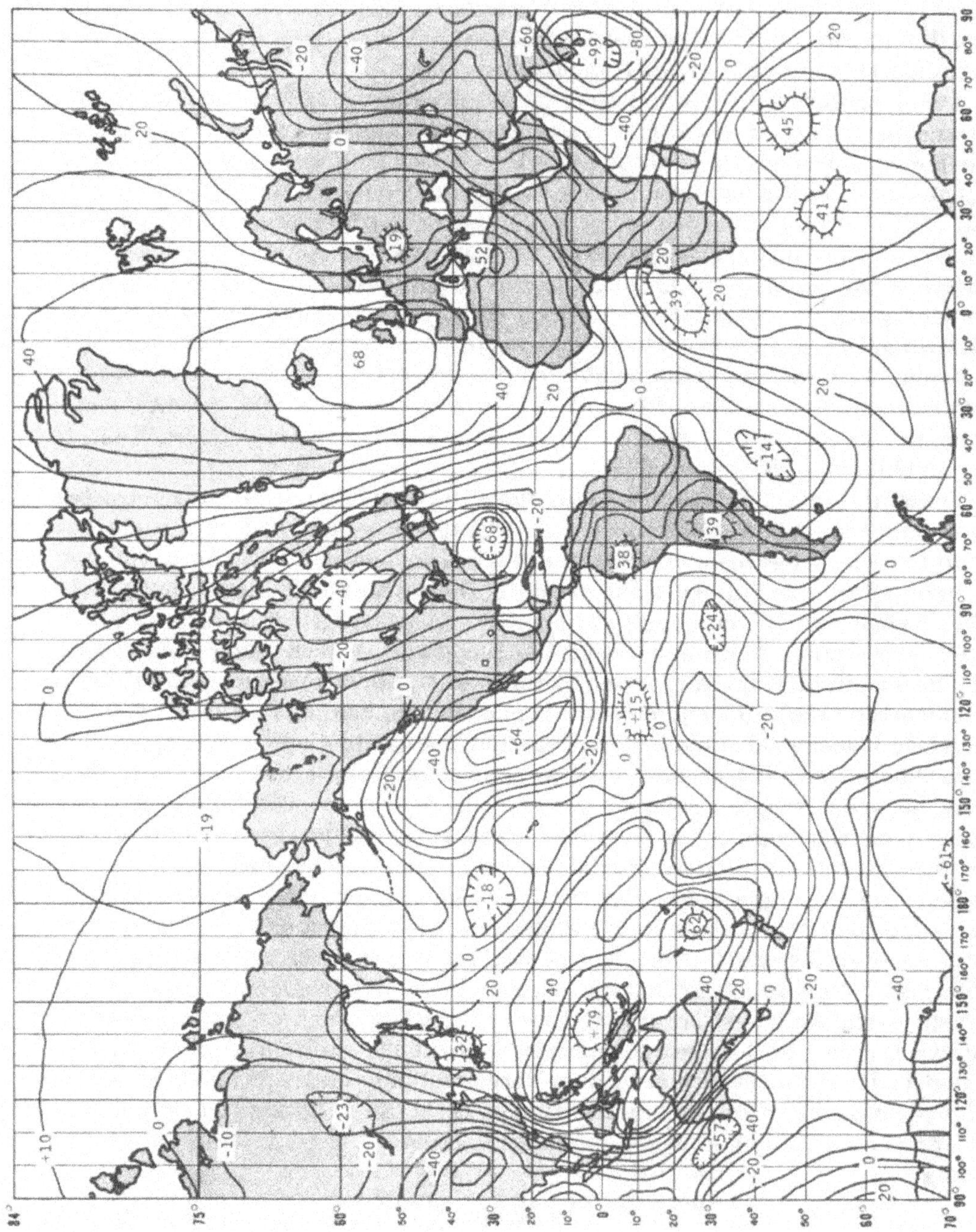

FIGURE 4429b.—Contour map of geoidal height (meters).

measured by a receiver whenever a transmitting navigation satellite passes within radio range, and is due to the combination of three effects:

1. motion of the satellite in its orbit,
2. motion of the navigator on the earth's surface, and
3. rotation of the earth (and therefore of the navigator) about the earth's axis.

It follows that the motion of the navigator must be properly measured or estimated and inserted into the computation if accurate fixes are to be obtained.

The integral of the doppler shift over a 2-minute interval (as measured by the AN/SRN-9 receiver, using a doppler frequency counter that is controlled by the 2-minute time markers received from the satellite) is a measure of how much the slant range from satellite to navigator has changed during this 2-minute interval. In order to derive his position, the navigator also needs to know the position of the satellite in its orbit every 2 minutes. The information required from computing a fix are as follows:

Data	*Obtained From*
Two-minute doppler frequency counts (integrated doppler).	AN/SRN-9 receiver, using self-contained frequency standard.
Satellite orbital position every 2 minutes	AN/SRN-9 receiver recognition of satellite data bits plus subsequent computer operation (tab. 4430).
Own ship's *estimated* position every 2 minutes. *Or* Own ship's course and speed plus one estimate of ship's position at a stated time.	Ship's navigational aids (with DR plot if necessary).
Own ship's antenna height above geoid	See figure 4429b.

Figures 4430a and 4430b illustrate how the measurements are made and how the navigation fix is computed after the satellite pass is over. In figure 4430a, the positions of the satellite in its orbit are shown for times t_1 through t_4, which are the even minutes at which the satellite transmits its synchronization signal. The positions of the navigator, P_1 through P_4, refer to the times at which his AN/SRN-9 receiver recognizes the satellite synchronization signal, i.e., times $t_1+\Delta t_1$ through $t_4+\Delta t_4$. Note that the times of reception are slightly later than the times of transmission because of the radio propagation time over the slant ranges S_1, S_2, S_3, and S_4. Also, since the propagation times Δt_1, Δt_2, Δt_3, and Δt_4 are not all equal, the time intervals over which the AN/SRN-9 makes its integral doppler measurements will differ slightly from the exact 2-minute value, being somewhat smaller prior to satellite closest approach (while S is decreasing). This fact, however, does not affect the result of the measurement because the number of RF cycles *transmitted* by the satellite between synchronization signals (exactly 2 minutes apart) must necessarily equal the number *received* by the navigator between receptions of the synchronization signals, since no RF cycles can be "lost" or "gained."

Word number	Symbol	Meaning	Units
56	t_p	Time of perigee.	Min.
62	$\dot{M}$	Rate of change of mean anomaly.	Deg/Min.
68	$\lvert\phi\rvert$	Argument of perigee at t_p.	Deg.
74	$\dot{\phi}$	Rate of change of argument of perigee (absolute value).	Deg/Min.
80	ϵ	Eccentricity.	

Word number	Symbol	Meaning	Units
86	A_o	Semimajor axis.	Km.
92	Ω_N	Right ascension ascending node at t_p.	Deg.
98	$\dot{\Omega}$	Rate of change of Ω_N.	Deg/Min.
104	$\cos\psi$	Cosine of orbit inclination.	
110	Ω_G	Right Ascension Greenwich.	Deg.
128	$\sin\psi$	Sine of orbit inclination.	

TABLE 4430.—Satellite orbital information.

The integral doppler measurements are simply the count N_{12} of the number of doppler cycles received between $t_1+\Delta t_1$ and $t_2+\Delta t_2$, the count N_{23} of the number of doppler cycles between $t_2+\Delta t_2$ and $t_3+\Delta t_3$, and so on for all 2-minute intervals during the satellite pass. These counts are a direct measure of the amount by which the slant range from satellite to navigator has changed (S_2-S_1, S_3-S_2, etc.) during the count intervals. This measure is quite accurate since each doppler count added (or subtracted) due to the relative motion means that S has decreased (or increased) by one wavelength, or by 0.75 meter at 400 MHz. Therefore, one of the required inputs to the fix computation (slant range increment over each 2-minute interval) is directly measured by the AN/SRN-9 integral doppler count, suitably scaled as indicated in figure 4430a. Note that the slant ranges S_1 through S_4 (and therefore their respective differences, or slant range increments) are all functions of the navigator's position (ϕ, λ). Since the satellite orbital positions can be calculated from the data recovered from the signal phase modulation by the AN/SRN-9 receiver, and since the navigator's *estimated* position every 2 minutes is available, values of *estimated* slant range from satellite to navigator can be computed. These estimated slant ranges are differenced to obtain *estimated* slant range increments, which then can be compared with the slant range increments *measured* by means of the integral doppler counts as already described.

Unless the navigator's estimate of his position happens to be exactly correct there will, of course, be a difference or residual when each *estimated* slant range increment is subtracted from the corresponding *measured* increment. The fix calculation then consists of changing the navigator's estimate of position (ϕ, λ) in small steps until the sum of the squares of the slant range residuals is minimized, at which point the closest achievable agreement exists between the (revised) estimates and the measures of slant range increment. The values of ϕ and λ so determined (i.e., the revised estimates that yield the smallest residual) are then the fix result, which is printed out at the end of the fix computation.

In practice, two factors complicate this simple explanation, and therefore represent extra computing steps in the fix computation:

1. The frequency of the satellite oscillator and also that of the reference oscillator used in the AN/SRN-9 receiver are constant but not precisely known to the navigator.

2. The process of minimizing the sum of the squares of the differences between the estimated and measured slant range differences calls for a number of different manipulations to be performed in the computer used to calculate the navigation fix and is, in fact, an iterative process wherein the same mathematical steps are successively repeated in the same sequence several times in order to get the final result.

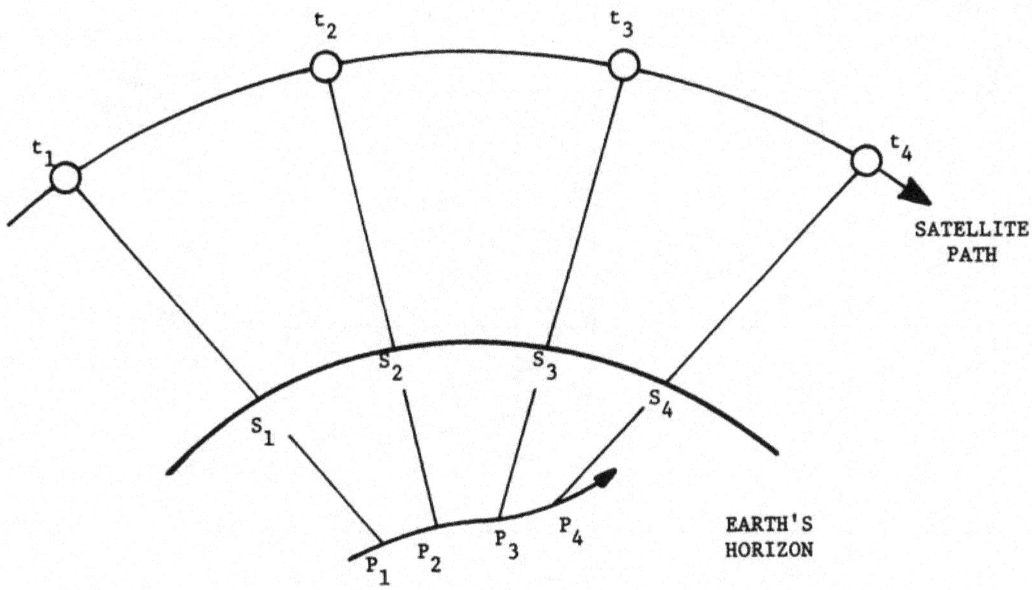

f_o = NOMINAL VALUE OF NAVIGATOR'S REFERENCE FREQUENCY

c = SPEED OF LIGHT

$\bar{f}$ = DIFFERENCE BETWEEN NAVIGATOR'S REFERENCE FREQUENCY AND SATELLITE TRANSMISSION FREQUENCY

T = 2 MINUTES (i.e., $t_2 - t_1$, $t_3 - t_2$, etc.)

ϕ = LATITUDE

λ = LONGITUDE

t = TIME OF TRANSMISSION OF TIMING MARK

$t + \Delta t$ = TIME OF RECEPTION OF TIMING MARK

N = DOPPLER COUNT

S = SLANT RANGE

$$N_{12} = \frac{f_o}{c}[S_2(\phi,\lambda) - S_1(\phi,\lambda)] + \Delta f \cdot T$$

$$N_{23} = \frac{f_o}{c}[S_3(\phi\ \lambda) - S_2(\phi,\lambda)] + \Delta f \cdot T$$

$$N_{34} = \frac{f_o}{c}[S_4(\phi,\lambda) - S_3(\phi,\lambda)] + \Delta f \cdot T$$

FIGURE 4430a.—Integrated doppler measurement.

The absolute values of satellite and navigator's oscillators are not required in the computation provided that they are constant—only their difference is of interest. This is the quantity $\bar{f}$ referred to in figure 4430b and printed out along with ϕ and λ in the fix results. Since $\bar{f}$ is now known to the navigator and cannot be directly estimated or measured and since its actual value affects the numbers obtained for the integral 2-minute doppler counts measured by the AN/SRN-9 receiver, it must be solved for (in addition to ϕ and λ) in the calculation of the navigation fix. Note that the value of $\bar{f}$ does not affect the *estimated* slant range increments—only the measured increments defined by the integral doppler counts

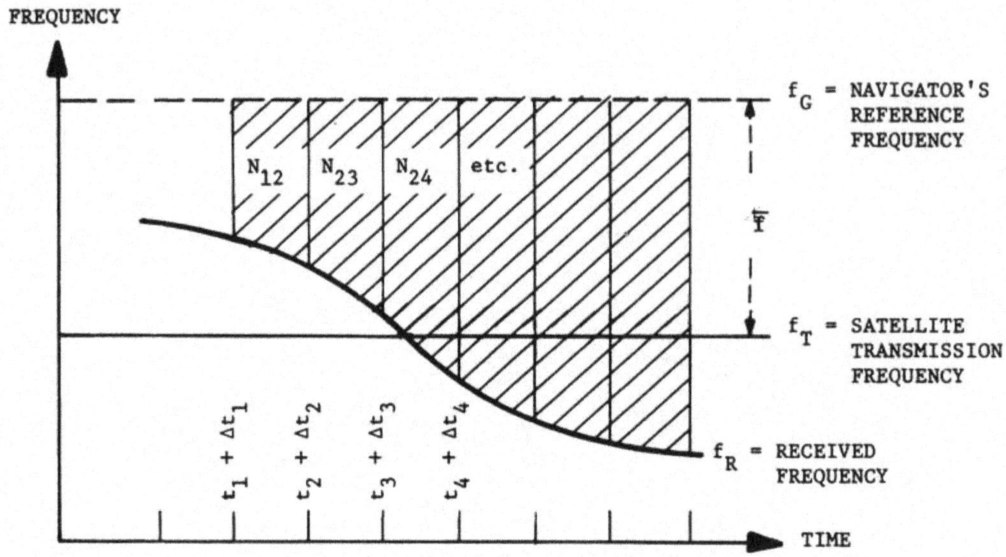

FIGURE 4430b.—Doppler frequency variation with time.

N_{12} through N_{34}. There are then three quantities to be determined by the computer: ϕ, λ, and $\bar{f}$, the last mentioned being of no immediate interest to the navigator but essential to the accurate determination of ϕ and λ. This means that integral doppler counts for at least *three* 2-minute intervals must be used (and preferably more than three) in order to provide three equations in the three unknowns, ϕ, λ, and $\bar{f}$.

That the integral doppler counts N_{12} and N_{34} are directly affected by $\bar{f}$ is illustrated in figure 4430b, wherein f_G is the (constant) frequency of the navigator's reference oscillator, f_T is the (constant) frequency of the satellite's transmitter, f_R is the received frequency containing the doppler component, and $\bar{f}=f_G-f_T$. The integral doppler counts, N_{12}, etc., are represented by the cross-hatched area in figure 4430b.

Since the values of three quantities (ϕ, λ, and $\bar{f}$) have to be simultaneously adjusted in minimizing the sum of the squares of the differences between estimated and measured slant range for three (or more) 2-minute intervals, the computations involve solution of a matrix whose general description can be illustrated as follows:

1. The *measured* slant range *increments* are calculated from the integral doppler counts for an assumed value of $\bar{f}$. Their rate of change as $\bar{f}$ changes is also determined.

2. The navigator's positions at the times t_1, t_2, etc., are calculated for an assumed initial position (ϕ, λ).

3. Using previously calculated satellite positions at t_1, t_2, etc., the estimated satellite-navigator slant range increments (for the assumed initial ϕ, λ are calculated).

4. The rate of change (partial derivative) of the estimated slant range increments (item 3 above) with respect to ϕ is determined.

5. The partial derivative of the estimated slant range increments with respect to λ is determined.

6. The differences (residuals) between *measured* slant range increments (item 1) and *estimated* slant range increments (item 3) are formed for each 2-minute interval.

7. Using the derivative of *measured* slant range increment with respect to $\bar{f}$, that of *estimated* slant range with respect to ϕ, and that of *estimated* slant range with respect to λ, new values of ϕ, λ, and $\bar{f}$ are calculated such that the sum of the squares of the residuals will be smaller than before these new values are used.

8. Steps 1 through 7 are repeated several times until the newly calculated values of $\bar{f}$, ϕ, and λ differ from the last values used by less than fixed threshold values. At this point the computing stops, and the last set of values of $\bar{f}$, ϕ, and λ is the final result.

4431. Computation algorithms.—In the computations the slant range is calculated for the beginning and end of each doppler counting interval based on the best estimates of the navigator's latitude, longitude, altitude above the geoid, and the transmitted satellite coordinates. Referring to figure 4430b, the doppler count N_{12} is repeated:

$$N_{12}=\int_{t_1+\Delta t_1}^{t_2+\Delta t_2}(f_G-f_R)dt. \tag{1}$$

It should be noted that the time $t^1+\Delta t_1$ is the time of *receipt* of the satellite time mark which was transmitted at time Δt_1. Therefore, Δt_m, $m=1, \ldots n$ (n=the total number of 2-minute intervals observed during one pass) represent the propagation time delay for the time mark to travel the distance S_m from the satellite to the receiver. The propagation time delay is defined by the slant range distance divided by the speed of light:

$$\Delta t_m=S_m/C. \tag{2}$$

Rearranging equation (1):

$$N_{12}=\int_{t_1+\Delta t_1}^{t_2+\Delta t_2} f_G dt-\int_{t_1+\Delta t_1}^{t_2+\Delta t_2} f_R dt. \tag{3}$$

The first integral in equation 3 is of a constant frequency and its integration is simple. The second integral is of the changing frequency f_R and represents the number of cycles received between the message receipt of two timing marks. It should be noted that the number of cycles received must identically equal the number of cycles transmitted:

$$\int_{t_1+\Delta t_1}^{t_2+\Delta t_2} f_R dt=\int_{t_1}^{t_2} f_T dt. \tag{4}$$

Substituting this expression into equation 3:

$$N_{12}=\int_{t_1+\Delta t_1}^{t_2+\Delta t_2} f_G dt-\int_{t_1}^{t_2} f_T dt. \tag{5}$$

Since the frequency functions, f_G *and* f_T, are assumed constant during a satellite pass, the integration of equation 5 gives:

$$N_{12}=f_G[(t_2-t_1)+(\Delta t_2-\Delta t_1)]-f_T(t_2-t_1). \tag{6}$$

Rearranging the terms in equation 6 gives:

$$N_{12}=(f_G-f_T)(t_2-t_1)+f_G(\Delta t_2-\Delta t_1). \tag{7}$$

The following comments apply to equation 7.

The quantity $f_G - f_T$ is assumed constant during a satellite pass:

$$f_G - f_T = \bar{f} \approx 32 \text{ kHz}, \tag{8}$$

and $f_G(\Delta t_2 - \Delta t_1)$ can be rewritten as:

$$f_G \frac{(S_2 - S_1)}{C}; \left(\Delta t_m = \frac{S_m}{C}\right), \tag{9}$$

defining L_G as the wavelength of the frequency f_G:

$$L_G = \frac{C}{f_G}; \; C = \text{speed of light in vacuum}, \tag{10}$$

and

$$\Delta T_{12} = t_2 - t_1.$$

Equation 7 becomes:

$$N_{12} = \bar{f}\Delta T_{12} + (1/L_G)(S_2 - S_1). \tag{11}$$

Defining $S_2 - S_1 = \hat{S}_{12}$ we have an expression for the *measured* slant range change in the kth counting interval:

$$\hat{S}_K = L_G N_K - L_G \bar{f} \Delta t_K. \tag{12}$$

The cartesian coordinates of a satellite are computed from the transmitted fixed and variable parameters of the satellite's position. Likewise, the navigator's cartesian coordinates are computed from his estimated position and geoidal height. The computed slant range at time t_K is then:

$$S_K = (X^2_K + Y^2_K + Z^2_K)^{1/2}, \tag{13}$$

where, with subscripts N and S for the navigator and satellite respectively:

$$X_K = (X_{SK} - Y_{NK}),\; Y_K = (Y_{SK} - Y_{NK}),\; Z_K = (Z_{SK} - Z_{NK}).$$

The computed slant range change between times t_K and t_{K-1} is therefore $S_K - S_{K-1}$. Defining a residual:

$$R(\phi, \lambda, \bar{f})_K = \hat{S}_K - (S_K - S_{K-1}).$$

If M doppler intervals are to be used in the computation, M such functions may be computed and $F(\phi, \lambda, \bar{f})$, the sum of the squares of the residuals, may be computed:

$$F(\phi, \lambda, \bar{f}) = \sum_{K=1}^{M} R(\phi, \lambda, \bar{f})^2_K. \tag{14}$$

Expanding the residuals in a Taylor series about the initial estimates $\bar{f}_K$, ϕ_K, and λ_K, then truncating the terms above first degree, this function becomes:

$$F = \sum_{K=1}^{M} \left[R_K - \frac{\partial(S_K - S_{K-1})}{\partial\phi} \Delta\phi - \frac{\partial(S_K - S_{K-1})}{\partial\lambda} \Delta\lambda + \frac{\partial}{\partial\bar{f}} \hat{S}_K \Delta\bar{f} \right]^2 \tag{15}$$

Since we have just shown that $\hat{S}_K$ is a function of $\bar{f}$, and S_K and S_{K-1} are both functions of ϕ and λ, or

$$F = \sum_{K=1}^{M} G^2_K.$$

where:

$$\Delta\bar{f} = \bar{f} - \bar{f}_K$$

$$\Delta\phi = \phi - \phi_K$$

$$\Delta\lambda = \lambda - \lambda_K$$

$$R_K = \hat{S}_K - (S_K - S_{K-1}),$$

and G_K symbolizes the bracketed function in equation 15.

To minimize $F(\phi, \lambda, \bar{f})$ with respect to $\phi, \lambda, \bar{f}$ equation 15 is differentiated partially with respect to $\phi, \lambda, \bar{f}$, the partials equated to zero, and the equations divided by 2.

$$\left.\begin{aligned}
\frac{\partial F}{\partial \phi} &= 0 = -\sum_{K=1}^{M} G_K \frac{\partial(S_K - S_{K-1})}{\partial \phi} \\
\frac{\partial F}{\partial \lambda} &= 0 = -\sum_{K=1}^{M} G_K \frac{\partial(S_K - S_{K-1})}{\partial \lambda} \\
\frac{\partial F}{\partial f} &= 0 = \sum_{K=1}^{M} G_K \frac{\partial \hat{S}_K}{\partial \bar{f}}
\end{aligned}\right\} \qquad (16)$$

Define the A matrix

$$\{A\} = \begin{bmatrix}
-\frac{\partial(S_1 - S_0)}{\partial \phi} & -\frac{\partial(S_1 - S_0)}{\partial \lambda} & \frac{\partial \hat{S}_1}{\partial \bar{f}} \\
-\frac{\partial(S_2 - S_1)}{\partial \phi} & -\frac{\partial(S_2 - S_1)}{\partial \lambda} & \frac{\partial \hat{S}_2}{\partial \bar{f}} \\
\vdots & \vdots & \vdots \\
-\frac{\partial(S_M - S_{M-1})}{\partial \phi} & -\frac{\partial(S_M - S_{M-1})}{\partial \lambda} & \frac{\partial \hat{S}_M}{\partial \bar{f}}
\end{bmatrix}$$

and the residual matrix

$$R = \begin{bmatrix} R_1 \\ \vdots \\ R_M \end{bmatrix}$$

and an error matrix,

$$E = \begin{bmatrix} \Delta\phi \\ \Delta\lambda \\ \Delta\bar{f} \end{bmatrix}$$

and rewrite the G_K functions in matrix form.

$$G_1 = R_1 + \frac{\partial}{\partial \bar{f}} \hat{S}_1 \Delta \bar{f} - \frac{\partial(S_1 - S_0)}{\partial \phi} \Delta\phi - \frac{\partial(S_1 - S_0)}{\partial \lambda} \Delta\lambda$$

$$G_2 = R_2 + \frac{\partial}{\partial \bar{f}} \hat{S}_2 \Delta \bar{f} - \frac{\partial(S_2 - S_1)}{\partial \phi} \Delta\phi - \frac{\partial(S_2 - S_1)}{\partial \lambda} \Delta\lambda$$

$$G_3 = R_3 + \frac{\partial}{\partial \bar{f}} \hat{S}_3 \Delta \bar{f} - \frac{\partial(S_3 - S_2)}{\partial \phi} \Delta\phi - \frac{\partial(S_2 - S_1)}{\partial \lambda} \Delta\lambda$$

$$\vdots \qquad \vdots \qquad \vdots$$

$$G_M = R_M + \frac{\partial}{\partial \bar{f}} \hat{S}_1 \Delta \bar{f} - \frac{\partial(S_M - S_{M-1})}{\partial \phi} \Delta\phi - \frac{\partial(S_M - S_{M-1})}{\partial \lambda} \Delta\lambda$$

$$G = R + AE.$$

Then, in matrix notation:

$$A^T G = A^T [R + AE] = [0]$$

or

$$E = -[A^T A]^{-1} A^T R.$$

Navstar Global Positioning System (GPS)

4432. Introduction.—The Navstar Global Positioning System (GPS) is a space-based, radio positioning, navigation, and time-transfer system that will become fully operational in the late 1980s. GPS has the potential for providing highly accurate three-dimensional position and velocity information along with Coordinated Universal Time (UTC) (art. 1805) to an unlimited number of suitably equipped users under all weather conditions, continuously, anywhere on or near the surface of the Earth.

GPS is being developed by the U.S. Government and is in the full-scale engineering development phase (1984). The system is managed by the U.S. Air Force Space Division as a joint program involving all the services, plus the Defense Mapping Agency, the Department of Transportation, NATO, and Australia. In this phase of the program the production baseline for the operational satellite will be established, the operational control segment will be completed, and the final determination on the configuration of user equipment to accommodate all classes of system users will be accomplished.

4433. System description.—The Navstar GPS navigation and time-transfer system operates on two L-band frequencies, L_1 (1575.4 MHz) and L_2 (1227.6 MHz). The system consists of three major segments: a *space segment,* satellites that transmit radio signals; a *control segment,* ground-based equipment to monitor the satellites and update their signals; and a *user segment,* equipment which passively receives and converts satellite signals into positioning and navigation information.

When fully operational at the end of 1988, the *space segment* will contain 18 satellites in six orbital planes, each having three satellites. The satellites will operate in circular 20,200-kilometer orbits at an inclination of 55 degrees with 12-hour periods. Precise spacing of the satellites in orbit will be arranged such that a minimum of four satellites will be in view to a user at any time on a worldwide basis. Each satellite will transmit an L_1 and an L_2 signal. Superimposed on these signals will be navigation and system data, including a predicted satellite ephemeris, atmospheric propagation correction data, satellite clock error information, and satellite health data.

The *control segment* includes a master control station (MCS) and a number of monitor stations and ground antennas located throughout the world. The MCS consists of equipment and facilities required for satellite monitoring, telemetry, tracking, commanding, and control, uploading, and navigation message generation. The monitor stations passively track the satellites, accumulate ranging data from their signals, and relay them to the MCS where they are processed to determine satellite position and signal data accuracy. The MCS updates the navigation message of each satellite and relays this information to the ground antennas which transmit it to the satellites. The ground antennas are also used for transmitting and receiving satellite control information.

The *user segment* includes low, medium, and high dynamic receivers designed to different requirements of various users. The user equipment is designed to receive and process signals from four orbiting satellites either simultaneously or sequentially. The processor in the receiver then converts these signals to three-dimensional navigation information in World Geodetic System (art. X 12) coordinates. Positioning data are presented on a display unit in any other coordinate system desired by the user. In addition, precise time information is available for use in the host vehicle.

4434. System capabilities.—GPS is being developed to provide multiple users with accurate continuous, worldwide, all-weather, common-grid, three-dimensional positioning and navigation information. Because of the unique capabilities of the system it is reversing the trend of proliferation of radio navigation systems within the U.S. Department of Defense.

To obtain a navigation solution of position and time (four unknowns), four satellites must be selected. The GPS user measures pseudorange and pseudorange rate by synchronizing and tracking the navigation signal from each of the four selected satellites. Pseudorange is the true distance between the satellite and the user plus an offset due to the user's clock bias. Pseudorange rate is the true slant range rate plus an offset due to the frequency of the user's clock. By decoding the ephemeris data and system timing information on each satellite's signal, the user's receiver/processor can convert the pseudorange and pseudorange rate to user three-dimensional position and velocity. Four measurements are necessary to solve for the three unknown components of position (or velocity) and the unknown user time bias (or frequency).

The navigation accuracy that can be achieved by any user depends primarily on the variability of the errors in making pseudorange measurements and the instantaneous geometry of the satellites as seen from the user's location on Earth.

4435. System status.—In December 1973 the Defense System Acquisition Review Council (DSARC) authorized a step-wise, design-to-cost development and test program which would lead, in successive phases, to an operational GPS. The system concept resulted from the integration of the best features of previous navigation satellite concepts being pursued independently by the U.S. Navy and the U.S. Air Force.

The Navstar program is being developed in three phases: Phase I, concept validation; Phase II, full-scale engineering and development; and Phase III, production. During Phase I prototype satellites were developed, a control segment was established, and advanced engineering models of user equipment were tested. Currently (1984), the program is in Phase II. The main tasks of this phase include developing the prototype operational satellite, the operational control system, the prototype user equipment, and the completion of the developmental test and evaluation/ initial operational test and evaluation (DT&E/IOT&E).

The current (1984) GPS constellation consists of a limited number of development satellites configured in two planes inclined at 63 degrees with respect to the equator. The current constellation is shown in figure 4435a. The plan is to maintain a minimum constellation of five satellites for the completion of the test and evaluation period for the system. It is currently projected that the last test satellites will be launched in late-1984 and in 1985 to maintain the test configuration.

The operational constellation is a 6-plane, uniform 18-satellite configuration plus three active spares. The spares are necessary to guarantee very high system availability for the 18-satellite constellation. Figure 4435b illustrates the operational constellation. The planes of the constellation are 60 degrees apart in longitude with each plane containing three satellites separated by 120 degrees. The phasing from plane to plane is 40 degrees so that a satellite in one plane has a satellite 40 degrees ahead (north) of it in the adjacent plane to the east. According to the current plan, satellites will be launched at a rate of eight or nine per year beginning in October 1986. Two-dimensional coverage should be available by the latter part of 1987 and three-dimensional coverage should be available by the end of 1988.

The final operational control system (OCS) will consist of a master control station, three ground antennas, and five monitor stations. The initial operational capability for the OCS is scheduled for October 1984. Details of the operational control system are shown in figure 4435c.

In 1984, contractors were developing a family of user equipment sets for demonstration and testing in the current phase of the GPS program. In order to enhance reliability and maintainability objectives of the user equipment, three main design features have been emphasized: the development and procurement of a small number of user equipment classes, the commonality/interchangeability of units across user classes, and the utilization of a higher order language. These design features prevent the proliferation of a large number of peculiar user sets, afford the advantage of large-lot buys and common spares, and permit more efficient programming and reprogramming of requirement changes. The primary objective is to design and build a family of GPS user sets at a minimum life cycle cost consistent with performance and functional capabilities. Testing continues on the host vehicles to collect data to support the source selection decision in January 1985 for the Phase III user equipment production contract. Since 1977 more than 800 test missions have been accomplished. In virtually all of these tests the Navstar GPS has exceeded specifications.

4436. System management.—The policies for use and operation of all radio navigation systems controlled by the U.S. Department of Defense (DoD) are reflected in the Federal Navigation Plan. The plan is a joint annual publication by the Departments of Defense and Transportation. The operational satellites will provide two unique and distinct navigation services, the *Precise Positioning Service* (PPS) and the *Standard Positioning Service* (SPS). The PPS will be available to military users and access will be controlled. The PPS is the service of highest accuracy which provides a positioning accuracy of 15 meters SEP, a velocity accuracy of 0.1 meter/sec (1 sigma), and a timing accuracy of better than 100 nanosec (1 sigma). The SPS will be made available to the worldwide community of civil users. Current policy (U.S. DoD, 1983) sets the accuracy of SPS at 100 meters with a 95% level of confidence.

The DoD policy is for Navstar GPS to become the primary DoD radionavigation system. As GPS becomes operational, DoD will terminate operations with other radionavigation systems.

4437. Applications.—Even though Navstar GPS is not fully operational, several potential civilian applications exist. GPS will provide a broad spectrum of civilian users with an accurate position, velocity, and time determination capability at a reasonable cost. Ultimately, it will replace many existing navigation systems.

There are many potential applications for GPS. Civilian users of airborne, land-based, and maritime vehicles will benefit from GPS for optimal course navigation, which will in turn reduce fuel costs and transportation time. Search-and-rescue techniques can be enhanced because of the precision position identification capability of GPS. The mineral exploration and geophysical survey communities will be able to accurately locate potential petroleum bearing areas, ore bodies, and active fault belts in a shorter period of time. The GPS common grid feature will enhance many land-vehicle operations. Airborne collision avoidance systems and maritime hazard systems are also potential users of the system. GPS time transfer applications include scientific, intra-system synchronization, and inter-system synchronization uses.

In addition to the potential applications given above, GPS has determined performance in space applications and baseline determinations. Continuous 10 to 15-meter positioning in space without the use of ground tracking systems has been demonstrated on LANDSAT-4 using a GPS receiver. Relative positions have been

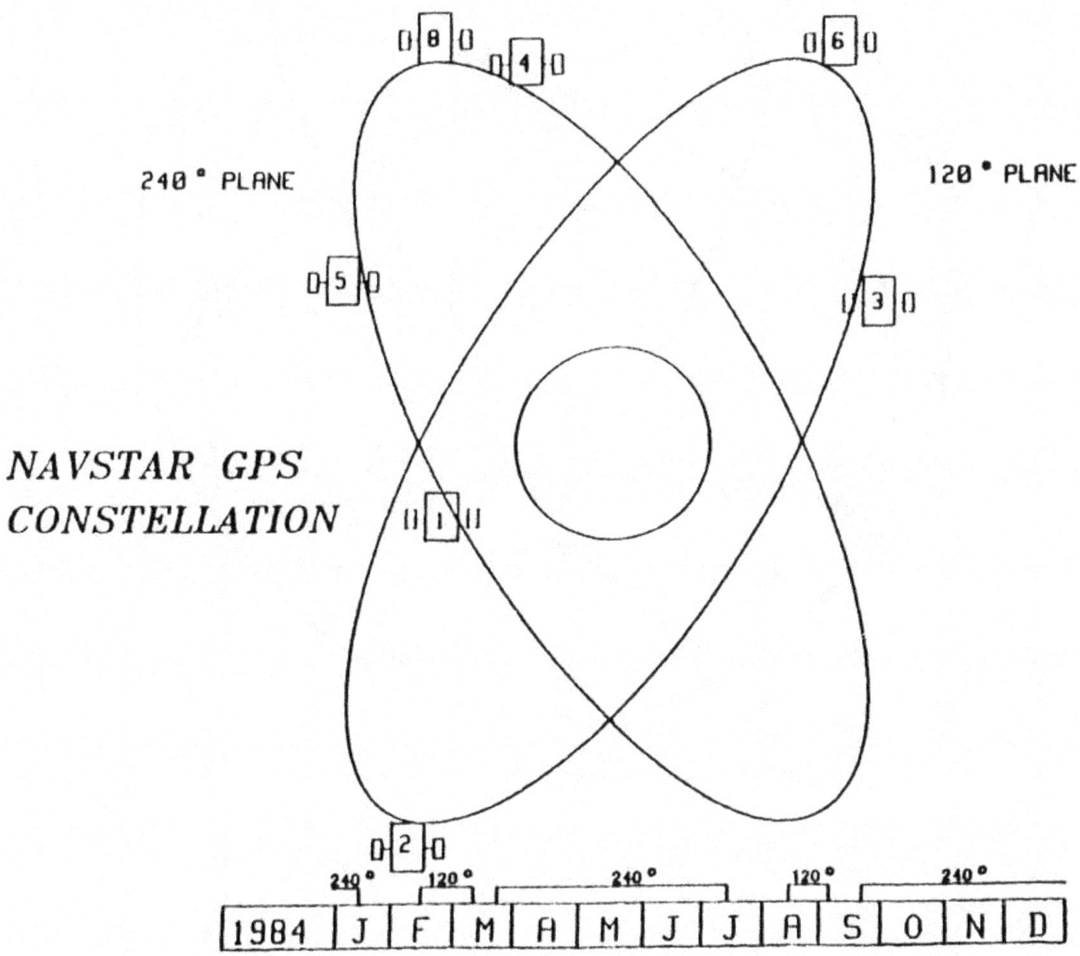

FIGURE 4435a.—Current (1984) GPS constellation.

demonstrated to be accurate to within several centimeters over baselines varying from hundreds of meters to tens of kilometers. The potential applications for GPS are boundless.

4438. Conclusions.—The Navstar GPS is evolving into a highly accurate, world-wide, all-weather navigation/positioning system applicable to the needs of both the military and civilian communities. Since 1977 extensive testing has taken place which verified the established accuracy goals and the system operation under various mission-oriented conditions. There are numerous applications for which GPS can be exploited. Even though the system is not fully operational (1984), it is being used by diverse civilian groups such as NASA for orbit determination onboard LANDSAT-4 and oil companies for geodetic positioning. As the system gains acceptance by the civilian community, more sophisticated uses for the system will be established.

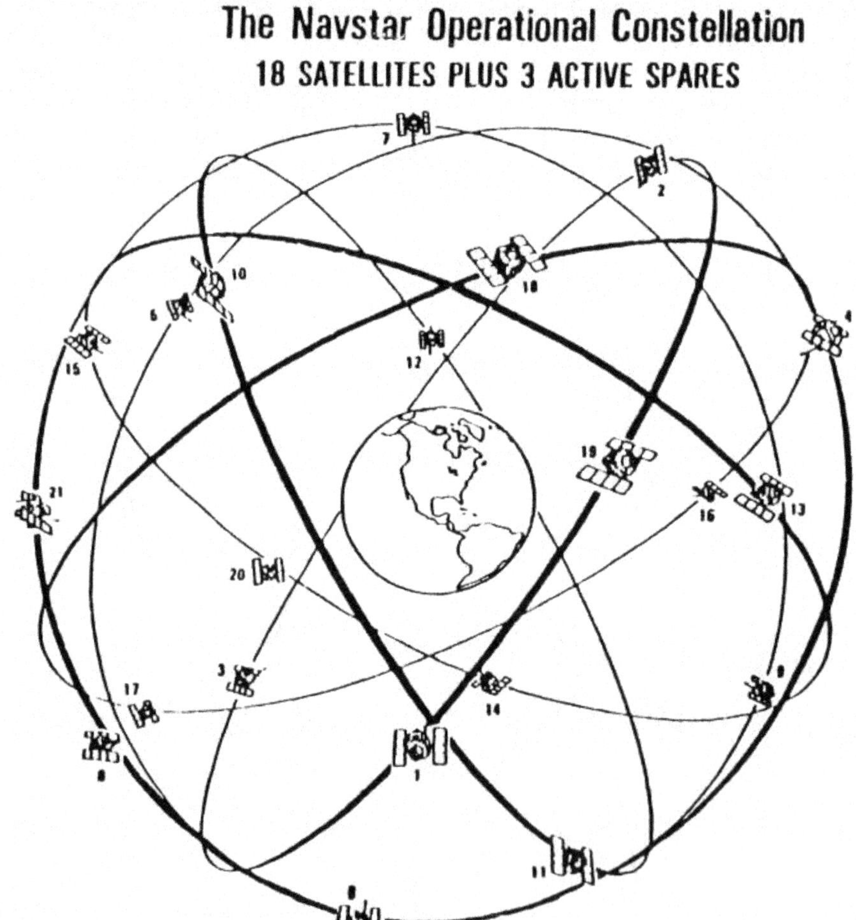

FIGURE 4435b.—Navstar operational constellation.

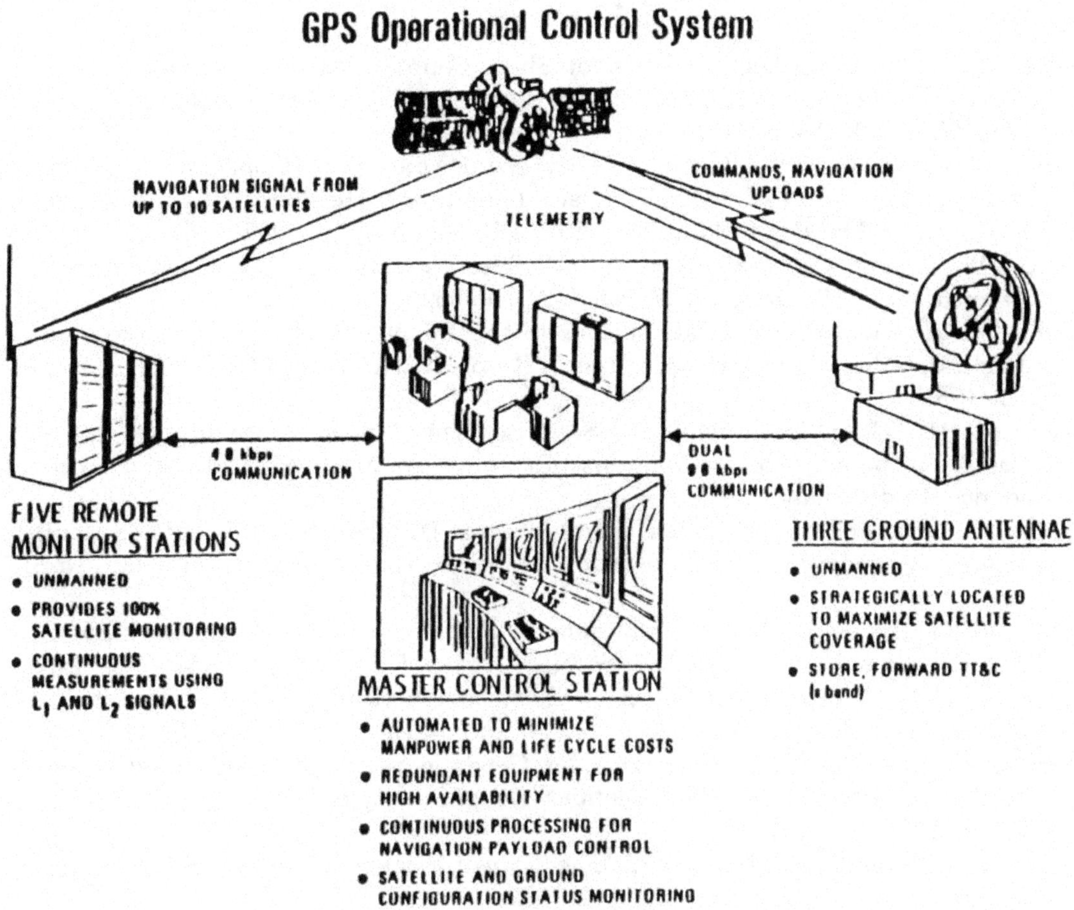

FIGURE 4435c.—GPS Operational Control System.

References

Dove, R. E., "Program Requirements for Short Count Integrated Doppler Satellite Navigation Solution," Applied Physics Laboratory, The Johns Hopkins University, TG-819-2, August 1974.

Guier, W. H. and G. C. Wieffenbach, "Theoretical Analysis of Doppler Radio Signals from Earth Satellites," Applied Physics Laboratory, The Johns Hopkins University, Laurel, Maryland, Bumblebee Series Report No. 276, April 1958.

Guier, W. H. and G. C. Wieffenbach, "A Satellite Doppler Navigation System," *Proceedings of the IRE,* vol. 48, No. 4, April 1960.

Guier, W. H., "Ionospheric Contributions to the Doppler Shift at VHF from Near-Earth Satellites," *Proceedings of the IRE,* vol. 49, No. 11, pp. 1680–1681, November 1961.

Guier, W. H., "Geodetic Problems and Satellite Orbits," Space Mathematics—part 2, vol. 6, Lectures in Applied Mathematics, American Mathematical Society, Providence, Rhode Island, 1966, p. 170 ff.

Guier, W. H., "Satellite Navigation using Integral Doppler Data: AN/SRN-9 Equipment," *Proceedings of the American Geophysical Union,* Washington, D.C., April 1965.

Hopfield, H. S., "The Effect of Tropospheric Refraction on the Doppler Shift of a Satellite Signal," *Journal of Geophysical Research,* vol. 68, pp. 5157–5168, 1963.

Hopfield, H. S., "Tropospheric Effect on Electromagnetically Measured Range: Prediction from Surface Weather Data," *Radio Science,* vol. 6, pp. 357–367, 1971.

Newton, R. R., "The U. S. Navy Doppler Geodetic System and Its Observational Accuracy," *Philosophical Transactions,* Royal Society of London, A, vol. 262, pp. 50–66, 1967.

Newton, R. R., and R. B. Kershner, "The Transit System," *Journal of the Institute of Navigation,* London, vol. 15, April 1962.

Yionoulis, S. M., F. T. Heuring, and W. H. Guier, "A Geopotential Model (APL 5.0–1967) Determined from Satellite Doppler Data at Seven Inclinations," *Journal of Geophysical Research,* vol. 77(20), 1972.

Yionoulis, S. M., "Determination of Coefficients Associated with the Geopotential Harmonics of Order Thirteen," *Journal of Geophysical Research,* vol. 71(6), 1966.

"Navy Navigation Satellite System User Equipment Handbook," prepared by the Space Development Department, Applied Physics Laboratory, The Johns Hopkins University, Laurel, Maryland, SDO-3100.

"The Navy Navigation Satellite System," U. S. Navy Astronautics Group, Point Mugu, California, January 1967.

NAVSTAR References

Birmingham, W. P., B. L. Miller, and W. L. Stein, 1983, "Experimental Results of Using the GPS for LANDSAT 4 On-board Navigation," Proceedings of the ION Aerospace Meeting, Arlington, Virginia.

Counselman, C. C., R. I. Abbot, S. A. Gourevitch, R. W. King, and A. R. Paradis, 1983, "Centimeter-Level Relative Positioning with GPS," *Journal of Surveying Engineering (ASCE),* Vol. 109, No. 2, pp. 81–89.

Milliken, R. J. and C. J. Zoller, 1980, "Principle of Operation of NAVSTAR and System Characteristics," in "Global Positioning System," Papers published in Navigation, Institute of Navigation.

Payne, C. R., 1982, "NAVSTAR Global Positioning System, 1982," *Proceedings of the Third International Geodetic Symposium on Satellite Doppler Positioning,* Vol. 2, pages 993–1021.

Senus, W. J. and R. W. Hill, 1981, "GPS Applications to Mapping, Charting, and Geodesy," *Navigation, the Journal of the Institute of Navigation,* Vol. 28, No. 2.

Van Dierendonck, A. J. and W. C. Melton, 1983, "Applications of Time Transfer Using NAVSTAR GPS," *Navigation,* Vol. 30, No. 2.

U.S. Department of Defense, 1983, Office of Assistant Secretary of Defense (Public Affairs), Washington, D.C., News Release, No. 328–83.

CHAPTER XLV

DOPPLER SONAR NAVIGATION

4501. Introduction.—The *doppler effect,* first described by Christian Johann Doppler in 1842, is observed as a frequency shift resulting from relative motion between a transmitter and receiver or reflector of acoustic or electromagnetic energy. The effect on ultrasonic energy is used in **doppler sonar speed logs** to measure the relative motion between the vessel and the reflective sea bottom (for bottom return mode) or suspended particulate matter in the seawater itself (for volume reverberation mode). The velocity so obtained and integrated with respect to time is used in **doppler sonar navigators** to determine position with respect to a start point. The doppler effect is also used in **docking aids** which provide precise speed measurements.

The maximum safe docking speed of vessels exceeding about 100,000 dead weight tons is of the order of 0.2 feet per second; the berthing facility may collapse when the vessel makes contact at a speed in the region of 1 foot per second. When anchoring these large vessels without the aid of tugs, a speed of the order of 1 foot per second over the ground can result in the loss of the anchor and chain. Anchoring should be effected at speeds of less than 0.5 feet per second. Therefore, precise determination of the speed of these large vessels with respect to the bottom is essential to safe operation. The required speed-measurement increments are of the order of 0.01 knot.

4502. The Janus configuration (fig. 4502) normally used with doppler sonar speed logs and navigators employs four beams of ultrasonic energy, displaced laterally 90° from each other, and each directed obliquely (30° from the vertical) at the ocean floor, to obtain true ground speed in the fore and aft and athwartship directions. These speeds are measured as doppler frequency shifts in the reflected beams.

Under ideal operating conditions (i.e. a calm flat sea), the speeds are expressed as follows:

$$V_{fa}=\frac{F_f-F_a}{\frac{2F_s}{C}}$$

$$V_{st-p}=\frac{F_{st}-F_p}{\frac{2F_s}{C}},$$

where:

1. V_{fa} is the vessel's speed over the ground in the fore and aft direction;
2. V_{st-p} is the vessel's speed over the ground in the athwartship (starboard to port) direction;
3. F_f is the reflected forward beam frequency;
4. F_a is the reflected after beam frequency;

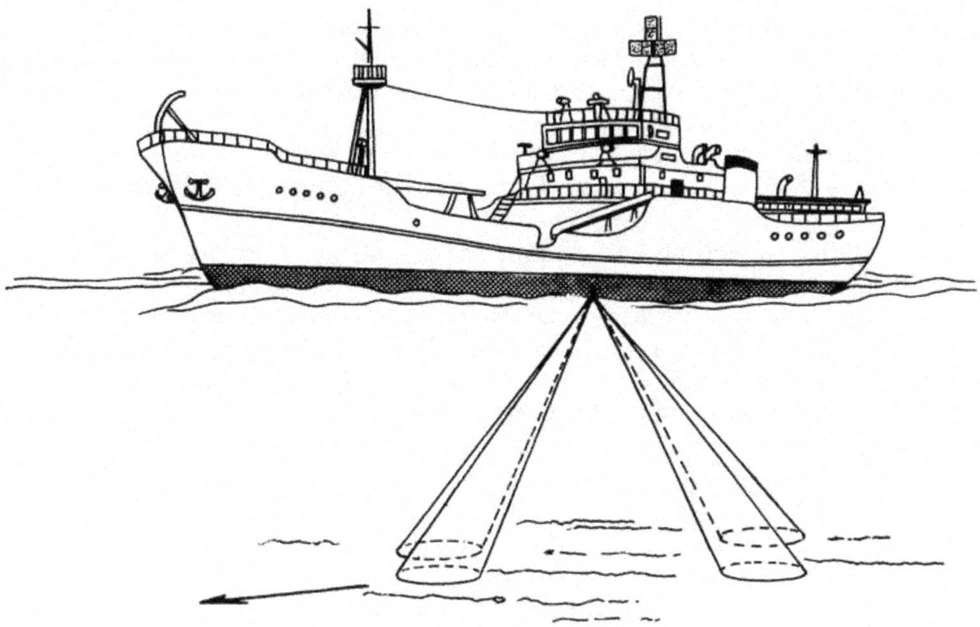

FIGURE 4502.—Janus configuration.

5. F_{st} is the reflected starboard beam frequency;
6. F_p is the reflected port beam frequency;
7. F_s is the transmitter operating frequency; and
8. C is the sound speed with respect to the water.

A development of the above simplified equations from the fundamental doppler equation for sound propagation is presented in article 4506.

4503. Operational errors.—Under actual operating conditions, the deviations from the ideal or assumed conditions introduce additional terms and nonlinearities not shown in the simplified equations. These effects may be classified as *transducer orientation errors, vessel motion induced errors, velocity of sound errors,* and *power loss errors.*

Transducer orientation errors. If the transducers are not properly aligned with the vessel's velocity vector, the speed signal is reduced by the factor, cos α, where α is the angle between the transducer plane and the velocity vector. For example, if the transducers are aligned when the vessel has no trim, and the vessel is trimmed to angle α, the speed signal is reduced by a factor of cosine α. This error is generally small; for example, a trim (or list) of 8° will reduce the fore and aft (or starboard to port) speed signal by 1 percent.

Vessel motion induced errors. Even though the speed of sound is large compared to the vessels' speed, the vessel may roll or pitch a degree or two during the time between transmission and reception of a particular wave front. This difference in transmission and reception angles will introduce significant errors in single-beam systems. However, the Janus configuration (art. 4502) causes the errors to very nearly cancel. For a 2° difference between the transmission and reception angle, the error induced, using a Janus configuration, is less than 0.1 percent of the speed signal. The forward speed of a vessel will also cause differences in transmission and reception angles, but the resulting speed errors also tend to cancel.

The most significant speed error is caused by the fact that, as a vessel rolls or pitches, the respective speed signal is reduced by the cosine of the instantaneous roll or pitch angle. Since this cosine factor reduces the speed signal for both positive and negative values of angles, this error does not cancel. Assuming sinusoidal rolling with a maximum roll of 25°, the average roll over one-half cycle is 16°; the cosine factor is 0.9613. The starboard to port speed signal is reduced by nearly 4 percent.

Another effect of rolling and pitching is the variation in the amplitude of the return signal. As the beam on the high side of a roll or pitch tilts toward the horizon, the path length increases and more sound energy is lost in the water. In addition, the beam reflection angle is decreasing and less energy is back-scattered along the beam. These effects can cause periodic loss of signal, especially when operating near the maximum depth in heavy seas.

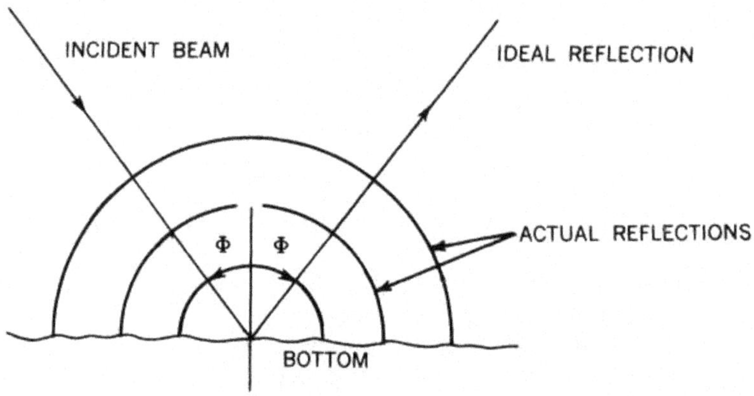

FIGURE 4503.—Bottom reflection.

The roll and pitch rates may also cause loss or attenuation of the return signals when narrow, sharply defined ultrasonic beams are used. Under normal operating conditions, the area of the bottom "illuminated" by the transmitted beam is also "seen" by the receiving transducer. However, if the roll (or pitch) rate is very large, the receiving transducer may be rotated sufficiently to reduce the time it takes a wave front to travel to the bottom and back, so that it no longer "sees" the area illuminated by the transmitted beam. In water 400 feet deep and a 3° beam width, the roll rate must be less than 16° per second if the transmitted and received beams are to overlap at the half-power points.

Velocity of sound errors. The doppler frequency shifts are dependent upon the velocity of sound in water in the immediate vicinity of the transducer. The actual frequency changes occur as the sound pressure waves move from the face of the transducer into water undisturbed by the vessel's hull. Similarly, a received beam undergoes a frequency shift as the sound travels from the undisturbed water to the face of the receiving transducer.

One empirical formula for the speed of sound in seawater (ch. XXXV) is

$$C=4422+11.25T-0.0450T^2+0.0182d+4.3\ (S-34),$$

where C is the velocity in feet per second, T is the temperature (°F), S is the salinity (parts per thousand), and d is the depth below the surface in feet. The velocity of sound in seawater is dependent primarily upon temperature which varies from about 28°F in polar regions to about 86°F in the Tropics. Over this temperature range, the velocity varies about 7 percent due to temperature alone. In the open sea, salinity varies from about 33 to 36 parts per thousand. However, the salinity in bays, inlets, and rivers varies from open-sea values to freshwater values. For example, as one goes from the Chesapeake Bay waters at Annapolis to the Atlantic Ocean, the salinity increases about 25 parts per thousand. This salinity change results in more than 1½ percent increase in velocity due to salinity alone.

Since the transmitter oscillator frequency is held constant, the calibration constant (art. 4506) should vary directly as the change in the velocity in seawater. In some doppler speed logs and navigators, the adjustment of the calibration constant may be limited to the effects of temperature change. In some installations a velocimeter may be used to determine the velocity of sound in the vicinity of the transducer.

Power loss errors. The speed information is contained in the frequency of the received signals. Therefore, the only power requirement is that the signal level be great enough for the receivers to separate it from the noise. If the speed signal is lost for very short periods at frequent intervals, there will be an erroneous speed indication. If the speed signal is lost for longer periods, the speed indication will drop to zero or the natural noise level. Commercial equipments have memory circuits that retain the last good speed signal until the next one is obtained to update the readout. This approach is not entirely satisfactory during tactical maneuvering.

Excluding internal system losses, there are two primary areas of power loss:

1. Power lost in the water during transit to and from the bottom.
2. Power lost at the bottom on reflection.

The losses in the water are caused by absorption, scatter, and normal inverse square spreading. The absorption and scatter losses are attenuation losses and are exponentially proportional to path length. The absorption loss is the energy lost to internal friction of the water. At the higher frequencies the loss is very nearly proportional to the square of the frequency but tends to become linear as the frequency decreases. The scattering loss is caused by the existence of suspended matter in the water that reflects or scatters the sound energy. This effect can be useful in that it may permit a doppler system to measure velocity with respect to water if the energy scattered back along the beam is sufficient to activate the receivers. Spreading losses are caused by the expansion or spreading of the wave front which then reduces the energy per unit area of the wave front. For spherical waves from a point source, the energy in a wave front, neglecting attenuation, is constant. Therefore, the energy density decreases inversely as the square of the distance from the source.

The losses on reflection from the bottom are caused by scatter and absorption. In general, the sound energy is reflected from the bottom in all directions, with the intensity of sound energy reflected in any particular direction being strongly dependent upon the angle of incidence and the character of the bottom.

For a perfectly smooth bottom, the incident sound beam would be reflected along the path labeled "ideal reflection" shown in figure 4503. This type of bottom would cause loss of a useful return. Actually, there is always a degree of back scatter. But for a very smooth mud bottom and a large angle of incidence, the back scatter along the incident beam is greatly reduced. The roughness of the bottom to a particular sound beam is dependent upon the wavelength of the sound. When the discontinuities of the bottom are much smaller than the wavelength, the bottom appears smooth and there is little back scatter.

The angle of incidence of the sound beam is dependent upon the slope of the bottom at the point of incidence. In general, the ocean bottom is similar to land areas with respect to valleys, hills, and similar terrain characteristics (art. 3022). Thus, the returned energy will vary greatly with the local slope of the bottom where the beam strikes it.

The sound energy absorbed on reflection depends largely on the material of the bottom. The average reflection coefficients for mud bottoms are approximately 0.16 to 0.17, whereas the reflection coefficients for sand bottoms are 0.51 to 0.82. Thus, when a system is operating near its maximum depth, a transition from a sand to a mud bottom could result in the loss of the signal reflected from the bottom.

This brief discussion of the factors affecting the power losses in underwater sound transmission and losses on reflection from the bottom indicates that the sound power level at the receiver will vary over many orders of magnitude due to environmental phenomena.

4504. Problems associated with beam geometry.—As shown in figure 4504, the side lobe reflects from the bottom at a more favorable angle with a shorter wave path than the main beam. It is evident from the development in article 4506 that if side lobe reception dominates over main beam reception, the result will be that the indicated speed will be lower than the actual speed. The Janus configuration does not eliminate this effect. This effect can be compensated for by transducer design tending to eliminate side-lobe effects. However, the beam width must be large enough (about 3° to 6°) to "illuminate" a sufficient area of the bottom to minimize the effects of bottom relief.

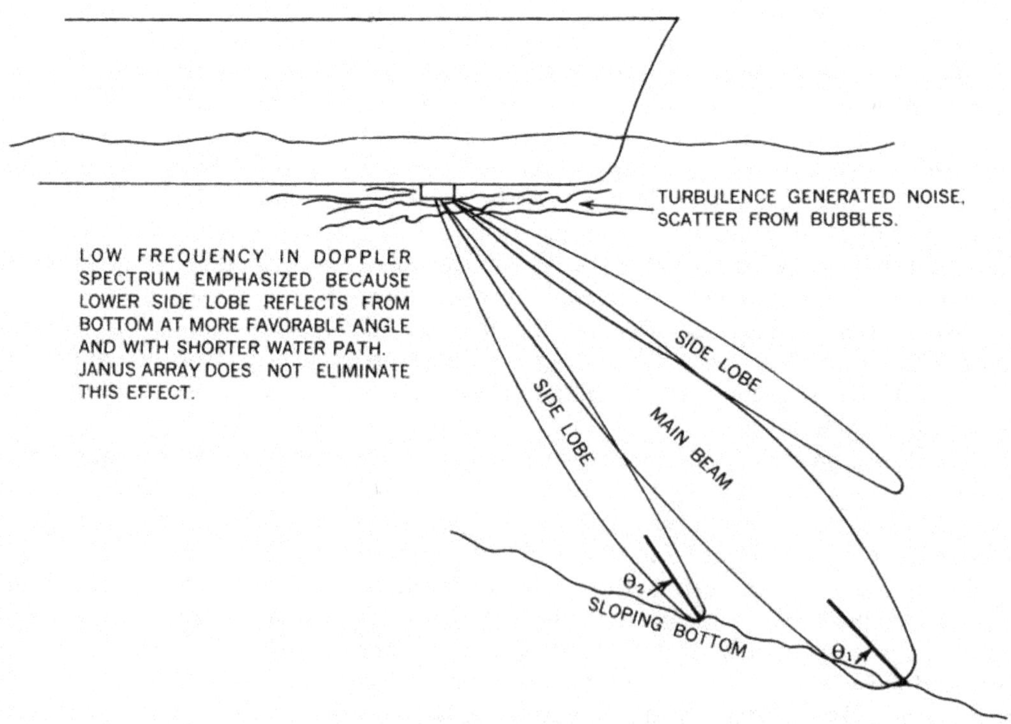

FIGURE 4504.—Side lobe reflection.

4505. Basic design considerations include the following:

Frequency. A high oscillator frequency is desirable for reducing transducer size and maximizing the doppler shift. However, at the higher frequencies the absorption loss (art. 4503) is very nearly proportional to the square of the frequency. As the frequency is decreased, the loss tends to become linear. As the frequency is reduced by the designer to obtain greater maximum depth of operation in the bottom return mode, he encounters transducer size problems. Improvement in operating depth diminishes near 100 kHz while transducer size has increased considerably. Also, as discussed in article 4503, the roughness of the bottom to a particular sound beam is dependent upon the wavelength of the sound. Therefore, the frequency selected by the designer is a trade off of desired maximum operating depth, transducer size, and bottom composition factors. The region between 150 and 600 kHz is finding the greatest application.

Sound transmission. The pulse transmission method permits the use of relatively high power for operating at greater depths in the bottom return mode. Continuous wave transmission, however, provides more precise velocity resolution permitting operation with very little clearance between the vessel's hull and the bottom. Therefore, a designer may select pulse transmission for a doppler sonar navigator but continuous wave transmission for a docking system to be used for precise speed measurements over a vessel operating with very little clearance.

4506. Development of expressions for the vessel's speed in the fore and aft and athwartship directions.—The doppler equation for sound propagation may be expressed as

$$F_o = F_s\left(\frac{C+V_m-V_o}{C+V_m-V_s}\right),$$

where F_o is the frequency detected by the observer, F_s is the frequency of the source, C is the speed of sound in the medium, V_m is the speed of the medium, V_o is the speed of the observer, and V_s is the speed of the source. All speeds are velocity components along the direction of sound propagation and are considered positive if moving in the same direction as the sound. Consider as shown in figure 4506a, the situation where the sound source (transmitter *A*) and observer (receiver *A*) are connected and moving at velocity V_A with respect to the ground. Similarly consider transmitter *B* and receiver *B* connected and moving at velocity V_B. V_W represents the water velocity with respect to the ground; *C* represents the sound velocity with respect to the water. If transmitter *A* has a frequency F_s, the frequency received by receiver *B* is given by

$$F_B = F_s\left(\frac{C+V_w-V_B}{C+V_w-V_A}\right).$$

Assuming that transmitter *B* always emits the same frequency received by receiver *B*, then receiver *A* receives a frequency given by

$$F_A = F_B\left(\frac{C+(-V_W)-(-V_A)}{C+(-V_W)-(-V_B)}\right).$$

The signs of V_w, V_A, and V_B are negative because they are directed opposite the sound propagation between transmitter *B* and receiver *A*. Thus, the frequency received by receiver *A* may be written as

$$F_A = F_s \left(\frac{C + V_W - V_B}{C + V_W - V_A} \right) \left(\frac{C - V_W + V_A}{C - V_W + V_B} \right).$$

If the transmitter-receiver unit, *B*, is stationary with respect to the ground, $V_B = 0$. Then, receiver *A* will receive a frequency given by

$$F_A = F_s \left(\frac{C + V_W}{C + V_W - V_A} \right) \left(\frac{C - V_W + V_A}{C - V_W} \right).$$

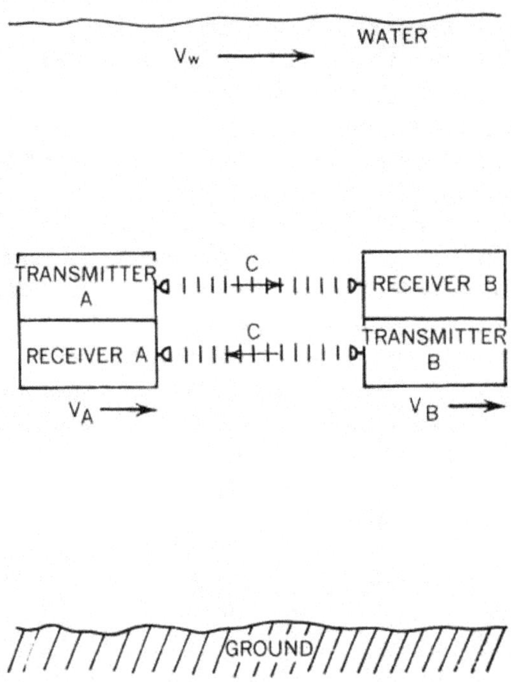

FIGURE 4506a.—Sound transmitters and receivers in a moving medium.

If unit A is attached to a vessel's hull, and a reflector, the ocean bottom, is considered to act in the same manner as unit B, then the vessel's speed over the ground can be derived from the doppler shift, $F_s - F_A$.

Because the ultrasonic beam is not usually directed along the vessel's velocity vector and only a component of velocity along the beam results in a doppler shift, the doppler equation as given must be modified.

Inspection of the Janus arrangement shown in figure 4506b reveals that the components of the vessel's velocity (V_{fa}) and the water's velocity (V_W) in the fore and aft plane which cause doppler shifts are cosine projections along the ultrasonic beam. These components for the forward beam are $V_{fa} \cos \theta$ and $V_W \cos \theta$, respectively.

For forward velocities, the reflected forward beam frequency, F_f, is increased; and the reflected after beam frequency, F_a, is decreased. If there were vertical velocity components, their sine projections along the beams would cause doppler frequency shifts. However, for vertical velocities, the frequencies of both the forward and after beams would be shifted in the same direction and by the same amount. No significant change of the difference frequency would be observed.

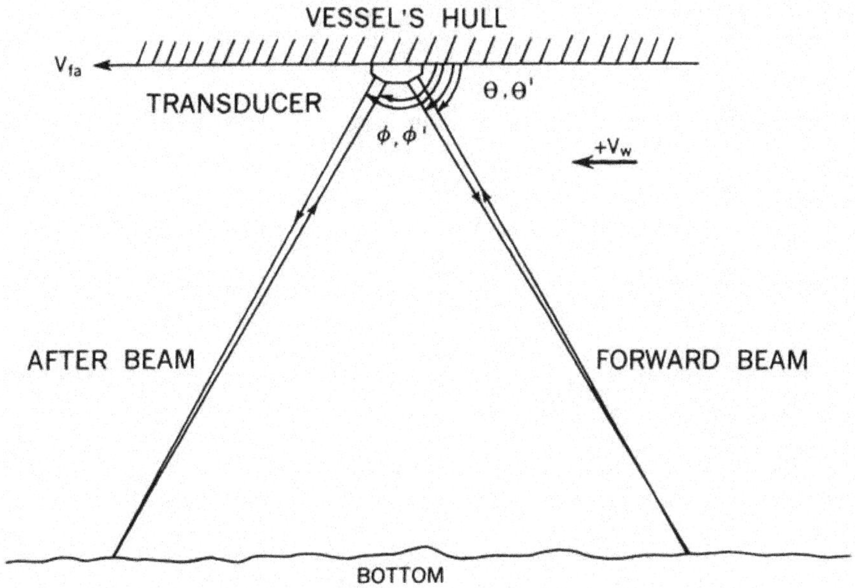

FIGURE 4506b.—Janus fore and aft beams.

If the vessel is pitching, the received signal will not return to the receiver at exactly the same angle at which it was transmitted. Thus, the reflected forward beam frequency is given by

$$F_f = F_s \left[\left(\frac{C + V_W \cos\theta}{C + V_W \cos\theta - V_{fa} \cos\theta} \right) \left(\frac{C - V_W \cos\theta' + V_{fa} \cos\theta'}{C - V_W \cos\theta'} \right) \right],$$

where θ is the beam angle with the vessel's fore and aft axis at transmission and θ' is the beam angle at reception.

The reflected after beam frequency is given by the same equation except that θ and θ' are replaced by ϕ and ϕ', respectively. The fact that the after beam is directed aft is accounted for in the cosine projections, i.e., cos ϕ and cos ϕ' are negative.

$$F_a = F_s \left[\left(\frac{C + V_W \cos\phi}{C + V_W \cos\phi - V_{fa} \cos\phi} \right) \left(\frac{C - V_W \cos\phi' + V_{fa} \cos\phi'}{C - V_W \cos\phi'} \right) \right].$$

Since the doppler shifts of the forward and after beams are in opposite directions, the difference between the frequencies of the two reflected beams is used to determine the vessel's fore and aft speed. If the sum of the frequencies were used, the doppler shifts would be cancelled. The difference frequency or speed signal is given by

$$F_f - F_a = F_s \left[\left(\frac{C + V_W \cos\theta}{C + V_W \cos\theta - V_{fa} \cos\theta} \right) \left(\frac{C - V_W \cos\theta' + V_{fa} \cos\theta'}{C - V_W \cos\theta'} \right) - \left(\frac{C + V_W \cos\phi}{C + V_W \cos\phi - V_{fa} \cos\phi} \right) \left(\frac{C - V_W \cos\phi' + V_{fa} \cos\phi'}{C - V_W \cos\phi'} \right) \right].$$

Several approximations can be made to simplify this equation. Multiplying the two factors in the numerator of the first term gives

$$(C+V_W\cos\theta)(C-V_W\cos\theta'+V_{fa}\cos\theta')=C^2-CV_W\cos\theta'$$
$$+CV_{fa}\cos\theta'+CV_W\cos\theta-V^2_W\cos\theta\cos\theta'+V_{fa}V_W\cos\theta\cos\theta'.$$

Multiplying the two factors in the denominator of the first term by $1/C^2$ gives

$$(C+V_W\cos\theta-V_{fa}\cos\theta)(C-V_W\cos\theta')=C^2-CV_W\cos\theta'$$
$$+CV_W\cos\theta-V^2_W\cos\theta\cos\theta'-CV_{fa}\cos\theta+V_{fa}V_W\cos\theta\cos\theta'.$$

By multiplying both the numerator and the denominator of the first term by $1/C^2$ we obtain:

$$\frac{1/C^2(C^2-CV_W\cos\theta'+CV_{fa}\cos\theta'+CV_W\cos\theta-V^2_W\cos\theta\cos\theta'+V_{fa}\cos\theta\cos\theta')}{1/C^2(C^2+CV_W\cos\theta-CV_{fa}\cos\theta-CV_W\cos\theta'-V^2_W\cos\theta\cos\theta'+V_{fa}V_W\cos\theta\cos\theta')}.$$

Since $C>>V_W$ or V_{fa}, terms containing V^2_W/C^2 and $V_{fa}V_W/C^2$, along with trigonometric functions can be neglected. The first term then becomes:

$$\frac{1/C^2(C^2-CV_W\cos\theta'+CV_{fa}\cos\theta'+CV_W\cos\theta')}{1C^2(C^2+CV_W\cos\theta'--CV_{fa}\cos\theta'-CV_W\cos\theta')}=\frac{C(C+V_{fa}\cos\theta')+CV_W\cos\theta'-\cos\theta')}{C(C-V_{fa}\cos\theta')+CV_W(\cos\theta'-\cos\theta')}$$
$$=\frac{(C+V_{fa}\cos\theta')+V_W(\cos\theta'-\cos\theta')}{(C-V_{fa}\cos\theta')+V_W(\cos\theta'-\cos\theta')}.$$

The second term of the equation can be simplified by similar approximations. The equation can then be written as:

$$F_f-F_a=F_s\left[\left\{\frac{(C+V_{fa}\cos\theta')+V_W(\cos\theta-\cos\theta')}{(C-V_{fa}\cos\theta)+V_W(\cos\theta-\cos\theta')}\right\}\right.$$
$$\left.-\left\{\frac{(C+V_{fa}\cos\phi')}{(C-V_{fa}\cos\phi)}+\frac{V_W(\cos\phi-\cos\phi')}{V_W(\cos\phi-\cos\phi')}\right\}\right].$$

Since the transmitted and received angles are very nearly the same, the cosine difference terms can also be neglected. Then,

$$F_f-F_a=F_s\left[\frac{(C+V_{fa}\cos\theta')}{(C-V_{fa}\cos\theta)}-\frac{(C+V_{fa}\cos\phi')}{(C-V_{fa}\cos\phi)}\right].$$

or

$$F_f-F_a=F_s\left[\left(1+\frac{V_{fa}}{C}\cos\theta'\right)\left(1-\frac{V_{fa}}{C}\cos\theta\right)^{-1}-\left(1+\frac{V_{fa}}{C}\cos\phi'\right)\left(1-\frac{V_{fa}}{C}\cos\phi\right)^{-1}\right].$$

Expanding the factors with the negative exponent in power series and multiplying gives:

$$F_f - F_a = F_s\left[\left(1+\frac{V_{fa}}{C}\cos\theta'\right)\left(1+\frac{V_{fa}}{C}\cos\theta+\frac{V^2_{fa}}{C^2}\cos^2\theta+\frac{V^3_{fa}}{C^3}\cos^3\theta+\ldots\right)\right.$$

$$\left.-\left(1+\frac{V_{fa}}{C}\cos\phi'\right)\left(1+\frac{V_{fa}}{C}\cos\phi+\frac{V^2_{fa}}{C^2}\cos^2\phi+\frac{V^3_{fa}}{C^3}\cos^3\phi+\ldots\right)\right].$$

Multiplying the factors and retaining only first order terms,

$$F_f - F_a = F_s\left[\frac{V_{fa}}{C}(\cos\theta+\cos\theta'-\cos\phi-\cos\phi')\right].$$

Assuming that $\theta \cong \theta'$ and $\phi \cong \phi'$, the cosines are very nearly equal. Then,

$$F_f - F_a = \frac{2F_s V_{fa}}{C}(\cos\theta - \cos\phi).$$

With $\theta = 60°$ and $\phi = 120°$,

$$F_f - F_a = \frac{2F_s}{C} V_{fa},$$

where V_{fa} is the vessel's velocity component in the fore and aft direction. A similar equation applies to the athwartship component:

$$F_{st} - F_p = \frac{2F_s}{C} \times V_{st-p}.$$

where F_{st} is the reflected starboard beam frequency, F_p is the reflected port beam frequency, and V_{st-p} is the athwartship (starboard to port) velocity.

The speed indication, S_i, is given by:

$$S_i = k\sqrt{(F_f - F_a)^2 + (F_{st} - F_p)^2}$$

$$= k\sqrt{\left(\frac{2F_s}{C}V_{fa}\right)^2 + \left(\frac{2F_s}{C}V_{st-p}\right)^2}$$

$$= k\frac{2F_s}{C}\sqrt{V^2_{fa} + V^2_{st-p}}$$

$$S_i = k\frac{2F_s}{C}V.$$

Thus, when the system calibration constant, k, has the value

$$k = \frac{C}{2F_s},$$

the speed indication is equal to the actual speed, V.

The drift angle

$$=\tan^{-1}\left(\frac{V_{st-p}}{V_{fa}}\right)$$

$$=\tan^{-1}\left(\frac{F_{st}-F_p}{F_f-F_a}\right).$$

References

Kritz, Jack and Morton J. Howard. "Channel Navigation and Docking of Supertankers." NAVIGATION: *Journal of the Institute of Navigation,* vol. 16, No. 1, April 1969.

Laster, D. R., and J. W. Fagan. "Exploratory Evaluation of Underwater Doppler Navigators." U. S. Navy Marine Engineering Laboratory Evaluation Report 8/65, October 1965.

CHAPTER XLVI

INTRODUCTION TO INERTIAL NAVIGATION

4601. Introduction.—*Inertial navigation* is the process of measuring a craft's velocity, attitude, and displacement from a known start point through sensing the accelerations acting on it in known directions by means of devices that mechanize Newton's laws of motion (art. 1407). Since these laws are expressed relative to inertial space (the fixed stars), the term "inertial" is applied to the process.

Inertial navigation is described as "self-contained," that is, independent of external aids to navigation. It is also described as "passive" because no energy is emitted to obtain information from an external source. Thus, inertial navigation is fundamentally different from other methods of navigation because it depends only on measurements made within the craft being navigated.

Inertial navigation is often referred to as a dead reckoning method because position is obtained by measuring displacements from a start point in accordance with the motion of the craft.

4602. The basic principle of inertial navigation is the measurement of the accelerations acting on a craft, other than those not associated with its orientation or motion with respect to the earth, and the double integration of these accelerations along known directions to obtain the displacement from the start point.

If the indicated acceleration of the craft from rest is constant, velocity and distance traveled can be found from the equations:

$$v=at,$$

and

$$s=\tfrac{1}{2}\,at^2,$$

where a is the acceleration, v is the velocity, s is the distance, and t is the time. But the acceleration must be constant. The equations cannot be used for varying accelerations unless very small time increments are used, and the increments of velocity or distance are integrated (art. 147, vol. II). The equations, using the calculus notation are

$$v=\int a\,dt$$

$$s=\int v\,dt,$$

where dt denotes a very small increment of time.

4603. Inertial sensors used in the mechanization of Newton's laws of motion, hereafter called the **inertial navigator** or **navigator,** are gyroscopes (art. 630) and accelerometers. The **gyroscopes** sense angular motions of the vessel. The **accelerometers** sense the vessel's linear accelerations which are changes in linear velocity. The inertial sensors are subject to all motions of the vessel in inertial space. Since some of these motions do not change the vessel's position or orientation on the earth and since the attitude and motion of the inertial navigator with respect to the earth is desired, it is necessary to apply certain corrections to the inertial motions (art. 4604) sensed in order to obtain the corresponding earth referenced values.

Although the gyroscope is discussed in some detail in article 630, the following discussion of this inertial sensor is written to meet the needs of an introductory treatment of inertial navigation.

The principle of **conservation of angular momentum** states that a system will maintain its angular momentum if no external forces are applied to it. Thus, the rotor (mass) of a gyroscope must maintain a constant angular momentum about its spin axis if no external forces are applied. Both the amplitude and direction of the angular momentum must be conserved. The spin axis, therefore, tends to maintain the same direction in inertial space. This property is called **gyroscopic inertia** or **rigidity in space.**

If a rotational or couple force is applied to the rotor through the gyroscope's supports, an additional angular momentum is introduced and the gyroscope's orientation will change to include this along with its original angular momentum about the spin axis and the axis of the applied force. This property of the gyroscope is called **precession.**

Precession causes the gyroscope to tend to align its spin axis with the axis of the applied torque (fig. 4603a). If the axis of applied torque is designated as the **input axis,** then the precession of the gyroscope can be determined by rotating the spin axis into the input axis through the smaller angle. Using this nomenclature the axis of precession is often called the **output axis.**

The rigidity of a gyroscope produces a phenomenon called **apparent precession.** This is due to the fact that as the gyroscope maintains its original orientation in space, the earth turns beneath it. As a result, to an observer on the earth, the gyroscope appears to turn or precess when actually it remains fixed in space. This apparent precession is shown in figure 4603b. If it is desired to have the gyroscope maintain an orientation referenced to the earth, it must be precessed (torqued) in such a manner as to remove the apparent precession.

The gyroscope described above is known as a **two-degree-of-freedom gyroscope.** It consists of an axially symmetric rotor supported in a case in such a way that one point on the axis is fixed relative to the case and the spin axis itself is free to assume any angular orientation relative to the case. A gimbal system supports the rotor within the case and provides angular freedom. An example of such an arrangement is shown in figure 4603c.

The spin axis has the ability to move about two axes with respect to the case (the enclosure not shown to which the gimbal system is attached). The gyroscope rotor is free to precess about either of the two supporting gimbal axes depending upon where the force is applied.

For practical application of the gyroscope to an inertial navigator it is more useful to design a gyroscope which is able to precess about only one axis called the output axis. Such a gyroscope will respond (precess) to rotations of the case about an axis known as the input axis. This allows the gyroscope to perform a control function with respect to an axis. With three such gyroscopes, mounted on a platform such that the input axes are mutually perpendicular, three-dimensional attitude control of the platform is obtained. Any component of rotation about the input axis causes a rotation about the output axis and can be sensed and used to control the orientation of the platform on which the gyroscope is mounted.

By attaching the outer gimbal of figure 4603c rigidly to the case such a configuration is obtained. The gyroscope now has just one axis of freedom with respect to the case about which it can precess. This is called a **single-degree-of-freedom gyroscope** (fig. 4603d).

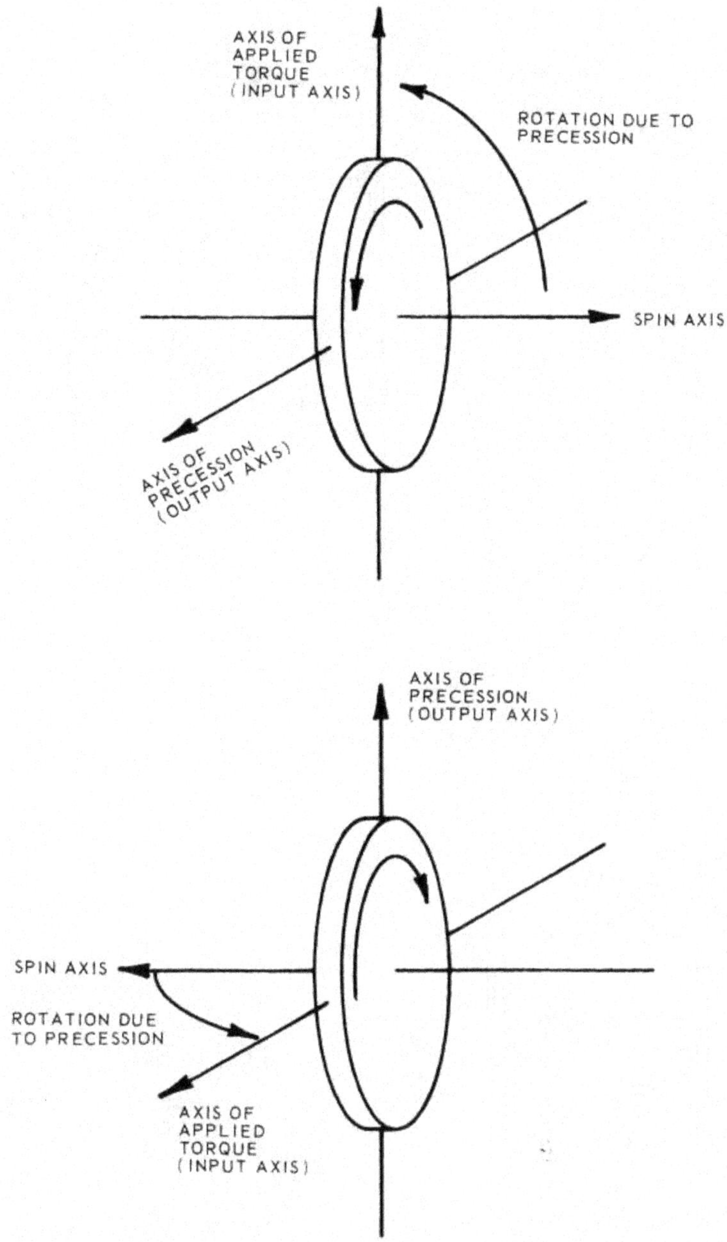

FIGURE 4603a.—Precession of a gyroscope.

Velocity is the linear rate of change of position. If the velocity is known it can be integrated with respect to time to determine the change in position. However, velocity cannot be sensed or measured with an inertial device since there is nothing inertially to distinguish one velocity from another.

A body at rest will remain at rest unless acted upon by an external force, and a body in motion will retain the motion unless acted upon by an external force. By measuring forces acting on a test mass, *changes* in velocity can be detected. This rate of change in velocity, called acceleration, is what accelerometers measure. The

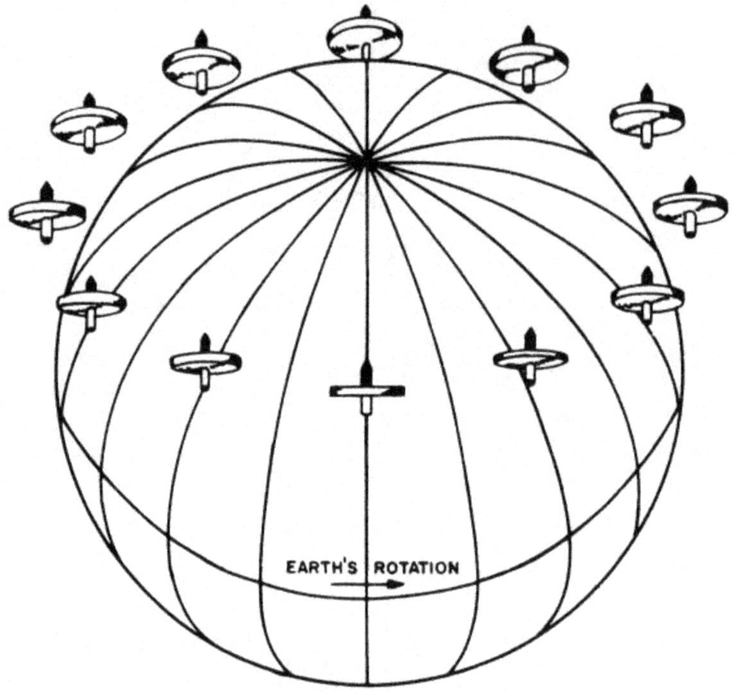

FIGURE 4603b.—Apparent precession.

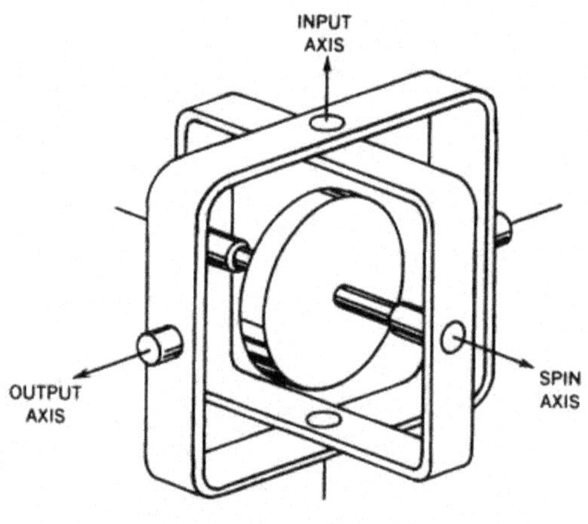

FIGURE 4603c.—Two-degree-of-freedom gyroscope.

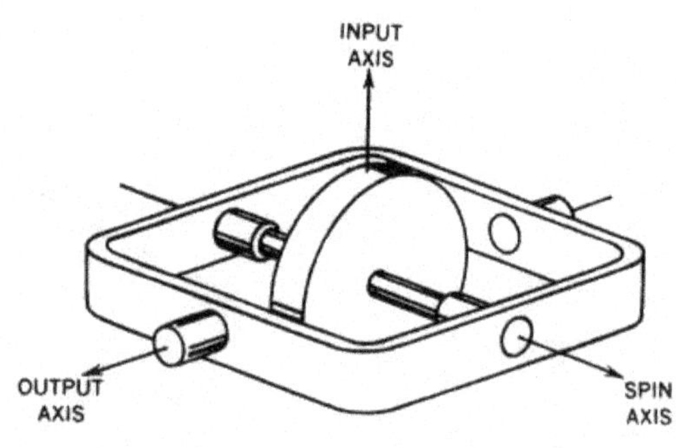

FIGURE 4603d.—Single-degree-of-freedom gyroscope.

measuring device is properly called an accelerometer; but, since the integration is a simple step, the term **velocity meter** often is used.

In its simplest form, the accelerometer consists of a test mass, as shown in figure 4603e, constrained to measure accelerations in a particular direction (the **sensitive axis**) with a scale, or other appropriate device, to indicate its output. If the frame is accelerated to the right (fig. 4603e) the test mass lags behind since the acceleration is applied to the frame, not the test mass. The test mass displaces enough for the constraining springs to apply a force proportional to the acceleration. The test mass then moves with the case maintaining its constant displacement. When the acceleration is removed from the frame, the constraining springs cause the test mass to move (with respect to the case) back to the neutral position. Thus, a body at rest or a body at constant velocity (zero acceleration) causes no displacement, providing the accelerometer is held horizontal. If the accelerometer is tilted or placed on end, the force of gravity causes the mass to move in the same way as does an actual acceleration, even though the frame is at rest.

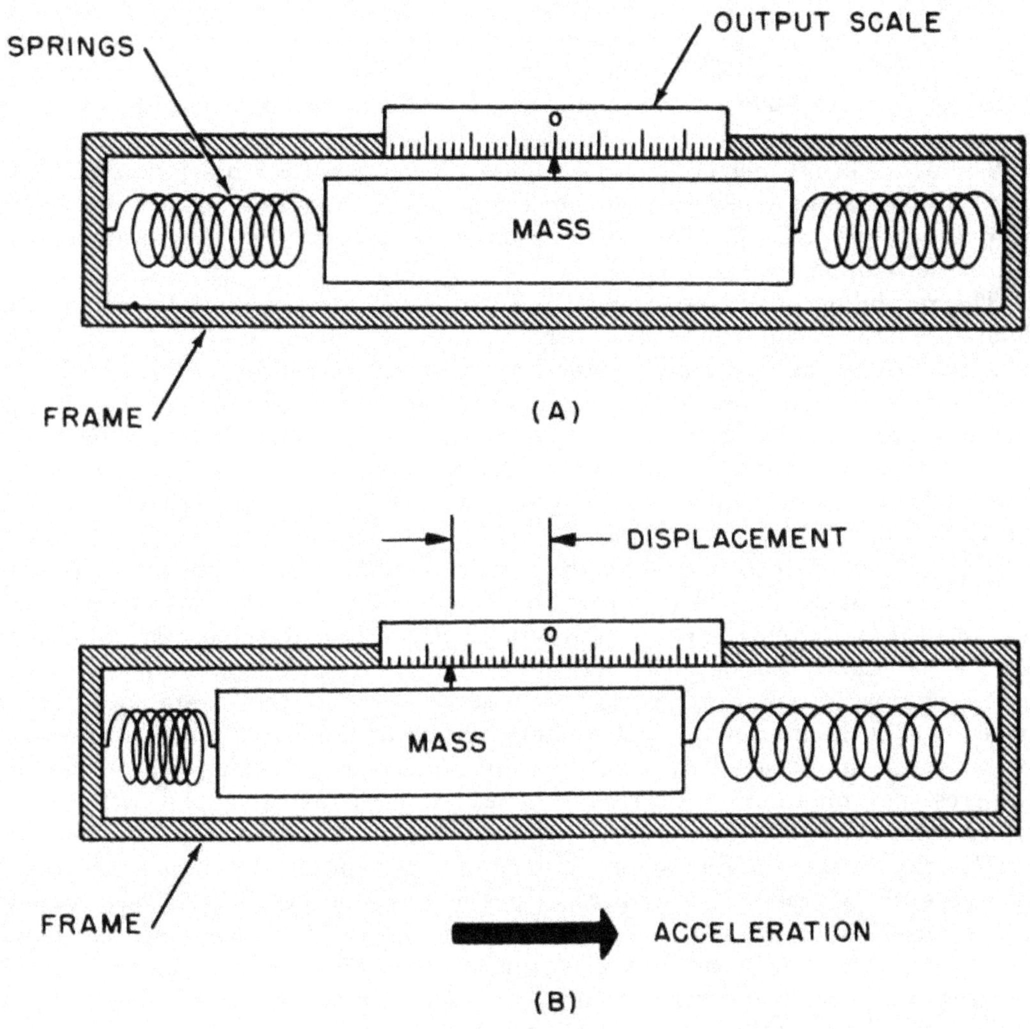

FIGURE 4603e.—Basic accelerometer.

This basic accelerometer demonstrates the principle of operation of inertial accelerometers. It must be kept in mind that these inertial accelerometers are sensitive to more than just the accelerations with respect to the earth. Since they are sensitive to accelerations in space, their output includes other inertial accelerations which are not due to travel over the earth's surface. A compensation must be made for these inertial accelerations so that the quantity left is the acceleration with respect to the earth. What these inertial accelerations are and how a compensation is made is discussed in article 4604.

4604. Inertial motions.—The motions affecting the inertial sensors may be divided into two categories: rotations and accelerations. The rotations are:

1. craft's roll, pitch, and yaw,

2. earth's rotation (earth rate),
3. changes in latitude and longitude.

The accelerations of concern are:

1. craft's acceleration with respect to the earth,
2. acceleration of gravity,
3. Coriolis acceleration.

Some of these motions are interrelated. For example, Coriolis acceleration results from the rotation of the earth. The rotation of the earth has an affect on gravity. There can be no latitude or longitude changes without acceleration with respect to the earth.

There are some inertial motions whose effects are negligible; that is, their effects are below the sensitivity level of the sensors. These motions are precession and nutation (art. 1419) and the acceleration of the earth in its orbit in accordance with Kepler's second law (art. 1407).

The revolution of the earth about the sun is included in the total earth rate, which equals the vector sum of the earth's rotation and revolution.

Since the inertial navigator deals with the earth referenced values of velocity, attitude, and position, and, since the gyroscopes maintain direction with respect to inertial space, it is necessary that the gyroscopes be controlled to maintain a reference with respect to the earth. In the discussion of the inertial navigator which follows, the earth reference used is the local vertical and an orientation with respect to true north.

The rotation of the earth causes the local vertical for a given position to change its direction in space. This change is not obvious to anyone on the earth because the local vertical maintains the same orientation with respect to the earth. To prevent the local vertical indication of the inertial navigator from being stationary in space due to rigidity in space (apparent precession), an earth rate torquing signal is applied so that the inertial navigator rotates about the earth's spin axis at the same rate that the earth does. As a result, the inertial navigator, which is controlled by the gyroscopes, maintains the desired orientation with respect to the earth as the earth rotates in inertial space.

The earth rate torquing signal is only one part of the total torquing signal going to each gyroscope. The change in position of the inertial navigator on the earth's surface also causes the local vertical to change direction in space. This is due to the fact that in going from one position to another the inertial navigator is changing from one local vertical to another. This is demonstrated in figure 4604a. As the inertial navigator travels over the earth's curved surface from position 1 to position 2, the "correct" orientation is shown by the solid line figure at position 2. The broken line figure represents the inertial navigator after the change in position without compensation for the change in the local vertical.

Each gyroscope must also be torqued because of its own internal drift with respect to inertial space. This precession can be caused by such factors as internal torques produced by friction and mass imbalance. The compensation applied for this drift is called **gyro bias.**

As shown in figure 4604a, the total angular change in the local vertical in moving from position 1 to position 2 is represented by θ. The value of θ, expressed in radians, is a function of the distance traveled, S, and the radius of the earth, R. Stated mathematically, this is:

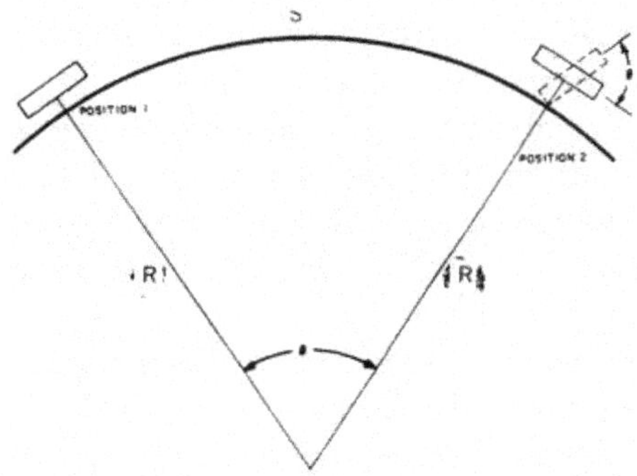

FIGURE 4604a.—Change in local vertical with movement over earth's surface.

$$\theta = \frac{S}{R}.$$

However, the quantity of interest in maintaining the correct orientation to the local vertical due to change in position is the rate of change of θ with respect to time. Assuming R to be constant,

$$\frac{d\theta}{dt} = \frac{1}{R}\frac{dS}{dt}.$$

Since the time rate of change of distance, S, is velocity, V,

$$\frac{d\theta}{dt} = \frac{V}{R}.$$

Thus, the appropriate gyroscopes must be torqued at this rate so that the vertical indication of the inertial navigator will remain correct as the craft moves over the curved surface of the earth. When the gyroscopes receive a properly calibrated signal to compensate for the change in the local vertical due to movement over the earth, the system is said to be *Schuler tuned.*

If the accelerometer is tilted, it will be affected by gravity. The output which would be produced would not be due to any acceleration of the inertial navigator with respect to the earth and so a compensation must be made. The simplest solution for elimination of the effect of gravity is to place the accelerometer so that its sensitive axis is perpendicular to gravity. Since all accelerations which result in a change in position are perpendicular to the local vertical, this orientation allows the accelerometer to measure these accelerations without being affected by gravity. Two accelerometers, then, are placed perpendicular to the local vertical and perpendicular to each other so that together they can measure any acceleration which will result in a change of position on the earth's surface. A third accelerometer is necessary to measure any accelerations of the navigator along the local vertical. This one also senses the total magnitude of gravity acceleration which is compensated for in the computer of the inertial navigator.

Another acceleration occurring in inertial space which is not an acceleration with respect to the earth is **Coriolis acceleration.** Since the inertially oriented accelerometers produce outputs due to Coriolis acceleration, a compensation must be made. This is not a difficult problem because the Coriolis acceleration can be calculated and therefore removed from the output of an accelerometer before that output is used to provide inertial navigator values.

A body experiences Coriolis acceleration when it travels over a rotating surface in a direction perpendicular to the axis of rotation. The acceleration is not within the reference frame of the rotating body, but within the reference frame in which the body is rotating. A merry-go-round can be used to illustrate the often misunderstood Coriolis acceleration phenomenon.

If as shown in figure 4604b a ticket-taker starts from the center with a velocity, v, toward horse A, he will reach the horse at some finite time, t, later. With respect to the merry-go-round he traveled at a constant velocity and therefore did not accelerate. However, when using the ground below the merry-go-round as the reference in which to make the measurements of his motion, it can be seen that in the time it takes for the man to get to the horse, the horse has moved to position A'. This is due to the rotation, ω, of the merry-go-round. The path of the man with respect to the ground is shown as the solid curve even though his velocity with respect to the merry-go-round is directed radially at all times as shown by the two representative vectors along the path.

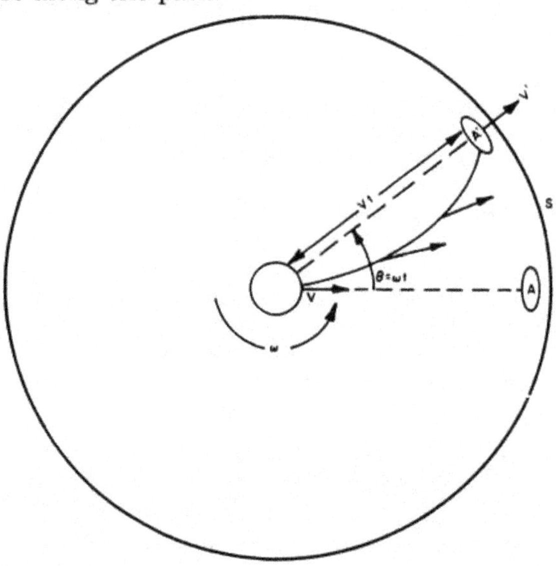

FIGURE 4604b.—Motion on a merry-go-round.

The Coriolis acceleration which causes the path over the ground to curve is determined as follows:

While going from the center to the horse the man was moved (with respect to the ground) counterclockwise, a distance S. This distance, written in terms of the acceleration which produced it, is:

$$S=1/2\ at^2.$$

The distance may also be expressed in terms of the angle θ (which equals the angular rate, ω, of the merry-go-round multiplied by the time, t, involved) and the radius (which equals the velocity, v, of the man with respect to the merry-go-round multiplied by the time, t, involved). This expression is:

$$S=\theta vt=\omega tvt=\omega vt^2.$$

By equating these two expressions and solving for a,

$$1/2\ at^2 = \omega v t^2,$$
$$a = 2v\omega.$$

In the case of the rotating earth in an inertial frame of reference, Coriolis is an apparent effect acting on a mass in motion. The principle of conservation of angular momentum states that a mass in motion will maintain its angular momentum if no external forces are applied. As a body in motion travels northerly in the Northern Hemisphere the radius of rotation of earth decreases with latitude causing the linear velocity of earth to slow down in an easterly direction while the angular momentum of the body is maintained. Hence, the earth's coordinate frame lags relative to the position of the moving body in inertial space causing the effect of the body to be deflected to the right. In the case of southerly motion of a body in the Northern Hemisphere, the radius of rotation of earth increases causing the earth's coordinate frame to speed up in an easterly direction relative to the position of the body resulting in the appearance of the body to lag the coordinate frame and be deflected to the right. The opposite effect occurs in the Southern Hemisphere causing the body in motion to appear to be deflected to the left. The change in the earth's radius of rotation is proportional to the convergence of the meridians, or the sine of the latitude. This apparent effect was first discovered by Gaspard Gustave de Coriolis shortly before the middle of the nineteenth century for whom it was named. It was given quantitative formulation by Ferrel. The general equation is:

$$D = 2\Omega v \sin lat.$$

where D is Coriolis displacement, $\Omega = 2\pi/86{,}164$ seconds in a sidereal day = 0.0000729 radians/sec,1852 meters = 1 knot, g = 9.8 meters/sec^2, sin *lat.* = convergence, 57.3×60 = conversion from radians to minutes, hence nautical miles.

$$D = \frac{2 \times 0.0000729 \times 1852\ v \sin lat.}{9.8 \times 3600} = 0.00000765 \text{ radians } v \sin lat.$$

$$0.00000765 \times 57.3 \times 60 = 0.0263\ v \sin lat. \text{ per knot} = 0.263\ v \sin lat. \text{ per 10 knots.}$$

4605. Stable platform.—The function of the accelerometers in the inertial navigator is to sense the accelerations of the craft with respect to the earth so that a **computer** can determine the craft's velocity, attitude, and position. The function of the gyroscopes is to maintain the accelerometers in the correct orientation to measure these accelerations. The function of the gimbals is to provide the physical support for the gyroscopes and accelerometers, and provide three degrees of freedom with respect to the craft. The gimbals, gyroscopes, and accelerometers together with associated electronics and gimbal torquer motors form a **stable platform.**

The function of the stable platform is to establish and maintain a reference system in which the measurements necessary to produce the navigator outputs are taken. The reference system most commonly used is based upon the local vertical and the direction of north. Both of these are referenced to the earth. As a result, "stable" in stable platform means fixed to or stable with respect to the earth. This means that any motion or force which might disturb the reference system must be accounted for and its effect must be eliminated. This would include motions in inertial space (art. 4604) as well as motions of the craft in which the navigator is mounted.

Since the craft in which the stable platform is mounted operates in three dimensional space, it has three degrees of freedom with respect to the earth. In order to maintain a reference fixed to the earth, the stable platform must have three degrees of freedom with respect to the craft. The stable platform contains three gimbals, one for each degree of freedom necessary for the stable platform. Each gimbal has rotational freedom about one axis with respect to its supporting element. The gimbal can be driven with respect to its support about the gimbal axis by a gimbal torquing motor.

In the configuration shown in figure 4605a, the gimbal which is supported by the craft is called the pitch gimbal. The pitch gimbal in turn supports the roll gimbal on an axis perpendicular to the pitch axis. The roll gimbal supports the heading or azimuth gimbal. The heading gimbal has one degree of freedom with respect to the roll gimbal, and rotational movement about the heading gimbal axis with respect to the roll gimbal is a change in heading. The heading gimbal through its own axis and the axes of the roll and pitch gimbals has three-degree or complete rotational freedom with respect to the craft in which the platform is mounted. The heading gimbal is able to assume any orientation with respect to the craft within the limits of the gimbal physical restraints.

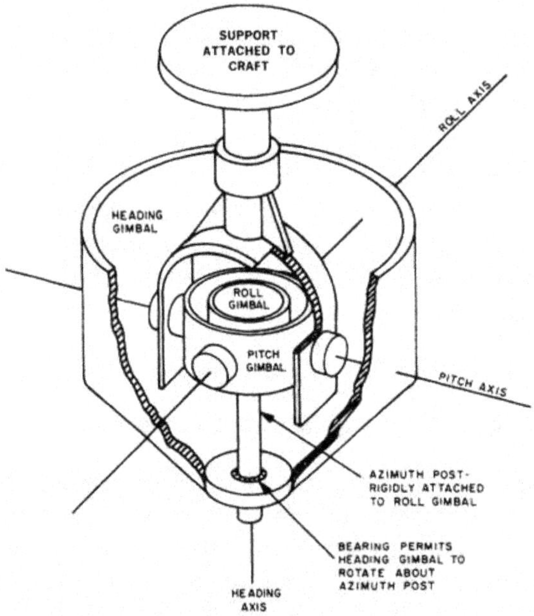

FIGURE 4605a.—Stable platform.

The heading gimbal, then, is the reference system which the stable platform maintains in some desired orientation with respect to the earth. Mounted on the heading gimbal are the accelerometers whose sensitive axes must be maintained within the desired reference system.

The following discussion of a single-degree-of-freedom stable platform (useful for training purposes only) is presented as an aid to understanding how the stable platform functions to maintain the reference system.

As shown in figure 4605b, the single-degree-of-freedom stable platform consists of a single-degree-of-freedom gyroscope, an accelerometer with integrator (velocity meter), a gimbal with rotational freedom about one axis with respect to the supporting craft, and a torquer motor which drives the gimbal with respect to the craft about the gimbal axis. The gyroscope input axis is aligned to the gimbal axis, and the input axis of the accelerometer is aligned perpendicular to the gimbal axis.

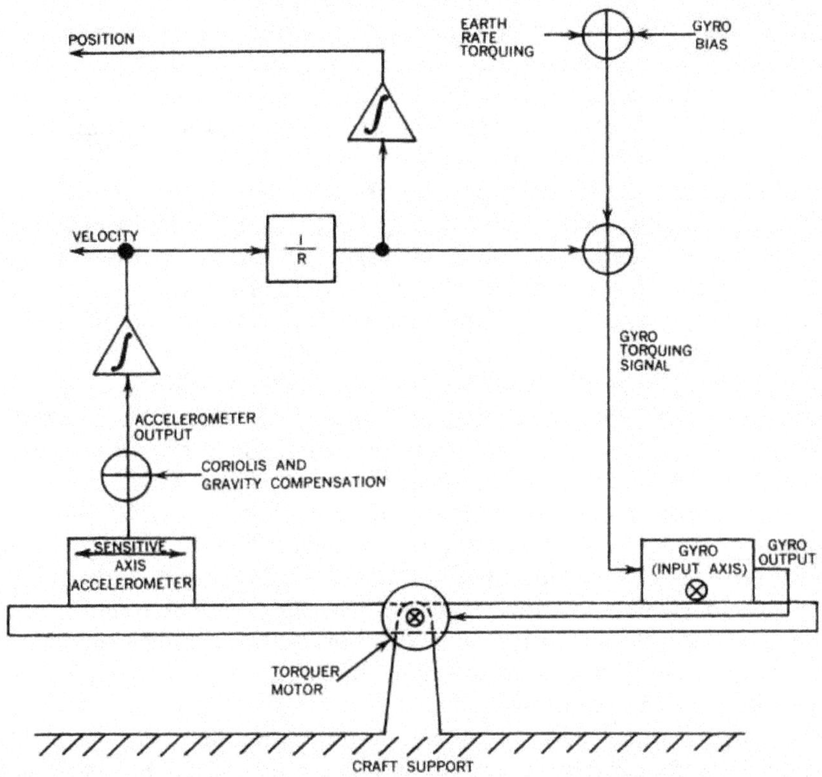

FIGURE 4605b.—Single-degree-of-freedom stable platform used for training purposes only.

As discussed in article 4604, there are several motions which can disturb the orientation of the heading gimbal. These are divided into rotations and linear accelerations. Some of these motions do not change the craft's position or orientation on the earth. The single-degree-of-freedom platform can be used to examine each motion.

If the supporting craft should undergo any base motion (roll, pitch, or yaw) about the gimbal axis, the platform through friction in the gimbal axis would also start to rotate about the axis with respect to the earth. The rotation rate of the gimbal with respect to the earth would not be nearly so great as the rotation rate of the craft about this axis because the axis is made as frictionless as possible. There could, however, be some displacement of the gimbal. The gyroscope would sense this instantaneously since its input axis is aligned with the axis of the platform. This disturbance about the gyroscope's input axis would cause an output from the gyroscope. This output would go to and excite the gimbal torquing motor which in turn would drive the gimbal with respect to the craft about the gimbal axis.

The gimbal, gyroscope, and torquer motor form a **closed system** or **closed loop.** The gyroscope senses any rotation of the platform with respect to the earth, and the gyroscope output drives the gimbal through the torquer motor at the exact rate necessary to maintain its orientation with respect to the earth. If the rate of platform disturbance increases or decreases, the gyroscope output signal increases or decreases to keep the gimbal in the same position relative to the earth. This keeps the accelerometer in the original position because the reaction of the gimbal, gyroscope, and torquer motor loop is instantaneous and linear.

The rotation of the earth (earth rate) is another motion that tends to disturb the heading gimbal. Figure 4605b shows the earth rate torquing signal as a part of the total gyroscope torquing signal. The value of this signal is a function of the orientation of the gyroscope's input axis with respect to the earth's axis of rotation. The effect of the torquing signal is to rotate the platform about its axis in space at the same rate that the earth is rotating about this axis so that the platform maintains its orientation with respect to the earth.

The single-degree-of-freedom stable platform is also compensated for any disturbing effects of Coriolis acceleration and gravity acceleration. As shown in figure 4605b, the usual procedure is to make a correction to the output of the velocity meter in the computer of the inertial navigator. (If these errors are not compensated, they become part of the Schuler tuning gyroscope torquing signal and cause this signal to be in error.)

As indicated in article 4604, the purpose of **Schuler tuning** is to insure that the vertical indication of the inertial navigator remains correct as the craft moves over the curved surface of the earth. Two Schuler tuned loops are needed to maintain the desired orientation of the local vertical. One, called the *X* Schuler loop, consists of the *X* gyroscope and *Y* velocity meter, and controls the vertical indication about the input axis of the *X* gyroscope which is positioned in the horizontal plane normally in a north-south alignment. The other, the *Y* Schuler loop, consists of the *Y* gyroscope and *X* velocity meter, and controls the vertical indication about the input axis of the *Y* gyroscope which is in the horizontal plane perpendicular to the *X* gyroscope input axis. Since these Schuler loops are identical in operation, only the *X* Schuler loop will be examined.

Figure 4605c is a block diagram of the *X* Schuler loop. This figure shows only those signals primarily concerned with that characteristic of the inertial navigator called Schuler tuning. Other signals such as earth rate, gimbal movements due to vehicle base motions, gyro bias and drift, and Coriolis acceleration are not represented. The inputs to the accelerometer are the acceleration of the vehicle in the *Y* direction (*AY*) and the gravity acceleration (*g·MX*) caused by a tilt about the *X* axis (*MX*). The actual value of gravity acceleration is *g* sin *MX* but *MX* is small enough to write this as *g·MX*. The inputs to the gyroscope are the accelerometer output scaled to be the Schuler gyro torquing signal and the rotation in space about the gyroscope's input axis due to the craft's movement over the surface of the earth.

Without tilt, no gravity acceleration is sensed by the accelerometer. The only input is an acceleration of the craft (*AY*) which produces the integrated output, velocity in the *Y* direction (*VY*). This is Schuler tuned by using the effective earth's radius as the scaling factor, and the signal that results (ρ_{XC}) is the computed value of angular rate of change of the vertical indication about the *X* axis. This should match the actual rate represented by ρ_X. When this happens there is no gyroscope output and therefore no tilt. Hence the system maintains its correct vertical indication and is properly Schuler tuned.

With the system Schuler tuned, if a tilt arises, the accelerometer has an output due to gravity (*g·MX*) which when integrated with time is represented as *VY*. This value is also scaled in the Schuler loop and becomes part of the gyro torquing signal. In fact, *the accelerometer is incapable of distinguishing between craft accelerations and gravity accelerations due to tilt.* This error in the gyro torquing signal causes the platform to turn at a rate different from its space angular rate due to

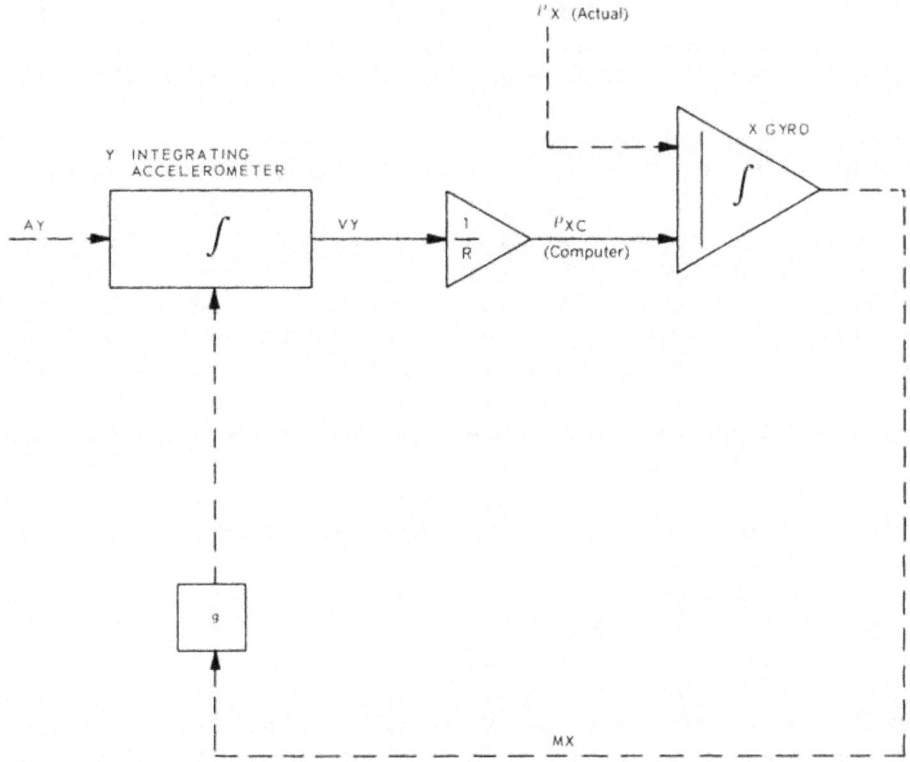

FIGURE 4605c.—Block diagram of Schuler tuned single-degree-of-freedom stable platform.

change in position. Fortunately, this error always drives the platform toward the proper vertical indication, decreasing *MX*.

However, as the tilt is decreasing, the gyroscope torquing signal causing the tilt to disappear is increasing because it is the integral of the tilt. As a result when $MX=0$, ρ_{XC} is a maximum and the platform continues to drive about the *X* axis causing a tilt to build up in the opposite sense. This $g\cdot(-MX)$ is integrated with time and causes the gyroscope torquing signal to decrease as the tilt builds up. The gyroscope torquing signal will be zero when the tilt has reached the original value with opposite sense. This tilt in turn causes the platform to drive back toward the vertical, and the process repeats itself in a mirror image. The process will continue to repeat itself, oscillating back and forth about the correct vertical indication.

This oscillation about the vertical is called **Schuler oscillation.** It has the characteristics of a pendulum whose length is equal to the earth's radius. This Schuler pendulum (art. 635) has a period of oscillation of about 84.4 minutes.

In the presence of such an oscillation, the platform would never settle to the vertical. Such a platform would yield erroneous values of ground velocity, and therefore, erroneous values of latitude, longitude, and heading. All of the data generated from the vertical loop would be varying about the true values by virtue of the oscillation. Provisions must be made for "damping" or settling this oscillation.

4606. Damping refers to the process of reducing the amplitude of the oscillation in the vertical. To reduce the amplitude of the oscillation, the velocity caused by the gravity acceleration must be caused to come to zero at an angle θ which is smaller each oscillation. After several oscillations, the angle θ and the gravity induced velocity are made to be zero at the same instant. The amount of velocity (due to tilt) is determined by the angle of tilt and the amount of time the platform stays off the vertical. The maximum angle of tilt is that angle at which the signal from the accelerometer begins to change in response to the gravity acceleration, and thereby starts the platform back to the vertical. The maximum rate of change of tilt occurs as the platform passes through the vertical. Since erection to the vertical is desired, maximum damping must occur as the tilt angle passes through zero. A change in the vertical loop gain such that the gain is minimum as the tilt angle passes through zero, and maximum when the tilt is maximum, yields the desired damping. A method of detecting the presence and magnitude of tilt, and of using a signal proportional to the *rate of change* of tilt to vary the gain of the vertical loop is necessary, then, to damp the vertical.

To detect the presence of rate of change of tilt, the change in velocity output of the accelerometer due to actual change in velocity of the craft must be distinguished from an output change due to gravity. To accomplish this differentiation, a secondary source of craft velocity unrelated to the accelerometers, and unaffected by gravity, must be used. By subtracting this secondary reference velocity from the accelerometer output, the actual craft velocity is cancelled out (if the secondary reference is accurate). The difference is the rate of change of tilt. This signal is maximum as θ passes through zero.

Since maximum damping is desired as θ passes through zero, a portion of this signal can be used for damping the vertical loop. It has the desired characteristic of being maximum in respect to the velocity signal when the platform is being driven through the vertical and minimum when the platform is a maximum tilt. Figure 4606 shows the block diagram of a damped vertical loop. The secondary velocity signal (VRL) from the electromagnetic log is resolved into the north and east components and subtracted from the inertial velocity. The velocity difference, ΔV, is multiplied by the damping constant (K) and applied as an input to the first integrator. This signal (ΔV) should represent the integral of the acceleration due to platform tilt, i.e., platform tilt rate. It is greatest when the platform crosses the vertical. Fed back in such a way as to subtract from and thereby reduce the velocity signal, it will settle or damp the vertical by reducing the amplitude of the maxima of the velocity oscillations.

Any velocity difference (ΔV) (inertial velocity in respect to secondary velocity) appears as a damping signal. A premium is placed on the secondary velocity source in this system because inaccurate response of the secondary velocity is interpreted by the damping circuitry as a tilt. When the reference velocity is noisy or inaccurate, a *forced error* is caused to appear in the vertical due to the damping circuitry. To prevent (or minimize) these errors, the value of K must be kept low (much less than unity), the secondary velocity source must be of good quality, *or* the damping circuit must be disabled during periods of error in the secondary source.

The electromagnetic log (art. 613) is subject to several errors. It does not measure the actual speed over the ground; in the presence of a current, there is a constant difference between the two velocities. The EM log lags the accelerometer in response to a change in velocity. The EM log is not usable in turbulent or shallow water, and its velocity components vary irregularly during ship maneuvers.

Figure 4606 shows a vertical loop which settles with a constant small tilt in the vertical when there is a constant ΔV. The constant tilt gives rise to an accelerometer output just equal and opposite to the product of ΔV and the damping constant. There is no error in the inertial velocity signal, however, and transient vertical

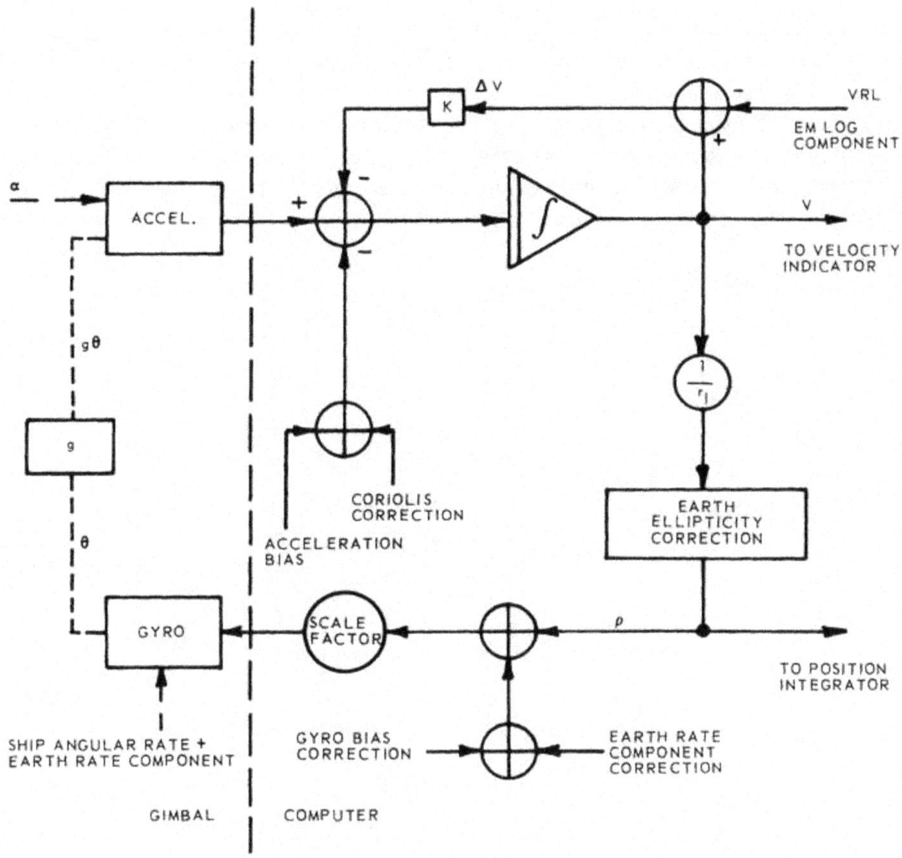

FIGURE 4606.—Block diagram of a damped vertical loop.

disturbances are settled out. Constant electromagnetic log errors can be compensated by a technique known as "third-order damping."

4607. Stabilization loop.—When going from the single-degree-of-freedom to the three-degree-of-freedom stable platform of a practical inertial navigator, provision must be made for the fact that a gimbal is rotated about a gyroscope's input axis in response to an output from that gyroscope, and the output from a particular gyroscope is not always routed to the same gimbal torquer in the stable platform. The function of the *stabilization loop* is to provide the means for the gyroscope outputs to be routed to the proper gimbal torquers.

Three gyroscopes are mounted on the stable platform of the inertial navigator so that the input axes of the *X* and *Y* gyroscopes are perpendicular to each other and in the horizontal plane. The input axis of the *Z* gyroscope is perpendicular to the input axes of both the *X* and *Y* gyroscopes and therefore in the vertical direction. By reference to figure 4605a it can be seen that with the gyroscopes on the heading gimbal (stable element):

1. The heading gimbal axis is always aligned with the input axis of the *Z* gyroscope.

2. The roll gimbal axis is aligned with the input axes of the *X* and *Y* gyroscopes as a function of heading.

3. The pitch gimbal axis is aligned with the input axes of the X and Y gyroscopes as a function of heading and roll, and with the input axis of the Z gyroscope as a function of roll.

Figure 4607a shows the resolution of the gyroscope output signals with heading and roll resolvers mounted on the inertial navigator gimbal system. This stabilization loop is simply a coordinate conversion unit converting signals from the coordinate system of the gyroscopes into the coordinate system of the gimbal axes.

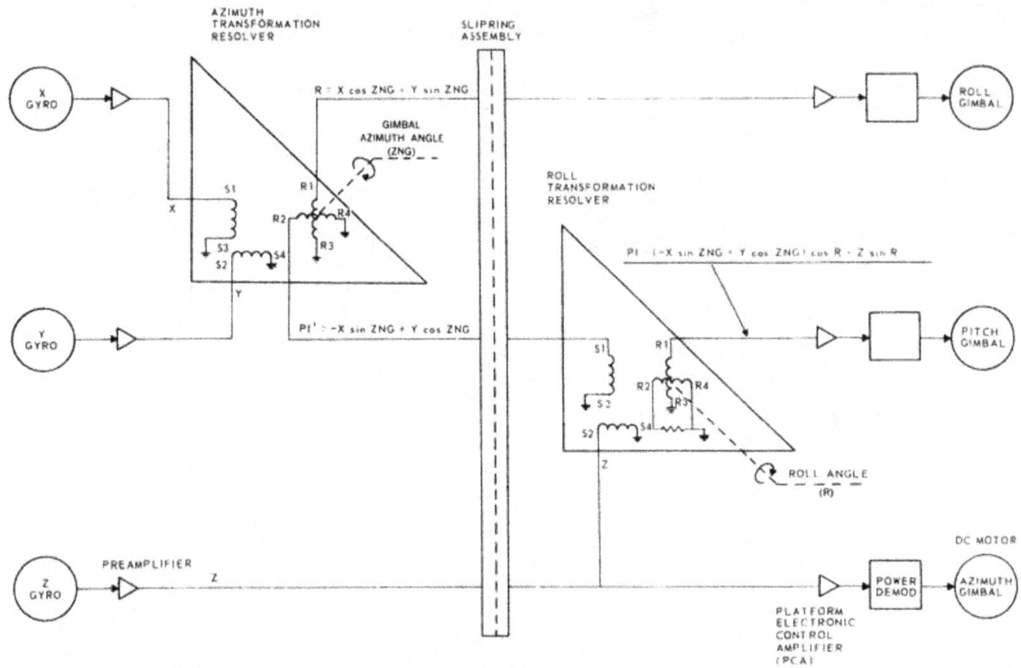

FIGURE 4607a.—Stabilization loop.

If the craft is heading true north, the gyroscopes are oriented on the stable platform as shown in figure 4607b. In this alignment any roll motion of the craft is sensed by the X gyroscope causing it to precess. The precession causes an output signal to appear across the pickoff (fig. 4607a). This output signal is amplified and applied to the $S1$–$S3$ stator winding of the azimuth transformer resolver. The rotor of the resolver is controlled by the azimuth gimbal and changes position in relation to the stator windings as the craft's heading changes.

The relative alignment of the windings for a true north heading is shown in figure 4607a. The signal across the $S1$–$S3$ stator winding of the azimuth transformation resolver produces a maximum output on rotor winding $R1$–$R3$. The signal passes through the slip ring assembly to the navigation console of this practical navigator. The signal is then amplified and modulated. The resulting DC signal to the roll DC gimbal torque motor drives the platform in the direction necessary to restore the gyroscope output to zero. Thus, the platform is stabilized about the craft's roll axis.

The Y gyroscope senses motion about the pitch axis and causes an amplified signal to be applied to the $S2$–$S4$ stator winding of the azimuth transformation resolver. With a heading of true north this causes a maximum signal at rotor winding $R2$–$R4$. The signal passes through the slip rings and is applied to the $S1$–$S3$ stator winding of the roll transformation resolver. This resolver is mounted between

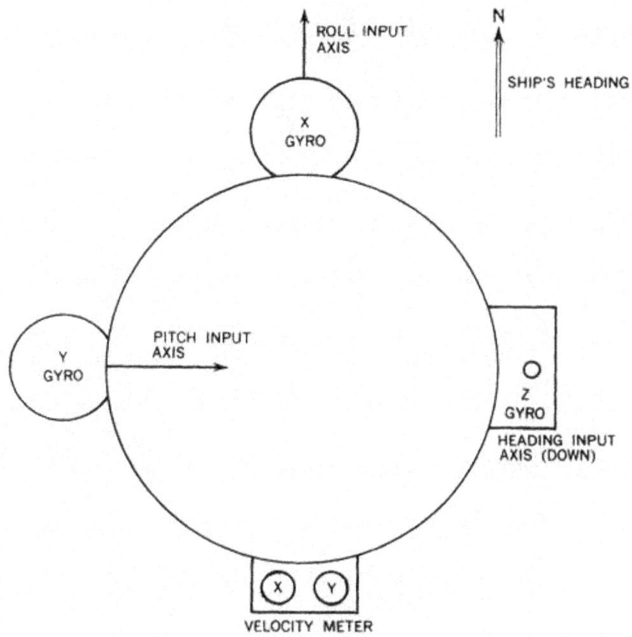

FIGURE 4607b.—Stable platform configuration and input axis for true north heading.

the roll and pitch gimbal and its rotor is positioned relative to the stator by the roll gimbal shaft. The output of the resolver is present at the *R*1-*R*3 winding of the rotor and is applied to the platform control amplifier and power demodulator. The pitch power demodulator DC output signal drives the pitch gimbal motor, which drives the gimbal in the correct direction to cancel the *Y* gyroscope output. Rotor winding *R*2-*R*4 of the roll transformation resolver is short circuited because its output is not used in the stabilization system.

Any change in heading introduces a torque about the input axis of the *Z* gyroscope and causes it to precess. The precession generates an output signal which is amplified by the azimuth gyroscope pickoff amplifier. The signal passes through the slip rings and branches off. One branch is applied to the *S*2-*S*4 stator winding of the roll transformation resolver. The other branch goes directly to the platform electronic control amplifier where the function is identical to that of the roll and pitch sections. The azimuth gimbal motor drives the azimuth gimbal in a direction to decrease the torque about the *Z* gyroscope. The orientation of the platform, therefore, remains unchanged in azimuth as the craft changes heading or experiences yaw.

If the craft changes heading to 090°, the motion sensed by the *Z* gyroscope drives the azimuth gimbal in the correct direction to cancel the torque. As a result the platform is oriented as shown in figure 4607c. The *X* gyroscope now senses pitch motions of the ship and the *Y* gyroscope senses roll motion. The operation of the stabilization loop, however, is not affected since the change in heading resulted in the azimuth gimbal being driven 90° in the opposite direction. The rotor of the azimuth transformation resolver is controlled by the azimuth gimbal shaft, and therefore, has rotated 90° with the change in craft's heading. The signal from the *X* gyroscope, applied to stator winding *S*1-*S*3, is induced on winding *R*2-*R*4 of the rotor and is used to drive the pitch gimbal. The signal from the *Y* gyroscope,

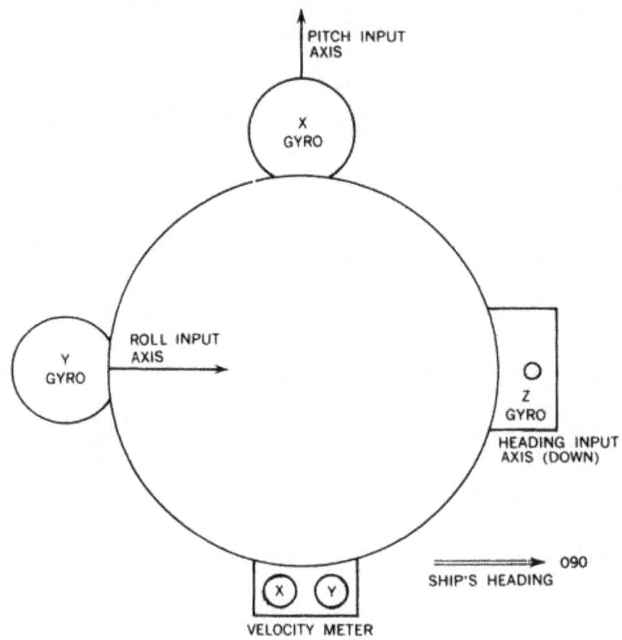

FIGURE 4607c.—Stable platform configuration and input axis for heading of 090°.

applied to stator *S*2–*S*4, induces a maximum signal on rotor winding *R*1–*R*3 and drives the roll gimbal.

At headings between 000° and 090°, outputs from the *X* or *Y* gyroscope result in a proportional mixture of the output signals applied to the pitch and roll gimbals. For example, at heading 045° an output signal from either gyroscope pickoffs will be applied equally to the roll and pitch gimbal drive motors. Thus, the azimuth transformation resolver causes the pickoff signal to drive the correct gimbal regardless of the heading. Similarly, if the craft should experience pitching motion while at some angle of roll, the pitch gimbal axis will not be in the level plane. The pitching motion will have a vertical component and the rotation of the stable platform about the pitch axis will be sensed not only by the *X* and *Y* gyroscopes, but also partly by the *Z* gyroscope. This results in a pickoff from the *Z* gyroscope being routed in the proper proportion through the roll transformation resolver to the pitch gimbal drive motor. The contributions from the level *X* and *Y* gyroscopes will be reduced appropriately by the same resolver.

4608. Coordinate systems.—Two orthogonal coordinate systems are used to relate the inherent errors (art. 4609) to the inertial navigator. One system consists of *X*, *Y*, and *Z* axes. The *X* and *Y* axes define the horizontal plane with the *X* axis directed northward and the *Y* axis directed eastward. The *Z* axis is directed downward along the local vertical. The relationship of this coordinate system to the second system, the equatorial coordinate system, is shown in figure 4608. The equatorial system consists of *P* (polar), *Q* (equatorial), and *E* (east) axes. The *P* axis is parallel to the earth's axis of rotation. The *Q* axis is parallel to the equatorial plane and is directed outward from the earth's axis of rotation. The *E* axis is coincident with the *Y* axis of the other coordinate system.

The first coordinate system is physically represented in the gyroscope arrangement on the stable platform of the inertial navigator. The computer of the inertial

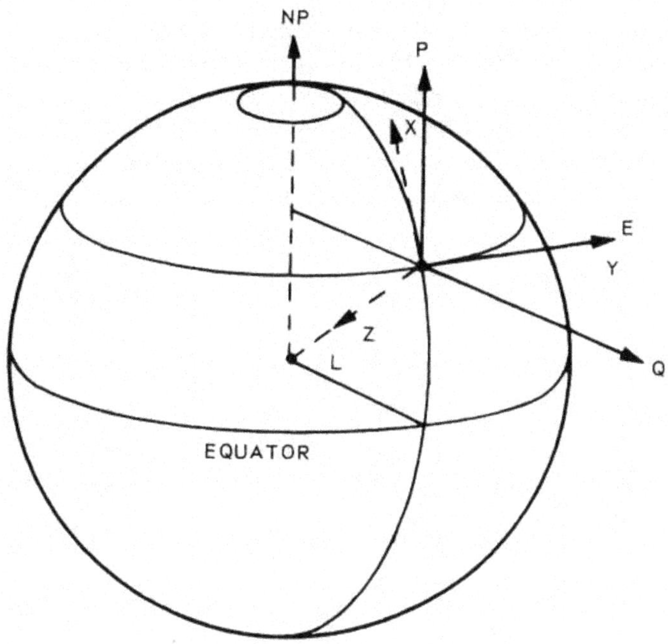

FIGURE 4608.—Orthogonal coordinate systems.

navigator maintains a mathematical representation of the equatorial coordinate system. The relationship of the two systems is expressed mathematically as:

$$X = P \cos L - Q \sin L$$

$$Y = E$$

$$Z = -P \sin L - Q \cos L.$$

4609. Inherent errors of an inertial navigator are the 84.4 minute Schuler oscillation and the 24-hour oscillations. The latter may have constant *standoffs* or *ramps* in the outputs associated with it.

As discussed in article 4607, the **84.4-minute Schuler oscillation** is an oscillation in the vertical loop resulting from initial tilts or velocity errors. It can also be caused by any transient disturbance in the system. The Schuler oscillations are continuously damped and limited in magnitude.

The **24-hour oscillation** is a property of a properly functioning practical inertial navigator which is caused by initial errors in position, heading, or gyro bias. An initial error in either latitude or heading causes both quantities to oscillate about their true values within a 24-hour period. A gyro bias error limits the ability of the inertial navigator to determine latitude and heading. A gyro bias error causes latitude or heading to oscillate about an offset value. A gyro bias error may also cause an ever increasing or *ramping* longitude error.

Although uncompensated gyro drift can cause a 24-hour oscillation, the following discussion is simplified by assuming properly biased gyros and therefore no uncompensated gyro drift. The discussion is also simplified by assuming that there is no Schuler oscillation.

Figure 4609a illustrates the 24-hour oscillation for an initial heading error. With the inertial navigator at the equator (point T_0), the displacement of the equatorial coordinate system (art. 4608) as indicated by the navigator from the true equatorial coordinate system due to the initial heading error is shown at A. This is a displacement about the Q axis and is represented by the angle δ_Q between E and E'. As shown, this is a positive heading error because the heading readout of the navigator will be larger than the actual heading by the amount δ_Q. No latitude or longitude error results because there is no displacement about either the E or the P axis.

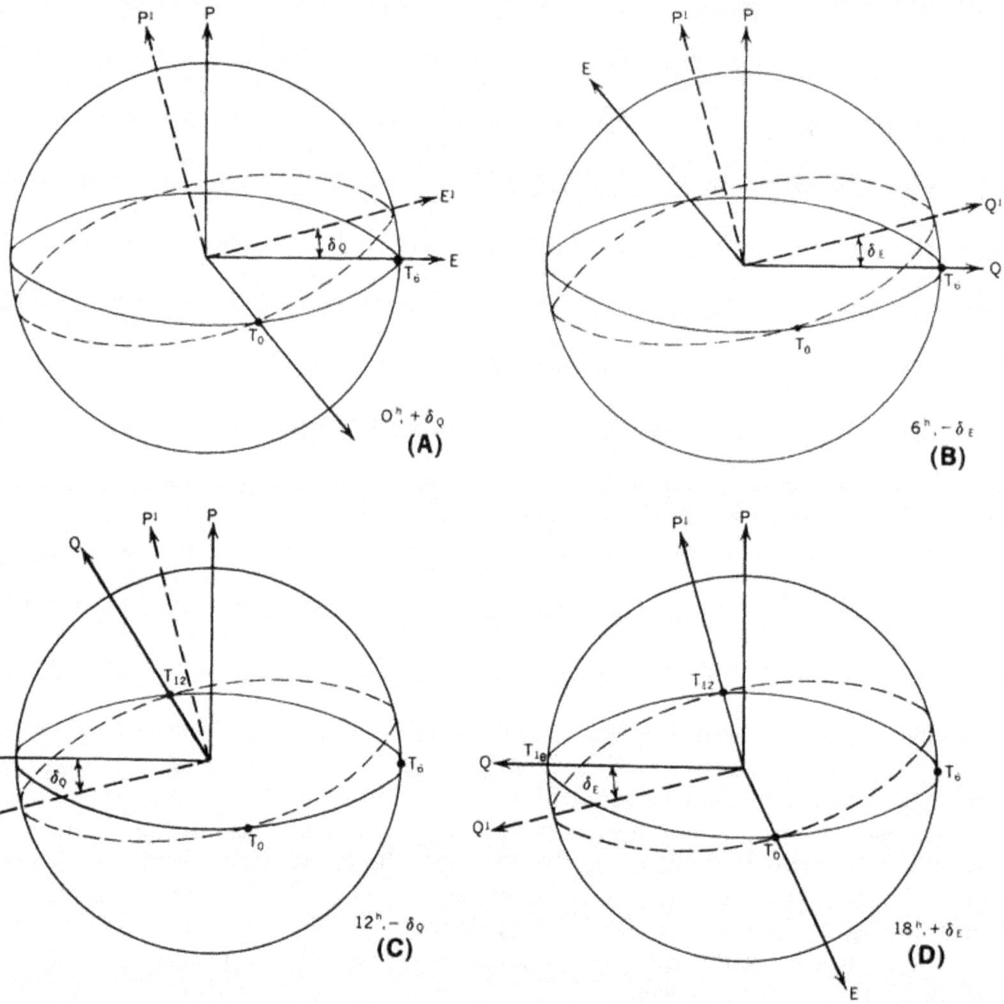

FIGURE 4609a.—The 24-hour oscillation for an initial heading error.

After this initial setup, the earth rotates about the P axis and the navigator rotates about the P' axis. In inertial space, the relationship between these axes remains the same because the navigator holds the P' axis, its instrumented polar axis, stationary in inertial space. However, with respect to the earth, it is seen at B that after 6 hours the navigator is at point T_6 due to earth's rotation, and the relationship between the coordinate systems has changed. The misalignment between the two systems is now about the E axis and is shown as the angle δ_E. The instrumented polar axis, P', is now in the meridian plane but no longer perpendicular to the vertical. This represents a negative latitude error in the navigator since its coordinates are displaced in a negative rotation about the E axis, and the heading error is now zero. Hence after 6 hours the heading error has been reduced to zero, but a negative latitude error has been produced.

As the earth continues to rotate, the navigator maintains the orientation of its instrumented polar axis in inertial space. At C the situation is shown after 6 more hours or 12 hours after the initial situation at A. Once again the misalignment of the navigator's reference system with respect to the earth is about Q axis, meaning an error in heading and no error in latitude. This time the heading error is negative but of the same magnitude as at T_0.

During the next 6 hours this negative heading error diminishes until it has decreased to zero as shown at D. At the same time the latitude error is building up in a positive direction to the value shown. At T_{18} the misalignment of the navigator is about the E axis again, which results in a positive latitude error and no heading error as stated. This positive latitude error has a magnitude equal to the negative latitude error at T_6.

View D of figure 4609a can be used to summarize the preceding action as follows: With the navigator on the equator at position T_0 set up with a positive heading error and no position error, the misalignments of its reference system with respect to the earth is about the Q axis. (It should be kept in mind that the Q axis goes through the position on the equator.) As position T_0 moves to position T_6 in space due to the rotation of the earth, the misalignment of the navigator with respect to the earth changes to be about the E axis. This means that the error in the navigator is latitude (negative at T_6) and there is no heading error. As the earth continues to rotate, the navigator reaches position T_{12} in space; and, because the instrumented polar axis has maintained its orientation in space, the navigator is now misaligned with respect to the earth about the Q axis, producing a negative heading error and no latitude error. The next 12 hours will be a mirror image of these first 12, completing the 24-hour oscillation.

If the errors shown in figure 4609a are plotted against time, the latitude error will be a negative sine wave with a 24-hour period since the oscillation started with a positive heading error. The heading error will be a cosine wave with a 24-hour period. The latitude error curve leads the heading error curve by a 90° phase relationship. At the equator, due to the geometry of the reference systems, the maximum value of δ_Q equals the maximum value of δ_E. At the equator the maximum latitude and heading errors due to the 24-hour oscillation are equal.

The relationship among navigator errors is a function of latitude. Figure 4609b represents an initial heading error as a vector along the local vertical. This vector represents a displacement of the navigator's reference system about this axis and is a positive heading error in this example. The components of this misalignment along the P axis and the Q axis are also shown. The component along the P axis is equal to $\delta_H \sin L$ and contributes to the longitude error of the navigator since it is a misalignment about the earth's spin axis or polar axis. This relationship shows that at the equator, the heading error would cause no longitude error and there would be no 24-hour oscillation in longitude. The other component along the Q axis is called δ_Q and has a value of $\delta_H \cos L$. This component determines the 24-hour oscillation due to an initial heading error in an inertial navigator with no uncompensated gyro drifts.

In the 24-hour oscillation, the maximum δ_Q occurs 6 hours after the maximum δ_E and the two have equal magnitude. The heading error is in phase with δ_Q but $\delta_H = \delta_Q \sec L$. Therefore, the 24-hour oscillation when present appears in the lati-

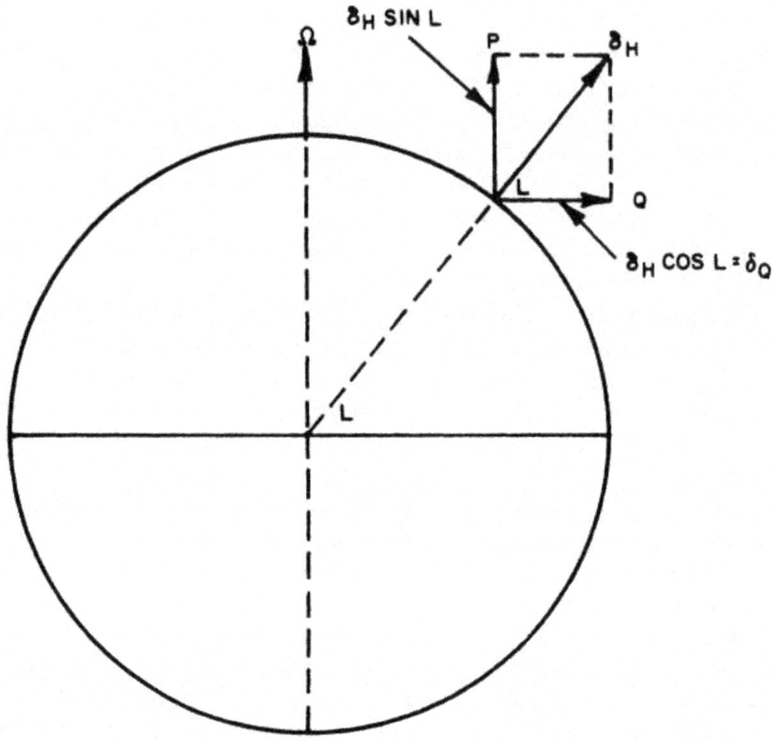

FIGURE 4609b.—Relationship among misalignment errors of inertial navigator.

tude, heading, and longitude outputs of the inertial navigator with the following relationships as shown in figure 4609c.

1. The latitude error equals δ_E.
2. The heading error equals δ_Q sec L.
3. The longitude error equals δ_H sin L.
4. Latitude, heading, and longitude errors oscillate as a sine wave with a 24-hour period.
5. The latitude error leads the heading error by 6 hours or 90°.
6. The heading error and longitude error oscillations are in phase.
7. The maximum heading error equals the maximum latitude error multiplied by the secant of the latitude position.
8. The longitude error equals the heading error times the sine of the latitude position.

Since heading error is a function of latitude, the usual practice is to use a normalized heading value for the plotting of this error. This is accomplished by multiplying the heading error by the cosine of the latitude position. This results in a plot of δ_Q. The δ_Q curve is equal in amplitude to the δ_L curve.

Gyro bias error. A gyro drift is an internal disturbance which causes an output signal from the gyroscope. The stabilization loop interprets it as a disturbance of the stable element's orientation and drives the gimbals accordingly. This causes a misorientation of the stable element and results in inertial navigator errors. Although the drift cannot be completely removed, it is possible to compensate for it by applying a gyro torquing signal called a **bias.** If the bias is proper, there is no gyroscope output due to drift. If the gyro bias is not correct or whenever the drift of

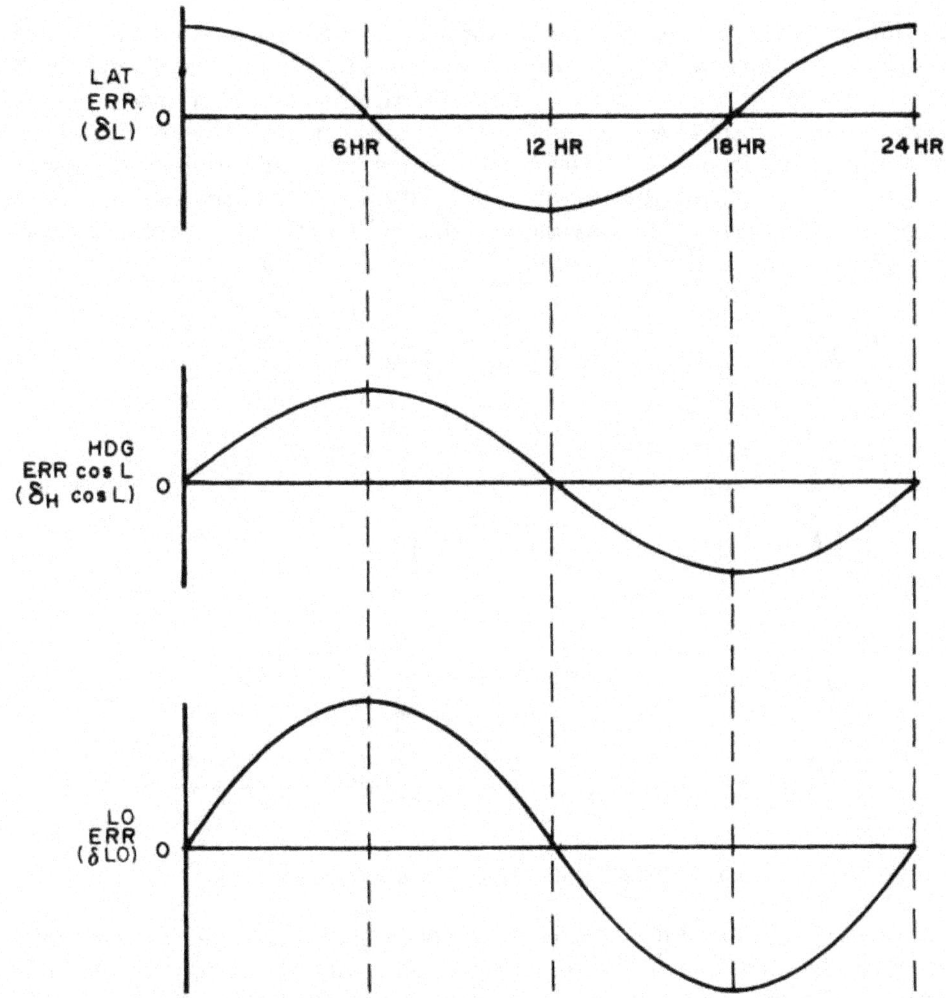

FIGURE 4609c.—Inertial navigator 24-hour oscillation error plots.

a gyroscope changes and a new bias is needed, then there is a gyro bias error or a gyroscope with uncompensated drift.

To analyze the effect of gyro bias errors on the navigator it is advantageous to examine a settled system. This is a special case in which there is no 24-hour oscillation in the inertial navigator. When the system is settled with gyro bias errors, there will be errors in navigator outputs. These errors will be a function of the gyro bias errors since they compensate for the gyro bias errors. In fact it is only when these errors in the navigator outputs exactly compensate for the gyro bias errors that the navigator will be settled.

Although the gyroscopes are placed physically in the inertial navigator's coordinate system (X, Y, Z), the effect of gyro bias errors is better described in terms of the equatorial coordinate system (P, Q, E). Using hypothetical gyroscopes in this coordinate system makes the analysis simpler. When completed the results can be transferred into the physical gyroscope coordinates by using the relationships given in article 4608. The X, Y, and Z gyroscopes have P, Q, and E uncompensated drift rate components (δ_{BP}, δ_{BQ}, and δ_{BE}).

Considering that the inertial navigator is a model of the earth and assuming that there is no movement with respect to the earth of the craft in which the navigator is installed, then the only motion of the navigator in space is about its polar axis at earth rate. However, if the equatorial gyroscope has a bias error, δ_{BQ}, then there is an additional rotation about this axis in space. The navigator will then rotate in space about the vector sum of these two rotations. This causes the instrumented polar axis of the navigator indicated as P'' to be displaced about the east axis as shown at *A* in figure 4609d.

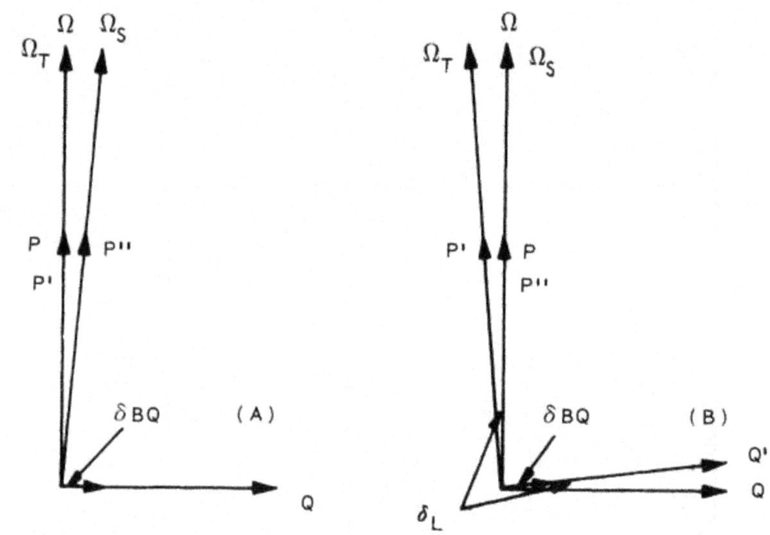

FIGURE 4609d.—Effect of equatorial gyro bias error.

Figure 4609d shows that the reference coordinates of the inertial navigator are aligned with the earth's coordinates. This means that earth rate torquing in the navigator (Ω_T) is applied about an axis parallel to the earth's spin axis. However the navigator is rotating at earth rate (shown by Ω_S, *S* for actual rotation of navigator) about P'' due to δ_{BQ}. If left this way the equatorial gyro bias error would cause a 24-hour oscillation. To achieve a settled system with the equatorial gyro bias error, δ_{BQ}, the instrumented polar axis, P'', must be made to coincide with the earth's spin axis. This is accomplished by a latitude error or δ_L as shown at *B* in figure 4609d. The amount of latitude error needed to settle the system is related to the equatorial gyro bias error by the following relationship:

$$\delta_E = \delta_L = \frac{\delta_{BQ}}{\Omega}.$$

View B of figure 4609d shows the earth rate torquing (Ω_T) of the navigator being applied about the displaced P' axis. However, the δ_{BQ} about the displaced *Q* axis (Q') causes the navigator to rotate at earth rate (Ω_S) about the P'' axis which is now aligned to the earth's spin axis (*P*). If there is an equatorial gyro bias error present but the navigator is not settled, then the 24-hour oscillation occurs not about zero error but about a latitude error given in the above relationship. As a result an equatorial gyro bias error results in a **stand-off** or constant error in latitude which may or may not have a 24-hour oscillation superimposed upon it. The equatorial gyro bias error defines the settling point for latitude.

In the case of an east gyro bias error, δ_{BE}, the instrumented polar axis would be displaced about the equatorial axis. Again, to settle the navigator with an east gyro bias error, the instrumented polar axis must be made coincident with the earth's spin axis. To do this the navigator coordinates must be misaligned about the equatorial axis an amount given by:

$$\delta_Q = \delta_H \cos L = \frac{\delta_{BE}}{\Omega}.$$

If the displacement about the equatorial axis is the above amount, the navigator would be settled. The δ_Q results in a heading error which varies as a function of latitude as discussed with respect to the 24-hour oscillation. If the navigator is not settled and has an east gyro bias error, the 24-hour oscillation in heading will be about a stand-off error defined by the above equation. If heading error times the cosine of latitude is plotted for a given east gyro bias error, this stand-off is constant and is a function of the magnitude of the gyro bias error.

If there is both an equatorial and east gyro bias error, there is a settling point for the navigator involving both a latitude error and a heading error. Figure 4609e illustrates the error propagation in an inertial navigator which had no errors previous to time 0 and then at time zero a bias error occurs in each of the gyros. View A shows that the latitude error oscillates about a value determined by the equatorial gyro bias error (the stand-off error in latitude due to an equatorial gyro bias error). The heading error as seen at view B oscillates about the stand-off error due to the east gyro bias error. The phase relationship is as discussed earlier and the magnitude of the 24-hour oscillation errors is a function of the initial conditions of the navigator. These are more easily studied in the circle plot.

The remaining gyro bias error is the polar gyro bias error. Since this bias error occurs about the earth's spin axis, it doesn't change the orientation of the inertial navigator's instrumented polar axis. Instead the polar gyro bias error causes the navigator to rotate about the polar axis at a rate different from the earth's rotation. The navigator interprets this as change of position on the earth about the polar axis which is longitude change or longitude rate. The bias error has a constant value and this results in a constant longitude rate error. The longitude rate is integrated with time in the navigator to produce an increasing longitude value. As a result the polar gyro bias error contributes a straight line function to the longitude error. The slope of this line equals the polar gyro bias error or longitude ramp.

View C of figure 4609e shows the longitude error at latitude 45°N. The longitude error starts at zero in this case because of the initial conditions previously set up. The longitude error oscillates relative to the polar gyro bias error in the same phase as the heading error oscillates about the latitude gyro bias error.

4610. The circle plot is a convenient method of representing the 24-hour oscillation. It combines both latitude and heading in one plot and shows all phase and amplitude relationships.

Latitude and heading errors oscillate with a 24-hour period. These oscillations are 90° out of phase. Also, the amplitude of the latitude oscillation equals the heading oscillation multiplied by cosine latitude. If latitude error is plotted against heading error times cosine latitude the result is a circle plot as shown in figure 4610a. Also shown are the latitude and heading cosine latitude curves which the circle represents. Note that the circle completely defines each of the component

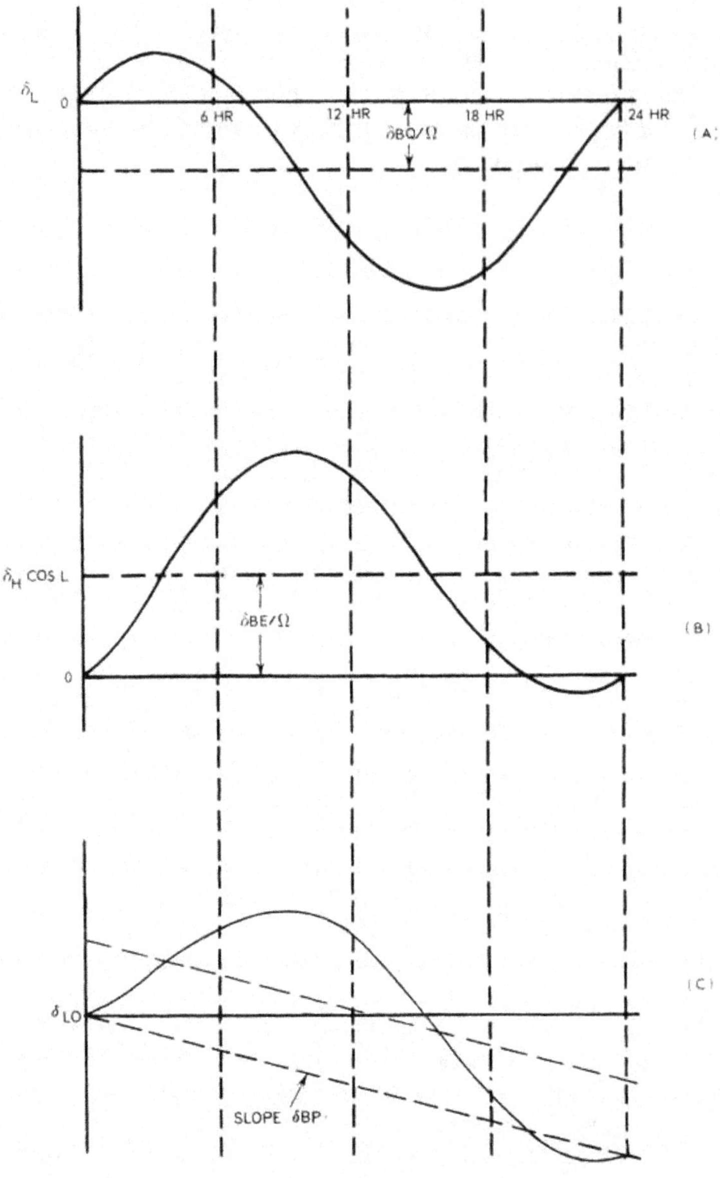

FIGURE 4609e.—Inertial navigator error plots.

oscillations. A point on the circle is known as the operating point. Its coordinates represent the latitude and heading errors at a particular time. As the operating point moves around the circle its projection on the vertical axis is the latitude oscillation. Its projection on the horizontal axis is the heading error times the cosine latitude oscillation. As the point goes through one rotation, it projects one cycle of a sine curve. Thus 360° on the circle represents 24 hours of time or 15° equals 1 hour. Because of the phase relationships involved, the circle always propagates in a clockwise direction.

The location of the center of the circle is a function of gyro bias errors as they represent the standoff values in latitude and heading cosine latitude. The center of the circle is known as the bias center. The circle diagram can be used to determine how the system will behave starting with given initial conditions. For example,

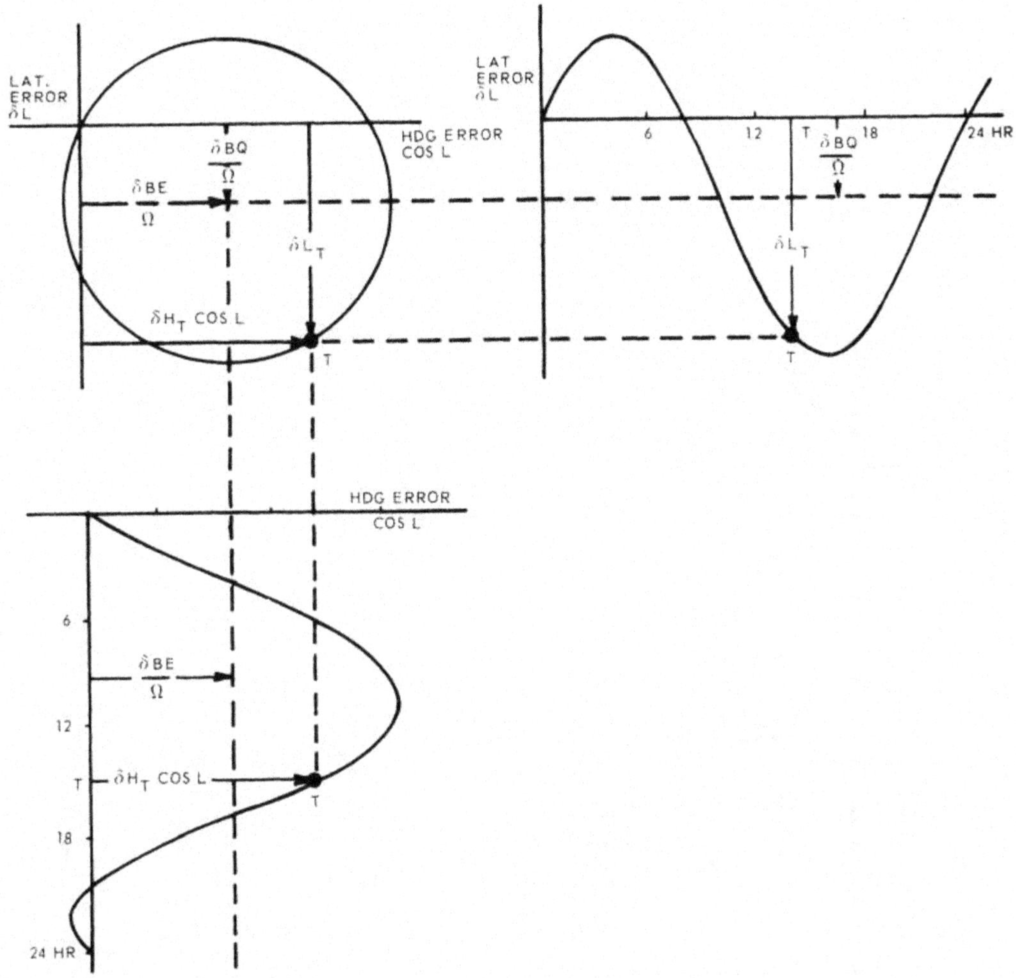

FIGURE 4610a.—Circle plot.

suppose the inertial navigator has zero initial latitude and heading error, but at time zero a bias error develops in the east gyro only. The circle plot will be used to determine the resulting oscillation.

The initial conditions are shown in figure 4610b at A. The operating point is at the origin and the bias center is along the heading cosine latitude axis. The circle propagates about the bias center in a clockwise direction as shown at B. To find the oscillations in latitude and heading, the operating point is projected on the vertical and horizontal axes as shown at C. Latitude error peaks at 6 hours, goes through zero at 12 hours, and reaches a negative peak at 18 hours. It is thus a sine curve oscillating about zero.

Heading error cosine latitude starts at zero, reaches a point of inflection at 6 hours, and a positive peak after 12 hours. It varies as a cosine function about an offset. The offset is due to the east gyro drift. The phase and amplitude relations may thus be determined from the circle plot directly. This can be done for any initial conditions or to see the effect of a reset or gyro drift change on the inertial navigator.

A longitude plot must be used in conjunction with the circle plot in order to describe the error propagation of the inertial navigator completely.

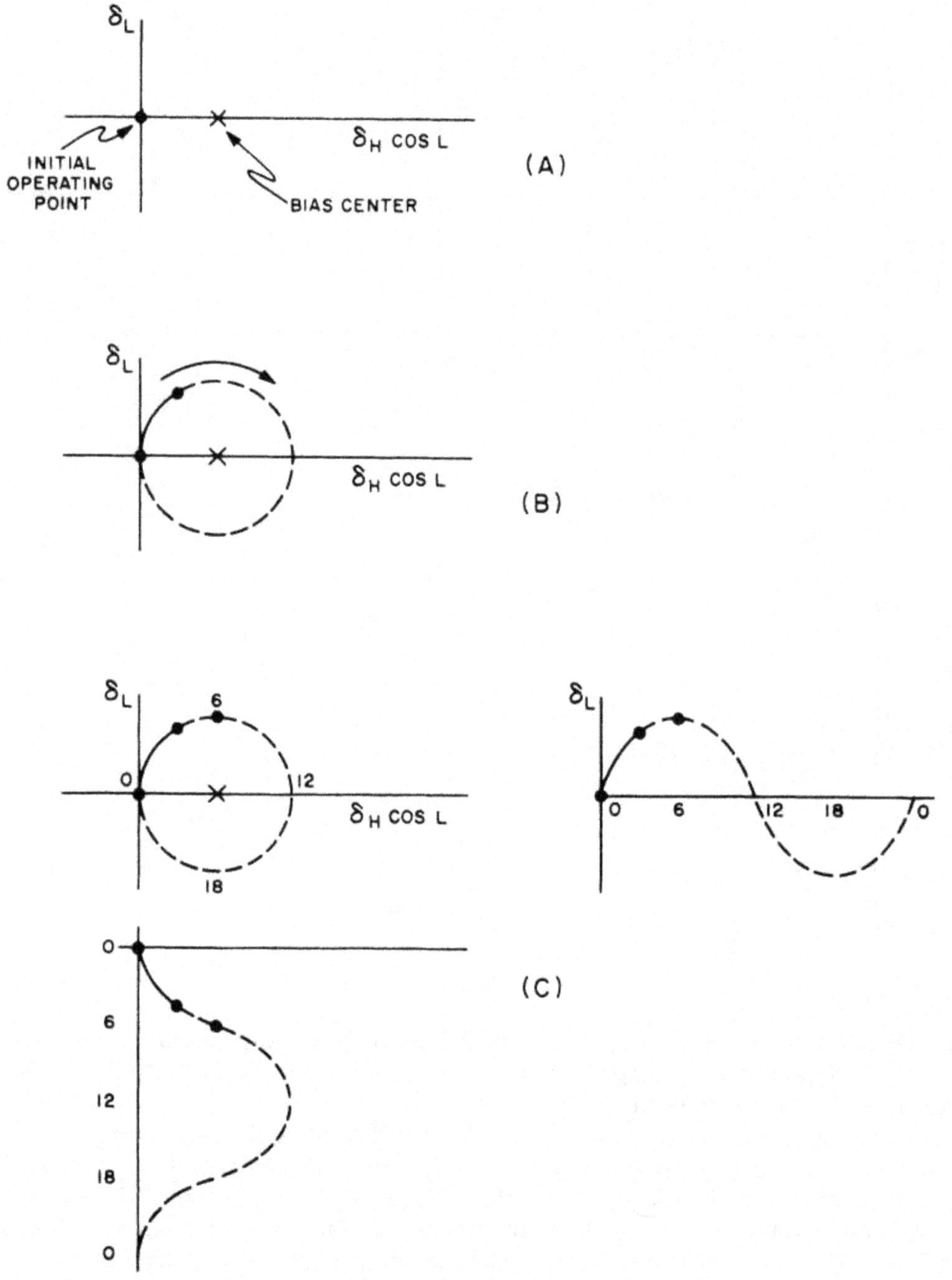

FIGURE 4610b.—Effect of east gyro bias error.

4611. Reset.—The characteristics of propagation of the 24-hour oscillation errors and the gyro bias error in the inertial navigator are well known. This knowledge is used with fix data to determine corrections which are entered into the navigator to eliminate the errors and to prevent the errors from occurring in the future. The process of entering these corrections into the navigator for this purpose is called *reset.*

To determine the error of the navigator, propagation plots such as those in 4609e could be made. Since the latitude error and heading error times the cosine of latitude are sine waves of known period, three points would be necessary to determine each curve. However, it is not always possible to determine points on the heading error plot due to lack of accurate heading information. If this happens it is still possible to determine the heading error oscillation from the latitude error plot because of their 90° relationship. In fact, it is just this relationship which allows these two plots to be combined into one plot for a more useful display of the errors.

It has been found advantageous in the construction of the circle plot to use bias "corrections" rather than bias "errors." For example, an equatorial bias error (δ_{BE}) would be plotted as an equatorial bias correction (BEC).

4612. Geodetic and geophysical errors.—Even if it were possible to make perfect inertial sensors, inertial navigators would still experience errors. These errors are due to uncertainties in knowledge of the physical environment.

A particularly serious error in precision marine inertial navigation systems is due to lack of knowledge of the gravitational environment (art. X4). Since an accelerometer cannot distinguish between a kinematic acceleration and a gravitational acceleration, any uncertainty in the gravitational environment manifests itself as a system error. In the case of marine inertial navigation in locally level coordinates, it is the horizontal components of gravity that cause significant errors. These are directly due to deflections of the vertical (art. X4), which are tilts of the actual (plumb bob) vertical vector relative to the presumed reference vertical.

Departures of the magnitude of the gravity vector from the presumed model value are called **gravity anomalies.**

Inertial navigation systems are mechanized in terms of a reference ellipsoid (art. X7). The reference ellipsoid is chosen as a good approximation to a surface called the geoid (art. X2), which is that surface to which the oceans would conform over the entire earth if free to adjust to the combined effect of the earth's mass attraction and the centrifugal force of the earth's rotation. It has the special property that, at every point, the direction to the geoid is given by the direction of a plumb line. The angle between the normal to the geoid and the normal to the reference ellipsoid is the deflection of the vertical referred to above.

The use of an electromagnetic log or similar speed log for damping the inertial navigator results in two types of errors. The first is due to the speed sensor itself. The second occurs because the speed indication is made relative to the water, which is itself moving relative to the earth.

APPENDICES

APPENDICES

		Page
APPENDIX A.	Abbreviations	1161
APPENDIX B.	Greek Alphabet	1166
APPENDIX C.	Symbols	1167
APPENDIX D.	Miscellaneous Data	1169
APPENDIX E.	Navigational Coordinates	1181
APPENDIX F.	Extracts from *Nautical Almanac*	1182
APPENDIX G.	Extracts from *Air Almanac*	1199
APPENDIX H.	Long-term Almanac	1218
APPENDIX I.	Identification of Navigational Stars	1224
APPENDIX J.	Navigational Stars and Planets	1232
APPENDIX K.	Constellations	1233
APPENDIX L.	Extracts from Tide Tables	1235
APPENDIX M.	Extracts from Tidal Current Tables	1238
APPENDIX N.	Extracts from Pub. No. 214	1243
APPENDIX O.	Extracts from Pub. No. 229	1250
APPENDIX P.	Extracts from Pub. No. 249	1256
APPENDIX Q.	Navigational Errors	1260
APPENDIX R.	GHA and Declination of the Sun for the Years 1981–2016	1295
APPENDIX S.	Charts and Publications of other Agencies	1297
APPENDIX T.	Hand-held Digital Calculators	1299
APPENDIX U.	Underwater Log Calibration Guidelines	1309
APPENDIX V.	Beaufort Wind Scale	1312
APPENDIX W.	Sea State	1313
APPENDIX X.	Geodesy for the Navigator	1340
APPENDIX Y.	Buoyage Systems	1355

APPENDIX A

ABBREVIATIONS

Abbreviations used on nautical charts are given in appendix Z.

A, amplitude; augmentation; away (altitude intercept).

a, altitude intercept (Ho∼Hc); altitude factor (change of altitude in 1 minute of time from meridian transit); assumed.

a_0, first Polaris correction.

a_1, second Polaris correction.

a_2, third Polaris correction.

AC, alternating current.

add'l, additional.

ADF, automatic radio direction finder.

AF, audio frequency.

***a*L**, assumed latitude.

AM, amplitude modulation.

am, ante meridian (before noon).

AMVER, Automated Mutual-assistance Vessel Rescue System.

antilog, antilogarithm.

AP, assumed position.

approx., approximate, approximately.

ASF, Additional Secondary Phase Factor.

AT, Atomic Time.

AU, astronomical unit.

AUSREP, Australian Ships Reporting System.

***a*λ**, assumed longitude.

B, atmospheric pressure correction (altitude); bearing, bearing angle.

B_A, difference between heading and apparent wind direction.

BIH, Bureau Internationale de l'Heure.

Brg., bearing (as distinguished from bearing angle).

B_{pgc}, bearing per gyrocompass.

B_T, difference between heading and true wind direction.

C, Celsius (centigrade); chronometer time; compass (direction); correction; course, course angle.

CB, compass bearing.

CC, compass course; chronometer correction.

CCIR, International Radio Consultative Committee.

CCZ, Coastal Confluence Zone.

CE, chronometer error; compass error.

cec, centicycle.

cel, centilane.

CEP, circular probable error.

CFR, Code of Federal Regulations.

CH, compass heading.

cm, centimeter, centimeters.

CMG, course made good.

Cn, course (as distinguished from course angle).

co-, the complement of (90° *minus).*

COA, course of advance.

COG, course over ground.

coL, colatitude.

colog, cologarithm.

corr., correction.

cos, cosine.

cot, cotangent.

cov, coversine.

CPE, circular probable error.

cps, cycles per second.

C_{pgc}, course per gyrocompass.

C_{psc}, course per standard compass.

$C_{p\ stg\ c}$, course per steering compass.

CRT, cathode-ray tube.

csc, cosecant.

CW, continuous wave.

CZn, compass azimuth.

D, deviation; dip (of horizon); distance.

d, declination (astronomical); altitude difference.

d, declination change in 1 hour.

DC, direct current.

D. Lat., difference of latitude.

Dec., declination.

Dec. Inc., declination increment.
Dep., departure.
Dev., deviation.
DG, degaussing.
diff., difference.
Dist., distance.
DLo, difference of longitude (arc units).
DMAHTC, Defense Mapping Agency Hydrographic/Topographic Center.
DR, dead reckoning, dead reckoning position.
DRE, dead reckoning equipment.
DRT, dead reckoning tracer.
D_S, dip short of horizon.
DSD, double second difference.
dur., duration.
dλ, difference of longitude (time units).
E, east.
e, base of Naperian logarithms.
e, eccentricity.
EDD, estimated date of departure.
EHF, extremely high frequency.
EM, electromagnetic (underwater log).
EP, estimated position.
Eq.T, equation of time.
ET, Ephemeris Time.
ETA, estimated time of arrival.
ETD, estimated time of departure.
F, Fahrenheit; fast; longitude factor; phase correction (altitude).
f, latitude factor.
f, flattening or ellipticity.
FM, frequency modulation.
ft., foot, feet.
G, Greenwich, Greenwich meridian (upper branch); grid (direction).
g, acceleration due to gravity; Greenwich meridian (lower branch).
GAT, Greenwich apparent time.
GB, grid bearing.
GC, grid course.
GE, gyro error.
GH, grid heading.
GHA, Greenwich hour angle.
GMT, Greenwich mean time.
GP, geographical position.
Gr., Greenwich.
GRI, group repetition interval.
GST, Greenwich sidereal time.
GV, grid variation.
GZn, grid azimuth.
h, altitude (astronomical); height above sea level.
ha, apparent altitude.
hav, haversine.
Hc, computed altitude.
Hdg., heading.
HE, heeling error; height of eye.
HF, high frequency.
h_f, height above sea level in feet.
HHW, higher high water.
HLW, higher low water.
hm, height above sea level in meters.
Ho, observed altitude.
HP, horizontal parallax.
Hp, precomputed altitude.
H_{pgc}, heading per gyrocompass.
H_{psc}, heading per standard compass.
$H_{p\ stg\ c}$, heading per steering compass.
hr, rectified (apparent) altitude.
hr., hour.
hrs., hours.
hs, sextant altitude.
ht, tabulated altitude.
HW, high water.
HWF & C, high water full and change.
I, instrument correction.
IALA, International Association of Lighthouse Authorities.
IAU, International Astronomical Union.
IC, index correction.
ICW, Intracoastal Waterway.
IHB, International Hydrographic Bureau.
IHO, International Hydrographic Organization.
IMO, International Maritime Organization.
in., inch, inches.
INM, International Nautical Mile.
INS, inertial navigation system.
int., interval.
ISLW, Indian spring low water.
ITU, International Telecommunications Union.
IUGG, International Union of Geodesy and Geophysics.
J, irradiation correction (altitude).
K, Kelvin (temperature).
kHz, kilohertz.
km, kilometer, kilometers.

kn, knot, knots.
L, latitude; lower limb correction for moon (from *Nautical Almanac*).
l, difference of latitude; logarithm, logarithmic.
LAN, local apparent noon.
LAT, local apparent time.
lat., latitude.
LF, low frequency.
LHA, local hour angle.
LHW, lower high water.
LL, lower limb.
LLW, lower low water.
Lm, middle latitude; mean latitude.
LMT, local mean time.
log, logarithm, logarithmic.
log_e, natural logarithm (to the base e).
log_{10}, common logarithm (to the base 10).
long., longitude.
LOP, line of position.
LST, local sidereal time.
LW, low water.
M, celestial body; meridian (upper branch); magnetic (direction); meridional parts; nautical mile, miles.
m, meridian (lower branch); meridional difference ($M_1 \sim M_2$); meter, meters; statute mile, miles.
mag., magnetic; magnitude.
MB, magnetic bearing.
mb, millibar, millibars.
MC, magnetic course.
mc, megacycle, megacycles; megacycles per second.
Mer. Pass., meridian passage.
MF, medium frequency.
MH, magnetic heading.
MHHW, mean higher high water.
MHW, mean high water.
MHWN, mean high water neaps.
MHWS, mean high water springs.
MHz, megahertz.
mi., mile, miles.
mid, middle.
min., minute, minutes.
MLLW, mean lower low water.
MLW, mean low water.
MLWN, mean low water neaps.
MLWS, mean low water springs.
mm, millimeter.
mo., month.
mos., months.
mph, miles (statute) per hour.
MPP, most probable position.
ms, millisecond, milliseconds.
MSL, mean sea level.
MZn, magnetic azimuth.
N, north.
n, natural (trigonometric function).
Na, nadir.
NASA, National Aeronautics and Space Administration.
NAVSAT, Navy Navigation Satellite System.
NBS, National Bureau of Standards.
NLT, not less than (used with danger bearing).
NM, nautical mile, miles.
n. mi., nautical mile, miles.
NMT, not more than (used with danger bearing).
NNSS, Navy Navigation Satellite System.
NOAA, National Oceanic and Atmospheric Administration.
NOS, National Ocean Service.
OTSR, Optimum Track Ship Routing.
P, atmospheric pressure; parallax; planet; pole.
p, departure, polar distance.
PC, personal correction.
PCA, polar cap absorption.
PCD, polar cap disturbance.
pgc, per gyrocompass.
P in A, parallax in altitude.
PM, pulse modulation.
PM, post meridian (after noon).
Pn, north pole; north celestial pole.
PPC, predicted propagation correction.
PPI, plan position indicator.
PRR, pulse repetition rate.
Ps, south pole; south celestial pole.
psc, per standard compass.
p stg c, per steering compass.
Pub., publication.
PV, prime vertical.
Q, Polaris correction (*Air Almanac*).
QQ', celestial equator.
R, Rankine (temperature); refraction.
RA, right ascension.
rad, radian, radians.

RB, relative bearing.
R Bn, radiobeacon.
RDF, radio direction finder.
rev., reversed.
RF, radio frequency.
R Fix, running fix.
RMS, root mean square.
RSS, root sum square.
RZn, relative azimuth.
S, sea-air temperature difference correction (altitude); slow; south; set; speed.
S_A, speed of apparent wind in units of ship's speed.
SAM, system area monitor.
SAR, Search and Rescue.
SD, semidiameter.
sec, secant.
sec., second, seconds.
semidur., semiduration.
SF, Secondary Phase Factor.
SH, ship's head (heading).
SHA, sidereal hour angle.
SHF, super high frequency.
SI, International System of Units.
SID, sudden ionospheric disturbance.
sin, sine.
SINS, Ships Inertial Navigation System.
SMG, speed made good.
SOA, speed of advance.
SOG, speed over ground.
SPA, sudden phase anomoly.
S_T, speed of true wind in units of ship's speed.
St M, statute mile, miles.
T, air temperature correction (altitude); table; temperature; time; toward (altitude intercept); true (direction).
t, dry-bulb temperature; elapsed time; meridian angle.
t′, wet-bulb temperature.
tab., table.
TAI, International Atomic Time scale.
tan, tangent.
TB, true bearing; turning bearing; air temperature-atmospheric pressure correction (altitude).
TC, true course.
TcHHW, tropic higher high water.
TcHLW, tropic higher low water.
TcLHW, tropic lower high water.
TcLLW, tropic lower low water.
TD, time difference (Loran-C).
T_G, time difference of groundwaves from master and secondary (slave) stations (Loran).
T_{GS}, time difference of groundwave from master and skywave from secondary (slave) station (Loran).
TH, true heading.
TMG, track made good.
TOD, time of day (clock).
TR, track.
Tr., transit.
T_S, time difference of skywaves from master and secondary (slave) stations (Loran).
T_{SG}, time difference of skywave from master and groundwave from secondary (slave) station (Loran).
TZn, true azimuth.
U, upper limb correction for moon (from *Nautical Almanac*).
UHF, ultra high frequency.
UL, upper limb.
UPS, Universal Polar Stereographic.
USWMS, Uniform State Waterway Marking System.
UT, Universal Time.
UTC, Coordinated Universal Time.
UT0, Universal Time 0.
UT1, Universal Time 1.
UT2, Universal Time 2.
UTM, Universal Transverse Mercator.
V, variation; vertex.
v, excess of GHA change from adopted value for 1 hour.
Var., variation.
ver, versine.
VHF, very high frequency.
VLF, very low frequency.
W, west.
WARC, World Administrative Radio Council.
WE, watch error.
WGS, World Geodetic System.
WMO, World Meteorological Organization.
WT, watch time.
X, parallactic angle.
yd., yard.

yds., yards.
yr., year.
yrs., years.
Z, azimuth angle; zenith.
z, zenith distance.
ZD, zone description.
Z Diff., azimuth angle difference.
Zn, azimuth (as distinguished from azimuth angle).
Zn_{pgc}, azimuth per gyrocompass.
ZT, zone time.
Δ, a small increment, or the change in one quantity corresponding to unit change in another.
λ, longitude; wavelength (radiant energy).
σ, standard deviation.
μ, index of refraction.
μS, microsecond.
π, ratio of circumference of circle to diameter$=3.14159+$.

APPENDIX B

GREEK ALPHABET

Α α a	Alpha	Ν ν	Nu
Β β	Beta	Ξ ξ	Xi
Γ γ	Gamma	Ο ο	Omicron
Δ δ	Delta	Π π	Pi
Ε ϵ	Epsilon	Ρ ρ	Rho
Ζ ζ	Zeta	Σ σ	Sigma
Η η	Eta	Τ τ	Tau
Θ θ ϑ	Theta	Υ υ	Upsilon
Ι ι	Iota	Φ φ	Phi
Κ κ	Kappa	Χ χ	Chi
Λ λ	Lambda	Ψ ψ	Psi
Μ μ	Mu	Ω ω	Omega

APPENDIX C

SYMBOLS

Positions

⌓ Dead reckoning position.
⊙ Fix.
⊡ Estimated position.
See Note.

△ Symbol used for one set of fixes when simultaneously fixing by two means, e.g. visual and radar; sometimes used for radionavigation fixes.

Mathematical Symbols

+ Plus (addition)
− Minus (subtraction)
± Plus or minus
∼ difference
|| Absolute value
· Multiplied by
∷ Is to; ratio
× Times (multiplication)
÷ Divided by (division)
√‾ Square root
∠ Angle
∴ Therefore
∵ Because
⋆ nth root
= Equals
≠ Not equal to

≅ Congruent to approximately equal
≡ Identical with
≢ Not identical with
≈ Nearly equal to
> Is greater than
< Is less than
⋜ Equal to or less than
⋝ Equal to or greater than
⋞ Is dominated by
∂ or δ Differential; variation
f Function
∫ Integral sign
Σ Summation of; sum; sigma
→ Approaches limit of
∞ Infinity
. . . Repeating decimal

Celestial Bodies

⊙ Sun
☾ Moon
☿ Mercury
♀ Venus
⊕ Earth
♂ Mars
♃ Jupiter
♄ Saturn
♅ Uranus
♆ Neptune
♇ Pluto
☆ Star
☆-P Star-planet altitude correction (altitude)

⊙̲ ☾̲ Lower limb
⊖ ☾ Center
⊙̄ ☾̄ Upper limb
● New moon
◐ Crescent moon
◐ First quarter
◐ Gibbous moon
○ Full moon
◑ Gibbous moon
◑ Last quarter
◑ Crescent moon

Signs of the Zodiac

♈ Aries (vernal equinox)	♎ Libra (autumnal equinox)
♉ Taurus	♏ Scorpius
♊ Gemini	♐ Sagittarius
♋ Cancer (summer solstice)	♑ Capricornus (winter solstice)
♌ Leo	♒ Aquarius
♍ Virgo	♓ Pisces

Miscellaneous Symbols

ʸ Years	✳ Interpolation impractical
ᵐ Months	° Degrees
ᵈ Days	′ Minutes of arc
ʰ Hours	″ Seconds of arc
ᵐ Minutes of time	☌ Conjunction
ˢ Seconds of time	☍ Opposition
■ Remains below horizon	□ Quadrature
▭ Remains above horizon	☊ Ascending node
//// Twilight all night	☋ Descending node

NOTE.—The digits indicating the times of fixes and estimated positions are printed horizontally on the chart or plotting sheet; the digits indicating the times of dead reckoning positions are not printed horizontally. The running fix is the only position indicated on the chart or plotting sheet by both symbol and label (R Fix).

APPENDIX D

MISCELLANEOUS DATA

Exact relationships shown by asterisk (*). See footnote on page 1179

Area

1 square inch	=6.4516 square centimeters*
1 square foot	=144 square inches*
	=0.09290304 square meter*
	=0.000022957 acre
1 square yard	=9 square feet*
	=0.83612736 square meter
1 square (statute) mile	=27,878,400 square feet*
	=640 acres*
	=2.589988110336 square kilometers*
1 square centimeter	=0.1550003 square inch
	=0.00107639 square foot
1 square meter	=10.76391 square feet
	=1.19599005 square yards
1 square kilometer	=247.1053815 acres
	=0.38610216 square statute mile
	=0.29155335 square nautical mile

Astronomy

1 mean solar unit	=1.00273791 sidereal units
1 sidereal unit	=0.99726957 mean solar unit
1 microsecond	=0.000001 second*
1 second	=1,000,000 microseconds*
	=0.01666667 minute
	=0.00027778 hour
	=0.00001157 day
1 minute	=60 seconds*
	=0.01666667 hour
	=0.00069444 day
1 hour	=3,600 seconds*
	=60 minutes*
	=0.04166667 day
1 mean solar day	=$24^{h}03^{m}56^{s}.55536$ of mean sidereal time
	=1 rotation of earth with respect to sun (mean)*
	=1.00273791 rotations of earth with respect to vernal equinox (mean)
	=1.0027378118868 rotations of earth with respect to stars (mean)
1 mean sidereal day	=$23^{h}56^{m}04^{s}.09054$ of mean solar time
1 sidereal month	=27.321661 days
	=$27^{d}07^{h}43^{m}11^{s}.5$
1 synodical month	=29.530588 days
	=$29^{d}12^{h}44^{m}02^{s}.8$
1 tropical (ordinary) year	=31,556,925.975 seconds
	=525,948.766 minutes
	=8,765.8128 hours
	=$365^{d}.24219879-0^{d}.0000000614(t-1900)$, where t=the year (date)
	=$365^{d}05^{h}48^{m}46^{s}$ (−) $0^{s}.0053t$

Astronomy—Continued

1 sidereal year	$=365^{\mathrm{d}}.25636042+0.0000000011(t-1900)$, where $t=$ the year (date)
	$=365^{\mathrm{d}}06^{\mathrm{h}}09^{\mathrm{m}}09^{\mathrm{s}}.5$ (+) $0^{\mathrm{s}}.0001\mathrm{t}$
1 calendar year (common)	=31,536,000 seconds*
	=525,600 minutes*
	=8,760 hours*
	=365 days*
1 calendar year (leap)	=31,622,400 seconds*
	=527,040 minutes*
	=8,784 hours*
	=366 days*
1 light-year	=9,460,000,000,000 kilometers
	=5,880,000,000,000 statute miles
	=5,110,000,000,000 nautical miles
	=63,240 astronomical units
	=0.3066 parsecs
1 parsec	=30,860,000,000,000 kilometers
	=19,170,000,000,000 statute miles
	=16,660,000,000,000 nautical miles
	=206,300 astronomical units
	=3.262 light years
1 astronomical unit	=149,600,000 kilometers
	=92,960,000 statute miles
	=80,780,000 nautical miles
	$=499^{\mathrm{s}}.012$ light-time
	=mean distance, earth to sun
Mean distance, earth to moon	=384,400 kilometers
	=238,855 statute miles
	=207,559 nautical miles
Mean distance, earth to sun	=149,600,000 kilometers
	=92,957,000 statute miles
	=80,780,000 nautical miles
	=1 astronomical unit
Sun's diameter	=1,392,000 kilometers
	=865,000 statute miles
	=752,000 nautical miles
Sun's mass	=1,987,000,000,000,000,000,000,000,000,000,000 grams
	=2,200,000,000,000,000,000,000,000,000 short tons
	=2,000,000,000,000,000,000,000,000,000 long tons
Speed of sun relative to neighboring stars	=19.4 kilometers per second
	=12.1 statute miles per second
	=10.5 nautical miles per second
Orbital speed of earth	=29.8 kilometers per second
	=18.5 statute miles per second
	=16.1 nautical miles per second
Obliquity of the ecliptic	$=23°27'08''.26-0''.4684(t-1900)$, where $t=$ the year (date)
General precession of the equinoxes	$=50''.2564+0''.000222(t-1900)$ per year, where $t=$ the year (date)
Precession of the equinoxes in right ascension	$=46''.0850+0''.000279(t-1900)$ per year, where $t=$ the year (date)
Precession of the equinoxes in declination	$=20''.0468-0''.000085(t-1900)$ per year, where $t=$ the year (date)
Magnitude ratio	=2.512
	$=\sqrt[5]{100}$*

Charts

Nautical miles per inch	=reciprocal of natural scale ÷ 72,913.39
Statute miles per inch	=reciprocal of natural scale ÷ 63,360*

Charts—Continued

Inches per nautical mile .... $=72{,}913.39\times$ natural scale
Inches per statute mile .... $=63{,}360\times$ natural scale*
Natural scale .... $=1{:}72{,}913.39\times$ nautical miles per inch
$=1{:}63{,}360\times$ statute miles per inch*

Earth

Acceleration due to gravity (standard) .... =980.665 centimeters per second per second
=32.1740 feet per second per second
Mass-ratio—Sun/Earth .... =332,958
Mass-ratio—Sun/(Earth & Moon) .... =328,912
Mass-ratio—Earth/Moon .... =81.30
Mean density .... =5.517 grams per cubic centimeter
Velocity of escape .... =6.94 statute miles per second
Curvature of surface .... =0.8 foot per nautical mile

Airy ellipsoid

Equatorial radius (a) .... =6,377,563.396 meters
=3,443.609 nautical miles
Polar radius (b) .... =6,356,256.91 meters
=3,432.104 nautical miles
Mean radius $(2a+b)/3$ .... =6,370,461.234 meters
=3,439.774 nautical miles
Flattening or ellipticity ($f=1-b/a$) .... =1/299.325
=0.00334085
Eccentricity ($e=(2f-f^2)^{1/2}$) .... =0.081673374
Eccentricity squared (e^2) .... =0.00667054

Australian National-South American ellipsoid of 1969

Equatorial radius (a) .... =6,378,160 meters
=3,443.931 nautical miles
Polar radius (b) .... =6,356,774.719 meters
=3,432.384 nautical miles
Mean radius $(2a+b)/3$ .... =6,371,031.573 meters
=3,440.082 nautical miles
Flattening or ellipticity ($f=1-b/a$) .... =1/298.25
=0.00335289
Eccentricity ($e=(2f-f^2)^{1/2}$) .... =0.0818202
Eccentricity squared (e^2) .... =0.00669454

Bessel ellipsoid

Equatorial radius (a) .... =6,377,397.155 meters
=3,443.52 nautical miles
Polar radius (b) .... =6,356,078.963 meters
=3,432.01 nautical miles
Mean radius $(2a+b)/3$ .... =6,370,291.091 meters
=3,439.682 nautical miles
Flattening or ellipticity ($f=1-b/a$) .... =1/299.1528
=0.00334277
Eccentricity ($e=(2f-f^2)^{1/2}$) .... =0.08169683
Eccentricity squared (e^2) .... =0.00667437

Clarke ellipsoid of 1866

Equatorial radius (a) .... =6,378,206.4 meters
=3,443.957 nautical miles
Polar radius (b) .... =6,356,583.8 meters
=3,432.281 nautical miles
Mean radius $(2a+b)/3$ .... =6,370,998.9 meters
=3,440.064 nautical miles
Flattening or ellipticity ($f=1-b/a$) .... =1/294.98
=0.00339008

Earth—Continued

Eccentricity ($e=(2f-f^2)^{1/2}$) = 0.08227185
Eccentricity squared (e^2) = 0.00676866

Clarke ellipsoid of 1880

Equatorial radius (a) = 6,378,249.145 meters
= 3,443.98 nautical miles
Polar radius (b) = 6,356,514.87 meters
= 3,432.245 nautical miles
Mean radius $(2a+b)/3$ = 6,371,004.387 meters
= 3,440.067 nautical miles
Flattening or ellipticity ($f=1-b/a$) = 1/293.465
= 0.00340756
Eccentricity ($e=(2f-f^2)^{1/2}$) = 0.0824834
Eccentricity squared (e^2) = 0.00680351

Everest ellipsoid

Equatorial radius (a) = 6,377,276.345 meters
= 3,443.454 nautical miles
Polar radius (b) = 6,356,075.413 meters
= 3,432,006 nautical miles
Mean radius $(2a+b)/3$ = 6,370,209.37 meters
= 3,439.638 nautical miles
Flattening or ellipticity ($f=1-b/a$) = 1/300.8017
= 0.00332445
Eccentricity ($e=(2f-f^2)^{1/2}$) = 0.08147298
Eccentricity squared (e^2) = 0.00663785

Fischer ellipsoid of 1960 (Mercury Datum)

Equatorial radius (a) = 6,378,166 meters
= 3,443.934 nautical miles
Polar radius (b) = 6,356,784.284 meters
= 3,432.389 nautical miles
Mean radius $(2a+b)/3$ = 6,371,038.761 meters
= 3,440.086 nautical miles
Flattening or ellipticity ($f=1-b/a$) = 1/298.3
= 0.00335233
Eccentricity ($e=(2f-f^2)^{1/2}$) = 0.081813334
Eccentricity squared (e^2) = 0.00669342

Fischer South Asia ellipsoid of 1960

Equatorial radius (a) = 6,378,155 meters
= 3,443.928 nautical miles
Polar radius (b) = 6,356,773.32 meters
= 3,432.383 nautical miles
Mean radius $(2a+b)/3$ = 6,371,027.773 meters
= 3,440.08 nautical miles
Flattening or ellipticity ($f=1-b/a$) = 1/298.3
= 0.00335233
Eccentricity ($e=(2f-f^2)^{1/2}$) = 0.081813334
Eccentricity squared (e^2) = 0.00669342

Fischer ellipsoid of 1968

Equatorial radius (a) = 6,378,150 meters
= 3,443.925 nautical miles
Polar radius (b) = 6,356,768.337 meters
= 3,432.381 nautical miles
Mean radius $(2a+b)/3$ = 6,371,022.985 meters
= 3,440.077 nautical miles
Flattening or ellipticity ($f=1-b/a$) = 1/298.3
= 0.00335233
Eccentricity ($e=(2f-f^2)^{1/2}$) = 0.08181333

Earth—Continued

Eccentricity squared (e^2) .... $=0.00669342$

Hough ellipsoid

Equatorial radius (a) .... $=6{,}378{,}270$ meters
$=3{,}443.99$ nautical miles

Polar radius (b) .... $=6{,}356{,}794.343$ meters
$=3{,}432.394$ nautical miles

Mean radius $(2a+b)/3$ .... $=6{,}371{,}111.448$ meters
$=3{,}440.125$ nautical miles

Flattening or ellipticity ($f=1-b/a$) .... $=1/297$
$=0.003367003$

Eccentricity ($e=(2f-f^2)^{1/2}$) .... $=0.08199189$

Eccentricity squared (e^2) .... $=0.00672267$

International ellipsoid

Equatorial radius (a) .... $=6{,}378{,}388$ meters
$=3{,}444.054$ nautical miles

Polar radius (b) .... $=6{,}356{,}911.946$ meters
$=3{,}432.459$ nautical miles

Mean radius $(2a+b)/3$ .... $=6{,}371{,}229.315$ meters
$=3{,}440.19$ nautical miles

Flattening or ellipticity ($f=1-b/a$) .... $=1/297$
$=0.003367003$

Eccentricity ($e=(2f-f^2)^{1/2}$) .... $=0.08199189$

Eccentricity squared (e^2) .... $=0.00672267$

International Astronomical Union figure of earth (1968)

Equatorial radius (a) .... $=6{,}378{,}160$ meters
$=3{,}443.931$ nautical miles

Polar radius (b) .... $=6{,}356{,}774.719$ meters
$=3{,}432.384$ nautical miles

Mean radius $(2a+b)/3$ .... $=6{,}371{,}031.573$ meters
$=3{,}440.082$ nautical miles

Flattening or ellipticity ($f=1-b/a$) .... $=1/298.25$
$=0.00335289$

Eccentricity ($e=(2f-f^2)^{1/2}$) .... $=0.0818202$

Eccentricity squared (e^2) .... $=0.00669454$

Krassovskiy ellipsoid

Equatorial radius (a) .... $=6{,}378{,}245$ meters
$=3{,}443.977$ nautical miles

Polar radius (b) .... $=6{,}356{,}863.019$ meters
$=3{,}432.43$ nautical miles

Mean radius $(2a+b)/3$ .... $=6{,}371{,}117.673$ meters
$=3{,}440.128$ nautical miles

Flattening or ellipticity ($f=1-b/a$) .... $=1/298.3$
$=0.00335233$

Eccentricity ($e=(2f-f^2)^{1/2}$) .... $=0.08181333$

Eccentricity squared (e^2) .... $=0.00669342$

World Geodetic System (WGS) ellipsoid of 1972

Equatorial radius (a) .... $=6{,}378{,}135$ meters
$=3{,}443.917$ nautical miles

Polar radius (b) .... $=6{,}356{,}750.52$ meters
$=3{,}432.371$ nautical miles

Mean radius $(2a+b)/3$ .... $=6{,}371{,}006.84$ meters
$=3{,}440.068$ nautical miles

Flattening or ellipticity ($f=1-b/a$) .... $=1/298.26$
$=0.00335278$

Eccentricity ($e=(2f-f^2)^{1/2}$) .... $=0.0818188$

Earth—Continued

Eccentricity squared (e^2)	=0.00669432

Length

1 inch	=25.4 millimeters*
	=2.54 centimeters*
1 foot (U.S.)	=12 inches*
	=1 British foot
	=⅓ yard*
	=0.3048 meter*
	=⅙ fathom*
1 foot (U.S. Survey)	=0.30480061 meter
1 yard	=36 inches*
	=3 feet*
	=0.9144 meter*
1 fathom	=6 feet*
	=2 yards*
	=1.8288 meters*
1 cable	=720 feet*
	=240 yards*
	=219.4560 meters*
1 cable (British)	=0.1 nautical mile
1 statute mile	=5,280 feet*
	=1,760 yards*
	=1,609.344 meters*
	=1.609344 kilometers*
	=0.86897624 nautical mile
1 nautical mile	=6,076.11548556 feet
	=2,025.37182852 yards
	=1,852 meters*
	=1.852 kilometers*
	=1.150779448 statute miles
1 meter	=100 centimeters*
	=39.370079 inches
	=3.28083990 feet
	=1.09361330 yards
	=0.54680665 fathom
	=0.00062137 statute mile
	=0.00053996 nautical mile
1 kilometer	=3,280.83990 feet
	=1,093.61330 yards
	=1,000 meters*
	=0.62137119 statute mile
	=0.53995680 nautical mile

Mass

1 ounce	=437.5 grains*
	=28.349523125 grams*
	=0.0625 pound*
	=0.028349523125 kilogram*
1 pound	=7,000 grains*
	=16 ounces*
	=0.45359237 kilogram*
1 short ton	=2,000 pounds*
	=907.18474 kilograms*
	=0.90718474 metric ton*
	=0.8928571 long ton

Mass—Continued

1 long ton	=2,240 pounds*
	=1,016.0469088 kilograms*
	=1.12 short tons*
	=1.0160469088 metric tons*
1 kilogram	=2.204623 pounds
	=0.00110231 short ton
	=0.0009842065 long ton
1 metric ton	=2,204.623 pounds
	=1,000 kilograms*
	=1.102311 short tons
	=0.9842065 long ton

Mathematics

π	=3.14159265358979323846264338327950288841971
π^2	=9.8696044011
$\sqrt{\pi}$	=1.7724538509
Base of Naperian logarithms (e)	=2.718281828459
Modulus of common logarithms ($\log_{10}e$)	=0.4342944819032518
1 radian	=206,264.″80625
	=3,437.′7467707849
	=57.°2957795131
	=57°17′44.″80625
1 circle	=1,296,000″*
	=21,600′*
	=360°*
	=2π radians*
180°	=π radians*
1°	=3600″*
	=60′*
	=0.0174532925199432957666 radian
1′	=60″*
	=0.000290888208665721596 radian
1″	=0.000004848136811095359933 radian
Sine of 1′	=0.00029088820456342460
Sine of 1″	=0.00000484813681107637

Meteorology

Atmosphere (dry air)

Nitrogen	=78.08%	99.99%
Oxygen	=20.95%	
Argon	=0.93%	
Carbon dioxide	=0.03%	
Neon	=0.0018%	
Helium	=0.000524%	
Krypton	=0.0001%	
Hydrogen	=0.00005%	
Xenon	=0.0000087%	
Ozone	=0 to 0.000007% (increasing with altitude)	
Radon	=0.000000000000000006% (decreasing with altitude)	

Meteorology—Continued

Standard atmospheric pressure at sea level	=1,013.250 dynes per square centimeter*
	=1,033.227 grams per square centimeter
	=1,033.227 centimeters of water
	=1,013.250 millibars*
	=760 millimeters of mercury
	=76 centimeters of mercury
	=33.8985 feet of water
	=29.92126 inches of mercury
	=14.6960 pounds per square inch
	=1.033227 kilograms per square centimeter
	=1.013250 bars*
Absolute zero	=(−) 273°.16 C
	=(−) 459°.69 F

Pressure

1 dyne per square centimeter	=0.001 millibar*
	=0.000001 bar*
1 gram per square centimeter	=1 centimeter of water
	=0.980665 millibar*
	=0.07355592 centimeter of mercury
	=0.0289590 inch of mercury
	=0.0142233 pound per square inch
	=0.001 kilogram per square centimeter*
	=0.000967841 atmosphere
1 millibar	=1,000 dynes per square centimeter*
	=1.01971621 grams per square centimeter
	=0.7500617 millimeter of mercury
	=0.03345526 foot of water
	=0.02952998 inch of mercury
	=0.01450377 pound per square inch
	=0.001 bar*
	=0.00098692 atmosphere
1 millimeter of mercury	=1.35951 grams per square centimeter
	=1.3332237 millibars
	=0.1 centimeter of mercury*
	=0.04460334 foot of water
	=0.039370079 inch of mercury
	=0.01933677 pound per square inch
	=0.001315790 atmosphere
1 centimeter of mercury	=10 millimeters of mercury*
1 inch of mercury	=34.53155 grams per square centimeter
	=33.86389 millibars
	=25.4 millimeters of mercury*
	=1.132925 feet of water
	=0.4911541 pound per square inch
	=0.03342106 atmosphere
1 centimeter of water	=1 gram per square centimeter
	=0.001 kilogram per square centimeter
1 foot of water	=30.48000 grams per square centimeter
	=29.89067 millibars
	=2.241985 centimeters of mercury
	=0.882671 inch of mercury
	=0.4335275 pound per square inch
	=0.02949980 atmosphere

Pressure—Continued

1 pound per square inch	=68,947.57 dynes per square centimeter
	=70.30696 grams per square centimeter
	=70.30696 centimeters of water
	=68.94757 millibars
	=51.71493 millimeters of mercury
	=5.171493 centimeters of mercury
	=2.306659 feet of water
	=2.036021 inches of mercury
	=0.07030696 kilogram per square centimeter
	=0.06894757 bar
	=0.06804596 atmosphere
1 kilogram per square centimeter	=1,000 grams per square centimeter*
	=1,000 centimeters of water
1 bar	=1,000,000 dynes per square centimeter*
	=1,000 millibars*

Speed

1 foot per minute	=0.01666667 foot per second
	=0.00508 meter per second*
1 yard per minute	=3 feet per minute*
	=0.05 foot per second*
	=0.03409091 statute mile per hour
	=0.02962419 knot
	=0.01524 meter per second*
1 foot per second	=60 feet per minute*
	=20 yards per minute*
	=1.09728 kilometers per hour*
	=0.68181818 statute mile per hour
	=0.59248380 knot
	=0.3048 meter per second*
1 statute mile per hour	=88 feet per minute*
	=29.33333333 yards per minute
	=1.609344 kilometers per hour*
	=1.46666667 feet per second
	=0.86897624 knot
	=0.44704 meter per second*
1 knot	=101.26859143 feet per minute
	=33.75619714 yards per minute
	=1.852 kilometers per hour*
	=1.68780986 feet per second
	=1.15077945 statute miles per hour
	=0.51444444 meter per second
1 kilometer per hour	=0.62137119 statute mile per hour
	=0.53995680 knot
1 meter per second	=196.85039340 feet per minute
	=65.6167978 yards per minute
	=3.6 kilometers per hour*
	=3.28083990 feet per second
	=2.23693632 statute miles per hour
	=1.94384449 knots
Light in vacuo	=299,792.5 kilometers per second
	=186,282 statute miles per second
	=161,875 nautical miles per second
	=983.570 feet per microsecond
Light in air	=299,708 kilometers per second
	=186,230 statute miles per second
	=161,829 nautical miles per second
	=983.294 feet per microsecond

Speed—Continued

Sound in dry air at 59°F or 15°C and standard sea level pressure.	=1,116.45 feet per second
	=761.22 statute miles per hour
	=661.48 knots
	=340.29 meters per second
Sound in 3.485 percent saltwater at 60°F	=4,945.37 feet per second
	=3,371.85 statute miles per hour
	=2,930.05 knots
	=1,507.35 meters per second

Volume

1 cubic inch	=16.387064 cubic centimeters*
	=0.016387064 liter*
	=0.004329004 gallon
1 cubic foot	=1,728 cubic inches*
	=28.316846592 liters*
	=7.480519 U.S. gallons
	=6.228822 imperial (British) gallons
	=0.028316846592 cubic meter*
1 cubic yard	=46,656 cubic inches*
	=764.554857984 liters*
	=201.974026 U.S. gallons
	=168.1782 imperial (British) gallons
	=27 cubic feet*
	=0.764554857984 cubic meter*
1 milliliter	=0.06102374 cubic inch
	=0.0002641721 U.S. gallon
	=0.00021997 imperial (British) gallon
1 cubic meter	=264.172035 U.S. gallons
	=219.96878 imperial (British) gallons
	=35.31467 cubic feet
	=1.307951 cubic yards
1 quart (U.S.)	=57.75 cubic inches*
	=32 fluid ounces*
	=2 pints*
	=0.9463529 liter
	=0.25 gallon*
1 gallon (U.S.)	=3,785.412 milliliters
	=231 cubic inches*
	=0.1336806 cubic foot
	=4 quarts*
	=3.785412 liters
	=0.8326725 imperial (British) gallon
1 liter	=1,000 milliliters
	=61.02374 cubic inches
	=1.056688 quarts
	=0.2641721 gallon
1 register ton	=100 cubic feet*
	=2.8316846592 cubic meters*
1 measurement ton	=40 cubic feet*
	=1 freight ton*
1 freight ton	=40 cubic feet*
	=1 measurement ton*

Volume-mass

1 cubic foot of seawater .... = 64 pounds
1 cubic foot of freshwater .... = 62.428 pounds at temperature of maximum density (4°C = 39°.2F)
1 cubic foot of ice .... = 56 pounds
1 displacement ton .... = 35 cubic feet of seawater*
= 1 long ton

Vessel Tonnage

The several kinds of *vessel tonnage* are as follows:

Gross tonnage, or *gross register tonnage*, is the total cubical capacity of a ship expressed in *register* tons of 100 cubic feet, or 2.83 cubic meters, less such space as hatchways, bakeries, galleys, etc., as are exempted from measurement by different governments. There is some lack of uniformity in the gross tonnages as given by different nations on account of lack of agreement on the spaces that are to be exempted.

Official merchant marine statistics of most countries are published in terms of the *gross register tonnage*. Press references to *ship tonnage* are usually to the *gross tonnage*.

The *net tonnage*, or *net register tonnage*, is the *gross tonnage* less the different spaces specified by maritime nations in their measurement rules and laws. The spaces that are deducted are those totally unavailable for carrying cargo, such as the engine room, coal bunkers, crews quarters, chart and instrument room, etc.

The *net tonnage* is used in computing the amount of cargo that can be loaded on a ship. It is used as the basis for wharfage and other similar charges.

The *register under-deck tonnage* is the cubical capacity of a ship under her tonnage deck expressed in register tons. In a vessel having more than one deck the tonnage deck is the second from the keel.

There are several variations of *displacement tonnage*.

The *dead weight tonnage* is the difference between the "loaded" and "light" *displacement tonnages* of a vessel. It is expressed in terms of the long ton of 2,240 pounds, or the metric ton of 2,204.6 pounds, and is the weight of fuel, passengers, and cargo that a vessel can carry when loaded to her maximum draft.

The second variety of tonnage, *cargo tonnage*, refers to the weight of the particular items making up the cargo. In overseas traffic it is usually expressed in long tons of 2,240 pounds or metric tons of 2,204.6 pounds. The short ton is only occasionally used. The *cargo tonnage* is therefore very distinct from *vessel tonnage*.

NOTE:—All values in this appendix are based on the following relationships:
1 inch = 2.54 centimeters*
1 yard = 0.9144 meter*
1 pound (avoirdupois) = 0.45359237 kilogram*
1 nautical mile = 1852 meters*
Absolute zero = (−)273°.16C = (−)459°.69F.

Decibel Scale

The **decibel (dB)** is 10 times the logarithm to the base 10 of the ratio of two amounts of power. The decibel scale is used to express conveniently the ratio between widely different powers.

The ratio between one power P_1 and a second power P_2 is expressed in dB's as:

$$10 \log_{10}\left(\frac{P_1}{P_2}\right).$$

Thus if P_1 is 1,000 times P_2, their ratio is expressed as

$$10 \log_{10} (10)=10\times3=30 \text{ dB}.$$

If P_2 is 1,000 times P_1, their ratio is expressed as

$$10 \log (10)=10\times(-3)=-30 \text{ dB}.$$

Power ratio	dB
1	0
2	3
4	6
10	10
100	20
1, 000	30
10, 000	40
100, 000	50
1, 000, 000	60

Prefixes to Form Decimal Multiples and Sub-Multiples of International System of Units (SI)

Multiplying factor	Prefix	Symbol
1 000 000 000 000 $=10^{12}$	tera	T
1 000 000 000 $=10^{9}$	giga	G
1 000 000 $=10^{6}$	mega	M
1 000 $=10^{3}$	kilo	k
100 $=10^{2}$	hecto	h
10 $=10^{1}$	deka	da
0. 1 $=10^{-1}$	deci	d
0. 01 $=10^{-2}$	centi	c
0. 001 $=10^{-3}$	milli	m
0. 000 001 $=10^{-6}$	micro	μ
0. 000 000 001 $=10^{-9}$	nano	n
0. 000 000 000 001 $=10^{-12}$	pico	p
0. 000 000 000 000 001 $=10^{-15}$	femto	f
0. 000 000 000 000 000 001 $=10^{-18}$	atto	a

APPENDIX E

NAVIGATIONAL COORDINATES

Coordinate	Symbol	Measured from	Measured along	Direction	Measured to	Units	Precision	Maximum value	Labels
latitude	L, lat.	equator	meridian	N, S	parallel	°, ′	0′.1	90°	N, S
colatitude	colat.	poles	meridian	S, N	parallel	°, ′	0′.1	90°	—
longitude	λ, long.	prime meridian	parallel	E, W	local meridian	°, ′	0′.1	180°	E, W
declination	d, dec.	celestial equator	hour circle	N, S	parallel of declination	°, ′	0′.1	90°	N, S
polar distance	p	elevated pole	hour circle	S, N	parallel of declination	°, ′	0′.1	180°	—
altitude	h	horizon	vertical circle	up	parallel of altitude	°, ′	0′.1	90°*	—
zenith distance	z	zenith	vertical circle	down	parallel of altitude	°, ′	0′.1	180°	—
azimuth	Zn	north	horizon	E	vertical circle	°	0°.1	360°	—
azimuth angle	Z	north, south	horizon	E, W	vertical circle	°	0°.1	180° or 90°	N, S ... E, W
amplitude	A	east, west	horizon	N, S	body	°	0°.1	90°	E, W ... N, S
Greenwich hour angle	GHA	Greenwich celestial meridian	parallel of declination	W	hour circle	°, ′	0′.1	360°	—
local hour angle	LHA	local celestial meridian	parallel of declination	W	hour circle	°, ′	0′.1	360°	—
meridian angle	t	local celestial meridian	parallel of declination	E, W	hour circle	°, ′	0′.1	180°	E, W
sidereal hour angle	SHA	hour circle of vernal equinox	parallel of declination	W	hour circle	°, ′	0′.1	360°	—
right ascension	RA	hour circle of vernal equinox	parallel of declination	E	hour circle	^{h}, ^{m}, ^{s}	1^s	24^h	—
Greenwich mean time	GMT	lower branch Greenwich celestial meridian	parallel of declination	W	hour circle mean sun	^{h}, ^{m}, ^{s}	1^s	24^h	—
local mean time	LMT	lower branch local celestial meridian	parallel of declination	W	hour circle mean sun	^{h}, ^{m}, ^{s}	1^s	24^h	—
zone time	ZT	lower branch zone celestial meridian	parallel of declination	W	hour circle mean sun	^{h}, ^{m}, ^{s}	1^s	24^h	—
Greenwich apparent time	GAT	lower branch Greenwich celestial meridian	parallel of declination	W	hour circle apparent sun	^{h}, ^{m}, ^{s}	1^s	24^h	—
local apparent time	LAT	lower branch local celestial meridian	parallel of declination	W	hour circle apparent sun	^{h}, ^{m}, ^{s}	1^s	24^h	—
Greenwich sidereal time	GST	Greenwich celestial meridian	parallel of declination	W	hour circle vernal equinox	^{h}, ^{m}, ^{s}	1^s	24^h	—
local sidereal time	LST	local celestial meridian	parallel of declination	W	hour circle vernal equinox	^{h}, ^{m}, ^{s}	1^s	24^h	—

*When measured from celestial horizon.

APPENDIX F

EXTRACTS FROM *NAUTICAL ALMANAC*

ALTITUDE CORRECTION TABLES 10°-90°—SUN, STARS, PLANETS

SUN (OCT.—MAR. / APR.—SEPT.) — each correction applies between the App. Alt. in its row and the next.

OCT.—MAR. App. Alt.	Lower Limb	Upper Limb	APR.—SEPT. App. Alt.	Lower Limb	Upper Limb
° ′	′	′	° ′	′	′
9 34	+10.8	−21.5	9 39	+10.6	−21.2
9 45	+10.9	−21.4	9 51	+10.7	−21.1
9 56	+11.0	−21.3	10 03	+10.8	−21.0
10 08	+11.1	−21.2	10 15	+10.9	−20.9
10 21	+11.2	−21.1	10 27	+11.0	−20.8
10 34	+11.3	−21.0	10 40	+11.1	−20.7
10 47	+11.4	−20.9	10 54	+11.2	−20.6
11 01	+11.5	−20.8	11 08	+11.3	−20.5
11 15	+11.6	−20.7	11 23	+11.4	−20.4
11 30	+11.7	−20.6	11 38	+11.5	−20.3
11 46	+11.8	−20.5	11 54	+11.6	−20.2
12 02	+11.9	−20.4	12 10	+11.7	−20.1
12 19	+12.0	−20.3	12 28	+11.8	−20.0
12 37	+12.1	−20.2	12 46	+11.9	−19.9
12 55	+12.2	−20.1	13 05	+12.0	−19.8
13 14	+12.3	−20.0	13 24	+12.1	−19.7
13 35	+12.4	−19.9	13 45	+12.2	−19.6
13 56	+12.5	−19.8	14 07	+12.3	−19.5
14 18	+12.6	−19.7	14 30	+12.4	−19.4
14 42	+12.7	−19.6	14 54	+12.5	−19.3
15 06	+12.8	−19.5	15 19	+12.6	−19.2
15 32	+12.9	−19.4	15 46	+12.7	−19.1
15 59	+13.0	−19.3	16 14	+12.8	−19.0
16 28	+13.1	−19.2	16 44	+12.9	−18.9
16 59	+13.2	−19.1	17 15	+13.0	−18.8
17 32	+13.3	−19.0	17 48	+13.1	−18.7
18 06	+13.4	−18.9	18 24	+13.2	−18.6
18 42	+13.5	−18.8	19 01	+13.3	−18.5
19 21	+13.6	−18.7	19 42	+13.4	−18.4
20 03	+13.7	−18.6	20 25	+13.5	−18.3
20 48	+13.8	−18.5	21 11	+13.6	−18.2
21 35	+13.9	−18.4	22 00	+13.7	−18.1
22 26	+14.0	−18.3	22 54	+13.8	−18.0
23 22	+14.1	−18.2	23 51	+13.9	−17.9
24 21	+14.2	−18.1	24 53	+14.0	−17.8
25 26	+14.3	−18.0	26 00	+14.1	−17.7
26 36	+14.4	−17.9	27 13	+14.2	−17.6
27 52	+14.5	−17.8	28 33	+14.3	−17.5
29 15	+14.6	−17.7	30 00	+14.4	−17.4
30 46	+14.7	−17.6	31 35	+14.5	−17.3
32 26	+14.8	−17.5	33 20	+14.6	−17.2
34 17	+14.9	−17.4	35 17	+14.7	−17.1
36 20	+15.0	−17.3	37 26	+14.8	−17.0
38 36	+15.1	−17.2	39 50	+14.9	−16.9
41 08	+15.2	−17.1	42 31	+15.0	−16.8
43 59	+15.3	−17.0	45 31	+15.1	−16.7
47 10	+15.4	−16.9	48 55	+15.2	−16.6
50 46	+15.5	−16.8	52 44	+15.3	−16.5
54 49	+15.6	−16.7	57 02	+15.4	−16.4
59 23	+15.7	−16.6	61 51	+15.5	−16.3
64 30	+15.8	−16.5	67 17	+15.6	−16.2
70 12	+15.9	−16.4	73 16	+15.7	−16.1
76 26	+16.0	−16.3	79 43	+15.8	−16.0
83 05	+16.1	−16.2	86 32	+15.9	−15.9
90 00			90 00		

STARS AND PLANETS

App. Alt.	Corr^n
° ′	′
9 56	−5.3
10 08	−5.2
10 20	−5.1
10 33	−5.0
10 46	−4.9
11 00	−4.8
11 14	−4.7
11 29	−4.6
11 45	−4.5
12 01	−4.4
12 18	−4.3
12 35	−4.2
12 54	−4.1
13 13	−4.0
13 33	−3.9
13 54	−3.8
14 16	−3.7
14 40	−3.6
15 04	−3.5
15 30	−3.4
15 57	−3.3
16 26	−3.2
16 56	−3.1
17 28	−3.0
18 02	−2.9
18 38	−2.8
19 17	−2.7
19 58	−2.6
20 42	−2.5
21 28	−2.4
22 19	−2.3
23 13	−2.2
24 11	−2.1
25 14	−2.0
26 22	−1.9
27 36	−1.8
28 56	−1.7
30 24	−1.6
32 00	−1.5
33 45	−1.4
35 40	−1.3
37 48	−1.2
40 08	−1.1
42 44	−1.0
45 36	−0.9
48 47	−0.8
52 18	−0.7
56 11	−0.6
60 28	−0.5
65 08	−0.4
70 11	−0.3
75 34	−0.2
81 13	−0.1
87 03	0.0
90 00	

App. Alt.	Additional Corr^n
1975	
VENUS	
Jan. 1 – June 7	
0°	+0.1′
42	
June 8 – July 21	
0°	+0.3′
46	
July 22 – Aug. 6	
0°	+0.4′
11	+0.5
41	
Aug. 7 – Aug. 15	
0°	+0.5′
6	+0.6
20	+0.7
31	
Aug. 16 – Sept. 10	
0°	+0.6′
4	+0.7
12	+0.8
22	
Sept. 11 – Sept. 19	
0°	+0.5′
6	+0.6
20	+0.7
31	
Sept. 20 – Oct. 5	
0°	+0.4′
11	+0.5
41	
Oct. 6 – Nov. 22	
0°	+0.3′
46	
Nov. 23 – Dec. 31	
0°	+0.1′
42	
MARS	
Jan. 1 – Sept. 8	
0°	+0.1′
60	
Sept. 9 – Nov. 22	
0°	+0.2′
41	+0.1
75	
Nov. 23 – Dec. 31	
0°	+0.3′
34	+0.2
60	+0.1
80	

DIP

Ht. of Eye	Corr^n	Ht. of Eye
m	′	ft.
2.4	−2.8	8.0
2.6	−2.9	8.6
2.8	−3.0	9.2
3.0	−3.1	9.8
3.2	−3.2	10.5
3.4	−3.3	11.2
3.6	−3.4	11.9
3.8	−3.5	12.6
4.0	−3.6	13.3
4.3	−3.7	14.1
4.5	−3.8	14.9
4.7	−3.9	15.7
5.0	−4.0	16.5
5.2	−4.1	17.4
5.5	−4.2	18.3
5.8	−4.3	19.1
6.1	−4.4	20.1
6.3	−4.5	21.0
6.6	−4.6	22.0
6.9	−4.7	22.9
7.2	−4.8	23.9
7.5	−4.9	24.9
7.9	−5.0	26.0
8.2	−5.1	27.1
8.5	−5.2	28.1
8.8	−5.3	29.2
9.2	−5.4	30.4
9.5	−5.5	31.5
9.9	−5.6	32.7
10.3	−5.7	33.9
10.6	−5.8	35.1
11.0	−5.9	36.3
11.4	−6.0	37.6
11.8	−6.1	38.9
12.2	−6.2	40.1
12.6	−6.3	41.5
13.0	−6.4	42.8
13.4	−6.5	44.2
13.8	−6.6	45.5
14.2	−6.7	46.9
14.7	−6.8	48.4
15.1	−6.9	49.8
15.5	−7.0	51.3
16.0	−7.1	52.8
16.5	−7.2	54.3
16.9	−7.3	55.8
17.4	−7.4	57.4
17.9	−7.5	58.9
18.4	−7.6	60.5
18.8	−7.7	62.1
19.3	−7.8	63.8
19.8	−7.9	65.4
20.4	−8.0	67.1
20.9	−8.1	68.8
21.4		70.5

Ht. of Eye	Corr^n
m	′
1.0	−1.8
1.5	−2.2
2.0	−2.5
2.5	−2.8
3.0	−3.0
See table ←	
m	′
20	−7.9
22	−8.3
24	−8.6
26	−9.0
28	−9.3
30	−9.6
32	−10.0
34	−10.3
36	−10.6
38	−10.8
40	−11.1
42	−11.4
44	−11.7
46	−11.9
48	−12.2
ft.	′
2	−1.4
4	−1.9
6	−2.4
8	−2.7
10	−3.1
See table ←	
ft.	′
70	−8.1
75	−8.4
80	−8.7
85	−8.9
90	−9.2
95	−9.5
100	−9.7
105	−9.9
110	−10.2
115	−10.4
120	−10.6
125	−10.8
130	−11.1
135	−11.3
140	−11.5
145	−11.7
150	−11.9
155	−12.1

App. Alt. = Apparent altitude = Sextant altitude corrected for index error and dip.

For daylight observations of Venus, see page 260.

ALTITUDE CORRECTION TABLES 0°–10°—SUN, STARS, PLANETS A3

App. Alt.	OCT.–MAR. SUN Lower Limb	OCT.–MAR. SUN Upper Limb	APR.–SEPT. SUN Lower Limb	APR.–SEPT. SUN Upper Limb	STARS PLANETS
° ′	′	′	′	′	′
0 00	−18·2	−50·5	−18·4	−50·2	−34·5
03	17·5	49·8	17·8	49·6	33·8
06	16·9	49·2	17·1	48·9	33·2
09	16·3	48·6	16·5	48·3	32·6
12	15·7	48·0	15·9	47·7	32·0
15	15·1	47·4	15·3	47·1	31·4
0 18	−14·5	−46·8	−14·8	−46·6	−30·8
21	14·0	46·3	14·2	46·0	30·3
24	13·5	45·8	13·7	45·5	29·8
27	12·9	45·2	13·2	45·0	29·2
30	12·4	44·7	12·7	44·5	28·7
33	11·9	44·2	12·2	44·0	28·2
0 36	−11·5	−43·8	−11·7	−43·5	−27·8
39	11·0	43·3	11·2	43·0	27·3
42	10·5	42·8	10·8	42·6	26·8
45	10·1	42·4	10·3	42·1	26·4
48	9·6	41·9	9·9	41·7	25·9
51	9·2	41·5	9·5	41·3	25·5
0 54	−8·8	−41·1	−9·1	−40·9	−25·1
0 57	8·4	40·7	8·7	40·5	24·7
1 00	8·0	40·3	8·3	40·1	24·3
03	7·7	40·0	7·9	39·7	24·0
06	7·3	39·6	7·5	39·3	23·6
09	6·9	39·2	7·2	39·0	23·2
1 12	−6·6	−38·9	−6·8	−38·6	−22·9
15	6·2	38·5	6·5	38·3	22·5
18	5·9	38·2	6·2	38·0	22·2
21	5·6	37·9	5·8	37·6	21·9
24	5·3	37·6	5·5	37·3	21·6
27	4·9	37·2	5·2	37·0	21·2
1 30	−4·6	−36·9	−4·9	−36·7	−20·9
35	4·2	36·5	4·4	36·2	20·5
40	3·7	36·0	4·0	35·8	20·0
45	3·2	35·5	3·5	35·3	19·5
50	2·8	35·1	3·1	34·9	19·1
1 55	2·4	34·7	2·6	34·4	18·7
2 00	−2·0	−34·3	−2·2	−34·0	−18·3
05	1·6	33·9	1·8	33·6	17·9
10	1·2	33·5	1·5	33·3	17·5
15	0·9	33·2	1·1	32·9	17·2
20	0·5	32·8	0·8	32·6	16·8
25	−0·2	32·5	0·4	32·2	16·5
2 30	+0·2	−32·1	−0·1	−31·9	−16·1
35	0·5	31·8	+0·2	31·6	15·8
40	0·8	31·5	0·5	31·3	15·5
45	1·1	31·2	0·8	31·0	15·2
50	1·4	30·9	1·1	30·7	14·9
2 55	1·6	30·7	1·4	30·4	14·7
3 00	+1·9	−30·4	+1·7	−30·1	−14·4
05	2·2	30·1	1·9	29·9	14·1
10	2·4	29·9	2·1	29·7	13·9
15	2·6	29·7	2·4	29·4	13·7
20	2·9	29·4	2·6	29·2	13·4
25	3·1	29·2	2·9	28·9	13·2
3 30	+3·3	−29·0	+3·1	−28·7	−13·0

App. Alt.	OCT.–MAR. SUN Lower Limb	OCT.–MAR. SUN Upper Limb	APR.–SEPT. SUN Lower Limb	APR.–SEPT. SUN Upper Limb	STARS PLANETS
° ′	′	′	′	′	′
3 30	+3·3	−29·0	+3·1	−28·7	−13·0
35	3·6	28·7	3·3	28·5	12·7
40	3·8	28·5	3·5	28·3	12·5
45	4·0	28·3	3·7	28·1	12·3
50	4·2	28·1	3·9	27·9	12·1
3 55	4·4	27·9	4·1	27·7	11·9
4 00	+4·5	−27·8	+4·3	−27·5	−11·8
05	4·7	27·6	4·5	27·3	11·6
10	4·9	27·4	4·6	27·2	11·4
15	5·1	27·2	4·8	27·0	11·2
20	5·2	27·1	5·0	26·8	11·1
25	5·4	26·9	5·1	26·7	10·9
4 30	+5·6	−26·7	+5·3	−26·5	−10·7
35	5·7	26·6	5·5	26·3	10·6
40	5·9	26·4	5·6	26·2	10·4
45	6·0	26·3	5·8	26·0	10·3
50	6·2	26·1	5·9	25·9	10·1
4 55	6·3	26·0	6·0	25·8	10·0
5 00	+6·4	−25·9	+6·2	−25·6	−9·9
05	6·6	25·7	6·3	25·5	9·7
10	6·7	25·6	6·4	25·4	9·6
15	6·8	25·5	6·6	25·2	9·5
20	6·9	25·4	6·7	25·1	9·4
25	7·1	25·2	6·8	25·0	9·2
5 30	+7·2	−25·1	+6·9	−24·9	−9·1
35	7·3	25·0	7·0	24·8	9·0
40	7·4	24·9	7·2	24·6	8·9
45	7·5	24·8	7·3	24·5	8·8
50	7·6	24·7	7·4	24·4	8·7
5 55	7·7	24·6	7·5	24·3	8·6
6 00	+7·8	−24·5	+7·6	−24·2	−8·5
10	8·0	24·3	7·8	24·0	8·3
20	8·2	24·1	8·0	23·8	8·1
30	8·4	23·9	8·1	23·7	7·9
40	8·6	23·7	8·3	23·5	7·7
6 50	8·7	23·6	8·5	23·3	7·6
7 00	+8·9	−23·4	+8·6	−23·2	−7·4
10	9·1	23·2	8·8	23·0	7·2
20	9·2	23·1	9·0	22·8	7·1
30	9·3	23·0	9·1	22·7	7·0
40	9·5	22·8	9·2	22·6	6·8
7 50	9·6	22·7	9·4	22·4	6·7
8 00	+9·7	−22·6	+9·5	−22·3	−6·6
10	9·9	22·4	9·6	22·2	6·4
20	10·0	22·3	9·7	22·1	6·3
30	10·1	22·2	9·8	22·0	6·2
40	10·2	22·1	10·0	21·8	6·1
8 50	10·3	22·0	10·1	21·7	6·0
9 00	+10·4	−21·9	+10·2	−21·6	−5·9
10	10·5	21·8	10·3	21·5	5·8
20	10·6	21·7	10·4	21·4	5·7
30	10·7	21·6	10·5	21·3	5·6
40	10·8	21·5	10·6	21·2	5·5
9 50	10·9	21·4	10·6	21·2	5·4
10 00	+11·0	−21·3	+10·7	−21·1	−5·3

Additional corrections for temperature and pressure are given on the following page.

For bubble sextant observations ignore dip and use the star corrections for Sun, planets, and stars.

ALTITUDE CORRECTION TABLES—ADDITIONAL CORRECTIONS

ADDITIONAL REFRACTION CORRECTIONS FOR NON-STANDARD CONDITIONS

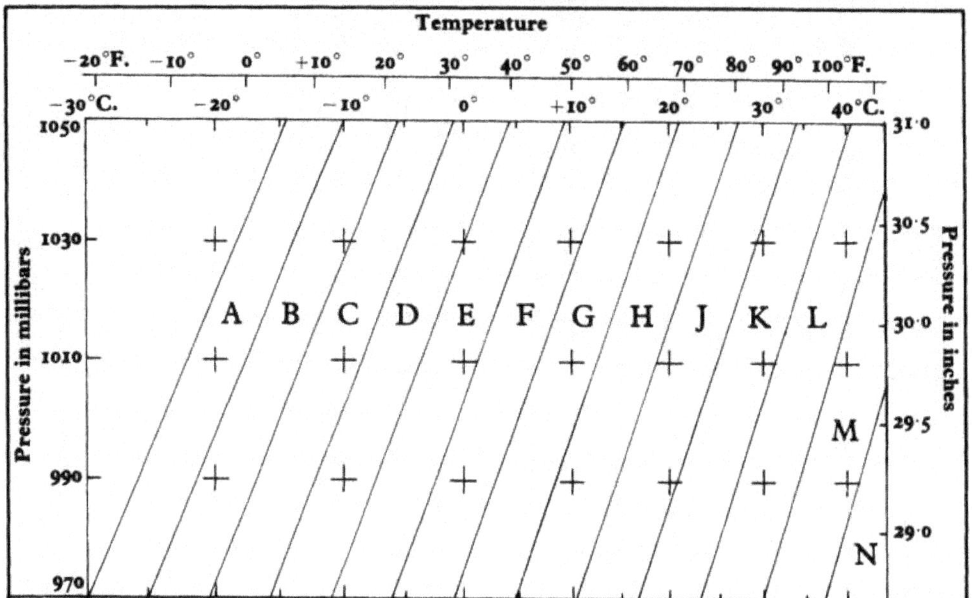

App. Alt.	A	B	C	D	E	F	G	H	J	K	L	M	N	App. Alt.
° ′	′	′	′	′	′	′	′	′	′	′	′	′	′	° ′
0 00	−6·9	−5·7	−4·6	−3·4	−2·3	−1·1	0·0	+1·1	+2·3	+3·4	+4·6	+5·7	+6·9	0 00
0 30	5·2	4·4	3·5	2·6	1·7	0·9	0·0	0·9	1·7	2·6	3·5	4·4	5·2	0 30
1 00	4·3	3·5	2·8	2·1	1·4	0·7	0·0	0·7	1·4	2·1	2·8	3·5	4·3	1 00
1 30	3·5	2·9	2·4	1·8	1·2	0·6	0·0	0·6	1·2	1·8	2·4	2·9	3·5	1 30
2 00	3·0	2·5	2·0	1·5	1·0	0·5	0·0	0·5	1·0	1·5	2·0	2·5	3·0	2 00
2 30	−2·5	−2·1	−1·6	−1·2	−0·8	−0·4	0·0	+0·4	+0·8	+1·2	+1·6	+2·1	+2·5	2 30
3 00	2·2	1·8	1·5	1·1	0·7	0·4	0·0	0·4	0·7	1·1	1·5	1·8	2·2	3 00
3 30	2·0	1·6	1·3	1·0	0·7	0·3	0·0	0·3	0·7	1·0	1·3	1·6	2·0	3 30
4 00	1·8	1·5	1·2	0·9	0·6	0·3	0·0	0·3	0·6	0·9	1·2	1·5	1·8	4 00
4 30	1·6	1·4	1·1	0·8	0·5	0·3	0·0	0·3	0·5	0·8	1·1	1·4	1·6	4 30
5 00	−1·5	−1·3	−1·0	−0·8	−0·5	−0·2	0·0	+0·2	+0·5	+0·8	+1·0	+1·3	+1·5	5 00
6	1·3	1·1	0·9	0·6	0·4	0·2	0·0	0·2	0·4	0·6	0·9	1·1	1·3	6
7	1·1	0·9	0·7	0·6	0·4	0·2	0·0	0·2	0·4	0·6	0·7	0·9	1·1	7
8	1·0	0·8	0·7	0·5	0·3	0·2	0·0	0·2	0·3	0·5	0·7	0·8	1·0	8
9	0·9	0·7	0·6	0·4	0·3	0·1	0·0	0·1	0·3	0·4	0·6	0·7	0·9	9
10 00	−0·8	−0·7	−0·5	−0·4	−0·3	−0·1	0·0	+0·1	+0·3	+0·4	+0·5	+0·7	+0·8	10 00
12	0·7	0·6	0·5	0·3	0·2	0·1	0·0	0·1	0·2	0·3	0·5	0·6	0·7	12
14	0·6	0·5	0·4	0·3	0·2	0·1	0·0	0·1	0·2	0·3	0·4	0·5	0·6	14
16	0·5	0·4	0·3	0·3	0·2	0·1	0·0	0·1	0·2	0·3	0·3	0·4	0·5	16
18	0·4	0·4	0·3	0·2	0·2	0·1	0·0	0·1	0·2	0·2	0·3	0·4	0·4	18
20 00	−0·4	−0·3	−0·3	−0·2	−0·1	−0·1	0·0	+0·1	+0·1	+0·2	+0·3	+0·3	+0·4	20 00
25	0·3	0·3	0·2	0·2	0·1	−0·1	0·0	+0·1	0·1	0·2	0·2	0·3	0·3	25
30	0·3	0·2	0·2	0·1	0·1	0·0	0·0	0·0	0·1	0·1	0·2	0·2	0·3	30
35	0·2	0·2	0·1	0·1	0·1	0·0	0·0	0·0	0·1	0·1	0·1	0·2	0·2	35
40	0·2	0·1	0·1	0·1	−0·1	0·0	0·0	0·0	+0·1	0·1	0·1	0·1	0·2	40
50 00	−0·1	−0·1	−0·1	−0·1	0·0	0·0	0·0	0·0	0·0	+0·1	+0·1	+0·1	+0·1	50 00

The graph is entered with arguments temperature and pressure to find a zone letter; using as arguments this zone letter and apparent altitude (sextant altitude corrected for dip), a correction is taken from the table. This correction is to be applied to the sextant altitude in addition to the corrections for standard conditions (for the Sun, planets and stars from the inside front cover and for the Moon from the inside back cover).

INDEX TO SELECTED STARS, 1975

Name	No.	Mag.	S.H.A.	Dec.
			°	°
Acamar	7	3·1	316	S. 40
Achernar	5	0·6	336	S. 57
Acrux	30	1·1	174	S. 63
Adhara	19	1·6	256	S. 29
Aldebaran	10	1·1	291	N. 16
Alioth	32	1·7	167	N. 56
Alkaid	34	1·9	153	N. 49
Al Na'ir	55	2·2	28	S. 47
Alnilam	15	1·8	276	S. 1
Alphard	25	2·2	218	S. 9
Alphecca	41	2·3	127	N. 27
Alpheratz	1	2·2	358	N. 29
Altair	51	0·9	63	N. 9
Ankaa	2	2·4	354	S. 42
Antares	42	1·2	113	S. 26
Arcturus	37	0·2	146	N. 19
Atria	43	1·9	108	S. 69
Avior	22	1·7	234	S. 59
Bellatrix	13	1·7	279	N. 6
Betelgeuse	16	Var.*	272	N. 7
Canopus	17	−0·9	264	S. 53
Capella	12	0·2	281	N. 46
Deneb	53	1·3	50	N. 45
Denebola	28	2·2	183	N. 15
Diphda	4	2·2	349	S. 18
Dubhe	27	2·0	194	N. 62
Elnath	14	1·8	279	N. 29
Eltanin	47	2·4	91	N. 51
Enif	54	2·5	34	N. 10
Fomalhaut	56	1·3	16	S. 30
Gacrux	31	1·6	173	S. 57
Gienah	29	2·8	176	S. 17
Hadar	35	0·9	149	S. 60
Hamal	6	2·2	329	N. 23
Kaus Australis	48	2·0	84	S. 34
Kochab	40	2·2	137	N. 74
Markab	57	2·6	14	N. 15
Menkar	8	2·8	315	N. 4
Menkent	36	2·3	149	S. 36
Miaplacidus	24	1·8	222	S. 70
Mirfak	9	1·9	309	N. 50
Nunki	50	2·1	77	S. 26
Peacock	52	2·1	54	S. 57
Pollux	21	1·2	244	N. 28
Procyon	20	0·5	245	N. 5
Rasalhague	46	2·1	97	N. 13
Regulus	26	1·3	208	N. 12
Rigel	11	0·3	282	S. 8
Rigil Kentaurus	38	0·1	141	S. 61
Sabik	44	2·6	103	S. 16
Schedar	3	2·5	350	N. 56
Shaula	45	1·7	97	S. 37
Sirius	18	−1·6	259	S. 17
Spica	33	1·2	159	S. 11
Suhail	23	2·2	223	S. 43
Vega	49	0·1	81	N. 39
Zubenelgenubi	39	2·9	138	S. 16

No.	Name	Mag.	S.H.A.	Dec.
			°	°
1	*Alpheratz*	2·2	358	N. 29
2	*Ankaa*	2·4	354	S. 42
3	*Schedar*	2·5	350	N. 56
4	*Diphda*	2·2	349	S. 18
5	*Achernar*	0·6	336	S. 57
6	*Hamal*	2·2	329	N. 23
7	*Acamar*	3·1	316	S. 40
8	*Menkar*	2·8	315	N. 4
9	*Mirfak*	1·9	309	N. 50
10	*Aldebaran*	1·1	291	N. 16
11	*Rigel*	0·3	282	S. 8
12	*Capella*	0·2	281	N. 46
13	*Bellatrix*	1·7	279	N. 6
14	*Elnath*	1·8	279	N. 29
15	*Alnilam*	1·8	276	S. 1
16	*Betelgeuse*	Var.*	272	N. 7
17	*Canopus*	−0·9	264	S. 53
18	*Sirius*	−1·6	259	S. 17
19	*Adhara*	1·6	256	S. 29
20	*Procyon*	0·5	245	N. 5
21	*Pollux*	1·2	244	N. 28
22	*Avior*	1·7	234	S. 59
23	*Suhail*	2·2	223	S. 43
24	*Miaplacidus*	1·8	222	S. 70
25	*Alphard*	2·2	218	S. 9
26	*Regulus*	1·3	208	N. 12
27	*Dubhe*	2·0	194	N. 62
28	*Denebola*	2·2	183	N. 15
29	*Gienah*	2·8	176	S. 17
30	*Acrux*	1·1	174	S. 63
31	*Gacrux*	1·6	173	S. 57
32	*Alioth*	1·7	167	N. 56
33	*Spica*	1·2	159	S. 11
34	*Alkaid*	1·9	153	N. 49
35	*Hadar*	0·9	149	S. 60
36	*Menkent*	2·3	149	S. 36
37	*Arcturus*	0·2	146	N. 19
38	*Rigil Kentaurus*	0·1	141	S. 61
39	*Zubenelgenubi*	2·9	138	S. 16
40	*Kochab*	2·2	137	N. 74
41	*Alphecca*	2·3	127	N. 27
42	*Antares*	1·2	113	S. 26
43	*Atria*	1·9	108	S. 69
44	*Sabik*	2·6	103	S. 16
45	*Shaula*	1·7	97	S. 37
46	*Rasalhague*	2·1	97	N. 13
47	*Eltanin*	2·4	91	N. 51
48	*Kaus Australis*	2·0	84	S. 34
49	*Vega*	0·1	81	N. 39
50	*Nunki*	2·1	77	S. 26
51	*Altair*	0·9	63	N. 9
52	*Peacock*	2·1	54	S. 57
53	*Deneb*	1·3	50	N. 45
54	*Enif*	2·5	34	N. 10
55	*Al Na'ir*	2·2	28	S. 47
56	*Fomalhaut*	1·3	16	S. 30
57	*Markab*	2·6	14	N. 15

* 0·1—1·2

1975 MAY 25, 26, 27 (SUN., MON., TUES.)

G.M.T.	ARIES	VENUS −3.8		MARS +1.0		JUPITER −1.7		SATURN +0.4		STARS		
	G.H.A.	G.H.A.	Dec.	G.H.A.	Dec.	G.H.A.	Dec.	G.H.A.	Dec.	Name	S.H.A.	Dec.
d h	° ′	° ′	° ′	° ′	° ′	° ′	° ′	° ′	° ′		° ′	° ′
25 00	241 57.5	133 22.6	N24 57.2	238 47.5	S 0 24.3	227 20.7	N 4 59.3	134 24.1	N22 20.6	Acamar	315 40.4	S40 24.1
01	257 00.0	148 22.1	56.9	253 48.2	23.5	242 22.7	59.5	149 26.2	20.5	Achernar	335 48.4	S57 21.5
02	272 02.4	163 21.6	56.5	268 49.0	22.8	257 24.7	59.7	164 28.4	20.5	Acrux	173 40.9	S62 58.1
03	287 04.9	178 21.1	·· 56.2	283 49.7	·· 22.1	272 26.7	4 59.9	179 30.6	·· 20.5	Adhara	255 35.3	S28 56.5
04	302 07.4	193 20.6	55.8	298 50.5	21.3	287 28.7	5 00.1	194 32.8	20.5	Aldebaran	291 22.5	N16 27.6
05	317 09.8	208 20.1	55.5	313 51.2	20.6	302 30.7	00.3	209 34.9	20.4			
06	332 12.3	223 19.6	N24 55.1	328 52.0	S 0 19.9	317 32.7	N 5 00.4	224 37.1	N22 20.4	Alioth	166 45.3	N56 05.7
07	347 14.7	238 19.1	54.8	343 52.7	19.1	332 34.7	00.6	239 39.3	20.4	Alkaid	153 20.9	N49 26.2
08	2 17.2	253 18.6	54.4	358 53.4	18.4	347 36.7	00.8	254 41.5	20.3	Al Na'ir	28 19.4	S47 04.5
S 09	17 19.7	268 18.1	·· 54.1	13 54.2	·· 17.6	2 38.7	·· 01.0	269 43.6	·· 20.3	Alnilam	276 15.7	S 1 13.1
U 10	32 22.1	283 17.6	53.7	28 54.9	16.9	17 40.7	01.2	284 45.8	20.3	Alphard	218 24.2	S 8 33.3
N 11	47 24.6	298 17.1	53.4	43 55.7	16.2	32 42.7	01.4	299 48.0	20.3			
D 12	62 27.1	313 16.6	N24 53.0	58 56.4	S 0 15.4	47 44.7	N 5 01.6	314 50.2	N22 20.2	Alphecca	126 34.8	N26 47.8
A 13	77 29.5	328 16.1	52.7	73 57.1	14.7	62 46.7	01.7	329 52.4	20.2	Alpheratz	358 13.2	N28 57.2
Y 14	92 32.0	343 15.7	52.3	88 57.9	14.0	77 48.7	01.9	344 54.5	20.2	Altair	62 35.8	N 8 48.2
15	107 34.5	358 15.2	·· 52.0	103 58.6	·· 13.2	92 50.7	·· 02.1	359 56.7	·· 20.2	Ankaa	353 44.0	S42 26.1
16	122 36.9	13 14.7	51.6	118 59.4	12.5	107 52.7	02.3	14 58.9	20.1	Antares	113 00.9	S26 22.7
17	137 39.4	28 14.2	51.3	134 00.1	11.8	122 54.7	02.5	30 01.1	20.1			
18	152 41.9	43 13.7	N24 50.9	149 00.9	S 0 11.0	137 56.7	N 5 02.7	45 03.2	N22 20.1	Arcturus	146 21.5	N19 18.6
19	167 44.3	58 13.2	50.5	164 01.6	10.3	152 58.7	02.9	60 05.4	20.0	Atria	108 27.9	S68 59.0
20	182 46.8	73 12.7	50.2	179 02.3	09.6	168 00.7	03.0	75 07.6	20.0	Avior	234 30.0	S59 26.2
21	197 49.2	88 12.3	·· 49.8	194 03.1	·· 08.8	183 02.7	·· 03.2	90 09.8	·· 20.0	Bellatrix	279 03.0	N 6 19.6
22	212 51.7	103 11.8	49.5	209 03.8	08.1	198 04.7	03.4	105 11.9	20.0	Betelgeuse	271 32.5	N 7 24.1
23	227 54.2	118 11.3	49.1	224 04.6	07.4	213 06.7	03.6	120 14.1	19.9			
26 00	242 56.6	133 10.8	N24 48.7	239 05.3	S 0 06.6	228 08.7	N 5 03.8	135 16.3	N22 19.9	Canopus	264 09.3	S52 41.2
01	257 59.1	148 10.3	48.4	254 06.1	05.9	243 10.7	04.0	150 18.5	19.9	Capella	281 17.1	N45 58.4
02	273 01.6	163 09.8	48.0	269 06.8	05.2	258 12.8	04.1	165 20.6	19.8	Deneb	49 50.7	N45 11.4
03	288 04.0	178 09.4	·· 47.6	284 07.5	·· 04.4	273 14.8	·· 04.3	180 22.8	·· 19.8	Denebola	183 02.6	N14 42.5
04	303 06.5	193 08.9	47.3	299 08.3	03.7	288 16.8	04.5	195 25.0	19.8	Diphda	349 24.7	S18 07.2
05	318 09.0	208 08.4	46.9	314 09.0	03.0	303 18.8	04.7	210 27.2	19.8			
06	333 11.4	223 07.9	N24 46.5	329 09.8	S 0 02.2	318 20.8	N 5 04.9	225 29.3	N22 19.7	Dubhe	194 26.4	N61 53.2
07	348 13.9	238 07.5	46.1	344 10.5	01.5	333 22.8	05.1	240 31.5	19.7	Elnath	278 49.1	N28 35.2
08	3 16.4	253 07.0	45.8	359 11.3	S 0 00.7	348 24.8	05.2	255 33.7	19.7	Eltanin	90 58.9	N51 29.4
M 09	18 18.8	268 06.5	·· 45.4	14 12.0	0 00.0	3 26.8	·· 05.4	270 35.9	·· 19.6	Enif	34 15.1	N 9 45.7
O 10	33 21.3	283 06.0	45.0	29 12.7	N 0 00.7	18 28.8	05.6	285 38.0	19.6	Fomalhaut	15 55.5	S29 45.0
N 11	48 23.7	298 05.6	44.6	44 13.5	01.5	33 30.8	05.8	300 40.2	19.6			
D 12	63 26.2	313 05.1	N24 44.3	59 14.2	N 0 02.2	48 32.8	N 5 06.0	315 42.4	N22 19.6	Gacrux	172 32.5	S56 58.9
A 13	78 28.7	328 04.6	43.9	74 15.0	02.9	63 34.8	06.2	330 44.6	19.5	Gienah	176 21.5	S17 24.6
Y 14	93 31.1	343 04.1	43.5	89 15.7	03.7	78 36.8	06.3	345 46.7	19.5	Hadar	149 27.9	S60 15.5
15	108 33.6	358 03.7	·· 43.1	104 16.5	·· 04.4	93 38.8	·· 06.5	0 48.9	·· 19.5	Hamal	328 33.3	N23 20.7
16	123 36.1	13 03.2	42.7	119 17.2	05.1	108 40.8	06.7	15 51.1	19.4	Kaus Aust.	84 21.3	S34 23.7
17	138 38.5	28 02.7	42.4	134 17.9	05.9	123 42.8	06.9	30 53.3	19.4			
18	153 41.0	43 02.3	N24 42.0	149 18.7	N 0 06.6	138 44.8	N 5 07.1	45 55.4	N22 19.4	Kochab	137 17.7	N74 15.5
19	168 43.5	58 01.8	41.6	164 19.4	07.3	153 46.8	07.3	60 57.6	19.4	Markab	14 06.8	N15 04.3
20	183 45.9	73 01.3	41.2	179 20.2	08.1	168 48.8	07.4	75 59.8	19.3	Menkar	314 45.2	N 3 59.6
21	198 48.4	88 00.9	·· 40.8	194 20.9	·· 08.8	183 50.8	·· 07.6	91 01.9	·· 19.3	Menkent	148 40.9	S36 15.2
22	213 50.9	103 00.4	40.4	209 21.7	09.5	198 52.9	07.8	106 04.1	19.3	Miaplacidus	221 46.0	S69 37.4
23	228 53.3	117 59.9	40.0	224 22.4	10.3	213 54.9	08.0	121 06.3	19.2			
27 00	243 55.8	132 59.5	N24 39.7	239 23.2	N 0 11.0	228 56.9	N 5 08.2	136 08.5	N22 19.2	Mirfak	309 21.8	N49 46.4
01	258 58.2	147 59.0	39.3	254 23.9	11.7	243 58.9	08.4	151 10.6	19.2	Nunki	76 33.4	S26 19.6
02	274 00.7	162 58.5	38.9	269 24.6	12.5	259 00.9	08.5	166 12.8	19.2	Peacock	54 03.8	S56 48.6
03	289 03.2	177 58.1	·· 38.5	284 25.4	·· 13.2	274 02.9	·· 08.7	181 15.0	·· 19.1	Pollux	244 02.8	N28 05.2
04	304 05.6	192 57.6	38.1	299 26.1	13.9	289 04.9	08.9	196 17.2	19.1	Procyon	245 29.8	N 5 17.2
05	319 08.1	207 57.2	37.7	314 26.9	14.7	304 06.9	09.1	211 19.3	19.1			
06	334 10.6	222 56.7	N24 37.3	329 27.6	N 0 15.4	319 08.9	N 5 09.3	226 21.5	N22 19.0	Rasalhague	96 32.6	N12 34.6
07	349 13.0	237 56.2	36.9	344 28.4	16.1	334 10.9	09.5	241 23.7	19.0	Regulus	208 13.9	N12 05.2
T 08	4 15.5	252 55.8	36.5	359 29.1	16.9	349 12.9	09.6	256 25.8	19.0	Rigel	281 39.8	S 8 13.9
U 09	19 18.0	267 55.3	·· 36.1	14 29.9	·· 17.6	4 14.9	·· 09.8	271 28.0	·· 18.9	Rigil Kent.	140 30.1	S60 44.2
E 10	34 20.4	282 54.9	35.7	29 30.6	18.3	19 16.9	10.0	286 30.2	18.9	Sabik	102 44.9	S15 41.7
S 11	49 22.9	297 54.4	35.3	44 31.3	19.1	34 19.0	10.2	301 32.4	18.9			
D 12	64 25.3	312 54.0	N24 34.9	59 32.1	N 0 19.8	49 21.0	N 5 10.4	316 34.5	N22 18.9	Schedar	350 13.5	N56 24.0
A 13	79 27.8	327 53.5	34.5	74 32.8	20.5	64 23.0	10.6	331 36.7	18.8	Shaula	97 00.3	S37 05.2
Y 14	94 30.3	342 53.1	34.1	89 33.6	21.3	79 25.0	10.7	346 38.9	18.8	Sirius	258 59.2	S16 41.1
15	109 32.7	357 52.6	·· 33.7	104 34.3	·· 22.0	94 27.0	·· 10.9	1 41.1	·· 18.8	Spica	159 01.1	S11 02.2
16	124 35.2	12 52.1	33.3	119 35.1	22.7	109 29.0	11.1	16 43.2	18.7	Suhail	223 13.6	S43 20.3
17	139 37.7	27 51.7	32.9	134 35.8	23.5	124 31.0	11.3	31 45.4	18.7			
18	154 40.1	42 51.2	N24 32.5	149 36.6	N 0 24.2	139 33.0	N 5 11.5	46 47.6	N22 18.7	Vega	80 57.9	N38 45.5
19	169 42.6	57 50.8	32.1	164 37.3	24.9	154 35.0	11.6	61 49.7	18.7	Zuben'ubi	137 36.7	S15 56.5
20	184 45.1	72 50.3	31.7	179 38.0	25.7	169 37.0	11.8	76 51.9	18.6		S.H.A.	Mer. Pass.
21	199 47.5	87 49.9	·· 31.2	194 38.8	·· 26.4	184 39.0	·· 12.0	91 54.1	·· 18.6		° ′	h m
22	214 50.0	102 49.5	30.8	209 39.5	27.1	199 41.1	12.2	106 56.2	18.6	Venus	250 14.2	15 08
23	229 52.5	117 49.0	30.4	224 40.3	27.9	214 43.1	12.4	121 58.4	18.5	Mars	356 08.7	8 03
										Jupiter	345 12.1	8 46
Mer. Pass.	h m 7 46.9	v −0.5	d 0.4	v 0.7	d 0.7	v 2.0	d 0.2	v 2.2	d 0.0	Saturn	252 19.6	14 57

1975 MAY 25, 26, 27 (SUN., MON., TUES.)

G.M.T.	SUN G.H.A.	SUN Dec.	MOON G.H.A.	v	MOON Dec.	d	H.P.
d h	° ′	° ′	° ′	′	° ′	′	′
25 00	180 48.4	N20 47.7	4 04.3	6.9	S20 14.7	4.1	58.3
01	195 48.4	48.2	18 30.2	7.0	20 18.8	4.0	58.3
02	210 48.3	48.7	32 56.2	6.9	20 22.8	3.8	58.3
03	225 48.2	·· 49.1	47 22.1	7.0	20 26.6	3.7	58.2
04	240 48.2	49.6	61 48.1	6.9	20 30.3	3.6	58.2
05	255 48.1	50.0	76 14.0	7.0	20 33.9	3.5	58.2
06	270 48.1	N20 50.5	90 40.0	6.9	S20 37.4	3.3	58.2
07	285 48.0	51.0	105 05.9	7.0	20 40.7	3.2	58.1
08	300 48.0	51.4	119 31.9	6.9	20 43.9	3.0	58.1
S 09	315 47.9	·· 51.9	133 57.8	7.0	20 46.9	3.0	58.1
U 10	330 47.8	52.3	148 23.8	7.0	20 49.9	2.7	58.1
N 11	345 47.8	52.8	162 49.8	7.0	20 52.6	2.7	58.0
D 12	0 47.7	N20 53.2	177 15.8	7.0	S20 55.3	2.5	58.0
A 13	15 47.7	53.7	191 41.8	7.0	20 57.8	2.4	58.0
Y 14	30 47.6	54.1	206 07.8	7.0	21 00.2	2.3	58.0
15	45 47.5	·· 54.6	220 33.8	7.1	21 02.5	2.1	57.9
16	60 47.5	55.0	234 59.9	7.0	21 04.6	2.0	57.9
17	75 47.4	55.5	249 25.9	7.1	21 06.6	1.9	57.9
18	90 47.4	N20 55.9	263 52.0	7.1	S21 08.5	1.8	57.8
19	105 47.3	56.4	278 18.1	7.2	21 10.3	1.6	57.8
20	120 47.2	56.8	292 44.3	7.1	21 11.9	1.5	57.8
21	135 47.2	·· 57.3	307 10.4	7.2	21 13.4	1.3	57.8
22	150 47.1	57.7	321 36.6	7.2	21 14.7	1.2	57.7
23	165 47.1	58.2	336 02.8	7.2	21 15.9	1.1	57.7
26 00	180 47.0	N20 58.6	350 29.0	7.2	S21 17.0	1.0	57.7
01	195 46.9	59.1	4 55.2	7.3	21 18.0	0.8	57.7
02	210 46.9	20 59.5	19 21.5	7.3	21 18.8	0.7	57.6
03	225 46.8	21 00.0	33 47.8	7.4	21 19.5	0.6	57.6
04	240 46.7	00.4	48 14.2	7.3	21 20.1	0.4	57.6
05	255 46.7	00.9	62 40.5	7.4	21 20.5	0.3	57.5
06	270 46.6	N21 01.3	77 06.9	7.5	S21 20.8	0.2	57.5
07	285 46.6	01.7	91 33.4	7.5	21 21.0	0.1	57.5
08	300 46.5	02.2	105 59.9	7.5	21 21.1	0.1	57.5
M 09	315 46.4	·· 02.6	120 26.4	7.5	21 21.0	0.2	57.4
O 10	330 46.4	03.1	134 52.9	7.6	21 20.8	0.3	57.4
N 11	345 46.3	03.5	149 19.5	7.6	21 20.5	0.5	57.4
D 12	0 46.2	N21 03.9	163 46.1	7.7	S21 20.0	0.6	57.3
A 13	15 46.2	04.4	178 12.8	7.7	21 19.4	0.7	57.3
Y 14	30 46.1	04.8	192 39.5	7.8	21 18.7	0.8	57.3
15	45 46.0	·· 05.3	207 06.3	7.8	21 17.9	1.0	57.3
16	60 46.0	05.7	221 33.1	7.8	21 16.9	1.0	57.2
17	75 45.9	06.1	235 59.9	7.9	21 15.9	1.2	57.2
18	90 45.8	N21 06.6	250 26.8	7.9	S21 14.7	1.4	57.2
19	105 45.8	07.0	264 53.7	8.0	21 13.3	1.4	57.1
20	120 45.7	07.4	279 20.7	8.1	21 11.9	1.6	57.1
21	135 45.7	·· 07.9	293 47.8	8.0	21 10.3	1.7	57.1
22	150 45.6	08.3	308 14.8	8.2	21 08.6	1.8	57.1
23	165 45.5	08.7	322 42.0	8.2	21 06.8	1.9	57.0
27 00	180 45.5	N21 09.2	337 09.2	8.2	S21 04.9	2.0	57.0
01	195 45.4	09.6	351 36.4	8.3	21 02.9	2.2	57.0
02	210 45.3	10.0	6 03.7	8.4	21 00.7	2.3	56.9
03	225 45.3	·· 10.4	20 31.1	8.4	20 58.4	2.4	56.9
04	240 45.2	10.9	34 58.5	8.5	20 56.0	2.5	56.9
05	255 45.1	11.3	49 26.0	8.5	20 53.5	2.6	56.9
06	270 45.0	N21 11.7	63 53.5	8.6	S20 50.9	2.8	56.8
07	285 45.0	12.2	78 21.1	8.6	20 48.1	2.8	56.8
08	300 44.9	12.6	92 48.7	8.7	20 45.3	3.0	56.8
T 09	315 44.8	·· 13.0	107 16.4	8.8	20 42.3	3.1	56.7
U 10	330 44.8	13.4	121 44.2	8.8	20 39.2	3.2	56.7
E 11	345 44.7	13.9	136 12.0	8.9	20 36.0	3.3	56.7
S 12	0 44.6	N21 14.3	150 39.9	9.0	S20 32.7	3.4	56.6
D 13	15 44.6	14.7	165 07.9	9.0	20 29.3	3.6	56.6
A 14	30 44.5	15.1	179 35.9	9.0	20 25.7	3.6	56.6
Y 15	45 44.4	·· 15.5	194 03.9	9.2	20 22.1	3.8	56.6
16	60 44.4	16.0	208 32.1	9.2	20 18.3	3.8	56.5
17	75 44.3	16.4	223 00.3	9.3	20 14.5	4.0	56.5
18	90 44.2	N21 16.8	237 28.6	9.3	S20 10.5	4.0	56.5
19	105 44.1	17.2	251 56.9	9.4	20 06.5	4.2	56.4
20	120 44.1	17.6	266 25.3	9.5	20 02.3	4.3	56.4
21	135 44.0	·· 18.1	280 53.8	9.5	19 58.0	4.4	56.4
22	150 43.9	18.5	295 22.3	9.7	19 53.6	4.4	56.4
23	165 43.9	18.9	309 51.0	9.6	19 49.2	4.6	56.3
	S.D. 15.8	d 0.4	S.D. 15.8		15.6		15.4

Lat.	Twilight Naut.	Twilight Civil	Sunrise	Moonrise 25	Moonrise 26	Moonrise 27	Moonrise 28
°	h m	h m	h m	h m	h m	h m	h m
N 72	▭	▭	▭	■	■	■	■
N 70	▭	▭	▭	■	■	■	01 46
68	////	////	00 29	23 42	24 27	00 27	00 36
66	////	////	01 41	22 42	23 33	24 00	00 00
64	////	////	02 16	22 07	23 01	23 33	23 53
62	////	01 03	02 41	21 43	22 36	23 13	23 37
60	////	01 46	03 01	21 23	22 17	22 56	23 24
N 58	////	02 15	03 17	21 07	22 01	22 42	23 12
56	00 56	02 36	03 30	20 53	21 48	22 30	23 02
54	01 37	02 53	03 42	20 41	21 36	22 19	22 53
52	02 03	03 08	03 52	20 30	21 26	22 10	22 45
50	02 23	03 20	04 02	20 21	21 16	22 02	22 38
45	03 00	03 46	04 21	20 01	20 57	21 44	22 23
N 40	03 26	04 05	04 37	19 45	20 41	21 29	22 10
35	03 46	04 21	04 50	19 32	20 27	21 17	22 00
30	04 02	04 35	05 01	19 20	20 16	21 06	21 50
20	04 28	04 57	05 21	19 00	19 55	20 47	21 34
N 10	04 49	05 15	05 38	18 42	19 38	20 31	21 19
0	05 05	05 31	05 53	18 26	19 22	20 15	21 06
S 10	05 20	05 46	06 09	18 10	19 05	20 00	20 53
20	05 35	06 02	06 25	17 53	18 48	19 44	20 38
30	05 49	06 18	06 44	17 33	18 28	19 25	20 22
35	05 56	06 27	06 55	17 21	18 16	19 14	20 12
40	06 04	06 37	07 07	17 08	18 03	19 01	20 01
45	06 13	06 49	07 22	16 52	17 47	18 47	19 48
S 50	06 22	07 03	07 40	16 33	17 27	18 28	19 33
52	06 27	07 09	07 48	16 24	17 18	18 20	19 25
54	06 31	07 16	07 58	16 13	17 08	18 10	19 17
56	06 36	07 24	08 08	16 02	16 56	17 59	19 08
58	06 42	07 32	08 20	15 49	16 43	17 47	18 57
S 60	06 48	07 42	08 34	15 33	16 27	17 32	18 45

Lat.	Sunset	Twilight Civil	Twilight Naut.	Moonset 25	Moonset 26	Moonset 27	Moonset 28
°	h m	h m	h m	h m	h m	h m	h m
N 72	▭	▭	▭	■	■	■	■
N 70	▭	▭	▭	■	■	■	03 18
68	23 43	////	////	01 16	01 36	02 46	04 27
66	22 17	////	////	01 59	02 36	03 40	05 03
64	21 40	////	////	02 28	03 10	04 12	05 29
62	21 15	22 57	////	02 50	03 35	04 36	05 49
60	20 55	22 10	////	03 08	03 55	04 55	06 05
N 58	20 38	21 41	////	03 23	04 11	05 11	06 19
56	20 24	21 20	23 02	03 36	04 25	05 24	06 31
54	20 13	21 02	22 19	03 47	04 37	05 36	06 41
52	20 02	20 47	21 53	03 57	04 48	05 46	06 50
50	19 53	20 34	21 32	04 06	04 57	05 55	06 58
45	19 33	20 09	20 55	04 25	05 17	06 15	07 16
N 40	19 18	19 49	20 29	04 40	05 33	06 30	07 30
35	19 04	19 33	20 08	04 53	05 47	06 44	07 42
30	18 53	19 19	19 52	05 04	05 59	06 55	07 52
20	18 33	18 57	19 26	05 24	06 19	07 15	08 10
N 10	18 16	18 39	19 05	05 41	06 37	07 32	08 26
0	18 01	18 23	18 48	05 57	06 53	07 48	08 40
S 10	17 45	18 07	18 33	06 12	07 09	08 04	08 55
20	17 28	17 52	18 19	06 29	07 27	08 21	09 10
30	17 10	17 36	18 05	06 49	07 47	08 40	09 27
35	16 59	17 26	17 57	07 00	07 59	08 51	09 38
40	16 46	17 16	17 49	07 13	08 12	09 04	09 49
45	16 32	17 04	17 41	07 29	08 28	09 20	10 03
S 50	16 14	16 51	17 31	07 48	08 48	09 38	10 19
52	16 05	16 44	17 27	07 57	08 57	09 47	10 27
54	15 56	16 37	17 22	08 07	09 08	09 57	10 36
56	15 45	16 30	17 17	08 18	09 19	10 08	10 45
58	15 33	16 21	17 12	08 31	09 33	10 21	10 56
S 60	15 19	16 12	17 06	08 46	09 49	10 36	11 09

Day	SUN Eqn. of Time 00ʰ	SUN Eqn. of Time 12ʰ	SUN Mer. Pass.	MOON Mer. Pass. Upper	MOON Mer. Pass. Lower	MOON Age	MOON Phase
	m s	m s	h m	h m	h m	d	
25	03 14	03 11	11 57	24 40	12 11	14	
26	03 08	03 05	11 57	00 40	13 07	15	○
27	03 02	02 59	11 57	01 35	14 02	16	

1975 MAY 31, JUNE 1, 2 (SAT., SUN., MON.)

G.M.T.		ARIES	VENUS −3.8		MARS +1.0		JUPITER −1.7		SATURN +0.4		STARS		
d	h	G.H.A.	G.H.A.	Dec.	G.H.A.	Dec.	G.H.A.	Dec.	G.H.A.	Dec.	Name	S.H.A.	Dec.
		° ′	° ′	° ′	° ′	° ′	° ′	° ′	° ′	° ′		° ′	° ′
31	00	247 52.3	132 18.6	N23 57.7	240 34.8	N 1 21.2	232 10.1	N 5 25.4	139 36.7	N22 16.3	Acamar	315 40.4	S40 24.1
	01	262 54.8	147 18.2	57.2	255 35.6	22.0	247 12.2	25.6	154 38.8	16.3	Achernar	335 48.4	S57 21.4
	02	277 57.3	162 17.8	56.7	270 36.3	22.7	262 14.2	25.8	169 41.0	16.3	Acrux	173 40.9	S62 58.2
	03	292 59.7	177 17.5	·· 56.3	285 37.1	·· 23.4	277 16.2	·· 25.9	184 43.1	·· 16.2	Adhara	255 35.3	S28 56.5
	04	308 02.2	192 17.1	55.8	300 37.8	24.2	292 18.2	26.1	199 45.3	16.2	Aldebaran	291 22.5	N16 27.6
	05	323 04.7	207 16.7	55.3	315 38.6	24.9	307 20.2	26.3	214 47.5	16.2			
	06	338 07.1	222 16.3	N23 54.8	330 39.3	N 1 25.6	322 22.3	N 5 26.5	229 49.6	N22 16.2	Alioth	166 45.3	N56 05.7
	07	353 09.6	237 15.9	54.3	345 40.1	26.3	337 24.3	26.7	244 51.8	16.1	Alkaid	153 21.0	N49 26.2
S	08	8 12.1	252 15.6	53.8	0 40.8	27.1	352 26.3	26.8	259 54.0	16.1	Al Na'ir	28 19 4	S47 04.5
A	09	23 14.5	267 15.2	·· 53.3	15 41.6	·· 27.8	7 28.3	·· 27.0	274 56.1	·· 16.1	Alnilam	276 15.6	S 1 13.1
T	10	38 17.0	282 14.8	52.8	30 42.3	28.5	22 30.3	27.2	289 58.3	16.0	Alphard	218 24.2	S 8 33.3
U	11	53 19.4	297 14.4	52.3	45 43.1	29.3	37 32.4	27.4	305 00.5	16.0			
R	12	68 21.9	312 14.0	N23 51.8	60 43.8	N 1 30.0	52 34.4	N 5 27.5	320 02.6	N22 16.0	Alphecca	126 34.8	N26 47.8
D	13	83 24.4	327 13.7	51.3	75 44.6	30.7	67 36.4	27.7	335 04.8	15.9	Alpheratz	358 13.2	N28 57.2
A	14	98 26.8	342 13.3	50.8	90 45.3	31.4	82 38.4	27.9	350 06.9	15.9	Altair	62 35.8	N 8 48.2
Y	15	113 29.3	357 12.9	·· 50.4	105 46.1	·· 32.2	97 40.5	·· 28.1	5 09.1	·· 15.9	Ankaa	353 44.0	S42 26.1
	16	128 31.8	12 12.5	49.9	120 46.8	32.9	112 42.5	28.2	20 11.3	15.8	Antares	113 00.9	S26 22.7
	17	143 34.2	27 12.2	49.4	135 47.6	33.6	127 44.5	28.4	35 13.4	15.8			
	18	158 36.7	42 11.8	N23 48.9	150 48.3	N 1 34.4	142 46.5	N 5 28.6	50 15.6	N22 15.8	Arcturus	146 21.5	N19 18.6
	19	173 39.2	57 11.4	48.4	165 49.1	35.1	157 48.5	28.8	65 17.8	15.8	Atria	108 27.8	S68 59.1
	20	188 41.6	72 11.1	47.9	180 49.8	35.8	172 50.6	28.9	80 19.9	15.7	Avior	234 30.1	S59 26.2
	21	203 44.1	87 10.7	·· 47.4	195 50.6	·· 36.5	187 52.6	·· 29.1	95 22.1	·· 15.7	Bellatrix	279 02.9	N 6 19.6
	22	218 46.5	102 10.3	46.8	210 51.3	37.3	202 54.6	29.3	110 24.2	15.7	Betelgeuse	271 32.5	N 7 24.1
	23	233 49.0	117 10.0	46.3	225 52.1	38.0	217 56.6	29.5	125 26.4	15.6			
1	00	248 51.5	132 09.6	N23 45.8	240 52.8	N 1 38.7	232 58.7	N 5 29.6	140 28.6	N22 15.6	Canopus	264 09.3	S52 41.1
	01	263 53.9	147 09.2	45.3	255 53.6	39.4	248 00.7	29.8	155 30.7	15.6	Capella	281 17.1	N45 58.4
	02	278 56.4	162 08.9	44.8	270 54.3	40.2	263 02.7	30.0	170 32.9	15.5	Deneb	49 50.7	N45 11.4
	03	293 58.9	177 08.5	·· 44.3	285 55.1	·· 40.9	278 04.7	·· 30.2	185 35.1	·· 15.5	Denebola	183 02.6	N14 42.5
	04	309 01.3	192 08.1	43.8	300 55.8	41.6	293 06.7	30.3	200 37.2	15.5	Diphda	349 24.7	S18 07.2
	05	324 03.8	207 07.8	43.3	315 56.6	42.4	308 08.8	30.5	215 39.4	15.4			
	06	339 06.3	222 07.4	N23 42.8	330 57.3	N 1 43.1	323 10.8	N 5 30.7	230 41.5	N22 15.4	Dubhe	194 26.5	N61 53.2
	07	354 08.7	237 07.1	42.3	345 58.1	43.8	338 12.8	30.9	245 43.7	15.4	Elnath	278 49.1	N28 35.2
	08	9 11.2	252 06.7	41.8	0 58.8	44.5	353 14.8	31.0	260 45.9	15.3	Eltanin	90 58.9	N51 29.4
S	09	24 13.7	267 06.3	·· 41.3	15 59.6	·· 45.3	8 16.9	·· 31.2	275 48.0	·· 15.3	Enif	34 15.0	N 9 45.7
U	10	39 16.1	282 06.0	40.7	31 00.3	46.0	23 18.9	31.4	290 50.2	15.3	Fomalhaut	15 55.4	S29 44.9
N	11	54 18.6	297 05.6	40.2	46 01.1	46.7	38 20.9	31.5	305 52.3	15.2			
D	12	69 21.0	312 05.3	N23 39.7	61 01.8	N 1 47.4	53 22.9	N 5 31.7	320 54.5	N22 15.2	Gacrux	172 32.5	S56 58.9
A	13	84 23.5	327 04.9	39.2	76 02.6	48.2	68 25.0	31.9	335 56.7	15.2	Gienah	176 21.5	S17 24.6
Y	14	99 26.0	342 04.6	38.7	91 03.3	48.9	83 27.0	32.1	350 58.8	15.2	Hadar	149 27.9	S60 15.6
	15	114 28.4	357 04.2	·· 38.2	106 04.1	·· 49.6	98 29.0	·· 32.2	6 01.0	·· 15.1	Hamal	328 33.3	N23 20.7
	16	129 30.9	12 03.9	37.6	121 04.8	50.4	113 31.0	32.4	21 03.2	15.1	Kaus Aust.	84 21.2	S34 23.7
	17	144 33.4	27 03.5	37.1	136 05.6	51.1	128 33.1	32.6	36 05.3	15.1			
	18	159 35.8	42 03.2	N23 36.6	151 06.3	N 1 51.8	143 35.1	N 5 32.8	51 07.5	N22 15.0	Kochab	137 17.7	N74 15.5
	19	174 38.3	57 02.8	36.1	166 07.1	52.5	158 37.1	32.9	66 09.6	15.0	Markab	14 06.8	N15 04.3
	20	189 40.8	72 02.5	35.5	181 07.8	53.3	173 39.1	33.1	81 11.8	15.0	Menkar	314 45.2	N 3 59.6
	21	204 43.2	87 02.1	·· 35.0	196 08.6	·· 54.0	188 41.2	·· 33.3	96 14.0	·· 14.9	Menkent	148 40.9	S36 15.2
	22	219 45.7	102 01.8	34.5	211 09.3	54.7	203 43.2	33.5	111 16.1	14.9	Miaplacidus	221 46.1	S69 37.4
	23	234 48.2	117 01.4	34.0	226 10.1	55.4	218 45.2	33.6	126 18.3	14.9			
2	00	249 50.6	132 01.1	N23 33.4	241 10.8	N 1 56.2	233 47.2	N 5 33.8	141 20.4	N22 14.8	Mirfak	309 21.7	N49 46.4
	01	264 53.1	147 00.7	32.9	256 11.6	56.9	248 49.3	34.0	156 22.6	14.8	Nunki	76 33.3	S26 19.6
	02	279 55.5	162 00.4	32.4	271 12.3	57.6	263 51.3	34.1	171 24.8	14.8	Peacock	54 03.7	S56 48.6
	03	294 58.0	177 00.0	·· 31.9	286 13.1	·· 58.3	278 53.3	·· 34.3	186 26.9	·· 14.7	Pollux	244 02.8	N28 05.2
	04	310 00.5	191 59.7	31.3	301 13.8	59.1	293 55.4	34.5	201 29.1	14.7	Procyon	245 29.8	N 5 17.2
	05	325 02.9	206 59.4	30.8	316 14.6	1 59.8	308 57.4	34.7	216 31.2	14.7			
	06	340 05.4	221 59.0	N23 30.3	331 15.3	N 2 00.5	323 59.4	N 5 34.8	231 33.4	N22 14.6	Rasalhague	96 32.6	N12 34.6
	07	355 07.9	236 58.7	29.7	346 16.1	01.2	339 01.4	35.0	246 35.6	14.6	Regulus	208 13.9	N12 05.2
	08	10 10.3	251 58.4	29.2	1 16.8	02.0	354 03.5	35.2	261 37.7	14.6	Rigel	281 39.8	S 8 13.9
M	09	25 12.8	266 58.0	·· 28.7	16 17.6	·· 02.7	9 05.5	·· 35.4	276 39.9	·· 14.6	Rigil Kent.	140 30.1	S60 44.2
O	10	40 15.3	281 57.7	28.1	31 18.3	03.4	24 07.5	35.5	291 42.0	14.5	Sabik	102 44.9	S15 41.7
N	11	55 17.7	296 57.3	27.6	46 19.1	04.1	39 09.5	35.7	306 44.2	14.5			
D	12	70 20.2	311 57.0	N23 27.1	61 19.8	N 2 04.9	54 11.6	N 5 35.9	321 46.4	N22 14.5	Schedar	350 13.4	N56 24.0
A	13	85 22.6	326 56.7	26.5	76 20.6	05.6	69 13.6	36.0	336 48.5	14.4	Shaula	97 00.2	S37 05.2
Y	14	100 25.1	341 56.3	26.0	91 21.3	06.3	84 15.6	36.2	351 50.7	14.4	Sirius	258 59.2	S16 41.1
	15	115 27.6	356 56.0	·· 25.4	106 22.1	·· 07.0	99 17.7	·· 36.4	6 52.8	·· 14.4	Spica	159 01.1	S11 02.2
	16	130 30.0	11 55.7	24.9	121 22.8	07.8	114 19.7	36.6	21 55.0	14.3	Suhail	223 13.6	S43 20.3
	17	145 32.5	26 55.4	24.4	136 23.6	08.5	129 21.7	36.7	36 57.1	14.3			
	18	160 35.0	41 55.0	N23 23.8	151 24.3	N 2 09.2	144 23.7	N 5 36.9	51 59.3	N22 14.3	Vega	80 57.9	N38 45.6
	19	175 37.4	56 54.7	23.3	166 25.1	09.9	159 25.8	37.1	67 01.5	14.2	Zuben'ubi	137 36.7	S15 56.5
	20	190 39.9	71 54.4	22.7	181 25.8	10.7	174 27.8	37.2	82 03.6	14.2		S.H.A.	Mer. Pass.
	21	205 42.4	86 54.0	·· 22.2	196 26.6	·· 11.4	189 29.8	·· 37.4	97 05.8	·· 14.2		° ′	h m
	22	220 44.8	101 53.7	21.6	211 27.3	12.1	204 31.9	37.6	112 07.9	14.1	Venus	243 18.1	15 12
	23	235 47.3	116 53.4	21.1	226 28.1	12.8	219 33.9	37.8	127 10.1	14.1	Mars	352 01.3	7 56
											Jupiter	344 07.2	8 27
Mer. Pass.		h m 7 23.4	v −0.4	d 0.5	v 0.8	d 0.7	v 2.0	d 0.2	v 2.2	d 0.0	Saturn	251 37.1	14 36

1975 MAY 31, JUNE 1, 2 (SAT., SUN., MON.)

G.M.T. d	h	SUN G.H.A.	SUN Dec.	MOON G.H.A.	v	MOON Dec.	d	H.P.
		° ′	° ′	° ′	′	° ′	′	′
31	00	180 38.1	N21 47.6	289 46.4	14.5	S10 46.8	9.6	54.7
	01	195 38.0	47.9	304 19.9	14.5	10 37.2	9.6	54.6
	02	210 37.9	48.3	318 53.4	14.5	10 27.6	9.7	54.6
	03	225 37.8	·· 48.7	333 26.9	14.6	10 17.9	9.7	54.6
	04	240 37.7	49.0	348 00.5	14.7	10 08.2	9.7	54.6
	05	255 37.6	49.4	2 34.2	14.6	9 58.5	9.8	54.6
	06	270 37.6	N21 49.8	17 07.8	14.8	S 9 48.7	9.8	54.6
	07	285 37.5	50.1	31 41.6	14.7	9 38.9	9.8	54.6
S	08	300 37.4	50.5	46 15.3	14.9	9 29.1	9.9	54.5
A	09	315 37.3	·· 50.9	60 49.2	14.8	9 19.2	9.9	54.5
T	10	330 37.2	51.2	75 23.0	14.9	9 09.3	9.9	54.5
U	11	345 37.1	51.6	89 56.9	15.0	8 59.4	10.0	54.5
R	12	0 37.0	N21 51.9	104 30.9	14.9	S 8 49.4	10.0	54.5
D	13	15 36.9	52.3	119 04.8	15.0	8 39.4	10.0	54.5
A	14	30 36.9	52.7	133 38.8	15.1	8 29.4	10.1	54.5
Y	15	45 36.8	·· 53.0	148 12.9	15.1	8 19.3	10.1	54.5
	16	60 36.7	53.4	162 47.0	15.1	8 09.2	10.1	54.4
	17	75 36.6	53.7	177 21.1	15.2	7 59.1	10.1	54.4
	18	90 36.5	N21 54.1	191 55.3	15.2	S 7 49.0	10.2	54.4
	19	105 36.4	54.5	206 29.5	15.2	7 38.8	10.2	54.4
	20	120 36.3	54.8	221 03.7	15.3	7 28.6	10.2	54.4
	21	135 36.2	·· 55.2	235 38.0	15.3	7 18.4	10.3	54.4
	22	150 36.1	55.5	250 12.3	15.3	7 08.1	10.2	54.4
	23	165 36.0	55.9	264 46.6	15.4	6 57.9	10.3	54.4
1	00	180 36.0	N21 56.2	279 21.0	15.4	S 6 47.6	10.4	54.4
	01	195 35.9	56.6	293 55.4	15.4	6 37.2	10.3	54.4
	02	210 35.8	56.9	308 29.8	15.4	6 26.9	10.4	54.4
	03	225 35.7	·· 57.3	323 04.2	15.5	6 16.5	10.4	54.3
	04	240 35.6	57.6	337 38.7	15.5	6 06.1	10.4	54.3
	05	255 35.5	58.0	352 13.2	15.5	5 55.7	10.4	54.3
	06	270 35.4	N21 58.3	6 47.7	15.6	S 5 45.3	10.4	54.3
	07	285 35.3	58.7	21 22.3	15.5	5 34.9	10.5	54.3
	08	300 35.2	59.0	35 56.8	15.6	5 24.4	10.5	54.3
S	09	315 35.1	·· 59.4	50 31.4	15.7	5 13.9	10.5	54.3
U	10	330 35.0	21 59.7	65 06.1	15.6	5 03.4	10.5	54.3
N	11	345 34.9	22 00.1	79 40.7	15.7	4 52.9	10.6	54.3
D	12	0 34.8	N22 00.4	94 15.4	15.6	S 4 42.3	10.5	54.3
A	13	15 34.8	00.8	108 50.0	15.7	4 31.8	10.6	54.3
Y	14	30 34.7	01.1	123 24.7	15.8	4 21.2	10.6	54.3
	15	45 34.6	·· 01.5	137 59.5	15.7	4 10.6	10.6	54.3
	16	60 34.5	01.8	152 34.2	15.7	4 00.0	10.6	54.3
	17	75 34.4	02.1	167 08.9	15.8	3 49.4	10.6	54.3
	18	90 34.3	N22 02.5	181 43.7	15.8	S 3 38.8	10.7	54.3
	19	105 34.2	02.8	196 18.5	15.8	3 28.1	10.6	54.3
	20	120 34.1	03.2	210 53.3	15.8	3 17.5	10.7	54.3
	21	135 34.0	·· 03.5	225 28.1	15.8	3 06.8	10.7	54.3
	22	150 33.9	03.8	240 02.9	15.8	2 56.1	10.7	54.3
	23	165 33.8	04.2	254 37.7	15.9	2 45.4	10.6	54.2
2	00	180 33.7	N22 04.5	269 12.6	15.8	S 2 34.8	10.8	54.2
	01	195 33.6	04.8	283 47.4	15.9	2 24.0	10.7	54.2
	02	210 33.5	05.2	298 22.3	15.8	2 13.3	10.7	54.2
	03	225 33.4	·· 05.5	312 57.1	15.9	2 02.6	10.7	54.2
	04	240 33.3	05.9	327 32.0	15.9	1 51.9	10.8	54.2
	05	255 33.2	06.2	342 06.9	15.8	1 41.1	10.7	54.2
	06	270 33.1	N22 06.5	356 41.7	15.9	S 1 30.4	10.8	54.2
	07	285 33.0	06.9	11 16.6	15.9	1 19.6	10.7	54.2
	08	300 32.9	07.2	25 51.5	15.9	1 08.9	10.8	54.2
M	09	315 32.9	·· 07.5	40 26.4	15.9	0 58.1	10.7	54.2
O	10	330 32.8	07.9	55 01.3	15.9	0 47.4	10.8	54.2
N	11	345 32.7	08.2	69 36.2	15.9	0 36.6	10.8	54.3
D	12	0 32.6	N22 08.5	84 11.1	15.9	S 0 25.8	10.8	54.3
A	13	15 32.5	08.8	98 46.0	15.8	0 15.0	10.7	54.3
Y	14	30 32.4	09.2	113 20.8	15.9	S 0 04.3	10.8	54.3
	15	45 32.3	·· 09.5	127 55.7	15.9	N 0 06.5	10.8	54.3
	16	60 32.2	09.8	142 30.6	15.9	0 17.3	10.8	54.3
	17	75 32.1	10.1	157 05.5	15.8	0 28.1	10.8	54.3
	18	90 32.0	N22 10.5	171 40.3	15.9	N 0 38.9	10.7	54.3
	19	105 31.9	10.8	186 15.2	15.9	0 49.6	10.8	54.3
	20	120 31.8	11.1	200 50.1	15.8	1 00.4	10.8	54.3
	21	135 31.7	·· 11.4	215 24.9	15.8	1 11.2	10.8	54.3
	22	150 31.6	11.8	229 59.7	15.9	1 22.0	10.8	54.3
	23	165 31.5	12.1	244 34.6	15.8	1 32.8	10.7	54.3
		S.D. 15.8	*d* 0.3	S.D. 14.8		14.8		14.8

Lat.	Twilight Naut.	Twilight Civil	Sunrise	Moonrise 31	Moonrise 1	Moonrise 2	Moonrise 3
°	h m	h m	h m	h m	h m	h m	h m
N 72	▭	▭	▭	01 05	00 49	00 36	00 23
N 70	▭	▭	▭	00 48	00 40	00 33	00 25
68	▭	▭	▭	00 34	00 32	00 30	00 27
66	////	////	01 15	00 23	00 26	00 27	00 29
64	////	////	02 00	00 14	00 20	00 25	00 30
62	////	00 15	02 28	00 06	00 15	00 24	00 31
60	////	01 28	02 50	24 11	00 11	00 22	00 32
N 58	////	02 02	03 08	24 07	00 07	00 21	00 33
56	00 13	02 26	03 23	24 04	00 04	00 19	00 34
54	01 20	02 45	03 35	24 01	00 01	00 18	00 35
52	01 51	03 00	03 46	23 58	24 17	00 17	00 36
50	02 14	03 14	03 56	23 56	24 16	00 16	00 36
45	02 53	03 41	04 17	23 50	24 14	00 14	00 38
N 40	03 21	04 01	04 33	23 46	24 13	00 13	00 39
35	03 43	04 18	04 47	23 42	24 11	00 11	00 40
30	04 00	04 32	04 59	23 38	24 10	00 10	00 41
20	04 27	04 56	05 20	23 32	24 07	00 07	00 42
N 10	04 48	05 15	05 38	23 27	24 05	00 05	00 44
0	05 06	05 32	05 54	23 22	24 04	00 04	00 45
S 10	05 22	05 48	06 10	23 17	24 02	00 02	00 46
20	05 37	06 04	06 27	23 11	24 00	00 00	00 48
30	05 52	06 21	06 47	23 05	23 58	24 50	00 50
35	06 00	06 31	06 59	23 02	23 56	24 50	00 50
40	06 08	06 42	07 12	22 58	23 55	24 52	00 52
45	06 17	06 54	07 27	22 53	23 53	24 53	00 53
S 50	06 28	07 09	07 46	22 47	23 51	24 54	00 54
52	06 33	07 16	07 55	22 45	23 50	24 55	00 55
54	06 38	07 23	08 05	22 42	23 49	24 56	00 56
56	06 43	07 31	08 17	22 39	23 48	24 57	00 57
58	06 49	07 40	08 30	22 35	23 47	24 58	00 58
S 60	06 55	07 51	08 45	22 31	23 45	24 59	00 59

Lat.	Sunset	Twilight Civil	Twilight Naut.	Moonset 31	Moonset 1	Moonset 2	Moonset 3
°	h m	h m	h m	h m	h m	h m	h m
N 72	▭	▭	▭	08 53	10 38	12 18	13 58
N 70	▭	▭	▭	09 09	10 45	12 19	13 52
68	▭	▭	▭	09 21	10 51	12 19	13 47
66	22 44	////	////	09 31	10 56	12 20	13 43
64	21 58	////	////	09 39	11 00	12 20	13 40
62	21 28	////	////	09 46	11 04	12 20	13 37
60	21 06	22 30	////	09 52	11 07	12 21	13 34
N 58	20 48	21 55	////	09 58	11 10	12 21	13 32
56	20 33	21 31	////	10 03	11 12	12 21	13 30
54	20 21	21 12	22 37	10 07	11 14	12 21	13 28
52	20 09	20 56	22 06	10 11	11 16	12 21	13 27
50	20 00	20 42	21 43	10 14	11 18	12 22	13 25
45	19 39	20 15	21 03	10 22	11 22	12 22	13 22
N 40	19 22	19 54	20 35	10 28	11 25	12 22	13 19
35	19 08	19 37	20 13	10 33	11 28	12 22	13 17
30	18 56	19 23	19 56	10 38	11 31	12 23	13 15
20	18 35	19 00	19 28	10 46	11 35	12 23	13 11
N 10	18 18	18 41	19 07	10 53	11 39	12 23	13 08
0	18 01	18 24	18 50	11 00	11 42	12 24	13 05
S 10	17 45	18 08	18 34	11 06	11 46	12 24	13 02
20	17 28	17 51	18 19	11 13	11 49	12 24	12 59
30	17 08	17 34	18 04	11 21	11 53	12 24	12 55
35	16 56	17 24	17 56	11 26	11 56	12 24	12 53
40	16 43	17 13	17 47	11 31	11 58	12 25	12 51
45	16 28	17 01	17 38	11 37	12 01	12 25	12 48
S 50	16 09	16 46	17 27	11 44	12 05	12 25	12 45
52	16 00	16 40	17 22	11 47	12 07	12 25	12 44
54	15 50	16 32	17 17	11 51	12 08	12 25	12 42
56	15 38	16 24	17 12	11 54	12 10	12 25	12 40
58	15 25	16 15	17 06	11 59	12 13	12 26	12 38
S 60	15 10	16 04	16 59	12 04	12 15	12 26	12 36

Day	SUN Eqn. of Time 00ʰ	SUN Eqn. of Time 12ʰ	SUN Mer. Pass.	MOON Mer. Pass. Upper	MOON Mer. Pass. Lower	Age	Phase
	m s	m s	h m	h m	h m	d	
31	02 32	02 28	11 58	04 49	17 11	20	◑
1	02 24	02 20	11 58	05 32	17 53	21	
2	02 15	02 10	11 58	06 14	18 34	22	

1975 JUNE 12, 13, 14 (THURS., FRI., SAT.)

G.M.T.	ARIES	VENUS −3.9		MARS +0.8		JUPITER −1.9		SATURN +0.4	
d h	G.H.A. ° ′	G.H.A. ° ′	Dec. ° ′	G.H.A. ° ′	Dec. ° ′	G.H.A. ° ′	Dec. ° ′	G.H.A. ° ′	Dec. ° ′
12 00	259 42.0	131 06.2	N21 02.9	244 12.2	N 4 47.9	241 58.0	N 6 13.1	149 56.7	N22 06.6
01	274 44.5	146 06.1	02.2	259 13.0	48.6	257 00.1	13.3	164 58.8	06.6
02	289 46.9	161 06.0	01.5	274 13.8	49.3	272 02.2	13.4	180 01.0	06.6
03	304 49.4	176 05.9	·· 00.8	289 14.5	·· 50.0	287 04.2	·· 13.6	195 03.1	·· 06.5
04	319 51.9	191 05.8	21 00.1	304 15.3	50.7	302 06.3	13.7	210 05.2	06.5
05	334 54.3	206 05.7	20 59.3	319 16.0	51.4	317 08.4	13.9	225 07.4	06.4
06	349 56.8	221 05.6	N20 58.6	334 16.8	N 4 52.1	332 10.4	N 6 14.0	240 09.5	N22 06.4
07	4 59.3	236 05.5	57.9	349 17.6	52.8	347 12.5	14.2	255 11.7	06.4
T 08	20 01.7	251 05.4	57.2	4 18.3	53.5	2 14.6	14.3	270 13.8	06.3
H 09	35 04.2	266 05.3	·· 56.4	19 19.1	·· 54.2	17 16.6	·· 14.5	285 16.0	·· 06.3
U 10	50 06.6	281 05.3	55.7	34 19.8	54.9	32 18.7	14.6	300 18.1	06.3
R 11	65 09.1	296 05.2	55.0	49 20.6	55.6	47 20.8	14.8	315 20.2	06.2
S 12	80 11.6	311 05.1	N20 54.3	64 21.4	N 4 56.3	62 22.8	N 6 15.0	330 22.4	N22 06.2
D 13	95 14.0	326 05.0	53.5	79 22.1	57.0	77 24.9	15.1	345 24.5	06.1
A 14	110 16.5	341 04.9	52.8	94 22.9	57.7	92 27.0	15.3	0 26.7	06.1
Y 15	125 19.0	356 04.8	·· 52.1	109 23.7	·· 58.4	107 29.0	·· 15.4	15 28.8	·· 06.1
16	140 21.4	11 04.7	51.3	124 24.4	59.1	122 31.1	15.6	30 30.9	06.0
17	155 23.9	26 04.6	50.6	139 25.2	4 59.8	137 33.2	15.7	45 33.1	06.0
18	170 26.4	41 04.5	N20 49.9	154 25.9	N 5 00.5	152 35.2	N 6 15.9	60 35.2	N22 06.0
19	185 28.8	56 04.5	49.2	169 26.7	01.2	167 37.3	16.0	75 37.4	05.9
20	200 31.3	71 04.4	48.4	184 27.5	01.9	182 39.4	16.2	90 39.5	05.9
21	215 33.8	86 04.3	·· 47.7	199 28.2	·· 02.6	197 41.4	·· 16.3	105 41.6	·· 05.9
22	230 36.2	101 04.2	47.0	214 29.0	03.3	212 43.5	16.5	120 43.8	05.8
23	245 38.7	116 04.1	46.2	229 29.7	04.0	227 45.6	16.6	135 45.9	05.8
13 00	260 41.1	131 04.1	N20 45.5	244 30.5	N 5 04.7	242 47.6	N 6 16.8	150 48.1	N22 05.7
01	275 43.6	146 04.0	44.8	259 31.3	05.4	257 49.7	16.9	165 50.2	05.7
02	290 46.1	161 03.9	44.0	274 32.0	06.1	272 51.8	17.1	180 52.4	05.7
03	305 48.5	176 03.8	·· 43.3	289 32.8	·· 06.8	287 53.8	·· 17.2	195 54.5	·· 05.6
04	320 51.0	191 03.8	42.6	304 33.6	07.5	302 55.9	17.4	210 56.6	05.6
05	335 53.5	206 03.7	41.8	319 34.3	08.2	317 58.0	17.5	225 58.8	05.6
06	350 55.9	221 03.6	N20 41.1	334 35.1	N 5 08.9	333 00.0	N 6 17.7	241 00.9	N22 05.5
07	5 58.4	236 03.5	40.3	349 35.8	09.6	348 02.1	17.9	256 03.1	05.5
08	21 00.9	251 03.5	39.6	4 36.6	10.3	3 04.2	18.0	271 05.2	05.4
F 09	36 03.3	266 03.4	·· 38.9	19 37.4	·· 11.0	18 06.3	·· 18.2	286 07.3	·· 05.4
R 10	51 05.8	281 03.3	38.1	34 38.1	11.7	33 08.3	18.3	301 09.5	05.4
I 11	66 08.3	296 03.3	37.4	49 38.9	12.4	48 10.4	18.5	316 11.6	05.3
D 12	81 10.7	311 03.2	N20 36.6	64 39.6	N 5 13.1	63 12.5	N 6 18.6	331 13.8	N22 05.3
A 13	96 13.2	326 03.1	35.9	79 40.4	13.8	78 14.5	18.8	346 15.9	05.3
Y 14	111 15.6	341 03.1	35.1	94 41.2	14.5	93 16.6	18.9	1 18.0	05.2
15	126 18.1	356 03.0	·· 34.4	109 41.9	·· 15.2	108 18.7	·· 19.1	16 20.2	·· 05.2
16	141 20.6	11 03.0	33.7	124 42.7	15.9	123 20.8	19.2	31 22.3	05.1
17	156 23.0	26 02.9	32.9	139 43.5	16.6	138 22.8	19.4	46 24.5	05.1
18	171 25.5	41 02.8	N20 32.2	154 44.2	N 5 17.3	153 24.9	N 6 19.5	61 26.6	N22 05.1
19	186 28.0	56 02.8	31.4	169 45.0	18.0	168 27.0	19.7	76 28.7	05.0
20	201 30.4	71 02.7	30.7	184 45.7	18.7	183 29.0	19.8	91 30.9	05.0
21	216 32.9	86 02.7	·· 29.9	199 46.5	·· 19.4	198 31.1	·· 20.0	106 33.0	·· 05.0
22	231 35.4	101 02.6	29.2	214 47.3	20.1	213 33.2	20.1	121 35.2	04.9
23	246 37.8	116 02.6	28.4	229 48.0	20.8	228 35.3	20.3	136 37.3	04.9
14 00	261 40.3	131 02.5	N20 27.7	244 48.8	N 5 21.5	243 37.3	N 6 20.4	151 39.4	N22 04.8
01	276 42.8	146 02.5	26.9	259 49.6	22.2	258 39.4	20.6	166 41.6	04.8
02	291 45.2	161 02.4	26.2	274 50.3	22.9	273 41.5	20.7	181 43.7	04.8
03	306 47.7	176 02.4	·· 25.4	289 51.1	·· 23.6	288 43.6	·· 20.9	196 45.9	·· 04.7
04	321 50.1	191 02.3	24.7	304 51.8	24.2	303 45.6	21.0	211 48.0	04.7
05	336 52.6	206 02.3	23.9	319 52.6	24.9	318 47.7	21.2	226 50.1	04.7
06	351 55.1	221 02.2	N20 23.2	334 53.4	N 5 25.6	333 49.8	N 6 21.3	241 52.3	N22 04.6
07	6 57.5	236 02.2	22.4	349 54.1	26.3	348 51.8	21.5	256 54.4	04.6
S 08	22 00.0	251 02.1	21.7	4 54.9	27.0	3 53.9	21.6	271 56.6	04.5
A 09	37 02.5	266 02.1	·· 20.9	19 55.7	·· 27.7	18 56.0	·· 21.8	286 58.7	·· 04.5
T 10	52 04.9	281 02.1	20.1	34 56.4	28.4	33 58.1	21.9	302 00.8	04.5
U 11	67 07.4	296 02.0	19.4	49 57.2	29.1	49 00.1	22.1	317 03.0	04.4
R 12	82 09.9	311 02.0	N20 18.6	64 58.0	N 5 29.8	64 02.2	N 6 22.2	332 05.1	N22 04.4
D 13	97 12.3	326 01.9	17.9	79 58.7	30.5	79 04.3	22.4	347 07.3	04.4
A 14	112 14.8	341 01.9	17.1	94 59.5	31.2	94 06.4	22.5	2 09.4	04.3
Y 15	127 17.2	356 01.9	·· 16.4	110 00.2	·· 31.9	109 08.4	·· 22.7	17 11.5	·· 04.3
16	142 19.7	11 01.8	15.6	125 01.0	32.6	124 10.5	22.8	32 13.7	04.2
17	157 22.2	26 01.8	14.8	140 01.8	33.3	139 12.6	23.0	47 15.8	04.2
18	172 24.6	41 01.8	N20 14.1	155 02.5	N 5 34.0	154 14.7	N 6 23.1	62 17.9	N22 04.2
19	187 27.1	56 01.8	13.3	170 03.3	34.7	169 16.7	23.3	77 20.1	04.1
20	202 29.6	71 01.7	12.6	185 04.1	35.4	184 18.8	23.4	92 22.2	04.1
21	217 32.0	86 01.7	·· 11.8	200 04.8	·· 36.1	199 20.9	·· 23.6	107 24.4	·· 04.1
22	232 34.5	101 01.7	11.0	215 05.6	36.8	214 23.0	23.7	122 26.5	04.0
23	247 37.0	116 01.6	10.3	230 06.3	37.5	229 25.1	23.9	137 28.6	04.0
Mer. Pass.	h m 6 36.2	*v* −0.1	*d* 0.7	*v* 0.8	*d* 0.7	*v* 2.1	*d* 0.2	*v* 2.1	*d* 0.0

STARS Name	S.H.A. ° ′	Dec. ° ′
Acamar	315 40.3	S40 24.0
Achernar	335 48.3	S57 21.4
Acrux	173 41.0	S62 58.2
Adhara	255 35.3	S28 56.5
Aldebaran	291 22.4	N16 27.6
Alioth	166 45.4	N56 05.7
Alkaid	153 21.0	N49 26.3
Al Na'ir	28 19.2	S47 04.5
Alnilam	276 15.6	S 1 13.1
Alphard	218 24.2	S 8 33.3
Alphecca	126 34.8	N26 47.9
Alpheratz	358 13.1	N28 57.2
Altair	62 35.7	N 8 48.2
Ankaa	353 43.9	S42 26.1
Antares	113 00.9	S26 22.7
Arcturus	146 21.5	N19 18.6
Atria	108 27.8	S68 59.1
Avior	234 30.1	S59 26.1
Bellatrix	279 02.9	N 6 19.6
Betelgeuse	271 32.5	N 7 24.1
Canopus	264 09.3	S52 41.1
Capella	281 17.1	N45 58.4
Deneb	49 50.6	N45 11.5
Denebola	183 02.7	N14 42.5
Diphda	349 24.6	S18 07.1
Dubhe	194 26.6	N61 53.2
Elnath	278 49.0	N28 35.2
Eltanin	90 58.8	N51 29.5
Enif	34 14.9	N 9 45.8
Fomalhaut	15 55.3	S29 44.9
Gacrux	172 32.6	S56 58.9
Gienah	176 21.6	S17 24.6
Hadar	149 28.0	S60 15.6
Hamal	328 33.2	N23 20.7
Kaus Aust.	84 21.2	S34 23.7
Kochab	137 17.9	N74 15.5
Markab	14 06.7	N15 04.4
Menkar	314 45.1	N 3 59.6
Menkent	148 41.0	S36 15.2
Miaplacidus	221 46.2	S69 37.4
Mirfak	309 21.7	N49 46.4
Nunki	76 33.2	S26 19.6
Peacock	54 03.6	S56 48.6
Pollux	244 02.8	N28 05.2
Procyon	245 29.8	N 5 17.2
Rasalhague	96 32.6	N12 34.7
Regulus	208 13.9	N12 05.2
Rigel	281 39.8	S 8 13.8
Rigil Kent.	140 30.1	S60 44.3
Sabik	102 44.9	S15 41.7
Schedar	350 13.3	N56 24.0
Shaula	97 00.2	S37 05.2
Sirius	258 59.2	S16 41.1
Spica	159 01.1	S11 02.2
Suhail	223 13.7	S43 20.3
Vega	80 57.8	N38 45.6
Zuben'ubi	137 36.7	S15 56.5

	S.H.A. ° ′	Mer. Pass h m
Venus	230 22.9	15 16
Mars	343 49.4	7 42
Jupiter	342 06.5	7 48
Saturn	250 06.9	13 55

1975 JUNE 12, 13, 14 (THURS., FRI., SAT.)

	G.M.T.	SUN G.H.A.	SUN Dec.	MOON G.H.A.	MOON v	MOON Dec.	MOON d	MOON H.P.
	d h	° ′	° ′	° ′	′	° ′	′	′
	12 00	180 06.5	N23 05.8	149 45.2	7.2	N18 13.9	7.0	58.9
	01	195 06.3	05.9	164 11.4	7.2	18 06.9	7.1	58.9
	02	210 06.2	06.1	178 37.6	7.2	17 59.8	7.2	58.9
	03	225 06.1	·· 06.3	193 03.8	7.3	17 52.6	7.3	59.0
	04	240 06.0	06.4	207 30.1	7.3	17 45.3	7.4	59.0
	05	255 05.8	06.6	221 56.4	7.3	17 37.9	7.5	59.0
	06	270 05.7	N23 06.8	236 22.7	7.4	N17 30.4	7.7	59.0
	07	285 05.6	06.9	250 49.1	7.4	17 22.7	7.7	59.0
T	08	300 05.4	07.1	265 15.5	7.4	17 15.0	7.9	59.0
H	09	315 05.3	·· 07.3	279 41.9	7.5	17 07.1	8.0	59.0
U	10	330 05.2	07.4	294 08.4	7.5	16 59.1	8.1	59.0
R	11	345 05.1	07.6	308 34.9	7.6	16 51.0	8.2	59.1
S	12	0 04.9	N23 07.8	323 01.5	7.6	N16 42.8	8.3	59.1
D	13	15 04.8	07.9	337 28.1	7.6	16 34.5	8.4	59.1
A	14	30 04.7	08.1	351 54.7	7.7	16 26.1	8.5	59.1
Y	15	45 04.5	·· 08.3	6 21.4	7.7	16 17.6	8.6	59.1
	16	60 04.4	08.4	20 48.1	7.8	16 09.0	8.7	59.1
	17	75 04.3	08.6	35 14.9	7.8	16 00.3	8.8	59.1
	18	90 04.2	N23 08.7	49 41.7	7.8	N15 51.5	8.9	59.1
	19	105 04.0	08.9	64 08.5	7.9	15 42.6	9.0	59.2
	20	120 03.9	09.1	78 35.4	7.9	15 33.6	9.1	59.2
	21	135 03.8	·· 09.2	93 02.3	8.0	15 24.5	9.2	59.2
	22	150 03.6	09.4	107 29.3	8.0	15 15.3	9.3	59.2
	23	165 03.5	09.5	121 56.3	8.1	15 06.0	9.4	59.2
	13 00	180 03.4	N23 09.7	136 23.4	8.0	N14 56.6	9.5	59.2
	01	195 03.3	09.8	150 50.4	8.2	14 47.1	9.6	59.2
	02	210 03.1	10.0	165 17.6	8.2	14 37.5	9.6	59.2
	03	225 03.0	·· 10.1	179 44.8	8.2	14 27.9	9.8	59.2
	04	240 02.9	10.3	194 12.0	8.3	14 18.1	9.8	59.2
	05	255 02.7	10.4	208 39.3	8.3	14 08.3	10.0	59.2
	06	270 02.6	N23 10.6	223 06.6	8.3	N13 58.3	10.0	59.3
	07	285 02.5	10.7	237 33.9	8.4	13 48.3	10.1	59.3
	08	300 02.3	10.9	252 01.3	8.5	13 38.2	10.2	59.3
F	09	315 02.2	·· 11.0	266 28.8	8.4	13 28.0	10.3	59.3
R	10	330 02.1	11.2	280 56.2	8.6	13 17.7	10.3	59.3
I	11	345 02.0	11.3	295 23.8	8.5	13 07.4	10.4	59.3
D	12	0 01.8	N23 11.5	309 51.3	8.7	N12 57.0	10.5	59.3
A	13	15 01.7	11.6	324 19.0	8.6	12 46.5	10.6	59.3
Y	14	30 01.6	11.8	338 46.6	8.7	12 35.9	10.7	59.3
	15	45 01.4	·· 11.9	353 14.3	8.7	12 25.2	10.7	59.3
	16	60 01.3	12.1	7 42.0	8.8	12 14.5	10.8	59.3
	17	75 01.2	12.2	22 09.8	8.9	12 03.7	10.9	59.3
	18	90 01.0	N23 12.3	36 37.7	8.8	N11 52.8	11.0	59.3
	19	105 00.9	12.5	51 05.5	8.9	11 41.8	11.0	59.3
	20	120 00.8	12.6	65 33.4	9.0	11 30.8	11.1	59.4
	21	135 00.6	·· 12.8	80 01.4	9.0	11 19.7	11.2	59.4
	22	150 00.5	12.9	94 29.4	9.0	11 08.5	11.2	59.4
	23	165 00.4	13.1	108 57.4	9.1	10 57.3	11.3	59.4
	14 00	180 00.3	N23 13.2	123 25.5	9.1	N10 46.0	11.4	59.4
	01	195 00.1	13.3	137 53.6	9.2	10 34.6	11.4	59.4
	02	210 00.0	13.5	152 21.8	9.2	10 23.2	11.5	59.4
	03	224 59.9	·· 13.6	166 50.0	9.2	10 11.7	11.5	59.4
	04	239 59.7	13.7	181 18.2	9.3	10 00.2	11.6	59.4
	05	254 59.6	13.9	195 46.5	9.3	9 48.6	11.7	59.4
	06	269 59.5	N23 14.0	210 14.8	9.3	N 9 36.9	11.7	59.4
	07	284 59.3	14.1	224 43.1	9.4	9 25.2	11.8	59.4
S	08	299 59.2	14.3	239 11.5	9.4	9 13.4	11.8	59.4
A	09	314 59.1	·· 14.4	253 39.9	9.5	9 01.6	11.9	59.4
T	10	329 58.9	14.5	268 08.4	9.5	8 49.7	11.9	59.4
U	11	344 58.8	14.7	282 36.9	9.5	8 37.8	12.0	59.4
R	12	359 58.7	N23 14.8	297 05.4	9.6	N 8 25.8	12.0	59.4
D	13	14 58.5	14.9	311 34.0	9.6	8 13.8	12.1	59.4
A	14	29 58.4	15.0	326 02.6	9.6	8 01.7	12.1	59.4
Y	15	44 58.3	·· 15.2	340 31.2	9.7	7 49.6	12.2	59.4
	16	59 58.1	15.3	354 59.9	9.7	7 37.4	12.2	59.4
	17	74 58.0	15.4	9 28.6	9.7	7 25.2	12.2	59.4
	18	89 57.9	N23 15.6	23 57.3	9.8	N 7 13.0	12.3	59.4
	19	104 57.7	15.7	38 26.1	9.8	7 00.7	12.3	59.4
	20	119 57.6	15.8	52 54.9	9.8	6 48.4	12.4	59.4
	21	134 57.5	·· 15.9	67 23.7	9.8	6 36.0	12.4	59.4
	22	149 57.4	16.0	81 52.5	9.9	6 23.6	12.5	59.4
	23	164 57.2	16.2	96 21.4	9.9	6 11.1	12.4	59.4
		S.D. 15.8	*d* 0.1	S.D. 16.1		16.2		16.2

Lat.	Twilight Naut.	Twilight Civil	Sunrise	Moonrise 12	Moonrise 13	Moonrise 14	Moonrise 15
°	h m	h m	h m	h m	h m	h m	h m
N 72	▭	▭	▭	02 46	05 50	08 06	10 12
N 70	▭	▭	▭	04 08	06 19	08 21	10 18
68	▭	▭	▭	04 46	06 41	08 33	10 22
66	////	////	00 08	05 13	06 57	08 42	10 26
64	////	////	01 36	05 34	07 11	08 50	10 29
62	////	////	02 12	05 50	07 23	08 57	10 31
60	////	00 58	02 37	06 04	07 32	09 03	10 34
N 58	////	01 44	02 57	06 15	07 41	09 08	10 36
56	////	02 12	03 14	06 26	07 48	09 13	10 37
54	00 53	02 34	03 28	06 35	07 55	09 17	10 39
52	01 35	02 51	03 40	06 42	08 01	09 21	10 41
50	02 02	03 06	03 50	06 50	08 06	09 24	10 42
45	02 46	03 36	04 13	07 05	08 18	09 31	10 45
N 40	03 16	03 58	04 30	07 17	08 27	09 37	10 47
35	03 39	04 16	04 45	07 28	08 35	09 42	10 49
30	03 58	04 31	04 58	07 37	08 42	09 47	10 51
20	04 26	04 55	05 20	07 53	08 55	09 55	10 54
N 10	04 49	05 16	05 39	08 07	09 05	10 02	10 57
0	05 08	05 34	05 56	08 20	09 15	10 08	11 00
S 10	05 24	05 51	06 13	08 33	09 25	10 15	11 03
20	05 40	06 08	06 32	08 47	09 36	10 22	11 05
30	05 57	06 26	06 53	09 03	09 48	10 30	11 09
35	06 05	06 37	07 05	09 12	09 55	10 34	11 11
40	06 14	06 48	07 19	09 23	10 03	10 39	11 13
45	06 25	07 02	07 35	09 35	10 12	10 45	11 15
S 50	06 36	07 17	07 56	09 50	10 23	10 52	11 18
52	06 41	07 25	08 06	09 57	10 28	10 55	11 19
54	06 47	07 33	08 17	10 04	10 34	10 59	11 21
56	06 53	07 42	08 29	10 13	10 40	11 03	11 22
58	07 00	07 52	08 44	10 22	10 47	11 07	11 24
S 60	07 07	08 04	09 01	10 33	10 55	11 12	11 26

Lat.	Sunset	Twilight Civil	Twilight Naut.	Moonset 12	Moonset 13	Moonset 14	Moonset 15
°	h m	h m	h m	h m	h m	h m	h m
N 72	▭	▭	▭	01 39	00 31	{00 07 / 23 50}	23 35
N 70	▭	▭	▭	00 16	{00 08 / 23 58}	23 42	23 34
68	▭	▭	▭	23 38	23 37	23 35	23 33
66	▭	▭	▭	23 20	23 26	23 29	23 32
64	22 25	////	////	23 05	23 16	23 25	23 32
62	21 48	////	////	22 53	23 08	23 20	23 31
60	21 23	23 04	////	22 42	23 01	23 17	23 30
N 58	21 03	22 17	////	22 33	22 55	23 13	23 30
56	20 46	21 48	////	22 25	22 50	23 11	23 30
54	20 32	21 26	23 09	22 17	22 45	23 08	23 29
52	20 20	21 09	22 25	22 11	22 40	23 06	23 29
50	20 09	20 54	21 58	22 05	22 36	23 03	23 29
45	19 47	20 24	21 14	21 52	22 27	22 59	23 28
N 40	19 29	20 02	20 44	21 42	22 20	22 55	23 27
35	19 14	19 44	20 21	21 33	22 13	22 51	23 27
30	19 01	19 29	20 02	21 24	22 08	22 48	23 26
20	18 40	19 04	19 33	21 11	21 58	22 42	23 26
N 10	18 21	18 44	19 11	20 58	21 49	22 38	23 25
0	18 03	18 26	18 52	20 47	21 41	22 33	23 24
S 10	17 46	18 09	18 35	20 35	21 32	22 28	23 23
20	17 28	17 52	18 19	20 23	21 23	22 23	23 23
30	17 07	17 33	18 03	20 08	21 13	22 18	23 22
35	16 55	17 23	17 54	20 00	21 07	22 14	23 21
40	16 41	17 11	17–45	19 50	21 00	22 11	23 20
45	16 24	16 58	17 35	19 39	20 52	22 06	23 20
S 50	16 04	16 42	17 23	19 25	20 43	22 01	23 19
52	15 54	16 35	17 18	19 19	20 38	21 58	23 18
54	15 43	16 26	17 13	19 12	20 33	21 56	23 18
56	15 30	16 17	17 06	19 04	20 28	21 53	23 17
58	15 16	16 07	17 00	18 55	20 22	21 49	23 17
S 60	14 59	15 56	16 52	18 44	20 15	21 46	23 16

Day	SUN Eqn. of Time 00ʰ	SUN Eqn. of Time 12ʰ	SUN Mer. Pass.	MOON Mer. Pass. Upper	MOON Mer. Pass. Lower	MOON Age	MOON Phase
	m s	m s	h m	h m	h m	d	
12	00 26	00 20	12 00	14 34	02 06	03	
13	00 14	00 08	12 00	15 28	03 01	04	
14	00 01	00 05	12 00	16 21	03 55	05	

STARS, 1975 JANUARY—JUNE

Mag.	Name and Number	S.H.A.							Dec.						
		°	JAN. ′	FEB. ′	MAR. ′	APR. ′	MAY ′	JUNE ′	°	JAN. ′	FEB. ′	MAR. ′	APR. ′	MAY ′	JUNE ′
3·1	γ Ursæ Minoris †	129	49·3	48·8	48·3	47·9	47·8	48·0	N. 71	55·0	55·0	55·0	55·1	55·3	55·5
3·1	γ Trianguli Aust.	130	51·5	50·9	50·4	50·0	49·8	49·7	S. 68	35·1	35·1	35·2	35·3	35·5	35·6
2·7	β Libræ	131	05·0	04·7	04·5	04·4	04·3	04·2	S. 9	17·5	17·6	17·7	17·7	17·7	17·6
2·8	β Lupi	135	46·5	46·2	45·9	45·7	45·6	45·6	S. 43	01·9	02·0	02·1	02·2	02·3	02·3
2·2	β Ursæ Minoris 40	137	19·2	18·6	18·0	17·7	17·6	17·9	N. 74	15·1	15·0	15·1	15·2	15·4	15·5
2·9	α Libræ 39	137	37·4	37·2	37·0	36·8	36·7	36·7	S. 15	56·3	56·4	56·5	56·5	56·5	56·5
2·6	ε Bootis	139	01·5	01·2	01·0	00·9	00·8	00·9	N. 27	10·5	10·4	10·4	10·5	10·6	10·7
2·9	α Lupi	139	55·9	55·5	55·3	55·1	54·9	55·0	S. 47	16·8	16·8	16·9	17·0	17·1	17·2
0·1	α Centauri 38	140	31·2	30·8	30·5	30·2	30·1	30·2	S. 60	43·7	43·8	43·9	44·0	44·2	44·3
2·6	η Centauri	141	31·0	30·7	30·5	30·3	30·2	30·2	S. 42	02·8	02·9	03·0	03·1	03·2	03·3
3·0	γ Bootis	142	13·9	13·6	13·4	13·2	13·2	13·2	N. 38	24·7	24·7	24·7	24·8	24·9	25·0
0·2	α Bootis 37	146	22·0	21·8	21·6	21·5	21·5	21·5	N. 19	18·5	18·4	18·4	18·5	18·5	18·6
2·3	θ Centauri 36	148	41·6	41·3	41·1	41·0	40·9	41·0	S. 36	14·8	14·9	15·0	15·1	15·2	15·2
0·9	β Centauri 35	149	28·9	28·5	28·2	28·0	27·9	28·0	S. 60	15·0	15·1	15·2	15·4	15·5	15·6
3·1	ζ Centauri	151	30·1	29·8	29·5	29·4	29·3	29·4	S. 47	09·8	09·9	10·0	10·2	10·3	10·3
2·8	η Bootis	151	37·4	37·1	37·0	36·9	36·9	36·9	N. 18	31·1	31·0	31·0	31·1	31·2	31·2
1·9	η Ursæ Majoris 34	153	21·5	21·2	21·0	20·9	20·9	21·0	N. 49	25·9	25·9	25·9	26·0	26·2	26·3
2·6	ε Centauri †	155	25·2	24·8	24·6	24·4	24·4	24·5	S. 53	20·2	20·3	20·5	20·6	20·7	20·8
1·2	α Virginis 33	159	01·6	01·4	01·2	01·1	01·1	01·1	S. 11	02·0	02·1	02·1	02·2	02·2	02·2
2·2	ζ Ursæ Majoris	159	16·0	15·7	15·5	15·4	15·4	15·6	N. 55	02·9	02·9	03·0	03·1	03·3	03·4
2·9	ι Centauri	160	11·8	11·6	11·4	11·3	11·3	11·3	S. 36	34·8	34·9	35·0	35·1	35·2	35·3
3·0	ε Virginis	164	45·7	45·5	45·4	45·3	45·3	45·4	N. 11	05·4	05·3	05·3	05·3	05·4	05·4
2·9	α Canum Venat.	166	16·8	16·6	16·4	16·4	16·4	16·5	N. 38	26·8	26·8	26·9	27·0	27·1	27·1
1·7	ε Ursæ Majoris 32	166	45·8	45·4	45·2	45·2	45·3	45·4	N. 56	05·3	05·3	05·4	05·5	05·7	05·7
1·5	β Crucis	168	25·7	25·3	25·1	25·0	25·1	25·3	S. 59	33·0	33·1	33·3	33·4	33·6	33·7
2·9	γ Virginis	169	53·8	53·6	53·4	53·4	53·4	53·5	S. 1	18·9	19·0	19·0	19·1	19·0	19·0
2·4	γ Centauri	169	57·6	57·3	57·1	57·0	57·1	57·2	S. 48	49·3	49·4	49·5	49·7	49·8	49·8
2·9	α Muscæ	171	04·1	03·7	03·4	03·3	03·5	03·7	S. 68	59·7	59·8	60·0	60·2	60·3	60·4
2·8	β Corvi	171	43·5	43·3	43·2	43·1	43·2	43·2	S. 23	15·6	15·7	15·8	15·9	15·9	15·9
1·6	γ Crucis 31	172	32·9	32·6	32·4	32·3	32·4	32·6	S. 56	58·3	58·4	58·6	58·7	58·9	58·9
1·1	α Crucis 30	173	41·4	41·0	40·8	40·7	40·8	41·0	S. 62	57·5	57·6	57·8	58·0	58·1	58·2
2·8	γ Corvi 29	176	21·8	21·6	21·5	21·5	21·5	21·6	S. 17	24·3	24·4	24·5	24·5	24·6	24·6
2·9	δ Centauri	178	13·6	13·3	13·2	13·2	13·3	13·4	S. 50	34·9	35·1	35·2	35·4	35·5	35·5
2·5	γ Ursæ Majoris	181	51·7	51·4	51·3	51·3	51·5	51·7	N. 53	49·6	49·6	49·7	49·9	50·0	50·0
2·2	β Leonis 28	183	02·8	02·6	02·5	02·5	02·6	02·7	N. 14	42·4	42·4	42·4	42·4	42·5	42·5
2·6	δ Leonis	191	47·8	47·6	47·6	47·6	47·7	47·8	N. 20	39·4	39·3	39·4	39·4	39·5	39·5
3·2	ψ Ursæ Majoris	192	55·5	55·3	55·2	55·3	55·4	55·6	N. 44	37·7	37·7	37·8	37·9	38·0	38·0
2·0	α Ursæ Majoris 27	194	26·4	26·1	26·0	26·1	26·3	26·6	N. 61	52·8	52·8	53·0	53·1	53·2	53·2
2·4	β Ursæ Majoris	194	54·2	54·0	53·9	54·0	54·2	54·4	N. 56	30·6	30·7	30·8	30·9	31·0	31·0
2·8	μ Velorum	198	34·0	33·8	33·8	33·9	34·0	34·2	S. 49	17·2	17·4	17·6	17·7	17·8	17·8
3·0	θ Carinæ †	199	28·2	28·0	28·0	28·1	28·4	28·7	S. 64	15·7	15·9	16·1	16·2	16·3	16·4
2·3	γ Leonis	205	20·5	20·4	20·3	20·4	20·5	20·6	N. 19	57·8	57·8	57·8	57·9	57·9	57·9
1·3	α Leonis 26	208	13·8	13·7	13·7	13·7	13·8	13·9	N. 12	05·2	05·1	05·1	05·1	05·2	05·2
3·1	ε Leonis	213	52·9	52·8	52·8	52·8	53·0	53·1	N. 23	53·1	53·1	53·2	53·2	53·2	53·3
3·0	*N* Velorum	217	22·3	22·2	22·3	22·5	22·7	23·0	S. 56	55·4	55·6	55·8	55·9	56·0	55·9
2·2	α Hydræ 25	218	24·0	24·0	24·0	24·1	24·2	24·2	S. 8	33·1	33·2	33·3	33·3	33·3	33·3
2·6	κ Velorum	219	39·2	39·1	39·2	39·4	39·6	39·8	S. 54	54·2	54·4	54·6	54·7	54·7	54·7
2·2	ι Carinæ	220	52·9	52·8	52·9	53·1	53·4	53·7	S. 59	10·2	10·4	10·6	10·7	10·7	10·7
1·8	β Carinæ 24	221	44·9	44·9	45·1	45·4	45·9	46·3	S. 69	36·9	37·1	37·2	37·3	37·4	37·3
2·2	λ Velorum 23	223	13·2	13·1	13·2	13·4	13·5	13·7	S. 43	19·9	20·1	20·3	20·3	20·4	20·3
3·1	ι Ursæ Majoris	225	36·6	36·6	36·6	36·8	37·0	37·1	N. 48	08·2	08·3	08·3	08·4	08·4	08·4
2·0	δ Velorum	228	59·0	59·0	59·1	59·4	59·6	59·8	S. 54	37·0	37·2	37·4	37·5	37·5	37·4
1·7	ε Carinæ 22	234	29·1	29·2	29·3	29·6	29·9	30·2	S. 59	25·8	26·0	26·1	26·2	26·2	26·1
1·9	γ Velorum	237	47·9	47·9	48·0	48·2	48·4	48·6	S. 47	15·9	16·0	16·1	16·2	16·2	16·1
2·9	ρ Puppis	238	22·2	22·2	22·3	22·4	22·5	22·6	S. 24	14·0	14·2	14·2	14·3	14·3	14·2
2·3	ζ Puppis	239	18·8	18·8	18·9	19·1	19·2	19·4	S. 39	56·1	56·2	56·3	56·4	56·4	56·3
1·2	β Geminorum 21	244	02·5	02·4	02·5	02·7	02·8	02·8	N. 28	05·1	05·1	05·1	05·2	05·2	05·2
0·5	α Canis Minoris 20	245	29·5	29·4	29·5	29·7	29·8	29·8	N. 5	17·2	17·2	17·2	17·2	17·2	17·2

† Not suitable for use with H.O. 214 (H.D. 486)

POLARIS (POLE STAR) TABLES, 1975

FOR DETERMINING LATITUDE FROM SEXTANT ALTITUDE AND FOR AZIMUTH

L.H.A. ARIES	240°– 249°	250°– 259°	260°– 269°	270°– 279°	280°– 289°	290°– 299°	300°– 309°	310°– 319°	320°– 329°	330°– 339°	340°– 349°	350°– 359°
	a_0	a_0	a_0	a_0	a_0	a_0	a_0	a_0	a_0	a_0	a_0	a_0
°	° ′	° ′	° ′	° ′	° ′	° ′	° ′	° ′	° ′	° ′	° ′	° ′
0	1 43·8	1 39·1	1 33·1	1 26·1	1 18·3	1 09·9	1 01·1	0 52·3	0 43·6	0 35·4	0 27·9	0 21·3
1	43·4	38·5	32·5	25·4	17·5	09·0	1 00·2	51·4	42·8	34·6	27·2	20·7
2	43·0	38·0	31·8	24·6	16·7	08·2	0 59·4	50·5	41·9	33·8	26·5	20·1
3	42·5	37·4	31·1	23·9	15·8	07·3	58·5	49·6	41·1	33·1	25·8	19·6
4	42·1	36·8	30·4	23·1	15·0	06·4	57·6	48·8	40·3	32·3	25·1	19·0
5	1 41·6	1 36·2	1 29·7	1 22·3	1 14·2	1 05·5	0 56·7	0 47·9	0 39·4	0 31·5	0 24·5	0 18·5
6	41·1	35·6	29·0	21·5	13·3	04·7	55·8	47·0	38·6	30·8	23·8	17·9
7	40·6	35·0	28·3	20·7	12·5	03·8	54·9	46·2	37·8	30·0	23·2	17·4
8	40·1	34·4	27·6	19·9	11·6	02·9	54·0	45·3	37·0	29·3	22·5	16·9
9	39·6	33·8	26·9	19·1	10·8	02·0	53·2	44·5	36·2	28·6	21·9	16·4
10	1 39·1	1 33·1	1 26·1	1 18·3	1 09·9	1 01·1	0 52·3	0 43·6	0 35·4	0 27·9	0 21·3	0 15·9
Lat.	a_1	a_1	a_1	a_1	a_1	a_1	a_1	a_1	a_1	a_1	a_1	a_1
°	′	′	′	′	′	′	′	′	′	′	′	′
0	0·5	0·4	0·3	0·2	0·2	0·2	0·2	0·2	0·2	0·3	0·4	0·4
10	·5	·4	·4	·3	·3	·2	·2	·2	·3	·3	·4	·5
20	·5	·5	·4	·4	·3	·3	·3	·3	·3	·4	·4	·5
30	·5	·5	·5	·4	·4	·4	·4	·4	·4	·4	·5	·5
40	0·6	0·5	0·5	0·5	0·5	0·5	0·5	0·5	0·5	0·5	0·5	0·6
45	·6	·6	·6	·5	·5	·5	·5	·5	·5	·5	·6	·6
50	·6	·6	·6	·6	·6	·6	·6	·6	·6	·6	·6	·6
55	·6	·6	·7	·7	·7	·7	·7	·7	·7	·7	·6	·6
60	·7	·7	·7	·8	·8	·8	·8	·8	·8	·7	·7	·7
62	0·7	0·7	0·8	0·8	0·8	0·9	0·9	0·8	0·8	0·8	0·7	0·7
64	·7	·7	·8	·9	0·9	0·9	0·9	0·9	·9	·8	·8	·7
66	·7	·8	·9	0·9	1·0	1·0	1·0	1·0	0·9	·9	·8	·7
68	0·7	0·8	0·9	1·0	1·0	1·1	1·1	1·1	1·0	0·9	0·9	0·8
Month	a_2	a_2	a_2	a_2	a_2	a_2	a_2	a_2	a_2	a_2	a_2	a_2
	′	′	′	′	′	′	′	′	′	′	′	′
Jan.	0·4	0·4	0·4	0·5	0·5	0·5	0·6	0·6	0·6	0·7	0·7	0·7
Feb.	·3	·3	·3	·3	·3	·4	·4	·4	·5	·5	·6	·6
Mar.	·3	·3	·3	·3	·3	·3	·3	·3	·3	·4	·4	·5
Apr.	0·4	0·4	0·3	0·3	0·3	0·2	0·2	0·2	0·2	0·3	0·3	0·3
May	·6	·5	·5	·4	·3	·3	·3	·2	·2	·2	·2	·2
June	·7	·7	·6	·5	·5	·4	·4	·3	·3	·3	·2	·2
July	0·8	0·8	0·8	0·7	0·6	0·6	0·5	0·5	0·4	0·4	0·3	0·3
Aug.	·9	·9	·9	·8	·8	·8	·7	·7	·6	·5	·5	·4
Sept.	·9	·9	·9	·9	·9	·9	·9	·8	·8	·7	·7	·6
Oct.	0·8	0·8	0·9	0·9	0·9	0·9	0·9	0·9	0·9	0·9	0·9	0·8
Nov.	·6	·7	·8	·8	·9	·9	1·0	1·0	1·0	1·0	1·0	1·0
Dec.	0·5	0·5	0·6	0·7	0·8	0·9	0·9	1·0	1·0	1·0	1·1	1·1
Lat.	AZIMUTH											
°	°	°	°	°	°	°	°	°	°	°	°	°
0	0·5	0·6	0·7	0·8	0·8	0·8	0·8	0·8	0·8	0·7	0·6	0·5
20	0·5	0·6	0·7	0·8	0·9	0·9	0·9	0·9	0·8	0·8	0·7	0·5
40	0·6	0·7	0·9	1·0	1·1	1·1	1·1	1·1	1·0	0·9	0·8	0·7
50	0·7	0·9	1·0	1·2	1·3	1·3	1·3	1·3	1·2	1·1	1·0	0·8
55	0·8	1·0	1·2	1·3	1·4	1·5	1·5	1·4	1·4	1·3	1·1	0·9
60	0·9	1·1	1·3	1·5	1·6	1·7	1·7	1·7	1·6	1·4	1·3	1·0
65	1·1	1·3	1·6	1·8	1·9	2·0	2·0	2·0	1·9	1·7	1·5	1·2

Latitude = Apparent altitude (corrected for refraction) $- 1° + a_0 + a_1 + a_2$

The table is entered with L.H.A. Aries to determine the column to be used; each column refers to a range of 10°. a_0 is taken, with mental interpolation, from the upper table with the units of L.H.A. Aries in degrees as argument; a_1, a_2 are taken, without interpolation, from the second and third tables with arguments latitude and month respectively. a_0, a_1, a_2 are always positive. The final table gives the azimuth of *Polaris*.

CONVERSION OF ARC TO TIME

0°–59°		60°–119°		120°–179°		180°–239°		240°–299°		300°–359°			0'·00	0'·25	0'·50	0'·75
°	h m	°	h m	°	h m	°	h m	°	h m	°	h m	'	m s	m s	m s	m s
0	0 00	60	4 00	120	8 00	180	12 00	240	16 00	300	20 00	0	0 00	0 01	0 02	0 03
1	0 04	61	4 04	121	8 04	181	12 04	241	16 04	301	20 04	1	0 04	0 05	0 06	0 07
2	0 08	62	4 08	122	8 08	182	12 08	242	16 08	302	20 08	2	0 08	0 09	0 10	0 11
3	0 12	63	4 12	123	8 12	183	12 12	243	16 12	303	20 12	3	0 12	0 13	0 14	0 15
4	0 16	64	4 16	124	8 16	184	12 16	244	16 16	304	20 16	4	0 16	0 17	0 18	0 19
5	0 20	65	4 20	125	8 20	185	12 20	245	16 20	305	20 20	5	0 20	0 21	0 22	0 23
6	0 24	66	4 24	126	8 24	186	12 24	246	16 24	306	20 24	6	0 24	0 25	0 26	0 27
7	0 28	67	4 28	127	8 28	187	12 28	247	16 28	307	20 28	7	0 28	0 29	0 30	0 31
8	0 32	68	4 32	128	8 32	188	12 32	248	16 32	308	20 32	8	0 32	0 33	0 34	0 35
9	0 36	69	4 36	129	8 36	189	12 36	249	16 36	309	20 36	9	0 36	0 37	0 38	0 39
10	0 40	70	4 40	130	8 40	190	12 40	250	16 40	310	20 40	10	0 40	0 41	0 42	0 43
11	0 44	71	4 44	131	8 44	191	12 44	251	16 44	311	20 44	11	0 44	0 45	0 46	0 47
12	0 48	72	4 48	132	8 48	192	12 48	252	16 48	312	20 48	12	0 48	0 49	0 50	0 51
13	0 52	73	4 52	133	8 52	193	12 52	253	16 52	313	20 52	13	0 52	0 53	0 54	0 55
14	0 56	74	4 56	134	8 56	194	12 56	254	16 56	314	20 56	14	0 56	0 57	0 58	0 59
15	1 00	75	5 00	135	9 00	195	13 00	255	17 00	315	21 00	15	1 00	1 01	1 02	1 03
16	1 04	76	5 04	136	9 04	196	13 04	256	17 04	316	21 04	16	1 04	1 05	1 06	1 07
17	1 08	77	5 08	137	9 08	197	13 08	257	17 08	317	21 08	17	1 08	1 09	1 10	1 11
18	1 12	78	5 12	138	9 12	198	13 12	258	17 12	318	21 12	18	1 12	1 13	1 14	1 15
19	1 16	79	5 16	139	9 16	199	13 16	259	17 16	319	21 16	19	1 16	1 17	1 18	1 19
20	1 20	80	5 20	140	9 20	200	13 20	260	17 20	320	21 20	20	1 20	1 21	1 22	1 23
21	1 24	81	5 24	141	9 24	201	13 24	261	17 24	321	21 24	21	1 24	1 25	1 26	1 27
22	1 28	82	5 28	142	9 28	202	13 28	262	17 28	322	21 28	22	1 28	1 29	1 30	1 31
23	1 32	83	5 32	143	9 32	203	13 32	263	17 32	323	21 32	23	1 32	1 33	1 34	1 35
24	1 36	84	5 36	144	9 36	204	13 36	264	17 36	324	21 36	24	1 36	1 37	1 38	1 39
25	1 40	85	5 40	145	9 40	205	13 40	265	17 40	325	21 40	25	1 40	1 41	1 42	1 43
26	1 44	86	5 44	146	9 44	206	13 44	266	17 44	326	21 44	26	1 44	1 45	1 46	1 47
27	1 48	87	5 48	147	9 48	207	13 48	267	17 48	327	21 48	27	1 48	1 49	1 50	1 51
28	1 52	88	5 52	148	9 52	208	13 52	268	17 52	328	21 52	28	1 52	1 53	1 54	1 55
29	1 56	89	5 56	149	9 56	209	13 56	269	17 56	329	21 56	29	1 56	1 57	1 58	1 59
30	2 00	90	6 00	150	10 00	210	14 00	270	18 00	330	22 00	30	2 00	2 01	2 02	2 03
31	2 04	91	6 04	151	10 04	211	14 04	271	18 04	331	22 04	31	2 04	2 05	2 06	2 07
32	2 08	92	6 08	152	10 08	212	14 08	272	18 08	332	22 08	32	2 08	2 09	2 10	2 11
33	2 12	93	6 12	153	10 12	213	14 12	273	18 12	333	22 12	33	2 12	2 13	2 14	2 15
34	2 16	94	6 16	154	10 16	214	14 16	274	18 16	334	22 16	34	2 16	2 17	2 18	2 19
35	2 20	95	6 20	155	10 20	215	14 20	275	18 20	335	22 20	35	2 20	2 21	2 22	2 23
36	2 24	96	6 24	156	10 24	216	14 24	276	18 24	336	22 24	36	2 24	2 25	2 26	2 27
37	2 28	97	6 28	157	10 28	217	14 28	277	18 28	337	22 28	37	2 28	2 29	2 30	2 31
38	2 32	98	6 32	158	10 32	218	14 32	278	18 32	338	22 32	38	2 32	2 33	2 34	2 35
39	2 36	99	6 36	159	10 36	219	14 36	279	18 36	339	22 36	39	2 36	2 37	2 38	2 39
40	2 40	100	6 40	160	10 40	220	14 40	280	18 40	340	22 40	40	2 40	2 41	2 42	2 43
41	2 44	101	6 44	161	10 44	221	14 44	281	18 44	341	22 44	41	2 44	2 45	2 46	2 47
42	2 48	102	6 48	162	10 48	222	14 48	282	18 48	342	22 48	42	2 48	2 49	2 50	2 51
43	2 52	103	6 52	163	10 52	223	14 52	283	18 52	343	22 52	43	2 52	2 53	2 54	2 55
44	2 56	104	6 56	164	10 56	224	14 56	284	18 56	344	22 56	44	2 56	2 57	2 58	2 59
45	3 00	105	7 00	165	11 00	225	15 00	285	19 00	345	23 00	45	3 00	3 01	3 02	3 03
46	3 04	106	7 04	166	11 04	226	15 04	286	19 04	346	23 04	46	3 04	3 05	3 06	3 07
47	3 08	107	7 08	167	11 08	227	15 08	287	19 08	347	23 08	47	3 08	3 09	3 10	3 11
48	3 12	108	7 12	168	11 12	228	15 12	288	19 12	348	23 12	48	3 12	3 13	3 14	3 15
49	3 16	109	7 16	169	11 16	229	15 16	289	19 16	349	23 16	49	3 16	3 17	3 18	3 19
50	3 20	110	7 20	170	11 20	230	15 20	290	19 20	350	23 20	50	3 20	3 21	3 22	3 23
51	3 24	111	7 24	171	11 24	231	15 24	291	19 24	351	23 24	51	3 24	3 25	3 26	3 27
52	3 28	112	7 28	172	11 28	232	15 28	292	19 28	352	23 28	52	3 28	3 29	3 30	3 31
53	3 32	113	7 32	173	11 32	233	15 32	293	19 32	353	23 32	53	3 32	3 33	3 34	3 35
54	3 36	114	7 36	174	11 36	234	15 36	294	19 36	354	23 36	54	3 36	3 37	3 38	3 39
55	3 40	115	7 40	175	11 40	235	15 40	295	19 40	355	23 40	55	3 40	3 41	3 42	3 43
56	3 44	116	7 44	176	11 44	236	15 44	296	19 44	356	23 44	56	3 44	3 45	3 46	3 47
57	3 48	117	7 48	177	11 48	237	15 48	297	19 48	357	23 48	57	3 48	3 49	3 50	3 51
58	3 52	118	7 52	178	11 52	238	15 52	298	19 52	358	23 52	58	3 52	3 53	3 54	3 55
59	3 56	119	7 56	179	11 56	239	15 56	299	19 56	359	23 56	59	3 56	3 57	3 58	3 59

The above table is for converting expressions in arc to their equivalent in time; its main use in this Almanac is for the conversion of longitude for application to L.M.T. (*added* if *west*, *subtracted* if *east*) to give G.M.T. or vice versa, particularly in the case of sunrise, sunset, etc.

24ᵐ INCREMENTS AND CORRECTIONS 25ᵐ

24ᵐ	SUN PLANETS	ARIES	MOON	v or d	Corrⁿ	v or d	Corrⁿ	v or d	Corrⁿ
s	° ′	° ′	° ′	′	′	′	′	′	′
00	6 00·0	6 01·0	5 43·6	0·0	0·0	6·0	2·5	12·0	4·9
01	6 00·3	6 01·2	5 43·8	0·1	0·0	6·1	2·5	12·1	4·9
02	6 00·5	6 01·5	5 44·1	0·2	0·1	6·2	2·5	12·2	5·0
03	6 00·8	6 01·7	5 44·3	0·3	0·1	6·3	2·6	12·3	5·0
04	6 01·0	6 02·0	5 44·6	0·4	0·2	6·4	2·6	12·4	5·1
05	6 01·3	6 02·2	5 44·8	0·5	0·2	6·5	2·7	12·5	5·1
06	6 01·5	6 02·5	5 45·0	0·6	0·2	6·6	2·7	12·6	5·1
07	6 01·8	6 02·7	5 45·3	0·7	0·3	6·7	2·7	12·7	5·2
08	6 02·0	6 03·0	5 45·5	0·8	0·3	6·8	2·8	12·8	5·2
09	6 02·3	6 03·2	5 45·7	0·9	0·4	6·9	2·8	12·9	5·3
10	6 02·5	6 03·5	5 46·0	1·0	0·4	7·0	2·9	13·0	5·3
11	6 02·8	6 03·7	5 46·2	1·1	0·4	7·1	2·9	13·1	5·3
12	6 03·0	6 04·0	5 46·5	1·2	0·5	7·2	2·9	13·2	5·4
13	6 03·3	6 04·2	5 46·7	1·3	0·5	7·3	3·0	13·3	5·4
14	6 03·5	6 04·5	5 46·9	1·4	0·6	7·4	3·0	13·4	5·5
15	6 03·8	6 04·7	5 47·2	1·5	0·6	7·5	3·1	13·5	5·5
16	6 04·0	6 05·0	5 47·4	1·6	0·7	7·6	3·1	13·6	5·6
17	6 04·3	6 05·2	5 47·7	1·7	0·7	7·7	3·1	13·7	5·6
18	6 04·5	6 05·5	5 47·9	1·8	0·7	7·8	3·2	13·8	5·6
19	6 04·8	6 05·7	5 48·1	1·9	0·8	7·9	3·2	13·9	5·7
20	6 05·0	6 06·0	5 48·4	2·0	0·8	8·0	3·3	14·0	5·7
21	6 05·3	6 06·3	5 48·6	2·1	0·9	8·1	3·3	14·1	5·8
22	6 05·5	6 06·5	5 48·8	2·2	0·9	8·2	3·3	14·2	5·8
23	6 05·8	6 06·8	5 49·1	2·3	0·9	8·3	3·4	14·3	5·8
24	6 06·0	6 07·0	5 49·3	2·4	1·0	8·4	3·4	14·4	5·9
25	6 06·3	6 07·3	5 49·6	2·5	1·0	8·5	3·5	14·5	5·9
26	6 06·5	6 07·5	5 49·8	2·6	1·1	8·6	3·5	14·6	6·0
27	6 06·8	6 07·8	5 50·0	2·7	1·1	8·7	3·6	14·7	6·0
28	6 07·0	6 08·0	5 50·3	2·8	1·1	8·8	3·6	14·8	6·0
29	6 07·3	6 08·3	5 50·5	2·9	1·2	8·9	3·6	14·9	6·1
30	6 07·5	6 08·5	5 50·8	3·0	1·2	9·0	3·7	15·0	6·1
31	6 07·8	6 08·8	5 51·0	3·1	1·3	9·1	3·7	15·1	6·2
32	6 08·0	6 09·0	5 51·2	3·2	1·3	9·2	3·8	15·2	6·2
33	6 08·3	6 09·3	5 51·5	3·3	1·3	9·3	3·8	15·3	6·2
34	6 08·5	6 09·5	5 51·7	3·4	1·4	9·4	3·8	15·4	6·3
35	6 08·8	6 09·8	5 52·0	3·5	1·4	9·5	3·9	15·5	6·3
36	6 09·0	6 10·0	5 52·2	3·6	1·5	9·6	3·9	15·6	6·4
37	6 09·3	6 10·3	5 52·4	3·7	1·5	9·7	4·0	15·7	6·4
38	6 09·5	6 10·5	5 52·7	3·8	1·6	9·8	4·0	15·8	6·5
39	6 09·8	6 10·8	5 52·9	3·9	1·6	9·9	4·0	15·9	6·5
40	6 10·0	6 11·0	5 53·1	4·0	1·6	10·0	4·1	16·0	6·5
41	6 10·3	6 11·3	5 53·4	4·1	1·7	10·1	4·1	16·1	6·6
42	6 10·5	6 11·5	5 53·6	4·2	1·7	10·2	4·2	16·2	6·6
43	6 10·8	6 11·8	5 53·9	4·3	1·8	10·3	4·2	16·3	6·7
44	6 11·0	6 12·0	5 54·1	4·4	1·8	10·4	4·2	16·4	6·7
45	6 11·3	6 12·3	5 54·3	4·5	1·8	10·5	4·3	16·5	6·7
46	6 11·5	6 12·5	5 54·6	4·6	1·9	10·6	4·3	16·6	6·8
47	6 11·8	6 12·8	5 54·8	4·7	1·9	10·7	4·4	16·7	6·8
48	6 12·0	6 13·0	5 55·1	4·8	2·0	10·8	4·4	16·8	6·9
49	6 12·3	6 13·3	5 55·3	4·9	2·0	10·9	4·5	16·9	6·9
50	6 12·5	6 13·5	5 55·5	5·0	2·0	11·0	4·5	17·0	6·9
51	6 12·8	6 13·8	5 55·8	5·1	2·1	11·1	4·5	17·1	7·0
52	6 13·0	6 14·0	5 56·0	5·2	2·1	11·2	4·6	17·2	7·0
53	6 13·3	6 14·3	5 56·2	5·3	2·2	11·3	4·6	17·3	7·1
54	6 13·5	6 14·5	5 56·5	5·4	2·2	11·4	4·7	17·4	7·1
55	6 13·8	6 14·8	5 56·7	5·5	2·2	11·5	4·7	17·5	7·1
56	6 14·0	6 15·0	5 57·0	5·6	2·3	11·6	4·7	17·6	7·2
57	6 14·3	6 15·3	5 57·2	5·7	2·3	11·7	4·8	17·7	7·2
58	6 14·5	6 15·5	5 57·4	5·8	2·4	11·8	4·8	17·8	7·3
59	6 14·8	6 15·8	5 57·7	5·9	2·4	11·9	4·9	17·9	7·3
60	6 15·0	6 16·0	5 57·9	6·0	2·5	12·0	4·9	18·0	7·4

25ᵐ	SUN PLANETS	ARIES	MOON	v or d	Corrⁿ	v or d	Corrⁿ	v or d	Corrⁿ
s	° ′	° ′	° ′	′	′	′	′	′	′
00	6 15·0	6 16·0	5 57·9	0·0	0·0	6·0	2·6	12·0	5·1
01	6 15·3	6 16·3	5 58·2	0·1	0·0	6·1	2·6	12·1	5·1
02	6 15·5	6 16·5	5 58·4	0·2	0·1	6·2	2·6	12·2	5·2
03	6 15·8	6 16·8	5 58·6	0·3	0·1	6·3	2·7	12·3	5·2
04	6 16·0	6 17·0	5 58·9	0·4	0·2	6·4	2·7	12·4	5·3
05	6 16·3	6 17·3	5 59·1	0·5	0·2	6·5	2·8	12·5	5·3
06	6 16·5	6 17·5	5 59·3	0·6	0·3	6·6	2·8	12·6	5·4
07	6 16·8	6 17·8	5 59·6	0·7	0·3	6·7	2·8	12·7	5·4
08	6 17·0	6 18·0	5 59·8	0·8	0·3	6·8	2·9	12·8	5·4
09	6 17·3	6 18·3	6 00·1	0·9	0·4	6·9	2·9	12·9	5·5
10	6 17·5	6 18·5	6 00·3	1·0	0·4	7·0	3·0	13·0	5·5
11	6 17·8	6 18·8	6 00·5	1·1	0·5	7·1	3·0	13·1	5·6
12	6 18·0	6 19·0	6 00·8	1·2	0·5	7·2	3·1	13·2	5·6
13	6 18·3	6 19·3	6 01·0	1·3	0·6	7·3	3·1	13·3	5·7
14	6 18·5	6 19·5	6 01·3	1·4	0·6	7·4	3·1	13·4	5·7
15	6 18·8	6 19·8	6 01·5	1·5	0·6	7·5	3·2	13·5	5·7
16	6 19·0	6 20·0	6 01·7	1·6	0·7	7·6	3·2	13·6	5·8
17	6 19·3	6 20·3	6 02·0	1·7	0·7	7·7	3·3	13·7	5·8
18	6 19·5	6 20·5	6 02·2	1·8	0·8	7·8	3·3	13·8	5·9
19	6 19·8	6 20·8	6 02·5	1·9	0·8	7·9	3·4	13·9	5·9
20	6 20·0	6 21·0	6 02·7	2·0	0·9	8·0	3·4	14·0	6·0
21	6 20·3	6 21·3	6 02·9	2·1	0·9	8·1	3·4	14·1	6·0
22	6 20·5	6 21·5	6 03·2	2·2	0·9	8·2	3·5	14·2	6·0
23	6 20·8	6 21·8	6 03·4	2·3	1·0	8·3	3·5	14·3	6·1
24	6 21·0	6 22·0	6 03·6	2·4	1·0	8·4	3·6	14·4	6·1
25	6 21·3	6 22·3	6 03·9	2·5	1·1	8·5	3·6	14·5	6·2
26	6 21·5	6 22·5	6 04·1	2·6	1·1	8·6	3·7	14·6	6·2
27	6 21·8	6 22·8	6 04·4	2·7	1·1	8·7	3·7	14·7	6·2
28	6 22·0	6 23·0	6 04·6	2·8	1·2	8·8	3·7	14·8	6·3
29	6 22·3	6 23·3	6 04·8	2·9	1·2	8·9	3·8	14·9	6·3
30	6 22·5	6 23·5	6 05·1	3·0	1·3	9·0	3·8	15·0	6·4
31	6 22·8	6 23·8	6 05·3	3·1	1·3	9·1	3·9	15·1	6·4
32	6 23·0	6 24·0	6 05·6	3·2	1·4	9·2	3·9	15·2	6·5
33	6 23·3	6 24·3	6 05·8	3·3	1·4	9·3	4·0	15·3	6·5
34	6 23·5	6 24·5	6 06·0	3·4	1·4	9·4	4·0	15·4	6·5
35	6 23·8	6 24·8	6 06·3	3·5	1·5	9·5	4·0	15·5	6·6
36	6 24·0	6 25·1	6 06·5	3·6	1·5	9·6	4·1	15·6	6·6
37	6 24·3	6 25·3	6 06·7	3·7	1·6	9·7	4·1	15·7	6·7
38	6 24·5	6 25·6	6 07·0	3·8	1·6	9·8	4·2	15·8	6·7
39	6 24·8	6 25·8	6 07·2	3·9	1·7	9·9	4·2	15·9	6·8
40	6 25·0	6 26·1	6 07·5	4·0	1·7	10·0	4·3	16·0	6·8
41	6 25·3	6 26·3	6 07·7	4·1	1·7	10·1	4·3	16·1	6·8
42	6 25·5	6 26·6	6 07·9	4·2	1·8	10·2	4·3	16·2	6·9
43	6 25·8	6 26·8	6 08·2	4·3	1·8	10·3	4·4	16·3	6·9
44	6 26·0	6 27·1	6 08·4	4·4	1·9	10·4	4·4	16·4	7·0
45	6 26·3	6 27·3	6 08·7	4·5	1·9	10·5	4·5	16·5	7·0
46	6 26·5	6 27·6	6 08·9	4·6	2·0	10·6	4·5	16·6	7·1
47	6 26·8	6 27·8	6 09·1	4·7	2·0	10·7	4·5	16·7	7·1
48	6 27·0	6 28·1	6 09·4	4·8	2·0	10·8	4·6	16·8	7·1
49	6 27·3	6 28·3	6 09·6	4·9	2·1	10·9	4·6	16·9	7·2
50	6 27·5	6 28·6	6 09·8	5·0	2·1	11·0	4·7	17·0	7·2
51	6 27·8	6 28·8	6 10·1	5·1	2·2	11·1	4·7	17·1	7·3
52	6 28·0	6 29·1	6 10·3	5·2	2·2	11·2	4·8	17·2	7·3
53	6 28·3	6 29·3	6 10·6	5·3	2·3	11·3	4·8	17·3	7·4
54	6 28·5	6 29·6	6 10·8	5·4	2·3	11·4	4·8	17·4	7·4
55	6 28·8	6 29·8	6 11·0	5·5	2·3	11·5	4·9	17·5	7·4
56	6 29·0	6 30·1	6 11·3	5·6	2·4	11·6	4·9	17·6	7·5
57	6 29·3	6 30·3	6 11·5	5·7	2·4	11·7	5·0	17·7	7·5
58	6 29·5	6 30·6	6 11·8	5·8	2·5	11·8	5·0	17·8	7·6
59	6 29·8	6 30·8	6 12·0	5·9	2·5	11·9	5·1	17·9	7·6
60	6 30·0	6 31·1	6 12·2	6·0	2·6	12·0	5·1	18·0	7·7

TABLES FOR INTERPOLATING SUNRISE, MOONRISE, ETC.

TABLE I—FOR LATITUDE

Tabular Interval			Difference between the times for consecutive latitudes															
10°	5°	2°	5m	10m	15m	20m	25m	30m	35m	40m	45m	50m	55m	60m	1h 05m	1h 10m	1h 15m	1h 20m
° ′	° ′	° ′	m	m	m	m	m	m	m	m	m	m	m	m	h m	h m	h m	h m
0 30	0 15	0 06	0	0	1	1	1	1	1	2	2	2	2	2	0 02	0 02	0 02	0 02
1 00	0 30	0 12	0	1	1	2	2	3	3	3	4	4	4	5	05	05	05	05
1 30	0 45	0 18	1	1	2	3	3	4	4	5	5	6	7	7	07	07	07	07
2 00	1 00	0 24	1	2	3	4	5	5	6	7	7	8	9	10	10	10	10	10
2 30	1 15	0 30	1	2	4	5	6	7	8	9	9	10	11	12	12	13	13	13
3 00	1 30	0 36	1	3	4	6	7	8	9	10	11	12	13	14	0 15	0 15	0 16	0 16
3 30	1 45	0 42	2	3	5	7	8	10	11	12	13	14	16	17	18	18	19	19
4 00	2 00	0 48	2	4	6	8	9	11	13	14	15	16	18	19	20	21	22	22
4 30	2 15	0 54	2	4	7	9	11	13	15	16	18	19	21	22	23	24	25	26
5 00	2 30	1 00	2	5	7	10	12	14	16	18	20	22	23	25	26	27	28	29
5 30	2 45	1 06	3	5	8	11	13	16	18	20	22	24	26	28	0 29	0 30	0 31	0 32
6 00	3 00	1 12	3	6	9	12	14	17	20	22	24	26	29	31	32	33	34	36
6 30	3 15	1 18	3	6	10	13	16	19	22	24	26	29	31	34	36	37	38	40
7 00	3 30	1 24	3	7	10	14	17	20	23	26	29	31	34	37	39	41	42	44
7 30	3 45	1 30	4	7	11	15	18	22	25	28	31	34	37	40	43	44	46	48
8 00	4 00	1 36	4	8	12	16	20	23	27	30	34	37	41	44	0 47	0 48	0 51	0 53
8 30	4 15	1 42	4	8	13	17	21	25	29	33	36	40	44	48	0 51	0 53	0 56	0 58
9 00	4 30	1 48	4	9	13	18	22	27	31	35	39	43	47	52	0 55	0 58	1 01	1 04
9 30	4 45	1 54	5	9	14	19	24	28	33	38	42	47	51	56	1 00	1 04	1 08	1 12
10 00	5 00	2 00	5	10	15	20	25	30	35	40	45	50	55	60	1 05	1 10	1 15	1 20

Table I is for interpolating the L.M.T. of sunrise, twilight, moonrise, etc. for latitude. It is to be noted that the interpolation is not linear, so that when using this table it is essential to take out the required phenomenon for the latitude *less* than the true latitude. The table is entered with the nearest value of the difference between the times for the tabular latitude and the next higher one, and, in the appropriate column, with the difference between true latitude and tabular latitude; the correction so obtained is applied to the time for the tabular latitude; the sign of the correction can be seen by inspection.

TABLE II—FOR LONGITUDE

Long. East or West	Difference between the times for given date and preceding date (for east longitude) or for given date and following date (for west longitude)																	
	10m	20m	30m	40m	50m	60m	1h + 10m	1h + 20m	1h + 30m	1h + 40m	1h + 50m	1h + 60m	2h 10m	2h 20m	2h 30m	2h 40m	2h 50m	3h 00m
°	m	m	m	m	m	m	m	m	m	m	m	m	h m	h m	h m	h m	h m	h m
0	0	0	0	0	0	0	0	0	0	0	0	0	0 00	0 00	0 00	0 00	0 00	0 00
10	0	1	1	1	1	2	2	2	2	3	3	3	04	04	04	04	05	05
20	1	1	2	2	3	3	4	4	5	6	6	7	07	08	08	09	09	10
30	1	2	2	3	4	5	6	7	7	8	9	10	11	12	12	13	14	15
40	1	2	3	4	6	7	8	9	10	11	12	13	14	16	17	18	19	20
50	1	3	4	6	7	8	10	11	12	14	15	17	0 18	0 19	0 21	0 22	0 24	0 25
60	2	3	5	7	8	10	12	13	15	17	18	20	22	23	25	27	28	30
70	2	4	6	8	10	12	14	16	17	19	21	23	25	27	29	31	33	35
80	2	4	7	9	11	13	16	18	20	22	24	27	29	31	33	36	38	40
90	2	5	7	10	12	15	17	20	22	25	27	30	32	35	37	40	42	45
100	3	6	8	11	14	17	19	22	25	28	31	33	0 36	0 39	0 42	0 44	0 47	0 50
110	3	6	9	12	15	18	21	24	27	31	34	37	40	43	46	49	0 52	0 55
120	3	7	10	13	17	20	23	27	30	33	37	40	43	47	50	53	0 57	1 00
130	4	7	11	14	18	22	25	29	32	36	40	43	47	51	54	0 58	1 01	1 05
140	4	8	12	16	19	23	27	31	35	39	43	47	51	54	0 58	1 02	1 06	1 10
150	4	8	13	17	21	25	29	33	38	42	46	50	0 54	0 58	1 03	1 07	1 11	1 15
160	4	9	13	18	22	27	31	36	40	44	49	53	0 58	1 02	1 07	1 11	1 16	1 20
170	5	9	14	19	24	28	33	38	42	47	52	57	1 01	1 06	1 11	1 16	1 20	1 25
180	5	10	15	20	25	30	35	40	45	50	55	60	1 05	1 10	1 15	1 20	1 25	1 30

Table II is for interpolating the L.M.T. of moonrise, moonset and the Moon's meridian passage for longitude. It is entered with longitude and with the difference between the times for the given date and for the preceding date (in east longitudes) or following date (in west longitudes). The correction is normally *added* for west longitudes and *subtracted* for east longitudes, but if, as occasionally happens, the times become earlier each day instead of later, the signs of the corrections must be reversed.

ALTITUDE CORRECTION TABLES 0°–35°—MOON

App. Alt.	0°–4° Corrⁿ	5°–9° Corrⁿ	10°–14° Corrⁿ	15°–19° Corrⁿ	20°–24° Corrⁿ	25°–29° Corrⁿ	30°–34° Corrⁿ	App. Alt.
′	° ′	° ′	° ′	° ′	° ′	° ′	° ′	′
00	0 33·8	5 58·2	10 62·1	15 62·8	20 62·2	25 60·8	30 58·9	00
10	35·9	58·5	62·2	62·8	62·1	60·8	58·8	10
20	37·8	58·7	62·2	62·8	62·1	60·7	58·8	20
30	39·6	58·9	62·3	62·8	62·1	60·7	58·7	30
40	41·2	59·1	62·3	62·8	62·0	60·6	58·6	40
50	42·6	59·3	62·4	62·7	62·0	60·6	58·5	50
00	1 44·0	6 59·5	11 62·4	16 62·7	21 62·0	26 60·5	31 58·5	00
10	45·2	59·7	62·4	62·7	61·9	60·4	58·4	10
20	46·3	59·9	62·5	62·7	61·9	60·4	58·3	20
30	47·3	60·0	62·5	62·7	61·9	60·3	58·2	30
40	48·3	60·2	62·5	62·7	61·8	60·3	58·2	40
50	49·2	60·3	62·6	62·7	61·8	60·2	58·1	50
00	2 50·0	7 60·5	12 62·6	17 62·7	22 61·7	27 60·1	32 58·0	00
10	50·8	60·6	62·6	62·6	61·7	60·1	57·9	10
20	51·4	60·7	62·6	62·6	61·6	60·0	57·8	20
30	52·1	60·9	62·7	62·6	61·6	59·9	57·8	30
40	52·7	61·0	62·7	62·6	61·5	59·9	57·7	40
50	53·3	61·1	62·7	62·6	61·5	59·8	57·6	50
00	3 53·8	8 61·2	13 62·7	18 62·5	23 61·5	28 59·7	33 57·5	00
10	54·3	61·3	62·7	62·5	61·4	59·7	57·4	10
20	54·8	61·4	62·7	62·5	61·4	59·6	57·4	20
30	55·2	61·5	62·8	62·5	61·3	59·6	57·3	30
40	55·6	61·6	62·8	62·4	61·3	59·5	57·2	40
50	56·0	61·6	62·8	62·4	61·2	59·4	57·1	50
00	4 56·4	9 61·7	14 62·8	19 62·4	24 61·2	29 59·3	34 57·0	00
10	56·7	61·8	62·8	62·3	61·1	59·3	56·9	10
20	57·1	61·9	62·8	62·3	61·1	59·2	56·9	20
30	57·4	61·9	62·8	62·3	61·0	59·1	56·8	30
40	57·7	62·0	62·8	62·2	60·9	59·1	56·7	40
50	57·9	62·1	62·8	62·2	60·9	59·0	56·6	50
H.P.	L U	L U	L U	L U	L U	L U	L U	H.P.
′	′ ′	′ ′	′ ′	′ ′	′ ′	′ ′	′ ′	′
54·0	0·3 0·9	0·3 0·9	0·4 1·0	0·5 1·1	0·6 1·2	0·7 1·3	0·9 1·5	54·0
54·3	0·7 1·1	0·7 1·2	0·7 1·2	0·8 1·3	0·9 1·4	1·1 1·5	1·2 1·7	54·3
54·6	1·1 1·4	1·1 1·4	1·1 1·4	1·2 1·5	1·3 1·6	1·4 1·7	1·5 1·8	54·6
54·9	1·4 1·6	1·5 1·6	1·5 1·6	1·6 1·7	1·6 1·8	1·8 1·9	1·9 2·0	54·9
55·2	1·8 1·8	1·8 1·8	1·9 1·9	1·9 1·9	2·0 2·0	2·1 2·1	2·2 2·2	55·2
55·5	2·2 2·0	2·2 2·0	2·3 2·1	2·3 2·1	2·4 2·2	2·4 2·3	2·5 2·4	55·5
55·8	2·6 2·2	2·6 2·2	2·6 2·3	2·7 2·3	2·7 2·4	2·8 2·4	2·9 2·5	55·8
56·1	3·0 2·4	3·0 2·5	3·0 2·5	3·0 2·5	3·1 2·6	3·1 2·6	3·2 2·7	56·1
56·4	3·4 2·7	3·4 2·7	3·4 2·7	3·4 2·7	3·4 2·8	3·5 2·8	3·5 2·9	56·4
56·7	3·7 2·9	3·7 2·9	3·8 2·9	3·8 2·9	3·8 3·0	3·8 3·0	3·9 3·0	56·7
57·0	4·1 3·1	4·1 3·1	4·1 3·1	4·1 3·1	4·2 3·1	4·2 3·2	4·2 3·2	57·0
57·3	4·5 3·3	4·5 3·3	4·5 3·3	4·5 3·3	4·5 3·3	4·5 3·4	4·6 3·4	57·3
57·6	4·9 3·5	4·9 3·5	4·9 3·5	4·9 3·5	4·9 3·5	4·9 3·5	4·9 3·6	57·6
57·9	5·3 3·8	5·3 3·8	5·2 3·8	5·2 3·7	5·2 3·7	5·2 3·7	5·2 3·7	57·9
58·2	5·6 4·0	5·6 4·0	5·6 4·0	5·6 4·0	5·6 3·9	5·6 3·9	5·6 3·9	58·2
58·5	6·0 4·2	6·0 4·2	6·0 4·2	6·0 4·2	6·0 4·1	5·9 4·1	5·9 4·1	58·5
58·8	6·4 4·4	6·4 4·4	6·4 4·4	6·3 4·4	6·3 4·3	6·3 4·3	6·2 4·2	58·8
59·1	6·8 4·6	6·8 4·6	6·7 4·6	6·7 4·6	6·7 4·5	6·6 4·5	6·6 4·4	59·1
59·4	7·2 4·8	7·1 4·8	7·1 4·8	7·1 4·8	7·0 4·7	7·0 4·7	6·9 4·6	59·4
59·7	7·5 5·1	7·5 5·0	7·5 5·0	7·5 5·0	7·4 4·9	7·3 4·8	7·2 4·7	59·7
60·0	7·9 5·3	7·9 5·3	7·9 5·2	7·8 5·2	7·8 5·1	7·7 5·0	7·6 4·9	60·0
60·3	8·3 5·5	8·3 5·5	8·2 5·4	8·2 5·4	8·1 5·3	8·0 5·2	7·9 5·1	60·3
60·6	8·7 5·7	8·7 5·7	8·6 5·7	8·6 5·6	8·5 5·5	8·4 5·4	8·2 5·3	60·6
60·9	9·1 5·9	9·0 5·9	9·0 5·9	8·9 5·8	8·8 5·7	8·7 5·6	8·6 5·4	60·9
61·2	9·5 6·2	9·4 6·1	9·4 6·1	9·3 6·0	9·2 5·9	9·1 5·8	8·9 5·6	61·2
61·5	9·8 6·4	9·8 6·3	9·7 6·3	9·7 6·2	9·5 6·1	9·4 5·9	9·2 5·8	61·5

DIP

Ht. of Eye	Corrⁿ	Ht. of Eye	Ht. of Eye	Corrⁿ	Ht. of Eye
m		ft.	m		ft.
2·4	′	8·0	9·5	′	31·5
	−2·8			−5·5	
2·6		8·6	9·9		32·7
	−2·9			−5·6	
2·8		9·2	10·3		33·9
	−3·0			−5·7	
3·0		9·8	10·6		35·1
	−3·1			−5·8	
3·2		10·5	11·0		36·3
	−3·2			−5·9	
3·4		11·2	11·4		37·6
	−3·3			−6·0	
3·6		11·9	11·8		38·9
	−3·4			−6·1	
3·8		12·6	12·2		40·1
	−3·5			−6·2	
4·0		13·3	12·6		41·5
	−3·6			6·3	
4·3		14·1	13·0		42·8
	−3·7			−6·4	
4·5		14·9	13·4		44·2
	−3·8			6·5	
4·7		15·7	13·8		45·5
	−3·9			6·6	
5·0		16·5	14·2		46·9
	−4·0			−6·7	
5·2		17·4	14·7		48·4
	−4·1			−6·8	
5·5		18·3	15·1		49·8
	−4·2			−6·9	
5·8		19·1	15·5		51·3
	−4·3			−7·0	
6·1		20·1	16·0		52·8
	−4·4			−7·1	
6·3		21·0	16·5		54·3
	−4·5			−7·2	
6·6		22·0	16·9		55·8
	−4·6			−7·3	
6·9		22·9	17·4		57·4
	−4·7			−7·4	
7·2		23·9	17·9		58·9
	−4·8			−7·5	
7·5		24·9	18·4		60·5
	−4·9			−7·6	
7·9		26·0	18·8		62·1
	−5·0			−7·7	
8·2		27·1	19·3		63·8
	−5·1			−7·8	
8·5		28·1	19·8		65·4
	−5·2			−7·9	
8·8		29·2	20·4		67·1
	−5·3			−8·0	
9·2		30·4	20·9		68·8
	−5·4			−8·1	
9·5		31·5	21·4		70·5

MOON CORRECTION TABLE

The correction is in two parts; the first correction is taken from the upper part of the table with argument apparent altitude, and the second from the lower part, with argument H.P., in the same column as that from which the first correction was taken. Separate corrections are given in the lower part for lower (L) and upper (U) limbs. All corrections are to be **added** to apparent altitude, *but* 30′ *is to be subtracted from the altitude of the upper limb.*

For corrections for pressure and temperature see page A4.

For bubble sextant observations ignore dip, take the mean of upper and lower limb corrections and subtract 15′ from the altitude.

App. Alt. = Apparent altitude = Sextant altitude corrected for index error and dip.

ALTITUDE CORRECTION TABLES 35°-90°—MOON

App. Alt.	35°–39° Corrⁿ	40°–44° Corrⁿ	45°–49° Corrⁿ	50°–54° Corrⁿ	55°–59° Corrⁿ	60°–64° Corrⁿ	65°–69° Corrⁿ	70°–74° Corrⁿ	75°–79° Corrⁿ	80°–84° Corrⁿ	85°–89° Corrⁿ	App. Alt.
′ 00	° 35 56·5′	° 40 53·7	° 45 50·5	° 50 46·9	° 55 43·1	° 60 38·9	° 65 34·6	° 70 30·1	° 75 25·3	° 80 20·5	° 85 15·6	′ 00
10	56·4	53·6	50·4	46·8	42·9	38·8	34·4	29·9	25·2	20·4	15·5	10
20	56·3	53·5	50·2	46·7	42·8	38·7	34·3	29·7	25·0	20·2	15·3	20
30	56·2	53·4	50·1	46·5	42·7	38·5	34·1	29·6	24·9	20·0	15·1	30
40	56·2	53·3	50·0	46·4	42·5	38·4	34·0	29·4	24·7	19·9	15·0	40
50	56·1	53·2	49·9	46·3	42·4	38·2	33·8	29·3	24·5	19·7	14·8	50
00	36 56·0	41 53·1	46 49·8	51 46·2	56 42·3	61 38·1	66 33·7	71 29·1	76 24·4	81 19·6	86 14·6	00
10	55·9	53·0	49·7	46·0	42·1	37·9	33·5	29·0	24·2	19·4	14·5	10
20	55·8	52·8	49·5	45·9	42·0	37·8	33·4	28·8	24·1	19·2	14·3	20
30	55·7	52·7	49·4	45·8	41·8	37·7	33·2	28·7	23·9	19·1	14·1	30
40	55·6	52·6	49·3	45·7	41·7	37·5	33·1	28·5	23·8	18·9	14·0	40
50	55·5	52·5	49·2	45·5	41·6	37·4	32·9	28·3	23·6	18·7	13·8	50
00	37 55·4	42 52·4	47 49·1	52 45·4	57 41·4	62 37·2	67 32·8	72 28·2	77 23·4	82 18·6	87 13·7	00
10	55·3	52·3	49·0	45·3	41·3	37·1	32·6	28·0	23·3	18·4	13·5	10
20	55·2	52·2	48·8	45·2	41·2	36·9	32·5	27·9	23·1	18·2	13·3	20
30	55·1	52·1	48·7	45·0	41·0	36·8	32·3	27·7	22·9	18·1	13·2	30
40	55·0	52·0	48·6	44·9	40·9	36·6	32·2	27·6	22·8	17·9	13·0	40
50	55·0	51·9	48·5	44·8	40·8	36·5	32·0	27·4	22·6	17·8	12·8	50
00	38 54·9	43 51·8	48 48·4	53 44·6	58 40·6	63 36·4	68 31·9	73 27·2	78 22·5	83 17·6	88 12·7	00
10	54·8	51·7	48·2	44·5	40·5	36·2	31·7	27·1	22·3	17·4	12·5	10
20	54·7	51·6	48·1	44·4	40·3	36·1	31·6	26·9	22·1	17·3	12·3	20
30	54·6	51·5	48·0	44·2	40·2	35·9	31·4	26·8	22·0	17·1	12·2	30
40	54·5	51·4	47·9	44·1	40·1	35·8	31·3	26·6	21·8	16·9	12·0	40
50	54·4	51·2	47·8	44·0	39·9	35·6	31·1	26·5	21·7	16·8	11·8	50
00	39 54·3	44 51·1	49 47·6	54 43·9	59 39·8	64 35·5	69 31·0	74 26·3	79 21·5	84 16·6	89 11·7	00
10	54·2	51·0	47·5	43·7	39·6	35·3	30·8	26·1	21·3	16·5	11·5	10
20	54·1	50·9	47·4	43·6	39·5	35·2	30·7	26·0	21·2	16·3	11·4	20
30	54·0	50·8	47·3	43·5	39·4	35·0	30·5	25·8	21·0	16·1	11·2	30
40	53·9	50·7	47·2	43·3	39·2	34·9	30·4	25·7	20·9	16·0	11·0	40
50	53·8	50·6	47·0	43·2	39·1	34·7	30·2	25·5	20·7	15·8	10·9	50
H.P.	L U	L U	L U	L U	L U	L U	L U	L U	L U	L U	L U	H.P.
′ 54·0	′ ′ 1·1 1·7	1·3 1·9	1·5 2·1	1·7 2·4	2·0 2·6	2·3 2·9	2·6 3·2	2·9 3·5	3·2 3·8	3·5 4·1	3·8 4·5	′ 54·0
54·3	1·4 1·8	1·6 2·0	1·8 2·2	2·0 2·5	2·3 2·7	2·5 3·0	2·8 3·2	3·0 3·5	3·3 3·8	3·6 4·1	3·9 4·4	54·3
54·6	1·7 2·0	1·9 2·2	2·1 2·4	2·3 2·6	2·5 2·8	2·7 3·0	3 0 3·3	3·2 3·5	3·5 3·8	3·7 4·1	4·0 4·3	54·6
54·9	2·0 2·2	2·2 2·3	2·3 2·5	2·5 2·7	2·7 2·9	2·9 3·1	3·2 3·3	3·4 3·5	3·6 3·8	3·9 4·0	4·1 4·3	54·9
55·2	2·3 2·3	2·5 2·4	2·6 2·6	2·8 2·8	3·0 2·9	3·2 3·1	3·4 3·3	3·6 3·5	3·8 3·7	4·0 4·0	4·2 4·2	55·2
55·5	2·7 2·5	2·8 2·6	2·9 2·7	3·1 2·9	3·2 3·0	3·4 3·2	3·6 3·4	3·7 3·5	3·9 3·7	4·1 3·9	4·3 4·1	55·5
55·8	3·0 2·6	3·1 2·7	3·2 2·8	3·3 3·0	3·5 3·1	3·6 3·3	3·8 3·4	3·9 3·6	4·1 3·7	4·2 3·9	4·4 4·0	55·8
56·1	3·3 2·8	3·4 2·9	3·5 3·0	3·6 3·1	3·7 3·2	3·8 3·3	4·0 3·4	4·1 3·6	4·2 3·7	4·4 3·8	4·5 4·0	56·1
56·4	3·6 2·9	3·7 3·0	3·8 3·1	3·9 3·2	3·9 3·3	4·0 3·4	4·1 3·5	4·3 3·6	4·4 3·7	4·5 3·8	4·6 3·9	56·4
56·7	3·9 3·1	4·0 3·1	4·1 3·2	4·1 3·3	4·2 3·3	4·3 3·4	4·3 3·5	4·4 3·6	4·5 3·7	4·6 3·8	4·7 3·8	56·7
57·0	4·3 3·2	4·3 3·3	4·3 3·3	4·4 3·4	4·4 3·4	4·5 3·5	4·5 3·5	4·6 3·6	4·7 3·6	4·7 3·7	4·8 3·8	57·0
57·3	4·6 3·4	4·6 3·4	4·6 3·4	4·6 3·5	4·7 3·5	4·7 3·5	4·7 3·6	4·8 3·6	4·8 3·6	4·8 3·7	4·9 3·7	57·3
57·6	4·9 3·6	4·9 3·6	4·9 3·6	4·9 3·6	4·9 3·6	4·9 3·6	4·9 3·6	4·9 3·6	5·0 3·6	5·0 3·6	5·0 3·6	57·6
57·9	5·2 3·7	5·2 3·7	5·2 3·7	5·2 3·7	5·2 3·7	5·1 3·6	5·1 3·6	5·1 3·6	5·1 3·6	5·1 3·6	5·1 3·6	57·9
58·2	5·5 3·9	5·5 3·8	5·5 3·8	5·4 3·8	5·4 3·7	5·4 3·7	5·3 3·7	5·3 3·6	5·2 3·6	5·2 3·5	5·2 3·5	58·2
58·5	5·9 4·0	5·8 4·0	5·8 3·9	5·7 3·9	5·6 3·8	5·6 3·8	5·5 3·7	5·5 3·6	5·4 3·6	5·3 3·5	5·3 3·4	58·5
58·8	6·2 4·2	6·1 4·1	6·0 4·1	6·0 4·0	5·9 3·9	5·8 3·8	5·7 3·7	5·6 3·6	5·5 3·5	5·4 3·5	5·3 3·4	58·8
59·1	6·5 4·3	6·4 4·3	6·3 4·2	6·2 4·1	6·1 4·0	6·0 3·9	5·9 3·8	5·8 3·6	5·7 3·5	5·6 3·4	5·4 3·3	59·1
59·4	6·8 4·5	6·7 4·4	6·6 4·3	6·5 4·2	6·4 4·1	6·2 3·9	6·1 3·8	6·0 3·7	5·8 3·5	5·7 3·4	5·5 3·2	59·4
59·7	7·1 4·6	7·0 4·5	6·9 4·4	6·8 4·3	6·6 4·1	6·5 4·0	6·3 3·8	6·2 3·7	6·0 3·5	5·8 3·3	5·6 3·2	59·7
60·0	7·5 4·8	7·3 4·7	7·2 4·5	7·0 4·4	6·9 4·2	6·7 4·0	6·5 3·9	6·3 3·7	6·1 3·5	5·9 3·3	5·7 3·1	60·0
60·3	7·8 5·0	7·6 4·8	7·5 4·7	7·3 4·5	7·1 4·3	6·9 4·1	6·7 3·9	6·5 3·7	6·3 3·5	6·0 3·2	5·8 3·0	60·3
60·6	8·1 5·1	7·9 5·0	7·7 4·8	7·6 4·6	7·3 4·4	7·1 4·2	6·9 3·9	6·7 3·7	6·4 3·4	6·2 3·2	5·9 2·9	60·6
60·9	8·4 5·3	8·2 5·1	8·0 4·9	7·8 4·7	7·6 4·5	7·3 4·2	7·1 4·0	6·8 3·7	6·6 3·4	6·3 3·2	6·0 2·9	60·9
61·2	8·7 5·4	8·5 5·2	8·3 5·0	8·1 4·8	7·8 4·5	7·6 4·3	7·3 4·0	7·0 3·7	6·7 3·4	6·4 3·1	6·1 2·8	61·2
61·5	9·1 5·6	8·8 5·4	8·6 5·1	8·3 4·9	8·1 4·6	7·8 4·3	7·5 4·0	7·2 3·7	6·9 3·4	6·5 3·1	6·2 2·7	61·5

APPENDIX G

EXTRACTS FROM *AIR ALMANAC*

STARS, MAY—AUG., 1975

No.	Name		Mag.	S.H.A.	Dec.
				° ′	° ′
7*	*Acamar*		3·1	315 40	S. 40 24
5*	*Achernar*		0·6	335 48	S. 57 21
30*	*Acrux*		1·1	173 41	S. 62 58
19	*Adhara*	†	1·6	255 35	S. 28 56
10*	*Aldebaran*	†	1·1	291 22	N. 16 28
32*	*Alioth*		1·7	166 46	N. 56 06
34*	*Alkaid*		1·9	153 21	N. 49 26
55	*Al Na'ir*		2·2	28 19	S. 47 05
15	*Alnilam*	†	1·8	276 16	S. 1 13
25*	*Alphard*	†	2·2	218 24	S. 8 33
41*	*Alphecca*	†	2·3	126 35	N. 26 48
1*	*Alpheratz*	†	2·2	358 13	N. 28 57
51*	*Altair*	†	0·9	62 36	N. 8 48
2	*Ankaa*		2·4	353 44	S. 42 26
42*	*Antares*	†	1·2	113 01	S. 26 23
37*	*Arcturus*	†	0·2	146 22	N. 19 19
43	*Atria*		1·9	108 28	S. 68 59
22	*Avior*		1·7	234 30	S. 59 26
13	*Bellatrix*	†	1·7	279 03	N. 6 20
16*	*Betelgeuse*	†	0·1–1·2	271 32	N. 7 24
17*	*Canopus*		−0·9	264 09	S. 52 41
12*	*Capella*		0·2	281 17	N. 45 58
53*	*Deneb*		1·3	49 51	N. 45 12
28*	*Denebola*	†	2·2	183 03	N. 14 43
4*	*Diphda*	†	2·2	349 24	S. 18 07
27*	*Dubhe*		2·0	194 27	N. 61 53
14	*Elnath*	†	1·8	278 49	N. 28 35
47	*Eltanin*		2·4	90 59	N. 51 30
54*	*Enif*	†	2·5	34 15	N. 9 46
56*	*Fomalhaut*	†	1·3	15 55	S. 29 45
31	*Gacrux*		1·6	172 33	S. 56 59
29*	*Gienah*	†	2·8	176 22	S. 17 25
35	*Hadar*		0·9	149 28	S. 60 16
6*	*Hamal*	†	2·2	328 33	N. 23 21
48	*Kaus Aust.*		2·0	84 21	S. 34 24
40*	*Kochab*		2·2	137 18	N. 74 16
57	*Markab*	†	2·6	14 07	N. 15 04
8*	*Menkar*	†	2·8	314 45	N. 4 00
36	*Menkent*		2·3	148 41	S. 36 15
24*	*Miaplacidus*		1·8	221 46	S. 69 37
9*	*Mirfak*		1·9	309 21	N. 49 46
50*	*Nunki*	†	2·1	76 33	S. 26 20
52*	*Peacock*		2·1	54 03	S. 56 49
21*	*Pollux*	†	1·2	244 03	N. 28 05
20*	*Procyon*	†	0·5	245 30	N. 5 17
46*	*Rasalhague*	†	2·1	96 33	N. 12 35
26*	*Regulus*	†	1·3	208 14	N. 12 05
11*	*Rigel*	†	0·3	281 40	S. 8 14
38*	*Rigil Kent.*		0·1	140 30	S. 60 44
44	*Sabik*	†	2·6	102 45	S. 15 42
3*	*Schedar*		2·5	350 13	N. 56 24
45*	*Shaula*		1·7	97 00	S. 37 05
18*	*Sirius*	†	−1·6	258 59	S. 16 41
33*	*Spica*	†	1·2	159 01	S. 11 02
23*	*Suhail*		2·2	223 14	S. 43 20
49*	*Vega*		0·1	80 58	N. 38 46
39	*Zuben'ubi*	†	2·9	137 37	S. 15 57

INTERPOLATION OF G.H.A.

Increment to be added for intervals of G.M.T. to G.H.A. of: Sun, Aries (♈) and planets; Moon

SUN, etc. m s	° ′	MOON m s	SUN, etc. m s	° ′	MOON m s	SUN, etc. m s	° ′	MOON m s
00 00	0 00	00 00	03 17	0 50	03 25	06 37	1 40	06 52
01	0 01	00 02	21	0 51	03 29	41	1 41	06 56
05	0 02	00 06	25	0 52	03 33	45	1 42	07 00
09	0 03	00 10	29	0 53	03 37	49	1 43	07 04
13	0 04	00 14	33	0 54	03 41	53	1 44	07 08
17	0 05	00 18	37	0 55	03 45	06 57	1 45	07 13
21	0 06	00 22	41	0 56	03 49	07 01	1 46	07 17
25	0 07	00 26	45	0 57	03 54	05	1 47	07 21
29	0 08	00 31	49	0 58	03 58	09	1 48	07 25
33	0 09	00 35	53	0 59	04 02	13	1 49	07 29
37	0 10	00 39	03 57	1 00	04 06	17	1 50	07 33
41	0 11	00 43	04 01	1 01	04 10	21	1 51	07 37
45	0 12	00 47	05	1 02	04 14	25	1 52	07 42
49	0 13	00 51	09	1 03	04 19	29	1 53	07 46
53	0 14	00 55	13	1 04	04 23	33	1 54	07 50
00 57	0 15	01 00	17	1 05	04 27	37	1 55	07 54
01 01	0 16	01 04	21	1 06	04 31	41	1 56	07 58
05	0 17	01 08	25	1 07	04 35	45	1 57	08 02
09	0 18	01 12	29	1 08	04 39	49	1 58	08 06
13	0 19	01 16	33	1 09	04 43	53	1 59	08 11
17	0 20	01 20	37	1 10	04 48	07 57	2 00	08 15
21	0 21	01 24	41	1 11	04 52	08 01	2 01	08 19
25	0 22	01 29	45	1 12	04 56	05	2 02	08 23
29	0 23	01 33	49	1 13	05 00	09	2 03	08 27
33	0 24	01 37	53	1 14	05 04	13	2 04	08 31
37	0 25	01 41	04 57	1 15	05 08	17	2 05	08 35
41	0 26	01 45	05 01	1 16	05 12	21	2 06	08 40
45	0 27	01 49	05	1 17	05 17	25	2 07	08 44
49	0 28	01 53	09	1 18	05 21	29	2 08	08 48
53	0 29	01 58	13	1 19	05 25	33	2 09	08 52
01 57	0 30	02 02	17	1 20	05 29	37	2 10	08 56
02 01	0 31	02 06	21	1 21	05 33	41	2 11	09 00
05	0 32	02 10	25	1 22	05 37	45	2 12	09 04
09	0 33	02 14	29	1 23	05 41	49	2 13	09 09
13	0 34	02 18	33	1 24	05 46	53	2 14	09 13
17	0 35	02 22	37	1 25	05 50	08 57	2 15	09 17
21	0 36	02 27	41	1 26	05 54	09 01	2 16	09 21
25	0 37	02 31	45	1 27	05 58	05	2 17	09 25
29	0 38	02 35	49	1 28	06 02	09	2 18	09 29
33	0 39	02 39	53	1 29	06 06	13	2 19	09 33
37	0 40	02 43	05 57	1 30	06 10	17	2 20	09 38
41	0 41	02 47	06 01	1 31	06 15	21	2 21	09 42
45	0 42	02 51	05	1 32	06 19	25	2 22	09 46
49	0 43	02 56	09	1 33	06 23	29	2 23	09 50
53	0 44	03 00	13	1 34	06 27	33	2 24	09 54
02 57	0 45	03 04	17	1 35	06 31	37	2 25	09 58
03 01	0 46	03 08	21	1 36	06 35	41	2 26	10 00
05	0 47	03 12	25	1 37	06 39	45	2 27	
09	0 48	03 16	29	1 38	06 44	49	2 28	
13	0 49	03 20	33	1 39	06 48	53	2 29	
17	0 50	03 25	37	1 40	06 52	09 57	2 30	
03 21		03 29	06 41		06 56	10 00		

*Stars used in H.O. 249 (A.P. 3270) Vol. 1.

†Stars that may be used with Vols. 2 and 3.

(DAY 152) GREENWICH A. M. 1975 JUNE 1 (SUNDAY)

GMT	☉ SUN GHA	SUN Dec.	ARIES GHA ♈	VENUS−3.8 GHA	VENUS Dec.	JUPITER−1.7 GHA	JUPITER Dec.	SATURN 0.4 GHA	SATURN Dec.	◑ MOON GHA	MOON Dec.
h m	° ′	° ′	° ′	° ′	° ′	° ′	° ′	° ′	° ′	° ′	° ′
00 00	180 36.0	N21 56.2	248 51.5	132 10	N23 46	232 59	N 5 30	140 29	N22 16	279 21	S 6 47
10	183 06.0	56.3	251 21.9	134 40		235 29		142 59		281 47	45
20	185 36.0	56.3	253 52.3	137 09		237 59		145 29		284 13	43
30	188 06.0	· 56.4	256 22.7	139 39 ·	·	240 30 ·	·	148 00 ·	·	286 39	· 42
40	190 35.9	56.5	258 53.1	142 09		243 00		150 30		289 04	40
50	193 05.9	56.5	261 23.5	144 39		245 30		153 00		291 30	38
01 00	195 35.9	N21 56.6	263 53.9	147 09	N23 45	248 01	N 5 30	155 31	N22 16	293 56	S 6 36
10	198 05.9	56.6	266 24.4	149 39		250 31		158 01		296 22	35
20	200 35.9	56.7	268 54.8	152 09		253 01		160 31		298 47	33
30	203 05.9	· 56.8	271 25.2	154 39 ·	·	255 32 ·	·	163 02 ·	·	301 13	· 31
40	205 35.8	56.8	273 55.6	157 09		258 02		165 32		303 39	29
50	208 05.8	56.9	276 26.0	159 39		260 32		168 03		306 05	28
02 00	210 35.8	N21 56.9	278 56.4	162 09	N23 45	263 03	N 5 30	170 33	N22 16	308 30	S 6 26
10	213 05.8	57.0	281 26.8	164 39		265 33		173 03		310 56	24
20	215 35.8	57.1	283 57.2	167 09		268 03		175 34		313 22	23
30	218 05.8	· 57.1	286 27.6	169 39 ·	·	270 34 ·	·	178 04 ·	·	315 47	· 21
40	220 35.8	57.2	288 58.0	172 09		273 04		180 34		318 13	19
50	223 05.7	57.2	291 28.5	174 39		275 34		183 05		320 39	17
03 00	225 35.7	N21 57.3	293 58.9	177 09	N23 44	278 05	N 5 30	185 35	N22 16	323 05	S 6 16
10	228 05.7	57.3	296 29.3	179 38		280 35		188 05		325 30	14
20	230 35.7	57.4	298 59.7	182 08		283 05		190 36		327 56	12
30	233 05.7	· 57.5	301 30.1	184 38 ·	·	285 36 ·	·	193 06 ·	·	330 22	· 10
40	235 35.7	57.5	304 00.5	187 08		288 06		195 37		332 48	09
50	238 05.6	57.6	306 30.9	189 38		290 36		198 07		335 13	07
04 00	240 35.6	N21 57.6	309 01.3	192 08	N23 44	293 07	N 5 30	200 37	N22 15	337 39	S 6 05
10	243 05.6	57.7	311 31.7	194 38		295 37		203 08		340 05	04
20	245 35.6	57.8	314 02.2	197 08		298 07		205 38		342 31	02
30	248 05.6	· 57.8	316 32.6	199 38 ·	·	300 38 ·	·	208 08 ·	·	344 56	6 00
40	250 35.6	57.9	319 03.0	202 08		303 08		210 39		347 22	5 58
50	253 05.6	57.9	321 33.4	204 38		305 38		213 09		349 48	57
05 00	255 35.5	N21 58.0	324 03.8	207 08	N23 43	308 09	N 5 31	215 39	N22 15	352 14	S 5 55
10	258 05.5	58.0	326 34.2	209 38		310 39		218 10		354 39	53
20	260 35.5	58.1	329 04.6	212 08		313 09		220 40		357 05	51
30	263 05.5	· 58.2	331 35.0	214 38 ·	·	315 40 ·	·	223 10 ·	·	359 31	· 50
40	265 35.5	58.2	334 05.4	217 08		318 10		225 41		1 57	48
50	268 05.5	58.3	336 35.9	219 37		320 40		228 11		4 22	46
06 00	270 35.5	N21 58.3	339 06.3	222 07	N23 43	323 11	N 5 31	230 42	N22 15	6 48	S 5 44
10	273 05.4	58.4	341 36.7	224 37		325 41		233 12		9 14	43
20	275 35.4	58.5	344 07.1	227 07		328 11		235 42		11 40	41
30	278 05.4	· 58.5	346 37.5	229 37 ·	·	330 42 ·	·	238 13 ·	·	14 05	· 39
40	280 35.4	58.6	349 07.9	232 07		333 12		240 43		16 31	38
50	283 05.4	58.6	351 38.3	234 37		335 42		243 13		18 57	36
07 00	285 35.4	N21 58.7	354 08.7	237 07	N23 42	338 13	N 5 31	245 44	N22 15	21 23	S 5 34
10	288 05.3	58.7	356 39.1	239 37		340 43		248 14		23 49	32
20	290 35.3	58.8	359 09.5	242 07		343 13		250 44		26 14	31
30	293 05.3	· 58.9	1 40.0	244 37 ·	·	345 44 ·	·	253 15 ·	·	28 40	· 29
40	295 35.3	58.9	4 10.4	247 07		348 14		255 45		31 06	27
50	298 05.3	59.0	6 40.8	249 37		350 44		258 16		33 32	25
08 00	300 35.3	N21 59.0	9 11.2	252 07	N23 42	353 15	N 5 31	260 46	N22 15	35 57	S 5 24
10	303 05.3	59.1	11 41.6	254 37		355 45		263 16		38 23	22
20	305 35.2	59.1	14 12.0	257 07		358 16		265 47		40 49	20
30	308 05.2	· 59.2	16 42.4	259 37 ·	·	0 46 ·	·	268 17 ·	·	43 15	· 18
40	310 35.2	59.3	19 12.8	262 06		3 16		270 47		45 40	17
50	313 05.2	59.3	21 43.2	264 36		5 47		273 18		48 06	15
09 00	315 35.2	N21 59.4	24 13.7	267 06	N23 41	8 17	N 5 31	275 48	N22 15	50 32	S 5 13
10	318 05.2	59.4	26 44.1	269 36		10 47		278 18		52 58	11
20	320 35.1	59.5	29 14.5	272 06		13 18		280 49		55 23	10
30	323 05.1	· 59.6	31 44.9	274 36 ·	·	15 48 ·	·	283 19 ·	·	57 49	· 08
40	325 35.1	59.6	34 15.3	277 06		18 18		285 49		60 15	06
50	328 05.1	59.7	36 45.7	279 36		20 49		288 20		62 41	04
10 00	330 35.1	N21 59.7	39 16.1	282 06	N23 40	23 19	N 5 31	290 50	N22 15	65 07	S 5 03
10	333 05.1	59.8	41 46.5	284 36		25 49		293 21		67 32	5 01
20	335 35.1	59.8	44 16.9	287 06		28 20		295 51		69 58	4 59
30	338 05.0	21 59.9	46 47.4	289 36 ·	·	30 50 ·	·	298 21 ·	·	72 24	· 57
40	340 35.0	22 00.0	49 17.8	292 06		33 20		300 52		74 50	56
50	343 05.0	00.0	51 48.2	294 36		35 51		303 22		77 15	54
11 00	345 35.0	N22 00.1	54 18.6	297 06	N23 40	38 21	N 5 32	305 52	N22 15	79 41	S 4 52
10	348 05.0	00.1	56 49.0	299 36		40 51		308 23		82 07	50
20	350 35.0	00.2	59 19.4	302 06		43 22		310 53		84 33	48
30	353 04.9	· 00.2	61 49.8	304 35 ·	·	45 52 ·	·	313 23 ·	·	86 59	· 47
40	355 34.9	00.3	64 20.2	307 05		48 22		315 54		89 24	45
50	358 04.9	00.4	66 50.6	309 35		50 53		318 24		91 50	43
Rate	14 59.9	N0 00.3		14 59.6	S0 00.5	15 02.0	N0 00.2	15 02.2	0 00.0	14 34.5	N0 10.4

Lat.	Moonrise	Diff.
N		
°	h m	m
72	00 49	−07
70	00 40	−04
68	00 32	−01
66	00 26	+01
64	00 20	03
62	00 15	05
60	00 11	06
58	00 07	07
56	00 04	08
54	00 01	09
52	24 17	10
50	24 16	10
45	24 14	12
40	24 13	13
35	24 11	15
30	24 10	16
20	24 07	18
10	24 05	19
0	24 04	21
10	24 02	22
20	24 00	24
30	23 58	26
35	23 56	27
40	23 55	29
45	23 53	30
50	23 51	32
52	23 50	33
54	23 49	34
56	23 48	35
58	23 47	36
60	23 45	37
S		

Moon's P. in A.

Alt.	Corr.	Alt.	Corr.
°	+ ′	°	+ ′
0		57	
	54		29
10		58	
	53		28
15		59	
	52		27
18		60	
	51		26
21		62	
	50		25
24		63	
	49		24
26		64	
	48		23
29		65	
	47		22
31		66	
	46		21
33		67	
	45		20
35		68	
	44		19
36		70	
	43		18
38		71	
	42		17
40		72	
	41		16
41		73	
	40		15
43		74	
	39		14
44		75	
	38		13
46		76	
	37		12
47		77	
	36		11
49		78	
	35		10
50		79	
	34		
51			
	33		
53			
	32		
54			
	31		
55			
	30		
57			
	29		
58			

Sun SD 15′.8

Moon SD 15′

Age 21d

(DAY 152) GREENWICH P. M. 1975 JUNE 1 (SUNDAY)

GMT	☉ SUN GHA	SUN Dec.	ARIES GHA ♈	VENUS −3.8 GHA	VENUS Dec.	JUPITER −1.7 GHA	JUPITER Dec.	SATURN 0.4 GHA	SATURN Dec.	◑ MOON GHA	MOON Dec.
h m	° ′	° ′	° ′	° ′	° ′	° ′	° ′	° ′	° ′	° ′	° ′
12 00	0 34.9	N22 00.4	69 21.0	312 05	N23 39	53 23	N 5 32	320 55	N22 15	94 16	S 4 41
10	3 04.9	00.5	71 51.5	314 35		55 53		323 25		96 42	40
20	5 34.9	00.5	74 21.9	317 05		58 24		325 55		99 07	38
30	8 04.8	· 00.6	76 52.3	319 35	· ·	60 54	· ·	328 26	· ·	101 33	· 36
40	10 34.8	00.6	79 22.7	322 05		63 24		330 56		103 59	34
50	13 04.8	00.7	81 53.1	324 35		65 55		333 26		106 25	33
13 00	15 34.8	N22 00.8	84 23.5	327 05	N23 39	68 25	N 5 32	335 57	N22 15	108 51	S 4 31
10	18 04.8	00.8	86 53.9	329 35		70 55		338 27		111 16	29
20	20 34.8	00.9	89 24.3	332 05		73 26		340 57		113 42	27
30	23 04.8	· 00.9	91 54.7	334 35	· ·	75 56	· ·	343 28	· ·	116 08	· 26
40	25 34.7	01.0	94 25.2	337 05		78 26		345 58		118 34	24
50	28 04.7	01.1	96 55.6	339 35		80 57		348 28		120 59	22
14 00	30 34.7	N22 01.1	99 26.0	342 05	N23 38	83 27	N 5 32	350 59	N22 15	123 25	S 4 20
10	33 04.7	01.2	101 56.4	344 35		85 57		353 29		125 51	19
20	35 34.7	01.2	104 26.8	347 04		88 28		356 00		128 17	17
30	38 04.7	· 01.3	106 57.2	349 34	· ·	90 58	· ·	358 30	· ·	130 43	· 15
40	40 34.6	01.3	109 27.6	352 04		93 28		1 00		133 08	13
50	43 04.6	01.4	111 58.0	354 34		95 59		3 31		135 34	11
15 00	45 34.6	N22 01.5	114 28.4	357 04	N23 38	98 29	N 5 32	6 01	N22 15	138 00	S 4 10
10	48 04.6	01.5	116 58.8	359 34		100 59		8 31		140 26	08
20	50 34.6	01.6	119 29.3	2 04		103 30		11 02		142 52	06
30	53 04.6	· 01.6	121 59.7	4 34	· ·	106 00	· ·	13 32	· ·	145 17	· 04
40	55 34.6	01.7	124 30.1	7 04		108 30		16 02		147 43	03
50	58 04.5	01.7	127 00.5	9 34		111 01		18 33		150 09	4 01
16 00	60 34.5	N22 01.8	129 30.9	12 04	N23 37	113 31	N 5 33	21 03	N22 15	152 35	S 3 59
10	63 04.5	01.9	132 01.3	14 34		116 01		23 34		155 00	57
20	65 34.5	01.9	134 31.7	17 04		118 32		26 04		157 26	56
30	68 04.5	· 02.0	137 02.1	19 34	· ·	121 02	· ·	28 34	· ·	159 52	· 54
40	70 34.5	02.0	139 32.5	22 04		123 32		31 05		162 18	52
50	73 04.4	02.1	142 03.0	24 34		126 03		33 35		164 44	50
17 00	75 34.4	N22 02.1	144 33.4	27 04	N23 37	128 33	N 5 33	36 05	N22 15	167 09	S 3 49
10	78 04.4	02.2	147 03.8	29 33		131 03		38 36		169 35	47
20	80 34.4	02.3	149 34.2	32 03		133 34		41 06		172 01	45
30	83 04.4	· 02.3	152 04.6	34 33	· ·	136 04	· ·	43 36	· ·	174 27	· 43
40	85 34.4	02.4	154 35.0	37 03		138 34		46 07		176 53	41
50	88 04.3	02.4	157 05.4	39 33		141 05		48 37		179 18	40
18 00	90 34.3	N22 02.5	159 35.8	42 03	N23 36	143 35	N 5 33	51 08	N22 15	181 44	S 3 38
10	93 04.3	02.5	162 06.2	44 33		146 05		53 38		184 10	36
20	95 34.3	02.6	164 36.7	47 03		148 36		56 08		186 36	34
30	98 04.3	· 02.6	167 07.1	49 33	· ·	151 06	· ·	58 39	· ·	189 02	· 33
40	100 34.3	02.7	169 37.5	52 03		153 36		61 09		191 27	31
50	103 04.3	02.8	172 07.9	54 33		156 07		63 39		193 53	29
19 00	105 34.2	N22 02.8	174 38.3	57 03	N23 36	158 37	N 5 33	66 10	N22 15	196 19	S 3 27
10	108 04.2	02.9	177 08.7	59 33		161 07		68 40		198 45	25
20	110 34.2	02.9	179 39.1	62 03		163 38		71 10		201 11	24
30	113 04.2	· 03.0	182 09.5	64 33	· ·	166 08	· ·	73 41	· ·	203 36	· 22
40	115 34.2	03.0	184 39.9	67 03		168 38		76 11		206 02	20
50	118 04.2	03.1	187 10.3	69 33		171 09		78 41		208 28	18
20 00	120 34.1	N22 03.2	189 40.8	72 03	N23 35	173 39	N 5 33	81 12	N22 15	210 54	S 3 17
10	123 04.1	03.2	192 11.2	74 32		176 09		83 42		213 20	15
20	125 34.1	03.3	194 41.6	77 02		178 40		86 13		215 45	13
30	128 04.1	· 03.3	197 12.0	79 32	· ·	181 10	· ·	88 43	· ·	218 11	· 11
40	130 34.1	03.4	199 42.4	82 02		183 41		91 13		220 37	09
50	133 04.1	03.4	202 12.8	84 32		186 11		93 44		223 03	08
21 00	135 34.0	N22 03.5	204 43.2	87 02	N23 35	188 41	N 5 33	96 14	N22 15	225 29	S 3 06
10	138 04.0	03.6	207 13.6	89 32		191 12		98 44		227 54	04
20	140 34.0	03.6	209 44.0	92 02		193 42		101 15		230 20	02
30	143 04.0	· 03.7	212 14.5	94 32	· ·	196 12	· ·	103 45	· ·	232 46	3 01
40	145 34.0	03.7	214 44.9	97 02		198 43		106 15		235 12	2 59
50	148 04.0	03.8	217 15.3	99 32		201 13		108 46		237 38	57
22 00	150 34.0	N22 03.8	219 45.7	102 02	N23 34	203 43	N 5 34	111 16	N22 15	240 03	S 2 55
10	153 03.9	03.9	222 16.1	104 32		206 14		113 46		242 29	53
20	155 33.9	03.9	224 46.5	107 02		208 44		116 17		244 55	52
30	158 03.9	· 04.0	227 16.9	109 32	· ·	211 14	· ·	118 47	· ·	247 21	· 50
40	160 33.9	04.1	229 47.3	112 02		213 45		121 18		249 47	48
50	163 03.9	04.1	232 17.7	114 31		216 15		123 48		252 12	46
23 00	165 33.9	N22 04.2	234 48.2	117 01	N23 34	218 45	N 5 34	126 18	N22 15	254 38	S 2 45
10	168 03.8	04.2	237 18.6	119 31		221 16		128 49		257 04	43
20	170 33.8	04.3	239 49.0	122 01		223 46		131 19		259 30	41
30	173 03.8	· 04.3	242 19.4	124 31	· ·	226 16	· ·	133 49	· ·	261 56	· 39
40	175 33.8	04.4	244 49.8	127 01		228 47		136 20		264 21	37
50	178 03.8	04.5	247 20.2	129 31		231 17		138 50		266 47	36
Rate	14 59.9	N0 00.3		14 59.7	S0 00.5	15 02.0	N0 00.2	15 02.2	0 00.0	14 34.8	N0 10.6

Lat.	Moon-set	Diff.
N		
°	h m	m
72	10 38	51
70	10 45	48
68	10 51	45
66	10 56	42
64	11 00	40
62	11 04	39
60	11 07	37
58	11 10	36
56	11 12	35
54	11 14	34
52	11 16	33
50	11 18	32
45	11 22	30
40	11 25	29
35	11 28	27
30	11 31	26
20	11 35	24
10	11 39	23
0	11 42	21
10	11 46	20
20	11 49	18
30	11 53	16
35	11 56	15
40	11 58	14
45	12 01	12
50	12 05	10
52	12 07	10
54	12 08	09
56	12 10	08
58	12 13	07
60	12 15	06
S		

Moon's P. in A.

Alt. °	Corr. + ′	Alt. °	Corr. + ′
0		57	
	54		29
9		58	
	53		28
14		59	
	52		27
18		60	
	51		26
21		61	
	50		25
24		63	
	49		24
26		64	
	48		23
28		65	
	47		22
31		66	
	46		21
33		67	
	45		20
34		68	
	44		19
36		70	
	43		18
38		71	
	42		17
40		72	
	41		16
41		73	
	40		15
43		74	
	39		14
44		75	
	38		13
46		76	
	37		12
47		77	
	36		11
49		78	
	35		10
50		79	
	34		
51			
	33		
53			
	32		
54			
	31		
55			
	30		
57			
	29		
58			

Sun SD 15′.8

Moon SD 15′

Age 21d

(DAY 153) GREENWICH A. M. 1975 JUNE 2 (MONDAY)

GMT	⊙ SUN GHA	SUN Dec.	ARIES GHA ♈	VENUS −3.8 GHA	VENUS Dec.	JUPITER −1.7 GHA	JUPITER Dec.	SATURN 0.4 GHA	SATURN Dec.	◑ MOON GHA	MOON Dec.
h m	° ′	° ′	° ′	° ′	° ′	° ′	° ′	° ′	° ′	° ′	° ′
00 00	180 33.8	N22 04.5	249 50.6	132 01	N23 33	233 47	N 5 34	141 20	N22 15	269 13	S 2 34
10	183 03.7	04.6	252 21.0	134 31		236 18		143 51		271 39	32
20	185 33.7	04.6	254 51.4	137 01		238 48		146 21		274 05	30
30	188 03.7	· 04.7	257 21.8	139 31	· ·	241 18	· ·	148 52	· ·	276 31	· 29
40	190 33.7	04.7	259 52.3	142 01		243 49		151 22		278 56	27
50	193 03.7	04.8	262 22.7	144 31		246 19		153 52		281 22	25
01 00	195 33.7	N22 04.8	264 53.1	147 01	N23 33	248 49	N 5 34	156 23	N22 15	283 48	S 2 23
10	198 03.7	04.9	267 23.5	149 31		251 20		158 53		286 14	21
20	200 33.6	05.0	269 53.9	152 01		253 50		161 23		288 40	20
30	203 03.6	· 05.0	272 24.3	154 31	· ·	256 20	· ·	163 54	· ·	291 05	· 18
40	205 33.6	05.1	274 54.7	157 01		258 51		166 24		293 31	16
50	208 03.6	05.1	277 25.1	159 30		261 21		168 54		295 57	14
02 00	210 33.6	N22 05.2	279 55.5	162 00	N23 32	263 51	N 5 34	171 25	N22 15	298 23	S 2 12
10	213 03.6	05.2	282 26.0	164 30		266 22		173 55		300 49	11
20	215 33.5	05.3	284 56.4	167 00		268 52		176 26		303 14	09
30	218 03.5	· 05.4	287 26.8	169 30	· ·	271 22	· ·	178 56	· ·	305 40	· 07
40	220 33.5	05.4	289 57.2	172 00		273 53		181 26		308 06	05
50	223 03.5	05.5	292 27.6	174 30		276 23		183 57		310 32	03
03 00	225 33.5	N22 05.5	294 58.0	177 00	N23 32	278 53	N 5 34	186 27	N22 15	312 58	S 2 02
10	228 03.5	05.6	297 28.4	179 30		281 24		188 57		315 23	2 00
20	230 33.4	05.6	299 58.8	182 00		283 54		191 28		317 49	1 58
30	233 03.4	· 05.7	302 29.2	184 30	· ·	286 24	· ·	193 58	· ·	320 15	· 56
40	235 33.4	05.7	304 59.7	187 00		288 55		196 28		322 41	55
50	238 03.4	05.8	307 30.1	189 30		291 25		198 59		325 07	53
04 00	240 33.4	N22 05.9	310 00.5	192 00	N23 31	293 55	N 5 35	201 29	N22 15	327 33	S 1 51
10	243 03.4	05.9	312 30.9	194 30		296 26		203 59		329 58	49
20	245 33.4	06.0	315 01.3	197 00		298 56		206 30		332 24	47
30	248 03.3	· 06.0	317 31.7	199 30	· ·	301 26	· ·	209 00	· ·	334 50	· 46
40	250 33.3	06.1	320 02.1	202 00		303 57		211 31		337 16	44
50	253 03.3	06.1	322 32.5	204 29		306 27		214 01		339 42	42
05 00	255 33.3	N22 06.2	325 02.9	206 59	N23 31	308 57	N 5 35	216 31	N22 15	342 07	S 1 40
10	258 03.3	06.2	327 33.3	209 29		311 28		219 02		344 33	38
20	260 33.3	06.3	330 03.8	211 59		313 58		221 32		346 59	37
30	263 03.2	· 06.4	332 34.2	214 29	· ·	316 28	· ·	224 02	· ·	349 25	· 35
40	265 33.2	06.4	335 04.6	216 59		318 59		226 33		351 51	33
50	268 03.2	06.5	337 35.0	219 29		321 29		229 03		354 16	31
06 00	270 33.2	N22 06.5	340 05.4	221 59	N23 30	323 59	N 5 35	231 33	N22 15	356 42	S 1 30
10	273 03.2	06.6	342 35.8	224 29		326 30		234 04		359 08	28
20	275 33.2	06.6	345 06.2	226 59		329 00		236 34		1 34	26
30	278 03.1	· 06.7	347 36.6	229 29	· ·	331 30	· ·	239 05	· ·	4 00	· 24
40	280 33.1	06.7	350 07.0	231 59		334 01		241 35		6 25	22
50	283 03.1	06.8	352 37.5	234 29		336 31		244 05		8 51	21
07 00	285 33.1	N22 06.9	355 07.9	236 59	N23 29	339 01	N 5 35	246 36	N22 15	11 17	S 1 19
10	288 03.1	06.9	357 38.3	239 29		341 32		249 06		13 43	17
20	290 33.1	07.0	0 08.7	241 59		344 02		251 36		16 09	15
30	293 03.0	· 07.0	2 39.1	244 29	· ·	346 32	· ·	254 07	· ·	18 35	· 13
40	295 33.0	07.1	5 09.5	246 59		349 03		256 37		21 00	12
50	298 03.0	07.1	7 39.9	249 28		351 33		259 07		23 26	10
08 00	300 33.0	N22 07.2	10 10.3	251 58	N23 29	354 04	N 5 35	261 38	N22 15	25 52	S 1 08
10	303 03.0	07.2	12 40.7	254 28		356 34		264 08		28 18	06
20	305 33.0	07.3	15 11.1	256 58		359 04		266 38		30 44	04
30	308 02.9	· 07.4	17 41.6	259 28	· ·	1 35	· ·	269 09	· ·	33 09	· 03
40	310 32.9	07.4	20 12.0	261 58		4 05		271 39		35 35	1 01
50	313 02.9	07.5	22 42.4	264 28		6 35		274 10		38 01	0 59
09 00	315 32.9	N22 07.5	25 12.8	266 58	N23 28	9 06	N 5 35	276 40	N22 15	40 27	S 0 57
10	318 02.9	07.6	27 43.2	269 28		11 36		279 10		42 53	55
20	320 32.9	07.6	30 13.6	271 58		14 06		281 41		45 19	54
30	323 02.9	· 07.7	32 44.0	274 28	· ·	16 37	· ·	284 11	· ·	47 44	· 52
40	325 32.8	07.7	35 14.4	276 58		19 07		286 41		50 10	50
50	328 02.8	07.8	37 44.8	279 28		21 37		289 12		52 36	48
10 00	330 32.8	N22 07.9	40 15.3	281 58	N23 28	24 08	N 5 36	291 42	N22 15	55 02	S 0 47
10	333 02.8	07.9	42 45.7	284 28		26 38		294 12		57 28	45
20	335 32.8	08.0	45 16.1	286 58		29 08		296 43		59 53	43
30	338 02.8	· 08.0	47 46.5	289 28	· ·	31 39	· ·	299 13	· ·	62 19	· 41
40	340 32.7	08.1	50 16.9	291 57		34 09		301 43		64 45	39
50	343 02.7	08.1	52 47.3	294 27		36 39		304 14		67 11	38
11 00	345 32.7	N22 08.2	55 17.7	296 57	N23 27	39 10	N 5 36	306 44	N22 15	69 37	S 0 36
10	348 02.7	08.2	57 48.1	299 27		41 40		309 15		72 03	34
20	350 32.7	08.3	60 18.5	301 57		44 10		311 45		74 28	32
30	353 02.7	· 08.3	62 49.0	304 27	· ·	46 41	· ·	314 15	· ·	76 54	· 30
40	355 32.6	08.4	65 19.4	306 57		49 11		316 46		79 20	29
50	358 02.6	08.5	67 49.8	309 27		51 41		319 16		81 46	27
Rate	14 59.9	N0 00.3		14 59.7	S0 00.5	15 02.0	N0 00.2	15 02.2	0 00.0	14 34.9	N0 10.8

Lat.	Moon-rise	Diff.
N		
°	h m	m
72	00 36	−07
70	00 33	−04
68	00 30	−01
66	00 27	+01
64	00 25	03
62	00 24	04
60	00 22	05
58	00 21	07
56	00 19	08
54	00 18	09
52	00 17	10
50	00 16	10
45	00 14	12
40	00 13	13
35	00 11	15
30	00 10	16
20	00 07	18
10	00 05	19
0	00 04	21
10	00 02	22
20	00 00	24
30	24 50	26
35	24 50	27
40	24 52	29
45	24 53	30
50	24 54	32
52	24 55	33
54	24 56	34
56	24 57	35
58	24 58	36
60	24 59	37
S		

Moon's P. in A.

Alt. °	Corr. ′ +	Alt. °	Corr. ′ +
0	54	57	29
9	53	58	28
14	52	59	27
18	51	60	26
21	50	61	25
24	49	63	24
26	48	64	23
28	47	65	22
31	46	66	21
33	45	67	20
34	44	68	19
36	43	70	18
38	42	71	17
40	41	72	16
41	40	73	15
43	39	74	14
44	38	75	13
46	37	76	12
47	36	77	11
49	35	78	10
50	34	79	
51	33		
53	32		
54	31		
55	30		
57	29		
58			

Sun SD 15′.8

Moon SD 15′

Age 22d

(DAY 153) GREENWICH P. M. 1975 JUNE 2 (MONDAY)

GMT	☉ SUN GHA	SUN Dec.	ARIES GHA ♈	VENUS−3.8 GHA	VENUS Dec.	JUPITER−1.7 GHA	JUPITER Dec.	SATURN 0.4 GHA	SATURN Dec.	◑ MOON GHA	MOON Dec.
h m	° ′	° ′	° ′	° ′	° ′	° ′	° ′	° ′	° ′	° ′	° ′
12 00	0 32.6	N22 08.5	70 20.2	311 57	N23 27	54 12	N 5 36	321 46	N22 14	84 12	S 0 25
10	3 02.6	08.6	72 50.6	314 27		56 42		324 17		86 37	23
20	5 32.6	08.6	75 21.0	316 57		59 12		326 47		89 03	21
30	8 02.6	· 08.7	77 51.4	319 27	· ·	61 43	·	329 17	· ·	91 29	· 20
40	10 32.5	08.7	80 21.8	321 57		64 13		331 48		93 55	18
50	13 02.5	08.8	82 52.2	324 27		66 43		334 18		96 21	16
13 00	15 32.5	N22 08.8	85 22.6	326 57	N23 26	69 14	N 5 36	336 49	N22 14	98 47	S 0 14
10	18 02.5	08.9	87 53.1	329 27		71 44		339 19		101 12	12
20	20 32.5	08.9	90 23.5	331 57		74 14		341 49		103 38	11
30	23 02.5	· 09.0	92 53.9	334 27	· ·	76 45	·	344 20	· ·	106 04	· 09
40	25 32.4	09.1	95 24.3	336 56		79 15		346 50		108 30	07
50	28 02.4	09.1	97 54.7	339 26		81 45		349 20		110 56	05
14 00	30 32.4	N22 09.2	100 25.1	341 56	N23 26	84 16	N 5 36	351 51	N22 14	113 21	S 0 03
10	33 02.4	09.2	102 55.5	344 26		86 46		354 21		115 47	S 0 02
20	35 32.4	09.3	105 25.9	346 56		89 16		356 51		118 13	0 00
30	38 02.4	· 09.3	107 56.3	349 26	· ·	91 47	·	359 22	· ·	120 39	N 0 02
40	40 32.3	09.4	110 26.8	351 56		94 17		1 52		123 05	04
50	43 02.3	09.4	112 57.2	354 26		96 47		4 22		125 30	06
15 00	45 32.3	N22 09.5	115 27.6	356 56	N23 25	99 18	N 5 37	6 53	N22 14	127 56	N 0 07
10	48 02.3	09.5	117 58.0	359 26		101 48		9 23		130 22	09
20	50 32.3	09.6	120 28.4	1 56		104 18		11 54		132 48	11
30	53 02.3	· 09.7	122 58.8	4 26	· ·	106 49	·	14 24	· ·	135 14	· 13
40	55 32.3	09.7	125 29.2	6 56		109 19		16 54		137 39	15
50	58 02.2	09.8	127 59.6	9 26		111 49		19 25		140 05	16
16 00	60 32.2	N22 09.8	130 30.0	11 56	N23 25	114 20	N 5 37	21 55	N22 14	142 31	N 0 18
10	63 02.2	09.9	133 00.5	14 26		116 50		24 25		144 57	20
20	65 32.2	09.9	135 30.9	16 56		119 20		26 56		147 23	22
30	68 02.2	· 10.0	138 01.3	19 26	· ·	121 51	·	29 26	· ·	149 49	· 24
40	70 32.2	10.0	140 31.7	21 56		124 21		31 56		152 14	25
50	73 02.1	10.1	143 02.1	24 25		126 51		34 27		154 40	27
17 00	75 32.1	N22 10.1	145 32.5	26 55	N23 24	129 22	N 5 37	36 57	N22 14	157 06	N 0 29
10	78 02.1	10.2	148 02.9	29 25		131 52		39 27		159 32	31
20	80 32.1	10.3	150 33.3	31 55		134 22		41 58		161 58	33
30	83 02.1	· 10.3	153 03.7	34 25	· ·	136 53	·	44 28	· ·	164 23	· 34
40	85 32.1	10.4	155 34.1	36 55		139 23		46 59		166 49	36
50	88 02.0	10.4	158 04.6	39 25		141 53		49 29		169 15	38
18 00	90 32.0	N22 10.5	160 35.0	41 55	N23 24	144 24	N 5 37	51 59	N22 14	171 41	N 0 40
10	93 02.0	10.5	163 05.4	44 25		146 54		54 30		174 07	42
20	95 32.0	10.6	165 35.8	46 55		149 24		57 00		176 32	43
30	98 02.0	· 10.6	168 06.2	49 25	· ·	151 55	·	59 30	· ·	178 58	· 45
40	100 32.0	10.7	170 36.6	51 55		154 25		62 01		181 24	47
50	103 01.9	10.7	173 07.0	54 25		156 55		64 31		183 50	49
19 00	105 31.9	N22 10.8	175 37.4	56 55	N23 23	159 26	N 5 37	67 02	N22 14	186 16	N 0 51
10	108 01.9	10.9	178 07.8	59 25		161 56		69 32		188 42	52
20	110 31.9	10.9	180 38.3	61 55		164 26		72 02		191 07	54
30	113 01.9	· 11.0	183 08.7	64 25	· ·	166 57	·	74 33	· ·	193 33	· 56
40	115 31.9	11.0	185 39.1	66 55		169 27		77 03		195 59	0 58
50	118 01.8	11.1	188 09.5	69 24		171 57		79 33		198 25	1 00
20 00	120 31.8	N22 11.1	190 39.9	71 54	N23 22	174 28	N 5 37	82 04	N22 14	200 51	N 1 01
10	123 01.8	11.2	193 10.3	74 24		176 58		84 34		203 16	03
20	125 31.8	11.2	195 40.7	76 54		179 28		87 04		205 42	05
30	128 01.8	· 11.3	198 11.1	79 24	· ·	181 59	·	89 35	· ·	208 08	· 07
40	130 31.8	11.3	200 41.5	81 54		184 29		92 05		210 34	09
50	133 01.7	11.4	203 12.0	84 24		186 59		94 35		213 00	10
21 00	135 31.7	N22 11.4	205 42.4	86 54	N23 22	189 30	N 5 38	97 06	N22 14	215 25	N 1 12
10	138 01.7	11.5	208 12.8	89 24		192 00		99 36		217 51	14
20	140 31.7	11.6	210 43.2	91 54		194 31		102 07		220 17	16
30	143 01.7	· 11.6	213 13.6	94 24	· ·	197 01	·	104 37	· ·	222 43	· 18
40	145 31.7	11.7	215 44.0	96 54		199 31		107 07		225 09	19
50	148 01.6	11.7	218 14.4	99 24		202 02		109 38		227 34	21
22 00	150 31.6	N22 11.8	220 44.8	101 54	N23 21	204 32	N 5 38	112 08	N22 14	230 00	N 1 23
10	153 01.6	11.8	223 15.2	104 24		207 02		114 38		232 26	25
20	155 31.6	11.9	225 45.6	106 54		209 33		117 09		234 52	27
30	158 01.6	· 11.9	228 16.1	109 24	· ·	212 03	·	119 39	· ·	237 18	· 28
40	160 31.6	12.0	230 46.5	111 54		214 33		122 09		239 43	30
50	163 01.5	12.0	233 16.9	114 23		217 04		124 40		242 09	32
23 00	165 31.5	N22 12.1	235 47.3	116 53	N23 21	219 34	N 5 38	127 10	N22 14	244 35	N 1 34
10	168 01.5	12.1	238 17.7	119 23		222 04		129 40		247 01	35
20	170 31.5	12.2	240 48.1	121 53		224 35		132 11		249 27	37
30	173 01.5	· 12.3	243 18.5	124 23	· ·	227 05	·	134 41	· ·	251 53	· 39
40	175 31.5	12.3	245 48.9	126 53		229 35		137 12		254 18	41
50	178 01.4	12.4	248 19.3	129 23		232 06		139 42		256 44	43
Rate	14 59.9	N0 00.3		14 59.7	S0 00.6	15 02.0	N0 00.2	15 02.2	0 00.0	14 34.9	N0 10.8

Lat.	Moon-set	Diff.
N		
°	h m	m
72	12 18	50
70	12 19	47
68	12 19	44
66	12 20	42
64	12 20	40
62	12 20	38
60	12 21	37
58	12 21	36
56	12 21	35
54	12 21	34
52	12 21	33
50	12 22	32
45	12 22	30
40	12 22	29
35	12 22	27
30	12 23	26
20	12 23	24
10	12 23	22
0	12 24	21
10	12 24	19
20	12 24	18
30	12 24	16
35	12 24	14
40	12 25	13
45	12 25	12
50	12 25	10
52	12 25	09
54	12 25	09
56	12 25	08
58	12 26	06
60	12 26	05
S		

Moon's P. in A.

Alt.	Corr.	Alt.	Corr.
°	+ ′	°	+ ′
0		57	
	54		29
9		58	
	53		28
14		59	
	52		27
18		60	
	51		26
21		61	
	50		25
24		63	
	49		24
26		64	
	48		23
29		65	
	47		22
31		66	
	46		21
33		67	
	45		20
34		68	
	44		19
36		70	
	43		18
38		71	
	42		17
40		72	
	41		16
41		73	
	40		15
43		74	
	39		14
44		75	
	38		13
46		76	
	37		12
47		77	
	36		11
49		78	
	35		10
50		79	
	34		
51			
	33		
53			
	32		
54			
	31		
55			
	30		
57			
	29		
58			

Sun SD 15′.8

Moon SD 15′

Age 22d

F4 INTERPOLATION OF MOONRISE, MOONSET

FOR LONGITUDE

Add if longitude *west*
Subtract if longitude *east*

Longi-tude	Diff.*					
	05	10	15	20	25	30
°	m	m	m	m	m	m
0	00	00	00	00	00	00
20	01	01	02	02	03	03
40	01	02	03	04	06	07
60	02	03	05	07	08	10
80	02	04	07	09	11	13
100	03	06	08	11	14	17
120	03	07	10	13	17	20
140	04	08	12	16	19	23
160	04	09	13	18	22	27
180	05	10	15	20	25	30

Longi-tude	Diff.*					
	35	40	45	50	55	60
°	m	m	m	m	m	m
0	00	00	00	00	00	00
15	03	03	04	04	05	05
30	06	07	08	08	09	10
45	09	10	11	12	14	15
60	12	13	15	17	18	20
75	15	17	19	21	23	25
90	18	20	22	25	28	30
105	20	23	26	29	32	35
120	23	27	30	33	37	40
135	26	30	34	38	41	45
150	29	33	38	42	46	50
165	32	37	41	46	50	55
180	35	40	45	50	55	60

Longi-tude	Diff.*					
	65	70	75	80	85	90
°	m	m	m	m	m	m
0	00	00	00	00	00	00
10	04	04	04	04	05	05
20	07	08	08	09	09	10
30	11	12	12	13	14	15
40	14	16	17	18	19	20
50	18	19	21	22	24	25
60	22	23	25	27	28	30
70	25	27	29	31	33	35
80	29	31	33	36	38	40
90	32	35	38	40	42	45
100	36	39	42	44	47	50
110	40	43	46	49	52	55
120	43	47	50	53	57	60
130	47	51	54	58	61	65
140	51	54	58	62	66	70
150	54	58	62	67	71	75
160	58	62	67	71	76	80
170	61	66	71	76	80	85
180	65	70	75	80	85	90

*When negative *subtract* correction if longitude *west*, and *add* if *east*.

STAR INDEX, MAY—AUG., 1975

No.	Name		Mag.	S.H.A.	Dec.
				° ′	° ′
1*	*Alpheratz*	†	2·2	358 13	N. 28 57
2	*Ankaa*		2·4	353 44	S. 42 26
3*	*Schedar*		2·5	350 13	N. 56 24
4*	*Diphda*	†	2·2	349 24	S. 18 07
5*	*Achernar*		0·6	335 48	S. 57 21
6*	*Hamal*	†	2·2	328 33	N. 23 21
7*	*Acamar*		3·1	315 40	S. 40 24
8*	*Menkar*	†	2·8	314 45	N. 4 00
9*	*Mirfak*		1·9	309 21	N. 49 46
10*	*Aldebaran*	†	1·1	291 22	N. 16 28
11*	*Rigel*	†	0·3	281 40	S. 8 14
12*	*Capella*		0·2	281 17	N. 45 58
13	*Bellatrix*	†	1·7	279 03	N. 6 20
14	*Elnath*	†	1·8	278 49	N. 28 35
15	*Alnilam*	†	1·8	276 16	S. 1 13
16*	*Betelgeuse*	†	0·1–1·2	271 32	N. 7 24
17*	*Canopus*		−0·9	264 09	S. 52 41
18*	*Sirius*	†	−1·6	258 59	S. 16 41
19	*Adhara*	†	1·6	255 35	S. 28 56
20*	*Procyon*	†	0·5	245 30	N. 5 17
21*	*Pollux*	†	1·2	244 03	N. 28 05
22	*Avior*		1·7	234 30	S. 59 26
23*	*Suhail*		2·2	223 14	S. 43 20
24*	*Miaplacidus*		1·8	221 46	S. 69 37
25*	*Alphard*	†	2·2	218 24	S. 8 33
26*	*Regulus*	†	1·3	208 14	N. 12 05
27*	*Dubhe*		2·0	194 27	N. 61 53
28*	*Denebola*	†	2·2	183 03	N. 14 43
29*	*Gienah*	†	2·8	176 22	S. 17 25
30*	*Acrux*		1·1	173 41	S. 62 58
31	*Gacrux*		1·6	172 33	S. 56 59
32*	*Alioth*		1·7	166 46	N. 56 06
33*	*Spica*	†	1·2	159 01	S. 11 02
34*	*Alkaid*		1·9	153 21	N. 49 26
35	*Hadar*		0·9	149 28	S. 60 16
36	*Menkent*		2·3	148 41	S. 36 15
37*	*Arcturus*	†	0·2	146 22	N. 19 19
38*	*Rigil Kentaurus*		0·1	140 30	S. 60 44
39	*Zubenelgenubi*	†	2·9	137 37	S. 15 57
40*	*Kochab*		2·2	137 18	N. 74 16
41*	*Alphecca*	†	2·3	126 35	N. 26 48
42*	*Antares*	†	1·2	113 01	S. 26 23
43	*Atria*		1·9	108 28	S. 68 59
44	*Sabik*	†	2·6	102 45	S. 15 42
45*	*Shaula*		1·7	97 00	S. 37 05
46*	*Rasalhague*	†	2·1	96 33	N. 12 35
47	*Eltanin*		2·4	90 59	N. 51 30
48	*Kaus Australis*		2·0	84 21	S. 34 24
49*	*Vega*		0·1	80 58	N. 38 46
50*	*Nunki*	†	2·1	76 33	S. 26 20
51*	*Altair*	†	0·9	62 36	N. 8 48
52*	*Peacock*		2·1	54 03	S. 56 49
53*	*Deneb*		1·3	49 51	N. 45 12
54*	*Enif*	†	2·5	34 15	N. 9 46
55	*Al Na'ir*		2·2	28 19	S. 47 05
56*	*Fomalhaut*	†	1·3	15 55	S. 29 45
57	*Markab*	†	2·6	14 07	N. 15 04

*Stars used in H.O. 249 (A.P. 3270) Vol. 1.
†Stars that may be used with Vols. 2 and 3.

SUNRISE

Lat.	Dec.	January										February				Lat.
	30	2	5	8	11	14	17	20	23	26	29	1	4	7	10	
°	h m	h m	h m	h m	h m	h m	h m	h m	h m	h m	h m	h m	h m	h m	h m	°
N 72	—	—	—	—	—	—	—	—	—	12 04	11 03	10 33	10 09	09 48	09 28	N 72
70	—	—	—	—	—	—	12 05	11 11	10 46	10 25	10 06	09 49	09 33	09 17	09 02	70
68	—	—	11 34	11 13	10 56	10 41	10 27	10 13	09 59	09 46	09 33	09 20	09 08	08 55	08 42	68
66	10 31	10 27	10 21	10 14	10 06	09 58	09 49	09 39	29	09 19	09 09	08 59	08 48	37	26	66
64	09 51	09 49	09 45	09 41	09 35	29	22	09 15	09 07	08 59	08 50	42	32	23	14	64
62	24	22	20	09 16	09 12	09 07	09 02	08 56	08 49	43	35	27	19	11	08 03	62
N 60	09 03	09 02	09 00	08 58	08 54	08 50	08 46	08 40	08 35	08 29	08 22	08 15	08 08	08 01	07 53	N 60
58	08 46	08 45	08 44	42	39	36	32	27	22	17	11	08 05	07 59	07 52	45	58
56	32	31	30	28	26	23	20	16	12	08 07	08 02	07 56	50	44	38	56
54	19	19	18	17	15	12	09	08 06	08 02	07 58	07 53	48	43	37	32	54
52	08 08	08 08	08 07	08 06	08 05	08 02	08 00	07 57	07 54	50	46	41	36	31	26	52
N 50	07 58	07 58	07 58	07 57	07 56	07 54	07 52	07 49	07 46	07 42	07 39	07 35	07 30	07 25	07 21	N 50
45	38	38	38	38	37	36	34	32	30	27	24	21	17	13	09	45
40	22	22	22	22	22	21	20	18	16	14	12	07 09	07 06	07 03	07 00	40
35	07 07	07 08	07 08	07 08	07 08	07 08	07 07	07 06	07 05	07 03	07 01	06 59	06 57	06 55	06 52	35
30	06 55	06 56	06 57	06 57	06 57	06 57	06 56	06 56	06 55	06 54	06 52	51	49	47	45	30
N 20	06 34	06 35	06 36	06 37	06 37	06 38	06 38	06 38	06 38	06 37	06 37	06 36	06 35	06 33	06 32	N 20
N 10	06 16	17	18	19	20	21	22	22	22	23	23	23	22	22	21	N 10
0	05 59	06 00	06 02	06 03	06 04	06 05	06 06	06 07	06 08	06 09	06 10	06 10	06 10	06 11	11	0
S 10	41	05 43	05 44	05 46	05 48	05 49	05 51	05 52	05 54	05 55	05 56	05 57	05 58	05 59	06 00	S 10
20	23	24	26	28	30	32	34	36	38	40	42	44	45	47	05 49	20
S 30	05 01	05 03	05 05	05 07	05 10	05 12	05 15	05 17	05 20	05 23	05 25	05 28	05 31	05 33	05 36	S 30
35	04 48	04 50	04 52	04 55	04 57	05 00	05 03	05 06	05 09	13	16	19	22	25	28	35
40	33	35	38	41	44	04 47	04 50	04 54	04 57	05 01	05 04	05 08	12	16	19	40
45	04 15	04 18	04 21	24	27	31	35	39	43	04 47	04 51	04 56	05 00	05 05	05 09	45
50	03 53	03 56	03 59	04 03	04 07	11	15	20	25	30	35	40	04 46	04 51	04 56	50
S 52	03 42	03 45	03 49	03 52	03 57	04 01	04 06	04 11	04 17	04 22	04 28	04 33	04 39	04 45	04 50	S 52
54	30	33	37	41	46	03 51	03 56	04 01	04 07	13	19	25	31	38	44	54
56	03 16	19	23	28	33	38	44	03 50	03 57	04 03	04 10	16	23	30	37	56
58	02 59	03 03	03 08	03 13	03 18	24	31	37	44	03 52	03 59	04 06	14	21	29	58
S 60	02 39	02 44	02 49	02 54	03 00	03 07	03 15	03 22	03 30	03 38	03 46	03 54	04 03	04 11	04 19	S 60

SUNSET

Lat.	Dec.	January										February				Lat.
	30	2	5	8	11	14	17	20	23	26	29	1	4	7	10	
°	h m	h m	h m	h m	h m	h m	h m	h m	h m	h m	h m	h m	h m	h m	h m	°
N 72	—	—	—	—	—	—	—	—	—	12 25	13 24	13 55	14 20	14 42	15 02	N 72
70	—	—	—	—	—	—	12 21	13 11	13 39	14 01	14 21	14 39	14 56	15 12	28	70
68	—	—	12 36	13 00	13 20	13 37	13 54	14 09	14 25	14 39	14 54	15 08	15 21	35	15 48	68
66	13 33	13 41	13 49	13 59	14 09	14 20	14 32	14 43	14 55	15 06	15 18	29	41	15 52	16 03	66
64	14 13	14 19	14 25	14 32	14 40	14 49	14 58	15 07	15 17	27	36	15 46	15 56	16 06	16	64
62	14 40	14 45	14 50	14 57	15 03	15 10	15 18	26	35	43	15 52	16 00	16 09	18	27	62
N 60	15 02	15 06	15 10	15 16	15 22	15 28	15 35	15 42	15 49	15 57	16 05	16 12	16 20	16 28	16 36	N 60
58	19	22	27	32	37	43	15 49	15 55	16 02	16 09	15	23	30	37	44	58
56	33	37	41	45	15 50	15 55	16 01	16 06	12	19	25	32	38	45	51	56
54	46	15 49	15 53	15 57	16 01	16 06	11	16	22	28	34	40	46	52	16 58	54
52	15 57	16 00	16 03	16 07	11	16	20	25	30	36	41	47	52	16 58	17 03	52
N 50	16 06	16 09	16 13	16 16	16 20	16 24	16 29	16 33	16 38	16 43	16 48	16 53	16 58	17 03	17 09	N 50
45	27	29	32	35	39	42	16 46	16 50	16 54	16 58	17 03	17 07	17 11	16	20	45
40	44	16 46	16 48	16 51	16 54	16 57	17 01	17 04	17 08	17 11	15	18	22	25	29	40
35	16 57	17 00	17 02	17 04	17 07	17 10	13	16	19	22	25	28	31	34	37	35
30	17 10	12	14	16	19	21	24	26	29	32	34	37	39	42	44	30
N 20	17 31	17 32	17 34	17 36	17 38	17 40	17 42	17 44	17 46	17 48	17 50	17 52	17 53	17 55	17 57	N 20
N 10	17 49	17 50	17 52	17 54	17 55	17 57	17 58	18 00	18 01	18 02	18 04	18 05	18 06	18 07	18 07	N 10
0	18 06	18 07	18 09	18 10	18 11	18 12	18 14	14	15	16	17	17	17	18	18	0
S 10	24	25	26	27	28	29	29	29	30	30	30	30	29	29	28	S 10
20	18 42	18 43	18 44	18 45	18 45	18 46	18 46	18 46	18 45	18 45	18 44	43	42	41	39	20
S 30	19 04	19 05	19 05	19 06	19 06	19 05	19 05	19 04	19 03	19 02	19 00	18 59	18 57	18 55	18 52	S 30
35	17	17	18	18	18	17	16	15	14	12	10	19 08	19 05	19 03	19 00	35
40	32	32	32	32	31	30	29	28	26	24	21	18	15	12	09	40
45	19 49	19 50	19 49	19 49	19 48	19 46	19 45	19 42	40	37	34	31	27	23	19	45
50	20 12	20 12	20 11	20 10	20 08	20 06	20 04	20 01	19 58	19 54	50	46	41	36	31	50
S 52	20 22	20 22	20 21	20 20	20 18	20 16	20 13	20 09	20 06	20 02	19 57	19 53	19 48	19 42	19 37	S 52
54	34	34	33	31	29	26	23	19	15	11	20 06	20 01	19 55	49	43	54
56	20 48	20.48	20 46	44	41	38	35	30	26	21	15	09	20 03	19 57	50	56
58	21 05	21 04	21 02	20 59	20 56	20 52	20 48	43	38	32	26	19	13	20 06	19 58	58
S 60	21 25	21 23	21 21	21 18	21 14	21 09	21 04	20 58	20 52	20 45	20 38	20 31	20 23	20 15	20 07	S 60

MORNING CIVIL TWILIGHT

Lat.	Dec.	January										February				Lat.
	30	2	5	8	11	14	17	20	23	26	29	1	4	7	10	
°	h m	h m	h m	h m	h m	h m	h m	h m	h m	h m	h m	h m	h m	h m	h m	°
N 72	10 49	10 41	10 32	10 21	10 10	09 58	09 46	09 33	09 21	09 08	08 55	08 41	08 28	08 14	08 01	N 72
70	09 52	09 49	09 44	09 38	09 30	09 22	09 14	09 04	08 54	08 44	33	22	08 11	08 00	07 48	70
68	09 19	09 16	09 13	09 08	09 03	08 57	08 50	08 42	34	26	17	08 07	07 58	07 48	37	68
66	08 54	08 53	08 50	08 47	08 42	37	32	25	18	08 11	08 03	07 55	46	38	28	66
64	35	34	32	29	26	21	17	08 11	08 05	07 59	07 52	45	37	29	21	64
62	19	18	17	15	12	08 08	08 04	07 59	07 54	48	42	36	29	22	14	62
N 60	08 06	08 05	08 04	08 02	08 00	07 57	07 53	07 49	07 44	07 39	07 34	07 28	07 22	07 15	07 08	N 60
58	07 54	07 54	07 53	07 51	07 49	47	44	40	36	31	26	21	15	09	07 03	58
56	44	44	44	42	41	38	35	32	29	24	20	15	10	04	06 58	56
54	36	35	35	34	32	30	28	25	22	18	14	09	05	07 00	54	54
52	27	28	27	26	25	23	21	18	15	12	08	04	07 00	06 55	50	52
N 50	07 20	07 20	07 20	07 19	07 18	07 17	07 15	07 13	07 10	07 07	07 03	07 00	06 56	06 51	06 47	N 50
45	07 04	07 05	07 05	07 04	07 04	07 03	07 01	07 00	06 58	06 55	06 52	06 49	46	43	39	45
40	06 51	06 52	06 52	06 52	06 52	06 51	06 50	06 49	47	45	43	41	38	35	32	40
35	39	40	40	41	41	40	40	39	38	36	35	33	31	28	26	35
30	29	30	30	31	31	31	31	30	29	28	27	26	24	22	20	30
N 20	06 10	06 11	06 12	06 13	06 14	06 14	06 14	06 14	06 14	06 14	06 13	06 13	06 12	06 11	06 10	N 20
N 10	05 53	05 54	05 55	05 56	05 57	05 58	05 59	06 00	06 00	06 01	06 01	06 01	06 00	06 00	06 00	N 10
0	36	38	39	40	42	43	44	05 45	05 46	05 47	05 48	05 48	05 49	05 49	05 49	0
S 10	05 18	20	22	23	25	27	28	30	31	33	34	35	36	38	38	S 10
20	04 58	05 00	05 02	05 04	05 06	05 08	05 10	05 12	05 14	05 16	05 18	20	22	24	26	20
S 30	04 33	04 35	04 38	04 40	04 43	04 45	04 48	04 51	04 54	04 56	04 59	05 02	05 05	05 08	05 10	S 30
35	18	20	23	26	28	31	35	38	41	44	48	04 51	04 54	04 58	05 01	35
40	04 00	04 03	04 05	04 08	04 12	04 15	04 19	22	26	30	34	38	42	46	04 50	40
45	03 38	03 41	03 44	03 47	03 51	03 55	03 59	04 04	04 08	04 13	04 18	22	27	32	37	45
50	03 09	03 12	15	19	24	29	34	03 39	03 45	03 51	03 56	04 02	04 08	14	20	50
S 52	02 54	02 57	03 01	03 05	03 10	03 16	03 21	03 27	03 34	03 40	03 46	03 53	03 59	04 06	04 12	S 52
54	36	40	02 44	02 49	02 55	03 01	03 07	03 14	20	27	34	42	49	03 56	04 03	54
56	02 14	02 19	02 24	29	36	02 43	02 50	02 57	03 05	03 13	21	29	37	45	03 53	56
58	01 45	01 50	01 57	02 04	02 12	02 20	28	37	02 46	02 56	03 05	03 14	23	32	41	58
S 60	00 58	01 07	01 16	01 27	01 38	01 49	02 00	02 12	02 23	02 34	02 45	02 56	03 06	03 17	03 27	S 60

EVENING CIVIL TWILIGHT

Lat.	Dec.	January										February				Lat.
	30	2	5	8	11	14	17	20	23	26	29	1	4	7	10	
°	h m	h m	h m	h m	h m	h m	h m	h m	h m	h m	h m	h m	h m	h m	h m	°
N 72	13 16	13 26	13 39	13 52	14 06	14 20	14 35	14 49	15 04	15 18	15 33	15 47	16 01	16 15	16 29	N 72
70	14 12	14 19	14 27	14 36	14 45	14 56	15 07	15 18	30	15 42	15 54	16 06	18	30	42	70
68	14 46	14 51	14 57	15 05	15 13	15 21	30	40	15 50	16 00	16 10	21	31	42	16 53	68
66	15 10	15 15	15 20	27	33	41	15 49	15 57	16 06	15	24	33	42	16 52	17 01	66
64	30	34	38	44	15 50	15 57	16 04	16 11	19	27	35	43	16 52	17 00	09	64
62	45	15 49	15 54	15 58	16 04	16 10	16	23	30	37	45	16 52	17 00	08	16	62
N 60	15 59	16 02	16 06	16 11	16 16	16 21	16 27	16 33	16 40	16 46	16 53	17 00	17 07	17 14	17 21	N 60
58	16 10	13	17	21	26	31	36	42	48	16 54	17 00	07	13	20	26	58
56	20	23	27	31	35	40	45	50	16 56	17 01	07	13	19	25	31	56
54	29	32	36	39	43	48	52	16 57	17 02	08	13	18	24	30	35	54
52	38	40	43	47	51	16 55	16 59	17 04	09	13	18	23	29	34	39	52
N 50	16 45	16 47	16 50	16 54	16 57	17 01	17 05	17 10	17 14	17 19	17 23	17 28	17 33	17 38	17 42	N 50
45	17 01	17 03	17 06	17 09	17 12	15	19	23	26	30	34	38	42	46	50	45
40	14	16	19	21	24	27	30	33	37	40	43	47	50	17 54	17 57	40
35	26	28	30	32	35	38	40	43	46	49	52	17 55	17 58	18 00	18 03	35
30	36	38	40	17 42	17 45	17 47	17 49	17 52	17 54	17 57	17 59	18 02	18 04	07	09	30
N 20	17 55	17 56	17 58	18 00	18 02	18 04	18 06	18 08	18 09	18 11	18 13	18 15	18 16	18 18	18 19	N 20
N 10	18 12	18 13	18 15	16	18	19	21	22	23	25	26	27	28	28	29	N 10
0	29	30	32	33	34	35	36	37	37	38	38	39	39	39	39	0
S 10	18 47	18 48	18 49	18 50	18 50	18 51	18 52	18 52	18 52	18 52	18 52	18 52	18 51	18 51	18 50	S 10
20	19 07	19 08	19 08	19 09	19 09	19 10	19 10	19 09	19 09	19 08	19 07	19 06	19 05	19 04	19 02	20
S 30	19 31	19 32	19 32	19 33	19 33	19 32	19 31	19 31	19 29	19 28	19 26	19 24	19 22	19 20	19 18	S 30
35	19 46	19 47	19 47	19 47	19 47	19 46	19 45	43	42	40	38	35	33	30	27	35
40	20 05	20 05	20 05	20 04	20 03	20 02	20 01	19 59	19 57	19 54	19 51	19 48	19 45	42	38	40
45	27	27	26	25	24	22	20	20 17	20 14	20 11	20 08	20 04	20 00	19 55	19 51	45
50	20 56	20 56	20 54	20 53	20 51	20 48	45	41	37	33	28	24	18	20 13	20 07	50
S 52	21 11	21 10	21 09	21 07	21 04	21 01	20 57	20 53	20 49	20 44	20 39	20 33	20 27	20 21	20 15	S 52
54	28	27	25	23	19	16	21 11	21 07	21 02	20 56	20 50	44	38	31	24	54
56	21 50	21 48	21 46	21 42	21 38	34	28	23	17	21 10	21 03	20 56	20 49	42	34	56
58	22 19	22 16	22 12	22 07	22 02	21 56	21 49	21 42	35	27	19	21 11	21 03	20 54	46	58
S 60	23 04	22 58	22 51	22 44	22 35	22 26	22 17	22 07	21 58	21 48	21 38	21 29	21 19	21 09	20 59	S 60

SUNRISE

Lat.	Apr.	May										June				Lat.
	29	2	5	8	11	14	17	20	23	26	29	1	4	7	10	
°	h m	h m	h m	h m	h m	h m	h m	h m	h m	h m	h m	h m	h m	h m	h m	°
N 72	02 14	01 50	01 22	00 41	▭	▭	▭	▭	▭	▭	▭	▭	▭	▭	▭	N 72
70	02 47	02 29	02 11	01 51	01 29	01 02	00 12	▭	▭	▭	▭	▭	▭	▭	▭	70
68	03 10	02 56	02 42	02 28	02 12	01 57	01 40	01 21	01 00	00 29	▭	▭	▭	▭	▭	68
66	28	03 17	03 05	02 53	02 41	02 29	02 17	02 05	01 53	01 41	01 28	01 15	01 02	00 48	00 33	66
64	43	33	23	03 12	03 02	02 53	02 43	34	02 25	02 16	02 07	02 00	01 52	01 46	01 40	64
62	03 55	46	37	28	20	03 11	03 03	02 55	02 48	02 41	34	28	02 23	02 19	02 15	62
N 60	04 05	03 57	03 49	03 41	03 34	03 26	03 19	03 13	03 06	03 01	02 55	02 50	02 46	02 43	02 40	N 60
58	15	04 07	04 00	03 53	46	39	33	27	22	17	03 12	03 08	03 04	03 01	02 59	58
56	22	16	09	04 02	03 56	03 50	45	40	35	30	26	23	20	17	03 15	56
54	29	23	17	11	04 06	04 00	03 55	03 50	46	42	39	35	33	31	29	54
52	36	30	24	19	14	09	04 04	04 00	03 56	03 52	49	46	44	42	41	52
N 50	04 41	04 36	04 31	04 26	04 21	04 17	04 12	04 09	04 05	04 02	03 59	03 56	03 54	03 53	03 51	N 50
45	04 54	04 49	45	41	37	33	30	27	24	21	04 19	04 17	04 15	04 14	04 13	45
40	05 04	05 00	04 56	04 53	04 49	46	44	41	39	37	35	33	32	31	31	40
35	12	09	05 06	05 03	05 00	04 58	04 56	04 53	04 52	04 50	04 49	47	47	46	46	35
30	20	17	14	12	10	05 08	05 06	05 04	05 03	05 01	05 00	04 59	04 59	04 58	04 58	30
N 20	05 32	05 31	05 29	05 27	05 26	05 25	05 23	05 22	05 22	05 21	05 20	05 20	05 20	05 20	05 20	N 20
N 10	44	43	42	41	40	39	39	38	38	38	38	38	38	38	38	N 10
0	05 54	05 54	05 53	05 53	05 53	05 53	05 53	05 53	05 53	05 53	05 54	05 54	05 55	05 55	05 56	0
S 10	06 04	06 04	06 05	06 05	06 06	06 06	06 07	06 07	06 08	06 09	06 10	06 10	06 11	06 12	06 13	S 10
20	15	16	17	18	19	20	21	23	24	25	26	27	29	30	31	20
S 30	06 27	06 29	06 31	06 33	06 35	06 37	06 38	06 40	06 42	06 44	06 46	06 47	06 49	06 50	06 51	S 30
35	34	37	39	41	44	46	48	06 50	06 53	06 55	06 57	06 59	07 00	07 02	07 03	35
40	42	45	48	06 51	06 54	06 57	06 59	07 02	07 05	07 07	07 09	07 12	14	16	17	40
45	06 51	06 55	06 59	07 02	07 06	07 09	07 12	16	19	22	25	27	30	32	34	45
50	07 02	07 07	07 11	16	20	24	28	32	36	40	43	46	49	07 52	07 54	50
S 52	07 08	07 12	07 17	07 22	07 27	07 31	07 36	07 40	07 44	07 48	07 52	07 55	07 58	08 01	08 04	S 52
54	13	19	24	29	34	39	44	49	07 53	07 58	08 02	08 05	08 09	12	14	54
56	20	25	31	37	42	48	07 53	07 58	08 03	08 08	12	17	20	24	27	56
58	26	33	39	46	07 52	07 58	08 04	08 10	15	20	25	30	34	38	41	58
S 60	07 34	07 41	07 49	07 56	08 03	08 09	08 16	08 22	08 29	08 34	08 40	08 45	08 50	08 54	08 58	S 60

SUNSET

Lat.	Apr.	May										June				Lat.
	29	2	5	8	11	14	17	20	23	26	29	1	4	7	10	
°	h m	h m	h m	h m	h m	h m	h m	h m	h m	h m	h m	h m	h m	h m	h m	°
N 72	21 47	22 11	22 41	23 35	▭	▭	▭	▭	▭	▭	▭	▭	▭	▭	▭	N 72
70	21 12	21 29	21 47	22 07	22 31	23 01	▭	▭	▭	▭	▭	▭	▭	▭	▭	70
68	20 48	21 01	21 15	21 30	21 45	22 01	22 18	22 37	23 01	23 43	▭	▭	▭	▭	▭	68
66	29	20 40	20 52	21 03	21 15	21 27	21 39	21 51	22 04	22 17	22 30	22 44	22 58	23 14	23 32	66
64	14	24	33	20 43	20 53	21 02	21 12	21 22	21 31	21 40	21 49	21 58	22 06	22 13	22 20	64
62	20 02	20 10	18	27	35	20 43	20 51	20 59	21 07	21 15	22	28	21 34	21 40	21 45	62
N 60	19 51	19 59	20 06	20 13	20 21	20 28	20 35	20 42	20 48	20 55	21 01	21 06	21 11	21 16	21 20	N 60
58	42	48	19 55	20 02	20 08	15	21	27	33	38	20 43	20 48	20 53	20 57	21 00	58
56	34	40	46	19 52	19 58	20 03	20 09	14	20	24	29	33	37	41	20 44	56
54	27	32	38	43	48	19 53	19 58	20 03	20 08	13	17	21	24	27	30	54
52	20	25	30	35	40	45	49	19 54	19 58	20 02	20 06	09	13	15	18	52
N 50	19 15	19 19	19 24	19 28	19 33	19 37	19 41	19 45	19 49	19 53	19 56	20 00	20 02	20 05	20 07	N 50
45	19 02	19 06	19 09	13	17	20	24	27	30	33	36	19 39	19 41	19 43	19 45	45
40	18 52	18 55	18 58	19 01	19 04	19 07	19 09	12	15	18	20	22	24	26	28	40
35	43	46	48	18 50	18 53	18 55	18 58	19 00	19 02	19 04	19 06	19 08	19 10	19 12	13	35
30	36	37	39	41	43	45	47	18 49	18 51	18 53	18 54	18 56	18 58	18 59	19 00	30
N 20	18 23	18 24	18 25	18 26	18 27	18 28	18 29	18 31	18 32	18 33	18 34	18 35	18 37	18 38	18 39	N 20
N 10	11	12	12	13	13	14	14	15	16	16	17	18	19	19	20	N 10
0	18 01	18 00	18 00	18 00	18 00	18 00	18 00	18 00	18 00	18 01	18 01	18 01	18 02	18 02	18 03	0
S 10	17 51	17 50	17 49	17 48	17 47	17 46	17 46	17 45	17 45	17 45	17 45	17 45	17 45	17 46	17 46	S 10
20	40	38	36	35	33	32	31	30	29	28	28	28	28	28	28	20
S 30	17 27	17 25	17 22	17 20	17 18	17 16	17 14	17 12	17 11	17 10	17 09	17 08	17 07	17 07	17 07	S 30
35	20	17	14	11	17 09	17 06	17 04	17 02	17 00	16 59	16 57	16 56	16 56	16 55	16 55	35
40	12	17 09	17 05	17 02	16 59	16 56	16 53	16 51	16 48	46	45	43	42	41	41	40
45	17 03	16 58	16 54	16 50	47	43	40	37	34	32	30	28	27	25	25	45
50	16 52	47	42	37	32	28	24	20	17	14	11	09	16 07	16 05	16 04	50
S 52	16 47	16 41	16 35	16 30	16 25	16 21	16 16	16 12	16 09	16 05	16 02	16 00	15 58	15 56	15 55	S 52
54	41	35	29	23	18	13	16 08	16 04	16 00	15 56	15 53	15 50	47	45	44	54
56	35	28	22	15	10	16 04	15 59	15 54	15 49	45	42	38	36	33	32	56
58	28	20	13	16 07	16 00	15 54	48	43	38	33	29	25	22	19	17	58
S 60	16 19	16 12	16 04	15 56	15 49	15 43	15 36	15 30	15 24	15 19	15 14	15 10	15 06	15 03	15 00	S 60

MORNING CIVIL TWILIGHT

Lat.	Apr. 29	May 2	May 5	May 8	May 11	May 14	May 17	May 20	May 23	May 26	May 29	June 1	June 4	June 7	June 10	Lat.
°	h m	h m	h m	h m	h m	h m	h m	h m	h m	h m	h m	h m	h m	h m	h m	°
N 72	////	////	////	////	▭	▭	▭	▭	▭	▭	▭	▭	▭	▭	▭	N 72
70	////	////	////	////	////	////	////	▭	▭	▭	▭	▭	▭	▭	▭	70
68	01 33	01 04	00 02	////	////	////	////	////	////	////	▭	▭	▭	▭	▭	68
66	02 12	01 54	01 34	01 10	00 37	////	////	////	////	////	////	////	////	////	////	66
64	38	02 24	02 10	01 54	01 38	01 20	00 59	00 27	////	////	////	////	////	////	////	64
62	02 58	02 46	35	02 23	02 11	01 59	01 46	01 32	01 18	01 03	00 44	00 15	////	////	////	62
N 60	03 14	03 04	02 54	02 44	02 35	02 25	02 15	02 05	01 56	01 46	01 37	01 28	01 20	01 11	01 04	N 60
58	27	19	03 10	03 02	02 53	02 45	37	29	02 22	02 15	02 08	02 02	01 56	01 51	01 47	58
56	39	31	23	16	03 09	03 01	02 55	02 48	42	36	31	26	02 21	02 18	02 14	56
54	49	42	35	28	22	15	03 09	03 04	02 58	02 53	02 49	02 45	41	38	36	54
52	03 58	51	45	39	33	27	22	17	03 12	03 08	03 04	03 00	02 57	02 55	02 53	52
N 50	04 05	03 59	03 54	03 48	03 43	03 38	03 33	03 28	03 24	03 20	03 17	03 14	03 11	03 09	03 07	N 50
45	22	04 17	04 12	04 07	04 03	03 59	03 55	03 52	03 49	03 46	03 43	03 41	03 39	37	36	45
40	35	31	27	23	19	04 16	04 13	04 10	04 07	04 05	04 03	04 01	04 00	03 59	03 58	40
35	45	42	39	36	33	30	27	25	23	21	20	18	17	04 16	04 16	35
30	04 55	04 52	04 49	04 46	04 44	04 42	04 40	38	36	35	33	32	32	31	31	30
N 20	05 10	05 08	05 06	05 04	05 03	05 01	05 00	04 59	04 58	04 57	04 56	04 56	04 55	04 55	04 55	N 20
N 10	22	21	20	19	18	17	16	05 16	05 15	05 15	05 15	05 15	05 15	05 15	05 15	N 10
0	33	32	32	31	31	31	31	31	31	31	31	32	32	33	33	0
S 10	43	43	43	43	44	44	44	45	05 46	05 46	05 47	05 48	05 48	05 49	05 50	S 10
20	05 52	05 53	05 54	05 55	05 56	05 57	05 58	05 59	06 00	06 02	06 03	06 04	06 05	06 06	06 07	20
S 30	06 03	06 04	06 06	06 08	06 10	06 11	06 13	06 15	06 16	06 18	06 20	06 21	06 23	06 24	06 25	S 30
35	08	10	12	15	17	19	21	23	25	27	29	31	32	34	36	35
40	14	17	20	22	25	27	30	33	35	37	40	42	44	45	06 47	40
45	21	24	28	31	34	37	40	43	46	06 49	06 52	06 54	06 56	06 58	07 00	45
50	28	33	37	41	45	49	52	06 56	06 59	07 03	07 06	07 09	07 11	07 14	16	50
S 52	06 32	06 37	06 41	06 45	06 50	06 54	06 58	07 02	07 06	07 09	07 12	07 16	07 18	07 21	07 23	S 52
54	36	41	46	50	06 55	06 59	07 04	08	12	16	20	23	26	29	31	54
56	40	45	51	06 56	07 01	07 06	10	15	19	24	27	31	34	37	40	56
58	44	50	06 56	07 02	07	13	18	23	28	32	36	40	44	47	07 50	58
S 60	06 49	06 56	07 02	07 08	07 14	07 20	07 26	07 31	07 37	07 42	07 46	07 51	07 55	07 58	08 01	S 60

EVENING CIVIL TWILIGHT

Lat.	Apr. 29	May 2	May 5	May 8	May 11	May 14	May 17	May 20	May 23	May 26	May 29	June 1	June 4	June 7	June 10	Lat.
°	h m	h m	h m	h m	h m	h m	h m	h m	h m	h m	h m	h m	h m	h m	h m	°
N 72	////	////	////	////	▭	▭	▭	▭	▭	▭	▭	▭	▭	▭	▭	N 72
70	////	////	////	////	////	////	▭	▭	▭	▭	▭	▭	▭	▭	▭	70
68	22 29	23 01	////	////	////	////	////	////	////	////	▭	▭	▭	▭	▭	68
66	21 48	22 06	22 26	22 51	23 33	////	////	////	////	////	////	////	////	////	////	66
64	21	21 34	21 48	22 03	22 20	22 39	23 02	23 50	////	////	////	////	////	////	////	64
62	21 00	21 11	22	21 33	21 45	21 58	22 11	22 25	22 40	22 57	23 18	////	////	////	////	62
N 60	20 43	20 53	21 02	21 11	21 21	21 30	21 40	21 50	22 00	22 10	22 20	22 30	22 39	22 48	22 56	N 60
58	30	38	20 45	20 54	21 02	21 10	18	26	21 34	21 41	21 48	21 55	22 02	22 08	22 13	58
56	18	25	32	39	20 46	20 53	21 00	21 07	21 13	20	26	31	21 36	21 41	21 45	56
54	20 08	14	20	26	33	39	20 45	20 51	20 56	21 02	21 07	21 12	16	20	23	54
52	19 59	20 04	10	16	21	27	32	37	42	20 47	20 52	20 56	21 00	21 03	21 06	52
N 50	19 51	19 56	20 01	20 06	20 11	20 16	20 21	20 26	20 30	20 34	20 38	20 42	20 46	20 49	20 51	N 50
45	34	38	19 42	19 46	19 50	19 54	19 58	20 02	20 05	20 09	20 12	20 15	20 18	20 20	23	45
40	21	24	28	31	34	37	40	19 43	19 46	19 49	19 52	19 54	19 57	19 59	20 00	40
35	10	13	15	18	21	23	26	28	31	33	35	37	39	41	19 43	35
30	19 01	19 03	19 05	19 07	19 09	19 11	19 13	19 15	19 17	19 19	19 21	23	25	26	28	30
N 20	18 46	18 47	18 48	18 49	18 50	18 52	18 53	18 54	18 56	18 57	18 58	19 00	19 01	19 02	19 03	N 20
N 10	33	33	34	34	35	36	36	37	38	39	40	18 41	18 41	18 42	18 43	N 10
0	22	22	22	22	22	22	22	22	22	23	23	24	24	25	25	0
S 10	12	11	18 10	18 09	18 09	18 08	18 08	18 08	18 07	18 07	18 07	18 08	18 08	18 08	18 08	S 10
20	18 02	18 01	17 59	17 58	17 56	17 55	17 54	17 53	17 53	17 52	17 52	17 51	17 51	17 51	17 52	20
S 30	17 52	17 49	17 47	17 45	17 43	17 41	17 39	17 38	17 37	17 36	17 35	17 34	17 33	17 33	17 33	S 30
35	46	43	41	38	35	33	31	29	28	26	25	24	24	23	23	35
40	40	37	34	30	27	25	22	20	18	16	15	13	12	17 12	17 11	40
45	33	29	25	22	18	15	12	17 09	17 07	17 04	17 03	17 01	17 00	16 59	16 58	45
50	26	21	16	12	07	17 03	17 00	16 57	16 54	16 51	16 48	16 46	16 45	43	43	50
S 52	17 22	17 17	17 12	17 07	17 02	16 58	16 54	16 51	16 47	16 44	16 42	16 40	16 38	16 36	16 35	S 52
54	18	13	07	17 02	16 57	52	48	44	41	37	34	32	30	28	27	54
56	14	08	17 02	16 56	51	46	42	37	33	30	27	24	22	20	18	56
58	10	17 03	16 57	51	45	39	34	29	25	21	18	15	12	16 10	16 08	58
S 60	17 05	16 57	16 50	16 44	16 38	16 32	16 26	16 21	16 16	16 12	16 08	16 04	16 01	15 59	15 57	S 60

RISING, SETTING AND DEPRESSION GRAPHS

TABLE 1—MERIDIAN PASSAGE AND DECLINATION OF THE SUN AT 12ʰ G.M.T.

Day	May Mer. Pass.	May Dec.	June Mer. Pass.	June Dec.	July Mer. Pass.	July Dec.	August Mer. Pass.	August Dec.
	h m	°	h m	°	h m	°	h m	°
1	11 57	N.15·0	11 58	N.22·0	12 04	N.23·1	12 06	N.18·1
2	11 57	15·3	11 58	22·1	12 04	23·1	12 06	17·9
3	11 57	15·6	11 58	22·3	12 04	23·0	12 06	17·6
4	11 57	15·9	11 58	22·4	12 04	22·9	12 06	17·3
5	11 57	16·2	11 58	22·5	12 04	22·8	12 06	17·1
6	11 57	N.16·4	11 59	N.22·6	12 05	N.22·7	12 06	N.16·8
7	11 57	16·7	11 59	22·7	12 05	22·6	12 06	16·5
8	11 56	17·0	11 59	22·8	12 05	22·5	12 06	16·2
9	11 56	17·3	11 59	22·9	12 05	22·4	12 06	16·0
10	11 56	17·5	11 59	23·0	12 05	22·3	12 05	15·7
11	11 56	N.17·8	11 59	N.23·1	12 05	N.22·2	12 05	N.15·4
12	11 56	18·0	12 00	23·1	12 06	22·0	12 05	15·1
13	11 56	18·3	12 00	23·2	12 06	21·9	12 05	14·8
14	11 56	18·5	12 00	23·2	12 06	21·7	12 05	14·5
15	11 56	18·8	12 00	23·3	12 06	21·6	12 05	14·2
16	11 56	N.19·0	12 01	N.23·3	12 06	N.21·4	12 04	N.13·8
17	11 56	19·3	12 01	23·4	12 06	21·3	12 04	13·5
18	11 56	19·5	12 01	23·4	12 06	21·1	12 04	13·2
19	11 56	19·7	12 01	23·4	12 06	20·9	12 04	12·9
20	11 56	19·9	12 01	23·4	12 06	20·7	12 03	12·6
21	11 57	N.20·1	12 02	N.23·4	12 06	N.20·5	12 03	N.12·2
22	11 57	20·3	12 02	23·4	12 06	20·3	12 03	11·9
23	11 57	20·5	12 02	23·4	12 06	20·1	12 03	11·6
24	11 57	20·7	12 02	23·4	12 06	19·9	12 02	11·2
25	11 57	20·9	12 02	23·4	12 06	19·7	12 02	10·9
26	11 57	N.21·1	12 03	N.23·4	12 06	N.19·5	12 02	N.10·5
27	11 57	21·2	12 03	23·3	12 06	19·3	12 02	10·2
28	11 57	21·4	12 03	23·3	12 06	19·1	12 01	9·8
29	11 57	21·6	12 03	23·2	12 06	18·8	12 01	9·5
30	11 57	21·7	12 03	N.23·2	12 06	18·6	12 01	9·1
31	11 58	N.21·9			12 06	N.18·4	12 00	N. 8·8

TABLES 2 and 3—DEPRESSION OF THE SUN AT VARIOUS HEIGHTS

Height	TABLE 2 AT SUNRISE AND SUNSET		TABLE 3 AT CIVIL TWILIGHT	
Feet	Depression	Diff. from 0°·8	Depression	Diff. from 0°·8
	°	°	°	°
0	0·8	—	6·0	5·2
500	1·3	0·5	6·0	5·2
1 000	1·5	0·7	6·0	5·2
2 000	1·7	0·9	6·1	5·3
3 000	1·9	1·1	6·1	5·3
4 000	2·1	1·3	6·1	5·3
5 000	2·2	1·4	6·2	5·4
6 000	2·4	1·6	6·2	5·4
7 000	2·5	1·7	6·2	5·4
8 000	2·6	1·8	6·3	5·5
9 000	2·7	1·9	6·3	5·5
10 000	2·8	2·0	6·3	5·5
15 000	3·2	2·4	6·5	5·7
20 000	3·6	2·8	6·6	5·8
25 000	3·9	3·1	6·8	6·0
30 000	4·2	3·4	6·9	6·1
35 000	4·4	3·6	7·1	6·3
40 000	4·7	3·9	7·2	6·4
45 000	4·9	4·1	7·3	6·5
50 000	5·1	4·3	7·5	6·7
55 000	5·3	4·5	7·6	6·8
60 000	5·5	4·7	7·7	6·9

An alternative method to those given on pages A12–A14 is to use the graphs to give the corrections to the tabulated times of sunrise and sunset at ground level; in this case it is adequate to use the graphs for the *nearest* tabular latitude and declination. The difference in hour angle is found between the hour angle for zero depression and the hour angle at the tabular depression minus 0°·8. The difference in hour angle so found is then applied to the time of sunrise or sunset. The result will be less than 5ᵐ in error if the declination curve cuts all the depression lines.

Example. To find the times of sunrise and sunset on 1975 August 23 in latitude N. 65° 17′, longitude W. 35° 15′, at a height of 37 000 feet. From Table **1**, Dec.= N. 11°·6; Table **2**, Depression diff. from 0°·8=3°·7.

	Sunrise	Sunset
	h m	h m
Page A68, N. 65° 17′ (August 24)	04 10	19 53
Page A71, Lat. 66°, Dec. 11° (same); diff. in H.A. from depression 0° to 3°·7	45	45
L.M.T.	03 25	20 38
Longitude W. 35° 15′	2 21	2 21
G.M.T.	05 46	22 59

SEMIDURATION OF SUNLIGHT

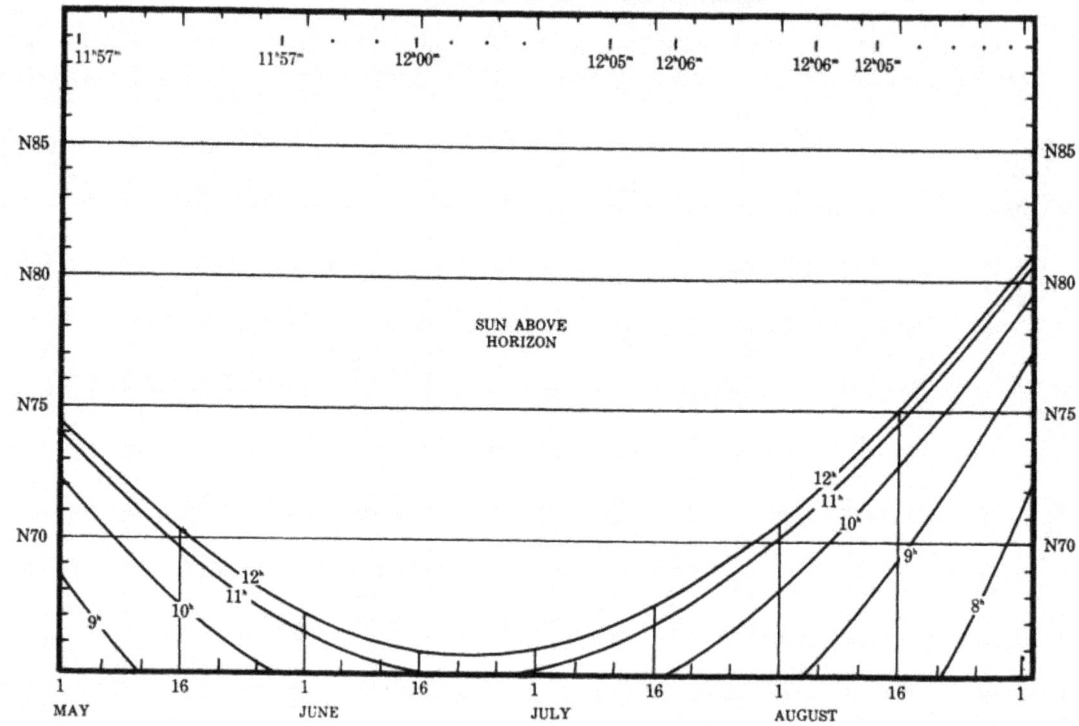

DURATION OF TWILIGHT

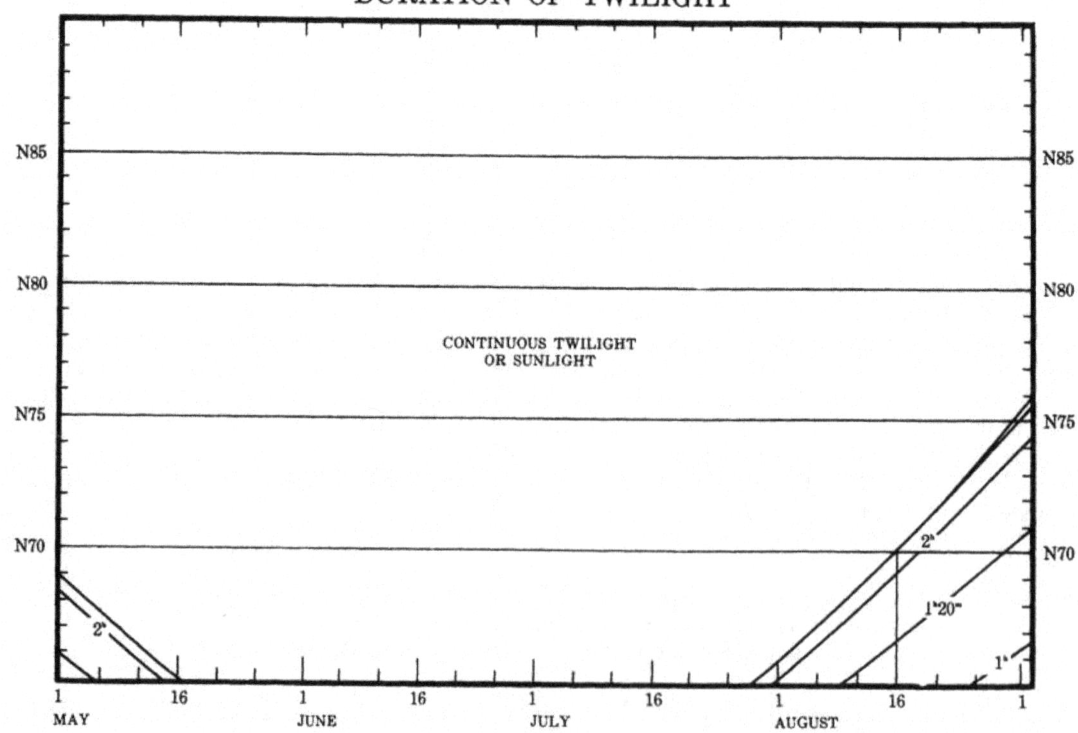

SEMIDURATION OF MOONLIGHT

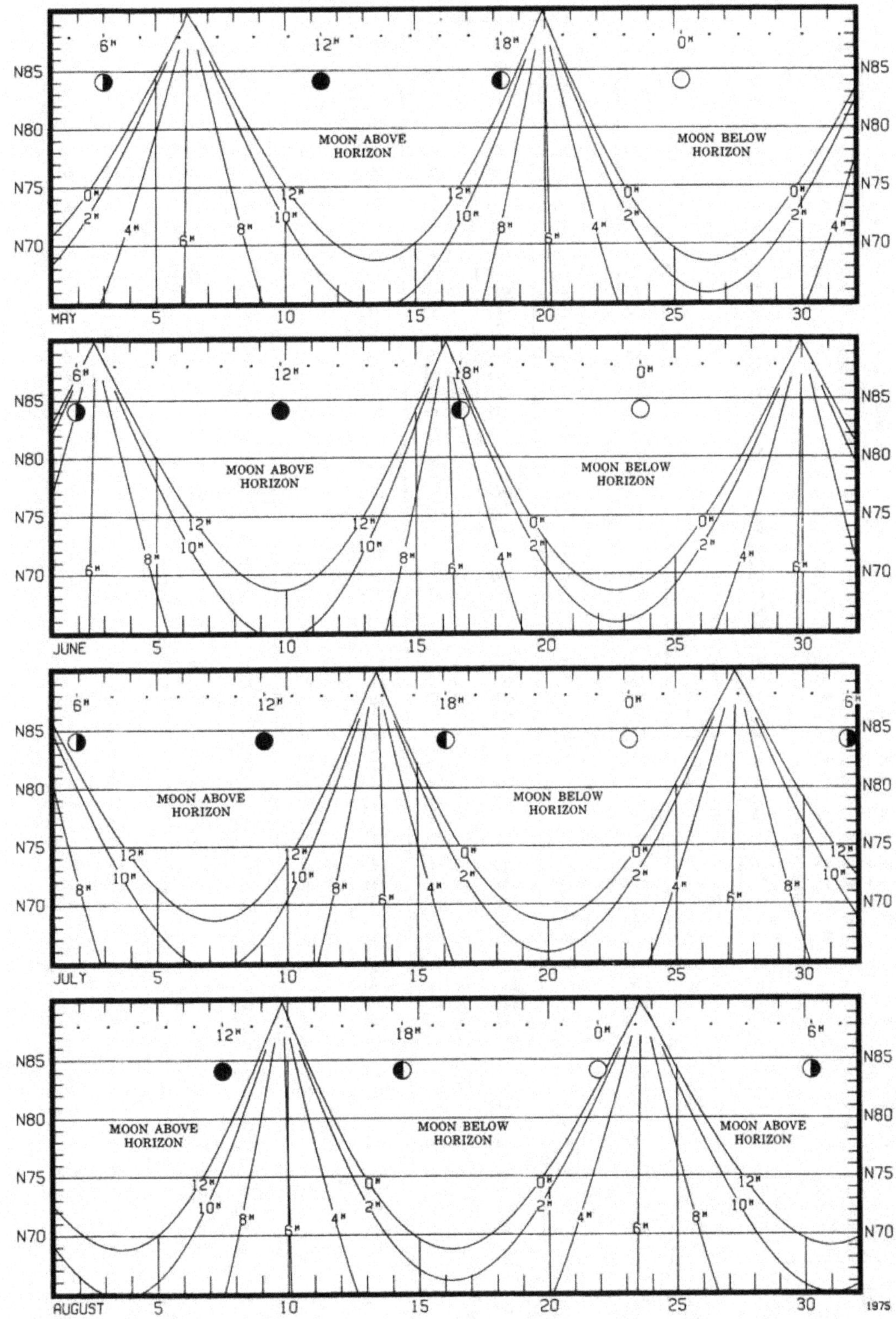

CONVERSION OF ARC TO TIME

°	h m	°	h m	°	h m	°	h m	°	h m	°	h m	′	m s
0	0 00	**60**	4 00	**120**	8 00	**180**	12 00	**240**	16 00	**300**	20 00	**0**	0 00
1	0 04	**61**	4 04	**121**	8 04	**181**	12 04	**241**	16 04	**301**	20 04	**1**	0 04
2	0 08	**62**	4 08	**122**	8 08	**182**	12 08	**242**	16 08	**302**	20 08	**2**	0 08
3	0 12	**63**	4 12	**123**	8 12	**183**	12 12	**243**	16 12	**303**	20 12	**3**	0 12
4	0 16	**64**	4 16	**124**	8 16	**184**	12 16	**244**	16 16	**304**	20 16	**4**	0 16
5	0 20	**65**	4 20	**125**	8 20	**185**	12 20	**245**	16 20	**305**	20 20	**5**	0 20
6	0 24	**66**	4 24	**126**	8 24	**186**	12 24	**246**	16 24	**306**	20 24	**6**	0 24
7	0 28	**67**	4 28	**127**	8 28	**187**	12 28	**247**	16 28	**307**	20 28	**7**	0 28
8	0 32	**68**	4 32	**128**	8 32	**188**	12 32	**248**	16 32	**308**	20 32	**8**	0 32
9	0 36	**69**	4 36	**129**	8 36	**189**	12 36	**249**	16 36	**309**	20 36	**9**	0 36
10	0 40	**70**	4 40	**130**	8 40	**190**	12 40	**250**	16 40	**310**	20 40	**10**	0 40
11	0 44	**71**	4 44	**131**	8 44	**191**	12 44	**251**	16 44	**311**	20 44	**11**	0 44
12	0 48	**72**	4 48	**132**	8 48	**192**	12 48	**252**	16 48	**312**	20 48	**12**	0 48
13	0 52	**73**	4 52	**133**	8 52	**193**	12 52	**253**	16 52	**313**	20 52	**13**	0 52
14	0 56	**74**	4 56	**134**	8 56	**194**	12 56	**254**	16 56	**314**	20 56	**14**	0 56
15	1 00	**75**	5 00	**135**	9 00	**195**	13 00	**255**	17 00	**315**	21 00	**15**	1 00
16	1 04	**76**	5 04	**136**	9 04	**196**	13 04	**256**	17 04	**316**	21 04	**16**	1 04
17	1 08	**77**	5 08	**137**	9 08	**197**	13 08	**257**	17 08	**317**	21 08	**17**	1 08
18	1 12	**78**	5 12	**138**	9 12	**198**	13 12	**258**	17 12	**318**	21 12	**18**	1 12
19	1 16	**79**	5 16	**139**	9 16	**199**	13 16	**259**	17 16	**319**	21 16	**19**	1 16
20	1 20	**80**	5 20	**140**	9 20	**200**	13 20	**260**	17 20	**320**	21 20	**20**	1 20
21	1 24	**81**	5 24	**141**	9 24	**201**	13 24	**261**	17 24	**321**	21 24	**21**	1 24
22	1 28	**82**	5 28	**142**	9 28	**202**	13 28	**262**	17 28	**322**	21 28	**22**	1 28
23	1 32	**83**	5 32	**143**	9 32	**203**	13 32	**263**	17 32	**323**	21 32	**23**	1 32
24	1 36	**84**	5 36	**144**	9 36	**204**	13 36	**264**	17 36	**324**	21 36	**24**	1 36
25	1 40	**85**	5 40	**145**	9 40	**205**	13 40	**265**	17 40	**325**	21 40	**25**	1 40
26	1 44	**86**	5 44	**146**	9 44	**206**	13 44	**266**	17 44	**326**	21 44	**26**	1 44
27	1 48	**87**	5 48	**147**	9 48	**207**	13 48	**267**	17 48	**327**	21 48	**27**	1 48
28	1 52	**88**	5 52	**148**	9 52	**208**	13 52	**268**	17 52	**328**	21 52	**28**	1 52
29	1 56	**89**	5 56	**149**	9 56	**209**	13 56	**269**	17 56	**329**	21 56	**29**	1 56
30	2 00	**90**	6 00	**150**	10 00	**210**	14 00	**270**	18 00	**330**	22 00	**30**	2 00
31	2 04	**91**	6 04	**151**	10 04	**211**	14 04	**271**	18 04	**331**	22 04	**31**	2 04
32	2 08	**92**	6 08	**152**	10 08	**212**	14 08	**272**	18 08	**332**	22 08	**32**	2 08
33	2 12	**93**	6 12	**153**	10 12	**213**	14 12	**273**	18 12	**333**	22 12	**33**	2 12
34	2 16	**94**	6 16	**154**	10 16	**214**	14 16	**274**	18 16	**334**	22 16	**34**	2 16
35	2 20	**95**	6 20	**155**	10 20	**215**	14 20	**275**	18 20	**335**	22 20	**35**	2 20
36	2 24	**96**	6 24	**156**	10 24	**216**	14 24	**276**	18 24	**336**	22 24	**36**	2 24
37	2 28	**97**	6 28	**157**	10 28	**217**	14 28	**277**	18 28	**337**	22 28	**37**	2 28
38	2 32	**98**	6 32	**158**	10 32	**218**	14 32	**278**	18 32	**338**	22 32	**38**	2 32
39	2 36	**99**	6 36	**159**	10 36	**219**	14 36	**279**	18 36	**339**	22 36	**39**	2 36
40	2 40	**100**	6 40	**160**	10 40	**220**	14 40	**280**	18 40	**340**	22 40	**40**	2 40
41	2 44	**101**	6 44	**161**	10 44	**221**	14 44	**281**	18 44	**341**	22 44	**41**	2 44
42	2 48	**102**	6 48	**162**	10 48	**222**	14 48	**282**	18 48	**342**	22 48	**42**	2 48
43	2 52	**103**	6 52	**163**	10 52	**223**	14 52	**283**	18 52	**343**	22 52	**43**	2 52
44	2 56	**104**	6 56	**164**	10 56	**224**	14 56	**284**	18 56	**344**	22 56	**44**	2 56
45	3 00	**105**	7 00	**165**	11 00	**225**	15 00	**285**	19 00	**345**	23 00	**45**	3 00
46	3 04	**106**	7 04	**166**	11 04	**226**	15 04	**286**	19 04	**346**	23 04	**46**	3 04
47	3 08	**107**	7 08	**167**	11 08	**227**	15 08	**287**	19 08	**347**	23 08	**47**	3 08
48	3 12	**108**	7 12	**168**	11 12	**228**	15 12	**288**	19 12	**348**	23 12	**48**	3 12
49	3 16	**109**	7 16	**169**	11 16	**229**	15 16	**289**	19 16	**349**	23 16	**49**	3 16
50	3 20	**110**	7 20	**170**	11 20	**230**	15 20	**290**	19 20	**350**	23 20	**50**	3 20
51	3 24	**111**	7 24	**171**	11 24	**231**	15 24	**291**	19 24	**351**	23 24	**51**	3 24
52	3 28	**112**	7 28	**172**	11 28	**232**	15 28	**292**	19 28	**352**	23 28	**52**	3 28
53	3 32	**113**	7 32	**173**	11 32	**233**	15 32	**293**	19 32	**353**	23 32	**53**	3 32
54	3 36	**114**	7 36	**174**	11 36	**234**	15 36	**294**	19 36	**354**	23 36	**54**	3 36
55	3 40	**115**	7 40	**175**	11 40	**235**	15 40	**295**	19 40	**355**	23 40	**55**	3 40
56	3 44	**116**	7 44	**176**	11 44	**236**	15 44	**296**	19 44	**356**	23 44	**56**	3 44
57	3 48	**117**	7 48	**177**	11 48	**237**	15 48	**297**	19 48	**357**	23 48	**57**	3 48
58	3 52	**118**	7 52	**178**	11 52	**238**	15 52	**298**	19 52	**358**	23 52	**58**	3 52
59	3 56	**119**	7 56	**179**	11 56	**239**	15 56	**299**	19 56	**359**	23 56	**59**	3 56

The above table is for converting expressions in arc to their equivalent in time; its main use in this Almanac is for the conversion of longitude for application to L.M.T. (*added* if *west*, *subtracted* if *east*) to give G.M.T., or vice versa, particularly in the case of sunrise, sunset, etc.

INTERPOLATION OF G.H.A. SUN

	0^m	1^m	2^m	3^m	4^m	5^m	6^m	7^m	8^m	9^m	
s	° ′	° ′	° ′	° ′	° ′	° ′	° ′	° ′	° ′	° ′	s
00	0 00·0	0 15·0	0 30·0	0 45·0	1 00·0	1 15·0	1 30·0	1 45·0	2 00·0	2 15·0	00
01	0 00·3	0 15·3	0 30·3	0 45·3	1 00·3	1 15·3	1 30·3	1 45·3	2 00·3	2 15·3	01
02	0 00·5	0 15·5	0 30·5	0 45·5	1 00·5	1 15·5	1 30·5	1 45·5	2 00·5	2 15·5	02
03	0 00·8	0 15·8	0 30·8	0 45·8	1 00·8	1 15·8	1 30·8	1 45·8	2 00·8	2 15·8	03
04	0 01·0	0 16·0	0 31·0	0 46·0	1 01·0	1 16·0	1 31·0	1 46·0	2 01·0	2 16·0	04
05	0 01·3	0 16·3	0 31·3	0 46·3	1 01·3	1 16·3	1 31·3	1 46·3	2 01·3	2 16·3	05
06	0 01·5	0 16·5	0 31·5	0 46·5	1 01·5	1 16·5	1 31·5	1 46·5	2 01·5	2 16·5	06
07	0 01·8	0 16·8	0 31·8	0 46·8	1 01·8	1 16·8	1 31·8	1 46·8	2 01·8	2 16·8	07
08	0 02·0	0 17·0	0 32·0	0 47·0	1 02·0	1 17·0	1 32·0	1 47·0	2 02·0	2 17·0	08
09	0 02·3	0 17·3	0 32·3	0 47·3	1 02·3	1 17·3	1 32·3	1 47·3	2 02·3	2 17·3	09
10	0 02·5	0 17·5	0 32·5	0 47·5	1 02·5	1 17·5	1 32·5	1 47·5	2 02·5	2 17·5	10
11	0 02·8	0 17·8	0 32·8	0 47·8	1 02·8	1 17·8	1 32·8	1 47·8	2 02·8	2 17·8	11
12	0 03·0	0 18·0	0 33·0	0 48·0	1 03·0	1 18·0	1 33·0	1 48·0	2 03·0	2 18·0	12
13	0 03·3	0 18·3	0 33·3	0 48·3	1 03·3	1 18·3	1 33·3	1 48·3	2 03·3	2 18·3	13
14	0 03·5	0 18·5	0 33·5	0 48·5	1 03·5	1 18·5	1 33·5	1 48·5	2 03·5	2 18·5	14
15	0 03·8	0 18·8	0 33·8	0 48·8	1 03·8	1 18·8	1 33·8	1 48·8	2 03·8	2 18·8	15
16	0 04·0	0 19·0	0 34·0	0 49·0	1 04·0	1 19·0	1 34·0	1 49·0	2 04·0	2 19·0	16
17	0 04·3	0 19·3	0 34·3	0 49·3	1 04·3	1 19·3	1 34·3	1 49·3	2 04·3	2 19·3	17
18	0 04·5	0 19·5	0 34·5	0 49·5	1 04·5	1 19·5	1 34·5	1 49·5	2 04·5	2 19·5	18
19	0 04·8	0 19·8	0 34·8	0 49·8	1 04·8	1 19·8	1 34·8	1 49·8	2 04·8	2 19·8	19
20	0 05·0	0 20·0	0 35·0	0 50·0	1 05·0	1 20·0	1 35·0	1 50·0	2 05·0	2 20·0	20
21	0 05·3	0 20·3	0 35·3	0 50·3	1 05·3	1 20·3	1 35·3	1 50·3	2 05·3	2 20·3	21
22	0 05·5	0 20·5	0 35·5	0 50·5	1 05·5	1 20·5	1 35·5	1 50·5	2 05·5	2 20·5	22
23	0 05·8	0 20·8	0 35·8	0 50·8	1 05·8	1 20·8	1 35·8	1 50·8	2 05·8	2 20·8	23
24	0 06·0	0 21·0	0 36·0	0 51·0	1 06·0	1 21·0	1 36·0	1 51·0	2 06·0	2 21·0	24
25	0 06·3	0 21·3	0 36·3	0 51·3	1 06·3	1 21·3	1 36·3	1 51·3	2 06·3	2 21·3	25
26	0 06·5	0 21·5	0 36·5	0 51·5	1 06·5	1 21·5	1 36·5	1 51·5	2 06·5	2 21·5	26
27	0 06·8	0 21·8	0 36·8	0 51·8	1 06·8	1 21·8	1 36·8	1 51·8	2 06·8	2 21·8	27
28	0 07·0	0 22·0	0 37·0	0 52·0	1 07·0	1 22·0	1 37·0	1 52·0	2 07·0	2 22·0	28
29	0 07·3	0 22·3	0 37·3	0 52·3	1 07·3	1 22·3	1 37·3	1 52·3	2 07·3	2 22·3	29
30	0 07·5	0 22·5	0 37·5	0 52·5	1 07·5	1 22·5	1 37·5	1 52·5	2 07·5	2 22·5	30
31	0 07·8	0 22·8	0 37·8	0 52·8	1 07·8	1 22·8	1 37·8	1 52·8	2 07·8	2 22·8	31
32	0 08·0	0 23·0	0 38·0	0 53·0	1 08·0	1 23·0	1 38·0	1 53·0	2 08·0	2 23·0	32
33	0 08·3	0 23·3	0 38·3	0 53·3	1 08·3	1 23·3	1 38·3	1 53·3	2 08·3	2 23·3	33
34	0 08·5	0 23·5	0 38·5	0 53·5	1 08·5	1 23·5	1 38·5	1 53·5	2 08·5	2 23·5	34
35	0 08·8	0 23·8	0 38·8	0 53·8	1 08·8	1 23·8	1 38·8	1 53·8	2 08·8	2 23·8	35
36	0 09·0	0 24·0	0 39·0	0 54·0	1 09·0	1 24·0	1 39·0	1 54·0	2 09·0	2 24·0	36
37	0 09·3	0 24·3	0 39·3	0 54·3	1 09·3	1 24·3	1 39·3	1 54·3	2 09·3	2 24·3	37
38	0 09·5	0 24·5	0 39·5	0 54·5	1 09·5	1 24·5	1 39·5	1 54·5	2 09·5	2 24·5	38
39	0 09·8	0 24·8	0 39·8	0 54·8	1 09·8	1 24·8	1 39·8	1 54·8	2 09·8	2 24·8	39
40	0 10·0	0 25·0	0 40·0	0 55·0	1 10·0	1 25·0	1 40·0	1 55·0	2 10·0	2 25·0	40
41	0 10·3	0 25·3	0 40·3	0 55·3	1 10·3	1 25·3	1 40·3	1 55·3	2 10·3	2 25·3	41
42	0 10·5	0 25·5	0 40·5	0 55·5	1 10·5	1 25·5	1 40·5	1 55·5	2 10·5	2 25·5	42
43	0 10·8	0 25·8	0 40·8	0 55·8	1 10·8	1 25·8	1 40·8	1 55·8	2 10·8	2 25·8	43
44	0 11·0	0 26·0	0 41·0	0 56·0	1 11·0	1 26·0	1 41·0	1 56·0	2 11·0	2 26·0	44
45	0 11·3	0 26·3	0 41·3	0 56·3	1 11·3	1 26·3	1 41·3	1 56·3	2 11·3	2 26·3	45
46	0 11·5	0 26·5	0 41·5	0 56·5	1 11·5	1 26·5	1 41·5	1 56·5	2 11·5	2 26·5	46
47	0 11·8	0 26·8	0 41·8	0 56·8	1 11·8	1 26·8	1 41·8	1 56·8	2 11·8	2 26·8	47
48	0 12·0	0 27·0	0 42·0	0 57·0	1 12·0	1 27·0	1 42·0	1 57·0	2 12·0	2 27·0	48
49	0 12·3	0 27·3	0 42·3	0 57·3	1 12·3	1 27·3	1 42·3	1 57·3	2 12·3	2 27·3	49
50	0 12·5	0 27·5	0 42·5	0 57·5	1 12·5	1 27·5	1 42·5	1 57·5	2 12·5	2 27·5	50
51	0 12·8	0 27·8	0 42·8	0 57·8	1 12·8	1 27·8	1 42·8	1 57·8	2 12·8	2 27·8	51
52	0 13·0	0 28·0	0 43·0	0 58·0	1 13·0	1 28·0	1 43·0	1 58·0	2 13·0	2 28·0	52
53	0 13·3	0 28·3	0 43·3	0 58·3	1 13·3	1 28·3	1 43·3	1 58·3	2 13·3	2 28·3	53
54	0 13·5	0 28·5	0 43·5	0 58·5	1 13·5	1 28·5	1 43·5	1 58·5	2 13·5	2 28·5	54
55	0 13·8	0 28·8	0 43·8	0 58·8	1 13·8	1 28·8	1 43·8	1 58·8	2 13·8	2 28·8	55
56	0 14·0	0 29·0	0 44·0	0 59·0	1 14·0	1 29·0	1 44·0	1 59·0	2 14·0	2 29·0	56
57	0 14·3	0 29·3	0 44·3	0 59·3	1 14·3	1 29·3	1 44·3	1 59·3	2 14·3	2 29·3	57
58	0 14·5	0 29·5	0 44·5	0 59·5	1 14·5	1 29·5	1 44·5	1 59·5	2 14·5	2 29·5	58
59	0 14·8	0 29·8	0 44·8	0 59·8	1 14·8	1 29·8	1 44·8	1 59·8	2 14·8	2 29·8	59
60	0 15·0	0 30·0	0 45·0	1 00·0	1 15·0	1 30·0	1 45·0	2 00·0	2 15·0	2 30·0	60

INTERPOLATION OF G.H.A. ARIES

	0^m	1^m	2^m	3^m	4^m	5^m	6^m	7^m	8^m	9^m	
s	° ′	° ′	° ′	° ′	° ′	° ′	° ′	° ′	° ′	° ′	s
00	0 00.0	0 15.0	0 30.1	0 45.1	1 00.2	1 15.2	1 30.2	1 45.3	2 00.3	2 15.4	00
01	0 00.3	0 15.3	0 30.3	0 45.4	1 00.4	1 15.5	1 30.5	1 45.5	2 00.6	2 15.6	01
02	0 00.5	0 15.5	0 30.6	0 45.6	1 00.7	1 15.7	1 30.7	1 45.8	2 00.8	2 15.9	02
03	0 00.8	0 15.8	0 30.8	0 45.9	1 00.9	1 16.0	1 31.0	1 46.0	2 01.1	2 16.1	03
04	0 01.0	0 16.0	0 31.1	0 46.1	1 01.2	1 16.2	1 31.2	1 46.3	2 01.3	2 16.4	04
05	0 01.3	0 16.3	0 31.3	0 46.4	1 01.4	1 16.5	1 31.5	1 46.5	2 01.6	2 16.6	05
06	0 01.5	0 16.5	0 31.6	0 46.6	1 01.7	1 16.7	1 31.8	1 46.8	2 01.8	2 16.9	06
07	0 01.8	0 16.8	0 31.8	0 46.9	1 01.9	1 17.0	1 32.0	1 47.0	2 02.1	2 17.1	07
08	0 02.0	0 17.0	0 32.1	0 47.1	1 02.2	1 17.2	1 32.3	1 47.3	2 02.3	2 17.4	08
09	0 02.3	0 17.3	0 32.3	0 47.4	1 02.4	1 17.5	1 32.5	1 47.5	2 02.6	2 17.6	09
10	0 02.5	0 17.5	0 32.6	0 47.6	1 02.7	1 17.7	1 32.8	1 47.8	2 02.8	2 17.9	10
11	0 02.8	0 17.8	0 32.8	0 47.9	1 02.9	1 18.0	1 33.0	1 48.0	2 03.1	2 18.1	11
12	0 03.0	0 18.0	0 33.1	0 48.1	1 03.2	1 18.2	1 33.3	1 48.3	2 03.3	2 18.4	12
13	0 03.3	0 18.3	0 33.3	0 48.4	1 03.4	1 18.5	1 33.5	1 48.5	2 03.6	2 18.6	13
14	0 03.5	0 18.6	0 33.6	0 48.6	1 03.7	1 18.7	1 33.8	1 48.8	2 03.8	2 18.9	14
15	0 03.8	0 18.8	0 33.8	0 48.9	1 03.9	1 19.0	1 34.0	1 49.0	2 04.1	2 19.1	15
16	0 04.0	0 19.1	0 34.1	0 49.1	1 04.2	1 19.2	1 34.3	1 49.3	2 04.3	2 19.4	16
17	0 04.3	0 19.3	0 34.3	0 49.4	1 04.4	1 19.5	1 34.5	1 49.5	2 04.6	2 19.6	17
18	0 04.5	0 19.6	0 34.6	0 49.6	1 04.7	1 19.7	1 34.8	1 49.8	2 04.8	2 19.9	18
19	0 04.8	0 19.8	0 34.8	0 49.9	1 04.9	1 20.0	1 35.0	1 50.1	2 05.1	2 20.1	19
20	0 05.0	0 20.1	0 35.1	0 50.1	1 05.2	1 20.2	1 35.3	1 50.3	2 05.3	2 20.4	20
21	0 05.3	0 20.3	0 35.3	0 50.4	1 05.4	1 20.5	1 35.5	1 50.6	2 05.6	2 20.6	21
22	0 05.5	0 20.6	0 35.6	0 50.6	1 05.7	1 20.7	1 35.8	1 50.8	2 05.8	2 20.9	22
23	0 05.8	0 20.8	0 35.8	0 50.9	1 05.9	1 21.0	1 36.0	1 51.1	2 06.1	2 21.1	23
24	0 06.0	0 21.1	0 36.1	0 51.1	1 06.2	1 21.2	1 36.3	1 51.3	2 06.3	2 21.4	24
25	0 06.3	0 21.3	0 36.3	0 51.4	1 06.4	1 21.5	1 36.5	1 51.6	2 06.6	2 21.6	25
26	0 06.5	0 21.6	0 36.6	0 51.6	1 06.7	1 21.7	1 36.8	1 51.8	2 06.8	2 21.9	26
27	0 06.8	0 21.8	0 36.9	0 51.9	1 06.9	1 22.0	1 37.0	1 52.1	2 07.1	2 22.1	27
28	0 07.0	0 22.1	0 37.1	0 52.1	1 07.2	1 22.2	1 37.3	1 52.3	2 07.3	2 22.4	28
29	0 07.3	0 22.3	0 37.4	0 52.4	1 07.4	1 22.5	1 37.5	1 52.6	2 07.6	2 22.6	29
30	0 07.5	0 22.6	0 37.6	0 52.6	1 07.7	1 22.7	1 37.8	1 52.8	2 07.8	2 22.9	30
31	0 07.8	0 22.8	0 37.9	0 52.9	1 07.9	1 23.0	1 38.0	1 53.1	2 08.1	2 23.1	31
32	0 08.0	0 23.1	0 38.1	0 53.1	1 08.2	1 23.2	1 38.3	1 53.3	2 08.4	2 23.4	32
33	0 08.3	0 23.3	0 38.4	0 53.4	1 08.4	1 23.5	1 38.5	1 53.6	2 08.6	2 23.6	33
34	0 08.5	0 23.6	0 38.6	0 53.6	1 08.7	1 23.7	1 38.8	1 53.8	2 08.9	2 23.9	34
35	0 08.8	0 23.8	0 38.9	0 53.9	1 08.9	1 24.0	1 39.0	1 54.1	2 09.1	2 24.1	35
36	0 09.0	0 24.1	0 39.1	0 54.1	1 09.2	1 24.2	1 39.3	1 54.3	2 09.4	2 24.4	36
37	0 09.3	0 24.3	0 39.4	0 54.4	1 09.4	1 24.5	1 39.5	1 54.6	2 09.6	2 24.6	37
38	0 09.5	0 24.6	0 39.6	0 54.6	1 09.7	1 24.7	1 39.8	1 54.8	2 09.9	2 24.9	38
39	0 09.8	0 24.8	0 39.9	0 54.9	1 09.9	1 25.0	1 40.0	1 55.1	2 10.1	2 25.1	39
40	0 10.0	0 25.1	0 40.1	0 55.2	1 10.2	1 25.2	1 40.3	1 55.3	2 10.4	2 25.4	40
41	0 10.3	0 25.3	0 40.4	0 55.4	1 10.4	1 25.5	1 40.5	1 55.6	2 10.6	2 25.6	41
42	0 10.5	0 25.6	0 40.6	0 55.7	1 10.7	1 25.7	1 40.8	1 55.8	2 10.9	2 25.9	42
43	0 10.8	0 25.8	0 40.9	0 55.9	1 10.9	1 26.0	1 41.0	1 56.1	2 11.1	2 26.1	43
44	0 11.0	0 26.1	0 41.1	0 56.2	1 11.2	1 26.2	1 41.3	1 56.3	2 11.4	2 26.4	44
45	0 11.3	0 26.3	0 41.4	0 56.4	1 11.4	1 26.5	1 41.5	1 56.6	2 11.6	2 26.7	45
46	0 11.5	0 26.6	0 41.6	0 56.7	1 11.7	1 26.7	1 41.8	1 56.8	2 11.9	2 26.9	46
47	0 11.8	0 26.8	0 41.9	0 56.9	1 11.9	1 27.0	1 42.0	1 57.1	2 12.1	2 27.2	47
48	0 12.0	0 27.1	0 42.1	0 57.2	1 12.2	1 27.2	1 42.3	1 57.3	2 12.4	2 27.4	48
49	0 12.3	0 27.3	0 42.4	0 57.4	1 12.4	1 27.5	1 42.5	1 57.6	2 12.6	2 27.7	49
50	0 12.5	0 27.6	0 42.6	0 57.7	1 12.7	1 27.7	1 42.8	1 57.8	2 12.9	2 27.9	50
51	0 12.8	0 27.8	0 42.9	0 57.9	1 12.9	1 28.0	1 43.0	1 58.1	2 13.1	2 28.2	51
52	0 13.0	0 28.1	0 43.1	0 58.2	1 13.2	1 28.2	1 43.3	1 58.3	2 13.4	2 28.4	52
53	0 13.3	0 28.3	0 43.4	0 58.4	1 13.5	1 28.5	1 43.5	1 58.6	2 13.6	2 28.7	53
54	0 13.5	0 28.6	0 43.6	0 58.7	1 13.7	1 28.7	1 43.8	1 58.8	2 13.9	2 28.9	54
55	0 13.8	0 28.8	0 43.9	0 58.9	1 14.0	1 29.0	1 44.0	1 59.1	2 14.1	2 29.2	55
56	0 14.0	0 29.1	0 44.1	0 59.2	1 14.2	1 29.2	1 44.3	1 59.3	2 14.4	2 29.4	56
57	0 14.3	0 29.3	0 44.4	0 59.4	1 14.5	1 29.5	1 44.5	1 59.6	2 14.6	2 29.7	57
58	0 14.5	0 29.6	0 44.6	0 59.7	1 14.7	1 29.7	1 44.8	1 59.8	2 14.9	2 29.9	58
59	0 14.8	0 29.8	0 44.9	0 59.9	1 15.0	1 30.0	1 45.0	2 00.1	2 15.1	2 30.2	59
60	0 15.0	0 30.1	0 45.1	1 00.2	1 15.2	1 30.2	1 45.3	2 00.3	2 15.4	2 30.4	60

POLARIS (POLE STAR) TABLE, 1975

FOR DETERMINING THE LATITUDE FROM A SEXTANT ALTITUDE

L.H.A. ♈	Q	L.H.A. ♈	Q	L.H.A. ♈	Q	L.H.A. ♈	Q	L.H.A. ♈	Q	L.H.A. ♈	Q	L.H.A. ♈	Q	L.H.A. ♈	Q
° ′	′	° ′	′	° ′	′	° ′	′	° ′	′	° ′	′	° ′	′	° ′	′
359 08	−43	80 32	−33	113 11	− 7	143 05	+19	183 04	+45	264 07	+31	296 18	+ 5	326 18	−21
1 14	−44	82 00	−32	114 20	− 6	144 18	+20	185 31	+46	265 33	+30	297 26	+ 4	327 32	−22
3 29	−45	83 27	−31	115 28	− 5	145 32	+21	188 12	+47	266 57	+29	298 34	+ 3	328 47	−23
5 54	−46	84 52	−30	116 36	− 4	146 46	+22	191 12	+48	268 20	+28	299 43	+ 2	330 03	−24
8 32	−47	86 15	−29	117 44	− 3	148 01	+23	194 41	+49	269 41	+27	300 50	+ 1	331 19	−25
11 30	−48	87 38	−28	118 51	− 2	149 18	+24	199 02	+50	271 01	+26	301 58	0	332 37	−26
14 56	−49	88 58	−27	119 59	− 1	150 35	+25	205 51	+51	272 20	+25	303 06	− 1	333 55	−27
19 13	−50	90 18	−26	121 07	0	151 53	+26	218 22	+50	273 38	+24	304 14	− 2	335 15	−28
25 56	−51	91 36	−25	122 15	+ 1	153 12	+27	225 11	+49	274 55	+23	305 22	− 3	336 35	−29
38 17	−50	92 54	−24	123 23	+ 2	154 32	+28	229 32	+48	276 12	+22	306 29	− 4	337 58	−30
45 00	−49	94 10	−23	124 30	+ 3	155 53	+29	233 01	+47	277 27	+21	307 37	− 5	339 21	−31
49 17	−48	95 26	−22	125 39	+ 4	157 16	+30	236 01	+46	278 41	+20	308 45	− 6	340 46	−32
52 43	−47	96 41	−21	126 47	+ 5	158 40	+31	238 42	+45	279 55	+19	309 53	− 7	342 13	−33
55 41	−46	97 55	−20	127 55	+ 6	160 06	+32	241 09	+44	281 08	+18	311 02	− 8	343 41	−34
58 19	−45	99 08	−19	129 03	+ 7	161 34	+33	243 25	+43	282 21	+17	312 10	− 9	345 12	−35
60 44	−44	100 21	−18	130 11	+ 8	163 03	+34	245 33	+42	283 33	+16	313 19	−10	346 45	−36
62 59	−43	101 33	−17	131 20	+ 9	164 35	+35	247 34	+41	284 44	+15	314 27	−11	348 20	−37
65 05	−42	102 45	−16	132 29	+10	166 09	+36	249 29	+40	285 55	+14	315 37	−12	349 59	−38
67 05	−41	103 56	−15	133 38	+11	167 45	+37	251 20	+39	287 06	+13	316 46	−13	351 40	−39
68 59	−40	105 07	−14	134 47	+12	169 24	+38	253 06	+38	288 16	+12	317 56	−14	353 25	−40
70 48	−39	106 17	−13	135 57	+13	171 07	+39	254 49	+37	289 26	+11	319 06	−15	355 14	−41
72 33	−38	107 27	−12	137 07	+14	172 53	+40	256 28	+36	290 35	+10	320 17	−16	357 08	−42
74 14	−37	108 36	−11	138 18	+15	174 44	+41	258 04	+35	291 44	+ 9	321 28	−17	359 08	−43
75 53	−36	109 46	−10	139 29	+16	176 39	+42	259 38	+34	292 53	+ 8	322 40	−18	1 14	−44
77 28	−35	110 54	− 9	140 40	+17	178 40	+43	261 10	+33	294 02	+ 7	323 52	−19	3 29	−45
79 01	−34	112 03	− 8	141 52	+18	180 48	+44	262 39	+32	295 10	+ 6	325 05	−20	5 54	−46
80 32		113 11		143 05		183 04		264 07		296 18		326 18		8 32	

Q, which does *not* include refraction, is to be applied to the corrected sextant altitude of *Polaris*.
Polaris: Mag. 2·1, S.H.A. 327° 53′, Dec. N. 89° 09′·2

STANDARD DOME REFRACTION

To be *subtracted* from sextant altitude when using sextant suspension in a perspex dome

Alt.	Refn.	Alt.	Refn.
°	′	°	′
10	8	50	4
20	7	60	4
30	6	70	3
40	5	80	3

This table must not be used if a calibration table is fitted to the dome, or if a flat glass plate is provided, or for non-standard domes.

BUBBLE SEXTANT ERROR

Sextant Number	Alt.	Corr.
	°	′

AZIMUTH OF *POLARIS*

L.H.A. ♈ 300°–120°	Latitude 0°	30°	50°	55°	60°	65°	70°	L.H.A. ♈ 120°–300°
°	°	°	°	°	°	°	°	°
300	0·8	1·0	1·3	1·5	1·7	2·0	2·5	300
310	0·8	1·0	1·3	1·4	1·7	2·0	2·4	290
320	0·8	0·9	1·2	1·4	1·6	1·9	2·3	280
330	0·7	0·8	1·1	1·2	1·4	1·7	2·1	270
340	0·6	0·7	1·0	1·1	1·3	1·5	1·8	260
350	0·5	0·6	0·8	0·9	1·0	1·2	1·5	250
0	0·4	0·5	0·6	0·7	0·8	0·9	1·2	240
10	0·3	0·3	0·4	0·5	0·5	0·6	0·8	230
20	0·1	0·1	0·2	0·2	0·2	0·3	0·3	220
30	0·0	0·0	0·0	0·1	0·1	0·1	0·1	210
40	359·9	359·9	359·8	359·8	359·8	359·7	359·7	200
50	359·7	359·7	359·6	359·5	359·5	359·4	359·2	190
60	359·6	359·5	359·4	359·3	359·2	359·1	358·8	180
70	359·5	359·4	359·2	359·1	359·0	358·8	358·5	170
80	359·4	359·3	359·0	358·9	358·7	358·5	358·2	160
90	359·3	359·2	358·9	358·8	358·6	358·3	357·9	150
100	359·2	359·1	358·8	358·6	358·4	358·1	357·7	140
110	359·2	359·0	358·7	358·6	358·3	358·0	357·6	130
120	359·2	359·0	358·7	358·5	358·3	358·0	357·5	120

When Cassiopeia is left (right), *Polaris* is west (east).

CORRECTIONS TO BE APPLIED TO SEXTANT ALTITUDE

REFRACTION

To be subtracted from sextant altitude (referred to as observed altitude in A.P. 3270).

R_0	Height above sea level in units of 1,000 ft. — 0	5	10	15	20	25	30	35	40	45	50	55	R_0	$R=R_0\times f$ — 0·9	1·0	1·1	1·2
	Sextant Altitude ° ′	° ′	° ′	° ′	° ′	° ′	° ′	° ′	° ′	° ′	° ′	° ′					
′	90	90	90	90	90	90	90	90	90	90	90	90	′	′	′	′	′
0													0	0	0	0	0
	63	59	55	51	46	41	36	31	26	20	17	13					
1													1	1	1	1	1
	33	29	26	22	19	16	14	11	9	7	6	4					
2													2	2	2	2	2
	21	19	16	14	12	10	8	7	5	4	2 40	1 40					
3													3	3	3	3	4
	16	14	12	10	8	7	6	5	3 10	2 20	1 30	0 40					
4													4	4	4	4	5
	12	11	9	8	7	5	4 00	3 10	2 10	1 30	0 39	+0 05					
5													5	5	5	5	6
	10	9	7	5 50	4 50	3 50	3 10	2 20	1 30	0 49	+0 11	−0 19					
6													6	5	6	7	7
	8 10	6 50	5 50	4 50	4 00	3 00	2 20	1 50	1 10	0 24	−0 11	−0 38					
7													7	6	7	8	8
	6 50	5 50	5 00	4 00	3 10	2 30	1 50	1 20	0 38	+0 04	−0 28	−0 54					
8													8	7	8	9	10
	6 00	5 10	4 10	3 20	2 40	2 00	1 30	1 00	0 19	−0 13	−0 42	−1 08					
9													9	8	9	10	11
	5 20	4 30	3 40	2 50	2 10	1 40	1 10	0 35	+0 03	−0 27	−0 53	−1 18					
10													10	9	10	11	12
	4 30	3 40	2 50	2 20	1 40	1 10	0 37	+0 11	−0 16	−0 43	−1 08	−1 31					
12													12	11	12	13	14
	3 30	2 50	2 10	1 40	1 10	0 34	+0 09	− 014	−0 37	−1 00	−1 23	−1 44					
14													14	13	14	15	17
	2 50	2 10	1 40	1 10	0 37	+0 10	−0 13	−0 34	−0 53	−1 14	−1 35	−1 56					
16													16	14	16	18	19
	2 20	1 40	1 20	0 43	+0 15	−0 08	−0 31	−0,52	−1 08	−1 27	−1 46	−2 05					
18													18	16	18	20	22
	1 50	1 20	0 49	+0 23	−0 02	−0 26	−0 46	−1 06	−1 22	−1 39	−1 57	−2 14					
20													20	18	20	22	24
	1 12	0 44	+0 19	−0 06	−0 28	−0 48	−1 09	−1 27	−1 42	−1 58	−2 14	−2 30					
25													25	22	25	28	30
	0 34	+0 10	−0 13	−0 36	−0 55	−1 14	−1 32	−1 51	−2 06	−2 21	−2 34	−2 49					
30													30	27	30	33	36
	+0 06	−0 16	−0 37	−0 59	−1 17	−1 33	−1 51	−2 07	−2 23	−2 37	−2 51	−3 04					
35													35	31	35	38	42
	−0 18	−0 37	−0 58	−1 16	−1 34	−1 49	−2 06	−2 22	−2 35	−2 49	−3 03	−3 16					
40													40	36	40	44	48
		−0 53	−1 14	−1 31	−1 47	−2 03	−2 18	−2 33	−2 47	−2 59	−3 13	−3 25					
45													45	40	45	50	54
		−1 10	−1 28	−1 44	−1 59	−2 15	−2 28	−2 43	−2 56	−3 08	−3 22	−3 33					
50													50	45	50	55	60
			−1 40	−1 53	−2 09	−2 24	−2 38	−2 52	−3 04	−3 17	−3 29	−3 41					
55													55	49	55	60	66
				−2 03	−2 18	−2 33	−2 46	−3 01	−3 12	−3 25	−3 37	−3 48					
60													60	54	60	66	72
							−2 53	−3 07	−3 19	−3 31	−3 42	−3 53					

f	Temperature in °C — 0	5	10	15	20	25	30	35	40	45	50	55	f
	+47	+36	+27	+18	+10	+ 3	− 5	−13					
0·9													0·9
	+26	+16	+ 6	− 4	−13	−22	−31	−40					
1·0													1·0
	+ 5	− 5	−15	−25	−36	−46	−57	−68					
1·1													1·1
	−16	−25	−36	−46	−58	−71	−83	−95					
1·2													1·2
	−37	−45	−56	−67	−81	−95							

For these heights no temperature correction is necessary: take $f=1·0$ and use $R=R_0$

When R_0 is less than 10′ or the height is greater than 35,000 ft. take $f=1·0$ and use $R=R_0$

Choose the column appropriate to height, in units of 1,000 ft., and find the range of altitude in which the sextant altitude lies; the corresponding value of R_0 is the refraction, to be subtracted from sextant altitude, unless conditions are extreme. In that case find f from the lower table, with critical argument temperature. Use the table on the right to form the refraction, $R=R_0\times f$.

CORIOLIS (Z) CORRECTION

To be applied by moving the position line a distance Z to starboard (right) of the track in northern latitudes and to port (left) in southern latitudes. The argument is given as T.A.S. (True Air Speed) in A.P. 3270.

G/S KNOTS	Latitude 0°	10°	20°	30°	40°	50°	60°	70°	80°	90°	G/S KNOTS	Latitude 0°	10°	20°	30°	40°	50°	60°	70°	80°	90°
	′	′	′	′	′	′	′	′	′	′		′	′	′	′	′	′	′	′	′	′
150	0	1	1	2	3	3	3	4	4	4	450	0	2	4	6	8	9	10	11	12	12
200	0	1	2	3	3	4	5	5	5	5	500	0	2	4	7	8	10	11	12	13	13
250	0	1	2	3	4	5	6	6	6	7	550	0	3	5	7	9	11	12	14	14	14
300	0	1	3	4	5	6	7	7	8	8	600	0	3	5	8	10	12	14	15	16	16
350	0	2	3	5	6	7	8	9	9	9	650	0	3	6	9	11	13	15	16	17	17
400	0	2	4	5	7	8	9	10	10	10	700	0	3	6	9	12	14	16	17	18	18

CORRECTIONS TO BE APPLIED TO MARINE SEXTANT ALTITUDES

MARINE SEXTANT ERROR	CORRECTIONS	CORRECTION FOR DIP OF THE HORIZON To be subtracted from sextant altitude									
Sextant Number	In addition to sextant error and dip, corrections are to be applied for:	Ht.	Dip	Ht.	Dip	Ht.	Dip	Ht.	Dip	Ht.	Dip
		Ft.	′	Ft.	′	Ft.	′	Ft.	′	Ft.	′
Index Error	Refraction	0		114		437		968		1 707	
	Semi-diameter (for the Sun and Moon)		1		11		21		31		41
	Parallax (for the Moon)	2		137		481		1 033		1 792	
	Dome refraction (if applicable)		2		12		22		32		42
		6		162		527		1 099		1 880	
			3		13		23		33		43
		12		189		575		1 168		1 970	
			4		14		24		34		44
		21		218		625		1 239		2 061	
			5		15		25		35		45
		31		250		677		1 311		2 155	
			6		16		26		36		46
		43		283		731		1 386		2 251	
			7		17		27		37		47
		58		318		787		1 463		2 349	
			8		18		28		38		48
		75		356		845		1 543		2 449	
			9		19		29		39		49
		93		395		906		1 624		2 551	
			10		20		30		40		50
		114		437		968		1 707		2 655	

LIST OF CONTENTS

Pages	Contents
Inside front cover	Star list (57 stars) and G.H.A. interpolation tables
Daily pages	Ephemerides of Sun, Moon, Aries and planets; moonrise and moonset
F1–F2 (flap)	Star chart
F3 (flap)	Star list (57 stars) and G.H.A. interpolation tables
F4 (flap)	Interpolation of moonrise and moonset for longitude, and star index
A1–A3	Title page, preface, etc.
A4–A17	Explanation
A18–A19	List of symbols and abbreviations
A20–A23	Standard times
A24–A73	Sky diagrams
A74–A75	Planet location diagram
A76–A79	Star recognition diagrams for periscopic sextants
A80–A87	Sunrise, sunset and civil twilight
A88–A93	Rising, setting and depression graphs
A94–A95	Semi-duration graphs of sunlight, twilight and moonlight, in high latitudes
A96	Index of star names
A97–A99	Star list, 173 stars (accuracy 0′·1)
A100–A101	Interpolation of G.H.A. Sun and G.H.A. Aries (accuracy 0′·1)
A102	Conversion of arc to time
A103	*Polaris* tables and dome refraction
A104	Corrections for (total) refraction and Coriolis (Z) table
Inside back cover	Corrections to marine sextant observations

APPENDIX H

LONG-TERM ALMANAC

This appendix is intended for use when a more complete almanac is not available. It is based principally upon the fact that approximately correct values for the Greenwich hour angle and declination of the sun, and the Greenwich hour angle of Aries, can be obtained from an almanac that is exactly four years out of date. The differences in these values at intervals of exactly four years can be largely removed by applying an average correction to the values obtained from the tables of this appendix. The maximum error in an altitude computed by means of this appendix should not exceed 2'.0 for the sun or 1'.3 for stars.

This four-year, or quadrennial, correction varies throughout the year for the GHA of the sun (between about plus and minus one-half of a minute) and for the declination of the sun (between about plus and minus three-fourths of a minute). For the GHA of Aries the quadrennial correction is a constant, (+)1'.84. The appropriate quadrennial correction is applied once for each full four years which has passed since the base year of the tabulation (1972 in this appendix).

The tabulated values for GHA−175° and declination of the sun and GHA of Aries are given in four columns, labeled 0, 1, 2, and 3. The "0" column contains the data for the leap year in each four-year cycle and the 1, 2, and 3 columns contain data for, respectively, the first, second, and third years following each leap year.

The GHA−175° and declination of the sun are given at intervals of three days throughout the four-year cycle, except for the final days of each month, when the interval varies between one and four days. Linear interpolation is made between entries to obtain data for a given day. Additional corrections to the GHA of the sun of 15° per hour, 15' per minute, and 15" per second are made to obtain the GHA at a given time. Declination of the sun is obtained to sufficient accuracy by linear interpolation alone.

The GHA of Aries is given for each month of the four-year cycle. Additional corrections of 0°59'.14 per day, 15°02'.5 per hour, 15' per minute, and 15" per second are made to obtain the GHA at a given time.

The SHA and declination of 38 navigational stars are given for the base year, 1972.0. Annual (not quadrennial) corrections are made to these data to obtain the values for a given year and tenth of a year.

A multiplication table is included as an aid in applying corrections to tabulated values.

Sun tables. 1. Subtract 1972 from the year and divide the difference by four, obtaining (*a*) a whole number, and (*b*) a remainder. Enter column indicated by remainder (*b*) and take out values on either side of given time and date.

2. Multiply quadrennial correction for each value by whole number (*a*) obtained in step 1 and apply to tabulated values plus 175°.

3. Divide difference between corrected values by number of days (usually three) between them to determine daily change.

4. Multiply daily change by number of days and tenths since 0^h GMT of earlier tabulated date, and mark correction plus (+) or minus (−) as appropriate.

5. (GHA only.) Enter multiplication table with hours, minutes, and seconds of GMT, and take out corrections *A*, *B*, and *C*, respectively. These are all positive.

6. Apply corrections of steps 4 and 5 to corrected *earlier* values of step 2.

Example.—Find GHA and declination of sun at GMT $17^h13^m49^s$ on July 18, 2002.

Solution.—Steps 1 *and* 2: (2002−1972)÷4=7, remainder 2. Use column 2, and multiply quadrennial corrections by 7. Corrected values: GHA, July 16, 178°31′.1+(7×0′.05)=178°31′.5; July 19, 178°27′.2+(7×0′.06)=178°27′.6. Dec., July 16, 21°27′.9N−(7×0′.41)=21°25′.0N; July 19, 20°57′.5N−(7×0′.44)=20°54′.4N.

	GHA	
July 16	178°31′.5	*Step 3*
July 19	178°27′.6	
3-day change	(−)3′.9	
daily change	(−)1′.3	*Step 4*
days and tenths	2.7	
corr.	(−)3′.5	
A	255°00′.0	*Step 5*
B	3°15′.0	
C	12′.1	
0^h July 16	178°31′.5	*Step 6*
GHA	76°55′.2	

	Declination	
July 16	21°25′.0 N	*Step 3*
July 19	20°54′.4 N	
3-day change	(−)30′.6	
daily change	(−)10′.2	*Step 4*
days and tenths	2.7	
corr.	(−)27′.5	
0^h July 16	21°25′.0 N	*Step 6*
Dec.	20°57′.5 N	

Aries table. 1. Subtract 1972 from the year and divide the difference by four, obtaining (*a*) a whole number, and (*b*) a remainder. Enter column indicated by remainder (*b*) and take out value for given month.

2. Enter multiplication table with whole number (*a*) of step 1, day of month, hours of GMT, minutes of GMT, and seconds of GMT, and take out corrections *D*, *E*, *F*, *G*, and *C*, respectively.

3. Add values of steps 1 and 2.

Example.—Find GHA ♈ at GMT $11^h06^m33^s$ on November 28, 1995.

Solution.—Step 1: (1995−1972)÷4=5, remainder 3. Use column 3.

	GHA ♈	
Nov.	38°40′.6	*Step 1*
D	9′.2	*Step 2*
E	27°35′.9	
F	165°27′.1	
G	1°30′.2	
C	8′.2	
GHA ♈	233°31′.2	*Step 3*

Stars table. 1. Enter table with star name, and take out tabulated values.

2. Subtract 1972.0 from given year and tenth, and multiply annual correction by difference. Apply as correction (+or−, as appropriate) to value of step 1.

Example.—Find SHA and declination of Spica on September 11, 2011.
Solution.—From decimal table, September 11, 2011 = 2011.7. 2011.7 − 1972.0 = 39.7.

	SHA			*Declination*	
1972.0	159°04′.3	*Step 1*	1972.0	11°01′.0 S	*Step 1*
39.7×(−)0′.79	(−)31′.4	*Step 2*	39.7×0.31	(+)12′.3	*Step 2*
SHA	158°32′.9		Dec.	11°13′.3 S	

To determine GHA of star, add GHA ♈ and SHA☆ for given time and date.

SUN

0		Quad. GHA Corr.	1		Date	2		Quad. Dec. Corr.	3	
GHA − 175°	Dec.		GHA − 175°	Dec.		GHA − 175°	Dec.		GHA − 175°	Dec.
JANUARY										
° ′	° ′	′	° ′	° ′		° ′	° ′	′	° ′	° ′
4 14.4	23 05.5 S	−0.11	4 09.0	23 02.0 S	1	4 10.8	23 03.1 S	−0.32	4 12.9	23 04.2 S
3 53.3	22 50.2 S	−0.13	3 48.0	22 45.6 S	4	3 49.8	22 47.0 S	−0.35	3 51.9	22 48.4 S
3 33.0	22 30.7 S	−0.12	3 27.8	22 25.1 S	7	3 29.8	22 26.8 S	−0.39	3 31.7	22 28.6 S
3 13.8	22 07.2 S	−0.09	3 08.8	22 00.6 S	10	3 10.8	22 02.7 S	−0.42	3 12.5	22 04.8 S
2 55.7	21 39.9 S	−0.04	2 51.1	21 32.2 S	13	2 53.0	21 34.7 S	−0.44	2 54.5	21 37.1 S
2 38.9	21 08.7 S	+0.03	2 34.8	21 00.2 S	16	2 36.6	21 02.9 S	−0.46	2 37.9	21 05.6 S
2 23.7	20 33.9 S	+0.09	2 20.1	20 24.5 S	19	2 21.7	20 27.5 S	−0.48	2 22.9	20 30.5 S
2 10.1	19 55.6 S	+0.13	2 07.0	19 45.4 S	22	2 08.4	19 48.7 S	−0.49	2 09.5	19 51.9 S
1 58.2	19 13.9 S	+0.15	1 55.7	19 02.9 S	25	1 56.8	19 06.4 S	−0.52	1 57.9	19 10.0 S
1 48.2	18 29.1 S	+0.15	1 46.1	18 17.3 S	28	1 47.0	18 21.1 S	−0.54	1 48.0	18 24.9 S
FEBRUARY										
1 37.7	17 24.7 S	+0.13	1 36.1	17 11.9 S	1	1 36.8	17 16.0 S	−0.57	1 37.8	17 20.2 S
1 32.0	16 33.2 S	+0.14	1 30.7	16 19.6 S	4	1 31.3	16 24.0 S	−0.59	1 32.1	16 28.4 S
1 28.0	15 39.0 S	+0.15	1 27.1	15 24.8 S	7	1 27.8	15 29.5 S	−0.60	1 28.3	15 34.0 S
1 25.9	14 42.5 S	+0.19	1 25.4	14 27.7 S	10	1 26.0	14 32.5 S	−0.61	1 26.3	14 37.2 S
1 25.4	13 43.6 S	+0.24	1 25.4	13 28.4 S	13	1 25.9	13 33.4 S	−0.60	1 26.0	13 38.2 S
1 26.7	12 42.8 S	+0.29	1 27.2	12 27.1 S	16	1 27.5	12 32.3 S	−0.60	1 27.4	12 37.2 S
1 29.6	11 40.1 S	+0.34	1 30.6	11 24.0 S	19	1 30.7	11 29.3 S	−0.59	1 30.4	11 34.4 S
1 34.0	10 35.8 S	+0.36	1 35.6	10 19.3 S	22	1 35.4	10 24.7 S	−0.59	1 35.1	10 30.0 S
1 40.0	9 30.0 S	+0.36	1 41.9	9 13.2 S	25	1 41.5	9 18.6 S	−0.60	1 41.2	9 24.1 S
1 47.4	8 23.0 S	+0.34	1 49.5	8 05.8 S	28	1 49.0	8 11.3 S	−0.61	1 48.7	8 17.0 S
MARCH										
1 52.9	7 37.6 S	+0.32	1 52.3	7 43.1 S	1	1 51.7	7 48.7 S	−0.62	1 51.4	7 54.3 S
2 02.3	6 28.8 S	+0.31	2 01.4	6 34.4 S	4	2 00.8	6 40.0 S	−0.62	2 00.4	6 45.7 S
2 12.6	5 19.2 S	+0.31	2 11.6	5 24.8 S	7	2 11.0	5 30.5 S	−0.62	2 10.5	5 36.3 S
2 23.8	4 08.9 S	+0.33	2 22.6	4 14.5 S	10	2 22.1	4 20.4 S	−0.61	2 21.4	4 26.1 S
2 35.7	2 58.2 S	+0.36	2 34.5	3 03.8 S	13	2 33.9	3 09.7 S	−0.59	2 33.1	3 15.5 S
2 48.2	1 47.1 S	+0.40	2 47.1	1 52.8 S	16	2 46.4	1 58.7 S	−0.56	2 45.4	2 04.5 S
3 01.1	0 35.9 S	+0.43	3 00.1	0 41.7 S	19	2 59.4	0 47.6 S	−0.54	2 58.3	0 53.3 S
3 14.5	0 35.2 N	+0.44	3 13.6	0 29.4 N	22	3 12.7	0 23.6 N	+0.53	3 11.6	0 17.8 N
3 28.1	1 46.1 N	+0.43	3 27.2	1 40.3 N	25	3 26.2	1 34.6 N	+0.52	3 25.2	1 28.8 N
3 41.9	2 56.6 N	+0.40	3 40.9	2 50.9 N	28	3 39.8	2 45.2 N	+0.52	3 38.9	2 39.4 N
APRIL										
4 00.0	4 29.8 N	+0.35	3 59.0	4 24.3 N	1	3 57.9	4 18.6 N	+0.52	3 57.1	4 12.8 N
4 13.4	5 38.9 N	+0.33	4 12.3	5 33.4 N	4	4 11.2	5 27.8 N	+0.51	4 10.5	5 22.1 N
4 26.4	6 47.1 N	+0.32	4 25.2	6 41.7 N	7	4 24.3	6 36.2 N	+0.49	4 23.5	6 30.6 N
4 38.8	7 54.3 N	+0.33	4 37.7	7 49.0 N	10	4 36.9	7 43.5 N	+0.47	4 36.0	7 38.0 N
4 50.6	9 00.3 N	+0.35	4 49.6	8 55.1 N	13	4 48.9	8 49.7 N	+0.44	4 48.0	8 44.3 N
5 01.7	10 05.0 N	+0.37	5 00.9	9 59.8 N	16	5 00.2	9 54.5 N	+0.41	4 59.2	9 49.3 N
5 12.0	11 08.1 N	+0.39	5 11.3	11 03.0 N	19	5 10.6	10 57.9 N	+0.39	5 09.7	10 52.8 N
5 21.4	12 09.6 N	+0.37	5 20.9	12 04.6 N	22	5 20.0	11 59.7 N	+0.37	5 19.3	11 54.7 N
5 29.8	13 09.2 N	+0.33	5 29.4	13 04.4 N	25	5 28.5	12 59.7 N	+0.36	5 28.0	12 54.8 N
5 37.2	14 06.9 N	+0.28	5 36.8	14 02.3 N	28	5 36.0	13 57.7 N	+0.34	5 35.6	13 53.0 N
MAY										
5 43.4	15 02.6 N	+0.24	5 43.0	14 58.1 N	1	5 42.3	14 53.7 N	+0.33	5 42.1	14 49.2 N
5 48.4	15 56.0 N	+0.20	5 48.0	15 51.7 N	4	5 47.5	15 47.5 N	+0.31	5 47.3	15 43.1 N
5 52.1	16 47.0 N	+0.19	5 51.7	16 43.0 N	7	5 51.4	16 38.9 N	+0.28	5 51.3	16 34.7 N
5 54.5	17 35.5 N	+0.19	5 54.2	17 31.7 N	10	5 54.1	17 27.8 N	+0.25	5 53.9	17 23.9 N
5 55.6	18 21.4 N	+0.21	5 55.5	18 17.7 N	13	5 55.5	18 14.1 N	+0.22	5 55.3	18 10.4 N
5 55.4	19 04.5 N	+0.23	5 55.5	19 01.0 N	16	5 55.5	18 57.6 N	+0.19	5 55.4	18 54.2 N
5 53.9	19 44.7 N	+0.23	5 54.3	19 41.4 N	19	5 54.3	19 38.3 N	+0.16	5 54.3	19 35.1 N
5 51.3	20 21.8 N	+0.20	5 51.8	20 18.8 N	22	5 51.7	20 15.9 N	+0.13	5 51.9	20 13.0 N
5 47.5	20 55.8 N	+0.16	5 48.1	20 53.1 N	25	5 48.0	20 50.5 N	+0.11	5 48.4	20 47.7 N
5 42.6	21 26.6 N	+0.11	5 43.2	21 24.2 N	28	5 43.2	21 21.8 N	+0.08	5 43.8	21 19.3 N
JUNE										
5 34.5	22 02.4 N	+0.04	5 35.1	22 00.4 N	1	5 35.2	21 58.3 N	+0.04	5 36.0	21 56.2 N
5 27.4	22 25.3 N	+0.02	5 27.9	22 23.5 N	4	5 28.2	22 21.7 N	+0.01	5 28.9	22 19.9 N
5 19.4	22 44.6 N	+0.02	5 19.9	22 43.1 N	7	5 20.4	22 41.6 N	−0.02	5 21.1	22 40.1 N
5 10.7	23 00.3 N	+0.04	5 11.3	22 59.1 N	10	5 11.9	22 57.9 N	−0.05	5 12.5	22 56.7 N
5 01.5	23 12.4 N	+0.07	5 02.3	23 11.5 N	13	5 02.9	23 10.6 N	−0.08	5 03.4	23 09.7 N
4 51.9	23 20.8 N	+0.09	4 52.8	23 20.2 N	16	4 53.4	23 19.6 N	−0.12	4 53.9	23 19.0 N
4 42.1	23 25.5 N	+0.08	4 43.2	23 25.3 N	19	4 43.6	23 24.9 N	−0.15	4 44.2	23 24.6 N
4 32.4	23 26.5 N	+0.06	4 33.4	23 26.6 N	22	4 33.7	23 26.5 N	−0.18	4 34.5	23 26.5 N
4 22.8	23 23.8 N	+0.01	4 23.8	23 24.1 N	25	4 24.0	23 24.4 N	−0.22	4 24.8	23 24.6 N
4 13.4	23 17.4 N	−0.03	4 14.3	23 18.0 N	28	4 14.5	23 18.5 N	−0.25	4 15.4	23 19.1 N

SUN										
0		Quad. GHA Corr.	1		Date	2		Quad. Dec. Corr.	3	
GHA − 175°	Dec.		GHA − 175°	Dec.		GHA − 175°	Dec.		GHA − 175°	Dec.
JULY										
° ′	° ′	′	° ′	° ′		° ′	° ′	′	° ′	° ′
4 04.5	23 07.3 N	−0.06	4 05.2	23 08.2 N	1	4 05.4	23 09.0 N	−0.28	4 06.3	23 09.9 N
3 56.1	22 53.5 N	−0.07	3 56.6	22 54.7 N	4	3 57.0	22 55.8 N	−0.31	3 57.7	22 57.0 N
3 48.4	22 36.2 N	−0.05	3 48.8	22 37.6 N	7	3 49.2	22 39.1 N	−0.34	3 49.8	22 40.5 N
3 41.5	22 15.3 N	−0.01	3 41.9	22 17.1 N	10	3 42.3	22 18.8 N	−0.37	3 42.7	22 20.5 N
3 35.5	21 51.0 N	+0.03	3 36.0	21 53.0 N	13	3 36.2	21 55.0 N	−0.39	3 36.4	21 57.0 N
3 30.6	21 23.3 N	+0.05	3 31.1	21 25.7 N	16	3 31.1	21 27.9 N	−0.41	3 31.3	21 30.1 N
3 26.9	20 52.4 N	+0.06	3 27.4	20 55.0 N	19	3 27.2	20 57.5 N	−0.44	3 27.4	21 00.0 N
3 24.5	20 18.3 N	+0.04	3 24.9	20 21.2 N	22	3 24.5	20 23.8 N	−0.47	3 24.7	20 26.6 N
3 23.4	19 41.1 N	+0.01	3 23.6	19 44.2 N	25	3 23.0	19 47.1 N	−0.50	3 23.3	19 50.2 N
3 23.6	19 01.0 N	−0.02	3 23.6	19 04.3 N	28	3 23.0	19 07.5 N	−0.54	3 23.1	19 10.8 N
AUGUST										
3 26.0	18 03.2 N	−0.03	3 25.6	18 06.8 N	1	3 25.1	18 10.3 N	−0.57	3 25.1	18 13.9 N
3 29.3	17 16.7 N	−0.02	3 28.8	17 20.5 N	4	3 28.2	17 24.2 N	−0.59	3 28.0	17 28.0 N
3 33.9	16 27.7 N	+0.02	3 33.4	16 31.6 N	7	3 32.8	16 35.6 N	−0.60	3 32.3	16 39.6 N
3 39.9	15 36.3 N	+0.06	3 39.3	15 40.4 N	10	3 38.6	15 44.6 N	−0.60	3 38.0	15 48.7 N
3 47.1	14 42.6 N	+0.09	3 46.5	14 47.0 N	13	3 45.7	14 51.3 N	−0.61	3 44.9	14 55.6 N
3 55.7	13 46.8 N	+0.11	3 55.0	13 51.4 N	16	3 54.0	13 55.8 N	−0.62	3 53.2	14 00.3 N
4 05.4	12 49.1 N	+0.11	4 04.7	12 53.9 N	19	4 03.4	12 58.4 N	−0.64	4 02.7	13 03.1 N
4 16.2	11 49.6 N	+0.09	4 15.4	11 54.5 N	22	4 14.0	11 59.1 N	−0.66	4 13.3	12 04.0 N
4 28.1	10 48.4 N	+0.06	4 27.2	10 53.3 N	25	4 25.7	10 58.1 N	−0.68	4 24.9	11 03.2 N
4 40.9	9 45.6 N	+0.03	4 39.8	9 50.6 N	28	4 38.3	9 55.6 N	−0.70	4 37.5	10 00.8 N
SEPTEMBER										
4 59.2	8 19.8 N	+0.02	4 57.9	8 24.9 N	1	4 56.5	8 30.0 N	−0.71	4 55.6	8 35.4 N
5 13.7	7 13.9 N	+0.04	5 12.3	7 19.2 N	4	5 10.9	7 24.4 N	−0.70	5 09.9	7 29.8 N
5 28.7	6 07.0 N	+0.08	5 27.3	6 12.4 N	7	5 26.0	6 17.8 N	−0.69	5 24.7	6 23.2 N
5 44.1	4 59.2 N	+0.11	5 42.8	5 04.7 N	10	5 41.4	5 10.1 N	−0.68	5 40.1	5 15.6 N
5 59.9	3 50.7 N	+0.13	5 58.6	3 56.2 N	13	5 57.1	4 01.7 N	−0.67	5 55.8	4 07.2 N
6 15.8	2 41.5 N	+0.13	6 14.7	2 47.1 N	16	6 13.0	2 52.6 N	−0.67	6 11.7	2 58.2 N
6 31.9	1 31.9 N	+0.11	6 30.7	1 37.5 N	19	6 29.0	1 43.0 N	−0.68	6 27.8	1 48.7 N
6 47.9	0 21.9 N	+0.08	6 46.6	0 27.6 N	22	6 44.9	0 33.1 N	−0.68	6 43.8	0 38.8 N
7 03.6	0 48.1 S	+0.04	7 02.3	0 42.5 S	25	7 00.6	0 37.0 S	+0.69	6 59.6	0 31.2 S
7 19.0	1 58.3 S	+0.01	7 17.6	1 52.7 S	28	7 16.1	1 47.1 S	+0.68	7 15.1	1 41.3 S
OCTOBER										
7 33.8	3 08.2 S	0.00	7 32.4	3 02.7 S	1	7 31.0	2 57.1 S	+0.67	7 30.0	2 51.4 S
7 48.0	4 17.9 S	+0.02	7 46.6	4 12.4 S	4	7 45.4	4 06.8 S	+0.65	7 44.3	4 01.2 S
8 01.3	5 27.2 S	+0.05	8 00.1	5 21.7 S	7	7 59.0	5 16.1 S	+0.62	7 57.9	5 10.5 S
8 13.7	6 35.8 S	+0.07	8 12.7	6 30.3 S	10	8 11.6	6 24.8 S	+0.60	8 10.5	6 19.3 S
8 25.1	7 43.6 S	+0.08	8 24.3	7 38.1 S	13	8 23.2	7 32.8 S	+0.57	8 22.2	7 27.3 S
8 35.4	8 50.5 S	+0.07	8 34.7	8 45.0 S	16	8 33.6	8 39.8 S	+0.56	8 32.8	8 34.4 S
8 44.4	9 56.2 S	+0.04	8 43.8	9 50.9 S	19	8 42.8	9 45.7 S	+0.54	8 42.2	9 40.3 S
8 52.0	11 00.6 S	−0.01	8 51.5	10 55.4 S	22	8 50.5	10 50.3 S	+0.53	8 50.1	10 45.0 S
8 58.1	12 03.5 S	−0.05	8 57.6	11 58.5 S	25	8 56.9	11 53.5 S	+0.52	8 56.6	11 48.3 S
9 02.6	13 04.7 S	−0.08	9 02.1	12 59.9 S	28	9 01.6	12 55.0 S	+0.50	9 01.5	12 50.0 S
NOVEMBER										
9 05.8	14 23.5 S	−0.09	9 05.5	14 18.9 S	1	9 05.4	14 14.2 S	+0.46	9 05.3	14 09.4 S
9 06.1	15 20.3 S	−0.07	9 06.0	15 15.8 S	4	9 06.1	15 11.2 S	+0.42	9 06.0	15 06.6 S
9 04.6	16 14.7 S	−0.04	9 04.7	16 10.4 S	7	9 05.0	16 06.0 S	+0.38	9 04.9	16 01.6 S
9 01.1	17 06.7 S	−0.03	9 01.6	17 02.5 S	10	9 01.9	16 58.4 S	+0.35	9 02.0	16 54.2 S
8 55.8	17 56.1 S	−0.03	8 56.5	17 52.1 S	13	8 56.9	17 48.2 S	+0.31	8 57.2	17 44.2 S
8 48.6	18 42.6 S	−0.05	8 49.5	18 38.8 S	16	8 49.9	18 35.2 S	+0.28	8 50.5	18 31.4 S
8 39.6	19 26.1 S	−0.09	8 40.6	19 22.7 S	19	8 41.1	19 19.2 S	+0.25	8 41.9	19 15.7 S
8 28.7	20 06.5 S	−0.14	8 29.8	20 03.3 S	22	8 30.5	20 00.1 S	+0.22	8 31.5	19 56.9 S
8 16.1	20 43.6 S	−0.18	8 17.2	20 40.7 S	25	8 18.1	20 37.8 S	+0.19	8 19.3	20 34.8 S
8 01.7	21 17.2 S	−0.20	8 02.8	21 14.6 S	28	8 04.1	21 11.9 S	+0.16	8 05.3	21 09.3 S
DECEMBER										
7 45.6	21 47.3 S	−0.18	7 46.9	21 44.9 S	1	7 48.4	21 42.5 S	+0.12	7 49.6	21 40.1 S
7 28.1	22 13.5 S	−0.15	7 29.6	22 11.5 S	4	7 31.3	22 09.4 S	+0.08	7 32.5	22 07.3 S
7 09.3	22 35.9 S	−0.11	7 11.0	22 34.2 S	7	7 12.8	22 32.4 S	+0.04	7 13.9	22 30.6 S
6 49.3	22 54.3 S	−0.09	6 51.3	22 52.9 S	10	6 53.0	22 51.4 S	0.00	6 54.3	22 50.0 S
6 28.5	23 08.6 S	−0.08	6 30.6	23 07.6 S	13	6 32.3	23 06.4 S	−0.04	6 33.7	23 05.3 S
6 06.9	23 18.8 S	−0.09	6 09.1	23 18.1 S	16	6 10.8	23 17.3 S	−0.08	6 12.4	23 16.5 S
5 44.9	23 24.8 S	−0.12	5 47.1	23 24.4 S	19	5 48.8	23 24.0 S	−0.12	5 50.5	23 23.6 S
5 22.6	23 26.6 S	−0.15	5 24.7	23 26.6 S	22	5 26.5	23 26.5 S	−0.16	5 28.3	23 26.4 S
5 00.2	23 24.1 S	−0.17	5 02.2	23 24.4 S	25	5 04.1	23 24.7 S	−0.19	5 05.9	23 25.0 S
4 38.0	23 17.5 S	−0.17	4 39.9	23 18.1 S	28	4 41.9	23 18.7 S	−0.23	4 43.6	23 19.3 S

STARS

SHA (1972.0)	Annual Corr.	Star	Dec. (1972.0)	Annual Corr.
° ′	′		° ′	′
315 42.0	−0.57	Acamar	40 25.0 S	−0.24
335 49.9	−0.56	Achernar	57 22.7 S	−0.30
173 44.6	−0.84	Acrux	62 56.6 S	+0.33
291 25.3	−0.86	Aldebaran	16 27.2 N	+0.12
153 23.4	−0.59	Alkaid	49 27.2 N	−0.30
218 26.8	−0.74	Alphard	8 32.2 S	+0.26
126 37.5	−0.64	Alphecca	26 48.5 N	−0.20
358 16.0	−0.78	Alpheratz	28 56.2 N	+0.33
62 38.7	−0.73	Altair	8 47.6 N	+0.16
113 04.7	−0.92	Antares	26 22.3 S	+0.13
146 24.2	−0.68	Arcturus	19 19.6 N	−0.31
108 34.7	−1.59	Atria	68 58.7 S	+0.11
271 35.2	−0.81	Betelgeuse	7 24.2 N	+0.01
264 10.0	−0.33	Canopus	52 40.8 S	+0.03
281 20.7	−1.11	Capella	45 58.3 N	+0.06
49 52.9	−0.51	Deneb	45 10.8 N	+0.22
183 05.5	−0.76	Denebola	14 43.7 N	−0.34
349 27.2	−0.75	Diphda	18 08.4 S	−0.33
194 29.8	−0.92	Dubhe	61 54.2 N	−0.32
34 17.8	−0.74	Enif	9 44.8 N	+0.28
15 58.4	−0.83	Fomalhaut	29 46.3 S	−0.32
328 36.2	−0.85	Hamal	23 19.8 N	+0.28
137 18.4	+0.04	Kochab	74 16.2 N	−0.25
148 44.6	−0.88	Menkent	36 14.0 S	+0.29
309 25.3	−1.07	Mirfak	49 45.8 N	+0.21
76 37.1	−0.93	Nunki	26 20.0 S	−0.08
54 08.4	−1.18	Peacock	56 49.6 S	−0.19
244 05.9	−0.92	Pollux	28 05.7 N	−0.15
245 32.4	−0.78	Procyon	5 17.9 N	−0.16
96 35.5	−0.70	Rasalhague	12 34.8 N	−0.04
208 16.8	−0.80	Regulus	12 06.3 N	−0.29
281 42.1	−0.72	Rigel	8 14.0 S	−0.07
140 34.8	−1.02	Rigil Kent.	60 43.2 S	+0.25
350 16.4	−0.86	Schedar	56 23.0 N	+0.33
259 01.3	−0.66	Sirius	16 40.6 S	+0.08
159 04.3	−0.79	Spica	11 01.0 S	+0.31
223 15.5	−0.55	Suhail	43 19.1 S	+0.24
81 00.2	−0.51	Vega	38 45.4 N	+0.06

MULTIPLICATION TABLE

No.	A	B	C	D	E	F	G
	°	° ′	′	′	° ′	° ′	° ′
1	15	0 15	0.2	1.8	0 59.1	15 02.5	0 15.0
2	30	0 30	0.5	3.7	1 58.3	30 04.9	0 30.1
3	45	0 45	0.8	5.5	2 57.4	45 07.4	0 45.1
4	60	1 00	1.0	7.4	3 56.6	60 09.9	1 00.2
5	75	1 15	1.2	9.2	4 55.7	75 12.3	1 15.2
6	90	1 30	1.5	11.0	5 54.8	90 14.8	1 30.2
7	105	1 45	1.8	12.9	6 54.0	105 17.2	1 45.3
8	120	2 00	2.0	14.7	7 53.1	120 19.7	2 00.3
9	135	2 15	2.2	16.6	8 52.3	135 22.2	2 15.4
10	150	2 30	2.5	18.4	9 51.4	150 24.6	2 30.4
11	165	2 45	2.8	20.2	10 50.5	165 27.1	2 45.5
12	180	3 00	3.0	22.1	11 49.7	180 29.6	3 00.5
13	195	3 15	3.2	23.9	12 48.8	195 32.0	3 15.5
14	210	3 30	3.5	25.8	13 48.0	210 34.5	3 30.6
15	225	3 45	3.8	27.6	14 47.1	225 37.0	3 45.6
16	240	4 00	4.0	29.4	15 46.2	240 39.4	4 00.7
17	255	4 15	4.2	31.3	16 45.4	255 41.9	4 15.7
18	270	4.30	4.5	33.1	17 44.5	270 44.4	4 30.7
19	285	4 45	4.8	35.0	18 43.7	285 46.8	4 45.8
20	300	5 00	5.0	36.8	19 42.8	300 49.3	5 00.8
21	315	5 15	5.2	38.6	20 41.9	315 51.7	5 15.9
22	330	5 30	5.5	40.5	21 41.1	330 54.2	5 30.9
23	345	5 45	5.8	42.3	22 40.2	345 56.7	5 45.9
24	360	6 00	6.0	44.2	23 39.4	360 59.1	6 01.0
25	—	6 15	6.2	46.0	24 38.5	—	6 16.0
26	—	6 30	6.5	47.8	25 37.6	—	6 31.1
27	—	6 45	6.8	49.7	26 36.8	—	6 46.1
28	—	7.00	7.0	51.5	27 35.9	—	7 01.1
29	—	7 15	7.2	53.4	28 35.1	—	7 16.2
30	—	7 30	7.5	55.2	29 34.2	—	7 31.2
31	—	7 45	7.8	57.0	30 33.3	—	7 46.3
32	—	8 00	8.0	58.9	—	—	8 01.3
33	—	8 15	8.2	60.7	—	—	8 16.4
34	—	8 30	8.5	62.6	—	—	8 31.4
35	—	8 45	8.8	64.4	—	—	8 46.4
36	—	9 00	9.0	66.2	—	—	9 01.5
37	—	9 15	9.2	68.1	—	—	9 16.5
38	—	9 30	9.5	69.9	—	—	9 31.6
39	—	9 45	9.8	71.8	—	—	9 46.6
40	—	10 00	10.0	73.6	—	—	10 01.6
41	—	10 15	10.2	75.4	—	—	10 16.7
42	—	10 30	10.5	77.3	—	—	10 31.7
43	—	10 45	10.8	79.1	—	—	10 46.8
44	—	11 00	11.0	81.0	—	—	11 01.8
45	—	11 15	11.2	82.8	—	—	11 16.8
46	—	11 30	11.5	84.6	—	—	11 31.9
47	—	11.45	11.8	86.5	—	—	11 46.9
48	—	12 00	12.0	88.3	—	—	12 02.0
49	—	12 15	12.2	90.2	—	—	12 17.0
50	—	12 30	12.5	92.0	—	—	12 32.1
51	—	12 45	12.8	93.8	—	—	12 47.1
52	—	13 00	13.0	95.7	—	—	13 02.1
53	—	13 15	13.2	97.5	—	—	13 17.2
54	—	13 30	13.5	99.4	—	—	13 32.2
55	—	13 45	13.8	—	—	—	13 47.3
56	—	14 00	14.0	—	—	—	14 02.3
57	—	14 15	14.2	—	—	—	14 17.3
58	—	14 30	14.5	—	—	—	14 32.4
59	—	14 45	14.8	—	—	—	14 47.4
60	—	15 00	15.0	—	—	—	15 02.5

ARIES (♈)

0	1	Month	2	3
° ′	° ′		° ′	° ′
98 46.2	99 31.0	Jan.	99 16.7	99 02.4
129 19.5	130 04.4	Feb.	129 50.1	129 35.7
157 54.5	157 40.3	Mar.	157 25.9	157 11.6
188 27.8	188 13.5	Apr.	187 59.2	187 44.9
218 02.0	217 47.7	May	217 33.4	217 19.0
248 35.3	248 21.0	June	248 06.7	247 52.3
278 09.5	277 55.2	July	277 40.9	277 26.5
308 42.8	308 28.5	Aug.	308 14.2	307 59.8
339 16.1	339 01.8	Sept.	338 47.5	338 33.1
8 50.3	8 36.0	Oct.	8 21.6	8 07.3
39 23.6	39 09.2	Nov.	38 54.9	38 40.6
68 57.7	68 43.4	Dec.	68 29.1	68 14.7

DECIMAL PARTS OF DAY AND YEAR

Decimal	0.0	0.1	0.2	0.3	0.4	0.5	0.6	0.7	0.8	0.9	1.0
Hour of Day	0000 to 0112	0112 to 0336	0336 to 0600	0600 to 0824	0824 to 1048	1048 to 1312	1312 to 1536	1536 to 1800	1800 to 2024	2024 to 2248	2248 to 2400
Day of Year	Jan. 1 to Jan. 18	Jan. 19 to Feb. 23	Feb. 24 to Apr. 1	Apr 2 to May 7	May 8 to June 13	June 14 to July 19	July 20 to Aug. 25	Aug. 26 to Sept 30	Oct. 1 to Nov. 6	Nov. 7 to Dec. 12	Dec. 13 to Dec. 31

APPENDIX I

IDENTIFICATION OF NAVIGATIONAL STARS

Introduction.—The following summary is not intended as a substitute for a star finder such as No. 2102-D, or of a knowledge of the heavens, but is given as a supplementary reference to assist in locating the 57 stars included in the main listing in the *Nautical Almanac,* plus Polaris. The observer is assumed to be at about the average latitude of the United States, unless another latitude is indicated. If a celestial body is said to be *east* of another, it is lower in the sky if both are rising and higher if both are setting. A body *north* of another is nearer the north celestial pole. Directions refer to great circles on the celestial sphere. Figures referred to are the star charts of chapter XXII, which should be of assistance in interpreting the descriptions given. It is assumed the reader is familiar with such well-known configurations as the Big Dipper and *Orion.* Constellation names are given in *italics.*

Acamar crosses the celestial meridian near the southern horizon during evening twilight in February, and during morning twilight in August. It is part of the constellation *Eridanus,* the river, which is not a striking configuration. It is the faintest star listed among the 57 in the almanac, but is the brightest in its immediate vicinity. The nearest bright star is Achernar, about 20° away in a generally southwesterly direction. Dec. 40°S, SHA 316°, mag. 3.1. Fig. 2205.

Achernar, at the southern end of the inconspicuous constellation *Eridanus,* the river, is one of the brightest stars of the Southern Hemisphere. It is not visible north of latitude 33°N. It crosses the celestial meridian during evening twilight in January, and during morning twilight in early August. Nearly a straight line is formed by Fomalhaut, about 40° WNW; Achernar; and Canopus, about the same distance in the opposite direction. However, since these stars are widely separated, the relationship is not striking. Achernar forms large triangles with Acamar and Ankaa, Ankaa and Al Na'ir, and with Al Na'ir and Peacock. Dec. 57°S, SHA 336°, mag. 0.6. Fig. 2205.

Acrux is the brightest and most southerly star in the famed Southern Cross. It is not visible north of latitude 27°N. It crosses the celestial meridian during evening twilight in early June and during morning twilight in January. It is about 15° WSW of first magnitude Hadar and Rigil Kentaurus. Dec. 63°S, SHA 174°, mag. 1.1. Fig. 2207.

Adhara. About 10°S and a little to the east of Sirius is a small, approximately equilateral triangle of three second magnitude stars. Adhara is the westernmost and brightest of the three. It crosses the celestial meridian to the south during evening twilight in March, and during morning twilight in October. Dec. 29°S, SHA 256°, mag. 1.6. Fig. 2206.

Aldebaran. If the line formed by the belt of *Orion,* the hunter, is extended about 20° to the northwestward, and curved somewhat toward the north, it leads to first magnitude Aldebaran in *Taurus,* the bull. This is a group of stars forming a V. A long, curving line starting at Sirius extends through Procyon, Pollux, Capella, and Aldebaran. Dec. 16°N, SHA 291°, mag. 1.1. Fig. 2206.

Alioth is the third star from the outer end of the handle of the Big Dipper, and the brightest star of the group. Dec. 56°N, SHA 167°, mag. 1.7. Fig. 2207.

Alkaid is the star at the outer end of the handle of the Big Dipper, farthest from the bowl. It is the second brightest star of the group. Dec. 49°N, SHA 153°, mag. 1.9. Fig. 2207.

Al Na'ir is the westernmost of two second magnitude stars of nearly equal brightness about midway between first magnitude Fomalhaut, approximately 20° to the northeast, and second magnitude Peacock, about the same distance in the opposite direction. A curved line extending eastward from the Southern Cross passes through Hadar and Rigil Kentaurus and, if extended with less curvature, leads first to Peacock and then to Al Na'ir. This star forms triangles with Fomalhaut and Ankaa, Ankaa and Achernar, and with Achernar and Peacock. It is not visible north of latitude 43°N. It crosses the celestial meridian during evening twilight early in December, and during morning twilight in June. Dec. 47°S, SHA 28°, mag. 2.2. Figs. 2205, 2208.

Alnilam is the middle star of the belt of *Orion,* the hunter. Dec. 1°S, SHA 276°, mag. 1.8. Fig. 2206.

Alphard, a second magnitude star, is the brightest in the inconspicuous constellation *Hydra,* the water monster. The nearest bright star is first magnitude Regulus, about 20° NNE. It is about midway between the horizon and zenith when it crosses the celestial meridian to the southward during evening twilight in late April, and during morning twilight in November. Dec. 9°S, SHA 218°, mag. 2.2. Fig. 2207.

Alphecca is the brightest star of *Corona Borealis,* the Northern Crown, about 20° ENE of first magnitude Arcturus. It forms a triangle with Arcturus and Alkaid. It crosses the celestial meridian near the zenith during evening twilight in July, and during morning twilight in February. Dec. 27°N, SHA 127°, mag. 2.3. Figs. 2207, 2208.

Alpheratz, a second magnitude star, is at the northeast corner of the great square of *Pegasus,* the winged horse, and is the brightest of the four stars forming the square. It crosses the celestial meridian near the zenith during evening twilight early in January, and during morning twilight in July. Dec. 29°N, SHA 358°, mag. 2.2. Fig. 2205.

Altair is at the southern vertex of a large, nearly right triangle which is a conspicuous feature of the evening sky in late summer and in autumn. The right angle is at Vega and the northern vertex is at Deneb. All three are first magnitude stars. Two fainter stars close to Altair, one on each side in a line through Vega, form a characteristic pattern making Altair one of the easiest stars to identify. It crosses the celestial meridian during evening twilight in October, and during morning twilight in May. Dec. 9°N, SHA 63°, mag. 0.9. Fig. 2208.

Ankaa, a second magnitude star, is the brightest star in inconspicuous *Phoenix.* It is surrounded by and forms a series of triangles with Diphda, Fomalhaut, Al Na'ir, Achernar, and Acamar. It crosses the celestial meridian low in the southern sky in January, and during morning twilight in July. Dec. 42°S, SHA 354°, mag. 2.4. Fig. 2205.

Antares is the brightest star in the conspicuous constellation *Scorpio,* the scorpion, which is low in the southern sky during evening twilight in late July, and morning twilight in late February. No other first magnitude star is within 40° of Antares and none toward the north is within 60°. It has a noticeable reddish hue and in appearance somewhat resembles Mars, which is occasionally near it in the sky. Dec. 26°S, SHA 113°, mag. 1.2. Fig. 2208.

Arcturus. The curved line along the stars forming the handle of the Big Dipper, if continued in a direction away from the bowl, passes through brilliant, first magnitude Arcturus. The distance from Alkaid, at the end of the Big Dipper, to Arcturus is a little more than the length of the dipper. Arcturus forms a large triangle with Alkaid and Alphecca. Dec. 19°N, SHA 146°, mag. 0.2. Figs. 2207, 2208.

Atria is the brightest of three stars forming a small triangle called *Triangulum Australe*, the southern triangle, not far from the south celestial pole. It is not seen north of latitude 21°N. A line through the east-west arm of the Southern Cross, if continued toward the east and curved somewhat toward the south, leads first to Hadar, then to Rigil Kentaurus, then, by curving more sharply, to the northernmost star of the triangle, and finally to Atria, only about 21° from the south celestial pole. Dec. 69°S, SHA 108°, mag 1.9. Fig. 2207.

Avior is the westernmost star of *Vela*, the sails, or false southern cross, about 30° WNW of the true Southern Cross, about 15° ESE of the brilliant Canopus, and nearly enclosed within a large triangle formed by Canopus, Suhail, and Miaplacidus. It is not visible north of latitude 31°N. Below this, it crosses the celestial meridian low in the southern sky during evening twilight in April, and morning twilight in early November. Dec. 59°S, SHA 234°, mag. 1.7. Figs. 2206, 2207.

Bellatrix is a second magnitude star north and a little west of the belt of *Orion*, the hunter. It is about equidistant from the belt and first magnitude, red Betelgeuse. Bellatrix is at the northwest corner of a box surrounding the belt of *Orion*. Dec. 6°N, SHA 279°, mag. 1.7. Fig. 2206.

Betelgeuse is a conspicuous, reddish star of variable brightness about 10° north and a little east of the belt of *Orion*, the hunter. A line through the center of the belt and perpendicular to it passes close to red Betelgeuse to the north and blue Rigel about the same distance south of the belt. Betelgeuse and Rigel are at opposite corners of a box surrounding the belt of *Orion*. Dec. 7°N., SHA 272°, mag. 0.1–1.2 (variable). Fig. 2206.

Canopus, second brightest star in the sky, is about 35° south of Sirius. A line extending eastward through the belt of *Orion* and curving toward the south passes first through Sirius, then through the small triangle of which Adhara is the brightest star, and finally to Canopus, which forms a large, almost equilateral triangle with Suhail and Miaplacidus. This triangle nearly encloses *Vela*, the sails or false southern cross, about 20° ESE of Canopus. Canopus is not visible north of latitude 37°N. It is on the edge of the Milky Way and while many relatively bright stars are nearby, none in the immediate vicinity of Canopus approaches it in brightness. Dec. 53°S, SHA 264°, mag. (−)0.9. Fig. 2206.

Capella is a brilliant star about 45° north of the belt of *Orion*, the hunter. A curved line starting at Sirius and extending through Procyon, Pollux, Capella, Aldebaran, the belt of *Orion*, and back to Sirius forms an inverted tear-drop figure with Capella at the top and the various parts being about equally spaced along the curve. Capella crosses the celestial meridian near the zenith during evening twilight in early March, and during morning twilight in late September. Dec. 46°N, SHA 281°, mag. 0.2. Fig. 2206.

Deneb is a bright star at the northeastern vertex of a large, nearly right triangle formed by Altair, Vega, and Deneb, the right angle being at Vega. These three stars are the brightest in the eastern sky during summer evenings. Deneb is not as bright as the other two, but is the brightest star in the constellation *Cygnus*, the swan. It crosses the celestial meridian near the zenith during evening twilight in November, and during morning twilight in late May. Dec. 45°N, SHA 50°, mag. 1.3. Fig. 2208.

Denebola, in *Leo,* the lion, is a second magnitude star at the opposite end of the constellation from Regulus. A straight line from Regulus, on the west, to Arcturus, on the east, passes close to Denebola, which is somewhat nearer Regulus. Denebola crosses the celestial meridian to the south during evening twilight in May, and during morning twilight in December. Dec. 15°N, SHA 183°, mag. 2.2. Fig. 2207.

Diphda. A line extending southward through the eastern side of the great square of *Pegasus,* the winged horse, and curving slightly toward the east, leads to second magnitude Diphda. The distance from the southern star of *Pegasus* to Diphda is about twice the length of one side of the square. Diphda is part of the inconspicuous constellation *Cetus,* the whale. The only nearby first magnitude star is Fomalhaut, about 25° in a generally southwest direction. Diphda, Fomalhaut, and Ankaa form a nearly equilateral triangle. Dec. 18°S, SHA 349°, mag. 2.2. Fig. 2205.

Dubhe forms the outer rim of the bowl of the Big Dipper. It and Merak (not one of the 57 navigational stars) are the two "pointers" used to locate Polaris, Dubhe being the one nearer the Pole Star. Dec. 62°N, SHA 194°, mag. 2.0. Fig. 2207.

Elnath is a second magnitude star between Capella, about 15° to the north, and Betelgeuse, about 20° to the south. It is a little north of a line connecting Aldebaran and Pollux. It is at the end of the northern fork of V-shaped *Taurus,* the bull. Aldebaran is the principal star at the closed end of the V. This constellation is approximately 25° NNW of *Orion,* the hunter. Dec. 29°N, SHA 279°, mag. 1.8. Fig. 2206.

Eltanin is the southernmost and brightest star in the inconspicuous constellation *Draco,* the dragon, south and somewhat east of the Little Dipper. A straight line extending northwestward through Altair and its two fainter companions passes first through brilliant Vega, and, about 15° beyond, to second magnitude Eltanin. Eltanin crosses the celestial meridian high in the sky toward the north during evening twilight in early September, and during morning twilight in late March. Dec. 51°N, SHA 91°, mag. 2.4. Fig. 2208.

Enif is a third magnitude star approximately midway between Altair, about 25° west, and Markab, about 20° ENE. From Markab, at the southwestern corner of the great square of *Pegasus,* the winged horse, a line extending in a generally west-southwesterly direction passes through two almost equally spaced fourth magnitude stars. From the second of these, a line about 5° long extending in a northwesterly direction leads to Enif. Enif crosses the celestial meridian to the south during evening twilight in November, and during morning twilight in June. Dec. 10°N, SHA 34°, mag. 2.5. Figs. 2205, 2208.

Fomalhaut is a first magnitude star well separated from stars of comparable brightness and from conspicuous configurations. A line through the western side of the great square of *Pegasus,* the winged horse, and extended about 45° toward the south passes close to Fomalhaut, which forms two large, nearly equilateral triangles with Diphda and Ankaa and with Ankaa and Al Na'ir. Dec. 30°S, SHA 16°, mag. 1.3. Fig. 2205.

Gacrux is the northernmost star of the Southern Cross. It is bright for a second magnitude star, but its brilliance is overshadowed by the brighter *β Crucis* (not listed among the 57 navigational stars) and Acrux, the two brightest stars of the Southern Cross, and by Hadar and Rigil Kentaurus, about 15° ESE. Gacrux crosses the celestial meridian during evening twilight in early June, and during morning twilight in late December, but is not visible north of latitude 33°N. Dec. 57°S, SHA 173°, mag. 1.6. Fig. 2207.

Gienah is a third magnitude star, the brightest in the constellation *Corvus,* the crow. A long, sweeping arc starting with the handle of the Big Dipper and extending successively through Arcturus and Spica leads to this relatively small, four-sided figure made up of third magnitude stars. Gienah is at the northwest corner. It crosses the celestial meridian during evening twilight in late May, and during morning twilight in December. Dec. 17°S, SHA 176°, mag. 2.8. Fig. 2207.

Hadar is a first magnitude star about 10° east of the Southern Cross, and about 5° west of Rigil Kentaurus, the brightest of several bright stars in this part of the sky. Dec. 60°S, SHA 149°, mag. 0.9. Fig. 2207.

Hamal is the brightest star of the inconspicuous constellation *Aries,* the ram. A line through the center of the great square of *Pegasus,* the winged horse, extended about 25° east, and curved slightly toward the north, leads to Hamal. It is over the meridian to the south during evening twilight in January, and during morning twilight in August. Dec. 23°N, SHA 329°, mag. 2.2. Fig. 2205.

Kaus Australis is near the southern end of a group of second and third magnitude stars forming the constellation *Sagittarius,* the archer, about 25° ESE of Antares, in *Scorpio,* the scorpion. It is about 10° SW of Nunki, also in *Sagittarius,* and about the same distance ENE of Shaula, in *Scorpio.* With Antares, Sabik, and Nunki, it forms a large, poorly defined box. It is over the meridian to the south during evening twilight in September and during morning twilight in April. Dec. 34°S, SHA 84°, mag. 2.0. Fig. 2208.

Kochab forms the outer rim of the bowl of the Little Dipper, at the opposite end from Polaris, about 15° north. It is directly above the pole during evening twilight in early July and during morning twilight in January; and directly below the pole, low in the northern sky, during evening twilight of early February and morning twilight of late August. Dec. 74°N, SHA 137°, mag. 2.2. Fig. 2208.

Markab is the star at the southwest corner of the great square of *Pegasus,* the winged horse, at the opposite corner from Alpheratz. It is over the celestial meridian to the south during evening twilight in December, and during morning twilight late in June. Dec. 15°N, SHA 14°, mag. 2.6. Fig. 2205.

Menkar is a third magnitude star at the eastern end of the inconspicuous constellation *Cetus,* the whale. No bright stars are nearby. A straight line from Aldebaran extending about 25° in the direction indicated by the point of the V of *Taurus,* the bull, leads to Menkar. A long, straight line from Fomalhaut east-northeastward through Diphda, and extended about 40°, leads to Menkar. It crosses the celestial meridian during evening twilight in February, and during morning twilight in August. Dec. 4°N, SHA 315°, mag. 2.8. Figs. 2205, 2206.

Menkent is a second magnitude star about 25° north of Hadar and about 30° northeast of the Southern Cross. A line from Gienah across the opposite corner of the small, four-sided *Corvus,* the crow, and then curving a little toward the east, leads to Menkent. A number of third magnitude stars are nearby, but they do not form a conspicuous configuration. With Antares and Rigil Kentaurus, Menkent forms a large triangle. It crosses the celestial meridian low in the southern sky during evening twilight in late June and during morning twilight in early January. Dec. 36°S, SHA 149°, mag. 2.3. Figs. 2207, 2208.

Miaplacidus is a second magnitude star about 10° south of the false southern cross. It is the nearest of the 57 navigational stars to the south celestial pole, about 20° away, and is not visible north of latitude 20°N. With Suhail and brilliant

Canopus it forms a large, nearly equilateral triangle almost enclosing the false southern cross. South of latitude 20°S, it does not set, but circles the south celestial pole in a clockwise direction, reaching its maximum altitude above the pole during evening twilight in early May and during morning twilight in November. Dec. 70°S, SHA 222°, mag. 1.8. Figs. 2206, 2207.

Mirfak is a second magnitude star at the northeastern end of a gently curving line extending in a northeasterly direction from Alpheratz at the northeastern corner of the great square of *Pegasus,* the winged horse, through two other second magnitude stars, Mirach and Almach, not included among the 57 navigational stars. Mirfak is about 25° east and a little south of *Cassiopeia,* and about 20° WNW of Capella. A line from Kochab through Polaris, and curved slightly toward the east, leads to Mirfak. Dec. 50°N, SHA 309°, mag. 1.9. Figs. 2205, 2206.

Nunki is the more northerly of the two brightest stars of a group of second and third magnitude stars forming the constellation *Sagittarius,* the archer, about 30° E of Antares. It is about 10° NE of Kaus Australis, also in *Sagittarius.* With Sabik, Antares, and Kaus Australis, it forms a large, poorly defined box. It is over the meridian to the south during evening twilight in early October and during morning twilight in April. Dec. 26°S, SHA 77°, mag. 2.1. Fig. 2208.

Peacock, the brightest star in the southern constellation of the same name, is not a part of a conspicuous configuration of stars. A curved line extending eastward from the Southern Cross passes through Hadar and Rigil Kentaurus and, if extended with less curvature, leads to Peacock, about 30° southeast of *Scorpio,* the scorpion, and about 20° southwest of Al Na'ir. With Al Na'ir and Achernar it forms a large, poorly defined triangle. It crosses the celestial meridian during evening twilight in early November, and during morning twilight in late May, but is not visible north of latitude 33°N. Dec. 57°S, SHA 54°, mag. 2.1. Figs. 2205, 2208.

Polaris is not listed among the 57 navigational stars, but is treated separately because it is less than 1° from the north celestial pole. It is about midway between the Big Dipper and *Cassiopeia.* A line through Dubhe and Merak (not one of the 57 navigational stars), the pointers forming the outer side of the bowl of the Big Dipper, if extended northward for about 30°, leads almost directly to Polaris. A line extending north from Alpheratz at the northwest corner of the great square of *Pegasus,* the winged horse, passes through Caph (not one of the 57 navigational stars) in *Cassiopeia* and then Polaris at about equal intervals. Dec. 89°N, SHA 328°, mag. 2.1. Figs. 2205–2208.

Pollux is the brighter of the "twins of *Gemini,*" two relatively bright stars about 45° NE of *Orion,* the hunter, and about 45° ENE of Aldebaran. A curved line starting at Sirius extends through Procyon, Pollux, and Capella, all first magnitude stars. Dec. 28°N, SHA 244°, mag. 1.2. Fig. 2206.

Procyon is a bright star about 30° east of *Orion,* the hunter. A curved line starting at Sirius extends through Procyon, Pollux, and Capella, all first magnitude stars. Dec. 5°N, SHA 245°, mag. 0.5. Fig. 2206.

Rasalhague forms a large, nearly equilateral triangle with Altair and Vega, Rasalhague being at the western vertex. Both of the other stars are considerably brighter than Rasalhague. It crosses the celestial meridian to the south during evening twilight in early September, and during morning twilight in late March. Dec. 13°N, SHA 97°, mag. 2.1. Fig. 2208.

Regulus is at the opposite end of *Leo,* the lion, from Denebola, and is the brightest star of the constellation. A line through Dubhe and Merak (not one of the 57 navigational stars), the pointers by which Polaris is usually identified, extended about 45° *southward,* and curved slightly toward the west, leads to Regulus, which forms the southern end of the handle of the Sickle, part of *Leo.* Dec. 12°N, SHA 208°, mag. 1.3. Fig. 2207.

Rigel is a brilliant bluish star about 10°S and a little to the west of the belt of *Orion,* the hunter. A line through the center of the belt and perpendicular to it passes close to blue Rigel to the south and red Betelgeuse about the same distance north of the belt. Rigel and Betelgeuse are at opposite corners of a box surrounding the belt of *Orion.* Dec. 8°S, SHA 282°, mag. 0.3. Fig. 2206.

Rigil Kentaurus is the brighter and more easterly of two first magnitude stars about 15° east of the Southern Cross. It is over the meridian during evening twilight in early July, and during morning twilight in late January, but is not visible north of latitude 29°N. Dec. 61°S, SHA 141°, mag. 0.1. Figs. 2207, 2208.

Sabik is part of the inconspicuous constellation *Ophiuchus,* the serpent holder, about 20° north of *Scorpio,* the scorpion. With Antares, Kaus Australis, and Nunki, it forms a large, poorly defined box in the southern sky on summer evenings. Sabik crosses the celestial meridian during evening twilight in August, and during morning twilight in March. Dec. 16°S, SHA 103°, mag. 2.6. Fig. 2208.

Schedar is the southernmost star of the W (or M) of *Cassiopeia,* on the opposite side of Polaris from the Big Dipper. It is the second star from the leading edge of this configuration as it circles the north celestial pole. Dec. 56°N, SHA 350°, mag. 2.5. Figs. 2205, 2206, 2208.

Shaula is a second magnitude star marking the end of the tail of *Scorpio,* the scorpion, at the opposite end from Antares. This constellation is low in the southern sky on summer evenings. Shaula is about 15° southeast of Antares and about 10° WSW of Kaus Australis. It crosses the celestial meridian during evening twilight in early September, and during morning twilight in March. Dec. 37°S, SHA 97°, mag. 1.7. Fig. 2208.

Sirius, the brightest star in the heavens, is in the constellation *Canis Major,* the "large dog" of *Orion,* the hunter. The line formed by the belt of *Orion,* if extended about 20° to the eastward and curved toward the south, leads to Sirius. Dec. 17°S, SHA 259°, mag. (−)1.6. Fig. 2206.

Spica is the brightest star of *Virgo,* the virgin, an inconspicuous constellation on the celestial equator to the south during evening twilight in early summer. The curved line along the stars forming the handle of the Big Dipper, if continued in a direction away from the pointers, passes through Arcturus and then Spica. The distance between Alkaid, at the end of the Big Dipper, and Arcturus is about the same as that between Arcturus and Spica, and is a little more than the length of the Big Dipper. Spica crosses the celestial meridian during evening twilight in June, and during morning twilight late in December. Dec. 11°S, SHA 159°, mag. 1.2. Fig. 2207.

Suhail is one of a number of second magnitude stars extending along the Milky Way between Sirius and the Southern Cross. It is about 10° north of the false southern cross, which is nearly enclosed by a large, nearly equilateral triangle formed by Suhail, Canopus, and Miaplacidus. Canopus and Suhail are on opposite edges of the Milky Way, with a number of second magnitude stars between them. A straight line extending eastward through the east-west arm of the Southern Cross leads to Suhail, about 35° away. In the southern United States, Suhail crosses the celestial meridian near the southern horizon during evening twilight in April, and during morning twilight in November. Dec. 43°S, SHA 223°, mag. 2.2. Figs. 2206, 2207.

Vega is the brightest star north of the celestial equator, and the third brightest in the entire sky. It is at the western vertex and the nearly right angle of a large triangle which is a conspicuous feature of the evening sky in late summer and in autumn. The other two stars of the triangle are Altair and Deneb, both of the first magnitude. Vega passes through the zenith approximately at latitude 38°45′N during evening twilight in September and during morning twilight in April. Dec. 39°N, SHA 81°, mag. 0.1. Fig. 2208.

Zubenelgenubi, a third magnitude star, is the southern (or western) basket of *Libra,* the balance. The boxlike *Libra* is about 25° WNW of Antares, in *Scorpio,* the scorpion. A long line extending eastward from Alphard, between Gienah and Spica, leads to Zubenelgenubi. Dec. 16°S, SHA 138°, mag. 2.9. Figs. 2207, 2208.

APPENDIX J

NAVIGATIONAL STARS AND THE PLANETS

Name	Pronunciation	Bayer name	Origin of name	Meaning of name	Distance*
Acamar	ā'kȧ·mär	θ Eridani	Arabic	another form of Achernar	120
Achernar	ā'kēr·när	α Eridani	Arabic	end of the river (Eridanus)	72
Acrux	ā'krŭks	α Crucis	Modern	coined from Bayer name	220
Adhara	ȧ·dā'rȧ	ε Canis Majoris	Arabic	the virgin(s)	350
Aldebaran	ăl dĕb'ȧ·rȧn	α Tauri	Arabic	follower (of the Pleiades)	64
Alioth	ăl'ĭ·ŏth	ε Ursa Majoris	Arabic	another form of Capella	49
Alkaid	ăl·kād'	η Ursa Majoris	Arabic	leader of the daughters of the bier	190
Al Na'ir	ăl·nār'	α Gruis	Arabic	bright one (of the fish's tail)	90
Alnilam	ăl'nĭ·lăm	ε Orionis	Arabic	string of pearls	410
Alphard	ăl'färd	α Hydrae	Arabic	solitary star of the serpent	200
Alphecca	ăl·fĕk'ȧ	α Corona Borealis	Arabic	feeble one (in the crown)	76
Alpheratz	ăl·fē'răts	α Andromeda	Arabic	the horse's navel	120
Altair	ăl·tār'	α Aquilae	Arabic	flying eagle or vulture	16
Ankaa	än'kä	α Phoenicis	Arabic	coined name	93
Antares	ăn·tā'rēz	α Scorpii	Greek	rival of Mars (in color)	250
Arcturus	ärk·tū'rŭs	α Bootis	Greek	the bear's guard	37
Atria	ăt'rĭ·ȧ	α Trianguli Australis	Modern	coined from Bayer name	130
Avior	ā'vĭ·ôr	ε Carinae	Modern	coined name	350
Bellatrix	bĕ·lā'trĭks.	γ Orionis	Latin	female warrior	250
Betelgeuse	bĕt'ĕl·jūz	α Orionis	Arabic	the arm pit (of Orion)	300
Canopus	kȧ·nō'pŭs	α Carinae	Greek	city of ancient Egypt	230
Capella	kȧ·pĕl'ȧ	α Aurigae	Latin	little she-goat	46
Deneb	dĕn'ĕb	α Cygni	Arabic	tail of the hen	600
Denebola	dĕ·nĕb'ō·lȧ	β Leonis	Arabic	tail of the lion	42
Diphda	dĭf'dȧ	β Ceti	Arabic	the second frog (Fomalhaut was once the first)	57
Dubhe	dŭb'ē	α Ursa Majoris	Arabic	the bear's back	100
Elnath	ĕl'năth	β Tauri	Arabic	one butting with horns	130
Eltanin	ĕl·tā'nĭn	γ Draconis	Arabic	head of the dragon	150
Enif	ĕn'ĭf	ε Pegasi	Arabic	nose of the horse	250
Fomalhaut	fō'mȧl·ôt	α Piscis Austrini	Arabic	mouth of the southern fish	23
Gacrux	gā'krŭks	γ Crucis	Modern	coined from Bayer name	72
Gienah	jē'nȧ	γ Corvi	Arabic	right wing of the raven	136
Hadar	hā'där	β Centauri	Modern	leg of the centaur	200
Hamal	hăm'ȧl	α Arietis	Arabic	full-grown lamb	76
Kaus Australis	kôs ôs·trā'lĭs	ε Sagittarii	Ar., L.	southern part of the bow	163
Kochab	kō'kăb	β Ursa Minoris	Arabic	shortened form of "north star" (named when it was that, c. 1500 BC-A D 300)	100
Markab	mär'kăb	α Pegasi	Arabic	saddle (of Pegasus)	100
Menkar	mĕn'kär	α Ceti	Arabic	nose (of the whale)	1,100
Menkent	mĕn'kĕnt	θ Centauri	Modern	shoulder of the centaur	55
Miaplacidus	mī'ȧ·plăs'ĭ·dŭs	β Carinae	Ar., L.	quiet or still waters	86
Mirfak	mĭr'făk	α Persei	Arabic	elbow of the Pleiades	130
Nunki	nŭn'kē	σ Sagittarii	Bab.	constellation of the holy city (Eridu)	150
Peacock	pē'kŏk	α Pavonis	Modern	coined from English name of constellation	250
Polaris	pō·lā'rĭs	α Ursa Minoris	Latin	the pole (star)	450
Pollux	pŏl'ŭks	β Geminorum	Latin	Zeus' other twin son (Castor, α Geminorum, is first twin)	33
Procyon	prō'sĭ·ŏn	α Canis Minoris	Greek	before the dog (rising before the dog star, Sirius)	11
Rasalhague	ràs'ăl·hā'gwē	α Ophiuchi	Arabic	head of the serpent charmer	67
Regulus	rĕg'ū·lŭs	α Leonis	Latin	the prince	67
Rigel	rī'jĕl	β Orionis	Arabic	foot (left foot of Orion)	500
Rigil Kentaurus	rī'jĭl kĕn·tô'rŭs	α Centauri	Arabic	foot of the centaur	4.3
Sabik	sā'bĭk	η Ophiuchi	Arabic	second winner or conqueror	69
Schedar	shĕd'ȧr	α Cassiopeiae	Arabic	the breast (of Cassiopeia)	360
Shaula	shô'lȧ	λ Scorpii	Arabic	cocked-up part of the scorpion's tail	200
Sirius	sĭr'ĭ·ŭs	α Canis Majoris	Greek	the scorching one (popularly, the dog star)	8.6
Spica	spī'kȧ	α Virginis	Latin	the ear of corn	155
Suhail	sōō·hāl'	λ Velorum	Arabic	shortened form of Al Suhail, one Arabic name for Canopus	200
Vega	vē'gȧ	α Lyrae	Arabic	the falling eagle or vulture	27
Zubenelgenubi	zōō·bĕn'ĕl·jĕ·nū'bē	α Librae	Arabic	southern claw (of the scorpion)	66

PLANETS

Name	Pronunciation	Origin of name	Meaning of name
Mercury	mûr'kū·rĭ	Latin	god of commerce and gain
Venus	vē'nŭs	Latin	goddess of love
Earth	ûrth	Mid. Eng.	—
Mars	märz	Latin	god of war
Jupiter	jōō'pĭ·tēr	Latin	god of the heavens, identified with the Greek Zeus, chief of the Olympian gods
Saturn	săt'ērn	Latin	god of seed-sowing
Uranus	ū'rȧ·nŭs	Greek	the personification of heaven
Neptune	nĕp'tūn	Latin	god of the sea
Pluto	plōō'tō	Greek	god of the lower world (Hades)

Guide to pronunciations:
fāte, ădd, finȧl, làst, ȧbound, ärm; bē, ĕnd, camĕl, readẽr; īce, bĭt, anĭmal; ōver, pȯetic, hŏt, lôrd, mōōn; tūbe, ůnite, tŭb, circŭs, ûrn

*Distances in light-years. One light-year equals approximately 63,300 AU, or 5,880,000,000,000 miles. Authorities differ on distances of the stars; the values given are representative.

APPENDIX K

CONSTELLATIONS

Name	Pronunciation	Genitive	Pronunciation	Meaning	Navigational stars or approximate position
Andromeda*	ăn·drŏm'·ē·dȧ	Andromedae	ăn·drŏm'ē·dē	Andromeda [the chained woman]†	Alpheratz
Antlia	ănt'lĭ·ȧ	Antliae	ănt'lĭ·ē	(air) pump††	d 35°S, SHA 210°
Apus	ā'pŭs	Apodis	ăp'ŏ·dĭs	bird of paradise	d 75°S, SHA 120°
Aquarius (=)*	ȧ·kwâr'ĭ·ŭs	Aquarii	ȧ·kwâr'ĭ·ī	water carrier	d 5°S, SHA 25°
Aquila*	ăk'wĭ·lȧ	Aquilae	ăk'wĭ·lē	eagle	Altair
Ara*	ā'rȧ	Arae	ā'rē	altar	d 55°S, SHA 100°
Aries (♈)*	ā'rĭ·ēz	Arietis	ȧ·rī'ē·tĭs	ram	Hamal
Auriga*	ô·rī'gȧ	Aurigae	ô·rī'jē	charioteer	Capella
Bootes*	bō·ō'tēz	Bootis	bō̇·ō'tĭs	herdsman	Arcturus
Caelum	sē'lŭm	Caeli	sē'lī	graving tool	d 40°S, SHA 290°
Camelopardalis	kȧ·mĕl'ō̇·pär'dȧ·lĭs	Camelopardalis	kȧ·mĕl'ō̇·pär'dȧ·lĭs	giraffe	d 70°N, SHA 275°
Cancer (♋)*	kăn'sēr	Cancri	kăng'krī	crab	d 20°N, SHA 230°
Canes Venatici	kā'nēz vē̇·năt'ĭ·sī	Canum Venaticorum	kā'nŭm vē̇·năt'ĭ·kō'rŭm	hunting dogs	d 40°N, SHA 165°
Canis Major*	kā'nĭs mā'jēr	Canis Majoris	kā'nĭs mȧ·jō'rĭs	larger dog	Adhara, Sirius
Canis Minor*	kā'nĭs mī'nēr	Canis Minoris	kā'nĭs mĭ·nō'rĭs	smaller dog	Procyon
Capricornus (♑)	kăp'rĭ·kôr'nŭs	Capricorni	kăp'rĭ·kôr'nī	horned goat	d 20°S, SHA 45°
Carina**	kȧ·rī'nȧ	Carinae	kȧ·rī'nē	keel	Avior, Canopus, Miaplacidus
Cassiopeia*	kăs'ĭ·ō̇·pē'yȧ	Cassiopeiae	kăs'ĭ·ō̇·pē'yē	Cassiopeia [the lady in the chair]†	Schedar
Centaurus*	sĕn·tô'rŭs	Centauri	sĕn·tô'rī	centaur	Hadar, Menkent, Rigil Kentaurus
Cepheus*	sē'fūs	Cephei	sē'fē̇·ī	Cepheus [the shepherd]†	d 75°N, SHA 15°
Cetus*	sē'tŭs	Ceti	sē'tī	whale	Diphda, Menkar
Chamaeleon	kȧ·mē'lē̇·ŭn	Chamaeleontis	kȧ·mē'lē̇·ŏn'tĭs	chameleon	d 80°S, SHA 200°
Circinus	sûr'sĭ·nŭs	Circini	sûr'sĭ·nī	pair of compasses	d 65°S, SHA 140°
Columba	kō̇·lŭm'bȧ	Columbae	kō̇·lŭm'bē	dove	d 35°S, SHA 275°
Coma Berenices	kō'mȧ bĕr'ē̇·nī'sēz	Comae Berenices	kō'mē bĕr'ē̇·nī'sēz	Berenice's hair	d 25°N, SHA 170°
Corona Australis*	kō̇·rō'nȧ ôs·trā'lĭs	Coronae Australis	kō̇·rō'nē ôs·trā'lĭs	southern crown	d 40°S, SHA 80°
Corona Borealis*	kō̇·rō'nȧ bō'rē̇·ā'lĭs	Coronae Borealis	kō̇·rō'nē bō'rē̇·ā'lĭs	northern crown	Alphecca
Corvus*	kôr'vŭs	Corvi	kôr'vī	crow	Gienah
Crater*	krā'tēr	Crateris	krȧ̇·tē'rĭs	cup	d 15°S, SHA 190°
Crux	krŭks	Crucis	kroo͞'sĭs	cross	Acrux, Gacrux
Cygnus*	sĭg'nŭs	Cygni	sĭg'nī	swan	Deneb
Delphinus*	dĕl·fī'nŭs	Delphini	dĕl·fī'nī	dolphin	d 15°N, SHA 50°
Dorado	dō̇·rä'dō	Doradus	dō̇·rä'dŭs	dorado [a fish]†	d 60°S, SHA 285°
Draco*	drā'kō	Draconis	drȧ̇·kō'nĭs	dragon	Eltanin
Equuleus*	ē̇·kwoo͞'lē̇·ŭs	Equulei	ē̇·kwoo͞'lē̇·ī	colt	d 10°N, SHA 40°
Eridanus*	ē̇·rĭd'ȧ·nŭs	Eridani	ē̇·rĭd'ȧ·nī	Eridanus [a river]†	Acamar, Achernar
Fornax	fôr'năks	Fornacis	fôr·nā'sĭs	furnace	d 30°S, SHA 320°
Gemini (II)*	jĕm'ĭ·nī	Geminorum	jĕm'ĭ·nō'rŭm	twins	Pollux
Grus	grŭs	Gruis	groo͞'ĭs	crane [a bird]†	Al Na'ir
Hercules*	hûr'kû·lēz	Herculis	hûr'kû·lĭs	Hercules [mythological hero]†	d 30°N, SHA 100°
Horologium	hŏr'ō̇·lō'jĭ·ŭm	Horologii	hŏr'ō̇·lō'jĭ·ī	clock	d 50°S, SHA 310°

Zodiacal constellations are given in bold type, with their symbols.
*One of the original constellations of Ptolemy.
**Part of the single constellation Argo Navis of Ptolemy.
†Parts within brackets are amplifications of the meanings of constellation names.
††Parts within parentheses are the meanings of words deleted from former, more complete constellation names.

Guide to pronunciations:

fāte, câre, hăt, finăl, ȧbound, sofȧ, ärm; bē, crēate, ĕnd, readēr; īce, bĭt; ōver, pōetic, hŏt, cŏnnect, lôrd, moo͞n; tūbe, ûnite, tŭb, circŭs, ûrn.

APPENDIX K
CONSTELLATIONS

Name	Pronunciation	Genitive	Pronunciation	Meaning	Navigational stars or approximate position
Hydra*	hī'drȧ	Hydrae	hī'drē	water monster	Alphard
Hydrus	hī'drŭs	Hydri	hī'drī	water snake	d 70°S, SHA 320°
Indus	ĭn'dŭs	Indi	ĭn'dī	Indian	d 60°S, SHA 35°
Lacerta	lȧ·sûr'tȧ	Lacertae	lȧ·sûr'tē	lizard	d 45°N, SHA 25°
Leo (♌)*	lē'ō	Leonis	lē̇·ō'nĭs	lion	Denebola, Regulus
Leo Minor	lē'ō mī'nẽr	Leonis Minoris	lē̇·ō'nĭs mĭ·nō'rĭs	smaller lion	d 35°N, SHA 205°
Lepus*	lē'pŭs	Leporis	lĕp'ō·rĭs	hare	d 20°S, SHA 275°
Libra (♎)*	lī'brȧ	Librae	lī'brē	balance [scales]†	Zubenelgenubi
Lupus*	lū'pŭs	Lupi	lū'pī	wolf	d 45°S, SHA 130°
Lynx	lĭngks	Lyncis	lĭn'sĭs	lynx	d 50°N, SHA 240°
Lyra*	lī'rȧ	Lyrae	lī'rē	lyre	Vega
Mensa	mĕn'sȧ	Mensae	mĕn'sē	table (mountain)††	d 75°S, SHA 275°
Microscopium	mī'krō̇·skō'pĭ·ŭm	Microscopii	mī'krō̇·skō'pĭ·ī	microscope	d 35°S, SHA 45°
Monoceros	mō̇·nŏs'ẽr·ŏs	Monocerotis	mō̇·nŏs'ẽr·ō'tĭs	unicorn	d 0°, SHA 255°
Musca	mŭs'kȧ	Muscae	mŭs'sē	fly	d 70°S, SHA 175°
Norma	nôr'mȧ	Normae	nôr'mē	square (and rule)††	d 50°S, SHA 120°
Octans	ŏk'tănz	Octantis	ŏk·tăn'tĭs	octant	d 85°S, SHA 40°
Ophiuchus*	ŏf'ĭ·ū'kŭs	Ophiuchi	ŏf'ĭ·ū'kī	serpent holder	Rasalhague, Sabik
Orion*	ō̇·rī'ŏn	Orionis	ō'rĭ·ō'nĭs	Orion [the hunter]†	Alnilam, Bellatrix, Betelgeuse, Rigel
Pavo	pā'vō	Pavonis	pȧ·vō'nĭs	peacock	Peacock
Pegasus*	pĕg'ȧ·sŭs	Pegasi	pĕg'ȧ·sī	Pegasus [winged horse]†	Enif, Markab
Perseus*	pûr'sūs	Persei	pûr'sē̇·ī	Perseus [mythological character]†	Mirfak
Phoenix	fē'nĭks	Phoenicis	fē̇·nī'sĭs	phoenix [the immortal bird]†	Ankaa
Pictor	pĭk'tẽr	Pictoris	pĭk·tō'rĭs	painter (easel of)††	d 55°S, SHA 275°
Pisces (♓)*	pĭs'ēz	Piscium	pĭsh'ĭ·ŭm	fishes	d 15°N, SHA 355°
Piscis Austrinus*	pĭs'ĭs ôs·trī'nŭs	Piscis Austrini	pĭs'ĭs ôs·trī'nī	southern fish	Fomalhaut
Puppis**	pŭp'ĭs	Puppis	pŭp'ĭs	stern [of ship]†	d 30°S, SHA 245°
Pyxis**	pĭk'sĭs	Pyxidis	pĭk'sĭ·dĭs	mariner's compass	d 25°S, SHA 230°
Reticulum	rē̇·tĭk'ū̇·lŭm	Reticuli	rē̇·tĭk'ū̇·lī	net	d 60°S, SHA 300°
Sagitta*	sȧ·jĭt'ȧ	Sagittae	sȧ·jĭt'ē	arrow	d 20°N, SHA 65°
Sagittarius (♐)*	săj'ĭ·tā'rĭ·ŭs	Sagittarii	săj'ĭ·tā'rĭ·ī	archer	Kaus Australis, Nunki
Scorpius (♏)*	skôr'pĭ·ŭs	Scorpii	skôr'pĭ·ī	scorpion	Antares, Shaula
Sculptor	skŭlp'tẽr	Sculptoris	skŭlp·tō'rĭs	sculptor (workshop of)††	d 30°S, SHA 355°
Scutum	skū'tŭm	Scuti	skū'tī	shield	d 10°S, SHA 80°
Serpens*	sûr'pĕnz	Serpentis	sẽr·pĕn'tĭs	serpent	d 10°N, SHA 125°
Sextans	sĕks'tănz	Sextantis	sĕks·tăn'tĭs	sextant	d 0°, SHA 205°
Taurus (♉)*	tô'rŭs	Tauri	tô'rī	bull	Aldebaran, Elnath
Telescopium	tĕl'ē̇·skō'pĭ·ŭm	Telescopii	tĕl'ē̇·skō'pĭ·ī	telescope	d 50°S, SHA 75°
Triangulum*	trī·ăng'gū̇·lŭm	Trianguli	trī·ăng'gū̇·lī	triangle	d 30°N, SHA 330°
Triangulum Australe	trī·ăng'gū̇·lŭm ôs·trā'lē	Trianguli Australis	trī·ăng'gū̇·lī ôs·trā'lĭs	southern triangle	Atria
Tucana	tū̇·kā'nȧ	Tucanae	tū̇·kā'nē	toucan [a bird]†	d 65°S, SHA 5°
Ursa Major*	ûr'sȧ mā'jẽr	Ursae Majoris	ûr'sē mȧ·jō'rĭs	larger bear	Alioth, Alkaid, Dubhe
Ursa Minor*	ûr'sȧ mī'nẽr	Ursae Minoris	ûr'sē mĭ·nō'rĭs	smaller bear	Kochab, Polaris
Vela**	vē'lȧ	Velorum	vē̇·lō'rŭm	sails	Suhail
Virgo (♍)*	vûr'gō	Virginis	vûr'jĭ·nĭs	virgin	Spica
Volans	vō'lănz	Volantis	vō̇·lăn'tĭs	flying (fish)††	d 70°S, SHA 240°
Vulpecula	vŭl·pĕk'ū̇·lȧ	Vulpeculae	vŭl·pĕk'ū̇·lē	little fox	d 25°N, SHA 60°

Zodiacal constellations are given in bold type, with their symbols.
*One of the original constellations of Ptolemy.
**Part of the single constellation Argo Navis of Ptolemy.
†Parts within brackets are amplifications of the meanings of constellation names.
††Parts within parentheses are the meanings of words deleted from former, more complete constellation names.

Guide to pronunciations:

fāte, câre, hăt, finăl, ȧbound, sofȧ, ärm; bē, crēate, ĕnd, readẽr; īce, bĭt; ōver, pōetic, hŏt, cŏnnect, lôrd, mōōn; tūbe, ūnite, tŭb, circŭs, ûrn.

APPENDIX L

EXTRACTS FROM TIDE TABLES

NEW YORK (THE BATTERY), N.Y., 1975

TIMES AND HEIGHTS OF HIGH AND LOW WATERS

JANUARY

DAY	TIME H.M.	HT. FT.	DAY	TIME H.M.	HT. FT.
1	0422	-0.8	16	0431	0.0
W	1043	5.1	TH	1045	4.1
	1659	-1.1		1656	-0.2
	2321	4.6		2319	3.8
2	0516	-0.6	17	0501	0.2
TH	1138	4.9	F	1120	3.9
	1749	-0.9		1722	0.0
				2357	3.7
3	0017	4.6	18	0533	0.4
F	0615	-0.4	SA	1154	3.7
	1234	4.6		1746	0.2
	1847	-0.6			
4	0111	4.6	19	0029	3.7
SA	0724	-0.1	SU	0613	0.6
	1329	4.3		1228	3.5
	1951	-0.4		1818	0.3
5	0207	4.6	20	0108	3.8
SU	0835	0.0	M	0727	0.7
	1428	4.0		1311	3.4
	2053	-0.3		1917	0.4
6	0307	4.5	21	0156	3.9
M	0939	-0.1	TU	0856	0.7
	1530	3.7		1404	3.3
	2153	-0.3		2050	0.4
7	0407	4.5	22	0255	4.0
TU	1038	-0.2	W	1001	0.4
	1635	3.6		1515	3.2
	2247	-0.3		2157	0.2
8	0508	4.6	23	0404	4.2
W	1131	-0.3	TH	1057	0.1
	1735	3.7		1637	3.4
	2339	-0.3		2256	0.0
9	0603	4.7	24	0510	4.5
TH	1222	-0.4	F	1150	-0.2
	1828	3.8		1742	3.7
				2352	-0.3
10	0028	-0.4	25	0609	4.9
F	0649	4.8	SA	1242	-0.6
	1310	-0.5		1838	4.0
	1915	3.9			
11	0117	-0.4	26	0048	-0.6
SA	0733	4.8	SU	0701	5.2
	1354	-0.6		1330	-1.0
	1959	3.9		1929	4.4
12	0202	-0.4	27	0140	-0.9
SU	0814	4.8	M	0749	5.4
	1436	-0.6		1419	-1.3
	2040	3.9		2019	4.7
13	0242	-0.4	28	0231	-1.1
M	0854	4.7	TU	0840	5.5
	1516	-0.6		1505	-1.4
	2122	3.9		2110	4.9
14	0321	-0.3	29	0321	-1.2
TU	0931	4.5	W	0931	5.4
	1552	-0.5		1550	-1.5
	2202	3.9		2203	5.0
15	0358	-0.1	30	0410	-1.2
W	1011	4.3	TH	1024	5.2
	1625	-0.4		1635	-1.3
	2242	3.8		2258	5.0
			31	0459	-0.9
			F	1120	4.9
				1725	-1.0
				2354	4.9

FEBRUARY

DAY	TIME H.M.	HT. FT.	DAY	TIME H.M.	HT. FT.
1	0554	-0.6	16	0503	0.2
SA	1213	4.5	SU	1111	3.8
	1818	-0.7		1704	0.1
				2338	4.0
2	0047	4.8	17	0535	0.3
SU	0659	-0.2	M	1149	3.6
	1309	4.1		1733	0.2
	1919	-0.3			
3	0142	4.6	18	0020	4.0
M	0808	0.0	TU	0621	0.5
	1405	3.8		1231	3.5
	2024	-0.1		1818	0.4
4	0240	4.4	19	0108	4.1
TU	0915	0.1	W	0759	0.6
	1508	3.6		1329	3.4
	2129	0.0		1932	0.5
5	0341	4.3	20	0211	4.1
W	1015	0.0	TH	0926	0.5
	1612	3.5		1439	3.3
	2225	0.0		2125	0.4
6	0444	4.3	21	0327	4.2
TH	1110	-0.1	F	1026	0.2
	1715	3.5		1607	3.5
	2320	0.0		2233	0.1
7	0542	4.4	22	0444	4.5
F	1200	-0.2	SA	1123	-0.2
	1811	3.7		1718	3.9
				2334	-0.3
8	0009	-0.1	23	0545	4.8
SA	0629	4.5	SU	1214	-0.6
	1246	-0.3		1816	4.4
	1856	3.9			
9	0057	-0.2	24	0030	-0.7
SU	0713	4.6	M	0640	5.2
	1329	-0.5		1306	-1.0
	1937	4.0		1909	4.8
10	0141	-0.3	25	0123	-1.0
M	0752	4.6	TU	0731	5.4
	1410	-0.5		1354	-1.2
	2015	4.1		1959	5.2
11	0222	-0.4	26	0215	-1.3
TU	0829	4.6	W	0820	5.4
	1449	-0.6		1441	-1.4
	2053	4.2		2048	5.4
12	0300	-0.4	27	0305	-1.4
W	0906	4.5	TH	0912	5.3
	1523	-0.5		1526	-1.4
	2129	4.2		2139	5.4
13	0335	-0.3	28	0352	-1.3
TH	0941	4.3	F	1003	5.1
	1555	-0.4		1612	-1.2
	2203	4.1		2232	5.3
14	0407	-0.2			
F	1011	4.1			
	1622	-0.2			
	2235	4.1			
15	0435	0.0			
SA	1043	3.9			
	1643	-0.1			
	2307	4.0			

MARCH

DAY	TIME H.M.	HT. FT.	DAY	TIME H.M.	HT. FT.
1	0442	-1.0	16	0416	-0.1
SA	1058	4.8	SU	1013	4.0
	1658	-0.9		1613	0.0
	2325	5.1		2223	4.4
2	0533	-0.6	17	0443	0.0
SU	1152	4.4	M	1043	3.8
	1749	-0.5		1637	0.2
				2259	4.4
3	0021	4.9	18	0517	0.2
M	0631	-0.2	TU	1124	3.7
	1247	4.1		1708	0.3
	1847	0.0		2346	4.3
4	0115	4.6	19	0602	0.4
TU	0739	0.1	W	1218	3.6
	1344	3.8		1752	0.5
	1956	0.3			
5	0212	4.3	20	0042	4.3
W	0848	0.3	TH	0724	0.5
	1443	3.6		1318	3.6
	2103	0.4		1906	0.6
6	0312	4.1	21	0146	4.3
TH	0949	0.3	F	0853	0.4
	1545	3.5		1429	3.6
	2204	0.4		2106	0.5
7	0415	4.1	22	0300	4.3
F	1041	0.2	SA	0959	0.2
	1649	3.6		1548	3.9
	2258	0.3		2217	0.2
8	0513	4.2	23	0417	4.5
SA	1131	0.0	SU	1056	-0.2
	1744	3.8		1658	4.3
	2347	0.1		2315	-0.2
9	0603	4.3	24	0524	4.8
SU	1216	-0.1	M	1147	-0.5
	1830	4.1		1755	4.8
10	0033	0.0	25	0011	-0.6
M	0645	4.4	TU	0620	5.0
	1259	-0.3		1238	-0.8
	1909	4.3		1848	5.2
11	0117	-0.2	26	0107	-1.0
TU	0726	4.5	W	0712	5.2
	1341	-0.4		1328	-1.1
	1946	4.4		1938	5.6
12	0158	-0.3	27	0157	-1.2
W	0803	4.5	TH	0802	5.3
	1418	-0.4		1416	-1.2
	2022	4.5		2025	5.7
13	0237	-0.4	28	0247	-1.3
TH	0837	4.5	F	0852	5.2
	1453	-0.4		1502	-1.1
	2054	4.5		2114	5.7
14	0313	-0.4	29	0334	-1.2
F	0912	4.3	SA	0942	4.9
	1525	-0.3		1547	-0.9
	2126	4.5		2206	5.5
15	0346	-0.3	30	0422	-0.9
SA	0941	4.2	SU	1037	4.7
	1551	-0.1		1632	-0.6
	2152	4.5		2259	5.3
			31	0512	-0.6
			M	1132	4.3
				1721	-0.2
				2352	4.9

TIME MERIDIAN 75° W. 0000 IS MIDNIGHT. 1200 IS NOON.
HEIGHTS ARE RECKONED FROM THE DATUM OF SOUNDINGS ON CHARTS OF THE LOCALITY WHICH IS MEAN LOW WATER.

TABLE 2.—TIDAL DIFFERENCES AND OTHER CONSTANTS

No.	PLACE	POSITION		DIFFERENCES				RANGES		Mean Tide Level
		Lat.	Long.	Time		Height		Mean	Spring	
				High water	Low water	High water	Low water			
		° ′	° ′	h. m.	h. m.	feet	feet	feet	feet	feet
	NEW YORK and NEW JERSEY — Continued									
	Hudson River‡	N.	W.	on **NEW YORK, p.56**						
	Time meridian, 75°W.									
1513	Jersey City, Pa. RR. Ferry, N. J----	40 43	74 02	+0 07	+0 07	-0.1	0.0	4.4	5.3	2.2
1515	New York, Desbrosses Street---------	40 43	74 01	+0 10	+0 10	-0.1	0.0	4.4	5.3	2.2
1517	New York, Chelsea Docks------------	40 45	74 01	+0 17	+0 16	-0.2	0.0	4.3	5.2	2.1
1519	Hoboken, Castle Point, N. J---------	40 45	74 01	+0 17	+0 16	-0.2	0.0	4.3	5.2	2.1
1521	Weehawken, Days Point, N. J---------	40 46	74 01	+0 24	+0 23	-0.3	0.0	4.2	5.0	2.1
1523	New York, Union Stock Yards---------	40 47	74 00	+0 27	+0 26	-0.3	0.0	4.2	5.0	2.1
1525	New York, 130th Street-------------	40 49	73 58	+0 37	+0 35	-0.5	0.0	4.0	4.8	2.0
1527	George Washington Bridge------------	40 51	73 57	+0 46	+0 43	-0.6	0.0	3.9	4.6	1.9
1529	Spuyten Duyvil, West of RR. bridge--	40 53	73 56	+0 58	+0 53	-0.7	0.0	3.8	4.5	1.9
1531	Yonkers----------------------------	40 56	73 54	+1 09	+1 10	-0.8	0.0	3.7	4.4	1.8
1533	Dobbs Ferry------------------------	41 01	73 53	+1 29	+1 40	-1.1	0.0	3.4	4.0	1.7
1535	Tarrytown--------------------------	41 05	73 52	+1 45	+1 54	-1.3	0.0	3.2	3.7	1.6
1537	Ossining---------------------------	41 10	73 52	+1 53	+2 14	-1.4	0.0	3.1	3.6	1.5
1539	Haverstraw-------------------------	41 12	73 58	+1 59	+2 25	-1.6	0.0	2.9	3.4	1.4
1541	Peekskill--------------------------	41 17	73 56	+2 24	+3 00	-1.3	+0.3	2.9	3.4	1.7
1543	West Point-------------------------	41 24	73 57	+3 16	+3 37	-1.5	+0.3	2.7	3.1	1.6
1545	Newburgh---------------------------	41 30	74 00	+3 42	+4 00	-1.5	+0.2	2.8	3.2	1.6
1547	New Hamburg------------------------	41 35	73 57	+4 00	+4 25	-1.5	+0.1	2.9	3.3	1.5
1549	Poughkeepsie-----------------------	41 42	73 57	+4 30	+4 43	-1.3	+0.1	3.1	3.5	1.6
1551	Hyde Park--------------------------	41 47	73 57	+4 56	+5 09	-1.3	0.0	3.2	3.6	1.6
1553	Kingston Point---------------------	41 56	73 58	+5 16	+5 31	-0.9	-0.1	3.7	4.2	1.7
1555	Tivoli-----------------------------	42 04	73 56	+5 46	+6 01	-0.8	-0.2	3.9	4.4	1.7
1557	Catskill---------------------------	42 13	73 51	+6 37	+6 55	-0.7	-0.3	4.1	4.6	1.7
1559	Hudson-----------------------------	42 15	73 48	+6 54	+7 09	-0.9	-0.4	4.0	4.4	1.6
				on **ALBANY, p.60**						
1561	Coxsackie--------------------------	42 21	73 48	-1 01	-1 38	-0.5	+0.2	3.9	4.3	2.1
1563	New Baltimore----------------------	42 27	73 47	-0 34	-0 56	-0.1	+0.4	4.1	4.5	2.4
1565	Castleton-on-Hudson----------------	42 32	73 46	-0 17	-0 29	-0.2	+0.1	4.3	4.7	2.2
1567	ALBANY-----------------------------	42 39	73 45	Daily predictions				4.6	5.0	2.5
1569	Troy-------------------------------	42 44	73 42	+0 08	+0 10	+0.1	0.0	4.7	5.1	2.3
	The Kills and Newark Bay			on **NEW YORK, p.56**						
	Kill Van Kull									
1571	Constable Hook---------------------	40 39	74 05	-0 34	-0 21	0.0	0.0	4.5	5.4	2.2
1573	New Brighton-----------------------	40 39	74 05	-0 12	-0 18	0.0	0.0	4.5	5.4	2.2
1575	Port Richmond----------------------	40 38	74 08	-0 03	+0 05	0.0	0.0	4.5	5.4	2.2
1577	Bergen Point-----------------------	40 39	74 08	+0 03	+0 03	+0.1	0.0	4.6	5.5	2.3
1579	Shooters Island--------------------	40 39	74 10	+0 06	+0 18	+0.1	0.0	4.6	5.5	2.3
1581	Port Newark Terminal---------------	40 41	74 08	-0 01	+0 18	+0.6	0.0	5.1	6.1	2.5
1583	Newark, Passaic River--------------	40 44	74 10	+0 22	+0 52	+0.6	0.0	5.1	6.1	2.5
1585	Passaic, Gregory Ave. bridge--------	40 51	74 07	+0 49	+1 57	+0.6	0.0	5.1	6.1	2.5
	Hackensack River									
1586	Kearny Point-----------------------	40 44	74 06	+0 09	+0 33	+0.5	0.0	5.0	6.0	2.5
1587	Secaucus---------------------------	40 48	74 04	+1 13	+1 09	+0.6	0.0	5.1	6.1	2.6
1588	Little Ferry-----------------------	40 51	74 02	+1 22	+1 14	+0.8	0.0	5.3	6.4	2.7
1589	Hackensack-------------------------	40 53	74 02	+1 33	+1 58	+0.8	0.0	5.3	6.4	2.6
				on **SANDY HOOK, p.64**						
	Arthur Kill									
1591	Elizabethport----------------------	40 39	74 11	+0 25	+0 39	+0.3	0.0	4.9	5.9	2.4
1593	Chelsea----------------------------	40 36	74 12	+0 24	+0 35	+0.4	0.0	5.0	6.0	2.5
1595	Carteret---------------------------	40 35	74 13	+0 23	+0 31	+0.5	0.0	5.1	6.2	2.6
1597	Rossville--------------------------	40 33	74 13	+0 17	+0 25	+0.7	0.0	5.3	6.4	2.6
1599	Tottenville------------------------	40 31	74 15	+0 03	+0 13	+0.7	0.0	5.3	6.4	2.6
1601	Perth Amboy------------------------	40 30	74 16	+0 13	+0 19	+0.6	0.0	5.2	6.3	2.6

‡Values for the Hudson River above the George Washington Bridge are based upon averages for the six months May to October, when the fresh-water discharge is a minimum.

TABLE 3.—HEIGHT OF TIDE AT ANY TIME

Duration of rise or fall, see footnote	Time from the nearest high water or low water														
h. m.	*h. m.*	*h. m.*	*h. m.*	*h. m.*	*h. m.*	*h. m.*	*h. m.*	*h. m.*	*h. m.*	*h. m.*	*h. m.*	*h. m.*	*h. m.*	*h. m.*	*h. m.*
4 00	0 08	0 16	0 24	0 32	0 40	0 48	0 56	1 04	1 12	1 20	1 28	1 36	1 44	1 52	2 00
4 20	0 09	0 17	0 26	0 35	0 43	0 52	1 01	1 09	1 18	1 27	1 35	1 44	1 53	2 01	2 10
4 40	0 09	0 19	0 28	0 37	0 47	0 56	1 05	1 15	1 24	1 33	1 43	1 52	2 01	2 11	2 20
5 00	0 10	0 20	0 30	0 40	0 50	1 00	1 10	1 20	1 30	1 40	1 50	2 00	2 10	2 20	2 30
5 20	0 11	0 21	0 32	0 43	0 53	1 04	1 15	1 25	1 36	1 47	1 57	2 08	2 19	2 29	2 40
5 40	0 11	0 23	0 34	0 45	0 57	1 08	1 19	1 31	1 42	1 53	2 05	2 16	2 27	2 39	2 50
6 00	0 12	0 24	0 36	0 48	1 00	1 12	1 24	1 36	1 48	2 00	2 12	2 24	2 36	2 48	3 00
6 20	0 13	0 25	0 38	0 51	1 03	1 16	1 29	1 41	1 54	2 07	2 19	2 32	2 45	2 57	3 10
6 40	0 13	0 27	0 40	0 53	1 07	1 20	1 33	1 47	2 00	2 13	2 27	2 40	2 53	3 07	3 20
7 00	0 14	0 28	0 42	0 56	1 10	1 24	1 38	1 52	2 06	2 20	2 34	2 48	3 02	3 16	3 30
7 20	0 15	0 29	0 44	0 59	1 13	1 28	1 43	1 57	2 12	2 27	2 41	2 56	3 11	3 25	3 40
7 40	0 15	0 31	0 46	1 01	1 17	1 32	1 47	2 03	2 18	2 33	2 49	3 04	3 19	3 35	3 50
8 00	0 16	0 32	0 48	1 04	1 20	1 36	1 52	2 08	2 24	2 40	2 56	3 12	3 28	3 44	4 00
8 20	0 17	0 33	0 50	1 07	1 23	1 40	1 57	2 13	2 30	2 47	3 03	3 20	3 37	3 53	4 10
8 40	0 17	0 35	0 52	1 09	1 27	1 44	2 01	2 19	2 36	2 53	3 11	3 28	3 45	4 03	4 20
9 00	0 18	0 36	0 54	1 12	1 30	1 48	2 06	2 24	2 42	3 00	3 18	3 36	3 54	4 12	4 30
9 20	0 19	0 37	0 56	1 15	1 33	1 52	2 11	2 29	2 48	3 07	3 25	3 44	4 03	4 21	4 40
9 40	0 19	0 39	0 58	1 17	1 37	1 56	2 15	2 35	2 54	3 13	3 33	3 52	4 11	4 31	4 50
10 00	0 20	0 40	1 00	1 20	1 40	2 00	2 20	2 40	3 00	3 20	3 40	4 00	4 20	4 40	5 00
10 20	0 21	0 41	1 02	1 23	1 43	2 04	2 25	2 45	3 06	3 27	3 47	4 08	4 29	4 49	5 10
10 40	0 21	0 43	1 04	1 25	1 47	2 08	2 29	2 51	3 12	3 33	3 55	4 16	4 37	4 59	5 20

Range of tide, see footnote	Correction to height														
Ft.	*Ft.*	*Ft.*	*Ft.*	*Ft.*	*Ft.*	*Ft.*	*Ft.*	*Ft.*	*Ft.*	*Ft.*	*Ft.*	*Ft.*	*Ft.*	*Ft.*	*Ft.*
0.5	0.0	0.0	0.0	0.0	0.0	0.0	0.1	0.1	0.1	0.1	0.1	0.2	0.2	0.2	0.2
1.0	0.0	0.0	0.0	0.0	0.1	0.1	0.1	0.2	0.2	0.2	0.3	0.3	0.4	0.4	0.5
1.5	0.0	0.0	0.0	0.1	0.1	0.1	0.2	0.2	0.3	0.4	0.4	0.5	0.6	0.7	0.8
2.0	0.0	0.0	0.0	0.1	0.1	0.2	0.3	0.3	0.4	0.5	0.6	0.7	0.8	0.9	1.0
2.5	0.0	0.0	0.1	0.1	0.2	0.2	0.3	0.4	0.5	0.6	0.7	0.9	1.0	1.1	1.2
3.0	0.0	0.0	0.1	0.1	0.2	0.3	0.4	0.5	0.6	0.8	0.9	1.0	1.2	1.3	1.5
3.5	0.0	0.0	0.1	0.2	0.2	0.3	0.4	0.6	0.7	0.9	1.0	1.2	1.4	1.6	1.8
4.0	0.0	0.0	0.1	0.2	0.3	0.4	0.5	0.7	0.8	1.0	1.2	1.4	1.6	1.8	2.0
4.5	0.0	0.0	0.1	0.2	0.3	0.4	0.6	0.7	0.9	1.1	1.3	1.6	1.8	2.0	2.2
5.0	0.0	0.1	0.1	0.2	0.3	0.5	0.6	0.8	1.0	1.2	1.5	1.7	2.0	2.2	2.5
5.5	0.0	0.1	0.1	0.2	0.4	0.5	0.7	0.9	1.1	1.4	1.6	1.9	2.2	2.5	2.8
6.0	0.0	0.1	0.1	0.3	0.4	0.6	0.8	1.0	1.2	1.5	1.8	2.1	2.4	2.7	3.0
6.5	0.0	0.1	0.2	0.3	0.4	0.6	0.8	1.1	1.3	1.6	1.9	2.2	2.6	2.9	3.2
7.0	0.0	0.1	0.2	0.3	0.5	0.7	0.9	1.2	1.4	1.8	2.1	2.4	2.8	3.1	3.5
7.5	0.0	0.1	0.2	0.3	0.5	0.7	1.0	1.2	1.5	1.9	2.2	2.6	3.0	3.4	3.8
8.0	0.0	0.1	0.2	0.3	0.5	0.8	1.0	1.3	1.6	2.0	2.4	2.8	3.2	3.6	4.0
8.5	0.0	0.1	0.2	0.4	0.6	0.8	1.1	1.4	1.8	2.1	2.5	2.9	3.4	3.8	4.2
9.0	0.0	0.1	0.2	0.4	0.6	0.9	1.2	1.5	1.9	2.2	2.7	3.1	6.6	4.0	4.5
9.5	0.0	0.1	0.2	0.4	0.6	0.9	1.2	1.6	2.0	2.4	2.8	3.3	3.8	4.3	4.8
10.0	0.0	0.1	0.2	0.4	0.7	1.0	1.3	1.7	2.1	2.5	3.0	3.5	4.0	4.5	5.0
10.5	0.0	0.1	0.3	0.5	0.7	1.0	1.3	1.7	2.2	2.6	3.1	3.6	4.2	4.7	5.2
11.0	0.0	0.1	0.3	0.5	0.7	1.1	1.4	1.8	2.3	2.8	3.3	3.8	4.4	4.9	5.5
11.5	0.0	0.1	0.3	0.5	0.8	1.1	1.5	1.9	2.4	2.9	3.4	4.0	4.6	5.1	5.8
12.0	0.0	0.1	0.3	0.5	0.8	1.1	1.5	2.0	2.5	3.0	3.6	4.1	4.8	5.4	6.0
12.5	0.0	0.1	0.3	0.5	0.8	1.2	1.6	2.1	2.6	3.1	3.7	4.3	5.0	5.6	6.2
13.0	0.0	0.1	0.3	0.6	0.9	1.2	1.7	2.2	2.7	3.2	3.9	4.5	5.1	5.8	6.5
13.5	0.0	0.1	0.3	0.6	0.9	1.3	1.7	2.2	2.8	3.4	4.0	4.7	5.3	6.0	6.8
14.0	0.0	0.2	0.3	0.6	0.9	1.3	1.8	2.3	2.9	3.5	4.2	4.8	5.5	6.3	7.0
14.5	0.0	0.2	0.4	0.6	1.0	1.4	1.9	2.4	3.0	3.6	4.3	5.0	5.7	6.5	7.2
15.0	0.0	0.2	0.4	0.6	1.0	1.4	1.9	2.5	3.1	3.8	4.4	5.2	5.9	6.7	7.5
15.5	0.0	0.2	0.4	0.7	1.0	1.5	2.0	2.6	3.2	3.9	4.6	5.4	6.1	6.9	7.8
16.0	0.0	0.2	0.4	0.7	1.1	1.5	2.1	2.6	3.3	4.0	4.7	5.5	6.3	7.2	8.0
16.5	0.0	0.2	0.4	0.7	1.1	1.6	2.1	2.7	3.4	4.1	4.9	5.7	6.5	7.4	8.2
17.0	0.0	0.2	0.4	0.7	1.1	1.6	2.2	2.8	3.5	4.2	5.0	5.9	6.7	7.6	8.5
17.5	0.0	0.2	0.4	0.8	1.2	1.7	2.2	2.9	3.6	4.4	5.2	6.0	6.9	7.8	8.8
18.0	0.0	0.2	0.4	0.8	1.2	1.7	2.3	3.0	3.7	4.5	5.3	6.2	7.1	8.1	9.0
18.5	0.1	0.2	0.5	0.8	1.2	1.8	2.4	3.1	3.8	4.6	5.5	6.4	7.3	8.3	9.2
19.0	0.1	0.2	0.5	0.8	1.3	1.8	2.4	3.1	3.9	4.8	5.6	6.6	7.5	8.5	9.5
19.5	0.1	0.2	0.5	0.8	1.3	1.9	2.5	3.2	4.0	4.9	5.8	6.7	7.7	8.7	9.8
20.0	0.1	0.2	0.5	0.9	1.3	1.9	2.6	3.3	4.1	5.0	5.9	6.9	7.9	9.0	10.0

Obtain from the predictions the high water and low water, one of which is before and the other after the time for which the height is required. The difference between the times of occurrence of these tides is the duration of rise or fall, and the difference between their heights is the range of tide for the above table. Find the difference between the nearest high or low water and the time for which the height is required.

Enter the table with the duration of rise or fall, printed in heavy-faced type, which most nearly agrees with the actual value, and on that horizontal line find the time from the nearest high or low water which agrees most nearly with the corresponding actual difference. The correction sought is in the column directly below, on the line with the range of tide.

When the nearest tide is high water, subtract the correction.

When the nearest tide is low water, add the correction.

APPENDIX M

EXTRACTS FROM TIDAL CURRENT TABLES

THE NARROWS, NEW YORK HARBOR, N.Y., 1975

F-FLOOD, DIR. 340° TRUE E-EBB, DIR. 160° TRUE

JANUARY

DAY	SLACK WATER TIME H.M.	MAXIMUM CURRENT TIME H.M.	MAXIMUM CURRENT VEL. KNOTS	DAY	SLACK WATER TIME H.M.	MAXIMUM CURRENT TIME H.M.	MAXIMUM CURRENT VEL. KNOTS
1		0254	2.4E	16	0002	0310	1.8E
W	0617	0857	2.2F	TH	0641	0907	1.6F
	1211	1526	2.5E		1217	1532	2.0E
	1901	2132	2.0F		1917	2139	1.5F
2	0037	0345	2.3E	17	0046	0351	1.7E
TH	0716	0952	2.1F	F	0730	0954	1.5F
	1301	1614	2.4E		1257	1611	1.9E
	1954	2227	2.0F		2000	2224	1.5F
3	0132	0440	2.2E	18	0131	0437	1.6E
F	0819	1049	1.9F	SA	0823	1043	1.4F
	1352	1708	2.3E		1339	1654	1.8E
	2050	2321	2.0F		2045	2309	1.5F
4	0230	0542	2.1E	19	0218	0530	1.6E
SA	0924	1147	1.7F	SU	0920	1130	1.3F
	1445	1807	2.1E		1424	1745	1.7E
	2146				2132	2358	1.5F
5		0018	1.9F	20	0310	0631	1.5E
SU	0331	0648	2.0E	M	1017	1220	1.2F
	1027	1244	1.5F		1514	1842	1.6E
	1543	1909	2.0E		2220		
	2242						
6		0119	1.9F	21		0047	1.6F
M	0434	0751	2.0E	TU	0407	0728	1.6E
	1129	1352	1.4F		1114	1313	1.1F
	1643	2007	2.0E		1610	1940	1.6E
	2337				2310		
7		0232	1.9F	22		0140	1.6F
TU	0537	0850	2.0E	W	0505	0823	1.7E
	1230	1515	1.3F		1211	1410	1.1F
	1744	2102	2.0E		1710	2033	1.7E
8	0033	0343	1.9F	23	0001	0237	1.7F
W	0635	0946	2.0E	TH	0602	0917	1.9E
	1328	1623	1.4F		1305	1509	1.2F
	1841	2154	1.9E		1808	2124	1.8E
9	0127	0442	2.0F	24	0055	0337	1.9F
TH	0727	1037	2.1E	F	0655	1007	2.0E
	1423	1714	1.4F		1357	1612	1.4F
	1932	2242	1.9E		1903	2215	2.0E
10	0219	0527	2.0F	25	0148	0434	2.1F
F	0814	1126	2.1E	SA	0745	1059	2.2E
	1511	1757	1.5F		1445	1703	1.6F
	2020	2336	1.9E		1955	2309	2.1F
11	0307	0609	2.0F	26	0240	0523	2.3F
SA	0857	1215	2.1E	SU	0834	1150	2.4E
	1556	1836	1.5F		1531	1751	1.8F
	2106				2046		
12		0023	1.9E	27		0003	2.3E
SU	0351	0638	2.0F	M	0330	0611	2.4F
	0938	1300	2.1E		0922	1241	2.5E
	1637	1906	1.5F		1615	1838	2.0F
	2150				2138		
13		0108	1.9E	28		0057	2.4E
M	0434	0709	1.9F	TU	0420	0658	2.5F
	1018	1340	2.2E		1011	1329	2.6E
	1717	1937	1.5F		1659	1925	2.2F
	2235				2231		
14		0151	1.9E	29		0148	2.5E
TU	0515	0743	1.8F	W	0510	0745	2.4F
	1058	1418	2.1E		1100	1416	2.7E
	1756	2010	1.5F		1745	2015	2.2F
	2318				2324		
15		0231	1.9E	30		0237	2.6E
W	0557	0822	1.7F	TH	0602	0838	2.3F
	1137	1455	2.1E		1149	1503	2.6E
	1836	2051	1.5F		1833	2106	2.2F
				31	0017	0327	2.5E
				F	0659	0932	2.1F
					1238	1551	2.5E
					1925	2201	2.1F

FEBRUARY

DAY	SLACK WATER TIME H.M.	MAXIMUM CURRENT TIME H.M.	MAXIMUM CURRENT VEL. KNOTS	DAY	SLACK WATER TIME H.M.	MAXIMUM CURRENT TIME H.M.	MAXIMUM CURRENT VEL. KNOTS
1	0111	0420	2.3E	16	0058	0404	1.8E
SA	0759	1027	1.9F	SU	0751	1011	1.4F
	1328	1643	2.3E		1307	1615	1.8E
	2021	2258	2.1F		1958	2236	1.6F
2	0207	0517	2.1E	17	0144	0451	1.7E
SU	0902	1124	1.7F	M	0846	1100	1.3F
	1420	1738	2.1E		1350	1656	1.7E
	2118	2355	2.0F		2046	2325	1.6F
3	0305	0621	2.0E	18	0233	0547	1.6E
M	1005	1222	1.5F	TU	0944	1151	1.2F
	1516	1839	1.9E		1439	1800	1.6E
	2216				2138		
4		0056	1.8F	19		0015	1.6F
TU	0408	0727	1.9E	W	0329	0650	1.6E
	1107	1331	1.3F		1041	1252	1.1F
	1617	1943	1.8E		1535	1903	1.6E
	2314				2234		
5		0207	1.7F	20		0108	1.6F
W	0511	0830	1.9E	TH	0429	0752	1.7E
	1208	1455	1.2F		1138	1337	1.1F
	1720	2039	1.8E		1638	2004	1.7E
					2331		
6	0011	0322	1.7F	21		0205	1.7F
TH	0611	0925	1.9E	F	0529	0848	1.9E
	1306	1602	1.3F		1233	1439	1.2F
	1820	2133	1.8E		1741	2100	1.9E
7	0107	0422	1.8F	22	0030	0308	1.8F
F	0704	1016	1.9E	SA	0627	0941	2.1E
	1359	1655	1.4F		1325	1544	1.5F
	1913	2226	1.8E		1840	2154	2.1E
8	0200	0511	1.8F	23	0127	0408	2.0F
SA	0751	1104	2.0E	SU	0720	1032	2.2E
	1447	1741	1.5F		1415	1639	1.7F
	2001	2313	1.8E		1935	2248	2.2E
9	0248	0552	1.9F	24	0222	0504	2.2F
SU	0833	1149	2.0E	M	0810	1123	2.4E
	1530	1818	1.5F		1502	1730	2.0F
	2045				2027	2343	2.4E
10		0000	1.9E	25	0315	0552	2.4F
M	0333	0625	1.9F	TU	0859	1214	2.5E
	0912	1232	2.1E		1547	1818	2.2F
	1609	1845	1.6F		2119		
	2128						
11		0044	1.9E	26		0037	2.6E
TU	0414	0650	1.8F	W	0405	0641	2.4F
	0951	1312	2.1E		0947	1304	2.6E
	1647	1911	1.6F		1632	1903	2.4F
	2209				2211		
12		0127	2.0E	27		0129	2.7E
W	0455	0718	1.8F	TH	0455	0726	2.4F
	1029	1350	2.1E		1036	1353	2.7E
	1723	1942	1.7F		1718	1952	2.4F
	2251				2303		
13		0207	2.0E	28		0218	2.7E
TH	0534	0757	1.7F	F	0547	0816	2.2F
	1108	1426	2.1E		1125	1439	2.6E
	1800	2019	1.7F		1805	2041	2.3F
	2332				2356		
14		0245	1.9E				
F	0616	0838	1.6F				
	1147	1501	2.0E				
	1836	2102	1.7F				
15	0015	0323	1.9E				
SA	0700	0923	1.5F				
	1226	1536	1.9E				
	1915	2149	1.6F				

TIME MERIDIAN 75° W. 0000 IS MIDNIGHT. 1200 IS NOON.

TABLE 2.—CURRENT DIFFERENCES AND OTHER CONSTANTS

No.	PLACE	POSITION		TIME DIFFERENCES		VELOCITY RATIOS		MAXIMUM CURRENTS			
								Flood		Ebb	
		Lat.	Long.	Slack water	Maximum current	Maximum flood	Maximum ebb	Direction (true)	Average velocity	Direction (true)	Average velocity
		° ′	° ′	h. m.	h. m.			deg.	knots	deg.	knots
	LONG ISLAND, South Coast—Continued	N.	W.	on THE NARROWS, p.52 *Time meridian, 75° W.*							
2250	Shinnecock Inlet	40 51	72 29	-0 20	-0 40	1.5	1.2	350	2.5	170	2.3
2255	Fire I. Inlet, 0.5 mi. S. of Oak Beach	40 38	73 18	+0 15	0 00	1.4	1.2	80	2.4	245	2.4
2260	Jones Inlet	40 35	73 34	-1 00	-0 55	1.8	1.3	35	3.1	215	2.6
2265	Long Beach, inside, between bridges	40 36	73 40	-0 10	+0 10	0.3	0.3	75	0.5	275	0.6
2270	East Rockaway Inlet	40 35	73 45	-1 25	-1 35	1.3	1.2	40	2.2	225	2.3
2275	Ambrose Light	40 27	73 49	See table 5.							
2280	Sandy Hook App. Lighted Horn Buoy 2A	40 27	73 55	See table 5.							
	JAMAICA BAY										
2285	Rockaway Inlet	40 34	73 56	-1 45	-2 15	1.1	1.3	85	1.8	245	2.7
2290	Barren Island, east of	40 35	73 53	-2 00	-2 25	0.7	0.9	5	1.2	190	1.7
2295	Canarsie (midchannel, off Pier)	40 38	73 53	-1 35	-1 50	0.3	0.3	45	0.5	220	0.7
2300	Beach Channel (bridge)	40 35	73 49	-1 20	-1 20	1.1	1.0	60	1.9	225	2.0
2305	Grass Hassock Channel	40 37	73 47	-1 10	-1 00	0.6	0.5	50	1.0	230	1.0
	NEW YORK HARBOR ENTRANCE										
2310	Ambrose Channel entrance	40 30	73 58	-1 10	-1 05	1.0	1.2	310	1.7	110	2.3
2315	Ambrose Channel, SE. of West Bank Lt	40 32	74 01	(1)	-0 25	0.8	0.9	310	1.3	170	1.8
2320	Coney Island Lt., 1.6 miles SSW. of	40 33	74 01	-0 10	(2)	0.5	0.8	330	0.8	145	1.5
2325	Ambrose Channel, north end	40 34	74 02	+0 05	+0 15	0.8	0.9	330	1.3	175	1.9
2330	Coney Island, 0.2 mile west of	40 35	74 01	-0 55	-0 55	0.9	1.0	330	1.5	170	2.0
2335	Ft. Lafayette, channel east of	40 36	74 02	(3)	(3)	0.6	0.5	345	1.1	195	0.9
2340	THE NARROWS, midchannel	40 37	74 03	Daily predictions				340	1.7	160	2.0
	NEW YORK HARBOR, Upper Bay										
2345	Tompkinsville	40 38	74 04	-0 10	+0 20	0.9	1.0	5	1.6	170	2.0
2350	Bay Ridge Channel	40 39	74 02	-0 35	-0 45	0.6	0.6	40	1.0	220	1.1
2355	Red Hook Channel	40 40	74 01	-0 35	-0 35	0.6	0.4	355	1.0	170	0.7
2360	Robbins Reef Light, east of	40 39	74 03	+0 10	+0 20	0.8	0.8	15	1.3	205	1.6
2365	Red Hook, 1 mile west of	40 41	74 02	+0 45	+1 00	0.8	1.2	25	1.3	205	2.3
2370	Statue of Liberty, east of	40 42	74 02	+0 55	+1 00	0.8	1.0	30	1.4	205	1.9
	HUDSON RIVER, Midchannel[4]										
2375	The Battery, northwest of	40 43	74 02	+1 30	+1 35	0.9	1.2	15	1.5	195	2.3
2380	Desbrosses Street	40 43	74 01	+1 35	+1 40	0.9	1.2	10	1.5	----	2.3
2385	Chelsea Docks	40 45	74 01	+1 30	+1 40	1.0	1.0	20	1.7	185	2.0
2390	Forty-second Street	40 46	74 00	+1 35	+1 45	1.0	1.2	30	1.7	----	2.3
2395	Ninety-sixth Street	40 48	73 59	+1 40	+1 50	1.0	1.2	30	1.7	----	2.3
2400	Grants Tomb, 123d Street	40 49	73 58	+1 45	+1 55	0.9	1.2	25	1.6	----	2.3
2405	George Washington Bridge	40 51	73 57	+1 45	+2 00	0.9	1.1	20	1.6	200	2.2
2410	Spuyten Duyvil	40 53	73 56	+2 00	+2 10	0.9	1.1	20	1.6	----	2.1
2415	Riverdale	40 54	73 55	+2 05	+2 20	0.8	1.0	15	1.4	200	2.0
2420	Dobbs Ferry	41 01	73 53	+2 25	+2 40	0.8	0.9	10	1.3	----	1.7
2425	Tarrytown	41 05	73 53	+2 40	+2 55	0.6	0.8	0	1.1	----	1.5
2430	Ossining	41 10	73 54	+2 55	+3 10	0.5	0.7	320	0.9	----	1.3
2435	Haverstraw	41 12	73 57	+3 05	+3 15	0.5	0.7	335	0.8	----	1.3
2440	Peekskill	41 17	73 57	+3 20	+3 35	0.5	0.6	0	0.8	----	1.2
2445	Bear Mountain Bridge	41 19	73 59	+3 25	+3 40	0.5	0.6	0	0.8	----	1.1
2450	Highland Falls	41 22	73 58	+3 35	+3 50	0.6	0.6	5	1.0	185	1.2
2455	West Point, off Duck Island	41 24	73 57	+3 40	+3 55	0.5	0.6	10	1.0	----	1.1

[1] Current is rotary, turning clockwise. Minimum current of 0.9 knot sets SW. about time of "Slack, flood begins" at The Narrows. Minimum current of 0.5 knot sets NE. about 1 hour before "Slack, ebb begins" at The Narrows.

[2] Maximum flood, $-0^h\ 50^m$; maximum ebb, $+0^h\ 55^m$.

[3] Flood begins, $-2^h\ 15^m$; maximum flood, $-0^h\ 05^m$; ebb begins, $+0^h\ 05^m$; maximum ebb, $-1^h\ 50^m$.

[4] The values for the Hudson River are for the summer months, when the fresh-water discharge is a minimum.

TABLE 3.—VELOCITY OF CURRENT AT ANY TIME

Table A

Interval between slack and desired time	Interval between slack and maximum current													
	h. m. 1 20	*h. m.* 1 40	*h. m.* 2 00	*h. m.* 2 20	*h. m.* 2 40	*h. m.* 3 00	*h. m.* 3 20	*h. m.* 3 40	*h. m.* 4 00	*h. m.* 4 20	*h. m.* 4 40	*h. m.* 5 00	*h. m.* 5 20	*h. m.* 5 40
h. m.	*f.*	*f.*	*f.*	*f.*	*f.*	*f.*	*f.*	*f.*	*f.*	*f.*	*f.*	*f.*	*f.*	*f.*
0 20	0.4	0.3	0.3	0.2	0.2	0.2	0.2	0.1	0.1	0.1	0.1	0.1	0.1	0.1
0 40	0.7	0.6	0.5	0.4	0.4	0.3	0.3	0.3	0.3	0.2	0.2	0.2	0.2	0.2
1 00	0.9	0.8	0.7	0.6	0.6	0.5	0.5	0.4	0.4	0.4	0.3	0.3	0.3	0.3
1 20	1.0	1.0	0.9	0.8	0.7	0.6	0.6	0.5	0.5	0.5	0.4	0.4	0.4	0.4
1 40	-----	1.0	1.0	0.9	0.8	0.8	0.7	0.7	0.6	0.6	0.5	0.5	0.5	0.4
2 00	-----	-----	1.0	1.0	0.9	0.9	0.8	0.8	0.7	0.7	0.6	0.6	0.6	0.5
2 20	-----	-----	-----	1.0	1.0	0.9	0.9	0.8	0.8	0.7	0.7	0.7	0.6	0.6
2 40	-----	-----	-----	-----	1.0	1.0	1.0	0.9	0.9	0.8	0.8	0.7	0.7	0.7
3 00	-----	-----	-----	-----	-----	1.0	1.0	1.0	0.9	0.9	0.8	0.8	0.8	0.7
3 20	-----	-----	-----	-----	-----	-----	1.0	1.0	1.0	0.9	0.9	0.9	0.8	0.8
3 40	-----	-----	-----	-----	-----	-----	-----	1.0	1.0	1.0	0.9	0.9	0.9	0.9
4 00	-----	-----	-----	-----	-----	-----	-----	-----	1.0	1.0	1.0	1.0	0.9	0.9
4 20	-----	-----	-----	-----	-----	-----	-----	-----	-----	1.0	1.0	1.0	1.0	0.9
4 40	-----	-----	-----	-----	-----	-----	-----	-----	-----	-----	1.0	1.0	1.0	1.0
5 00	-----	-----	-----	-----	-----	-----	-----	-----	-----	-----	-----	1.0	1.0	1.0
5 20	-----	-----	-----	-----	-----	-----	-----	-----	-----	-----	-----	-----	1.0	1.0
5 40	-----	-----	-----	-----	-----	-----	-----	-----	-----	-----	-----	-----	-----	1.0

Table B

Interval between slack and desired time	Interval between slack and maximum current													
	h. m. 1 20	*h. m.* 1 40	*h. m.* 2 00	*h. m.* 2 20	*h. m.* 2 40	*h. m.* 3 00	*h. m.* 3 20	*h. m.* 3 40	*h. m.* 4 00	*h. m.* 4 20	*h. m.* 4 40	*h. m.* 5 00	*h. m.* 5 20	*h. m.* 5 40
h. m.	*f.*	*f.*	*f.*	*f.*	*f.*	*f.*	*f.*	*f.*	*f.*	*f.*	*f.*	*f.*	*f.*	*f.*
0 20	0.5	0.4	0.4	0.3	0.3	0.3	0.3	0.3	0.2	0.2	0.2	0.2	0.2	0.2
0 40	0.8	0.7	0.6	0.5	0.5	0.5	0.4	0.4	0.4	0.4	0.3	0.3	0.3	0.3
1 00	0.9	0.8	0.8	0.7	0.7	0.6	0.6	0.5	0.5	0.5	0.4	0.4	0.4	0.4
1 20	1.0	1.0	0.9	0.8	0.8	0.7	0.7	0.6	0.6	0.6	0.5	0.5	0.5	0.5
1 40	-----	1.0	1.0	0.9	0.9	0.8	0.8	0.7	0.7	0.7	0.6	0.6	0.6	0.6
2 00	-----	-----	1.0	1.0	0.9	0.9	0.9	0.8	0.8	0.7	0.7	0.7	0.7	0.6
2 20	-----	-----	-----	1.0	1.0	1.0	0.9	0.9	0.8	0.8	0.8	0.7	0.7	0.7
2 40	-----	-----	-----	-----	1.0	1.0	1.0	0.9	0.9	0.9	0.8	0.8	0.8	0.7
3 00	-----	-----	-----	-----	-----	1.0	1.0	1.0	0.9	0.9	0.9	0.9	0.8	0.8
3 20	-----	-----	-----	-----	-----	-----	1.0	1.0	1.0	1.0	0.9	0.9	0.9	0.8
3 40	-----	-----	-----	-----	-----	-----	-----	1.0	1.0	1.0	1.0	0.9	0.9	0.9
4 00	-----	-----	-----	-----	-----	-----	-----	-----	1.0	1.0	1.0	1.0	0.9	0.9
4 20	-----	-----	-----	-----	-----	-----	-----	-----	-----	1.0	1.0	1.0	1.0	0.9
4 40	-----	-----	-----	-----	-----	-----	-----	-----	-----	-----	1.0	1.0	1.0	1.0
5 00	-----	-----	-----	-----	-----	-----	-----	-----	-----	-----	-----	1.0	1.0	1.0
5 20	-----	-----	-----	-----	-----	-----	-----	-----	-----	-----	-----	-----	1.0	1.0
5 40	-----	-----	-----	-----	-----	-----	-----	-----	-----	-----	-----	-----	-----	1.0

Use Table A for all places except those listed below for Table B.

Use Table B for Cape Cod Canal, Hell Gate, Chesapeake and Delaware Canal and all stations in Table 2 which are referred to them.

1. From predictions find the time of slack water and the time and velocity of maximum current (flood or ebb), one of which is immediately before and the other after the time for which the velocity is desired.
2. Find the interval of time between the above slack and maximum current, and enter the top of Table A or B with the interval which most nearly agrees with this value.
3. Find the interval of time between the above slack and the time desired, and enter the side of Table A or B with the interval which most nearly agrees with this value.
4. Find, in the table, the factor corresponding to the above two intervals and multiply the maximum velocity by this factor. The result will be the approximate velocity at the time desired.

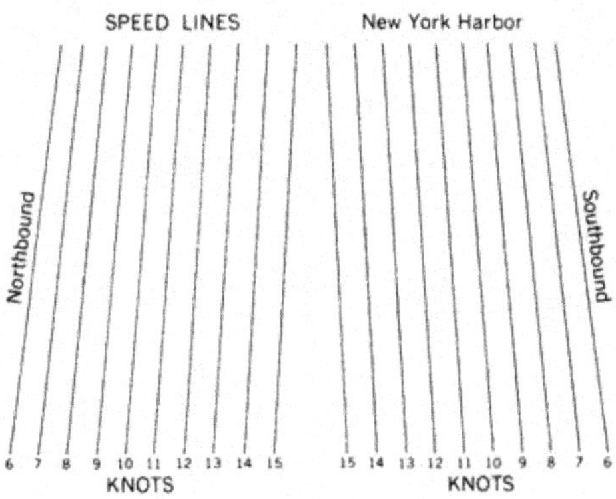
SPEED LINES
New York Harbor
Northbound
Southbound
KNOTS
KNOTS

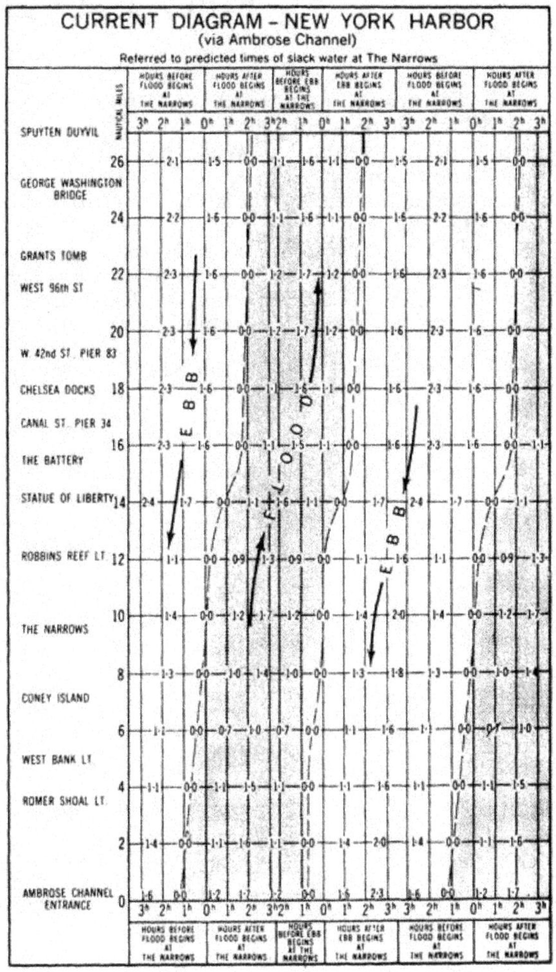
CURRENT DIAGRAM – NEW YORK HARBOR
(via Ambrose Channel)
Referred to predicted times of slack water at The Narrows
SPUYTEN DUYVIL
GEORGE WASHINGTON BRIDGE
GRANTS TOMB
WEST 96th ST
W. 42nd ST. PIER 83
CHELSEA DOCKS
CANAL ST. PIER 34
THE BATTERY
STATUE OF LIBERTY
ROBBINS REEF LT.
THE NARROWS
CONEY ISLAND
WEST BANK LT.
ROMER SHOAL LT.
AMBROSE CHANNEL ENTRANCE
EBB
FLOOD
EBB

TABLE 4.—DURATION OF SLACK

The predicted times of slack water given in this publication indicate the instant of zero velocity, which is only momentary. There is a period each side of slack water, however, during which the current is so weak that for practical purposes it may be considered as negligible.

The following tables give, for various maximum currents, the approximate period of time during which weak currents not exceeding 0.1 to 0.5 knot will be encountered. This duration includes the last of the flood or ebb and the beginning of the following ebb or flood, that is, half of the duration will be before and half after the time of slack water.

Table A should be used for all places *except* those listed below for table B.

Table B should be used for **Cape Cod Canal, Hell Gate, Chesapeake and Delaware Canal,** and all stations in table 2 which are referred to them.

Duration of weak current near time of slack water

TABLE A

Maximum current	Period with a velocity not more than—				
	0.1 knot	0.2 knot	0.3 knot	0.4 knot	0.5 knot
Knots	*Minutes*	*Minutes*	*Minutes*	*Minutes*	*Minutes*
1.0	23	46	70	94	120
1.5	15	31	46	62	78
2.0	11	23	35	46	58
3.0	8	15	23	31	38
4.0	6	11	17	23	29
5.0	5	9	14	18	23
6.0	4	8	11	15	19
7.0	3	7	10	13	16
8.0	3	6	9	11	14
9.0	3	5	8	10	13
10.0	2	5	7	9	11

TABLE B

Maximum current	Period with a velocity not more than—				
	0.1 knot	0.2 knot	0.3 knot	0.4 knot	0.5 knot
Knots	*Minutes*	*Minutes*	*Minutes*	*Minutes*	*Minutes*
1.0	13	28	46	66	89
1.5	8	18	28	39	52
2.0	6	13	20	28	36
3.0	4	8	13	18	22
4.0	3	6	9	13	17
5.0	3	5	8	10	13

When there is a difference between the velocities of the maximum flood and ebb preceding and following the slack for which the duration is desired, it will be sufficiently accurate for practical purposes to find a separate duration for each maximum velocity and take the average of the two as the duration of the weak current.

APPENDIX N

EXTRACTS FROM PUB. NO. 214

Lat. 41°

DECLINATION SAME NAME AS LATITUDE

H.A.	19° 00′ Alt.	Δd Δt	Az.	19° 30′ Alt.	Δd Δt	Az.	20° 00′ Alt.	Δd Δt	Az.	20° 30′ Alt.	Δd Δt	Az.
°	° ′		°	° ′		°	° ′		°	° ′		°
00	68 00.0	1.0 02	180.0	68 30.0	1.0 02	180.0	69 00.0	1.0 02	180.0	69 30.0	1.0 02	180.0
1	67 59.0	1.0 05	177.5	68 29.0	1.0 05	177.4	68 59.0	1.0 05	177.4	69 28.9	1.0 05	177.3
2	67 56.0	1.0 08	175.0	68 25.9	1.0 08	174.9	68 55.9	1.0 09	174.8	69 25.8	1.0 09	174.7
3	67 51.1	99 12	172.5	68 20.9	99 12	172.3	68 50.7	99 12	172.2	69 20.5	99 12	172.0
4	67 44.1	99 15	170.0	68 13.8	99 15	169.8	68 43.5	99 15	169.6	69 13.2	99 16	169.4
05	67 35.3	99 18	167.5	68 04.8	98 18	167.3	68 34.4	98 19	167.0	69 03.9	98 19	166.8
6	67 24.6	98 21	165.1	67 53.9	98 21	164.8	68 23.2	98 22	164.5	68 52.5	98 22	164.2
7	67 12.0	97 24	162.7	67 41.1	97 24	162.4	68 10.2	97 25	162.1	68 39.3	97 25	161.7
8	66 57.7	96 27	160.4	67 26.5	96 27	160.0	67 55.4	96 28	159.6	68 24.2	96 28	159.3
9	66 41.6	95 30	158.0	67 10.2	95 30	157.7	67 38.7	95 31	157.3	68 07.2	95 31	156.8
10	66 23.8	94 32	155.8	66 52.1	94 33	155.4	67 20.4	94 33	154.9	67 48.6	94 34	154.5
1	66 04.5	93 35	153.6	66 32.5	93 35	153.1	67 00.4	93 36	152.7	67 28.2	93 36	152.2
2	65 43.6	92 37	151.4	66 11.2	92 38	151.0	66 38.8	92 38	150.5	67 06.3	92 39	150.0
3	65 21.2	91 40	149.3	65 48.5	91 40	148.8	66 15.8	91 41	148.3	66 42.9	90 41	147.8
4	64 57.4	90 42	147.3	65 24.4	90 42	146.8	65 51.3	89 43	146.2	66 18.1	89 44	145.7
30	56 15.7	73 65	121.7	56 37.6	73 65	121.0	56 59.3	72 65	120.4	57 20.9	71 66	119.8
1	55 36.9	73 65	120.4	55 58.6	72 66	119.8	56 20.0	71 66	119.2	56 41.3	71 67	118.5
2	54 57.6	72 66	119.2	55 19.0	71 67	118.6	55 40.3	71 67	118.0	56 01.3	70 67	117.4
3	54 17.9	71 67	118.1	54 39.1	70 67	117.5	55 00.1	70 68	116.8	55 20.9	69 68	116.2
4	53 37.7	70 68	116.9	53 58.7	70 68	116.3	54 19.5	69 68	115.7	54 40.1	68 69	115.1
35	52 57.1	70 68	115.8	53 17.9	69 69	115.2	53 38.5	68 69	114.6	53 58.9	68 69	114.0
6	52 16.2	69 69	114.7	52 36.8	68 69	114.1	52 57.1	68 69	113.5	53 17.4	67 70	112.9
7	51 34.9	68 69	113.7	51 55.3	68 70	113.1	52 15.5	67 70	112.5	52 35.5	66 70	111.9
8	50 53.3	68 70	112.7	51 13.5	67 70	112.1	51 33.5	66 70	111.5	51 53.3	66 71	110.9
9	50 11.3	67 70	111.7	50 31.4	66 71	111.1	50 51.2	66 71	110.5	51 10.9	65 71	109.9
60	34 46.6	61 75	94.5	35 04.7	60 75	94.0	35 22.7	60 75	93.5	35 40.6	59 75	93.0
1	34 01.5	60 75	93.8	34 19.6	60 75	93.3	34 37.5	60 75	92.8	34 55.4	59 75	92.3
2	33 16.3	60 75	93.1	33 34.3	60 75	92.6	33 52.3	60 75	92.2	34 10.1	59 75	91.7
3	32 31.0	60 75	92.4	32 49.1	60 75	92.0	33 07.0	60 75	91.5	33 24.9	59 75	91.0
4	31 45.8	60 75	91.8	32 03.8	60 75	91.3	32 21.8	60 75	90.8	32 39.6	59 75	90.3
80	19 43.5	61 75	81.6	20 01.9	61 75	81.1	20 20.2	61 74	80.7	20 38.5	61 74	80.3
1	18 58.7	62 74	81.0	19 17.2	61 74	80.5	19 35.5	61 74	80.1	19 53.9	61 74	79.7
2	18 14.1	62 74	80.3	18 32.5	62 74	79.9	18 51.0	61 74	79.5	19 09.4	61 74	79.1
3	17 29.5	62 74	79.7	17 48.0	62 74	79.3	18 06.5	62 74	78.9	18 24.9	61 74	78.5
4	16 45.0	62 74	79.1	17 03.6	62 74	78.7	17 22.1	62 74	78.3	17 40.6	62 74	77.9
85	16 00.5	62 74	78.5	16 19.2	62 74	78.1	16 37.8	62 74	77.7	16 56.4	62 74	77.3
6	15 16.2	63 74	77.9	15 34.9	62 74	77.5	15 53.6	62 73	77.1	16 12.3	62 73	76.7
7	14 32.0	63 74	77.3	14 50.8	63 73	76.9	15 09.6	62 73	76.5	15 28.3	62 73	76.1
8	13 47.9	63 73	76.7	14 06.7	63 73	76.3	14 25.6	63 73	75.9	14 44.4	63 73	75.5
9	13 03.9	63 73	76.0	13 22.8	63 73	75.6	13 41.7	63 73	75.3	14 00.6	63 73	74.8
90	12 20.0	63 73	75.4	12 39.0	63 73	75.0	12 58.0	63 73	74.6	13 17.0	63 73	74.2

H.A.	21° 00′ Alt.	Δd Δt	Az.	21° 30′ Alt.	Δd Δt	Az.	22° 00′ Alt.	Δd Δt	Az.	22° 30′ Alt.	Δd Δt	Az.
°	° ′		°	° ′		°	° ′		°	° ′		°
00	70 00.0	1.0 02	180.0	70 30.0	1.0 02	180.0	71 00.0	1.0 02	180.0	71 30.0	1.0 02	180.0
1	69 58.9	1.0 05	177.3	70 28.9	1.0 06	177.2	70 58.9	1.0 06	177.2	71 28.8	1.0 06	177.1
2	69 55.7	1.0 09	174.6	70 25.6	1.0 09	174.4	70 55.5	1.0 09	174.3	71 25.4	1.0 10	174.2
3	69 50.3	99 12	171.9	70 20.1	99 13	171.7	70 49.9	99 13	171.5	71 19.7	99 13	171.3
4	69 42.9	99 16	169.2	70 12.5	99 16	168.9	70 42.1	99 17	168.7	71 11.7	99 17	168.5
05	69 33.3	98 19	166.5	70 02.8	98 20	166.3	70 32.2	98 20	166.0	71 01.6	98 20	165.7
6	69 21.8	97 22	163.9	69 51.0	97 23	163.6	70 20.2	97 23	163.3	70 49.3	97 24	162.9
7	69 08.3	97 26	161.4	69 37.2	96 26	161.0	70 06.1	96 27	160.6	70 35.0	96 27	160.2
8	68 52.9	96 29	158.9	69 21.5	95 29	158.4	69 50.2	95 30	158.0	70 18.7	95 30	157.6
9	68 35.6	95 32	156.4	69 04.0	94 32	156.0	69 32.3	94 33	155.5	70 00.5	94 33	155.0
10	68 16.7	94 34	154.0	68 44.7	93 35	153.5	69 12.6	93 36	153.0	69 40.4	93 36	152.5
1	67 56.0	92 37	151.7	68 23.7	92 38	151.2	68 51.2	92 38	150.6	69 18.7	91 39	150.1
2	67 33.7	91 40	149.4	68 01.0	91 40	148.9	68 28.2	90 41	148.3	68 55.3	90 42	147.7
3	67 10.0	90 42	147.2	67 36.9	90 43	146.7	68 03.7	89 43	146.1	68 30.3	89 44	145.4
4	66 44.8	89 44	145.1	67 11.3	88 45	144.5	67 37.7	88 46	143.9	68 03.9	87 46	143.2
30	57 42.2	71 66	119.1	58 03.4	70 67	118.4	58 24.3	69 67	117.8	58 45.0	69 68	117.1
1	57 02.4	70 67	117.9	57 23.3	69 67	117.2	57 44.0	69 68	116.6	58 04.4	68 68	115.9
2	56 22.2	69 68	116.7	56 42.8	68 68	116.1	57 03.3	68 68	115.4	57 23.5	67 69	114.7
3	55 41.5	68 68	115.6	56 02.0	68 69	114.9	56 22.2	67 69	114.2	56 42.2	66 69	113.6
4	55 00.5	68 69	114.4	55 20.7	67 69	113.8	55 40.7	66 70	113.1	56 00.5	66 70	112.5
35	54 19.1	67 70	113.4	54 39.1	66 70	112.7	54 58.9	66 70	112.1	55 18.5	65 71	111.4
6	53 37.4	66 70	112.3	53 57.2	66 70	111.7	54 16.8	65 71	111.0	54 36.2	64 71	110.4
7	52 55.3	66 71	111.3	53 14.9	65 71	110.6	53 34.4	64 71	110.0	53 53.6	64 71	109.3
8	52 13.0	65 71	110.3	52 32.4	65 71	109.6	52 51.7	64 72	109.0	53 10.7	63 72	108.4
9	51 30.4	65 71	109.3	51 49.7	64 72	108.7	52 08.8	63 72	108.0	52 27.6	63 72	107.4
60	35 58.4	59 75	92.5	36 16.0	59 75	92.0	36 33.5	58 75	91.5	36 50.9	58 75	91.0
1	35 13.1	59 75	91.9	35 30.7	59 75	91.4	35 48.2	58 75	90.9	36 05.6	58 75	90.4
2	34 27.9	59 75	91.2	34 45.5	58 75	90.7	35 02.9	58 75	90.2	35 20.3	58 75	89.7
3	33 42.6	59 75	90.5	34 00.2	58 75	90.0	34 17.7	58 75	89.5	34 35.0	58 75	89.1
4	32 57.3	59 75	89.9	33 14.9	58 75	89.4	33 32.4	58 75	88.9	33 49.8	58 75	88.4
80	20 56.7	61 74	79.9	21 14.8	60 74	79.5	21 32.9	60 74	79.0	21 50.9	60 74	78.6
1	20 12.1	61 74	79.3	20 30.3	61 74	78.9	20 48.5	60 74	78.4	21 06.5	60 74	78.0
2	19 27.7	61 74	78.7	19 45.9	61 74	78.2	20 04.1	61 74	77.8	20 22.3	60 74	77.4
3	18 43.3	61 74	78.1	19 01.7	61 74	77.6	19 19.9	61 74	77.2	19 38.1	61 73	76.8
4	17 59.1	61 74	77.5	18 17.5	61 73	77.0	18 35.8	61 73	76.6	18 54.1	61 73	76.2
85	17 14.9	62 73	76.9	17 33.4	61 73	76.4	17 51.8	61 73	76.0	18 10.2	61 73	75.6
6	16 30.9	62 73	76.3	16 49.4	62 73	75.8	17 07.9	62 73	75.4	17 26.4	61 73	75.0
7	15 46.9	62 73	75.7	16 05.6	62 73	75.2	16 24.2	62 73	74.8	16 42.7	62 73	74.4
8	15 03.1	62 73	75.1	15 21.8	62 73	74.6	15 40.5	62 73	74.2	15 59.1	62 72	73.8
9	14 19.4	63 73	74.4	14 38.2	63 72	74.0	14 57.0	62 72	73.6	15 15.7	62 72	73.2
90	13 35.9	63 72	73.8	13 54.8	63 72	73.4	14 13.6	63 72	73.0	14 32.4	63 72	72.6

Lat. 41°

DECLINATION CONTRARY NAME TO LATITUDE

H.A.	19° 00′ Alt.	Δd	Δt	Az.	19° 30′ Alt.	Δd	Δt	Az.	20° 00′ Alt.	Δd	Δt	Az.	20° 30′ Alt.	Δd	Δt	Az.
°	° ′			°	° ′			°	° ′			°	° ′			°
00	30 00.0	1.0	01	180.0	29 30.0	1.0	01	180.0	29 00.0	1.0	01	180.0	28 30.0	1.0	01	180.0
1	29 59.6	1.0	02	178.9	29 29.6	1.0	02	178.9	28 59.6	1.0	02	178.9	28 29.6	1.0	02	178.9
2	29 58.3	1.0	04	177.8	29 28.3	1.0	04	177.8	28 58.3	1.0	04	177.9	28 28.3	1.0	03	177.9
3	29 56.1	1.0	05	176.7	29 26.2	1.0	05	176.8	28 56.2	1.0	05	176.8	28 26.2	1.0	05	176.8
4	29 53.1	1.0	06	175.6	29 23.2	1.0	06	175.7	28 53.2	1.0	06	175.7	28 23.3	1.0	06	175.7
05	29 49.2	1.0	08	174.5	29 19.3	1.0	08	174.6	28 49.4	1.0	08	174.6	28 19.5	1.0	08	174.7
6	29 44.5	1.0	09	173.5	29 14.6	1.0	09	173.5	28 44.7	1.0	09	173.6	28 14.9	1.0	09	173.6
7	29 38.9	99	11	172.4	29 09.1	99	11	172.4	28 39.3	99	11	172.5	28 09.4	99	10	172.6
8	29 32.5	99	12	171.3	29 02.7	99	12	171.4	28 32.9	99	12	171.4	28 03.1	99	12	171.5
9	29 25.2	99	14	170.2	28 55.5	99	13	170.3	28 25.8	99	13	170.4	27 56.0	99	13	170.5
10	29 17.1	99	15	169.1	28 47.5	99	15	169.2	28 17.8	99	15	169.3	27 48.1	99	15	169.4
1	29 08.2	99	16	168.1	28 38.6	99	16	168.2	28 09.0	99	16	168.3	27 39.4	99	16	168.4
2	28 58.4	98	18	167.0	28 28.9	98	18	167.1	27 59.4	98	17	167.2	27 29.9	98	17	167.3
3	28 47.8	98	19	166.0	28 18.4	98	19	166.1	27 49.0	98	19	166.2	27 19.5	98	19	166.3
4	28 36.4	98	20	164.9	28 07.1	98	20	165.0	27 37.7	98	20	165.1	27 08.4	98	20	165.2
15	28 24.2	98	22	163.8	27 55.0	98	22	164.0	27 25.7	98	21	164.1	26 56.5	98	21	164.2
6	28 11.2	97	23	162.8	27 42.1	97	23	162.9	27 12.9	97	23	163.1	26 43.8	97	22	163.2
7	27 57.5	97	24	161.8	27 28.4	97	24	161.9	26 59.3	97	24	162.0	26 30.3	97	24	162.2
8	27 42.9	96	26	160.7	27 14.0	97	25	160.9	26 45.0	97	25	161.0	26 16.0	97	25	161.2
9	27 27.6	96	27	159.7	26 58.7	96	27	159.9	26 29.9	96	26	160.0	26 01.0	96	26	160.2
20	27 11.5	96	28	158.7	26 42.8	96	28	158.8	26 14.0	96	28	159.0	25 45.3	96	27	159.2
1	26 54.7	95	29	157.7	26 26.1	95	29	157.8	25 57.5	95	29	158.0	25 28.8	95	29	158.2
2	26 37.1	95	31	156.7	26 08.6	95	30	156.8	25 40.1	95	30	157.0	25 11.6	95	30	157.2
3	26 18.8	94	32	155.7	25 50.4	94	31	155.8	25 22.1	95	31	156.0	24 53.7	95	31	156.2
4	25 59.8	94	33	154.7	25 31.6	94	33	154.9	25 03.3	94	32	155.0	24 35.1	94	32	155.2
25	25 40.0	94	34	153.7	25 12.0	94	34	153.9	24 43.9	94	34	154.1	24 15.8	94	33	154.3
6	25 19.6	93	35	152.7	24 51.7	93	35	152.9	24 23.7	93	35	153.1	23 55.8	93	34	153.3
7	24 58.5	93	36	151.7	24 30.7	93	36	151.9	24 02.9	93	36	152.2	23 35.1	93	36	152.4
8	24 36.7	92	37	150.8	24 09.1	92	37	151.0	23 41.4	92	37	151.2	23 13.8	92	37	151.4
9	24 14.3	92	38	149.8	23 46.8	92	38	150.0	23 19.3	92	38	150.3	22 51.8	92	38	150.5
55	11 17.2	78	60	127.8	10 53.9	78	60	128.2	10 30.5	78	59	128.5	10 07.2	78	59	128.8
6	10 41.2	77	60	127.1	10 18.1	77	60	127.4	9 54.9	77	60	127.7	9 31.7	77	60	128.1
7	10 04.9	77	61	126.3	9 41.9	77	61	126.7	9 18.9	77	61	127.0	8 55.9	77	60	127.3
8	9 28.3	76	62	125.6	9 05.4	76	61	125.9	8 42.6	76	61	126.3	8 19.7	76	61	126.6
9	8 51.3	76	62	124.9	8 28.6	76	62	125.2	8 05.9	76	62	125.6	7 43.2	76	61	125.9
60	8 14.0	75	63	124.2	7 51.5	75	62	124.5	7 28.9	75	62	124.8	7 06.3	75	62	125.2
1	7 36.4	75	63	123.5	7 14.0	75	63	123.8	6 51.6	75	63	124.1	6 29.2	75	62	124.5
2	6 58.4	74	64	122.7	6 36.2	74	63	123.1	6 13.9	74	63	123.4	5 51.7	74	63	123.8
3	6 20.2	74	64	122.0	5 58.1	74	64	122.4	5 36.0	74	64	122.7	5 13.9	74	63	123.1
4	5 41.7	73	65	121.3	5 19.7	73	64	121.7								

21° 00′ Alt.	Δd	Δt	Az.	21° 30′ Alt.	Δd	Δt	Az.	22° 00′ Alt.	Δd	Δt	Az.	22° 30′ Alt.	Δd	Δt	Az.	H.A.
° ′			°	° ′			°	° ′			°	° ′			°	°
28 00.0	1.0	01	180.0	27 30.0	1.0	01	180.0	27 00.0	1.0	01	180.0	26 30.0	1.0	01	180.0	00
27 59.6	1.0	02	178.9	27 29.6	1.0	02	179.0	26 59.6	1.0	02	179.0	26 29.6	1.0	02	179.0	1
27 58.3	1.0	04	177.9	27 28.3	1.0	03	177.9	26 58.4	1.0	03	177.9	26 28.4	1.0	03	177.9	2
27 56.2	1.0	05	176.8	27 26.3	1.0	05	176.9	26 56.3	1.0	05	176.9	26 26.3	1.0	05	176.9	3
27 53.3	1.0	06	175.8	27 23.4	1.0	06	175.8	26 53.4	1.0	06	175.8	26 23.5	1.0	06	175.9	4
27 49.6	1.0	08	174.7	27 19.7	1.0	08	174.8	26 49.7	1.0	08	174.8	26 19.8	1.0	07	174.8	05
27 45.0	1.0	09	173.7	27 15.1	1.0	09	173.7	26 45.2	1.0	09	173.8	26 15.3	1.0	09	173.8	6
27 39.6	99	10	172.6	27 09.7	99	10	172.7	26 39.9	99	10	172.7	26 10.1	99	10	172.8	7
27 33.4	99	12	171.6	27 03.6	99	12	171.6	26 33.8	99	12	171.7	26 04.0	99	11	171.8	8
27 26.3	99	13	170.5	26 56.6	99	13	170.6	26 26.8	99	13	170.7	25 57.1	99	13	170.8	9
27 18.5	99	14	169.5	26 48.8	99	14	169.6	26 19.1	99	14	169.7	25 49.4	99	14	169.7	10
27 09.8	99	16	168.5	26 40.2	99	16	168.5	26 10.6	99	16	168.6	25 41.0	99	15	168.7	1
27 00.3	98	17	167.4	26 30.8	98	17	167.5	26 01.3	98	17	167.6	25 31.7	98	17	167.7	2
26 50.1	98	18	166.4	26 20.6	98	18	166.5	25 51.2	98	18	166.6	25 21.7	98	18	166.7	3
26 39.0	98	20	165.4	26 09.6	98	20	165.5	25 40.3	98	19	165.6	25 10.9	98	19	165.7	4
26 27.2	98	21	164.3	25 57.9	98	21	164.5	25 28.6	98	21	164.6	24 59.3	98	21	164.7	15
26 14.6	97	22	163.3	25 45.4	97	22	163.5	25 16.2	97	22	163.6	24 47.0	97	22	163.7	6
26 01.2	97	24	162.3	25 32.1	97	23	162.5	25 03.0	97	23	162.6	24 33.9	97	23	162.7	7
25 47.1	97	25	161.3	25 18.1	97	25	161.5	24 49.1	97	24	161.6	24 20.1	97	24	161.7	8
25 32.2	96	26	160.3	25 03.3	96	26	160.5	24 34.4	96	26	160.6	24 05.5	96	25	160.8	9
25 16.6	96	27	159.3	24 47.8	96	27	159.5	24 19.0	96	27	159.6	23 50.3	96	27	159.8	20
25 00.2	95	28	158.3	24 31.6	95	28	158.5	24 02.9	96	28	158.7	23 34.3	96	28	158.8	1
24 43.1	95	30	157.4	24 14.6	95	29	157.5	23 46.1	95	29	157.7	23 17.6	95	29	157.9	2
24 25.3	95	31	156.4	23 57.0	95	31	156.6	23 28.6	95	30	156.7	23 00.1	95	30	156.9	3
24 06.9	94	32	155.4	23 38.6	94	32	155.6	23 10.3	94	32	155.8	22 42.0	94	31	156.0	4
23 47.7	94	33	154.5	23 19.5	94	33	154.6	22 51.4	94	33	154.8	22 23.3	94	32	155.0	25
23 27.8	93	34	153.5	22 59.8	93	34	153.7	22 31.8	93	34	153.9	22 03.8	93	34	154.1	6
23 07.3	93	35	152.6	22 39.4	93	35	152.8	22 11.6	93	35	153.0	21 43.7	93	35	153.2	7
22 46.1	92	36	151.6	22 18.4	92	36	151.8	21 50.6	92	36	152.0	21 22.9	92	36	152.2	8
22 24.2	92	38	150.7	21 56.7	92	37	150.9	21 29.1	92	37	151.1	21 01.5	92	37	151.3	9
9 43.9	78	59	129.1	9 20.5	78	59	129.4	8 57.1	78	58	129.7	8 33.7	78	58	130.1	55
9 08.5	77	59	128.4	8 45.3	77	59	128.7	8 22.1	77	59	129.0	7 58.8	78	59	129.3	6
8 32.9	77	60	127.7	8 09.8	77	60	128.0	7 46.7	77	60	128.3	7 23.6	77	59	128.6	7
7 56.8	76	61	126.9	7 33.9	76	60	127.3	7 11.0	76	60	127.6	6 48.1	76	60	127.9	8
7 20.5	76	61	126.2	6 57.7	76	61	126.5	6 35.0	76	61	126.9	6 12.2	76	60	127.2	9
6 43.8	75	62	125.5	6 21.2	75	61	125.8	5 58.6	75	61	126.2	5 35.9	75	61	126.5	60
6 06.7	75	62	124.8	5 44.3	75	62	125.1	5 21.8	75	62	125.5					1
5 29.4	74	63	124.1	5 07.1	74	62	124.4									2
																3
																4

Lat. 42°

DECLINATION **CONTRARY** NAME TO LATITUDE

H.A.	19° 00′			19° 30′			20° 00′			20° 30′			21° 00′			21° 30′			22° 00′			22° 30′			H.A.
	Alt.		Az.	Alt.		Az.	Alt.		Az.	Alt.		Az.	Alt.		Az.	Alt.		Az.	Alt.		Az.	Alt.		Az.	
°	° ′	Δd Δt	°	° ′	Δd Δt	°	° ′	Δd Δt	°	° ′	Δd Δt	°	° ′	Δd Δt	°	° ′	Δd Δt	°	° ′	Δd Δt	°	° ′	Δd Δt	°	°
00	29 00.0	1.0 01	180.0	28 30.0	1.0 01	180.0	28 00.0	1.0 01	180.0	27 30.0	1.0 01	180.0	27 00.0	1.0 01	180.0	26 30.0	1.0 01	180.0	26 00.0	1.0 01	180.0	25 30.0	1.0 01	180.0	*00*
1	28 59.6	1.0 02	178.9	28 29.6	1.0 02	178.9	27 59.6	1.0 02	178.9	27 29.6	1.0 02	178.9	26 59.6	1.0 02	179.0	26 29.6	1.0 02	179.0	25 59.6	1.0 02	179.0	25 29.6	1.0 02	179.0	*1*
2	28 58.3	1.0 04	177.8	28 28.3	1.0 03	177.9	27 58.3	1.0 03	177.9	27 28.4	1.0 03	177.9	26 58.4	1.0 03	177.9	26 28.4	1.0 03	177.9	25 58.4	1.0 03	177.9	25 28.4	1.0 03	178.0	*2*
3	28 56.2	1.0 05	176.8	28 26.2	1.0 05	176.8	27 56.3	1.0 05	176.8	27 26.3	1.0 05	176.8	26 56.3	1.0 05	176.9	26 26.4	1.0 05	176.9	25 56.4	1.0 05	176.9	25 26.4	1.0 05	176.9	*3*
4	28 53.3	1.0 06	175.7	28 23.3	1.0 06	175.7	27 53.4	1.0 06	175.7	27 23.4	1.0 06	175.8	26 53.5	1.0 06	175.8	26 23.5	1.0 06	175.8	25 53.6	1.0 06	175.9	25 23.6	1.0 06	175.9	*4*
05	28 49.5	1.0 08	174.6	28 19.6	1.0 08	174.6	27 49.7	1.0 08	174.7	27 19.7	1.0 08	174.7	26 49.8	1.0 07	174.8	26 19.9	1.0 07	174.8	25 50.0	1.0 07	174.8	25 20.1	1.0 07	174.9	*05*
6	28 44.9	1.0 09	173.5	28 15.0	1.0 09	173.6	27 45.1	1.0 09	173.6	27 15.2	1.0 09	173.7	26 45.4	1.0 09	173.7	26 15.5	1.0 09	173.8	25 45.6	1.0 09	173.8	25 15.7	1.0 09	173.9	*6*
7	28 39.4	99 10	172.5	28 09.6	99 10	172.5	27 39.8	99 10	172.6	27 09.9	99 10	172.6	26 40.1	99 10	172.7	26 10.2	99 10	172.7	25 40.4	99 10	172.8	25 10.5	99 10	172.9	*7*
8	28 33.2	99 12	171.4	28 03.4	99 12	171.5	27 33.6	99 12	171.5	27 03.8	99 12	171.6	26 34.0	99 11	171.6	26 04.2	99 11	171.7	25 34.4	99 11	171.8	25 04.6	99 11	171.8	*8*
9	28 26.1	99 13	170.3	27 56.4	99 13	170.4	27 26.6	99 13	170.5	26 56.9	99 13	170.5	26 27.1	99 13	170.6	25 57.4	99 13	170.7	25 27.6	99 13	170.8	24 57.9	99 12	170.8	*9*
10	28 18.2	99 15	169.3	27 48.5	99 14	169.3	27 18.8	99 14	169.4	26 49.1	99 14	169.5	26 19.5	99 14	169.6	25 49.8	99 14	169.7	25 20.1	99 14	169.7	24 50.4	99 14	169.8	*10*
1	28 09.5	99 16	168.2	27 39.9	99 16	168.3	27 10.2	99 16	168.4	26 40.6	99 16	168.5	26 11.0	99 15	168.6	25 41.4	99 15	168.6	25 11.7	99 15	168.7	24 42.1	99 15	168.8	*1*
2	27 59.9	98 17	167.1	27 30.4	98 17	167.2	27 00.9	98 17	167.3	26 31.3	98 17	167.4	26 01.8	99 17	167.5	25 32.2	99 17	167.6	25 02.6	99 16	167.7	24 33.1	99 16	167.8	*2*
3	27 49.6	98 19	166.1	27 20.1	98 18	166.2	26 50.7	98 18	166.3	26 21.2	98 18	166.4	25 51.7	98 18	166.5	25 22.3	98 18	166.6	24 52.8	98 18	166.7	24 23.3	98 18	166.8	*3*
4	27 38.5	98 20	165.0	27 09.1	98 20	165.2	26 39.7	98 20	165.3	26 10.3	98 19	165.4	25 40.9	98 19	165.5	25 11.5	98 19	165.6	24 42.1	98 19	165.7	24 12.7	98 19	165.8	*4*
15	27 26.6	98 21	164.0	26 57.3	98 21	164.1	26 28.0	98 21	164.2	25 58.7	98 21	164.4	25 29.4	98 21	164.5	25 00.1	98 20	164.6	24 30.8	98 20	164.7	24 01.4	98 20	164.8	*15*
6	27 13.9	97 22	163.0	26 44.7	97 22	163.1	26 15.5	97 22	163.2	25 46.3	97 22	163.3	25 17.1	97 22	163.5	24 47.9	97 22	163.6	24 18.6	97 21	163.7	23 49.4	97 21	163.8	*6*
7	27 00.4	97 24	161.9	26 31.4	97 24	162.1	26 02.2	97 23	162.2	25 33.1	97 23	162.3	25 04.0	97 23	162.5	24 34.9	97 23	162.6	24 05.7	97 23	162.7	23 36.6	97 22	162.9	*7*
8	26 46.2	97 25	160.9	26 17.2	97 25	161.0	25 48.2	97 25	161.2	25 19.2	97 24	161.3	24 50.2	97 24	161.5	24 21.2	97 24	161.6	23 52.1	97 24	161.7	23 23.1	97 24	161.9	*8*
9	26 31.3	96 26	159 9	26 02.4	96 26	160.0	25 33.5	96 26	160.2	25 04.6	96 26	160.3	24 35.7	96 25	160.5	24 06.7	96 25	160.6	23 37.8	96 25	160.8	23 08.9	96 25	160.9	*9*
20	26 15.6	96 27	158.9	25 46.8	96 27	159.0	25 18.0	96 27	159.2	24 49.2	96 27	159.3	24 20.4	96 27	159.5	23 51.6	96 26	159.6	23 22.8	96 26	159.8	22 53.9	96 26	159.9	*20*
1	25 59.1	96 29	157.9	25 30.5	96 28	158.0	25 01.8	96 28	158.2	24 33.1	96 28	158.3	24 04.4	96 28	158.5	23 35.7	96 28	158.7	23 07.0	96 27	158.8	22 38.3	96 27	159.0	*1*
2	25 41.9	95 30	156.9	25 13.4	95 30	157.0	24 44.9	95 29	157.2	24 16.3	95 29	157.4	23 47.7	95 29	157.5	23 19.1	95 29	157.7	22 50.5	95 29	157.9	22 21.9	95 28	158.0	*2*
3	25 24.1	95 31	155.9	24 55.7	95 31	156.0	24 27.2	95 31	156.2	23 58.8	95 30	156.4	23 30.3	95 30	156.6	23 01.9	95 30	156.7	22 33.4	95 30	156.9	22 04.9	95 30	157.1	*3*
4	25 05.5	94 32	154.9	24 37.2	94 32	155.1	24 08.9	94 32	155.2	23 40.6	94 32	155.4	23 12.3	94 31	155.6	22 43.9	94 31	155.8	22 15.6	95 31	156.0	21 47.2	95 31	156.1	*4*
25	24 46.2	94 33	153.9	24 18.0	94 33	154.1	23 49.9	94 33	154.3	23 21.7	94 33	154.5	22 53.5	94 32	154.6	22 25.3	94 32	154.8	21 57.1	94 32	155.0	21 28.8	94 32	155.2	*25*
6	24 26.2	93 34	152.9	23 58.2	93 34	153.1	23 30.2	93 34	153.3	23 02.1	94 34	153.5	22 34.1	94 34	153.7	22 06.0	94 33	153.9	21 37.9	94 33	154.1	21 09.8	94 33	154.3	*6*
7	24 05.6	93 36	152.0	23 37.7	93 35	152.2	23 09.8	93 35	152.4	22 41.9	93 35	152.6	22 14.0	93 35	152.8	21 46.0	93 34	152.9	21 18.1	93 34	153.1	20 50.1	93 34	153.3	*7*
8	23 44.3	92 37	151.0	23 16.6	93 36	151.2	22 48.8	93 36	151.4	22 21.0	·93 36	151.6	21 53.2	93 36	151.8	21 25.4	93 35	152.0	20 57.6	93 35	152.2	20 29.8	93 35	152.4	*8*
9	23 22.4	92 38	150.0	22 54.8	92 37	150.3	22 27.1	92 37	150.5	21 59.5	92 37	150.7	21 31.9	92 37	150.9	21 04.2	92 36	151.1	20 36.5	92 36	151.3	20 08.8	92 36	151.5	*9*
55	10 40.3	78 59	128.0	10 16.7	79 59	128.3	9 53.2	79 58	128.6	9 29.6	79 58	128.9	9 06.0	79 58	129.2	8 42.3	79 58	129.6	8 18.7	79 57	129.9	7 55.0	79 57	130.2	*55*
6	10 05.0	78 59	127.2	9 41.6	78 59	127.6	9 18.1	78 59	127.9	8 54.7	78 59	128.2	8 31.2	78 58	128.5	8 07.8	78 58	128.8	7 44.3	78 58	129.1	7 20.8	78 58	129.4	*6*
7	9 29.3	78 60	126.5	9 06.0	78 60	126.8	8 42.8	78 60	127.1	8 19.5	78 59	127.4	7 56.2	78 59	127.8	7 32.8	78 59	128.1	7 09.5	78 59	128.4	6 46.2	78 58	128.7	*7*
8	8 53.3	77 61	125.7	8 30.2	77 60	126.1	8 07.0	77 60	126.4	7 43.9	77 60	126.7	7 20.7	77 60	127.0	6 57.6	77 59	127.4	6 34.4	77 59	127.7	6 11.2	77 59	128.0	*8*
9	8 16.9	77 61	125.0	7 54.0	77 61	125.3	7 31.0	77 61	125.7	7 08.0	77 60	126.0	6 45.0	77 60	126.3	6 22.0	77 60	126.6	5 58.9	77 60	127.0	5 35.9	77 59	127.3	*9*
60	7 40.2	76 62	124.3	7 17.4	76 61	124.6	6 54.6	76 61	124.9	6 31.7	76 61	125.3	6 08.9	76 61	125.6	5 46.0	76 60	125.9	5 23.1	76 60	126.2	5 00.2	76 60	126.6	*60*
1	7 03.2	76 62	123.6	6 40.6	76 62	123.9	6 17.9	76 62	124.2	5 55.2	76 61	124.6	5 32.5	76 61	124.9	5 09.7	76 61	125.2							*1*
2	6 25.9	75 63	122.8	6 03.4	75 62	123.2	5 40.9	75 62	123.5	5 18.3	75 62	123.8													*2*
3	5 48.3	75 63	122.1	5 25.9	75 63	122.5	5 03.5	75 63	122.8																*3*
4	5 10.4	74 64	121.4																						*4*

ALTITUDE CORRECTION FOR D. R. LATITUDE

Az.		LATITUDE DIFFERENCE (minutes of arc) 1′	2′	3′	4′	5′	6′	7′	8′	9′	10′	11′	12′	13′	14′	15′	Az.		LAT. DIFF. (tenths of minutes of arc) 0.1′	0.2′	0.3′	0.4′	0.5′	0.6′	0.7′	0.8′	0.9′	Az.	
°	°																°	°										°	°
0	**180**	1.0	2.0	3.0	4.0	5.0	6.0	7.0	8.0	9.0	10.0	11.0	12.0	13.0	14.0	15.0	**0**	**180**	0. 1	0.2	0.3	0.4	0.5	0.6	0.7	0.8	0.9	**0**	**180**
1	179	1.0	2.0	3.0	4.0	5.0	6.0	7.0	8.0	9.0	10.0	11.0	12.0	13.0	14.0	15.0	1	179										1	179
2	178	1.0	2.0	3.0	4.0	5.0	6.0	7.0	8.0	9.0	10.0	11.0	12.0	13.0	14.0	15.0	2	178										2	178
3	177	1.0	2.0	3.0	4.0	5.0	6.0	7.0	8.0	9.0	10.0	11.0	12.0	13.0	14.0	15.0	3	177										3	177
4	176	1.0	2.0	3.0	4.0	5.0	6.0	7.0	8.0	9.0	10.0	11.0	12.0	13.0	14.0	15.0	4	176										4	176
5	**175**	1.0	2.0	3.0	4.0	5.0	6.0	7.0	8.0	9.0	10.0	11.0	12.0	13.0	13.9	14 9	**5**	**175**	0.1	0.2	0.3	0.4	0.5	0.6	0.7	0.8	0.9	**5**	**175**
6	174	1.0	2.0	3.0	4.0	5.0	6.0	7.0	8.0	9.0	9.9	10.9	11.9	12.9	13.9	14.9	6	174										6	174
7	173	1.0	2.0	3.0	4.0	5.0	6.0	6.9	7.9	8.9	9.9	10.9	11.9	12.9	13.9	14.9	7	173										7	173
8	172	1.0	2.0	3.0	4.0	5.0	5.9	6.9	7.9	8.9	9.9	10.9	11.9	12.9	13.9	14.9	8	172										8	172
9	171	1.0	2.0	3.0	4.0	4.9	5.9	6.9	7.9	8.9	9.9	10.9	11.9	12.8	13.8	14.8	9	171										9	171
10	**170**	1.0	2.0	3.0	3.9	4.9	5.9	6.9	7.9	8.9	9.8	10.8	11.8	12.8	13.8	14.8	**10**	**170**	0.1	0.2	0.3	0.4	0.5	0.6	0.7	0.8	0.9	**10**	**170**
11	169	1.0	2.0	2.9	3.9	4.9	5.9	6.9	7.9	8.8	9.8	10.8	11.8	12.8	13.7	14.7	11	169										11	169
12	168	1.0	2.0	2.9	3.9	4.9	5.9	6.8	7.8	8.8	9.8	10.8	11.7	12.7	13.7	14.7	12	168										12	168
13	167	1.0	1.9	2.9	3.9	4.9	5.8	6.8	7.8	8.8	9.7	10.7	11.7	12.6	13.6	14.6	13	167										13	167
14	166	1.0	1.9	2.9	3.9	4.9	5.8	6.8	7.8	8.7	9.7	10.7	11.6	12.6	13.6	14.6	14	166										14	166
15	**165**	1.0	1.9	2.9	3.9	4.8	5.8	6.8	7.7	8.7	9.7	10.6	11.6	12.6	13.5	14.5	**15**	**165**	0.1	0.2	0.3	0.4	0.5	0.6	0.7	0.8	0.9	**15**	**165**
16	164	1.0	1.9	2.9	3.8	4.8	5.8	6.7	7.7	8.7	9.6	10.6	11.5	12.5	13.5	14.4	16	164										16	164
17	163	1.0	1.9	2.9	3.8	4.8	5.7	6.7	7.7	8.6	9.6	10.5	11.5	12.4	13.4	14.3	17	163										17	163
18	162	1.0	1.9	2.9	3.8	4.8	5.7	6.7	7.6	8.6	9.5	10.5	11.4	12.4	13.3	14.3	18	162										18	162
19	161	0.9	1.9	2.8	3.8	4.7	5.7	6.6	7.6	8.5	9.5	10.4	11.3	12.3	13.2	14.2	19	161									0.9	19	161
20	**160**	0.9	1.9	2.8	3.8	4.7	5.6	6.6	7.5	8.5	9.4	10.3	11.3	12.2	13.2	14.1	**20**	**160**	0.1	0.2	0.3	0.4	0.5	0.6	0.7	0.8	0.8	**20**	**160**
21	159	0.9	1.9	2.8	3.7	4.7	5.6	6.5	7.5	8.4	9.3	10.3	11.2	12.1	13.1	14.0	21	159							0.7	0.7		21	159
22	158	0.9	1.9	2.8	3.7	4.6	5.6	6.5	7.4	8.3	9.3	10.2	11.1	12.1	13.0	13.9	22	158							0.6			22	158
23	157	0.9	1.8	2.8	3.7	4.6	5.5	6.4	7.4	8.3	9.2	10.1	11.0	12.0	12.9	13.8	23	157						0.6				23	157
24	156	0.9	1.8	2.7	3.7	4.6	5.5	6.4	7.3	8.2	9.1	10.0	11.0	11.9	12.8	13.7	24	156						0.5				24	156
25	**155**	0.9	1.8	2.7	3.6	4.5	5.4	6.3	7.3	8.2	9.1	10.0	10.9	11.8	12.7	13.6	**25**	**155**	0.1	0.2	0.3	0.4	0.5	0.5	0.6	0.7	0.8	**25**	**155**
26	154	0.9	1.8	2.7	3.6	4.5	5.4	6.3	7.2	8.1	9.0	9.9	10.8	11.7	12.6	13.5	26	154					0.4					26	154
27	153	0.9	1.8	2.7	3.6	4.5	5.3	6.2	7.1	8.0	8.9	9.8	10.7	11.6	12.5	13.4	27	153										27	153
28	152	0.9	1.8	2.6	3.5	4.4	5.3	6.2	7.1	7.9	8.8	9.7	10.6	11.5	12.4	13.2	28	152				0.4						28	152
29	151	0.9	1.7	2.6	3.5	4.4	5.2	6.1	7.0	7.9	8.7	9.6	10.5	11.4	12.2	13.1	29	151				0.3						29	151
60	**120**	0.5	1.0	1.5	2.0	2.5	3.0	3.5	4.0	4.5	5.0	5.5	6.0	6.5	7.0	7.5	**60**	**120**	0.0	0.1	0.2	0.2	0.2	0.3	0.4	0.4	0.4	**60**	**120**
61	119	0.5	1.0	1.5	1.9	2.4	2.9	3.4	3.9	4.4	4.8	5.3	5.8	6.3	6.8	7.3	61	119			0.1				0.3			61	119
62	118	0.5	0.9	1.4	1.9	2.3	2.8	3.3	3.8	4.2	4.7	5.2	5.6	6.1	6.6	7.0	62	118										62	118
63	117	0.5	0.9	1.4	1.8	2.3	2.7	3.2	3.6	4.1	4.5	5.0	5.4	5.9	6.4	6.8	63	117										63	117
64	116	0.4	0.9	1.3	1.8	2.2	2.6	3.1	3.5	3.9	4.4	4.8	5.3	5.7	6.1	6.6	64	116								0.4		64	116
75	**105**	0.3	0.5	0.8	1.0	1.3	1.6	1.8	2.1	2.3	2.6	2.8	3.1	3.4	3.6	3.9	**75**	**105**	0.0	0.1	0.1	0.1	0.1	0.2	0.2	0.2	0.2	**75**	**105**
76	104	0.2	0.5	0.7	1.0	1.2	1.5	1.7	1.9	2.2	2.4	2.7	2.9	3.1	3.4	3.6	76	104		0.0				0.1				76	104
77	103	0.2	0.4	0.7	0.9	1.1	1.3	1.6	1.8	2.0	2.2	2.5	2.7	2.9	3.1	3.4	77	103							0.2			77	103
78	102	0.2	0.4	0.6	0.8	1.0	1.2	1.5	1.7	1.9	2.1	2.3	2.5	2.7	2.9	3.1	78	102							0.1			78	102
79	101	0.2	0.4	0.6	0.8	1.0	1.1	1.3	1.5	1.7	1.9	2.1	2.3	2.5	2.7	2.9	79	101								0.2		79	101
80	**100**	0.2	0.3	0.5	0.7	0.9	1.0	1.2	1.4	1.6	1.7	1.9	2.1	2.3	2.4	2.6	**80**	**100**	0.0	0.0	0.1	0.1	0.1	0.1	0.1	0.1	0.2	**80**	**100**
81	99	0.2	0.3	0.5	0.6	0.8	0.9	1.1	1.3	1.4	1.6	1.7	1.9	2.0	2.2	2.3	81	99			0.0						0.1	81	99
82	98	0.1	0.3	0.4	0.6	0.7	0.8	1.0	1.1	1.3	1.4	1.5	1.7	1.8	1.9	2.1	82	98				0.1						82	98
83	97	0.1	0.2	0.4	0.5	0.6	0.7	0.9	1.0	1.1	1.2	1.3	1.5	1.6	1.7	1.8	83	97				0.0						83	97
84	96	0.1	0.2	0.3	0.4	0.5	0.6	0.7	0.8	0.9	1.0	1.1	1.3	1.4	1.5	1.6	84	96					0.1					84	96
Az.		1′	2′	3′	4′	5′	6′	7′	8′	9′	10′	11′	12′	13′	14′	15′	Az.		0.1′	0.2′	0.3′	0.4′	0.5′	0.6′	0.7′	0.8′	0.9′	Az.	

Azimuth angle greater than 90°:
If DR latitude is greater than selected tabulated latitude, ΔL correction is *minus;* but for DR latitude less than selected tabulated latitude, the correction is *plus.*
Azimuth angle less than 90°:
If DR latitude is greater than selected tabulated latitude, ΔL correction is *plus;* but for DR latitude less than selected tabulated latitude, the correction is *minus.*

ALTITUDE CORRECTION FOR D. R. LATITUDE

		LATITUDE DIFFERENCE (minutes of arc)																	LAT. DIFF. (tenths of minutes of arc)										
Az.		16′	17′	18′	19′	20′	21′	22′	23′	24′	25′	26′	27′	28′	29′	30′	Az.		0.1′	0.2′	0.3′	0.4′	0.5′	0.6′	0.7′	0.8′	0.9′	Az.	
°	°																°	°										°	°
0	**180**	16.0	17.0	18.0	19.0	20.0	21.0	22.0	23.0	24.0	25.0	26.0	27.0	28.0	29.0	30.0	**0**	**180**	0.1	0.2	0.3	0.4	0.5	0.6	0.7	0.8	0.9	**0**	**180**
1	179	16.0	17.0	18.0	19.0	20.0	21.0	22.0	23.0	24.0	25.0	26.0	27.0	28.0	29.0	30.0	1	179										1	179
2	178	16.0	17.0	18.0	19.0	20.0	21.0	22.0	23.0	24.0	25.0	26.0	27.0	28.0	29.0	30.0	2	178										2	178
3	177	16.0	17.0	18.0	19.0	20.0	21.0	22.0	23.0	24.0	25.0	26.0	27.0	28.0	29.0	30.0	3	177										3	177
4	176	16.0	17.0	18.0	19.0	20.0	20.9	21.9	22.9	23.9	24.9	25.9	26.9	27.9	28.9	29.9	4	176										4	176
5	**175**	15.9	16.9	17.9	18.9	19.9	20.9	21.9	22.9	23.9	24.9	25.9	26.9	27.9	28.9	29.9	**5**	**175**	0.1	0.2	0.3	0.4	0.5	0.6	0.7	0.8	0.9	**5**	**175**
6	174	15.9	16.9	17.9	18.9	19.9	20.9	21.9	22.9	23.9	24.9	25.9	26.9	27.8	28.8	29.8	6	174										6	174
7	173	15.9	16.9	17.9	18.9	19.9	20.8	21.8	22.8	23.8	24.8	25.8	26.8	27.8	28.8	29.8	7	173										7	173
8	172	15.8	16.8	17.8	18.8	19.8	20.8	21.8	22.8	23.8	24.8	25.7	26.7	27.7	28.7	29.7	8	172										8	172
9	171	15.8	16.8	17.8	18.8	19.8	20.7	21.7	22.7	23.7	24.7	25.7	26.7	27.7	28.6	29.6	9	171										9	171
10	**170**	15.8	16.7	17.7	18.7	19.7	20.7	21.7	22.7	23.6	24.6	25.6	26.6	27.6	28.6	29.5	**10**	**170**	0.1	0.2	0.3	0.4	0.5	0.6	0.7	0.8	0.9	**10**	**170**
11	169	15.7	16.7	17.7	18.7	19.6	20.6	21.6	22.6	23.6	24.5	25.5	26.5	27.5	28.5	29.4	11	169										11	169
12	168	15.7	16.6	17.6	18.6	19.6	20.5	21.5	22.5	23.5	24.5	25.4	26.4	27.4	28.4	29.3	12	168										12	168
13	167	15.6	16.6	17.5	18.5	19.5	20.5	21.4	22.4	23.4	24.4	25.3	26.3	27.3	28.3	29.2	13	167										13	167
14	166	15.5	16.5	17.5	18.4	19.4	20.4	21.3	22.3	23.3	24.3	25.2	26.2	27.2	28.1	29.1	14	166										14	166
15	**165**	15.5	16.4	17.4	18.4	19.3	20.3	21.3	22.2	23.2	24.1	25.1	26.1	27.0	28.0	29.0	**15**	**165**	0.1	0.2	0.3	0.4	0.5	0.6	0.7	0.8	0.9	**15**	**165**
16	164	15.4	16.3	17.3	18.3	19.2	20.2	21.1	22.1	23.1	24.0	25.0	26.0	26.9	27.9	28.8	16	164										16	164
17	163	15.3	16.3	17.2	18.2	19.1	20.1	21.0	22.0	23.0	23.9	24.9	25.8	26.8	27.7	28.7	17	163										17	163
18	162	15.2	16.2	17.1	18.1	19.0	20.0	20.9	21.9	22.8	23.8	24.7	25.7	26.6	27.6	28.5	18	162										18	162
19	161	15.1	16.1	17.0	18.0	18.9	19.9	20.8	21.7	22.7	23.6	24.6	25.5	26.5	27.4	28.4	19	161									0.9	19	161
20	**160**	15.0	16.0	16.9	17.9	18.8	19.7	20.7	21.6	22.6	23.5	24.4	25.4	26.3	27.3	28.2	**20**	**160**	0.1	0.2	0.3	0.4	0.5	0.6	0.7	0.8	0.8	**20**	**160**
21	159	14.9	15.9	16.8	17.7	18.7	19.6	20.5	21.5	22.4	23.3	24.3	25.2	26.1	27.1	28.0	21	159							0.7	0.7		21	159
22	158	14.8	15.8	16.7	17.6	18.5	19.5	20.4	21.3	22.3	23.2	24.1	25.0	26.0	26.9	27.8	22	158							0.6			22	158
23	157	14.7	15.6	16.6	17.5	18.4	19.3	20.3	21.2	22.1	23.0	23.9	24.9	25.8	26.7	27.6	23	157						0.6				23	157
24	156	14.6	15.5	16.4	17.4	18.3	19.2	20.1	21.0	21.9	22.8	23.8	24.7	25.6	26.5	27.4	24	156						0.5				24	156
25	**155**	14.5	15.4	16.3	17.2	18.1	19.0	19.9	20.8	21.8	22.7	23.6	24.5	25.4	26.3	27.2	**25**	**155**	0.1	0.2	0.3	0.4	0.5	0.5	0.6	0.7	0.8	**25**	**155**
26	154	14.4	15.3	16.2	17.1	18.0	18.9	19.8	20.7	21.6	22.5	23.4	24.3	25.2	26.1	27.0	26	154					0.4					26	154
27	153	14.3	15.1	16.0	16.9	17.8	18.7	19.6	20.5	21.4	22.3	23.2	24.1	24.9	25.8	26.7	27	153										27	153
28	152	14.1	15.0	15.9	16.8	17.7	18.5	19.4	20.3	21.2	22.1	23.0	23.8	24.7	25.6	26.5	28	152				0.4						28	152
29	151	14.0	14.9	15.7	16.6	17.5	18.4	19.2	20.1	21.0	21.9	22.7	23.6	24.5	25.4	26.2	29	151				0.3						29	151
60	**120**	8.0	8.5	9.0	9.5	10.0	10.5	11.0	11.5	12.0	12.5	13.0	13.5	14.0	14.5	15.0	**60**	**120**	0.0	0.1	0.2	0.2	0.2	0.3	0.4	0.4	0.4	**60**	**120**
61	119	7.8	8.2	8.7	9.2	9.7	10.2	10.7	11.2	11.6	12.1	12.6	13.1	13.6	14.1	14.5	61	119			0.1				0.3			61	119
62	118	7.5	8.0	8.5	8.9	9.4	9.9	10.3	10.8	11.3	11.7	12.2	12.7	13.1	13.6	14.1	62	118										62	118
63	117	7.3	7.7	8.2	8.6	9.1	9.5	10.0	10.4	10.9	11.3	11.8	12.3	12.7	13.2	13.6	63	117										63	117
64	116	7.0	7.5	7.9	8.3	8.8	9.2	9.6	10.1	10.5	11.0	11.4	11.8	12.3	12.7	13.2	64	116								0.4		64	116
75	**105**	4.1	4.4	4.7	4.9	5.2	5.4	5.7	6.0	6.2	6.5	6.7	7.0	7.2	7.5	7.8	**75**	**105**	0.0	0.1	0.1	0.1	0.1	0.2	0.2	0.2	0.2	**75**	**105**
76	104	3.9	4.1	4.4	4.6	4.8	5.1	5.3	5.6	5.8	6.0	6.3	6.5	6.8	7.0	7.3	76	104		0.0				0.1				76	104
77	103	3.6	3.8	4.0	4.3	4.5	4.7	5.0	5.2	5.4	5.6	5.8	6.1	6.3	6.5	6.7	77	103							0.2			77	103
78	102	3.3	3.5	3.7	4.0	4.2	4.4	4.6	4.8	5.0	5.2	5.4	5.6	5.8	6.0	6.2	78	102							0.1			78	102
79	101	3.0	3.2	3.4	3.6	3.8	4.0	4.2	4.4	4.6	4.8	5.0	5.2	5.3	5.5	5.7	79	101								0.2		79	101
80	**100**	2.8	3.0	3.1	3.3	3.5	3.6	3.8	4.0	4.2	4.3	4.5	4.7	4.9	5.0	5.2	**80**	**100**	0.0	0.0	0.1	0.1	0.1	0.1	0.1	0.1	0.2	**80**	**100**
81	99	2.5	2.7	2.8	3.0	3.1	3.3	3.4	3.6	3.8	3.9	4.1	4.2	4.4	4.5	4.7	81	99			0.0						0.1	81	99
82	98	2.2	2.4	2.5	2.6	2.8	2.9	3.1	3.2	3.3	3.5	3.6	3.8	3.9	4.0	4.2	82	98				0.1						82	98
83	97	1.9	2.1	2.2	2.3	2.4	2.6	2.7	2.8	2.9	3.0	3.2	3.3	3.4	3.5	3.7	83	97				0.0						83	97
84	96	1.7	1.8	1.9	2.0	2.1	2.2	2.3	2.4	2.5	2.6	2.7	2.8	2.9	3.0	3.1	84	96					0.1					84	96
Az.		16′	17′	18′	19′	20′	21′	22′	23′	24′	25′	26′	27′	28′	29′	30′	Az.		0.1′	0.2′	0.3′	0.4′	0.5′	0.6′	0.7′	0.8′	0.9′	Az.	

Azimuth angle greater than 90°:
If DR latitude is greater than selected tabulated latitude, ΔL correction is *minus;* but for DR latitude less than selected tabulated latitude, the correction is *plus.*

Azimuth angle less than 90°:
If DR latitude is greater than selected tabulated latitude, ΔL correction is *plus;* but for DR latitude less than selected tabulated latitude, the correction is *minus.*

MULTIPLICATION TABLE

	DEC. DIFF. OR H. A. DIFF. (minutes of arc)																DEC. DIFF. OR H. A. DIFF. (tenths of minutes)									
Δ	1′	2′	3′	4′	5′	6′	7′	8′	9′	10′	11′	12′	13′	14′	15′	Δ	0.1′	0.2′	0.3′	0.4′	0.5′	0.6′	0.7′	0.8′	0.9′	Δ
01	0.0	0.0	0.0	0.0	0.1	0.1	0.1	0.1	0.1	0.1	0.1	0.1	0.1	0.1	0.2	**01**	0.0	0.0	0.0	0.0	0.0	0.0	0.0	0.0	0.0	**01**
2	.0	.0	.1	.1	.1	.1	.1	.2	.2	.2	.2	.2	.3	.3	.3	2										2
3	.0	.1	.1	.1	.2	.2	.2	.2	.3	.3	.3	.4	.4	.4	.5	3										3
4	.0	.1	.1	.2	.2	.2	.3	.3	.4	.4	.4	.5	.5	.6	.6	4										4
25	0.3	0.5	0.8	1.0	1.3	1.5	1.8	2.0	2.3	2.5	2.8	3.0	3.3	3.5	3.8	**25**	0.0	0.1	0.1	0.1	0.1	0.2	0.2	0.2	0.2	**25**
6	.3	.5	.8	1.0	1.3	1.6	1.8	2.1	2.3	2.6	2.9	3.1	3.4	3.6	3.9	6										6
7	.3	.5	.8	1.1	1.4	1.6	1.9	2.2	2.4	2.7	3.0	3.2	3.5	3.8	4.1	7									0.2	7
8	.3	.6	.8	1.1	1.4	1.7	2.0	2.2	2.5	2.8	3.1	3.4	3.6	3.9	4.2	8									0.3	8
9	.3	.6	.9	1.2	1.5	1.7	2.0	2.3	2.6	2.9	3.2	3.5	3.8	4.1	4.4	9					0.1					9
30	0.3	0.6	0.9	1.2	1.5	1.8	2.1	2.4	2.7	3.0	3.3	3.6	3.9	4.2	4.5	**30**	0.0	0.1	0.1	0.1	0.2	0.2	0.2	0.2	0.3	**30**
1	.3	.6	.9	1.2	1.6	1.9	2.2	2.5	2.8	3.1	3.4	3.7	4.0	4.3	4.7	1								0.2		1
2	.3	.6	1.0	1.3	1.6	1.9	2.2	2.6	2.9	3.2	3.5	3.8	4.2	4.5	4.8	2								0.3		2
3	.3	.7	1.0	1.3	1.7	2.0	2.3	2.6	3.0	3.3	3.6	4.0	4.3	4.6	5.0	3										3
4	.3	.7	1.0	1.4	1.7	2.0	2.4	2.7	3.1	3.4	3.7	4.1	4.4	4.8	5.1	4										4
60	0.6	1.2	1.8	2.4	3.0	3.6	4.2	4.8	5.4	6.0	6.6	7.2	7.8	8.4	9.0	**60**	0.1	0.1	0.2	0.2	0.3	0.4	0.4	0.5	0.5	**60**
1	.6	1.2	1.8	2.4	3.1	3.7	4.3	4.9	5.5	6.1	6.7	7.3	7.9	8.5	9.2	1									0.5	1
2	.6	1.2	1.9	2.5	3.1	3.7	4.3	5.0	5.6	6.2	6.8	7.4	8.1	8.7	9.3	2				0.2					0.6	2
3	.6	1.3	1.9	2.5	3.2	3.8	4.4	5.0	5.7	6.3	6.9	7.6	8.2	8.8	9.5	3				0.3						3
4	.6	1.3	1.9	2.6	3.2	3.8	4.5	5.1	5.8	6.4	7.0	7.7	8.3	9.0	9.6	4							0.4			4
65	0.7	1.3	2.0	2.6	3.3	3.9	4.6	5.2	5.9	6.5	7.2	7.8	8.5	9.1	9.8	**65**	0.1	0.1	0.2	0.3	0.3	0.4	0.5	0.5	0.6	**65**
6	.7	1.3	2.0	2.6	3.3	4.0	4.6	5.3	5.9	6.6	7.3	7.9	8.6	9.2	9.9	6										6
7	.7	1.3	2.0	2.7	3.4	4.0	4.7	5.4	6.0	6.7	7.4	8.0	8.7	9.4	10.1	7										7
8	.7	1.4	2.0	2.7	3.4	4.1	4.8	5.4	6.1	6.8	7.5	8.2	8.8	9.5	10.2	8								0.5		8
9	.7	1.4	2.1	2.8	3.5	4.1	4.8	5.5	6.2	6.9	7.6	8.3	9.0	9.7	10.4	9					0.3			0.6		9
70	0.7	1.4	2.1	2.8	3.5	4.2	4.9	5.6	6.3	7.0	7.7	8.4	9.1	9.8	10.5	**70**	0.1	0.1	0.2	0.3	0.4	0.4	0.5	0.6	0.6	**70**
1	.7	1.4	2.1	2.8	3.6	4.3	5.0	5.7	6.4	7.1	7.8	8.5	9.2	9.9	10.7	1										1
2	.7	1.4	2.2	2.9	3.6	4.3	5.0	5.8	6.5	7.2	7.9	8.6	9.4	10.1	10.8	2									0.6	2
3	.7	1.5	2.2	2.9	3.7	4.4	5.1	5.8	6.6	7.3	8.0	8.8	9.5	10.2	11.0	3									0.7	3
4	.7	1.5	2.2	3.0	3.7	4.4	5.2	5.9	6.7	7.4	8.1	8.9	9.6	10.4	11.1	4		0.1				0.4				4
75	0.8	1.5	2.3	3.0	3.8	4.5	5.3	6.0	6.8	7.5	8.3	9.0	9.8	10.5	11.3	**75**	0.1	0.2	0.2	0.3	0.4	0.5	0.5	0.6	0.7	**75**
6	.8	1.5	2.3	3.0	3.8	4.6	5.3	6.1	6.8	7.6	8.4	9.1	9.9	10.6	11.4	6										6
7	.8	1.5	2.3	3.1	3.9	4.6	5.4	6.2	6.9	7.7	8.5	9.2	10.0	10.8	11.6	7										7
8	.8	1.6	2.3	3.1	3.9	4.7	5.5	6.2	7.0	7.8	8.6	9.4	10.1	10.9	11.7	8							0.5			8
9	.8	1.6	2.4	3.2	4.0	4.7	5.5	6.3	7.1	7.9	8.7	9.5	10.3	11.1	11.9	9							0.6			9
90	0.9	1.8	2.7	3.6	4.5	5.4	6.3	7.2	8.1	9.0	9.9	10.8	11.7	12.6	13.5	**90**	0.1	0.2	0.3	0.4	0.5	0.5	0.6	0.7	0.8	**90**
1	.9	1.8	2.7	3.6	4.6	5.5	6.4	7.3	8.2	9.1	10.0	10.9	11.8	12.7	13.7	1						0.5				1
2	.9	1.8	2.8	3.7	4.6	5.5	6.4	7.4	8.3	9.2	10.1	11.0	12.0	12.9	13.8	2						0.6	0.6			2
3	.9	1.9	2.8	3.7	4.7	5.6	6.5	7.4	8.4	9.3	10.2	11.2	12.1	13.0	14.0	3							0.7	0.7		3
4	.9	1.9	2.8	3.8	4.7	5.6	6.6	7.5	8.5	9.4	10.3	11.3	12.2	13.2	14.1	4								0.8	0.8	4
95	1.0	1.9	2.9	3.8	4.8	5.7	6.7	7.6	8.6	9.5	10.5	11.4	12.4	13.3	14.3	**95**	0.1	0.2	0.3	0.4	0.5	0.6	0.7	0.8	0.9	**95**
6	1.0	1.9	2.9	3.8	4.8	5.8	6.7	7.7	8.6	9.6	10.6	11.5	12.5	13.4	14.4	6										6
7	1.0	1.9	2.9	3.9	4.9	5.8	6.8	7.8	8.7	9.7	10.7	11.6	12.6	13.6	14.6	7										7
8	1.0	2.0	2.9	3.9	4.9	5.9	6.9	7.8	8.8	9.8	10.8	11.8	12.7	13.7	14.7	8										8
99	1.0	2.0	3.0	4.0	5.0	5.9	6.9	7.9	8.9	9.9	10.9	11.9	12.9	13.9	14.9	**99**										**99**
Δ	1′	2′	3′	4′	5′	6′	7′	8′	9′	10′	11′	12′	13′	14′	15′	Δ	0.1′	0.2′	0.3′	0.4′	0.5′	0.6′	0.7′	0.8′	0.9′	Δ

Δd or Δt

Δd or Δt

Lat. 41°

STAR IDENTIFICATION TABLE

ALTITUDE

AZ.	4°		8°		12°		16°		20°		24°		28°		32°		36°		40°		44°		AZ.
	Dec.	H.A.	Dec.	H.A.	Dec.	H.A.	Dec.	H.A.	Dec.	H.A.	Dec.	H.A.	Dec.	H.A.	Dec.	H.A.	Dec.	H.A.	Dec.	H.A.	Dec.	H.A.	
°	°	°	°	°	°	°	°	°	°	°	°	°	°	°	°	°	°	°	°	°	°	°	°
80	10	94	13	90	15	87	18	84	20	81	23	77	25	74	27	70	29	66	31	62	33	58	80
84	07	91	10	88	12	85	15	82	17	78	20	75	22	72	24	68	27	64	29	60	31	56	84
88	04	88	07	85	09	82	12	79	14	76	17	73	19	69	22	66	24	62	26	59	28	55	88
92	01	86	04	83	06	80	09	77	12	73	14	70	17	67	19	64	21	60	24	57	26	53	92
96	*02*	83	01	80	03	77	06	74	09	71	11	68	14	65	16	61	19	58	21	55	24	51	96
100	*05*	80	*02*	77	00	74	03	72	06	68	08	65	11	62	14	59	16	56	19	53	21	49	100
104	*08*	78	*05*	75	*02*	72	00	69	03	66	06	63	08	60	11	57	14	54	16	51	19	48	104
108	*11*	75	*08*	72	*05*	69	*02*	66	00	63	03	60	06	58	09	55	11	52	14	49	17	46	108
112	*14*	72	*11*	69	*08*	66	*05*	64	*02*	61	00	58	03	55	06	52	09	49	12	47	15	44	112
116	*17*	69	*14*	66	*11*	64	*08*	61	*05*	58	*02*	55	01	53	04	50	07	47	10	44	13	41	116
120	*19*	66	*16*	63	*13*	61	*10*	58	*07*	55	*04*	53	*01*	50	02	47	05	45	08	42	11	39	120
124	*22*	63	*19*	60	*16*	58	*13*	55	*10*	52	*07*	50	*04*	47	*01*	45	03	42	06	40	09	37	124
128	*25*	60	*22*	57	*19*	54	*15*	52	*12*	49	*09*	47	*06*	44	*03*	42	01	40	04	37	07	35	128
132	*27*	57	*24*	54	*21*	51	*18*	49	*14*	46	*11*	44	*08*	41	*05*	39	*01*	37	02	35	05	32	132
136	*30*	53	*27*	50	*23*	48	*20*	45	*17*	43	*13*	41	*10*	39	*06*	36	*03*	34	00	32	04	30	136
140	*32*	49	*29*	47	*25*	44	*22*	42	*19*	40	*15*	37	*12*	35	*08*	33	*05*	31	*01*	30	02	28	140
144	*34*	45	*31*	43	*27*	40	*24*	38	*20*	36	*17*	34	*13*	32	*10*	30	*06*	29	*03*	27	01	25	144
148	*36*	41	*33*	39	*29*	36	*26*	34	*22*	33	*19*	31	*15*	29	*11*	27	*08*	26	*04*	24	*00*	22	148
152	*38*	37	*35*	34	*31*	32	*27*	31	*24*	29	*20*	27	*16*	26	*13*	24	*09*	23	*05*	21	*01*	20	152
156	*40*	32	*36*	30	*33*	28	*29*	27	*25*	25	*21*	24	*18*	22	*14*	21	*10*	20	*06*	18	*02*	17	156
	4°		8°		12°		16°		20°		24°		28°		32°		36°		40°		44°		

FIGURES IN ITALICS INDICATE THAT DECLINATION IS OF **CONTRARY** NAME TO LATITUDE

APPENDIX O

EXTRACTS FROM PUB. NO. 229

INTERPOLATION TABLE

Dec. Inc.	Tens 10'	20'	30'	40'	50'	Decimals	Units 0'	1'	2'	3'	4'	5'	6'	7'	8'	9'	Double Second Diff. and Corr.
0.0	0.0	0.0	0.0	0.0	0.0	.0	0.0	0.0	0.0	0.0	0.0	0.0	0.0	0.1	0.1	0.1	0.0 [0.0] 48.2
0.1	0.0	0.0	0.0	0.0	0.1	.1	0.0	0.0	0.0	0.0	0.0	0.0	0.1	0.1	0.1	0.1	
0.2	0.0	0.0	0.1	0.1	0.1	.2	0.0	0.0	0.0	0.0	0.0	0.0	0.1	0.1	0.1	0.1	
0.3	0.0	0.1	0.1	0.2	0.2	.3	0.0	0.0	0.0	0.0	0.0	0.0	0.1	0.1	0.1	0.1	
0.4	0.1	0.1	0.2	0.3	0.3	.4	0.0	0.0	0.0	0.0	0.0	0.0	0.1	0.1	0.1	0.1	
0.5	0.1	0.2	0.3	0.3	0.4	.5	0.0	0.0	0.0	0.0	0.0	0.0	0.1	0.1	0.1	0.1	16.2 [0.1] 48.6
0.6	0.1	0.2	0.3	0.4	0.5	.6	0.0	0.0	0.0	0.0	0.0	0.0	0.1	0.1	0.1	0.1	
0.7	0.1	0.3	0.4	0.5	0.6	.7	0.0	0.0	0.0	0.0	0.0	0.0	0.1	0.1	0.1	0.1	
0.8	0.2	0.3	0.4	0.6	0.7	.8	0.0	0.0	0.0	0.0	0.0	0.0	0.1	0.1	0.1	0.1	
0.9	0.2	0.3	0.5	0.6	0.8	.9	0.0	0.0	0.0	0.0	0.0	0.0	0.1	0.1	0.1	0.1	
1.0	0.1	0.3	0.5	0.6	0.8	.0	0.0	0.0	0.0	0.1	0.1	0.1	0.1	0.2	0.2	0.2	8.2 [0.1] 24.6 [0.2] 41.0
1.1	0.2	0.3	0.5	0.7	0.9	.1	0.0	0.0	0.1	0.1	0.1	0.1	0.2	0.2	0.2	0.2	
1.2	0.2	0.4	0.6	0.8	1.0	.2	0.0	0.0	0.1	0.1	0.1	0.1	0.2	0.2	0.2	0.2	
1.3	0.2	0.4	0.6	0.9	1.1	.3	0.0	0.0	0.1	0.1	0.1	0.1	0.2	0.2	0.2	0.2	
1.4	0.2	0.5	0.7	0.9	1.2	.4	0.0	0.0	0.1	0.1	0.1	0.1	0.2	0.2	0.2	0.2	
1.5	0.3	0.5	0.8	1.0	1.3	.5	0.0	0.0	0.1	0.1	0.1	0.1	0.2	0.2	0.2	0.2	
1.6	0.3	0.5	0.8	1.1	1.3	.6	0.0	0.0	0.1	0.1	0.1	0.1	0.2	0.2	0.2	0.2	
1.7	0.3	0.6	0.9	1.2	1.4	.7	0.0	0.0	0.1	0.1	0.1	0.1	0.2	0.2	0.2	0.2	
1.8	0.3	0.6	0.9	1.2	1.5	.8	0.0	0.0	0.1	0.1	0.1	0.1	0.2	0.2	0.2	0.2	
1.9	0.4	0.7	1.0	1.3	1.6	.9	0.0	0.0	0.1	0.1	0.1	0.1	0.2	0.2	0.2	0.2	
2.0	0.3	0.6	1.0	1.3	1.6	.0	0.0	0.0	0.1	0.1	0.2	0.2	0.2	0.3	0.3	0.4	5.0 [0.1] 15.0 [0.2] 25.0 [0.3] 35.1
2.1	0.3	0.7	1.0	1.4	1.7	.1	0.0	0.0	0.1	0.1	0.2	0.2	0.3	0.3	0.3	0.4	
2.2	0.3	0.7	1.1	1.4	1.8	.2	0.0	0.0	0.1	0.1	0.2	0.2	0.3	0.3	0.3	0.4	
2.3	0.4	0.8	1.1	1.5	1.9	.3	0.0	0.1	0.1	0.1	0.2	0.2	0.3	0.3	0.3	0.4	
2.4	0.4	0.8	1.2	1.6	2.0	.4	0.0	0.1	0.1	0.1	0.2	0.2	0.3	0.3	0.3	0.4	
2.5	0.4	0.8	1.3	1.7	2.1	.5	0.0	0.1	0.1	0.1	0.2	0.2	0.3	0.3	0.4	0.4	
2.6	0.4	0.9	1.3	1.7	2.2	.6	0.0	0.1	0.1	0.1	0.2	0.2	0.3	0.3	0.4	0.4	
2.7	0.5	0.9	1.4	1.8	2.3	.7	0.0	0.1	0.1	0.2	0.2	0.2	0.3	0.3	0.4	0.4	
2.8	0.5	1.0	1.4	1.9	2.4	.8	0.0	0.1	0.1	0.2	0.2	0.2	0.3	0.3	0.4	0.4	
2.9	0.5	1.0	1.5	2.0	2.5	.9	0.0	0.1	0.1	0.2	0.2	0.2	0.3	0.3	0.4	0.4	
3.0	0.5	1.0	1.5	2.0	2.5	.0	0.0	0.1	0.1	0.2	0.2	0.3	0.3	0.4	0.5	0.5	3.6 [0.1] 10.9 [0.2] 18.2 [0.3] 25.5 [0.4] 32.8 [0.5] 40.1
3.1	0.5	1.0	1.5	2.0	2.6	.1	0.0	0.1	0.1	0.2	0.2	0.3	0.4	0.4	0.5	0.5	
3.2	0.5	1.0	1.6	2.1	2.6	.2	0.0	0.1	0.1	0.2	0.2	0.3	0.4	0.4	0.5	0.5	
3.3	0.5	1.1	1.6	2.2	2.7	.3	0.0	0.1	0.1	0.2	0.3	0.3	0.4	0.4	0.5	0.5	
3.4	0.6	1.1	1.7	2.3	2.8	.4	0.0	0.1	0.1	0.2	0.3	0.3	0.4	0.4	0.5	0.5	
3.5	0.6	1.2	1.8	2.3	2.9	.5	0.0	0.1	0.1	0.2	0.3	0.3	0.4	0.4	0.5	0.6	
3.6	0.6	1.2	1.8	2.4	3.0	.6	0.0	0.1	0.2	0.2	0.3	0.3	0.4	0.4	0.5	0.6	
3.7	0.6	1.3	1.9	2.5	3.1	.7	0.0	0.1	0.2	0.2	0.3	0.3	0.4	0.4	0.5	0.6	
3.8	0.7	1.3	1.9	2.6	3.2	.8	0.0	0.1	0.2	0.2	0.3	0.3	0.4	0.5	0.5	0.6	
3.9	0.7	1.3	2.0	2.6	3.3	.9	0.1	0.1	0.2	0.2	0.3	0.3	0.4	0.5	0.5	0.6	
4.0	0.6	1.3	2.0	2.6	3.3	.0	0.0	0.1	0.1	0.2	0.3	0.4	0.4	0.5	0.6	0.7	2.9 [0.1] 8.6 [0.2] 14.4 [0.3] 20.2 [0.4] 25.9 [0.5] 31.7 [0.6] 37.5
4.1	0.7	1.3	2.0	2.7	3.4	.1	0.0	0.1	0.2	0.2	0.3	0.4	0.5	0.5	0.6	0.7	
4.2	0.7	1.4	2.1	2.8	3.5	.2	0.0	0.1	0.2	0.2	0.3	0.4	0.5	0.5	0.6	0.7	
4.3	0.7	1.4	2.1	2.9	3.6	.3	0.0	0.1	0.2	0.2	0.3	0.4	0.5	0.5	0.6	0.7	
4.4	0.7	1.5	2.2	2.9	3.7	.4	0.0	0.1	0.2	0.3	0.3	0.4	0.5	0.6	0.6	0.7	
4.5	0.8	1.5	2.3	3.0	3.8	.5	0.0	0.1	0.2	0.3	0.3	0.4	0.5	0.6	0.6	0.7	
4.6	0.8	1.5	2.3	3.1	3.8	.6	0.0	0.1	0.2	0.3	0.3	0.4	0.5	0.6	0.6	0.7	
4.7	0.8	1.6	2.4	3.2	3.9	.7	0.1	0.1	0.2	0.3	0.4	0.4	0.5	0.6	0.7	0.7	
4.8	0.8	1.6	2.4	3.2	4.0	.8	0.1	0.1	0.2	0.3	0.4	0.4	0.5	0.6	0.7	0.7	
4.9	0.9	1.7	2.5	3.3	4.1	.9	0.1	0.1	0.2	0.3	0.4	0.4	0.5	0.6	0.7	0.7	
5.0	0.8	1.6	2.5	3.3	4.1	.0	0.0	0.1	0.2	0.3	0.4	0.5	0.5	0.6	0.7	0.8	2.4 [0.1] 7.2 [0.2] 12.0 [0.3] 16.8 [0.4] 21.6 [0.5] 26.4 [0.6] 31.2 [0.7] 36.0
5.1	0.8	1.7	2.5	3.4	4.2	.1	0.0	0.1	0.2	0.3	0.4	0.5	0.6	0.7	0.7	0.8	
5.2	0.8	1.7	2.6	3.4	4.3	.2	0.0	0.1	0.2	0.3	0.4	0.5	0.6	0.7	0.8	0.8	
5.3	0.9	1.8	2.6	3.5	4.4	.3	0.0	0.1	0.2	0.3	0.4	0.5	0.6	0.7	0.8	0.9	
5.4	0.9	1.8	2.7	3.6	4.5	.4	0.0	0.1	0.2	0.3	0.4	0.5	0.6	0.7	0.8	0.9	
5.5	0.9	1.8	2.8	3.7	4.6	.5	0.0	0.1	0.2	0.3	0.4	0.5	0.6	0.7	0.8	0.9	
5.6	0.9	1.9	2.8	3.7	4.7	.6	0.1	0.1	0.2	0.3	0.4	0.5	0.6	0.7	0.8	0.9	
5.7	1.0	1.9	2.9	3.8	4.8	.7	0.1	0.2	0.2	0.3	0.4	0.5	0.6	0.7	0.8	0.9	
5.8	1.0	2.0	2.9	3.9	4.9	.8	0.1	0.2	0.3	0.3	0.4	0.5	0.6	0.7	0.8	0.9	
5.9	1.0	2.0	3.0	4.0	5.0	.9	0.1	0.2	0.3	0.4	0.4	0.5	0.6	0.7	0.8	0.9	
6.0	1.0	2.0	3.0	4.0	5.0	.0	0.0	0.1	0.2	0.3	0.4	0.5	0.6	0.8	0.9	1.0	2.1 [0.1] 6.2 [0.2] 10.4 [0.3] 14.5 [0.4] 18.6 [0.5] 22.8 [0.6] 26.9 [0.7] 31.1 [0.8] 35.2
6.1	1.0	2.0	3.0	4.0	5.1	.1	0.0	0.1	0.2	0.3	0.4	0.6	0.7	0.8	0.9	1.0	
6.2	1.0	2.0	3.1	4.1	5.1	.2	0.0	0.1	0.2	0.3	0.5	0.6	0.7	0.8	0.9	1.0	
6.3	1.0	2.1	3.1	4.2	5.2	.3	0.0	0.1	0.2	0.4	0.5	0.6	0.7	0.8	0.9	1.0	
6.4	1.1	2.1	3.2	4.3	5.3	.4	0.0	0.2	0.3	0.4	0.5	0.6	0.7	0.8	0.9	1.0	
6.5	1.1	2.2	3.3	4.3	5.4	.5	0.1	0.2	0.3	0.4	0.5	0.6	0.7	0.8	0.9	1.0	
6.6	1.1	2.2	3.3	4.4	5.5	.6	0.1	0.2	0.3	0.4	0.5	0.6	0.7	0.8	0.9	1.0	
6.7	1.1	2.3	3.4	4.5	5.6	.7	0.1	0.2	0.3	0.4	0.5	0.6	0.7	0.8	0.9	1.1	
6.8	1.2	2.3	3.4	4.6	5.7	.8	0.1	0.2	0.3	0.4	0.5	0.6	0.7	0.8	1.0	1.1	
6.9	1.2	2.3	3.5	4.6	5.8	.9	0.1	0.2	0.3	0.4	0.5	0.6	0.7	0.9	1.0	1.1	
7.0	1.1	2.3	3.5	4.6	5.8	.0	0.0	0.1	0.2	0.4	0.5	0.6	0.7	0.9	1.0	1.1	1.8 [0.1] 5.5 [0.2] 9.1 [0.3] 12.8 [0.4] 16.5 [0.5] 20.1 [0.6] 23.8 [0.7] 27.4 [0.8] 31.1 [0.9] 34.7
7.1	1.2	2.3	3.5	4.7	5.9	.1	0.0	0.1	0.3	0.4	0.5	0.6	0.8	0.9	1.0	1.1	
7.2	1.2	2.4	3.6	4.8	6.0	.2	0.0	0.1	0.3	0.4	0.5	0.6	0.8	0.9	1.0	1.1	
7.3	1.2	2.4	3.6	4.9	6.1	.3	0.0	0.2	0.3	0.4	0.5	0.7	0.8	0.9	1.0	1.2	
7.4	1.2	2.5	3.7	4.9	6.2	.4	0.0	0.2	0.3	0.4	0.5	0.7	0.8	0.9	1.0	1.2	
7.5	1.3	2.5	3.8	5.0	6.3	.5	0.1	0.2	0.3	0.4	0.6	0.7	0.8	0.9	1.1	1.2	
7.6	1.3	2.5	3.8	5.1	6.3	.6	0.1	0.2	0.3	0.4	0.6	0.7	0.8	0.9	1.1	1.2	
7.7	1.3	2.6	3.9	5.2	6.4	.7	0.1	0.2	0.3	0.5	0.6	0.7	0.8	1.0	1.1	1.2	
7.8	1.3	2.6	3.9	5.2	6.5	.8	0.1	0.2	0.3	0.5	0.6	0.7	0.8	1.0	1.1	1.2	
7.9	1.4	2.7	4.0	5.3	6.6	.9	0.1	0.2	0.4	0.5	0.6	0.7	0.9	1.0	1.1	1.2	
	10'	20'	30'	40'	50'		0'	1'	2'	3'	4'	5'	6'	7'	8'	9'	

Dec. Inc.	Tens 10'	20'	30'	40'	50'	Decimals	Units 0'	1'	2'	3'	4'	5'	6'	7'	8'	9'	Double Second Diff. and Corr.
8.0	1.3	2.6	4.0	5.3	6.6	.0	0.0	0.1	0.3	0.4	0.6	0.7	0.8	1.0	1.1	1.3	1.6 [0.1] 4.8 [0.2] 8.0 [0.3] 11.2 [0.4] 14.5 [0.5] 17.7 [0.6] 20.9 [0.7] 24.1 [0.8] 27.3 [0.9] 30.5 [1.0] 33.7 [1.1] 36.9
8.1	1.3	2.7	4.0	5.4	6.7	.1	0.0	0.2	0.3	0.4	0.6	0.7	0.9	1.0	1.1	1.3	
8.2	1.3	2.7	4.1	5.4	6.8	.2	0.0	0.2	0.3	0.5	0.6	0.7	0.9	1.0	1.2	1.3	
8.3	1.4	2.8	4.1	5.5	6.9	.3	0.0	0.2	0.3	0.5	0.6	0.8	0.9	1.0	1.2	1.3	
8.4	1.4	2.8	4.2	5.6	7.0	.4	0.1	0.2	0.3	0.5	0.6	0.8	0.9	1.0	1.2	1.3	
8.5	1.4	2.8	4.3	5.7	7.1	.5	0.1	0.2	0.4	0.5	0.6	0.8	0.9	1.1	1.2	1.3	
8.6	1.4	2.9	4.3	5.7	7.2	.6	0.1	0.2	0.4	0.5	0.7	0.8	0.9	1.1	1.2	1.4	
8.7	1.5	2.9	4.4	5.8	7.3	.7	0.1	0.2	0.4	0.5	0.7	0.8	0.9	1.1	1.2	1.4	
8.8	1.5	3.0	4.4	5.9	7.4	.8	0.1	0.3	0.4	0.5	0.7	0.8	1.0	1.1	1.2	1.4	
8.9	1.5	3.0	4.5	6.0	7.5	.9	0.1	0.3	0.4	0.6	0.7	0.8	1.0	1.1	1.3	1.4	
9.0	1.5	3.0	4.5	6.0	7.5	.0	0.0	0.2	0.3	0.5	0.6	0.8	0.9	1.1	1.3	1.4	
9.1	1.5	3.0	4.5	6.0	7.6	.1	0.0	0.2	0.3	0.5	0.6	0.8	1.0	1.1	1.3	1.4	
9.2	1.5	3.0	4.6	6.1	7.6	.2	0.0	0.2	0.3	0.5	0.7	0.8	1.0	1.1	1.3	1.5	
9.3	1.5	3.1	4.6	6.2	7.7	.3	0.0	0.2	0.4	0.5	0.7	0.8	1.0	1.2	1.3	1.5	
9.4	1.6	3.1	4.7	6.3	7.8	.4	0.1	0.2	0.4	0.5	0.7	0.9	1.0	1.2	1.3	1.5	
9.5	1.6	3.2	4.8	6.3	7.9	.5	0.1	0.2	0.4	0.6	0.7	0.9	1.0	1.2	1.3	1.5	1.4 [0.1] 4.2 [0.2] 7.1 [0.3] 9.9 [0.4] 12.7 [0.5] 15.5 [0.6] 18.4 [0.7] 21.2 [0.8] 24.0 [0.9] 26.8 [1.0] 29.7 [1.1] 32.5 [1.2] 35.3
9.6	1.6	3.2	4.8	6.4	8.0	.6	0.1	0.3	0.4	0.6	0.7	0.9	1.0	1.2	1.4	1.5	
9.7	1.6	3.3	4.9	6.5	8.1	.7	0.1	0.3	0.4	0.6	0.7	0.9	1.1	1.2	1.4	1.5	
9.8	1.7	3.3	4.9	6.6	8.2	.8	0.1	0.3	0.4	0.6	0.8	0.9	1.1	1.2	1.4	1.6	
9.9	1.7	3.3	5.0	6.6	8.3	.9	0.1	0.3	0.5	0.6	0.8	0.9	1.1	1.3	1.4	1.6	
10.0	1.6	3.3	5.0	6.6	8.3	.0	0.0	0.2	0.3	0.5	0.7	0.9	1.0	1.2	1.4	1.6	
10.1	1.7	3.3	5.0	6.7	8.4	.1	0.0	0.2	0.4	0.5	0.7	0.9	1.1	1.2	1.4	1.6	
10.2	1.7	3.4	5.1	6.8	8.5	.2	0.0	0.2	0.4	0.6	0.7	0.9	1.1	1.3	1.4	1.6	
10.3	1.7	3.4	5.1	6.9	8.6	.3	0.1	0.2	0.4	0.6	0.8	0.9	1.1	1.3	1.5	1.6	
10.4	1.7	3.5	5.2	6.9	8.7	.4	0.1	0.2	0.4	0.6	0.8	0.9	1.1	1.3	1.5	1.6	
10.5	1.8	3.5	5.3	7.0	8.8	.5	0.1	0.3	0.4	0.6	0.8	1.0	1.1	1.3	1.5	1.7	
10.6	1.8	3.5	5.3	7.1	8.8	.6	0.1	0.3	0.5	0.6	0.8	1.0	1.2	1.3	1.5	1.7	
10.7	1.8	3.6	5.4	7.2	8.9	.7	0.1	0.3	0.5	0.6	0.8	1.0	1.2	1.3	1.5	1.7	
10.8	1.8	3.6	5.4	7.2	9.0	.8	0.1	0.3	0.5	0.7	0.8	1.0	1.2	1.4	1.5	1.7	
10.9	1.9	3.7	5.5	7.3	9.1	.9	0.2	0.3	0.5	0.7	0.9	1.0	1.2	1.4	1.6	1.7	
11.0	1.8	3.6	5.5	7.3	9.1	.0	0.0	0.2	0.4	0.6	0.8	1.0	1.1	1.3	1.5	1.7	1.3 [0.1] 3.8 [0.2] 6.3 [0.3] 8.9 [0.4] 11.4 [0.5] 14.0 [0.6] 16.5 [0.7] 19.0 [0.8] 21.6 [0.9] 24.1 [1.0] 26.7 [1.1] 29.2 [1.2] 31.7 [1.3] 34.3
11.1	1.8	3.7	5.5	7.4	9.2	.1	0.0	0.2	0.4	0.6	0.8	1.0	1.2	1.4	1.6	1.7	
11.2	1.8	3.7	5.6	7.4	9.3	.2	0.0	0.2	0.4	0.6	0.8	1.0	1.2	1.4	1.6	1.8	
11.3	1.9	3.8	5.6	7.5	9.4	.3	0.1	0.2	0.4	0.6	0.8	1.0	1.2	1.4	1.6	1.8	
11.4	1.9	3.8	5.7	7.6	9.5	.4	0.1	0.3	0.5	0.7	0.8	1.0	1.2	1.4	1.6	1.8	
11.5	1.9	3.8	5.8	7.7	9.6	.5	0.1	0.3	0.5	0.7	0.9	1.1	1.2	1.4	1.6	1.8	
11.6	1.9	3.9	5.8	7.7	9.7	.6	0.1	0.3	0.5	0.7	0.9	1.1	1.3	1.5	1.6	1.8	
11.7	2.0	3.9	5.9	7.8	9.8	.7	0.1	0.3	0.5	0.7	0.9	1.1	1.3	1.5	1.7	1.9	
11.8	2.0	4.0	5.9	7.9	9.9	.8	0.2	0.3	0.5	0.7	0.9	1.1	1.3	1.5	1.7	1.9	
11.9	2.0	4.0	6.0	8.0	10.0	.9	0.2	0.4	0.6	0.7	0.9	1.1	1.3	1.5	1.7	1.9	
12.0	2.0	4.0	6.0	8.0	10.0	.0	0.0	0.2	0.4	0.6	0.8	1.0	1.2	1.5	1.7	1.9	
12.1	2.0	4.0	6.0	8.0	10.1	.1	0.0	0.2	0.4	0.6	0.9	1.1	1.3	1.5	1.7	1.9	
12.2	2.0	4.0	6.1	8.1	10.1	.2	0.0	0.2	0.5	0.7	0.9	1.1	1.3	1.5	1.7	1.9	
12.3	2.0	4.1	6.1	8.2	10.2	.3	0.1	0.3	0.5	0.7	0.9	1.1	1.3	1.5	1.7	1.9	
12.4	2.1	4.1	6.2	8.3	10.3	.4	0.1	0.3	0.5	0.7	0.9	1.1	1.3	1.5	1.7	2.0	
12.5	2.1	4.2	6.3	8.3	10.4	.5	0.1	0.3	0.5	0.7	0.9	1.1	1.4	1.6	1.8	2.0	1.2 [0.1] 3.5 [0.2] 5.8 [0.3] 8.1 [0.4] 10.5 [0.5] 12.8 [0.6] 15.1 [0.7] 17.4 [0.8] 19.8 [0.9] 22.1 [1.0] 24.4 [1.1] 26.7 [1.2] 29.1 [1.3] 31.4 [1.4] 33.7 [1.5] 36.0
12.6	2.1	4.2	6.3	8.4	10.5	.6	0.1	0.3	0.5	0.7	1.0	1.2	1.4	1.6	1.8	2.0	
12.7	2.1	4.3	6.4	8.5	10.6	.7	0.1	0.4	0.6	0.8	1.0	1.2	1.4	1.6	1.8	2.0	
12.8	2.2	4.3	6.4	8.6	10.7	.8	0.2	0.4	0.6	0.8	1.0	1.2	1.4	1.6	1.8	2.0	
12.9	2.2	4.3	6.5	8.6	10.8	.9	0.2	0.4	0.6	0.8	1.0	1.2	1.4	1.6	1.9	2.1	
13.0	2.1	4.3	6.5	8.6	10.8	.0	0.0	0.2	0.4	0.7	0.9	1.1	1.3	1.6	1.8	2.0	
13.1	2.2	4.3	6.5	8.7	10.9	.1	0.0	0.2	0.5	0.7	0.9	1.1	1.4	1.6	1.8	2.0	
13.2	2.2	4.4	6.6	8.8	11.0	.2	0.0	0.3	0.5	0.7	0.9	1.2	1.4	1.6	1.8	2.1	
13.3	2.2	4.4	6.6	8.9	11.1	.3	0.1	0.3	0.5	0.7	1.0	1.2	1.4	1.6	1.9	2.1	
13.4	2.2	4.5	6.7	8.9	11.2	.4	0.1	0.3	0.5	0.8	1.0	1.2	1.4	1.7	1.9	2.1	
13.5	2.3	4.5	6.8	9.0	11.3	.5	0.1	0.3	0.6	0.8	1.0	1.2	1.5	1.7	1.9	2.1	
13.6	2.3	4.5	6.8	9.1	11.3	.6	0.1	0.4	0.6	0.8	1.0	1.3	1.5	1.7	1.9	2.2	
13.7	2.3	4.6	6.9	9.2	11.4	.7	0.2	0.4	0.6	0.8	1.1	1.3	1.5	1.7	2.0	2.2	
13.8	2.3	4.6	6.9	9.2	11.5	.8	0.2	0.4	0.6	0.9	1.1	1.3	1.5	1.8	2.0	2.2	
13.9	2.4	4.7	7.0	9.3	11.6	.9	0.2	0.4	0.7	0.9	1.1	1.3	1.6	1.8	2.0	2.2	
14.0	2.3	4.6	7.0	9.3	11.6	.0	0.0	0.2	0.5	0.7	1.0	1.2	1.4	1.7	1.9	2.2	1.1 [0.1] 3.2 [0.2] 5.3 [0.3] 7.5 [0.4] 9.6 [0.5] 11.7 [0.6] 13.9 [0.7] 16.0 [0.8] 18.1 [0.9] 20.3 [1.0] 22.4 [1.1] 24.5 [1.2] 26.7 [1.3] 28.8 [1.4] 30.9 [1.5] 33.1 [1.6] 35.2
14.1	2.3	4.7	7.0	9.4	11.7	.1	0.0	0.3	0.5	0.7	1.0	1.2	1.5	1.7	2.0	2.2	
14.2	2.3	4.7	7.1	9.4	11.8	.2	0.0	0.3	0.5	0.8	1.0	1.3	1.5	1.7	2.0	2.2	
14.3	2.4	4.8	7.1	9.5	11.9	.3	0.1	0.3	0.6	0.8	1.0	1.3	1.5	1.8	2.0	2.2	
14.4	2.4	4.8	7.2	9.6	12.0	.4	0.1	0.3	0.6	0.8	1.1	1.3	1.5	1.8	2.0	2.3	
14.5	2.4	4.8	7.3	9.7	12.1	.5	0.1	0.4	0.6	0.8	1.1	1.3	1.6	1.8	2.1	2.3	
14.6	2.4	4.9	7.3	9.7	12.2	.6	0.1	0.4	0.6	0.9	1.1	1.4	1.6	1.8	2.1	2.3	
14.7	2.5	4.9	7.4	9.8	12.3	.7	0.2	0.4	0.7	0.9	1.1	1.4	1.6	1.9	2.1	2.3	
14.8	2.5	5.0	7.4	9.9	12.4	.8	0.2	0.4	0.7	0.9	1.2	1.4	1.6	1.9	2.1	2.4	
14.9	2.5	5.0	7.5	10.0	12.5	.9	0.2	0.5	0.7	0.9	1.2	1.4	1.7	1.9	2.2	2.4	
15.0	2.5	5.0	7.5	10.0	12.5	.0	0.0	0.3	0.5	0.8	1.0	1.3	1.5	1.8	2.1	2.3	
15.1	2.5	5.0	7.5	10.0	12.6	.1	0.0	0.3	0.5	0.8	1.1	1.3	1.6	1.8	2.1	2.4	
15.2	2.5	5.0	7.6	10.1	12.6	.2	0.1	0.3	0.6	0.8	1.1	1.3	1.6	1.9	2.1	2.4	
15.3	2.5	5.1	7.6	10.2	12.7	.3	0.1	0.3	0.6	0.9	1.1	1.4	1.6	1.9	2.1	2.4	
15.4	2.6	5.1	7.7	10.3	12.8	.4	0.1	0.4	0.6	0.9	1.1	1.4	1.7	1.9	2.2	2.4	
15.5	2.6	5.2	7.8	10.3	12.9	.5	0.1	0.4	0.6	0.9	1.2	1.4	1.7	1.9	2.2	2.5	
15.6	2.6	5.2	7.8	10.4	13.0	.6	0.2	0.4	0.7	0.9	1.2	1.4	1.7	2.0	2.2	2.5	
15.7	2.6	5.3	7.9	10.5	13.1	.7	0.2	0.4	0.7	1.0	1.2	1.5	1.7	2.0	2.2	2.5	
15.8	2.7	5.3	7.9	10.6	13.2	.8	0.2	0.5	0.7	1.0	1.2	1.5	1.8	2.0	2.3	2.5	
15.9	2.7	5.3	8.0	10.6	13.3	.9	0.2	0.5	0.7	1.0	1.3	1.5	1.8	2.0	2.3	2.6	
	10'	20'	30'	40'	50'		0'	1'	2'	3'	4'	5'	6'	7'	8'	9'	

The Double-Second-Difference correction (Corr.) is always to be added to the tabulated altitude.

INTERPOLATION TABLE

Dec. Inc.	Tens 10	20	30	40	50	Decimals	Units 0	1	2	3	4	5	6	7	8	9	Double Second Diff and Corr
16.0	2.6	5.3	8.0	10.6	13.3	.0	0.0	0.3	0.5	0.8	1.1	1.4	1.6	1.9	2.2	2.5	
16.1	2.7	5.3	8.0	10.7	13.4	.1	0.0	0.3	0.6	0.9	1.1	1.4	1.7	2.0	2.2	2.5	
16.2	2.7	5.4	8.1	10.8	13.5	.2	0.1	0.3	0.6	0.9	1.2	1.4	1.7	2.0	2.3	2.5	1.0 / 0.1
16.3	2.7	5.4	8.1	10.9	13.6	.3	0.1	0.4	0.6	0.9	1.2	1.5	1.7	2.0	2.3	2.6	3.0 / 0.2
16.4	2.7	5.5	8.2	10.9	13.7	.4	0.1	0.4	0.7	0.9	1.2	1.5	1.8	2.0	2.3	2.6	4.9 / 0.3
16.5	2.8	5.5	8.3	11.0	13.8	.5	0.1	0.4	0.7	1.0	1.2	1.5	1.8	2.1	2.3	2.6	6.9 / 0.4
16.6	2.8	5.5	8.3	11.1	13.8	.6	0.2	0.4	0.7	1.0	1.3	1.5	1.8	2.1	2.4	2.6	8.9 / 0.5
16.7	2.8	5.6	8.4	11.2	13.9	.7	0.2	0.5	0.7	1.0	1.3	1.6	1.8	2.1	2.4	2.7	10.8 / 0.6
16.8	2.8	5.6	8.4	11.2	14.0	.8	0.2	0.5	0.8	1.0	1.3	1.6	1.9	2.1	2.4	2.7	12.8 / 0.7
16.9	2.9	5.7	8.5	11.3	14.1	.9	0.2	0.5	0.8	1.1	1.3	1.6	1.9	2.2	2.4	2.7	14.8 / 0.8 / 16.7 / 0.9
17.0	2.8	5.6	8.5	11.3	14.1	.0	0.0	0.3	0.6	0.9	1.2	1.5	1.7	2.0	2.3	2.6	18.7 / 1.0
17.1	2.8	5.7	8.5	11.4	14.2	.1	0.0	0.3	0.6	0.9	1.2	1.5	1.8	2.1	2.4	2.7	20.7 / 1.1
17.2	2.8	5.7	8.6	11.4	14.3	.2	0.1	0.3	0.6	0.9	1.2	1.5	1.8	2.1	2.4	2.7	22.7 / 1.2
17.3	2.9	5.8	8.6	11.5	14.4	.3	0.1	0.4	0.7	1.0	1.3	1.5	1.8	2.1	2.4	2.7	24.6 / 1.3
17.4	2.9	5.8	8.7	11.6	14.5	.4	0.1	0.4	0.7	1.0	1.3	1.6	1.9	2.2	2.4	2.7	26.6 / 1.4
17.5	2.9	5.8	8.8	11.7	14.6	.5	0.1	0.4	0.7	1.0	1.3	1.6	1.9	2.2	2.5	2.8	28.6 / 1.5
17.6	2.9	5.9	8.8	11.7	14.7	.6	0.2	0.5	0.8	1.0	1.3	1.6	1.9	2.2	2.5	2.8	30.5 / 1.6
17.7	3.0	5.9	8.9	11.8	14.8	.7	0.2	0.5	0.8	1.1	1.4	1.7	2.0	2.2	2.5	2.8	32.5 / 1.7
17.8	3.0	6.0	8.9	11.9	14.9	.8	0.2	0.5	0.8	1.1	1.4	1.7	2.0	2.3	2.6	2.9	34.5
17.9	3.0	6.0	9.0	12.0	15.0	.9	0.3	0.6	0.8	1.1	1.4	1.7	2.0	2.3	2.6	2.9	
18.0	3.0	6.0	9.0	12.0	15.0	.0	0.0	0.3	0.6	0.9	1.2	1.5	1.8	2.2	2.5	2.8	
18.1	3.0	6.0	9.0	12.0	15.1	.1	0.0	0.3	0.6	1.0	1.3	1.6	1.9	2.2	2.5	2.8	0.9 / 0.1
18.2	3.0	6.0	9.1	12.1	15.1	.2	0.1	0.4	0.7	1.0	1.3	1.6	1.9	2.2	2.5	2.8	2.8 / 0.2
18.3	3.0	6.1	9.1	12.2	15.2	.3	0.1	0.4	0.7	1.0	1.3	1.6	1.9	2.3	2.6	2.9	4.6 / 0.3
18.4	3.1	6.1	9.2	12.3	15.3	.4	0.1	0.4	0.7	1.0	1.4	1.7	2.0	2.3	2.6	2.9	6.5 / 0.4
18.5	3.1	6.7	9.3	12.3	15.4	.5	0.2	0.5	0.8	1.1	1.4	1.7	2.0	2.3	2.6	2.9	8.3 / 0.5
18.6	3.1	6.2	9.3	12.4	15.5	.6	0.2	0.5	0.8	1.1	1.4	1.7	2.0	2.3	2.7	3.0	10.2 / 0.6
18.7	3.1	6.3	9.4	12.5	15.6	.7	0.2	0.5	0.8	1.1	1.4	1.8	2.1	2.4	2.7	3.0	12.0 / 0.7
18.8	3.2	6.3	9.4	12.6	15.7	.8	0.2	0.6	0.9	1.2	1.5	1.8	2.1	2.4	2.7	3.0	13.9 / 0.8
18.9	3.2	6.3	9.5	12.6	15.8	.9	0.3	0.6	0.9	1.2	1.5	1.8	2.1	2.4	2.7	3.1	15.7 / 0.9 / 17.6 / 1.0
19.0	3.1	6.3	9.5	12.6	15.8	.0	0.0	0.3	0.6	1.0	1.3	1.6	1.9	2.3	2.6	2.9	19.4 / 1.1
19.1	3.2	6.3	9.5	12.7	15.9	.1	0.0	0.4	0.7	1.0	1.3	1.7	2.0	2.3	2.6	3.0	21.3 / 1.2
19.2	3.2	6.4	9.6	12.8	16.0	.2	0.1	0.4	0.7	1.0	1.4	1.7	2.0	2.3	2.7	3.0	23.1 / 1.3
19.3	3.2	6.4	9.6	12.9	16.1	.3	0.1	0.4	0.7	1.1	1.4	1.7	2.0	2.4	2.7	3.0	25.0 / 1.4
19.4	3.2	6.5	9.7	12.9	16.2	.4	0.1	0.5	0.8	1.1	1.4	1.8	2.1	2.4	2.7	3.1	26.8 / 1.5
19.5	3.3	6.5	9.8	13.0	16.3	.5	0.2	0.5	0.8	1.1	1.5	1.8	2.1	2.4	2.8	3.1	28.7 / 1.6
19.6	3.3	6.5	9.8	13.1	16.3	.6	0.2	0.5	0.8	1.2	1.5	1.8	2.1	2.5	2.8	3.1	30.5 / 1.7
19.7	3.3	6.6	9.9	13.2	16.4	.7	0.2	0.6	0.9	1.2	1.5	1.9	2.2	2.5	2.8	3.2	32.3 / 1.8
19.8	3.3	6.6	9.9	13.2	16.5	.8	0.3	0.6	0.9	1.2	1.6	1.9	2.2	2.5	2.9	3.2	34.2
19.9	3.4	6.7	10.0	13.3	16.6	.9	0.3	0.6	0.9	1.3	1.6	1.9	2.2	2.6	2.9	3.2	
20.0	3.3	6.6	10.0	13.3	16.6	.0	0.0	0.3	0.7	1.0	1.4	1.7	2.0	2.4	2.7	3.1	
20.1	3.3	6.7	10.0	13.4	16.7	.1	0.0	0.4	0.7	1.1	1.4	1.7	2.1	2.4	2.8	3.1	0.9 / 0.1
20.2	3.3	6.7	10.1	13.4	16.8	.2	0.1	0.4	0.8	1.1	1.4	1.8	2.1	2.5	2.8	3.1	2.6 / 0.2
20.3	3.4	6.8	10.1	13.5	16.9	.3	0.1	0.4	0.8	1.1	1.5	1.8	2.2	2.5	2.8	3.2	4.4 / 0.3
20.4	3.4	6.8	10.2	13.6	17.0	.4	0.1	0.5	0.8	1.2	1.5	1.8	2.2	2.5	2.9	3.2	6.2 / 0.4
20.5	3.4	6.8	10.3	13.7	17.1	.5	0.2	0.5	0.9	1.2	1.5	1.9	2.2	2.6	2.9	3.2	7.9 / 0.5
20.6	3.4	6.9	10.3	13.7	17.2	.6	0.2	0.5	0.9	1.2	1.6	1.9	2.3	2.6	2.9	3.3	9.7 / 0.6
20.7	3.5	6.9	10.4	13.8	17.3	.7	0.2	0.6	0.9	1.3	1.6	1.9	2.3	2.6	3.0	3.3	11.4 / 0.7
20.8	3.5	7.0	10.4	13.9	17.4	.8	0.3	0.6	1.0	1.3	1.6	2.0	2.3	2.7	3.0	3.3	13.2 / 0.8
20.9	3.5	7.0	10.5	14.0	17.5	.9	0.3	0.6	1.0	1.3	1.7	2.0	2.4	2.7	3.0	3.4	14.9 / 0.9 / 16.7 / 1.0
21.0	3.5	7.0	10.5	14.0	17.5	.0	0.0	0.4	0.7	1.1	1.4	1.8	2.1	2.5	2.9	3.2	18.5 / 1.1
21.1	3.5	7.0	10.5	14.0	17.6	.1	0.0	0.4	0.8	1.1	1.5	1.8	2.2	2.5	2.9	3.3	20.2 / 1.2
21.2	3.5	7.0	10.6	14.1	17.6	.2	0.1	0.4	0.8	1.1	1.5	1.9	2.2	2.6	2.9	3.3	22.0 / 1.3
21.3	3.5	7.1	10.6	14.2	17.7	.3	0.1	0.5	0.8	1.2	1.5	1.9	2.3	2.6	3.0	3.3	23.7 / 1.4
21.4	3.6	7.1	10.7	14.3	17.8	.4	0.1	0.5	0.9	1.2	1.6	1.9	2.3	2.7	3.0	3.4	25.5 / 1.5 / 27.3 / 1.6
21.5	3.6	7.2	10.8	14.3	17.9	.5	0.2	0.5	0.9	1.3	1.6	2.0	2.3	2.7	3.0	3.4	29.0 / 1.7
21.6	3.6	7.2	10.8	14.4	18.0	.6	0.2	0.6	0.9	1.3	1.6	2.0	2.4	2.7	3.1	3.4	30.8 / 1.8
21.7	3.6	7.3	10.9	14.5	18.1	.7	0.3	0.6	1.0	1.3	1.7	2.0	2.4	2.8	3.1	3.5	32.5 / 1.9
21.8	3.7	7.3	10.9	14.6	18.2	.8	0.3	0.6	1.0	1.4	1.7	2.1	2.4	2.8	3.2	3.5	34.3
21.9	3.7	7.3	11.0	14.6	18.3	.9	0.3	0.7	1.0	1.4	1.8	2.1	2.5	2.8	3.2	3.5	
22.0	3.6	7.3	11.0	14.6	18.3	.0	0.0	0.4	0.7	1.1	1.5	1.9	2.2	2.6	3.0	3.4	0.8 / 0.1
22.1	3.7	7.3	11.0	14.7	18.4	.1	0.0	0.4	0.8	1.2	1.5	1.9	2.3	2.7	3.0	3.4	2.5 / 0.2
22.2	3.7	7.4	11.1	14.8	18.5	.2	0.1	0.4	0.8	1.2	1.6	1.9	2.3	2.7	3.1	3.4	4.2 / 0.3
22.3	3.7	7.4	11.1	14.9	18.6	.3	0.1	0.5	0.9	1.2	1.6	2.0	2.4	2.7	3.1	3.5	5.9 / 0.4
22.4	3.7	7.5	11.2	14.9	18.7	.4	0.1	0.5	0.9	1.3	1.6	2.0	2.4	2.8	3.1	3.5	7.6 / 0.5
22.5	3.8	7.5	11.3	15.0	18.8	.5	0.2	0.6	0.9	1.3	1.7	2.1	2.4	2.8	3.2	3.6	9.3 / 0.6
22.6	3.8	7.5	11.3	15.1	18.8	.6	0.2	0.6	1.0	1.3	1.7	2.1	2.5	2.8	3.2	3.6	11.0 / 0.7
22.7	3.8	7.6	11.4	15.2	18.9	.7	0.3	0.6	1.0	1.4	1.8	2.1	2.5	2.9	3.3	3.6	12.7 / 0.8
22.8	3.8	7.6	11.4	15.2	19.0	.8	0.3	0.7	1.0	1.4	1.8	2.2	2.5	2.9	3.3	3.7	14.4 / 0.9
22.9	3.9	7.7	11.5	15.3	19.1	.9	0.3	0.7	1.1	1.5	1.8	2.2	2.6	3.0	3.3	3.7	16.1 / 1.0 / 17.8 / 1.1
23.0	3.8	7.6	11.5	15.3	19.1	.0	0.0	0.4	0.8	1.2	1.6	2.0	2.3	2.7	3.1	3.5	19.5 / 1.2
23.1	3.8	7.7	11.5	15.4	19.2	.1	0.0	0.4	0.8	1.2	1.6	2.0	2.4	2.8	3.2	3.6	21.2 / 1.3
23.2	3.8	7.7	11.6	15.4	19.3	.2	0.1	0.5	0.9	1.3	1.6	2.0	2.4	2.8	3.2	3.6	22.8 / 1.4
23.3	3.9	7.8	11.6	15.5	19.4	.3	0.1	0.5	0.9	1.3	1.7	2.1	2.5	2.9	3.3	3.6	24.5 / 1.5
23.4	3.9	7.8	11.7	15.6	19.5	.4	0.2	0.5	0.9	1.3	1.7	2.1	2.5	2.9	3.3	3.7	26.2 / 1.6 / 27.9 / 1.7
23.5	3.9	7.8	11.8	15.7	19.6	.5	0.2	0.6	1.0	1.4	1.8	2.2	2.5	2.9	3.3	3.7	29.6 / 1.8
23.6	3.9	7.9	11.8	15.7	19.7	.6	0.2	0.6	1.0	1.4	1.8	2.2	2.6	3.0	3.4	3.8	31.3 / 1.9
23.7	4.0	7.9	11.9	15.8	19.8	.7	0.3	0.7	1.1	1.4	1.8	2.2	2.6	3.0	3.4	3.8	33.0 / 2.0
23.8	4.0	8.0	11.9	15.9	19.9	.8	0.3	0.7	1.1	1.5	1.9	2.3	2.7	3.1	3.4	3.8	34.7
23.9	4.0	8.0	12.0	16.0	20.0	.9	0.4	0.7	1.1	1.5	1.9	2.3	2.7	3.1	3.5	3.9	
	10	20	30	40	50		0	1	2	3	4	5	6	7	8	9	

Dec. Inc.	Tens 10	20	30	40	50	Decimals	Units 0	1	2	3	4	5	6	7	8	9	Double Second Diff and Corr
24.0	4.0	8.0	12.0	16.0	20.0	.0	0.0	0.4	0.8	1.2	1.6	2.0	2.4	2.9	3.3	3.7	0.8 / 0.1
24.1	4.0	8.0	12.0	16.0	20.1	.1	0.0	0.4	0.9	1.3	1.7	2.1	2.5	2.9	3.3	3.7	2.5 / 0.2
24.2	4.0	8.0	12.1	16.1	20.1	.2	0.1	0.5	0.9	1.3	1.7	2.1	2.5	2.9	3.3	3.8	4.1 / 0.3
24.3	4.0	8.1	12.1	16.2	20.2	.3	0.1	0.5	0.9	1.3	1.8	2.2	2.6	3.0	3.4	3.8	5.8 / 0.4
24.4	4.1	8.1	12.2	16.3	20.3	.4	0.2	0.6	1.0	1.4	1.6	2.2	2.6	3.0	3.4	3.8	7.4 / 0.5 / 9.1 / 0.6
24.5	4.1	8.2	12.3	16.3	20.4	.5	0.2	0.6	1.0	1.4	1.8	2.2	2.7	3.1	3.5	3.9	10.7 / 0.7
24.6	4.1	8.2	12.3	16.4	20.5	.6	0.2	0.7	1.1	1.5	1.9	2.3	2.7	3.1	3.5	3.9	12.3 / 0.8
24.7	4.1	8.3	12.4	16.5	20.6	.7	0.3	0.7	1.1	1.5	1.9	2.3	2.7	3.1	3.6	4.0	14.0 / 0.9
24.8	4.2	8.3	12.4	16.6	20.7	.8	0.3	0.7	1.1	1.6	2.0	2.4	2.8	3.2	3.6	4.0	15.6 / 1.0
24.9	4.2	8.3	12.5	16.6	20.8	.9	0.4	0.8	1.2	1.6	2.0	2.4	2.8	3.2	3.6	4.0	17.3 / 1.1 / 18.9 / 1.2
25.0	4.1	8.3	12.5	16.6	20.8	.0	0.0	0.4	0.8	1.3	1.7	2.1	2.5	3.0	3.4	3.8	20.6 / 1.3
25.1	4.2	8.3	12.5	16.7	20.9	.1	0.0	0.5	0.9	1.3	1.7	2.2	2.6	3.0	3.4	3.9	22.2 / 1.4
25.2	4.2	8.4	12.6	16.8	21.0	.2	0.1	0.5	0.9	1.4	1.8	2.2	2.6	3.1	3.5	3.9	23.9 / 1.5
25.3	4.2	8.4	12.6	16.9	21.1	.3	0.1	0.6	1.0	1.4	1.8	2.3	2.7	3.1	3.5	4.0	25.5 / 1.6
25.4	4.2	8.5	12.7	16.9	21.2	.4	0.2	0.6	1.0	1.4	1.9	2.3	2.7	3.1	3.6	4.0	27.2 / 1.7
25.5	4.3	8.5	12.8	17.0	21.3	.5	0.2	0.6	1.1	1.5	1.9	2.3	2.8	3.2	3.6	4.0	28.8 / 1.8
25.6	4.3	8.5	12.8	17.1	21.3	.6	0.3	0.7	1.1	1.5	2.0	2.4	2.8	3.2	3.7	4.1	30.4 / 1.9
25.7	4.3	8.6	12.9	17.2	21.4	.7	0.3	0.7	1.1	1.6	2.0	2.4	2.8	3.3	3.7	4.1	32.1 / 2.0
25.8	4.3	8.6	12.9	17.2	21.5	.8	0.3	0.8	1.2	1.6	2.0	2.5	2.9	3.3	3.7	4.2	33.7 / 2.1
25.9	4.4	8.7	13.0	17.3	21.6	.9	0.4	0.8	1.2	1.7	2.1	2.5	2.9	3.4	3.8	4.2	35.4
26.0	4.3	8.6	13.0	17.3	21.6	.0	0.0	0.4	0.9	1.3	1.8	2.2	2.6	3.1	3.5	4.0	0.8 / 0.1
26.1	4.3	8.7	13.0	17.4	21.7	.1	0.0	0.5	0.9	1.4	1.8	2.3	2.7	3.1	3.6	4.0	2.4 / 0.2
26.2	4.3	8.7	13.1	17.4	21.8	.2	0.1	0.5	1.0	1.4	1.9	2.3	2.7	3.2	3.6	4.1	4.0 / 0.3
26.3	4.4	8.8	13.1	17.5	21.9	.3	0.1	0.6	1.0	1.5	1.9	2.3	2.8	3.2	3.7	4.1	5.7 / 0.4
26.4	4.4	8.8	13.2	17.6	22.0	.4	0.2	0.6	1.1	1.5	1.9	2.4	2.8	3.3	3.7	4.2	7.3 / 0.5 / 8.9 / 0.6
26.5	4.4	8.8	13.3	17.7	22.1	.5	0.2	0.7	1.1	1.5	2.0	2.4	2.9	3.3	3.8	4.2	10.5 / 0.7
26.6	4.4	8.9	13.3	17.7	22.2	.6	0.3	0.7	1.1	1.6	2.0	2.5	2.9	3.4	3.8	4.2	12.1 / 0.8
26.7	4.5	8.9	13.4	17.8	22.3	.7	0.3	0.8	1.2	1.6	2.1	2.5	3.0	3.4	3.8	4.3	13.7 / 0.9
26.8	4.5	9.0	13.4	17.9	22.4	.8	0.4	0.8	1.2	1.7	2.1	2.6	3.0	3.4	3.9	4.3	15.4 / 1.0
26.9	4.5	9.0	13.5	18.0	22.5	.9	0.4	0.8	1.3	1.7	2.2	2.6	3.0	3.5	3.9	4.4	17.0 / 1.1 / 18.6 / 1.2
27.0	4.5	9.0	13.5	18.0	22.5	.0	0.0	0.5	0.9	1.4	1.8	2.3	2.7	3.2	3.7	4.1	20.2 / 1.3
27.1	4.5	9.0	13.5	18.0	22.6	.1	0.0	0.5	1.0	1.4	1.9	2.3	2.8	3.3	3.7	4.2	21.8 / 1.4
27.2	4.5	9.0	13.6	18.1	22.6	.2	0.1	0.5	1.0	1.5	1.9	2.4	2.8	3.3	3.8	4.2	23.4 / 1.5
27.3	4.5	9.1	13.6	18.2	22.7	.3	0.1	0.6	1.1	1.5	2.0	2.4	2.9	3.3	3.8	4.3	25.1 / 1.6
27.4	4.6	9.1	13.7	18.3	22.8	.4	0.2	0.6	1.1	1.6	2.0	2.5	2.9	3.4	3.8	4.3	26.7 / 1.7
27.5	4.6	9.2	13.8	18.3	22.9	.5	0.2	0.7	1.1	1.6	2.1	2.5	3.0	3.4	3.9	4.4	28.3 / 1.8
27.6	4.6	9.2	13.8	18.4	23.0	.6	0.3	0.7	1.2	1.6	2.1	2.6	3.0	3.5	3.9	4.4	29.9 / 1.9
27.7	4.6	9.3	13.9	18.5	23.1	.7	0.3	0.8	1.2	1.7	2.2	2.6	3.1	3.5	4.0	4.4	31.5 / 2.0
27.8	4.7	9.3	13.9	18.6	23.2	.8	0.4	0.8	1.3	1.7	2.2	2.7	3.1	3.6	4.0	4.5	33.1 / 2.1
27.9	4.7	9.3	14.0	18.6	23.3	.9	0.4	0.9	1.3	1.8	2.2	2.7	3.2	3.6	4.1	4.5	34.7
28.0	4.6	9.3	14.0	18.6	23.3	.0	0.0	0.5	0.9	1.4	1.9	2.4	2.8	3.3	3.8	4.3	0.8 / 0.1
28.1	4.7	9.3	14.0	18.7	23.4	.1	0.0	0.5	1.0	1.5	1.9	2.4	2.9	3.4	3.8	4.3	2.4 / 0.2
28.2	4.7	9.4	14.1	18.8	23.5	.2	0.1	0.6	1.0	1.5	2.0	2.5	2.9	3.4	3.9	4.4	4.0 / 0.3
28.3	4.7	9.4	14.1	18.9	23.6	.3	0.1	0.6	1.1	1.6	2.0	2.5	3.0	3.5	3.9	4.4	5.6 / 0.4
28.4	4.7	9.5	14.2	18.9	23.7	.4	0.2	0.7	1.1	1.6	2.1	2.6	3.0	3.5	4.0	4.5	7.2 / 0.5 / 8.8 / 0.6
28.5	4.8	9.5	14.3	19.0	23.8	.5	0.2	0.7	1.2	1.7	2.1	2.6	3.1	3.6	4.0	4.5	10.4 / 0.7
28.6	4.8	9.5	14.3	19.1	23.8	.6	0.3	0.8	1.2	1.7	2.2	2.7	3.1	3.6	4.1	4.6	12.0 / 0.8
28.7	4.8	9.6	14.4	19.2	23.9	.7	0.3	0.8	1.3	1.8	2.2	2.7	3.2	3.7	4.1	4.6	13.6 / 0.9
28.8	4.8	9.6	14.4	19.2	24.0	.8	0.4	0.9	1.3	1.8	2.3	2.8	3.2	3.7	4.2	4.7	15.2 / 1.0
28.9	4.9	9.7	14.5	19.3	24.1	.9	0.4	0.9	1.4	1.9	2.3	2.8	3.3	3.8	4.2	4.7	16.8 / 1.1 / 18.4 / 1.2
29.0	4.8	9.6	14.5	19.3	24.1	.0	0.0	0.5	1.0	1.5	2.0	2.5	2.9	3.4	3.9	4.4	20.0 / 1.3
29.1	4.8	9.7	14.5	19.4	24.2	.1	0.0	0.5	1.0	1.5	2.0	2.5	3.0	3.5	4.0	4.5	21.6 / 1.4
29.2	4.8	9.7	14.6	19.4	24.3	.2	0.1	0.6	1.1	1.6	2.1	2.6	3.0	3.5	4.0	4.5	23.2 / 1.5
29.3	4.9	9.8	14.6	19.5	24.4	.3	0.1	0.6	1.1	1.6	2.1	2.6	3.1	3.6	4.1	4.6	24.8 / 1.6
29.4	4.9	9.8	14.7	19.6	24.5	.4	0.2	0.7	1.2	1.7	2.2	2.7	3.1	3.6	4.1	4.6	26.4 / 1.7
29.5	4.9	9.8	14.8	19.7	24.6	.5	0.2	0.7	1.2	1.7	2.2	2.7	3.2	3.7	4.2	4.7	28.0 / 1.8
29.6	4.9	9.9	14.8	19.7	24.7	.6	0.3	0.8	1.3	1.8	2.3	2.8	3.2	3.7	4.2	4.7	29.6 / 1.9
29.7	5.0	9.9	14.9	19.8	24.8	.7	0.3	0.8	1.3	1.8	2.3	2.8	3.3	3.8	4.3	4.8	31.2 / 2.0
29.8	5.0	10.0	14.9	19.9	24.9	.8	0.4	0.9	1.4	1.9	2.4	2.9	3.3	3.8	4.3	4.8	32.8 / 2.1
29.9	5.0	10.0	15.0	20.0	25.0	.9	0.4	0.9	1.4	1.9	2.4	2.9	3.4	3.9	4.4	4.9	34.4
30.0	5.0	10.0	15.0	20.0	25.0	.0	0.0	0.5	1.0	1.5	2.0	2.5	3.0	3.6	4.1	4.6	0.8 / 0.1
30.1	5.0	10.0	15.0	20.0	25.1	.1	0.1	0.6	1.1	1.6	2.1	2.6	3.1	3.6	4.1	4.6	2.4 / 0.2
30.2	5.0	10.0	15.1	20.1	25.1	.2	0.1	0.6	1.1	1.6	2.1	2.6	3.2	3.7	4.2	4.7	4.0 / 0.3
30.3	5.0	10.1	15.1	20.2	25.2	.3	0.2	0.7	1.2	1.7	2.2	2.7	3.2	3.7	4.2	4.7	5.6 / 0.4
30.4	5.1	10.1	15.2	20.3	25.3	.4	0.2	0.7	1.2	1.7	2.2	2.7	3.3	3.8	4.3	4.8	7.2 / 0.5 / 8.8 / 0.6
30.5	5.1	10.2	15.3	20.3	25.4	.5	0.3	0.8	1.3	1.8	2.3	2.8	3.3	3.8	4.3	4.8	10.4 / 0.7
30.6	5.1	10.2	15.3	20.4	25.5	.6	0.3	0.8	1.3	1.8	2.3	2.8	3.4	3.9	4.4	4.9	12.0 / 0.8
30.7	5.1	10.3	15.4	20.5	25.6	.7	0.4	0.9	1.4	1.9	2.4	2.9	3.4	3.9	4.4	4.9	13.6 / 0.9
30.8	5.2	10.3	15.4	20.6	25.7	.8	0.4	0.9	1.4	1.9	2.4	2.9	3.5	4.0	4.5	5.0	15.2 / 1.0
30.9	5.2	10.3	15.5	20.6	25.8	.9	0.5	1.0	1.5	2.0	2.5	3.0	3.5	4.0	4.5	5.0	16.8 / 1.1 / 18.4
31.0	5.1	10.3	15.5	20.6	25.8	.0	0.0	0.5	1.0	1.6	2.1	2.6	3.1	3.7	4.2	4.7	20.0
31.1	5.2	10.3	15.5	20.7	25.9	.1	0.1	0.6	1.1	1.6	2.2	2.7	3.2	3.7	4.3	4.8	21.6
31.2	5.2	10.4	15.6	20.8	26.0	.2	0.1	0.6	1.2	1.7	2.2	2.7	3.3	3.8	4.3	4.8	23.2
31.3	5.2	10.4	15.6	20.9	26.1	.3	0.2	0.7	1.2	1.7	2.3	2.8	3.3	3.8	4.4	4.9	24.8
31.4	5.2	10.5	15.7	20.9	26.2	.4	0.2	0.7	1.3	1.8	2.3	2.8	3.4	3.9	4.4	4.9	26.4
31.5	5.3	10.5	15.8	21.0	26.3	.5	0.3	0.8	1.3	1.8	2.4	2.9	3.4	3.9	4.5	5.0	28.0
31.6	5.3	10.5	15.8	21.1	26.3	.6	0.3	0.8	1.4	1.9	2.4	2.9	3.5	4.0	4.5	5.0	29.6
31.7	5.3	10.6	15.9	21.2	26.4	.7	0.4	0.9	1.4	1.9	2.5	3.0	3.5	4.0	4.6	5.1	31.2
31.8	5.3	10.6	15.9	21.2	26.5	.8	0.4	0.9	1.5	2.0	2.5	3.0	3.6	4.1	4.6	5.1	32.8
31.9	5.4	10.7	16.0	21.3	26.6	.9	0.5	1.0	1.5	2.0	2.6	3.1	3.6	4.1	4.7	5.2	34.4
	10	20	30	40	50		0	1	2	3	4	5	6	7	8	9	

The Double-Second-Difference correction (Corr.) is always to be added to the tabulated altitude

60°, 300° L.H.A. LATITUDE SAME NAME AS DECLINATION

N. Lat. { L.H.A. greater than 180° ...... Zn=Z; L.H.A. less than 180° ...... Zn=360°−Z }

Dec.	38° Hc	38° d	38° Z	39° Hc	39° d	39° Z	40° Hc	40° d	40° Z	41° Hc	41° d	41° Z
°	° ′	′	°	° ′	′	°	° ′	′	°	° ′	′	°
0	23 12.2	+40.1	109.6	22 51.9	+40.9	110.0	22 31.3	+41.6	110.4	22 10.2	+42.4	110.7
1	23 52.3	39.8	108.8	23 32.8	40.6	109.2	23 12.9	41.4	109.6	22 52.6	42.2	110.0
2	24 32.1	39.5	107.9	24 13.4	40.4	108.4	23 54.3	41.2	108.8	23 34.8	41.9	109.2
3	25 11.6	39.3	107.1	24 53.8	40.1	107.6	24 35.5	40.9	108.0	24 16.7	41.7	108.4
4	25 50.9	39.0	106.3	25 33.9	39.8	106.7	25 16.4	40.6	107.2	24 58.4	41.5	107.6
5	26 29.9	+38.6	105.4	26 13.7	+39.5	105.9	25 57.0	+40.4	106.4	25 39.9	+41.1	106.8
6	27 08.5	38.4	104.6	26 53.2	39.2	105.1	26 37.4	40.0	105.5	26 21.0	40.9	106.0
7	27 46.9	38.0	103.7	27 32.4	38.9	104.2	27 17.4	39.8	104.7	27 01.9	40.7	105.2
8	28 24.9	37.7	102.8	28 11.3	38.6	103.3	27 57.2	39.5	103.9	27 42.6	40.3	104.4
9	29 02.6	37.3	101.9	28 49.9	38.2	102.5	28 36.7	39.1	103.0	28 22.9	40.0	103.5
10	29 39.9	+37.0	101.0	29 28.1	+37.9	101.6	29 15.8	+38.8	102.1	29 02.9	+39.7	102.7
11	30 16.9	36.5	100.1	30 06.0	37.5	100.7	29 54.6	38.4	101.3	29 42.6	39.3	101.8
12	30 53.4	36.2	99.2	30 43.5	37.2	99.8	30 33.0	38.1	100.4	30 21.9	39.0	101.0
13	31 29.6	35.8	98.3	31 20.7	36.7	98.9	31 11.1	37.7	99.5	31 00.9	38.6	100.1
14	32 05.4	35.3	97.3	31 57.4	36.3	97.9	31 48.8	37.3	98.6	31 39.5	38.3	99.2
15	32 40.7	+34.9	96.4	32 33.7	+35.9	97.0	32 26.1	+36.9	97.6	32 17.8	+37.8	98.3
16	33 15.6	34.5	95.4	33 09.6	35.5	96.1	33 03.0	36.4	96.7	32 55.6	37.5	97.3
17	33 50.1	33.9	94.4	33 45.1	35.0	95.1	33 39.4	36.0	95.7	33 33.1	37.0	96.4
18	34 24.0	33.5	93.4	34 20.1	34.5	94.1	34 15.4	35.6	94.8	34 10.1	36.5	95.5
19	34 57.5	33.0	92.4	34 54.6	34.0	93.1	34 51.0	35.1	93.8	34 46.6	36.1	94.5
20	35 30.5	+32.4	91.4	35 28.6	+33.6	92.1	35 26.1	+34.5	92.8	35 22.7	+35.7	93.5
21	36 02.9	31.9	90.4	36 02.2	32.9	91.1	36 00.6	34.1	91.8	35 58.4	35.1	92.5
22	36 34.8	31.3	89.3	36 35.1	32.5	90.0	36 34.7	33.6	90.8	36 33.5	34.6	91.5
23	37 06.1	30.8	88.2	37 07.6	31.9	89.0	37 08.3	32.9	89.8	37 08.1	34.1	90.5
24	37 36.9	30.2	87.2	37 39.5	31.3	87.9	37 41.2	32.5	88.7	37 42.2	33.5	89.5
25	38 07.1	+29.5	86.1	38 10.8	+30.7	86.8	38 13.7	+31.8	87.6	38 15.7	+33.0	88.4
26	38 36.6	28.9	85.0	38 41.5	30.0	85.7	38 45.5	31.2	86.5	38 48.7	32.4	87.4
27	39 05.5	28.3	83.8	39 11.5	29.5	84.6	39 16.7	30.6	85.4	39 21.1	31.7	86.3
28	39 33.8	27.5	82.7	39 41.0	28.7	83.5	39 47.3	30.0	84.3	39 52.8	31.1	85.2
29	40 01.3	26.9	81.5	40 09.7	28.1	82.4	40 17.3	29.2	83.2	40 23.9	30.5	84.0
30	40 28.2	+26.1	80.4	40 37.8	+27.4	81.2	40 46.5	+28.6	82.1	40 54.4	+29.8	82.9
31	40 54.3	25.5	79.2	41 05.2	26.6	80.0	41 15.1	27.9	80.9	41 24.2	29.1	81.8
32	41 19.8	24.6	78.0	41 31.8	25.9	78.8	41 43.0	27.1	79.7	41 53.3	28.3	80.6
33	41 44.4	23.9	76.8	41 57.7	25.1	77.6	42 10.1	26.4	78.5	42 21.6	27.6	79.4
34	42 08.3	23.0	75.5	42 22.8	24.4	76.4	42 36.5	25.6	77.3	42 49.2	26.9	78.2
35	42 31.3	+22.3	74.3	42 47.2	+23.5	75.2	43 02.1	+24.8	76.1	43 16.1	+26.0	77.0
36	42 53.6	21.4	73.0	43 10.7	22.7	73.9	43 26.9	23.9	74.8	43 42.1	25.3	75.7
37	43 15.0	20.5	71.7	43 33.4	21.8	72.6	43 50.8	23.1	73.5	44 07.4	24.3	74.5
38	43 35.5	19.7	70.4	43 55.2	20.9	71.3	44 13.9	22.2	72.3	44 31.7	23.6	73.2
39	43 55.2	18.7	69.1	44 16.1	20.0	70.0	44 36.1	21.4	71.0	44 55.3	22.6	71.9
40	44 13.9	+17.8	67.8	44 36.1	+19.2	68.7	44 57.5	+20.4	69.6	45 17.9	+21.7	70.6
41	44 31.7	16.9	66.5	44 55.3	18.1	67.4	45 17.9	19.5	68.3	45 39.6	20.8	69.3
42	44 48.6	15.9	65.1	45 13.4	17.2	66.0	45 37.4	18.5	67.0	46 00.4	19.8	67.9
43	45 04.5	15.0	63.8	45 30.6	16.3	64.7	45 55.9	17.5	65.6	46 20.2	18.8	66.5
44	45 19.5	13.9	62.4	45 46.9	15.2	63.3	46 13.4	16.5	64.2	46 39.0	17.9	65.2
45	45 33.4	+12.9	61.0	46 02.1	+14.2	61.9	46 29.9	+15.5	62.8	46 56.9	+16.8	63.8
46	45 46.3	11.9	59.6	46 16.3	13.1	60.5	46 45.4	14.5	61.4	47 13.7	15.7	62.4
47	45 58.2	10.9	58.2	46 29.4	12.1	59.1	46 59.9	13.3	60.0	47 29.4	14.7	60.9
48	46 09.1	9.7	56.8	46 41.5	11.1	57.7	47 13.2	12.4	58.6	47 44.1	13.6	59.5
49	46 18.8	8.8	55.3	46 52.6	9.9	56.2	47 25.6	11.2	57.1	47 57.7	12.5	58.0
50	46 27.6	+7.6	53.9	47 02.5	+8.9	54.8	47 36.8	+10.1	55.7	48 10.2	+11.4	56.6
51	46 35.2	6.5	52.5	47 11.4	7.7	53.3	47 46.9	8.9	54.2	48 21.6	10.2	55.1
52	46 41.7	5.5	51.0	47 19.1	6.7	51.9	47 55.8	7.9	52.7	48 31.8	9.1	53.6
53	46 47.2	4.3	49.6	47 25.8	5.5	50.4	48 03.7	6.7	51.2	48 40.9	7.9	52.1
54	46 51.5	3.3	48.1	47 31.3	4.3	48.9	48 10.4	5.5	49.8	48 48.8	6.7	50.6
55	46 54.8	+2.1	46.6	47 35.6	+3.3	47.4	48 15.9	+4.4	48.3	48 55.5	+5.6	49.1
56	46 56.9	+1.0	45.2	47 38.9	2.0	46.0	48 20.3	3.2	46.8	49 01.1	4.3	47.6
57	46 57.9	0.2	43.7	47 40.9	+1.0	44.5	48 23.5	2.0	45.3	49 05.4	3.1	46.1
58	46 57.7	1.2	42.3	47 41.9	0.2	43.0	48 25.5	+0.8	43.8	49 08.5	2.0	44.5
59	46 56.5	2.4	40.8	47 41.7	1.4	41.5	48 26.3	0.3	42.2	49 10.5	+0.7	43.0
60	46 54.1	3.5	39.3	47 40.3	−2.5	40.0	48 26.0	1.5	40.7	49 11.2	−0.5	41.5
61	46 50.6	4.6	37.9	47 37.8	3.7	38.5	48 24.5	2.7	39.2	49 10.7	1.7	40.0
62	46 46.0	5.7	36.4	47 34.1	4.8	37.1	48 21.8	3.9	37.7	49 09.0	2.9	38.4
63	46 40.3	6.8	35.0	47 29.3	6.0	35.6	48 17.9	5.1	36.2	49 06.1	4.1	36.9
64	46 33.5	7.9	33.5	47 23.3	7.0	34.1	48 12.8	6.2	34.7	49 02.0	5.4	35.4
65	46 25.6	9.0	32.1	47 16.3	8.2	32.6	48 06.6	7.3	33.2	48 56.6	−6.5	33.9
66	46 16.6	10.0	30.6	47 08.1	9.3	31.2	47 59.3	8.5	31.8	48 50.1	7.7	32.4
67	46 06.6	11.1	29.2	46 58.8	10.4	29.7	47 50.8	9.6	30.3	48 42.4	8.8	30.8
68	45 55.5	12.1	27.8	46 48.4	11.4	28.3	47 41.2	10.8	28.8	48 33.6	10.0	29.4
69	45 43.4	13.2	26.4	46 37.0	12.5	26.9	47 30.4	11.8	27.4	48 23.6	11.2	27.9
70	45 30.2	14.2	25.0	46 24.5	13.6	25.4	47 18.6	13.0	25.9	48 12.4	12.3	26.4
71	45 16.0	15.1	23.6	46 10.9	14.6	24.0	47 05.6	14.0	24.5	48 00.1	13.4	24.9
72	45 00.9	16.2	22.2	45 56.3	15.6	22.6	46 51.6	15.0	23.0	47 46.7	14.4	23.5
73	44 44.7	17.1	20.9	45 40.7	16.6	21.2	46 36.6	16.1	21.6	47 32.3	15.6	22.0
74	44 27.6	18.0	19.5	45 24.1	17.6	19.9	46 20.5	17.1	20.2	47 16.7	16.6	20.6
75	44 09.6	19.0	18.2	45 06.5	18.5	18.5	46 03.4	18.1	18.8	47 00.1	−17.7	19.2
76	43 50.6	19.8	16.9	44 48.0	19.5	17.2	45 45.3	19.1	17.5	46 42.4	18.6	17.8
77	43 30.8	20.8	15.6	44 28.5	20.4	15.8	45 26.2	20.0	16.1	46 23.8	19.6	16.4
78	43 10.0	21.6	14.3	44 08.1	21.3	14.5	45 06.2	21.0	14.8	46 04.2	20.6	15.0
79	42 48.4	22.4	13.0	43 46.8	22.1	13.2	44 45.2	21.8	13.5	45 43.6	21.6	13.7
80	42 26.0	23.3	11.8	43 24.7	23.0	11.9	44 23.4	22.8	12.1	45 22.0	−22.5	12.4
81	42 02.7	24.0	10.5	43 01.7	23.8	10.7	44 00.6	23.6	10.9	44 59.5	23.3	11.0
82	41 38.7	24.9	9.3	42 37.9	24.7	9.4	43 37.0	24.4	9.6	44 36.2	24.2	9.7
83	41 13.8	25.6	8.1	42 13.2	25.4	8.2	43 12.6	25.2	8.3	44 12.0	25.1	8.5
84	40 48.2	26.3	6.9	41 47.8	26.2	7.0	42 47.4	26.1	7.1	43 46.9	25.9	7.2
85	40 21.9	27.0	5.7	41 21.6	26.9	5.8	42 21.3	26.8	5.9	43 21.0	26.7	6.0
86	39 54.9	27.7	4.5	40 54.7	27.7	4.6	41 54.5	27.6	4.7	42 54.3	27.5	4.7
87	39 27.2	28.5	3.4	40 27.0	28.3	3.4	41 26.9	28.2	3.5	42 26.8	28.2	3.5
88	38 58.7	29.0	2.2	39 58.7	29.0	2.3	40 58.7	29.0	2.3	41 58.6	28.9	2.3
89	38 29.7	29.7	1.1	39 29.7	29.7	1.1	40 29.7	29.7	1.1	41 29.7	29.7	1.2
90	38 00.0	30.3	0.0	39 00.0	30.3	0.0	40 00.0	−30.3	0.0	41 00.0	−30.3	0.0

Dec.	42° Hc	42° d	42° Z	43° Hc	43° d	43° Z	44° Hc	44° d	44° Z	45° Hc	45° d	45° Z
°	° ′	′	°	° ′	′	°	° ′	′	°	° ′	′	°
0	21 48.8	+43.1	111.1	21 27.0	+43.8	111.5	21 04.8	+44.6	111.9	20 42.3	+45.3	112.2
1	22 31.9	42.9	110.4	22 10.8	43.7	110.8	21 49.4	44.4	111.1	21 27.6	45.0	111.5
2	23 14.8	42.7	109.6	22 54.5	43.4	110.0	22 33.8	44.2	110.4	22 12.6	45.0	110.8
3	23 57.5	42.5	108.8	23 37.9	43.3	109.3	23 18.0	43.9	109.7	22 57.6	44.7	110.1
4	24 40.0	42.3	108.1	24 21.2	43.0	108.5	24 01.9	43.8	108.9	23 42.3	44.5	109.3
5	25 22.3	+41.9	107.3	25 04.2	+42.8	107.7	24 45.7	+43.5	108.2	24 26.8	+44.3	108.6
6	26 04.2	41.8	106.5	25 47.0	42.5	107.0	25 29.2	43.3	107.4	25 11.1	44.0	107.9
7	26 46.0	41.4	105.7	26 29.5	42.3	106.2	26 12.5	43.1	106.7	25 55.1	43.8	107.1
8	27 27.4	41.2	104.9	27 11.8	41.9	105.4	26 55.6	42.8	105.9	26 38.9	43.6	106.4
9	28 08.6	40.8	104.1	27 53.7	41.7	104.6	27 38.4	42.5	105.1	27 22.5	43.3	105.6
10	28 49.4	+40.6	103.2	28 35.4	+41.4	103.8	28 20.9	+42.2	104.3	28 05.8	+43.1	104.8
11	29 30.0	40.2	102.4	29 16.8	41.1	102.9	29 03.1	42.0	103.5	28 48.9	42.8	104.0
12	30 10.2	39.9	101.5	29 57.9	40.8	102.1	29 45.1	41.6	102.7	29 31.7	42.4	103.2
13	30 50.1	39.6	100.7	30 38.7	40.5	101.2	30 26.7	41.3	101.8	30 14.1	42.2	102.4
14	31 29.7	39.1	99.8	31 19.2	40.0	100.4	31 08.0	41.0	101.0	30 56.3	41.9	101.6
15	32 08.8	+38.8	98.9	31 59.2	+39.8	99.5	31 49.0	+40.6	100.1	31 38.2	+41.5	100.7
16	32 47.6	38.4	98.0	32 39.0	39.3	98.6	32 29.6	40.3	99.3	32 19.7	41.1	99.9
17	33 26.0	38.0	97.1	33 18.3	38.9	97.7	33 09.9	39.9	98.4	33 00.8	40.8	99.0
18	34 04.0	37.6	96.1	33 57.2	38.6	96.8	33 49.8	39.5	97.5	33 41.6	40.5	98.1
19	34 41.6	37.1	95.2	34 35.8	38.1	95.9	34 29.3	39.1	96.6	34 22.1	40.0	97.2
20	35 18.7	+36.6	94.2	35 13.9	+37.7	94.9	35 08.4	+38.6	95.6	35 02.1	+39.6	96.3
21	35 55.3	36.2	93.3	35 51.6	37.2	94.0	35 47.0	38.2	94.7	35 41.7	39.2	95.4
22	36 31.5	35.7	92.3	36 28.8	36.7	93.0	36 25.2	37.8	93.7	36 20.9	38.8	94.5
23	37 07.2	35.2	91.3	37 05.5	36.2	92.0	37 03.0	37.2	92.8	36 59.7	38.2	93.5
24	37 42.4	34.6	90.2	37 41.7	35.7	91.0	37 40.2	36.8	91.8	37 37.9	37.8	92.6
25	38 17.0	+34.0	89.2	38 17.4	+35.2	90.0	38 17.0	+36.2	90.8	38 15.7	+37.3	91.6
26	38 51.0	33.5	88.2	38 52.6	34.6	89.0	38 53.2	35.7	89.8	38 53.0	36.8	90.6
27	39 24.5	32.9	87.1	39 27.2	34.0	87.9	39 28.9	35.2	88.7	39 29.8	36.3	89.6
28	39 57.4	32.3	86.0	40 01.2	33.4	86.8	40 04.1	34.5	87.7	40 06.1	35.6	88.5
29	40 29.7	31.7	84.9	40 34.6	32.8	85.7	40 38.6	34.0	86.6	40 41.7	35.1	87.5
30	41 01.4	+30.9	83.8	41 07.4	+32.2	84.6	41 12.6	+33.3	85.5	41 16.8	+34.5	86.4
31	41 32.3	30.3	82.6	41 39.6	31.4	83.5	41 45.9	32.6	84.4	41 51.3	33.8	85.3
32	42 02.6	29.6	81.5	42 11.0	30.8	82.4	42 18.5	32.0	83.3	42 25.1	33.2	84.2
33	42 32.2	28.8	80.3	42 41.8	30.1	81.2	42 50.5	31.3	82.1	42 58.3	32.5	83.0
34	43 01.0	28.1	79.1	43 11.9	29.3	80.0	43 21.8	30.6	81.0	43 30.8	31.8	81.9
35	43 29.1	+27.3	77.9	43 41.2	+28.6	78.8	43 52.4	+29.8	79.8	44 02.6	+31.0	80.7
36	43 56.4	26.5	76.7	44 09.8	27.8	77.6	44 22.2	29.0	78.6	44 33.6	30.3	79.5
37	44 22.9	25.7	75.4	44 37.6	26.9	76.4	44 51.2	28.3	77.3	45 03.9	29.5	78.3
38	44 48.6	24.8	74.1	45 04.5	26.1	75.1	45 19.5	27.4	76.1	45 33.4	28.7	77.1
39	45 13.4	24.0	72.9	45 30.6	25.3	73.8	45 46.9	26.5	74.8	46 02.1	27.8	75.8
40	45 37.4	+23.0	71.5	45 55.9	+24.3	72.5	46 13.4	+25.6	73.5	46 29.9	+27.0	74.5
41	46 00.4	22.1	70.2	46 20.2	23.4	71.2	46 39.0	24.8	72.2	46 56.9	26.0	73.2
42	46 22.5	21.1	68.9	46 43.6	22.5	69.9	47 03.8	23.8	70.9	47 22.9	25.2	71.9
43	46 43.6	20.2	67.5	47 06.1	21.5	68.5	47 27.6	22.8	69.5	47 48.1	24.1	70.5
44	47 03.8	19.1	66.1	47 27.6	20.5	67.1	47 50.4	21.8	68.1	48 12.2	23.2	69.2
45	47 22.9	+18.2	64.7	47 48.1	+19.4	65.7	48 12.2	+20.8	66.8	48 35.4	+22.2	67.8
46	47 41.1	17.0	63.3	48 07.5	18.4	64.3	48 33.0	19.8	65.3	48 57.6	21.1	66.4
47	47 58.1	16.0	61.9	48 25.9	17.4	62.9	48 52.8	18.7	63.9	49 18.7	20.0	65.0
48	48 14.1	15.0	60.5	48 43.3	16.2	61.4	49 11.5	17.6	62.5	49 38.7	19.0	63.5
49	48 29.1	13.7	59.0	48 59.5	15.1	60.0	49 29.1	16.4	61.0	49 57.7	17.8	62.0
50	48 42.8	+12.7	57.5	49 14.6	+14.0	58.5	49 45.5	+15.3	59.5	50 15.5	+16.7	60.5
51	48 55.5	11.5	56.0	49 28.6	12.8	57.0	50 00.8	14.2	58.0	50 32.2	15.4	59.0
52	49 07.0	10.3	54.5	49 41.4	11.6	55.5	50 15.0	12.9	56.5	50 47.6	14.3	57.5
53	49 17.3	9.2	53.0	49 53.0	10.4	54.0	50 27.9	11.7	55.0	51 01.9	13.0	56.0
54	49 26.5	7.9	51.5	50 03.4	9.2	52.5	50 39.6	10.5	53.4	51 14.9	11.8	54.4
55	49 34.4	+6.8	50.0	50 12.6	+8.0	50.9	50 50.1	+9.2	51.9	51 26.7	+10.6	52.8
56	49 41.2	5.5	48.5	50 20.6	6.7	49.4	50 59.3	8.0	50.3	51 37.3	9.2	51.3
57	49 46.7	4.3	46.9	50 27.3	5.5	47.8	51 07.3	6.7	48.7	51 46.5	7.9	49.7
58	49 51.0	3.1	45.4	50 32.8	4.2	46.2	51 14.0	5.4	47.1	51 54.4	6.7	48.1
59	49 54.1	1.8	43.8	50 37.0	3.0	44.7	51 19.4	4.1	45.5	52 01.1	5.3	46.4
60	49 55.9	+0.5	42.3	50 40.0	+1.6	43.1	51 23.5	+2.8	43.9	52 06.4	+3.9	44.8
61	49 56.4	−0.6	40.7	50 41.6	+0.4	41.5	51 26.3	1.5	42.3	52 10.3	2.6	43.2
62	49 55.8	2.0	39.2	50 42.0	0.9	39.9	51 27.8	+0.1	40.7	52 12.9	+1.3	41.6
63	49 53.8	3.1	37.6	50 41.1	2.1	38.4	51 27.9	1.1	39.1	52 14.2	0.1	39.9
64	49 50.7	4.4	36.1	50 39.0	3.5	36.8	51 26.8	2.5	37.5	52 14.1	1.4	38.3
65	49 46.3	5.7	34.5	50 35.5	−4.7	35.2	51 24.3	−3.8	35.9	52 12.7	2.8	36.7
66	49 40.6	6.8	33.0	50 30.8	6.0	33.6	51 20.5	5.0	34.3	52 09.9	4.1	35.0
67	49 33.8	8.1	31.4	50 24.8	7.2	32.1	51 15.5	6.4	32.7	52 05.8	5.5	33.4
68	49 25.7	9.2	29.9	50 17.6	8.5	30.5	51 09.1	7.6	31.1	52 00.3	6.8	31.8
69	49 16.5	10.5	28.4	50 09.1	9.7	29.0	51 01.5	9.0	29.6	51 53.5	8.1	30.2
70	49 06.0	11.6	26.9	49 59.4	10.9	27.4	50 52.5	10.1	28.0	51 45.4	−9.4	28.6
71	48 54.4	12.7	25.4	49 48.5	12.1	25.9	50 42.4	11.4	26.4	51 36.0	10.7	27.0
72	48 41.7	13.9	23.9	49 36.4	13.2	24.4	50 31.0	12.7	24.9	51 25.3	12.0	25.4
73	48 27.8	15.0	22.4	49 23.2	14.5	22.9	50 18.3	13.8	23.4	51 13.3	13.2	23.8
74	48 12.8	16.1	21.0	49 08.7	15.5	21.4	50 04.5	15.0	21.8	51 00.1	14.4	22.3
75	47 56.7	−17.2	19.5	48 53.2	16.7	19.9	49 49.5	16.2	20.3	50 45.7	−15.6	20.8
76	47 39.5	18.2	18.1	48 36.5	17.8	18.5	49 33.3	17.3	18.8	50 30.1	16.9	19.2
77	47 21.3	19.2	16.7	48 18.7	18.8	17.0	49 16.0	18.4	17.4	50 13.2	17.9	17.7
78	47 02.1	20.3	15.3	47 59.9	19.9	15.6	48 57.6	19.5	15.9	49 55.3	19.1	16.2
79	46 41.8	21.2	13.9	47 40.0	20.9	14.2	48 38.1	20.5	14.5	49 36.2	20.2	14.8
80	46 20.6	−22.2	12.6	47 19.1	21.9	12.8	48 17.6	21.6	13.1	49 16.0	−21.2	13.3
81	45 58.4	23.1	11.2	46 57.2	22.8	11.4	47 56.0	22.6	11.7	48 54.8	22.3	11.9
82	45 35.3	24.0	9.9	46 34.4	23.8	10.1	47 33.4	23.5	10.3	48 32.5	23.3	10.5
83	45 11.3	24.9	8.6	46 10.6	24.7	8.8	47 09.9	24.5	8.9	48 09.2	24.3	9.1
84	44 46.4	25.7	7.3	45 45.9	25.6	7.5	46 45.4	25.4	7.6	47 44.9	25.3	7.7
85	44 20.7	−26.6	6.1	45 20.3	−26.4	6.2	46 20.0	26.3	6.3	47 19.6	26.2	6.4
86	43 54.1	27.4	4.8	44 53.9	27.3	4.9	45 53.7	27.2	5.0	46 53.4	27.1	5.1
87	43 26.7	28.1	3.6	44 26.6	28.1	3.6	45 26.5	28.1	3.7	46 26.3	27.9	3.8
88	42 58.6	29.0	2.4	43 58.5	28.9	2.4	44 58.4	28.8	2.4	45 58.4	28.8	2.5
89	42 29.6	29.6	1.2	43 29.6	29.6	1.2	44 29.6	29.6	1.2	45 29.6	29.6	1.2
90	42 00.0	−30.3	0.0	43 00.0	30.4	0.0	44 00.0	30.4	0.0	45 00.0	−30.4	0.0

60°, 300° L.H.A. LATITUDE SAME NAME AS DECLINATION

LATITUDE CONTRARY NAME TO DECLINATION L.H.A. 60°, 300°

Dec.	38°			39°			40°			41°			42°			43°			44°			45°			Dec.
	Hc	d	Z	Hc	d	Z	Hc	d	Z	Hc	d	Z	Hc	d	Z	Hc	d	Z	Hc	d	Z	Hc	d	Z	
°	° ′	′	°	° ′	′	°	° ′	′	°	° ′	′	°	° ′	′	°	° ′	′	°	° ′	′	°	° ′	′	°	°
0	23 12.2	-40.2	109.6	22 51.9	-41.1	110.0	22 31.3	41.9	110.4	22 10.2	42.6	110.7	21 48.8	43.4	111.1	21 27.0	-44.1	111.5	21 04.8	44.8	111.9	20 42.3	-45.5	112.2	0
1	22 31.9	40.5	110.4	22 10.8	41.3	110.8	21 49.4	42.1	111.1	21 27.6	42.8	111.5	21 05.4	43.5	111.9	20 42.9	44.2	112.2	20 20.0	44.9	112.6	19 56.8	45.6	112.9	1
2	21 51.4	40.8	111.2	21 29.5	41.5	111.5	21 07.3	42.3	111.9	20 44.8	43.0	112.3	20 21.9	43.7	112.6	19 58.7	44.5	112.9	19 35.1	45.1	113.3	19 11.2	45.7	113.6	2
3	21 10.6	41.0	112.0	20 48.0	41.8	112.3	20 25.0	42.4	112.7	20 01.8	43.2	113.0	19 38.2	43.9	113.3	19 14.2	44.5	113.7	18 50.0	45.2	114.0	18 25.5	45.9	114.3	3
4	20 29.6	41.2	112.7	20 06.2	41.9	113.1	19 42.6	42.7	113.4	19 18.6	43.4	113.7	18 54.3	44.1	114.1	18 29.7	44.8	114.4	18 04.8	45.4	114.7	17 39.6	46.0	115.0	4
5	19 48.4	-41.4	113.5	19 24.3	-42.1	113.8	18 59.9	42.8	114.2	18 35.2	-43.5	114.5	18 10.2	-44.2	114.8	17 44.9	44.9	115.1	17 19.4	-45.6	115.3	16 53.6	-46.2	115.6	5
6	19 07.0	41.6	114.3	18 42.2	42.3	114.6	18 17.1	43.0	114.9	17 51.7	43.7	115.2	17 26.0	44.4	115.5	17 00.0	45.0	115.8	16 33.8	45.6	116.0	16 07.4	46.3	116.3	6
7	18 25.4	41.8	115.0	17 59.9	42.5	115.3	17 34.1	43.2	115.6	17 08.0	43.9	115.9	16 41.6	44.5	116.2	16 15.0	45.1	116.4	15 48.2	45.8	116.7	15 21.1	46.4	117.0	7
8	17 43.6	41.9	115.8	17 17.4	42.6	116.1	16 50.9	43.3	116.4	16 24.1	44.0	116.6	15 57.1	44.6	116.9	15 29.9	45.3	117.1	15 02.4	45.9	117.4	14 34.7	46.5	117.6	8
9	17 01.7	42.1	116.5	16 34.8	42.8	116.8	16 07.6	43.5	117.1	15 40.1	44.1	117.3	15 12.5	44.8	117.6	14 44.6	45.4	117.8	14 16.5	46.0	118.0	13 48.2	46.6	118.3	9
10	16 19.6	-42.3	117.3	15 52.0	-43.0	117.5	15 24.1	-43.6	117.8	14 56.0	-44.2	118.0	14 27.7	-44.9	118.3	13 59.2	-45.5	118.5	13 30.5	-46.1	118.7	13 01.6	-46.7	118.9	10
11	15 37.3	42.4	118.0	15 09.0	43.1	118.3	14 40.5	43.7	118.5	14 11.8	44.4	118.7	13 42.8	45.0	118.9	13 13.7	45.6	119.2	12 44.4	46.3	119.4	12 14.9	46.9	119.6	11
12	14 54.9	42.5	118.8	14 25.9	43.2	119.0	13 56.8	43.9	119.2	13 27.4	44.5	119.4	12 57.8	45.1	119.6	12 28.1	45.7	119.8	11 58.1	46.3	120.0	11 28.0	46.8	120.2	12
13	14 12.4	42.7	119.5	13 42.7	43.3	119.7	13 12.9	44.0	119.9	12 42.9	44.6	120.1	12 12.7	45.2	120.3	11 42.4	45.8	120.5	11 11.8	46.4	120.7	10 41.2	47.0	120.8	13
14	13 29.7	42.8	120.2	12 59.4	43.4	120.4	12 28.9	44.0	120.6	11 58.3	44.7	120.8	11 27.5	45.3	121.0	10 56.6	45.9	121.1	10 25.4	46.4	121.3	9 54.2	47.0	121.5	14
15	12 46.9	-42.9	120.9	12 16.0	-43.6	121.1	11 44.9	-44.2	121.3	11 13.6	44.8	121.5	10 42.2	-45.4	121.6	10 10.7	-46.0	121.8	9 39.0	-46.6	121.9	9 07.2	-47.1	122.1	15
16	12 04.0	43.1	121.6	11 32.4	43.6	121.8	11 00.7	44.3	122.0	10 28.8	44.8	122.2	9 56.8	45.4	122.3	9 24.7	46.0	122.5	8 52.4	46.6	122.6	8 20.1	47.2	122.7	16
17	11 20.9	43.1	122.4	10 48.8	43.8	122.5	10 16.4	44.3	122.7	9 44.0	45.0	122.8	9 11.4	45.6	123.0	8 38.7	46.1	123.1	8 05.8	46.6	123.2	7 32.9	47.2	123.3	17
18	10 37.8	43.2	123.1	10 05.0	43.8	123.2	9 32.1	44.5	123.4	8 59.0	45.0	123.5	8 25.8	45.6	123.6	7 52.6	46.2	123.7	7 19.2	46.7	123.9	6 45.7	47.3	124.0	18
19	9 54.6	43.3	123.8	9 21.2	43.9	123.9	8 47.6	44.5	124.0	8 14.0	45.1	124.2	7 40.2	45.6	124.3	7 06.4	46.2	124.4	6 32.5	45.8	124.5	5 58.4	47.2	124.6	19
20	9 11.3	-43.4	124.5	8 37.3	-44.0	124.6	8 03.1	-44.5	124.7	7 28.9	-45.1	124.8	6 54.6	-45.7	124.9	6 20.2	-46.3	125.0	5 45.7	-46.8	125.1	5 11.2	-47.4	125.2	20
21	8 27.9	43.5	125.2	7 53.3	44.1	125.3	7 18.6	44.7	125.4	6 43.8	45.2	125.5	6 08.9	45.8	125.6	5 33.9	46.3	125.7	4 58.9	46.8	125.8	4 23.8	47.3	125.8	21
22	7 44.4	43.5	125.9	7 09.2	44.1	126.0	6 33.9	44.7	126.1	5 58.6	45.3	126.2	5 23.1	45.8	126.2	4 47.6	46.3	126.3	4 12.1	46.9	126.4	3 36.5	47.4	126.4	22
23	7 00.9	43.6	126.6	6 25.1	44.2	126.7	5 49.2	44.7	126.7	5 13.3	45.3	126.8	4 37.3	45.8	126.9	4 01.3	46.4	127.0	3 25.2	46.9	127.0	2 49.1	47.4	127.0	23
24	6 17.3	43.7	127.3	5 40.9	44.2	127.3	5 04.5	44.8	127.4	4 28.0	45.3	127.5	3 51.5	45.9	127.5	3 14.9	46.4	127.6	2 38.3	46.9	127.6	2 01.7	47.5	127.7	24
25	5 33.6	-43.7	127.9	4 56.7	-44.2	128.0	4 19.7	-44.8	128.1	3 42.7	-45.3	128.1	3 05.6	-45.9	128.2	2 28.5	-46.4	128.2	1 51.4	-46.9	128.3	1 14.2	47.4	128.3	25
26	4 49.9	43.7	128.6	4 12.5	44.3	128.7	3 34.9	44.8	128.7	2 57.4	45.4	128.8	2 19.7	45.9	128.8	1 42.1	46.4	128.9	1 04.5	47.0	128.9	0 26.8	47.4	128.9	26
27	4 06.2	43.8	129.3	3 28.2	44.4	129.4	2 50.1	44.9	129.4	2 12.0	45.4	129.4	1 33.8	45.9	129.5	0 55.7	46.4	129.5	0 17.5	-46.9	129.5	0 20.6	-47.5	50.5	27
28	3 22.4	43.8	130.0	2 43.8	44.3	130.0	2 05.2	44.9	130.1	1 26.6	45.4	130.1	0 47.9	45.9	130.1	0 09.3	-46.5	130.1	0 29.4	+46.9	49.9	1 08.1	47.4	49.9	28
29	2 38.6	43.8	130.7	1 59.5	44.4	130.7	1 20.3	44.8	130.7	0 41.2	-45.4	130.8	0 02.0	-45.9	130.8	0 37.2	+46.4	49.2	1 16.3	47.0	49.3	1 55.5	47.4	49.3	29
30	1 54.8	-43.8	131.4	1 15.1	-44.3	131.4	0 35.5	-44.9	131.4	0 04.2	+45.4	48.6	0 43.9	+45.9	48.6	1 23.6	+46.4	48.6	2 03.3	+46.9	48.6	2 42.9	+47.4	48.7	30
31	1 11.0	43.9	132.1	0 30.8	-44.4	132.1	0 09.4	+44.9	47.9	0 49.6	45.4	47.9	1 29.8	45.9	48.0	2 10.0	46.4	48.0	2 50.2	46.9	48.0	3 30.3	47.4	48.0	31
32	0 27.1	-43.9	132.7	0 13.6	+44.4	47.3	0 54.3	44.9	47.3	1 35.0	45.4	47.3	2 15.7	45.9	47.3	2 56.4	46.4	47.3	3 37.1	46.8	47.4	4 17.7	47.3	47.4	32
33	0 16.8	+43.8	46.6	0 58.0	44.4	46.6	1 39.2	44.9	46.6	2 20.4	45.4	46.6	3 01.6	45.9	46.7	3 42.8	46.3	46.7	4 23.9	46.8	46.8	5 05.0	47.3	46.8	33
34	1 00.6	43.8	45.9	1 42.4	44.3	45.9	2 24.1	44.8	45.9	3 05.8	45.4	46.0	3 47.5	45.8	46.0	4 29.1	46.4	46.1	5 10.7	46.8	46.1	5 52.3	47.3	46.2	34
35	1 44.4	+43.9	45.2	2 26.7	+44.3	45.2	3 08.9	+44.9	45.3	3 51.2	+45.3	45.3	4 33.3	+45.8	45.4	5 15.5	+46.2	45.4	5 57.5	+46.8	45.5	6 39.6	+47.2	45.6	35
36	2 28.3	43.8	44.5	3 11.0	44.3	44.6	3 53.8	44.8	44.6	4 36.5	45.3	44.7	5 19.1	45.8	44.7	6 01.7	46.3	44.8	6 44.3	46.7	44.9	7 26.8	47.1	45.0	36
37	3 12.1	43.8	43.8	3 55.3	44.3	43.9	4 38.6	44.7	43.9	5 21.8	45.2	44.0	6 04.9	45.7	44.1	6 48.0	46.1	44.1	7 31.0	46.6	44.2	8 13.9	47.1	44.3	37
38	3 55.9	43.7	43.2	4 39.6	44.3	43.2	5 23.3	44.7	43.3	6 07.0	45.2	43.3	6 50.6	45.7	43.4	7 34.1	46.2	43.5	8 17.6	46.6	43.6	9 01.0	47.1	43.7	38
39	4 39.6	43.7	42.5	5 23.9	44.1	42.5	6 08.0	44.7	42.6	6 52.2	45.1	42.7	7 36.3	45.6	42.8	8 20.3	46.0	42.9	9 04.2	46.5	43.0	9 48.1	47.0	43.1	39
40	5 23.3	+43.7	41.8	6 08.0	+44.2	41.9	6 52.7	+44.6	41.9	7 37.3	+45.1	42.0	8 21.9	+45.5	42.1	9 06.3	+46.0	42.2	9 50.7	+46.5	42.3	10 35.1	+46.9	42.4	40
41	6 07.0	43.6	41.1	6 52.2	44.1	41.2	7 37.3	44.6	41.3	8 22.4	45.0	41.3	9 07.4	45.5	41.5	9 52.3	46.0	41.6	10 37.2	46.4	41.7	11 22.0	46.8	41.8	41
42	6 50.6	43.5	40.4	7 36.3	44.0	40.5	8 21.9	44.4	40.6	9 07.4	44.9	40.7	9 52.9	45.4	40.8	10 38.3	45.8	40.9	11 23.6	46.2	41.0	12 08.8	46.7	41.2	42
43	7 34.1	43.5	39.7	8 20.3	43.9	39.8	9 06.3	44.4	39.9	9 52.3	44.9	40.0	10 38.3	45.3	40.1	11 24.1	45.7	40.2	12 09.8	46.2	40.4	12 55.5	46.6	40.5	43
44	8 17.6	43.4	39.0	9 04.2	43.9	39.1	9 50.7	44.4	39.2	10 37.2	44.8	39.3	11 23.6	45.2	39.5	12 09.8	45.7	39.6	12 56.0	46.1	39.7	13 42.1	46.5	39.9	44
45	9 01.0	+43.4	38.3	9 48.1	+43.8	38.4	10 35.1	+44.2	38.5	11 22.0	+44.6	38.7	12 08.8	+45.1	38.8	12 55.5	+45.5	38.9	13 42.1	+46.0	39.1	14 28.7	+46.4	39.2	45
46	9 44.4	43.2	37.6	10 31.9	43.7	37.7	11 19.3	44.1	37.8	12 06.6	44.6	38.0	12 53.9	45.0	38.1	13 41.0	45.5	38.3	14 28.1	45.9	38.4	15 15.1	46.3	38.6	46
47	10 27.6	43.2	36.9	11 15.6	43.6	37.0	12 03.4	44.0	37.2	12 51.2	44.5	37.3	13 38.9	44.9	37.4	14 26.5	45.3	37.6	15 14.0	45.7	37.7	16 01.4	46.2	37.9	47
48	11 10.8	43.0	36.2	11 59.2	43.4	36.3	12 47.4	44.0	36.5	13 35.7	44.3	36.6	14 23.8	44.8	36.7	15 11.8	45.2	36.9	15 59.7	45.7	37.1	16 47.6	46.0	37.3	48
49	11 53.8	43.0	35.5	12 42.6	43.4	35.6	13 31.4	43.8	35.8	14 20.0	44.2	35.9	15 08.6	44.6	36.1	15 57.0	45.1	36.2	16 45.4	45.5	36.4	17 33.6	45.9	36.6	49
50	12 36.8	+42.8	34.8	13 26.0	+43.3	34.9	14 15.2	+43.7	35.1	15 04.2	+44.1	35.2	15 53.2	+44.5	35.4	16 42.1	+44.9	35.5	17 30.9	+45.3	35.7	18 19.5	+45.8	35.9	50
51	13 19.6	42.7	34.1	14 09.3	43.1	34.2	14 58.9	43.5	34.3	15 48.3	44.0	34.5	16 37.7	44.4	34.7	17 27.0	44.8	34.8	18 16.2	45.2	35.0	19 05.3	45.6	35.2	51
52	14 02.3	42.6	33.3	14 52.4	43.0	33.5	15 42.4	43.4	33.6	16 32.3	43.8	33.8	17 22.1	44.3	34.0	18 11.8	44.7	34.1	19 01.4	45.1	34.3	19 50.9	45.5	34.5	52
53	14 44.9	42.4	32.6	15 35.4	42.8	32.8	16 25.8	43.3	32.9	17 16.1	43.7	33.1	18 06.4	44.0	33.3	18 56.5	44.5	33.4	19 46.5	44.9	33.6	20 36.4	45.3	33.8	53
54	15 27.3	42.3	31.9	16 18.2	42.7	32.0	17 09.1	43.1	32.2	17 59.8	43.5	32.4	18 50.4	43.9	32.5	19 41.0	44.3	32.7	20 31.4	44.7	32.9	21 21.7	45.1	33.1	54
55	16 09.6	+42.2	31.1	17 00.9	+42.6	31.3	17 52.2	+42.9	31.5	18 43.3	+43.3	31.6	19 34.3	+43.8	31.8	20 25.3	+44.1	32.0	21 16.1	+44.5	32.2	22 06.8	+44.9	32.4	55
56	16 51.8	42.0	30.4	17 43.5	42.4	30.6	18 35.1	42.8	30.7	19 26.6	43.2	30.9	20 18.1	43.5	31.1	21 09.4	43.9	31.3	22 00.6	44.4	31.5	22 51.7	44.8	31.7	56
57	17 33.8	41.8	29.7	18 25.9	42.2	29.8	19 17.9	42.6	30.0	20 09.8	43.0	30.2	21 01.6	43.4	30.4	21 53.3	43.8	30.6	22 45.0	44.1	30.8	23 36.5	44.5	31.0	57
58	18 15.6	41.6	28.9	19 08.1	42.0	29.1	20 00.5	42.3	29.2	20 52.8	42.7	29.4	21 45.0	43.1	29.6	22 37.1	43.5	29.8	23 29.1	43.9	30.0	24 21.0	44.3	30.2	58
59	18 57.2	41.4	28.1	19 50.1	41.8	28.3	20 42.8	42.2	28.5	21 35.5	42.6	28.7	22 28.1	43.0	28.9	23 20.6	43.4	29.1	24 13.0	43.7	29.3	25 05.3	44.1	29.5	59
60	19 38.6	+41.3	27.4	20 31.9	+41.6	27.5	21 25.0	+42.0	27.7	22 18.1	+42.4	27.9	23 11.1	+42.7	28.1	24 04.0	+43.1	28.3	24 56.7	+43.5	28.5	25 49.4	+43.8	28.8	60
61	20 19.9	41.0	26.6	21 13.5	41.4	26.8	22 07.0	41.8	26.9	23 00.5	42.1	27.1	23 53.8	42.5	27.3	24 47.1	42.8	27.5	25 40.2	43.3	27.8	26 33.2	43.7	28.0	61
62	21 00.9	40.9	25.8	21 54.9	41.2	26.0	22 48.8	41.5	26.2	23 42.6	41.9	26.4	24 36.3	42.3	26.6	25 29.9	42.7	26.8	26 23.5	42.9	27.0	27 16.9	43.3	27.2	62
63	21 41.8	40.6	25.0	22 36.1	40.9	25.2	23 30.3	41.3	25.4	24 24.5	41.7	25.6	25 18.6	42.0	25.8	26 12.6	42.3	26.0	27 06.4	42.8	26.2	28 00.2	43.1	26.4	63
64	22 22.4	40.3	24.2	23 17.0	40.8	24.4	24 11.6	41.1	24.6	25 06.2	41.4	24.8	26 00.6	41.7	25.0	26 54.9	42.1	25.2	27 49.2	42.4	25.4	28 43.3	42.8	25.7	64
65	23 02.7	+40.2	23.4	23 57.8	+40.4	23.6	24 52.7	+40.8	23.8	25 47.6	+41.1	24.0	26 42.3	+41.5	24.2	27 37.0	+41.8	24.4	28 31.6	+42.2	24.6	29 26.1	+42.5	24.9	65
66	23 42.9	39.8	22.6	24 38.2	40.2	22.8	25 33.5	40.5	23.0	26 28.7	40.8	23.2	27 23.8	41.2	23.4	28 18.8	41.6	23.6	29 13.8	41.8	23.8	30 08.6	42.2	24.0	66
67	24 22.7	39.6	21.8	25 18.4	39.9	22.0	26 14.0	40.3	22.2	27 09.5	40.6	22.4	28 05.0	40.9	22.6	29 00.4	41.2	22.8	29 55.6	41.6	23.0	30 50.8	41.9	23.2	67
68	25 02.3	39.4	21.0	25 58.3	39.7	21.2	26 54.3	39.9	21.3	27 50.1	40.3	21.5	28 45.9	40.6	21.7	29 41.6	40.9	21.9	30 37.2	41.2	22.1	31 32.7	41.6	22.4	68
69	25 41.7	39.0	20.1	26 38.0	39.3	20.3	27 34.2	39.6	20.5	28 30.4	39.9	20.7	29 26.5	40.2	20.9	30 22.5	40.6	21.1	31 18.4	40.9	21.3	32 14.3	41.2	21.5	69
70	26 20.7	+38.7	19.3	27 17.3	+39.0	19.5	28 13.8	+39.4	19.6	29 10.3	+39.6	19.8	30 06.7	+39.9	20.0	31 03.1	+40.2	20.2	31 59.3	+40.6	20.4	32 55.5	+40.9	20.7	70
71	26 59.4	38.5	18.4	27 56.3	38.7	18.6	28 53.2	38.9	18.8	29 49.9	39.3	19.0	30 46.6	39.6	19.2	31 43.3	39.9	19.4	32 39.9	40.1	19.6	33 36.4	40.4	19.8	71
72	27 37.9	38.1	17.6	28 35.0	38.4	17.7	29 32.1	38.7	17.9	30 29.2	38.9	18.1	31 26.2	39.2	18.3	32 23.2	39.4	18.5	33 20.0	39.8	18.7	34 16.8	40.1	18.9	72
73	28 16.0	37.7	16.7	29 13.4	38.0	16.9	30 10.8	38.3	17.0	31 08.1	38.6	17.2	32 05.4	38.8	17.4	33 02.6	39.2	17.6	33 59.8	39.4	17.8	34 56.9	39.7	18.0	73
74	28 53.7	37.4	15.8	29 51.4	37.7	16.0	30 49.1	37.9	16.1	31 46.7	38.2	16.3	32 44.2	38.5	16.5	33 41.8	38.7	16.7	34 39.2	39.0	16.9	35 36.6	39.2	17.1	74
75	29 31.1	+37.1	14.9	30 29.1	+37.3	15.1	31 27.0	+37.5	15.2	32 24.9	+37.7	15.4	33 22.7	+38.0	15.6	34 20.5	+38.2	15.8	35 18.2	+38.5	15.9	36 15.8	+38.8	16.1	75
76	30 08.2	36.6	14.0	31 06.4	36.9	14.2	32 04.5	37.1	14.3	33 02.6	37.4	14.5	34 00.7	37.6	14.6	34 58.7	37.9	14.8	35 56.7	38.1	15.0	36 54.6	38.4	15.2	76
77	30 44.8	36.3	13.1	31 43.3	36.4	13.2	32 41.6	36.7	13.4	33 40.0	36.9	13.5	34 38.3	37.2	13.7	35 36.6	37.4	13.9	36 34.8	37.6	14.0	37 33.0	37.9	14.2	77
78	31 21.1	35.9	12.2	32 19.7	36.1	12.3	33 18.3	36.3	12.4	34 16.9	36.5	12.6	35 15.5	36.7	12.7	36 14.0	36.9	12.9	37 12.4	37.2	13.1	38 10.9	37.3	13.2	78
79	31 57.0	35.4	11.2	32 55.8	35.6	11.4	33 54.6	35.8	11.5	34 53.4	36.0	11.6	35 52.2	36.2	11.8	36 50.9	36.4	11.9	37 49.6	36.6	12.1	38 48.2	36.9	12.2	79
80	32 32.4	+35.0	10.3	33 31.4	+35.2	10.4	34 30.4	+35.4	10.5	35 29.4	+35.6	10.6	36 28.4	+35.7	10.8	37 27.3	+35.9	10.9	38 26.2	+36.1	11.1	39 25.1	+36.3	11.2	80
81	33 07.4	34.6	9.3	34 06.6	34.7	9.4	35 05.8	34.9	9.5	36 05.0	35.0	9.7	37 04.1	35.2	9.8	38 03.2	35.4	9.9	39 02.3	35.6	10.0	40 01.4	35.7	10.2	81
82	33 42.0	34.0	8.3	34 41.3	34.2	8.4	35 40.7	34.3	8.5	36 40.0	34.5	8.6	37 39.3	34.7	8.8	38 38.6	34.8	8.9	39 37.9	35.0	9.0	40 37.1	35.2	9.1	82
83	34 16.0	33.6	7.3	35 15.5	33.8	7.4	36 15.0	33.9	7.5	37 14.5	34.0	7.6	38 14.0	34.1	7.7	39 13.4	34.3	7.8	40 12.9	34.4	7.9	41 12.3	34.5	8.1	83
84	34 49.6	33.1	6.3	35 49.3	33.2	6.4	36 48.9	33.3	6.5	37 48.5	33.4	6.6	38 48.1	33.5	6.7	39 47.7	33.6	6.8	40 47.3	33.7	6.9	41 46.8	33.9	7.0	84
85	35 22.7	+32.6	5.3	36 22.5	+32.6	5.4	37 22.2	+32.8	5.4	38 21.9	+32.9	5.5	39 21.6	+33.0	5.6	40 21.3	+33.1	5.7	41 21.0	+33.2	5.8	42 20.7	+33.3	5.9	85
86	35 55.3	32.0	4.3	36 55.1	32.1	4.3	37 55.0	32.1	4.4	38 54.8	32.2	4.5	39 54.6	32.3	4.5	40 54.4	32.4	4.6	41 54.2	32.5	4.7	42 54.0	32.6	4.7	86
87	36 27.3	31.5	3.2	37 27.2	31.6	3.3	38 27.1	31.6	3.3	39 27.0	31.7	3.4	40 26.9	31.7	3.4	41 26.8	31.8	3.5	42 26.7	31.8	3.5	43 26.6	31.9	3.6	87
88	36 58.8	30.9	2.2	37 58.8	30.9	2.2	38 58.7	31.0	2.2	39 58.7	31.0	2.3	40 58.6	31.1	2.3	41 58.6	31.0	2.3	42 58.5	31.1	2.4	43 58.5	31.1	2.4	88
89	37 29.7	30.3	1.1	38 29.7	30.3	1.1	39 29.7	30.3	1.1	40 29.7	30.3	1.1	41 29.7	30.3	1.2	42 29.6	30.4	1.2	43 29.6	30.4	1.2	44 29.6	30.4	1.2	89
90	38 00.0	+29.7	0.0	39 00.0	+29.7	0.0	40 00.0	+29.7	0.0	41 00.0	+29.7	0.0	42 00.0	+29.6	0.0	43 00.0	+29.6	0.0	44 00.0	+29.6	0.0	45 00.0	+29.6	0.0	90
	38°			39°			40°			41°			42°			43°			44°			45°			

S. Lat. { L.H.A. greater than 180°......Zn=180°−Z; L.H.A. less than 180°..........Zn=180°+Z }

LATITUDE SAME NAME AS DECLINATION L.H.A. 120°, 240°

INTERPOLATION TABLE

Dec. Inc.	Tens 10	20	30	40	50	Decimals	Units 0	1	2	3	4	5	6	7	8	9	Double Second Diff. and Corr.
28.0	4.6	9.3	14.0	18.6	23.3	.0	0.0	0.5	0.9	1.4	1.9	2.4	2.8	3.3	3.8	4.3	0.8 (0.1) 2.4 (0.2) 4.0 (0.3) 5.6 (0.4) 7.2 (0.5) 8.8 (0.6) 10.4 (0.7) 12.0 (0.8) 13.6 (0.9) 15.2 (1.0) 16.8 (1.1) 18.4 (1.2) 20.0 (1.3) 21.6 (1.4) 23.2 (1.5) 24.8 (1.6) 26.4 (1.7) 28.0 (1.8) 29.6 (1.9) 31.2 (2.0) 32.8 (2.1) 34.4
28.1	4.7	9.3	14.0	18.7	23.4	.1	0.0	0.5	1.0	1.5	1.9	2.4	2.9	3.4	3.8	4.3	
28.2	4.7	9.4	14.1	18.8	23.5	.2	0.1	0.6	1.0	1.5	2.0	2.5	2.9	3.4	3.9	4.4	
28.3	4.7	9.4	14.1	18.9	23.6	.3	0.1	0.6	1.1	1.6	2.0	2.5	3.0	3.5	3.9	4.4	
28.4	4.7	9.5	14.2	18.9	23.7	.4	0.2	0.7	1.1	1.6	2.1	2.6	3.0	3.5	4.0	4.5	
28.5	4.8	9.5	14.3	19.0	23.8	.5	0.2	0.7	1.2	1.7	2.1	2.6	3.1	3.6	4.0	4.5	
28.6	4.8	9.5	14.3	19.1	23.8	.6	0.3	0.8	1.2	1.7	2.2	2.7	3.1	3.6	4.1	4.6	
28.7	4.8	9.6	14.4	19.2	23.9	.7	0.3	0.8	1.3	1.8	2.2	2.7	3.2	3.7	4.1	4.6	
28.8	4.8	9.6	14.4	19.2	24.0	.8	0.4	0.9	1.3	1.8	2.3	2.8	3.2	3.7	4.2	4.7	
28.9	4.9	9.7	14.5	19.3	24.1	.9	0.4	0.9	1.4	1.9	2.3	2.8	3.3	3.8	4.2	4.7	
29.0	4.8	9.6	14.5	19.3	24.1	.0	0.0	0.5	1.0	1.5	2.0	2.5	2.9	3.4	3.9	4.4	
29.1	4.8	9.7	14.5	19.4	24.2	.1	0.0	0.5	1.0	1.5	2.0	2.5	3.0	3.5	4.0	4.5	
29.2	4.8	9.7	14.6	19.4	24.3	.2	0.1	0.6	1.1	1.6	2.1	2.6	3.0	3.5	4.0	4.5	
29.3	4.9	9.8	14.6	19.5	24.4	.3	0.1	0.6	1.1	1.6	2.1	2.6	3.1	3.6	4.1	4.6	
29.4	4.9	9.8	14.7	19.6	24.5	.4	0.2	0.7	1.2	1.7	2.2	2.7	3.1	3.6	4.1	4.6	
29.5	4.9	9.8	14.8	19.7	24.6	.5	0.2	0.7	1.2	1.7	2.2	2.7	3.2	3.7	4.2	4.7	
29.6	4.9	9.9	14.8	19.7	24.7	.6	0.3	0.8	1.3	1.8	2.3	2.8	3.2	3.7	4.2	4.7	
29.7	5.0	9.9	14.9	19.8	24.8	.7	0.3	0.8	1.3	1.8	2.3	2.8	3.3	3.8	4.3	4.8	
29.8	5.0	10.0	14.9	19.9	24.9	.8	0.4	0.9	1.4	1.9	2.4	2.9	3.3	3.8	4.3	4.8	
29.9	5.0	10.0	15.0	20.0	25.0	.9	0.4	0.9	1.4	1.9	2.4	2.9	3.4	3.9	4.4	4.9	
30.0	5.0	10.0	15.0	20.0	25.0	.0	0.0	0.5	1.0	1.5	2.0	2.5	3.0	3.6	4.1	4.6	0.8 (0.1) 2.4 (0.2) 4.0 (0.3) 5.6 (0.4) 7.2 (0.5) 8.8 (0.6) 10.4 (0.7) 12.0 (0.8) 13.6 (0.9) 15.2 (1.0) 16.8 (1.1) 18.4 (1.2) 20.0 (1.3) 21.6 (1.4) 23.2 (1.5) 24.8 (1.6) 26.4 (1.7) 28.0 (1.8) 29.6 (1.9) 31.2 (2.0) 32.8 (2.1) 34.4
30.1	5.0	10.0	15.0	20.0	25.1	.1	0.1	0.6	1.1	1.6	2.1	2.6	3.1	3.6	4.1	4.6	
30.2	5.0	10.0	15.1	20.1	25.1	.2	0.1	0.6	1.1	1.6	2.1	2.6	3.2	3.7	4.2	4.7	
30.3	5.0	10.1	15.1	20.2	25.2	.3	0.2	0.7	1.2	1.7	2.2	2.7	3.2	3.7	4.2	4.7	
30.4	5.1	10.1	15.2	20.3	25.3	.4	0.2	0.7	1.2	1.7	2.2	2.7	3.3	3.8	4.3	4.8	
30.5	5.1	10.2	15.3	20.3	25.4	.5	0.3	0.8	1.3	1.8	2.3	2.8	3.3	3.8	4.3	4.8	
30.6	5.1	10.2	15.3	20.4	25.5	.6	0.3	0.8	1.3	1.8	2.3	2.8	3.4	3.9	4.4	4.9	
30.7	5.1	10.3	15.4	20.5	25.6	.7	0.4	0.9	1.4	1.9	2.4	2.9	3.4	3.9	4.4	4.9	
30.8	5.2	10.3	15.4	20.6	25.7	.8	0.4	0.9	1.4	1.9	2.4	2.9	3.5	4.0	4.5	5.0	
30.9	5.2	10.3	15.5	20.6	25.8	.9	0.5	1.0	1.5	2.0	2.5	3.0	3.5	4.0	4.5	5.0	
31.0	5.1	10.3	15.5	20.6	25.8	.0	0.0	0.5	1.0	1.6	2.1	2.6	3.1	3.7	4.2	4.7	
31.1	5.2	10.3	15.5	20.7	25.9	.1	0.1	0.6	1.1	1.6	2.2	2.7	3.2	3.7	4.3	4.8	
31.2	5.2	10.4	15.6	20.8	26.0	.2	0.1	0.6	1.2	1.7	2.2	2.7	3.3	3.8	4.3	4.8	
31.3	5.2	10.4	15.6	20.9	26.1	.3	0.2	0.7	1.2	1.7	2.3	2.8	3.3	3.8	4.4	4.9	
31.4	5.2	10.5	15.7	20.9	26.2	.4	0.2	0.7	1.3	1.8	2.3	2.8	3.4	3.9	4.4	4.9	
31.5	5.3	10.5	15.8	21.0	26.3	.5	0.3	0.8	1.3	1.8	2.4	2.9	3.4	3.9	4.5	5.0	
31.6	5.3	10.5	15.8	21.1	26.3	.6	0.3	0.8	1.4	1.9	2.4	2.9	3.5	4.0	4.5	5.0	
31.7	5.3	10.6	15.9	21.2	26.4	.7	0.4	0.9	1.4	1.9	2.5	3.0	3.5	4.0	4.6	5.1	
31.8	5.3	10.6	15.9	21.2	26.5	.8	0.4	0.9	1.5	2.0	2.5	3.0	3.6	4.1	4.6	5.1	
31.9	5.4	10.7	16.0	21.3	26.6	.9	0.5	1.0	1.5	2.0	2.6	3.1	3.6	4.1	4.7	5.2	
32.0	5.3	10.6	16.0	21.3	26.6	.0	0.0	0.5	1.1	1.6	2.2	2.7	3.2	3.8	4.3	4.9	0.8 (0.1) 2.4 (0.2) 4.0 (0.3) 5.7 (0.4) 7.3 (0.5) 8.9 (0.6) 10.5 (0.7) 12.1 (0.8) 13.7 (0.9) 15.4 (1.0) 17.0 (1.1) 18.6 (1.2) 20.2 (1.3) 21.8 (1.4) 23.4 (1.5) 25.1 (1.6) 26.7 (1.7) 28.3 (1.8) 29.9 (1.9) 31.5 (2.0) 33.1 (2.1) 34.7
32.1	5.3	10.7	16.0	21.4	26.7	.1	0.1	0.6	1.1	1.7	2.2	2.8	3.3	3.8	4.4	4.9	
32.2	5.3	10.7	16.1	21.4	26.8	.2	0.1	0.6	1.2	1.7	2.3	2.8	3.4	3.9	4.4	5.0	
32.3	5.4	10.8	16.1	21.5	26.9	.3	0.2	0.7	1.2	1.8	2.3	2.9	3.4	4.0	4.5	5.0	
32.4	5.4	10.8	16.2	21.6	27.0	.4	0.2	0.8	1.3	1.8	2.4	2.9	3.5	4.0	4.5	5.1	
32.5	5.4	10.8	16.3	21.7	27.1	.5	0.3	0.8	1.4	1.9	2.4	3.0	3.5	4.1	4.6	5.1	
32.6	5.4	10.9	16.3	21.7	27.2	.6	0.3	0.9	1.4	1.9	2.5	3.0	3.6	4.1	4.7	5.2	
32.7	5.5	10.9	16.4	21.8	27.3	.7	0.4	0.9	1.5	2.0	2.5	3.1	3.6	4.2	4.7	5.3	
32.8	5.5	11.0	16.4	21.9	27.4	.8	0.4	1.0	1.5	2.1	2.6	3.1	3.7	4.2	4.8	5.3	
32.9	5.5	11.0	16.5	22.0	27.5	.9	0.5	1.0	1.6	2.1	2.7	3.2	3.7	4.3	4.8	5.4	
33.0	5.5	11.0	16.5	22.0	27.5	.0	0.0	0.6	1.1	1.7	2.2	2.8	3.3	3.9	4.5	5.0	
33.1	5.5	11.0	16.5	22.0	27.6	.1	0.1	0.6	1.2	1.7	2.3	2.8	3.4	4.0	4.5	5.1	
33.2	5.5	11.0	16.6	22.1	27.6	.2	0.1	0.7	1.2	1.8	2.3	2.9	3.5	4.0	4.6	5.1	
33.3	5.5	11.1	16.6	22.2	27.7	.3	0.2	0.7	1.3	1.8	2.4	3.0	3.5	4.1	4.6	5.2	
33.4	5.6	11.1	16.7	22.3	27.8	.4	0.2	0.8	1.3	1.9	2.5	3.0	3.6	4.1	4.7	5.2	
33.5	5.6	11.2	16.8	22.3	27.9	.5	0.3	0.8	1.4	2.0	2.5	3.1	3.6	4.2	4.7	5.3	
33.6	5.6	11.2	16.8	22.4	28.0	.6	0.3	0.9	1.5	2.0	2.6	3.1	3.7	4.2	4.8	5.4	
33.7	5.6	11.3	16.9	22.5	28.1	.7	0.4	0.9	1.5	2.1	2.6	3.2	3.7	4.3	4.9	5.4	
33.8	5.7	11.3	16.9	22.6	28.2	.8	0.4	1.0	1.6	2.1	2.7	3.2	3.8	4.4	4.9	5.5	
33.9	5.7	11.3	17.0	22.6	28.3	.9	0.5	1.1	1.6	2.2	2.7	3.3	3.9	4.4	5.0	5.5	
34.0	5.6	11.3	17.0	22.6	28.3	.0	0.0	0.6	1.1	1.7	2.3	2.9	3.4	4.0	4.6	5.2	0.8 (0.1) 2.5 (0.2) 4.1 (0.3) 5.8 (0.4) 7.4 (0.5) 9.1 (0.6) 10.7 (0.7) 12.3 (0.8) 14.0 (0.9) 15.6 (1.0) 17.3 (1.1) 18.9 (1.2) 20.6 (1.3) 22.2 (1.4) 23.9 (1.5) 25.5 (1.6) 27.2 (1.7) 28.8 (1.8) 30.4 (1.9) 32.1 (2.0) 33.7 (2.1) 35.4
34.1	5.7	11.3	17.0	22.7	28.4	.1	0.1	0.6	1.2	1.8	2.4	2.9	3.5	4.1	4.7	5.2	
34.2	5.7	11.4	17.1	22.8	28.5	.2	0.1	0.7	1.3	1.8	2.4	3.0	3.6	4.1	4.7	5.3	
34.3	5.7	11.4	17.1	22.9	28.6	.3	0.2	0.7	1.3	1.9	2.5	3.0	3.6	4.2	4.8	5.3	
34.4	5.7	11.5	17.2	22.9	28.7	.4	0.2	0.8	1.4	2.0	2.5	3.1	3.7	4.3	4.8	5.4	
34.5	5.8	11.5	17.3	23.0	28.8	.5	0.3	0.9	1.4	2.0	2.6	3.2	3.7	4.3	4.9	5.5	
34.6	5.8	11.5	17.3	23.1	28.8	.6	0.3	0.9	1.5	2.1	2.6	3.2	3.8	4.4	4.9	5.5	
34.7	5.8	11.6	17.4	23.2	28.9	.7	0.4	1.0	1.6	2.1	2.7	3.3	3.9	4.4	5.0	5.6	
34.8	5.8	11.6	17.4	23.2	29.0	.8	0.5	1.0	1.6	2.2	2.8	3.3	3.9	4.5	5.1	5.6	
34.9	5.9	11.7	17.5	23.3	29.1	.9	0.5	1.1	1.7	2.2	2.8	3.4	4.0	4.5	5.1	5.7	
35.0	5.8	11.6	17.5	23.3	29.1	.0	0.0	0.6	1.2	1.8	2.4	3.0	3.5	4.1	4.7	5.3	
35.1	5.8	11.7	17.5	23.4	29.2	.1	0.1	0.7	1.2	1.8	2.4	3.0	3.6	4.2	4.8	5.4	
35.2	5.8	11.7	17.6	23.4	29.3	.2	0.1	0.7	1.3	1.9	2.5	3.1	3.7	4.3	4.9	5.4	
35.3	5.9	11.8	17.6	23.5	29.4	.3	0.2	0.8	1.4	2.0	2.5	3.1	3.7	4.3	4.9	5.5	
35.4	5.9	11.8	17.7	23.6	29.5	.4	0.2	0.8	1.4	2.0	2.6	3.2	3.8	4.4	5.0	5.6	
35.5	5.9	11.8	17.8	23.7	29.6	.5	0.3	0.9	1.5	2.1	2.7	3.3	3.8	4.4	5.0	5.6	
35.6	5.9	11.9	17.8	23.7	29.7	.6	0.4	0.9	1.5	2.1	2.7	3.3	3.9	4.5	5.1	5.7	
35.7	6.0	11.9	17.9	23.8	29.8	.7	0.4	1.0	1.6	2.2	2.8	3.4	4.0	4.6	5.1	5.7	
35.8	6.0	12.0	17.9	23.9	29.9	.8	0.5	1.1	1.7	2.2	2.8	3.4	4.0	4.6	5.2	5.8	
35.9	6.0	12.0	18.0	24.0	30.0	.9	0.5	1.1	1.7	2.3	2.9	3.5	4.1	4.7	5.3	5.9	
	10	20	30	40	50		0	1	2	3	4	5	6	7	8	9	

Dec. Inc.	Tens 10	20	30	40	50	Decimals	Units 0	1	2	3	4	5	6	7	8	9	Double Second Diff. and Corr.
36.0	6.0	12.0	18.0	24.0	30.0	.0	0.0	0.6	1.2	1.8	2.4	3.0	3.6	4.3	4.9	5.5	0.8 (0.1) 2.5 (0.2) 4.2 (0.3) 5.9 (0.4) 7.6 (0.5) 9.3 (0.6) 11.0 (0.7) 12.7 (0.8) 14.4 (0.9) 16.1 (1.0) 17.8 (1.1) 19.5 (1.2) 21.2 (1.3) 22.8 (1.4) 24.5 (1.5) 26.2 (1.6) 27.9 (1.7) 29.6 (1.8) 31.3 (1.9) 33.0 (2.0) 34.7
36.1	6.0	12.0	18.0	24.0	30.1	.1	0.1	0.7	1.3	1.9	2.5	3.1	3.7	4.3	4.9	5.5	
36.2	6.0	12.0	18.1	24.1	30.1	.2	0.1	0.7	1.3	1.9	2.6	3.2	3.8	4.4	5.0	5.6	
36.3	6.0	12.1	18.1	24.2	30.2	.3	0.2	0.8	1.4	2.0	2.6	3.2	3.8	4.4	5.0	5.7	
36.4	6.1	12.1	18.2	24.3	30.3	.4	0.2	0.9	1.5	2.1	2.7	3.3	3.9	4.5	5.1	5.7	
36.5	6.1	12.2	18.3	24.3	30.4	.5	0.3	0.9	1.5	2.1	2.7	3.3	4.0	4.6	5.2	5.8	
36.6	6.1	12.2	18.3	24.4	30.5	.6	0.4	1.0	1.6	2.2	2.8	3.4	4.0	4.6	5.2	5.8	
36.7	6.1	12.3	18.4	24.5	30.6	.7	0.4	1.0	1.6	2.3	2.9	3.5	4.1	4.7	5.3	5.9	
36.8	6.2	12.3	18.4	24.6	30.7	.8	0.5	1.1	1.7	2.3	2.9	3.5	4.1	4.7	5.4	6.0	
36.9	6.2	12.3	18.5	24.6	30.8	.9	0.5	1.2	1.8	2.4	3.0	3.6	4.2	4.8	5.4	6.0	
37.0	6.1	12.3	18.5	24.6	30.8	.0	0.0	0.6	1.2	1.9	2.5	3.1	3.7	4.4	5.0	5.6	
37.1	6.2	12.3	18.5	24.7	30.9	.1	0.1	0.7	1.3	1.9	2.6	3.2	3.8	4.4	5.1	5.7	
37.2	6.2	12.4	18.6	24.8	31.0	.2	0.1	0.7	1.4	2.0	2.6	3.2	3.9	4.5	5.1	5.7	
37.3	6.2	12.4	18.6	24.9	31.1	.3	0.2	0.8	1.4	2.1	2.7	3.3	3.9	4.6	5.2	5.8	
37.4	6.2	12.5	18.7	24.9	31.2	.4	0.2	0.9	1.5	2.1	2.7	3.4	4.0	4.6	5.2	5.9	
37.5	6.3	12.5	18.8	25.0	31.3	.5	0.3	0.9	1.6	2.2	2.8	3.4	4.1	4.7	5.3	5.9	
37.6	6.3	12.5	18.8	25.1	31.3	.6	0.4	1.0	1.6	2.2	2.9	3.5	4.1	4.7	5.4	6.0	
37.7	6.3	12.6	18.9	25.2	31.4	.7	0.4	1.1	1.7	2.3	2.9	3.6	4.2	4.8	5.4	6.1	
37.8	6.3	12.6	18.9	25.2	31.5	.8	0.5	1.1	1.7	2.4	3.0	3.6	4.2	4.9	5.5	6.1	
37.9	6.4	12.7	19.0	25.3	31.6	.9	0.6	1.2	1.8	2.4	3.1	3.7	4.3	4.9	5.6	6.2	
38.0	6.3	12.6	19.0	25.3	31.6	.0	0.0	0.6	1.3	1.9	2.6	3.2	3.8	4.5	5.1	5.8	0.9 (0.1) 2.6 (0.2) 4.4 (0.3) 6.2 (0.4) 7.9 (0.5) 9.7 (0.6) 11.4 (0.7) 13.2 (0.8) 14.9 (0.9) 16.7 (1.0) 18.5 (1.1) 20.2 (1.2) 22.0 (1.3) 23.7 (1.4) 25.5 (1.5) 27.3 (1.6) 29.0 (1.7) 30.8 (1.8) 32.5 (1.9) 34.3
38.1	6.3	12.7	19.0	25.4	31.7	.1	0.1	0.7	1.3	2.0	2.6	3.3	3.9	4.6	5.2	5.8	
38.2	6.3	12.7	19.1	25.4	31.8	.2	0.1	0.8	1.4	2.1	2.7	3.3	4.0	4.6	5.3	5.9	
38.3	6.4	12.8	19.1	25.5	31.9	.3	0.2	0.8	1.5	2.1	2.8	3.4	4.0	4.7	5.3	6.0	
38.4	6.4	12.8	19.2	25.6	32.0	.4	0.3	0.9	1.5	2.2	2.8	3.5	4.1	4.7	5.4	6.0	
38.5	6.4	12.8	19.3	25.7	32.1	.5	0.3	1.0	1.6	2.2	2.9	3.5	4.2	4.8	5.5	6.1	
38.6	6.4	12.9	19.3	25.7	32.2	.6	0.4	1.0	1.7	2.3	3.0	3.6	4.2	4.9	5.5	6.2	
38.7	6.5	12.9	19.4	25.8	32.3	.7	0.4	1.1	1.7	2.4	3.0	3.7	4.3	4.9	5.6	6.2	
38.8	6.5	13.0	19.4	25.9	32.4	.8	0.5	1.2	1.8	2.4	3.1	3.7	4.4	5.0	5.6	6.3	
38.9	6.5	13.0	19.5	26.0	32.5	.9	0.6	1.2	1.9	2.5	3.1	3.8	4.4	5.1	5.7	6.4	
39.0	6.5	13.0	19.5	26.0	32.5	.0	0.0	0.7	1.3	2.0	2.6	3.3	3.9	4.6	5.3	5.9	
39.1	6.5	13.0	19.5	26.0	32.6	.1	0.1	0.7	1.4	2.0	2.7	3.4	4.0	4.7	5.3	6.0	
39.2	6.5	13.0	19.6	26.1	32.6	.2	0.1	0.8	1.4	2.1	2.8	3.4	4.1	4.7	5.4	6.1	
39.3	6.5	13.1	19.6	26.2	32.7	.3	0.2	0.9	1.5	2.2	2.8	3.5	4.1	4.8	5.5	6.1	
39.4	6.6	13.1	19.7	26.3	32.8	.4	0.3	0.9	1.6	2.2	2.9	3.6	4.2	4.9	5.5	6.2	
39.5	6.6	13.2	19.8	26.3	32.9	.5	0.3	1.0	1.6	2.3	3.0	3.6	4.3	4.9	5.6	6.3	
39.6	6.6	13.2	19.8	26.4	33.0	.6	0.4	1.1	1.7	2.4	3.0	3.7	4.3	5.0	5.7	6.3	
39.7	6.6	13.3	19.9	26.5	33.1	.7	0.5	1.1	1.8	2.4	3.1	3.8	4.4	5.1	5.7	6.4	
39.8	6.7	13.3	19.9	26.6	33.2	.8	0.5	1.2	1.8	2.5	3.2	3.8	4.5	5.1	5.8	6.5	
39.9	6.7	13.3	20.0	26.6	33.3	.9	0.6	1.3	1.9	2.6	3.2	3.9	4.5	5.2	5.9	6.5	
40.0	6.6	13.3	20.0	26.6	33.3	.0	0.0	0.7	1.3	2.0	2.7	3.4	4.0	4.7	5.4	6.1	0.9 (0.1) 2.8 (0.2) 4.6 (0.3) 6.5 (0.4) 8.3 (0.5) 10.2 (0.6) 12.0 (0.7) 13.9 (0.8) 15.7 (0.9) 17.6 (1.0) 19.4 (1.1) 21.3 (1.2) 23.1 (1.3) 25.0 (1.4) 26.8 (1.5) 28.7 (1.6) 30.5 (1.7) 32.3 (1.8) 34.2
40.1	6.7	13.3	20.0	26.7	33.4	.1	0.1	0.7	1.4	2.1	2.8	3.4	4.1	4.8	5.5	6.1	
40.2	6.7	13.4	20.1	26.8	33.5	.2	0.1	0.8	1.5	2.2	2.8	3.5	4.2	4.9	5.5	6.2	
40.3	6.7	13.4	20.1	26.9	33.6	.3	0.2	0.9	1.6	2.2	2.9	3.6	4.3	4.9	5.6	6.3	
40.4	6.7	13.5	20.2	26.9	33.7	.4	0.3	0.9	1.6	2.3	3.0	3.6	4.3	5.0	5.7	6.3	
40.5	6.8	13.5	20.3	27.0	33.8	.5	0.3	1.0	1.7	2.4	3.0	3.7	4.4	5.1	5.7	6.4	
40.6	6.8	13.5	20.3	27.1	33.8	.6	0.4	1.1	1.8	2.4	3.1	3.8	4.5	5.1	5.8	6.5	
40.7	6.8	13.6	20.4	27.2	33.9	.7	0.5	1.1	1.8	2.5	3.2	3.8	4.5	5.2	5.9	6.5	
40.8	6.8	13.6	20.4	27.2	34.0	.8	0.5	1.2	1.9	2.6	3.2	3.9	4.6	5.3	5.9	6.6	
40.9	6.9	13.7	20.5	27.3	34.1	.9	0.6	1.3	2.0	2.6	3.3	4.0	4.7	5.3	6.0	6.7	
41.0	6.8	13.6	20.5	27.3	34.1	.0	0.0	0.7	1.4	2.1	2.8	3.5	4.1	4.8	5.5	6.2	
41.1	6.8	13.7	20.5	27.4	34.2	.1	0.1	0.8	1.5	2.1	2.8	3.5	4.2	4.9	5.6	6.3	
41.2	6.8	13.7	20.6	27.4	34.3	.2	0.1	0.8	1.5	2.2	2.9	3.6	4.3	5.0	5.7	6.4	
41.3	6.9	13.8	20.6	27.5	34.4	.3	0.2	0.9	1.6	2.3	3.0	3.7	4.4	5.0	5.7	6.4	
41.4	6.9	13.8	20.7	27.6	34.5	.4	0.3	1.0	1.7	2.4	3.0	3.7	4.4	5.1	5.8	6.5	
41.5	6.9	13.8	20.8	27.7	34.6	.5	0.3	1.0	1.7	2.4	3.1	3.8	4.5	5.2	5.9	6.6	
41.6	6.9	13.9	20.8	27.7	34.7	.6	0.4	1.1	1.8	2.5	3.2	3.9	4.6	5.3	5.9	6.6	
41.7	7.0	13.9	20.9	27.8	34.8	.7	0.5	1.2	1.9	2.6	3.3	3.9	4.6	5.3	6.0	6.7	
41.8	7.0	14.0	20.9	27.9	34.9	.8	0.6	1.2	1.9	2.6	3.3	4.0	4.7	5.4	6.1	6.8	
41.9	7.0	14.0	21.0	28.0	35.0	.9	0.6	1.3	2.0	2.7	3.4	4.1	4.8	5.5	6.2	6.8	
42.0	7.0	14.0	21.0	28.0	35.0	.0	0.0	0.7	1.4	2.1	2.8	3.5	4.2	5.0	5.7	6.4	1.0 (0.1) 3.0 (0.2) 4.9 (0.3) 6.9 (0.4) 8.9 (0.5) 10.8 (0.6) 12.8 (0.7) 14.8 (0.8) 16.7 (0.9) 18.7 (1.0) 20.7 (1.1) 22.7 (1.2) 24.6 (1.3) 26.6 (1.4) 28.6 (1.5) 30.5 (1.6) 32.5 (1.7) 34.5
42.1	7.0	14.0	21.0	28.0	35.1	.1	0.1	0.8	1.5	2.2	2.9	3.6	4.3	5.0	5.7	6.4	
42.2	7.0	14.0	21.1	28.1	35.1	.2	0.1	0.8	1.6	2.3	3.0	3.7	4.4	5.1	5.8	6.5	
42.3	7.0	14.1	21.1	28.2	35.2	.3	0.2	0.9	1.6	2.3	3.0	3.8	4.5	5.2	5.9	6.6	
42.4	7.1	14.1	21.2	28.3	35.3	.4	0.3	1.0	1.7	2.4	3.1	3.8	4.5	5.2	5.9	6.7	
42.5	7.1	14.2	21.3	28.3	35.4	.5	0.4	1.1	1.8	2.5	3.2	3.9	4.6	5.3	6.0	6.7	
42.6	7.1	14.2	21.3	28.4	35.5	.6	0.4	1.1	1.8	2.5	3.3	4.0	4.7	5.4	6.1	6.8	
42.7	7.1	14.3	21.4	28.5	35.6	.7	0.5	1.2	1.9	2.6	3.3	4.0	4.7	5.5	6.2	6.9	
42.8	7.2	14.3	21.4	28.6	35.7	.8	0.6	1.3	2.0	2.7	3.4	4.1	4.8	5.5	6.2	6.9	
42.9	7.2	14.3	21.5	28.6	35.8	.9	0.6	1.3	2.1	2.8	3.5	4.2	4.9	5.6	6.3	7.0	
43.0	7.1	14.3	21.5	28.6	35.8	.0	0.0	0.7	1.4	2.2	2.9	3.6	4.3	5.1	5.8	6.5	
43.1	7.2	14.3	21.5	28.7	35.9	.1	0.1	0.8	1.5	2.2	3.0	3.7	4.4	5.1	5.9	6.6	
43.2	7.2	14.4	21.6	28.8	36.0	.2	0.1	0.9	1.6	2.3	3.0	3.8	4.5	5.2	5.9	6.7	
43.3	7.2	14.4	21.6	28.9	36.1	.3	0.2	0.9	1.7	2.4	3.1	3.8	4.6	5.3	6.0	6.7	
43.4	7.2	14.5	21.7	28.9	36.2	.4	0.3	1.0	1.7	2.5	3.2	3.9	4.6	5.4	6.1	6.8	
43.5	7.3	14.5	21.8	29.0	36.3	.5	0.4	1.1	1.8	2.5	3.3	4.0	4.7	5.4	6.2	6.9	
43.6	7.3	14.5	21.8	29.1	36.3	.6	0.4	1.2	1.9	2.6	3.3	4.1	4.8	5.5	6.2	7.0	
43.7	7.3	14.6	21.9	29.2	36.4	.7	0.5	1.2	2.0	2.7	3.4	4.1	4.9	5.6	6.3	7.0	
43.8	7.3	14.6	21.9	29.2	36.5	.8	0.6	1.3	2.0	2.8	3.5	4.2	4.9	5.7	6.4	7.1	
43.9	7.4	14.7	22.0	29.3	36.6	.9	0.7	1.4	2.1	2.8	3.6	4.3	5.0	5.7	6.5	7.2	
	10	20	30	40	50		0	1	2	3	4	5	6	7	8	9	

The Double-Second-Difference correction (Corr.) is always to be added to the tabulated altitude.

INTERPOLATION TABLE

Dec. Inc.	Tens 10'	20'	30'	40'	50'	Decimals	Units 0'	1'	2'	3'	4'	5'	6'	7'	8'	9'	Double Second Diff. and Corr.
44.0	7.3	14.6	22.0	29.3	36.6	.0	0.0	0.7	1.5	2.2	3.0	3.7	4.4	5.2	5.9	6.7	1.1 0.1 3.2 0.2 5.3 0.3 7.5 0.4 9.6 0.5 11.7 0.6 13.9 0.7 16.0 0.8 18.1 0.9 20.3 1.0 22.4 1.1 24.5 1.2 26.7 1.3 28.8 1.4 30.9 1.5 33.1 1.6 35.2
44.1	7.3	14.7	22.0	29.4	36.7	.1	0.1	0.8	1.6	2.3	3.0	3.8	4.5	5.3	6.0	6.7	
44.2	7.3	14.7	22.1	29.4	36.8	.2	0.1	0.9	1.6	2.4	3.1	3.9	4.6	5.3	6.1	6.8	
44.3	7.4	14.8	22.1	29.5	36.9	.3	0.2	1.0	1.7	2.4	3.2	3.9	4.7	5.4	6.2	6.9	
44.4	7.4	14.8	22.2	29.6	37.0	.4	0.3	1.0	1.8	2.5	3.3	4.0	4.7	5.5	6.2	7.0	
44.5	7.4	14.8	22.3	29.7	37.1	.5	0.4	1.1	1.9	2.6	3.3	4.1	4.8	5.6	6.3	7.0	
44.6	7.4	14.9	22.3	29.7	37.2	.6	0.4	1.2	1.9	2.7	3.4	4.2	4.9	5.6	6.4	7.1	
44.7	7.5	14.9	22.4	29.8	37.3	.7	0.5	1.3	2.0	2.7	3.5	4.2	5.0	5.7	6.5	7.2	
44.8	7.5	15.0	22.4	29.9	37.4	.8	0.6	1.3	2.1	2.8	3.6	4.3	5.0	5.8	6.5	7.3	
44.9	7.5	15.0	22.5	30.0	37.5	.9	0.7	1.4	2.2	2.9	3.6	4.4	5.1	5.9	6.6	7.3	
45.0	7.5	15.0	22.5	30.0	37.5	.0	0.0	0.8	1.5	2.3	3.0	3.8	4.5	5.3	6.1	6.8	
45.1	7.5	15.0	22.5	30.0	37.6	.1	0.1	0.8	1.6	2.4	3.1	3.9	4.6	5.4	6.1	6.9	
45.2	7.5	15.0	22.6	30.1	37.6	.2	0.2	0.9	1.7	2.4	3.2	3.9	4.7	5.5	6.2	7.0	
45.3	7.5	15.1	22.6	30.2	37.7	.3	0.2	1.0	1.7	2.5	3.3	4.0	4.8	5.5	6.3	7.1	
45.4	7.6	15.1	22.7	30.3	37.8	.4	0.3	1.1	1.8	2.6	3.3	4.1	4.9	5.6	6.4	7.1	
45.5	7.6	15.2	22.8	30.3	37.9	.5	0.4	1.1	1.9	2.7	3.4	4.2	4.9	5.7	6.4	7.2	
45.6	7.6	15.2	22.8	30.4	38.0	.6	0.5	1.2	2.0	2.7	3.5	4.2	5.0	5.8	6.5	7.3	
45.7	7.6	15.3	22.9	30.5	38.1	.7	0.5	1.3	2.0	2.8	3.6	4.3	5.1	5.8	6.6	7.4	
45.8	7.7	15.3	22.9	30.6	38.2	.8	0.6	1.4	2.1	2.9	3.6	4.4	5.2	5.9	6.7	7.4	
45.9	7.7	15.3	23.0	30.6	38.3	.9	0.7	1.4	2.2	3.0	3.7	4.5	5.2	6.0	6.7	7.5	
46.0	7.6	15.3	23.0	30.6	38.3	.0	0.0	0.8	1.5	2.3	3.1	3.9	4.6	5.4	6.2	7.0	1.2 0.1 3.5 0.2 5.8 0.3 8.1 0.4 10.5 0.5 12.8 0.6 15.1 0.7 17.4 0.8 19.8 0.9 22.1 1.0 24.4 1.1 26.7 1.2 29.1 1.3 31.4 1.4 33.7 1.5 36.0
46.1	7.7	15.3	23.0	30.7	38.4	.1	0.1	0.9	1.6	2.4	3.2	4.0	4.7	5.5	6.3	7.1	
46.2	7.7	15.4	23.1	30.8	38.5	.2	0.2	0.9	1.7	2.5	3.3	4.0	4.8	5.6	6.4	7.1	
46.3	7.7	15.4	23.1	30.9	38.6	.3	0.2	1.0	1.8	2.6	3.3	4.1	4.9	5.7	6.4	7.2	
46.4	7.7	15.5	23.2	30.9	38.7	.4	0.3	1.1	1.9	2.6	3.4	4.2	5.0	5.7	6.5	7.3	
46.5	7.8	15.5	23.3	31.0	38.8	.5	0.4	1.2	1.9	2.7	3.5	4.3	5.0	5.8	6.6	7.4	
46.6	7.8	15.5	23.3	31.1	38.8	.6	0.5	1.2	2.0	2.8	3.6	4.3	5.1	5.9	6.7	7.4	
46.7	7.8	15.6	23.4	31.2	38.9	.7	0.5	1.3	2.1	2.9	3.6	4.4	5.2	6.0	6.7	7.5	
46.8	7.8	15.6	23.4	31.2	39.0	.8	0.6	1.4	2.2	2.9	3.7	4.5	5.3	6.0	6.8	7.6	
46.9	7.9	15.7	23.5	31.3	39.1	.9	0.7	1.5	2.2	3.0	3.8	4.6	5.3	6.1	6.9	7.7	
47.0	7.8	15.6	23.5	31.3	39.1	.0	0.0	0.8	1.6	2.4	3.2	4.0	4.7	5.5	6.3	7.1	
47.1	7.8	15.7	23.5	31.4	39.2	.1	0.1	0.9	1.7	2.5	3.2	4.0	4.8	5.6	6.4	7.2	
47.2	7.8	15.7	23.6	31.4	39.3	.2	0.2	0.9	1.7	2.5	3.3	4.1	4.9	5.7	6.5	7.3	
47.3	7.9	15.8	23.6	31.5	39.4	.3	0.2	1.0	1.8	2.6	3.4	4.2	5.0	5.8	6.6	7.4	
47.4	7.9	15.8	23.7	31.6	39.5	.4	0.3	1.1	1.9	2.7	3.5	4.3	5.1	5.9	6.6	7.4	
47.5	7.9	15.8	23.8	31.7	39.6	.5	0.4	1.2	2.0	2.8	3.6	4.4	5.1	5.9	6.7	7.5	1.3 0.1 3.8 0.2 6.3 0.3 8.9 0.4 11.4 0.5 14.0 0.6 16.5 0.7 19.0 0.8 21.6 0.9 24.1 1.0 26.7 1.1 29.2 1.2 31.7 1.3 34.3
47.6	7.9	15.9	23.8	31.7	39.7	.6	0.5	1.3	2.1	2.8	3.6	4.4	5.2	6.0	6.8	7.6	
47.7	8.0	15.9	23.9	31.8	39.8	.7	0.6	1.3	2.1	2.9	3.7	4.5	5.3	6.1	6.9	7.7	
47.8	8.0	16.0	23.9	31.9	39.9	.8	0.6	1.4	2.2	3.0	3.8	4.6	5.4	6.2	7.0	7.8	
47.9	8.0	16.0	24.0	32.0	40.0	.9	0.7	1.5	2.3	3.1	3.9	4.7	5.5	6.3	7.0	7.8	
48.0	8.0	16.0	24.0	32.0	40.0	.0	0.0	0.8	1.6	2.4	3.2	4.0	4.8	5.7	6.5	7.3	
48.1	8.0	16.0	24.0	32.0	40.1	.1	0.1	0.9	1.7	2.5	3.3	4.1	4.9	5.7	6.5	7.4	
48.2	8.0	16.0	24.1	32.1	40.1	.2	0.2	1.0	1.8	2.6	3.4	4.2	5.0	5.8	6.6	7.4	
48.3	8.0	16.1	24.1	32.2	40.2	.3	0.2	1.1	1.9	2.7	3.5	4.3	5.1	5.9	6.7	7.5	
48.4	8.1	16.1	24.2	32.3	40.3	.4	0.3	1.1	1.9	2.7	3.6	4.4	5.2	6.0	6.8	7.6	
48.5	8.1	16.2	24.3	32.3	40.4	.5	0.4	1.2	2.0	2.8	3.6	4.4	5.3	6.1	6.9	7.7	
48.6	8.1	16.2	24.3	32.4	40.5	.6	0.5	1.3	2.1	2.9	3.7	4.5	5.3	6.1	7.0	7.8	
48.7	8.1	16.3	24.4	32.5	40.6	.7	0.6	1.4	2.2	3.0	3.8	4.6	5.4	6.2	7.0	7.8	
48.8	8.2	16.3	24.4	32.6	40.7	.8	0.6	1.5	2.3	3.1	3.9	4.7	5.5	6.3	7.1	7.9	
48.9	8.2	16.3	24.5	32.6	40.8	.9	0.7	1.5	2.3	3.2	4.0	4.8	5.6	6.4	7.2	8.0	
49.0	8.1	16.3	24.5	32.6	40.8	.0	0.0	0.8	1.6	2.5	3.3	4.1	4.9	5.8	6.6	7.4	1.4 0.1 4.2 0.2 7.1 0.3 9.9 0.4 12.7 0.5 15.5 0.6 18.4 0.7 21.2 0.8 24.0 0.9 26.8 1.0 29.7 1.1 32.5 1.2 35.3
49.1	8.2	16.3	24.5	32.7	40.9	.1	0.1	0.9	1.7	2.6	3.4	4.2	5.0	5.9	6.7	7.5	
49.2	8.2	16.4	24.6	32.8	41.0	.2	0.2	1.0	1.8	2.6	3.5	4.3	5.1	5.9	6.8	7.6	
49.3	8.2	16.4	24.6	32.9	41.1	.3	0.2	1.1	1.9	2.7	3.5	4.4	5.2	6.0	6.8	7.7	
49.4	8.2	16.5	24.7	32.9	41.2	.4	0.3	1.2	2.0	2.8	3.6	4.5	5.3	6.1	6.9	7.8	
49.5	8.3	16.5	24.8	33.0	41.3	.5	0.4	1.2	2.1	2.9	3.7	4.5	5.4	6.2	7.0	7.8	
49.6	8.3	16.5	24.8	33.1	41.3	.6	0.5	1.3	2.1	3.0	3.8	4.6	5.4	6.3	7.1	7.9	
49.7	8.3	16.6	24.9	33.2	41.4	.7	0.6	1.4	2.2	3.1	3.9	4.7	5.5	6.4	7.2	8.0	
49.8	8.3	16.6	24.9	33.2	41.5	.8	0.7	1.5	2.3	3.1	4.0	4.8	5.6	6.4	7.3	8.1	
49.9	8.4	16.7	25.0	33.3	41.6	.9	0.7	1.6	2.4	3.2	4.0	4.9	5.7	6.5	7.3	8.2	
50.0	8.3	16.6	25.0	33.3	41.6	.0	0.0	0.8	1.7	2.5	3.4	4.2	5.0	5.9	6.7	7.6	
50.1	8.3	16.7	25.0	33.4	41.7	.1	0.1	0.9	1.8	2.6	3.5	4.3	5.1	6.0	6.8	7.7	
50.2	8.3	16.7	25.1	33.4	41.8	.2	0.2	1.0	1.9	2.7	3.5	4.4	5.2	6.1	6.9	7.7	
50.3	8.4	16.8	25.1	33.5	41.9	.3	0.3	1.1	1.9	2.8	3.6	4.5	5.3	6.1	7.0	7.8	
50.4	8.4	16.8	25.2	33.6	42.0	.4	0.3	1.2	2.0	2.9	3.7	4.5	5.4	6.2	7.1	7.9	
50.5	8.4	16.8	25.3	33.7	42.1	.5	0.4	1.3	2.1	2.9	3.8	4.6	5.5	6.3	7.2	8.0	1.6 0.1 4.8 0.2 8.0 0.3 11.2 0.4 14.5 0.5 17.7 0.6 20.9 0.7 24.1 0.8 27.3 0.9 30.5 1.0 33.7 1.1 36.9
50.6	8.4	16.9	25.3	33.7	42.2	.6	0.5	1.3	2.2	3.0	3.9	4.7	5.6	6.4	7.2	8.1	
50.7	8.5	16.9	25.4	33.8	42.3	.7	0.6	1.4	2.3	3.1	4.0	4.8	5.6	6.5	7.3	8.2	
50.8	8.5	17.0	25.4	33.9	42.4	.8	0.7	1.5	2.4	3.2	4.0	4.9	5.7	6.6	7.4	8.2	
50.9	8.5	17.0	25.5	34.0	42.5	.9	0.8	1.6	2.4	3.3	4.1	5.0	5.8	6.6	7.5	8.3	
51.0	8.5	17.0	25.5	34.0	42.5	.0	0.0	0.9	1.7	2.6	3.4	4.3	5.1	6.0	6.9	7.7	
51.1	8.5	17.0	25.5	34.0	42.6	.1	0.1	0.9	1.8	2.7	3.5	4.4	5.2	6.1	7.0	7.8	
51.2	8.5	17.0	25.6	34.1	42.6	.2	0.2	1.0	1.9	2.7	3.6	4.5	5.3	6.2	7.0	7.9	
51.3	8.5	17.1	25.6	34.2	42.7	.3	0.3	1.1	2.0	2.8	3.7	4.5	5.4	6.3	7.1	8.0	
51.4	8.6	17.1	25.7	34.3	42.8	.4	0.3	1.2	2.1	2.9	3.8	4.6	5.5	6.4	7.2	8.1	
51.5	8.6	17.2	25.8	34.3	42.9	.5	0.4	1.3	2.1	3.0	3.9	4.7	5.6	6.4	7.3	8.2	
51.6	8.6	17.2	25.8	34.4	43.0	.6	0.5	1.4	2.2	3.1	3.9	4.8	5.7	6.5	7.4	8.2	
51.7	8.6	17.3	25.9	34.5	43.1	.7	0.6	1.5	2.3	3.2	4.0	4.9	5.8	6.6	7.5	8.3	
51.8	8.7	17.3	25.9	34.6	43.2	.8	0.7	1.5	2.4	3.3	4.1	5.0	5.8	6.7	7.6	8.4	
51.9	8.7	17.3	26.0	34.6	43.3	.9	0.8	1.6	2.5	3.3	4.2	5.1	5.9	6.8	7.6	8.5	
	10'	20'	30'	40'	50'		0'	1'	2'	3'	4'	5'	6'	7'	8'	9'	

Dec. Inc.	Tens 10'	20'	30'	40'	50'	Decimals	Units 0'	1'	2'	3'	4'	5'	6'	7'	8'	9'	Double Second Diff. and Corr.
52.0	8.6	17.3	26.0	34.6	43.3	.0	0.0	0.9	1.7	2.6	3.5	4.4	5.2	6.1	7.0	7.9	1.8 0.1 5.5 0.2 9.1 0.3 12.8 0.4 16.5 0.5 20.1 0.6 23.8 0.7 27.4 0.8 31.1 0.9 34.7
52.1	8.7	17.3	26.0	34.7	43.4	.1	0.1	1.0	1.8	2.7	3.6	4.5	5.3	6.2	7.1	8.0	
52.2	8.7	17.4	26.1	34.8	43.5	.2	0.2	1.0	1.9	2.8	3.7	4.5	5.4	6.3	7.2	8.0	
52.3	8.7	17.4	26.1	34.9	43.6	.3	0.3	1.1	2.0	2.9	3.8	4.6	5.5	6.4	7.3	8.1	
52.4	8.7	17.5	26.2	34.9	43.7	.4	0.3	1.2	2.1	3.0	3.8	4.7	5.6	6.5	7.3	8.2	
52.5	8.8	17.5	26.3	35.0	43.8	.5	0.4	1.3	2.2	3.1	3.9	4.8	5.7	6.6	7.4	8.3	
52.6	8.8	17.5	26.3	35.1	43.8	.6	0.5	1.4	2.3	3.1	4.0	4.9	5.8	6.6	7.5	8.4	
52.7	8.8	17.6	26.4	35.2	43.9	.7	0.6	1.5	2.4	3.2	4.1	5.0	5.9	6.7	7.6	8.5	
52.8	8.8	17.6	26.4	35.2	44.0	.8	0.7	1.6	2.4	3.3	4.2	5.1	5.9	6.8	7.7	8.6	
52.9	8.9	17.7	26.5	35.3	44.1	.9	0.8	1.7	2.5	3.4	4.3	5.2	6.0	6.9	7.8	8.7	
53.0	8.8	17.6	26.5	35.3	44.1	.0	0.0	0.9	1.8	2.7	3.6	4.5	5.3	6.2	7.1	8.0	2.1 0.1 6.2 0.2 10.4 0.3 14.5 0.4 18.6 0.5 22.8 0.6 26.9 0.7 31.1 0.8 35.2
53.1	8.8	17.7	26.5	35.4	44.2	.1	0.1	1.0	1.9	2.8	3.7	4.5	5.4	6.3	7.2	8.1	
53.2	8.8	17.7	26.6	35.4	44.3	.2	0.2	1.1	2.0	2.9	3.7	4.6	5.5	6.4	7.3	8.2	
53.3	8.9	17.8	26.6	35.5	44.4	.3	0.3	1.2	2.1	2.9	3.8	4.7	5.6	6.5	7.4	8.3	
53.4	8.9	17.8	26.7	35.6	44.5	.4	0.4	1.2	2.1	3.0	3.9	4.8	5.7	6.6	7.5	8.4	
53.5	8.9	17.8	26.8	35.7	44.6	.5	0.4	1.3	2.2	3.1	4.0	4.9	5.8	6.7	7.6	8.5	
53.6	8.9	17.9	26.8	35.7	44.7	.6	0.5	1.4	2.3	3.2	4.1	5.0	5.9	6.8	7.7	8.6	
53.7	9.0	17.9	26.9	35.8	44.8	.7	0.6	1.5	2.4	3.3	4.2	5.1	6.0	6.9	7.8	8.6	
53.8	9.0	18.0	26.9	35.9	44.9	.8	0.7	1.6	2.5	3.4	4.3	5.2	6.1	7.0	7.8	8.7	
53.9	9.0	18.0	27.0	36.0	45.0	.9	0.8	1.7	2.6	3.5	4.4	5.3	6.2	7.0	7.9	8.8	
54.0	9.0	18.0	27.0	36.0	45.0	.0	0.0	0.9	1.8	2.7	3.6	4.5	5.4	6.4	7.3	8.2	2.4 0.1 7.2 0.2 12.0 0.3 16.8 0.4 21.6 0.5 26.4 0.6 31.2 0.7 36.0
54.1	9.0	18.0	27.0	36.0	45.1	.1	0.1	1.0	1.9	2.8	3.7	4.6	5.5	6.4	7.4	8.3	
54.2	9.0	18.0	27.1	36.1	45.1	.2	0.2	1.1	2.0	2.9	3.8	4.7	5.6	6.5	7.4	8.4	
54.3	9.0	18.1	27.1	36.2	45.2	.3	0.3	1.2	2.1	3.0	3.9	4.8	5.7	6.6	7.5	8.4	
54.4	9.1	18.1	27.2	36.3	45.3	.4	0.4	1.3	2.2	3.1	4.0	4.9	5.8	6.7	7.6	8.5	
54.5	9.1	18.2	27.3	36.3	45.4	.5	0.5	1.4	2.3	3.2	4.1	5.0	5.9	6.8	7.7	8.6	
54.6	9.1	18.2	27.3	36.4	45.5	.6	0.5	1.5	2.4	3.3	4.2	5.1	6.0	6.9	7.8	8.7	
54.7	9.1	18.3	27.4	36.5	45.6	.7	0.6	1.5	2.5	3.4	4.3	5.2	6.1	7.0	7.9	8.8	
54.8	9.2	18.3	27.4	36.6	45.7	.8	0.7	1.6	2.5	3.5	4.4	5.3	6.2	7.1	8.0	8.9	
54.9	9.2	18.3	27.5	36.6	45.8	.9	0.8	1.7	2.6	3.5	4.5	5.4	6.3	7.2	8.1	9.0	
55.0	9.1	18.3	27.5	36.6	45.8	.0	0.0	0.9	1.8	2.8	3.7	4.6	5.5	6.5	7.4	8.3	2.9 0.1 8.6 0.2 14.4 0.3 20.2 0.4 25.9 0.5 31.7 0.6 37.5
55.1	9.2	18.3	27.5	36.7	45.9	.1	0.1	1.0	1.9	2.9	3.8	4.7	5.6	6.6	7.5	8.4	
55.2	9.2	18.4	27.6	36.8	46.0	.2	0.2	1.1	2.0	3.0	3.9	4.8	5.7	6.7	7.6	8.5	
55.3	9.2	18.4	27.6	36.9	46.1	.3	0.3	1.2	2.1	3.1	4.0	4.9	5.8	6.8	7.7	8.6	
55.4	9.2	18.5	27.7	36.9	46.2	.4	0.4	1.3	2.2	3.1	4.1	5.0	5.9	6.8	7.8	8.7	
55.5	9.3	18.5	27.8	37.0	46.3	.5	0.5	1.4	2.3	3.2	4.2	5.1	6.0	6.9	7.9	8.8	
55.6	9.3	18.5	27.8	37.1	46.3	.6	0.6	1.5	2.4	3.3	4.3	5.2	6.1	7.0	8.0	8.9	
55.7	9.3	18.6	27.9	37.2	46.4	.7	0.6	1.6	2.5	3.4	4.3	5.3	6.2	7.1	8.0	9.0	
55.8	9.3	16.6	27.9	37.2	46.5	.8	0.7	1.7	2.6	3.5	4.4	5.4	6.3	7.2	8.1	9.1	
55.9	9.4	18.7	28.0	37.3	46.6	.9	0.8	1.8	2.7	3.6	4.5	5.5	6.4	7.3	8.2	9.2	
56.0	9.3	18.6	28.0	37.3	46.6	.0	0.0	0.9	1.9	2.8	3.8	4.7	5.6	6.6	7.5	8.5	3.6 0.1 10.9 0.2 18.2 0.3 25.5 0.4 32.8 0.5 40.1
56.1	9.3	18.7	28.0	37.4	46.7	.1	0.1	1.0	2.0	2.9	3.9	4.8	5.7	6.7	7.6	8.6	
56.2	9.3	18.7	28.1	37.4	46.8	.2	0.2	1.1	2.1	3.0	4.0	4.9	5.8	6.8	7.7	8.7	
56.3	9.4	18.8	28.1	37.5	46.9	.3	0.3	1.2	2.2	3.1	4.0	5.0	5.9	6.9	7.8	8.8	
56.4	9.4	18.8	28.2	37.6	47.0	.4	0.4	1.3	2.3	3.2	4.1	5.1	6.0	7.0	7.9	8.9	
56.5	9.4	18.8	28.3	37.7	47.1	.5	0.5	1.4	2.4	3.3	4.2	5.2	6.1	7.1	8.0	8.9	
56.6	9.4	18.9	28.3	37.7	47.2	.6	0.6	1.5	2.4	3.4	4.3	5.3	6.2	7.2	8.1	9.0	
56.7	9.5	18.9	28.4	37.8	47.3	.7	0.7	1.6	2.5	3.5	4.4	5.4	6.3	7.3	8.2	9.1	
56.8	9.5	19.0	28.4	37.9	47.4	.8	0.8	1.7	2.6	3.6	4.5	5.5	6.4	7.3	8.3	9.2	
56.9	9.5	19.0	28.5	38.0	47.5	.9	0.8	1.8	2.7	3.7	4.6	5.6	6.5	7.4	8.4	9.3	
57.0	9.5	19.0	28.5	38.0	47.5	.0	0.0	1.0	1.9	2.9	3.8	4.8	5.7	6.7	7.7	8.6	5.0 0.1 15.0 0.2 25.0 0.3 35.1
57.1	9.5	19.0	28.5	38.0	47.6	.1	0.1	1.1	2.0	3.0	3.9	4.9	5.8	6.8	7.8	8.7	
57.2	9.5	19.0	28.6	38.1	47.6	.2	0.2	1.1	2.1	3.1	4.0	5.0	5.9	6.9	7.9	8.8	
57.3	9.5	19.1	28.6	38.2	47.7	.3	0.3	1.2	2.2	3.2	4.1	5.1	6.0	7.0	8.0	8.9	
57.4	9.6	19.1	28.7	38.3	47.8	.4	0.4	1.3	2.3	3.3	4.2	5.2	6.1	7.1	8.0	9.0	
57.5	9.6	19.2	28.8	38.3	47.9	.5	0.5	1.4	2.4	3.4	4.3	5.3	6.2	7.2	8.1	9.1	
57.6	9.6	19.2	28.8	38.4	48.0	.6	0.6	1.5	2.5	3.4	4.4	5.4	6.3	7.3	8.2	9.2	
57.7	9.6	19.3	28.9	38.5	48.1	.7	0.7	1.6	2.6	3.5	4.5	5.5	6.4	7.4	8.3	9.3	
57.8	9.7	19.3	28.9	38.6	48.2	.8	0.8	1.7	2.7	3.6	4.6	5.6	6.5	7.5	8.4	9.4	
57.9	9.7	19.3	29.0	38.6	48.3	.9	0.9	1.8	2.8	3.7	4.7	5.7	6.6	7.6	8.5	9.5	
58.0	9.6	19.3	29.0	38.6	48.3	.0	0.0	1.0	1.9	2.9	3.9	4.9	5.8	6.8	7.8	8.8	8.2 0.1 24.6 0.2 41.0
58.1	9.7	19.3	29.0	38.7	48.4	.1	0.1	1.1	2.0	3.0	4.0	5.0	5.9	6.9	7.9	8.9	
58.2	9.7	19.4	29.1	38.8	48.5	.2	0.2	1.2	2.1	3.1	4.1	5.1	6.0	7.0	8.0	9.0	
58.3	9.7	19.4	29.1	38.9	48.6	.3	0.3	1.3	2.2	3.2	4.2	5.2	6.1	7.1	8.1	9.1	
58.4	9.7	19.5	29.2	38.9	48.7	.4	0.4	1.4	2.3	3.3	4.3	5.3	6.2	7.2	8.2	9.2	
58.5	9.8	19.5	29.3	39.0	48.8	.5	0.5	1.5	2.4	3.4	4.4	5.4	6.3	7.3	8.3	9.3	
58.6	9.8	19.5	29.3	39.1	48.8	.6	0.6	1.6	2.5	3.5	4.5	5.5	6.4	7.4	8.4	9.4	
58.7	9.8	19.6	29.4	39.2	48.9	.7	0.7	1.7	2.6	3.6	4.6	5.6	6.5	7.5	8.5	9.5	
58.8	9.8	19.6	29.4	39.2	49.0	.8	0.8	1.8	2.7	3.7	4.7	5.7	6.6	7.6	8.6	9.6	
58.9	9.9	19.7	29.5	39.3	49.1	.9	0.9	1.9	2.8	3.8	4.8	5.8	6.7	7.7	8.7	9.7	
59.0	9.8	19.6	29.5	39.3	49.1	.0	0.0	1.0	2.0	3.0	4.0	5.0	5.9	6.9	7.9	8.9	16.2 0.1 48.6
59.1	9.8	19.7	29.5	39.4	49.2	.1	0.1	1.1	2.1	3.1	4.1	5.1	6.0	7.0	8.0	9.0	
59.2	9.8	19.7	29.6	39.4	49.3	.2	0.2	1.2	2.2	3.2	4.2	5.2	6.1	7.1	8.1	9.1	
59.3	9.9	19.8	29.6	39.5	49.4	.3	0.3	1.3	2.3	3.3	4.3	5.3	6.2	7.2	8.2	9.2	
59.4	9.9	19.8	29.7	39.6	49.5	.4	0.4	1.4	2.4	3.4	4.4	5.4	6.3	7.3	8.3	9.3	
59.5	9.9	19.8	29.8	39.7	49.6	.5	0.5	1.5	2.5	3.5	4.5	5.5	6.4	7.4	8.4	9.4	0.0 0.0 48.2
59.6	9.9	19.9	29.8	39.7	49.7	.6	0.6	1.6	2.6	3.6	4.6	5.6	6.5	7.5	8.5	9.5	
59.7	10.0	19.9	29.9	39.8	49.8	.7	0.7	1.7	2.7	3.7	4.7	5.7	6.6	7.6	8.6	9.6	
59.8	10.0	20.0	29.9	39.9	49.9	.8	0.8	1.8	2.8	3.8	4.8	5.8	6.7	7.7	8.7	9.7	
59.9	10.0	20.0	30.0	40.0	50.0	.9	0.9	1.9	2.9	3.9	4.9	5.9	6.8	7.8	8.8	9.8	
	10'	20'	30'	40'	50'		0'	1'	2'	3'	4'	5'	6'	7'	8'	9'	

The Double-Second-Difference correction (Corr.) is always to be added to the tabulated altitude.

APPENDIX P

EXTRACTS FROM PUB. NO. 249

LAT 41°S

LHA ♈	Hc Zn	Hc Zn	Hc Zn	Hc Zn	Hc Zn	Hc Zn	Hc Zn
	Alpheratz	Hamal	•RIGEL	CANOPUS	•RIGIL KENT.	Peacock	•Enif
0	20 02 002	19 21 031	14 11 089	28 22 137	16 43 199	52 19 226	30 14 320
1	20 03 001	19 43 030	14 56 088	28 53 136	16 28 198	51 46 226	29 45 319
2	20 03 000	20 05 029	15 41 087	29 25 136	16 14 198	51 13 226	29 15 318
3	20 02 359	20 27 028	16 26 087	29 56 136	16 00 198	50 40 226	28 44 317
4	20 01 358	20 48 027	17 12 086	30 28 135	15 46 197	50 08 226	28 13 316
5	19 59 357	21 08 026	17 57 085	31 00 135	15 33 197	49 35 226	27 41 315
6	19 56 356	21 27 025	18 42 085	31 32 135	15 20 196	49 02 226	27 09 314
7	19 53 355	21 46 024	19 27 084	32 04 134	15 08 196	48 29 226	26 37 313
8	19 49 354	22 04 023	20 12 083	32 37 134	14 56 195	47 57 226	26 03 312
9	19 44 353	22 22 022	20 57 083	33 09 134	14 44 195	47 24 226	25 30 312
10	19 38 352	22 39 021	21 42 082	33 42 133	14 33 194	46 51 226	24 56 311
11	19 32 351	22 55 020	22 27 081	34 15 133	14 21 194	46 19 226	24 21 310
12	19 25 351	23 10 019	23 11 080	34 48 133	14 11 193	45 46 226	23 46 309
13	19 17 350	23 25 018	23 56 080	35 22 132	14 00 193	45 14 226	23 11 308
14	19 08 349	23 39 017	24 40 079	35 55 132	13 50 193	44 41 226	22 35 307
	•Hamal	RIGEL	•CANOPUS	RIGIL KENT.	•Peacock	FOMALHAUT	Alpheratz
15	23 52 017	25 25 078	36 29 132	13 41 192	44 09 225	62 35 284	18 59 348
16	24 05 016	26 09 078	37 02 132	13 32 192	43 37 225	61 51 283	18 49 347
17	24 16 015	26 53 077	37 36 131	13 23 191	43 05 225	61 06 282	18 38 346
18	24 27 014	27 37 076	38 10 131	13 14 191	42 33 225	60 22 281	18 27 345
19	24 38 013	28 21 075	38 45 131	13 06 190	42 01 225	59 38 281	18 15 344
20	24 47 012	29 05 075	39 19 131	12 58 190	41 29 225	58 53 280	18 02 343
21	24 56 011	29 48 074	39 53 130	12 51 189	40 57 224	58 08 279	17 49 342
22	25 04 010	30 32 073	40 28 130	12 44 189	40 25 224	57 24 278	17 35 341
23	25 11 009	31 15 072	41 03 130	12 37 188	39 54 224	56 39 277	17 20 341
24	25 17 008	31 58 071	41 38 130	12 31 188	39 22 224	55 54 276	17 05 340
25	25 23 007	32 41 071	42 13 129	12 25 187	38 51 224	55 09 276	16 49 339
26	25 27 006	33 23 070	42 48 129	12 20 187	38 20 223	54 24 275	16 32 338
27	25 31 005	34 06 069	43 23 129	12 15 186	37 49 223	53 38 274	16 15 337
28	25 35 004	34 48 068	43 58 129	12 10 186	37 18 223	52 53 273	15 57 336
29	25 37 002	35 30 067	44 33 129	12 06 185	36 47 223	52 08 273	15 39 335
	•Hamal	RIGEL	SIRIUS	•CANOPUS	RIGIL KENT.	•Peacock	FOMALHAUT
30	25 38 001	36 11 066	25 04 091	45 09 128	12 02 185	36 17 222	51 23 272
31	25 39 000	36 53 065	25 49 090	45 45 128	11 58 184	35 46 222	50 38 271
32	25 39 359	37 34 065	26 34 089	46 20 128	11 55 184	35 16 222	49 52 271
33	25 38 358	38 15 064	27 19 089	46 56 128	11 52 183	34 45 222	49 07 270
34	25 37 357	38 55 063	28 05 088	47 32 128	11 50 183	34 15 221	48 22 269
35	25 34 356	39 35 062	28 50 087	48 08 128	11 48 182	33 46 221	47 36 269
36	25 31 355	40 15 061	29 35 087	48 44 127	11 46 182	33 16 221	46 51 268
37	25 27 354	40 54 060	30 20 086	49 20 127	11 45 181	32 46 221	46 06 267
38	25 22 353	41 33 059	31 05 085	49 56 127	11 44 181	32 17 220	45 21 267
39	25 16 352	42 12 058	31 51 085	50 32 127	11 44 180	31 48 220	44 35 266
40	25 10 351	42 50 057	32 36 084	51 08 127	11 44 180	31 19 220	43 50 266
41	25 03 350	43 27 056	33 21 083	51 44 127	11 44 179	30 50 219	43 05 265
42	24 55 349	44 05 055	34 06 083	52 20 127	11 45 179	30 21 219	42 20 264
43	24 46 348	44 41 054	34 50 082	52 57 127	11 46 178	29 53 219	41 35 264
44	24 36 347	45 18 053	35 35 081	53 33 127	11 48 178	29 24 218	40 50 263
	•RIGEL	SIRIUS	CANOPUS	•RIGIL KENT.	Peacock	•FOMALHAUT	Hamal
45	45 53 051	36 20 080	54 0[illegible] 127	11 50 177	28 56 218	40 05 263	24 26 346
46	46 28 050	37 05 080	54 46 127	11 52 177	28 28 218	39 20 262	24 15 345
47	47 03 049	37 49 079	55 22 127	11 55 176	28 01 217	38 35 262	24 03 344
48	47 37 048	38 33 078	55 59 127	11 58 176	27 33 217	37 51 261	23 50 343
49	48 10 047	39 18 077	56 35 127	12 02 175	27 06 217	37 06 260	23 37 342
50	48 43 045	40 02 076	57 11 127	12 06 175	26 39 216	36 21 260	23 23 341
51	49 14 044	40 46 076	57 48 127	12 10 174	26 12 216	35 37 259	23 08 340
52	49 46 043	41 29 075	58 24 127	12 15 174	25 46 216	34 52 259	22 53 340
53	50 16 041	42 13 074	59 00 127	12 20 173	25 19 215	34 08 258	22 37 339
54	50 46 040	42 57 073	59 36 127	12 25 173	24 53 215	33 24 258	22 20 338
55	51 14 039	43 40 072	60 13 127	12 31 172	24 28 215	32 40 257	22 02 337
56	51 42 037	44 23 071	60 49 127	12 38 172	24 02 214	31 55 257	21 44 336
57	52 09 036	45 06 071	61 25 127	12 44 171	23 37 214	31 12 256	21 25 335
58	52 35 034	45 48 070	62 01 128	12 51 171	23 11 213	30 28 255	21 05 334
59	53 00 033	46 31 069	62 36 128	12 59 170	22 47 213	29 44 255	20 45 333
	ALDEBARAN	•SIRIUS	Suhail	•RIGIL KENT	ACHERNAR	•Diphda	Hamal
60	31 59 010	47 13 068	35 10 120	13 06 170	61 57 222	42 07 283	20 24 332
61	32 06 009	47 54 067	35 49 120	13 15 169	61 27 223	41 23 283	20 03 331
62	32 13 008	48 36 066	36 28 119	13 23 169	60 56 223	40 39 282	19 41 330
63	32 18 006	49 17 065	37 08 119	13 32 168	60 25 223	39 54 281	19 18 329
64	32 23 005	49 58 064	37 48 118	13 41 168	59 54 224	39 10 280	18 54 329
65	32 26 004	50 38 063	38 27 118	13 51 167	59 23 224	38 25 279	18 31 328
66	32 29 003	51 18 062	39 08 118	14 01 167	58 51 224	37 41 279	18 06 327
67	32 31 002	51 58 060	39 48 117	14 11 167	58 20 224	36 56 278	17 41 326
68	32 32 001	52 37 059	40 28 117	14 22 166	57 48 224	36 11 277	17 15 325
69	32 32 000	53 15 058	41 08 117	14 33 166	57 16 225	35 26 276	16 49 324
70	32 32 358	53 54 057	41 49 116	14 45 165	56 45 225	34 41 276	16 22 323
71	32 30 357	54 31 056	42 30 116	14 56 165	56 13 225	33 56 275	15 55 323
72	32 27 356	55 08 054	43 10 116	15 09 164	55 41 225	33 11 274	15 27 322
73	32 24 355	55 45 053	43 51 115	15 21 164	55 08 225	32 26 274	14 59 321
74	32 19 354	56 21 052	44 32 115	15 34 163	54 36 225	31 40 273	14 30 320
	BETELGEUSE	•SIRIUS	Suhail	•RIGIL KENT.	ACHERNAR	•Diphda	ALDEBARAN
75	40 02 018	56 56 050	45 14 115	15 47 163	54 04 225	30 55 272	32 14 353
76	40 16 016	57 30 049	45 55 114	16 01 162	53 32 225	30 10 272	32 08 352
77	40 28 015	58 04 047	46 36 114	16 15 162	52 59 226	29 25 271	32 01 351
78	40 39 014	58 37 046	47 18 114	16 29 161	52 27 226	28 39 270	31 53 349
79	40 49 012	59 09 044	47 59 113	16 43 161	51 55 226	27 54 270	31 44 348
80	40 59 011	59 41 043	48 41 113	16 58 161	51 22 226	27 09 269	31 35 347
81	41 07 010	60 11 041	49 23 113	17 13 160	50 50 226	26 23 268	31 24 346
82	41 14 009	60 40 040	50 05 112	17 29 160	50 18 226	25 38 268	31 13 345
83	41 20 007	61 09 038	50 46 112	17 45 159	49 45 226	24 53 267	31 01 344
84	41 25 006	61 36 036	51 29 112	18 01 159	49 13 226	24 08 266	30 48 343
85	41 30 005	62 02 034	52 11 111	18 17 158	48 41 225	23 23 266	30 34 342
86	41 33 003	62 27 032	52 53 111	18 34 158	48 08 225	22 37 265	30 19 341
87	41 35 002	62 51 031	53 35 111	18 51 158	47 36 225	21 52 265	30 04 340
88	41 36 001	63 13 029	54 18 110	19 09 157	47 04 225	21 07 264	29 48 338
89	41 36 359	63 34 027	55 00 110	19 26 157	46 32 225	20 22 263	29 31 337

LHA ♈	Hc Zn	Hc Zn	Hc Zn	Hc Zn	Hc Zn	Hc Zn	Hc Zn
	PROCYON	Alphard	•Suhail	RIGIL KENT.	•ACHERNAR	RIGEL	•BETELGEUSE
90	38 34 032	34 08 069	55 43 110	19 44 156	46 00 225	55 38 339	41 35 358
91	38 57 031	34 51 069	56 25 110	20 03 156	45 28 225	55 21 338	41 32 357
92	39 20 030	35 33 068	57 08 109	20 21 156	44 56 225	55 03 336	41 29 355
93	39 42 028	36 14 067	57 51 109	20 40 155	44 24 225	54 44 334	41 25 354
94	40 03 027	36 56 066	58 34 109	20 59 155	43 52 225	54 24 333	41 20 353
95	40 23 026	37 37 065	59 17 108	21 19 154	43 20 224	54 03 331	41 13 351
96	40 42 025	38 18 064	60 00 108	21 39 154	42 49 224	53 40 330	41 06 350
97	41 01 023	38 59 063	60 43 108	21 59 154	42 17 224	53 17 328	40 58 349
98	41 18 022	39 39 062	61 26 108	22 19 153	41 46 224	52 52 327	40 48 347
99	41 35 021	40 19 061	62 09 108	22 39 153	41 14 224	52 27 325	40 38 346
100	41 51 020	40 58 060	62 52 107	23 00 152	40 43 224	52 00 324	40 27 345
101	42 05 018	41 38 059	63 35 107	23 21 152	40 12 223	51 33 322	40 14 344
102	42 19 017	42 16 058	64 19 107	23 43 152	39 41 223	51 05 321	40 01 342
103	42 32 016	42 55 057	65 02 107	24 04 151	39 10 223	50 36 319	39 47 341
104	42 43 014	43 33 056	65 45 106	24 26 151	38 39 223	50 06 318	39 32 340
	PROCYON	REGULUS	•Gienah	RIGIL KENT.	•ACHERNAR	RIGEL	•BETELGEUSE
105	42 54 013	21 36 050	19 46 096	24 48 151	38 08 223	49 35 317	39 16 339
106	43 04 012	22 11 049	20 31 096	25 10 150	37 38 222	49 04 315	38 59 337
107	43 12 010	22 45 048	21 16 095	25 33 150	37 07 222	48 32 314	38 41 336
108	43 20 009	23 18 047	22 01 094	25 56 150	36 37 222	47 59 313	38 22 335
109	43 26 008	23 51 047	22 46 094	26 19 149	36 07 222	47 25 312	38 03 334
110	43 32 006	24 24 046	23 32 093	26 42 149	35 37 221	46 51 311	37 42 333
111	43 36 005	24 56 045	24 17 092	27 05 149	35 07 221	46 17 309	37 21 331
112	43 39 003	25 28 044	25 02 092	27 29 148	34 37 221	45 41 308	36 59 330
113	43 41 002	25 59 043	25 47 091	27 53 148	34 07 221	45 06 307	36 36 329
114	43 42 001	26 29 042	26 33 091	28 17 148	33 38 220	44 29 306	36 12 328
115	43 42 359	26 59 041	27 18 090	28 42 147	33 09 220	43 52 305	35 48 327
116	43 41 358	27 29 040	28 03 089	29 06 147	32 40 220	43 15 304	35 23 326
117	43 39 357	27 58 039	28 49 089	29 31 147	32 11 220	42 37 303	34 57 325
118	43 36 355	28 26 038	29 34 088	29 56 146	31 42 219	41 59 302	34 31 324
119	43 32 354	28 54 037	30 19 087	30 21 146	31 13 219	41 20 301	34 03 323
	REGULUS	•SPICA	RIGIL KENT.	•ACHERNAR	RIGEL	BETELGEUSE	•PROCYON
120	29 21 036	14 00 093	30 46 146	30 45 219	40 41 300	33 36 321	43 26 352
121	29 47 035	14 45 092	31 12 146	30 17 218	40 02 299	33 07 320	43 20 351
122	30 13 034	15 30 091	31 37 145	29 49 218	39 22 298	32 38 319	43 12 350
123	30 38 033	16 15 091	32 03 145	29 21 218	38 42 297	32 08 318	43 04 348
124	31 02 032	17 01 090	32 29 145	28 54 217	38 01 296	31 38 317	42 54 347
125	31 26 031	17 46 089	32 56 145	28 26 217	37 20 295	31 07 316	42 43 346
126	31 49 030	18 31 089	33 22 144	27 59 217	36 39 294	30 35 315	42 32 344
127	32 11 029	19 17 088	33 48 144	27 32 216	35 58 293	30 03 314	42 19 343
128	32 33 028	20 02 087	34 15 144	27 05 216	35 16 292	29 31 313	42 05 342
129	32 54 027	20 47 087	34 42 144	26 39 216	34 34 292	28 58 313	41 50 340
130	33 14 026	21 32 086	35 09 143	26 12 215	33 52 291	28 24 312	41 35 339
131	33 33 025	22 17 085	35 36 143	25 46 215	33 09 290	27 50 311	41 18 338
132	33 51 023	23 02 085	36 03 143	25 21 215	32 27 289	27 15 310	41 01 337
133	34 09 022	23 47 084	36 31 143	24 55 214	31 44 288	26 40 309	40 42 335
134	34 26 021	24 32 083	36 58 142	24 30 214	31 01 288	26 05 308	40 23 334
	REGULUS	•SPICA	RIGIL KENT.	•ACHERNAR	CANOPUS	SIRIUS	•PROCYON
135	34 42 020	25 17 083	37 26 142	24 04 214	61 14 233	51 59 300	40 03 333
136	34 57 019	26 02 082	37 54 142	23 40 213	60 38 233	51 19 299	39 42 332
137	35 11 018	26 47 081	38 22 142	23 15 213	60 02 233	50 39 297	39 20 330
138	35 24 017	27 32 080	38 50 142	22 51 212	59 25 233	49 59 296	38 57 329
139	35 37 015	28 16 080	39 18 141	22 26 212	58 49 233	49 18 295	38 34 328
140	35 48 014	29 01 079	39 46 141	22 03 212	58 13 233	48 37 294	38 09 327
141	35 59 013	29 45 078	40 15 141	21 39 211	57 37 233	47 55 293	37 44 326
142	36 09 012	30 29 077	40 43 141	21 16 211	57 00 233	47 14 292	37 18 325
143	36 18 011	31 14 077	41 12 141	20 53 210	56 24 233	46 31 291	36 52 324
144	36 26 009	31 58 076	41 40 141	20 30 210	55 48 233	45 49 290	36 25 322
145	36 33 008	32 41 075	42 09 141	20 07 210	55 11 233	45 07 289	35 57 321
146	36 39 007	33 25 074	42 38 140	19 45 209	54 35 233	44 24 289	35 28 320
147	36 44 006	34 09 074	43 07 140	19 23 209	53 58 233	43 41 288	34 59 319
148	36 48 005	34 52 073	43 36 140	19 02 208	53 22 233	42 58 287	34 29 318
149	36 51 003	35 35 072	44 05 140	18 40 208	52 46 233	42 14 286	33 58 317
	REGULUS	•SPICA	RIGIL KENT.	•ACHERNAR	CANOPUS	SIRIUS	•PROCYON
150	36 53 002	36 18 071	44 34 140	18 19 207	52 09 233	41 30 285	33 27 316
151	36 54 001	37 01 070	45 03 140	17 59 207	51 33 233	40 47 284	32 56 315
152	36 55 000	37 43 069	45 32 140	17 38 207	50 57 233	40 03 284	32 23 314
153	36 54 358	38 26 069	46 01 140	17 18 206	50 21 233	39 19 283	31 51 313
154	36 52 357	39 08 068	46 31 140	16 58 206	49 45 233	38 34 282	31 17 312
155	36 50 356	39 49 067	47 00 140	16 39 205	49 09 233	37 50 281	30 44 311
156	36 46 355	40 31 066	47 29 140	16 20 205	48 33 233	37 06 280	30 09 310
157	36 41 354	41 12 065	47 58 140	16 01 204	47 57 232	36 21 280	29 34 309
158	36 36 352	41 53 064	48 28 140	15 42 204	47 21 232	35 36 279	28 59 308
159	36 29 351	42 33 063	48 57 140	15 24 203	46 45 232	34 51 278	28 24 307
160	36 22 350	43 14 062	49 26 140	15 07 203	46 09 232	34 07 277	27 47 307
161	36 14 349	43 53 061	49 56 140	14 49 202	45 34 232	33 22 277	27 11 306
162	36 04 348	44 33 060	50 25 140	14 32 202	44 58 232	32 37 276	26 34 305
163	35 54 346	45 12 059	50 54 140	14 15 201	44 23 231	31 52 275	25 56 304
164	35 43 345	45 50 058	51 24 140	13 59 201	43 47 231	31 07 275	25 19 303
	SPICA	•ANTARES	RIGIL KENT.	ACHERNAR	•CANOPUS	SIRIUS	•REGULUS
165	46 28 057	22 42 106	51 53 140	13 43 201	43 12 231	30 21 274	35 31 344
166	47 06 056	23 26 105	52 22 140	13 27 200	42 37 231	29 36 273	35 18 343
167	47 43 055	24 10 105	52 51 140	13 12 200	42 02 231	28 51 273	35 04 342
168	48 20 053	24 53 104	53 20 140	12 57 199	41 27 230	28 06 272	34 50 341
169	48 56 052	25 37 104	53 49 140	12 42 199	40 52 230	27 20 271	34 34 339
170	49 32 051	26 21 103	54 18 140	12 28 198	40 18 230	26 35 271	34 18 338
171	50 07 050	27 06 103	54 47 141	12 14 198	39 43 230	25 50 270	34 01 337
172	50 41 049	27 50 102	55 15 141	12 00 197	39 09 229	25 05 269	33 43 336
173	51 14 047	28 34 101	55 44 141	11 47 197	38 34 229	24 19 269	33 24 335
174	51 47 046	29 19 101	56 12 141	11 35 196	38 00 229	23 34 268	33 04 334
175	52 20 045	30 03 100	56 41 141	11 22 196	37 26 229	22 49 267	32 44 333
176	52 51 043	30 48 100	57 09 142	11 10 19[illegible]	36 52 228	22 04 267	32 23 332
177	53 22 042	31 32 099	57 37 142	10 59 195	36 19 228	21 18 266	32 01 331
178	53 52 040	32 17 098	58 05 142	10 48 194	35 45 228	20 33 265	31 38 329
179	54 21 039	33 02 098	58 32 143	10 37 194	35 12 227	19 48 265	31 15 328

242

TABLE 5.—Precession and Nutation Correction

L.H.A. ♈	North latitudes								South latitudes							L.H.A. ♈
	N. 89°	N. 80°	N. 70°	N. 60°	N. 50°	N. 40°	N. 20°	0°	S. 20°	S. 40°	S. 50°	S. 60°	S. 70°	S. 80°	S. 89°	
								1976								
°	mi. °	mi. °	mi. °	mi. °	mi. °	mi. °	mi. °	mi. °	mi. °	mi. °	mi. °	mi. °	mi. °	mi. °	mi. °	°
0	1 020	1 030	1 050	1 050	1 060	1 060	1 070	1 070	1 060	1 060	1 050	1 040	1 030	1 010	1 350	0
30	1 040	1 060	1 060	1 070	1 070	1 070	1 070	1 070	1 070	1 060	1 050	1 040	0 —	0 —	1 320	30
60	1 070	1 080	1 080	1 080	1 080	1 080	1 080	1 080	1 080	1 080	0 —	0 —	0 —	0 —	1 290	60
90	1 100	1 100	1 100	1 100	1 100	1 100	1 100	1 100	1 100	1 100	0 —	0 —	0 —	0 —	1 260	90
120	1 130	1 120	1 120	1 110	1 110	1 110	1 110	1 110	1 110	1 120	1 130	0 —	0 —	0 —	1 230	120
150	1 160	1 140	1 130	1 120	1 120	1 120	1 110	1 110	1 120	1 120	1 130	1 140	1 150	1 170	1 200	150
180	1 190	1 170	1 150	1 140	1 130	1 120	1 120	1 110	1 110	1 120	1 120	1 130	1 140	1 150	1 170	180
210	1 220	0 —	0 —	1 140	1 130	1 120	1 110	1 110	1 110	1 110	1 110	1 110	1 120	1 130	1 140	210
240	1 250	0 —	0 —	0 —	0 —	1 110	1 100	1 100	1 100	1 100	1 100	1 100	1 100	1 100	1 110	240
270	1 280	0 —	0 —	0 —	0 —	1 080	1 080	1 090	1 090	1 090	1 080	1 080	1 080	1 080	1 080	270
300	1 310	0 —	0 —	0 —	1 050	1 060	1 070	1 070	1 070	1 070	1 070	1 070	1 070	1 060	1 050	300
330	1 350	1 010	1 030	1 040	1 050	1 060	1 070	1 070	1 070	1 060	1 060	1 060	1 050	1 040	1 020	330
360	1 020	1 030	1 050	1 050	1 060	1 060	1 070	1 070	1 060	1 060	1 050	1 040	1 030	1 010	1 350	360
								1977								
0	1 010	1 030	1 040	1 050	2 060	2 060	2 070	2 070	2 070	2 060	1 050	1 050	1 030	1 010	1 350	0
30	1 040	1 050	1 060	2 070	2 070	2 070	2 070	2 070	2 070	1 060	1 050	1 040	1 010	1 340	1 320	30
60	1 070	1 080	2 080	2 080	2 080	2 080	2 080	2 080	2 080	1 070	1 070	0 —	0 —	1 300	1 290	60
90	1 100	1 100	2 100	2 100	2 090	2 090	2 090	2 090	2 100	1 100	1 100	0 —	0 —	1 250	1 260	90
120	1 130	1 120	1 110	2 110	2 110	2 110	2 110	2 110	2 110	1 120	1 130	1 140	1 180	1 210	1 230	120
150	1 160	1 140	1 130	2 120	2 120	2 120	2 110	2 110	2 120	2 120	1 130	1 140	1 150	1 180	1 200	150
180	1 190	1 170	1 150	1 140	1 130	2 120	2 120	2 110	2 110	2 120	2 120	1 130	1 140	1 150	1 170	180
210	1 220	1 200	1 170	1 140	1 130	1 120	2 110	2 110	2 110	2 110	2 110	2 110	1 120	1 130	1 140	210
240	1 250	1 240	0 —	0 —	1 120	1 110	2 100	2 100	2 100	2 100	2 100	2 100	2 100	1 110	1 110	240
270	1 280	1 290	0 —	0 —	1 080	1 080	2 080	2 090	2 090	2 090	2 090	2 090	2 080	1 080	1 080	270
300	1 310	1 330	1 010	1 040	1 050	1 060	2 070	2 070	2 080	2 070	2 070	2 070	1 070	1 060	1 050	300
330	1 340	1 000	1 030	1 040	1 050	2 060	2 070	2 070	2 070	2 060	2 060	2 060	1 050	1 040	1 020	330
360	1 010	1 030	1 040	1 050	2 060	2 060	2 070	2 070	2 070	2 060	1 050	1 050	1 030	1 010	1 350	360
								1978								
0	1 010	1 030	2 040	2 050	2 060	2 060	3 070	3 070	3 070	2 060	2 050	2 050	1 030	1 020	1 350	0
30	1 040	2 050	2 060	2 070	2 070	3 070	3 070	3 070	2 070	2 060	2 050	1 040	1 010	1 350	1 320	30
60	1 070	2 070	2 080	2 080	3 080	3 080	3 080	3 080	2 080	1 070	1 060	1 040	0 —	1 310	1 290	60
90	1 100	2 100	2 100	2 090	3 090	3 090	3 090	3 090	2 090	1 100	1 100	0 —	0 —	1 260	1 260	90
120	1 130	2 120	2 110	2 110	3 110	3 110	3 100	3 110	2 110	2 120	1 130	1 140	1 180	1 210	1 230	120
150	1 160	1 140	2 130	2 120	2 120	3 120	3 110	3 110	3 120	2 120	2 130	1 140	1 160	1 180	1 200	150
180	1 190	1 170	1 150	2 130	2 130	2 120	3 120	3 110	3 110	2 120	2 120	2 130	2 140	1 150	1 170	180
210	1 220	1 200	1 170	1 140	2 130	2 120	2 110	3 110	3 110	3 110	2 110	2 120	2 120	2 130	1 140	210
240	1 250	1 230	0 —	1 140	1 120	1 110	2 100	3 100	3 100	3 100	3 100	2 100	2 100	2 110	1 110	240
270	1 280	1 280	0 —	0 —	1 080	1 080	2 090	3 090	3 090	3 090	3 090	2 090	2 090	2 080	1 080	270
300	1 310	1 330	1 000	1 040	1 060	2 060	2 070	3 080	3 080	3 080	3 070	2 070	2 070	2 060	1 050	300
330	1 340	1 000	1 030	1 040	2 050	2 060	3 070	3 070	3 070	3 070	2 060	2 060	2 050	1 040	1 020	330
360	1 010	1 030	2 040	2 050	2 060	2 060	3 070	3 070	3 070	2 060	2 050	2 050	1 030	1 020	1 350	360
								1979								
0	1 010	2 030	2 040	2 050	3 060	3 060	4 070	4 070	3 070	3 060	3 050	2 050	2 040	2 020	1 000	0
30	2 040	2 050	2 060	3 060	3 070	3 070	4 070	4 070	3 070	2 060	2 050	2 040	1 020	1 350	1 330	30
60	2 070	2 070	2 080	3 080	3 080	4 080	4 080	3 080	3 080	2 070	1 060	1 040	1 350	1 310	1 300	60
90	2 100	2 090	3 090	3 090	3 090	4 090	4 090	3 090	3 090	2 100	1 100	0 —	0 —	1 260	1 260	90
120	2 130	2 120	2 110	3 110	3 110	3 100	4 100	3 100	3 110	2 120	2 130	1 140	1 180	1 210	1 230	120
150	2 150	2 140	2 130	3 120	3 120	3 110	4 110	4 110	3 110	3 120	2 130	2 140	1 160	1 180	1 200	150
180	1 180	2 160	2 150	2 130	3 130	3 120	3 120	4 110	4 110	3 120	3 120	2 130	2 140	2 150	1 170	180
210	1 210	1 190	1 160	2 140	2 130	2 120	3 110	4 110	4 110	3 110	3 110	3 120	2 120	2 130	2 140	210
240	1 250	1 230	1 190	1 140	1 120	2 110	3 100	3 100	4 100	4 100	3 100	3 100	2 100	2 110	2 110	240
270	1 280	1 280	0 —	0 —	1 080	2 090	3 090	3 090	4 090	4 090	3 090	3 090	3 090	2 090	2 080	270
300	1 310	1 330	1 000	1 040	2 060	2 070	3 070	3 080	4 080	3 080	3 070	3 070	2 070	2 060	2 060	300
300	1 340	1 000	1 020	2 040	2 050	3 060	3 070	4 070	4 070	3 070	3 060	3 060	2 050	2 040	2 030	330
360	1 010	2 030	2 040	2 050	3 060	3 060	4 070	4 070	3 070	3 060	3 050	2 050	2 040	2 020	1 000	360

LAT 41°

N. Lat. {LHA greater than 180°........ Zn=Z; LHA less than 180°........ Zn=360−Z}

DECLINATION (15°-29°) CONTRARY NAME TO LATITUDE

LHA	15° Hc	15° d	15° Z	16° Hc	16° d	16° Z	17° Hc	17° d	17° Z	18° Hc	18° d	18° Z	19° Hc	19° d	19° Z
	° ′	′	°	° ′	′	°	° ′	′	°	° ′	′	°	° ′	′	°
69	05 15	43	115	04 32	42	116	03 50	43	117	03 07	42	117	02 25	43	118
68	05 56	43	116	05 13	43	117	04 30	42	117	03 48	43	118	03 05	43	119
67	06 36	43	117	05 53	42	117	05 11	44	118	04 27	43	119	03 44	43	119
66	07 17	43	117	06 34	44	118	05 50	43	119	05 07	43	119	04 24	44	120
65	07 57	−43	118	07 14	−44	119	06 30	−43	119	05 47	−44	120	05 03	−44	121
64	08 37	44	119	07 53	44	119	07 09	43	120	06 26	44	121	05 42	44	121
63	09 16	43	119	08 33	44	120	07 49	45	121	07 04	44	121	06 20	44	122
62	09 56	44	120	09 12	45	121	08 27	44	121	07 43	45	122	06 58	44	123
61	10 35	45	121	09 50	44	121	09 06	45	122	08 21	45	123	07 36	44	124
60	11 14	−45	122	10 29	−45	122	09 44	−45	123	08 59	−45	124	08 14	−45	124
59	11 52	45	122	11 07	45	123	10 22	45	124	09 37	46	124	08 51	45	125
58	12 30	45	123	11 45	46	124	10 59	45	124	10 14	46	125	09 28	45	126
57	13 08	46	124	12 22	45	124	11 37	46	125	10 51	46	126	10 05	46	126
56	13 46	46	125	13 00	46	125	12 14	47	126	11 27	46	126	10 41	46	127
55	14 23	−46	125	13 37	−47	126	12 50	−46	127	12 04	−47	127	11 17	−46	128
54	15 00	47	126	14 13	47	127	13 26	46	127	12 40	47	128	11 53	47	129
53	15 36	47	127	14 49	47	127	14 02	47	128	13 15	47	129	12 28	47	129
52	16 12	47	128	15 25	47	128	14 38	48	129	13 50	47	130	13 03	48	130
51	16 48	48	128	16 00	47	129	15 13	48	130	14 25	48	130	13 37	48	131
50	17 23	−48	129	16 35	−48	130	15 47	−48	130	14 59	−48	131	14 11	−48	132
49	17 58	48	130	17 10	48	131	16 22	49	131	15 33	48	132	14 45	49	132
48	18 33	49	131	17 44	48	131	16 56	49	132	16 07	49	133	15 18	49	133
47	19 07	49	132	18 18	49	132	17 29	49	133	16 40	49	133	15 51	49	134
46	19 40	49	132	18 51	49	133	18 02	49	134	17 13	50	134	16 23	49	135
45	20 13	−49	133	19 24	−50	134	18 34	−49	135	17 45	−50	135	16 55	−50	136
44	20 46	50	134	19 56	49	135	19 07	50	135	18 17	50	136	17 27	51	137
43	21 18	50	135	20 28	50	136	19 38	50	136	18 48	51	137	17 57	50	137
42	21 50	50	136	21 00	51	137	20 09	50	137	19 19	51	138	18 28	51	138
41	22 21	50	137	21 31	51	137	20 40	51	138	19 49	51	139	18 58	51	139
40	22 52	−51	138	22 01	−51	138	21 10	−51	139	20 19	−52	139	19 27	−51	140
39	23 22	51	139	22 31	51	139	21 40	52	140	20 48	52	140	19 56	52	141
38	23 52	52	139	23 00	51	140	22 09	52	141	21 17	52	141	20 25	53	142
37	24 21	52	140	23 29	52	141	22 37	52	141	21 45	53	142	20 52	52	143
36	24 50	52	141	23 58	53	142	23 05	53	142	22 12	52	143	21 20	53	143
35	25 18	−53	142	24 25	−53	143	23 32	−52	143	22 40	−54	144	21 46	−53	144
34	25 45	53	143	24 52	53	144	23 59	53	144	23 06	53	145	22 13	54	145
33	26 12	53	144	25 19	54	145	24 25	53	145	23 32	54	146	22 38	54	146
32	26 39	54	145	25 45	54	146	24 51	54	146	23 57	54	147	23 03	54	147
31	27 04	54	146	26 10	54	147	25 16	54	147	24 22	54	148	23 28	55	148
30	27 29	−54	147	26 35	−55	148	25 40	−54	148	24 46	−55	148	23 51	−54	149
29	27 54	55	148	26 59	55	149	26 04	55	149	25 09	55	149	24 14	55	150
28	28 17	55	149	27 22	55	150	26 27	55	150	25 32	55	150	24 37	56	151
27	28 40	55	150	27 45	55	151	26 50	56	151	25 54	55	151	24 59	56	152
26	29 02	55	151	28 07	56	152	27 11	56	152	26 15	55	152	25 20	56	153
25	29 24	−56	152	28 28	−56	153	27 32	−56	153	26 36	−56	153	25 40	−56	154
24	29 45	56	153	28 49	57	154	27 52	56	154	26 56	56	154	26 00	57	155
23	30 05	56	154	29 09	57	155	28 12	57	155	27 15	56	155	26 19	57	156
22	30 24	56	155	29 28	57	156	28 31	57	156	27 34	57	156	26 37	57	157
21	30 43	57	156	29 46	57	157	28 49	57	157	27 52	57	157	26 55	57	158
20	31 01	−57	157	30 04	−58	158	29 06	−57	158	28 09	−57	158	27 12	−58	159
19	31 18	58	158	30 20	57	159	29 23	58	159	28 25	57	159	27 28	58	160
18	31 34	58	160	30 36	57	160	29 39	58	160	28 41	58	160	27 43	58	161
17	31 50	58	161	30 52	58	161	29 54	58	161	28 56	58	162	27 58	59	162
16	32 04	58	162	31 06	58	162	30 08	58	162	29 10	59	163	28 11	58	163
15	32 18	−58	163	31 20	−59	163	30 21	−58	163	29 23	−59	164	28 24	−58	164
14	32 31	59	164	31 32	58	164	30 34	59	164	29 35	59	165	28 36	58	165
13	32 43	59	165	31 44	58	165	30 46	59	166	29 47	59	166	28 48	59	166
12	32 54	59	166	31 55	59	166	30 56	59	167	29 57	59	167	28 58	59	167
11	33 05	59	167	32 06	59	168	31 07	60	168	30 07	59	168	29 08	59	168
10	33 14	−59	168	32 15	−59	169	31 16	−60	169	30 16	−59	169	29 17	−59	169
9	33 23	59	170	32 24	60	170	31 24	59	170	30 25	60	170	29 25	59	170
8	33 31	60	171	32 31	59	171	31 32	60	171	30 32	59	171	29 33	60	171
7	33 38	60	172	32 38	60	172	31 38	59	172	30 39	60	172	29 39	60	172
6	33 44	60	173	32 44	60	173	31 44	60	173	30 44	59	173	29 45	60	174
5	33 49	−60	174	32 49	−60	174	31 49	−60	174	30 49	−60	175	29 49	−60	175
4	33 53	60	175	32 53	60	175	31 53	60	176	30 53	60	176	29 53	60	176
3	33 56	60	177	32 56	60	177	31 56	60	177	30 56	60	177	29 56	60	177
2	33 58	60	178	32 58	60	178	31 58	60	178	30 58	60	178	29 58	60	178
1	34 00	60	179	33 00	60	179	32 00	60	179	31 00	60	179	30 00	60	179
0	34 00	−60	180	33 00	−60	180	32 00	−60	180	31 00	−60	180	30 00	−60	180

LHA	20° Hc	20° d	20° Z	21° Hc	21° d	21° Z	22° Hc	22° d	22° Z	23° Hc	23° d	23° Z	24° Hc	24° d	24° Z
	° ′	′	°	° ′	′	°	° ′	′	°	° ′	′	°	° ′	′	°
69	01 42	42	119	01 00	43	119	00 17	42	120	−0 25	43	121	−1 08	43	121
68	02 22	43	119	01 39	43	120	00 56	43	121	00 13	42	121	−0 29	43	122
67	03 01	43	120	02 18	43	121	01 35	43	121	00 52	43	122	00 09	43	123
66	03 40	43	121	02 57	43	121	02 14	44	122	01 30	43	123	00 47	44	123
65	04 19	−43	121	03 36	−44	122	02 52	−44	123	02 08	−44	123	01 24	−43	124
64	04 58	44	122	04 14	44	123	03 30	44	123	02 46	44	124	02 02	44	125
63	05 36	44	123	04 52	45	123	04 07	44	124	03 23	44	125	02 39	45	125
62	06 14	45	123	05 29	44	124	04 45	45	125	04 00	44	125	03 16	45	126
61	06 52	45	124	06 07	45	125	05 22	45	126	04 37	45	126	03 52	45	127
60	07 29	−45	125	06 44	−45	126	05 59	−46	126	05 13	−45	127	04 28	−45	127
59	08 06	45	126	07 21	46	126	06 35	46	127	05 49	45	128	05 04	46	128
58	08 43	46	126	07 57	46	127	07 11	46	128	06 25	46	128	05 39	46	129
57	09 19	46	127	08 33	46	128	07 47	46	128	07 01	47	129	06 14	46	130
56	09 55	46	128	09 09	47	128	08 22	46	129	07 36	47	130	06 49	47	130
55	10 31	−47	129	09 44	−47	129	08 57	−47	130	08 10	−47	130	07 23	−47	131
54	11 06	47	129	10 19	47	130	09 32	47	131	08 45	48	131	07 57	47	132
53	11 41	48	130	10 53	47	131	10 06	47	131	09 19	48	132	08 31	48	133
52	12 15	47	131	11 28	48	131	10 40	48	132	09 52	48	133	09 04	48	133
51	12 49	48	132	12 01	48	132	11 13	48	133	10 25	48	133	09 37	48	134
50	13 23	−48	132	12 35	−49	133	11 46	−48	134	10 58	−49	134	10 09	−48	135
49	13 56	48	133	13 08	49	134	12 19	49	134	11 30	49	135	10 41	49	135
48	14 29	49	134	13 40	49	134	12 51	49	135	12 02	49	136	11 13	49	136
47	15 02	50	135	14 12	49	135	13 23	49	136	12 34	50	136	11 44	50	137
46	15 34	50	135	14 44	50	136	13 54	49	137	13 05	50	137	12 15	50	138
45	16 05	−50	136	15 15	−50	137	14 25	−50	137	13 35	−50	138	12 45	−50	139
44	16 36	50	137	15 46	50	138	14 56	51	138	14 05	50	139	13 15	51	139
43	17 07	51	138	16 16	50	139	15 26	51	139	14 35	51	140	13 44	51	140
42	17 37	51	139	16 46	51	139	15 55	51	140	15 04	51	140	14 13	51	141
41	18 07	52	140	17 15	51	140	16 24	51	141	15 33	52	141	14 41	51	142
40	18 36	−52	140	17 44	−51	141	16 53	−52	142	16 01	−52	142	15 09	−52	143
39	19 04	52	141	18 12	52	142	17 20	52	142	16 28	52	143	15 36	52	143
38	19 32	52	142	18 40	52	143	17 48	53	143	16 55	52	144	16 03	53	144
37	20 00	53	143	19 07	52	144	18 15	53	144	17 22	53	145	16 29	53	145
36	20 27	53	144	19 34	53	144	18 41	53	145	17 48	53	145	16 55	53	146
35	20 53	−53	145	20 00	−53	145	19 07	−54	146	18 13	−53	146	17 20	−53	147
34	21 19	53	146	20 26	54	146	19 32	54	147	18 38	53	147	17 45	54	148
33	21 44	53	147	20 51	54	147	19 57	54	148	19 03	54	148	18 09	55	148
32	22 09	54	148	21 15	54	148	20 21	55	148	19 26	54	149	18 32	54	149
31	22 33	54	148	21 39	55	149	20 44	54	149	19 50	55	150	18 55	55	150
30	22 57	−55	149	22 02	−55	150	21 07	−55	150	20 12	−55	151	19 17	−55	151
29	23 19	55	150	22 24	55	151	21 29	55	151	20 34	55	152	19 39	56	152
28	23 41	55	151	22 46	55	152	21 51	56	152	20 55	55	152	20 00	56	153
27	24 03	56	152	23 07	55	153	22 12	56	153	21 16	56	153	20 20	56	154
26	24 24	56	153	23 28	56	154	22 32	56	154	21 36	56	154	20 40	56	155
25	24 44	−56	154	23 48	−57	155	22 51	−56	155	21 55	−56	155	20 59	−57	156
24	25 03	56	155	24 07	57	155	23 10	56	156	22 14	57	156	21 17	57	157
23	25 22	57	156	24 25	56	156	23 29	57	157	22 32	57	157	21 35	57	157
22	25 40	57	157	24 43	57	157	23 46	57	158	22 49	57	158	21 52	57	158
21	25 58	58	158	25 00	57	158	24 03	57	159	23 06	58	159	22 08	57	159
20	26 14	−57	159	25 17	−58	159	24 19	−57	160	23 22	−58	160	22 24	−58	160
19	26 30	58	160	25 32	58	160	24 34	57	161	23 37	58	161	22 39	58	161
18	26 45	59	161	25 47	58	161	24 49	58	162	23 51	58	162	22 53	58	162
17	26 59	58	162	26 01	58	162	25 03	58	163	24 05	58	163	23 07	59	163
16	27 13	58	163	26 15	59	163	25 16	58	164	24 18	59	164	23 19	58	164
15	27 26	−59	164	26 27	−58	164	25 29	−59	165	24 30	−59	165	23 31	−58	165
14	27 38	59	165	26 39	59	165	25 40	58	166	24 42	59	166	23 43	59	166
13	27 49	59	166	26 50	59	166	25 51	59	167	24 52	59	167	23 53	59	167
12	27 59	59	167	27 00	59	167	26 01	59	168	25 02	59	168	24 03	59	168
11	28 09	59	168	27 10	59	169	26 11	60	169	25 11	59	169	24 12	59	169
10	28 18	−59	169	27 19	−60	170	26 19	−59	170	25 20	−60	170	24 20	−59	170
9	28 26	60	170	27 26	59	171	26 27	60	171	25 27	59	171	24 28	60	171
8	28 33	60	171	27 33	59	172	26 34	60	172	25 34	59	172	24 35	60	172
7	28 39	59	173	27 40	60	173	26 40	60	173	25 40	59	173	24 41	60	173
6	28 45	60	174	27 45	60	174	26 45	59	174	25 46	60	174	24 46	60	174
5	28 49	−59	175	27 50	−60	175	26 50	−60	175	25 50	−60	175	24 50	−60	175
4	28 53	60	176	27 53	60	176	26 53	59	176	25 54	60	176	24 54	60	176
3	28 56	60	177	27 56	60	177	26 56	60	177	25 56	60	177	24 56	59	177
2	28 58	60	178	27 58	60	178	26 58	60	178	25 58	60	178	24 58	60	178
1	29 00	60	179	28 00	60	179	27 00	60	179	26 00	60	179	25 00	60	179
0	29 00	−60	180	28 00	−60	180	27 00	−60	180	26 00	−60	180	25 00	−60	180

LHA	25° Hc	25° d	25° Z	26° Hc	26° d	26° Z	27° Hc	27° d	27° Z	28° Hc	28° d	28° Z	29° Hc	29° d	29° Z	LHA
	° ′	′	°	° ′	′	°	° ′	′	°	° ′	′	°	° ′	′	°	
69	−1 51	42	122	−2 33	43	123	−3 16	42	124	−3 58	43	124	−4 41	42	125	291
68	−1 12	43	123	−1 55	43	124	−2 38	43	124	−3 21	43	125	−4 04	42	126	292
67	−0 34	44	123	−1 18	43	124	−2 01	43	125	−2 44	43	126	−3 27	43	126	293
66	00 03	43	124	−0 40	44	125	−1 24	43	125	−2 07	44	126	−2 51	43	127	294
65	00 41	−44	125	−0 03	−44	125	−0 47	−44	126	−1 31	−43	127	−2 14	−44	127	295
64	01 18	44	125	00 34	45	126	−0 11	44	127	−0 55	44	127	−1 39	44	128	296
63	01 54	44	126	01 10	44	127	00 26	45	127	−0 19	44	128	−1 03	45	129	297
62	02 31	45	127	01 46	45	127	01 01	44	128	00 17	45	129	−0 28	45	129	298
61	03 07	45	127	02 22	45	128	01 37	45	129	00 52	45	129	00 07	45	130	299
60	03 43	−46	128	02 57	−45	129	02 12	−45	129	01 27	−46	130	00 41	−45	131	300
59	04 18	46	129	03 32	45	129	02 47	46	130	02 01	46	131	01 15	45	131	301
58	04 53	46	130	04 07	46	130	03 21	46	131	02 35	46	131	01 49	46	132	302
57	05 28	46	130	04 42	47	131	03 55	46	132	03 09	46	132	02 23	47	133	303
56	06 02	46	131	05 16	47	132	04 29	47	132	03 42	46	133	02 56	47	133	304
55	06 36	−47	132	05 49	−47	132	05 02	−47	133	04 15	−47	134	03 28	−47	134	305
54	07 10	47	132	06 23	48	133	05 35	47	134	04 48	47	134	04 01	48	135	306
53	07 43	47	133	06 56	48	134	06 08	48	134	05 20	47	135	04 33	48	136	307
52	08 16	48	134	07 28	48	134	06 40	48	135	05 52	48	136	05 04	48	136	308
51	08 49	49	135	08 00	48	135	07 12	48	136	06 24	49	136	05 35	48	137	309
50	09 21	−49	135	08 32	−49	136	07 43	−48	137	06 55	−49	137	06 06	−49	138	310
49	09 52	48	136	09 04	50	137	08 14	49	137	07 25	49	138	06 36	49	138	311
48	10 24	50	137	09 34	49	137	08 45	49	138	07 56	50	139	07 06	49	139	312
47	10 54	49	138	10 05	50	138	09 15	50	139	08 25	49	139	07 36	50	140	313
46	11 25	50	138	10 35	50	139	09 45	50	139	08 55	50	140	08 05	51	141	314
45	11 55	−51	139	11 04	−50	140	10 14	−50	140	09 24	−51	141	08 33	−50	141	315
44	12 24	51	140	11 33	50	140	10 43	51	141	09 52	51	142	09 01	51	142	316
43	12 53	51	141	12 02	51	141	11 11	51	142	10 20	51	142	09 29	51	143	317
42	13 22	52	141	12 30	51	142	11 39	52	143	10 47	51	143	09 56	52	144	318
41	13 50	52	142	12 58	52	143	12 06	52	143	11 14	51	144	10 23	52	144	319
40	14 17	−52	143	13 25	−52	144	12 33	−52	144	11 41	−52	145	10 49	−52	145	320
39	14 44	52	144	13 52	53	144	12 59	52	145	12 07	53	145	11 14	52	146	321
38	15 10	52	145	14 18	53	145	13 25	53	146	12 32	52	146	11 40	53	147	322
37	15 36	53	146	14 43	53	146	13 50	53	147	12 57	53	147	12 04	53	147	323
36	16 02	54	146	15 08	53	147	14 15	53	147	13 22	54	148	12 28	53	148	324
35	16 27	−54	147	15 33	−54	148	14 39	−53	148	13 46	−54	149	12 52	−54	149	325
34	16 51	54	148	15 57	54	149	15 03	54	149	14 09	54	149	13 15	54	150	326
33	17 14	54	149	16 20	54	149	15 26	54	150	14 32	55	150	13 37	54	151	327
32	17 38	55	150	16 43	54	150	15 49	55	151	14 54	55	151	13 59	54	152	328
31	18 00	55	151	17 05	54	151	16 11	55	152	15 16	55	152	14 21	55	152	329
30	18 22	−55	152	17 27	−55	152	16 32	−55	152	15 37	−56	153	14 41	−55	153	330
29	18 43	55	152	17 48	55	153	16 53	56	153	15 57	55	154	15 02	56	154	331
28	19 04	56	153	18 08	55	154	17 13	56	154	16 17	56	154	15 21	56	155	332
27	19 24	56	154	18 28	56	155	17 32	56	155	16 36	56	155	15 40	56	156	333
26	19 44	57	155	18 47	56	155	17 51	56	156	16 55	56	156	15 59	57	157	334
25	20 02	−56	156	19 06	−57	156	18 09	−56	157	17 13	−57	157	16 16	−56	157	335
24	20 20	56	157	19 24	57	157	18 27	57	158	17 30	57	158	16 33	56	158	336
23	20 38	57	158	19 41	57	158	18 44	57	158	17 47	57	159	16 50	57	159	337
22	20 55	57	159	19 58	58	159	19 00	57	159	18 03	57	160	17 06	58	160	338
21	21 11	58	160	20 13	57	160	19 16	58	160	18 18	57	161	17 21	58	161	339
20	21 26	−57	161	20 29	−58	161	19 31	−58	161	18 33	−58	161	17 35	−57	162	340
19	21 41	58	162	20 43	58	162	19 45	58	162	18 47	58	162	17 49	58	163	341
18	21 55	58	162	20 57	58	163	19 59	58	163	19 01	58	163	18 03	59	164	342
17	22 08	58	163	21 10	58	164	20 12	59	164	19 13	58	164	18 15	58	164	343
16	22 21	59	164	21 22	58	165	20 24	59	165	19 25	58	165	18 27	59	165	344
15	22 33	−59	165	21 34	−58	166	20 36	−59	166	19 37	−59	166	18 38	−59	166	345
14	22 44	59	166	21 45	59	167	20 46	59	167	19 47	58	167	18 49	59	167	346
13	22 54	59	167	21 55	59	167	20 56	59	168	19 57	59	168	18 58	59	168	347
12	23 04	59	168	22 05	59	168	21 06	59	169	20 07	60	169	19 07	59	169	348
11	23 13	59	169	22 14	60	169	21 14	59	170	20 15	59	170	19 16	60	170	349
10	23 21	−59	170	22 22	−60	170	21 22	−59	170	20 23	−60	171	19 23	−59	171	350
9	23 28	59	171	22 29	60	171	21 29	59	171	20 30	60	172	19 30	59	172	351
8	23 35	60	172	22 35	59	172	21 36	60	172	20 36	59	173	19 37	60	173	352
7	23 41	60	173	22 41	60	173	21 41	59	173	20 42	60	173	19 42	60	174	353
6	23 46	60	174	22 46	60	174	21 46	59	174	20 47	60	174	19 47	60	174	354
5	23 50	−60	175	22 50	−59	175	21 51	−60	175	20 51	−60	175	19 51	−60	175	355
4	23 54	60	176	22 54	60	176	21 54	60	176	20 54	60	176	19 54	60	176	356
3	23 57	60	177	22 57	60	177	21 57	60	177	20 57	60	177	19 57	60	177	357
2	23 58	59	178	22 59	60	178	21 59	60	178	20 59	60	178	19 59	60	178	358
1	24 00	60	179	23 00	60	179	22 00	60	179	21 00	60	179	20 00	60	179	359
0	24 00	−60	180	23 00	−60	180	22 00	−60	180	21 00	−60	180	20 00	−60	180	360

S. Lat. {LHA greater than 180°........ Zn=180−Z; LHA less than 180°........ Zn=180+Z}

DECLINATION (15°-29°) CONTRARY NAME TO LATITUDE

LAT 41°

TABLE 5.—Correction to Tabulated Altitude for Minutes of Declination

d / ′	1	2	3	4	5	6	7	8	9	10	11	12	13	14	15	16	17	18	19	20	21	22	23	24	25	26	27	28	29	30	31	32	33	34	35	36	37	38	39	40	41	42	43	44	45	46	47	48	49	50	51	52	53	54	55	56	57	58	59	60	d / ′
0	0	0	0	0	0	0	0	0	0	0	0	0	0	0	0	0	0	0	0	0	0	0	0	0	0	0	0	0	0	0	0	0	0	0	0	0	0	0	0	0	0	0	0	0	0	0	0	0	0	0	0	0	0	0	0	0	0	0	0	0	0
1	0	0	0	0	0	0	0	0	0	0	0	0	0	0	0	0	0	0	0	0	0	0	0	0	0	0	0	0	0	0	1	1	1	1	1	1	1	1	1	1	1	1	1	1	1	1	1	1	1	1	1	1	1	1	1	1	1	1	1	1	1
2	0	0	0	0	0	0	0	0	0	0	0	0	0	0	0	1	1	1	1	1	1	1	1	1	1	1	1	1	1	1	1	1	1	1	1	1	1	1	1	1	1	1	1	1	2	2	2	2	2	2	2	2	2	2	2	2	2	2	2	2	2
3	0	0	0	0	0	0	0	0	0	0	1	1	1	1	1	1	1	1	1	1	1	1	1	1	1	1	1	1	1	2	2	2	2	2	2	2	2	2	2	2	2	2	2	2	2	2	2	2	2	2	3	3	3	3	3	3	3	3	3	3	3
4	0	0	0	0	0	0	0	1	1	1	1	1	1	1	1	1	1	1	1	1	1	1	2	2	2	2	2	2	2	2	2	2	2	2	2	2	2	3	3	3	3	3	3	3	3	3	3	3	3	3	3	3	4	4	4	4	4	4	4	4	4
5	0	0	0	0	0	0	1	1	1	1	1	1	1	1	1	1	1	2	2	2	2	2	2	2	2	2	2	2	2	2	3	3	3	3	3	3	3	3	3	3	3	4	4	4	4	4	4	4	4	4	4	4	4	4	5	5	5	5	5	5	5
6	0	0	0	0	0	1	1	1	1	1	1	1	1	1	2	2	2	2	2	2	2	2	2	2	2	3	3	3	3	3	3	3	3	3	4	4	4	4	4	4	4	4	4	4	4	5	5	5	5	5	5	5	5	5	6	6	6	6	6	6	6
7	0	0	0	0	1	1	1	1	1	1	1	1	2	2	2	2	2	2	2	2	2	3	3	3	3	3	3	3	3	4	4	4	4	4	4	4	4	4	5	5	5	5	5	5	5	5	5	6	6	6	6	6	6	6	6	7	7	7	7	7	7
8	0	0	0	1	1	1	1	1	1	1	1	2	2	2	2	2	2	2	3	3	3	3	3	3	3	3	4	4	4	4	4	4	4	5	5	5	5	5	5	5	5	6	6	6	6	6	6	6	7	7	7	7	7	7	7	7	8	8	8	8	8
9	0	0	0	1	1	1	1	1	1	2	2	2	2	2	2	2	3	3	3	3	3	3	3	4	4	4	4	4	4	4	5	5	5	5	5	5	6	6	6	6	6	6	6	7	7	7	7	7	7	8	8	8	8	8	8	8	9	9	9	9	9
10	0	0	0	1	1	1	1	1	2	2	2	2	2	2	2	3	3	3	3	3	4	4	4	4	4	4	4	5	5	5	5	5	6	6	6	6	6	6	6	7	7	7	7	7	8	8	8	8	8	8	8	9	9	9	9	9	10	10	10	10	10
11	0	0	1	1	1	1	1	1	2	2	2	2	2	3	3	3	3	3	3	4	4	4	4	4	5	5	5	5	5	6	6	6	6	6	6	7	7	7	7	7	8	8	8	8	8	8	9	9	9	9	9	10	10	10	10	10	10	11	11	11	11
12	0	0	1	1	1	1	1	2	2	2	2	2	3	3	3	3	3	4	4	4	4	4	5	5	5	5	5	6	6	6	6	6	7	7	7	7	7	8	8	8	8	8	9	9	9	9	9	10	10	10	10	10	11	11	11	11	11	12	12	12	12
13	0	0	1	1	1	1	2	2	2	2	2	3	3	3	3	3	4	4	4	4	5	5	5	5	5	6	6	6	6	6	7	7	7	7	8	8	8	8	8	9	9	9	9	10	10	10	10	10	11	11	11	11	11	12	12	12	12	13	13	13	13
14	0	0	1	1	1	1	2	2	2	2	3	3	3	3	4	4	4	4	4	5	5	5	5	6	6	6	6	7	7	7	7	7	8	8	8	8	9	9	9	9	10	10	10	10	10	11	11	11	11	12	12	12	12	13	13	13	13	14	14	14	14
15	0	0	1	1	1	2	2	2	2	2	3	3	3	4	4	4	4	4	5	5	5	6	6	6	6	6	7	7	7	8	8	8	8	8	9	9	9	10	10	10	10	10	11	11	11	12	12	12	12	12	13	13	13	14	14	14	14	14	15	15	15
16	0	1	1	1	1	2	2	2	2	3	3	3	3	4	4	4	5	5	5	5	6	6	6	6	7	7	7	7	8	8	8	9	9	9	9	10	10	10	10	11	11	11	11	12	12	12	13	13	13	13	14	14	14	14	15	15	15	15	16	16	16
17	0	1	1	1	1	2	2	2	3	3	3	3	4	4	4	5	5	5	5	6	6	6	7	7	7	7	8	8	8	8	9	9	9	10	10	10	10	11	11	11	12	12	12	12	13	13	13	14	14	14	14	15	15	15	16	16	16	16	17	17	17
18	0	1	1	1	2	2	2	2	3	3	3	4	4	4	4	5	5	5	6	6	6	7	7	7	8	8	8	8	9	9	9	10	10	10	10	11	11	11	12	12	12	13	13	13	14	14	14	14	15	15	15	16	16	16	16	17	17	17	18	18	18
19	0	1	1	1	2	2	2	3	3	3	3	4	4	4	5	5	5	6	6	6	7	7	7	8	8	8	9	9	9	10	10	10	10	11	11	11	12	12	12	13	13	13	14	14	14	15	15	15	16	16	16	16	17	17	17	18	18	18	19	19	19
20	0	1	1	1	2	2	2	3	3	3	4	4	4	5	5	5	6	6	6	7	7	7	8	8	8	9	9	9	10	10	10	11	11	11	12	12	12	13	13	13	14	14	14	15	15	15	16	16	16	17	17	17	18	18	18	19	19	19	20	20	20
21	0	1	1	1	2	2	2	3	3	4	4	4	5	5	5	6	6	6	7	7	7	8	8	8	9	9	9	10	10	10	11	11	12	12	12	13	13	13	14	14	14	15	15	15	16	16	16	17	17	18	18	18	19	19	19	20	20	20	21	21	21
22	0	1	1	1	2	2	3	3	3	4	4	4	5	5	6	6	6	7	7	7	8	8	8	9	9	10	10	10	11	11	11	12	12	12	13	13	14	14	14	15	15	15	16	16	16	17	17	18	18	18	19	19	19	20	20	21	21	21	22	22	22
23	0	1	1	2	2	2	3	3	3	4	4	5	5	5	6	6	7	7	7	8	8	8	9	9	10	10	10	11	11	12	12	12	13	13	13	14	14	15	15	15	16	16	16	17	17	18	18	18	19	19	20	20	20	21	21	21	22	22	23	23	23
24	0	1	1	2	2	2	3	3	4	4	4	5	5	6	6	6	7	7	8	8	8	9	9	10	10	10	11	11	12	12	12	13	13	14	14	14	15	15	16	16	16	17	17	18	18	18	19	19	20	20	20	21	21	22	22	22	23	23	24	24	24
25	0	1	1	2	2	2	3	3	4	4	5	5	5	6	6	7	7	8	8	8	9	9	10	10	10	11	11	12	12	12	13	13	14	14	15	15	15	16	16	17	17	18	18	18	19	19	20	20	20	21	21	22	22	22	23	23	24	24	25	25	25
26	0	1	1	2	2	3	3	3	4	4	5	5	6	6	6	7	7	8	8	9	9	10	10	10	11	11	12	12	13	13	13	14	14	15	15	16	16	16	17	17	18	18	19	19	20	20	20	21	21	22	22	23	23	23	24	24	25	25	26	26	26
27	0	1	1	2	2	3	3	4	4	4	5	5	6	6	7	7	8	8	9	9	9	10	10	11	11	12	12	13	13	14	14	14	15	15	16	16	17	17	18	18	18	19	19	20	20	21	21	22	22	22	23	23	24	24	25	25	26	26	27	27	27
28	0	1	1	2	2	3	3	4	4	5	5	6	6	7	7	7	8	8	9	9	10	10	11	11	12	12	13	13	14	14	14	15	15	16	16	17	17	18	18	19	19	20	20	21	21	21	22	22	23	23	24	24	25	25	26	26	27	27	28	28	28
29	0	1	1	2	2	3	3	4	4	5	5	6	6	7	7	8	8	9	9	10	10	11	11	12	12	13	13	14	14	14	15	15	16	16	17	17	18	18	19	19	20	20	21	21	22	22	23	23	24	24	25	25	26	26	27	27	28	28	29	29	29
30	0	1	2	2	2	3	4	4	4	5	6	6	6	7	8	8	8	9	10	10	10	11	12	12	12	13	14	14	14	15	16	16	16	17	18	18	18	19	20	20	20	21	22	22	22	23	24	24	24	25	26	26	26	27	28	28	28	29	30	30	30
31	1	1	2	2	3	3	4	4	5	5	6	6	7	7	8	8	9	9	10	10	11	11	12	12	13	13	14	14	15	16	16	17	17	18	18	19	19	20	20	21	21	22	22	23	23	24	24	25	25	26	26	27	27	28	28	29	29	30	30	31	31
32	1	1	2	2	3	3	4	4	5	5	6	6	7	7	8	9	9	10	10	11	11	12	12	13	13	14	14	15	15	16	17	17	18	18	19	19	20	20	21	21	22	22	23	23	24	25	25	26	26	27	27	28	28	29	29	30	30	31	31	32	32
33	1	1	2	2	3	3	4	4	5	6	6	7	7	8	8	9	9	10	10	11	12	12	13	13	14	14	15	15	16	16	17	18	18	19	19	20	20	21	21	22	23	23	24	24	25	25	26	26	27	28	28	29	29	30	30	31	31	32	32	33	33
34	1	1	2	2	3	3	4	5	5	6	6	7	7	8	8	9	10	10	11	11	12	12	13	14	14	15	15	16	16	17	18	18	19	19	20	20	21	22	22	23	23	24	24	25	26	26	27	27	28	28	29	29	30	31	31	32	32	33	33	34	34
35	1	1	2	2	3	4	4	5	5	6	6	7	8	8	9	9	10	10	11	12	12	13	13	14	15	15	16	16	17	18	18	19	19	20	20	21	22	22	23	23	24	24	25	26	26	27	27	28	29	29	30	30	31	32	32	33	33	34	34	35	35
36	1	1	2	2	3	4	4	5	5	6	7	7	8	8	9	10	10	11	11	12	13	13	14	14	15	16	16	17	17	18	19	19	20	20	21	22	22	23	23	24	25	25	26	26	27	28	28	29	29	30	31	31	32	32	33	34	34	35	35	36	36
37	1	1	2	2	3	4	4	5	6	6	7	7	8	9	9	10	10	11	12	12	13	14	14	15	15	16	17	17	18	18	19	20	20	21	22	22	23	23	24	25	25	26	27	27	28	28	29	30	30	31	31	32	33	33	34	35	35	36	36	37	37
38	1	1	2	3	3	4	4	5	6	6	7	8	8	9	10	10	11	11	12	13	13	14	15	15	16	16	17	18	18	19	20	20	21	22	22	23	23	24	25	25	26	27	27	28	28	29	30	30	31	32	32	33	34	34	35	35	36	37	37	38	38
39	1	1	2	3	3	4	5	5	6	6	7	8	8	9	10	10	11	12	12	13	14	14	15	16	16	17	18	18	19	20	20	21	21	22	23	23	24	25	25	26	27	27	28	29	29	30	31	31	32	32	33	34	34	35	36	36	37	38	38	39	39
40	1	1	2	3	3	4	5	5	6	7	7	8	9	9	10	11	11	12	13	13	14	15	15	16	17	17	18	19	19	20	21	21	22	23	23	24	25	25	26	27	27	28	29	29	30	31	31	32	33	33	34	35	35	36	37	37	38	39	39	40	40
41	1	1	2	3	3	4	5	5	6	7	8	8	9	10	10	11	12	12	13	14	14	15	16	16	17	18	18	19	20	20	21	22	23	23	24	25	25	26	27	27	28	29	29	30	31	31	32	33	33	34	35	36	36	37	38	38	39	40	40	41	41
42	1	1	2	3	4	4	5	6	6	7	8	8	9	10	10	11	12	13	13	14	15	15	16	17	18	18	19	20	20	21	22	22	23	24	24	25	26	27	27	28	29	29	30	31	32	32	33	34	34	35	36	36	37	38	38	39	40	41	41	42	42
43	1	1	2	3	4	4	5	6	6	7	8	9	9	10	11	11	12	13	14	14	15	16	16	17	18	19	19	20	21	22	22	23	24	24	25	26	27	27	28	29	29	30	31	32	32	33	34	34	35	36	37	37	38	39	39	40	41	42	42	43	43
44	1	1	2	3	4	4	5	6	7	7	8	9	10	10	11	12	12	13	14	15	15	16	17	18	18	19	20	21	21	22	23	23	24	25	26	26	27	28	29	29	30	31	32	32	33	34	34	35	36	37	37	38	39	40	40	41	42	43	43	44	44
45	1	2	2	3	4	4	5	6	7	8	8	9	10	10	11	12	13	14	14	15	16	16	17	18	19	20	20	21	22	22	23	24	25	26	26	27	28	28	29	30	31	32	32	33	34	34	35	36	37	38	38	39	40	40	41	42	43	44	44	45	45
46	1	2	2	3	4	5	5	6	7	8	8	9	10	11	12	12	13	14	15	15	16	17	18	18	19	20	21	21	22	23	24	25	25	26	27	28	28	29	30	31	31	32	33	34	34	35	36	37	38	38	39	40	41	41	42	43	44	44	45	46	46
47	1	2	2	3	4	5	5	6	7	8	9	9	10	11	12	13	13	14	15	16	16	17	18	19	20	20	21	22	23	24	24	25	26	27	27	28	29	30	31	31	32	33	34	34	35	36	37	38	38	39	40	41	42	42	43	44	45	45	46	47	47
48	1	2	2	3	4	5	6	6	7	8	9	10	10	11	12	13	14	14	15	16	17	18	18	19	20	21	22	22	23	24	25	26	26	27	28	29	30	30	31	32	33	34	34	35	36	37	38	38	39	40	41	42	42	43	44	45	46	46	47	48	48
49	1	2	2	3	4	5	6	7	7	8	9	10	11	11	12	13	14	15	16	16	17	18	19	20	20	21	22	23	24	24	25	26	27	28	29	29	30	31	32	33	33	34	35	36	37	38	38	39	40	41	42	42	43	44	45	46	47	47	48	49	49
50	1	2	2	3	4	5	6	7	8	8	9	10	11	12	12	13	14	15	16	17	18	18	19	20	21	22	22	23	24	25	26	27	28	28	29	30	31	32	32	33	34	35	36	37	38	38	39	40	41	42	42	43	44	45	46	47	48	48	49	50	50
51	1	2	3	3	4	5	6	7	8	8	9	10	11	12	13	14	14	15	16	17	18	19	20	20	21	22	23	24	25	26	26	27	28	29	30	31	31	32	33	34	35	36	37	37	38	39	40	41	42	42	43	44	45	46	47	48	48	49	50	51	51
52	1	2	3	3	4	5	6	7	8	9	10	10	11	12	13	14	15	16	16	17	18	19	20	21	22	23	23	24	25	26	27	28	29	29	30	31	32	33	34	35	36	36	37	38	39	40	41	42	42	43	44	45	46	47	48	49	49	50	51	52	52
53	1	2	3	4	4	5	6	7	8	9	10	11	11	12	13	14	15	16	17	18	19	19	20	21	22	23	24	25	26	26	27	28	29	30	31	32	33	34	34	35	36	37	38	39	40	41	42	42	43	44	45	46	47	48	49	49	50	51	52	53	53
54	1	2	3	4	4	5	6	7	8	9	10	11	12	13	14	14	15	16	17	18	19	20	21	22	22	23	24	25	26	27	28	29	30	31	32	32	33	34	35	36	37	38	39	40	40	41	42	43	44	45	46	47	48	49	50	50	51	52	53	54	54
55	1	2	3	4	5	6	6	7	8	9	10	11	12	13	14	15	16	16	17	18	19	20	21	22	23	24	25	26	27	28	28	29	30	31	32	33	34	35	36	37	38	38	39	40	41	42	43	44	45	46	47	48	49	50	50	51	52	53	54	55	55
56	1	2	3	4	5	6	7	7	8	9	10	11	12	13	14	15	16	17	18	19	20	21	21	22	23	24	25	26	27	28	29	30	31	32	33	34	35	35	36	37	38	39	40	41	42	43	44	45	46	47	48	49	49	50	51	52	53	54	55	56	56
57	1	2	3	4	5	6	7	8	9	10	10	11	12	13	14	15	16	17	18	19	20	21	22	23	24	25	26	27	28	28	29	30	31	32	33	34	35	36	37	38	39	40	41	42	43	44	45	46	47	48	48	49	50	51	52	53	54	55	56	57	57
58	1	2	3	4	5	6	7	8	9	10	11	12	13	14	14	15	16	17	18	19	20	21	22	23	24	25	26	27	28	29	30	31	32	33	34	35	36	37	38	39	40	41	42	43	44	44	45	46	47	48	49	50	51	52	53	54	55	56	57	58	58
59	1	2	3	4	5	6	7	8	9	10	11	12	13	14	15	16	17	18	19	20	21	22	23	24	25	26	27	28	29	30	30	31	32	33	34	35	36	37	38	39	40	41	42	43	44	45	46	47	48	49	50	51	52	53	54	55	56	57	58	59	59

APPENDIX Q

NAVIGATIONAL ERRORS

Q1. Introduction.—As commonly practiced, navigation is not an exact science. A number of approximations which would be unacceptable in careful scientific work are used by the navigator, because greater accuracy may not be consistent with the requirements or time available, or because there is no alternative.

Thus, when the navigator uses his latitude graduations as a mile scale, or computes a great-circle course and distance, he neglects the flattening of the earth at the poles, a practice that is not acceptable to the geodetic surveyor. When the navigator plots a visual bearing, or an azimuth line for a celestial line of position, on a Mercator chart, he uses a rhumb line to represent a great circle. When he plots the celestial line of position, he substitutes a rhumb line for a small circle. When he interpolates in sight reduction or lattice tables, he assumes a linear (constant-rate) change between tabulated values. When he measures distance by radar, or depth by echo sounder, he assumes that the radio- or sound-wave has constant speed under all conditions. When he applies dip and refraction corrections to his sextant altitude, he generally assumes standard atmospheric conditions.

These are only a few of the approximations commonly applied by a navigator. There are so many that there is a natural tendency for some of them to cancel others. Thus, under favorable conditions, a position at sea, determined from celestial observation by an experienced observer, should seldom be in error by more than 2 miles. However, if the various small errors in a particular observation all have the same sign (all plus or all minus), the error might be several times this amount, without any mistake having been made by the navigator.

Greater accuracy could be attained, but at a price. The navigator is a practical individual. In the course of ordinary navigation, he would rather spend 10 minutes determining a position having a probable error of plus or minus 2 miles, than to spend several hours learning where he *was* to an accuracy of a few meters. But if he can determine a recent or present position to greater accuracy, the decrease in error is attractive to him. The various navigational aids have been designed with this in mind. Greater accuracy in plotting could be achieved by increasing the scale of the chart or plotting sheet. This has been done for confined waters where a higher degree of accuracy is needed, but a large-scale plotting sheet would be a nuisance at sea. The hand-held marine sextant is not sufficiently accurate for use in determining an astronomical position in a geodetic survey. But it is much more satisfactory at sea than the surveyor's astrolabe or theodolite, which require stable platforms if their potential accuracy is to be realized.

An understanding of the kinds of errors involved in navigation, and of the elementary principles of probability, should be of assistance to a navigator in interpreting his results.

Q2. Definitions.—The following definitions apply to the discussions of this appendix:

Error is the difference between a specific value and the correct or standard value. As used here, it does not include mistakes, but is related to lack of perfection. Thus, an altitude determined by marine sextant is corrected for a standard amount of refraction, but if the actual refraction at the time of observation varies from the standard, the value taken from the table is in error by the difference between standard and actual refraction. This error will be compounded with others in the observed altitude. Similarly, depth determined by echo sounder is in error, among other things, by the difference between the actual speed of sound waves in the water and the speed used for calibration of the instrument. The depth will also be in error if an echo is returned from a phantom bottom instead of from the actual bottom. This appendix is concerned primarily with the deviation from standards. Thus, while variation of the compass is an error when referred to true directions, the difference between the assumed variation and that actually existing is an error with reference to magnetic direction. Corrections can be applied for standard values of error. It is the deviation from standard, as well as mistakes, that produce inaccurate results in navigation. Various kinds of errors are discussed in the following articles.

Mistake is a blunder, such as an incorrect reading of an instrument, the taking of a wrong value from a table, or the plotting of a reciprocal bearing. The mistake is discussed in more detail in article Q10.

Standard is something established by custom, agreement, or authority as a basis for comparison. It is customary to use nautical miles for measuring distances between ports. By international agreement the nautical mile is defined as exactly 1852 meters. By authority of various countries which are parties to the agreement, this length is translated to the linear units adopted by that country. It is the fact of establishment or general acceptance that determines whether a given quantity or condition has become a standard of measure or quality. Thus, in 1960, the standard unit of length agreed upon at the Eleventh General (International) Conference on Weights and Measures to redefine the meter was 1,650,763.73 wavelengths of the orange-red radiation in vacuum of krypton 86 corresponding to the unperturbed transition between the $2p_{10}$ and $5d_5$ levels. Where accepted, this established standard of length now serves as a basis for measurement of any physical magnitude, as the length of the meridian, rather than the reverse, which was originally proposed. Multiples and submultiples of a standard are exact. In 1959, the U.S. adopted the exact relationships of 1 yard as equal to 0.9144 meter and 1 inch as equal to 2.54 centimeters. Hence, 39.37 U.S. inches are approximately equal to 1 meter. Because 1 foot equals 12 inches by definition, and the international nautical mile has been defined as 1852 meters, the international nautical mile is equal to 6,076.11549 U.S. feet (approximately). The previous U.S. foot (6,076.10333 . . . feet equals 1 nautical mile) has been redesignated as the U.S. survey foot.

Frequently, a standard is so chosen that it serves as a model which approximates a mean or average condition. However, the distinction between the standard value and the actual value at any time should not be forgotten. Thus, a standard atmosphere has been established in which the temperature, pressure, density, etc., are *precisely* specified for each altitude. Actual conditions, however, are generally different from those defined by the standard atmosphere. Similarly, the values for dip given in the almanacs are considered standard by those who use them, but actual dip may be appreciably different from that tabulated.

Accuracy is the degree of conformance with the correct value, while **precision** is the degree of refinement of a value. Thus, an altitude determined by marine sextant might be stated to the nearest 0′.1, and yet be accurate only to the nearest 1′ if the horizon is indistinct.

Q3. Systematic errors are those which follow some law by which they can be predicted. The accuracy with which a systematic error can be predicted depends upon the accuracy with which the governing law is understood. An error which can be predicted can be eliminated, or compensation can be made for it.

The simplest form of systematic error is one of unchanging magnitude and sign. This is called a **constant error.** Examples are the index error of a marine sextant, watch error, or the error resulting from a lubber's line not being accurately aligned with the longitudinal axis of the craft. In each of these cases, all readings are in error by a constant amount *as long as the adjustment remains unchanged,* and can be removed by applying a correction of equal magnitude and opposite sign. Index error and watch error can be removed by adjustment of the instrument. Lubber's line error can be removed by aligning the lubber's line with the longitudinal axis of the craft.

Another type of systematic error results from a nonstandard rate. If a watch is gaining 4 seconds per day, its readings will be in error by 1 second after an interval of 6 hours, 8 seconds at the end of 2 days, etc. This principle is used in establishing a chronometer rate (art. 1815) for determination of chronometer error between comparisons of the chronometer with time signals. It can be eliminated by adjusting the rate. If a current is running and no allowance for it is made in the dead reckoning, the DR position is in error by an amount proportional to elapsed time. The error introduced by maintaining heading by means of an inaccurate compass is proportional to distance, as is the lateral error in a line of position plotted from an inaccurate bearing.

One of the causes of equation of time (art. 1809) is the fact that the ecliptic, around which annual motion occurs, is not parallel to the celestial equator, around or parallel to which apparent daily motion takes place. The same type systematic error is involved in other measurements. Consider the measurement of bearing with a tilted compass card. Bearing is measured by a system of uniform graduations (degrees) of a circle (such as a compass card) in the horizontal plane. If the card is tilted, and its graduations are projected onto the horizontal plane, the circle becomes an ellipse with the graduations unequally spaced. Along the axis of tilt and a line perpendicular to it, directions are correct. But near the axis of tilt the graduations are too close together, and near the perpendicular they are too widely spaced. The error thus introduced is similar to that which would arise if a watch face were tilted but the motion of the hands remained horizontal. If it were tilted around the "3–9" line, it would appear to run slow near the hour and half hour, and fast near the quarter and three-quarter hours. If the direction to be observed is of an object above or below the horizontal, as the azimuth of a celestial body, measurement is made to the foot of the perpendicular through the object. The sight vanes of a compass move in a plane perpendicular to the compass card. Hence, if the card is tilted, measurement is made to the foot of a perpendicular to the card, rather than to the foot of a perpendicular to the horizontal, introducing an error which increases with the angle of tilt and also with the angle of elevation (or depression) of the object. This error is greatest along the axis of tilt, and zero along the perpendicular to it. Both of these tilt errors can be corrected by leveling the compass card.

A different type of tilt error occurs when a reflection takes place from a tilted surface, such as the ionosphere, the error being proportional to the angle of tilt. In some respects, this error is similar to coastal refraction of a radio wave.

Additional examples of systematic error are uncorrected deviation of the compass, error due to a position in a pattern of hyperbolas, error due to incorrect location of a Loran transmitter, uncorrected parallax, and uncorrected personal error.

Q4. Random errors are chance errors, unpredictable in magnitude or sign. They are governed by the laws of probability. If the altitude of a celestial body is observed, the reading may be (1) too great, (2) correct, or (3) too small. If a number of observations are made, and there is no systematic error, the probability of a positive error is exactly equal to the probability of a negative error. This does not mean that every second observation having an error will be too great. However, the greater the number of observations, the greater is the probability that the percentage of positive errors will equal the percentage of negative ones, and that their magnitudes will correspond.

Error	No. of obs.	Percent of obs.
−10′	0	0. 0
− 9′	1	0. 2
− 8′	2	0. 4
− 7′	4	0. 8
− 6′	9	1. 8
− 5′	17	3. 4
− 4′	28	5. 6
− 3′	40	8. 0
− 2′	53	10. 6
− 1′	63	12. 6
0	66	13. 2
+ 1′	63	12. 6
+ 2′	53	10. 6
+ 3′	40	8. 0
+ 4′	28	5. 6
+ 5′	17	3. 4
+ 6′	9	1. 8
+ 7′	4	0. 8
+ 8′	2	0. 4
+ 9′	1	0. 2
+10′	0	0. 0
0	500	100. 0

TABLE Q4.—Normal distribution of random errors.

Suppose that 500 observations are made, with the results shown in table Q4. A close approximation of the plot of these errors is shown in figure Q4a. The plot has been modified slightly to constitute the **normal curve** of random errors, which is the same as the actual curve except that the normal curve *approaches* zero as the error increases, while the actual curve *reaches* zero at (+)10′ and (−)10′. The height of the curve at any point represents the percentage of observations that can be expected to have the error indicated at that point. The probability of any similar observation having any given error is the proportion of the number of observations having this error to the total number of observations, or the percentage expressed as a decimal. Thus, the probability of an observation having an error of (−)3′ is

$$\frac{40}{500}=\frac{1}{12.5}=0.08\ (8\%).$$

If the area under the curve represents 100 percent of the observations, half the area (the shaded portion of figure Q4a) represents 50 percent of the observations. The value of the error at the limits of this shaded portion is often called the "50 percent error," or **probable error,** meaning that 50 percent of the observations can be expected to have less error, and 50 percent greater error. Similarly, the limits which contain the central 95 percent of the area denote the 95 percent error. The percentage of error is found mathematically. For a normal curve, each error is squared, the sum of the squares is divided by one less than the number of observations, and the square root of the quotient is determined. This value is called the **standard deviation** or **standard error** (σ, the Greek letter sigma). In the illustration, the standard deviation is the square root of $0\times(-)10^2+1\times(-)9^2+2\times(-)8^2+4\times(-)7^2+9\times(-)6^2$, etc., divided by 499 or

$$\sqrt{\frac{4474}{499}}=\sqrt{8.966}=2.99 \text{ (about 3)}.$$

The standard deviation is the 68.27 percent error. The probability of the occurrence of an error of or less than a specific magnitude may be approximately determined by the following relationship (with the answers for the illustration given):

$$
\begin{aligned}
50\%\ \text{error} &= \tfrac{2}{3}\times\sigma = 2' \text{ (approx.)}\\
68\%\ \text{error} &= 1\ \ \times\sigma = 3' \text{ (approx.)}\\
95\%\ \text{error} &= 2\ \ \times\sigma = 6' \text{ (approx.)}\\
99\%\ \text{error} &= 2\tfrac{2}{3}\times\sigma = 8' \text{ (approx.)}\\
99.9\%\ \text{error} &= 3\tfrac{1}{3}\times\sigma = 10' \text{ (approx.)}
\end{aligned}
$$

Many of the errors of navigation do not follow the normal distribution discussed above. Pub. No. 229 values of altitude can be taken only to the nearest 0.′1. The error in tabular altitude might have any value from (+) 0.′05 to (−) 0.′05, and any value within these limits is as likely to occur as any other of the same precision. The same is true of a sextant that cannot be read more precisely than 0.′1, and of a time-difference that cannot be measured more precisely than 1 μs. These values refer to the single errors indicated, and not to the total error that might be involved. This is a **rectangular error,** so called because of the shape of its plot, as shown in figure Q4b. The 100 percent error is half the difference between readings. The 50 percent error is half this amount, the 95 percent error is 0.95 times this amount, etc. In some cases it may be more meaningful to refer to the rectangular error as the **resolution error.**

Still another type random error is encountered in navigation. If a compass is fluctuating periodically due to yaw of a ship, its motion slows as the end of a swing is approached, when the error approaches maximum value. If readings were taken continuously or at equal intervals of time, the interval being a small percentage of the total period of oscillation, the curve of errors would have a characteristic U-shape, as shown in figure Q4c. The same type error is involved in measurement of altitude of a celestial body from a wing of the bridge of a heavily rolling vessel, when the roll causes large changes in the height of eye. This type of error is called a **periodic error.** The effect is accentuated by the tendency of the observer to make readings near one of the extreme values because the instrument appears steadiest at this time. If it is impractical to make a reading at the center of the period, the error can be eliminated or reduced by averaging readings taken continuously or at short intervals, as indicated above. This is the method used in averaging type artificial-horizon sextants. Generally, better results can be obtained by taking maximum positive and maximum negative readings, and averaging the results.

The curve of any type of random error is symmetrical about the line representing zero error. This means that in the ideal plot every point on one side of the curve is exactly matched by one on the other side, or for every positive error there is a negative error of the same magnitude. The average of all readings, considering signs, is zero. The larger the number of readings, the greater the probability of the errors fitting the ideal curve. Another way of stating this is that as the number of readings increases, the error of the average can be expected to decrease.

Q5. Combinations of errors.—Many of the results obtained in navigation are subject to more than one error. Chapter XVI lists 19 errors applicable to sextant altitudes. Some of these have several components. A number of possible errors are involved in the determination of computed altitude and azimuth. A rectangular error is possible in finding the altitude difference. Several additional errors may affect the accuracy of plotting. Thus, the line of position as finally plotted may include 30 errors or more. Corrections are applied for some of the larger ones, so that in each of these cases the applicable error is the difference between the applied correction and the actual error. Thus, a dip correction may be applied for a height of eye of 30 feet, while the actual height at the moment of observation may be 31 feet 6 inches. Even if the height of eye is exactly 30 feet, a rectangular error may be involved in taking the dip correction from the table.

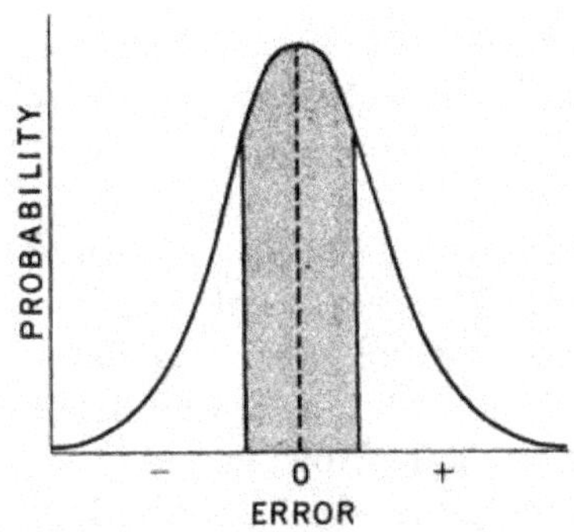

FIGURE Q4a.—Normal curve of random error with 50 percent of area shaded. Limits of shaded area indicate probable error.

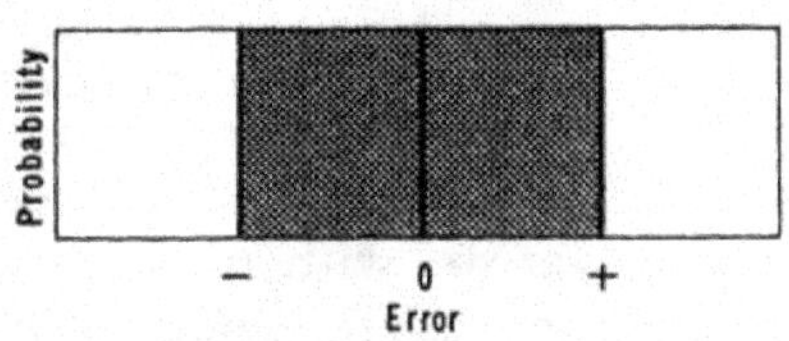

FIGURE Q4b.—Rectangular error, with 50 percent area shaded.

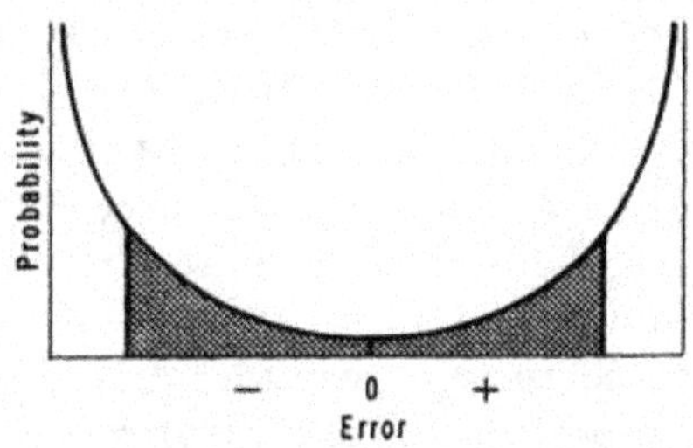

FIGURE Q4c.—Periodic error, with 50 percent area shaded.

Corrections which might be random as far as an individual observation is concerned may be systematic for a series of observations. Thus, if the average or standard conditions upon which a correction is based do not exist at the time of observation the value at any given time is as likely to be greater as it is to be less than the standard amount. But if a number of observations are taken in quick succession, the error will be about the same for each.

If two or more errors are applicable to a given result, the total error is equal to the algebraic sums of all errors. Thus, if a given number is subject to errors of (+) 4, (−) 2, (−) 1, (+) 3, (+) 2, 0, and (−) 2, the total error is (+) 4. Systematic errors can be combined by adding the curves of individual errors. Thus, a magnetic compass may have a quadrantal error as shown by the top curve of figure Q5, and a semicircular error as shown by the second curve. The sum of these two errors is shown in the bottom curve. If, in addition, the compass has a constant error, the

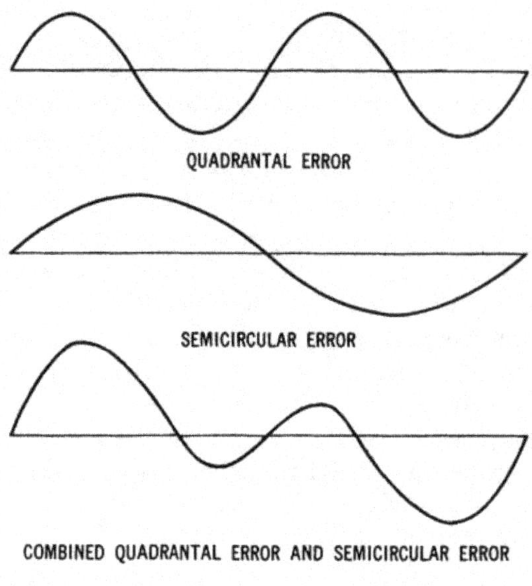

FIGURE Q5.—Combining systematic errors.

bottom curve is moved vertically upward or downward by the amount of the constant error, without undergoing a change of form. If the constant error is greater than the maximum value of the combined curves, all errors are positive or all are negative, but of varying magnitude.

If a number of random errors are combined, the result tends to follow a normal curve regardless of the shape of the individual errors, and the greater the number, the more nearly the result can be expected to approach the normal curve (fig. Q4a). If a given result is subject to errors of plus or minus 3, 2, 1, 2, 4, 2, 1, 8, 1, and 2, the total error *could* be as much as 26 if all errors had the same sign. However, if these are truly random, the probability of them all having the same sign is only 1 in 1024. This is so because the chance of any one being positive (or negative) is ½. That is, of a large number of results, approximately half will have any one particular correction positive (or negative). By the same reasoning, approximately half of the positive (or negative) results will have any one particular additional correction positive (or negative). Thus, the probability of any two particular corrections having a positive (or negative) sign is $\frac{1}{2} \times \frac{1}{2} = (\frac{1}{2})^2 = \frac{1}{4}$. The probability of all 10 corrections having a positive (or negative) sign is $(\frac{1}{2})^{10} = \frac{1}{1024}.$ If there were 20 corrections, the probability of all having a positive (or negative) sign would be $(\frac{1}{2})^{20} = \frac{1}{1{,}048{,}576}.$

When both systematic and random errors are present in a process, both effects are present. An increase in the number of readings decreases the residual random error, but regardless of the number of readings, a systematic error is present in its entirety. Thus, if a number of phase-difference readings are made at a fixed point, the average should be a good approximation of the true value if there is no systematic error. But if the equipment is out of adjustment to the extent that the lane is incorrectly identified, no number of readings will correct this error. In this illustration, a constant error is combined with a normal random error. The normal curve has the correct shape, but is offset from the zero value.

Under some conditions, systematic errors can be eliminated from the results even when the magnitude is not determined. Thus, if two celestial bodies differ in azimuth by 180°, and the altitude of each is observed, the line midway between the lines of position resulting from these observations is free from any *constant* error in the *altitude* (such as abnormal refraction or dip, or incorrect IC). It would *not* be free from such a constant error as one in time (unless the bodies were on the celestial meridian). Similarly, a fix obtained by observations of three stars differing in azimuth by 120°, or four stars differing by 90° is free from constant error in the altitude, if the center of the figure made by the lines of position is used. The center of the figure formed by circles of position from distances of objects equally spaced in azimuth is free from a constant error in range. A constant error in bearing lines does not introduce an error in the fix if the objects are equally spaced in azimuth. In all of these examples, the correct position is *outside* the figure formed by the lines of position if all objects observed are on the same side of the observer (that is, if they lie within an arc of less than 180°).

Q6. Navigation accuracy is normally expressed in terms of the probability of being within a specified distance of a desired point during the navigation process.

If the accuracy of only a single line of position is being considered, the specified distance may be stated as the standard deviation (art. Q4) or some multiple thereof, assuming that the errors of the line of position follow a **single-axis normal distribution.** The distance as stated for the standard deviation of a line of position is measured from the arithmetic mean of the positions which could be established from a large number of observations at a given place and time. Therefore, this distance does not indicate the separation between the line of position and the observer's actual position, except by chance. If the error is stated as 1σ, 68.27 percent of the cases should result in line of position displacements from the arithmetic mean in any direction not exceeding the distance specified for 1σ. If the error is stated as 2σ, 95.45 percent of the lines of position should not be displaced from the arithmetic mean in any direction by more than the distance specified for 2σ. If the error is stated as the **probable error,** 50 percent of the lines of position should not be displaced from the arithmetic mean in any direction by more than the distance specified for 0.6745σ.

The standard deviation is also employed in developing expressions for the probability of a fix position being within a specified distance of the mean of the positions which could be established from a large number of observations at a given place and time by means of the system used to establish the fix.

In the following discussion, the fix is established by the intersection of two lines of position, each of which may be in error. The lines of position (fig. Q6a) are range measurements from two points at the extremities of a baseline of known length. Because of inaccuracies in measurement, the actual ranges differ from the measured values and may lie somewhere between the limits shown as additional arcs either side of the measured arc.

The intersection of the two lines of position together with the standard deviations associated with each are drawn to an expanded scale in figure Q6b. It can be shown that the *contours of equal probability density* about such an intersection are ellipses with their center at the intersection. Thus, the ellipse shown in figure Q6b might be the 75 percent probability ellipse, meaning that there are three chances in four that a fix will lie within such ellipse centered upon the mean of the positions which would be established from a large number of observations at a given place and time by means of the system used to establish the fix.

For simplicity in this discussion of navigation accuracy, the following assumptions are made:

1. All constant errors or **bias errors** have been removed, leaving only the random errors. Thus, the mean or average error is assumed to be zero.

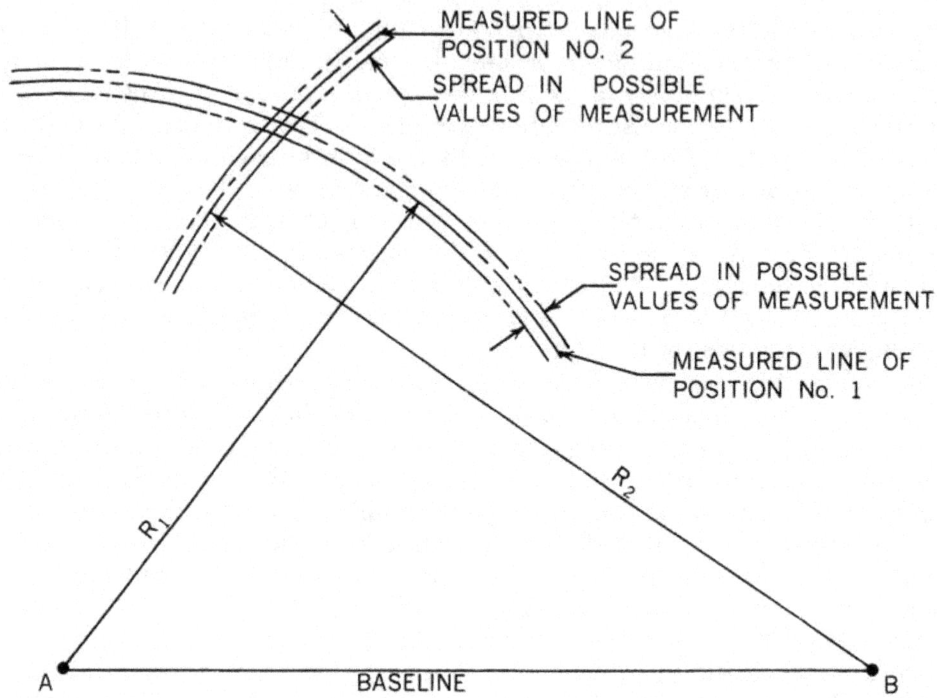

FIGURE Q6a.—Fix established at intersection of two lines of position having different values of error.

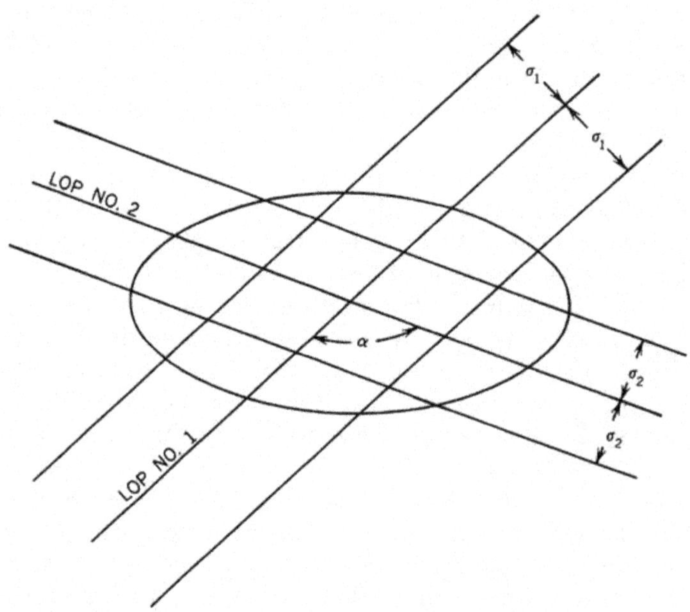

FIGURE Q6b.—Expanded view of intersection of two lines of position.

2. These random errors are assumed to be normally distributed.

3. The errors associated with the two intersecting lines of position are assumed to be independent. This assumption implies that a change in the error of one line of position has no effect upon the other.

4. The lines of position are assumed to be straight lines in the small area in the immediate vicinity of their intersection. This assumption is valid so long as the standard deviation is small compared to the radius of curvature of the line of position.

5. Errors of position are limited to the two-dimensional case.

As shown in figure Q6b, the general case of the intersection of two lines of position at any angle of cut and with different values of error associated with each line of position results in an elliptical error figure. Figure Q6c shows the ellipse simplified to geometrical terms.

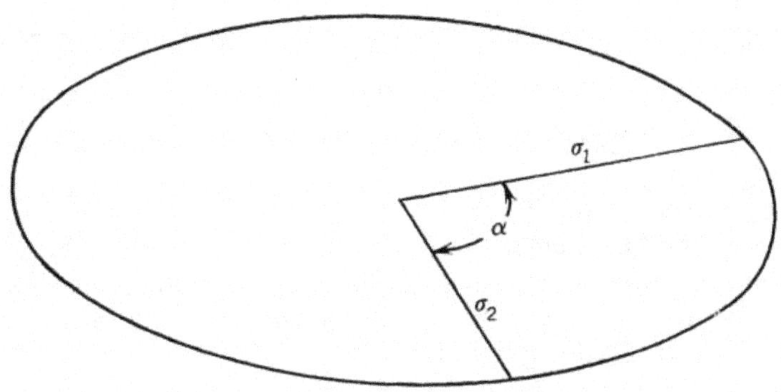

FIGURE Q6c.—Basic error ellipse.

One may readily surmise from figure Q6c that the exact shape of the error figure varies with the magnitudes of the two one-dimensional input errors, σ_1 and σ_2 as well as with the angle of cut, α. The angle α is also the angle between the two values of sigma because the standard deviations are mutually perpendicular to their corresponding lines of position. These variations can be calculated to provide the probability that a point is located within a circle of stated radius. *When this is done, the error is stated in terms more meaningful to the practicing navigator.* The basis of this concept may best be seen by first considering the special case when the two errors are equal, and the angle of intersection of the lines of position is a right angle. In this case, *and in this case alone,* the error figure becomes a circle and is described by the circular normal distribution. A plot of this special function is given in figure Q6d. In this plot, the horizontal axis is measured in terms of R/σ, R being the stated radius of the circle and σ being the measure of error. The error measure is given simply as σ, for in this circular case $\sigma_1 = \sigma_2$. To illustrate, a measurement system gives a circular error figure and has a value of $\sigma = 100$ meters; the probability of actually being located within a circle of 100 meters radius when $R/\sigma = 1.0$ may be read from the vertical axis to be 39.3 percent. To obtain the radius of a circle within which a 50 percent probability results, the corresponding value of R/σ is

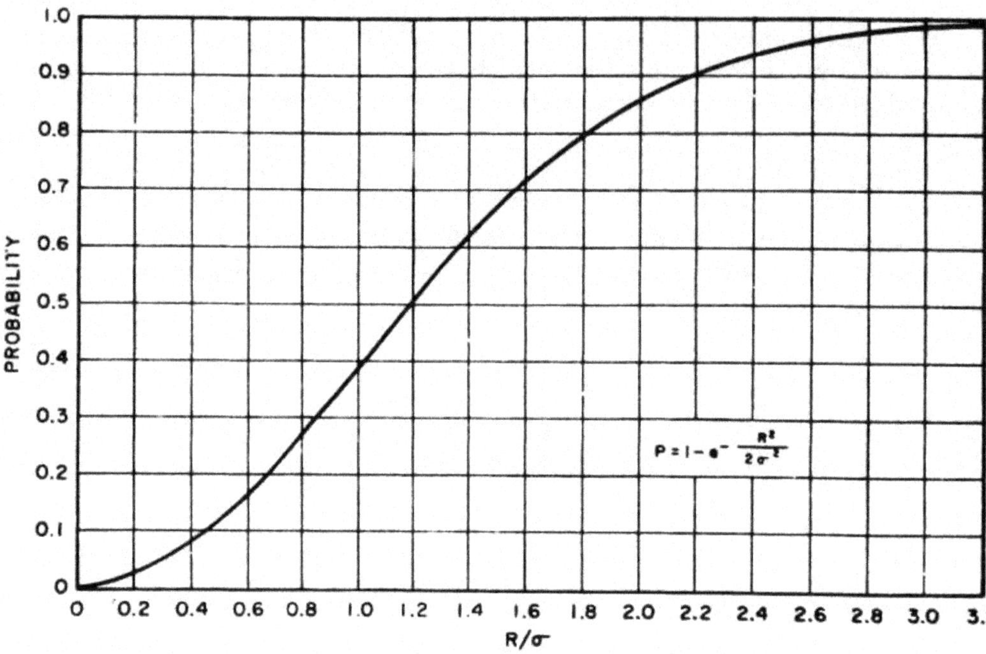

FIGURE Q6d.—Circular normal distribution.

seen to be 1.18 from the graph. Thus, for this example, the **circular probable error (CPE** or **CEP)** would be 118 meters.

In one method of using error ellipses to obtain the radii of **circles of equivalent probability**, new values of σ are found along the major and minor axes of the ellipse (fig. Q6e) using the following equations:

$$\sigma_x^2 = \frac{1}{2\sin^2\alpha}\left[\sigma_1^2+\sigma_2^2+\sqrt{(\sigma_1^2+\sigma_2^2)^2-4\sin^2\alpha\,\sigma_1^2\sigma_2^2}\right]$$

$$\sigma_y^2 = \frac{1}{2\sin^2\alpha}\left[\sigma_1^2+\sigma_2^2-\sqrt{(\sigma_1^2+\sigma_2^2)^2-4\sin^2\alpha\,\sigma_1^2\sigma_2^2}\right].$$

Then the ratio $c=\frac{\sigma_y}{\sigma_x}$, where σ_x is the larger of the two new standard deviations, is used in entering table Q6a which relates ellipses of varying values of ellipticity to the radii of circles of equivalent probability.

For a numerical example to illustrate the method of calculation, assume that the angle of cut α is 50°, σ_1 is 15 meters, and σ_2 is 20 meters to determine the probability of location within a circle of 30 meters radius.

For the computation the following numbers are needed:

$$\sigma_1^2=225$$

$$\sigma_2^2=400$$

$$\sin^2\alpha=0.5868.$$

Substituting in the equations for σ_x^2 and σ_y^2, σ_x and σ_y are calculated as 29.9 meters and 13.1 meters, respectively. Since the function K multiplied by the larger

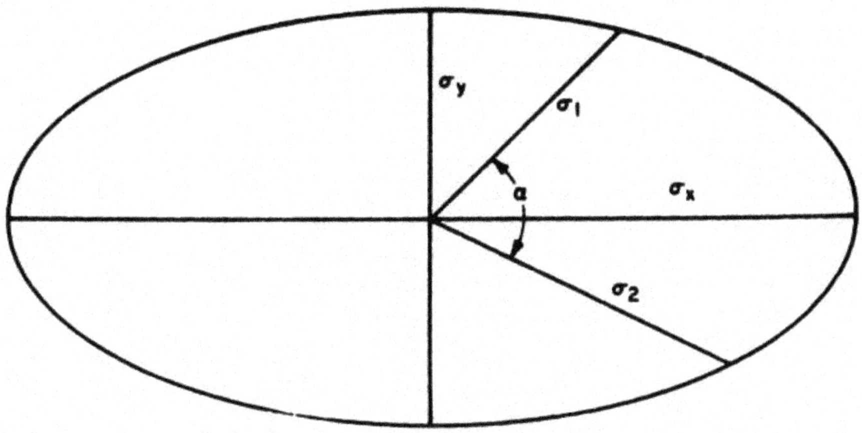

FIGURE Q6e.—Transformation to standard deviations along ellipse axes.

of the two standard deviations obtained by the transformation method gives the value of the radius of the circle of the corresponding value of probability shown in table Q6a, $K=1.003$. On entering table Q6a with $K=1.0$ and $c=0.44$, the probability is found to be 62 percent.

Table Q6b and figure Q6f provide ready information about the sizes of circles of specific probability value associated with ellipses of varying eccentricities.

In another method, fictitious values of sigma of identical value, indicated by σ^*, are assumed to replace the two unequal values originally given (σ_1 and σ_2). A fictitious angle of cut α^* is also assumed to replace the angle of cut (α) originally given (fig. Q6g).

The method utilizes a set of probability curves, with a separate curve for each value of angle of cut (fig. Q6h). These curves can be used only when the two error measures are equal, hence the need for making the transformation to the fictitious σ^*.

The values of σ^* and α^* needed to utilize the probability curves may either be determined from figures Q6i and Q6j or by means of the following equations:

$$\sigma^* = \frac{\sin \beta \sqrt{\sigma_1^2 + \sigma_2^2}}{\sqrt{2}}$$

$$\alpha^* = \text{arc sin} (\sin 2\beta \sin \alpha)$$

where

$$\beta = \text{arc tan} (\sigma_1/\sigma_2).$$

Thus,

$$\sin 2\beta = \frac{2\sigma_1\sigma_2}{\sigma_1^2 + \sigma_2^2}.$$

To use the curve and nomogram for obtaining σ^* and α^*, one must first calculate the ratio σ_2/σ_1. The value σ_1, is always taken as the larger of the two in the ratio so that its value is always less than 1.0. With this ratio, enter the curve of figure Q6i and obtain the σ^* factor. Multiply σ_1 by this factor to obtain the fictitious function σ^*. The nomogram of figure Q6j is entered with the same ratio to obtain the fictitious angle of cut α^*.

K \ c	0.0	0.1	0.2	0.3	0.4	0.5	0.6	0.7	0.8	0.9	1.0
0.1	.0796557	.0443987	.0242119	.0164176	.0123875	.0099377	.0082940	.0071157	.0062299	.0055400	.0049875
0.2	.1585194	.1339783	.0884533	.0628396	.0482413	.0390193	.0327123	.0281415	.0246824	.0219757	.0198013
0.3	.2358228	.2213804	.1739300	.1318281	.1039193	.0851535	.0719102	.0621386	.0546598	.0487639	.0440025
0.4	.3108435	.3010228	.2635181	.2139084	.1742045	.1451808	.1237982	.1076237	.0950495	.0850326	.0768837
0.5	.3829249	.3755884	.3481790	.3003001	.2532953	.2152886	.1857448	.1626829	.1443941	.1296286	.1175031
0.6	.4514938	.4457708	.·255605	.3846374	.3357384	.2914682	.2548177	.2251114	.2009797	.1811783	.1647298
0.7	.5160727	.5115048	.4960683	.4633258	.4170862	.3699305	.3280302	.2925654	.2629373	.2381583	.2172955
0.8	.5762892	.5725957	.5604457	.5349387	.4941882	.4474207	.4025628	.3627122	.3283453	.2989700	.2738510
0.9	.6318797	.6288721	.6191354	.5993140	.5651564	.5213998	.4759375	.4333628	.3953279	.3620135	.3330232
1.0	.6826895	.6802325	.6723586	.6568242	.6291249	.5900953	.5461319	.5025790	.4621421	.4257553	.3934693
1.1	.7286679	.7266597	.7202682	.7079681	.6859367	.6524489	.6116316	.5687467	.5272462	.4887873	.4539256
1.2	.7698607	.7682215	.7630305	.7532175	.7359558	.7079973	.6714269	.6306168	.5893494	.5498736	.5132477
1.3	.8063990	.8050648	.8008554	.7929968	.7793550	.7567265	.7249673	.6873122	.6474394	.6079822	.5704426
1.4	.8384867	.8374049	.8340018	.8277048	.8169851	.7989288	.7720889	.7383089	.7007900	.6623035	.6216889
1.5	.8663856	.8655127	.8627728	.8577362	.8493071	.8350816	.8129287	.7833962	.7489500	.7122546	.6753475
1.6	.8904014	.8897008	.8875060	.8834914	.8768644	.8657559	.8478393	.8226246	.7917194	.7574708	.7219627
1.7	.9108691	.9103102	.9085619	.9053766	.9001746	.8915536	.8773116	.8562471	.8291137	.7977882	.7462539
1.8	.9281394	.9276964	.9263125	.9237989	.9197275	.9130680	.9019110	.8846624	.8613238	.8332175	.8021013
1.9	.9425669	.9422182	.9411299	.9391586	.9359855	.9308615	.9222277	.9083609	.8886731	.8639149	.8355255
2.0	.9544997	.9542272	.9533775	.9518415	.9493815	.9454546	.9388418	.9278799	.9115762	.8901495	.8646647
2.1	.9642712	.9640598	.9634011	.9622127	.9603170	.9573205	.9522999	.9437668	.9305013	.9122714	.8897495
2.2	.9721931	.9720304	.9715237	.9706109	.9691597	.9668845	.9631017	.9565522	.9459386	.9306821	.9110784
2.3	.9785518	.9784275	.9780408	.9773450	.9762419	.9745239	.9716934	.9667306	.9583739	.9458085	.9289946
2.4	.9836049	.9835108	.9832180	.9826918	.9818594	.9805703	.9784661	.9747495	.9682698	.9580804	.9438652
2.5	.9875807	.9875100	.9872900	.9868953	.9862720	.9853112	.9837569	.9810035	.9760522	.9679136	.9560631
2.6	.9906776	.9906249	.9904612	.9901674	.9897045	.9889934	.9878527	.9858331	.9821023	.9756969	.9659525
2.7	.9930661	.9930271	.9929062	.9926894	.9923483	.9918260	.9909944	.9895268	.9867530	.9817837	.9738786
2.8	.9948897	.9948612	.9947727	.9946141	.9943649	.9939842	.9933821	.9923249	.9902888	.9864876	.9801589
2.9	.9962684	.9962477	.9961834	.9960684	.9958878	.9956126	.9951798	.9944246	.9929482	.9900803	.9850792
3.0	.9973002	.9972853	.9972391	.9971564	.9970266	.9968294	.9965205	.9959854	.9949274	.9927925	.9888910
3.1	.9980648	.9980542	.9980212	.9979622	.9978699	.9977296	.9975109	.9971348	.9963851	.9948168	.9918113
3.2	.9986257	.9986182	.9985949	.9985533	.9984880	.9983892	.9982356	.9979733	.9974478	.9963105	.9940240
3.3	.9990332	.9990279	.9990116	.9989824	.9989368	.9988677	.9987607	.9985792	.9982147	.9974004	.9956822
3.4	.9993261	.9993225	.9993112	.9992909	.9992593	.9992115	.9991376	.9990129	.9987626	.9981868	.9969113
3.5	.9995347	.9995323	.9995245	.9995105	.9994888	.9994559	.9994053	.9993204	.9991502	.9987480	.9978125
3.6	.9996818	.9996801	.9996748	.9996653	.9996505	.9996281	.9995938	.9995364	.9994218	.9991442	.9984662
3.7	.9997844	.9997832	.9997797	.9997733	.9997633	.9997482	.9997251	.9996867	.9996102	.9994208	.9989352
3.8	.9998553	.9998545	.9998522	.9998478	.9998412	.9998311	.9998157	.9997902	.9997396	.9996119	.9992682
3.9	.9999038	.9999033	.9999018	.9998989	.9998945	.9998878	.9998776	.9998608	.9998276	.9997426	.9995020
4.0	.9999367	.9999363	.9999353	.9999334	.9999305	.9999261	.9999195	.9999085	.9998870	.9998309	.9996645
4.1	.9999587	.9999585	.9999578	.9999566	.9999547	.9999519	.9999475	.9999404	.9999266	.9998900	.9997763
4.2	.9999733	.9999732	.9999727	.9999720	.9999707	.9999689	.9999661	.9999616	.9999527	.9999292	.9998523
4.3	.9999829	.9999828	.9999826	.9999821	.9999813	.9999801	.9999783	.9999754	.9999698	.9999548	.9999034
4.4	.9999892	.9999891	.9999889	.9999886	.9999881	.9999874	.9999863	.9999845	.9999809	.9999715	.9999375
4.5	.9999932	.9999932	.9999931	.9999929	.9999925	.9999921	.9999914	.9999902	.9999881	.9999822	.9999599
4.6	.9999958	.9999957	.9999957	.9999955	.9999954	.9999951	.9999947	.9999939	.9999926	.9999889	.9999746
4.7	.9999974	.9999974	.9999973	.9999973	.9999971	.9999970	.9999967	.9999963	.9999955	.9999932	.9999840
4.8	.9999984	.9999984	.9999984	.9999983	.9999983	.9999982	.9999980	.9999977	.9999972	.9999959	.9999901
4.9	.9999990	.9999990	.9999990	.9999990	.9999990	.9999989	.9999988	.9999986	.9999983	.9999975	.9999939
5.0	.9999994	.9999994	.9999994	.9999994	.9999994	.9999993	.9999993	.9999992	.9999990	.9999985	.9999963
5.1	.9999997	.9999997	.9999997	.9999996	.9999996	.9999996	.9999996	.9999995	.9999994	.9999991	.9999978
5.2	.9999998	.9999998	.9999998	.9999998	.9999998	.9999998	.9999998	.9999997	.9999997	.9999995	.9999987
5.3	.9999999	.9999999	.9999999	.9999999	.9999999	.9999999	.9999999	.9999998	.9999998	.9999997	.9999992
5.4	.9999999	.9999999	.9999999	.9999999	.9999999	.9999999	.9999999	.9999999	.9999999	.9999998	.9999995
5.5	1.0000000	1.0000000	1.0000000	1.0000000	1.0000000	1.0000000	1.0000000	.9999999	.9999999	.9999999	.9999997
5.6								1.0000000	1.0000000	.9999999	.9999998
5.7										1.0000000	.9999999
5.8											1.0000000
5.9											
6.0											

TABLE Q6a.—Circular error probabilities. Argument c is the ratio of the smaller standard deviation to the larger standard deviation. For the arguments c and K, the table provides the probability that a point lies within a circle whose center is at the origin and whose radius is K times the larger standard deviation.

P \ c	0.0	0.1	0.2	0.3	0.4	0.5	0.6	0.7	0.8	0.9	1.0
.5000	0.67449	0.68199	0.70585	0.74993	0.80785	0.87042	0.93365	0.99621	1.05769	1.11807	1.17741
.7500	1.15035	1.15473	1.16825	1.19246	1.23100	1.28534	1.35143	1.42471	1.50231	1.58271	1.66511
.9000	1.64485	1.64791	1.65731	1.67383	1.69918	1.73708	1.79152	1.86253	1.94761	2.04236	2.14597
.9500	1.95996	1.96253	1.97041	1.98420	2.00514	2.03586	2.08130	2.14598	2.23029	2.33180	2.44775
.9750	2.24140	2.24365	2.25053	2.26255	2.28073	2.30707	2.34581	2.40356	2.48494	2.58999	2.71620
.9900	2.57583	2.57778	2.58377	2.59421	2.60995	2.63257	2.66533	2.71515	2.79069	2.89743	3.03485
.9950	2.80703	2.80883	2.81432	2.83289	2.83830	2.85894	2.88859	2.93347	3.00431	3.11073	3.25525
.9975	3.02334	2.02500	3.03010	3.03898	3.05234	3.07144	3.09871	3.13969	3.20586	3.31099	3.46164
.9990	3.29053	3.29206	3.29673	3.30489	3.31715	3.33464	3.35949	3.39647	3.45698	3.55939	3.71692

TABLE Q6b.—Factors for conversion of probability ellipse to circle of equivalent probability.

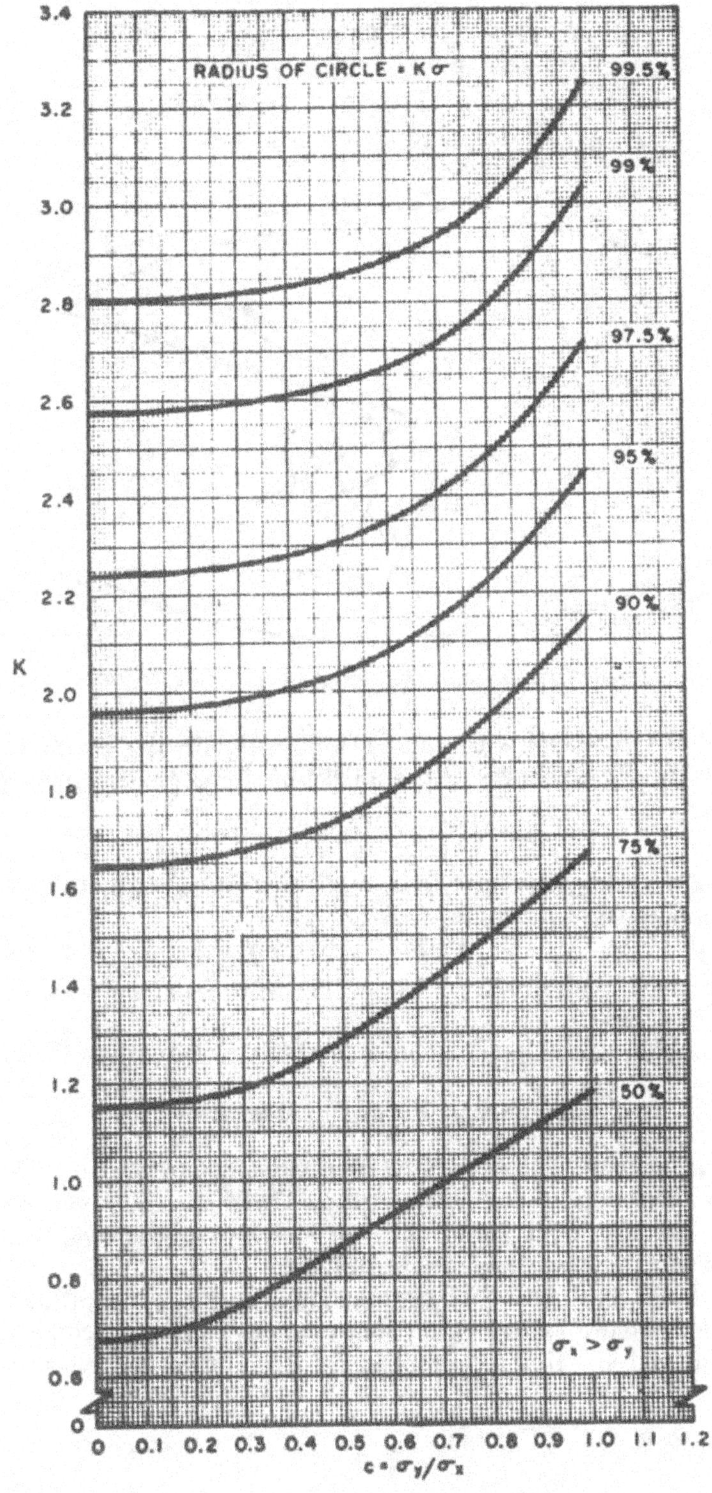

FIGURE Q6f.—Factors for conversion of probability ellipse to circle of equivalent probability.

For a numerical example to illustrate the method of calculation, assume that the angle of cut of 50°, σ_1, is 20 meters, and σ_2 is 15 meters to determine the probability of location within a circle of 30 meters radius.

Calculate the ratio $\sigma_2/\sigma_1 = \frac{15}{20} = 0.75$.

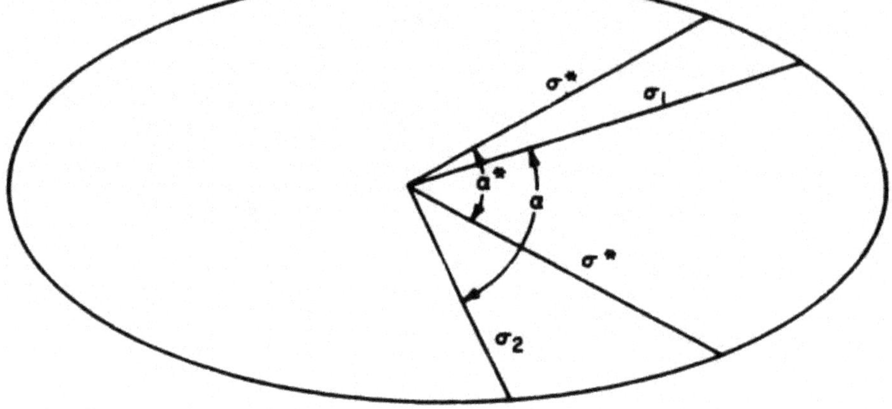

FIGURE Q6g.—Transformed parameters of error ellipse.

Enter the curve of figure Q6i with this ratio and obtain the σ^* factor (0.845). Multiply this factor by σ_1 to obtain σ^* equals 16.9 meters. Calculate the ratio

$$R/\sigma^* = 30/16.9 = 1.78.$$

Enter the nomogram of figure Q6j with the ratio σ_2/σ_1, and with the given angle α to obtain the fictitious angle of cut $\alpha^* = 47°$.

The values $R/\sigma^* = 1.78$ and $\alpha^* = 47°$ are then used to enter the probability curves of figure Q6h to obtain P=0.62 or 62 percent, interpolating between the 40° and 50° curves for $\alpha^* = 47°$.

Table Q6c presents the same data as figure Q6h in numerical form.

Geometrical Error Considerations

From the information that can be derived using the two methods of transformation of elliptical error data, one can develop curves which show for constant values of initial error that the size of a circle of fixed value of probability varies as a function of the angle of cut of the lines of position.

To simplify the investigation of geometrical factors, it is initially desirable to consider the special case of $\sigma_1 = \sigma_2 = \sigma$. Under this special condition, the long equations for σ_x and σ_y can be simplified to facilitate computation as follows:

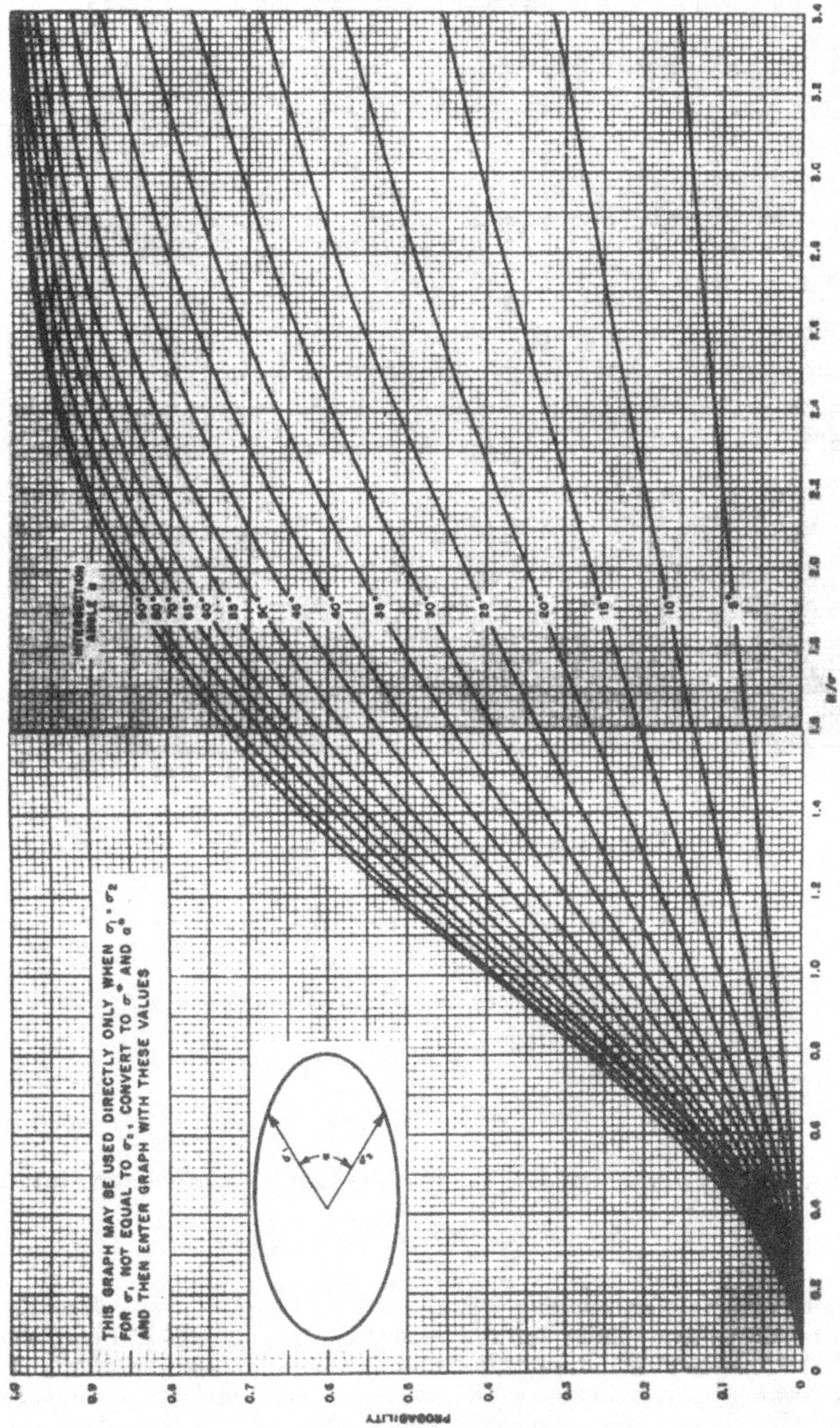

FIGURE Q6h.—Probability versus R/σ and α for elliptical bivariate distributions with two equal standard deviations.

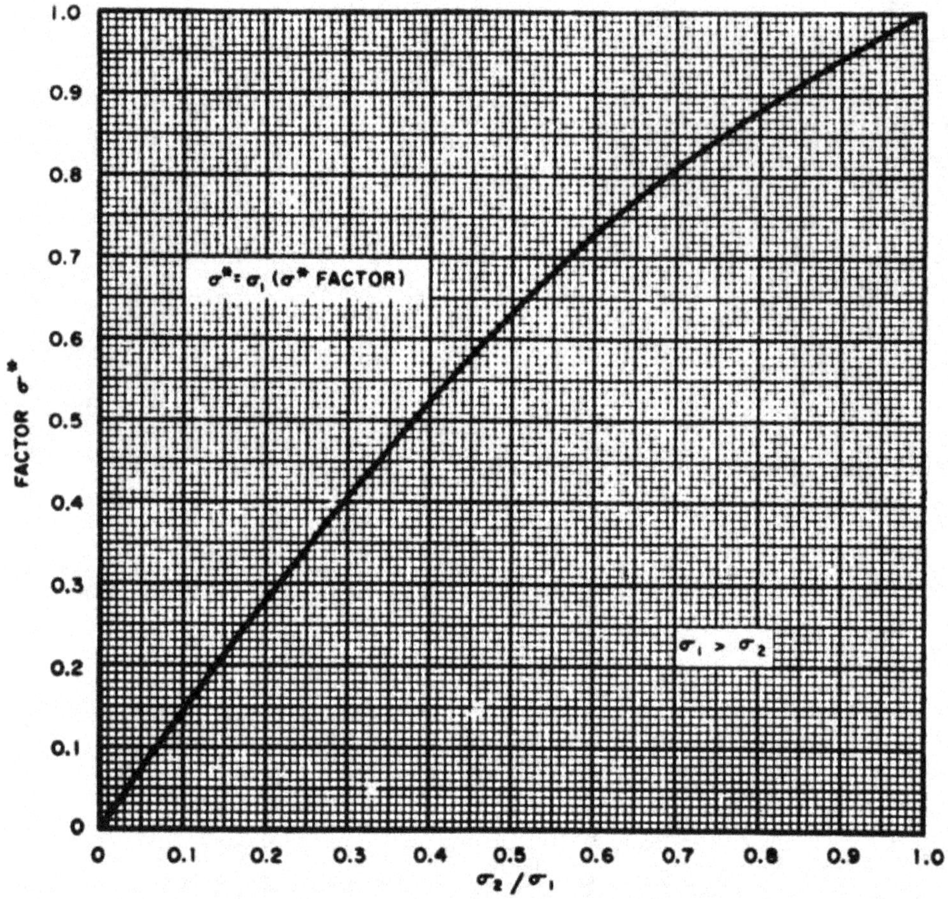

FIGURE Q6i.—σ^* factor versus σ_2/σ_1 ratio.

$$\sigma_x = \frac{\sqrt{2}}{2 \sin \frac{1}{2} \alpha} \sigma \qquad (\sigma_1 = \sigma_2)$$

$$\sigma_y = \frac{\sqrt{2}}{2 \cos \frac{1}{2} \alpha} \sigma \qquad (\sigma_1 = \sigma_2)$$

Taking the ratio of these two values, a simple equation is found for the ratio c.

$$c = \frac{\sigma_y}{\sigma_x} = \tan \frac{1}{2} \alpha.$$

Utilizing these simplified equations, significant parameters of error ellipses are tabulated in table Q6d as a function of the angle of cut α. Using the CEP curve of figure Q6f, values of the CEP are calculated for each angle, showing that the CEP increases as the angle of cut decreases. The last column in the table gives the factor by which the CEP for angles less than 90° is greater than the CEP for a right angle. This magnification of error curve is plotted in figure Q6k. The curve for the 90 percent probability circle has a slightly differing shape from the CEP curve as

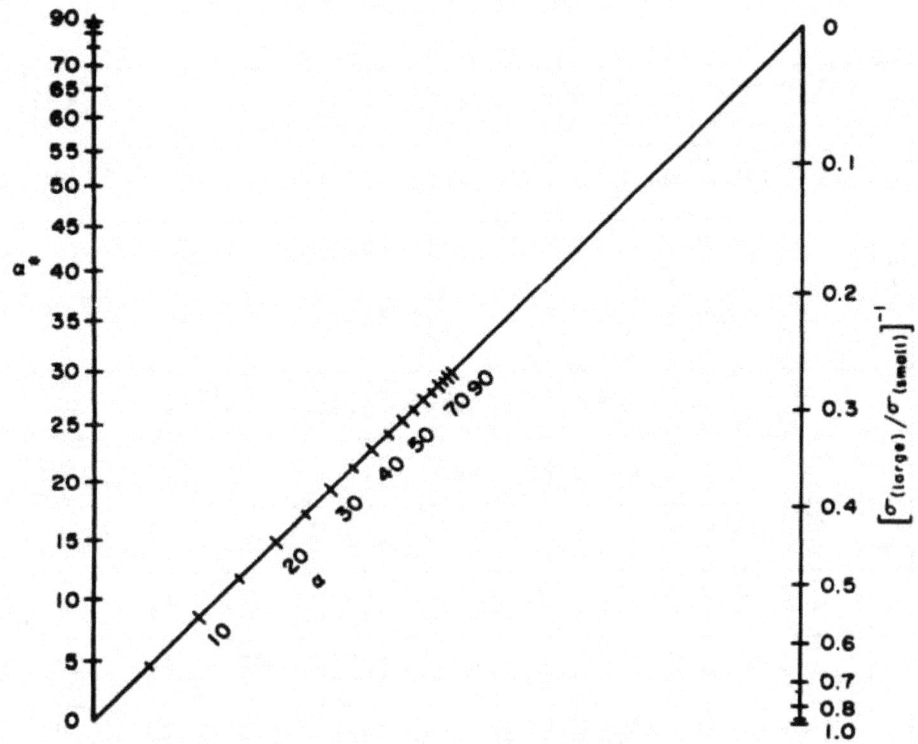

FIGURE Q6j.—Nomogram to obtain α^*.

shown in figure Q6k. Values for the 90 percent probability circle are given in table Q6e. Figure Q6k indicates the magnitude of the growth of error as the angle of cut varies from 90°.

It is also of interest to consider what values of probability result if the radius of the circle is held constant at the minimum value corresponding to that obtained for the 90° angle of cut. These values may be obtained from the probability versus angle of cut curves in figure Q6h.

Along the ordinate $R/\sigma=1.177$ which corresponds to the CEP for the circular case, one may read the lesser values of probability corresponding to the various angles of cut. Likewise, one may also obtain the probability values corresponding to holding a circle the size of the 90 percent probability circle for the circular case by using the ordinate $R/\sigma=2.15$ (also equivalent to 1.82 times the CEP). These two curves are plotted in figure Q6l and the numerical values are given in table Q6f. It is to be noted that the probability values are not inversely related to the error factors plotted in the preceding curves. The geometric error factor is a simple trigonometric function; the probability curves are exponential functions.

Q7. Clarification of terminology.—The following discussion is presented to insure that there is no misunderstanding with respect to the use of terms having one meaning when discussing one-dimensional errors and another when discussing two-dimensional errors.

Although the basic problem of position location is concerned with the two dimensions necessary to describe an area, one-dimensional error measures are commonly applied to each of the two dimensions involved. As demonstrated in

R/σ \ α	5°	10°	15°	20°	25°	30°	35°	40°	45°	50°	55°	60°	65°	70°	75°	80°	85°	90°
0.0																		.000
0.1																		.005
0.2	.001	.004	.006	.007	.009	.011	.011	.012	.014	.015	.016	.017	.018	.018	.019	.019	.019	.020
0.3	.005	.009	.012	.015	.019	.022	.025	.028	.031	.034	.036	.038	.040	.041	.042	.043	.044	.044
0.4	.008	.014	.021	.027	.033	.039	.045	.050	.055	.059	.063	.067	.070	.072	.074	.076	.077	.077
0.5	.009	.019	.029	.039	.048	.057	.066	.074	.083	.090	.096	.102	.106	.110	.114	.116	.117	.118
0.6	.014	.029	.043	.057	.070	.082	.095	.106	.117	.127	.135	.143	.150	.156	.159	.162	.164	.165
0.7	.020	.039	.057	.075	.093	.110	.126	.141	.155	.168	.179	.189	.198	.205	.211	.215	.217	.217
0.8	.025	.048	.072	.095	.117	.138	.158	.177	.195	.211	.226	.238	.249	.258	.265	.270	.273	.274
0.9	.031	.060	.088	.116	.143	.169	.196	.217	.238	.257	.275	.290	.303	.314	.322	.328	.332	.333
1.0	.035	.070	.104	.137	.169	.199	.228	.256	.281	.305	.326	.343	.358	.371	.381	.388	.392	.393
1.1	.040	.080	.120	.158	.196	.231	.265	.296	.325	.352	.366	.396	.414	.428	.440	.448	.452	.454
1.2	.046	.092	.136	.181	.224	.264	.302	.337	.370	.400	.426	.449	.469	.485	.497	.506	.512	.513
1.3	.053	.104	.155	.204	.251	.296	.339	.378	.414	.446	.475	.501	.522	.540	.553	.563	.569	.570
1.4	.058	.115	.171	.226	.278	.327	.374	.417	.456	.491	.523	.550	.573	.582	.606	.617	.623	.625
1.5	.063	.126	.188	.247	.304	.358	.408	.455	.497	.535	.568	.597	.621	.641	.656	.667	.673	.675
1.6	.069	.138	.204	.269	.330	.388	.442	.491	.536	.576	.611	.640	.666	.686	.702	.713	.720	.722
1.7	.074	.148	.220	.289	.355	.417	.471	.526	.573	.615	.650	.682	.707	.728	.744	.756	.762	.764
1.8	.079	.159	.236	.310	.380	.445	.505	.559	.608	.651	.688	.779	.745	.766	.782	.793	.800	.802
1.9	.084	.169	.251	.329	.403	.471	.536	.591	.641	.685	.722	.754	.780	.800	.816	.827	.833	.836
2.0	.089	.179	.266	.349	.426	.498	.563	.621	.672	.716	.753	.785	.810	.832	.846	.856	.862	.865
2.1	.093	.189	.280	.367	.448	.523	.589	.649	[illegible]	.744	.780	.812	.837	.857	.871	.881	.887	.890
2.2	.099	.199	.295	.386	.470	.546	.615	.675	.727	.771	.808	.838	.861	.880	.894	.903	.909	.911
2.3	.104	.209	.309	.404	.490	.570	.639	.700	.752	.795	.831	.860	.883	.900	.913	.922	.927	.929
2.4	.109	.219	.324	.422	.512	.592	.663	.724	.775	.818	.852	.880	.901	.918	.930	.938	.942	.944
2.5	.115	.229	.338	.440	.532	.614	.686	.746	.797	.838	.871	.897	.918	.933	.943	.951	.955	.956
2.6	.119	.239	.352	.457	.551	.634	.706	.766	.816	.856	.888	.912	.931	.945	.954	.961	.964	.966
2.7	.124	.248	.366	.474	.570	.654	.726	.786	.835	.873	.903	.926	.943	.955	.964	.969	.973	.974
2.8	.129	.258	.379	.490	.589	.674	.746	.805	.851	.888	.916	.937	.953	.963	.971	.976	.979	.980
2.9	.134	.268	.393	.507	.607	.692	.763	.821	.867	.902	.928	.947	.961	.971	.978	.982	.984	.985
3.0	.139	.277	.406	.522	.624	.710	.781	.837	.880	.914	.938	.956	.968	.976	.982	.986	.988	.989
3.1	.144	.287	.419	.538	.641	.727	.797	.852	.893	.924	.947	.963	.974	.982	.986	.989	.991	.992
3.2	.149	.296	.432	.553	.657	.743	.812	.865	.905	.934	.955	.969	.979	.985	.990	.992	.993	.994
3.3	.154	.306	.445	.568	.673	.759	.826	.878	.916	.943	.962	.975	.983	.989	.992	.994	.995	.996
3.4	.159	.315	.458	.583	.688	.773	.840	.889	.925	.951	.968	.979	.986	.991	.994	.996	.997	.997
3.5	.164	.324	.470	.597	.703	.788	.852	.900	.934	.957	.973	.983	.989	.993	.996	.997	.998	.998
3.6	.169	.333	.482	.611	.717	.801	.864	.910	.942	.963	.977	.986	.991	.995	.997	.998	.998	.999
3.7	.174	.343	.494	.625	.731	.814	.875	.919	.949	.968	.981	.990	.993	.996	.998	.998	.999	.999
3.8	.179	.352	.506	.638	.744	.826	.886	.927	.955	.973	.984	.990	.995	.997	.998	.999	.999	.999
3.9	.183	.360	.518	.651	.757	.838	.895	.935	.961	.977	.987	.993	.996	.998	.999	.999	.999	1.000
4.0	.189	.370	.529	.664	.770	.848	.904	.942	.966	.980	.989	.994	.997	.998	.999	.999	1.000	1.000
4.1	.193	.378	.541	.676	.781	.859	.913	.948	.970	.983	.991	.995	.998	.999	.999	1.000	1.000	1.000
4.2	.198	.387	.552	.688	.793	.869	.920	.954	.974	.986	.993	.996	.998	.999	.999	1.000	1.000	1.000
4.3	.203	.396	.564	.700	.804	.878	.927	.958	.978	.988	.994	.997	.999	.999	1.000	1.000	1.000	1.000
4.4																		1.000

TABLE Q6c.—Probability versus R/σ and α for elliptical bivariate distributions with two equal standard deviations.

a	σ_x	σ_y	c	K	CEP	ERROR FACTOR
90	1.0	1.0	1.0	1.177	1.177	1.00
80	1.10	0.924	0.839	1.078	1.186	1.01
70	1.234	0.865	0.700	0.996	1.228	1.042
60	1.414	0.817	0.577	0.914	1.292	1.099
50	1.672	0.782	0.466	0.847	1.420	1.206
45	1.847	0.766	0.414	0.815	1.508	1.281
40	2.06	0.753	0.364	0.783	1.620	1.376
30	2.74	0.733	0.268	0.734	2.01	1.710
20	4.06	0.718	0.176	0.700	2.85	2.42
10	8.11	0.710	0.087	0.680	5.52	4.69

TABLE Q6d.—Significant parameters of error ellipses when $\sigma_1 = \sigma_2$.

a	c	K	90% R	ERROR FACTOR
90	1.0	2 145	2.145	1.00
80	0.839	1.98	2.18	1.015
70	0.700	1.86	2.30	1.07
60	0.577	1.775	2.51	1.17
50	0.466	1.72	2.88	1.34
45	0.414	1.702	3.15	1.47
40	0.364	1.687	3.47	1.615
30	0.268	1.665	4.53	2.11
20	0.176	1.652	6.72	3.13
10	0.087	1.645	13.35	6.22

TABLE Q6e.—90 percent error factor.

a	P	P
90	50	90
80	49.4	89.2
70	47.5	86.9
60	44.0	82.4
50	39.5	76
40	37	66
30	25	53
20	17	37
10	8	19

TABLE Q6f.—Probability decrease with decreasing angle of cut for a circle of constant radius ($\sigma_1 = \sigma_2$).

article Q6, the use of the one-dimensional standard deviation of each line of position permitted a general approach to the consideration of the error ellipse.

One-Dimensional Errors

The terms **standard deviation, sigma** (σ), and **root mean square (RMS) error** have the same meaning in reference to one-dimensional errors. The basic equation of the normal (Gaussian) distribution indicates the use of the Greek letter sigma from which its use for standard deviation arises:

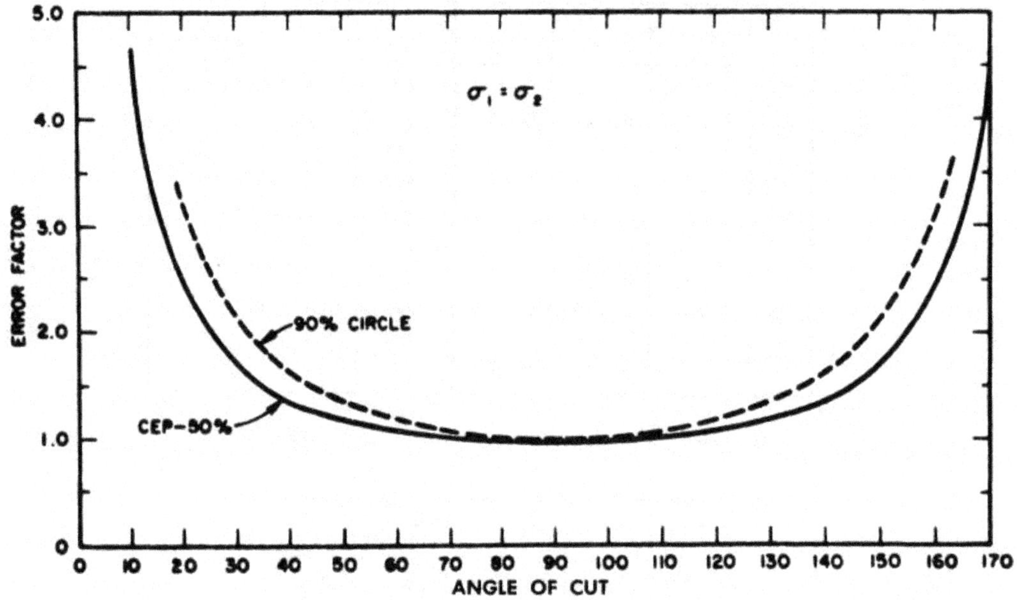

FIGURE Q6k.—CEP magnification versus angle of cut.

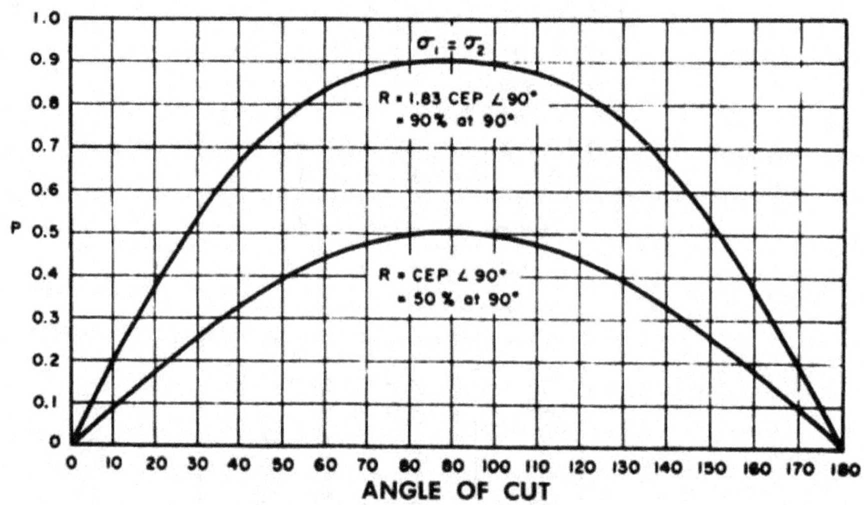

FIGURE Q6l.—Decrease in probability for a circle of constant radius angle of cut.

$$f(x)=\frac{1}{\sigma\sqrt{2\pi}}\,e^{-\frac{(x-a)^2}{2\sigma^2}} \qquad -\infty<x<\infty$$

where a is the mean of the distribution.

Standard deviation of a measurement system is a property that may be determined experimentally. If a large number of measurements of the same quantity—a length for example—are made and compared with a standard, the standard deviation is the square root of the sum of the squares of the deviations from the mean or average value divided by one less than the number of measurements taken. Symbolically this operation is represented as:

$$\sigma = \sqrt{\frac{\sum_{i=1}^{n} (x_i - a)^2}{n-1}}.$$

The term root-mean-square (RMS) error comes from this latter method of computation.

Numerically, one sigma corresponds to 68.27 percent of the distribution. That is, if a large number of measurements were made of a given quantity, 68.27 percent of the errors would be no greater than the value of one standard deviation. Likewise 2σ corresponds to 95.45 percent of the total errors and 3σ corresponds to 99.73 percent of the total errors.

The term **probable error** is identical *in concept* to standard deviation. The term differs from standard deviation in that it refers to the median error; that is, no more than half the errors in the measurement sample are greater than the value of the probable error. Linear probable error is related to standard deviation by a multiplication factor (tab. Q7a). One probable error equals 0.6745 times one standard deviation.

From \ To	50.00%	68.27%	95.00%	99.73%
50. 00%	1. 0000	1. 4826	2. 9059	4. 4475
68. 27%	0. 6745	1. 0000	1. 9600	3. 0000
95. 00%	0. 3441	0. 5102	1. 0000	1. 5307
99. 73%	0. 2248	0. 3333	0. 6533	1. 0000

TABLE Q7a.—Linear error conversion factors.

The term **variance** is met most frequently in detailed mathematical discussions. The term refers to a square of a standard deviation. It is useful in simplifying the algebra of some complex mathematical derivations.

Two-Dimensional Errors

Terms similar or identical in words to those used for one-dimensional error descriptions are also used with two-dimensional or **bivariate error** descriptions. However, in the two-dimensional case, not all of these terms have the same meaning as before; considerable care is needed to avoid confusion.

Standard deviation or sigma has a definable meaning only in the specific case of the circular normal distribution where $\sigma_x = \sigma_y$:

$$P_R = 1 - e^{-\frac{R^2}{2\sigma^2}}.$$

In the case of the circular normal distribution, the standard deviation σ is equivalent to the standard deviation along both orthogonal axes. Because of concern with a radial distribution, the total distribution of errors involves numbers different from those of the linear case (tabs. Q7a and Q7b). In the circular case, 1σ error indicates that 39.35 percent of the errors would not exceed the value of the 1σ error; 86.47 percent would not exceed the 2σ error; 98.89 percent would not exceed the 3σ error; and 99.78 percent would not exceed the 3.5σ error.

From \ To	39. 35%	50. 00%	63. 21%	95. 00%	99. 78%
39. 35%	1. 0000	1. 1774	1. 4142	2. 4477	3. 5000
50. 00%	0. 8493	1. 0000	1. 2011	2. 0789	2. 9726
63. 21%	0. 7071	0. 8325	1. 0000	1. 7308	2. 4749
95. 00%	0. 4085	0. 4810	0. 5778	1. 0000	1. 4299
99. 78%	0. 2857	0. 3364	0. 4040	0. 6993	1. 0000

TABLE Q7b.—Circular error conversion factors.

Because the usual case where there are two-dimensional distributions is that the standard deviations are different, resulting in an elliptical distribution, the circular standard deviation is less useful than the linear standard deviation. It is more common to describe two-dimensional distributions by the two separate one-dimensional standard deviations associated with each error axis. References, however, often do not make this distinction, referring to the position accuracy of a system as 600 feet (2σ), for example. Such a description should leave the reader wondering whether the measure is circular error, in which case the numbers describe the 86 percent probability circle, or whether the numbers are to be interpreted as one-dimensional sigmas along each axis, in which case the 95 percent probability circle is indicated (assuming the distribution to be circular, which actually it may not be).

The term **RMS (root mean square) error** when applied to two-dimensional errors does not have the same meaning as standard deviation. The term has the same meaning as radial error or d_{rms}, discussed later. Such use of the term is deprecated.

In a circular normal distribution, the term **circular probable error (CPE)** or **circular error probable (CEP)** refers to the radius of the circle inside of which there is a 50 percent probability of being located.

The term CEP is also used to indicate the radius of a circle inside of which there is a 50 percent probability of being located, even though the actual error figure (fig. Q7a) is an ellipse. Article Q6 describes one of the methods of obtaining such CEP equivalents when given ellipses of varying eccentricities. Curves and tables are available for performing this calculation. Despite the availability of these curves and tables, approximations are often made for this calculation of a CEP when the actual error distribution is elliptical. Several of these approximations are indicated and plotted for comparison with the exact curve in figure Q7b. Of the various approximations shown, the top curve, the one which diverges the most rapidly, appears to be the most commonly used.

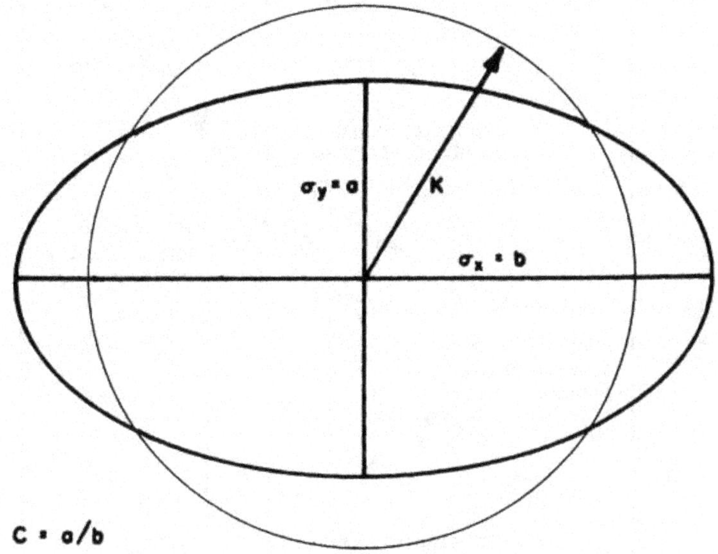

FIGURE Q7a.—Error ellipse and circle of equivalent probability.

Another factor of interest concerning the relationship of the CEP to various ellipses is that the area of the CEP circle is always greater than the basic ellipse. Table Q7c indicates that the divergence between the actual area of the ellipse of interest and the circle of equivalent probability increases as the ellipse becomes thinner and more elongated.

The value of the CEP may be related to the radius of other values of probability circles analytically for the case of the circular normal distribution by solving the basic equation for various values of probability. For this special case of the circular normal distribution, these relationships are shown drawn to scale in figure Q7c with the associated values tabulated in table Q7d.

The derivation of these values is shown in the following analysis. First, the factor relating the CEP to the circular sigma is derived, then, as a second example, the relationship between the 75 percent probability circle and the circular sigma is derived. The ratio of these two values is then the value shown in table Q7d for the 75 percent value.

The circular normal distribution equation is:

$$P_R = 1 - e^{-\frac{R^2}{2\sigma^2}},$$

and

$$\text{CEP} \equiv P(R) = 0.5$$

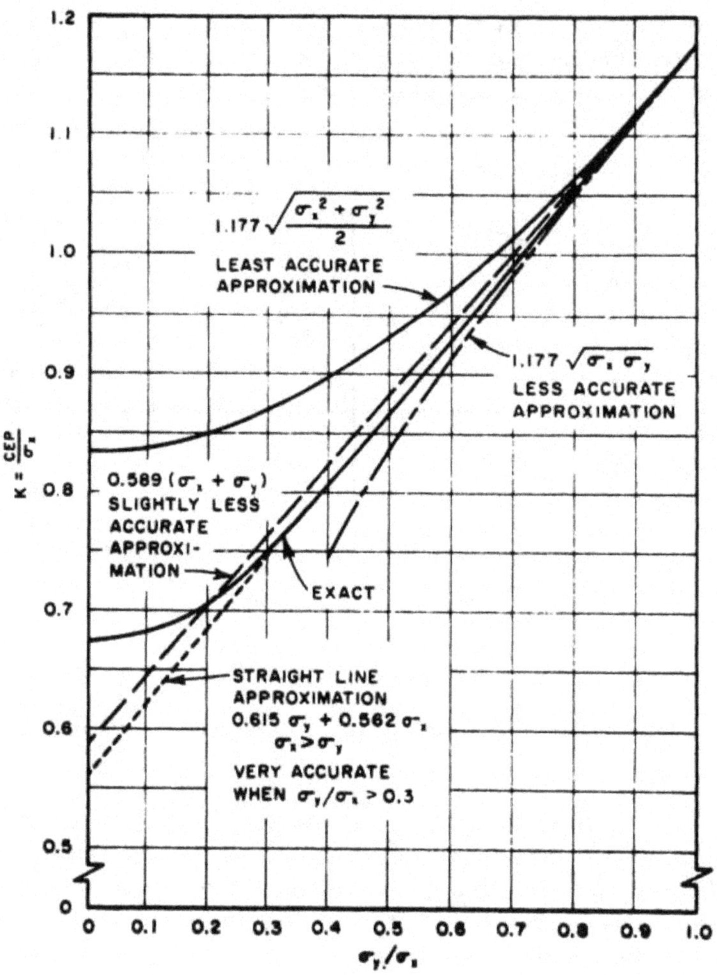

FIGURE Q7b.—CEP for elliptical error distribution approximations.

$$1-e^{-\frac{R^2}{2\sigma^2}}=0.5$$

$$e^{-\frac{R^2}{2\sigma^2}}=0.5.$$

Take natural logarithm of both sides

$$\ln\left(e^{-\frac{R^2}{2\sigma^2}}\right)=\ln 0.5$$

$$\frac{R^2}{2\sigma^2}=\ln 2 \qquad (\ln 0.5=-\ln 2)$$

$$R=1.1774\sigma.$$

$C=a/b$	Area of 50% ellipse	Area of equivalent circle
0.0	0	1. 43
0.1	0. 437	1. 46
0.2	0. 874	1. 56
0.3	1. 31	1. 76
0.4	1. 75	2. 06
0.5	2. 08	2. 37
0.6	2. 62	2. 74
0.7	3. 06	3. 12
0.8	3. 49	3. 52
0.9	3. 93	3. 94
1.0	4. 37	4. 37

TABLE Q7c.—Comparison of areas of 50 percent ellipses of varying eccentricities with areas of circles of equivalent probability.

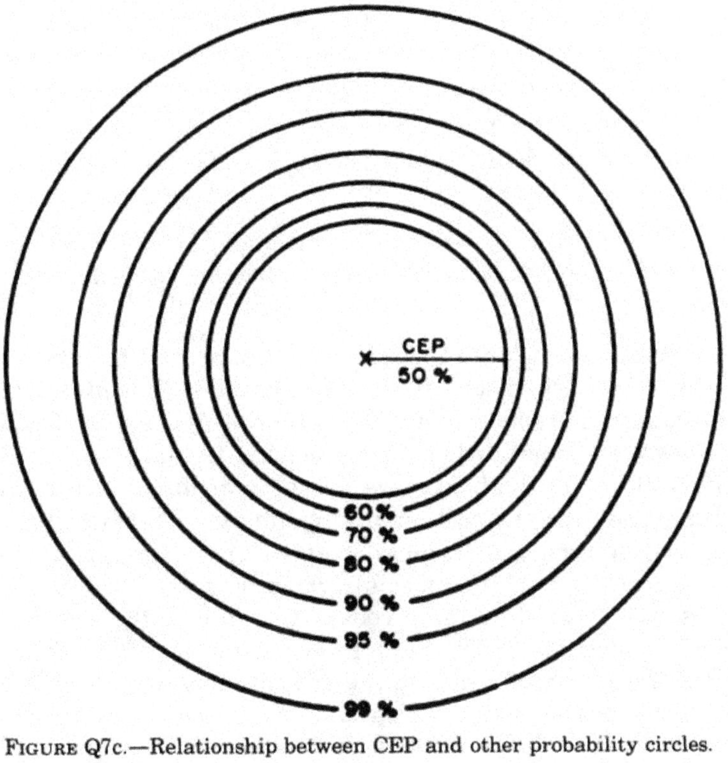

FIGURE Q7c.—Relationship between CEP and other probability circles.

For the 75 percent probability circle,

$$1-e^{-\frac{R^2}{2\sigma^2}}=0.75$$

$$e^{-\frac{R^2}{2\sigma^2}}=0.25$$

$$\ln\left(e^{-\frac{R^2}{2\sigma^2}}\right)=\ln 0.25$$

$$\frac{R^2}{2\sigma^2}=\ln 4$$

$$R=1.665\sigma$$

$$\frac{R(75\%)}{R(50\%)}=\frac{1.665\sigma}{1.177\sigma}=1.414.$$

Multiply value of CEP by	To obtain radius of circle of probability
1. 150	60%
1. 318	70%
1. 414	75%
1. 524	80%
1. 655	85%
1. 823	90%
2. 079	95%
2. 578	99%

TABLE Q7d.—Relationship between CEP and radii of other probability circles of the circular normal distribution.

The factors tabulated in table Q7d are sometimes used to relate varying probability circles when the basic distribution is not circular, but elliptical. That such a procedure is inaccurate may be seen by the curves of figure Q7d. It can be seen that the errors involved are small when the eccentricities are small. But the errors increase significantly when both high values of probability are desired and when the ellipticity increases in the direction of long, narrow distributions.

The terms **radial error, root mean square error,** and d_{rms} are identical in meaning when applied to two-dimensional errors. Figure Q7e illustrates the definition of d_{rms}. It is seen to be the square root of the sum of the square of the 1 sigma error components along the major and minor axes of a probability ellipse. The figure details the definition of 1 d_{rms}. Similarly, other values of d_{rms} can be derived by using the corresponding values of sigma. The measure d_{rms} is not equal to the square root of the sum of the squares of σ_1 and σ_2 that are the basic errors associated with the lines of position of a particular measuring system. The procedures described in art. Q6 must first be utilized to obtain the values shown as σ_x and σ_y.

The three terms (radial error, root-mean-square error, and d_{rms}) used as a measure of error are somewhat confusing because they do not correspond to a fixed value of probability for a given value of the error measure. The terms can be conveniently related to other error measures only when $\sigma_x = \sigma_y$, and the probability figure is a circle. In the more common elliptical cases, the probability associated with a fixed value of d_{rms} varies as a function of the eccentricity of the ellipse. One d_{rms} is defined as the radius of the circle obtained when $\sigma_x = 1$, in figure Q7e, and σ_y varies from 0 to 1. Likewise, 2 d_{rms} is the radius of the circle obtained when $\sigma_x = 2$, and σ_y varies from 0 to 2. Values of the length of the radius d_{rms} can be calculated as shown in table Q7e. From these values the associated probabilities can be determined from the tables of article Q6. The variations of probability associated with the values of 1 d_{rms} and 2 d_{rms} are shown in the curves of figures Q7f and Q7g. Figure Q7h shows the lack of a constant relationship in a slightly different way. Here the ratio d_{rms}/CEP is plotted against the same measure of ellipticity. The three figures show graphically that there is not a constant value of probability associated with a single value of d_{rms}.

Figure Q7i shows the substitution of the circular form for elliptical error distributions. When σ_x and σ_y are equal, the probability represented by 1 d_{rms} is 63.21 percent. When σ_x and σ_y are unequal (σ_x being the greater value), the probability varies from 64 percent when $\sigma_y / \sigma_x = 0.8$ to 68 percent when $\sigma_y / \sigma_x = 0.3$.

Q8. Navigation system accuracy.—In a navigation system, **predictability** is the measure of the accuracy with which the system can define the position in terms of geographical coordinates; **repeatability** is the measure of the accuracy with which the system permits the user to return to a position as defined only in terms of the coordinates peculiar to that system. **Predictable accuracy**, therefore, is the accuracy of positioning with respect to geographical coordinates; **repeatable accuracy** is the accuracy with which the user can return to a position whose coordinates have been measured previously with the same system. For example, the distance specified for the repeatable accuracy of a system such as Loran-C is the distance between two Loran-C positions established using the same stations and time-difference readings at different times. The correlation between the geographical coordinates and the system coordinates may or may not be known.

Relative accuracy is the accuracy with which a user can determine his position relative to that of another user of the same navigation system at the same time. Hence, a system with high relative accuracy provides good rendezvous capability for the users of the system. The correlation between the geographical coordinates and the system coordinates is not relevant.

Q9. Most probable position.—Some navigators, particularly those of little experience, have been led by the simplified definitions and explanations usually given in texts to conclude that the line of position is infallible, and that a fix is without error, overlooking the frequent incompatibility of these two notions. Too often the idea has prevailed that information is either all right or all wrong. An example is the practice of establishing an estimated position at the foot of the perpendicular from a dead reckoning position to a line of position. The assumption is that the vessel *must* be somewhere on the line of position. The limitations of this often valuable practice are not understood by these inexperienced navigators.

A more realistic concept is that of the **most probable position (MPP)**, which recognizes the probability of error in *all* navigational information, and determines position by an evaluation of all available information, using the principles of errors.

Suppose a vessel were to start from a completely accurate position and proceed on dead reckoning. If course and speed over the bottom were of equal accuracy, the uncertainty of dead reckoning positions would increase equally in all directions with either distance or elapsed time (for any one speed these would be directly

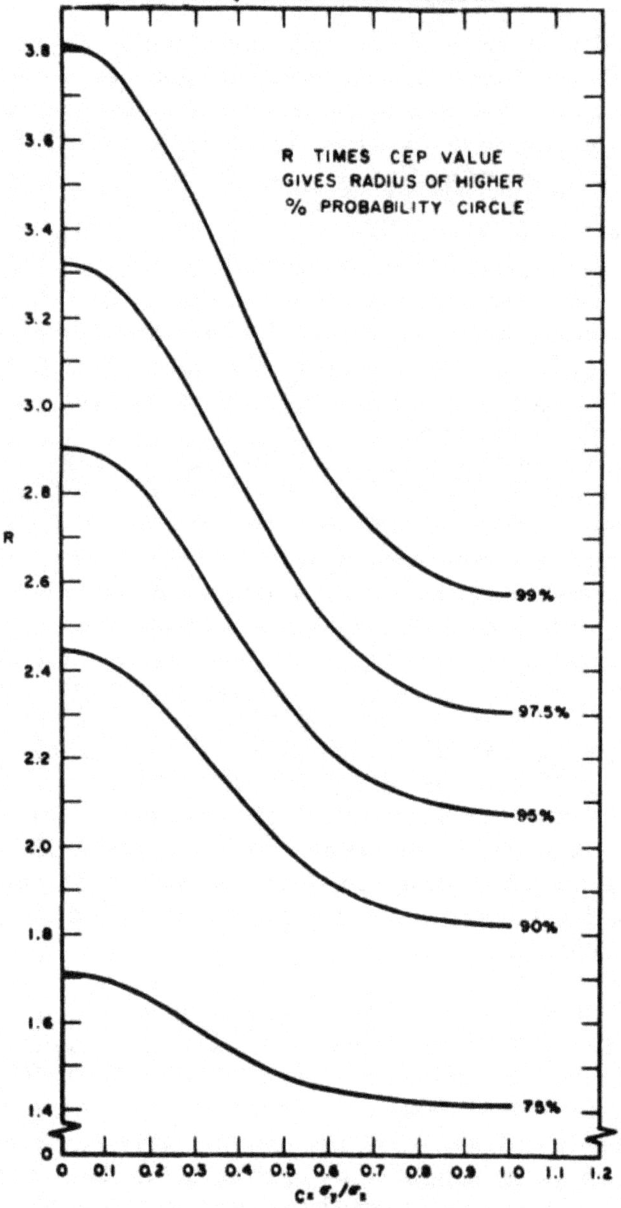

FIGURE Q7d.—Relation of probability circles to CEP versus ellipticity.

proportional and therefore either could be used). Therefore, a circle of uncertainty would grow around the dead reckoning position as the vessel proceeded. If the navigator had full knowledge of the distribution and nature of the errors of course and speed, and the necessary knowledge of statistical analysis, he could compute the radius of the circle of uncertainty, using the 50 percent, 95 percent, or other probabilities.

In ordinary navigation, this is not practicable, but based upon his experience and judgment, the navigator might estimate at any time the likely error of his dead reckoning or estimated position. With practice, he might acquire considerable skill in making this estimate. He would take into account, too, the fact that the area of uncertainty might better be represented by an ellipse than a circle, the major axis

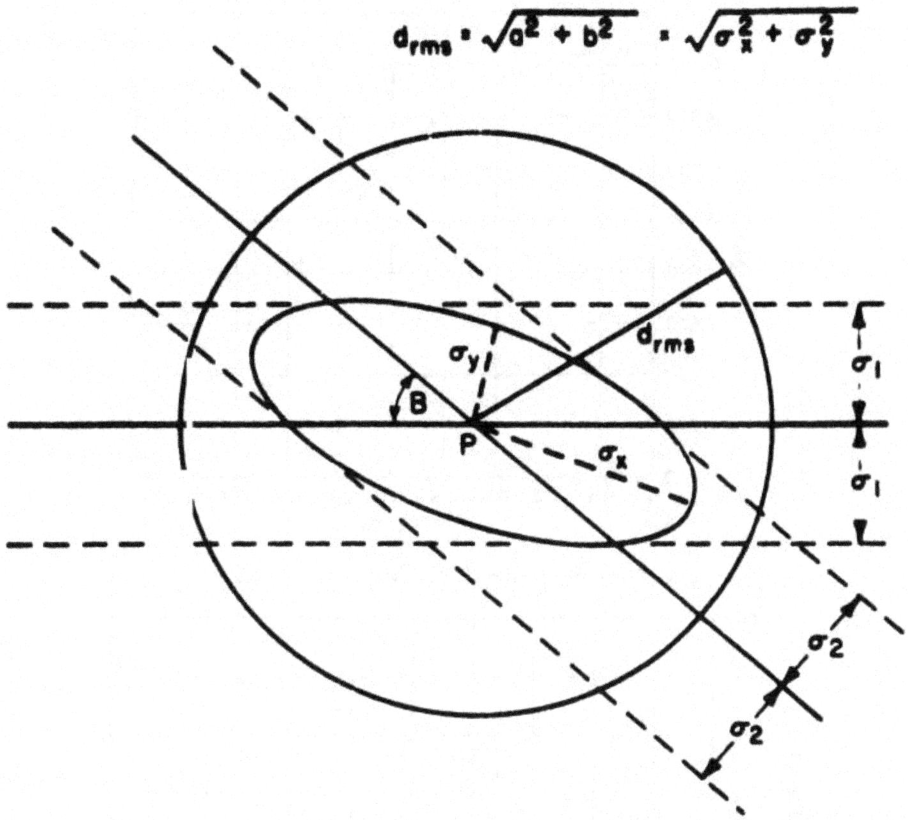

FIGURE Q7e.—Illustration of radial error or d_{rms}.

σ_y	σ_x	LENGTH OF 1 d_{rms}	PROBABILITY	
			1 d_{rms}	2 d_{rms}
0.0	1.0	1.000	0.683	0.954
0.1	1.0	1.005	0.682	0.955
0.2	1.0	1.020	0.682	0.957
0.3	1.0	1.042	0.676	0.961
0.4	1.0	1.077	0.671	0.966
0.5	1.0	1.118	0.662	0.969
0.6	1.0	1.166	0.650	0.9"3
0.7	1.0	1.220	0.641	0.977
0.8	1.0	1.280	0.635	0.980
0.9	1.0	1.345	0.632	0.981
1.0	1.0	1.414	0.632	0.982

$d_{rms} = \sqrt{\sigma_x^2 + \sigma_y^2}$ when σ_x and σ_y are at right angles to each other.

TABLE Q7e.—Calculation of d_{rms}.

being along the course line if the estimated error of the speed were greater than that of the course, and the minor axis being along the course line if the estimated error of the course were greater. He would recognize, too, that the size of the area of uncertainty would not grow in direct proportion to the distance or elapsed time, because disturbing factors such as wind and current could not be expected to

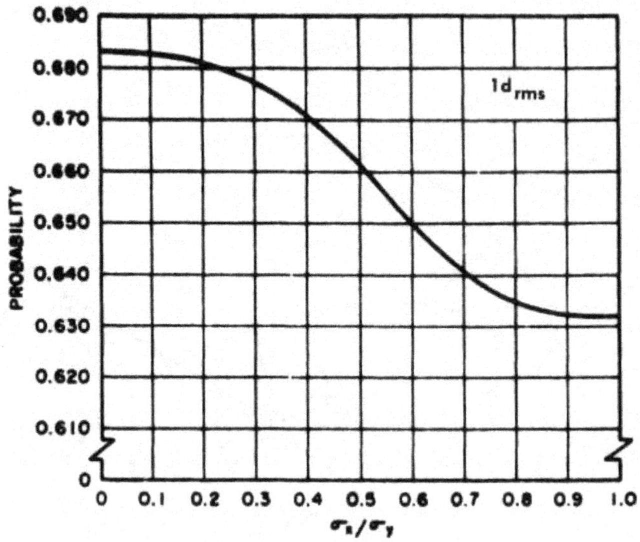

FIGURE Q7f.—Variation in d_{rms} with ellipticity (1 d_{rms}).

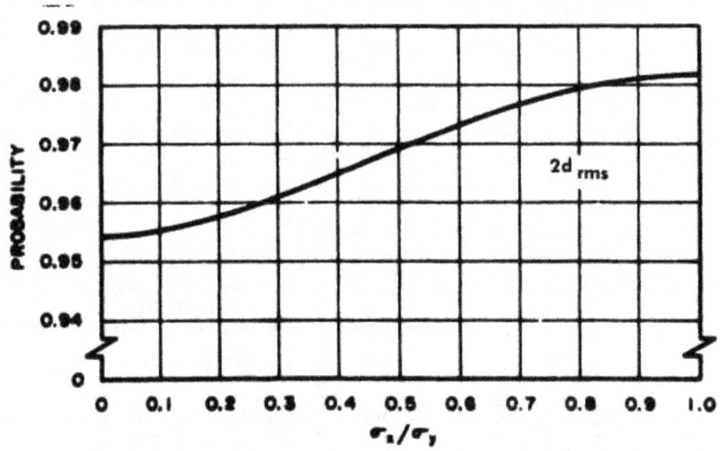

FIGURE Q7g.—Variation in d_{rms} with ellipticity (2 d_{rms}).

remain of constant magnitude and direction. Also, he would know that the starting point of the dead reckoning would not be completely free from error.

At some future time additional positional information would be obtained. This might be a line of position from a celestial observation or by Omega. This, too, would be accompanied by an estimated error which might be computed for a certain probability if the necessary information and knowledge were available. If the dead reckoning had started from a good position obtained by means of landmarks, the likely error of the initial position would be very small. At first the dead reckoning or estimated position would probably be more reliable than a line of position obtained by celestial observation or Omega. But at *some* distance the two would be equal, and beyond this the line of position might be more accurate.

However, the determination of most probable position does not depend upon determination of *which* information is most accurate. In figure Q9a a dead reckoning position is shown surrounded by a circle of uncertainty. A line of position is also shown, with its area of uncertainty. The most probable position is within the overlapping area, and if the uncertainty of the dead reckoning position and that of

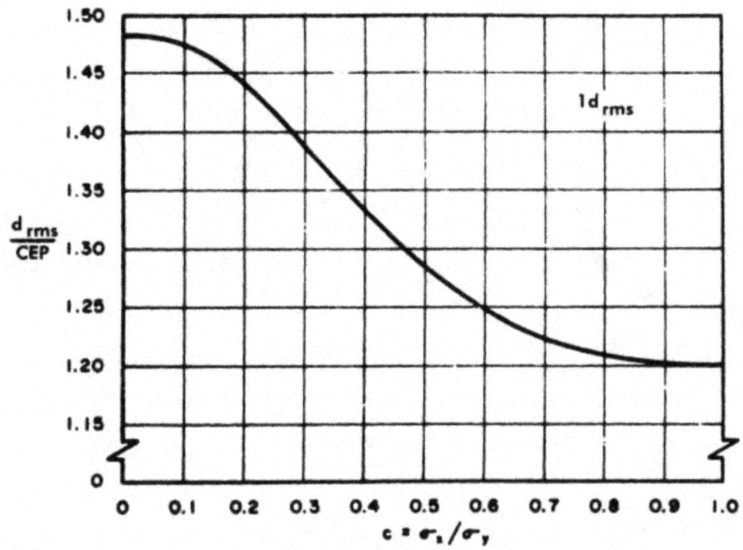

FIGURE Q7h.—Ellipticity versus d_{rms}/CEP (1 d_{rms}).

the line of position are about equal, it might be taken at the center of the area. If the overall errors are considered normal, and they are probably approximately so, *the effect of each error is proportional to its square.* Thus, if the likely error of a dead reckoning position is 3 miles, and that of a line of position is 2 miles, the most probable position is nearer the line of position, being at a distance equal to $\frac{2^2}{3^2}=\frac{4}{9}$ that from the dead reckoning position (or $\frac{4}{13}$ of the perpendicular distance from the dead reckoning position to the line of position).

If a fix is obtained from two lines of position, the area of uncertainty is a circle if the lines are perpendicular, have equal likely errors, and these errors can be considered normal. If one is considered more accurate than the other, the area is an ellipse, the two axes being proportional to the standard deviations of the two lines of position. As shown in figure Q9b, it is also an ellipse if the likely error of each is equal and the lines cross at an oblique angle. If the errors are unequal, the major axis of the ellipse is more nearly in line with the line of position having the smaller likely error.

If a fix is obtained from three or more lines of position spread in azimuth by more than 180°, and the error of each line is normal and equal to that of the others, the most probable position is the center of the figure. By "center" is meant that point within the figure which is equidistant from the sides. If the lines are of unequal likely error, the distance of the most probable position from each line of position is proportional to the *square* of the likely error of that line times the sine of the angle formed by the other two lines.

In the discussion of most probable position from lines of position, it has been assumed that no other positional information is available. Usually, this is an incorrect assumption, for there is nearly always a dead reckoning or estimated position. This can be considered in any of several ways. The square of its likely error can be used in the same manner as the square of the likely error of each line of position. A most probable position based upon the dead reckoning or estimated position and the most reliable line of position might be determined as explained above, and that line of position replaced with a new one parallel to it but passing through the most probable position just determined. This adjusted line of position

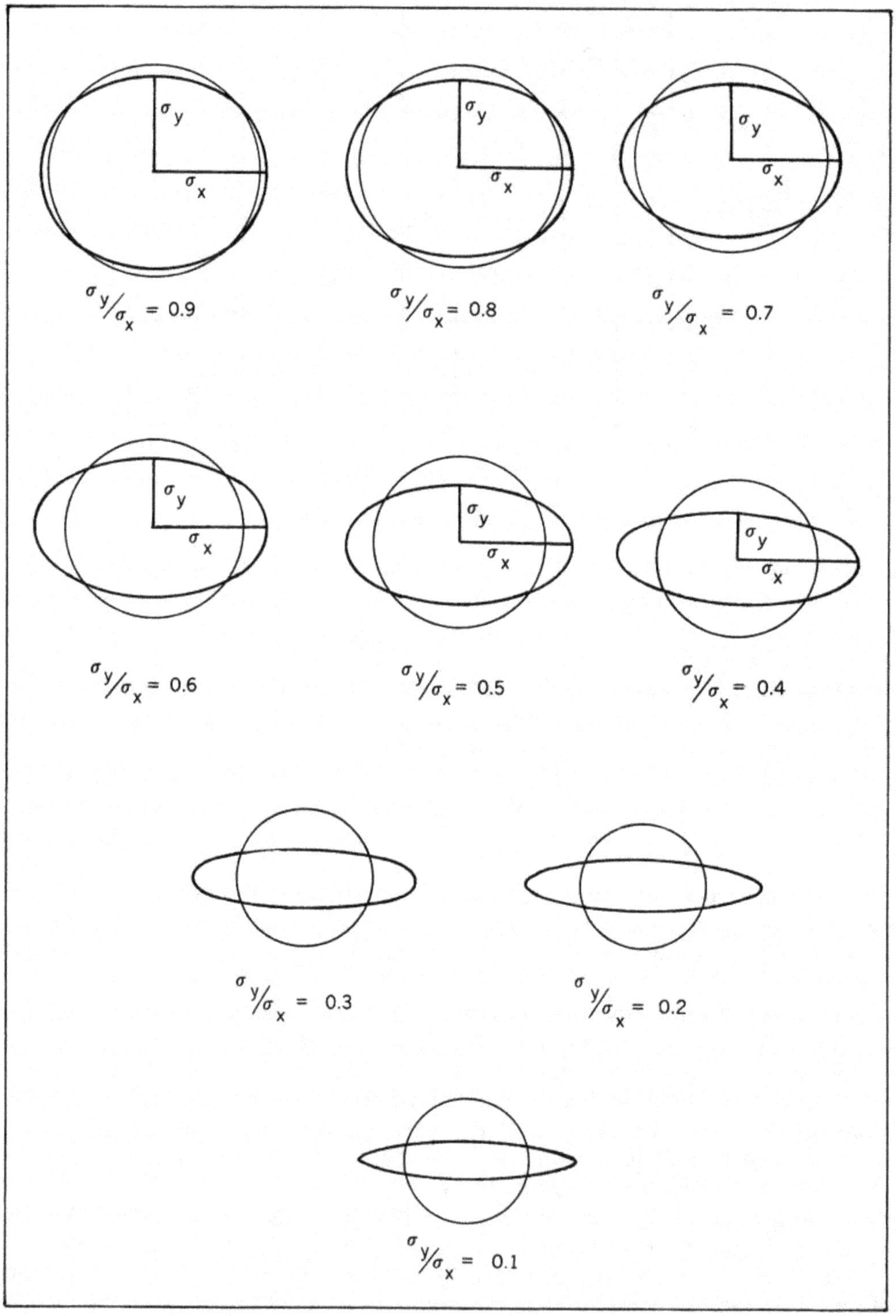

FIGURE Q7i.—Substitution of the circular form for elliptical error distributions.

can then be assigned a smaller likely error and used with the other lines of position to determine the overall most probable position. A third way is to establish a likely error for the fix, and consider the most probable position as that point along the straight line joining the fix and the dead reckoning or estimated position, the relative distances being equal to the square of the likely error of each position.

The value of the most probable position determined as suggested above depends upon the degree to which the various errors are in fact normal, and the accuracy with which the likely error of each is established. From a practical standpoint, the second factor is largely a matter of judgment based upon experience. It might seem that interpretation of results and establishment of most probable position is a matter of judgment anyway, and that the procedure outlined above is not needed. If a person will follow this procedure while gaining experience, and evaluate his results, the judgment he develops should be more reliable than if developed without benefit of a knowledge of the principles involved. The important point to remember is that the relative effects of normal random errors in any one direction are proportional to their *squares.*

Systematic errors are treated differently. Generally, an attempt is made to discover the errors and eliminate them or compensate for them. In the case of a position determined by three or more lines of position resulting from readings with constant error, the error might be eliminated by finding and applying that correction (including sign) which will bring all lines through a common point.

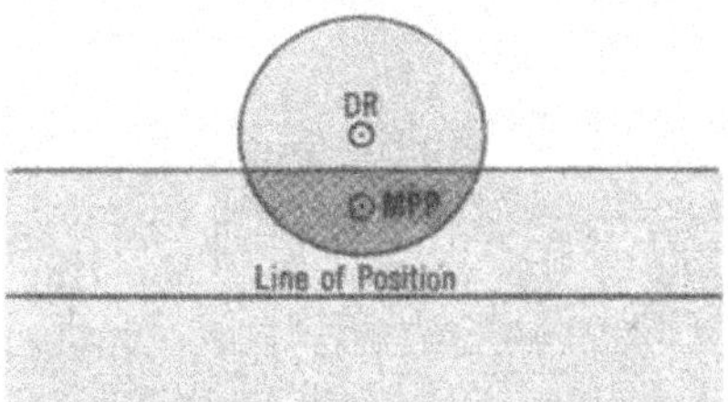

FIGURE Q9a.—A most probable position based upon a dead reckoning position and line of position having equal probable errors.

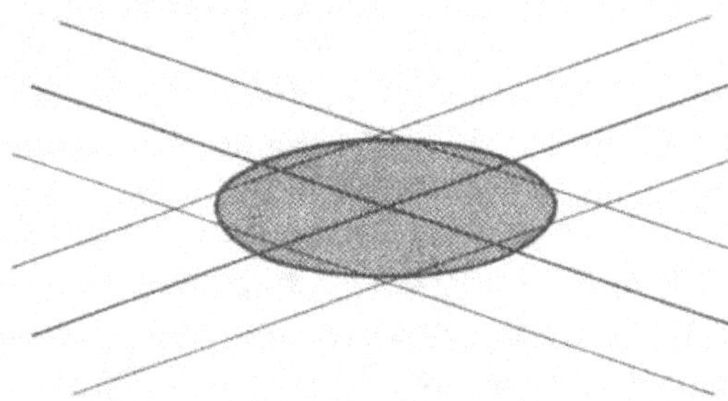

FIGURE Q9b.—Ellipse of uncertainty with lines of positions of equal probable errors crossing at an oblique angle.

Q10. Mistakes.—The recognition of a mistake, as contrasted with an error (art. Q2), is not always easy, since a mistake may have any magnitude, and may be either positive or negative. A large mistake should be readily apparent if the navigator is alert and has an understanding of the size of error to be reasonably expected. A small mistake is usually not detected unless the work is checked.

If results by two methods are compared, as a dead reckoning position and a line of position, exact agreement is not to be expected. But if the discrepancy is unreasonably large, a mistake is logically suspected. The definition of "unreasonably large" is a matter of opinion. If the 99.9 percent areas of the two results just touch, it is *possible* that no mistake has been made. However, the *probability* of either one having so great an error is remote if the errors are normal. The probability of both having 99.9 percent error of opposite sign at the same instant is very small indeed. Perhaps a reasonable standard is that unless the most accurate result lies within the 95 percent area of the least accurate result, the possibility of a mistake should be investigated. Thus, if the areas of uncertainty shown in figure Q9a represent the 95 percent areas, it is probable that a mistake has been made.

As in other matters pertaining to navigation, judgment is important. The use to be made of the results is certainly a consideration. In the middle of an ocean passage a mistake is usually not serious, and will undoubtedly be corrected before it jeopardizes the safety of the vessel. But if landfall is soon to be made, or if search and rescue operations are to be based upon the position, almost any mistake is intolerable.

Q11. Conclusion.—The correct identification of the nature of an error is important if the error is to be handled intelligently. Thus, the statement is sometimes made that a radio bearing need not be corrected if the receiver is within 50 miles of the transmitter. The need for a correction arises from the fact that radio waves are assumed to follow great circles, and if radio bearings are to be plotted on a Mercator chart, the equivalent rhumb line is needed. The statement regarding 50 miles implies that the size of the correction is proportional to distance only. It overlooks the fact that latitude and direction of the bearing line are also important factors, and is therefore a dangerous statement unless its limitations are understood.

The recognition of the type of error is also important. A systematic error has quite a different effect than a random error, and cannot be reduced by additional readings unless some method or procedure is instituted which will cause the errors to cancel each other.

The errors for various percentage probabilities are usually of greater interest than the "average" value. The average of a large number of normal errors approaches zero, but the probable (50 percent) error might be quite large.

A person who understands the nature of errors avoids many pitfalls. Thus, the magnitude of the errors of individual lines of position is not a reliable indication of the size of the error of the fix obtained from them. The size of the triangle formed by three lines of position has often been used as a guide to the accuracy of the fix, although a large triangle might be the result of a large constant error if the objects observed are equally spaced in azimuth. On the other hand, two lines of position with small errors might produce a fix having a much larger error if the lines cross at a small angle.

References

Burt, W. A., D. J. Kaplan, and R. R. Keenly, et al. "Mathematical Considerations Pertaining to the Accuracy of Position Location and Navigation Systems." Naval Warfare Research Center Research Memorandum NWRC—RM 34, Stanford Research Institute, Menlo Park, California, November 1965.

Greenwalt, C. R. and M. E. Shultz. "Principles of Error Theory and Cartographic Applications." Aeronautical Chart and Information Center Technical Report No. 96, St. Louis, Missouri, February 1962.

APPENDIX R

GHA and Declination of the Sun for the Years 1981–2016—Argument "Orbit Time"

OT	JAN			FEB			MAR			APR			MAY			JUN			JUL			AUG			SEP			OCT			NOV			DEC			OT
00h	E	Dec.		E	Dec.		E	Dec.		E	Dec.		E	Dec.		E	Dec.		E	Dec.		E	Dec.		E	Dec.		E	Dec.		E	Dec.		E	Dec.		00h
d	° ′	° ′		° ′	° ′		° ′	° ′		° ′	° ′		° ′	° ′		° ′	° ′		° ′	° ′		° ′	° ′		° ′	° ′		° ′	° ′		° ′	° ′		° ′	° ′		d
1	4 12	S 23 03	5	1 38	S 17 16	17	1 52	S 7 49	23	3 58	N 4 19	23	5 42	N 14 54	18	5 35	N 21 58	8	4 05	N 23 09	4	3 24	N 18 10	15	4 56	N 8 30	22	7 31	S 2 57	24	9 06	S 14 14	19	7 49	S 21 42	10	1
2	4 05	22 58	5	1 36	16 59	18	1 55	7 26	23	4 02	4 42	23	5 44	15 12	18	5 32	22 06	8	4 02	23 05	5	3 25	17 55	16	5 01	8 08	22	7 36	3 21	23	9 06	14 33	19	7 44	21 52	9	2
3	3 58	22 53	6	1 34	16 41	17	1 58	7 03	23	4 07	5 05	23	5 46	15 30	18	5 30	22 14	8	3 59	23 00	5	3 26	17 39	15	5 06	7 46	22	7 41	3 44	23	9 06	14 52	19	7 38	22 01	8	3
4	3 51	22 47	7	1 32	16 24	18	2 01	6 40	23	4 11	5 28	23	5 47	15 48	17	5 28	22 22	7	3 56	22 55	5	3 28	17 24	16	5 11	7 24	22	7 45	4 07	23	9 06	15 11	19	7 32	22 09	8	4
5	3 44	22 40	6	1 31	16 06	18	2 05	6 17	23	4 16	5 51	23	5 49	16 05	17	5 25	22 29	6	3 54	22 50	5	3 29	17 08	17	5 16	7 02	22	7 50	4 30	23	9 06	15 30	18	7 26	22 17	8	5
6	3 37	S 22 34	7	1 30	S 15 48	19	2 08	S 5 54	24	4 20	N 6 14	22	5 50	N 16 22	17	5 22	N 22 35	6	3 51	N 22 45	6	3 31	N 16 51	16	5 21	N 6 40	23	7 55	S 4 53	23	9 06	S 15 48	18	7 20	S 22 25	7	6
7	3 31	22 27	8	1 29	15 29	18	2 11	5 30	23	4 24	6 36	23	5 51	16 39	17	5 20	22 41	6	3 48	22 39	7	3 32	16 35	17	5 26	6 17	22	7 59	5 16	23	9 05	16 06	18	7 14	22 32	7	7
8	3 24	22 19	8	1 28	15 11	19	2 15	5 07	23	4 29	6 59	22	5 52	16 56	16	5 17	22 47	6	3 46	22 32	7	3 34	16 18	17	5 31	5 55	23	8 03	5 39	23	9 05	16 24	17	7 07	22 39	6	8
9	3 18	22 11	9	1 27	14 52	20	2 19	4 44	24	4 33	7 21	23	5 53	17 12	16	5 14	22 53	5	3 44	22 25	7	3 36	16 01	17	5 36	5 32	22	8 08	6 02	23	9 04	16 41	17	7 01	22 45	6	9
10	3 12	22 02	8	1 27	14 32	19	2 22	4 20	23	4 37	7 44	22	5 54	17 28	16	5 11	22 58	4	3 42	22 18	7	3 38	15 44	17	5 41	5 10	23	8 12	6 25	23	9 02	16 58	17	6 54	22 51	6	10
11	3 06	S 21 54	10	1 26	S 14 13	20	2 26	S 3 57	24	4 41	N 8 06	22	5 54	N 17 44	15	5 08	N 23 02	5	3 39	N 22 11	8	3 40	N 15 27	18	5 46	N 4 47	23	8 16	S 6 48	22	9 01	S 17 15	17	6 47	S 22 57	5	11
12	3 00	21 44	10	1 26	13 53	20	2 30	3 33	23	4 45	8 28	22	5 55	17 59	15	5 05	23 07	3	3 37	22 03	8	3 43	15 09	18	5 52	4 24	23	8 20	7 10	23	8 59	17 32	16	6 40	23 02	4	12
13	2 54	21 34	10	1 27	13 33	20	2 34	3 10	24	4 49	8 50	22	5 55	18 14	15	5 02	23 10	4	3 36	21 55	9	3 45	14 51	19	5 57	4 01	23	8 23	7 33	22	8 58	17 48	16	6 33	23 06	4	13
14	2 48	21 24	10	1 27	13 13	20	2 38	2 46	24	4 53	9 12	21	5 55	18 29	14	4 59	23 14	3	3 34	21 46	9	3 48	14 32	18	6 02	3 38	23	8 27	7 55	23	8 55	18 04	16	6 26	23 10	4	14
15	2 43	21 14	11	1 28	12 53	21	2 43	2 22	23	4 56	9 33	22	5 55	18 43	15	4 56	23 17	2	3 32	21 37	10	3 51	14 14	19	6 08	3 15	23	8 31	8 18	22	8 53	18 20	15	6 19	23 14	3	15
16	2 38	S 21 03	12	1 28	S 12 32	21	2 47	S 1 59	24	5 00	N 9 55	21	5 55	N 18 58	13	4 53	N 23 19	3	3 31	N 21 27	9	3 54	N 13 55	19	6 13	N 2 52	23	8 34	S 8 40	22	8 51	S 18 35	15	6 12	S 23 17	3	16
17	2 32	20 51	12	1 29	12 11	21	2 51	1 35	24	5 04	10 16	21	5 55	19 11	14	4 50	23 22	1	3 29	21 18	11	3 57	13 36	19	6 18	2 29	23	8 37	9 02	22	8 48	18 50	15	6 05	23 20	2	17
18	2 28	20 39	12	1 30	11 50	21	2 55	1 11	24	5 07	10 37	21	5 54	19 25	13	4 46	23 23	2	3 28	21 07	10	4 00	13 17	19	6 24	2 06	23	8 40	9 24	22	8 45	19 05	14	5 57	23 22	2	18
19	2 23	20 27	12	1 31	11 29	21	3 00	0 47	23	5 10	10 58	21	5 54	19 38	13	4 43	23 25	1	3 27	20 57	11	4 03	12 58	20	6 29	1 43	24	8 43	9 46	21	8 42	19 19	14	5 50	23 24	1	19
20	2 18	20 15	13	1 33	11 08	22	3 04	S 0 24	24	5 14	11 19	20	5 53	19 51	13	4 40	23 26	0	3 26	20 46	11	4 07	12 38	19	6 34	1 19	23	8 46	10 07	22	8 39	19 33	14	5 42	23 25	1	20
21	2 14	S 20 02	14	1 34	S 10 46	21	3 09	0 00	24	5 17	N 11 39	21	5 52	N 20 04	12	4 36	N 23 26	0	3 25	N 20 35	12	4 10	N 12 19	20	6 40	N 0 56	23	8 48	S 10 29	21	8 35	S 19 47	13	5 35	S 23 26	0	21
22	2 09	19 48	13	1 36	10 25	22	3 13	N 0 24	23	5 20	12 00	20	5 51	20 16	12	4 33	23 26	0	3 24	20 23	12	4 14	11 59	21	6 45	0 33	24	8 51	10 50	22	8 31	20 00	13	5 28	23 26	0	22
23	2 05	19 35	14	1 38	10 03	22	3 17	0 47	24	5 23	12 20	20	5 50	20 28	11	4 30	23 26	1	3 23	20 11	12	4 18	11 38	20	6 50	N 0 09	23	8 53	11 12	21	8 27	20 13	12	5 20	23 26	0	23
24	2 02	19 21	15	1 40	9 41	22	3 22	1 11	24	5 26	12 40	20	5 49	20 39	11	4 27	23 25	1	3 23	19 59	12	4 21	11 18	20	6 55	S 0 14	23	8 55	11 33	20	8 23	20 25	13	5 13	23 26	2	24
25	1 58	19 06	15	1 42	9 19	23	3 26	1 35	23	5 28	13 00	19	5 48	20 50	11	4 23	23 24	1	3 23	19 47	13	4 25	10 58	21	7 01	0 37	24	8 57	11 53	21	8 19	20 38	11	5 05	23 24	1	25
26	1 54	S 18 51	15	1 45	S 8 56	22	3 31	N 1 58	24	5 31	N 13 19	20	5 46	N 21 01	11	4 20	N 23 23	2	3 22	N 19 34	13	4 29	N 10 37	21	7 06	S 1 01	23	8 59	S 12 14	21	8 14	S 20 49	12	4 58	S 23 23	2	26
27	1 51	18 36	15	1 47	8 34	23	3 36	2 22	23	5 33	13 39	19	5 44	21 12	10	4 17	23 21	3	3 22	19 21	14	4 34	10 16	21	7 11	1 24	23	9 01	12 35	20	8 10	21 01	11	4 50	23 21	3	27
28	1 48	18 21	16	1 50	8 11	22	3 40	2 45	24	5 36	13 58	19	5 43	21 22	9	4 14	23 18	3	3 22	19 07	14	4 38	9 55	21	7 16	1 47	24	9 02	12 55	20	8 05	21 12	10	4 43	23 18	2	28
29	1 45	18 05	16	1 52	7 49	23	3 45	3 09	23	5 38	14 17	18	5 41	21 31	10	4 11	23 15	3	3 23	18 53	14	4 42	9 34	21	7 21	2 11	23	9 03	13 15	20	8 00	21 22	11	4 36	23 16	4	29
30	1 42	17 49	16				3 49	3 32	23	5 40	14 35	19	5 39	21 41	9	4 08	23 12	3	3 23	18 39	15	4 47	9 13	22	7 26	2 34	23	9 04	13 35	20	7 55	21 33	9	4 28	23 12	4	30
31	1 40	S 17 33	17				3 54	N 3 55	24				5 37	N 21 50	8				3 24	N 18 24	14	4 51	N 8 51	21				9 05	S 13 55	19				4 21	S 23 8	4	31

a. Corr. to GMT

Year	Corr.	Year	Corr.	Year	Corr.
	h		h		h
1981	+ 7	1993	+ 9	2005	+11
1982	+ 1	1994	+ 4	2006	+ 6
1983	− 5	1995	− 2	2007	0
1984	−10	1996	− 8	2008	− 6
"	+14*	"	+16*	"	+18*
1985	+ 8	1997	+10	2009	+12
1986	+ 2	1998	+ 4	2010	+ 6
1987	− 4	1999	− 2	2011	+ 1
1988	−10	2000	− 8	2012	− 5
"	+14*	"	+16*	"	+19*
1989	+ 9	2001	+10	2013	+13
1990	+ 3	2002	+ 5	2014	+ 7
1991	− 3	2003	− 1	2015	+ 1
1992	− 9	2004	− 7	2016	− 5
"	+15*	"	+17*	"	+19*

* After Feb. 29

b. Interpolation for Hours of OT

Diff.	1	2	3	4	5	6	7	8	9	10	11	12	13	14	15	16	17	18	19	20	21	22	23	24
h	′	′	′	′	′	′	′	′	′	′	′	′	′	′	′	′	′	′	′	′	′	′	′	′
0	0	0	0	0	0	0	0	0	0	0	0	0	0	0	0	0	0	0	0	0	0	0	0	0
1	0	0	0	0	0	0	0	0	0	0	0	1	1	1	1	1	1	1	1	1	1	1	1	1
2	0	0	0	0	0	1	1	1	1	1	1	1	1	1	1	1	1	2	2	2	2	2	2	2
3	0	0	0	1	1	1	1	1	1	1	1	2	2	2	2	2	2	2	2	3	3	3	3	3
4	0	0	1	1	1	1	1	1	2	2	2	2	2	2	3	3	3	3	3	3	4	4	4	4
5	0	0	1	1	1	1	1	2	2	2	2	3	3	3	3	3	4	4	4	4	4	5	5	5
6	0	1	1	1	1	2	2	2	2	3	3	3	3	4	4	4	4	5	5	5	5	6	6	6
7	0	1	1	1	1	2	2	2	3	3	3	4	4	4	4	5	5	5	6	6	6	6	7	7
8	0	1	1	1	2	2	2	3	3	3	4	4	4	5	5	5	6	6	6	7	7	7	8	8
9	0	1	1	2	2	2	3	3	3	4	4	5	5	5	6	6	6	7	7	8	8	8	9	9
10	0	1	1	2	2	3	3	3	4	4	5	5	5	6	6	7	7	8	8	8	9	9	10	10
11	0	1	1	2	2	3	3	4	4	5	5	6	6	6	7	7	8	8	9	9	10	10	11	11

Diff.	1	2	3	4	5	6	7	8	9	10	11	12	13	14	15	16	17	18	19	20	21	22	23	24
h	′	′	′	′	′	′	′	′	′	′	′	′	′	′	′	′	′	′	′	′	′	′	′	′
12	1	1	2	2	3	3	4	4	5	5	6	6	7	7	8	8	9	9	10	10	11	11	12	12
13	1	1	2	2	3	3	4	4	5	5	6	7	7	8	8	9	9	10	10	11	11	12	12	13
14	1	1	2	2	3	4	4	5	5	6	6	7	8	8	9	9	10	11	11	12	12	13	13	14
15	1	1	2	3	3	4	4	5	6	6	7	8	8	9	9	10	11	11	12	13	13	14	14	15
16	1	1	2	3	3	4	5	5	6	7	7	8	9	9	10	11	11	12	13	13	14	15	15	16
17	1	1	2	3	4	4	5	6	6	7	8	9	9	10	11	11	12	13	13	14	15	16	16	17
18	1	2	2	3	4	5	5	6	7	8	8	9	10	11	11	12	13	14	14	15	16	17	17	18
19	1	2	2	3	4	5	6	6	7	8	9	10	10	11	12	13	13	14	15	16	17	17	18	19
20	1	2	3	3	4	5	6	7	8	8	9	10	11	12	13	13	14	15	16	17	18	18	19	20
21	1	2	3	4	4	5	6	7	8	9	10	11	11	12	13	14	15	16	17	18	18	19	20	21
22	1	2	3	4	5	6	6	7	8	9	10	11	12	13	14	15	16	17	17	18	19	20	21	22
23	1	2	3	4	5	6	7	8	9	10	11	12	12	13	14	15	16	17	18	19	20	21	22	23

GHA and Declination of the Sun for the Years 1981–2016—Argument "Orbit Time"—Continued

c. Hours and Tens of minutes of GMT

	00m	10m	20m	30m	40m	50m
h	° ′	° ′	° ′	° ′	° ′	° ′
00	175 00	177 30	180 00	182 30	185 00	187 30
01	190 00	192 30	195 00	197 30	200 00	202 30
02	205 00	207 30	210 00	212 30	215 00	217 30
03	220 00	222 30	225 00	227 30	230 00	232 30
04	235 00	237 30	240 00	242 30	245 00	247 30
05	250 00	252 30	255 00	257 30	260 00	262 30
06	265 00	267 30	270 00	272 30	275 00	277 30
07	280 00	282 30	285 00	287 30	290 00	292 30
08	295 00	297 30	300 00	302 30	305 00	307 30
09	310 00	312 30	315 00	317 30	320 00	322 30
10	325 00	327 30	330 00	332 30	335 00	337 30
11	340 00	342 30	345 00	347 30	350 00	352 30
12	355 00	357 30	0 00	2 30	5 00	7 30
13	10 00	12 30	15 00	17 30	20 00	22 30
14	25 00	27 30	30 00	32 30	35 00	37 30
15	40 00	42 30	45 00	47 30	50 00	52 30
16	55 00	57 30	60 00	62 30	65 00	67 30
17	70 00	72 30	75 00	77 30	80 00	82 30
18	85 00	87 30	90 00	92 30	95 00	97 30
19	100 00	102 30	105 00	107 30	110 00	112 30
20	115 00	117 30	120 00	122 30	125 00	127 30
21	130 00	132 30	135 00	137 30	140 00	142 30
22	145 00	147 30	150 00	152 30	155 00	157 30
23	160 00	162 30	165 00	167 30	170 00	172 30

d. Minutes and Seconds of GMT (in critical cases ascend)

m s	° ′	m s	° ′	m s	° ′	m s	° ′	m s	° ′	m s	° ′
00 00	0 00	01 37	0 25	03 17	0 50	04 57	1 15	06 37	1 40	08 17	2 05
01	0 01	41	0 26	21	0 51	05 01	1 16	41	1 41	21	2 06
05	0 02	45	0 27	25	0 52	05	1 17	45	1 42	25	2 07
09	0 03	49	0 28	29	0 53	09	1 18	49	1 43	29	2 08
13	0 04	53	0 29	33	0 54	13	1 19	53	1 44	33	2 09
17	0 05	01 57	0 30	37	0 55	17	1 20	06 57	1 45	37	2 10
21	0 06	02 01	0 31	41	0 56	21	1 21	07 01	1 46	41	2 11
25	0 07	05	0 32	45	0 57	25	1 22	05	1 47	45	2 12
29	0 08	09	0 33	49	0 58	29	1 23	09	1 48	49	2 13
33	0 09	13	0 34	53	0 59	33	1 24	13	1 49	53	2 14
37	0 10	17	0 35	03 57	1 00	37	1 25	17	1 50	08 57	2 15
41	0 11	21	0 36	04 01	1 01	41	1 26	21	1 51	09 01	2 16
45	0 12	25	0 37	05	1 02	45	1 27	25	1 52	05	2 17
49	0 13	29	0 38	09	1 03	49	1 28	29	1 53	09	2 18
53	0 14	33	0 39	13	1 04	53	1 29	33	1 54	13	2 19
00 57	0 15	37	0 40	17	1 05	05 57	1 30	37	1 55	17	2 20
01 01	0 16	41	0 41	21	1 06	06 01	1 31	41	1 56	21	2 21
05	0 17	45	0 42	25	1 07	05	1 32	45	1 57	25	2 22
09	0 18	49	0 43	29	1 08	09	1 33	49	1 58	29	2 23
13	0 19	53	0 44	33	1 09	13	1 34	53	1 59	33	2 24
17	0 20	02 57	0 45	37	1 10	17	1 35	07 57	2 00	37	2 25
21	0 21	03 01	0 46	41	1 11	21	1 36	08 01	2 01	41	2 26
25	0 22	05	0 47	45	1 12	25	1 37	05	2 02	45	2 27
29	0 23	09	0 48	49	1 13	29	1 38	09	2 03	49	2 28
33	0 24	13	0 49	53	1 14	33	1 39	13	2 04	53	2 29
37	0 25	17	0 50	04 57	1 15	37	1 40	17	2 05	09 57	2 30
01 41		03 21		05 01		06 41		08 21		10 00	

EXPLANATION

Table 4 and supplementary tables a, b, c, d, make possible the determination of the GHA and declination of the Sun for any time during the years 1981–2016. The main table gives E (5° + Equation of Time) and declination of the Sun for the argument "Orbit Time" O T, the latter is formed by applying the h correction from Table a to the nearest integral hour of GMT. In leap years the upper value of the correction is to be used for January and February and the lower value for the rest of the year. Thus, O T's corresponding to 1996 February $29^d\ 16^h\ 31^m$ GMT and 1996 March $1^d\ 05^h\ 29^m$ GMT are February $29^d\ 09^h\ 00^m$ and March $1^d\ 21^h\ 00^m$ respectively.

Corrections to E and declination for O T are determined by entering Table b with the differences between consecutive values of E and of declination respectively as the horizontal argument, and with the number of hours of O T as the vertical argument. The declination differences are given in the main table.

The GHA is obtained by adding to the corrected E the value of the diurnal arc obtained from Tables c and d. The latter two tables must be entered with argument GMT.

Example. To find the GHA and declination of the Sun on 1996 January 18 at $03^h\ 30^m\ 35^s$ GMT.

O T = GMT (nearest integral hour) + Corr. (Table a).
= Jan. $18^d\ 04^h - 8^h$ = Jan. $17^d\ 20^h$.

		° ′	Diff.		° ′	Diff.
Main Table, Jan. 17^d O T,	E	2 32	(−4)	Dec. S	20 51	(−12)
Table b for 20^h O T		−3			−10	
Jan. $17^d\ 20^h$ O T, corrected	E	2 29		Dec. S	20 41	
Table c for $03^h\ 30^m$ GMT		227 30				
Table d for $00^m\ 35^s$ GMT		0 09				
Sum	GHA Sun =	230 08				

APPENDIX S

CHARTS AND PUBLICATIONS OF OTHER AGENCIES

The following items are publications and charts about which the Defense Mapping Agency Hydrographic/Topographic Center frequently receives inquiries. The Defense Mapping Agency Office of Distribution Services does not sell the charts and publications listed below. Application should be made to the authorities named.

Charts

United States waters and possessions.—Published by National Ocean Service. Charts and related publications may be ordered from Distribution Branch, CG33, National Ocean Service, Riverdale, Maryland 20737 (Counter Sales: 6501 Lafayette Avenue, Riverdale, Maryland—Building 1, 6001 Executive Boulevard, Rockville, Maryland—Atlantic Marine Service, NOS, 439 West York Street, Norfolk, Virginia) or from authorized sales agents.

Orders mailed to Riverdale, Maryland should be accompanied by a check or money order payable to NOS, Department of Commerce. Remittance from outside of the United States should be made either by an International Money Order or by a check payable on a United States Bank.

Canadian coastal and Great Lakes waters.—Charts may be ordered from Hydrographic Chart Distribution Office, Department of the Environment, 1675 Russel Road, P.O. Box 8080, Ottawa, Ontario, KIG 3H6, Canada. Pacific Coast and Western Arctic charts may be ordered from Canadian Hydrographic Service, Department of the Environment, 512 Federal Building, Victoria, British Columbia. Payment in the form of a money order or bankable remittance payable in Canadian funds to the Receiver General of Canada must accompany an order. C.O.D. orders and postage stamps are not accepted.

Lower Mississippi River.—Charts and maps of the Mississippi River from the Gulf of Mexico to the Ohio River (also St. Francis, White, Big Sunflower, and other rivers) may be ordered from U. S. Army Corps of Engineers, Vicksburg District, P.O. Box 60, Vicksburg, Mississippi 39180.

Middle and Upper Mississippi River and Illinois Waterway to Lake Michigan.—Charts may be ordered from U. S. Army Corps of Engineers, Chicago District, 219 South Dearborn Street, Chicago, Illinois, 60604.

Missouri River.—Charts may be ordered from U. S. Army Corps of Engineers, Omaha District, U. S. Post Office and Courthouse, 215 North 17th Street, Omaha, Nebraska 68102.

Ohio River.—Charts may be ordered from U. S. Army Corps of Engineers, Ohio River Division, P.O. Box 1159, Cincinnati, Ohio 45201.

Tennessee River and tributaries.—Charts may be ordered from Tennessee Valley Authority, Map Sales, 400 Commerce Avenue (WP A3), Knoxville, Tennessee 37902.

Black Warrior River, Alabama River, Tombigbee River, Apalachicola River, Chattahooche River, Flint River, and Pearl River.—Charts may be ordered from

U.S. Army Corps of Engineers, Mobile District, P.O. Box 2288, Mobile Alabama 36628 (Attention SAMEN-DI).

Gulf Intracoastal Waterway.—A *booklet* made from National Ocean Service nautical charts and showing depths of water, channels, navigation markers, and types of bottom from New Orleans, Louisiana to St. Marks, Florida may be ordered from U.S. Army Engineer District, Mobile, P.O. Box 2288, Mobile, Alabama 36628 (Attention: SAMEN-DI). *Nautical charts* for the Gulf Intracoastal Waterway from Brownsville, Texas to New Orleans, Louisiana, and from St. Marks Florida to the Florida Keys may be obtained from Distribution Branch, CG33, National Ocean Service, Riverdale, Maryland 20737.

National Ocean Service Publications

Tide Tables, Tidal Current Tables, and *United States Coast Pilots* are sold by the Distribution Branch, CG33, National Ocean Service, Riverdale, Maryland 20737, National Ocean Service, and its agents.

U.S. Naval Observatory Publications

Air Almanac, Nautical Almanac, and *Astronomical Almanac* are for sale by Superintendent of Documents, Government Printing Office, Washington, D.C. 20402.

U.S. Coast Guard Publications

Light Lists, United States coasts and possessions, are for sale by Coast Guard Sales Agents and by the Superintendent of Documents, Government Printing Office, Washington, D.C. 20402.

Navigation Rules, International—Inland, Commandant Instruction Ml6672.2, may be obtained from U.S. Coast Guard Marine Inspection Offices or the Commandant (G-N), U.S. Coast Guard Headquarters, Washington, D.C. 20593.

Local Notice to Mariners may be obtained free from the Commander of the local Coast Guard District.

National Weather Service Publications

Selected Worldwide Marine Weather Broadcasts, published jointly by the Naval Oceanographic Command and the National Weather Service of the National Oceanic and Atmospheric Administration, is for sale by the Distribution Branch, CG33, National Ocean Service, Riverdale, Maryland 20737, Sales Agents of the National Ocean Service, and by the Superintendent of Documents, Government Printing Office, Washington, D.C. 20402.

APPENDIX T

HAND-HELD DIGITAL CALCULATORS

T1. Introduction.—Many hand-held digital calculators are available for solving calculations in the practice of navigation. Any advantage obtained over manual or tabular methods through use of any one of these calculators is largely dependent upon the sophistication of the calculator, the user's familiarity with it, and the navigational problem involved. However, even the simplest calculator can be used to save time and increase reliability and accuracy in some common calculations. Since the sophistication of calculators is so diverse, they will be discussed in three general categories described here as **basic, intermediate,** and **advanced** or **programmable calculators.**

Basic Calculators

T2. Basic Calculators can be used to add, subtract, multiply, and divide. Some calculators in this category can be used for obtaining square roots, reciprocals, and the logarithms of numbers, but they cannot be used for obtaining trigonometric functions or the inverse trigonometric functions. The basic calculators lack the addressable storage features of the more advanced calculators which can be used to facilitate arithmetic operations.

The navigational applications of the basic calculators include: (1) linear interpolation, (2) solution of basic formulas requiring only routine arithmetic operations, (3) use with tables of trigonometric functions for solutions of basic formulas involving trigonometric functions, and (4) other routine arithmetic operations.

T3. Linear interpolation.—The use of the basic calculator for linear interpolation is simple, accurate, reliable, and time saving. The calculator can be used to expedite the linear interpolations normally associated with the use of sight reduction tables. Note that the interpolation table of Pub. No. 229 does not always provide interpolations to the nearest tenth of a minute.

T4. Basic calculator with square root capability.—The basic calculators that provide the square roots of numbers can be used for the solution of several common navigation problems.

Distance to the visible horizon (art. 724, vol. II) in *nautical miles* can be calculated using the formula:

$$d=1.17\sqrt{h_f}$$

or

$$d=2.11\sqrt{h_m},$$

depending upon whether the height of eye of the observer above sea level is in *feet* (h_f) or in *meters* (h_m).

Dip of the visible horizon (art. 725, vol. II) in *minutes of arc* can be calculated using the formula:

$$D=0'.97\sqrt{h_f}$$

or

$$D=1'.76\sqrt{h_m},$$

depending upon whether the height of eye of the observer above sea level is in *feet* (h_f) or in *meters* (h_m).

Distance to the radar horizon (art. 506, vol. II), in *nautical miles* can be calculated using the formula:

$$d = 1.22\sqrt{h_f}$$

or

$$d = 2.21\sqrt{h_m},$$

depending upon whether the height of the antenna above sea level is in *feet* (h_f) or in *meters* (h_m).

T5. Dip of the sea short of the horizon.—The basic calculator can be used instead of table 22 for the solution of the dip of the sea short of the horizon (art. 726, vol. II) by means of the formula:

$$D_s = 60 \tan^{-1}\left(\frac{h_f}{6076.1\ d_s} + \frac{\beta_o\ d_s}{2r_o}\right),$$

where D_s is the dip short of the horizon in minutes of arc; h_f is the height of eye of the observer above sea level in feet and must be entered in the formula in nautical miles or parts thereof; β_o is a parameter (0.8321) which characterizes terrestrial refraction; r_o is the mean radius of the earth, 3440.1 nautical miles; and d_s is the distance to waterline of obstruction in nautical miles.

The formula simplifies to:

$$D_s = 60 \tan^{-1}\left(\frac{h_f}{6076.1\ d_s} + \frac{d_s}{8268}\right).$$

T6. Use with tables of trigonometric functions.—The basic calculator can be used with tables of natural trigonometric functions (table 31) to avoid the use of logarithmic functions which might otherwise be necessary.

The solutions of the examples given in article 706 of volume II can be effected with the basic calculator if the natural trigonometric functions are extracted from table 31 or other tables of trigonometric functions. The solutions by this method are generally more laborious and time consuming than the use of the intermediate calculator (art. T7) or modern sight reduction tables.

Intermediate Calculators

T7. Intermediate calculators provide the trigonometric functions and inverse trigonometric functions lacking in the basic calculator, but these calculators cannot be programmed. The intermediate calculators normally contain many of the special features which enhance the use of the calculator for solving navigational problems. In addition to the special features which the basic calculator may have, the intermediate calculator may have the capability for conversion of degrees and minutes to degrees and decimal degree, and for rectangular to polar coordinate conversion. These intermediate calculators usually have additional working storage registers and addressable storage registers for expeditious solution of the more complex arithmetic operations.

The intermediate calculators can be used for the solutions of most, if not all, of the problems normally encountered in marine navigation. Some of the applications of these intermediate calculators are discussed under the following categories: (1) sight reduction, (2) azimuth and amplitude solutions, and (3) the sailings.

T8. Sight reduction by basic formulas permits the use of the normally nonintegral values of latitude of the observer, and LHA and declination of the celestial body. The reductions are effected without need for the interpolation normally associated with the use of sight reduction tables of the inspection type. Solving the formulas by means of the intermediate calculator, the sights can be readily reduced from the observer's most probable position or any assumed position of his choice. Simultaneous, or nearly simultaneous, observations can be reduced using a single assumed position. The use of the most probable position for the assumed position instead of that position which would provide integral values of latitude and LHA to avoid interpolation for these quantities generally tends to provide a shorter straight line LOP, and thus a better representation of the circle of equal altitude (art. 1701).

The computed altitude is calculated using the basic formula for solution of the undivided navigational triangle (art. 706, vol. II):

$$\sin h = \sin L \sin d + \cos L \cos d \cos \text{LHA},$$

in which h is the altitude to be computed (Hc), L is the latitude of the assumed position for the reduction, d is the declination of the celestial body, and LHA is the local hour angle of the body. Meridian angle (t) can be substituted for LHA in the basic formula.

For use with the intermediate calculator, the basic formula is restated in terms of the inverse trigonometric function (art. 141, vol. II):

$$\text{Hc} = \sin^{-1}[(\sin L \sin d) + (\cos L \cos d \cos \text{LHA})].$$

When latitude and declination are of contrary name, declination is treated as a negative quantity. No special sign convention is required for the local hour angle as in the following azimuth angle calculations.

The azimuth angle (Z) can be calculated using the *altitude azimuth* formula (art. 709, vol. II) if the altitude is known. The formula stated in terms of the inverse trigonometric function is

$$\text{Z} = \cos^{-1}\left[\frac{\sin d - (\sin L \sin \text{Hc})}{(\cos L \cos \text{Hc})}\right].$$

If the altitude is unknown or a solution independent of altitude is required, the azimuth angle can be calculated using the *time azimuth* formula (art. 707, vol. II). The formula stated in terms of the inverse trigonometric function is

$$\text{Z} = \tan^{-1}\left[\frac{\sin \text{LHA}}{(\cos L \tan d) - (\sin L \cos \text{LHA})}\right].$$

The sign conventions used in the calculations of both azimuth formulas are as follows: (1) If latitude and declination are of contrary name, declination is treated as a negative quantity; (2) If the local hour angle is greater than 180°, it is treated as a negative quantity.

If the azimuth angle as calculated is negative, it is necessary to add 180° to obtain the desired value.

The calculation of altitude and azimuth angle by means of the intermediate calculator requires many individual key strokes. Error-free accomplishment of a large number of strokes is not easy. Skill can be obtained by practice using values differing by only a few minutes from the integral entering arguments of Pub. No. 229. The tabulated values in the tables provide a convenient means of checking the solution by calculator for gross error.

T9. Azimuths.—When checking the error of the compass by an azimuth observation of a celestial body, it is necessary to compute the azimuth for the time and place of observation for comparison with the observed azimuth. When tables are used for the purpose (art. 719, vol. II) tedious triple interpolation is usually required. When calculators are used, solution can be effected by one of several formulas, including the **time and altitude azimuth** formula:

$$\sin Z = \frac{\sin t \cos d}{\cos h}.$$

This formula is not generally recommended because of ambiguity with respect to quadrant as discussed in article 710 of volume II.

When the altitude is known and a solution independent of the altitude is not desired, solution can be effected by the **altitude azimuth** formula or by the **time azimuth** formula given in article T8.

Example.—In DR lat. 41°25′.9S, the azimuth of the sun is observed as 016°.0 pgc. At the time of the observation, the declination of the sun is 22°19′.6N; the local hour angle of the sun is 342°37′.6.

Required.—The gyro error by calculation of

$$Z = \tan^{-1}\left[\frac{\sin \text{LHA}}{(\cos L \tan d) - (\sin L \cos \text{LHA})}\right],$$

using the intermediate calculator.

Preliminary.—(1) Convert each known quantity to degrees and decimal degree:

Latitude 41°25′.9S	=41°.432
Declination 22°19′.6N	=(−) 22°.327
LHA 342°37′.6	=(−) 342°.627

(2) Prepare form on which to record results obtained in the several procedural steps of the calculations.

Solution.—(1) Procedure varies according to calculator design and the degree to which the user employs the features of the design enabling more expeditious solutions.

(2) In this example, only the initial step of substituting the given quantities in the formula, in accordance with the sign conventions, is given before the azimuth angle as obtained by the calculator is stated.

$$Z = \tan^{-1}\left[\frac{\sin \text{LHA}}{(\cos L \tan d) - (\sin L \cos \text{LHA})}\right]$$

$$= \tan^{-1}\left[\frac{\sin (-)\, 342°.627}{(\cos 41°.432 \times \tan (-)\, 22°.327) - (\sin 41°.432 \times \cos (-)\, 342°.627)}\right]$$

$$Z = (-) 17°.6.$$

(3) Since Z as calculated is a negative angle (−17°.6), 180° is added to obtain the desired azimuth angle, 162°.4.

Z	S162°.4E
Zn	017°.6
Zn	pgc 016°.0

Answer.—Gyro Error $1\overset{\circ}{.}6$E.

T10. Amplitudes (art. 720, vol. II) can be easily computed by means of the intermediate calculator. The basic formula is

$$A=\sin^{-1}(\sin d \sec L),$$

which can be stated as

$$A=\sin^{-1}\left(\frac{\sin d}{\cos L}\right).$$

Where A is the arc of the horizon between the prime vertical and the body, L is the latitude at the point of observation, and d is the declination of the celestial body, usually the sun.

Example.—The DR latitude of a ship is 51°24′.6N, at a time when the declination of the sun is 19°40′.4N.

Required.—The amplitude (A), when the center of the setting sun is on the celestial horizon, by calculation of

$$A=\sin^{-1}\left(\frac{\sin d}{\cos L}\right)$$

using the intermediate calculator.

Preliminary.—(1) Convert each known quantity to degrees and decimal degree.

Latitude 51°24′.6N = 51°.410
Declination 19°40′.4N = 19°.673

(2) Prepare form on which to record results obtained in the several procedural steps of the calculations.

Rules.—(1) All terms are treated as positive quantities, whether or not latitude and declination are of contrary name.

(2) The amplitude is given the prefix E if the body is rising and W if it is setting; it is given the suffix N when the declination is north and the suffix S when the declination is south.

Solution.—(1) Procedure varies according to calculator design and the degree to which the user employs the features of the design enabling more expeditious solutions.

(2) In this example, only the initial step of substituting the given quantities in the formula, in accordance with the rules, is given before the amplitude as obtained by the calculator is stated.

$$A=\sin^{-1}\left(\frac{\sin d}{\cos L}\right)$$

$$=\sin^{-1}(\sin 19\overset{\circ}{.}673 \div \cos 51\overset{\circ}{.}410)$$

$$A=32\overset{\circ}{.}7$$

Answer.—A W32°.7N.

Interconversion of amplitude and azimuth is similar to that of azimuth angle and azimuth. Thus, if the amplitude is E15°S, the body is 15° south of east or 90° + 15° = Zn 105°.

T11. Great-circle solutions for distance and initial course angle can be calculated from the formulas:

$$D=\cos^{-1}[(\sin L_1 \sin L_2)+(\cos L_1 \cos L_2 \cos \text{DLo})]$$

$$C=\tan^{-1}\left[\frac{\sin \text{DLo}}{(\cos L_1 \tan L_2)-(\sin L_1 \cos \text{DLo})}\right]$$

where D is the great-circle distance, C is the initial great-circle course angle, L_1 is the latitude of the point of departure, L_2 is the latitude of the destination, and DLo is the difference of longitude of the points of departure and destination. If the name of the latitude of the destination is contrary to that of the point of departure, it is treated as a negative quantity.

Example 1.—A ship is proceeding from Manila to Los Angeles. The captain desires to use great-circle sailing from lat. 12°45.′2N, long. 124°20.′1E, off the entrance to San Bernardino Strait, to lat. 33°48.′8N, long. 120°07.′1W, 5 miles south of Santa Rosa Island.

Required.—(1) The great-circle distance and
(2) initial great-circle course.

Preliminary.—(1) Convert each known quantity to degrees and decimal degree:

L_1 12°45.′2N = 12.°753N

L_2 33°48.′8N = 33.°813N

DLo 115°32.′8E = 115.°547E

(2) Prepare form on which to record results obtained in the several procedural steps of the calculations.

Rule.—When the latitudes of departure and destination are of contrary name, the latitude of the destination is treated as a negative quantity.

Solution.—(1) Procedure varies according to calculator design and the degree to which the user employs the features of the design enabling more expeditious solutions.

(2) In this example, only the initial step of substituting the given quantities in the formula, in accordance with the sign convention or rule, is given before the distance and course angle as obtained by the calculator are stated.

$$D=\cos^{-1}[(\sin L_1 \sin L_2)+(\cos L_1 \cos L_2 \cos \text{DLo})]$$

$$=\cos^{-1}[(\sin 12.^\circ753\times\sin 33.^\circ813)+(\cos 12.^\circ753\times\cos 33.^\circ813\times\cos 115.^\circ547)]$$

$$D=103.^\circ099$$

$$C=\tan^{-1}\left[\frac{\sin \text{DLo}}{(\cos L_1 \tan L_2)-(\sin L_1 \cos \text{DLo})}\right]$$

$$=\tan^{-1}\left[\frac{\sin 115.^\circ547}{(\cos 12.^\circ753\times\tan 33.^\circ813)-(\sin 12.^\circ753\times\cos 115.^\circ547)}\right]$$

$$C=50.^\circ322$$

Answers.—(1) D 6,186 nautical miles.
(2) C N50.°3E, Cn 050.°3.

The Vertex

The latitude of the vertex, L_v, is always numerically equal to or greater than L_1 or L_2. If initial course angle, C, is less than 90°, the nearer vertex is toward L_2; but if C is greater than 90°, the nearer vertex is in the opposite direction. The vertex nearer L_1 has the same name as L_1.

The latitude of the vertex can be calculated from the formula:

$$L_v = \cos^{-1}(\cos L_1 \sin C).$$

The difference of longitude of the vertex and the point of departure (DLo_v) can be calculated from the formula:

$$DLo_v = \sin^{-1}\left(\frac{\cos C}{\sin L_v}\right).$$

The distance from the point of departure to the vertex (D_v) can be calculated from the formula:

$$D_v = \sin^{-1}(\cos L_1 \sin DLo_v).$$

Example 2.—The situation is the same as in example 1.

Required.—(1) The latitude of the vertex.

(2) The longitude of the vertex.

(3) The distance from the point of departure to the vertex.

Solution.—(1) Procedure varies according to calculator design and the degree to which the user employs the design features enabling more expeditious solutions.

(2) In this example, only the initial steps of substituting the given quantities in the formulas are given before the latitude of the vertex, difference of longitude of the vertex, and the distance from the point of departure to the vertex are stated.

$$L_v = \cos^{-1}[\cos L_1 \sin C]$$

$$= \cos^{-1}[\cos 12^\circ.753 \times \sin 50^\circ.322]$$

Answers.—(1) $L_v = 41°21'.1N.$

$$DLo_v = \sin^{-1}\left[\frac{\cos C}{\sin L_v}\right]$$

$$= \sin^{-1}[\cos 50^\circ.322 \div \sin 41^\circ.353]$$

$$DLo_v = 75°05'.8.$$

(2) $\lambda_v = 160°34'.1W.$

$$D_v = \sin^{-1}[\cos L_1 \sin DLo_v]$$

$$= \sin^{-1}[\cos 12^\circ.753 \times \sin 75^\circ.097]$$

$$D_v = 70^\circ.480.$$

(3) 4,229 nautical miles.

Points on the Great-Circle Track

DLo_v and D_v of the nearer vertex are never greater than 90°. However, when L_1 and L_2 are of contrary name, the other vertex, 180° away, may be the better one to use in the solution for points on the great-circle track if it is nearer the mid point of the track.

The latitudes of points on the great-circle track can be determined for equal DLo intervals each side of the vertex (DLo_{vx}) using the formula:

$$L_x = \tan^{-1}(\cos DLo_{vx} \tan L_v).$$

Example 3.—Same situation as example 1.

Required.—Points on the great-circle track for equal DLo intervals each side of the vertex (DLo_{vx}).

Preliminary.—Select DLo interval appropriate for latitude and speed of advance, 12°.

Solution.—(1) Procedure varies according to calculator design and the degree to which the user employs the design features enabling more expeditious solutions.

(2) In this example, only the initial steps of substituting the given quantities in the formulas are given before the latitudes of points on the great-circle track are stated.

$$L_x = \tan^{-1}[\cos DLo_{vx} \tan L_v]$$

$$= \tan^{-1}[\cos 12° \times \tan 41.°353]$$

$$L_x = 40.°726.$$

Answers.—(1) $L_x = 40°43.'6.$

$\lambda_x = 172°34.'1W(12°W \text{ of } \lambda_v).$

$\lambda_x = 148°34.'1W(12°E \text{ of } \lambda_v).$

$L_x = \tan^{-1}[\cos 24° \times \tan 41.°353]$

$L_x = 38.°802$

(2) $L_x = 38°48.'1.$

$\lambda_x = 175°25.'6E(24°W \text{ of } \lambda_v).$

$\lambda_x = 136°34.'1W(24°E \text{ of } \lambda_v).$

Other points are similarly found.

Alternative Solution for Points on Track

The method of selecting the longitude (or DLo_{vx}) and determining the latitude at which the great-circle crosses the selected meridian provides shorter legs in higher latitudes and longer legs in lower latitudes. Points at desired distances or desired equal intervals of distance on the great-circle from the vertex can be calculated using the formulas:

$$L_x = \sin^{-1}[\sin L_v \cos D_{vx}]$$

$$DLo_{vx} = \sin^{-1}\left[\frac{\sin D_{vx}}{\cos L_x}\right].$$

Example 4.—Same situation as example 1.

Required.—Points on the great-circle track at equal intervals of distance from the vertex.

Preliminary.—Select distance intervals appropriate for speed of advance, 300 nautical miles (5°).

Solution.—(1) Procedure varies according to calculator design and the degree to which the user employs the design features enabling more expeditious solutions.

(2) In this example, only the initial steps of substituting the given quantities in the formulas are given before the latitude and the difference of longitude of the point at the desired distance from the vertex are stated.

$$L_x = \sin^{-1}[\sin L_v \cos D_{vx}]$$

$$= \sin^{-1}[\sin 41.^\circ353 \times \cos 5^\circ]$$

$$L_x = 41.^\circ161$$

$$L_x = 41^\circ 09.'7\text{N}.$$

$$DLo_{vx} = \sin^{-1}\left[\frac{\sin D_{vx}}{\cos L_x}\right]$$

$$= \sin^{-1}[\sin 5^\circ \div \cos 41.^\circ353]$$

$$DLo_{vx} = 6.^\circ667$$

$$DLo_{vx} = 6^\circ 40.'0.$$

Answer.—(1) $L_x = 41^\circ 09.'7\text{N}.$

$\lambda_x = 153^\circ 54.'1\text{W}.$

$\lambda_x = 167^\circ 14.'1\text{W}.$

$$L_x = \sin^{-1}[\sin L_v \cos D_{vx}]$$

$$= \sin^{-1}[\sin 41.^\circ353 \times \cos 10^\circ]$$

$$L_x = 40.^\circ591$$

$$L_x = 40^\circ 35.'5\text{N}.$$

$$DLo_{vx} = \sin^{-1}\left[\frac{\sin D_{vx}}{\cos L_x}\right]$$

$$= \sin^{-1}[\sin 10^\circ \div \cos 41.^\circ353]$$

$$DLo_{vx} = 13.^\circ375$$

$$DLo_{vx} = 13^\circ 22.'5.$$

(2) $L_x = 40°35'.55$N.

$\lambda_x = 147°11'.6$W(300 miles W of λ_v).

$\lambda_x = 173°56'.6$W(300 miles E of λ_v).

Other points are similarly found.

T12. Plane sailings.—A calculator having the rectangular to polar coordinate conversion feature affords expeditious solutions of plane sailings. However, the user must keep in mind whether the *difference of latitude corresponds to the calculator's X-coordinate or to its Y-coordinate.*

Example 1.—A vessel steams 188 miles on course 005°.

Required.—(1) Difference of latitude, (2) departure.

Solution.—(1) This problem involves converting polar coordinates (188 miles, 005°) to rectangular coordinates.

(2) Procedure varies according to calculator design.

X-coordinate 187.285
Y-coordinate 16.385

Answer.—(1) l 187'.3 N.
(2) p 16.4 mi. E.

Advanced Calculators

T13. Advanced or programmable calculators have the potential for significantly reducing the quantity and complexity of the navigator's computational workload. Properly programmed calculators now available have the capability for "memorizing" and executing specific keystroke sequences required in the solution of a problem. This feature eliminates the requirement for calculating each step of a formula when using the basic or intermediate calculator. The need for format conversion (i.e., degrees and minutes to degrees and decimal degree) and calculating the formula in a specific order is no longer necessary. Thus, computation time is greatly reduced along with chances for human error.

Sight reduction speed is further enhanced when the "memory" of the calculator is stored with almanac data.

Other advantages to be derived from an advanced calculator include:

1. Relatively error-free reductions of a large number of celestial observations to improve position information.

2. Enables navigators to concentrate more on analysis and evaluation by freeing them from time-consuming and error prone repetitious tasks.

APPENDIX U

UNDERWATER LOG CALIBRATION GUIDELINES

U1. Introduction.—During the calibration of the underwater log on the measured mile course, the applicable requirements and procedures specified for the ship's standardization trial should be followed as closely as possible. The calibration should not be conducted when the effects of wind and sea can materially affect the accuracy of the results. Preferably, the measured mile should be run in a calm sea. The following guidelines do not include the computational procedures and equipment adjustments which may be found in the manufacturer's manual for the log being calibrated.

Guidelines

U2. Limiting sea and wind conditions.—The recommended maximum sea and wind conditions are sea state 3 and wind force 5 on the Beaufort scale (app. V). No more than 3° of rudder movement should be required to steer a steady course while on the measured mile. Excessive use of the rudder will reduce the accuracy of the calibration.

U3. Minimum depths.—The runs should be made on measured miles where the water is of such depth that shallow-water effects will not be experienced. A reasonable estimate of the required minimum depth in feet in the approach, measured mile, and turn around area is three times the square root of the product of the ship's beam and draft in feet, or 0.3 times the square of the ship's speed in knots, whichever is greater.

U4. Tidal current.—Each series of runs of the measured mile at the same speed must be made while the tidal current is running in the same direction. The method of averaging which compensates for water current will produce correct ship's speed only if the speed of the tidal current is increasing or decreasing uniformly with time. The time interval between runs of a series should be kept reasonably constant. If the first run of a series starts, for example, at 0900 and the second at 0920, then the third run should start at about 0940.

U5. Draft and trim.—Unless the log is to be calibrated for different loadings or different displacements, the runs should be made at normal trim and at approximately average draft.

U6. Propeller RPM.—During the run of the measured mile at a specific speed, RPM must be maintained constant within a tolerance of plus or minus two RPM's. If this tolerance is exceeded during any run, the data for the run should be discarded; the run should be repeated. To insure that the ship has reached its terminal speed for the RPM during a run, the ship must be steady on course at the required RPM when at least 1 mile from the range of the first set of markers. The speed at the various RPM's is obtained in addition to the log calibration data.

U7. Runs required.—A series of three consecutive runs, alternating in direction, should be made for each speed for which the log is to be calibrated. If any run must be discarded, additional runs must be made to obtain three consecutive runs alternating in direction. The effects of current and sea conditions being additive in one direction and subtractive in the other direction, the data for the second run of a series is used as a hypothetical fourth run in calculating the average speed through the water. The mathematical derivation is referred to as a mean of means. The correct mean value for speed and the RPM is the sum of the data for the first and third runs plus two times the data for the second run, and this sum total divided by four. This can be stated in the following formula:

$$v_m = \frac{v_1 + 2v_2 + v_3}{4}.$$

After a run has been completed in one direction, the ship should be taken well beyond the measured mile course and turned without excessive use of rudder so that the same track will be repeated on the following run. The limit on the rudder angle is to limit the loss of speed during the turn. The ship should be taken well away from the measured mile course to insure sufficient time to reach terminal speed for the RPM prior to passing the range of the first set of markers on the following run. The ship should be steady on course at the required RPM when at least 1 mile from the first set of markers.

In calibrating an electromagnetic log, each series of runs must be made at a speed as specified in the manufacturer's instruction manual. Thus, the number of runs and speeds during each series of runs are governed by equipment design. As specified in the manufacturer's instruction manual, a series of runs may have to be discarded if the speeds ascertained are not within certain limits.

Advance Preparations

U8. Advance preparations.—Complete calibration of the underwater log may require a full working day in an area distant from the area of the ship's normal operations. Advance preparations should include equipment tests in accordance with the manufacturer's instruction manual, and practice runs on a simulated measured mile to insure that the log is functioning properly (except for calibration) and that personnel are familiar with the procedures to be followed.

U9. Verification.—Since the markers for measured-mile courses are not U. S. Coast Guard maintained aids to navigation, advance preparation should include verification that the markers are in place.

Duties of Observers

U10. Duties of observers.—Two observers on the navigation bridge must make independent observations of the time required to run the measured mile. One bridge observer acts as coordinator; the other insures that the course is held steady during the run and that excessive rudder is not used. Approximately 3 minutes before the first set of markers are "in range" (or "in transit") the coordinator calls out "Get ready." A few seconds before the first set of markers are in range, the coordinator calls out "Stand by." At the instant the first set of markers are in range by his observation, the coordinator calls out "Mark" and simultaneously starts his stopwatch. The other bridge observer starts his stopwatch at the instant of his independent observation that the first set of markers are in range.

The engine room observer records the shaft RPM at the coordinator's "Mark" and at each 15-second interval thereafter during the run. The engine room observer insures that the tolerance of + or − two RPM's is not exceeded during the run.

The log observer determines the time required for the distance counter to increase 1 mile. He does not necessarily record the mileage on the counter and start his stopwatch at the instant of the coordinator's "Mark." Because of the difficulty of reading intermediate values on the counter, the normal procedure is to delay the log distance timing until such time after the coordinator's "Mark" that the bottom of the hundredths digit just appears in the window of the counter.

An assistant log observer may be employed for independent timing of the distance counter and for recording of data.

The coordinator must insure that speed and course are held steady until such time that the log counter has registered an increase of 1 mile. An initial high negative error in the log may extend the run several seconds beyond the time the second set of markers are in range. By comparing the time required to run the measured mile with the time required for the log counter to increase exactly 1 mile the percentage error of the log can be determined.

While running the measured mile for calibration of an electromagnetic log, the instantaneous readings of the speed dial can be ignored.

APPENDIX V
BEAUFORT WIND SCALE
WITH CORRESPONDING SEA STATE CODES

Beaufort number or force	Wind speed: knots	Wind speed: mph	Wind speed: meters per second	Wind speed: km per hour	World Meteorological Organization (1964)	Estimating wind speed: Effects observed far from land	Estimating wind speed: Effects observed near coast	Estimating wind speed: Effects observed on land	Sea State: Term and height of waves, in meters	Sea State: Code
0	under 1	under 1	0.0–0.2	under 1	Calm	Sea like mirror.	Calm.	Calm; smoke rises vertically.	Calm, glassy, 0	0
1	1–3	1–3	0.3–1.5	1–5	Light air	Ripples with appearance of scales; no foam crests.	Fishing smack just has steerage way.	Smoke drift indicates wind direction; vanes do not move.		
2	4–6	4–7	1.6–3.3	6–11	Light breeze	Small wavelets; crests of glassy appearance, not breaking.	Wind fills the sails of smacks which then travel at about 1–2 miles per hour.	Wind felt on face; leaves rustle; vanes begin to move.	Calm, rippled, 0–0.1	1
3	7–10	8–12	3.4–5.4	12–19	Gentle breeze	Large wavelets; crests begin to break; scattered whitecaps.	Smacks begin to careen and travel about 3–4 miles per hour.	Leaves, small twigs in constant motion; light flags extended.	Smooth, wavelets, 0.1–0.5	2
4	11–16	13–18	5.5–7.9	20–28	Moderate breeze	Small waves, becoming longer; numerous whitecaps.	Good working breeze, smacks carry all canvas with good list.	Dust, leaves, and loose paper raised up; small branches move.	Slight, 0.5–1.25	3
5	17–21	19–24	8.0–10.7	29–38	Fresh breeze	Moderate waves, taking longer form; many whitecaps; some spray.	Smacks shorten sail.	Small trees in leaf begin to sway.	Moderate, 1.25–2.5	4
6	22–27	25–31	10.8–13.8	39–49	Strong breeze	Larger waves forming; whitecaps everywhere; more spray.	Smacks have doubled reef in mainsail; care required when fishing.	Larger branches of trees in motion; whistling heard in wires.	Rough, 2.5–4	5
7	28–33	32–38	13.9–17.1	50–61	Near gale	Sea heaps up; white foam from breaking waves begins to be blown in streaks.	Smacks remain in harbor and those at sea lie-to.	Whole trees in motion; resistance felt in walking against wind.	Very rough, 4–6	6
8	34–40	39–46	17.2–20.7	62–74	Gale	Moderately high waves of greater length; edges of crests begin to break into spindrift; foam is blown in well-marked streaks.	All smacks make for harbor, if near.	Twigs and small branches broken off trees; progress generally impeded.		
9	41–47	47–54	20.8–24.4	75–88	Strong gale	High waves; sea begins to roll; dense streaks of foam; spray may reduce visibility.		Slight structural damage occurs; slate blown from roofs.		
10	48–55	55–63	24.5–28.4	89–102	Storm	Very high waves with overhanging crests; sea takes white appearance as foam is blown in very dense streaks; rolling is heavy and visibility reduced.		Seldom experienced on land; trees broken or uprooted; considerable structural damage occurs.	High, 6–9	7
11	56–63	64–72	28.5–32.6	103–117	Violent storm	Exceptionally high waves; sea covered with white foam patches; visibility still more reduced.		Very rarely experienced on land; usually accompanied by widespread damage.	Very high, 9–14	8
12	64 and over	73 and over	32.7 and over	118 and over	Hurricane	Air filled with foam; sea completely white with driving spray; visibility greatly reduced.			Phenomenal, over 14	9

Note: Since January 1, 1955, weather map symbols have been based upon wind speed in knots, at five-knot intervals, rather than upon Beaufort number.

APPENDIX W

SEA STATE

This appendix provides, by means of representative photographs illustrating the effects of the wind on the sea surface, a pictorial guide to mariners for estimating the wind speed at sea.

The photographs and associated text are taken from *State of Sea Photographs for the Beaufort Wind Scale,* Crown Copyright, Ottawa, 1975. The material is reprinted with minor changes through permission of the Atmospheric Environment Service, Department of the Environment, Canada.

State of Sea Photographs for the Beaufort Wind Scale

W1. Introduction.—This appendix presents the results of a project carried out on board the Canadian Ocean Weather Ships C.C.G.S. *St. Catharines* and C.C.G.S. *Stonetown.* The aim of the project was to collect photographs of the sea surface as it appears under the influence of the various ranges of wind speed defined by The Beaufort Scale of Wind Force (app. V). Word descriptions of the appearance of the sea for each Beaufort Force have been available for many years, but it was felt that photographs illustrating the conditions associated with each force would be of some assistance to ships' officers in estimating wind speed. Sea photographs taken by low-flying aircraft of the United States Navy were published by the Meteorological Branch under a circular memorandum dated December 20, 1957. However, these aerial photographs, while good, do not depict the aspect of the sea as viewed from the bridge of a ship. The apparent lack of good photographs of this nature prompted this project on the Ocean Weather Ships. A selection of the best photographs resulting from the project are presented.

W2. Estimating the wind at sea.—Observers on board ships at sea usually determine the speed of the wind by estimating its Beaufort Force, as merchant ships are not normally equipped with wind measuring instruments. Through experience, ships' officers have developed various methods of estimating this force. The effect of the wind on the observer himself, the ship's rigging, flags, etc., is used as a criterion; but, estimates based on these indications give the relative wind which must be corrected for the motion of the ship before an estimate of the true wind speed can be obtained.

The most common method involves the appearance of the sea surface. The state of the sea disturbance, i.e. the dimensions of the waves, the presence of white caps, foam or spray, depends principally on three factors:

1. *The wind speed.* The higher the speed of the wind, the greater is the sea disturbance.

2. *The duration of the wind.* At any point on the sea, the disturbance will increase the longer the wind blows at a given speed, until a maximum state of disturbance is reached.

3. *The fetch.* This is the length of the stretch of water over which the wind acts on the sea surface from the same direction. For a given wind speed and duration, the longer the fetch, the greater is the sea disturbance. If the fetch is short, say a

few miles, the disturbance will be relatively small no matter how great the wind speed is or how long it has been blowing.

There are other factors which can modify the appearance of the sea surface caused by wind alone. These are strong currents, shallow water, swell, precipitation, ice, and wind shifts. Their affects will be described later.

A wind of a given Beaufort Force will, therefore, produce a characteristic appearance of the sea surface provided that it has been blowing for a sufficient length of time, and over a sufficiently long fetch. The effects of currents, shallow water, swell, precipitation, etc., should also be absent. The "Sea Criteria" associated with each Beaufort Force from force 0 to force 12 were agreed upon and drawn up by the International Meteorological Committee in 1939. These word descriptions of the state of the sea are known as the Sea Criterion of the Beaufort Scale of Wind Force.

The use of the sea criterion has the advantage that the speed of the ship need not be considered. In practice, the mariner observes the sea surface, noting the size of the waves, the white caps, spindrift, etc., and then finds the criterion (app. V) which best describes the sea surface as he saw it. This criterion is associated with a Beaufort number, for which a corresponding mean wind speed and range in knots are given. Since meteorological reports require that wind speeds be reported in knots, the mean speed for the Beaufort number may be reported, or an experienced observer may judge that the sea disturbance is such that a higher or lower speed within the range for the force is more accurate.

This method, while it appears simple, should be used with caution however. It should be borne in mind that the sea conditions described for each Beaufort Force (app. V) are "steady-state" conditions; i.e. the conditions which result when the wind has been blowing for a relatively long time, and over a great stretch of water. At any particular time at sea, though, the duration of the wind or the fetch, or both, may not have been great enough to produce these "steady-state" conditions. When a high wind springs up suddenly after previously calm or near calm conditions, it will require some hours, depending on the strength of the wind, to generate waves of maximum height. The height of the waves increases rapidly in the first few hours after the commencement of the blow, but increases at a much slower rate later on. Considering the effect of fetch, if the observer could start at the beginning of the fetch (say at a coastline when the wind is offshore) after the wind has been blowing for a long time, and proceed downwind, he would notice that the waves were quite small at the beginning, and increased in height rapidly over the first 50 miles or so of the fetch. Farther along he would notice that the rate of increase in height with distance would slow down, and after 500 miles or so from the beginning of the fetch there would be little or no increase in height.

To illustrate the duration of winds and the length of fetches required for various wind forces to build seas to 50 percent, 75 percent, and 90 percent of their theoretical maximum heights, table W2 is of interest.

The theoretical maximum wave heights represent the average heights of the highest third of the waves, as these waves are of the most practical significance.

It will be seen that winds of force 5 or less can build seas to 90 percent of their maximum height in less than 12 hours, provided the fetch is long enough. Higher winds require a much greater time—force 11 winds requiring 32 hours to build waves to 90 percent of their maximum height. The times given in table W2 represent those required to build waves starting from initially calm sea conditions. If waves are already present at the onset of the blow, the times would be somewhat less depending on the initial wave heights and their direction relative to the direction of the wind which has sprung up.

The first consideration when using the sea criterion to estimate wind speed, therefore, is to decide whether the wind has been blowing long enough from the same direction to produce a steady state sea condition. If not, then it is possible that the wind speed may be underestimated. For example, if a wind with an actual speed of force 9 has been blowing for only 7 hours, it may have generated a sea condition which would perhaps correspond to a steady state for force 6. If, in this case, the sea criterion was used blindly without considering the short duration, and force 6 was reported, then the wind would be underestimated by three Beaufort forces, or approximately 20 knots. This is an extreme example however, as it is very unlikely that even an inexperienced seaman could not distinguish a force 6 from a force 9 wind.

Beaufort force of wind.	Theoretical maximum wave height (ft) unlimited duration and fetch.	Duration of winds, (hours), with unlimited fetch, to produce percent of maximum wave height indicated.			Fetch (nautical miles), with unlimited duration of blow, to produce percent of maximum wave height indicated.		
		50%	75%	90%	50%	75%	90%
3	2	1.5	5	8	3	13	25
5	8	3.5	8	12	10	30	60
7	20	5.5	12	21	22	75	150
9	40	7	16	25	55	150	280
11	70	9	19	32	85	200	450

TABLE—W2.

Experience has shown that the appearance of white caps, foam, spindrift, etc., reaches a steady state condition before the height of the waves attain their maximum value. It is a safe assumption that the appearance of the sea (as regards white caps, etc.) will reach a steady state in the time required to build the waves to 50–75 percent of their maximum height. Thus, from table W2, it is seen that a force 5 wind could require 8 hours at most to produce a characteristic appearance of the sea surface.

A second consideration when using the sea criterion is the length of the fetch over which the wind has been blowing to produce the present state of the sea. On the open sea, unless the mariner has a copy of the latest synoptic weather map available, he will not know the length of the fetch. It will be seen from table W2, though, that only relatively short fetches are required for the lower wind forces to generate their characteristic seas. On the open sea, the fetches associated with most storms and other weather systems are usually long enough so that even winds up to force 9 can build seas up to 90 percent or more of their maximum height, providing the wind blows from the same direction long enough.

When navigating close to a coast or in restricted waters, however, it may be necessary to make allowances for the shorter stretches of water over which the wind blows. For example, referring to table W2, if the ship is 22 miles from a coast and an offshore wind with an actual speed of force 7 is blowing, the waves at the ship will never attain more than 50 percent of their maximum height for this speed no matter how long the wind blows. Hence, if the sea criterion were used under these conditions without consideration of the short fetch, the wind speed would be underestimated. With an offshore wind, the sea criterion may be used with confidence if the distance to the coast is greater than the values given in the extreme right-hand column of table W2; again, provided that the wind has been blowing offshore for a sufficient length of time.

Other factors aside from the duration of the blow and the fetch affect the appearance of the sea surface, and should be considered if they are present.

W3. Tides and Currents.—A wind blowing against a tide or strong current causes a greater sea disturbance than normal, which may result in an overestimate of the wind speed. On the other hand, a wind blowing in the same direction as a tide or strong current causes less sea disturbance than normal, and may result in an underestimate of the wind speed.

W4. Shallow Water.—Waves running into shallow water increase in steepness, and hence, their tendency to break. With an onshore wind there will, therefore, be more white caps over the shallow waters than over the deeper water farther offshore. It is only over relatively deep water that the sea criterion can be used with confidence.

W5. Swell.—Swell is the name given to waves, generally of considerable length, which were raised in some distant area by winds blowing there, and which have moved into the vicinity of the ship; or to waves raised nearby and which continue to advance after the wind at the ship has abated or changed direction. The direction of swell waves is usually different from the direction of the wind and the sea waves. *Swell waves are not to be considered when estimating wind speed and direction. Only those waves raised by the wind blowing at the time are of any significance.* The wind-driven waves show a greater tendency to break when superimposed on the crests of swell, and hence more white caps may be formed than if the swell were absent. Under these conditions the use of the sea criterion may result in a slight overestimate of the wind speed.

W6. Precipitation.—Heavy rain has a damping or smoothing effect on the sea surface which must be mechanical in character. Since the sea surface will therefore appear less disturbed than would be the case without the rain, the wind speed may be underestimated unless the smoothing effect is taken into account.

W7. Ice.—Even small concentrations of ice floating on the sea surface will dampen waves considerably, and concentrations greater than about seven tenths average will eliminate waves altogether. Young sea ice, which in the early stages of formation has a thick soupy consistency, and later takes on a rubbery appearance, is very effective in dampening waves. Consequently, the sea criterion cannot be used with any degree of confidence when sea ice is present. In higher latitudes, the presence of an ice field some distance to windward of the ship may be suspected if, when the ship is not close to any coast, the wind is relatively strong but the seas abnormally underdeveloped. The edge of the ice field acts like a coastline, and the short fetch between the ice and the ship is not sufficient for the wind to develop the seas fully.

W8. Wind shifts.—Following a rapid change in the direction of the wind, as occurs at the passage of a cold front, the new wind will flatten out to a great extent the waves which were present before the wind shift. This is so because the direction of the wind after the shift may differ by 90° or more from the direction of the waves, which does not change. Hence, the wind may oppose the progress of the waves and dampen them out quickly. At the same time the new wind begins to generate its own waves on top of this dissipating swell, and it is not long before the cross pattern of waves gives the sea a "choppy" or confused appearance. It is during the first few hours following the wind shift that the appearance of the sea surface may not provide a reliable indication of the wind speed. The wind is normally stronger than the sea would indicate, as old waves are being flattened out, and new waves are just beginning to be developed.

W9. Night Observations.—On a dark night, when it is impossible to see the sea clearly, the observer may estimate the apparent wind from its effect on the ship's rigging, flags, etc., or simply the "feel" of the wind.

The State of Sea Photographs

The photographs were taken by the Meteorological Branch personnel of the Canadian Ocean Weather Ships C.C.G.S. *St. Catharines* and C.C.G.S. *Stonetown.*

The photographs were taken between March, 1960 and May, 1961. In this period a total of 247 photographs were obtained. Zeiss Ikon ($2\frac{1}{4} \times 3\frac{1}{4}$) cameras and Kodak Verichrome Pan film were used.

Of the 247 pictures available, 2 pictures were chosen to illustrate conditions associated with each Beaufort force from force 1 to force 10. Only one picture was considered acceptable to illustrate force 0. No representative photographs were available to illustrate force 11 conditions, and no photographs were made of force 12 conditions.

In selecting the pictures for presentation here, it was considered that they should meet two requirements. Firstly, a picture illustrating the effects of a given wind force should conform as closely as possible to the Sea Criterion for that force. Secondly, the wind prior to the time of the picture should be relatively steady both in direction and at the given force over many hours to ensure that near steady-state sea conditions for that force at the time of the picture existed. A large percentage of the photographs available were rejected because the wind at the time of the picture had not been blowing long enough to produce a disturbance of near steady-state proportion.

The pictures which follow were judged to fulfill best the requirements stated above. Opposite each picture is the accompanying technical and other data appropriate to each. In addition to the wind at the time of the photograph, the wind at 3-hourly intervals over the previous 24 hours is also included. The height of the ships' anemometers above the sea surface was approximately 60 feet. The synoptic weather situation at the time of the picture is described briefly, and other comments are given when warranted.

BEAUFORT FORCE 0

Wind speed less than 1 knot

Sea Criterion: Sea like a mirror.

Date/Time of photograph: June 5, 1960, 2340 GMT.

Height of Camera above sea: 35 Feet.

Time (GMT)	Direction (°T)	Wind speed (kn)
2340	—	calm
2100	190	01
1800	080	03
1500	040	07
1200	040	09
0900	030	10
0600	030	11
0300	030	11
0000	040	10

Waves at time of picture

	Direction (°T)	Period (sec.)	Height (ft.)
Sea Waves	—	—	—
Swell	100°	5	2

Synoptic Situation

High pressure area centered at 49°N, 144°W, or approximately 50 nautical miles southeast of ship at time of picture. Center of high moving southeastward.

Remarks

Calm winds, and hence calm sea conditions, are rare at Station "P," occurring only about 1.5 percent of the time during the course of a year. Although the wind was essentially calm during the 3 hours prior to this photograph, it averaged 4 knots (force 2) over the previous 12 hours, and 7 knots (force 3) in the previous 24 hours. Air temperature 52°2F, sea temperature 45°7F.

BEAUFORT FORCE 0 SEA STATE 0

BEAUFORT FORCE 1

Wind Speed 1 to 3 knots, mean 2 knots

Sea Criterion: Ripples with the appearance of scales are formed, but without foam crests.

Date/Time of Photograph: May 22, 1960, 2000 GMT.

Height of camera above sea: 35 feet.

Time (GMT)	Direction (°T)	Wind speed (kn)
2000	290	02
1800	—	calm
1500	220	04
1200	260	08
0900	280	02
0600	—	calm
0300	290	10
0000	290	12
2100	290	13

Waves at time of picture

	Direction (°T)	Period (sec.)	Height (ft.)
Sea Waves	—	—	—
Swell	290	10	3

Synoptic situation

The station was near the center of a narrow ridge of high pressure which extended northward into the Gulf of Alaska from a high centered near 30°N, 147°W. A low pressure area, elongated in a north-south direction, lay just off the coast of Canada and the United States. Another low was centered in the eastern Bering Sea.

Remarks

Winds averaged less than 3 knots (force 1) over the previous 14 hours, and less than 6 knots (force 2) over the previous 24 hours. Air temperature 44°F, sea temperature 44°F.

BEAUFORT FORCE 1 SEA STATE 0

BEAUFORT FORCE 2

Wind speed 4 to 6 knots, mean 5 knots

Sea criterion: Small wavelets, still short but more pronounced—crests have a glassy appearance and do not break.

Date/Time of photograph: May 26, 1961, 1700 GMT.

Height of camera above sea: 45 ft.

Time (GMT)	Direction (°T)	Wind speed (kn)
1700	120	05
1500	100	03
1200	—	calm
0900	340	06
0600	010	07
0300	340	03
0000	320	04
2100	310	04
1800	310	03

Waves at time of picture

	Direction (°T)	Period (sec.)	Height (ft.)
Sea waves	120	—	—
Swell	050	6	1

Synoptic situation

A weak trough of low pressure, lying in a general northwest-southeast direction, was immediately west of the ship and passed by during the next hour. The centerline of the following ridge of high pressure was approximately 400 nautical miles west of the ship.

Remarks

This picture, taken from the C.C.G.S. *ST. CATHARINES*, shows her sister ship the C.C.G.S. *STONETOWN*, and was taken at 49°.8N, 142°.5W, or approximately 100 nautical miles east of the Station "P" position. The wind speed averaged close to 4 knots (force 2) during the previous 24 hours. Air temperature 46°.5F; sea temperature 45°.0F.

BEAUFORT FORCE 2 SEA STATE 1

BEAUFORT FORCE 3

Wind speed 7 to 10 knots, mean 9 knots

Sea criterion: Large wavelets. Crests begin to break. Foam of a glassy appearance. Perhaps scattered white caps.

Date/Time of photograph: Feb. 19, 1961, 2000 GMT.

Height of camera above sea: 45 ft.

Time (GMT)	Direction (°T)	Wind speed (kn)
2000	190	03
1800	160	09
1500	180	10
1200	190	11
0900	260	08
0600	300	10
0300	320	10
0000	010	16
2100	090	10

Waves at time of picture

	Direction (°T)	Period (sec.)	Height (ft.)
Sea waves	—	—	—
Swell	180	7	8

Synoptic situation

Small low pressure area centered approximately 170 nautical miles west of station. The ship was approximately 40 nautical miles north of a warm front that extended eastward from the low center.

Remarks

Although the wind speed at the time of this picture was only 3 knots (force 1), it had decreased from 9 knots 2 hours previously, and had averaged almost 10 knots over the previous 24 hours. The direction had been constant from the south for 8 hours. Air temperature 44°.2F; sea temperature 40°.8F.

BEAUFORT FORCE 3 SEA STATE 2

BEAUFORT FORCE 4

Wind speed 11 to 16 knots, mean 13 knots

Sea criterion: Small waves, becoming longer, fairly frequent white caps.

Date/Time of photograph: July 3, 1960, 2240 GMT.

Height of camera above sea: 35 ft.

Time (GMT)	Direction (°T)	Wind speed (kn)
2240	320	16
2100	320	16
1800	340	11
1500	270	05
1200	270	08
0900	140	08
0600	230	09
0300	260	09
0000	250	13

NOTE:—Between 0000 and 2240 GMT, the ship had traveled eastward from 49°3N, 133°7W to 48°8N, 128°2W, a distance of approximately 220 nautical miles.

Waves at time of picture

	Direction (°T)	Period (sec.)	Height (ft.)
Sea waves	310	5	3
Swell	—	—	—

Synoptic situation

High pressure area centered at 40°N, 137°W, with ridge line extending northeastward to 100 nautical miles west of the ship, then northward to the northern section of the British Columbia coast. Low pressure area centered at 55°N, 150°W, in the Gulf of Alaska.

Remarks

This picture was made at 48°8N, 128°2W, or approximately 200 nautical miles west of Vancouver while the ship was returning from Station "P." The ship was moving into the area of northwesterly circulation east of the ridge of high pressure, and had passed the ridge line approximately 11 hours before the time of this picture. It is likely that force 4 winds had been acting on the area longer than would appear from the wind data, as the ship was moving eastward into the area. Air temperature 55°6F; sea temperature 52°7F.

BEAUFORT FORCE 4 SEA STATE 3

BEAUFORT FORCE 5

Wind speed 17 to 21 knots, mean 19 knots

Sea criterion: Moderate waves taking a more pronounced long form; many white caps are formed. (Chance of some spray.)

Date/Time of photograph: Apr. 7, 1961, 2315 GMT.

Height of camera above sea: 35 ft.

Time (GMT)	Direction (°T)	Wind speed (kn)
2315	280	23
2100	270	19
1800	240	20
1500	220	21
1200	220	24
0900	200	26
0600	190	25
0300	180	27
0000	180	30

Waves at time of picture

	Direction (°T)	Period (sec.)	Height (ft.)
Sea waves	280	6	7
Swell	240	8	6

Synoptic situation

High pressure area centered 550 nautical miles southeast of the station, which was in a northwesterly circulation 200 nautical miles northeast of the ridge line extending northwestward from the high. Low pressure area centered near 56°N, 136°W, in the Gulf of Alaska.

Remarks

The wind speed at the time of the picture was 23 knots (force 6), but had averaged less than 21 knots (force 5) over the previous 8 hours. In this period, the wind veered from southwest to west. Air temperature 41°0F; sea temperature 41°2F.

BEAUFORT FORCE 5 SEA STATE 4

BEAUFORT FORCE 6

Wind speed 22 to 27 knots, mean 24 knots

Sea criterion: Large waves begin to form; the white foam crests are more extensive everywhere. (Probably some spray.)

Date/Time of photograph: Feb. 10, 1961, 2115 GMT.

Height of camera above sea: 20 ft.

Time (GMT)	Direction (°T)	Wind speed (kn)
2115	280	25
1800	300	26
1500	300	14
1200	280	23
0900	280	26
0600	310	23
0300	300	26
0000	280	27

Waves at time of picture

	Direction (°T)	Period (sec.)	Height (ft.)
Sea waves	280	6	11
Swell	—	—	—

Synoptic situation

Station in a northwesterly flow south of a low pressure area centered at 56°N, 144°W, or 380 nautical miles north of the ship. Trough of low pressure extending southeastward from low to second low lying off west coast of the United States. Ridge of pressure 300 nautical miles west of the ship.

Remarks

Except for a period between 3 and 9 hours ago, the wind was fairly constant, in both speed and duration, over the previous 24 hours. The fetch of the wind at the time of the picture was at least 300 nautical miles. Air temperature 37.4°F; sea temperature 41.0°F.

BEAUFORT FORCE 6 SEA STATE 5

BEAUFORT FORCE 7

Wind speed 28 to 33 knots, mean 30 knots

Sea criterion: Sea heaps up and white foam from breaking waves begins to be blown in streaks along the direction of the wind.

Date/Time of photograph: Feb 28, 1961, 1900 GMT.

Height of camera above sea: 45 ft.

Time (GMT)	Direction (°T)	Wind speed (kn)
1900	300	30
1800	300	30
1500	260	35
1200	260	35
0900	270	30
0600	250	38
0300	220	39
0000	210	39
2100	180	30

Waves at time of picture

	Direction (°T)	Period (sec.)	Height (ft.)
Sea waves	300	6	13
Swell	250	9	10

Synoptic situation

A deep low pressure area centered at 60°N, 144°W dominated the northeastern Pacific east of 160°W and north of 45°N. Station in northwesterly circulation approximately 575 nautical miles south of the low center.

Remarks

Although the wind speed was force 7 at the time of the picture, it had decreased from low force 8 between 1 and 4 hours previously. The wind veered from south to northwest during the previous 24 hours as the low center passed west of the ship while moving northward into the Gulf of Alaska. The fetch of the wind at the time of the picture was approximately 350 nautical miles. Air temperature 38°.4F; sea temperature 40°.7F.

BEAUFORT FORCE 7 SEA STATE 6

BEAUFORT FORCE 8

Wind speed 34 to 40 knots, mean 37 knots

Sea criterion: Moderately high waves of greater length; edges of crests begin to break into spindrift. The foam is blown in well-marked streaks along the direction of the wind.

Date/Time of photograph: Jan. 15, 1961, 1955 GMT.

Height of camera above sea: 35 ft.

Time (GMT)	Direction (°T)	Wind speed (kn)
1955	250	45
1800	240	38
1500	240	38
1200	250	37
0900	250	35
0600	250	35
0300	250	43
0000	250	40
2100	240	30

Waves at time of picture

	Direction (°T)	Period (sec.)	Height (ft.)
Sea waves	260	7	18
Swell	—	—	—

Synoptic situation

Low pressure area centered at 58°N, 142°W, in the Gulf of Alaska. Station in southwesterly circulation 500 nautical miles south-southwest of the low center.

Remarks

Although the wind speed at the time of the picture was 45 knots (force 9), it had been force 8 for 12 hours previously, and had increased in speed less than 2 hours prior to the picture. The wind direction was remarkably constant at 240° to 250° over the previous 24 hours. Air temperature 39°.0F.; sea temperature 41°.2F.

BEAUFORT FORCE 8 SEA STATE 6

BEAUFORT FORCE 9

Wind speed 41 to 47 knots, mean 44 knots

Sea criterion: High waves. Dense streaks of foam along the direction of the wind. Crests of waves begin to topple, tumble, and roll over. Spray may affect visibility.

Date/Time of photograph: Jan. 17, 1961, 2130 GMT.

Height of camera above sea: 35 ft.

Time (GMT)	Direction (°T)	Wind speed (kn)
2130	120	50
2100	120	48
1800	120	46
1500	110	44
1200	110	30
0900	120	17
0600	140	08
0300	—	calm
0000	250	10

Waves at time of picture

	Direction (°T)	Period (sec.)	Height (ft.)
Sea waves	120	7	20
Swell	—	—	—

Synoptic situation

Station in strong southeasterly circulation, approximately 600 nautical miles east-northeast of a deep low, centered about 48°N, 160°W.

Remarks

Although the wind speed was 50 knots (force 10) at the time of the picture, the average speed over the previous 9½ hours was force 9. The wind had increased from calm conditions 18 hours ago. The appearance of the sea surface conforms well to the force 9 specification, but the reported wave heights are only 50 percent of the maximum theoretical height for force 9. It would be necessary for the wind to blow another 15 to 20 hours at force 9 before the waves approached their maximum height. Air temperature 42°.1F; sea temperature 41°.6F.

BEAUFORT FORCE 9 SEA STATE 6

BEAUFORT FORCE 10

Wind speed 48 to 55 knots, mean 52 knots

Sea criterion: Very high waves with long overhanging crests. The resulting foam, in great patches, is blown in dense white streaks along the direction of the wind. On the whole, the surface of the sea takes on a white appearance. The tumbling of the sea becomes heavy and shock-like. Visibility affected.

Date/Time of photograph: Mar. 14, 1961, 2330 GMT.

Height of camera above sea: 15 ft.

Time (GMT)	Direction (°T)	Wind speed (kn)
2330	340	55
2100	350	52
1800	000	55
1500	010	52
1200	010	44
0900	030	43
0600	030	29
0300	040	17
0000	050	12

Waves at time of picture

	Direction (°T)	Period (sec.)	Height (ft.)
Sea waves	340	9	22
Swell	—	—	—

Synoptic situation

Deep low pressure area centered at 49°N, 138°W, moving northeastward. Station in strong northerly circulation 250 nautical miles west-northwest of the low center.

Remarks

The wind had been blowing at force 10 for the past 8½ hours, the direction backing from 010° to 340°. Because of the relatively short duration of the force 10 winds, the wave heights have only attained roughly 50 percent of their maximum height for this force. The fetch of the wind at the time of the picture was estimated to lie between 150 and 200 nautical miles. Air temperature 37°6F; sea temperature 40°0F.

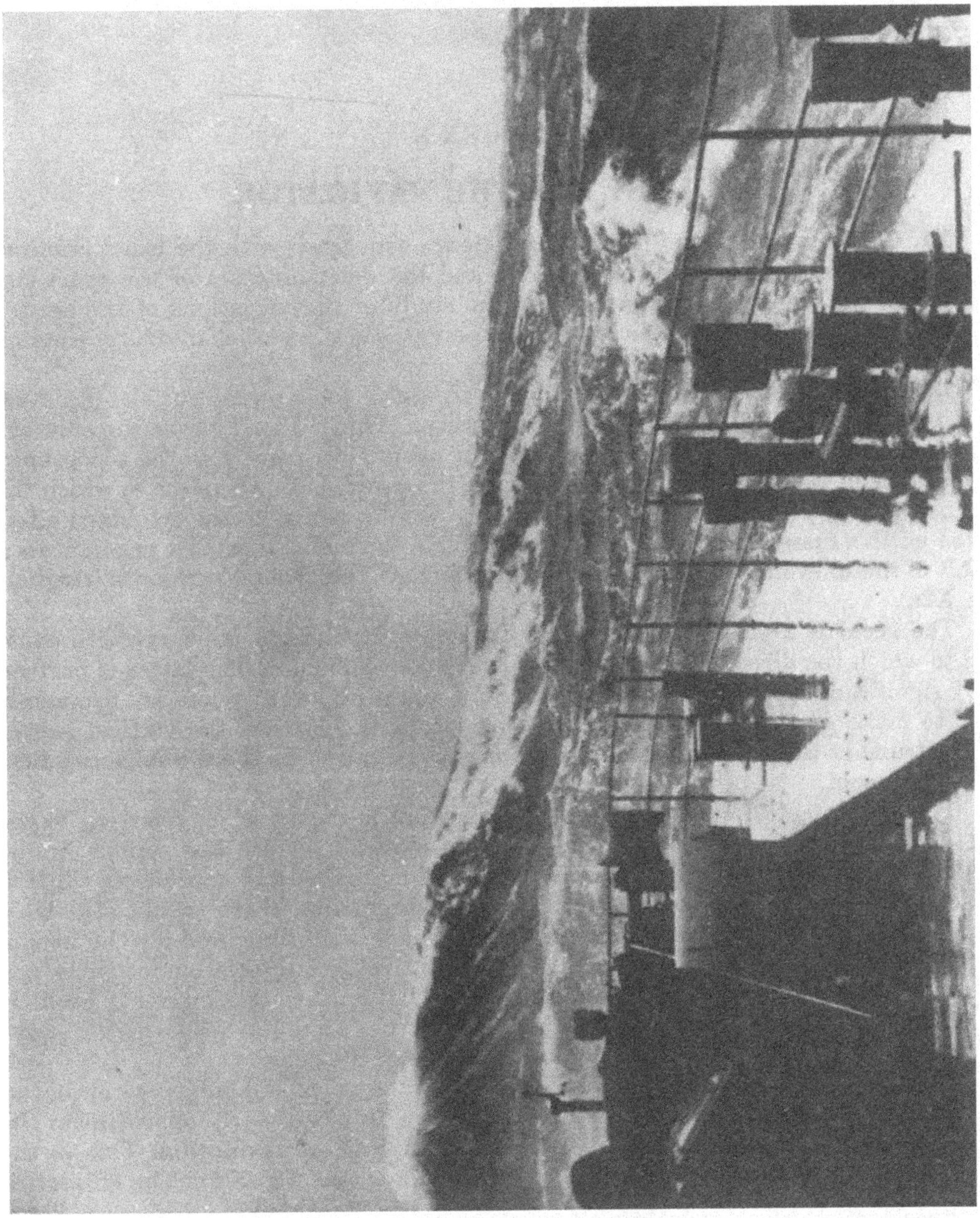

BEAUFORT FORCE 10 SEA STATE 7

APPENDIX X

GEODESY FOR THE NAVIGATOR

X1. Geodesy may be defined as that science concerned with the exact positioning of points on the surface of the earth, and the determination of the exact size and shape of the earth. It also involves the study of the variations of the earth's gravity (art. X4), and the application of these variations to exact measurements on the earth.

X2. Shape of the earth.—Although the irregular topographic surface (fig. X2a) is most apparent and is, in fact, the surface on which actual measurements are made, references to the earth's shape are not made with respect to the topography but to a mean sea-level surface. This surface, the **geoid**, is that surface to which the oceans would conform over the entire earth if free to adjust to the combined effect of the earth's mass attraction and the centrifugal force of the earth's rotation. As a result of the uneven distribution of the earth's mass, the geoidal surface is irregular (fig. X2a).

The geoid is a surface along which the gravity potential is everywhere equal and to which the direction of gravity is always perpendicular. The latter is particularly significant because optical instruments containing level devices are commonly used to make geodetic measurements. When properly adjusted, the vertical axis of the instrument coincides with the direction of gravity and is, therefore, perpendicular to the geoid.

The **equipotential surface** or the geoid is what is meant when referring to the size and shape of the earth, but such "potato-shaped" surface has serious limitations as an earth model: (1) It has no complete mathematical expression; (2) If it could be defined by an infinite series of measurements, there would still be a problem because of small variations in surface shape with time; and (3) The irregularity of the surface would necessitate a prohibitive amount of computations and complicate the problem of representation on the flat surface of a map. For geodetic and mapping purposes, it is therefore necessary to use a regular or geometric shape which provides a close approximation to the shape of the geoid.

Since the earth is in fact flattened slightly at the poles and bulges somewhat at the equator, the geometrical figure used in geodesy to most nearly approximate the shape of the earth is the oblate spheroid or **ellipsoid of revolution.** This is the figure obtained by rotating an ellipse about its minor axis (fig. X2b). The ellipsoidal earth model has its minor axis parallel to the earth's polar axis.

An ellipsoid of revolution is uniquely defined by specifying two dimensions. Geodesists, by convention, use the semimajor axis and flattening. The size is represented by the radius at the equator, the semimajor axis. The shape of the ellipsoid is given by the flattening, which indicates how closely an ellipsoid approaches a spherical shape. The **flattening** is the ratio of the difference between the semimajor and semiminor axes of the ellipsoid and the semimajor axis. If a and b represent the semimajor and semiminor axes, respectively, of the ellipsoid, and f is the flattening,

$$f=\frac{a-b}{a}.$$

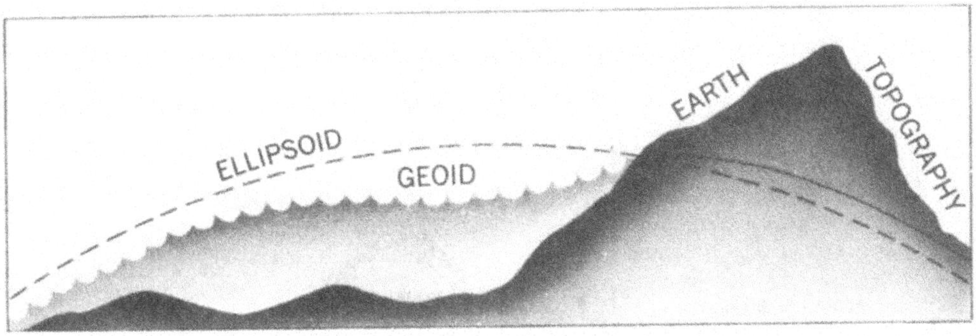

FIGURE X2a.—Surfaces of the earth, geoid, and ellipsoid.

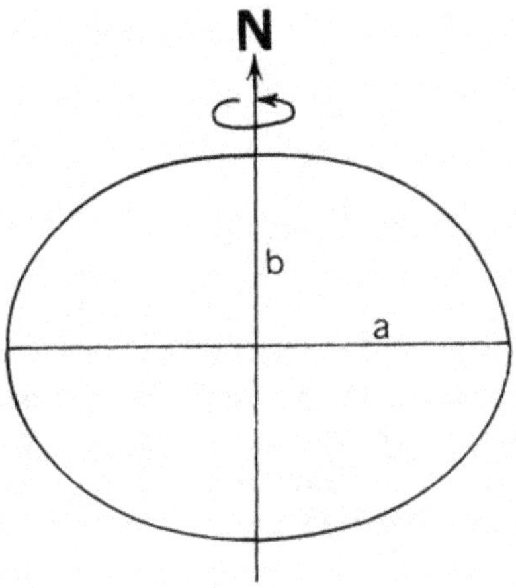

FIGURE X2b.—The ellipsoid of revolution.

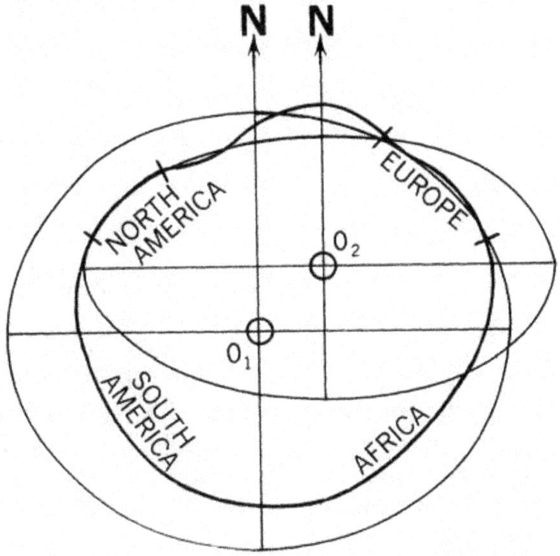

FIGURE X3.—The geoid and two ellipsoids.

As shown in figure X2a, the surface of the geoid tends to rise under mountains and to dip above ocean basins. The separations of the geoid and ellipsoid are called **geoidal heights, geoidal undulations,** or **geoidal separations.**

X3. Ellipsoids and the geoid.—Since the surface of the geoid is irregular and the surface of the ellipsoid is regular, no one ellipsoid can provide other than an approximation of part of the geoidal surface. As shown in figure X3, the ellipsoid that fits well in North America does not fit well in Europe.

X4. Deflection of the vertical.—Gravitation (art. 1407) is the mutual attraction between masses of matter. In geodesy, **gravitation** is the mutual attraction between the earth and bodies on or near its surface. **Gravity** is that force which tends to pull bodies toward the earth. It is the resultant of two opposing forces: gravitation and the centrifugal force due to the rotation of the earth.

The irregularities in density and heights of the material making up the surface crust of the earth result in slight alterations of the direction of gravity. These alterations are reflected in the irregular shape of the geoid, the surface of which is perpendicular to the plumb line.

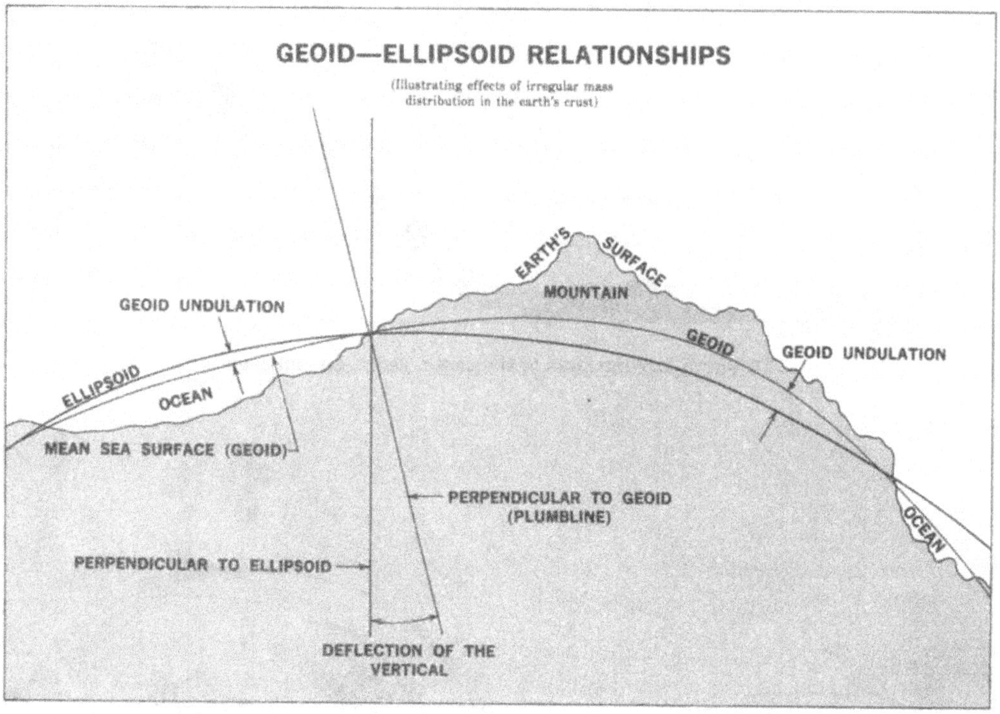

FIGURE X4.—Deflection of the vertical in the plane of the meridian.

The deflection of the vertical is the angle at a point on the geoid between the vertical (direction of the plumb line) and the direction of the normal to the ellipsoid of reference. For simplicity, the deflection of the vertical is shown in figure X4 at a point where the two normals coincide. Deflection of the vertical is also called **station error.** This deflection is usually resolved into two components, one in the plane of the meridian (art. 203) and the other in the plane of the prime vertical (art. 1430).

X5. Coordinates.—The **astronomic latitude** is the angle between the plumb line at a station and the plane of the celestial equator (fig. X5a). It is the latitude which results directly from observations of celestial bodies, uncorrected for deflection of the vertical which, in the United States, may amount to as much as 25″. Astronomic latitude applies only to positions on the earth, and is reckoned from the astronomic equator (0°), north and south through 90°.

The **astronomic longitude** is the angle between the plane of the celestial meridian at a station and the plane of the celestial meridian at Greenwich. It is the longitude which results directly from observations of celestial bodies, uncorrected for deflection of the vertical, the prime vertical component of which, in the United States, may amount to more than 18″.

Astronomic observations are made by optical instruments—theodolite, zenith camera, prismatic astrolabe—which all contain leveling devices. When properly adjusted, the vertical axis of the instrument coincides with the direction of gravity and is, therefore, perpendicular to the geoid. Thus, astronomic positions are referenced to the geoid. Since the geoid is an irregular, nonmathematical surface, astronomic positions are wholly independent of each other.

The **geodetic latitude** is the angle which the normal to the ellipsoid at a station makes with the plane of the geodetic equator (fig. X5b). In recording a geodetic position, it is essential that the *geodetic datum* (art. X8) on which it is based be also stated. A geodetic latitude differs from the corresponding astronomic latitude by the amount of the meridional component of the local deflection of the vertical.

Geodetic longitude is the angle between the plane of the geodetic meridian at a station and the plane of the geodetic meridian at Greenwich. A geodetic longitude differs from the corresponding astronomic longitude by the amount of the prime-vertical component of the local deflection of the vertical divided by the cosine of the latitude.

The geodetic coordinates are the ones used for mapping.

Geographic latitude is a general term applying alike to astronomic latitudes and geodetic latitudes. **Geographic longitude** is a general term applying alike to astronomic longitudes and geodetic longitudes.

Geocentric latitude is the angle at the center of the ellipsoid (used to represent the earth) between the plane of its equator and a straight line (or radius vector) to a point on the surface of the ellipsoid. This differs from geodetic latitude because the earth is a spheroid rather than a sphere, and the meridians are ellipses. Since the parallels of latitude are considered to be circles, geodetic longitude is geocentric, and a separate expression is not used. The difference between geocentric and geodetic latitudes has a maximum of about 11$'$.6 at latitude 45°.

Because of the oblate shape of the ellipsoid, the length of a degree of geodetic latitude is not everywhere the same, increasing from about 59.7 nautical miles at the equator to about 60.3 nautical miles at the poles, as shown by table 6.

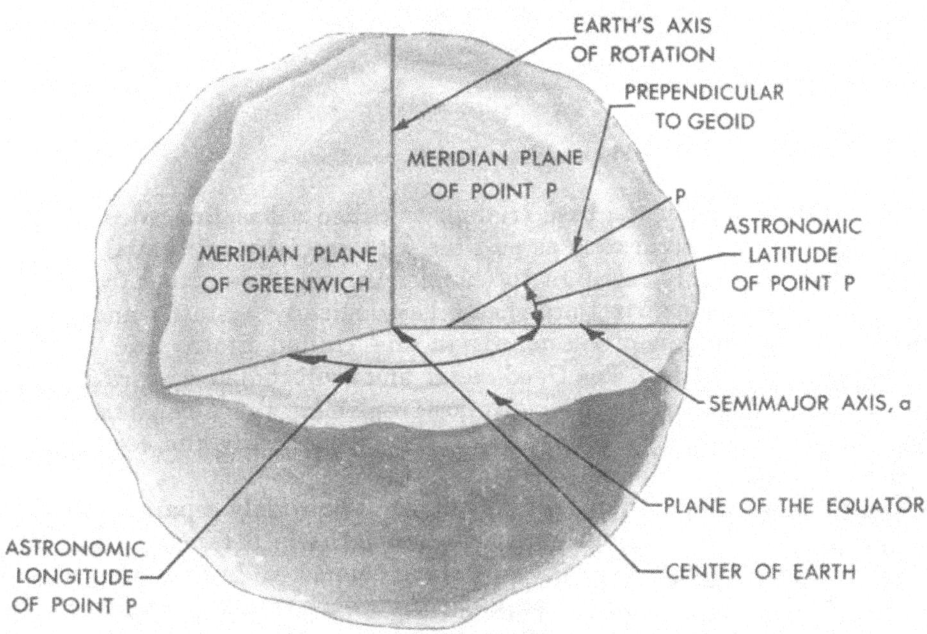

FIGURE X5a.—Astronomic coordinates.

X6. Geodetic surveys.—The most common type of geodetic survey is known as **triangulation** (fig. X6a). Basically, triangulation consists of the measurement of the angles of a series of triangles. The principle of triangulation is based on simple trigonometric procedures. If the distance along one side of the triangle and the angles at each end are accurately measured, the other two sides and the remaining angle can be computed. In practice all of the angles of every triangle are measured to provide exact data for use in computing the precision of the measurements. Also, the latitude and longitude of one end of the measured side along with the length and direction (azimuth) of the side provide sufficient data to compute the latitude and longitude of the other end of the side.

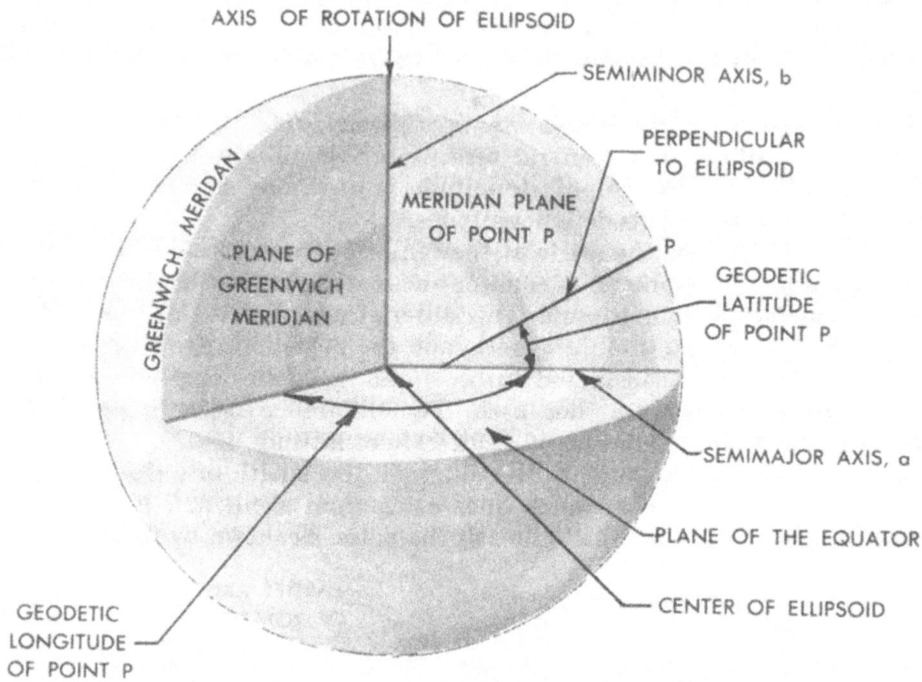

FIGURE X5b.—Geodetic coordinates.

The measured side of the base triangle is called a **baseline.** Measurements are made as carefully and accurately as possible with specially calibrated tapes or wires of Invar, an alloy, highly resistant to changes in length resulting from changes in temperature. The tape or wires are checked periodically against standard measures of length (at the Bureau of Standards in the United States and corresponding agencies in other countries). The Geodimeter and Tellurometer, operating on optical and electronic principles, respectively, are replacing the older methods of base measurement since the work can be completed more rapidly and economically than with wire or tape.

To establish an arc of triangulation between two widely separated locations, the baseline may be measured and longitude and latitude determined for the initial points at each location. The lines are then connected by a series of adjoining triangles forming quadrilaterals extending from each end. All angles of the triangles are observed repeatedly to reduce errors. With the longitude, latitude, and azimuth of the initial points, similar data is computed for each vertex of the triangles, thereby establishing triangulation stations or **geodetic control stations.** The coordinates of each of the stations are defined as geodetic coordinates.

Triangulation is extended over large areas by connecting and extending series of arcs to form a network or triangulation system. The network is adjusted in a manner which reduces the effect of observational errors to a minimum. A denser distribution of geodetic control is achieved in a system by subdividing or filling in with other surveys.

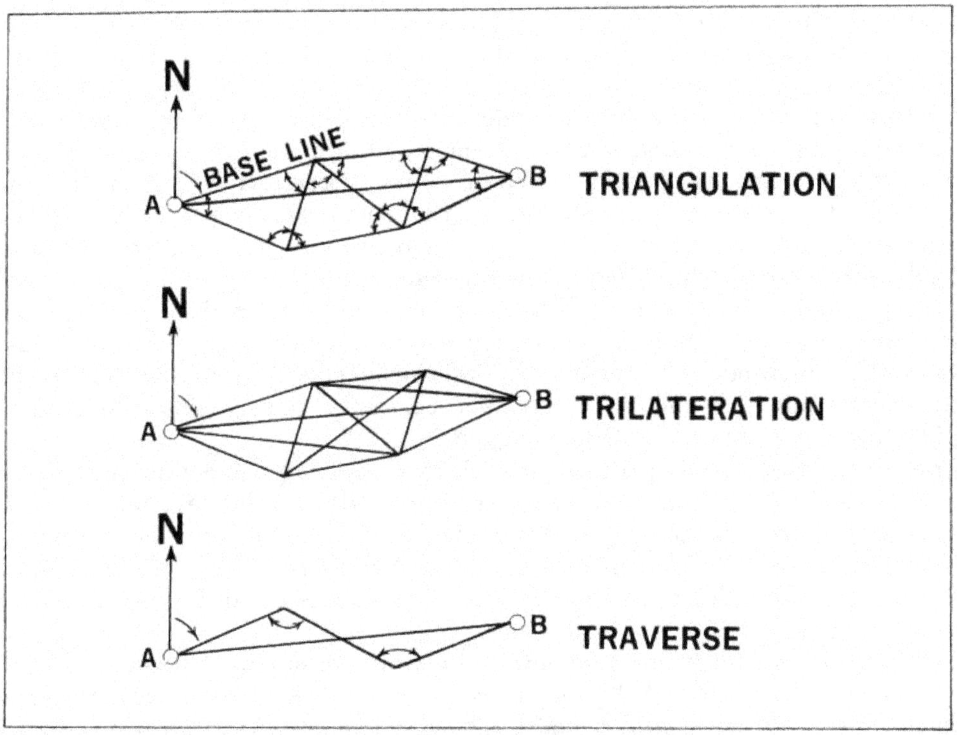

FIGURE X6a.—Triangulation, trilateration, and traverse.

There are four general classes or orders of triangulation. **First-order (primary) triangulation** is the most precise and exact type. It uses the most accurate instruments and rigorous computation methods. It is costly and time-consuming and is usually used to provide the basic framework of control data for an area and the determination of the figure of the earth. No precaution is ignored in making the linear and angular measurements, or in the computations. The most accurate first-order surveys furnish control points which can be interrelated with an accuracy ranging from 1 part in 25,000 over short distances to approximately 1 part in 100,000 for long distances.

Second-order triangulation furnishes points closer together than in the primary network. While second-order surveys may cover quite extensive areas, they are usually tied to a primary system where possible. The procedures are less exacting and the proportional error is 1 part in 10,000.

Third-order triangulation is run between points in a secondary survey. It is used to densify local control nets and thereby position the topographic and hydrographic detail of the area. Triangle error can amount to 1 part in 5,000 in third-order triangulation.

The sole accuracy requirement for **fourth-order triangulation** is that the positions be located without any appreciable error on maps compiled on the basis of the control. Fourth-order control is executed primarily as mapping control.

Trilateration involves the measuring of the sides of a chain of triangles or other polygons. From them, the distance and direction from *A* to *B* (fig. X6a) can be computed.

Traverse involves measuring distances, and the angles between them, without triangles for the purpose of computing the distance and direction from *A* to *B* (fig. X6a).

Vertical surveying is the process of determining elevations above the mean sea-level surface. In geodetic surveys executed primarily for mapping purposes, there is no problem in the fact that geodetic positions are referred to an ellipsoid, and the elevations of the positions are referred to the geoid. However, for certain other purposes such as satellite geodesy, the geoidal heights must be taken into account to establish the correct height above the ellipsoid.

Precise geodetic leveling is used to establish a basic network of vertical control points. From these, the height of other positions in the survey can be determined by supplementary methods. The mean sea-level surface used as a reference (vertical datum) is determined by obtaining an average of the hourly water heights for a specified period of time at specified tide gages.

There are three leveling techniques—differential, trigonometric, and barometric—which yield information of varying accuracy. **Differential leveling** is the most accurate of the three methods. With the instrument locked in position, readings are made on two calibrated staffs held in an upright position ahead of and behind the instrument (fig. X6b). The difference between readings is the difference in elevation between the points.

The optical instrument used for leveling contains a bubble tube to adjust it in a position parallel to the geoid. When properly "set up" at a point, the telescope is locked in a perfectly horizontal (level) position so that it will rotate through a 360° arc. The exact elevation of at least one point in a leveling line must be known and the rest computed from it.

Trigonometric leveling involves measuring a vertical angle from a known distance with a theodolite and computing the elevation of the point. With this method, vertical measurement can be made at the same time horizontal angles are measured for triangulation. It is, therefore, a somewhat more economical method but less accurate than differential leveling. It is often the only practical method of establishing accurate elevation control in mountainous areas.

In **barometric leveling**, differences in height are determined by measuring the differences in atmospheric pressure at various elevations. Air pressure is measured by mercurial or aneroid barometer, or a boiling point thermometer. Although the degree of accuracy possible with this method is not as great as either of the other two, it is a method which obtains relative heights very rapidly at points which are fairly far apart. It is widely used in reconnaissance and exploratory surveys where more exacting measurements will be made later or are not required.

X7. Reference ellipsoids.—A number of reference ellipsoids are used in geodesy and mapping because an ellipsoid is mathematically simpler than the geoid. Some of these ellipsoids and areas where used are as follows:

Clarke 1866 (North and Central America, Greenland)
International 1924 (Hayford 1909) (Europe, various countries in South America)
Modified Clarke 1880 (Africa)
Everest 1830 (India, Southeast Asia, Indonesia)
Bessell 1841 (China, Korea, Japan)
Krasovskiy 1942 (U.S.S.R. and adjacent countries)
Australian National 1965 (Australia)
South American 1969 (South America).

Reference ellipsoid constants are given in appendix D.

In the late 1950's, an earth model fitting the world as a whole was needed for manned space flights. The National Aeronautics and Space Administration, and the Department of Defense selected the Fischer Ellipsoid 1960 (Mercury Datum) for this program. A modification of the Mercury Datum in 1968 produced the Fischer Ellipsoid 1968, having a semimajor axis of 6,378,150 meters and a flattening of 1/298.3. The geoidal heights are shown in figure X7. These heights range from about 80 meters below the ellipsoid to about 60 meters above.

X8. Geodetic datum.—A *datum* is defined as any numerical or geometrical quantity or set of such quantities which may serve as a reference or base for other quantities.

In geodesy two types of datums must be considered: a *horizontal datum* which forms the basis for computations of horizontal control surveys in which the curvature of the earth is considered, and a *vertical datum* to which elevations are referred. In other words, the coordinates for points in specific geodetic surveys and triangulation networks are computed from certain initial quantities (datums).

A **horizontal geodetic datum** consists of the astronomic and geodetic latitude, and astronomic and geodetic longitude of an initial point (origin); an azimuth of a line (direction); the parameters (radius and flattening) of the ellipsoid selected for the computations; and the geoidal separation at the origin. A change in any of these quantities affects every point on the datum. For this reason, while positions within a system are directly and accurately relatable, those from different datums must be transformed to a common datum for consistency.

Just as horizontal surveys are referred to specific original conditions (datums), vertical surveys are also related to an initial quantity or datum. Elevations are referred to the geoid because the instruments used either for differential or trigonometric leveling (art. X6) are adjusted with the vertical axis coincident to the local vertical. As with horizontal datums, there are many discrepancies among **vertical datums.** There is never more than 2 meters variance between leveling nets based on different mean sea-level datums; however, elevations in some areas are related to surfaces other than the geoid; and barometrically determined heights are usually relative.

X9. Orientation of ellipsoid to geoid.—The selection of the reference ellipsoid provides two quantities of the geodetic datum: semimajor axis and flattening of the ellipsoid. The simplest means of obtaining the remaining quantities to establish the geodetic datum is to select a first-order triangulation station, preferably one located near the center of a triangulation network, to serve as the datum origin. Then the astronomical coordinates of the station and the astronomical azimuth of a line from the station to another control station are observed. The observed astronomical coordinates and azimuth are adopted without any correction as the geodetic coordinates and azimuth of the datum origin on the reference ellipsoid. Further, the geoid and ellipsoid are assumed to coincide at that point. This means that the deflection of the vertical and separation between the ellipsoid and geoid are defined as zero at the origin. By using this **single astronomical station datum orientation,** the normal to the ellipsoid is arbitrarily made to coincide with the plumb line at the datum origin (fig. X9).

Although the computed positions will be correct with respect to each other in this type of orientation, the entire net will be shifted with respect to the axis of the earth. This is not significant for local use of the positions but may introduce large systematic errors as the survey is expanded.

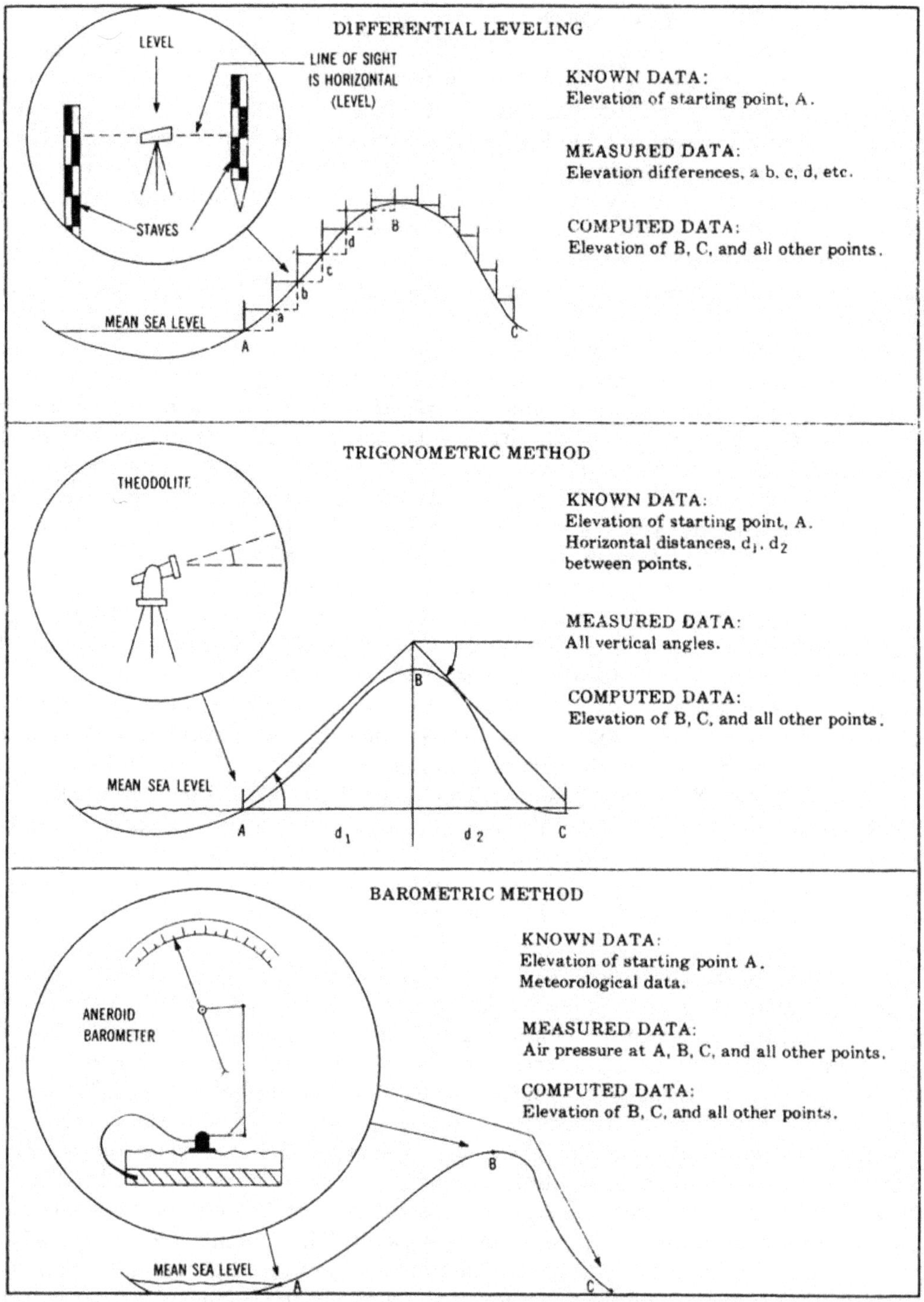

FIGURE X6b.—Methods of elevation determination.

It should be noted that although the deflection of the vertical and the geoidal height are defined as zero at the origin, deflections will occur at other positions within the network. Therefore, when comparing the geodetic latitude and longitude of any other point in the net with the corresponding astronomic latitude and

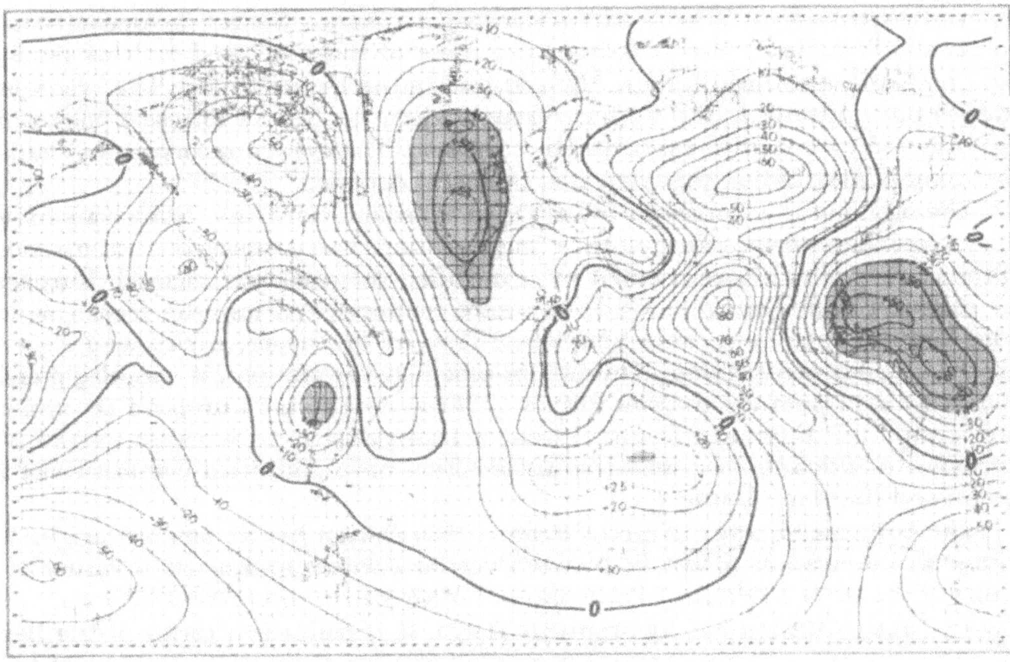

FIGURE X7.—Geoidal heights.

longitude of that point, systematic discrepancies will appear between the two sets of values.

A datum oriented by a single astronomical point may produce large systematic geoidal separations. The ellipsoid is not earth-centered and its rotational axis is not coincident with the axis of the earth. The inconvenience of such an orientation is that the positions derived from different astronomically oriented datums are not directly comparable to each other in any geodetic computation.

In the **astrogeodetic datum orientation,** the geoid and ellipsoid are oriented so that the sum of the squares of geoidal separations and several deflections of the vertical selected throughout the geodetic network is made as small as possible. Astrogeodetic datums are better suited over larger areas than those oriented by a single astronomic position because the separation between ellipsoid and geoid is minimized for a best fit.

X10. Datum connection.—In areas of overlapping geodetic triangulation networks, each computed on a different datum, the coordinates of the points given with respect to one datum will differ from those given with respect to the other. The differences can be used to derive transformation formulas.

Datums are connected by developing transformation formulas at common points, either between overlapping control networks or by satellite connections.

X11. Preferred datums.—Different countries developed their own geodetic datums which usually differed from those of their neighbors. Accordingly, national maps did not agree along the borders with those of the neighboring countries.

As military distance requirements increased, positioning information of local or even national scope became unsatisfactory. The capabilities of the various weapon systems increased until datums of at least continental limits were required. The best solution was the establishment of a preferred datum for an area and adjusting all local systems to it. The North American, Ordnance Survey of Great Britain, European, Tokyo, and Indian Datums are some of those selected for this purpose.

The **North American Datum, 1927,** is used in the United States. The origin is at Meades Ranch, Kansas. The datum is computed on the Clarke Ellipsoid 1866 which was oriented by a modified astrogeodetic method. The system incorporates Canada, Mexico, the West Indies, Greenland, and Central America.

The origin of the **European Datum** is at Potsdam, Germany. Numerous national systems have been joined into a large datum based upon the International Ellipsoid 1924 which was oriented by a modified astrogeodetic method. European, African, and Asian triangulation chains were connected. African are measurements from Cairo to Cape Town were completed. Thus all of Europe, Africa, and Asia are molded into one great system. Through common survey stations, it was also possible to convert data from the Russian Pulkova 1932 system to the European Datum, and as a result the European Datum includes triangulation as far east as the 84th meridian. Additional ties across the Middle East have permitted connection of the Indian and European Datums.

The Ordnance Survey of Great Britain 1936 Datum has no point of origin. The datum was derived as a best fit between retriangulation and original values of 11 points of the earlier Principal Triangulation of Great Britain (1783–1853).

The **Tokyo Datum** has its origin in Tokyo. It is defined in terms of the Bessel Ellipsoid and oriented by means of a single astronomic station. By means of triangulation ties through Korea, the Japanese datum is connected with the Manchurian datum. Unfortunately, Tokyo is situated on a steep slope on the geoid, and the single-station orientation has resulted in large systematic geoidal separations as the system is extended from its initial point.

The **Indian Datum** is accepted as the preferred datum for India and several adjacent countries in Southeast Asia. It is computed on the Everest Ellipsoid with its origin at Kalianpur in Central India.

Derived in 1830, the Everest Ellipsoid is one of the oldest of the ellipsoids in common use and is much too small. As a result, the datum cannot be extended too far from the origin or very large geoidal separations will occur. For this reason and the fact that the ties between local triangulation in Southeast Asia are typically weak, the Indian Datum is probably the least satisfactory of the preferred datums.

X12. World Geodetic System.—The Department of Defense (DOD) in the late 1950's generated a *geocentric* reference system to which different geodetic networks could be referred, and compatibility established between the coordinates of sites of interest. Efforts of the Army, Navy, and Air Force were combined leading to the development of the DOD World Geodetic System 1960 (WGS 60). In accomplishing WGS 60, a combination of available surface gravity data, astrogeodetic data, and results from Hiran and Canadian Shoran surveys were used to obtain a best-fitting ellipsoid for the major datum areas. The sole contribution of satellite data to the development of WGS 60 was the value for the ellipsoid flattening ($1/298.3 \pm 0.1$), which was obtained from the nodal motion of Satellite 1958β. The semimajor axis of the WGS 60 Ellipsoid was determined as $6{,}378{,}165 \pm 50$ meters.

In January 1966, a World Geodetic System Committee was charged with the responsibility for developing an improved WGS needed to satisfy mapping, charting, and geodetic requirements. Additional surface gravity observations, results from the extension of triangulation and trilateration networks, and large amounts of doppler and optical satellite data had become available since the development of WGS 60. Using the additional data and improved techniques, the Committee produced WGS

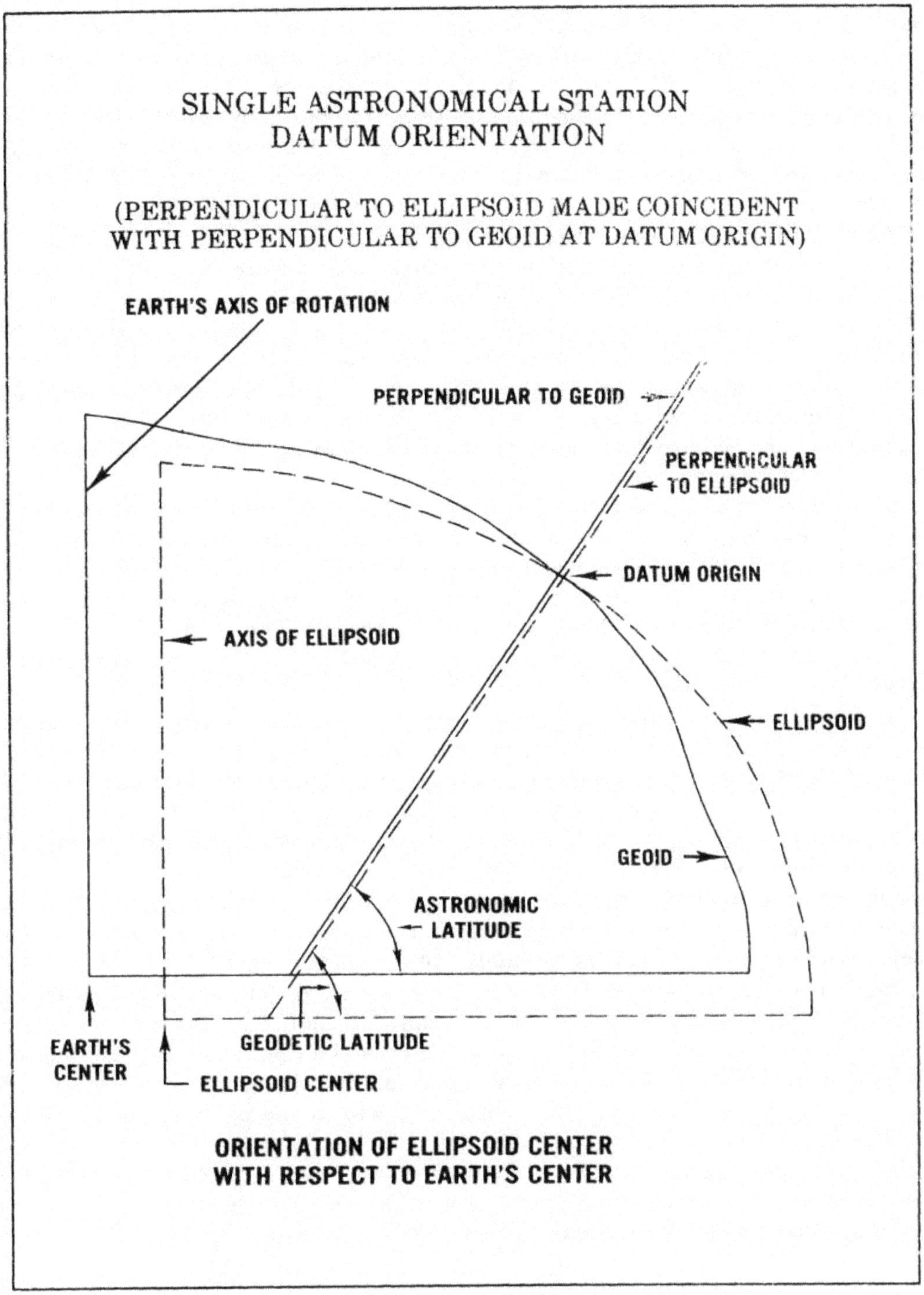

FIGURE X9.—Single astronomical station datum orientation.

66 which served DOD needs following its implementation in 1967. The defining parameters of the WGS 66 Ellipsoid were the flattening (1/298.25±0.02), determined from satellite data, and the semimajor axis (6,378,145±20 meters), determined from a combination of doppler satellite and astrogeodetic data involving a

geoid-match technique. A 24th degree and order geopotential coefficient set derived from a harmonic analysis of a worldwide $5° \times 5°$ mean free air gravity anomaly field was selected as the WGS 66 Gravitational Model. This geopotential coefficient set was also used in a spherical harmonic expansion to obtain the Worldwide WGS 66 Geoid. Also, a geoid referenced to the WGS 66 Ellipsoid, providing a detailed representation for limited land areas, was derived from available astrogeodetic data. Datum shift constants for the North American Datum 1927, European Datum, and Tokyo Datum were obtained for each datum.

The same World Geodetic System Committee began work in 1970 to develop a replacement for WGS 66. Since the development of WGS 66, large quantities of additional data had become available from both doppler and optical satellites, surface gravity surveys, triangulation and trilateration surveys, high precision traverses, and astronomic surveys. In addition, greater capabilities had been developed in both computers and computer software. Further, continued research in improved computational procedures and error analyses had produced better methods and a greater facility for handling and combining data.

After an extensive effort extending over a period of approximately 3 years, the Committee completed the development of the Department of Defense World Geodetic System 1972 (WGS 72). Selected satellite, surface gravity, and astrogeodetic data available through 1972 from both DOD and non-DOD sources was used in a Unified WGS Solution (a large scale least squares adjustment). The results of the adjustment consists of corrections to initial station coordinates and geopotential coefficients.

In determining the WGS 72 Ellipsoid and associated parameters, the Committee decided quite early to closely adhere to the approach used by the **International Union of Geodesy and Geophysics (IUGG)** in establishing the Geodetic Reference System 1967 (GRS 67). Accordingly, an equipotential ellipsoid of revolution was taken as the form for the WGS 72 Ellipsoid. An **equipotential ellipsoid** is simply an ellipsoid defined to be an equipotential surface; i.e., a surface on which all values of the potential are equal. Given an ellipsoid of revolution, it can be made an equipotential surface of a certain potential function, the normal gravity potential, *U*. This normal gravity potential can be uniquely determined, independent of the density distribution within the ellipsoid, by using any system of four independent parameters as the defining constants of the ellipsoid. To determine the normal gravity potential without resorting to the use of a mass distribution model for the ellipsoid, *U* can be expanded into a series of zonal ellipsoidal harmonics of linear eccentricity in $(a^2-b^2)^{1/2}$. The coefficients in the series are determined by using the condition that the ellipsoid is an equipotential surface (U=constant). Since all the zonal coefficients vanish, except the two of degree zero and two, a closed finite expression is obtained for *U*. Normal gravity (γ), the gradient of *U*, is given at the surface of the ellipsoid by the closed formula:

$$\gamma = \frac{a\gamma_e \cos^2\phi + b\gamma_p \sin^2\phi}{(a^2\cos^2\phi + b^2\sin^2\phi)^{1/2}}$$

where a is the semimajor axis, b is the semiminor axis, γ_e is the normal gravity at the equator, γ_p is the normal gravity at the poles, and ϕ is the geodetic latitude.

Thus the equipotential ellipsoid serves not only as the reference surface or geometric figure of the earth but leads to a closed formula for normal gravity at the ellipsoidal surface, a formula easily modified for spatial applications.

Consistent with the IUGG definition of GRS 67, the Committee took the four defining parameters of the WGS 72 Ellipsoid to be the semimajor axis (a), the earth's gravitational constant (GM) and angular velocity (ω), and the second degree zonal harmonic coefficient of the geopotential ($\bar{C}_{2,0}$). Other parameters associated with the ellipsoid, such as the semiminor axis (b) and the flattening (f), including the normal gravity formula, are calculated using the defining parameters. These and other parameters associated with the ellipsoid are given in table X12.

Parameters	*Notation*	*Magnitude*	*Standard error (68.27%)*
Gravitational constant	GM	398600.5 km^3/sec^2	± 0.4
Second degree zonal	$\bar{C}_{2,0}$	-484.1605×10^{-6}	—
Angular velocity	ω	$0.7292115147 \times 10^{-4}$ rad/sec	$\pm 0.1 \times 10^{-13}$
Semimajor axis	a	6378135 meters	± 5
Flattening	f	1/298.26	$\pm 0.6 \times 10^{-7}$
Equatorial gravity (Absolute system).	γ_e	978033.26 mgal	± 1.8
Gravitational constant (Mass of earth's atmosphere included).	GM'	398600.8 km^3/sec^2	± 0.4

TABLE X12.—Geodetic and geophysical parameters of World Geodetic System 1972 ellipsoid.

APPENDIX Y

BUOYAGE SYSTEMS

With modifications, two systems of buoyage have been in general use throughout the world from 1936 to 1977. These are the **lateral system** and the **cardinal system.**

The lateral system is best suited for well-defined channels. The location of each buoy indicates the direction of the danger it marks relative to the course which should normally be followed. Thus, a buoy which should be kept on the port hand lies between the vessel and the danger when the buoy is abeam to port, approximately.

In principle, the positions of marks in the lateral system are determined by the general direction taken by the mariner when approaching a harbor, river, estuary, or other waterway from seaward, and may also be determined with reference to the main stream of flood current. The application of this principle is defined, as required, by nautical documents such as sailing directions.

The cardinal system is best suited for coasts with numerous rocks, shoals, and islands, and for dangers in the open sea. The location of each buoy indicates the approximate true bearing of the danger it marks. Thus, an eastern quadrant buoy marks a danger, such as a shoal, which lies to the west of the buoy, approximately.

Although almost all of the major maritime nations have used either the lateral or the cardinal system for many years, details such as the shapes and colors of the buoys, and the characteristics and colors of lighted aids generally have varied from country to country. With the passage of time and the increase in maritime communication between countries, the desirability of a uniform system of buoyage has become increasingly apparent. Consequently, over the past century a number of attempts have been made to standardize the various systems of buoyage. International conferences have been held on the subject and recommendations have been made. These recommendations have often been conflicting, however, and although the differences in the various methods as applied to the cardinal system are comparatively slight, two distinct methods of applying the lateral system have evolved. The major discrepancy has been in the colors of the buoys and of their lights.

In 1889, the *International Marine Conference* held in Washington, D.C., recommended that in the lateral system starboard hand buoys be painted red and port hand buoys black. With the introduction of lighted aids to navigation, these recommendations logically led to the use, by nations which had accepted the recommendation, of red or white lights on the starboard side and green or white lights on the port side.

In 1936, a League of Nations subcommittee recommended a coloring system diametrically opposed to the 1889 proposal. This is part of the **Uniform System,** and it provides for black buoys with green or white lights on the starboard side and red buoys with red or white lights on the port side.

Most maritime countries using the lateral system have adopted one of these two systems, usually with small variations. Until 1977 it could be said that, *very generally,* European countries followed the Uniform System of 1936 and most other countries followed the system proposed in 1889.

In 1973, new terms of reference were given to the Technical Committee of the International Association of Lighthouse Authorities (IALA), which had been studying various projects, including buoyage, for the previous 8 years. IALA, a nongovernmental body which brings together representatives from the aids to navigation services in order to exchange information and recommend improvements to aids to navigation based on the latest technology, has decided that a single worldwide system of buoyage cannot be achieved at present but considers that the use of only two alternative systems is practicable. The two systems are termed: IALA Maritime Buoyage System 'A'—Combined Cardinal and Lateral System (Red to Port); IALA Maritime Buoyage System 'B'—Lateral System only (Red to Starboard).

The rules for System A which included both Cardinal and Lateral marks were completed in 1976 and agreed by the Inter-Governmental Maritime Consultative Organization (IMCO) which was renamed the International Maritime Organization (IMO) on 22 May 1982. The introduction of the System began in 1977 and its use has gradually spread throughout Europe, Australia, New Zealand, Africa, the Gulf and some Asian countries.

The rules for System B were completed in early 1980 and these were felt to be suitable for application in the countries of North, Central and South America, Japan, Korea and the Philippines.

United States System

In 1982, the United States agreed to make modifications to incorporate the *International Association of Lighthouse Authorities (IALA) Maritime Buoyage System* for Region B. The new system is called the *Modified U.S. Aids to Navigation System.* The *IALA Maritime Buoyage System* for Regions A and B and the *Modified U.S. Aids to Navigation System* are illustrated in this appendix. Until the conversion program is completed mariners should be familiar with both systems—the old U.S. lateral system of buoyage and the new *Modified U.S. Aids to Navigation System.*

The U.S. lateral system is not materially changed in the new system. A majority of U.S. aids to navigation already conform—red buoys and daymarks to starboard, green daymarks to port. The non conforming aids—chiefly black buoys and safe water buoys—will be converted on a gradual basis as they are routinely replaced or serviced. All changes to aids to navigation will be published in the U.S. Coast Guard *Local Notice to Mariners* and, where appropriate, the DMAHTC weekly *Notice to Mariners.*

The United States System and the *Modified U.S. Aids to Navigation System* are discussed in more detail in chapter IV.

The lateral system along the coast, as described in chapter IV, refers only to the side of the vessel on which buoys are kept, as indicated by color, shape, and light, if any; there is no numerical continuity between coast buoys. In fairways and channels, however, buoys are numbered consecutively from seaward.

Uniform System

As recommended by the League of Nations in 1936, a country uses the *Uniform Lateral System* or the *Uniform Cardinal System*, or both, according to its requirements or preference. When both are used, the transition from one to the other must be clearly indicated in appropriate publications, such as sailing directions, or by suitable buoyage marks.

Both the Uniform Lateral System and the Uniform Cardinal System employ **topmarks** as an additional means of identification. Unless otherwise stated in this appendix, a topmark is painted the darker of the colors used on the buoy. They are optional in every case except on wreck buoys in the Uniform Cardinal System. Topmarks are not used in the United States System.

In both the Uniform Lateral System and the Uniform Cardinal System, lighted buoys have the same shape as the unlighted buoys shown.

In both the Uniform Lateral System and the Uniform Cardinal System, a quick flashing light is regarded as a single flashing light.

The numbering or lettering of fairway and channel buoys is an optional feature of the Uniform Lateral System. In the United States System these buoys are always numbered, commencing from seaward.

IALA MARITIME BUOYAGE SYSTEM

General

The International Association of Lighthouse Authorities (IALA) is a non-governmental body which brings together representatives of the worldwide community of aids to navigation services to promote information exchange as well as recommend improvements.

In 1980 with the assistance of IMO and the IHO, the lighthouse authorities from 50 countries and representatives of 9 international organizations concerned with aids to navigation met and adopted the rules of the IALA Maritime Buoyage System and established the two regions, Region A and Region B. A graphic depicting the separation of the world in Regions is included in this appendix. Color plates illustrating the IALA Maritime Buoyage System are also included.

Lateral marks differ within Regions A and B. Lateral marks in Region A use red and green colors by day and night to indicate port and starboard sides of channels respectively. In Region B these colors are reversed with red to starboard and green to port.

Scope.—The IALA Maritime Buoyage System applies to all fixed and floating marks, other than lighthouses, sector lights, leading lights and marks, lightships and large navigational buoys (lighthouse buoys), and serves to indicate:

1. the sides and centerlines of navigable channels;
2. natural dangers and other obstructions, such as wrecks;
3. areas in which navigation may be subject to regulation;
4. other features of importance to the mariner.

Fixed Marks.—It should be understood that most lighted and unlighted beacons, other than leading marks, are included in the system. In general, beacon topmarks will have the same shape and colors as those used on buoys. (Because of the variety of beacon structures, the accompanying diagrams show mainly buoy shapes.)

Types of Marks.—The system provides *five types of marks* which may be used in any combination:

1. **lateral** marks indicate the port and starboard hand sides of channels;
2. **cardinal** marks, used in conjunction with the compass, indicate that the navigable water lies to the named side of the mark;
3. **isolated danger** marks erected on, or moored directly on or over, dangers of limited extent;
4. **safe water** marks, such as mid-channel buoys; and
5. **special** marks, the purpose of which is apparent from reference to the chart or other nautical documents.

Superseded Marks.—Certain marks were superseded by the introduction of the IALA System, including those which indicate wrecks, middle grounds, secondary channels, bifurcations, and junctions; there are no special "landfall" or "transition" marks in the System. There is no differentiation between the marks for such special features as spoil grounds, anchorages, cable areas, and military exercise areas, all of which will be marked by yellow buoys which may, in addition, carry lettering to indicate the purpose of the buoy.

Characteristics of Marks.—The significance of a mark depends on one or more features:

1. *by day*—color, shape, and topmark;
2. *by night*—light color and phase characteristics.

Colors of Marks.—The colors **red** and **green** are reserved for lateral marks, and **yellow** for special marks. The other types of marks have **black and yellow** or **black and red** horizontal bands or **red and white** vertical stripes, as described later.

Shapes of Marks.—There are *five basic buoy shapes,* namely, can, conical, spherical, pillar, and spar. In the case of can, conical, and spherical, the shape indicates the correct side to pass. With pillar and spar buoys, the shape has no such special significance.

The term "pillar" is used to describe any buoy which is smaller than a "lighthouse buoy" and which has a tall, central structure on a broad base; it includes beacon buoys, high focal plane buoys, and others (except spar buoys) whose body shape does not indicate the correct side to pass.

It must be understood that much existing equipment will be used in the new system including, for example, light-floats. Variations on the basic shapes will therefore be fairly common but, by day, the colors and topmarks should prevent ambiguity.

Topmarks.—The IALA System makes use of can, conical, spherical, and X-shaped topmarks only. Topmarks on pillar and spar buoys are particularly important and will be used wherever practicable, but ice or other severe conditions may occasionally prevent their use.

Colors of Lights.—Where marks are lighted, red and green lights are reserved for lateral marks, and yellow for special marks. The other types of mark have a white light, distinguished one from another by phase characteristic.

Rhythms of Lights.—Red and green lights may have any phase characteristic, as the color alone is sufficient to show on which side they should be passed. Special marks, when lighted, have a yellow light with any phase characteristic not reserved for white lights of the system. The other types of mark have clearly specified phase characteristics of white light: various quick flashing phase characteristics for cardinal marks, group flashing (2) for isolated danger marks, and relatively long periods of light for safe water marks.

Some shore lights, specifically excluded from the IALA System, may, by coincidence, have characteristics corresponding to those approved for use with the new marks. Care is needed to ensure that, on sight, such lights are not misinterpreted.

Lateral Marks

Lateral Marks—Direction of Buoyage Marks.—Lateral marks are generally used for well-defined channels; they indicate the port and starboard hand sides of the route to be followed, and are used in conjunction with a *conventional direction of buoyage.*

This direction is defined in one of two ways:

1. **local direction of buoyage**—the direction taken by the mariner when approaching a harbor, river estuary, or other waterway from seaward;

2. **general direction of buoyage**—in other areas, a direction determined by the buoyage authorities, following a clockwise direction around continental land masses, given in sailing directions, and, if necessary, indicated on charts by a symbol.

In some places, particularly straits (being open at both ends), the local direction of buoyage may be over-ridden by the general direction.

Cardinal Marks

Cardinal Marks.—Names of Marks.—A cardinal mark is used in conjunction with the compass to indicate where the mariner may find the best navigable water. It is placed in one of the four quadrants (north, east, south, and west), bounded by the true bearings NW-NE, NE-SE, SE-SW, and SW-NW, taken from the point of interest. A cardinal mark takes its name from the quadrant in which it is placed.

The mariner is safe if he passes north of a north mark, east of an east mark, south of a south mark, and west of a west mark.

Uses.—A cardinal mark may be used to:

1. indicate that the deepest water in an area is on the named side of the mark;

2. indicate the safe side on which to pass a danger; and

3. draw attention to a feature in a channel such as a bend, junction, bifurcation, or end of a shoal.

Topmarks.—Black double-cone topmarks are the most important feature, by day, of cardinal marks; *the arrangement of the cones must be memorized.* More difficult to remember than north (two cones points up) and south (two cones points down) are the east (one cone point up, one cone point down) and west (two cones point to point) topmarks: 'W for Wineglass' may help.

Cardinal marks carry topmarks whenever practicable, with the cones as large as possible and clearly separated.

Colors.—Black and yellow horizontal bands are used to color a cardinal mark. The position of the black band, or bands, is related to the points of the black topmarks, thus:

North—Points up—Black band above yellow band.
South—Points down—Black band below yellow band.
West—Points inward—Black band with yellow bands above and below.
East—Points outward—Black bands above and below yellow band.

Shape.—The shape of a cardinal mark is not significant, but in the case of a buoy will be pillar or spar.

Lights.—When lighted, a cardinal mark exhibits a white light; its characteristics are based on a group of quick or very quick flashes which distinguish it as a cardinal mark and indicate its quadrant.

The distinguishing quick or very quick flashes are:

North—Uninterrupted.
East—three flashes in a group.
South—six flashes in a group followed by a long flash.
West—nine flashes in a group.

As a memory aid, the number of flashes in each group can be associated with a clock face (3 o'clock—E, 6 o'clock—S, and 9 o'clock—W).

The long flash (of not less than 2 seconds duration), immediately following the group of flashes of a south cardinal mark, is to ensure that its six flashes cannot be mistaken for three or nine.

The periods of the east, south, and west lights are, respectively, 10, 15, and 15 seconds if quick flashing; and 5, 10, and 10 seconds if very quick flashing.

Quick flashing lights flash at a rate between 50 and 79 flashes per minute, usually either 50 or 60. Very quick flashing lights flash at a rate between 80 and 159 flashes per minute, usually either 100 or 120.

It is necessary to have a choice of quick flashing or very quick flashing lights in order to avoid confusion if, for example, two north buoys are placed near enough to each other for one to be mistaken for the other.

Isolated Danger Marks

Isolated Danger Marks—Use.—An isolated danger mark is erected on, or moored on or above, an isolated danger of limited extent which has navigable water all around it. The extent of the surrounding navigable water is immaterial: such a mark can, for example, indicate either a shoal which is well offshore or an islet separated by a narrow channel from the coast.

Charted Position.—On a chart, the position of a danger is the center of the symbol or sounding indicating it; an isolated danger buoy will inevitably therefore be slightly displaced on the chart.

Topmark.—A black double-sphere topmark is, by day, the most important feature of an isolated danger mark and, whenever practicable, this topmark will be carried, with the spheres as large as possible, disposed vertically, and clearly separated.

Color.—Black with one or more red horizontal bands are the colors used for isolated danger marks.

Shape.—The shape of an isolated danger mark is not significant, but in the case of a buoy will be pillar or spar.

Light.—When lighted, a white flashing light showing a group of two flashes is used to denote an isolated danger mark. The association of two flashes and two spheres in the topmark may be a help in remembering these characteristics.

Safe Water Marks

Safe Water Marks—Use.—A safe water mark is used to indicate that there is navigable water all around the mark. Such a mark may be used as a center line, mid-channel, or landfall buoy.

Color.—Red and white vertical stripes are used for safe water marks, and distinguish them from the black-banded, danger-marking marks.

Shape.—Spherical, pillar, or spar buoys may be used as safe water marks.

Topmark.—A single red sphere topmark will be carried, whenever practicable, by a pillar or spar buoy used as a safe water mark.

Light.—When lighted, safe water marks exhibit a **white** light, occulting, or equal interval (isophase), or showing a single long flash, or Morse "A." If a long flash (i.e. a flash of not less than 2 seconds) is used, the period of the light will be 10 seconds.

Special Marks

Special Marks—Use.—A special mark may be used to indicate to the mariner a special area or feature, the nature of which is apparent from reference to a chart, sailing directions, or notices to mariners. Uses include:

1. Ocean Data Acquisition System (ODAS), i.e. buoys carrying oceanographic or meteorological sensors;
2. traffic separation marks;
3. spoil ground marks;
4. military exercise zone marks;
5. cable or pipeline marks, including outfall pipes; and
6. recreation zone marks.

Another function of a special mark is to define a channel within a channel. For example, a channel for deep draft vessels in a wide estuary, where the limits of the channel for normal navigation are marked by red and green lateral buoys, may have the boundaries of the deep channel indicated by yellow buoys of the appropriate lateral shapes, or its center line marked by yellow spherical buoys.

Color.—Yellow is the color used for special marks.

Shape.—The shape of a special mark is optional, but must not conflict with that used for a lateral or a safe water mark. For example, an outfall buoy on the port hand side of a channel could be can-shaped but not conical.

Topmark.—When a topmark is carried it takes the form of a single yellow X.

Light.—When a light is exhibited it is yellow; the phase characteristic may be any, other than those used for the white lights of cardinal, isolated danger, and safe water marks, i.e.:

Group-occulting, single-flashing, group-flashing with a group of four, five, or (exceptionally) six flashes, composite group-flashing and morse code light. In the case of ODAS buoys, the phase characteristic used is group-flashing with a group of five flashes every 20 seconds.

New Dangers

New Dangers—Definition.—A newly discovered hazard to navigation not yet shown on charts, or included in sailing directions, or sufficiently promulgated by notices to mariners, is termed a **new danger.** The term covers naturally occurring obstructions such as sandbanks, rocks, or man-made dangers such as wrecks.

Marking.—A new danger is marked by one or more cardinal or lateral marks in accordance with the System rules. If the danger is especially grave, at least one of the marks will be duplicated as soon as practicable by an identical mark until the danger has been sufficiently promulgated.

Lights.—If a lighted mark is used for a new danger, it must exhibit a quick flashing or very quick flashing light. If it is a cardinal mark, it must exhibit a white light; if a lateral mark, a red or green light.

Racons.—The duplicate mark may carry a Racon, coded D(—··), showing a signal length of 1 nautical mile on a radar display.

Chart Symbols and Abbreviations

Changes.—New symbols and abbreviations, and altered ones, are being incorporated in DMAHTC charts when they are corrected or reprinted for use with the IALA Buoyage System. Symbols and abbreviations shown on charts to represent older systems of buoyage will remain unchanged until the new System is introduced into those areas.

Conventional Direction of Buoyage.—Where the conventional direction of buoyage may be open to doubt it is indicated on charts by a magenta symbol.

Pillar buoys.—The various forms of buoy termed 'pillar buoy' are indicated by the symbol introduced in 1976 for this purpose.

Spar buoys and Beacons.—The symbol for a spar buoy is also used to indicate a spindle buoy. In accordance with standard practice, spar buoy symbols are sloped to distinguish them from beacon symbols which are upright.

Colors.—The shading of buoy symbols formerly used to indicate the colors of buoys is omitted. A black (i.e. filled-in) symbol is used for green marks and for all spar buoys and beacons; an open symbol is used for all other colored buoys and beacon towers.

The abbreviated description of the color, or colors, of a buoy is given under the symbol.

Where a buoy is colored in bands, the colors are indicated in sequence from the top, e.g. east buoy—black with yellow band—BYB. If the sequence of the bands is not known, or if the buoy is striped, the colors are indicated with the darker color first, e.g. safe water buoy—red and white stripes—RW.

Topmarks.—Topmarks are charted boldly. Topmark symbols are inserted in solid black except when the topmark is red.

Lights.—The period of the light of a cardinal mark is determined by its quadrant and by whether the light is a quick light or a very quick light; the period is less important than its phase characteristic. Where space on charts is limited, and on second and smaller scale charts, the period may be omitted.

Light-stars.—Light-star symbols, formerly inserted above buoy symbols (and below the topmarks, if fitted), are omitted. This enables the topmark symbol to stand out more clearly, and avoids confusion with the X-shaped topmarks used on some special marks.

Light-flares.—Magenta light-flares are inserted with their points adjacent to the position circles at the base of the symbols: this avoids the light-flares obscuring the topmark symbols, and is in line with international chart practice.

Radar reflectors.—Radar reflectors are not affected by the IALA buoyage rules, but in 1976 their general significance was reconsidered in the study initiated by the need for new symbols. It was decided not to chart them on the introduction of the new buoyage for several reasons: it can be assumed that most major buoys are fitted with radar reflectors (some nations have already ceased to chart them on these grounds); it is necessary to reduce the size and complexity of buoy symbols and associated legends; and it is understood that, in the case of cardinal buoys, buoyage authorities site the reflector so that it cannot be mistaken for a topmark.

Promulgation of details of System 'A' symbols and abbreviations.—The symbols and abbreviations representative of IALA Maritime Buoyage System may be found in U.S. Chart No. 1, Nautical Chart Symbols and Abbreviations.

Implementation of the IALA Maritime Buoyage System.—Introduction of the new buoyage began in the Dover Strait in 1977. By the end of 1982 the process will have been completed in northwest Europe, including the Baltic Sea; in west Europe and in much of the Mediterranean Sea and the Red Sea; also in Singapore, Australia, Hong Kong, and parts of Africa, the Gulf, New Zealand, Malaysia, and Indonesia. Programs for the other areas have been drawn up.

In January of each year, the latest information about the progress to date, the program for the coming 12 months, and the long term schedule, is given in U.S. Notices to Mariners.

UNITED STATES SYSTEM

Fairways and Channels

PORT HAND

BUOY:

MARKING: Odd numbers, commencing from seaward.

LIGHTED BUOY: *White* or *green*, flashing or occulting; or, when marking important turns, quick flashing.

STARBOARD HAND

BUOY:

MARKING: Even numbers, commencing from seaward.

LIGHTED BUOY: *White* or *red*, flashing or occulting; or, when marking important turns, quick flashing.

Middle Grounds

MAIN CHANNEL TO RIGHT

BUOY:

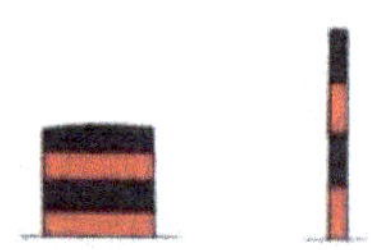

MARKING: May be lettered.

LIGHTED BUOY: *White* or *green*, interrupted quick flashing.

MAIN CHANNEL TO LEFT

BUOY:

MARKING: May be lettered.

LIGHTED BUOY: *White* or *red*, interrupted quick flashing.

Where channels are of equal importance, either of the above buoys is used, without regard to the uppermost band.

Mid Channel

BUOY:

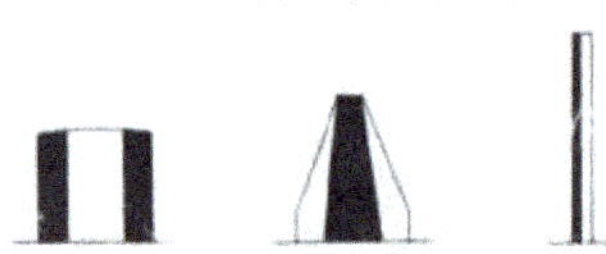

MARKING: May be lettered.

LIGHTED BUOY: *White*, short-long flashing.

Wrecks or Other Obstructions

TO BE PASSED ON PORT HAND

BUOY:

MARKING: Usually lettered "WR."

LIGHTED BUOY: *White* or *green*, quick flashing.

TO BE PASSED ON STARBOARD HAND

BUOY:

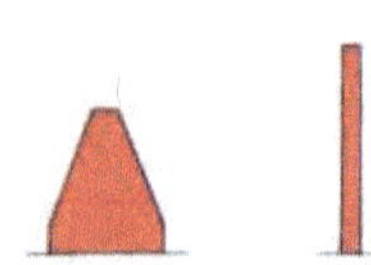

MARKING: Usually lettered "WR."

LIGHTED BUOY: *White* or *red*, quick flashing.

Where wrecks or other obstructions may be passed on either hand, either Middle Ground *buoy is used, without regard to the uppermost band.*

UNITED STATES SYSTEM

Miscellaneous

SHAPE: Optional.

COLOR: **Quarantine**—*Yellow.*
Anchorage—*White.*
Fish Nets—*Black-and-white* horizontal bands.
Dredging—*White* with *green* top.
Seadromes—*Yellow-and-black* vertical stripes.
Special Purpose—*White-and-international orange* horizontal or vertical bands.

MARKING: May be lettered.

LIGHTED BUOY: Any color except *red* or *green*; fixed, occulting, or slow flashing.

UNIFORM LATERAL SYSTEM

Fairways and Channels

	PORT HAND	STARBOARD HAND
TOPMARK:		
	"T"-shaped topmark not used, at channel entrance.	*Diamond-shaped topmark not used at channel entrance.*
BUOY:		

In secondary channels only, yellow *may be substituted for* white *in checkered buoys.*

	PORT HAND	STARBOARD HAND
MARKING:	Even numbers, commencing from seaward.	Odd numbers, commencing from seaward.
LIGHT:	*Red,* single flashing or occulting or group flashing or occulting, with a number of flashes or occultations up to four; or *white,* group flashing or occulting (2 or 4); both *red and white* with above characteristics.	*White,* single flashing or occulting, or group flashing or occulting (3); or *green,* of a different character from wreck markings; or both *white and green* with the above characteristics.

UNIFORM LATERAL SYSTEM

Middle Grounds

	MAIN CHANNEL TO RIGHT	CHANNELS OF EQUAL IMPORTANCE	MAIN CHANNEL TO LEFT
TOPMARK: *Bifurcation*			
Junction			
BUOY:			

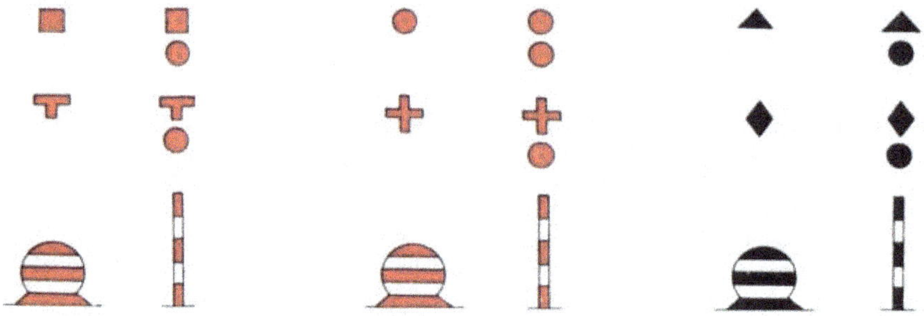

LIGHT:	Distinctive where possible.	Distinctive where possible.	Distinctive where possible.

Mid Channel

TOPMARK: Shape optional, but not conical, cylindrical, or spherical.

BUOY: Shape optional, but not conical, cylindrical, or spherical.

COLOR: *Red-and-white* or *black-and-white* vertical stripes; topmark *red* or *black* to conform with buoy.

LIGHT: Different from neighboring lights.

Marking of Wrecks

	TO BE PASSED ON PORT HAND	TO BE PASSED ON EITHER HAND	TO BE PASSED ON STARBOARD HAND

By Buoys

TOPMARK:			
BUOY:			

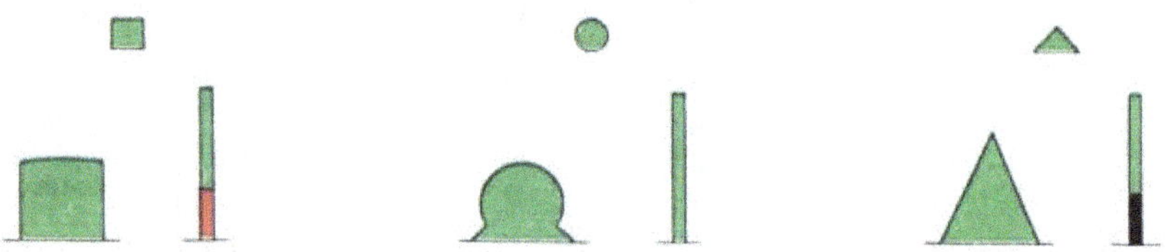

MARKING:	"W" in *white*.	"W" in *white*.	"W" in *white*.
LIGHT:	*Green*, group flashing (2).	*Green*, single occulting.	*Green*, group flashing (3).

By Vessels

VESSEL:

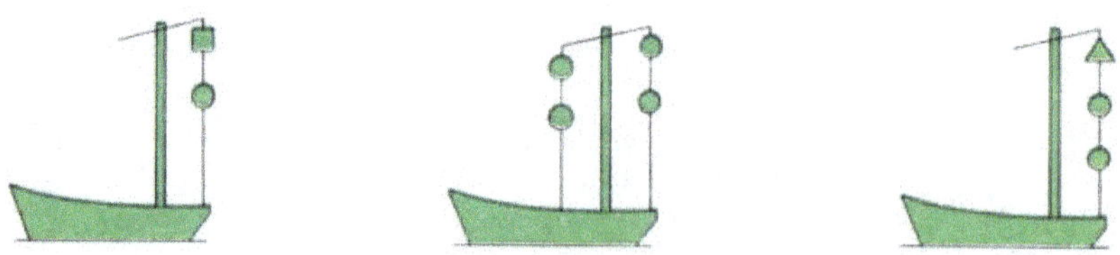

MARKING:	"W" or "WRECK" in *white*.	"W" or "WRECK" in *white*.	"W" or "WRECK" in *white*.
LIGHT:	Fixed *green*, corresponding in number and arrangement to shapes displayed by day.		
BELL:	Two strokes at intervals of not more than 30 seconds.	Four strokes at intervals of not more than 30 seconds.	Three strokes at intervals of not more than 30 seconds.

UNIFORM CARDINAL SYSTEM

Danger Markings

NW NE

TOPMARK:

BUOY:

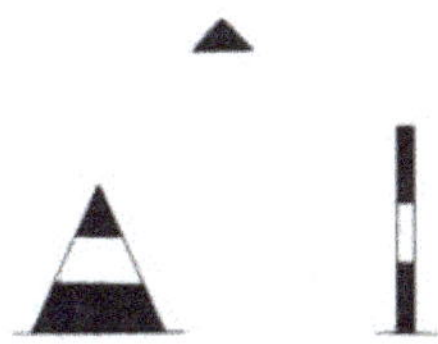

LIGHT: *White.* Preferably flashing or group flashing, with odd number of flashes; or occulting or group occulting, with odd number of occultations.

TOPMARK:

BUOY:

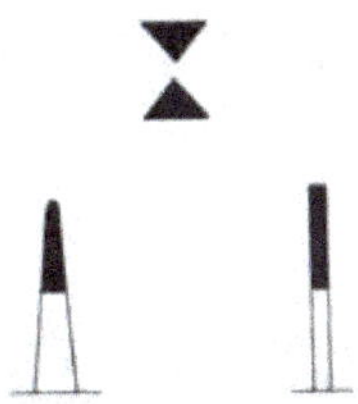

LIGHT: *White.* Preferably group flashing with even number of flashes, or group occulting with even number of occultations.

Danger

TOPMARK:

BUOY:

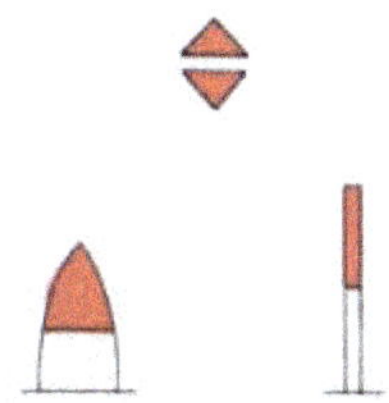

LIGHT: *Red,* preferably, or *white.* Flashing or group flashing, preferably, with odd number of flashes; or occulting or group occulting with odd number of occultations.

TOPMARK:

BUOY:

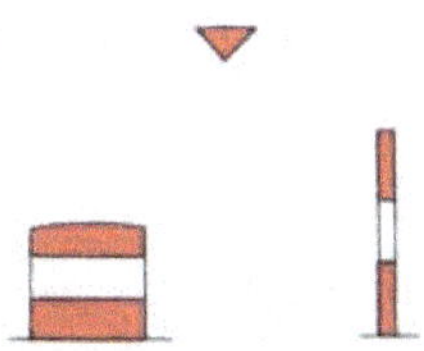

LIGHT: *Red,* preferably, or *white.* Group flashing, preferably, with even number of flashes; or group occulting with even number of occultations.

SW SE

Variations in Danger Markings

Northern Quadrant	*Eastern Quadrant*	*Southern Quadrant*	*Western Quadrant*	*Northern Quadrant*	*Eastern Quadrant*

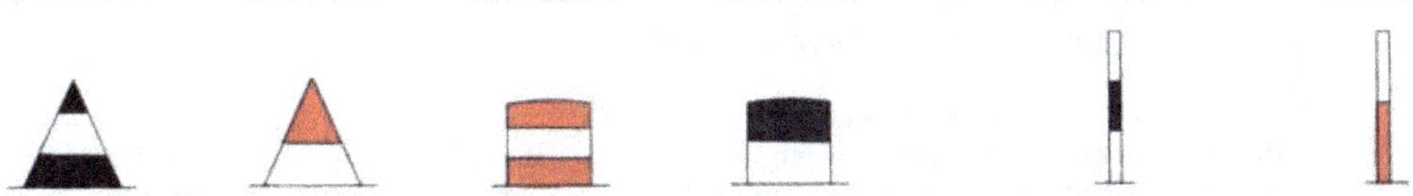

Note: The number of characteristic shapes employed for the buoy itself may be limited to two, the conical shape being employed in the northern and eastern quadrants and the cylindrical shape in the southern and western quadrants, as shown above.

Note: When spars only are used, it may be advantageous in the northern and eastern quadrants to reverse the positions of the dark colors, as shown above.

UNIFORM CARDINAL SYSTEM

Marking of Wrecks

	WESTERN QUADRANT	EASTERN QUADRANT
TOPMARK: / BUOY:		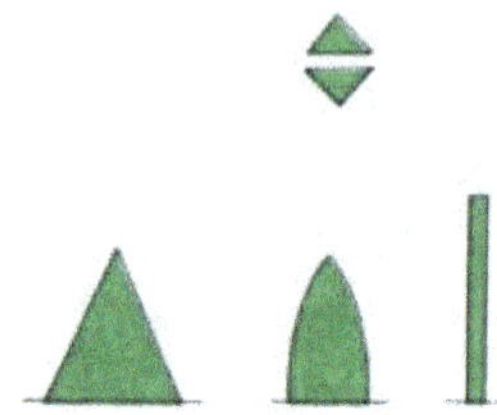
MARKING:	"W" in *white*, if possible.	"W" in *white*, if possible.
LIGHT:	*Green*, quick flashing.	*Green*, interrupted quick flashing.

In the Uniform Cardinal System, wreck buoys are not used in the northern or southern quadrants.

UNIFORM SYSTEM—LATERAL AND CARDINAL (Common To Both)

Isolated Dangers

TOPMARK:

BUOY:

LIGHT: *White* or *red*, rhythmic.

Miscellaneous

TOPMARK: **Landfall**—Shape optional, but not misleading.
Transition—Shape optional, but not misleading.
Others—None.

BUOY: Shape optional, but not misleading.

COLOR: **Landfall**—*Black-and-white* or *red-and-white* vertical stripes.
Transition—*Red-and-white* or *black-and-white* spiral bands.
Quarantine—*Yellow*.
Outfall—*Yellow* above and *black* below.
Military Practice Area—*White*, with two *blue* stripes rising from the waterline and intersecting at right angles on top of the buoy, and, optionally, lettering in the national language indicating a danger area (*e.g.*, in English, "D.A.").

LIGHT: **Landfall**—Rhythmic.
Outfall—Optional, with due regard to other lights in the area.
Others—None.

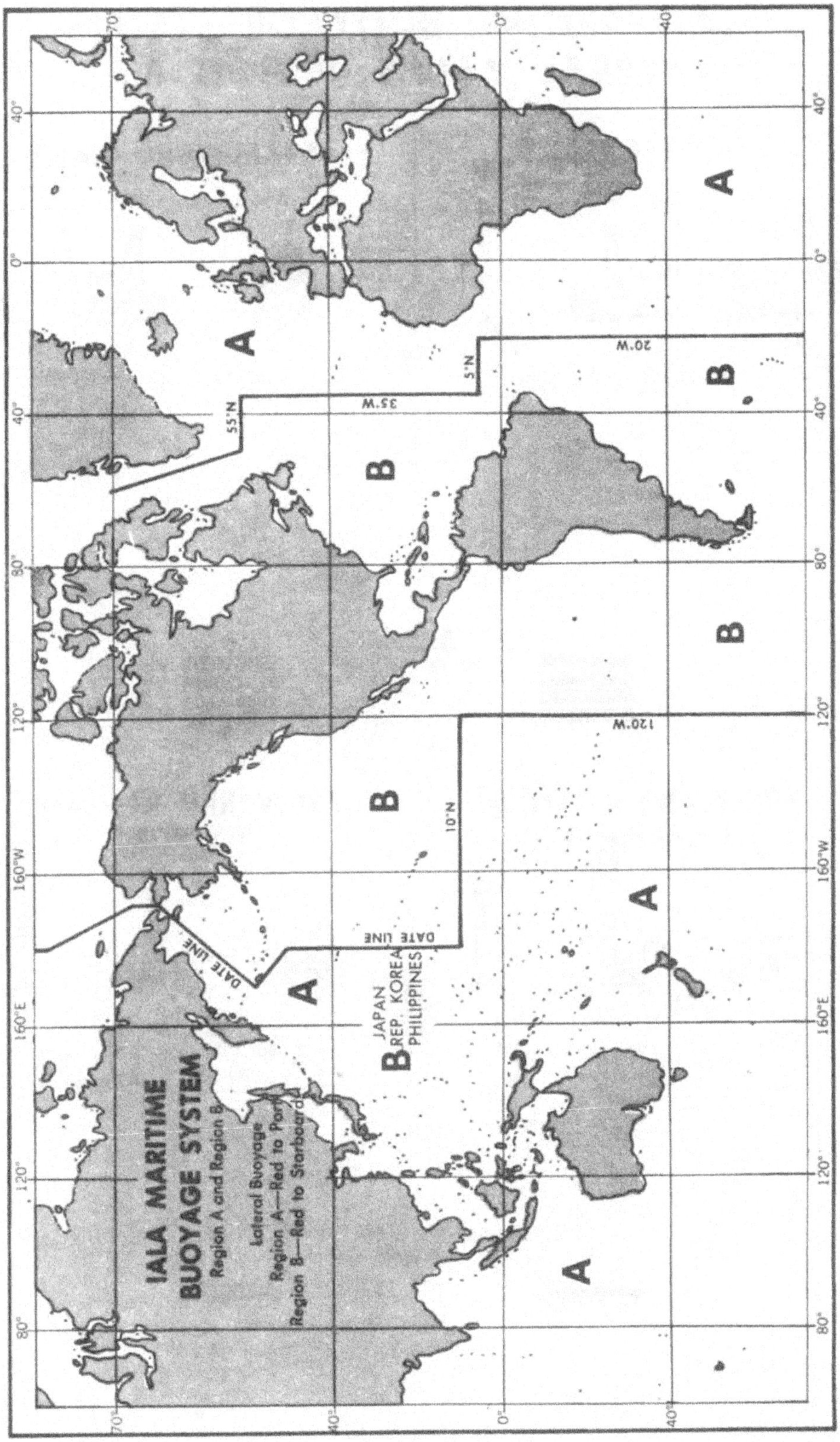
IALA MARITIME
BUOYAGE SYSTEM
Region A and Region B
Lateral Buoyage
Region A—Red to Port
Region B—Red to Starboard
JAPAN
REP. KOREA
PHILIPPINES
B
A
DATE LINE
10°N
120°W
55°N
35°W
5°N
20°W

IALA MARITIME BUOYAGE SYSTEM LATERAL MARKS REGION A

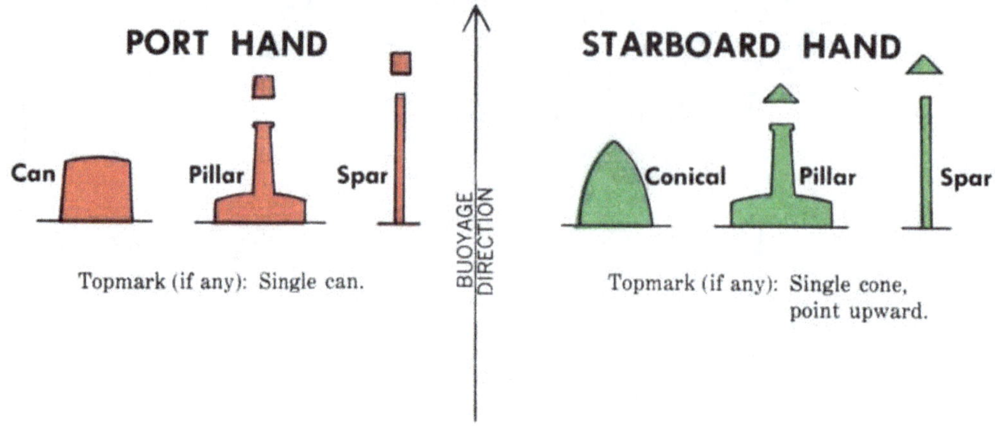

Lights, when fitted, may have any phase characteristic other than that used for preferred channels.

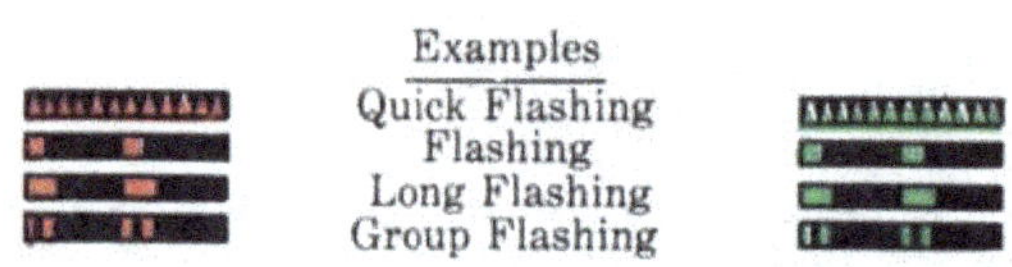

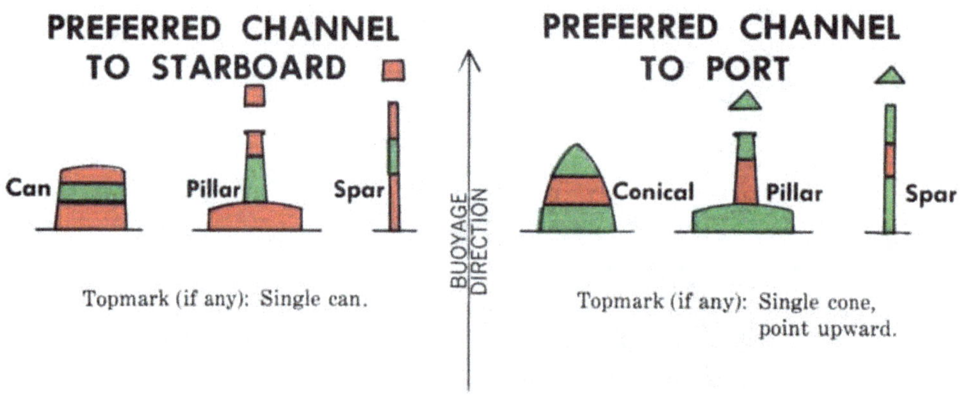

Lights, when fitted, are composite group flashing Fl (2 + 1).

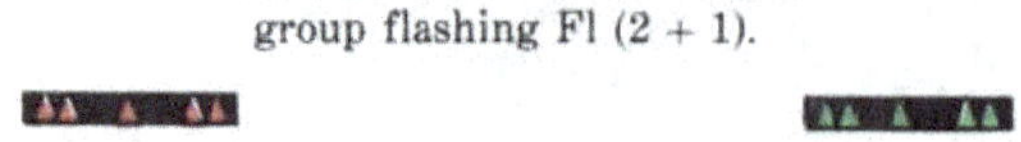

IALA MARITIME BUOYAGE SYSTEM LATERAL MARKS REGION B

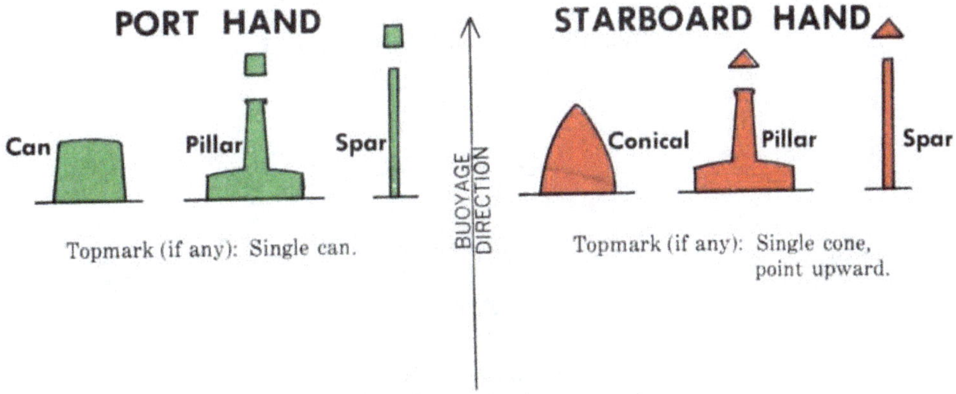

Topmark (if any): Single can.

Topmark (if any): Single cone, point upward.

Lights, when fitted, may have any phase characteristic other than that used for preferred channels.

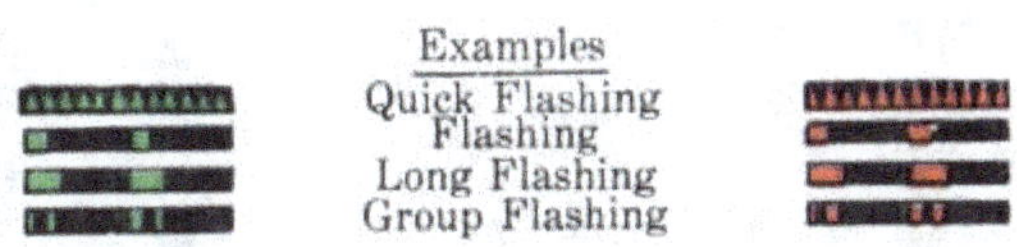

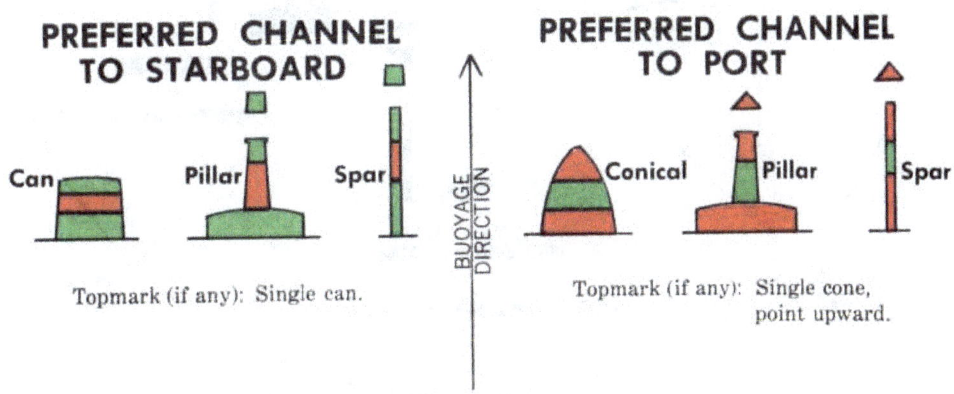

Topmark (if any): Single can.

Topmark (if any): Single cone, point upward.

Lights, when fitted, are composite group flashing Fl (2+1).

IALA MARITIME BUOYAGE SYSTEM CARDINAL MARKS REGIONS A AND B

Topmarks are always fitted (when practicable).
Buoy shapes are pillar or spar.

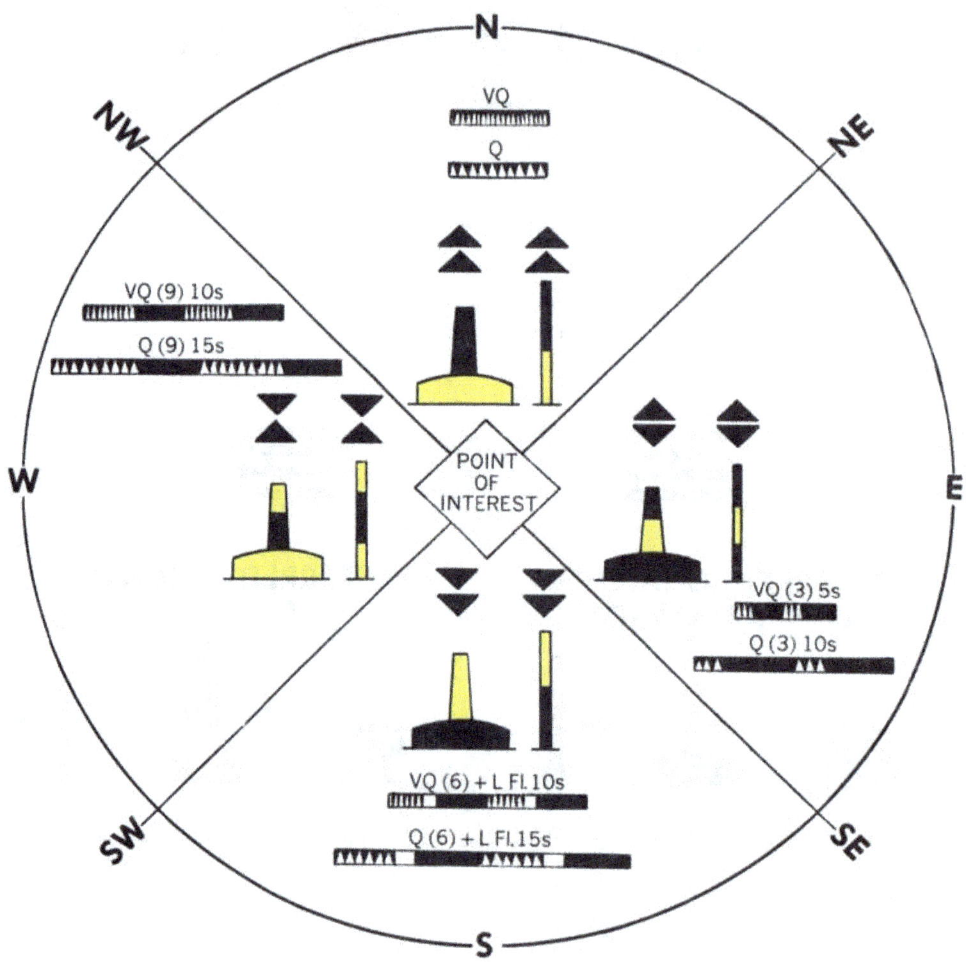

Lights, when fitted, are **white**. Very Quick Flashing or Quick Flashing; a South mark also has a Long Flash immediately following the quick flashes.

IALA MARITIME BUOYAGE SYSTEM
REGIONS A AND B

ISOLATED DANGER MARKS

Topmarks are always fitted (when practicable).

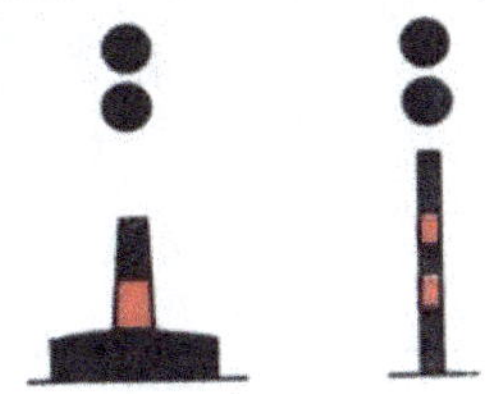

Light, when fitted, is **white** Group Flashing (2)

Fl (2)

Shape: Optional, but not conflicting with lateral marks; pillar or spar preferred.

SAFE WATER MARKS

Topmark (if any): Single sphere.

Light, when fitted, is **white** Isophase or Occulting, or one Long Flash every 10 seconds or Morse "A".

Iso

Occ

L Fl. 10s

Morse "A"

Shape: Spherical or pillar or spar.

SPECIAL MARKS

Topmark (if any): Single X shape.

Light (when fitted) is **yellow** and may have any phase characteristic not used for white lights.

Examples

Fl Y

Fl (4) Y

Shape: Optional, but not conflicting with navigational marks.

MODIFIED U.S. AID SYSTEM

LATERAL AIDS AS SEEN ENTERING FROM SEAWARD

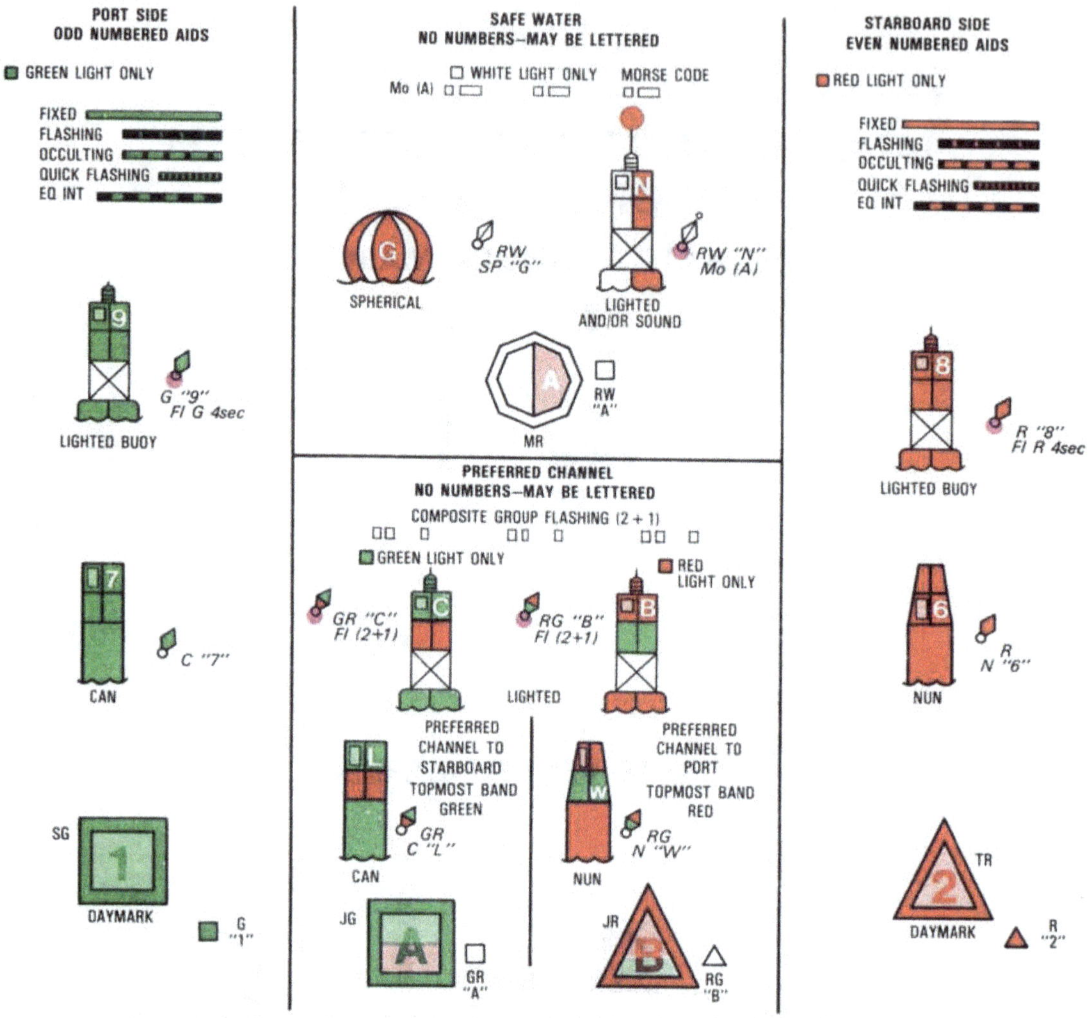

NOTE: WHEN USED ON THE INTRACOASTAL WATERWAY, THESE AIDS ARE ALSO EQUIPPED WITH SPECIAL YELLOW STRIPS, TRIANGLES, OR SQUARES. WHEN USED ON THE WESTERN RIVERS (MISSISSIPPI RIVER SYSTEM), THESE AIDS ARE NOT NUMBERED.

SPECIAL AIDS TO NAVIGATION

SHAPE: OPTIONAL—BUT SELECTED TO BE APPROPRIATE FOR THE POSITION OF THE MARK IN RELATION TO THE NAVIGABLE WATERWAY AND THE DIRECTION OF BUOYAGE.

YELLOW LIGHT ONLY
LIGHT RHYTHM—ANY RHYTHM <u>EXCEPT</u> THOSE NOTED BELOW

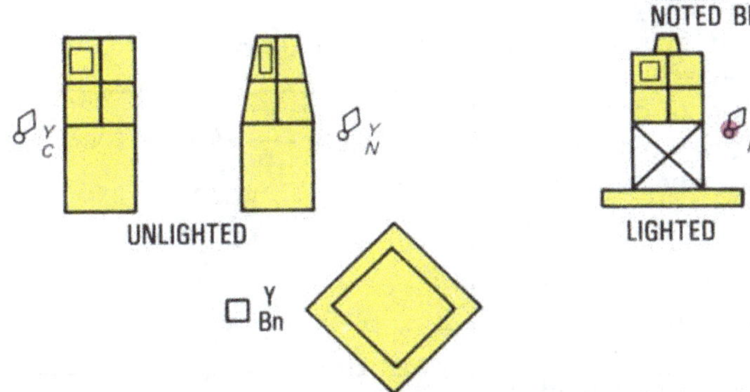

SPECIAL AIDS TO NAVIGATION MAY EXHIBIT ANY LIGHT RHYTHM <u>EXCEPT</u> MORSE "A", MORSE "U", GROUP FLASHING (2), ISOPHASE, OCCULTING, LONG FLASH, QUICK FLASHING, VERY QUICK FLASHING, GROUP QUICK FLASHING, GROUP VERY QUICK FLASHING, GROUP QUICK FLASHING PLUS LONG FLASH, GROUP VERY QUICK FLASHING PLUS LONG FLASH.

CROSSING DAYMARKS FOUND ON THE WESTERN RIVERS

(MISSISSIPPI RIVER SYSTEM)

RANGE DAYMARKS—MAY BE LETTERED

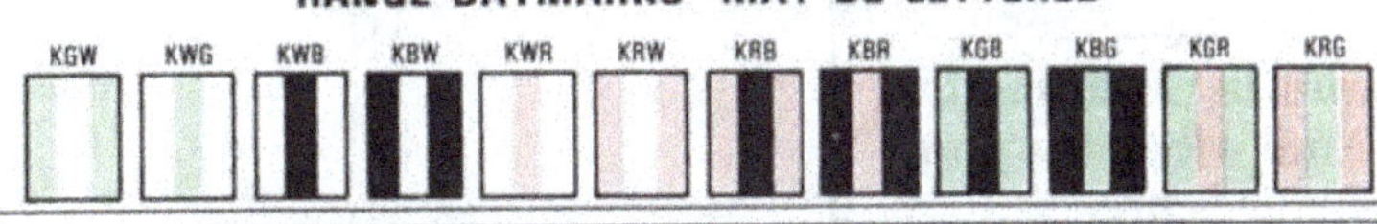

TYPICAL INFORMATION AND REGULATORY MARKS

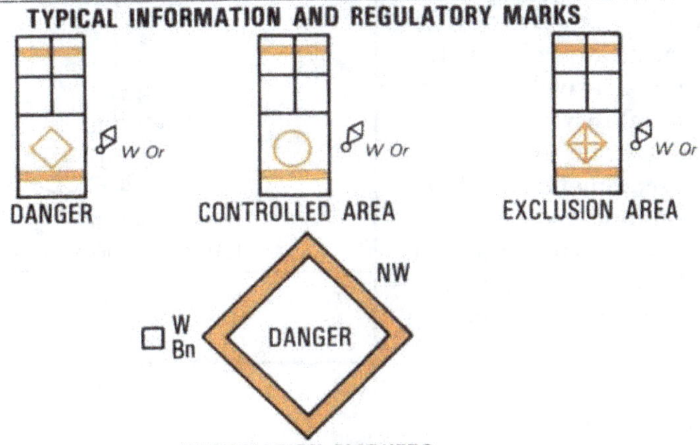

INFORMATION MARKERS

WHEN LIGHTED, INFORMATION AND REGULATORY MARKS WILL DISPLAY YELLOW LIGHTS.

The Articulated Light.—A new concept in the design and construction of offshore aids to navigation is slowly gaining acceptance among maritime nations of the world in the form of a uniquely pivoted structure termed the articulated light. Modified from an earlier experimental type of single point mooring, this novel structure has recently made its appearance in waters of the United States where it has been found particularly advantageous in marking navigable channels that require an increased measure of position integrity.

Essentially, the articulated light consists of three elements: on the sea floor, a weighted sinker; attached directly to this by a variety of devices incorporating the principles of the universal joint, an elongated pipe-like structure of sufficient length and diameter; and lastly, at a submerged, predetermined distance along the pipe, a buoyancy chamber. Its principle of operation is such that the buoyancy chamber seeks always to maintain the structure in a vertical position directly over the sinker. Should the structure heel over under the effects of wind, waves and/or current, the righting tendency of the chamber, becoming greater with deeper submergence, automatically forces the structure back to its normal stance. Thus the articulated light has a far more reduced watch circle than the traditionally moored buoy. And while its position is necessarily less precise than that of a fixed beacon, it nevertheless remains accurate enough that initial cost, comparatively easy deployment and ability to survive collision, among other advantages, increasingly recommend the structure as an economical yet efficient aid to navigation, particulary in sheltered waters and along narrow ship channels with little margin for error.

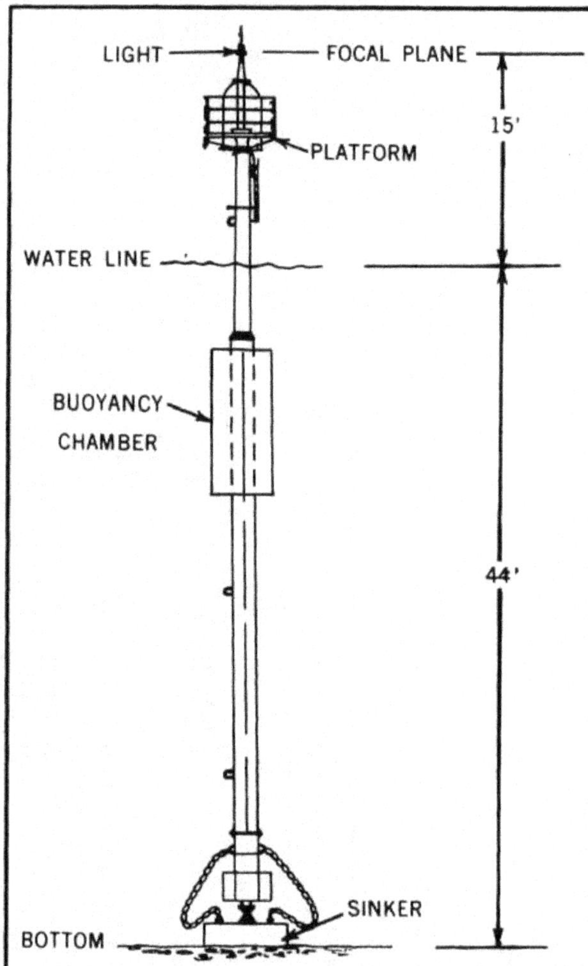

ARTICULATED LIGHT (FLOATING LIGHT)

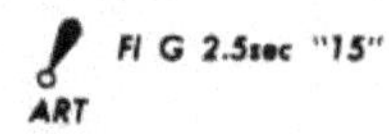

INDEX

Page

"A, B, C" *Tables for Azimuth, Great Circle Sailing, and Reduction to the Meridian;* of Blackburne 641
AMVER Bulletin, information in 766
AMVER Bulletin, publication 766
AMVER Users Manual, availability of 768
AN/BRN-3 Radionavigation Set, of satellite navigation, discussed 1075
AN/PRR-14 Geoceiver, accuracy of 1093
design characteristics 1093
discussed 1093
functional description 1093
AN/SRN-9 integral doppler navigation, discussed 1101
AN/SRN-9 Radionavigation Set, of satellite navigation, accuracy of 1083
development of 1079
AN/SRN-19 Radionavigation Set, of satellite navigation, accuracy of 1091
discussed 1087
functional description 1088
AN/WRN-5 Radionavigation Set, of satellite navigation, accuracy of 1085
design characteristics 1085
discussed 1085
functional description 1085
A-scope, of radar 983
Abandoning ship, before lowering boats 726
estimate of situation 727
Abbreviations 1161
of IALA system 1362
Abeam, defined 284
Aberration, discovery of 27
explained 373
Absolute motion, of celestial bodies, discussed 355
defined 355
Absolute temperature, defined 891
Absolute zero, defined 891
Absorption, and scattering, of radio waves 964
Accelerating anode, of cathode-ray tube 970
Acceleration error, of sextant 430
Accelerometer(s), in inertial navigation, defined 1129
sensitive axis of 1132
Accuracy, of charts 113
of Consol 1061
of Decca fix 1058
defined 1261
of Loran-C fix 1051
Additional Secondary Phase Factor, of Loran-C signals 1051
Admiralty List of Radio Signals (vol. 5) 511
Consol and Consolan procedures in 1060

Page

Admiralty List of Radio Signals (vol. 5)—Continued
Decca information in 1058
Advance, of line of position 468
of vessel in turn, defined 305
Advanced calculator(s) 1299
use of 1308
Advection fog, formation of 919
Aerial and Marine Navigation Tables, of Gingrich 599
Aeromarine radiobeacon(s) 972
Aeronautical light(s), on charts 125
discussed 93
Aeronautical radiobeacon(s) 972
Ageton, Arthur A.; sight reduction tables of 600
Agonic line, defined 128, 206
Agulhas Current 822
Aid(s) to navigation, chart symbols for 121
defined 92
history of 15
private, discussed 110
protection by law, discussed 110
visual and audible, introduction to 92
Air Almanac, of Weems 42
Air Almanac, The; altitude corrections by 449
arc to time table in 475
augmentation correction in 443
described 522
dip correction in 437, 448
equation of time by 505
extracts from 1199
finding GHA and declination, of moon 525
of planet 526
of star 528
of sun 524
finding time, of moonrise and moonset 532
of sunrise and sunset 529
of twilight 530
low altitude corrections 453
parallax correction in 446
planet location diagram in 656
Polaris correction table in 577
publication of 520
refraction correction in 440
semidiameter correction in 442
sky diagrams in 664
star chart in 645
tabulation of GHA and declination 523
Airmass, defined 911
types of 911
Air navigation, defined 59

Page

Air temperature, altitude correction for, discussed 441
Alaska Current 821
Alessio, A.; azimuth diagram of 642
Aleutian Current 821
Alfonsine Tables 24, 41
Algae, in seawater 782
Almagest, of Ptolemy 13, 24, 26, 37, 41
Almanac(s), history of 40
introduction to 520
for lifeboat navigation 725
use of 522
Almanach Perpetuum, of Zacuto 41
Almucantar, defined 395
Alpha, Beta, Gamma Navigation Tables, The; of Goodwin 588
Alt-Azimuth Tables, of Davis 604
Alternating current, defined 957
Alternating fixed and flashing light, *Light List* entry for 96
Alternating light, defined 96
Altitude, of celestial body, defined 395
computation of 540
determination of, by graph 627
by map projection 631
by mechanical methods 629
locating point on celestial sphere by 390
measurement of, in lifeboat 734
Altitude and Azimuth Almanac, of Japanese Hydrographic Office 610
Altitude and Azimuth Tables, of Aquino 601
Altitude azimuth, defined 639
by intermediate calculator 1302
Altitude, Azimuth, and Geographical Position; of Littlehales 45
Altitude, Azimuth, and Hour Angle Diagram; of Littlehales 627
Altitude difference, defined 463, 539
Altitude factor, defined 574
discussed 584
Altitude intercept, defined 463, 539, 590
finding of 436
Altitude intercept method, for line of position 590
Altitude method(s), for line of position 590
perpendicular from body, formulas for solution of 600
perpendicular from zenith, formulas for solution of 595
of sight reduction 583
triangle not divided, formulas for solution of 591
without use of navigational triangle 603
Altitude or Position Line Tables, of Ball 592
Altitude Tables for Mariners and Aviators, of Tillman 603
Alto, meaning of 895
Altocumulus clouds, described 895
Altostratus clouds, described 897
Ambiguity, of loop antenna 974
American Air Almanac, publishing of first 42

Page

American Ephemeris and Nautical Almanac, historical 42
replacement of 520
American Nautical Almanac, The; development of 42
lunar distance tables, deletion of 44
Amplifier, radio 969
Amplitude(s) in azimuth observations 582
for checking compass 639
defined 582
by intermediate calculator 1303
of radio wave, defined 958
Amplitude modulation, of carrier wave 968
Anabatic wind, defined 918
Anchor ice, defined 844
Anchorage(s), on charts 128
description of in *Sailing Directions (Enroute)* 344
holding quality of bottom 715
in polar regions 715
adequacy of room for swing 715
availability of, position determination 716
suitable exit 716
protection from wind and sea 716
Anchoring, piloting procedure 308
Anemometer 887
Aneroid barometer, described 883
discussed 884
Aneroid capsule, of aneroid barometer 884
Angle(s), of incidence 408
illustrated 439
of reflection 408
of refraction, illustrated 439
Angular distance, defined 160
Angular momentum, conservation of 1130
Angular relationship, correct, as desirable property of projection 67
Annular eclipse, of moon, defined 386
Anomalistic month, and tidal cycles 803
Anomalistic year, length of 374
Antarctic, features of 700
Antarctic airmass, source region of 911
Antarctic Circle, defined 377
Antarctic whiteout, described 704
Antenna(s), characteristics of 965
height of, in radarscope interpretation 996
of radio transmitter 969
Anthelion, defined 922
Anticrepuscular rays, described 923
Anticyclone(s), defined 916
description of 916
Antiselene, defined 922
Aphelion, defined 359
Apogean current 810
Apogean neap tide(s) 801
Apogean tide(s) 802
Apogee, defined 359
of moon, defined 801
distance at 365
Apparent (rectified) altitude, obtaining of 447
Apparent horizon, defined 393

	Page
Apparent motion, of celestial bodies, discussed	355
due to, earth revolution	372
earth rotation	371
movement of other bodies	374
defined	355
Apparent precession, of gyroscope, defined	1130
Apparent sidereal time, defined	482
Apparent solar day, defined	358, 486
Apparent solar time	378, 489
measurement of	486
Apparent sun	504
defined	378
Apparent time, finding	504
Apparent wind, defined	887
Appendages, as cause of speed sensor errors	149
Approach chart, classification of	114
Apron, defined	783
Apsides, line of, defined	359
Aquino, Radler de; sight reduction tables of	601
Arc, of marine sextant	409
and time, interconversion of	493
Arcano del Mare, of Dudley	21
Archipelagic apron, defined	783
Arctic, features of	700
Arctic airmass, source region of	911
Arctic Azimuth Tables (H.O. Pub. No. 66)	640
Arctic Circle, defined	377
Arctic Ocean, features of	700
Arctic paths, of Omega signals	1027
Arctic sea smoke, described	919
Arctic whiteout, described	704
AREA SURPIC, example of	769
Arithmetical Navigation, of Addison	21
Armed lead	156
Armillary sphere, origin of	37
Arte de Navegar, of Pedro de Medina	20
Articulated light	1376
Artificial horizon, altitude corrections for	447
defined	428
description and use of	428
Artificial horizon sextant(s), altitude corrections for	448
described	429
Ashton, Philip; sight reduction tables of	612
Asian charts	5
Aspect, of radar target, effect on range	993
Assumed latitude, selection of	463, 538
Assumed longitude, selection of	464, 538
Assumed position, in navigational triangle	403
selection of	538
in sight reduction	463, 539
Asteroid(s), in solar system	357
Astro-Navigation Tables for the Common Tangent Method, of Benest and Timberlake	600
Astrogeodetic datum orientation	1349
Astrograph-time Star Tables for Air Navigation, of Ashton	612
Astrolabe, invention of ancient	28
Astrology, pseudo science of	378
Astronomic latitude, defined	1342
illustrated	387
Astronomic longitude, defined	386, 1342
Astronomical day, defined	492
Astronomical Almanac, The; equatorial horizontal parallax in	365
Julian day listed in	493
parallax correction in	446
publication and contents of	520
semidiameter correction in	442
stars listed in	644
Astronomical distance, units of	355
Astronomical mean sun	487
Astronomical Navigation Tables (British Air Pub. 1618) (H.O. Pub. No. 218)	604
Astronomical observatories, history of	37
Astronomical refraction, defined	439
Astronomical triangle, defined	403
Astronomical twilight, defined	372, 530
Astronomical unit, defined	356
Astronomical year, length of	373
Astronomy, defined	355
history of	21
modern, history of	26
Atherton, J. W.; azimuth and great circle tables of	640
Athwartship degaussing coil, described	252
Atlantic Neptune	9
Atlantic Ocean, currents in	816
Atmosphere, circulation of	904
of earth	903
and radio waves	924
Atmospheric electricity, types of	924
Atmospheric noise, in radio, defined	965
Atmospheric pressure, defined	883
measurement of	883
Atmospheric pressure correction, of sextant altitude, discussed	441
Atmospheric refraction; defined	439
Atmospherics, in radio, defined	965
Atoll, coral, appearance on radarscope	998
Atom(s), defined	957
Atomic clock, basic principle of	483
Atomic second, defined	483
Atomic Time, basis of	483
Attenuation, of ocean waves	824
of radar beam	991
of radio waves, defined	965
Audio frequency, defined	959
Augmentation, altitude correction for, discussed	443
defined	366, 443
Aureole, defined	923
Aurora(s), described	924
due to sun spots	361
Aurora australis	702
Aurora borealis	702
Auroral zone(s), effect on Omega signals	1027

Page

Automated Mutual-assistance Vessel Rescue System (AMVER), communications network, discussed ... 766
discussed ... 765
participation in ... 767
plot information ... 769
uses of ... 770
reporting required ... 768
Automated weather stations and buoy systems, discussed ... 900
Automatic dead reckoning, discussed ... 267
Automatic radio direction finder ... 973
Autumn (season) ... 374
Autumnal equinox, defined ... 374
Average motion error, described ... 149
Away, sight reduction, defined ... 464, 539
Axis, of rotation, of earth ... 60
Azimut-Tabellen, of Ebsen ... 641
Azimutdiagramme, of German Navy ... 643
Azimuth, and bearing ... 65
computation of ... 540
defined ... 395
determination of ... 639
by intermediate calculator ... 1302
by tables ... 580
locating point on celestial sphere by ... 390
Azimuth and Hour Angle for Latitude and Declination, of Shortrede ... 640
Azimuth angle, defined ... 395
determination of, by graph ... 627
by map projection ... 631
by mechanical methods ... 629
Azimuth circle, description and use of ... 168
Azimuth by Logs, of Purey-Cust ... 642
Azimuth Diagram, of Weir ... 642
Azimuth diagram(s), discussed ... 642
Azimuth instrument ... 168
Azimuth method(s), for line of position ... 615
Azimuth table(s), discussed ... 639
Azimuthal equidistant projection, defined ... 83
features of ... 60, 83
Azimuthal orthomorphic projection, defined ... 82
Azimuthal projection(s), bearings on ... 80
defined ... 68
features of ... 80
Azimuthal radionavigation system ... 1012
Azimuths of Celestial Bodies (Pub. No. 261) ... 641
Azimuths of the Sun (Pub. No. 260) ... 641
Azimuts (Table 902), French azimuth table ... 642
Azimuty Svetil, of Yustchenko ... 642

Back sight(s), altitude corrections for ... 456
defined ... 456
method of taking ... 415
Backing wind ... 948
Backrush, of waves ... 835
Backstaff, invention of ... 29
Backstays of the sun, defined ... 923
Backwash, of waves ... 835
Baguio, in western North Pacific ... 932
Baily's Beads, defined ... 385

Page

Baker, T. Y.; sight reduction method of ... 635
Ball, Frederick; sight reduction tables of ... 592
Ballistic damping error, of gyrocompass, discussed ... 185
Ballistic deflection, defined ... 184
Ballistic deflection error, of gyrocompass, discussed ... 184
Band(s), of radio spectrum ... 959
Bank, defined ... 783
Bar, of tropical cyclone ... 944
Barograph, described ... 885
Barometer, atmospheric pressure measuring device ... 883
Barometer readings, adjustment of ... 886
Barometric leveling, in geodetic surveys ... 1346
Barometric pressure, defined ... 883
Baseline, in distance-difference measurement ... 1013
in geodetic surveys ... 1344
Baseline extension, in distance-difference measurement ... 1013
Basic calculator, use of ... 1299
Basin, of ocean ... 783
defined ... 783
Bathymetric chart(s), described ... 118
Bathymetric Data Requirements (chart 5103) ... 763
Bathythermograph, defined ... 778
Bayer's name, for stars ... 644
Beach, appearance of on radarscope ... 997
Beach sediments, discussed ... 837
Beaching of lifeboat ... 743
Beacon(s), discussed ... 105
Beam compasses and dividers, defined ... 141
Beam width, of radar ... 996
Bearing(s), and azimuth ... 65
on charts ... 127
defined ... 65, 283
discussed ... 283
in polar regions ... 714
Bearing angle, defined ... 65
Bearing bar, compass sighting device ... 168
Bearing circle, description and errors of ... 168
Bearing line, defined ... 284
Beaufort scale, of wind speeds ... 891, 1312
Bedell, A. L.; sight reduction instrument of. 639
Beehler, W. H.; "Solarometer" of ... 638
Beij, K. Hilding; sight reduction diagram of 633
Bell(s), sound signal ... 167
Bench, defined ... 784
Bending, of ice ... 847
Benest, E. E.; sight reduction tables of ... 600
Benguela Current ... 820
Bergy bit, defined ... 846
Bertin, Charles; sight reduction tables of ... 579
Bertin, Maurice; sight reduction diagram of 628
slide rule of ... 630
Beset, danger of, in polar regions ... 715
defined ... 871
Bias, gyro, defined ... 1150
Bias errors ... 1267

	Page
Binaural hearing	838
Binnacle, described	162, 219
Birds, as sign of land	741
Bivariate error, in two-dimensional errors	1281
Blackburne, H. S.; azimuth tables of	641
sight reduction tables of	587
Blackout, of radio	962
Blind sector, on radarscope, defined	1003
Blink, of Loran-C signals	1037
Blizzard, defined	918
Blue moon, occurrence of	921
Board of Longitude, establishment of	34
Boat compass	161
Bobbing a light, process of	351
Bolide, meteor as	368
Bombing practice area(s), on charts	129
Bora, defined	918
Borderland, defined	783
Bore, tidal	800
defined	832
Bottom contour chart(s), described	118
Bottom profile, by echo sounder	782
use of in piloting	304
Bottom relief, of oceans	782
Boundary layer, of water along the hull	149
Boundary waves, creation of	832
Bow and beam bearings, defined	302
Bowditch, Nathaniel; biography of	iii
Boxing the compass, defined	162
Braga, Romêo; sight reduction tables of	593
Brazil Current	820
Breaker(s), defined	828
features of	835
formation of	835
introduction to	834
Breaking, of waves	828
Breve de la Spera y de la Arte de Navegar, of Cortes	20
Bridge(s), marking of	94
Brief Celestial Navigation Table (Japanese H.O. Pub. No. 602)	603
Brieve Compendio del Arte del Navigar, of Rossa	8
Brill, Alfred; sight reduction device of	636
British Admiralty, *Astronomical Navigation Tables* (British Air Pub. 1618)	604
British thermal unit, defined	891
Brown-Nassau, "Navigational Computer"	632
Building(s), chart definition of	126
Bummock(s)	872
Buoy(s), as aid to radar navigation	1004
on charts	124
described	102
fallibility of	102
Buoy sounder	14
Buoy systems, and automated weather stations, discussed	900
Buoyage, systems discussed	104
modified U.S. Aids System	104, 1356
Burdwood, John; azimuth tables of	640
Buys Ballot's Law, statement of	916
Bygrave, slide rule of	630
CP-827 computer, of satellite navigation, functional description	1083
C-S Line, in Pub. No. 229	544
Cálculo del Punto, of Ménéclier and Chevalier	611
Calendar, time divisions by	373
Calibration, of radio direction finder	976
Calibration radiobeacon(s)	973
Calibration table, for radio bearings	975
California Current	820
Calorie, of heat, defined	891
Calving, of icebergs	846
Can buoy, shape of	102
Canary Current	819
Canyon, defined	783
Cape Horn Current	820
Cardinal mark(s), defined	1358
use of	1359
Cardinal points, of compass	162
Cardinal system of buoyage	1355
Cargo considerations, of ship weather routing	686
Carrier wave, of radio transmission, defined	968
Casa de Contratación	6
Cassini, M.; sight reduction tables of	586
Cathode, of cathode-ray tube, defined	970
Cathode-ray tube, principles of	970
Cat's paw, defined	918
Celestial body (bodies), identification of, introduction to	644
Celestial coordinates, in lifeboat navigation	732
Celestial equator, formation of	387
Celestial equator system of coordinates, described	390
Celestial horizon, illustrated	393
Celestial latitude, as coordinate of ecliptic system	390
defined	396
Celestial longitude, as coordinate of ecliptic system	390
defined	396
Celestial mechanics, history of	23
Celestial meridian, formation of	387
Celestial navigation	1
and dead reckoning	472
defined	59
in high latitudes	717
modern methods of	45
Celestial Navigation Computation Tables (Japanese H.O. Pub. No. 601)	593
Celestial observation(s), lines of position from	465
in polar regions	718
Celestial Navigation Observation Table (Japanese H.O. Pub. No. 351)	605
Celestial Navigation Tables, of Hoehne	605
Celestial Navigator for Aviators, of True	633
Celestial pole(s), location of	387

Page

Celestial sphere, defined and discussed
elements of ... 387
Celestial triangle, defined ... 402
Celsius temperature, defined ... 891
Centering error, of sextant ... 427
Centerline, in distance-difference measurement ... 1013
Centerline controlling depth, defined ... 753
Centigrade temperature, defined ... 891
Centimeter wave band, radio wave propagation of ... 968
Central computing center, of navigation satellites ... 1069
Ceres, described ... 364
Certaine Errors in Navigation Detected and Corrected, of Wright ... 8, 17, 21
Cesium beam atomic clock, frequency of ... 483
Chain, Decca, discussed ... 1054
numbers of ... 1058
Loran-C, discussed ... 1037
of radionavigation stations ... 1015
Channels, use of in radio ... 968
Chart(s), advance preparation of ... 136
classification by scale ... 113
dates, on, discussed ... 116
first edition ... 116
revised print ... 116
defined ... 67
history of ... 4
for lifeboat navigation ... 725
lighting of ... 137
maintaining of ... 137
new editions of ... 116
of other agencies ... 1297
projection, defined ... 67
reprint of ... 116
stowage of ... 137
use of ... 137
Chart datum sounding, defined ... 120
Chart Information in *Sailing Directions (Enroute)* ... 342
Chart numbering system, described ... 131
Chart/Publication Correction Record Card (DMAHTC 8660/9) ... 137
Chart sounding datum ... 804
Chart symbols, discussed ... 116
of IALA system ... 1362
Chevalier, Roberto; sight reduction tables of ... 611
Chiesa, Stefano; sight reduction tables of ... 594
Chimney, chart definition of ... 126
Chinook, defined ... 918
Chip log ... 11
in lifeboat navigation ... 731
speed measurement device ... 148
Chlorinity, of seawater ... 777
Chronometer, history of ... 34
Chronometer error, defined ... 432, 498
Chronometer rate, defined ... 432, 498
Chronometer Tables, of Percy L. H. Davis ... 47, 587, 604

Page

Chronometer Tables—Continued
Chronometer time ... 432
discussed ... 498
use of AM and PM ... 492
Circle(s), altitude measurement by ... 734
of equal altitude, as circle of position ... 460
discussed ... 460
utilization of ... 461
of equal angle, in three-point fix ... 314
of equivalent probability ... 1270
of latitude, defined ... 396
of longitude, defined ... 396
of position, as circle of equal altitude ... 460
of earth ... 60
Circular error probable, in two dimensional errors ... 1282
Circular probable error ... 1270
in two-dimensional errors ... 1282
Circular radiobeacon(s), defined ... 972
Circulation, of atmosphere, modifications of ... 910
Circumhorizontal arc, defined ... 922
Circumpolar celestial bodies, defined ... 371
Circumscribed halo, defined ... 922
Circumzenithal arc, defined ... 922
Cirripeda, defined ... 120
Cirrocumulus clouds, described ... 895
Cirrostratus clouds, described ... 895
Cirrus, meaning of ... 895
Cirrus clouds, described ... 895
Cities, on charts ... 126
Civil day ... 326
Civil twilight, defined ... 372, 529
Clamp screw, use of ... 411
Clamp screw vernier sextant, described ... 411
Climate, defined ... 903
and ocean currents ... 823
Clocks, and timekeeping ... 479
Close pack, concentration of ice ... 848
Cloud(s), height measurement of ... 899
types of, described ... 894
Clutter, on radarscope ... 996
Coaltitude, illustrated ... 404
Coarse lane, Omega, defined ... 1021
identification of ... 1021
Coast chart, description of ... 113
Coast pilot(s), and currents ... 813
Coast Pilots, correction of ... 344
projects depths in ... 753
subjects included ... 344
unverified information in ... 754
Coastal current, defined ... 815
Coastal Features, information on in *Sailing Directions (Enroute)* ... 343
Coastal refraction, of radio waves ... 961, 975
Coastal Winds and Currents, as depicted in *Sailing Directions (Enroute)* ... 342
Cocked hat fix, formation of ... 298
Code of Federal Regulations ... 754
time zones in ... 507
Code of Federal Regulations (33 CFR 70), protection of aids to navigation ... 110
Codeclination, illustrated ... 404

Page

Coefficient(s), *A*, defined ... 218
B, defined ... 218
C, defined ... 218
of compass deviation, approximate values of ... 234
discussed ... 218
D, defined ... 218
E, defined ... 219
J, defined ... 219
Coercive force, defined ... 202
Coercivity, defined ... 202
Colatitude, illustrated ... 404
Cold airmass ... 911
Cold front, defined ... 913
Cold low, defined ... 914
Cold wall, of Gulf Stream ... 818
Collins, Elmer B.; sight reduction tables of ... 599
Collins, Oliver C.; sight reduction method of ... 626
Color, of seawater ... 782
Coma (head), of comet, formation of ... 367
Combined Time and Altitude Azimuth Table, of Johnson ... 641
Comet(s), described ... 367
and meteors, discussed ... 367
in solar system ... 367
Command, establishing, in lifeboat ... 726
Commercial radio broadcasting station(s), on charts ... 127
Common establishment ... 804
Common log (speed indicator) ... 11
Common quadrant ... 28
Common year, defined ... 486
length of ... 373
Communications, in ship weather routing ... 698
Compact pack, concentration of ice ... 848
Comparing watch, described ... 433
Comparison frequency, of Decca signals ... 1056
Compass, history of ... 9
kinds of ... 161
for lifeboat navigation ... 726
location of, effect on deviation ... 219
U.S. Navy 5-inch, described ... 163
U.S. Navy 7½-inch, described ... 163
variation of, latitude determination by ... 740
longitude determination by ... 741
Compass adjustment, defined ... 216
discussed ... 216
procedure for, preliminary steps ... 238
underway ... 239
purpose of ... 248
Compass bearing, defined ... 283
Compass card, described ... 162
history of ... 10
of magnetic compass ... 161
Compass compensation, defined ... 216
degaussing ... 250
Compass course ... 64
Compass directions, defined ... 210
Compass error, defined ... 209
Compass north, defined ... 209

Page

Compass rose(s), on charts ... 128
on position plotting sheet ... 88
use of in plotting ... 262
Compasses (plotting instrument), defined ... 141
description and use of ... 141
Complete Coastal Navigator, The; of Charles H. Cotter ... 299
Composite group flashing light, *Light List* entry for ... 99
Composite radionavigation system ... 1012
Composite sailing, defined ... 270
discussed ... 280
Composition, of radar target, effect on range ... 994
Compressibility, of seawater ... 780
Computation algorithms, of satellite navigation ... 1107
Computed altitude, of celestial body ... 436
in sight reduction ... 539
use of ... 463
Computed and observed altitudes, comparison of ... 539
Comrie, L. J.; sight reduction tables of ... 599
Concentration, of pack ice ... 848
Conductor, electrical ... 957
Cone, defined ... 783
Conformal projection, characteristic of ... 67
Conformality, of polar projection ... 86
Conic projection(s), features of ... 75
with two standard parallels, defined ... 77
Conjunction, of planet, explained ... 382
Connaissance des Temps, first official almanac ... 42
Conning aid, in horizontal angles ... 316
Consol ... 56
discussed ... 1059
method of use ... 1060
operational range of ... 1064
in polar regions ... 717
signal of ... 1060
Consolan ... 1059
Consolidated pack, concentration of ice ... 848
Constan, P.; azimuth graph of ... 642
Constant error, defined ... 288, 1262
Constant of the cone, defined ... 707
Constellation(s), defined ... 369
names, meanings, positions of ... 1233
Continental borderland, defined ... 783
Continental drift ... 776
Continental margin, defined ... 783
Continental rise, defined ... 783
Continental shelf, defined ... 678, 783, 784
Continental slope, defined ... 678, 783, 784
Continuous radionavigation system, Decca as ... 1054
defined ... 1012
Continuous wave, of radio transmission ... 968
Contours, on charts ... 125
Contrary name, of declination and latitude, defined ... 401
Contrary-Same Line, in Pub. No. 229 ... 544

Page

Control center, of navigation satellites 1069
Controlling depth(s), defined 753
determination of 119
discussed 753
tabulations of, on charts 129
Convention on Safety of Life at Sea 765
Convergence, of wave energy 834
Convergence zone, defined 842
Convergency, defined 707
Conversion angle, correction to radio bearings 978
Coordinate(s), determination of from chart.. 261
in geodesy, discussed 1342
geographical, defined 61
Coordinate systems, on celestial sphere 390
summary of 396
Coordinated Universal Time, origin of 484
Cordillera, defined 783
Coriolis acceleration 1135
Coriolis force, defined 905
Corner reflector, as aid to radar navigation 1004
Cornet; azimuth graph of 642
Corona, defined 922
of sun, defined 385
of sun or moon, discussed 922
Corposant (St. Elmo's fire) 924
Corrected establishment 804
Correcting, of compass error, defined 212
Cosine-haversine formula, historical use of.. 47
publication of 591
Cosmographia, of Ptolemy 5, 24
Cotter, Charles H.; *The Complete Coastal Navigator* of 299
Countercurrent, defined 815
Counterglow, defined 368
Course(s), of advance, defined 64
on charts 127
and course angle 65
defined 64, 263
and heading 65
over ground, defined 64, 264
symbol used for 64
Course angle, and course 65
defined 64
symbol used for 64
Course line, celestial, defined 465
defined 64, 263
labeling of 263
plotting of 263
Course recorder, described 198, 267
Crepuscular rays, described 923
Crescent, of moon, explained 383
Crest, of radio wave, defined 958
of water wave, defined 824
Cross-section paper, in lifeboat navigation... 725
Cross-staff, description of 28
Crossed loop antenna 975
Crosstalk, in radio receiver 970
Cubit, length unit, origin of 12
Cugle, Charles H.; azimuth tables of 642
Cugle's Two-Minute Azimuths 642

Page

Culmination, defined 392
Cumbelić, Petar; sight reduction table of 600
Cumulonimbus clouds, described 899
Cumulus, meaning of 895
Cumulus clouds, described 898
Cupola, chart definition of 126
Current(s), on charts 128
defined 264, 807, 815
discussed 264, 789
effect on sea state 1316
observations of 813
in polar regions 713
prediction of 325
in surf zone 836
variation of, across estuary 812
with depth 812
Current diagrams, use of 335
Current rose (ellipse) 807
Cursor, of plan position indicator 985
Cusps, of moon, defined 383
Cyclan, development of 52
Cycle, defined 958
Cycloid, wave as 824
Cyclone(s) defined 916
description of 916
in South Indian Ocean 932
in Southwest Pacific and Australia 932
Cyclonic storm, in North Indian Ocean 932
Cylindrical projections, features of 68
Cytac 53

D layer, of ionosphere 962
Daily Memorandum, discussed 747
chart maintenance items 137
special warnings in 750
Daily rate, defined 432, 498
Damp haze, defined 919
Danger(s), chart symbols for, described 120
Danger angle 303
use of 322
Danger bearing, use of 302
Danger sounding, use of 303
Dangerous semicircle, defined 947
Dark trace tube 971
Date line, defined 496
origin of 496
use of 496
Datum(s), geodetic, preferred 1349
Datum connection, geodetic, discussed 1349
Davidson Current 821
Davies, Thomas D.; sight reduction tables of 614
Davis, John E.; azimuth tables of 640
Davis, Percy L.H.; Alt-Azimuth Tables of..... 592
azimuth tables of 641
Chronometer Tables of 587
Requisite Tables of 590
Dawn orbit, of navigation satellite 1072
Day, defined 378, 491
length of, latitude determination by 739

	Page
Daybeacon(s), in buoyage system	105
on charts	123
defined	105
Daylight saving time	491, 498
with tidal predictions	330
Daymark, described	106
Day's work, described	674
De Colong deflector, described	249
De Jonge, Joost H. Kiewiet; sight reduction method of	616
De Principiis Astronomiae, of Frisius	20
De Revolutionibus Orbium Coelestium, of Nicholaus Copernicus	25
Dead reckoning	1
and celestial navigation	472
defined	59, 260
importance of, in lifeboat navigation	728
introduction to	260
keeping of, in polar regions	714
by plot	263
in polar regions	709
Dead Reckoning Altitude and Azimuth Table (H.O. Pub. No. 211), of Ageton	602
Dead reckoning equipment, discussed	267
functions of	198
Dead reckoning position, defined	59, 260
determination of	465
Dead reckoning tracer, principles of	199, 267
Dead water	832
Decante; azimuth tables of	641
Decca	56
Decca Navigator Operating Instructions and Marine Data Sheets, The; of Decca Navigator Company	1054, 1058
accuracy and coverage diagrams in	1058
Decca Navigator System, coverage of	1058
interchain fixing	1058
introduction to	1054
principles of	1054
Decca Track Plotter, discussed	1059
December solstice, defined	377
Decibar, unit of pressure measurement	779
Deck ice, discussed	926
Deck log, in lifeboat navigation	729
Declination, defined	388, 390
as latitude on celestial sphere	391
magnetic, defined	205
of moon, finding by almanac	525
of planet, finding by almanac	526
of star, finding by almanac	527
of sun, finding by almanac	524
Declination increment, defined	546
Decometer(s) reading of	1057
Deduced reckoning, defined	59
Deep, on lead line	156
in ocean, defined	784
Deep scattering layer, in echo sounding	159
of seawater	840
Deep sea lead, description of	155
origin of	14
Deep sea sounding lines, discussed	760
Deep sound channel, defined	842
Deep-water layer, of seawater	839
Defense Mapping Agency Hydrographic/Topographic Center, establishment of	18
Defense Mapping Agency Hydrographic/Topographic Center publication(s), Pub. No(s).:	
229, *Sight Reduction Tables for Marine Navigation*	604
249, *Sight Reduction Tables for Air Navigation*	611
260, *Azimuths of the Sun* (Schroeder and Southerland)	641
261, *Azimuths of Celestial Bodies*	641
Deflection of the vertical, cause of	387
discussed	1341
Deflector, compass adjustment by	248
principles involved	248
defined	248
Degaussing, defined	250
principles of	249
Degaussing coils, types of	251
Degaussing compensation, of compass	253
defined	250
Degaussing system, securing the	253
Degree-of-freedom, of gyroscope, defined	177
Deimos, described	364
Delrac	56
Demodulation, of radio signals	968
Density, of seawater	779
Density differences, currents related to	816
Departure, defined	63
position of, defined	264
Departure time, adjustment of, in ship weather routing	696
Deperming, described	253
Depot des Cartes, Plans, Journaux et Memoirs Relatifs a la Navigation; origin of	17
Depressed pole, defined	394
Depth(s), on charts	120
minimum for underwater log calibration	1309
units of	12
Depth contour(s), defined	118
Depth curves, defined	114
Depth measurement, importance of	155
Depth of water, defined	324
Desert devil, defined	918
Deviation, adjustment for, induced magnetism in asymmetrical horizontal soft iron	230
induced magnetism in symmetrical horizontal soft iron	228
induced magnetism in vertical soft iron	223
permanent magnetism	220
analysis of	234
nature of	234
purpose of	234
application of	212
defined	209

Page

Deviation—Continued
finding the, placing vessel on desired magnetic heading 241
by azimuth of celestial body 243
by bearing of distant object 242
by gyrocompass 242
by magnetic compass 241
history of 10
by magnetic bearing or azimuth 244
by magnetic headings 244
by a range 246
by reciprocal bearings 247
Deviation table, use of 210
Dew, defined 892
Dew point, explained 892
Diagram, on plane of celestial meridian, discussed 396
illustrated 398
Diagrammi Altazimutali, of Alessio 642
Diaphone(s), sound signal 107
Diaphragm horn(s), sound signal 107
Diatom, defined 119
Difference, of latitude, explained 61
of longitude, explained 63
Differential leveling, in geodetic surveys 1346
Differential Omega 1028
Diffraction, of radar beam 990
of radio waves 964
Dip, abnormal, in polar regions 720
of horizon, altitude correction for 437
from lifeboat 738
of sea short of horizon, by basic calculator 1300
Dip needle, as vertical force instrument 233
Dipole antenna 966
Dipsey lead, description of 155
Direct current, defined 957
Direct motion, of planets, defined 382
Direct wave(s), defined 960
Direction, on earth 63
in ice 713
in lifeboat navigation 729
measurement of, on chart 261
at radio receiver 973
by sound 838
Direction finder error, defined 976
Directional light, described 93
Directional radiobeacon(s), defined 972
Dispersive ionosphere, defined 1099
Disposal areas, defined 752
Distance, by vertical angle 322
defined 63
in ice 713
in lifeboat navigation 731
measurement of, by sound 838
on chart 262
to an object 145
of an object, by two bearings 301
on earth 63
to new course, turning method 306
units of 12

Page

Distance—Continued
use of in piloting 285
Distance-difference measurement, in radionavigation systems 1013
Distance finding station(s), chart symbol for 125
use of 979
Distance off, determination of, from lifeboat 742
Distance, speed, and time; relationship of 145
Distance traveled, measurement of 146
Diurnal circle, defined 371, 388
Diurnal current 808
Diurnal libration, of moon 365
Diurnal tide, defined 799
Divergence, of waves 834
Diversion, in ship weather routing 696
Dividers, defined 141
description and use of 141
D'Ocagne, Maurice; altitude-azimuth graph of 627
Docking aid(s), doppler as 1118
Doldrums, weather in 905
Dome, chart definition of 126
Doniol, R.; sight reduction tables of 595
Doppler dead reckoning, described 267
Doppler effect, defined 1118
Doppler sonar navigation, basic design considerations 1123
development of expressions for the vessel's speed in the fore and aft and athwartship directions 1123
introduction to 1118
operational errors in 1119
problems associated with beam geometry. 1122
Doppler sonar navigator(s) 1118
Doppler sonar speed log(s) 1118
Double altitudes, longitude determination by 615
Double high water 801
Double low water 801
Double-second difference, defined 549
Double star, defined 369
Doubling angle on bow, defined 302
Dozier, Charles T.; sight reduction method of 617
Draft, ship's, in underwater log calibration. 1309
Drafting machine, description and use of 143
Drag circle, construction of 679
Dreisonstok, J. Y.; sight reduction tables of. 589
Drift, of current, in celestial navigation 473
defined 264, 815
discussed 808
Drift angle, of vessel in turn, defined 305
Drift current, defined 815
Drizzle 919
Dry-bulb temperature, indication of 893
Dry-bulb thermometer 893
Dry haze, defined 919

	Page
Duct, of radio waves	960
Dummy log	150
Dumping grounds, defined	752
Dusinberre, H. W.; sight reduction method of	637
Dust devil, defined	918
Dust whirl, defined	918
Dutchman's log, history of	11
in lifeboat navigation	731
speed measurement device	148
Dynamic pressure, of water	152
Dynamic water pressure, speed measurement by	152
Dynamical mean sun, defined	487
Dynamical time, defined	483
Dyne, unit of force, defined	884
E layer, of ionosphere	961
E-link, described	230
EM log, described	150
Earth, magnetic field of, effect on Omega signals	1026
measurement of	206
as a planet	362
shape of	60
discussed	1340
historical	23
size determination, by Eratosthenes	4
Earth rate, of gyro	173
Earthshine, on moon, defined	383
East, as direction of measurement	62
East Australia Current	821
East component, earth's magnetic field, defined	205
East Greenland Current	819, 822
Easterly wave, of tropical cyclone	939
Eastern North Pacific, tropical cyclones in	932
origin, season, frequency	932
Eastern standard time	498
Ebb, of current, discussed	808
strength of, defined	807
Ebsen, Julius; azimuth tables of	641
Eccentricity, of earth's orbit	359
Eccentricity component, defined	487
Echo, radar, recognition of unwanted	999
Echo sounder, description of	156
development of	48
in polar regions	717
Echo sounding, origin of	14
Echogram, use of, for records	762
Eclipse(s), of lights	96
longitude by	33
of sun and moon, explained	384
Eclipse year, length of	386
Ecliptic, discussed	374
as primary great circle	396
obliquity of, defined	374
Ecliptic poles, defined	396
Ecliptic system of coordinates, described	396
Eddy current, defined	815
Eddy viscosity, of seawater	780
Eighteenth century navigation, history of	3
Ekman spiral, defined	816
Ekman wind current, defined	815
Electric current, defined	957
Electric psychrometer, described	893
Electrical conductivity, of seawater	781
Electricity, history of	47
Electromagnetic radiation	958
Electromagnetic spectrum, defined	959
Electromagnetic underwater log, described	150
Electron(s)	957
Electron gun, of cathode-ray tube, defined	970
Electronic aids, other, in polar regions	717
Electronic interference, appearance on radarscope	1002
Electronic navigation, defined	59
Electronics, application of, to navigation	48
defined	60
history of	48
in piloting	310
Elements of Navigation, of Robertson	21
Elevated pole, defined	394
Ellipsoid(s), and geoid, fit of	1341
orientation of	1347
of revolution	1340
earth as	362
Elliptical orbits, Kepler's laws of	26
Elongation, of planet, explained	381
Endless tangent screw vernier sextant, described	412
Engine revolution counter	155
origin of	12
speed measurement by	154
Ephemeris second, defined	483
Ephemeris Time, defined	483, 505
Epicenter, of seismic sea waves	830
Epicycles, plan of	24
Epistola de Magnete, of Peregrinus de Maricourt	9
Epitome of Navigation, of Norie	21
Equal area(s), as desirable property of projection	67
Kepler's law of	26
Equal interval light, defined	96
Equation of time	504
defined	378, 487
discussed	487
Equator, defined	61
Equatorial airmass, source region of	912
Equatorial Azimuth-Table, An; of Goodwin	642
Equatorial Countercurrent, in Atlantic Ocean	818
in Indian Ocean	822
in Pacific Ocean	820
Equatorial current, in oceanic circulation	816
tidal	810
Equatorial horizontal parallax, defined	365
of moon, tabulation of	365
Equatorial projection, defined	67
Equatorial tide	801
Equinoctial, formation of	387
Equinoctial year, length of	373

Page

Equipotential ellipsoid, defined 1352
Equipotential surface, of geoid, defined 1340
Equisignal, of Consol 1059
Error(s), adjustable of sextant 425
of collimation 427
combinations of 1265
defined 1260
of inertial navigator, eighty-four and four-tenths minute Schuler oscillation 1147
gyro bias 1150
stand-off 1152
twenty-four hour oscillation 1147
nonadjustable of sextant 424
nonrandom sources, in satellite navigation 1099
random measurement, effect of, on satellite navigation 1096
of three-point fix, causes of 319
Escarpment, defined 783
Establishment, of the port 804
Estimated position, defined 59, 260, 466
as equivalent of most probable position 467
establishment of 465
in piloting 304
Eulerian motion, defined 373
European Datum, geodetic 1350
Evasion, in ship weather routing 697
Evening star, defined 382
Evening twilight, defined 372
Ex-meridian altitude(s) 575
defined and discussed 584
Ex-meridian observation 575
discussed 584
of sun, in latitude determination 32
Exact coefficients, defined 219
Excelsior Azimuth and Position Finding Table, The; of Blackburne 587
Exercises, of Blundeville 21
Expendable bathythermograph, described 778
Experimental Air Navigation Tables, development of 611
Heard's modification of 612
Explanatory Supplement to the Astronomical Ephemeris and to the American Ephemeris and Nautical Almanac, publication and contents of 520
Extragalactic nebula, defined 369
Extratropical anticyclone(s), defined 917
Extratropical cyclone(s), defined 917
Extremely high frequency, radio wave propagation of 968
Eye, of hurricane, described 939

F layer, of ionosphere 961
F-Tafel, of German Navy 599
Face, of cathode-ray tube 970
Facom, development of 53
Fading, of radio waves 963
Fahrenheit temperature, defined 891
Falkland Current 820
Fall wind, defined 695, 918

Page

False echo, radar, cause of 999
Fan, defined 783
Fast ice, defined 849
Fata morgana, defined 920
Fathom 12
origin of 13
Fault lines, in earth's surface 776
Favé, L.; sight reduction diagram of 627
sight reduction method of 635
Federal project depth, defined 753
Feeder current 837
Fetch, effect on sea state 1313
of wave, defined 824
Fictitious equator, of transverse and oblique Mercator projections 73
Fictitious graticule, of transverse and oblique Mercator projections 73
Fictitious meridian (parallel), of transverse and oblique Mercator projections 73
Fictitious parallel(s), of transverse and oblique Mercator projections 73
Fictitious pole(s), of transverse and oblique Mercator projections 73
Fictitious rhumb line, on polar grid 706
on transverse and oblique Mercator projections 73
Fidelity, of radio receiver, defined 970
Field, of magnetism 957
Final diameter, of vessel in turn, defined 305
Fine lane, Omega, defined 1021
Fireball, meteor as 368
Firing danger areas, discussed 754
Firing practice area(s), on chart 129
First differences, Pub. No. 229 interpolation, defined 546
First estimate, of zone time 576
First-order (primary) triangulation, in geodetic surveys 1345
First point of Aries, defined 377
as reference, hour circle 389
First quarter, of moon, defined 383
First-year ice 847
Fish havens, defined 752
Fishtrap areas, defined 752
Five-digit numbers, of chart numbering system 133
Fix, adjustment of for constant error 289
celestial, plotting of, discussed 467
formation of 286
labeling of 263
by nonsimultaneous observations 292
types of, discussed 286
Fixed structures, lights on, on charts 122
Fixed marks 1357
Flagpole (staff, tower), chart definition of 126
Flamsteed's number, for stars 645
Flashing, described 253
Flashing light, defined 96
Flashlight, in lifeboat navigation 726
Flattening, of ellipsoid, ratio of 1340

	Page
Flinders bar, description of	224
determination of length	225
origin of	10
Floating aid(s), buoys as	102
Floating ballistic, of gyrocompass	187
Flood, of current, defined	808
strength of, defined	807
Florida Current	818
Flyback, of cathode-ray tube	971
Foam line, of rip current	837
Focusing anode, of cathode-ray tube	970
Foehn, defined	918
Fog, defined	894
navigation in	681
types of	918
Fogbow, defined	921
Fontura da Costa, A.; sight reduction table of	603
Foot, length unit, origin of	12
Foraminifera, defined	119
Forecastle degaussing coil, described	252
Force log, description of	154
Foreign charts, use of	754
Form lines, on charts	125
Four-digit numbers, of chart numbering system	133
Fourth-order triangulation, in geodetic surveys	1345
Fractional scale, of charts	112
Fracture(s), in earth's surface	776
of ice	849
Fracture zone, defined	783
Frame, of marine sextant	409
Franklin Piloting Technique, discussed	290
Frazil ice, described	846
Free gyro	173
Free gyroscope, described	172
Frequency, in doppler sonar navigation	1123
radio, regulation of use	968
of radio waves, units of	958
and radio wave propagation	966
use of in radarscope interpretation	996
Frequency modulation, of radio signals	969
Frequency range, of radio receiver, defined	970
Friendly ice, defined	873
Frigid Zones, defined	377
Front(s), defined	912
discussed	912
Frontal surface, defined	912
Frost, defined	892
Frost degree day, defined	850
Frost smoke, defined	703
described	919
Fucus, defined	120
Full moon, explained	384
Fuss, V. E.; sight reduction, tables of	601
GHA and Declination of Sun for 1981–2016	1295
Galactic latitude, coordinate of galactic system	390
Galactic longitude, coordinate of galactic system	390
Galactic nebula, defined	369
Galactic system of coordinates	396
Galaxy (galaxies), described	370
Gamma, measurement of intensity, defined	205
Gap, defined	783
Gauss, magnetic unit, defined	250
Gaussin error, defined	239
Gee, development of	50
Gegenschein, defined	368
General chart, description of	113
General precession, of equinoxes, defined	377
Geocentric latitude, defined	1343
illustrated	387
Geocentric parallax, defined	445
of moon, defined	365
Geodesic, defined	60
Geodesy, defined	1340
Geodetic control stations	1344
Geodetic datum, discussed	1347
horizontal	1347
vertical	1347
Geodetic errors, in inertial navigation	1157
Geodetic latitude, defined	1342
illustrated	387
Geodetic longitude, defined	387, 1343
Geodetic surveys, discussed	1343
Geographic latitude	387
defined	1343
Geographic longitude	387
defined	1343
Geographic range, of a light, defined	347
Geographical mile, defined	63
Geographical position, of celestial body, in navigational triangle	403
defined	460
Geoid, defined	1340
and ellipsoid, fit of	1341
orientation of	1347
Geoidal heights, defined	1341
Geoidal horizon, illustrated	393
Geoidal separations, defined	1341
Geoidal undulations, defined	1341
Geomagnetic latitude, defined	205
Geomagnetic longitude, measurement of	205
Geomagnetic meridians, defined	205
Geomagnetic midnight, time of	206
Geomagnetic noon, time of	205
Geomagnetic poles, defined	205
Geomagnetic time, defined	206
Geometric dilution	1016
Geometric projection, defined	67
Geometrical horizon, illustrated	393
Geophysical errors, in inertial navigation	1157
Geostrophic wind, defined	905
German Navy, azimuth diagram of	643
Ghost(s), in radar	996
Gibbous, phase of moon, defined	384
Gigahertz, defined	958
Gimballing error, of gyrocompass, discussed	185
Gingrich, John E.; sight reduction tables of	599

Page

Glacier, described ... 844
Global Positioning System ... 1110
Global radionavigation system, defined ... 1012
Globigerina, defined ... 119
Globigerina ooze, defined ... 119
Globular cluster, of stars, defined ... 369
Glory ... 922
Gnomonic projection, defined ... 68, 80
  features of ... 80
  of Thales ... 4
Godfrey, Hugh; azimuth diagram of ... 642
Goniometer, direction measurement device ... 975
Goodwin, H. B.; azimuth table of ... 642
  sight reduction tables of ... 588
Gradient(s), defined ... 441
  in hyperbolic radionavigation systems ... 1015
Gradient wind, defined ... 905
Graduation error, of sextant ... 424
Graphic scale, of charts ... 112
Graphique d'Azimut, of Cornet ... 642
Graticule, defined ... 68
Gravitation, defined ... 1341
  universal law of ... 27
Gravity, defined ... 1341
Gravity anomoly, defined ... 1157
Gravity error, of mercurial barometer ... 886
Gravity waves, formation of ... 824
Grazing angle, of radio waves ... 959
Grease ice, formation of ... 846
Great circle(s), defined ... 60, 271
  as desirable property of projection ... 67
  on polar projections ... 86
  in sailing ... 270
  solutions of, by intermediate calculator ... 1304
Great-circle chart(s), gnomonic projection as ... 81, 273
Great-circle distance, defined ... 63
Great-circle sailing, by chart ... 272
  defined ... 270
  discussed ... 271
  history of ... 17
  by Pub. No. 229 ... 273
  by table ... 273
Great-circle track, altering to avoid obstructions ... 280
  defined ... 65, 271
Great Lakes ice, discussed ... 873
Great Lakes Pilot, publication, discussed ... 344
Greatest elongation east, of planet, defined. 381
Greatest elongation west, of planet, defined 381
Greek alphabet ... 1166
Green flash, described ... 923
Green moon, occurrence of ... 921
Greenland Current ... 819
Greenwich apparent time, finding ... 504
Greenwich hour angle ... 506
  as angular distance ... 389
  defined ... 391
  of moon, finding by almanac ... 525
  of planet, finding by almanac ... 526

Page

Greenwich hour angle—Continued
  of star, finding by almanac ... 527
  of sun, finding by almanac ... 524
Greenwich mean time ... 491
  finding ... 496
  UT 1 as ... 483
Greenwich Royal Observatory, history of ... 40
Greenwich sidereal time, defined ... 505
Greenwich time ... 491
  reference for ... 380
Gregorian calendar ... 486
  defined ... 373
Grey ice, described ... 847
Grey-white ice, described ... 847
Grid(s), defined ... 91
  purpose of ... 91
  types of ... 91
Grid bearing, defined ... 283
Grid course ... 64
Grid direction, defined ... 706
Grid variation, defined ... 205, 706
Grivation, defined ... 205, 706
Ground, as part of radio transmitter ... 969
Ground conductivity, effect on Omega signals ... 1026
Ground log, described ... 148
Ground swell, effect of upon ship ... 828
Groundwave, defined ... 962
Group flashing light, *Light List* entry for ... 99
Group repetition interval, of Loran-C ... 1037
Group velocity, of waves ... 827
Growler, defined ... 846
Guide to Marine Observing and Reporting (Pub. 606), publication ... 758
Guide to Meteorological Instruments and Observing Practices (WMO No. 8, T.P. 31). ... 879
Guinea Current ... 819
Gulf Stream ... 818
Gust, of wind, defined ... 918
Guyot, defined ... 304, 784
Gyro ... 172
Gyro bias, defined ... 1134
Gyro error, defined ... 180
Gyro repeaters, accessories for ... 195
Gyrocompass, desirable characteristics of ... 189
  discussed ... 178
  errors of ... 179
  history of ... 10
  models of ... 190
  in polar regions ... 712
  repeaters for ... 194
  undesirable characteristics of ... 189
  use of in polar regions ... 187
Gyroscope, as inertial navigator ... 1129
  principles of ... 170
Gyroscopic compass ... 161
Gyroscopic inertia, defined ... 170, 1130
Gyroscopic precession, defined ... 170
Gyrosphere, illustrated ... 175

Page
Hagner, Fred; "Position Finder" of ... 639
Half-power point, of radar, defined ... 987
Half-tide level ... 805
Halo(s), cause of ... 922
types of ... 922
Hamilton, W.; sight reduction tables of ... 599
Hand-held digital calculator(s), introduction to ... 1299
Hand lead, description of ... 155
history of ... 14
Handle, of marine sextant ... 411
Harbor chart, description of ... 113
Hard iron, magnetic characteristics of ... 201
Harmonic analysis, of tide observations ... 806
Harmonic constants, of tide predictions ... 806
Harmonic tide plane ... 805
Harvest moon, defined ... 384
Haze, described ... 919
Head, of feeder current ... 837
Heading, and course ... 65
discussed ... 65
Heard, John F.; sight reduction tables of ... 612
Heat lightning, defined ... 924
Heeling adjuster, described ... 233
Heeling error, cause of ... 231
defined ... 231
Heeling magnet, positioning of ... 232
Height(s), on charts ... 120
of radar target, effect on range ... 993
of wave, from ship ... 829
Height error, of barometer ... 886
Height of eye correction, defined ... 437
Height of tide, defined ... 324
Heliocentric parallax, defined ... 356, 369
of stars, defined ... 445
Hertz, defined ... 479, 958
Hestia, defined ... 24
Hevelian halo, defined ... 992
Hickerson, Thomas F.; sight reduction tables of ... 603
Hidalgo, orbit of ... 364
High, migratory anticyclone ... 917
High altitude observations, reduction methods for ... 567
High clouds, defined ... 894
types of ... 895
High frequency, radio wave propagation of. 967
High Latitude Celestial Navigation Tables (H.O. Pub. No. 230), of Goetz ... 611
High-latitude effects, on navigation ... 701
High tide, defined ... 798
High water ... 324, 798
High water datums ... 805
High water full and change ... 325, 804
Hill, defined ... 783
Hilly ground, appearance on radarscope ... 998
Hilsenrath, Joseph; azimuth device of ... 643
Hiltner, W. F.; sight reduction method of ... 638
Hoehne, George G.; sight reduction tables of ... 605
Hohentafeln nach Sternzeit, sight reduction table ... 610

Page
Hole, defined ... 783
Homing, on radio bearing ... 978
Hommey, Louis; sight reduction tables of ... 586
Horary tables, defined ... 586
Horizon(s), as primary great circle ... 392
types of ... 392
Horizon glass, on marine sextant ... 410
Horizon system of coordinates, described ... 394
Horizontal axis, of gyroscope ... 170
Horizontal beam width ... 985
Horizontal danger angle, described and discussed ... 322
Horizontal deflection plates, of cathode-ray tube ... 971
Horizontal earth rate, of gyro, described ... 173
Horizontal geodetic datum shift(s), on charts ... 128
Horizontal intensity, of earth's magnetic field ... 205
Horizontal parallax, defined ... 445
of moon, defined ... 365
Horizontal sextant angle(s), corrections to measurements of ... 457
Horse latitudes, defined ... 909
weather in ... 909
Hostile ice, defined ... 873
Hour(s), unit of time ... 493
Hour angle, and time ... 506
Hour circle ... 392
defined ... 388
Hour-glass effect, appearance on radarscope ... 1003
House, chart definition of ... 126
Hughes' Tables for Sea and Air Navigation, of Comrie ... 599
Hugon, P.; azimuth diagram of ... 643
sight reduction tables of ... 594
Hum, in radio ... 965
Humboldt Current ... 821
Humidity, measurement of ... 892
Hummock, of ice ... 848
Hummocked ice, defined ... 848
Hunt, of compass, defined ... 160
Hunter's moon, defined ... 384
Hurricane(s), defined ... 931
in eastern North Pacific ... 932
formation of ... 936
in North Atlantic ... 932
portrait of ... 937
Huygens' principle ... 964
Hyatt, Delwyn; "Celestial Coordinator" and "Coordinate Transformer" of ... 632
Hydraulic current, defined ... 808

	Page
Hydrographic Office (U.S.) publication(s), H.O. Pub. No(s).: 66, *Arctic Azimuth Tables* (Schroeder and Wainwright)...	640
201, *Simultaneous Altitudes and Azimuths of Celestial Bodies*	592
203, 204, *The Sumner Line of Position of Celestial Bodies* (Littlehales)	588
208, *Navigation Tables for Mariners and Aviators* (Dreisonstok)	598
209, *Position Tables for Aerial and Surface Navigation* (Pierce)	602
211, *Dead Reckoning Altitude, and Azimuth Tables* (Ageton)	602
218, *Astronomical Navigation Tables*	604
230, *High Latitude Celestial Navigation Tables* (Goetz)	611
Hydrographic offices, history of	17
Hydrographic reports, introduction to	756
Hygrometer, described	893
Hyperbola, in distance-difference measurement	1013
Hyperbolic radio aids, development of	49
Hyperbolic radionavigation systems	1013
discussed	1013
IALA Maritime Buoyage System	1357
I.C.S. Altitude and Azimuth Tables for Air and Sea Navigation, of Collins	599
Ice, density of	852
drift of	853
effect on sea state	1316
extent of in the sea	857
formation of	843
of land origin	844
and the navigator	843
in the North Atlantic	857
observing, reporting, and forecasting	875
operations in	867
in polar regions	722
salinity of	852
Ice accretion	926
Ice blink, described	865
Ice cake, defined	847
Ice canopy, defined	872
Ice cap(s)	700
defined	844
Ice cliff, defined	846
Ice detection, discussed	863
Ice field, defined	847
Ice floes, defined	847
Ice fog, defined	918
Ice front, defined	846
Ice island, defined	849
Ice keel(s)	872
Ice Observers Manual	879
Ice rind, described	847
Ice shelf, defined	849
described	700
Ice wall, defined	846
Iceberg(s), described	844
Iceberg drift	855
Impeller log, description of	154
development of	12
Inches of mercury, unit of measurement	883
Inclination, magnetic, defined	205
Index arm, on marine sextant	409
Index correction, of sextant	427, 738
of sextant altitude, need for	436
Index error, of sextant, cause of	426
Index mirror, on marine sextant	410
Index of refraction, defined	439
Indian Datum, geodetic	1350
Indian Ocean, currents in	822
Indian spring low water, defined	805
Indian tide plane, defined	805
Indirect echo, radar, cause of	999
Induced current	957
Induced magnetism, defined	201
parameters of	218
Inertial dead reckoning, described	268
Inertial motions, discussed	1133
Inertial navigation, basic principle of	1129
circle plot	1153
closed loop in	1139
closed system in	1139
coordinate systems	1146
damping	1141
defined	1129
development of	57
inherent errors of	1147
introduction to	1129
stabilization loop	1143
stable platform in	1137
Inertial navigator	1129
reset of	1156
Inertial sensors, discussed	1129
Inferior conjunction, defined	380
Inferior mirage, defined	920
Inferior planets, defined	361
Injection station, of satellite navigation data	1068
Inner planets, defined	361
Inshore current, defined	815
Instrucion Nauthica, of Diego Garcia de Palacio	21
Instrument(s), for celestial navigation, other	434
kinds of for plotting and dead reckoning	141
Instrument correction, of sextant	425
of sextant altitude, need for	436
Instrument error, of barometer	886
of sextant	425
Insulator, defined	957
Intended track, defined	64
plotting of	264
Intercardinal points, of compass	162
Intercardinal rolling error, of gyrocompass, discussed	186
Interference, radio	965
Intermediate calculator(s)	1300
use of	1300
Internal waves, creation of	832

Page

nternational Atomic Time 484
scale, discussed 484
nternational chart(s), discussed 131
nternational Convention for the Safety of Life at Sea (1960) 758, 861
nternational Hydrographic Bureau, establishment of 18
nternational Hydrographic Organization, establishment of 18
nternational Ice Patrol, discussed 754, 861
nternational Nautical Mile, origin of 63
nternational Regulations for Preventing Collisions at Sea, as part of Traffic Separation Schemes 750
nternational System of Units, time in 483
nternational Union of Geodesy and Geophysics 1352
nterpolation, in Omega lattice tables 1035
ntracoastal Waterway aids to navigation, discussed 108
nverse orthomorphic projection, defined 68, 74
nverse Mercator projection, defined 68, 74
on, defined 961
onosphere, defined 961
layers of 961
and radio waves 962
onospheric disturbance(s), sudden, effect on Omega signals 1026
rminger Current 819
rradiation, defined 442
rradiation correction, of sextant altitude, discussed 442
sland(s), long chains of, appearance on radarscope 998
low, appearance on radarscope 998
wooded, appearance on radarscope 998
sland shelf, defined 678, 783
sland slope, defined 678, 783
sobar(s), on synoptic charts 883
soclinal line, defined 206
sodynamic line, defined 206
sogonic line, defined 128, 206
sogriv, defined 707
solated danger mark(s), defined 1358
use of 1360
somagnetic charts, described 206
points of annual change on 209
sophase light, defined 96
sopor, on isomagnetic charts 209
sopycnic line, defined 816

Jamming, of radio 965
Janus configuration, with doppler sonar speed log and navigator 1118
Japan Current (Kuroshio) 820
Japanese Hydrographic Office, H.O. Pub. No(s).: 351, *Celestial Navigation Observation Table* 605
603, *Simplified Celestial Observation Table* 610
sight reduction tables of 593, 603

Page

Japanese Navy, sight reduction instrument of 638
Jernaes, Leiv; "Nauticator" of 628
Jet stream, defined 904
Johnson, A.C.; azimuth tables of 641
Julian day, defined 493
June solstice, defined 377
Jupiter, described 364

Kahn, Louis; sight reduction method of 637
Katabatic wind, defined 918
Kelvin deflector, described 249
Kelvin temperature, defined 891
Kelvin's rule for improving compass adjustment 228
Kepler's laws 359
Kilohertz, defined 958
Knoll, defined 783
Knot, defined 63
speed unit, origin of 12
Kortazzi; azimuth tables of 640
Kotlarić, Stjepo M.; sight reduction methods of 603, 618
sight reduction tables of 613
Star Finder and Identifier of 662
Kuroshio (Japan Current) 820

LF Loran, development of 51
Labrador Current 819
Labrosse, F.; azimuth tables of 640
Lagging the tides 798
Lagoon, appearance on radarscope 998
Lake, appearance on radarscope 998
Lalande, Jerome; sight reduction tables of 586
Lambert, Johann Heinrich; chart projections of 8
Lambert conformal projection, defined 68, 78
features of 77
origin of 8
Lamplough, F. E.; sight reduction method of 634
Land, appearance on radarscope 998
signs of 741
Land areas, chart symbols for 125
Land breeze, described 917
Land effect, of radio waves 961
on radio waves 975
Land mile, defined 63
origin and length of 13
Land navigation 1
defined 59
Land sky, described 865
Landfall, discussed 678
Landmark(s), on nautical charts 126
Lane, of Decca lattice 1057
of Decca system, identification of 1057
Lane slip, in Decca system 1058
Large-scale chart, defined 113
Last quarter, of moon, explained 384
Lateral mark(s) defined 1358
use of 1359

	Page
Lateral system of buoyage	1355
Latitude, from celestial observations	615
on celestial sphere	930
defined	61
determination of, by diagram	633
history of	32
from lifeboat	739
by Polaris	584
historical	32
effect of, on magnetic compass	217
on Omega signals	1026
graphical interpolation of altitude for	561
types of	386
Latitude and longitude by computation, sight reduction method	623
Latitude line, defined	465
Latitude, Longitude, and Azimuth by the Sun or Stars, of Hickerson	603
Latitude methods, for determination of latitude	583
Lattice, of Decca system	1057
in distance-difference measurement	1013
Lattice tables, of Omega Navigation System	1033
Lava, defined	119
LeSort; computing device of	630
Lead, depth measurement device, described	155
Lead(s), of ice	849
Lead line, origin of	14
Leading lights, defined	285
Leading line, defined	285
Leading marks, defined	285
Leap seconds	484, 515
adjustment of	515
Leap year, defined	486
length of	373
Lecky, S.T.S.; azimuth tables of	641
Leeway, defined and discussed	267
Left (dangerous) semicircle, avoiding storm center, Southern Hemisphere	950
Left (navigable) semicircle, avoiding storm center, Northern Hemisphere	950
Leick, A.; sight reduction diagram of	635
Length, of wave, from ship	830
Lettering, on nautical charts	117
Levee, defined	783
Libration, in latitude, of moon	365
in longitude, of moon	365
of moon	365
Lifeboat navigation, ashore after	743
defined	59
equipment for	724
introduction to	724
Lifeboats, keeping together	728
Light(s), as aids to navigation	92
in buoyage system	105
colors of	1358
rhythms of	1358
phase characteristics of	96
characteristics of, discussed	96
factors affecting	99
phase characteristics of	96
Light List—Continued	
Light List, of U.S. Coast Guard, contents of	346
emergency sound signals in	108
entry for, alternating fixed and flashing light	96
composite group flashing light	99
group flashing light	99
Luminous Range Diagram in	348
number of volumes	346
offshore light stations in	93
private aids to navigation in	110
radio direction finder information in	977
sound signals in	107
Light lists, information concerning	346
introduction to	340
Light sector, on charts	125
Light station, defined	93
Light-year, defined	356
Lighted buoy(s), described	102
Lighted sound buoy, described	102
Lighthouse(s), on charts	122
purpose of	93
Lightning, described	924
Lightship(s), on charts	123
Limb, of marine sextant	409
Line(s), of force, of magnetic field	202
of maxima, defined	988
of minima, defined	988
of nodes, of moon's orbit, defined	386
of soundings, in piloting	304
Line of position, advancing	293
celestial, advancing and retiring	467
determination of	463
circular	460
discovery of	44
Omega	1022
in piloting, defined	283
discussed	283
plot of	540
retiring of	293
Linear interpolation, by basic calculator	1299
List of Lights, of Defense Mapping Agency Hydrographic/Topographic Center, aeronautical lights in	93
contents of	346
number of volumes	346
Littlehales, George W.; altitude and azimuth angle by map projection	631
sight reduction method of	636
Local apparent noon	505, 576
Local apparent time, finding	504
Local attraction, defined	128
Local disturbances, defined	204
Local hour angle	506
as angular distance	389
defined	391
finding the	539
graphical interpolation of altitude for	561
Local knowledge, in piloting	282
Local mean time, finding	503
Local Notice to Mariners, discussed	748

	Page
Local sidereal time, finding	505
Local time, defined	490
reference for	380
Local warnings, radio navigational	749
Local winds, types and names of	917
Location vessel, example of	770
Lodestone, described	201
Log(s), history of	11
speed measurement instrument	148
Log board	16
Log chip	11
Log glass	11
Log line	11
of taffrail log	147
Log reel	11
Log ship	11
Logarithmic time-speed-distance nomogram, on charts	129
Long-range radionavigation aids, in polar regions	717
Long-range radionavigation system, defined	1013
Long-range radionavigational warnings, discussed	749
Long-term almanac	1218
Longitude(s), from celestial observations	615
on celestial sphere	390
on charts	128
defined	62
determination of, by diagram	633
history of	32
from lifeboat	740
and time	496
types of	386
Longitude Found, The; of Bond	33
Longitude line, defined	465
Longitude method(s), for determination of longitude	583
for line of position	585
Longitude Not Found, The; of Blackborrow	33
Longitudinal degaussing coil, described	252
Longley, C.D.N.; sight reduction method of	634
Longshore current, defined	815
described	836
Lookout, in lifeboat	728
Lookout station, chart definition of	127
Looming, of mirage	919
Loran-A, development of	51
Loran-C, development of	54
discussed	1036
lattice tables of	1043
lattices on charts	1050
propagation prediction	1051
signal format of	1037
synchronization control	1040
system calibration	1040
Loran-C Correction Tables	1051
Loran-C interference filters	1042
Loran-C receiver(s), characteristics of	1042
signal acquisition by	1043
Loran-C Table (Pub. 221), of DMAHTC	1043
Loran-D	1051

	Page
Low, migratory cyclone	917
Low altitude observations, corrections for	452
in polar regions	719
reduction methods for	566
Low clouds, defined	895
types of	897
Low frequency, radio wave propagation of	966
Low tide, defined	798
Low water	324, 798
Low water datum	805
Low water springs, defined	804
Lower branch, of celestial meridian, defined	388
of meridian of earth	61
Lower limb	414
Lower low water datum	805
Lower transit, of celestial meridian, defined	392
Lowest low water, defined	805
Lowest normal low water, defined	805
Loxodrome, defined	63
Loxodromic curve, defined	63
Lubber's line, of magnetic compass	161
Luminous range, of a light, defined	347
Luminous Range Diagram, explained	348
Lunar day, length of	380
reference for	489
and tidal cycles	804
and tidal day	803
Lunar distance(s), longitude by	33
solving by spherical triangle	43
Lunar eclipse, defined	386
Lunar Ephemeris for Aviators, origin of	42
Lunar month, and tidal cycles	803
Lunar time, basis for	489
reference for	380
Lunicurrent interval, defined	325
Lunitidal interval	804
Lynn Azimuth Tables	586
Lynn Horary Tables	587
Lynn, Thomas; *Lynn Azimuth Tables* of	640
Lynn Horary Tables of	586
Madrepore, defined	119
Magnet	201
Magnetic anomalies, defined	204
Magnetic azimuth(s), curve of	243
Magnetic bearing, defined	283
Magnetic compass	161
accessories for	168
described	161
history of	9
limitations of	166
other types of	166
in polar regions	710
Magnetic compass error, discussed	209
Magnetic compass rose, location of on isomagnetic charts	206
Magnetic course	64
Magnetic dip, defined	205
Magnetic direction	209

Page

Magnetic equator, defined 206
intensity at 205
Magnetic field, of earth, diurnal change 204
Magnetic force, of moving electric charge 957
Magnetic Inclanation or Dip (chart 30), described 206
Magnetic information, on charts 128
Magnetic latitude, defined 205
Magnetic materials, defined 201
Magnetic meridian, location of 205
Magnetic north 209
Magnetic observatories, number of 206
Magnetic pole(s), discussed 202
Magnetic signature of vessel 250
Magnetic Silencing, principles of 249
Magnetic treatment of vessel 253
Magnetic variation, use of 205
Magnetic Variation (chart 42), described 206
Magnetism, domains of 201
of steel vessel 213
theory of 201
Magnetometer, airborne 206
Magnitude, of celestial bodies, discussed 357
Magnitude ratio, defined 357
Main degaussing coil, described 251
Major light, defined 92
Major planets, named 362
Major ports, information on in *Sailing Directions (Enroute)* 344
Maneuvering board(s), in lifeboat navigation 725
Maneuvering errors, defined 150
Mangrove, appearance on radarscope 997
Manual for Ship's Surface Weather Observations 900
Manual on Codes (WMO No. 306) 879
Manuscript Tables, of Weems 599
Map(s), defined 67
Map projection, defined 67
March equinox, defined 377
Marcq St.-Hilaire method, for line of position 590
Margetts' Horary Tables 627
Marine chart, defined 67
Marine geology, defined 782
Marine microbarograph, described 885
Marine navigation 1
defined 59
practice of, advance preparation of 671
entering port 678
getting underway 672
introduction to 671
leaving port 673
taking departure 674
Marine radiobeacon(s) 972
Marine reports, discussed 756
Marine sediments, discussed 784
Marine sextant, altitude corrections for 447
defined 408
Marine Surface Observations 900
Marine weather broadcast, and tropical cyclones 941

Page

Mariner's Compass Rectified, The; of Wakeley 640
Mark, on lead line 156
Mark 27 Gyrocompass, described 193
Mark 227 Gyrocompass, described 194
Marks, discussed 1358
characteristics of 1358
colors of 1358
shapes of 1358
Marl, defined 119
Mars, described 363
Marsh, appearance on radarscope 997
Martelli, G. F.; sight reduction tables of 586
Master compass 195
Master gyrocompass 195
Master station, of radionavigation systems 1013
Matte, defined 120
Maximum range, of radar, limitations on 992
McKee, Merlin A.; sight reduction method of 626
McMillen, D. A.; sight reduction method of 638
Mean establishment 804
Mean high water, defined 805
Mean high water lunitidal interval 325, 804
Mean high water springs, defined 805
Mean higher high water, defined 805
Mean low water, as chart datum 120
defined 804
Mean low water springs, as chart datum 120
defined 804
Mean lower low water, as chart datum 120
defined 805
Mean lower low water springs, defined 805
Mean refraction, defined 440
Mean sea level, defined 805
Mean sidereal time, defined 482, 505
Mean solar time 378, 482, 489
Mean sun 489
defined 487
discussed 504
fictitious 378
Mean time 487
Measured mile, on charts 127
in speed measurements 155
Mécanique Céleste; of Laplace 26
Medium frequency, radio wave propagation of 967
Medium range radionavigation system, defined 1012
Megahertz, defined 958
Ménéclier, Victor; sight reduction tables of 611
Mercator chart, construction of 70
Mercator projection 68, 69
features of 69
origin of 8
Mercator sailing, defined 270
origin of 17
Merchant Ship Search and Rescue Manual 768
Mercurial barometer, atmospheric pressure measuring device 883
invention and construction of 884

Page

Mercury (planet), described 363
Mercury ballistic, of gyrocompass 178
Meridian(s), defined 60
of earth 60
on polar projections 86
Meridian altitude(s), latitude by 569
in lifeboat 739
method for 583
Meridian angle, defined 391
discussed 506
finding the 539
as measure of hour angle 389
Meridian transit, defined 392
direction from, in lifeboat 729
finding time of 575
of sun, in latitude determination 32
Meridional difference, use of in chart construction 70
Meridional parts, explanation of 70
Messenger, in water sampling 777
Meteor(s), and comet(s), discussed 367
described 368
in solar system 357
Meteor showers, occurrence of 368
Meteor swarms, defined 368
Meteorite, defined 368
Meteorological equator, defined 906
Meter, length unit, origin of 12
Micrometer drum, on marine sextant 409
Micrometer drum sextant, described 409
Microphone, radio 969
Microphonic noise, in radio 965
Microsecond, defined 493
Microwave(s), of radio 959
Microwave band, radio wave propagation of 968
Midchannel controlling depth, defined 753
Middle clouds, defined 894
types of 895
Middle (mid) latitude, defined 62
Middle- (mid-) latitude sailing, defined 270
origin of 16
Midnight, defined 491
Midnight sun, land of, defined 372
Military grids 91
Milky Way, sun as part of 370
Millibar, unit of pressure, defined 779, 884
Millimeters of mercury, unit of measurement 883
Millimicrosecond, defined 493
Millisecond, defined 493
Miniature Navigation Table for Altitude and Azimuth, of Doniol 595
Minor light(s), in buoyage system 105
defined 93
Minor planets, features of 364
orbits of 362
in solar system 357
Minute(s), unit of time 493
Mirage(s), defined 919
types of 919

Page

Mirifici Logarithmorum Canonis Descriptio, of Napier 21
Miscellaneous data, area 1169
astronomy 1169
charts 1170
decibel scale 1179
earth 1171
length 1174
mass 1174
mathematics 1175
meteorology 1175
pressure 1176
speed 1177
vessel tonnage 1179
volume 1178
volume-mass 1179
Mississippi River system aids to navigation, discussed 109
Mist, defined 919
Mistake(s), defined 1261
navigational, discussed 1293
Mistral 2
defined 918
Mixed current 808
Mixed layer, of seawater 839
Mixed tide, defined 799
Moat, defined 783
Mock moon, defined 922
Mock sun, defined 922
Modification des Tables d'Azimuth de Thomson, of Kortazzi 640
Modified facsimile chart(s), discussed 129
Modified Lambert conformal projection, defined 85
Modified U.S. Aid System 104, 1356
Modulating wave, radio 968
Modulation, of radio transmission 968
Modulator, radio 969
Molfino; azimuth diagram of 642
Monsoon, features of 910
Monsoon airmass 912
Monsoon Current 822
Moon, age of, defined 384
altitude corrections by *Nautical Almanac.* 449
altitude corrections of 450
discussed 365
observation of with sextant 414
Moon dog, defined 922
Moon pillar, described 922
Moonbow, defined 922
Moonrise, finding time of, by almanac 531
Moonset, finding time of, by almanac 531
Morning star, defined 382
Morning twilight, defined 372
Most probable position, defined 1287
discussed 1287
distinguished from estimated position 467
in piloting 304
Motion(s), of solar system bodies, discussed 357
Motive power, in lifeboat navigation 731

	Page
Mount Wilson Observatory, founding of	40
Mountains, defined	784
Mud flat, appearance on radarscope	997
Multiple echoes, appearance on radarscope	1002
Multiple star, defined	369
Multiple-trace echoes, appearance on radarscope	1002
Multiyear ice, defined	847
Muskeg, described	700
Myerscough, W.; sight reduction tables of	599
NBS Frequency and Time Broadcast Services (Special Pub. 236)	512
Nadir, defined	388
Name(s), for stars	644
Nanosecond, defined	493
Nansen bottle	777
Napier diagram, construction and use of	210
National Flags and Ensigns (chart 5), and chart numbering system	132
National Ocean Service, founding of	17
National Plan for Navigation, of Department of Transportation	1051
National Tidal Datum Epoch	806
Natural magnet, lodestone as	201
Natural range, formation of	285
Natural scale, of charts	112
Natural year, length of	373
Nautical Almanac, first British	37, 42
Nautical Almanac, The; air temperature correction in	441
altitude corrections by	448
arc to time table in	495
atmospheric pressure correction in	441
augmentation correction in	443
described	521
dip correction in	437
equation of time in	504
equatorial horizontal parallax in	365
extracts from	1182
finding GHA and declination, of moon	525
of planet	526
of star	527
of sun	524
finding time, of moonrise and moonset	531
of sunrise and sunset	529
of twilight	530
historical	520
irradiation correction in	442
for lifeboat navigation	725
low altitude corrections by	452
parallax correction in	446
phase correction in	443
planet diagram in	656
Polaris correction tables in	578
publication of	520
refraction correction in	440
semidiameter correction in	442
star charts in	645
stars listed in	644
sunrise, sunset, and twilight tables in	528
tabulation of GHA and declination	523

	Page
Nautical Almanac—Continued	
time of meridian transit by	575
v correction in	523
Nautical astronomy, defined	355
Nautical Astronomy, with New Tables; of Symonds	642
Nautical chart(s), defined	67
introduction to	112
miscellaneous features of, discussed	127
Nautical chart abbreviations, discussed	754
Nautical chart symbols, discussed	754
Nautical Chart Symbols and Abbreviations (Chart No. 1), and chart numbering system	132
publication	116
Nautical day, defined	492
historical	589
Nautical mile, defined	63
origin of	13
Nautical twilight, defined	372, 529
Nautičke Tablice, of Čumbelić	600
Nautische Tafeln, of Rose	595
Navaglobe, development of	52
NAVAREA Warnings, discussed	749
Navarho, development of	53
Navegação Moderna, of Newton and Pinto	597
Navigable semicircle, defined	947
Navigate, derivation of	59
Navigation, defined	59
history of, from art to science	1
background	1
conclusion	58
early celestial discoveries	25
epic voyages	1
piloting and dead reckoning, background	4
obstructions to, artificial	752
in polar regions, discussed	701
at sea, discussed	674
of small craft	684
Navigation accuracy, discussed	1267
Navigation kit, emergency	724
Navigation log, use of for records	763
Navigation manuals, history of	26
after sixteenth century	21
Navigation system accuracy, discussed	1287
Navigation Tables for Mariners and Aviators (H.O. Pub. No. 208), of Dreisonstok	598
Navigational astronomy, defined	355
introduction to	355
Navigational buoys, large, described	101
Navigational compass, desirable characteristics of	160
Navigational coordinates, table of	1181
Navigational errors, clarification of terminology	1277
conclusion to	1294
definitions of	1260
introduction to	1260
Navigational Handbook with Tables, of Hickerson	603

	Page
Navigational planets, named	361
Navigational safety, introduction to	747
Navigational stars, identification of	1224
names of, pronunciations and meanings	1232
Navigational triangle, components of	402
defined	403
description of	42
Navigator, inertial	1129
Navigator's chart, discussed	67
Navigator's notebook	261
NAVSTAR Global Positioning System	1110
Navy Navigation Satellite System, configuration of	1070
development of	56
discussed	1066
equipments of	1075
launching of	1070
memory organization	1073
packaging techniques	1073
power system and thermal design	1073
stabilization of	1070
timing of	1075
Neap current	810
Neap tide(s)	801
defined	798
Neatline on charts	70
Nebula, defined	369
Neck, of feeder current	837
Negative altitude, of celestial body, defined.	395
Neptune, described	364
Net bottom clearance(s), discussed	753
Neutron(s)	957
Névé, defined	853
New Altitude and Azimuth Tables, of Ogura	597
New American Practical Navigator, The; of Bowditch, first publication of	21
purchase of copyright of	18
simplification of lunar-distance method in	34, 44
New and Accurate Method of Finding a Ship's Position at Sea by Projection on Mercator's Chart, A; of Sumner	45
New danger(s), defined	1361
discussed	1361
New danger, defined	1355
New Line of Position Tables, of Weems	600
New moon, defined	383
New sea ice, formation of	846
Newton, J. A.; sight reduction tables of	597
Newton's laws of motion	359
Ney's projection, defined	85
Night effect, on radio bearings	975
on radio waves	964
Night order book, contents of	678
Nilas, ice development	846
Nimbostratus clouds, described	898
Nimbus, meaning of	895
Nipped, defined	877
Nocturnal, description of	29
Nodal period of moon, as tidal cycle	804
Node(s), of moon's orbit, defined	386
Noise—Continued	
Noise, radio, causes of	965
defined	965
Nominal range, of a light, defined	347
Nomograma degli Azimut del Sole, of Molfino	642
Non-saturable radionavigation system, Decca as	1054
defined	1012
Nontidal current, defined	807
discussed	807
effect of	810
Noon orbit, of navigation satellite	1072
North, as direction of measurement	61
North American Datum, 1927; geodetic	1350
North Atlantic, tropical cyclones in	932
origin, season, frequency	932
North Atlantic Current	819
North Cape Current	819
North component, earth's magnetic field defined	205
North Equatorial Current, in Atlantic Ocean	816
in Indian Ocean	822
in Pacific Ocean	820
North Indian Ocean, tropical cyclones in	932
origin, season, frequency	935
North magnetic pole, of earth	203
North Pacific Current	820
North pole, of magnet	202
Northeast Drift Current, in Atlantic Ocean	819
Northeast trades	906
Northeasterlies, of Arctic	909
Norway Current	819
Note(s), on charts	128
Notebook, for lifeboat navigation	724
Notice to Mariners, chart, new edition date in	116
reprint date in	116
controlling depths in	753
corrections to, charts by	116
Coast Pilots by	344
Light List and *List of Lights* by	346
danger areas in	129, 755
date of latest	137
discussed	747
NAVAREA Warnings in	750
oil-well information in	752
special warnings in	750
submarine cables and pipelines in	752
use of, in advance preparation	672
Nouvelles Tables Destinées à Abréger les Calculs Nautiques, of Perrin	640
Nouvelles Tables Pour le Calcul de la Droite de Hauteur a Partir du Point Estimé, of Hugon	594
Nova, defined	369
Nucleus, of atom	957
of comet, described	367
Null	974
Numerical scale, of charts	112
Nun buoy, shape of	102

Page

Nutation, defined ... 373
discovery of ... 27

Oblate spheroid, defined ... 60
earth as ... 362
Oblique conic projection, defined ... 75
Oblique orthomorphic projection, defined... 68, 73, 74
Oblique Mercator projection, defined ... 68, 73, 74
discussed ... 73
use of ... 74
Oblique projection, defined ... 67
Oblique sphere, illustrated ... 373
Observed altitude, of celestial body, defined 436
defined ... 463
in sight reduction ... 539
Observed latitude, defined ... 465
Observed longitude, defined ... 465
Occluded front ... 914
Occultation, of planet by sun ... 380
Occulting light, defined ... 96
Ocean(s), introduction to ... 775
origin of ... 776
Ocean current(s), causes of ... 815
and climate ... 823
defined ... 815
introduction to ... 815
Ocean waves, introduction to ... 824
Oceanic circulation, discussed ... 816
Oceanic soundings, introduction to ... 756
Oceanographic chemistry, discussed ... 777
Oceanography, defined ... 775
exploration ... 775
history of ... 775
Octant (instrument) ... 428
of Hadley ... 31
Odor(s), as sign of land ... 741
Oersted, measurement of intensity, defined. 205
Offshore current, defined ... 815
Offshore light station(s), described ... 93
Offshore oil-well structures, discussed ... 751
Ogura, Sinkiti; sight reduction tables of ... 597
Okta(s), defined ... 848
Old ice, defined ... 847
Omega lane, defined ... 1019
Omega Navigation System, characteristics of, propagation ... 1022
transmission ... 1022
geophysical parameters of ... 1026
identification of signal ... 1024
introduction to ... 1016
lane identification ... 1021
lanes in ... 1019
principles of ... 1017
signal format ... 1022
stability of propagation ... 1028
station locations and coverages ... 1022
Omega plotting chart(s), discussed ... 1032
Omega Propagation Correction Tables ... 1028, 1032
discussed ... 1028
Omnidirectional beacon ... 972

Page

One-digit number(s), of chart numbering system ... 132
On Finding the Latitude and Longitude in Cloudy Weather, of Johnson ... 47
Ooze, defined ... 119
Open cluster, of stars, defined ... 369
Open pack, concentration of ice ... 848
Open water, concentration of pack ice ... 848
Opposition, of planet, explained ... 382
Optimum Track Ship Routing, described ... 685
Orbit, of celestial body ... 357
Ordnance Survey of Great Britain 1936 Datum, geodetic ... 1350
Orthographic projection, defined ... 68, 82
features of ... 82
origin of ... 5
Orthomorphic projection, characteristics of. 67
Oscillating circuit ... 957
Oscillator, radio ... 969
Oscillatory errors, defined ... 149
Oscillatory wave, defined ... 827
Oscilloscope (scope), cathode-ray tube as ... 970
Outer Dangers, description of in *Sailing Directions (Enroute)* ... 342
Outer planets, defined ... 362
Overhead power cable, appearance on radarscope ... 1003
Overlapping charts, note on ... 128
Oyashio ... 821

Pace, length unit, origin of ... 12
Pacific Coast Pilot, of 1889 ... 17
Pacific Ocean, currents in ... 820
Pack ice, defined ... 857
Padrón Real ... 6
Palm trees, appearance on radarscope ... 998
Palomar Mountain Observatory, founding of ... 40
Pampero, defined ... 918
Pancake ice, described ... 847, 848
Parabola, comet's orbit as ... 368
Parallactic angle, defined ... 404
Parallax, altitude correction for, discussed .. 445
from lifeboat ... 738
of moon, variance of ... 366
Parallax in altitude, defined ... 366, 445
Parallel(s), of altitude, defined ... 395
of declination, defined ... 388, 391
defined ... 61
of latitude, defined ... 61, 396
Parallel rulers, description and use of ... 142
Parallel sailing, defined ... 270
early importance of ... 33
origin of ... 16
Parallel sphere, illustrated ... 371
Parallelism, of sextant mirror and horizon glass, adjustment for ... 426
of sextant telescope, adjustment for ... 427
Parameter(s), defined ... 218
P, defined ... 218
Q, defined ... 218

Page

Parameter(s)—Continued
R, defined ... 218
of ship's magnetic field, discussed ... 217
Paranthelion, defined ... 922
Parantiselene, defined ... 922
Paraselene, defined ... 922
Paraselenic circle, defined ... 922
Parhelic circle, defined ... 922
Parhelion, defined ... 922
Parsec, defined ... 356
Partial eclipse, of sun, explained ... 384
Passive radionavigation system, Decca as ... 1054
defined ... 1012
Peak, of radio wave, defined ... 958
undersea feature, defined ... 784
Pelorus, description and use of ... 196
Pelorus, pilot of Hannibal ... 9
Penteado, F.; sight reduction table of ... 603
Perigean current ... 810
Perigean spring tide(s) ... 802
Perigean tide(s) ... 801
Perigee, defined ... 359
of moon, defined ... 801
distance at ... 365
Perihelion, defined ... 359
Period, of wave, from ship ... 830
Periodic current, defined ... 815
Periodic error, defined ... 1264
Periplus, of Scylax ... 2, 8
Permafrost, defined ... 700
Permanent current, defined ... 815
Permanent magnet ... 202
Permanent magnetism, defined ... 201
parameters of ... 218
Permeability, magnetic, defined ... 202
Perpendicular lines of position, sight reduction method ... 622
Perpendicularity, of sextant horizon glass, adjustment for ... 425
of sextant index mirror, adjustment for ... 425
Perrin; azimuth tables of ... 640
Personal correction, to sextant altitude, need for ... 436
Personal equation, in sextant observations ... 420
Personal error, in sextant observations ... 420
Perspective projection, defined ... 67
Peru Current ... 821
Phantom, of gyrosphere ... 174
Phantom bottom, in echo sounding ... 159, 840
Phase(s), of moon, defined ... 382
discussed ... 382
of radio wave ... 959
Phase anomoly, sudden, defined ... 1027
Phase comparison system(s) ... 1011
Phase correction, for sextant altitudes, discussed ... 443
Phase locking, of Decca signals ... 1055
Phase synchronized, defined ... 1017
Phobus, described ... 363
Phosphor, on cathode-ray tube face ... 970
Phosphorescence, of cathode-ray tube ... 970

Page

Physical libration, of moon ... 365
Picture tube, cathode-ray tube as ... 970
Pierce, M. R.; sight reduction diagrams of ... 633
sight reduction method of ... 626
Pilot, defined ... 282
Pilot waters, defined ... 282
Pilotage ... 282
defined ... 59
Piloting ... 1, 282
defined ... 59, 282
electronics in ... 310
in high latitudes ... 714
introduction to ... 282
practical ... 311
preparation for ... 283
safely without a fix ... 302
use of sextant in, evaluation ... 323
introduction ... 313
Pinto, J. C.; sight reduction tables of ... 597
Pipelines, danger from ... 752
Pitch, of ship's screw ... 154
Pitot pressure, of water, defined ... 153
Pitot-static log, description of ... 153
development of ... 12
Pitot tube, dynamic pressure measuring device ... 153
Pivot point, of vessel in turn, defined ... 305
Plain, defined ... 784
Plan position indicator, principles of ... 984
Plane sailing(s), described ... 270
history of ... 16
by intermediate calculator ... 1308
Planet(s), altitude corrections by *Nautical Almanac* ... 448
altitude corrections of ... 451
defined ... 361
features of ... 361
names of, pronunciations and meanings ... 1232
observation of with sextant ... 414
other and minor, described ... 363
in solar system ... 357
Planet diagram, discussed ... 656
Planet location diagram, in *Air Almanac* ... 656
Planetary configurations, defined ... 382
discussed ... 380
Planning route, defined ... 696
Plateau, defined ... 784
Plotter(s), description of ... 142
for lifeboat navigation ... 725
Plotting, of celestial line of position, in polar regions ... 721
on polar charts ... 708
of position ... 261
Plotting equipment, in lifeboat navigation ... 725
Plotting materials, for lifeboat navigation ... 725
Plotting sheet(s), definition and use of ... 88
for lifeboat navigation ... 725
Plunging breaker, defined ... 835
Pluto, described ... 364
Point(s), of compass, defined ... 162
Polar airmass, source region of ... 911

	Page
Polar axis	60
of earth	60
Polar cap absorption, defined	1027
Polar cap disturbance, effect on Omega signals	1027
Polar charts, adequacy of	705
coverage of	706
inaccuracy of	706
lack of detail on	705
Polar circles, defined	377
Polar currents	822
Polar distance, defined	391
Polar front, defined	909
Polar geography, discussed	700
Polar grid, description and use of	706
Polar horizontal parallax, defined	445
Polar jet stream winds, defined	904
Polar motion, defined	373, 482
Polar navigation, defined	59
planning of	722
Polar Operations, of MacDonald	873
Polar projection(s), defined	67
features of	84
selection of	86
Polar regions, introduction to	700
knowledge of	722
meteorological effects in	702
miscellaneous information	704
natural landmarks in	714
summary of conditions in	704
Polaris, direction from, in lifeboat	729
latitude by	577
in lifeboat	739
Polaris correction table, in *Air Almanac*	579
Polarization, of electromagnetic field	959
of radio waves	959
Polarization error, of radio bearings	975
of radio waves	964
Polyconic projection, defined	68, 79
features of	79
Polynya(s), in ice, defined	849
Polyzoa, defined	120
Poor, Charles L.; slide rule of	630
Portolan charts	5
Portolano Rizo	8
Ports Around the World, of Pytheas	8
Position(s), dead reckoning, in lifeboat navigation	732
on earth, discussed	61
labeling of	263
from observation of two or more bodies	616
plotting of, on chart	261
of ship, in lifeboat navigation	726
from single-body observation	621
Position angle, defined	404
Position Line Tables, of Smart and Shearme	597
Position plotting sheet, defined	88
Position reporting systems, introduction to	765
Position Tables for Aerial and Surface Navigation (H.O. Pub. No. 209), of Pierce	602
Potential flow field, of water velocity	148
Power loss error(s), in doppler sonar navigation	1121
Power supply, radio	969
Practical Celestial Air Navigation Tables, of Hoehne	605
Practical Tables for Navigators and Aviators, of Rust	587
Pre-Christian navigation, history of	1
Precession, in declination, defined	377
of the equinoxes, defined	377
discovery of	26
gyroscopic, defined	1130
input axis of	1130
output axis of	1130
in right ascension, defined	377
Precipitation, effect on sea state	1316
Precision, defined	1261
Precomputation, defined	565
for sight reduction	565
Precomputed altitude, defined	565
Predictability, defined	1287
Predictable accuracy, defined	1287
Pressure, of seawater	779
Pressure gradient, defined	905
Pressure gradient force, defined	905
Pressure ice, formation of	848
Pressure tendency, defined	913
Prevailing westerlies, defined and discussed	909
Primary, of celestial body, defined	357
Primary radar	983
Primary seacoast light, defined	92
Prime meridian, defined	61
establishment of	37
Prime vertical, defined	395
direction from, in lifeboat	729
finding time on	589
Prime vertical circle, defined	395
Prime vertical observation, sight reduction method	621
Priming the tides	798
Principal vertical circle, defined	394
Prismatic error, of sextant	424
Pritchard, H. C.; sight reduction method of	634
Private aid(s), on charts	125
Probable error, defined	1263, 1267
in one-dimensional errors	1281
Programmable calculator(s)	1299
use of	1308
Progressive waves	832
Project depth, on charts	118
Projection(s), for nautical charts	112
polar	705
selection of	67
types of	67
Propagation, of radio waves, in polar regions	716
Propagation correction(s), of Omega signals	1030
Propagation prediction, of Omega signals	1028

Page

Propeller speed, in underwater log calibration 1309
Proper motion, of celestial bodies, defined ... 374
discovery of 28
of stars, defined 369
Proton(s) 957
Protractor, defined 142
description and use of 142
Province, defined 784
Psychrometer, described 893
Psychrometric tables, use of 893
Pteropod, defined 119
Ptolemy, Claudius; world map of 5
Publication(s), of other agencies 1297
Puddles, freshwater from 852
Pulsating current, defined 957
Pulse duration, defined 983
Pulse length, defined 983
Pulse modulation, of radio waves 969, 983
Pulse repetition (recurrence) rate(s) 983
Pulse width, defined 983
Pumice, defined 119
Pumping, of mercurial barometer 884, 943
Purey-Cust, H. E.; azimuth tables of 642

Quadrant (instrument) 428
of Godfrey 31
Quadrantal correctors, defined 228
Quadrantal deviation, defined 218
Quadrantal error, of gyrocompass, discussed 186
of radio bearings 975
Quadrature, of planet, defined 382
Quarter-power point, of radar, defined 987
Quarterdeck degaussing coil, described 252
Quartz crystal marine chronometer(s), described 433
Quintant (instrument) 428

Racon 1004
Radar, development of 49
minimum and maximum range 992
in polar regions 717
principles of 983
storm detection by 900
Radar beacon(s), as aid to navigation 1004
Radar beam, principles of 985
Radar horizon, distance to 990
by basic calculator 1300
Radar navigation 1005
aids to 1004
defined 59, 983
introduction to 983
Radar Navigation Manual (Pub. No. 1310) .. 983
Radar shadow 999
Radarscope interpretation 994
Radial error, in two-dimensional errors 1286
Radial motion, of stars, defined 369
Radiation fog, formation of 918
Radio, for lifeboat navigation 726
Radio bearing(s), accuracy of 975
errors of 975

Page

Radio bearing(s)—Continued
use of 978
Radio direction finder 972
first utilization of 48
in polar regions 717
Radio direction finder station(s), function of 979
Radio direction finding, introduction to 972
Radio horizon, discussed 960
Radio Navigational Aids (Pubs. Nos. 117A, 117B) 346, 511, 515
aeronautical radiobeacons in 973
Consol and Consolan procedures in 1060
Decca information in 1058
general information in 749
ice patrol information in 754
radio direction finder information in 977
sector of calibration in 979
Radio propagation factors, discussed 510
Radio spectrum, defined 959
Radio stars, defined 368
discovery of 28
Radio tower (mast), chart symbol for, defined 126
Radio transmission, kinds of 968
Radio wave(s), and the atmosphere 924
defined 958
and the ionosphere 962
source of 957
terminology of 958
Radio wave propagation, and frequency 966
Radioactivity, of seawater 781
Radiobeacon(s), on charts 123
discussed 972
factors affecting maximum range 976
Radiolaria, defined 119
Radionavigation, defined 59
Radionavigation systems, introduction to. 1011
Radius SURPIC, example of 769
Radius vector, of planetary orbits 359
Radux, development of 54
Radux-Omega, development of 55
Rafting, of ice 848
Rainbows, types of 921
Ram(s), of icebergs, defined 846
Ramark 1005
Random errors, discussed 1263
normal curve of 1263
Range(s), bearing, defined 285
on charts 125
of daybeacons 105
defined 246
in piloting 282
radio signal 966
of tide, defined 799
in piloting 324
variations in 801
undersea feature, defined 784
Range light(s), described 93
Range line, defined 285
Range-Range, in hyperbolic radionavigation systems 1016

	Page
Ranging mode, of hyperbolic radionavigation systems	1016
of Loran-C	1051
Ranging radionavigation system	1012
Rankine temperature, defined	891
Rapid Navigation Tables, of Myerscough and Hamilton	599
Rapid Navigation Tables for Mariners, of Myerscough and Hamilton	600
Rational horizon, celestial horizon as	393
Receiver, radio, defined	970
discussed	970
Reciprocal bearings, treatment of	975
Recommended trackline(s), on charts	129
Recovery time, of radio receiver	992
Rectangular error, defined	1264
Rectangular projection, defined	75
use of	75
Reduction to the meridian, process of	575, 584
Reef, defined	784
Reference direction(s), discussed	160
true north as	209
Reference ellipsoids, in geodesy and mapping, listed	1346
Reference station, defined	806
of tide tables	326
Reflected wave, radio	959
Reflection, of radio waves	959
Reflection binnacles	166
Refraction, abnormal, in polar regions	720
altitude correction for, from lifeboat	738
defined	438
discussed	438
of ocean waves	828, 834
of radio waves	960
Refractive index, of seawater	781
Regiment of the Pole Star, The; of Wakeley	640
Regimento do estrolabio e do quadrante	20
Regression of the nodes, moon's, defined	382
Regulatory buoy(s), color of	110
Relative accuracy, defined	1287
Relative bearing, defined	65, 283
Relative humidity, defined	892
Release, of tangent screw, on marine sextant	409
Relief, on charts	125
Remanence, defined	202
Repeatability, defined	1287
Repeatable accuracy, defined	1287
Repeater, of radarscope	985
Requisite Tables, of Davis	47, 593
Residual deviation, determination of	210
of properly adjusted compass	216
Residual magnetism, defined	201
Resolution, in bearing, defined	995
explained	994
in range, defined	995
explained	994
Resolution error, defined	1264
Retentivity, defined	202
Retire, line of position	468
Retrace, of cathode-ray tube	971
Retrograde motion, of planets, defined	382
Reversing current	807
Reversing thermometer, on Nansen bottle	778
Revolution, of celestial body, defined	357
Revolver, in horizontal sextant angles, defined	321
Rho-Rho, in hyperbolic radionavigation systems	1016
Rhumb line(s), defined	63, 270
as desirable property of projection	67
on nautical chart	262
in sailing	270
Richer, Jean Francisco; mechanical device of	630
Ridge, of ice, defined	848
undersea feature, defined	784
Ridging, of ice	848
Right ascension, defined	389
Right (dangerous) semicircle, avoiding storm center, Northern Hemisphere	950
Right (navigable) semicircle, avoiding storm center, Southern Hemisphere	951
Right sphere, defined	371
Right triangle, altitude measurement by	736
Rigidity in space, defined	1130
described	170
Rip current	837
Ripples, defined	824
Rise, undersea feature, defined	784
Rising(s), direction from, in lifeboat	730
of sun and moon, almanac time of	528
at moving craft	537
Roads, on charts	126
Roaring forties	909
Rock(s), appearance on radarscope	998
Rocking the sextant, defined	412
Rollet de l'Isle; sight reduction diagram of	627
Roman mile, length unit	12
Romanovsky, A. A.; azimuth nomogram of	643
Root mean square error, in one-dimensional errors	1279
in two-dimensional errors	1282, 1286
Rose, G.; sight reduction tables of	595
Rossel Current	822
Rotary current, described	807
Rotating loop antenna	975
Rotating loop radiobeacon, description of	980
use of	981
Rotation, of celestial body, defined	357
Rotator, of taffrail log	147
Rotten ice, described	852
Rough log, origin of	16
Round of sights, defined	468
Route, selection of, in lifeboat	728
Rudolphine Tables, of Kepler et al	41
Rule of Sixty, stated	287
Rules for Ice Strengthening of Ships, of the Board of Navigation	869
Rules of the road, on charts	127

Page

Running fix, from celestial observations 470
defined.. 292, 296
discussed 296
Rust, Armistead; azimuth diagram of 642
sight reduction tables of.............................. 588
Rutters.. 8

Saddle, defined 784
Safe water mark(s), defined 1358
use of.. 1360
Sailing(s), history of 16
introduction to.. 270
various kinds of.. 270
Sailing chart, description of 113
Sailing direction(s), and currents 813
foreign charts referred to............................ 754
format for 340
history of 8
introduction to.. 340
modern 9
for polar regions.. 716
during Renaissance 8
unverified information in............................ 754
Sailing Directions (Enroute), of Defense Mapping Agency Hydrographic/ Topographic Center 340
contents of.. 342
Sailing Directions (Planning Guide), of Defense Mapping Agency Hydrographic/Topographic Center 340
contents of.. 340
study of, in advance preparations 671
Salinity, of seawater 777
Salinometer, conductivity measuring instrument 777
Same name, of declination and latitude, defined 400
Sand dune, appearance on radarscope 997
Sandspits, appearance on radarscope 997
Santa Ana (wind), defined................................ 918
Sargasso Sea 819
Saros, defined 386
Satellite(s), in solar system 357
Satellite navigation, basic doppler navigation principle 1095
defined 59
development of.. 56
introduction to.. 1066
Satellite navigation system, design considerations 1095
Saturable radionavigation system, defined .. 1012
Saturated magnetic field.................................. 201
Saturn, described 364
Satyricon, of Copella 25
Scale, of chart, defined 112
constant, as desirable property of projection.. 67
of nautical charts, discussed........................ 112
representative fraction of............................ 112
variation of, on polar projections................ 86
Scarp, defined.. 783

Page

Scattering, and absorption, of radio waves 964
of radio waves, defined 965
Schist, defined 119
Schuler oscillation, described 1141
Schuler tuned, defined...................................... 1135
Schuler tuning, purpose of................................ 1140
Scintillation, of star 644
Scoria, defined.. 119
Sea 824
color of, as sign of land................................ 741
conditions for underwater log calibration. 1309
and wind.. 890
Sea and Air Navigation Tables, of Aquino... 601
Sea breeze, described 917
Sea ice, discussed.. 846, 849
problem of 843
thickness of 849
Sea-Ice Nomenclature (WMO No. 259. TP. 145) 879
Sea level pressure, defined 886
Sea-man's Practice, The; of Norwood 21
Sea quadrant, invention of 29
Sea return, defined.. 994
effect on radar range 992
Sea smoke, defined 703
Sea state, effect of night observation............ 1317
introduction to.. 1313
Sea tangle, defined 119
Seachannel, defined 784
Seaman's Secrets, The; of Davis.......... 21, 29, 32, 37
Seamount, defined.. 784
Seaquake, waves from 830
Seasonal current, defined.................................. 815
Seasons, cause of.. 374
Seawater, physical properties of...................... 777
Secant cone, defined.. 68
Secant conic projection, defined 77
Secchi disk, defined.. 782
Second(s), unit of time 493
Second differences, Pub. No. 229 interpolation, defined 546
Second estimate, of zone time 576
Second-order triangulation, in geodetic surveys.. 1345
Second-trace echoes, appearance on radarscope 1002
Second-year ice, defined 847
Secondary coding delay, of Loran-C.............. 1037
Secondary light, defined.................................... 92
Secondary Phase Factor, of Loran-C signals 1051
Secondary radar.. 983
Secondary station, of radionavigation systems 1012
Sector(s), of calibration, of radio direction finder station 979
in lights, discussed.. 99
of uncertain calibration, of radio direction finder station 979
Sectoring, appearance on radarscope 1003
Secular change, of magnetic field.................... 204

Page
Seiche(s), and standing waves 832
Seismic sea wave 830, 953
Selected Worldwide Marine Weather Broadcasts 941
Selectivity, of direction finder receiver 976
of radio receiver, defined 970
Semicircular deviation, defined 218
Semidiameter, altitude correction for, discussed 442
from lifeboat 738
Semidiurnal current 808
Semidiurnal tide, defined 326, 799
Semipermanent highs and lows 910
Sense antenna 974
Sensible horizon, illustrated 393
Sensitivity of direction finder receiver 976
of radio receiver, defined 970
September equinox, defined 377
Sequenced radiobeacon(s), described 972
Serrated range rings 1003
Set, of current, in celestial navigation 473
defined 264, 808, 815
Setting(s), direction from, in lifeboat 730
of sun and moon, almanac time of 528
at moving craft 537
Sextant(s), adjustments 424
altitude measuring device 436
care of 421
history of 28
invention of 30
for lifeboat navigation 725
observational skill, developing 418
optical principles of, discussed 408
reading the 416
selection of 428
use of 412
Sextant altitude, defined 414
in sight reduction 539
Sextant altitude correction(s) 436
in lifeboat navigation 737
need for 436
order of applying 447
summary of 446
Shade error, of sextant 424
Shade glasses, on marine sextant 410
Shadow, of radio waves 964
Shadow sector, on radarscope, defined 1003
Shallow water, effect on sea state 1316
and waves 828
Shallow water effect, described 148
Shape, of radar target, effect on range 993
true, as desirable property of projection ... 67
Shchetkin, N. O.; sight reduction method of 626
Shearme, F. N.; sight reduction tables of 597
Shelf, defined 784
Shingle, defined 119
Ship, appearance on radarscope 998
Ship considerations, of ship weather routing 686
Ship Inertial Navigation System, development of 58

Page
Ship log 11
Ship weather routing, benefits of 698
conclusion 698
defined 684
development of 685
environmental factors 687
introduction to 684
recommendations and advisories 696
special weather and environmental considerations 695
synoptic weather considerations 689
Shipboard Wind Plotter, use of 888
Ship's course indicator 195
Shoal, defined 784
Shoaling, defined 828
Shooting stars, described 368
Shoreline, on nautical charts 117
Short, Accurate, and Comprehensive Altitude-Azimuth, Tables, of Johnson 641
Short range radionavigation system, defined 1012
Short range warnings, radio navigational 749
Shortrede, Robert; azimuth tables of 640
Shuga, ice formation 846
Side error, of sextant, defined 425
Side-lobe echoes, appearance on radarscope. 1001
effect on radar range 992
Sidereal, rotation and revolution classification 358
Sidereal day, beginning of 492
defined 482, 486
reference for 379
Sidereal hour angle, defined 389
as longitude on celestial sphere 390
Sidereal month, duration of 365
Sidereal noon, defined 482, 492
Sidereal time, defined 482, 489
discussed 505
reference for 379
Sidereal year, length of 373
Sight reduction, by combined sextant and computer 638
comparison of various methods, introduction to 583
summary of 643
defined 538
graphical and mechanical solutions 626
by intermediate calculator 1300
introduction to 538
kinds of methods 583
in lifeboat 738
modern inspection tables 604
in polar regions 720
unique situations, use of 626
Sight reduction tables, defined 538
Sight Reduction Tables for Air Navigation (Pub. No. 249) 611
description and solutions using 553
extracts from 1256
star identification by 664

	Page
Sight Reduction Tables for Marine Navigation (Pub. No. 229)	604
description and use of	541
extracts from	1250
interpolation	546
near the horizon	563
of negative altitudes	564
when second differences are required	549
near the zenith	564
Interpolation Table, described	546
and *Nautical Almanac*, complete solution using	551
triple interpolation of azimuth angle	580
Sight Reduction Tables for Sun, Moon and Planets: Assumed Altitude Method of Celestial Navigation, of Davies	614
Sighting vanes, described	168
Sigma, in one-dimensional errors	1279
Sill, defined	784
Simple conic projection, defined	68, 75
discussed	75
Simplified Celestial Observation Table (Japanese H.O. Pub. No. 603)	610
Simultaneous Altitudes and Azimuths of Celestial Bodies (H.O. Pub. No. 201)	592
Single astronomical station datum orientation	1347
Single-axis normal distribution	1267
Single-degree-of-freedom gyroscope, described	1130
Sinking, of mirage	920
Siren(s), sound signal	107
Sirocco	2
Sixteenth century, navigation manuals of	20
Sixteenth century navigation, history of	2
Size of radar target, effect on range	993
Skip distance, of radio waves	962
Skip zone, of radio waves	962
Sky, appearance of, as sign of land	741
coloring of	920
Sky diagrams, in *Air Almanac*, described	664
Sky map, described	864
Skylight(s), defined	873
Skywave(s), avoiding contamination in Loran-C	1038
defined	962
Skywave Long Baseline Loran, development of	51
Skywave Synchronized Loran, development of	51
Slack water, defined	807
in piloting	324
Slave station, of radionavigation systems	1013
Slide bar technique, discussed	308
Sling psychrometer	893
Slip, of ship's screw	155
Slope, undersea feature, defined	784
Slush, ice formation	847
Small area investigation, parallel method	760
radial method	760
Small area plotting sheet, construction of	88
Small boat(s), as aid to radar navigation	1004
Small circle, defined	60
Small craft charts, use of	138
Small-scale chart, defined	113
Small's light, historical significance of	45
Smart, W. M.; sight reduction tables of	597
Smog, defined	919
Smooth log, origin of	16
Snapper, bottom sampler	787
Snell, Willebrord; law of	439
Snow blink, described	864
Soft iron, magnetic characteristics of	201
magnetism of	203
Soft iron correctors and nearby magnets, in compass adjustment	233
Solar activity, effect on Omega signals	1026
Solar day, defined	378
Solar eclipse, defined	386
Solar parallax, defined	445
Solar prominences, illustrated	385
Solar system, bodies of	357, 644
Solar tide	800
Solar time, defined	489
discussed	485
Solar year, length of	373
Solar zenith angle, effect on Omega signals	1026
Sonar, in polar regions	717
Sonne	56
Souillagouet, F.; sight reduction tables of	596
Soule, C. C.; sight reduction tables of	589
Sound, attenuation of	842
sources of, in the ocean	838
speed of, in seawater	838
Sound signals, discussed	107
limitations of	107
Sound signal apparatus, on charts	125
Sound transmission, in doppler sonar navigation	1123
Sound waves, reflection of, underwater	840
refraction of, underwater	841
Sounding(s), discussed	758
equipment for	759
evaluation of results	759
history of	14
investigation of small areas	760
in polar regions	715
use of in piloting	303
Sounding datum, on charts, discussed	120
Sounding machine, origin of	14
Sounding reports, records of	761
South, as direction of measurement	61
South Atlantic Current	820
South Equatorial Current, in Atlantic Ocean	816
in Indian Ocean	822
in Pacific Ocean	820
South Indian Ocean, tropical cyclones in	932
origin, season, frequency	936
South magnetic pole, of earth	203
South pole, of magnet	202
Southeast Drift Current, in Atlantic Ocean	819

Page

Southeast trades ... 906
Southeasterlies, of Antarctic ... 909
Southwest Pacific and Australian area, tropical cyclones in ... 932
origin, season, frequency ... 936
Southern Hemisphere, ship weather routing in ... 697
Space motion, of celestial bodies ... 374
defined ... 357
of stars, defined ... 369
Span, length unit ... 12
Spatial and dispersive correlation, effect on Omega signals ... 1028
Special mark(s), in buoyage system, defined 1358
use of ... 1361
Special Notice to Mariners Paragraphs, information in ... 748
Special purpose buoy(s), in buoyage system. 105
Special Warnings, radio navigational ... 750
Specific gravity, of seawater ... 779
Specific heat, of seawater ... 780
Spectrum, of colors ... 921
Speed, of advance, adjustment of, in ship weather routing ... 696
defined ... 63
of current, relationship to range of tide ... 812
defined ... 63
made good, defined ... 63, 264
over ground, defined ... 63, 269
measurement of ... 148
of wave, from ship ... 830
Speed-course-latitude error, of gyrocompass, discussed ... 181
Speed error, of gyrocompass, discussed ... 181
Speed line, celestial, defined ... 465
Sphere, developable property of ... 67
sight reduction by ... 637
undevelopable property of ... 67
Spherical hyperbola, in distance-difference measurement ... 1013
Spherical triangle, defined ... 402
Spherical Triangle Nomogram, of Wimperis ... 627
Spheroidal hyperbola, in distance-difference measurement ... 1013
Spicules, of ice, described ... 846
of marine sediment ... 119
Spieghel der Zeevaerdt (The Mariner's Mirror), of Waghenaer ... 9
Spilling breaker, defined ... 835
Spillover, in radio receiver ... 970
Spin axis, of gyroscope ... 170
Spire, chart definition of ... 126
Spitsbergen Current ... 819
Split fix, defined ... 316
in horizontal sextant angle observations .. 316
Spoil area(s), on charts ... 129
defined ... 752
Spoking, appearance on radarscope ... 1003
Spot elevations, on charts ... 125
Spring (season) ... 375

Page

Spring current ... 810
Spring-driven marine chronometer, described ... 432
Spring speed(s), of tidal current ... 335
Spring tide(s) ... 802
defined ... 798
Spur, defined ... 784
Sputnik I ... 56
Squall, defined ... 914
Squall line, defined ... 914
Square root, by basic calculator ... 1299
Squat, correction for ... 753
St. Elmo's fire, described ... 924, 965
Stability, of radio receiver, defined ... 970
Stabilization loop, in inertial navigation, function of ... 1143
Stabilized scope presentation ... 985
Stable platform, defined ... 1137
Stack, chart definition of ... 126
Stadimeter, description and use of ... 145
Stand, of tide ... 324
defined ... 799
Standard, defined ... 1261
Standard atmosphere, defined ... 904, 991
structure of ... 903
Standard compass ... 161
Standard deviation, defined ... 1264
in one-dimensional errors ... 1279
in two-dimensional errors ... 1281
Standard error, defined ... 1264
Standard lapse rate, defined ... 904
Standard Loran, development of ... 51
Standard meridian ... 490
Standard parallel, defined ... 75
Standard time ... 498
legal basis of ... 507
time meridian for ... 491
Standard time broadcasts, discussed ... 511
Standing waves, and seiches ... 832
Standpipe, chart definition of ... 127
Star(s), altitude corrections by *Nautical Almanac* ... 448
altitude corrections of ... 452
described ... 368
location of, in vicinity of Cygnus ... 654
in vicinity of Orion ... 650
in vicinity of Pegasus ... 647
in vicinity of Ursa Major ... 652
navigational ... 644
observation of with sextant ... 414
Star Air Navigation Tables, of Hoehne ... 605
Star Altitude Curves, of Weems ... 46, 634
Star catalog number ... 645
Star charts, discussed ... 645
Star clouds, defined ... 370
Star finder(s) ... 415
discussed ... 657
use of astrolabe as ... 28
Star globe ... 645
Star identification, by computation ... 665

Page

Star Sight Reduction Tables for 42 Stars: Assumed Altitude Method of Celestial Navigation, of Davies 614
Star Tables for Air Navigation 611
State of Sea Photographs for the Beaufort Wind Scale 1317
Static, in radio, defined 965
Static electricity 957
effect of on radarscope 1003
Static pressure, of water 152
Station buoy(s), in buoyage system 105
discussed 101
Station error, defined 1342
Station pressure, defined 886
Statute mile, defined 63
origin and length of 13
Steering compass 161
Stellar parallax, defined 445
Stereographic projection, defined 68, 82
features of 82
origin of 5
Stooping, of mirage 920
Storm center, maneuvering to avoid 947
Storm surges 831
Storm tide(s), described 831, 953
Storm track; on, ahead of center, avoiding in Northern Hemisphere 950
avoiding in Southern Hemisphere 951
on, behind center, avoiding in Northern Hemisphere 950
avoiding in Southern Hemisphere 951
Storm wave, defined 953
Straight line of position, adjustment of 561
Stratocumulus clouds, described 859
Stratosphere 903
Stratus, meaning of 895
Stratus clouds, described 898
Stray line 11
Stream current, defined 815
Sub-refraction, effect on radar rays 991
Subastral point 403
Sublunar point 403
Submarine cables, danger from 752
Subordinate station, defined 806
of tide tables 326
Subpolar regions, defined 700
Subsatellite point 403
Subsolar point 403
Substellar point 403
Subsurface current, defined 815
Subtropical anticyclone(s) 917
Suma de Geographia, of Encisco 20
Summary of Corrections, of Defense Mapping Agency Hydrographic/Topographic Center, information in 748
publication of 137
Summer (season), described 374
Summer solstice, defined 374
Summer time 491, 498
Sumner line, defined 464

Page

Sumner Line of Position of Celestial Bodies, The (H.O. Pubs. Nos. 203, 204); of Littlehales 588
Sun, altitude corrections by *Nautical Almanac* 448
altitude corrections of 449
features of 360
observation of with sextant 412
in solar system 357
Sun cross, described 922
Sun dog, defined 922
Sun drawing water, defined 923
Sun pillar, described 922
Sun spots, discussed 360
Sundial time 487
Sunlight, Moonlight, and Twilight for Antarctica (Circular No. 147); of U.S. Naval Observatory 529
Sunrise, finding time of, by almanac 529
longitude by 740
in polar regions 722
Sun's True Bearing or Azimuth Table, of Davis and Davis 47
Sun's way, defined 361
Sunset, finding time of, by almanac 529
longitude by 740
in polar regions 722
Super high frequency, radio wave propagation of 968
Super-refraction, effect on radar rays 991
of radio waves 961
Superior airmass 912
Superior conjunction, defined 380
Superior mirage, defined 920
Superior planets, defined 361
Supernova, defined 369
Supersaturation, of air 893
Superseded marks 1358
Supplement to A. E., 1968, The 520
Surf, breaking on reef, appearance on radarscope 998
defined 828
features of 835
introduction to 834
sound of, as sign of land 741
Surf zone 829
Surface current, defined 815
Surface navigation, defined 59
Surface radio duct, effect on radar rays 991
Surface tension, of seawater 782
Surge, of tropical cyclone 952
Surging breaker, defined 835
Sweat, defined 892
Sweep, of cathode-ray tube 971
Swell(s), appearance of, as sign of land 741
defined 824
effect on sea state 1316
Swing circle 681
Swing ship, procedure for 240

	Page
Swinger, in horizontal sextant angles, avoidance of	321
defined	321
Swinging the arc, defined	412
Sylphon cell, of aneroid barometer	884
Symbols, of celestial bodies	1167
mathematical	1167
miscellaneous	1168
for navigational positions	1167
zodiacal	1168
Symonds, W. P.; azimuth tables of	642
Synodic, rotation and revolution classification	358
Synodical month, duration of	365
and tidal cycles	803
Synoptic chart, of weather	927
Synoptic oceanography, discussed	787
System area monitor, of Loran-C	1040
System errors, in speed measurement	148
Systematic errors, discussed	1262
Table(s), for lifeboat navigation	725
Table d'angles horaires, of Hommey	586
Table des Azimuts du Soleil, of Labrosse	640
Table du Cadran Solaire Azimutal, of Decante	641
Table 902, *Azimuts;* French azimuth table	642
Tablemount, defined	784
Tables du Point Auxiliare, of Souillagouet	596
Tables for Calculating Altitude and Azimuth of Celestial Bodies, of Yonemura	593
Tables for Facilitating Sumner's Method at Sea, of Thomson (Lord Kelvin)	587
Tables for Finding Azimuth at Sea, of Shortrede	640
Tables Graphiques d'Azimut, of Constan	642
Tables K 1, of Kotlarić	603
Tables K 11, Two-Star Fix Without Use of Altitude Difference Method, of Kotlarić	620
Tables K 21, Computed Results of Altitudes and Azimuths for all Latitudes and all Celestial bodies, of Kotlarić	613
Tables of Azimuth of the Sun, of Zhdanko	641
Tables of Computed Altitude and Azimuth (Pub. No. 214)	604
azimuth determination	581
Δd solution	558
Δd and Δt solution	559
Δd, Δt, and ΔL solution	559
description of	556
extracts from	1243
Tables of Logarithms, of Martelli	586
Tables of Sunrise, Sunset, and Twilight	529, 536
Tables of Sun's True Bearing or Azimuth, of Burdwood	640
Tables of the Sun, of Newcomb	483
Tables to Facilitate the Practice of Great Circle Sailing, of Towson and Atherton	640
Tables to Find Altitudes and Azimuths, of Fuss	601
Tablette de Point Sphérique, of Bertin	597
Tàboas de Alturas, of Braga	593
Tabuas de Altura e Azimute, of Fontoura da Costa and Penteado	603
Tabulae Prutenicae, of Reinhold	41
Tabular iceberg, defined	849
Tachometer, speed measurement by	155
Tactical diameter, of vessel in turn, defined	305
Taffrail log, description and use of	147
Tail, of comet, formation of	367
Taking departure, defined	264, 674
Tangent arc, defined	922
Tangent latitude error, of gyrocompass, discussed	183
Tangent screw, on marine sextant	409
Tank (gas, oil); chart definition of	127
Target, height of, in radarscope interpretation	996
reflecting quality of, in radarscope interpretation	996
Tavole H (I.I. 3113), of Italian Navy Hydrographic Institute	600
Tavole nautiche e Tavole dei Semisenoversi, of Chiesa	594
Tehuantepecer, defined	918
Telescope, invention of	39
on marine sextant	410
Telescopic alidade, described	195
Temperate zones, defined	377
Temperature, defined	891
measurement of	892
in the ocean	777
Temperature error, of barometer	886
Temperature inversion	918
Temporary Geographic Grid	91
Tenting, of ice	848
Terminator, defined	384
Terra Australis Nondum Cognita	7
Terrace, defined	784
Terrestrial magnetism, discussed	203
Terrestrial refraction, defined	439
Texture, of radar target, effect on range	994
Thawhole(s), defined	850
Theatrum Orbis Terra	7
Thermal conductivity, of seawater	781
Thermal expansion, of seawater	781
Thermistor, defined	778
Thermocline, of seawater	778, 839
Thermometer, temperature measuring device	892
Thick growth, appearance on radarscope	998
Third-order triangulation, in geodetic surveys	1345
Thomson, William (Lord Kelvin); sight reduction tables of	587
Three-arm protractor, description and use of	142
Three-bearing plot, discussed	289
Three-digit number(s), of chart numbering system	132
Three-point fix, strength of	317

Page

Three-point problem, of horizontal sextant angles 313
solution of 313
without three-arm protractor 313
Three Star Position Tables for Aerial Navigation, of de Jonge 616
Thunder, described 924
Tidal box, on charts 129
Tidal current 324
defined 789, 807
discussed 807
features of 807
finding speed of by table 333
types of 808
in underwater log calibration 1309
variations and cycles 809
Tidal current charts 813
discussed 335
Tidal Current Diagrams 813
described 335
Tidal current predictions, for reference stations 330
for subordinate stations 332
Tidal current tables, contents of 330, 813
extracts from 1238
and other sources of information 813
Tidal datum(s) defined 804
discussed 804
Tidal day 326, 804
defined 489
Tidal effects, discussed 324
Tidal phenomenon, discussed 789
Tidal stream 789
Tidal wave(s) 831
described 830, 953
Tide(s) 324
age of 801
causes of 789
cycles of 803
defined 789, 798
discussed 789
effect on sea state 1316
finding height of, by table 327
general features of 798
meteorological effects on 806
observation of 806
in polar regions 713
prediction of 325, 806
special effects of 800
time of 804
types of 799
Tide predicting machine 806
Tide predictions, for reference stations 326
for subordinate stations 326
Tide rip, defined 325
Tide tables, extracts from 1235
publication and contents of 326, 806
tabulation of sunrise and sunset 536
Tide waves 832
Tillman, E.; sight reduction table of 603
Timberlake, E. M.; sight reduction tables of 600

Page

Time, and arc, interconversion of 493
basic concepts of 481
discussed 378
expressing 489
fundamental kinds of 482
and hour angle 506
introduction to 479
and longitude 496
of tide and tidal current, relationship between 812
Time and altitude azimuth, formulas for 639
Time azimuth, defined 639
by intermediate calculator 1302
Time Azimuth Diagram, of Godfrey 642
Time Azimuth Diagram, of Weir 642
Time ball 36
Time difference measuring system(s) 1011
Time interval, and time scales 484
Time meridian 490
defined 496
Time scale(s), defined and discussed 481
and time intervals 484
Time segment, of Omega signal 1024
Time Service Announcement Series 1, No. 2; of U.S. Naval Observatory 511
Time sight, formulas for 585
history of 44
longitude by 741
Time signal(s), CCIR standard format 514
codes for transmission of DUT 1 514
discussed 511
dissemination systems 509
characteristics of 509
first transmission of 48
history of 36
Morse code method 514
USSR extended format 514
Time tick, defined 511
Time zone, defined 496
Timekeeping, and clocks 479
Timepiece, for lifeboat navigation 725
Title, of chart 129
Tokyo Datum, geodetic 1350
Toledan Tables 37, 41
Topmark, on buoys, ... 1358
Tornado 925
Torque, defined 171
Torrid Zone, defined 376
Total eclipse, of sun, explained 384
Total intensity, of earth's magnetic field 205
Total pressure, of water, defined 153
Toward, in sight reduction 464, 539
Tower (water), chart definition of 127
Towering, of mirage 919
Towson, John Thomas; azimuth and great circle tables of 640
Trace, of cathode-ray tube 971
Track, defined 64
made food, defined 64, 264
made good, on Decca Track Plotter 1059
Tracking, process of 199

Page

Tracking station(s), of navigation satellites . 1068
Trackline, defined 64
Trackline SURPIC, example of 770
Tractive force(s), defined 793
Trade winds, discussed 906
Traffic separation schemes, on charts 129
discussed 750
Traité de Nomographie, of d'Ocagne 627
Transducer, of echo sounder 159
Transducer orientation error(s), in doppler sonar navigation 1119
Transfer, of vessel in turn, defined 305
Transit, of sun by planet 380
Mercury 363
time of, longitude by 740
Transmissometer 900
Transmitter, radio, components of 969
discussed 969
Transparency, of seawater 782
Transponder, defined 970
Transverse conic projection, defined 75
Transverse orthomorphic projection, defined 68, 73, 74
Transverse Mercator projection, defined... 68, 73, 74
discussed 73
special features of 74
Tratado da Sphera, of Nunes 17, 20
Travas, origin of 16
Traverse, in geodetic surveys 1346
Traverse sailing, described 270
history of 16
Traverse table(s), in lifeboat navigation 726
origin of 16
Trench, defined 784
Triangle of error, formation of 289
Triangulation, described 1343
Trigonometric functions, use of basic calculator with 1300
Trigonometric leveling, in geodetic surveys. 1346
Trilateration, in geodetic surveys 1345
Trim, ship's, in underwater log calibration.. 1309
Tropic current 810
Tropic higher high water 805
Tropic lower low water 805
Tropic of Cancer, defined 376
Tropic of Capricorn, defined 376
Tropic tide 801
Tropical airmass, source region of 911
Tropical cyclone(s) 917
areas of occurrence 932
defined 931
definitions of 931
effects of 951
frequency of occurrence 932
introduction to 931
life of 939
locating the center 947
origin of 931
passage of at sea 942
season of 932

Page

Tropical depression, defined 931
Tropical disturbance, defined 931
Tropical month, and tidal cycles 803
Tropical storm, defined 931
Tropical year, length of 373
Tropopause 903
Troposphere 903
Trough, of radio wave, defined 958
underwater feature, defined 784
of water wave 824
True, Clarence H.; sight reduction diagram of 633
True bearing, defined 283
True course 64
True direction, defined 209
True north, as reference direction 283
True wind, defined 887
Tsunamis, defined 830
Tsushima Current 820
Tufa, defined 119
Tundra, described 700
Turbidity current, described 787
Turning bearing, discussed 308
Turning characteristics of vessel, allowance for 305
Turning circle, of vessel, illustrated 306
Twentieth century navigation, history of 3
Twilight, defined 372
finding time of, by almanac 529
in polar regions 722
of sun and moon, almanac time of 528
at moving craft 537
Twinkling, of star 644
Two-bearing plot, discussed 288
Two-degree-of-freedom gyroscope, described 1130
Two-digit numbers, of chart numbering system 132
Typhoon, defined 931
in western North Pacific 932
Typhoon Havens Handbook for the Western Pacific and Indian Oceans, of Naval Environmental Prediction Research Facility 951

U.S. Coastal Confluence Zone, Loran-C in 1057
U.S. long range radio navigational warning system, discussed 749
U.S. Naval Observatory, establishment of 40
U.S. Navy Marine Climatic Atlas of the World 689
U.S. Navy Training Device Center, sight reduction method of 638
USSR tables, for sight reduction 603
Ultra high frequency, radio wave propagation of 967
Uncharted objects, cutting in, with horizontal sextant angles 321
Uncorrecting, of compass error, defined 212
Under-keel clearances, discussed 753
Underwater log calibration, advance preparations 1310
duties of observers 1310

Page

Underwater log calibration—Continued
introduction to ... 1309
runs required ... 1309
verification of markers ... 1310
Underwater navigation, defined ... 59
Underwater sound, and the navigator ... 838
Uniform Cardinal System of buoyage ... 1357
Uniform Lateral System of buoyage ... 1357
Uniform State Waterway Marking System, discussed ... 109
Uniform System of buoyage ... 1355
United States Coast Pilots, publications, discussed ... 344
United States of America Nautical Chart Symbols and Abbreviations (Chart No. 1) ... 754
United States system, of buoyage ... 1356
of time signals ... 512
Units of measurement, distance and speed ... 144
Universal law of gravitation, Newton's ... 359
Universal plotting sheet, defined ... 90
in lifeboat navigation ... 725
Universal Polar Stereographic grid ... 91
graticule for ... 82
Universal Time, defined ... 378, 496
discussed ... 482
reference for ... 380
Universal Transverse Mercator grid ... 73, 91
Unlighted buoy(s), shapes of ... 102
Unlighted sound buoy, described ... 102
Unstabilized scope presentation ... 985
Unverified information, discussed ... 754
Upper air, observations of ... 900
Upper branch, of celestial meridian, defined ... 388
of meridian, defined ... 60
Upper front ... 914
Upper limb ... 414
Upper transit, of celestial meridian, defined ... 392
Uprush, of waves ... 835
Uranus, described ... 364
Urgent reports, by radio ... 758
Utility, limits of, on polar projections ... 87

Valley, defined ... 784
Variable range marker, on PPI ... 984
Variable stars, defined ... 369
Variability, in speed measurement ... 150
Variance, defined ... 1281
Variation, application of ... 212
of compass, longitude determination by ... 33
defined ... 205, 209
history of ... 10
of latitude and longitude, due to eulerian motion ... 373
Veater, altitude and azimuth by map projection ... 632
Veering wind ... 947
Velocity meter, defined ... 1132
Velocity of sound error(s), in doppler sonar navigation ... 1120
Venus, described ... 363

Page

Vernal equinox, defined ... 375, 486
as reference hour circle ... 389
Vernier, on marine sextant ... 409
Vernier sextant, discussed ... 411
Vertex, of great circle, defined ... 271
Vertical axis, of gyroscope ... 170
Vertical beam width ... 985
effect on radar range ... 992
Vertical circle(s), defined ... 399
Vertical danger angle(s), described and discussed ... 322
Vertical deflection plates, of cathode-ray tube ... 971
Vertical earth rate, of gyro, described ... 173
Vertical force instrument, described ... 233
Vertical intensity, earth's magnetic field, defined ... 205
Vertical-plane coverage diagram ... 988
Vertical ring, of gyrosphere ... 174
Vertical surveying, process of ... 1346
Very close pack, concentration of ice ... 848
Very high frequency, radio wave propagation of ... 967
Very low frequency, radio wave propagation of ... 966
Very open pack, concentration of ice ... 848
Vessel motion induced error(s), in doppler sonar navigation ... 1119
Viscosity, of seawater ... 780
Visibility, defined ... 900
measurement of ... 900
Visible horizon, defined ... 393
dip of, by basic calculator ... 1299
distance to, by basic calculator ... 1299
Visual range, of lights, discussed ... 347
factors affecting ... 99
Voigt, "Orion" sight reduction instrument ... 636
Vucetic, Tamara; "Toposcope" sight reduction device ... 636
Vulgar establishment ... 804

Waggoner ... 8
Wakeley, Andrew; azimuth tables of ... 640
Waller, George W. D.; sight reduction method of ... 593
Wandering of the poles, defined ... 373
Waning, of moon ... 384
War time ... 498
Warm airmass ... 911
Warm front, defined ... 913
Warm sector ... 913
Watch, use of ... 433
Watch buoys, discussed ... 101
Watch error ... 434
defined ... 501
Watch rate ... 434, 503
Watch time, discussed ... 501
use of AM and PM ... 492
Watch tower, chart definition of ... 127
Water areas, on nautical charts ... 117
Water sky, described ... 865

Page
Water-speed measurement, problems of 148
Waterspout(s), described 924
Wave(s), causes of 824
characteristics of 824
defined 824
effect of, currents on 828
ice on 928
upon ship 832
energy of 829
measurement of, from ship 829
path of water particles in 827
and shallow water 828
use of oil on 833
Wave front, of radio wave, defined 958
Wave height, defined 824
Wave interference, defined 960
Wave period, defined 824
Wavelength, defined 824
of radiowave, units of 958
Waxing, of moon 384
Weather, defined 903
forecasting of 926
influencing the 929
information on, dissemination of 927
interpretation of 928
Weather advisory, defined 697
Weather elements, introduction to 903
Weather map 927
Weather observations, introduction to 936
recording of 900
Weems, Philip van Horn; *Line of Position Book* of 598
New Line of Position Tables of 599
"Polar Computer" of 635
Secant Time Sight of 588
Star Altitude Curves of 634
Weir, Patrick, azimuth diagram of 642
West, as direction of measurement 62
West Australia Current 822
West Greenland Current 819
West Wind Drift 820, 823
Western North Pacific, tropical cyclones in. 932
origin, season, frequency 932
Wet-bulb temperature, indication of 892
Wet-bulb thermometer 893
Whistle(s), sound signal 107
Whistle buoy, described 102
Willis, Edward J.; sight reduction method of 621, 638
Willy-Willy, in southwest Pacific and Australia 932
Wind(s), conditions for underwater log calibration 1309
defined 903

Page
Wind(s)—Continued
duration of, effect on sea state 1313
estimating at sea 1313
measurement of 887
of polar regions 713, 909
and the sea 890
true and apparent 887
Wind and Current Chart of the North Atlantic, of Maury 18
Wind current(s), defined 815
in oceans 815
Wind rose, Portolan 6
Wind shift(s) effect on sea state 1317
Wind-shift line, defined 914
Wind speed, effect on sea state 1313
Wind vane 887
Winter (season) 374
Winter solstice, defined 374
World Atlas of Sea Surface Temperatures (Pub. No. 225) 777
World Geodetic System, discussed 1350
World Geographic Referencing System (Georef) 91
World map(s), of Middle Ages 6
World Port Index (Pub. No. 150) 340
Worldwide radionavigation aids, in polar regions 717
Worldwide radionavigation system, defined. 1012
Wreck buoy(s), in buoyage system 105
"Wrinkles" in Practical Navigation, of Lecky 21

Year, astronomical significance of 378
defined 486
length of 373
Yonemura, S.; sight reduction tables of 593
Young ice, described 847
Yustchenko, A.; azimuth table of 642

Zenith, body in, latitude by 740
defined 388
Zenith distance, defined 395
Zenithal projection, bearings on 80
defined 68
Zerbee, Louis J.; sight reduction device of.... 638
Zhdanko, M.; azimuth tables of 641
Zodiac, described 378
Zodiacal light, defined 368
Zone, of Decca system 1057
Zone description, defined 496
Zone meridian 490
defined 496
Zone time, reference for 380
time meridian for 491, 496
use of 496

★ U.S. GOVERNMENT PRINTING OFFICE : 1984 O - 417-814 QL 3

www.ingramcontent.com/pod-product-compliance
Lightning Source LLC
LaVergne TN
LVHW081946200926
848484LV00009B/2515